A Textbook of
REFRIGERATION AND AIR-CONDITIONING

3rd Edition

Er. R.K. RAJPUT

A TEXTBOOK OF
REFRIGERATION AND AIR-CONDITIONING

For B.E./B.Tech., GATE, UPSC (Engg. & Civil Services), and Section-'B' of AMIE (India) Examinations

in

S.I. UNITS

By

Er. R.K. RAJPUT

M.E. (Heat Power Engg.) Hons., Gold Medallist, Grad. (Mech. Engg. & Elec. Engg.)
M.I.E. (India); M.S.E.S.I.; M.I.S.T.E.; C.E. (India)
Recipient of: "Best Teacher Award (Academic)"
"Distinguished Author Award"
"Jawahar Lal Nehru Memorial Gold Medal"
for an outstanding research paper
(Institution of Engineers—India)

Principal ***(Formerly):***

- Thapar Polytechnic College;
- Punjab College of Information Technology,

PATIALA (Punjab)

S.K. KATARIA & SONS®

Publisher of Engineering Books

4885/109, Prakash Mahal, Dr. Subhash Bhargav Lane,
Opposite Delhi Medical Association, Daryaganj, New Delhi-110002 (INDIA)
Phones: +91-11-43551243, Mobile: +91-9871775858, 7678425209
e-mail: katariabook@gmail.com; katariabooks@yahoo.com
Website: www.skkatariaandsons.com

(e-books, subscriptions and bulk buying also available on our website)

Published by:

S.K. Kataria & Sons®

4885/109, Prakash Mahal, Dr. Subhash Bhargav Lane,
Opposite Delhi Medical Association, Daryaganj, New Delhi-110002
Phones: +91-11-23243489, 43551243; Mobile: +91-9871775858
e-mail: katariabook@gmail.com; katariabooks@yahoo.com
Website: www.skkatariaandsons.com

First Edition: 2004; Reprint: 2005 (with Corrections); **Reprint: 2006** (with Corrections);
Reprint: 2007; 2008; 2009; 2010; 2011; Second Edition: 2012 (Two-coloured); Reprint: 2014
Third Edition: 2015 Reprint: 2026

PRINTED AT AL-AZIZ BOOKS, SAHIBABAD

Laser Typesetting by:
SARA Assignments

Preface to the Third Edition

I take pleasure in presenting the **"Third Edition"** of this book on **"Refrigeration and Air-Conditioning"**. The warm reception which the previous editions and reprints have received is a matter of much satisfaction to me.

In this edition the book has been thoroughly revised and a **new chapter 25** with the following **"contents"** has been added to make this book a complete and a comprehensive unit in all respects:

Chapter 25. "GATE" and "UPSC" Examinations' Questions (Latest-Selected) with Answers/Solutions

A. Conventional Questions with Answers/Solutions

B. Multiple-Choice Quesions with Answers and "Explanations"

Any suggestions for improvement of the book are most welcome.

Er. R.K. Rajput
(Author)

Preface to the First Edition

The main objective of writing this book on the subject **"Refrigeration and Air-conditioning"** is to present to the student-community, a book which should contain comprehensive treatment of the subject matter in simple, lucid and direct language and envelope a large number of solved problems properly graded, including typical examples, from examination point of view.

The book consists of 23 chapters and is divided into *three parts*, namely: ***Part-I: Refrigeration; Part-II: Air-conditioning; Part-III: Miscellaney.*** All chapters of the book are saturated with much needed text supported by simple and self-explanatory figures and large number of Worked Examples including Typical Examples (from competitive examinations). At the end of each chapter *Highlights, Objective Type Questions, Theoretical Questions* and *Unsolved Examples* have been added to make the book a comprehensive and complete unit in all respects.

The book will prove to be a boon to the students preparing for engineering undergraduate, AMIE Section-B (India) and competitive examinations.

The author's thanks are due to his wife Ramesh Rajput for extending all cooperation during preparation of the manuscript.

Although every care has been taken to make the book free of errors both in the text as well as solved examples, yet the author shall feel obliged if errors present are brought to his notice. Constructive criticism of the book will be warmly received.

Er. R.K. Rajput
(Author)

NOMENCLATURE

A	area
b	steady-flow availability function
C	velocity, mass concentration
$°C$	temperature on the celsius (or centigrade) scale
c	specific heat
c_p	specific heat at constant pressure
c_v	specific heat at constant volume
C_p	molar heat at constant pressure
C_v	molar heat at constant volume
D, d	bore; diameter
E	emissive power; total energy
e	base of natural logarithms
g	gravitational acceleration
H	enthalpy
h	specific enthalpy; heat transfer co-efficient
h_f	specific enthalpy of saturated liquid (fluid)
h_{fg}	latent heat
h_g	specific enthalpy of saturated vapour; gases
K	temperature on kelvin scale (*i.e.* celsius absolute, compressibility)
k	thermal conductivity, blade velocity co-efficient
L	stroke
M	molecular weight
m	mass
$\dot{m}$	rate of mass flow
N	rotational speed, mass flux of species
n	polytropic index, number of moles; number of cylinders
P	power
p	absolute pressure
p_m	mean effective pressure
p_{mi}	indicated mean effective pressure
p_{mb}	brake mean effective pressure, back pressure
Q	heat, rate of heat transfer
q	rate of heat transfer per unit area
R	gas constant; thermal resistance; radius; total expansion ratio in compound steam engines
R_0	universal gas constant
r	radius, expansion ratio, compression ratio

S entropy

s specific entropy

T absolute temperature; torque

t temperature

U internal energy; overall heat transfer co-efficient

u specific internal energy

V volume

v specific volume

W work; rate of work transfer; brake load; weight

w specific weight; velocity of whirl

x dryness fraction; length

Greek Symbols

α absorptivity

γ ratio of specific heats, c_p/c_v

$\in$ emissivity; effectiveness

η efficiency

θ temperature difference, angle

ρ density

σ Stefan-Boltzmann constant

ϕ relative humidity, angle.

Contents

Introduction to S.I. Units and Conversion Factors

A. INTRODUCTION TO S.I. UNITS

S.I., the international system of units is divided into three classes:

1. Base unit
2. Derived units
3. Supplementary units.

From the scientific point of view division of S.I. unit into these classes is to a certain extent arbitrary, because it is not essential to the physics of the subject. Nevertheless the General Conference, considering the advantages of a single, practical, world-wide system for international relations, for teaching and for scientific work, decided to base the international system on choice of six well-defined units given in Table 1 below:

Table 1. S.I. Base Units

Qaunitity	*Name*	*Symbol*
length	metre	m
mass	kilogram	kg
time	second	s
electric current	ampere	A
thermodynamic temperature	kelvin	K
luminous intensity	candela	cd
amount of substance	mole	mol

The second class of S.I. units contains derived units, i.e., units which can be formed by combining bases units according to the algebraic relations linking the corresponding quantities. Several of these algebric expressions in terms of base units can be replaced by special names and symbols can themselves be used to form other derived units.

Derived units may, therefore be classifed under three headings. Some of them are given in Tables 2, 3 and 4.

Table 2. Examples of S.I. Drived Units expressed in terms of Base Units

Quantity	*S.I. Unit*	
	Name	*Symbol*
area	square metre	m^2
volume	cubic metre	m^3
speed, velocity	metre per second	m/s
acceleration	metre per second squared	m/s^2
density, mass density	kilogram per cubic metre	kg/m^3
concentration (of amount of substance)	mole per cubic metre	mol/m^3
activity (radioactive)	1 per second	s^{-1}
specific volume	cubic metre per kilogram	m^3/kg
luminance	candela per square metre	cd/m^2

Table 3. S.I. Derived Units with Special Names

Quantity	*S.I. Unit*			
	Name	*Symbol*	*Expression in terms of other units*	*Expression in terms of S.I. bases units*
frequencey	hertz	Hz	—	s^{-1}
force	newton	N	—	$m.kg.s^{-2}$
pressure	pascal	Pa	N/m^2	$m^{-1}.kg.s^{-2}$
energy, work, quantity of heat power	joule	J	N.m	$m^2.kg.s^{-2}$
radiant flux, quantity of electricity	watt	W	J/s	$m^2.kg.s^{-1}$
electric charge	coloumb	C	A.s.	s.A
electric tension, electric potential	volt	V	W/A	$m^2.kg.s^{-3}.A^{-1}$
capacitance	farad	F	C/V	$m^{-2}.k^{g-1}.s^4$
electric resistance	ohm	Ω	V/A	$m^{-2}.kgs^{-3}.A^{-2}$
conductance	siemens	S	A/V	$m^{-2}.kg^{-1}.s^3.A^{-1}$
magnetic flux	weber	Wb	V.S	$m^2.kg.s^{-2}.A^{-1}$
magnetic flux density	tesla	T	Wb/m^2	$kg.s^{-2}.A^{-1}$
inductance	henry	H	Wb/A	$m^2kg.s^{-2}.A^{-2}$
iuminous flux	lumen	lm	—	cd.sr
illuminance	lux	lx	—	$m^{-2}.cd.sr$

Table 4. Examples of S.I. Derived Units expressed by means of Special Names

Quantity	*S.I. Unit*		
	Name	*Symbol*	*Expression in terms of S.I. base units*
dynamic viscosity	pascal second	Pa-s	$m^{-1}.kgs^{-1}$
moment of force	metre newton	N.m	$m^{-2}.kgs^{-2}$
surface tension	newton per metre	N/m	$kg.s^{-2}$
heat flux density, irradiance	watt per square metre	W/m^2	$kg.s^{-2}$
heat capacity, entropy	joule per kelvin	J/K	$m^{-2}.kg.s^{-2}.K^{-1}$
specific heat capacity, specific entropy	joule per kilogram kelvin	J/(kg.K)	$m^{-2}.s^{-2}.K^{-1}$
specific energy	joule per kilogram	J/kg	$m^{-2}.s^{-2}$
thermal conductivity	watt per metre kelvin	W/(m.K)	$m.kg.s^{-3}.K^{-1}$
energy density	joule per cubic metre	J/m^3	$m^{-1}.kg.s^{-2}$
electric field strength	volt per metre	V/m	$m.kg.s^{-3}.A^{-1}$
electric charge density	coloumb per cubic metre	C/m^3	$m^{-3}.s.A$
electric flux density	coloumb per square metre	C/m^2	$m^{-2}.s.A$
permitivity	farad per metre	F/m	$m^{-3}.kg^{-1}.s^4A^4$
current density	ampere per square metre	A/m^2	—
magnetic field strength	ampere per metre	A/m	—
permeability	henry per metre	H/m	$m.kg.s^{-2}.A^{-2}$
molar energy	joule per mole	J/mol	$m^{-2}.kg.s^{-2}mol^{-1}$
molar heat capacity	joule per mole kelvin	J/(mol.K)	$m^{-2}.kg.s^{-2}.K^{-1}mol^{-1}$

The S.I. units assigned to third class called "supplementary units" may be regarded either as base units or as derived units. Refer Table 5 and Table 6.

Table 5. S.I. Supplementary Units

Quantity	*S.I. Units*	
	Name	*Symbol*
plane angle	radian	rad
solid angle	steradian	sr

Table 6. Example of S.I. Derived Units formed by Using Supplementary Units

Quantity	*S.I. Units*	
	Name	*Symbol*
angular velocity	radian per second square	rad/s
angular acceleration	radian per second squared	rad/s^2
radiant intensity	watt per steradian	W/sr
radiance	watt per square metre steradian	$W\text{-}m^{-2}.sr^{-1}$

Table 7. S.I. Prefixes

Factor	*Prefix*	*Symbol*	*Factor*	*Prefix*	*Symbol*
10^{12}	tera	T	10^{-1}	deci	d
10^{9}	giga	G	10^{-2}	centi	c
10^{6}	mega	M	10^{-3}	milli	m
10^{3}	kilo	k	10^{-6}	micro	μ
10^{2}	hecto	h	10^{-9}	nano	n
10^{1}	deca	da	10^{-12}	pico	p
			10^{-15}	fasnot	f
			10^{-18}	atto	a

B. DEFINITIONS

The S.I. seven base units and two supplementary units are defined below:

(*i*) **Metre.** The metre is the length equal to 1, 650, 76373 wavelengths in vacuum of the radiation corresponding to the transition between the levels $2p_{10}$ and $5d_5$ of the krypton −86 atom.

(*ii*) **Kilogram.** One kilogram is equal to the mass of the international prototype of the kilogram.

(*iii*) **Second.** The second is the duration of 9, 192, 631, 770 periods of the radiation corresponding to the transition between the two hyperfine levels of the ground state of the cesium-133 atom.

(*iv*) **Ampere.** One ampere is that constant current which, if maintained in two straight parallel conductors of infinite length, of negligible circular cross-section and placed one meter apart in vacuum, would produce between these conductors a force equal to 2×10^{-7} newton per metre of length.

(*v*) **Kelvin.** The kelvin is the fraction $\frac{1}{273.16}$ of thermodynamic temperature of the triple point of water.

(*vi*) **Candela.** The candela is the luminous intensity, in the perpendicular direction, of a surface of a $\frac{1}{600,000}$ square metre of a black body at a temperature of freezing platinum under a pressure of 101, 325 newton per square metre.

(*vii*) **Mole.** The mole is the amount of substance of a system which contains as many elementary entities as there are atoms in 0.012 kg, carbon 12. When the mole is used, the elementary entities must be specified and may be atoms, molecules, ions, electrons, other particles or specified groups of such particles.

(*viii*) **Radian.** The radian is the plane angle between two radii of a circle that cut off on the circle an arc equal in length to the radius.

(*ix*) **Steradian.** The steradian is the solid angle which having its vertex in centre of a sphere, cuts off an area of the surface of the sphere equal to that of a square with sides of length equal to the radius of the sphere.

Some Other Definitions

Newton. The newton (N) is a derived unit of force and is defined as the unit of force which when acting on a mass of 1 kilogram gives it an acceleration of one metre per second per second. Since accelertion due to gravity equals 9.81 m/s^2, kilogram force equals 9.81 newtons.

Joule. The joule (J) is a derived unit of energy, work or quantity of heat and is defined as the work done when a force of one newton acts so as to cause a displacement of one metre. Energy is defined as the capacity to do work. A unit of energy in nuclear physics is the electron volt (eV) which is defined as the energy gained by an electron in rising through a potential difference of one volt.

$$1 \text{ eV} = 1.6021 \times 10^{-19} \text{ J.}$$

Watt. The watt (W) is a unit of power (i.e., rate of doing work)

$$\text{Power in watts} = \frac{\text{work (or energy) in joules}}{\text{time in seconds}}$$

Thus 1 watt equal 1 joule/sec.

$$\begin{aligned} 1 \text{ kilo watt-hour (kWh)} &= 1000 \text{ watt-hours} \\ &= 3600000 \text{ joules.} \end{aligned}$$

Coulomb. The coulomb (C) is the derived unit of charge. It is defined as *the quantity of electricity passing a given point in a circuit when a current of 1 A is maintained for 1 second.*

$$Q = I \, . \, t$$

Where Q = charge in coulombs,
I = current in amperes, and
t = time in second.

1 coulomb represents 6.24×10^{18} electrons.

Ohm. The ohm (Ω) is the unit of electric resistance and is defined as *the resistance in which a constant current of 1 A generates heat at the rate of 1 watt.*

Siemen. The siemen is a unit of electric conductance (i.e., reciprocal of resistance). If a circuit has a resistance of 5 ohms, its conductance is 0.2 siemens. A more commoly used name for siemen is mho (Ω).

Volt. The volt is a unit of potential difference and electomotive force. It is defined as *the difference of potential across a resistance of 1 ohm carrying a current of 1 ampere.*

Hertz. The hertz (Hz) is a unit of frequency. 1 Hz = 1 cycle per second.

Horse-power. It is a practical unit of mechanical output. BHP (British horse power or brake horse power) equals 746 watts. The metric horse power equals 735.5 watts. To avoid confusion between BHP and metric horse power, the mechnical output of machines in S.I. units, is expressed in watts or kilowatts.

C. SALIENT FEATURES OF S.I. UNITS

The salient features of S.I. units are as follows :

1. It is a coherent system of units, i.e., product or quotient of any two base quantities results in a unit resultant quantity. For example, unit length divided by unit time gives unit velocity.

2. It is a rationalised system of units, applicable to both, magnetism and electricity.
3. It is a non-gravitational system of units. It clearly distinguishes between the units of mass and weight (force) which are kilogram and newton respectively.
4. All the units of the system can be derived from the base and supplementary units.
5. The decimal relationship between units of same quantity makes possible to express any small or large quantity as a power of 10.
6. For any quantity there is one and only one S.I. unit. For example, joule is the unit of energy of all forms such as mechanical, heat, chemical, electrical and nuclear. However, kWh will also continue to be used as unit of electrical energy.

Advantages of S.I. Units:

1. Units for may different quantities are related through a series of simple and basic relationship.
2. Being an absolute system, it avoids the use of factor 'g' i.e. acceleration due to gravitiy in several expressions in physics and engineering which had been a nuisance in all numericals in physics and engineering.
3. Being a rationalised system, it ensures all the advantages of rationalised MKSA system in the fields of electricity, magnetism, electrical engineering and electronics.
4. Joule is the only sole unit of energy of all forms and watt is the sole unit of power hence a lot of labour is saved in calculations.
5. It is a coherent system of units and involves only decimal co-efficients. Hence it is very convenient and quick system for calculations.
6. In electricity, all the practical units like vol, ohm, ampere, henry, farad, coulomb, joule and watt accepted in industry and laboratories all over the worked for well over a century have become absolute in their own right in the S.I. system, without the need for any more practical units.

Disadvantages:

1. The non-S.I. time units 'minute' and 'hour' will still continue to be used until the clocks and watches are all changed to kilo seconds and mega seconds etc.
2. The base unit kilogram (kg) includes a prefix, which creates an ambiguity in the use of multipliers with gram.
3. S.I. units for energy, power and pressure (i.e., joule, watt and pascal) are too small to be expressed in science and technology, and, therefore, in such cases the use of larger units, such as MJ, kW, kPa will have to be made.
4. There are difficulties with regard to developing new S.I. units for apparent and reactive energy while joule is the accepted unit for active energy in S.I. systems.

D. CONVERSION FACTORS

1. Force:

$$1 \text{ newton} = \text{kg-m/sec}^2 = 0.012 \text{ kgf}$$
$$1 \text{ kgf} = 9.81 \text{ N}$$

2. Pressure:

$$1 \text{ bar} = 750.06 \text{ mm Hg} = 0.9869 \text{ atm} = 10^5 \text{ N/m}^2$$
$$= 103 \text{ kg/m-sec}^2$$
$$1 \text{ N/m}^2 = 1 \text{ pascal} = 10^{-5} \text{ bar} = 10^{-2} \text{ kg/m-sec}^2$$
$$= 1.01325 \times 10^5 \text{ N/m}^2$$

3. Work, Energy or Heat:

$$1 \text{ joule} = 1 \text{ newton metre} = 1 \text{ watt-sec}$$
$$= 2.7778 \times 10^{-7} \text{ kW-h} = 0.239 \text{ cal}$$
$$= 0.239 \times 10^{-3} \text{ kcal}$$
$$1 \text{ cal} = 4.184 \text{ joule} = 1.1622 \times 10^{-6} \text{ kWh}$$
$$1 \text{ kcal} = 4.184 \times 10^3 \text{ joule} = 427 \text{ kgf m}$$
$$= 1.1622 \times 10^{-3} \text{ kWh}$$
$$1 \text{ kwh} = 8.6042 \times 10^5 \text{ cal} = 860.42 \text{ kcal}$$
$$= 3.6 \times 10^6 \text{ joule}$$
$$1 \text{ kgf.m} = \left(\frac{1}{427}\right) \text{ kcal} = 9.81. \text{ joules}$$

4. Power:

$$1 \text{ watt} = 1 \text{ Joule/sec} = 0.86 \text{ kcal/h}$$
$$1 \text{ h.p.} = 75\text{m kgf/sec} = 0.1757 \text{ kcal/sec}$$
$$= 735.5 \text{ watt}$$
$$1\text{kW} = 1000 \text{ watts} = 860 \text{ kcal/h}$$

5. Specific beat:

$$1 \text{ kcal/kg-°K} = 0.4184 \text{ joules/kg-K}$$

6. Thermal conductivity:

$$1 \text{ watt/m-K} = 0.8598 \text{ kcal/h-m-°C}$$
$$1 \text{ kcal/h-m-°C} = 1.16123 \text{ watt/m-K}$$
$$= 1.16123 \text{ joules/s-m-K}$$

7. Heat transfer co-efficient:

$$1 \text{ watt/m}^2\text{-K} = 0.86 \text{ kcal/m}^2\text{-h-°C}$$
$$1 \text{ kcal/m}^2\text{-h-°C} = 1.163 \text{ watt/m}^2\text{-K}$$

The following conversion factors may be used to convert the quantities in non-SI. units into S.I. units.

To convert	*To*	*Multiply by*
angstroms	m	10^{-10}
atmospheres	kg/m^2	10332
bars	kg/m^2	1.02×10^4
Btu	joules	1054.8
Btu	kWh	2.928×10^{-4}
circular mils	m^2	5.067×10^{-10}
cubic feet	m^3	0.02831
dynes	newtones	10^{-5}
ergs	joules	10^{-7}
ergs	kWh	0.2778×10^{-13}
feet	m	0.3048
foot-pounds	joules	1.356
foot-pounds	kg-m	0.1383
gauss	tesla	10^{-4}
grams (force)	newton	9.807×10^{-3}
horse power (metric)	watts	735.5
lines/sq. inch	tesla	1.55×10^{-5}
Maxwell	webers	10^{-8}
mho	siemens	1
micron	metre	10^{-6}
miles	km	1.609
mils	cm	2.54×10^{-3}
poundals	newton	0.1383
pounds	kilogram	0.454
pounds (force)	newtons	0.448
pounds/sq. ft.	N/m^2	47.878
pounds/sq. inch.	N/m^2	6894.43

E. IMPORTANT ENGINEERING CONSTANTS AND EXPRESSIONS IN S.I. UNITS

Engineering constants and expressions	*M.K.S. system*	*S-I units*
1. Value of g_0	9.81 kg-m/kgf-sec^2	1 kg-m/N-sec^2
2. Universal gas constant	848 kgf-m/kg mole-°K	$848 \times 9.81 = 8341$ J/kg-mole-°C (∵ 1 kgf-m = 9.81 joules)
3. Gas constant (R)	29.27 kgf m/kg-°K	$\frac{8314}{29} = 287$ joules/kg-K
4. Specific heat (for air)	for air $c_v = 0.17$ kcal/kg-°K $c_v = 0.24$ kcal/kg-°K	for air $c_v = 0.17 \times 4.184 = 0.71128$ kJ/kg-K $c_p = 0.24 \times 4.184 = 1$ kJ/kg-K
5. Flow through nozzle-Exit velocity (C_2)	91.5 U where U is in kcal	$44.7 \sqrt{U}$ where U is in kJ
6. Refrigeration 1 ton	= 50 kcal/min	= 210 kJ/min
7. Heat transfer The Stefan Boltzmann Law is given by :	$Q = \sigma T^{-1}$ kcal/m^2-h where $\sigma = 4.9 \times 10^{-8}$ kcal/h-m^2-°K^4	$Q = \sigma T^4$ kcal/m^2-h where $\sigma = 5.67 \times 10^{-8}$ w/m^{2}K^4

F. DIMENSIONS OF QUANTITIES

Different units can be represented dimensionally in terms of units of length L, mass M, time T and current 1. The dimentions can be derived as under:

1. Velocity = length/time = L/T = LT^{-1}
2. Acceleration = velocity/time = $LT^{-1}/T = LT^{-2}$
3. Force = mass × acceleration = MLT^{-2}
4. Charge (coluomb) = current × time = IT
5. Work or energy = force × distance = ML^2T^{-2}
6. EMF or potential = work/charge = $ML^2T^{-2}/IT = ML^2I^{-1}T^{-3}$
7. Power = work/time = $ML^2T^{-2}/T = ML^2T^{-3}$
8. Current density = current/area = $I/L^2 = IL^{-2}$
9. Resistance = emf/current = $ML^2I^{-1}T^{-3}/I = ML^2I^{-2}T^{-3}$
10. Electric flux density = electric flux or charge/area = $IT/L^2 = ITL^{-2}$
11. MMF = current × number of turns = I
12. Conductance = 1/resistance = $1/ML^2I^{-2}T^{-3} = I^2T^3m^{-1}L^{-2}$
13. Electric field intensity = vol/metre = $ML^2I^{-1}T^{-3}/L = MLI^{-1}T^{-3}$
14. Resistivity = $\frac{\text{resistant} \times \text{area}}{\text{length}} = (ML^2I^{-2}T^{-3})\,(L^2)/L = ML^3I^{-2}T^{-3}$
15. Magnetic field intensity (H) = MMF/length = $I/L = IL^{-1}$
16. Magnetic flux = emf × time = $(ML^2I^{-1}T^{-3})\,(T) = ML^2I^{-1}T^{-2}$
17. Magnetic flux intensity = magnetic flux/area = $(ML^2I^{-1}T^{-2})/L^2 = MI^{-1}T^{-2}$
18. Impedence = emf/current = $ML^2I^{-2}T^{-3}$
19. Admittance = 1/impedence = $I^2T^3M^{-1}L^{-2}$
20. Inductance = magnetic flux/current = $ML^2T^{-2}I^{-1}/I = ML^2T^{-2}I^{-2}$
21. Capacitance = electric charge/potential = $IT/ML^2T^{-3}I^{-1} = M^{-1}L^{-2}T^4I^2$

❑❑❑

PART I: REFRIGERATION

Chapters:

Chapter 1

Fundamentals of Refrigeration

1.1. INTRODUCTION TO "REFRIGERATION"

- **Refrigeration** *is the science of the producing and maintaining temperatures below that of the surrounding atmosphere.* This means the removing of heat from a substance to be cooled. Heat always passes downhill, from a warm body to a cooler one, until both bodies are at the same temperature. Not only perishables today many human work spaces in offices and factory buildings are air-conditioned and a refrigeration unit is the heart of the system.
- Before the advent of mechanical refrigeration water was kept cool by storing it in semiporous jugs so that the water could seep through and evaporate. The evaporation carried away heat and cooled the water. This system was used by the Egyptians and by Indians in the Southwest. Natural ice from lakes and rivers was often cut during winter and stored in caves, straw-lined pits, and later in sawdust insulated buildings to be used as required. The Romans carried pack trains of snow from Alps to Rome for cooling the Emperor's drinks. Though these methods of cooling all make use of natural phenomena, they were used to maintain a lower temperature in a space or product and may properly be called refrigeration.
- In simple, **refrigeration** *means the cooling or removal of heat from a system.* The equipment employed to maintain the system at a low temperature is termed as **refrigerating system** and the system which is *kept at lower temperature is called* **refrigerated system**.
- Refrigeration is generally produced in one on the following *three ways*:

  (*i*) By melting of a solid.

  (*ii*) By sublimation of a solid, and

  (*iii*) By evaporation of a liquid.

- Most of the commercial refrigeration is produced by the evaporation of a liquid called *refrigerant*. **Mechanical refrigeration** depends upon the evaporation of liquid refrigerant and its circuit includes the equipments naming evaporator, compressor, condenser and expansion valve. It is used for preservation of food, manufacture of ice, solid carbon-dioxide and control of air temperature and humidity in the air-conditioning system.

1.2 APPLICATIONS OF REFRIGERATION

Important **refrigeration applications** are given below:

1. Ice making
2. Transportation of foods above and below freezing
3. Industrial air-conditioning
4. Comfort air-conditioning
5. Chemical and related industries
6. Medical and surgical aids
7. Processing food products and beverages
8. Oil refining and synthetic rubber manufacturing
9. Manufacturing and treatment of metals
10. Freezing food products
11. Miscellaneous applications:
    (*i*) Extremely low temperatures
    (*ii*) Plumbing
    (*iii*) Building construction etc.

1.3 ELEMENTS OF REFRIGERATION SYSTEMS

All refrigeration systems must include atleast **four basic** units as given below :

(*i*) A low temperature thermal "sink" to which heat will flow from the space to be cooled.

(*ii*) Means of extracting energy from the sink, raising the temperature level of this energy, and delivering it to a heat receiver.

(*iii*) A receiver to which heat will be transferred from the high temperature high pressure refrigerant.

(*iv*) Means of reducing pressure and temperature of the refrigerant as it returns from the receiver to the "sink".

1.4 REFRIGERATION SYSTEMS

The various refrigeration systems may be enumerated as below:

1. Ice refrigeration
2. Air-refrigeration system
3. Vapour compression refrigeration system
4. Vapour absorption refrigeration system

5. Special refrigeration systems:
   (*i*) Absorption refrigeration system
   (*ii*) Cascade refrigeration system
   (*iii*) Mixed refrigeration system
   (*iv*) Vortex tube refrigeration system
   (*v*) Thermoelectric refrigeration
   (*vi*) Steam jet refrigeration system.

1.5 CO-EFFICIENT OF PERFORMANCE (C.O.P.)

The performance of a refrigeration system is expressed by a term known as the **Co-efficient of performance** *which is defined as the ratio of heat absorbed by the refrigerant while passing through the evaporator to the work input required to compress the refrigerant in the compressor; in short it is the ratio between heat extracted and work done (in heat units).*

If, R_n = net refrigerating effect, and W = work expended in by the machine during the same interval to time. Then

$$\text{C.O.P.} = \frac{R_n}{W}$$

$$\textbf{Relative C.O.P.} = \frac{\text{Actual C.O.P.}}{\text{Theoretical C.O.P.}}$$

where, actual C.O.P. = ratio of R_n and W actually measured during a test and, theoretical C.O.P. = ratio of theoretical values of R_n and W obtained applying laws of thermodynamics to the refrigeration cycle.

1.6 STANDARD RATING OF A REFRIGERATION MACHINE

The rating of a refrigerating machine is obtained by refrigerating effect or amount of heat extracted in a given time from a body. The rating of the refrigeration machine is given by a unit of refrigeration known as **standard commercial tonne of refrigeration** *which is defined as the refrigerating effect (RE) produced by the melting of 1 tonne of ice from and at 0°C in 24 hours.* Since the latent heat of fusion of ice is 336 kJ/kg, the refrigerating effect of 336 × 1000 kJ/kg in 24 hours is rated as **one tonne**, i.e.,

$$\textbf{A tonne of refrigeration} = \frac{336\times1000}{24} = 14000 \text{ kJ/h.}$$

> **Note. Ton of refrigeration (TR).** A ton of refrigeration is basically an American unit of RE. It originated from the rate at which heat is required to be removed to freeze *one ton of water from and at 0°C.* Using American units this is equal to removal of 200 BTU of heat per minute, and in MKS units it is adopted as 50 kcal/min or 3000 kcal/hour. In S.I. units its conversion is rounded off to 3.5 kJ/s (kW) or 210 kJ/min.

In this book we shall be using *1 tonne of refrigeration* = 14000 kJ/h.

(1 ton of refrigeration may be taken equal to 0.9 tonne of refrigeration)

1.7 ICE REFRIGERATION

- For several years, melting ice has been used successfully as refrigerant. Not too many years ago ice was the only cooling agent available for use in domestic and small commercial refrigeration. Fig. 1.1. shows a typical ice refrigerator. In this refrigerator the heat entering the refrigerated space from all various sources reaches the melting ice primarily by convection current set up in the air of refrigerated convection space. The air in contact with the warm product and walls of the space is heated by heat conducted to it from these materials. As the air is warmed it expands and rises to the top of the space carrying the heat with it to the ice compartment. In passing over the ice the air is cooled as heat is conducted from the air to the ice. On cooling, the air becomes more dense and falls back into the storage space, where upon it absorbs more heat and the cycling continues. This air acts as a heat transfer agent in carrying the heat from the walls and stored product to the melting ice.

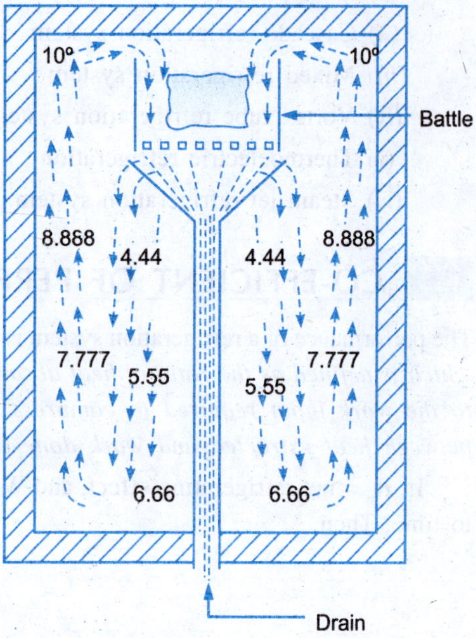

Fig. 1.1. Ice refrigerator.

In order that there may be adequate air circulation within the refrigerated space, the ice should be located near the top of the refrigerator and proper baffling should be installed to provide direct and unrestricted paths of air flow. To collect water which results from the melting, a *drip pan must be located beneath the ice.*

Ice entails certain **disadvantages** *which tend to limit its usefulness as a refrigerant.* These are as follows:

1. With ice it is not possible to obtain low temperatures required in many refrigeration applications. Ordinarily, 0°C is the minimum temperature obtainable through the melting of ice alone. In some cases, the inlet temperature of ice can be lowered to approximately –18°C by adding sodium chloride or calcium chloride to produce a freezing mixture.
2. There is necessity of frequently replenishing the supply; a practice which is neither convenient nor economical. In addition there is a problem of disposing of water resulting from the melting.
3. A difficulty is experienced in controlling the rate of refrigeration, which in turn makes it difficult to maintain the desired low temperature level within the refrigerated space.

Inspite of the fact ice has several disadvantages ice is preferable to *mechanical refrigeration in the following applications*:

1. Fresh vegetables, fish and poultry are often packed and shipped in cracked ice to prevent dehydration and to preserve appearance.
2. Ice has tremendous eye appeal and can be used to considerable advantages in the displaying and serving of certain foods such as salads, cocktails, etc. and in chilling beverages.

Note. Other refrigeration systems are dealt in subsequent chapters.

HIGHLIGHTS

1. *Refrigeration* is the science of producing and maintaining temperatures below that of the surrounding atmosphere.
2. Refrigeration is generally produced in one of the following *three ways*:
   (*i*) By melting a solid;
   (*ii*) By sublimation of a solid, and
   (*iii*) By evaporation of liquid.
3. *Co-efficient of performance (C.O.P.)* is defined as the ratio of heat absorbed by the refrigerant while passing through the evaporator to the work input required to compress the refrigerant in the compressor; in short it is the ratio between heat extracted and work done (in heat units).
4. *Relative C.O.P.* $= \dfrac{\text{Actual C.O.P.}}{\text{Theoretical C.O.P.}}$
5. The refrigerating effect of 14000 kJ/h is called *standard rating of a refrigeration machine.*

OBJECTIVE TYPE QUESTIONS

Fill in the Blanks

1. .... means the cooling of or removal of heat from a system.
2. Most of the commercial refrigeration is produced by the evaporation of a liquid..........
3. .... is the ratio between the heat extracted and the work done.
4. .... $= \dfrac{\text{Actual C.O.P.}}{\text{Theoretical C.O.P.}}$

ANSWERS

1. Refrigeration	2. Refrigerant	3. C.O.P.	4. Relative C.O.P.

THEORETICAL QUESTIONS

1. Define the following:
   (*i*) Refrigeration,
   (*ii*) Refrigeration system, and
   (*iii*) Refrigerated system.
2. Enumerate different ways of producing refrigeration.
3. Enumerate important refrigeration applications.
4. State elements of refrigeration systems.
5. Enumerate systems of refrigeration.
6. Define the following:
   (*i*) Actual C.O.P.
   (*ii*) Theoretical C.O.P., and
   (*iii*) Relative C.O.P.
7. What is standard rating of a refrigeration machine?
8. Write short note on ice refrigeration.

Chapter 2

Air-Refrigeration Cycles and Systems

2.1 INTRODUCTION

Air cycle refrigeration is one of the earliest methods of cooling developed. It became obsolete for several years because of its *low* coefficient of performance (C.O.P.). It has, however, found its application again in *aircraft refrigeration systems,* where with low equipment weight, it can utilise a portion of the cabin air according to the supercharger capacity.

The *main characteristic feature* of air-refrigeration system is that *throughout the cycle the refrigerant remains in gaseous state.*

The air-refrigeration system can be divided in *two systems:*

1. Closed system.
2. Open system.

1. Closed system (or Dense air machine). In this system the *air refrigerant is contained within the piping or components/parts of the system at all times and with refrigerator usually pressure is above atmospheric pressure.*

A closed system has the following *thermodynamic advantages:*

(*i*) It can work at suction pressure higher than atmospheric. This reduces the volume handled by the compressor and expander.

(*ii*) The operating pressure ratio can be reduced, resulting in higher C.O.P.

2. Open system. In this system the refrigerator is replaced by the actual space to be cooled with the air expanded to atmospheric pressure, circulated through the cold room and then compressed to the cooler pressure. The pressure of operation in this system is *inherently limited to operation at atmospheric pressure in the refrigerator.*

In this system, the refrigeration is obtained by three basic steps *viz. compression, cooling and expansion*, accompanied by expraction of work. The air after expansion is directly led to the conditioned space. It is therefore necessary that air be expanded to one atmosphere pressure. This requires larger volumes to be handled. Notwithstanding this, the sizes of the compressor and expander, now a days, are not affected significantly since both of them are turbo-type. This system, therefore, has an *advantage over closed system*, in respect that it *does not require a heat exchange for refrigeration process. This saves the weight and cost of the equipment.*

Advantages of closed system over open system:

1. In a *closed system* the sunction to compressor may be at high pressure. The sizes of expander and compressor can be kept within reasonable limits by using dense air.
2. In *open air system* the *air picks up moisture from the products kept in the refrigerated chamber, the moisture may freeze during expansion and is likely to choke the valves. Thus a drier in the circuit is required whereas it does not happen in closed system.*
3. C.O.P. is higher.
4. In open system, the expansion of the refrigerant can be varied only upto atmospheric pressure prevailing in the cold chamber but for a *closed system there is no such restriction.*

2.2 AIR-REFRIGERATOR WORKING ON A REVERSED CARNOT CYCLE

If a machine working on reversed Carnot cycle is driven from an external source, it will work or function as a **refrigerator**. The production of such a machine has not been possible practically *because the adiabatic portion of the stroke would need a high speed while during isothermal portion of the stroke a very low speed will be necessary. This variation of speed during the stroke, however is not practicable.*

p-V and T-*s* diagrams of reversed Carnot cycle are shown in Fig. 2.1 (*a* and *b*).

- Starting from point, the clearance space of the cylinder is full of air, the air is then *expanded adiabatically* to point *p* during which its temperature falls from T_1 to T_2, the cylinder is put in contact with a cold body at temperature T_2.
- The air is then *expanded isothermally* to the point *n*, as a result of which *heat is extracted* from the cold body at temperature T_2.
- Now the *cold body is removed*; from *n* to *m* air undergoes *adiabatic compression* with the assistance of the some external power and temperature rises to T_1.
- A hot body at temperature T_1 is put in contact with the cylinder.
- Finally the air is *compressed isothermally* during which process *heat is rejected to the hot body.*

Refer to Fig. 2.1 (*b*)

Heat abstracted from the cold body = area '*npqs*' = $T_2 \times pn$

Work done per cycle = area '*lpnm*' = $(T_1 - T_2) \times pn$

Co-efficient of performance,

$$\text{C.O.P.} = \frac{\text{Heat extracted from the cold body}}{\text{Work done per cycle}} = \frac{\text{Area ' }npqs\text{'}}{\text{Area ' }lpnm\text{'}}$$

$$= \frac{T_2 \times pn}{(T_1 - T_2) \times pn} = \frac{T_2}{T_1 - T_2} \quad \text{...(2.1)}$$

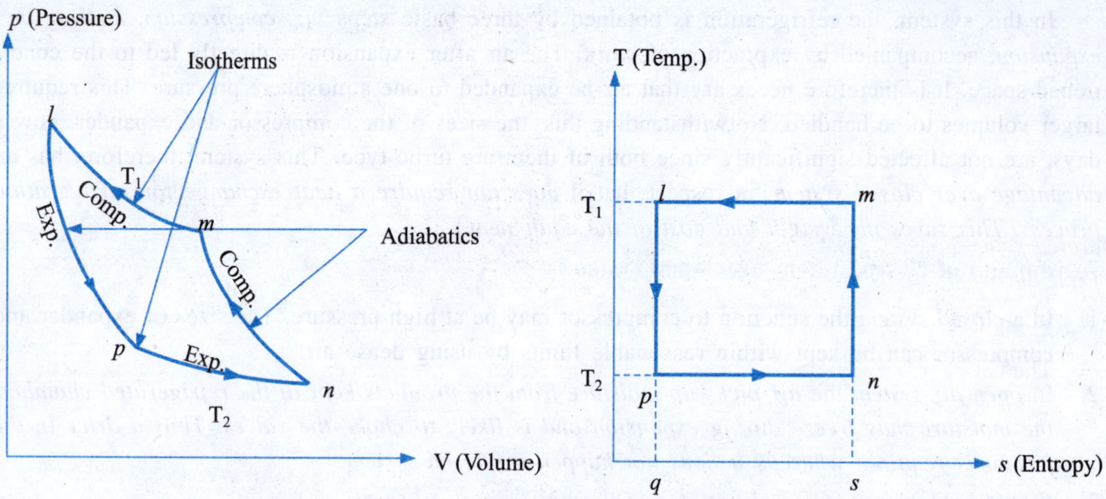

(a) p-V diagram (b) T-s diagram

Fig. 2.1. Reversed Carnot cycle.

Since the co-efficient of performance (C.O.P.) means the *ratio of the desired effect in kJ/kg to the energy supplied in kJ/kg*, therefore, C.O.P. in case of Carnot cycle run either as a refrigerating machine or a heat pump or as a heat engine will be as given below:

(i) For a reversed Carnot cycle 'refrigerating machine':

$$(\text{C.O.P.})_{ref.} = \frac{\text{Heat extracted from the cold body / cycle}}{\text{Work done per cycle}}$$

$$= \frac{T_2 \times pn}{(T_1 - T_2) \times pn} = \frac{T_2}{T_1 - T_2} \quad \text{...(2.2)}$$

(ii) For a Carnot cycle 'heat pump':

$$(\text{C.O.P.})_{heat\ pump} = \frac{\text{Heat rejected to the hot body / cycle}}{\text{Work done per cycle}}$$

[C.O.P. for heat pump is also called **E.P.R.** *i.e.*, ***Energy performance ratio***]

$$= \frac{T_1 \times lm}{(T_1 - T_2) \times pn} = \frac{T_1 \times pn}{(T_1 - T_2) \times pn} \quad (\because lm = pn)$$

$$= \frac{T_1}{T_1 - T_2} \quad \text{...(2.3)}$$

$$= 1 + \frac{T_2}{T_1 - T_2} \quad [\text{i.e., } 1 + (\text{C.O.P.})_{ref.}] \quad \text{...(2.4)}$$

This indicates that C.O.P. of heat pump is grater than that of a refrigerator working on reversed Carnot cycle between the same temperature limits T_1 and T_2 by **unity.**

(iii) For a Carnot cycle 'heat engine':

$$(\text{C.O.P.})\ \text{heat engine} = \frac{\text{Work obtained / cycle}}{\text{Heat supplied / cycle}} = \frac{(T_1 - T_2) \times pn}{T_1 \times lm}$$

$$= \frac{(T_1 - T_2) \times pn}{T_1 \times pn} \qquad (\because lm = pn)$$

$$= \frac{T_1 - T_2}{T_1} \qquad ...(2.5)$$

Example 2.1. *An inventor claims to have developed a refrigerating unit which maintains the refrigerated space at –5°C when operating in a room where temperature is 26°C and has a C.O.P. = 8.4. Find out whether his claim is correct or not.*

Solution: *Given:* $T_2 = 273 + (-5) = 268$ K; $T_1 = 273 + 26 = 299$ K; C.O.P. = 8.4.

The C.O.P. *of the refrigerating unit will be maximum when it is working on reserved Carnot cycle.*

$$\therefore \quad \text{Maximum possible C.O.P.} = \frac{T_2}{T_1 - T_2} = \frac{268}{299 - 268} = 8.64$$

Since the C.O.P. claimed by the invertor is *more* than the maximum possible C.O.P., hence the claim of the inventor is **incorrect.**

Example 2.2. *A certain machine works on reversed Carnot cycle between temperature limits of –10°C and 27°C. Find its C.O.P., when working as:*

(i) A refrigerating machine.

(ii) An heat pump.

(iii) An heat engine.

Solution: *Given:* $T_2 = 273 + (-10) = 263$ K; $T_1 = 273 + 27 = 300$ K;

C.O.P.:

(i) ***As a refrigerating machine:***

$$\text{C.O.P.} = \frac{T_2}{T_1 - T_2} = \frac{263}{300 - 263} = \mathbf{7.11} \quad \textbf{(Ans.)}$$

(ii) ***As a heat pump:***

$$\text{C.O.P.} = \frac{T_1}{T_1 - T_2} = \frac{300}{300 - 263} = \mathbf{8.11} \quad \textbf{(Ans.)}$$

(iii) ***As an heat engine:***

$$\text{C.O.P.} = \frac{T_1 - T_2}{T_1} = \frac{300 - 263}{300} = \mathbf{0.123} \quad \textbf{(Ans.)}$$

Example 2.3. *A reserved Carnot cycle is used for heating and cooling purpose. If the work supplied is 9.5 kW and C.O.P. is 3.6 for cooling, find:*

(i) T_2/T_1.

(ii) The refrigerating effect in tonnes of refrigeration.

(iii) C.O.P. for heating.

Solution: *Given:* Work supplied, W = 9.5 kW; C.O.P. = 3.6

(*i*) $\mathbf{T_2/T_1}$:

We know that, $$(\text{C.O.P.})_{cooling} = \frac{T_2}{T_1 - T_2} \quad \text{or} \quad \frac{T_1 - T_2}{T_2} = \frac{1}{\text{C.O.P.}} = \frac{1}{3.6}$$

$$\text{or} \quad \frac{T_1}{T_2} - 1 = \frac{1}{3.6} \quad \text{or} \quad \frac{T_1}{T_2} = 1 + \frac{1}{3.6} = 1.278$$

$$\therefore \quad \frac{T_2}{T_1} = \mathbf{0.782} \quad \textbf{(Ans.)}$$

(*ii*) **Refrigerating effect, R_n:**

$$\text{C.O.P.} = \frac{R_n}{W} \quad \text{or} \quad 3.6 = \frac{R_n}{9.5}$$

$$\therefore \quad R_n = 3.6 \times 9.5 \text{ kJ/s} = 3.6 \times 9.5 \times 3600 \text{ kJ/h}$$

$$= \frac{3.6 \times 9.5 \times 3600}{14000} = \mathbf{8.794\ tonnes} \quad \textbf{(Ans.)}$$

(*iii*) $$(\text{C.O.P.})_{heating} = \frac{T_1}{T_1 - T_2} = \frac{1}{1 - \frac{T_2}{T_1}} = \frac{1}{1 - 0.782} = \mathbf{4.587} \quad \textbf{(Ans.)}$$

Example 2.4. *A refrigerating machine working on reversed Carnot cycle consumes 5.5 kW for producing refrigerating effect of 940 kJ/min for maintaining a region at –38°C. Determine:*

(*i*) *C.O.P. of refrigerating machine;*

(*ii*) *Higher temperature of cycle;*

(*iii*) *Amount of heat delivered in kJ/min. when this device is used as an heat pump.*

Solution: *Given:* W = 5.5 kW; R_n = 940 kJ/min; t_2 = – 38°C

(*i*) **C.O.P. of the machine:**

$$T_2 = 273 + (-38) = 235 \text{ K}$$

$$\text{C.O.P. of the machine} = \frac{R_n \text{ (net refrigerating effect)}}{W \text{ (work done)}} = \frac{(940/60)}{5.5} = \mathbf{2.85} \quad \textbf{(Ans.)}$$

(*ii*) **Higher temperature of the cycle:**

$$\text{Also, C.O.P.} = \frac{T_2}{T_1 - T_2}$$

$$\therefore \quad 2.85 = \frac{235}{T_1 - 235} \quad \text{or} \quad T_1 = \frac{235}{2.85} + 235 = \mathbf{317.5\ K\ (or\ 44.5°C)} \quad \textbf{(Ans.)}$$

(*iii*) **Heat delivered as heat pump:**

Heat delivered = Heat absorbed + Work done

= 940 + 5.5 (kJ/s) × 60 = **1270 kJ/min** **(Ans.)**

Example 2.5. *A Carnot refrigerator requires 1.3 kW per tonne of refrigeration to maintain a region at low temperature of –38°C. Determine:*

(i) C.O.P. of Carnot refrigerator;

(ii) Higher temperature of the cycle;

(iii) The heat delivered and C.O.P. when this device is used as heat pump.

Solution: *Given:* $T_2 = 273 - 38 = 235$ K; Power required per tonne of refrigeration = 1.3 kW

(*i*) **C.O.P. of Carnot refrigerator:**

$$\text{C.O.P.}_{\text{(Carnot ref.)}} = \frac{\text{Heat absorbed}}{\text{Work done}} = \frac{1 \text{ tonne}}{1.3} = \frac{14000 \text{ (kJ/h)}}{1.3 \times 60 \times 60 \text{ (kJ/h)}} = \mathbf{2.99} \quad \textbf{(Ans.)}$$

(*ii*) **Higher temperature of the cycle, T_1:**

$$\text{C.O.P.}_{\text{(Carnot ref.)}} = \frac{T_2}{T_1 - T_2}$$

i.e.,
$$2.99 = \frac{235}{T_1 - 235}$$

$$\therefore \quad T_1 = \frac{235}{2.99} + 235$$

$$= 313.6 \text{ K}$$

$$= 313.6 - 273$$

$$= \mathbf{40.6°C} \quad \textbf{(Ans.)}$$

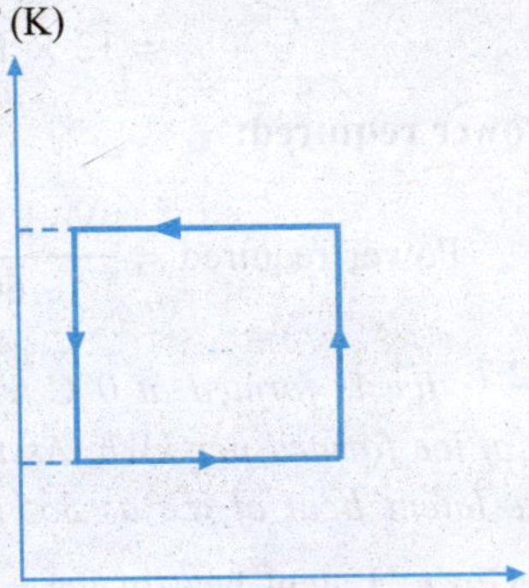

Fig. 2.2

(*iii*) **Heat delivered as heat pump:**

Heat delivered as heat pump = Heat absorbed + Work done

$$= \frac{14000}{60} + 1.3 \times 60 = \mathbf{311.3 \text{ kJ/min}} \quad \textbf{(Ans.)}$$

$$\text{C.O.P.}_{\text{(heat pump)}} = \frac{\text{Heat delivered}}{\text{Work done}} = \frac{311.3}{1.3 \times 60} = \mathbf{3.99} \quad \textbf{(Ans.)}$$

Example 2.6. *A refrigerating system operates on the reversed carnot cycle. The higher temperature of the refrigerant in the system is 35°C and the lower temperature is –15°C. The capacity is to be 12 tonnes. Neglect all losses. Determine:*

(i) Co-efficient of performance;

(ii) Heat rejected from the system per hour;

(iii) Power required.

Solution: *Given:* $T_1 = 273 + 35 = 308$ K; $T_2 = 273 - 15 = 258$ K;

Capacity = 12 tonnes.

(*i*) **C.O.P.:**

$$\text{C.O.P.} = \frac{T_2}{T_1 - T_2} = \frac{258}{308 - 258} = \mathbf{5.16} \quad \textbf{(Ans.)}$$

(*ii*) **Heat rejected from the system per hour:**

$$\text{C.O.P.} = \frac{\text{Refrigerating effect}}{\text{Work input}}$$

$$5.16 = \frac{12 \times 14000 \text{ kJ/h}}{\text{Work input}}$$

$$\therefore \quad \text{Work input} = \frac{12 \times 14000}{5.16} = 32558 \text{ kJ/h}$$

Thus, **heat rejected/hour**

$$= \text{Refrigerating effect/hour} + \text{Work input/hour}$$

$$= 12 \times 14000 + 32558 = \mathbf{200558\ kJ/h} \quad \textbf{(Ans.)}$$

(*iii*) **Power required:**

$$\text{Power required} = \frac{\text{Work input / hour}}{60 \times 60} = \frac{32558}{60 \times 60} \; \mathbf{9.04\ kW} \quad \textbf{(Ans.)}$$

Example 2.7. *Ice is formed at 0°C from water at 20°C. The temperature of the brine is –8°C. Find out the kg of ice formed per kWh. Assume that the refrigeration cycle used is perfect reversed Carnot cycle. Take latent heat of ice as 335 kJ/kg.*

Solution: *Given:* Latent heat of ice = 335 kJ/kg; $T_1 = 20 + 273 = 293$ K; $T_2 = -8 + 273 = 265$ K

$$\text{C.O.P.} = \frac{T_2}{T_1 - T_2} = \frac{265}{293 - 265} = 9.46$$

Heat to be extracted per kg of water (to form ice at 0°C i.e., 273 K),

$$R_n = 1 \times c_{pw} \times (293 - 273) + \text{Latent heat of ice}$$

$$= 1 \times 4.18 \times 20 + 335 = 418.6 \text{ kJ/kg}$$

Also, $1 \text{ kWh} = 1 \times 3600 = 3600 \text{ kJ}$

Also, $$\text{C.O.P.} = \frac{R_n}{W} = \frac{\text{Refrigerating effect in kJ / kg}}{\text{Work done in kJ}}$$

$$\therefore \quad 9.46 = \frac{m_{ice} \times 418.6}{3600}$$

$$\text{i.e., } m_{ice} = \frac{9.46 \times 3600}{418.6} = 81.35 \text{ kg}$$

Hence ice formed per kWh = **81.35 kg** **(Ans.)**

Example 2.8. *The capacity of a refrigerator is 450 tonnes when working between –15°C and 30°C. Find out the mass of ice produced within 24 hours when water is supplied at 20°C. Also, find out the minimum power required.*

Assume the machine to be working on Carnot cycle. Take latent heat of ice as 335 kJ/kg.

Solution: *Given:* Capacity of refrigerator, R_n = 450 tonnes.

$$T_2 = 273 + (-15) = 258 \text{ K}; \; t_1 = 273 + 30 = 303 \text{ K}; \; t_{water} = 20°\text{C}.$$

Mass of ice produced, m_{ice}:

Heat to be extracted per kg of water (to form ice at 0°C)

$$= 1 \times c_{pw} \times (20 - 0) + \text{Latent heat of ice}$$

$$= 1 \times 4.18 \times (20 - 0) + 335 = 418.6 \text{ kJ}$$

But, capacity of the plant, $R_n = 450 \times 14000$ kJ/h

∴ Mass of ice produced in 24 hours,

$$m_{ice} = \frac{(450 \times 14000) \times 24}{418.6 \times 1000} = \mathbf{361.2 \text{ tonnes}} \quad \textbf{(Ans.)}$$

Power required:

$$\text{C.O.P.} = \frac{T_2}{T_1 - T_2} = \frac{258}{303 - 258} = 5.733$$

Also, $$\text{C.O.P.} = \frac{R_n}{W}, \quad \text{or,} \quad 5.733 = \frac{450 \times 14000}{W}$$

or, $$W = \frac{450 \times 14000}{5.733} = 1098901 \text{ kJ/h, or, } 305.25 \text{ kJ/s}$$

∴ Power required = **305.25 kW (Ans.)**

Example 2.9. *Find the least power of a perfect reversed heat engine that makes 400 kg of ice per hour at –8°C from feed water at 18°C. Assume specific heat of ice as 2.09 kJ/kg K and latent heat 334 kJ/kg.*

Solution: *Given:* $T_1 = 18 + 273 = 291$ K; $T_2 = -8 + 273 = 265$ K

$$\text{C.O.P.} = \frac{T_2}{T_1 - T_2} = \frac{265}{291 - 265} = 10.19$$

Heat absorbed per kg of water (to form ice at – 8°C i.e,, 265 K),

$$R_n = 1 \times 4.18\,(291 - 273) + 334 + 1 \times 2.09 \times (273 - 265)$$

$$= \mathbf{425.96 \text{ kJ/kg}}$$

Also, $$\text{C.O.P.} = \frac{\text{Net refrigerating effect}}{\text{Work done}} = \frac{R_n}{W}$$

i.e., $$10.19 = \frac{425.96 \times 400}{W}$$

∴ $$W = \frac{425.96 \times 400}{10.19} = 16720.7 \text{ kJ/h} = 4.64 \text{ kJ/s, or, } 4.64 \text{ kW}$$

Hence least power = **4.64 kW (Ans.)**

Example 2.10. *The capacity of the refrigerator, working on reversed Carnot cycle, is 280 tonnes when operating between –10°C and 25°C. Determine:*

(i) Quantity of ice produced within 24 hours when water is supplied at 20°C;

(ii) Minimum power (in kW) required.

Assume latent heat of ice as 335 kJ/kg.

Solution: (*i*) **Quantity of ice produced:**

Heat to be extracted per kg of water (to form ice at 0°C)

$$= 4.18 \times 20 + 335 = 418.6 \text{ kJ/kg}$$

Heat extraction capacity of the refrigerator = 280 tonnes = 280 × 14000 = 3920000 kJ/h

∴ Quantity of ice produced in 24 hours

$$m_{ice} = \frac{3920000 \times 24}{418.6 \times 1000} = \mathbf{224.75\ tonnes} \quad \textbf{(Ans.)}$$

(*ii*) **Minimum power required:**

$$T_1 = 25 + 273 = 298 \text{ K}; \quad T_2 = -10 + 273 = 263 \text{ K}$$

$$\text{C.O.P.} = \frac{T_2}{T_1 - T_2} = \frac{263}{298 - 263} = 7.51$$

$$\text{Also,} \quad \text{C.O.P.} = \frac{\text{Net refrigerating effect}}{\text{Work done}} = \frac{R_n}{W}$$

$$\text{i.e.,} \quad 7.51 = \frac{3920000}{W}$$

$$\therefore \quad W = \frac{3920000}{7.51} \text{ kJ/h} = \frac{3920000}{7.51 \times 3600} \text{ kJ/s} = 145 \text{ kJ/s}$$

∴ Power required = **145 kW** **(Ans.)**

Example 2.11. *Find the least power of a perfect reversed heat engine that makes 500 kg of ice per hour at –4°C from feed water at 15°C. Take the specific heat of ice as 2.095 and latent heat as 335 kJ/kg.*

Solution: *Given:* m_{ice} = 500 kg/hour; $T_2 = 273 + (-4) = 269$ K; $T_1 = 273 + 15 = 288$ K;

Specific heat of ice = 2.095; Latent heat of ice = 335 kJ/kg

Least power of the heat engine:

$$\text{C.O.P.} = \frac{T_2}{T_1 - T_2} = \frac{269}{288 - 269} = 14.16$$

Heat to be extracted per kg of water (to form ice at – 4°C)

$$= 1 \times 4.18 \times (15 - 0) + 335 + 1 \times 2.095 \times [0 - (-4)]$$

$$= 62.7 + 335 + 8.38 = 406.1 \text{ kJ/kg}$$

$$\text{Also,} \quad \text{C.O.P.} = \frac{\text{Net refrigerating effect}}{\text{Work done}} = \frac{R_n}{W}$$

$$\text{i.e.,} \quad 14.16 = \frac{m_{ice} \times 406.1}{W}$$

$$\text{or,} \quad W = \frac{m_{ice} \times 406.1}{14.16} = \frac{5000 \times 406.1}{14.16} = 14339.7 \text{ kJ/h}$$

$$= \frac{14339.7}{3600} = 3.98 \text{ kJ/s or } 3.98 \text{ kW}$$

Hence least power = **3.98 kW** **(Ans.)**

Example 2.12. *An ice plant produces 12 tonnes of ice per day at 0°C, using water at room temperature of 15°C. If the C.O.P. of the plant is 2.6 and overall electro-mechanical efficiency is 87% determine the power rating of the compressor motor. Take specific heat of water as 4.18 kJ/kg and latent heat of ice 335 kJ/kg.*

Solution: *Given:* Mass of ice produced, m_{ice} = 12 tonnes; t_{water} = 15°C; C.O.P. = 2.6; η = 87%;

c_{pw} = 4.18 kJ/kg; Latent heat of ice = 335 kJ/kg.

Power rating of compressor motor:

Heat to be extracted per kg of water (to produce ice at 0°C)

$$= 1 \times 4.18\ (15 - 0) + 335 = 397.7 \text{ kJ/kg}$$

Heat to be extracted for producing 12 tonnes of ice per day

$$= 12 \times 1000 \times 397.7 \text{ kJ}$$

$$\text{Refrigerating effect,} \quad R_n = \frac{12 \times 1000 \times 397.7}{24 \times 3600} = 55.24 \text{ kJ/s}$$

$$\text{Also,} \quad \text{C.O.P.} = \frac{R_n}{W}$$

$$\text{i.e.,} \quad 2.6 = \frac{55.24}{W}, \quad \text{or,} \quad W = \frac{55.24}{2.6} = 21.2 \text{ kJ/s } (= 21.2 \text{ kW})$$

∴ Power rating of the compressor motor

$$= \frac{W}{\eta} = \frac{21.2}{0.87} = \mathbf{24.37\ kW} \quad \textbf{(Ans.)}$$

Example 2.13. *A refrigerating system operates on the reversed Carnot cycle between temperature limits of 25°C and –10°C. The capacity is to be 8 tonnes. Determine:*

(*i*) *Coefficient of performance;*

(*ii*) *Power rating of the compressor motor if the overall electro-mechanical efficiency is 85%;*

(*iii*) *Heat rejected from the system per min.*

Solution: *Given:* T_1 = 273 + 25 = 298 K; T_2 = 273 + (– 10) = 263 K; Capacity = 8 tonnes; η = 85%

(*i*) **Co-efficient of performance, C.O.P:**

$$\text{C.O.P.} = \frac{T_2}{T_1 - T_2} = \frac{263}{298 - 263} = \mathbf{7.51} \quad \textbf{(Ans.)}$$

(*ii*) **Power rating of the compressor motor:**

$$\text{We know that,} \quad \text{C.O.P.} = \frac{R_n \text{ (refrigerating effect)}}{W \text{ (work done)}}$$

$$\textit{i.e.,} \quad 7.51 = \frac{8 \times 14000 \text{ (kJ/h)}}{W}$$

$$\text{or,} \quad W = \frac{8 \times 14000}{7.51} \text{ kJ/h} = \frac{8 \times 14000}{7.51 \times 3600} \text{ kJ/s} = 4.14 \text{ kJ/s} = 4.14 \text{ kW}$$

∴ Power rating of the compressor motor

$$= \frac{4.14}{0.85} = \mathbf{4.87\ kW} \quad \textbf{(Ans.)}$$

(*iii*) **Heat rejected from the system:**

$$= \text{Refrigerating effect} + \text{Work done}$$

$$= \frac{8 \times 14000}{60} \text{ (kJ/min.)} + 4.14 \times 60 \text{ (kJ/min.)}$$

$$= 1866.67 + 248.4 = \mathbf{2115.07\ kJ/min} \quad \textbf{(Ans.)}$$

Example 2.14. *A cold storage plant is required to store 20 tonnes of fish. The temperature of the fish when supplied = 25°C; storage temperature of fish required = – 8°C; specific heat of fish above freezing point = 2.93 kJ/kg°C; specific heat of fish below freezing point = 1.25 kJ/kg°C; freezing point of fish = – 3°C. Latent heat of fish = 232 kJ/kg.*

If the cooling is achieved within 8 hours find out:

(*i*) *Capacity of the refrigerating plant;*

(*ii*) *Carnot cycle C.O.P. between this temperature range;*

(*iii*) *If the actual C.O.P. is 1/3rd of the Carnot C.O.P. find out the power required to run the plant.*

Solution: Heat removed in 8 hours from each kg of fish

$$= 1 \times 2.93 \times [25 - (-3)] + 232 + 1 \times 1.25\ [-3 - (-8)]$$

$$= 82.04 + 232 + 6.25 = 320.29 \text{ kJ/kg}$$

Heat removed by the plant/hour

$$= \frac{320.29 \times 20 \times 1000}{8} = 800725 \text{ kJ/h}$$

(*i*) **Capacity of the refrigerating plant:**

$$\text{Capacity of the refrigerating plant} = \frac{800725}{14000} = \mathbf{57.19\ tonnes} \quad \textbf{(Ans.)}$$

$$T_1 = 25 + 273 = 298 \text{ K}; \quad T_2 = -8 + 273 = 265 \text{ K}$$

(*ii*) **C.O.P. of reversed Carnot cycle:**

$$\text{Carnot C.O.P.} = \frac{T_2}{T_1 - T_2} = \frac{265}{298 - 265} = \mathbf{8.03} \quad \textbf{(Ans.)}$$

(*iii*) **Power required:**

$$\text{Actual C.O.P.} = \frac{1}{3} \times \text{Carnot C.O.P.} = \frac{1}{3} \times 8.03 = 2.67$$

$$\text{But, actual C.O.P.} = \frac{\text{Net refrigerating effect}}{\text{Work done}} = \frac{R_n}{W}$$

$$2.67 = \frac{800725}{W} \text{ kJ/h}$$

$$\therefore \quad W = \frac{800725}{2.67} = 299897 \text{ kJ/h} = 83.3 \text{ kJ/s}$$

$\therefore$ Power required to run the plant = **83.3 kW (Ans.)**

Example 2.15. *A Carnot refrigerator extracts 500 kJ of heat per minute from a cold room which is maintaind at – 10°C and it is discharged to atmosphere which is at 35°C. Find the ideal power required to run the refrigerator.*

Solution: Refer to Fig. 2.3. For a reversed Carnot cycle,

$$\frac{Q_1}{T_1} = \frac{Q_2}{T_2}$$

But, $T_1 = 35 + 273 = 308$ K;

$T_2 = -10 + 273 = 263$ K; $Q_2 = 500$ kJ/min

$$\therefore \quad \frac{Q_1}{308} = \frac{500}{263}$$

$$\text{or,} \quad Q_1 = \frac{500 \times 308}{263} = 585.5 \text{ kJ/min}$$

Work done, $W = Q_1 - Q_2$

$= 585.5 - 500 = 85.5$ kJ/min

$$\therefore \text{ Ideal power required} = \frac{85.5}{60} = \mathbf{1.425\ kW} \quad \textbf{(Ans.)}$$

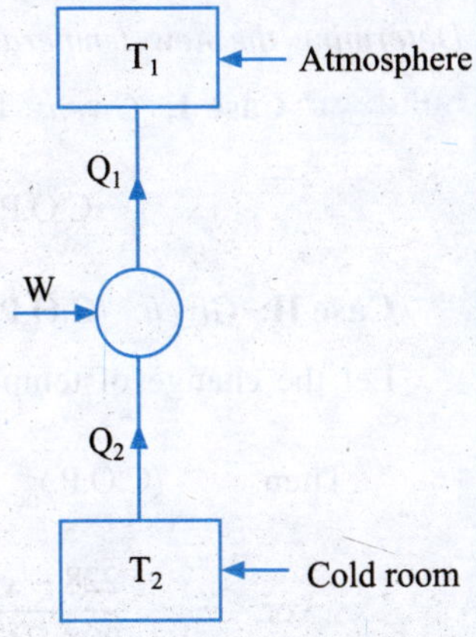

Fig. 2.3

Example 2.16. *A cold storage is to be maintained at –4°C while the surroundings are at 34°C. The heat leakage from the surroundings into the cold storage is estimated to be 30 kW. The actual C.O.P. of the refrigeration plant is one-third of an ideal plant working between the same temperatures. Find the power required to drive the plant.*

Solution: *Given*: $T_2 = -4 + 273 = 269$ K; $T_1 = 34 + 273 = 307$ K;

$Q_1 = 30$ kW;

$$(\text{C.O.P.})_{actual} = \frac{1}{3}(\text{C.O.P.})_{ideal}$$

Power required to drive the plant W:

$$(\text{C.O.P.})_{ideal} = \frac{T_2}{T_1 - T_2}$$

$$= \frac{269}{307 - 269} = 7.08$$

$$\therefore \quad (\text{C.O.P.})_{actual} = \frac{1}{3} \times 7.08$$

$$= 2.36$$

$$\text{Also, } (\text{C.O.P.})_{actual} = \frac{Q_2}{W}$$

$$= \frac{30}{W} = 2.36$$

$$\therefore W = \frac{30}{2.36} = \mathbf{12.71\ kW} \quad \textbf{(Ans.)}$$

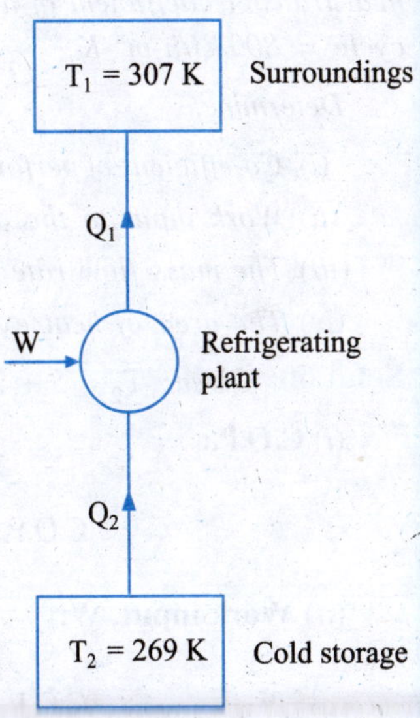

Fig. 2.4

Example 2.17. *A carnot refrigerator operates between temperatures of – 45ºC and 45ºC. Determine C.O.P. It is desired to make the C.O.P. equal to 3.5 by changing temperatures. The increase (or decrease) in upper temperature is to be equal to the decrease (or increase) in the lower temperature. Determine the new temperatures.*

Solution: Case I: *Given:* $T_2 = 273 + (-45) = 228$ K; $T_1 = 273 + 45 = 308$ K

$$\text{C.O.P.} = \frac{T_2}{T_1 - T_2} = \frac{228}{(318 - 228)} = \mathbf{2.533} \quad \textbf{(Ans.)}$$

Case II: *Given:* $(\text{C.O.P.})_{new} = 3.5$

Let the change of temperature be x.

Then, $$(\text{C.O.P.})_{new} = \frac{(228 - x)}{(318 + x) - (228 - x)} = 3.5$$

or, $$\frac{228 - x}{90 + 2x} = 3.5, \quad \text{or,} \quad (228 - x) = 3.5\,(90 + 2x)$$

or, $$228 - x = 315 + 7x, \quad \text{or,} \quad x = -10.875\text{ºC}$$

$$\therefore \quad T_2 = 228 - (-10.875) = \mathbf{238.875\ K} \quad \textbf{(Ans.)}$$

$$T_1 = 318 + (-10.875) = \mathbf{307.125\ K} \quad \textbf{(Ans.)}$$

$$\left[(\text{C.O.P.})_{new} = \frac{T_2}{T_1 - T_2} = \frac{238.875}{(307.125 - 238.875)} = 3.5 \textit{ confirmed}\right]$$

Example 2.18. *The following data relate to an air-conditioner working on the reversed Carnot cycle:*

Capacity = 2 tonnes; Temperature of heat absorption in cooling coil heat exchanger = 5ºC; Temperature of heat rejection of condenser heat exchanger = 60ºC: Temperature of air entering heat rejection heat exchanger = 40ºC; Temperature of air leaving heat rejection exchanger = 50ºC; Overall heat transfer coefficient of heat rejection of heat exchanger between air and working substance in the cycle = 800 kJ/h-m²-K

Determine:

(*i*) *Co-efficient of performance (C.O.P.);*

(*ii*) *Work input to the air-conditioner;*

(*iii*) *The mass flow rate of entering air;*

(*iv*) *The area of heat exchanger.*

Solution: *Given:* $T_2 = 5 + 273 = 278$ K; $T_1 = 60 + 273 = 333$ K; $U = 800$ kJ/h-m²-K

(*i*) **C.O.P.:**

$$\text{C.O.P.} = \frac{T_2}{T_1 - T_2} = \frac{278}{333 - 278} = \mathbf{5.05} \quad \textbf{(Ans.)}$$

(*ii*) **Work input, W:**

$$\text{C.O.P.} = \frac{R_n}{W}, \quad \text{or,} \quad 5.05 = \frac{2 \times 14000}{W} \qquad (\because 1\text{ tonne} = 1400\text{ kJ/h})$$

or, $$W = \frac{2 \times 14000}{5.05} = 5544.6\text{ kJ/h} = \frac{5544.6}{3600}\text{ kW} = \mathbf{1.54\ kW} \quad \textbf{(Ans.)}$$

(*iii*) **The mass flow rate of air, $\dot{m}$:**

Heat rejected, Q_R = Refrigerating effect/hour + Work input/hour

$$= 2 \times 14000 + 5554.6 = 33544.6 \text{ kJ/h} = 9.318 \text{ kJ/s}$$

Now, $Q_R = \dot{m} \times c_p \times \Delta t = 9.318$, or, $\dot{m} = \dfrac{9.318}{c_p \times \Delta t}$

or, $\dot{m} = \dfrac{9.318}{1 \times (50 - 40)} = $ **0.9318 kg/s (Ans.)**

(*iv*) **The area of heat exchanger, A:**

We know that, $Q_R = UA\,\Delta t_m$

Also, $\Delta t_m = \dfrac{\Delta t_1 - \Delta t_2}{\ln(\Delta t_1 / \Delta t_2)}$ (Assume counterflow)

$$= \frac{(60-40)-(60-50)}{\ln[(60-40)/(60-50)]} = \frac{20-10}{\ln(20/10)} = 14.43°\text{C}$$

$$\therefore \quad A = \frac{Q_R}{U\Delta t_m} = \frac{33544.6}{800 \times 14.43} = \mathbf{2.906\ m^2} \quad \textbf{(Ans.)}$$

Example 2.19. *A reversible heat engine operates between two reservoirs at temperatures 700°C and 50°C. The engine drives a reversible refrigerator which operators between reservoirs at temperatures of 50°C and –25°C. The heat transfer to the engine is 2500 kJ and the net work output of the combined engine refrigerator plant is 400 kJ.*

(*i*) *Determine the net heat transfer to the reservoir at 50°C;*

(*ii*) *Reconsider (i) given that the efficiency of the heat engine and the C.O.P. of the refrigerator are each 45 per cent of their maximum possible values.*

Solution: Refer to Fig. 2.5

Temperature, $T_1 = 700 + 273 = 973$ K

Temperature, $T_2 = 50 + 273 = 323$ K

Temperature, $T_3 = -25 + 273 = 248$ K

The heat transfer to the heat engine, $Q_1 = 2500$ kJ

The net work output of the combined engine refrigerator plant,

$$W = W_1 - W_2 = 400 \text{ kJ}$$

(*i*) Maximum efficiency of the heat engine cycle is given by

$$\eta_{max} = 1 - \frac{T_2}{T_1}$$

$$= 1 - \frac{323}{973} = 0.668$$

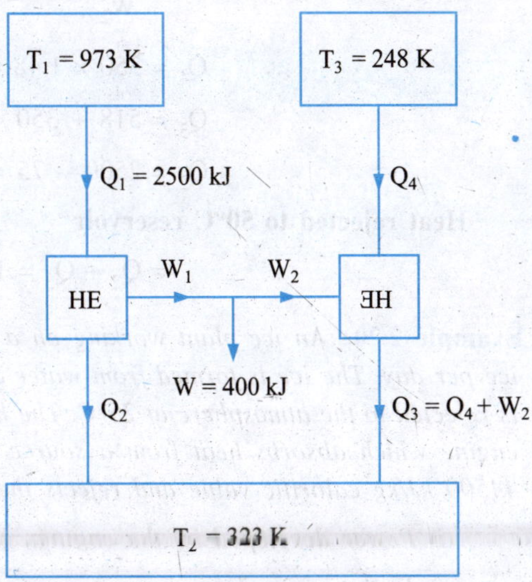

Fig. 2.5

Again, $\frac{W_1}{Q_1} = 0.668$

$\therefore \quad W_1 = 0.668 \times 2500 = 1670 \text{ kJ}$

$$(\text{C.O.P.})_{max} = \frac{T_3}{T_2 - T_3} = \frac{248}{323 - 248} = 3.306$$

Also, $\text{C.O.P.} = \frac{Q_4}{W_2} = 3.306$

Since, $W_1 - W_2 = W = 400 \text{ kJ}$

$W_2 = W_1 - W = 1670 - 400 = 1270 \text{ kJ}$

$\therefore \quad Q_4 = 3.306 \times 1270 = 4198.6 \text{ kJ}$

$Q_3 = Q_4 + W_2 = 4198.6 + 1270 = 5468.6 \text{ kJ}$

$Q_2 = Q_1 - W_1 = 2500 - 1670 = 830 \text{ kJ}$

Heat rejected to the 50°C reservoir

$= Q_2 + Q_3 = 830 + 5468.6 =$ **6298.6 kJ (Ans.)**

(*ii*) Efficiency of actual heat engine cycle,

$\eta = 0.45\eta_{max} = 0.45 \times 0.668 = 0.3$

$\therefore \quad W_1 = \eta \times Q_1 = 0.3 \times 2500 = 750 \text{ kJ}$

$\therefore \quad W_2 = 750 - 400 = 350 \text{ kJ}$

C.O.P. of the actual refrigerator cycle,

$$\text{C.O.P.} = \frac{Q_4}{W_2} = 0.45 \times 3.306 = 1.48$$

$\therefore \quad Q_4 = 350 \times 1.48 =$ **518 kJ (Ans.)**

$Q_3 = 518 + 350 = 868 \text{ kJ}$

$Q_2 = 2500 - 75 = 1750 \text{ kJ}$

Heat rejected to 50°C reservoir

$= Q_2 + Q_3 = 1750 + 868 =$ **2618 kJ (Ans.)**

Example 2.20. *An ice plant working on a reversed Carnot cycle heat pump produces 15 tonnes of ice per day. The ice is formed from water at 0°C and the formed ice is maintained at 0°C. The heat is rejected to the atmosphere at 25°C. The heat pump used to run the ice plant is coupled to a Carnot engine which absorbs heat from a source which is maintained at 220°C by burning liquid fuel of 44500 kJ/kg calorific value and rejects the heat to the atmosphere. Determine:*

(i) Power developed by the engine;

(ii) Fuel consumed per hour.

Take enthalpy of fusion of ice = 334.5 kJ/kg.

Solution: Fig. 2.6 shows the arrangement of the system.

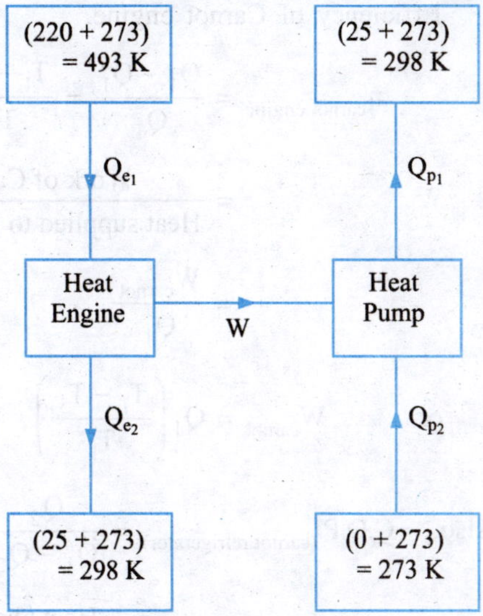

Fig. 2.6

(i) Power developed by the engine:

Amount of ice reproduced per day = 15 tonnes.

∴ The amount of heat removed by the heat pump,

$$Q_{P_2} = \frac{15 \times 1000 \times 334.5}{24 \times 60}$$

$$= 3484.4 \text{ kJ/min}$$

$$\text{C.O.P of the heat pump} = \frac{Q_{P_2}}{W} = \frac{273}{298 - 273}$$

$$\therefore \quad W = Q_{P_2} \times \frac{298 - 273}{273}$$

$$= 3484.4 \times \frac{25}{273} = 319.08 \text{ kJ/min}$$

This work must be developed by the Carnot engine,

$$W = \frac{319.08}{60} = 5.3 \text{ kJ/s} = 5.3 \text{ kW}$$

Thus power developed by the engine = **5.3 kW (Ans.)**

(ii) Fuel consumed per hour:

The efficiency of Carnot engine is given by:

$$\eta_{carnot} = \frac{W}{Q_{e_1}} = 1 - \frac{298}{493} = 0.396$$

$$\therefore \quad Q_{e_1} = \frac{W}{0.396} = \frac{5.3}{0.396} = 13.38 \text{ kJ/s}$$

$$\therefore \quad Q_{e_{1(\text{per hour})}} = 13.38 \times 60 \times 60 = 48168 \text{ kJ}$$

∴ Quantity of fuel consumed/hour

$$= \frac{48168}{44500} = \mathbf{1.082 \ kg/h} \quad \textbf{(Ans.)}$$

Example 2.21. *A Carnot heat engine draws heat from a reservoir at temperature T_1 and rejects heat to another reservoir at temperature T_3. The Carnot forward cycle engine drives a Carnot reversed cycle engine or Carnot refrigerator which absorbs heat from reservoir at temperature T_2 and rejects heat to a reservoir at temperature T_3. If the high temperature T_1 = 600 K and low temperature T_2 = 300 K, determine:*

(i) The temperature T_3 such that heat supplied to engine Q_1 is equal to the heat absorbed by refrigerator Q_2;

(ii) The efficiency of Carnot engine and C.O.P. of Carnot refrigerator.

Solution: Refer to Fig. 2.7

Temperature, $T_1 = 600$ K

Temperature, $T_2 = 300$ K

Efficiency of Carnot engine,

$$\eta_{\text{carnot engine}} = \frac{Q_1 - Q_1'}{Q_1} = \frac{T_1 - T_3}{T_1}$$

$$= \frac{\text{Work of Carnot engine}}{\text{Heat supplied to the Carnot engine}}$$

$$= \frac{W_{\text{carnot}}}{Q_1}$$

or, $$W_{\text{carnot}} = Q_1\left(\frac{T_1 - T_3}{T_1}\right) \quad ...(i)$$

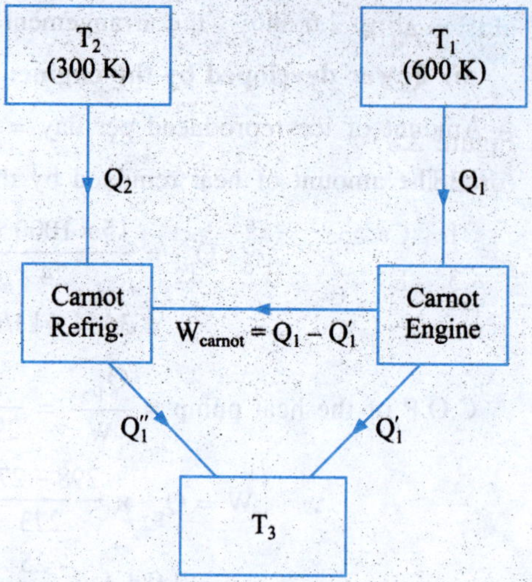

Fig. 2.7

Also, $$\text{C.O.P.}_{\text{(carnot refrigerator)}} = \frac{Q_2}{Q_1'' - Q_2} = \frac{T_2}{T_3 - T_2}$$

$$= \frac{\text{Heat absorbed}}{W_{\text{carnot}}} = \frac{Q_2}{W_{\text{carnot}}}$$

or, $$W_{\text{carnot}} = Q_2\left(\frac{T_3 - T_2}{T_2}\right) \quad ...(ii)$$

(*i*) **Temperature, T_3:**

From eqns. (*i*) and (*ii*), we get:

$$Q_1\left(\frac{T_1 - T_3}{T_1}\right) = Q_2\left(\frac{T_3 - T_2}{T_2}\right)$$

$$\therefore \quad \frac{Q_2}{Q_1} = \frac{T_2}{T_1}\left(\frac{T_1 - T_3}{T_3 - T_2}\right)$$

or, $$\frac{Q_2}{Q_1} = 1 = \frac{300}{600}\left(\frac{600 - T_3}{T_3 - 300}\right)$$

or, $$600 - T_3 = 2(T_3 - 300)$$

$$600 - T_3 = 2T_3 - 600$$

or, $$T_3 = 400 \text{ K}$$

Hence, temperature, T_3 = **400 K** **(Ans.)**

(*ii*) **Efficiency of carnot engine and C.O.P of carnot refrigerator:**

$$\eta_{\text{carnot engine}} = \frac{T_1 - T_3}{T_1} = \frac{600 - 400}{600} = 0.3333 = \mathbf{33.33\%} \quad \textbf{(Ans.)}$$

$$\text{C.O.P.}_{\text{refrigerator}} = \frac{T_2}{T_3 - T_2} = \frac{300}{400 - 300} = \mathbf{3} \quad \textbf{(Ans.)}$$

2.3 AIR-REFRIGERATION SYSTEM WORKING ON REVERSED BRAYTON CYCLE (OR BELL-COLEMAN OR JOULE CYCLE)

Figure 2.8 shows a schematic diagram of an air-refrigeration system working on the revered Brayton (or Bell-Coleman or Joule) cycle. Elements of this system are:

1. Compressor
2. Cooler (heat exchanger)
3. Expander
4. Refrigerator.

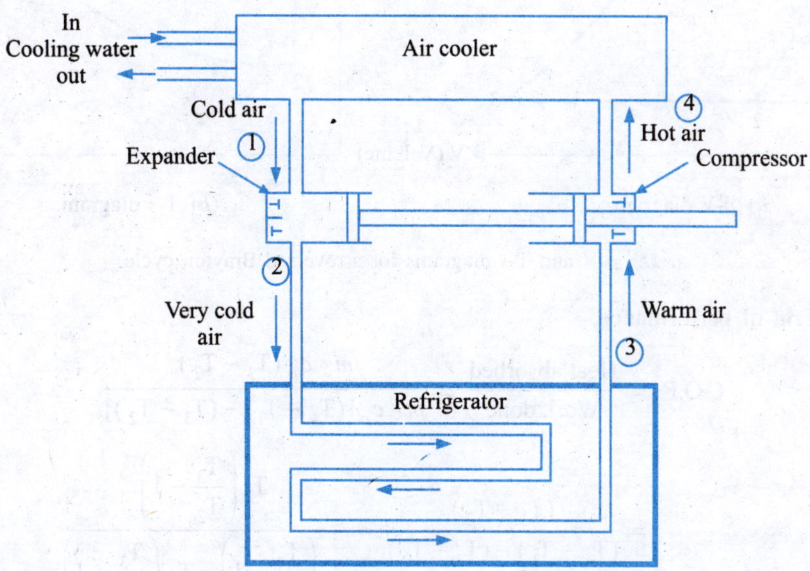

Fig. 2.8. Air-refrigeration system working on reversed Brayton or Bell-Coleman cycle.

In this system, work gained from expander is employed for compresstion of air, consequently less external work is needed for operation of the system. In practice it may not be done e.g., in some aircraft refrigeration systems which employ air-refrigeration cycle the expansion work may be used for *driving other devices.*

This system uses ***reversed Brayton cycle*** which is described below:

Fig. 2.9 (*a* and *b*) shows *p*-V and T-*s* diagrams for a reversed Brayton cycle. Here it is assumed that (*i*) absorption and rejection of heat are *constant pressure processes* and (*ii*) Compression and expansion are *isentropic processes.*

Consider *m* kg of air.

Heat absorbed in refrigerator $= Q_{added} = m \times c_p \times (T_3 - T_2)$

Heat rejected in cooler, $Q_{rejected} = m \times c_p (T_4 - T_1)$

Work done during the cycle = Heat rejected – Heat absorbed

$$= m \times c_p \times (T_4 - T_1) - m \times c_p \times (T_3 - T_2)$$

$$= m \times c_p [(T_4 - T_1) - (T_3 - T_2)].$$

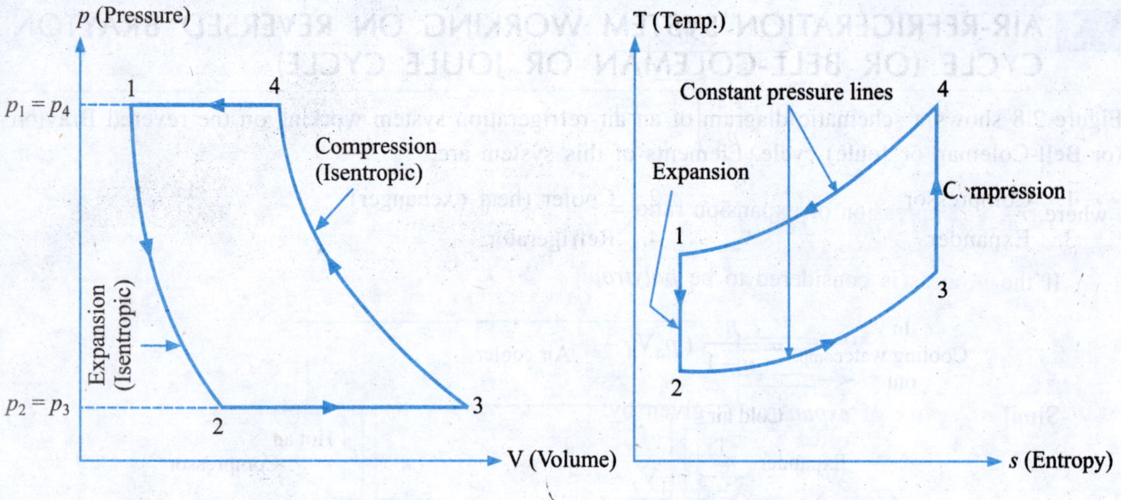

(a) p-V diagram. (b) T-s diagram.

Fig. 2.9. p-V and T-s diagrams for a reversed Brayton cycle.

∴ Coefficient of performance,

$$\text{C.O.P.} = \frac{\text{Heat absorbed}}{\text{Work done}} = \frac{m \times c_p (T_3 - T_2)}{m \times c_p [(T_4 - T_1) - (T_3 - T_2)]}$$

$$= \frac{(T_3 - T_2)}{(T_4 - T_1) - (T_3 - T_2)} = \frac{T_2\left(\frac{T_3}{T_2} - 1\right)}{T_1\left(\frac{T_4}{T_1} - 1\right) - T_2\left(\frac{T_3}{T_2} - 1\right)} \quad ...(i)$$

We know that for isontropic compression process 3-4,

$$\frac{T_4}{T_3} = \left(\frac{p_4}{p_3}\right)^{\frac{\gamma - 1}{\gamma}} \quad ...(ii)$$

Similarly, for isentropic expansion process 1–2,

$$\frac{T_1}{T_2} = \left(\frac{p_1}{p_2}\right)^{\frac{\gamma - 1}{\gamma}} \quad ...(iii)$$

Since $p_1 = p_4$ and $p_2 = p_3$, therefore from eqns. (ii) and (iii), we have

$$\text{C.O.P.} = \frac{T_4}{T_3} = \frac{T_1}{T_2}, \quad \text{or,} \quad \frac{T_4}{T_1} = \frac{T_3}{T_2}$$

Now substitutions these values in eqn. (i), we get:

$$\text{C.O.P.} = \frac{T_2}{T_1 - T_2} = \frac{1}{\frac{T_1}{T_2} - 1} = \frac{1}{\left(\frac{p_1}{p_2}\right)^{\frac{\gamma - 1}{\gamma}} - 1}$$

$$= \frac{1}{\left(\frac{p_4}{p_3}\right)^{\frac{\gamma-1}{\gamma}} - 1} = \frac{1}{(r_p)^{\frac{\gamma-1}{\gamma}} - 1} \quad ...(2.6)$$

where, r_p = compression or expansion ratio = $\frac{p_1}{p_2} = \frac{p_4}{p_3}$

If the process is considered to be *polytropic, the steady flow work of compression is given by*:

$$W_{comp.} = \frac{n}{n-1}(p_4V_4 - p_3V_3) \quad ...(2.7)$$

Similarly work of *expansion* is given by:

$$W_{exp.} = \frac{n}{n-1}(p_1V_1 - p_2V_2) \quad ...(2.8)$$

Equations (2.7) and (2.8) may easily be reduced to the theoretical isentropic process shown in Fig. 2.9 (*b*) by substituting γ = *n* and the known relationship:

$$R = c_p\left(\frac{\gamma-1}{\gamma}\right)J$$

The net external work required for operation of the cycle.

= Steady flow work of compression − Steady flow work of expansion

$$= W_{comp.} - W_{exp.}$$

$$= \left(\frac{n}{n-1}\right)(p_4V_4 - p_3V_3 - p_1V_1 + p_2V_2) \quad \left[\begin{array}{l} \because p_1V_1 = mRT_1 \\ p_2V_2 = mRT_2 \\ p_3V_3 = mRT_3 \\ p_4V_4 = mRT_4 \end{array}\right]$$

$$= \left(\frac{n}{n-1}\right)mR(T_4 - T_3 - T_1 + T_2)$$

$$= \left(\frac{n}{n-1}\right)\frac{mR}{J}(T_4 - T_3 - T_1 + T_2) \text{ in heat units.}$$

But, $R = c_p\left(\frac{\gamma-1}{\gamma}\right)J$ (where J = 1 in S.I. units)

$$\therefore \quad W_{comp.} - W_{exp.} = \left(\frac{n}{n-1}\right)\left(\frac{\gamma-1}{\gamma}\right)mc_p(T_4 - T_3 + T_2 - T_1) \quad ...(2.9)$$

For isentropic compression and expansion,

$$W_{net} = mc_p(T_4 - T_3 + T_2 - T_1)$$

According to law of conservation of energy the net work on the gas must be equivalent to the net heat rejected, therefore,

$$\text{C.O.P.} = \frac{Q_{added}}{Q_{rejected} - Q_{added}} = \frac{Q_{added}\text{ (kJ)}}{W_{net}\text{ (kJ)}}$$

For the air cycle assuming polytropic compression and expansion, co-efficient of performance,

$$\text{C.O.P.} = \frac{m \times c_p \times (T_3 - T_2)}{\left(\frac{n}{n-1}\right)\left(\frac{\gamma - 1}{\gamma}\right) m \times c_p \times (T_4 - T_3 + T_2 - T_1)}$$

$$= \frac{(T_3 - T_2)}{\left(\frac{n}{n-1}\right)\left(\frac{\gamma - 1}{\gamma}\right)(T_4 - T_3 + T_2 - T_1)} \quad ...(2.10)$$

Note: *The reversed Brayton cycle is same as the Bell-Coleman cycle. Conventionally Bell-Coleman cycle refers to closed cycle with expansion taking place in reciprocating expander and compressor respectively, and heat rejection and heat absorption taking place in condenser and evaporator respectively.*

With the development of efficient centrifugal compressors and gas turbines, the processes of compression and expansion can be carried out in centrifugal compressors and gas turbines respectively. Thus the shortcoming encountered with conventional reciprocating expander and compressor is overcome. Reversed Brayton cycle finds its application for air-conditioning of aeroplanes where air is used as refrigerant.

Merits and Demerits of Air-Refrigeration System

Merits:

1. Since air is non-flammable, therefore there is no risk of fire as in the machine using NH_3 as the refrigerant.
2. It is cheaper as air is easily available as compared to the other refrigerants.
3. As compared to the other refrigeration systems the weight of *air-refrigeration system per tonne of refrigeration is quite low, because of this reason this system is employed in aircrafts.*

Demerits:

1. The C.O.P. of this system is *very low* in comparison to other systems.
2. The weight of air required to be circulated is more compared with refrigerants used in other systems. This is due to the fact that heat is carried by air in the form of *sensible heat.*

Example 2.22. *One kg of air at a pressure of 1.05 bar and a temperature of 20°C is compressed to 6 bar. It is then cooled to 27°C in the cooler before entering the expansion cylinder. Assuming compression and expansion as isentropic processes, determine:*

(i) Refrigerating effect per kg of air;

(ii) Theoretical C.O.P.

Take $c_p = 1.0$ *and* $\gamma = 1.4$.

Solution: Fig. 2.10 shows the working cycle.

$$T_3 = 20 + 273 = 293 \text{ K};$$

$$T_1 = 27 + 273 = 300 \text{ K};$$

$$p_2 (= p_3) = 1.05 \text{ bar}$$
$$p_1 (= p_4) = 6 \text{ bar}$$

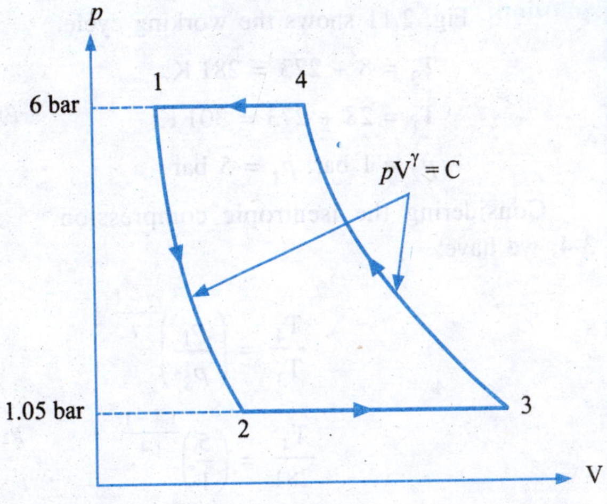

Fig. 2.10

(*i*) **Refrigerating effect per kg of air:**

Considering the isentropic compression 3-4, we have:

$$\frac{T_4}{T_3} = \left(\frac{p_4}{p_3}\right)^{\frac{\gamma-1}{\gamma}}$$

or,
$$\frac{T_4}{293} = \left(\frac{6}{1.05}\right)^{\frac{1.4-1}{1.4}} = 1.645$$

$$\therefore \quad T_4 = 293 \times 1.645 = 482 \text{ K}$$

Similarly considering isentropic expansion 1-2, we get:

$$\frac{T_2}{T_1} = \left(\frac{p_2}{p_1}\right)^{\frac{\gamma-1}{\gamma}}$$

or,
$$\frac{T_2}{300} = \left(\frac{1.05}{6}\right)^{\frac{1.4-1}{1.4}} = 0.608$$

$$\therefore \quad T_2 = 300 \times 0.608 = 182.4 \text{ K}$$

Now refrigerating effect per kg of air

$$= 1 \times c_p (T_3 - T_2) = 1 \times 1 \times (293 - 182.4)$$
$$= \mathbf{110.6 \text{ kJ/kg}} \quad \textbf{(Ans.)}$$

(*ii*) **Theoretical C.O.P.:**

Heat rejected to the cooler per kg of air

$$= 1 \times c_p (T_4 - T_1) = 1 \times 1 \times (482 - 300) = 182 \text{ kJ/kg}$$

$\therefore$ Work done per kg of air

$$= \text{Heat rejected} - \text{Heat abstracted}$$
$$= 182 - 110.6 = 71.4 \text{ kJ/kg}$$

$$\therefore \quad \text{Theoretical C.O.P} = \frac{\text{Heat abstracted}}{\text{Work done}} = \frac{110.6}{71.4} = \mathbf{1.55} \quad \textbf{(Ans.)}$$

Example 2.23. *500 kg of atmospheric air is circulated per hour in an open type of refrigeration installation. The air is drawn from the cold chamber at temperature 8°C and 1 bar, and then compressed isentropically to 5 bar abs. It is cooled at this pressure to 28°C and then led to the expander where it expands isentropically down to atmospheric pressure and is discharge to cold chamber. Determine:*

(*i*) *Heat extracted from cold chamber per hour;*

(*ii*) *Heat rejected to cooling water per hour;*

(*iii*) *C.O.P. of the system.*

For air, take $\gamma = 1.4$ *and* $c_p = 1.003$ *kJ/kg K.*

Solution: Fig. 2.11 shows the working cycle.

$$T_3 = 8 + 273 = 281 \text{ K};$$

$$T_1 = 28 + 273 = 301 \text{ K};$$

$$p_2 = 1 \text{ bar}; \; p_1 = 5 \text{ bar}$$

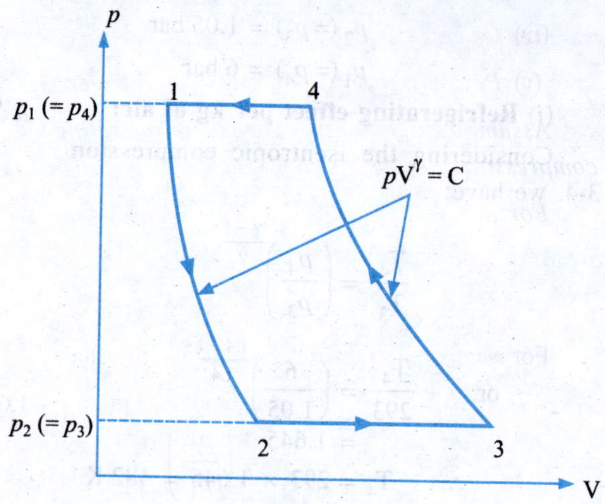

Fig. 2.11

Considering the isentropic compression 3-4, we have:

$$\frac{T_4}{T_3} = \left(\frac{p_1}{p_2}\right)^{\frac{\gamma-1}{\gamma}}$$

$$\frac{T_4}{281} = \left(\frac{5}{1}\right)^{\frac{1.4-1}{1.4}}$$

$$= (5)^{0.286} = 1.58$$

$$\therefore \quad T_4 = 445.2 \text{ K}$$

Similarly, considering isentripic expansion 1-2, we get:

$$\frac{T_2}{T_1} = \left(\frac{p_2}{p_1}\right)^{\frac{\gamma-1}{\gamma}} = \left(\frac{1}{5}\right)^{\frac{1.4-1}{1.4}} = \left(\frac{1}{5}\right)^{0.286} = 0.63$$

$$\therefore \quad \frac{T_2}{301} = 0.63 \text{ , or, } \quad T_2 = 301 \times 0.63 = 189.6 \text{ K}$$

(*i*) **Heat extracted from the cold chamber/hour**

$$= m\,c_p\,(T_3 - T_2) = 500 \times 1.003\,(281 - 189.6)$$

$$= \mathbf{45837 \text{ kJ/h}} \quad \textbf{(Ans.)}$$

(*ii*) **Heat rejected in the cooling chamber (or to cooling water)**

$$= m\,c_p\,(T_4 - T_1)$$

$$= 500 \times 1.003\,(445.2 - 301) = \mathbf{72316 \text{ kJ/h}} \quad \textbf{(Ans.)}$$

Since *both compression and expansion are reversible adiabatic*, therefore,

$$\text{Work done} = \text{Heat rejected} - \text{Heat abstracted}$$

$$= 72316 - 45837 = 26479 \text{ kJ/h}$$

$$\text{C.O.P.} = \frac{\text{Heat abstracted in cold chamber}}{\text{Work done}} = \frac{45837}{26479} = \mathbf{1.73} \quad \textbf{(Ans.)}$$

Example 2.24. *An open air cycle operated by air-refrigeration system is required to produce 6 tonnes of refrigerating effect with a cooler pressure of 11 bar abs. and a refrigerated space or region at a pressure of 1.05 bar. The temperature of air leaving the cooler is 38°C and the air leaving the room is 16°C. Calculate:*

(*i*) *Mass of air circulated per minute;*

(*ii*) *Compressor displacement required per minute;*

(*iii*) *Expander displacement required per minute;*

(*iv*) *C.O.P.;*

(*v*) *Power required per tonne of refrigeration.*

Assume that theoretical cycle is operating with isentropic compression and expansion with no compressor clearance and no losses.

For air, take $\gamma = 1.41$ *and* $c_p = 1.003$ *kJ/kg K.*

Solution: Refer to Fig. 2.12

$$T_1 = 38 + 273 = 311 \text{ K}$$

For isentropic expansion process 1-2, we have:

$$\frac{T_2}{T_1} = \left(\frac{p_2}{p_1}\right)^{\frac{\gamma-1}{\gamma}}$$

$$= \left(\frac{1.05}{11}\right)^{\frac{1.41-1}{1.41}}$$

$$= 0.505$$

or, $$\frac{T_2}{311} = 0.505$$

or, $$T_2 = 311 \times 0.505 = 157 \text{ K}$$

Also, $$T_3 = 16 + 273 = 289 \text{ K}$$

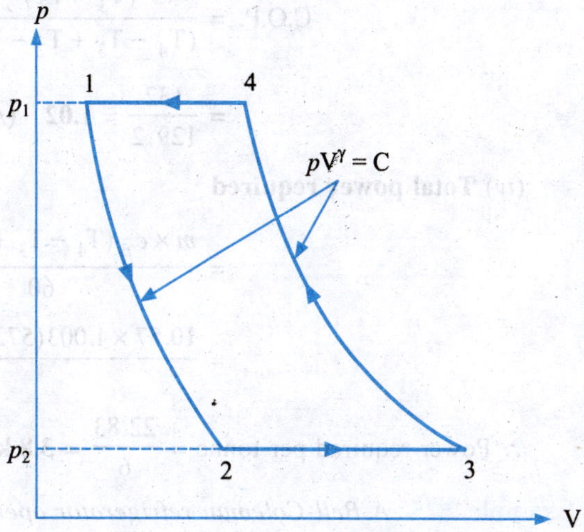

Fig. 2.12

For isentropic compression process 3-4, we have:

$$\frac{T_4}{T_3} = \left(\frac{p_1}{p_2}\right)^{\frac{\gamma-1}{\gamma}}$$

$$\frac{T_4}{289} = \left(\frac{11}{1.05}\right)^{\frac{1.41-1}{1.41}} = 1.98$$

$$T_4 = 289 \times 1.98 = 572.2 \text{ K}$$

Refrigerating effect per kg of air circulated

$$= c_p (T_3 - T_2) = 1.003 (289 - 157) = 132.4 \text{ kJ/kg}$$

(*i*) **Mass of air circulated per min,**

$$m = \frac{6 \times 14000}{132.4 \times 60} = \mathbf{10.57 \text{ kg}} \quad \textbf{(Ans.)}$$

(*ii*) **Compressor piston displacement,**

$$V_3 = \frac{mRT_3}{p_2} = \frac{10.57 \times (0.287 \times 1000) \times 289}{1.05 \times 10^5}$$

$$= \mathbf{8.35 \text{ m}^3\text{/min}} \quad \textbf{(Ans.)}$$

(*iii*) **Expander displacement,**

$$V_2 = \frac{mRT_2}{p_2} = \frac{10.57 \times (0.287 \times 1000) \times 157}{1.05 \times 10^5} = \mathbf{4.54\ m^3/min} \quad \textbf{(Ans.)}$$

For isentropic compression and expansion,

$$\text{C.O.P.} = \frac{(T_3 - T_2)}{(T_4 - T_3 + T_2 - T_1)} = \frac{289 - 157}{572.2 - 289 + 157 - 311}$$

$$= \frac{132}{129.2} = \mathbf{1.02} \quad \textbf{(Ans.)}$$

(*iv*) **Total power required**

$$= \frac{m \times c_p (T_4 - T_3 + T_2 - T_1)}{60} \text{kW}$$

$$= \frac{10.57 \times 1.003 (572.2 - 289 + 157 - 311)}{60} = 22.83 \text{ kW}$$

$$\therefore \text{ Power required per tonne} = \frac{22.83}{6} = \mathbf{3.8\ kW} \quad \textbf{(Ans.)}$$

Example 2.25. *A Bell-Coleman refrigerator operates between pressure limits of 1 bar and 8 bar. Air is drawn from the cold chamber at 9°C, compressed and then it is cooled to 29°C before entering the expansion cylinder. Expansion and compression follow the law $pv^{1.35}$ = constant. Calculate the theoretical C.O.P. of the system. For air take $\gamma = 1.4$, $c_p = 1.003$ kJ/kg K.*

Solution: Fig. 2.13 shows the working cycle of the refrigerator.

$$p_2 = 1 \text{ bar};\ p_1 = 8 \text{ bar};\ T_3 = 9 + 273 = 282 \text{ K};\ T_4 = 29 + 273 = 302 \text{ K}.$$

Considering polytropic compression 3-4, we get:

$$\frac{T_4}{T_3} = \left(\frac{p_1}{p_2}\right)^{\frac{\gamma - 1}{\gamma}}$$

$$= \left(\frac{8}{1}\right)^{\frac{1.35-1}{1.35}}$$

$$= (8)^{0.259} = 1.71$$

or $\quad T_4 = T_3 \times 1.71$

$$= 282 \times 1.71 = 482.2 \text{ K}$$

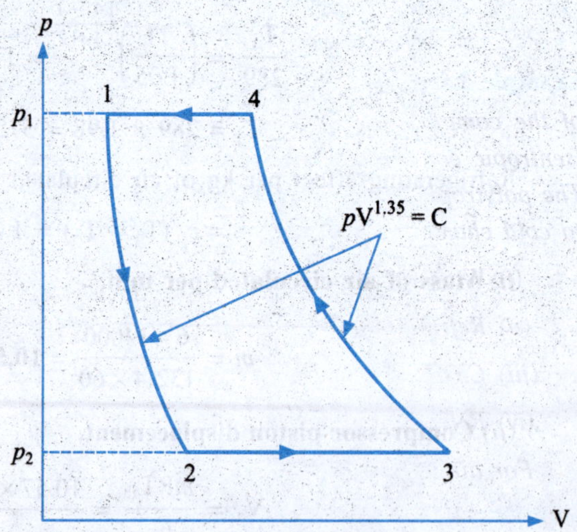

Fig. 2.13

Considering polytropic expansion 1-2, we have:

$$\frac{T_1}{T_2} = \left(\frac{p_1}{p_2}\right)^{\frac{\gamma - 1}{\gamma}}$$

$$= \left(\frac{8}{1}\right)^{\frac{1.35-1}{1.35}} = 1.71$$

$$T_2 = \frac{T_1}{1.71} = \frac{302}{1.71} = 176.6 \text{ K}$$

Heat extracted from the cold chamber per kg of air

$$= c_p (T_3 - T_2) = 1.003 \, (282 - 176.6) = 105.7 \text{ kJ/kg.}$$

Heat rejected in the cooling chamber per kg of air

$$= c_p (T_4 - T_1) = 1.003 \, (482.2 - 302) = 180.7 \text{ kJ/kg.}$$

Since the compression and expansion are not isentropic, difference between heat rejected and heat absorbed is not equal to the work done because there are heat transfers to the surroundings and from the surroundings during compression and expansion.

To find the work done, the area of the diagram '1234' is to be considered:

$$\text{Work done} = \frac{n}{n-1}(p_4 V_4 - p_3 V_3) - \frac{n}{n-1}(p_1 V_1 - p_2 V_2)$$

$$= \frac{n}{n-1} R[(T_4 - T_3) - (T_1 - T_2)] \qquad (m = 1 \text{ kg})$$

The value of R can be calculated as follows:

$$\frac{c_p}{c_v} = \gamma$$

$$\therefore \quad c_v = \frac{c_p}{\gamma} = \frac{1.003}{1.4} = 0.716 \text{ kJ/kg K}$$

$$R = (c_p - c_v) = 1.003 - 0.716 = 0.287 \text{ kJ/kg K}$$

$$\therefore \quad \text{Work done} = \frac{1.35}{0.35} \times 0.287 \, [(482.2 - 282) - (302 - 176.6)] = 82.8 \text{ kJ/kg}$$

$$\therefore \quad \text{C.O.P.} = \frac{\text{Heat abstracted}}{\text{Work done}} = \frac{105.7}{82.8} = \mathbf{1.27} \quad \textbf{(Ans.)}$$

Example 2.26. *In a refrigerator working on Bell-Coleman cycle, the air is drawn into the cylinder of the compressor from the cold chamber at a pressure of 1.03 bar and temperature 12°C. After isentropic compression to 5.5 bar, the air is cooled at constant pressure to a temperature of 22°C. The polytropic expansion* $pV^{1.25}$ = *constant then follows and the air expanded to 1.03 bar is passed to cold chamber. Determine:*

(*i*) *Work done per kg of air flow;*

(*ii*) *Refrigerating effect per kg of air flow;*

(*iii*) *C.O.P.;*

(*iv*) *Refrigerating capacity of the plant in tonnes for a mass flow rate of 90 kg/h.*

For air, take $\gamma = 1.4$ *and* $c_p = 1.003$ *kJ/kg K.*

Solution: Refer to Fig. 2.14

$T_3 = 12 + 273 = 285$ K; $T_1 = 22 + 273 = 295$ K; $p_1 (= p_4) = 5.5$ bar; $p_2 (= p_3) = 1.03$ bar.

Considering the isentropic compression 3-4, we have:

$$\frac{T_4}{T_3} = \left(\frac{p_4}{p_3}\right)^{\frac{\gamma-1}{\gamma}}$$

$$= \left(\frac{5.5}{1.03}\right)^{\frac{1.4-1}{1.4}}$$

$$= 1.614$$

$$\therefore \quad T_4 = T_3 \times 1.614$$

$$= 285 \times 1.614$$

$$= 460 \text{ K}$$

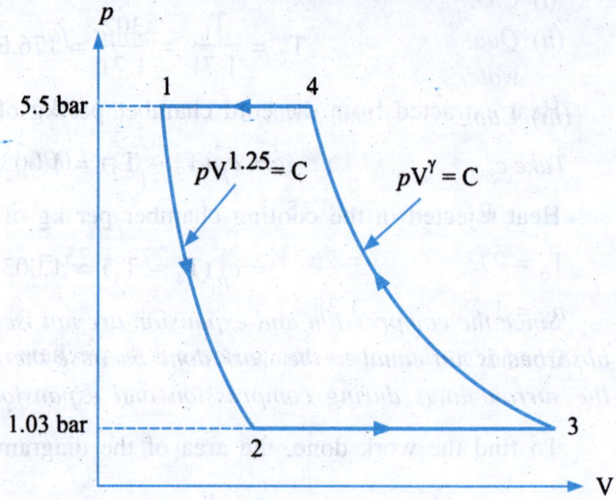

Fig. 2.14

Similarly considering isentropic expansion 1-2, we get:

$$\frac{T_2}{T_1} = \left(\frac{p_2}{p_1}\right)^{\frac{1.25-1}{1.25}}$$

$$= \left(\frac{1.03}{5.5}\right)^{0.2} = 0.715$$

$$\therefore \quad T_2 = T_1 \times 0.715 = 295 \times 0.715 = 210.9 \text{ K}$$

(i) **Work done per kg of air flow:**

Work done per kg of air = $W_{comp.} - W_{exp.}$

$$= \frac{\gamma}{\gamma-1} \times \frac{R}{J}(T_4 - T_3) - \frac{n}{n-1} \times \frac{R}{J}(T_1 - T_2)$$

$$= \frac{1.4}{1.4-1} \times \frac{0.287}{1}(460 - 285) - \frac{1.25}{1.25-1} \times \frac{0.287}{1}(295 - 210.9)$$

$$= 175.78 - 120.68 = \mathbf{55.1 \text{ kJ/kg}} \quad \textbf{(Ans.)}$$

(ii) **Refrigerating effect per kg of air flow:**

Refrigerating effect $= 1 \times c_p (T_3 - T_2) = 1 \times 1.003\ (285 - 210.9) = \mathbf{74.32 \text{ kJ/kg}}$ **(Ans.)**

(iii) **C.O.P.**

$$\text{C.O.P.} = \frac{\text{Refrigerating effect}}{\text{Work done}} = \frac{74.32}{55.1} = \mathbf{1.349} \quad \textbf{(Ans.)}$$

(iv) **Refrigerating capacity of the plant:**

$$\text{Refrigerating capacity} = \frac{\text{Refrigerating effect / hour}}{14000} = \frac{74.32 \times 90}{14000} = \mathbf{0.478 \text{ tonne}} \quad \textbf{(Ans.)}$$

Example 2.27. *An air-refrigeration system operating on Bell-Coleman cycle, takes in air from cold room at –6°C and compresses it from 1.04 bar to 6.2 bar. The index of compression being 1.26. The compressed air is cooled to 25°C. The ambient temperature is 18°C. Air expands in an expander where index of compression is 1.36. Determine:*

(i) *C.O.P. of the system;*

(ii) *Quantity of air circulated per minute for production of 1500 kg of ice per day at 0°C from water at 15°C;*

(iii) *Capacity of the plant.*

Take c_{pw} *= 4.18 kJ/kg K (for water);* c_{pa} *= 1.005 kJ/kg K (for air) and latent heat of ice = 335 kJ/kg.*

Solution: Refer to Fig. 2.15

$T_3 = 273 + (-6) = 267$ K; $T_1 = 273 + 25 = 298$ K; $p_1 (= p_4) = 6.2$ bar; $p_2 (= p_3) = 1.04$ bar;

$n_1 = 1.28$; $n_2 = 1.38$.

Considering compression process 3-4, we have:

$$\frac{T_4}{T_3} = \left(\frac{p_4}{p_3}\right)^{\frac{n_1-1}{n_1}}$$

$$= \left(\frac{6.2}{1.04}\right)^{\frac{1.28-1}{1.28}} = 1.477$$

$$\therefore \quad T_4 = T_3 \times 1.477$$

$$= 267 \times 1.477 = 394.4 \text{ K}$$

Similarly considering expansion process 1-2, we get:

$$\frac{T_2}{T_1} = \left(\frac{p_2}{p_1}\right)^{\frac{n_2-1}{n_2}}$$

$$= \left(\frac{1.04}{6.2}\right)^{\frac{1.38-1}{1.38}}$$

$$= 0.611$$

$$\therefore \quad T_2 = T_1 \times 0.611$$

$$= 298 \times 0.611$$

$$= 182.1 \text{ K}$$

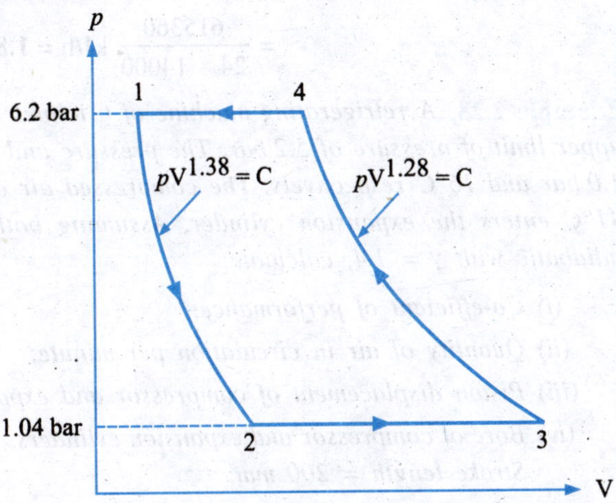

Fig. 2.15

(i) **C.O.P. of the system:**

Work done on the compressor per kg of air,

$$W_{comp.} = \frac{n_1}{n_1 - 1} R (T_4 - T_3)$$

$$= \frac{1.28}{1.28 - 1} \times 0.287 (394.4 - 267) = 167.15 \text{ kJ/kg}$$

Work done in expander per kg of air,

$$W_{exp.} = \frac{n_2}{n_2 - 1} R (T_1 - T_2)$$

$$= \frac{1.38}{1.38 - 1} \times 0.287 (298 - 182.1) = 120.8 \text{ kJ/kg}$$

$$\text{Net work done} = W_{comp.} - W_{exp.}$$
$$= 167.15 - 120.8 = 46.35 \text{ kJ/kg}$$

Refrigerating effect per kg of air

$$= c_{pa} (T_3 - T_2) = 1.003\,(267 - 182.1) = 85.15 \text{ kJ/kg}$$

$$\therefore \quad \text{C.O.P.} = \frac{\text{Refrigerating effect}}{\text{Net work done}} = \frac{85.15}{46.35} = \mathbf{1.837} \quad \textbf{(Ans.)}$$

(*ii*) **Quantity of air circulated per minute, m_{air}:**

Total heat to be abstracted for producing 1500 kg of ice per day

$$= 1500\,[1 \times c_{pw}\,(18 - 0) + \text{latent heat}]$$
$$= 1500\,[1 \times 4.18 \times 18 + 335] = 615360 \text{ kJ/day}$$

$$\therefore \quad m_{air} = \frac{615360}{\text{Refrigerating effect per day} \times 24 \times 60}$$

$$= \frac{615360}{85.15 \times 24 \times 60}\ \mathbf{5.02\ kg/min} \quad \textbf{(Ans.)}$$

(*iii*) **Capacity of the plant:**

Refrigerating capacity of the plant

$$= \frac{615360}{24 \times 14000} \text{ kJ/h} = \mathbf{1.83\ tonnes} \quad \textbf{(Ans.)}$$

Example 2.28. *A refrigerating machine of 6 tonnes capacity working on Bell-Coleman cycle has an upper limit of pressure of 5.2 bar. The pressure and temperature at the start of the compression are 1.0 bar and 16°C respectively. The compressed air cooled at constant pressure at a temperature of 41°C enters the expansion cylinder. Assuming both expansion and compression processes to be adiabatic with γ = 1.4, calculate:*

(*i*) *Co-efficient of performance;*

(*ii*) *Quantity of air in circulation per minute;*

(*iii*) *Piston displacement of compressor and expander;*

(*iv*) *Bore of compressor and expansion cylinders. The unit runs at 240 r.p.m. and is double-acting. Stroke length = 200 mm;*

(*v*) *Power required to drive the unit.*

For air take $\gamma = 1.4$ and $c_p = 1.003$ kJ/kg K.

Solution: Refer to Fig. 2.16

$T_3 = 16 + 273 = 298\text{K}$; $T_1 = 41 + 273 = 314$ K; $p_1 = 5.2$ bar; $p_2 = 1.0$ bar.

Considering the adiabatic compression 3-4, we get:

$$\frac{T_4}{T_3} = \left(\frac{p_1}{p_2}\right)^{\frac{\gamma - 1}{\gamma}} = \left(\frac{5.2}{1}\right)^{\frac{1.4 - 1}{1.4}} = (5.2)^{0.286} = 1.6$$

$$\therefore \quad T_4 = 1.6 \times T_3$$
$$= 1.6 \times 289$$
$$= 462.4 \text{ K}$$

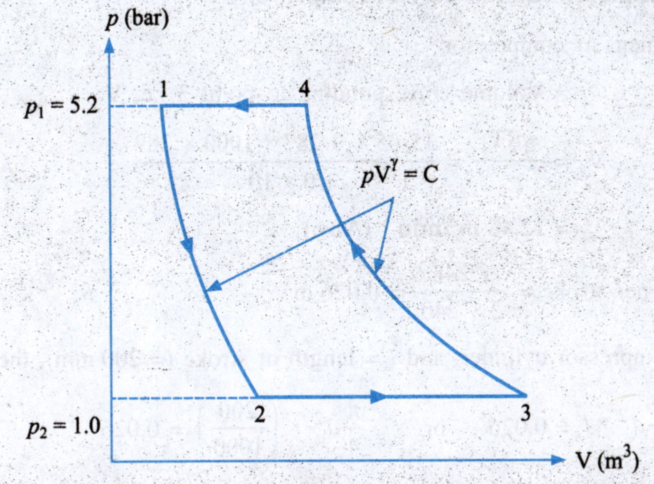

Fig. 2.16

Considering the adiabatic expansion 1-2, we have:

$$\frac{T_1}{T_2} = \left(\frac{p_1}{p_2}\right)^{\frac{\gamma - 1}{\gamma}}$$

$$\frac{314}{T_2} = \left(\frac{5.2}{1}\right)^{\frac{0.4}{1.4}} = 1.6$$

$$\therefore \quad T_2 = \frac{314}{1.6} = 196.25 \text{ K}$$

(*i*) **C.O.P.:**

Since both the compression and expansion processes are isentropic/adiabatic reversible, therefore

$$\text{C.O.P. of the machine} = \frac{T_2}{T_1 - T_2} = \frac{196.25}{314 - 196.25} = \mathbf{1.67} \quad \textbf{(Ans.)}$$

(*ii*) **Mass of air in circulation:**

Refrigerating effect per kg of air

$$= c_p (T_3 - T_2) = 1.003\ (289 - 196.25) = 93.03 \text{ kJ/kg}$$

Refrigerating effect produced by the refrigerating machine

$$= 6 \times 14000 = 84000 \text{ kJ/h}$$

Hence mass of air in circulation

$$= \frac{84000}{93.03 \times 60} = \mathbf{15.05\ kg/min} \quad \textbf{(Ans.)}$$

(*iii*), (*iv*) **Piston displacement and bore of compressor:**

Position displacement of compressor

$$= \text{Volume corresponding to point 3 i.e, } V_3$$

$$\therefore \quad V_3 = \frac{mRT_3}{p_2} = \frac{15.05 \times 0.287 \times 1000 \times 289}{1.0 \times 10^5}$$

$$= \mathbf{12.48\ m^3/min} \quad \textbf{(Ans.)}$$

$$\therefore \text{ Swept volume per stroke} = \frac{12.48}{2 \times 240} = 0.026 \text{ m}^3$$

If d_c = Dia. of compressor cylinder, and l = length of stroke (= 200 mm), then,

$$\frac{\pi}{4} d_c^2 \times l = 0.026, \quad \text{or,} \quad \frac{\pi}{4} d_c^2 \times \left(\frac{200}{1000}\right) = 0.026$$

$$d_c = \left(\frac{0.026 \times 1000 \times 4}{\pi \times 200}\right)^{1/2} = 0.407 \text{ m or } 407 \text{ mm}$$

i.e., Diameter or bore of the compressor cylinder = **407 mm (Ans.)**

Piston displacement and bore of expander:

Piston displacement of expander = Volume corresponding to point 2 i.e., V_2

$$\therefore \quad V_2 = \frac{mRT_2}{p_2} = \frac{15.05 \times 0.287 \times 1000 \times 196.25}{1.0 \times 10^5}$$

$$= \mathbf{8.476\ m^3/min.} \quad \textbf{(Ans.)}$$

$$\therefore \quad \text{Swept volume per stroke} = \frac{8.476}{2 \times 240} = 0.0176 \text{ m}^3$$

If d_e = Dia. of the expander, and l = length of stroke (= 200 mm), then,

$$\frac{\pi}{4} d_e^2 \times \left(\frac{200}{1000}\right) = 0.0176$$

$$d_e = \left(\frac{0.0176 \times 1000 \times 4}{\pi \times 200}\right)^{1/2} = 0.335 \text{ m or } 335 \text{ mm}$$

i.e, Diameter or bore of the expander cylinder = **335 mm (Ans.)**

(*v*) **Power required to drive the unit:**

$$\text{C.O.P.} = \frac{\text{Refrigerating effect}}{\text{Work done}} = \frac{R_n}{W}$$

$$1.67 = \frac{6 \times 14000}{W}$$

$$W = \frac{6 \times 14000}{1.67} = 50299.4 \text{ kJ/h} = 13.97 \text{ kJ/s}$$

Hence power required = **13.97 kW (Ans.)**

Example 2.29. *A dense air refrigeration cycle operates between 5 bar and 20 bar. The air temperature after heat rejection to surroundings is 37°C and air temperature at exit of refrigerator is 7°C. The isentropic efficiencies of compressor and turbine are 0.84 and 0.82 respectively. Determine:*

(i) Compressor and turbine work per tonne of refrigeration;

(ii) Co-efficient of performance;

(iii) Power per tonne of refrigeration.

Take $\gamma = 1.4$ *and* $c_p = 1.005$ *kJ/kg K.*

Solution: *Given:* $p_1(=p_4) = 20$ bar; $p_2(=p_3) = 5$ bar; $T_1 = 37 + 273 = 310$ K; $T_3 = 7 + 273 = 280$ K; $\eta_{comp.} = 0.84$, $\eta_{turbine} = 0.82$; $\gamma = 1.4$; $c_p = 1.005$ kJ/kg K.

Let, T_4 = Temperature of air at the end of isentropic compression in the compressor,

T_4' = Actual temperature of air leaving the compressor,

T_2 = Temperature of air at the end of isentropic expansion in the turbine, and

T_2' = Actual temperature of air leaving the turbine.

For isentropic compression process 3-4, we have:

$$\frac{T_4}{T_3} = \left(\frac{p_4}{p_3}\right)^{\frac{\gamma-1}{\gamma}} = \left(\frac{p_1}{p_2}\right)^{\frac{\gamma-1}{\gamma}} = \left(\frac{20}{5}\right)^{\frac{1.4-1}{1.4}} = 1.486$$

$$\therefore \quad T_4 = T_3 \times 1.486 = 280 \times 1.486 = 416 \text{ K}$$

Similarly, for isentropic expansion process 1-2, we have

$$\frac{T_1}{T_2} = \left(\frac{p_1}{p_2}\right)^{\frac{\gamma-1}{\gamma}} = \left(\frac{20}{5}\right)^{\frac{1.4-1}{1.4}} = 1.486$$

$$\therefore \quad T_2 = \frac{T_1}{1.486} = \frac{310}{1.486} = 208.6 \text{ K}$$

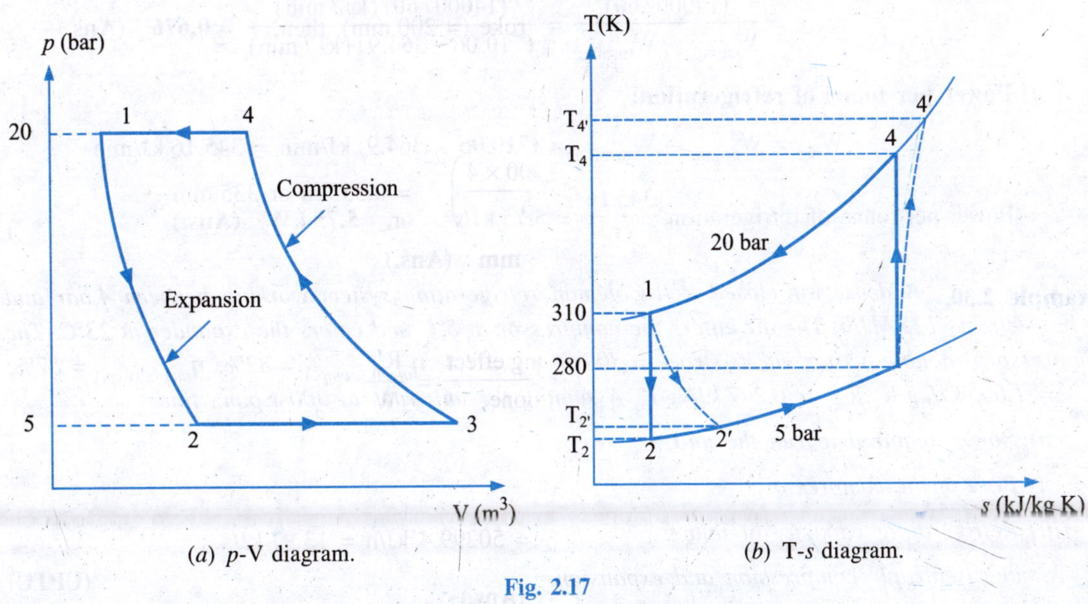

(*a*) *p*-V diagram. (*b*) T-*s* diagram.

Fig. 2.17

Now, Isentropic efficiency of the compressor,

$$\eta_{comp.} = \frac{T_4 - T_3}{T_{4'} - T_3}, \quad \text{or,} \quad 0.84 = \frac{416 - 280}{T_4 - 280}$$

$$\text{or,} \quad T_4 = \left(\frac{416 - 280}{0.84}\right) + 280 = 441.9 \text{ K}$$

and, isentropic efficiency of the turbine,

$$\eta_{turbine} = \frac{T_1 - T_{2'}}{T_1 - T_2}, \quad \text{or,} \quad 0.82 = \frac{310 - T_{2'}}{310 - 208.6}$$

$$\therefore \quad T_2' = 310 - 0.82\,(310 - 208.6) = 226.8 \text{ K}$$

Heat extracted from the refrigerator or refrigerating effect per kg of air

$$= c_p\,(T_3 - T_2') = 1.005\,(280 - 226.8)$$

$$= 53.47 \text{ kJ/kg,}$$

and mass of air per tonne of refrigeration,

$$m_a = \frac{14000\,(\text{kJ/h})}{53.47\,(\text{kJ/kg})} = 261.83 \text{ kg/h}, \quad \text{or,} \quad 4.364 \text{ kJ/min}$$

(*i*) **Compressor and turbine work per tonne of refrigeration, $W_{comp.}$, $W_{turbine}$:**

$$W_{comp.} = m_a\,c_p\,(T_{4'} - T_3) = 4.364 \times 1.005\,(441.9 - 280) = \mathbf{710.06 \text{ kJ/min}} \quad \textbf{(Ans.)}$$

$$W_{turbine} = m_a\,c_p\,(T_1 - T_{2'}) = 4.364 \times 1.005\,(310 - 226.8) = \mathbf{364.9 \text{ kJ/min}} \quad \textbf{(Ans.)}$$

(*ii*) **Co-efficient of performance (C.O.P.):**

$$\text{C.O.P.} = \frac{\text{Heat extracted per min}}{\text{Net work done}}$$

$$= \frac{(14000/60)}{W_{comp.} - W_{turbine}} = \frac{(14000/60)\,(\text{kJ/min})}{(710.06 - 364.9)\,(\text{kJ/min})} = \mathbf{0.676} \quad \textbf{(Ans.)}$$

(*iii*) **Power per tonne of refrigeration:**

$$W_{net} = W_{comp.} - W_{turbine} = (710.06 - 364.9) \text{ kJ/min} = 345.16 \text{ kJ/min}$$

$$\therefore \quad \text{Power per tonne of refrigeration } \frac{345.16}{60} = 5.75 \text{ kJ/s}, \quad \text{or,} \quad \mathbf{5.75 \text{ kW}} \quad \textbf{(Ans.)}$$

Example 2.30. *A dense air closed Bell-Coleman refrigeration system working between 4 bar and 16 bar extracts 125 MJ/h. The air enters the compressor at 5°C and enters the expander at 23°C. The compressor is double-acting and its stroke = 30 cm; γ_{air} = 1.4; $\eta_{mech.(comp.)}$ = 82%; $\eta_{mech.\,(exp.)}$ = 87%; $c_{p_{air}}$ = 1.005 kJ/kg K; R_{air} = 0.287 kJ/kg K. Assuming the unit runs at 300 r.p.m., find:*

(*i*) *Power required to run the unit;*

(*ii*) *Bore of the compressor;*

(*iii*) *Refrigerating capacity in tonnes.*

Assume isentropic compression and expansion. **(UPTU)**

Solution: *Given:* $p_1 = p_4 = 16$ bar; $p_2 = p_3 = 4$ bar; Heat extracted = 125 MJ/h;

$T_3 = 5 + 273 = 278$ K; $T_1 = 23 + 273 = 296$ K; N = 300 r.p.m.; l = 30 cm = 0.3 m; $\gamma_{air} = 1.4$; $\eta_{mech.(comp.)} = 82\%$;

$\eta_{mech.(exp.)} = 87\%$; $c_{p_{air}} = 1.005$ kJ/kg K; $R_{air} = 0.287$ kJ/kg K; N = 300 r.p.m.

(*i*) **Power required to run the unit, P:**

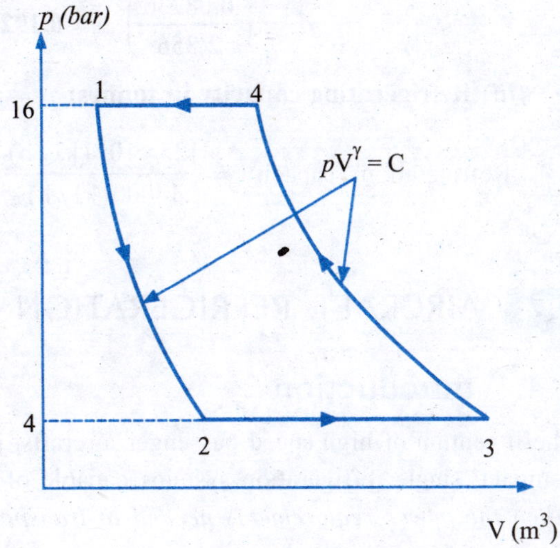

Fig. 2.18

$$\frac{T_2}{T_1} = \left(\frac{p_2}{p_1}\right)^{\frac{\gamma-1}{\gamma}} = \left(\frac{4}{16}\right)^{\frac{1.4-1}{1.4}} = 0.673$$

$$\therefore \quad T_2 = 296 \times 0.673 = 199.2 \text{ K}$$

$$\frac{T_4}{T_3} = \left(\frac{p_4}{p_3}\right)^{\frac{\gamma-1}{\gamma}}$$

$$= \left(\frac{16}{4}\right)^{\frac{1.4-1}{1.4}} = 1.486$$

$$\therefore \quad T_4 = 278 \times 1.486$$

$$= 413.1 \text{ K}$$

Refrigerations effect per kg $= c_p (T_3 - T_2)$

$$= 1.005 (278 - 199.2)$$

$$= 79.2 \text{ kJ/kg}$$

Mass of air in circulation,

$$m = \frac{125 \times 10^3}{79.2 \times 3600}$$

$$= 0.438 \text{ kg/s}$$

$$W_{comp.} = \frac{m \times c_p \times (T_4 - T_3)}{\eta_{mech.(comp.)}}$$

$$= \frac{0.438 \times 1.005 \times (413.1 - 278)}{0.82} = 72.5 \text{ kJ/s}$$

$$W_{exp.} = m \times c_p \times (T_1 - T_2) \times \eta_{mech.(exp.)}$$

$$= 0.438 \times 1.005 \times (296 - 199.2) \times 0.87 = 37.1 \text{ kJ/s}$$

Power required, **P** $= W_{comp.} - W_{exp.} = 72.5 - 37.1 =$ **35.4 kW** **(Ans.)**

(*ii*) **Bore of the compressor, d_c:** Now volume of air handled by the compressor (V_3) is found from the relation,

$$p_3 V_3 = mRT_3, \quad \text{or,} \quad V_3 = \frac{mRT_3}{p_3} = \frac{0.438 \times 0.287 \times 278}{4 \times 10^2} = 0.08736 \text{ m}^3\text{/s}$$

But, $V_3 = \frac{\pi}{4} d_c^2 \times l \times (2 \times N/60)$ ... since the compressor is double-acting

$$0.08736 = \frac{\pi}{4} d_c^2 \times 0.3 \times (2 \times 300/60) = 2.356\ d_c^2$$

$$\therefore \quad d_c = \left(\frac{0.08736}{2.356}\right)^{1/2} = \mathbf{0.1926\ m}, \quad \text{or,} \quad \mathbf{19.26\ cm} \quad \textbf{(Ans.)}$$

(*iii*) **Refrigerating capacity in tonnes:**

$$\text{Refrigerations capacity} = \frac{125 \times 10^3\ (\text{kJ/h})}{14000\ (\text{kJ/h})} = \mathbf{8.93\ tonnes} \quad \textbf{(Ans.)}$$

2.4 AIRCRAFT REFRIGERATION SYSTEMS

2.4.1 Introduction

The invention of high speed passenger aircrafts, jet aircrafts, and missiles has introduced the need for compact, single refrigeration systems capable of high capacity with minimum reduction of pay load. *When the power requirements needed to transport the additional weight of the refrigerating systems are taken into account, air cycle systems usually prove to be most efficient.*

In an aircraft application, the main consideration involved in order of importance are: *Weight, space* and *operating power.*

Although the power per tonne of refrigeration is considerably more for air-cycle refrigeration than for a vapour-compression system, yet the bulk and weight advantages of air cycle system, due to the heat exchanger at the cold end and a common turbo-compressor for both the engine and refrigeration plant, result in a greater overall power saving in the aircraft.

2.4.2 Advantages of Air Cycle for Aircraft Refrigeration

Following are the *advantages* of air cycle with regard to its application in aircraft refrigeration:

1. With air as the refrigerant, *leakage in small amounts is tolerable.*
2. *Quite cheap* (No cost of refrigerant in involved since air is used as a cooling medium).
3. The *pressure in the whole system is quite low*, thus, the piping, ducting, etc., are quite simple to design, fabricate and maintain.
4. In an air cycle (in its simplest form) with an open system, only *one heat exchanger is required.*
5. *Low maintenance cost* (since no complicated extra parts are involved).
6. *Light in weight comparatively.*
7. *No. additional vibrational problem* (as envisaged from the use of a vapour-compression system).
8. *Cabin pressurisation and air-conditioning can be carried out in one operation.*
9. *The ram effect* (due to high kinetic energy of the ambient air relative to the aircraft) *accomplishes the initial compression of the air.* The necessary power required for this, however, is derived from the aircraft engine.

10. As the main compressor is employed for compressed air source, there is *no problem for extra space, extra fixation for the compressor etc.* In some cases, however, separate compressors are used due to various seasons, since the air gets contaminated due to undesirable gases etc.
11. *The chilled air is directly used for cooling, thereby eliminating the costs of separator evaporator and its weight.*
12. Air is non-toxic and non-inflammable (desirable properties of refrigerant).
13. Refrigerant's availability in mid air is also an important consideration.

2.4.3 Dry Air Rated Temperature (DART)

- The **dry air rated temperature** *is the temperature in terms of which the comparison of different types of air-refrigeration systems is made.*
- *This is the temperature of the discharged air from the expander if there is no condensed moisture present and is calculated by the usual procedure* (assuming, c_p for air = 1.005 kJ/kg K). Thus DART will be essentially that which results from a calculation which assumes that air at a humidity so low that the expanded discharge is unsatured as it enters the system.
- The rating of the aircraft units is given in terms of kg of air per unit time at the design DART. Thus the machine's capacity giving *m* kg/s of air at a DART of t_0 to maintain a cabin at a temperature t_i is given by: $Q = \dot{m}\ c_p\ (t_0 - t_i)$.

2.4.4 Necessity of Cooling the Aeroplane

In case of an aircraft it is a common assumption there is no need of cooling or air-conditioning since the temperature is low at higher altitude as compared to ground. But, on the contrary, there are several external and internal heat sources which add heat in the passenger cabin resulting in the increase of air temperature in the cabin. Normally, the temperature of air should not exceed 22ºC in the passenger compartment and the temperature to the tune of –5ºC is required for preserving drinks and foods.

The following sources add heat in the compartment:

1. External heat sources:

(*i*) Heat from the *sun*.

(*ii*) The temperature of air entering the cabin is higher than the ambient temperature (due to *compression of air* from ambient pressure to the cabin pressure since the pressure of the atmosphere decreases with increase in altitude).

(*iii*) The increase in temperature of air, due to *skin friction*, when the fast moving aircraft passes through air.

- The actual rise in temperature will be 80 to 90% of the rise calculated theoretically because of heat transfer from hot compressed air to the colder surroundings.
- An aeroplane moving with 1000 km/h will experience 50ºC rise in temperature of the surface.

2. Internal sources:

(*i*) Heat flow from the occupants (Human body continuously generates heat at a rate of 400 kJ/h at rest).

(*ii*) Heat generated by the electrical and electronic equipments.

(*iii*) Heat flow from the engine parts of the aircraft.

- Air cycle refrigeration systems are commonly used now-a-days in air-conditioning of both commercial and military aircraft principally because of *extremely light weight, compact equipment* which is typical of air cycle system. The air cycle, air-conditioning has not been found economical in residential and commercial buildings because higher power is required to operate an air cycle system than a conventional vapour cycle system.
- In designing an aeroplane the following loads are considered:

  (*i*) Engine load (known as alive load).

  (*ii*) Fuel load.

  (*iii*) Dead load (or pay load).

In the passenger plane, the pay load is human load and aeroplane structural load.

2.4.5 Types of Aircraft Refrigeration Systems

Following are the various types of air-refrigeration systems employed for aircrafts these days:

1. Basic or simple air-refrigeration system

   (*a*) Without evaporative cooling, (*b*) With evaporative cooling.

2. Boot strap air-refrigeration system

   (*a*) Without evaporative cooling, (*b*) With evaporative cooling.

3. Regenerative air-refrigeration system.
4. Reduced ambient air-refrigeration system.

2.4.5.1 Basic or Simple air-refrigeration system

(*a*) ***Without evaporative cooling:***

The *main components* of this system are:

(*i*) Main compressor (driven by a gas turbine); (*ii*) A heat exchanger;

(*iii*) A cooling turbine; (*iv*) An exhaust fan.

Fig. 2.19 shows the arrangement of components used in simple *air-refrigeration system.* A portion of the compressed air from *the main compressor is passed through a heat exchanger*. The compressed air is cooled in the heat exchanger by passing the low *temperature ram air*. The air coming out from the heat exchanger expands in the cooling turbine and temperature of the air coming out from the cooling turbine is far below the surrounding air *temperatutre*. This cold air is directly passed through the freezing chamber if it is required for food preservation and it is mixed with the ram air and then passed into the cabin if it used for *air-conditioning the cabin*. The work developed by the cooling turbine is used to run the exhaust fan.

- *This system is good for ground surface cooling and for low flight speeds.*

The *various processes* (without pressure and friction losses) for the cycle are: (Refer to Fig. 2.20)

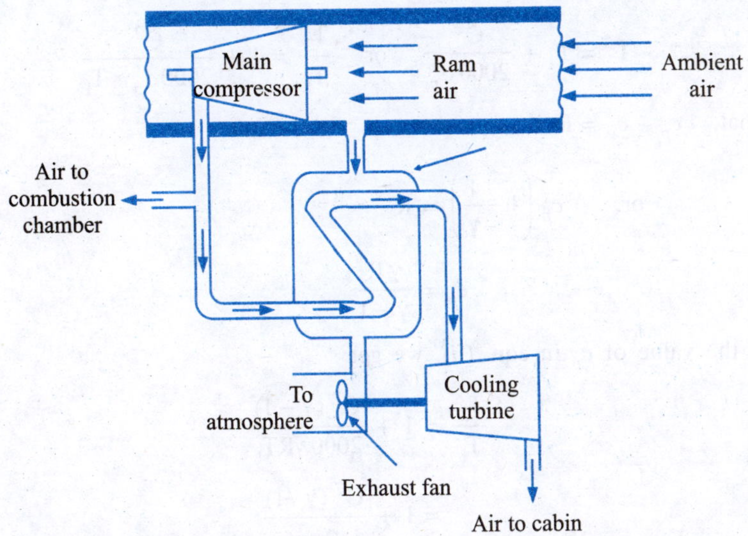

Fig. 2.19. Basic or simple air-refrigeration system (without evaporative cooling).

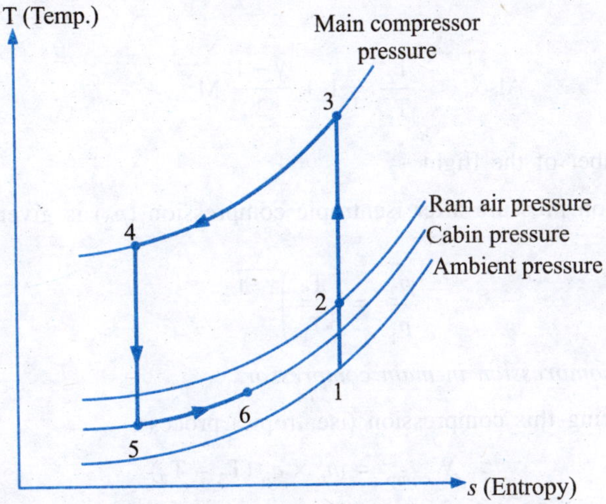

Fig. 2.20. T-*s* diagram for simple air-refrigeration system.

(*i*) **Process 1-2:** *Ramming of air.*

If C is the aircraft velocity or velocity of air relative to the aircraft in metres per second, then kinetic energy (K.E.) of outside air relative to aircraft is,

$$\text{K.E.} = \frac{C^2}{2} \text{ J/kg}, \quad \text{or,} \quad \frac{C^2}{2000} \text{ kJ/kg} \qquad ...(i)$$

From energy equation, we have:

$$h_2 - h_1 = \frac{C^2}{2000}$$

$$c_p T_2 - c_p T_1 = \frac{C^2}{2000}$$

$$\therefore \quad T_2 = T_1 + \frac{C^2}{2000\, c_p}, \quad \text{or,} \quad \frac{T_2}{T_1} = 1 + \frac{C^2}{2000\, c_p \times T_1} \qquad ...(ii)$$

We know that, $c_p - c_v = R$

$$\text{or,} \quad c_p\left(1 - \frac{1}{\gamma}\right) = R \qquad \left(\because \frac{c_p}{c_v} = \gamma\right)$$

$$\therefore \quad c_p = \frac{\gamma R}{\gamma - 1}$$

Substituting the value of c_p in eqn. (*ii*), we get:

$$\frac{T_2}{T_1} = 1 + \frac{C^2(\gamma - 1)}{2000\,\gamma R T_1} \qquad ...(iii)$$

$$= 1 + \frac{C^2(\gamma - 1)}{2a^2} \qquad ...(iv)$$

where a = *local sonic or acoustic velocity* at the ambient air condition = $\sqrt{\gamma R T_1}$, where R is in kJ/kg K.

$$\text{Also,} \quad \frac{T_2}{T_1} = 1 + \frac{\gamma - 1}{2} M^2$$

where, M = Mach number of the flight.

Again, the stagnation pressure after isentropic compression (p_2) is given by:

$$\frac{p_2}{p_1} = \left(\frac{T_2}{T_1}\right)^{\frac{\gamma}{\gamma - 1}}$$

(*ii*) **Process 2-3:** *Compression in main compressor.*

The work done during this compression (isentropic) process,

$$W_{comp.} = m_a \times c_p (T_3 - T_2) \qquad ...(v)$$

where m_a = Mass of air bled from the air compressor for refrigeration purposes.

(*iii*) **Process 3-4:** *Cooling of ram air in heat exchanger.*

The heat rejected in the heat exchanger during the cooling process is given by:

$$Q_{rej.} = m_a \times c_p \times (T_3 - T_4) \qquad ...(vi)$$

(*iv*) **Process 4-5:** *Expansion in turbine.*

The work done by the cooling turbine during this expansion process is given by:

$$W_{exp.} = m_a \times c_p \times (T_4 - T_5) \qquad ...(vii)$$

The work of this turbine is used to drive the exhaust fan which draws the cooling air from the heat exchanger.

(*v*) **Process 5-6:** *Air getting heated to cabin temperature.*

The refrigerating effects produced or heat absorbed is given by:

$$Q_{ref.} = m_a \times c_p \times (T_6 - T_5) \quad ...(viii)$$

where, T_6 = Inside temperature of the cabin.

Now, *C.O.P. of air cycle*

$$= \frac{\text{Refrigerating effect produced}}{\text{Work done}} = \frac{m_a \times c_p \times (T_6 - T_5)}{m_a \times c_p \times (T_3 - T_2)} = \frac{T_6 - T_5}{T_3 - T_2} \quad ...(ix)$$

If Q tonnes of refrigeration is the cooling load in the cabin, then the *air required for the refrigeration purpose,*

$$m_a = \frac{14000 \times Q}{c_p \times (T_6 - T_5)} \text{ kg/h} \quad ...(2.11)$$

Power required for the refrigeration system,

$$P = \frac{m_a \times c_p \times (T_3 - T_2)}{3600} \text{ kW} \quad ...(2.12)$$

C.O.P. of the refrigerating system

$$= \frac{14000\ Q}{m_a \times c_p \times (T_3 - T_2)} = \frac{14000\ Q}{P \times 3600} \quad ...(2.13)$$

Note: The value of c_p for air is taken as 1 kJ/kg K.

(*b*) ***With evaporative cooling:***

In case of *evaporative cooling* (Fig. 2.21) the compressed air from heat exchanger is passed through *evaporator*, for cooling by vaporisation of the low boiling point fluid which may be a refrigerant, ether or liquid nitrogen, etc. Thus air is further cooled to state 4′ as depicted in Fig. 2.22. The temperature of air at state 4 (without evaporative cooling) is further lowered due to vaporisation of the low boiling point fluid as shown by the state 4′. The magnitude by which the temperature T_4 is lowered, depends upon particular requirement.

- If Q tonnes of refrigeration is the cooling load in the cabin, then the *air required for the refrigeration purpose,*

$$m_a = \frac{14000\ Q}{c_p \times (T_6 - T_{5'})} \text{ kg/h} \quad ...(2.14)$$

- *Power required for the refrigerating system,*

$$P = \frac{m_a \times c_p \times (T_3 - T_2)}{3600} \text{ kW} \quad ...(2.15)$$

- *C.O.P. of the refrigerating system,*

$$\text{C.O.P.} = \frac{14000Q}{m_a \times c_p \times (T_3 - T_2)} = \frac{14000Q}{P \times 3600} \quad ...(2.16)$$

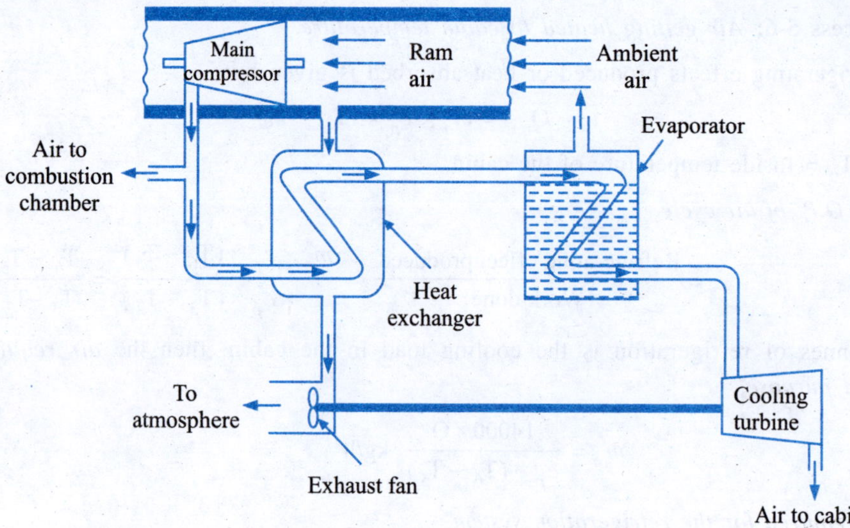

Fig. 2.21. Simple air-refrigeration system (with evaporative cooling).

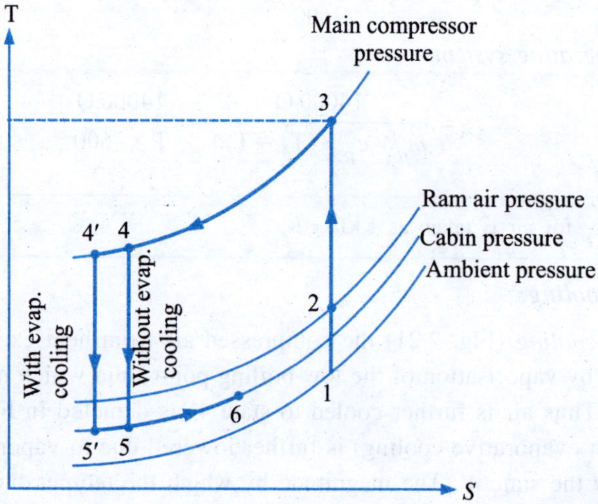

Fig. 2.22. T-*s* diagram for simple air-refrigeration system (with evaporative cooling).

- *The initial mass of evaporant* (m_e) *required to be carried for the given flight time is given by:*

$$m_e = \frac{Q_e \times t}{h_{fg}} \qquad ...(2.17)$$

where, Q_e = Heat to be removed in evaporation in kJ/h,

t = Flight time in hours, and

h_{fg} = Latent heat of evaporation in kJ/kg.

Note: When cooling of 45 minutes duration or less is required, it may be advantageous to use evaporative cooling alone.

Example 2.31. *A simple air cooled system is used for an aeroplane having a load of 9 tonnes. The atmospheric pressure and temperature are 0.9 bar and 10ºC respectively. During ramming pressure increases to 1.013 bar. In the heat exchanger, the temperature of air is reduced by 55ºC. The pressure in the cabin is 1.01 bar and the temperature of air leaving the cabin is 25ºC. Determine:*

(i) *Power required to take the load of cooling in the cabin;*

(ii) *C.O.P. of the system.*

Assume that all the expansions and compressions are isentropic. The pressure of the compressed air is 4 bar.

Solution: Refer to Fig. 2.23

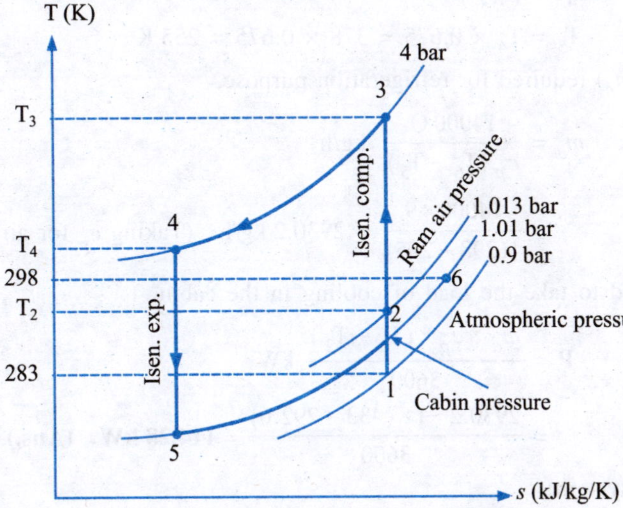

Fig. 2.23. T-*s* diagram.

Given: Cooling load Q = 9 tonnes;

$p_1 = 0.9$ bar; $p_2 = 1.013$ bar;

$p_3 = p_4 = 4$ bar; $p_5 = p_6 = 1.01$ bar

$T_1 = 10 + 273 = 283$ K; $T_6 = 25 + 273 = 298$ K;

Let, T_2 = Temperature of air at the end of ramming or entering the main compressor,

T_3 = Temperature of air leaving the compressor,

T_4 = Temperature of air leaving the heat exchanger, and

T_5 = Temperature of air leaving the cooling turbine.

We know that,

$$\frac{T_2}{T_1} = \left(\frac{p_2}{p_1}\right)^{\frac{\gamma-1}{\gamma}} = \left(\frac{1.013}{0.9}\right)^{\frac{1.4-1}{1.4}} = 1.034$$

$$\therefore \quad T_2 = T_1 \times 1.034 = 283 \times 1.034 = 292.6 \text{ K}$$

Similarly, $$\frac{T_3}{T_2} = \left(\frac{p_3}{p_2}\right)^{\frac{\gamma-1}{\gamma}} = \left(\frac{4}{1.013}\right)^{\frac{1.4-1}{1.4}} = 1.48$$

$$\therefore \quad T_3 = T_2 \times 1.48 = 292.6 \times 1.48 = 433 \text{ K} = 160°\text{C}$$

Since the temperature of air is reduced by 55ºC in the heat exchanger, therefore temperature of air leaving the heat exchanger,

$$T_4 = (160 - 55) + 273 = 378 \text{ K}.$$

We know that, $$\frac{T_5}{T_4} = \left(\frac{p_5}{p_4}\right)^{\frac{\gamma-1}{\gamma}} = \left(\frac{1.01}{4}\right)^{\frac{1.4-1}{1.4}} = 0.675$$

$$\therefore \quad T_5 = T_4 \times 0.675 = 378 \times 0.675 = 255 \text{ K}$$

The mass of air (m_a) required for refrigeration purpose,

$$m_a = \frac{14000\,Q}{c_p\,(T_6 - T_5)} \text{ kg/h} \qquad \text{...[Eqn. (2.11)]}$$

$$= \frac{14000 \times 9}{1(298 - 255)} = 2930.2 \text{ kg/h} \quad (\text{Taking } c_p \text{ for air} = 1 \text{ kJ/kg K})$$

∴ Power required to take the load of cooling in the cabin,

$$\mathbf{P} = \frac{m_a \times c_p \times (T_3 - T_2)}{3600} \text{ kW} \qquad \text{...(Eqn. 2.12)}$$

$$= \frac{2930.2 \times 1 \times (433 - 292.6)}{3600} = \mathbf{114.28\ kW} \quad \textbf{(Ans.)}$$

(*ii*) **C.O.P. of the system:**

$$\text{C.O.P.} = \frac{14000\,Q}{P \times 3600}$$

$$= \frac{14000 \times 9}{114.28 \times 3600} = \mathbf{0.306} \quad \textbf{(Ans.)} \qquad \text{...(Eqn. 2.16)}$$

Example 2.32. *An aircraft refrigeration plant has to handle a cabin load of 25 tonnes. The atmospheric temperature is 16ºC. The atmospheric air is compressed to a pressure of 0.96 bar and temperature of 29ºC due to ram action. This air is then further compressed in a compressor to 4.8 bar, cooled in a heat exchanger to 66ºC, expanded in a turbine to 1 bar pressure and supplied to the cabin. The air leaves the cabin at a temperature of 26ºC. The isentropic efficincies of both compressor and turbine are 0.9. Calculate:*

(*i*) *The mass of air circulated per minute;*

(*ii*) *C.O.P.*

Take for air: c_p *= 1.005 kJ/kg K and* γ *= 1.4.*

Solution: Refer to Fig. 2.24

Given: Q = 25 tonnes; $T_1 = 16 + 273 = 289$ K; $p_2 = 0.96$ bar; $T_2 = 29 + 273 = 302$ K;

$p_3 = p_{3'} = p_4 = 4.8$ bar; $T_4 = 66 + 273 = 339$ K; $p_5 = p_{5'} = 1$ bar; $T_6 = 26 + 273 = 299$ K;

$\eta_C = \eta_T = 0.9$; $c_p = 1.005$ kJ/kg K; $\gamma = 1.4$.

Let, T_3 = Temperature of air after isentropic compression in the compressor,

T_3 = Actual temperature of the air leaving the compressor,

T_5 = Temperature of the air leaving the turbine after isentropic expansion, and

$T_{5'}$ = Actual temperature of the air leaving the turbine.

For the isentropic compression process 2-3, we have:

$$\frac{T_3}{T_2} = \left(\frac{p_3}{p_2}\right)^{\frac{\gamma-1}{\gamma}} = \left(\frac{4.8}{0.96}\right)^{\frac{1.4-1}{1.4}} = 1.584$$

$$\therefore \quad T_3 = T_2 \times 1.584 = 302 \times 1.584 = 478.4 \text{ K}$$

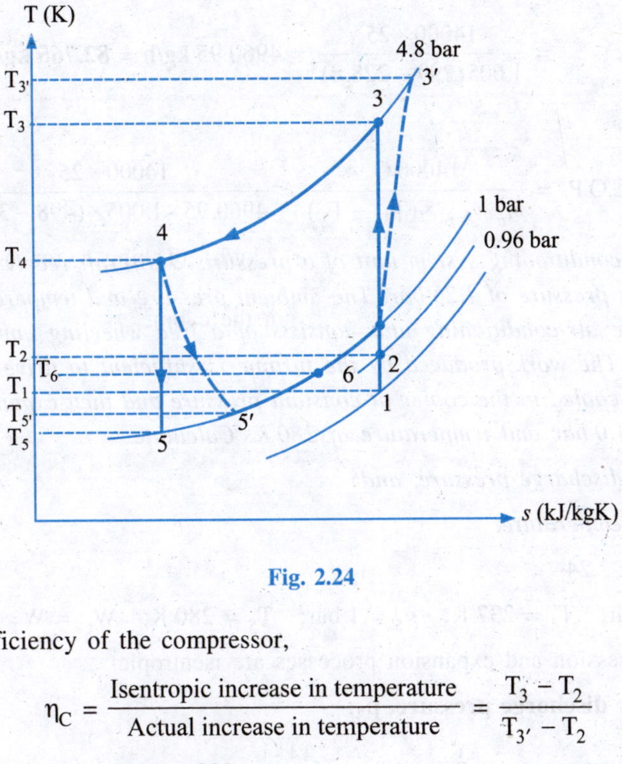

Fig. 2.24

The isentropic efficiency of the compressor,

$$\eta_C = \frac{\text{Isentropic increase in temperature}}{\text{Actual increase in temperature}} = \frac{T_3 - T_2}{T_{3'} - T_2}$$

$$\text{or,} \quad 0.9 = \frac{478.4 - 302}{T_{3'} - 302}$$

$$\therefore \quad T_{3'} = \frac{(478.4 - 302)}{0.9} + 302 = 498 \text{ K}$$

Now, for isentropic expansion process 4-5, we have:

$$\frac{T_4}{T_5} = \left(\frac{p_4}{p_5}\right)^{\frac{\gamma-1}{\gamma}} = \left(\frac{4.8}{1}\right)^{\frac{1.4-1}{1.4}} = 1.565$$

$$\therefore \quad T_5 = \frac{T_4}{1.565} = \frac{339}{1.565} = 216.6 \text{ K}$$

The isentropic efficiency of the turbine,

$$\eta_T = \frac{\text{Actual increase in temperature}}{\text{Isentropic increase in temperature}} = \frac{T_4 - T_{5'}}{T_4 - T_5}$$

or, $$0.9 = \frac{339 - T_{5'}}{339 - 216.6}, \quad \text{or,} \quad T_{5'} = 339 - 0.9 \times (339 - 216.6) = 228.8 \text{ K}$$

(*i*) Mass of air circulated for minute, m_a:

Mass of air circulated,

$$m_a = \frac{14000\,Q}{c_p(T_6 - T_{5'})} \text{ kg/h}$$

$$= \frac{14000 \times 25}{1.005(299 - 228.8)} = 4960.95 \text{ kg/h} = \mathbf{82.765\ kg/min} \quad \textbf{(Ans.)}$$

(*ii*) C.O.P.

$$\text{C.O.P.} = \frac{14000\,Q}{m_a \times c_p \times (T_{3'} - T_2)} = \frac{14000 \times 25}{4960.95 \times 1.005 \times (498 - 302)} = \mathbf{0.358} \quad \textbf{(Ans.)}$$

Example 2.33. *An air-conditioning system unit of a pressurised aircraft receives its air from the jet engine compressor at a pressure of 1.25 bar. The ambient pressure and temperature are 0.2 bar and 237 K respectively. The air-conditioning unit consists of a free wheeling compressor and turbine mounted on one shaft. The work produced by the turbine is sufficient to drive the compressor. The compressed air is then cooled in the cooler at constant pressure and then expanded in the turbine to the cabin pressure of 1.0 bar and temperature of 280 K. Calculate:*

(*i*) *The compressor discharge pressure, and*

(*ii*) *The cooler exit temperature.* **[AMIE]**

Solution: Refer to Fig. 2.24

Given: $p_1 = 0.2$ bar; $T_1 = 237$ K; $p_5 = 1$ bar; $T_5 = 280$ K; $W_C = W_T$

Assume that compression and expansion processes are isentropic.

(*i*) The compressor discharge pressure, p_3:

$$\frac{T_2}{T_1} = \left(\frac{p_2}{p_1}\right)^{\frac{\gamma-1}{\gamma}} = \left(\frac{1.25}{0.2}\right)^{\frac{1.4-1}{1.4}} = 1.688$$

$$\therefore \quad T_2 = T_1 \times 1.688 = 237 \times 1.688 = 400 \text{ K}$$

Also, $$W_C = W_T, \quad \text{or,} \quad T_3 - T_2 = T_4 - T_5$$

$$T_2\left(\frac{T_3}{T_2} - 1\right) = T_3\left(\frac{T_4}{T_5} - 1\right)$$

or, $$T_2\left[\left(\frac{p_3}{p_2}\right)^{\frac{\gamma-1}{\gamma}} - 1\right] = T_5\left[\left(\frac{p_4}{p_5}\right)^{\frac{\gamma-1}{\gamma}} - 1\right]$$

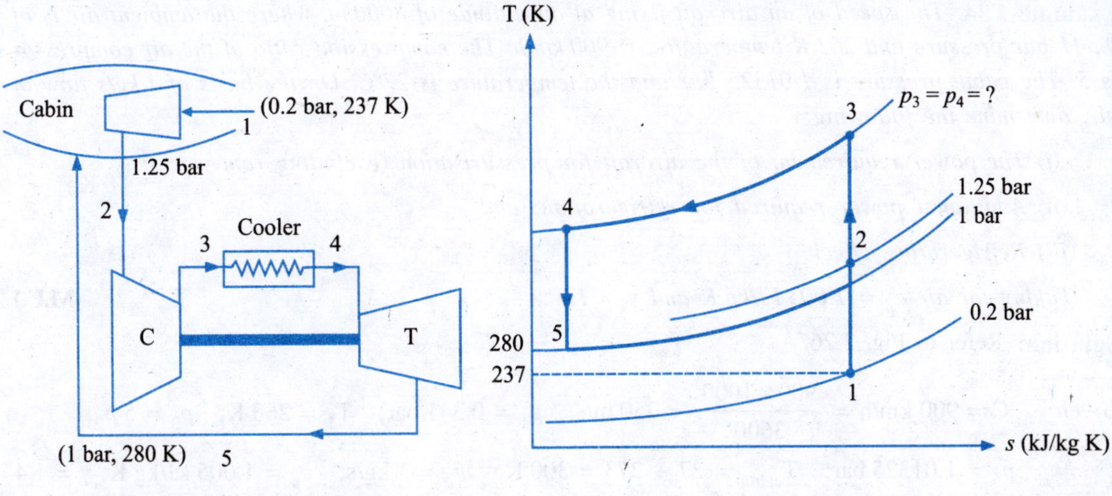

(a) Schematic arrangement. (b) T-s diagram.

Fig. 2.25

$$\text{or,} \quad \frac{T_2}{T_5}\left[\left(\frac{p_3}{p_2}\right)^{\frac{\gamma-1}{\gamma}} - 1\right] = \left[\left(\frac{p_4}{p_5}\right)^{\frac{\gamma-1}{\gamma}} - 1\right]$$

$$\frac{400}{280}\left[\left(\frac{p_3}{1.25}\right)^{\frac{1.4-1}{1.4}} - 1\right] = \left(\frac{p_3}{1}\right)^{\frac{1.4-1}{1.4}} - 1 \qquad [\because p_4 = p_3]$$

$$1.429 \times \left(\frac{p_3}{1.25}\right)^{0.2857} - 1.429 = \left(\frac{p_3}{1}\right)^{0.2857} - 1$$

$$1.429 \times \frac{(p_3)^{0.2857}}{(1.25)^{0.2857}} - 1.429 = (p_3)^{0.2857} - 1$$

$$1.34\ (p_3)^{0.2857} = (p_3)^{0.2857} + 0.429$$

$$(p_3)^{0.2857} = \frac{0.429}{0.34} = 1.26$$

$$\therefore \quad p_3 = (1.26)^{1/0.2857} = \mathbf{2.24\ bar} \quad \textbf{(Ans.)}$$

(*ii*) **The cooler exit temperature, T_4:**

$$\frac{T_4}{T_5} = \left(\frac{p_4}{p_5}\right)^{\frac{\gamma-1}{\gamma}} = \left(\frac{2.24}{1}\right)^{\frac{1.4-1}{1.4}} = 1.259$$

$$\therefore \quad T_4 = T_5 \times 1.259 = 280 \times 1.259 = \mathbf{352.5\ K} \quad \textbf{(Ans.)}$$

Example 2.34. *The speed of an aircraft flying at an altitude of 8000 m, where the ambient air is at 0.341 bar pressure and 263 K temperature, is 900 km/h. The compression ratio of the air compressor is 5. The cabin pressure is 1.01325 bar and the temperature is 27°C. On the basis of 1 kg/s flow of air, determine the following:*

(i) The power requirement of the aircraft for pressurisation (excluding ram work);

(ii) Additional power required for refrigeration; and

(iii) Refrigerating effect.

Taking for air c_p = 1.005 kJ/kg K and γ = 1.4 **(M.U.)**

Solution: Refer to Fig. 2.26.

Given: $C = 900 \text{ km/h} = \dfrac{900 \times 1000}{3600} = 250 \text{ m/s}$; $p_1 = 0.341$ bar; $T_1 = 263$ K; $p_3 = 5\,p_2$;

$p_5 = 1.01325$ bar; $T_{cabin} = 27 + 273 = 300$ K; $m_a = 1$ kg/s; $c_p = 1.005$ kJ/kg K; $\gamma = 1.4$

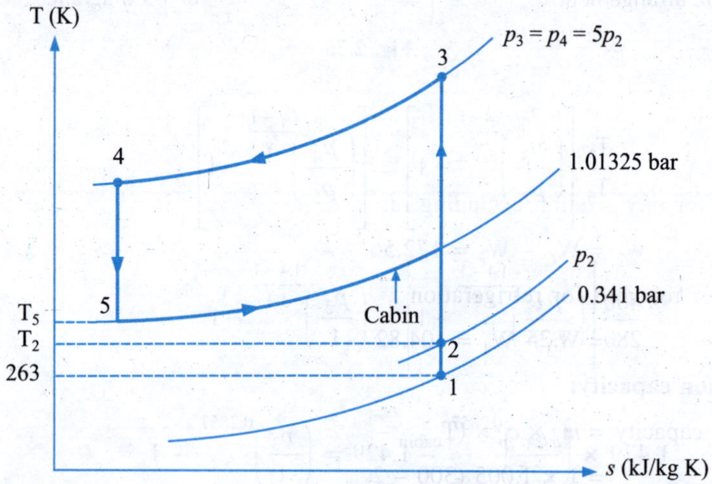

Fig. 2.26

(*i*) The power requirement of the aircraft for pressurisation (excluding ram work):

We know that,
$$T_2 = T_1 + \frac{C^2}{2000c_p} = 263 + \frac{(250)^2}{2000 \times 1.005} = 294.1 \text{ K}$$

Also,
$$\frac{p_2}{p_1} = \left(\frac{T_2}{T_1}\right)^{\frac{\gamma}{\gamma - 1}} = \left(\frac{294.1}{263}\right)^{\frac{1.4}{1.4-1}} = 1.479$$

$$\therefore \quad p_2 = p_1 \times 1.479 = 0.341 \times 1.479 = 0.504 \text{ bar}$$

Power required for pressurisation (excluding ram work),

$$W_1 = m_a c_p T_2 \left[\left(\frac{p_{cabin}}{p_2}\right)^{\frac{\gamma-1}{\gamma}} - 1 \right]$$

$$= 1 \times 1.005 \times 291.4 \left[\left(\frac{1.01325}{0.504} \right)^{\frac{1.4-1}{1.4}} - 1 \right] = \mathbf{64.67\ kW} \quad \textbf{(Ans.)}$$

(*ii*) **Additional power required for refrigeration:**

Compressor discharge temperature,

$$T_3 = T_2 \left(\frac{p_3}{p_2} \right)^{\frac{\gamma - 1}{\gamma}} = 294.1 \times (5)^{\frac{1.4-1}{1.4}} = 465.8 \text{ K}$$

$$p_3 = p_4 = 5p_2 = 5 \times 0.504 = 2.52 \text{ bar}$$

Expander exit temperature T_5 (assuming $T_4 = T_2$):

$$\frac{T_5}{T_4} = \left(\frac{p_5}{p_4} \right)^{\frac{\gamma - 1}{\gamma}} = \left(\frac{1.01325}{2.52} \right)^{\frac{1.4-1}{1.4}} = 0.7708$$

$$\therefore \quad T_5 = T_4 \times 0.7708 = 294.1 \times 0.7708 = 226.7 \text{ K}$$

Compressor work, $W_C = \dot{m}_a \times c_p (T_3 - T_2) = 1 \times 1.005 (465.8 - 294.1) = 172.56$ kW

Expander work, $W_e = \dot{m}_a \times c_p \times (T_4 - T_5) = 1 \times 1.005 \times (294.1 - 226.7) = 67.74$ kW

Power required for refrigeration excluding ram work,

$$W_2 = W_C - W_e = 172.56 - 67.74 = 104.82 \text{ kW}$$

Additional power required for refrigeration

$$= W_2 - W_1 = 104.82 - 64.67 = \mathbf{40.15\ kW} \quad \textbf{(Ans.)}$$

(*iii*) **Refrigeration capacity:**

$$\text{Refrigeration capacity} = m_a \times c_p \times (T_{\text{cabin}} - T_5)$$
$$= 1 \times 1.005 (300 - 226.7) = \mathbf{73.67\ kW} \quad \textbf{(Ans.)}$$

Example 2.35. *Ambient conditions for an aircraft cruising at 1000 km/h are 0.35 bar and –15°C. The cabin temperature is 25°C and turbine exit pressure is 1.06 bar. The pressure ratio of compressor is 3.0. Assuming 100 per cent efficiency for ram effect, compressor and turbine and ideal heat exchanger, determine for simple gas refrigeration cycle of 18 tonnes capacity:*

(*i*) *Temperatures and pressures at all points of cycle;*

(*ii*) *Mass flow rate and volume flow rates at compressor inlet and turbine outlet;*

(*iii*) *Work requirement;*

(*iv*) *C.O.P.*

Assume: $c_p = 1.005$ *kJ/kg K;* $R_{air} = 0.286$ *kJ/kg K;* $\gamma = 1.4$.

Solution: Refer to Fig. 2.27.

Given: $C = 1000 \text{ km/h} = \frac{1000 \times 1000}{60 \times 60} = 277.78$ m/s; $p_1 = 0.35$ bar, $T_1 = -15 + 273 = 258$ K;

$T_6 = 25 + 273 = 298$ K; $p_5 = 1.06$ bar; $p_3 = 3p_2$; Capacity, Q = 18 tonnes;

$c_p = 1.005$ kJ/kg K; $R_{air} = 0.286$ kJ/kg K; $\gamma = 1.4$.

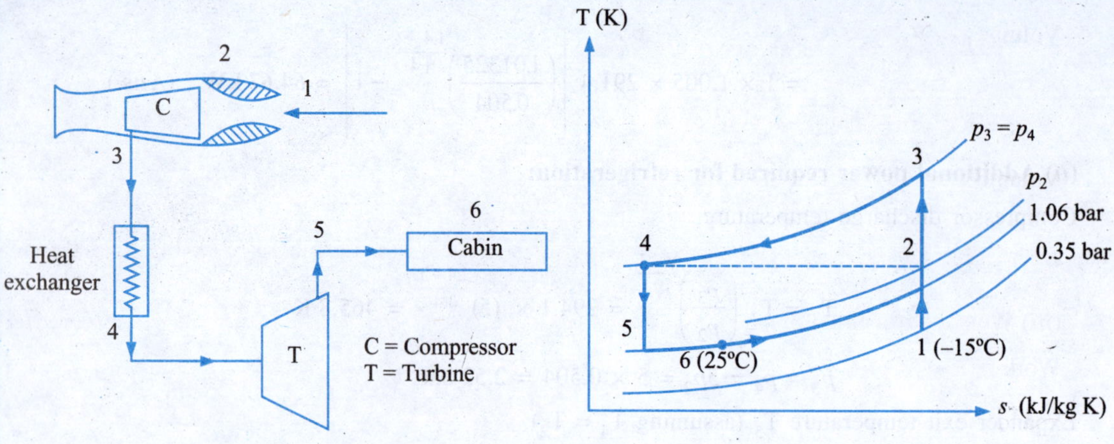

(*a*) Schematic arrangement. (*b*) T-*s* diagram.

Fig. 2.27

(*i*) **Temperatures and pressures at all points of cycle:**

$$\mathbf{T_2} = T_1 + \frac{C^2}{2000\,c_p} = 258 + \frac{(277.78)^2}{2000 \times 1.005} = 296.4 \text{ K} \quad \textbf{(Ans.)}$$

$$\frac{p_2}{p_1} = \left(\frac{T_2}{T_1}\right)^{\frac{\gamma}{\gamma-1}} = \left(\frac{296.4}{258}\right)^{\frac{1.4}{1.4-1}} = 1.625$$

$$\therefore \quad \mathbf{p_2} = p_1 \times 1.625 = 0.35 \times 1.625 = \mathbf{0.569\ bar} \quad \textbf{(Ans.)}$$

$$\mathbf{p_3} = 3\,p_2 = 3 \times 0.569 = \mathbf{1.707\ bar} \quad \textbf{(Ans.)}$$

$$\frac{T_3}{T_2} = \left(\frac{p_3}{p_2}\right)^{\frac{\gamma-1}{\gamma}} = (3)^{\frac{1.4-1}{1.4}} = 1.369$$

$$\therefore \quad \mathbf{T_3} = 296.4 \times 1.369 = \mathbf{405.8\ K} \quad \textbf{(Ans.)}$$

$$\mathbf{p_4 = p_3 = 1.707\ bar}; \quad \mathbf{T_4 = T_2 = 296.4\ K} \quad \textbf{(Ans.)}$$

$$\frac{T_5}{T_4} = \left(\frac{p_5}{p_4}\right)^{\frac{\gamma-1}{\gamma}} = \left(\frac{1.06}{1.707}\right)^{\frac{1.4-1}{1.4}} = 0.8727$$

$$\therefore \quad \mathbf{T_5} = T_4 \times 0.8727 = 296.4 \times 0.8727 = \mathbf{258.7\ K} \quad \textbf{(Ans.)}$$

Thus the temperatures and pressures at all points in the cycle are determined.

(*ii*) **Mass flow rate, volume flow rates at compressor inlet and turbine outlet:**

$$\text{Mass flow rate,} \quad \mathbf{m_a} = \frac{14000\,Q}{c_p\,(T_6 - T_5)} \text{ kg/h}$$

$$= \frac{14000 \times 18}{1.005(298 - 258.7)} = 6380.3 \text{ kg/h}, \quad \text{or,} \quad \mathbf{1.77\ kg/s} \quad \textbf{(Ans.)}$$

Volume flow rate at compressor inlet,

$$\mathbf{V_2} = \frac{m_a RT_2}{p_2} = \frac{1.77 \times (0.286 \times 10^3) \times 296.4}{0.569 \times 10^5} = \mathbf{2.64\ m^3/s} \quad \textbf{(Ans.)}$$

Volume flow rate at turbine outlet,

$$\mathbf{V_5} = \frac{m_a RT_5}{p_5} = \frac{1.77 \times (0.286 \times 10^3) \times 258.7}{1.06 \times 10^5} = \mathbf{1.235\ m^3/s} \quad \textbf{(Ans.)}$$

(*iii*) **Work requirement:**

Work requirement $= m_a \times c_p \times (T_3 - T_2) = 1.77 \times 1.005 \times (405.8 - 296.4) =$ **194.6 kW** **(Ans.)**

(*iv*) **C.O.P.:**

$$\text{C.O.P.} = \frac{\text{Refrigerating effect}}{\text{Work required}} = \frac{(14000 \times 18/3600)}{194.66} = \mathbf{0.3596} \quad \textbf{(Ans.)}$$

Example 2.36. *An aircraft moving with speed of 200 km/h uses simple gas refrigeration cycle for air-conditioning. The ambient pressure and temperature are 0.36 bar and –12°C respectively. The pressure ratio of compressor is 5. The heat exchanger effectiveness is 0.94. The isentropic efficiencies of compressor and expander are 0.82 each. The cabin pressure and temperatures are 1.05 bar and 27°C respectively. Determine:*

(*i*) *The temperatures and pressures at all the points of the cycle;*

(*ii*) *The volume flow rates through the compressor inlet and expander outlet for 90 tonnes refrigerator.*

Take for air: $c_p = 1.005$ *kJ/kg K;* $R = 0.287$ *kJ/kg K and* $\gamma = 1.4$.

Solution: Refer to Fig. 2.28.

Given: $C = 900 \text{ km/h} = \dfrac{900 \times 1000}{60 \times 60} = 250 \text{ m/s};\ p_1 = 0.36 \text{ bar};\ T_1 = -12 + 273 = 261 \text{ K};\ p_3/p_2 = 5;$

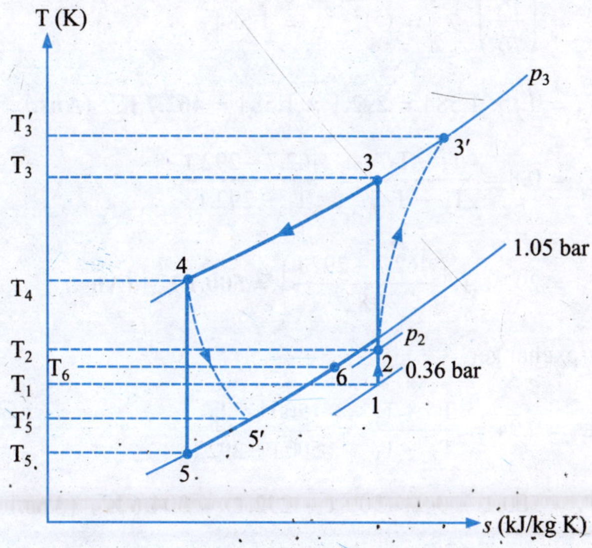

Fig. 2.28. T-*s* diagram

$\in = 0.94$; $\eta_C = \eta_T = 0.82$; Q = 90 tonnes $p_5 = p_6 = 1.05$ bar; $T_6 = 27 + 273 = 300$ K;

$c_p = 1.005$ kJ/kg K; R = 0.287 kJ/kg K; $\gamma = 1.4$.

Let, T_2 and p_2 = Stagnation temperature and pressure of the ambient air entering the compressor,

T_3 and p_3 = Temperature and pressure of the air leaving the compressor after isentropic compression,

$T_{3'}$ = Actual temperature of the air leaving the compressor,

T_4 = Temperature of the air leaving the heat exchanger or entering the expander,

p_4 = pressure of the air leaving the heat exchanger or entering the expander = $p_3 = p_{3'}$,

T_5 = Temperature of the air leaving the expander after isentropic expansion, and

$T_{5'}$ = Actual temperature of the air leaving the expander.

(*i*) **The temperatures and pressures at all the points of the cycle:**

$$T_2 = T_1 + \frac{C^2}{2000 \times c_p}$$

$$\text{or,} \quad \mathbf{T_2} = 261 + \frac{(250)^2}{2000 \times 1.005} = \mathbf{292.1\ K} \quad \textbf{(Ans.)}$$

$$\frac{p_2}{p_1} = \left(\frac{T_2}{T_1}\right)^{\frac{\gamma}{\gamma - 1}} = \left(\frac{292.1}{261}\right)^{\frac{1.4}{1.4 - 1}} = 1.483$$

$$\therefore \quad \mathbf{p_2} = p_1 \times 1.483 = 0.36 \times 1.483 = \mathbf{0.544\ bar} \quad \textbf{(Ans.)}$$

$$\mathbf{p_3} = 5p_2 = 5 \times 0.534 = \mathbf{2.67\ bar} \quad \textbf{(Ans.)}$$

For isentropic compression 2-3, we have:

$$\frac{T_3}{T_2} = \left(\frac{p_3}{p_2}\right)^{\frac{\gamma - 1}{\gamma}} = (5)^{\frac{1.4 - 1}{1.4}} = 1.584$$

$$\therefore \quad \mathbf{T_3} = T_2 \times 1.584 = 292.1 \times 1.584 = \mathbf{462.7\ K} \quad \textbf{(Ans.)}$$

$$\text{Also,} \quad \eta_C = 0.8 = \frac{T_3 - T_2}{T_{3'} - T_2} = \frac{462.7 - 292.1}{T_{3'} - 292.1}$$

$$\therefore \quad T_{3'} = 292.1 + \left(\frac{462.7 - 292.1}{0.82}\right) = \mathbf{500.1\ K} \quad \textbf{(Ans.)}$$

Effectiveness of heat exchanger,

$$\in = 0.94 = \frac{T_{3'} - T_4}{T_{3'} - T_2} = \frac{500.1 - T_4}{500.1 - 292.1}$$

$$\therefore \quad \mathbf{T_4} = 500.1 - 0.94\ (500.1 - 292.1) = \mathbf{304.6\ K} \quad \textbf{(Ans.)}$$

$$\mathbf{p_4} = \mathbf{p_3} = \mathbf{2.67\ bar} \quad \textbf{(Ans.)}$$

For isentropic expansion 4-5, we have:

$$\frac{T_5}{T_4} = \left(\frac{p_5}{p_4}\right)^{\frac{\gamma-1}{\gamma}} = \left(\frac{1.05}{2.67}\right)^{\frac{1.4-1}{1.4}} = 0.766$$

$$\therefore \quad \mathbf{T_5} = T_4 \times 0.766 = 304.6 \times 0.766 = \mathbf{233.3\ K} \quad \textbf{(Ans.)}$$

$$\text{Now,} \quad \eta_E = 0.82 = \frac{T_4 - T_{5'}}{T_4 - T_5} = \frac{304.6 - T_{5'}}{304.6 - 233.3}$$

$$\therefore \quad \mathbf{T_{5'}} = 304.6 - 0.82\,(304.6 - 233.3) = \mathbf{246.1\ K} \quad \textbf{(Ans.)}$$

(*ii*) The volume flow rates through the compressor inlet and expander outlet for 90 tonnes refrigeration:

$$\text{Mass flow rate,} \quad m_a = \frac{14000\,Q}{c_p\,(T_6 - T_{5'})} \times \frac{1}{60}\ \text{kg/min}$$

$$= \frac{14000 \times 90}{1.005(300 - 246)} \times \frac{1}{60} = 386.95\ \text{kg/min}$$

$$\text{and,} \quad p_2V_2 = m_aRT_2$$

$$\therefore \quad \mathbf{V_2} = \frac{m_aRT_2}{p_2} = \frac{386.95 \times (0.287 \times 10^3) \times 292.1}{0.534 \times 10^5}$$

$$= \mathbf{607.5\ m^3/min} \quad \textbf{(Ans.)}$$

$$\text{Similarly,} \quad p_{5'}V_{5'} = m_aRT_{5'}$$

$$\mathbf{V_{5'}} = \frac{m_aRT_{5'}}{p_{5'}} = \frac{386.95 \times (0.287 \times 10^3) \times 246.1}{1.05 \times 10^5}$$

$$= \mathbf{260.3\ m^3/min} \quad \textbf{(Ans.)}$$

Example 2.37. *The cock-pit of a jet plane flying at a speed of 1180 km/h is to be cooled by a simple air cooling system. The cock-pit is to maintained at 24°C and the pressure in the cock-pit is 1 bar. The ambient air pressure and temperature are 0.86 bar and 27°C respectively. The other data is as follows:*

Cock-pit cooling load ... 9 tonnes of refrigeration;

Main compressor pressure ratio ... 4.5;

Ram efficiency ... 88 per cent;

Temperature of air leaving the heat exchanger and entering the cooling turbine ... 58°C;

Pressure drop in the heat exchanger ... 0.51 bar;

Pressure loss between the cooler turbine and cock-pit ... 0.2 bar;

Isentropic efficiencies of main compressor and cooling turbine ... 82 per cent.

Determine the following:

(i) The quality of air passed through the cooling turbine, and

(ii) C.O.P. of the system.

Take for air: c_p = 1 kJ/kg K and γ = 1.4.

Solution: Refer to Fig. 2.29.

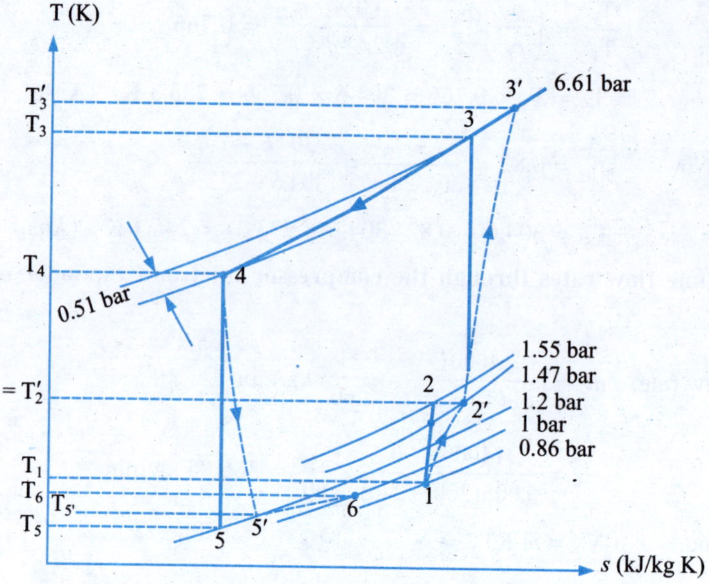

Fig. 2.29. T-*s* diagram.

Given: $C = 1180 \text{ km/h} = \dfrac{1180 \times 1000}{60 \times 60} = 327.8 \text{ m/s}$; $T_6 = 24 + 273 = 297$ K; $p_6 = 1$ bar;

$p_1 = 0.86$ bar; $T_1 = 27 + 273 = 300$ K; $Q = 9$ tonnes of refrigeration; $p_3/p_2 = 4.5$;

$\eta_{Ram} = 88\%$; $T_4 = 58 + 273 = 331$ K; $p_4 = (p_{3'} - 0.5)$ bar;

$p_5 = p_{5'} = p_6 + 0.2 = 1 + 0.2 = 1.2$ bar; $\eta_C = \eta_T = 82\%$.

Let, $T_{2'}$ = Stagnation temperature of the ambient air entering the main compressor,

p_2 = Pressure of air after isentropic ramming, and

$p_{2'}$ = Stagnation pressure of air entering the main compressor.

We know that, $$T_2 = T_1 + \frac{C^2}{2000 c_p} = 300 + \frac{(327.8)^2}{2000 \times 1} = 353.7 \text{ K}$$

and, $$\frac{p_2}{p_1} = \left(\frac{T_2}{T_1}\right)^{\frac{\gamma}{\gamma - 1}} = \left(\frac{353.7}{300}\right)^{\frac{1.4}{1.4-1}} = 1.8$$

$$\therefore \quad p_2 = p_1 \times 1.8 = 0.86 \times 1.8 = 1.55 \text{ bar}$$

Also, Ram efficiency, $$\eta_{Ram} = \frac{\text{Actual pressure rise}}{\text{Isentropic pressure rise}} = \frac{p_{2'} - p_1}{p_2 - p_1}, \quad \text{or,} \quad 0.88 = \frac{p_{2'} - 0.86}{1.55 - 0.86}$$

$$\therefore \quad p_{2'} = 0.86 + 0.88\,(1.55 - 0.86) \simeq 1.47 \text{ bar}$$

For isentropic process 2′-3, we have:

$$\frac{T_3}{T_{2'}} = \left(\frac{p_3}{p_{2'}}\right)^{\frac{\gamma-1}{\gamma}} = (4.5)^{\frac{1.4-1}{1.4}} = 1.537$$

$$\therefore \quad T_3 = T_{2'} \times 1.537 = 353.7 \times 1.537 = 543.6 \text{ K}$$

$$\text{Again,} \quad \eta_C = 0.82 = \frac{T_3 - T_{2'}}{T_{3'} - T_{2'}} = \frac{543.6 - 353.7}{T_{3'} - 353.7}$$

$$\therefore \quad T_{3'} = 353.7 + \frac{543.6 - 353.7}{0.82} = 585.3 \text{ K}$$

$$\text{Now,} \quad p_3 = p_{3'} = 4.5\, p_{2'} = 4.5 \times 1.47 = 6.61 \text{ bar}$$

Pressure drop in the heat exchanger = 0.51 bar

∴ Pressure of air passing through the heat exchanger or at entrance to the cooling turbine,

$$p_4 = p_{3'} - 0.51 = 6.61 - 0.51 = 6.1 \text{ bar}$$

Also there is a pressure loss of 0.2 bar between the cooling turbine and the cockpit. Therefore pressure of air leaving the cooling turbine,

$$p_5 = p_6 + 0.2 = 1 + 0.2 = 1.2 \text{ bar}$$

For isentropic process 4-5, we have:

$$\frac{T_4}{T_5} = \left(\frac{p_4}{p_5}\right)^{\frac{\gamma-1}{\gamma}} = \left(\frac{6.1}{1.2}\right)^{\frac{1.4-1}{1.4}} = 1.59$$

$$\therefore \quad T_5 = \frac{T_4}{1.59} = \frac{331}{1.59} = 208.2 \text{ K}$$

$$\eta_T = 0.82 = \frac{T_4 - T_{5'}}{T_4 - T_5} = \frac{331 - T_{5'}}{331 - 208.2}$$

$$\therefore \quad T_{5'} = 331 - 0.82\,(331 - 208.2) = 230.3 \text{ K}$$

(*i*) The quantity of air passed through the cooling turbine, m_a:

$$m_a = \frac{14000Q}{c_p\,(T_6 - T_{5'})} \text{ kg/h}$$

$$= \frac{14000 \times 9}{1(297 - 230.3)} = \mathbf{1889.06\ kg/h\ (or\ 31.48\ kg/min)} \quad \textbf{(Ans.)}$$

(*ii*) C.O.P. of the system:

$$\text{C.O.P.} = \frac{14000Q}{m_a\, c_p\,(T_{3'} - T_{2'})} = \frac{14000 \times 9}{1889.06 \times (585.3 - 353.7)} = \mathbf{0.288} \quad \textbf{(Ans.)}$$

Example 2.38. *A simple air-refrigeration system is employed in an aeroplane. The main compressor delivers the air at 5.2 bar and 210°C. The air taken from compressor is passed through a heat exchanger, cooled with the help of ram air so that the temperature of air leaving the heat exchanger is 50°C and pressure is 4.7 bar. The cooling turbine drives the exhaust fan which is used to force the ram air through the heat exchanger. The air leaving the heat exchanger passes through the cooling* *turbine and then supplied to cabin at 1 bar. The pressure loss between the cooling turbine and cabin is 0.2 bar. If the rate of flow of air through the cooling turbine is 18 kg/min, determine the following:*

(i) The temperature of the air leaving the expander,

(ii) The power delivered to the ram air which is passed through the heat exchanger, and

(iii) The refrigeration load in tonnes when the temperature of the air leaving the cabin is limited to 24°C.

Assume that the isentropic efficiency of the cooling turbine is 76 per cent and there is no loss of heat from air between the cooling turbine and cabin. Take c_p *= 1 kJ/kg K;* γ *= 1.4.*

Solution: *Given:* $p_3 = 5.2$ bar;

$T_3 = 210 + 273 = 483$ K;

$T_4 = 50 + 273 = 323$ K;

$p_4 = 4.7$ bar;

$T_6 = 24 + 273 = 297$ K;

$p_6 = 1$ bar;

$p_5 = 1.2$ bar ; $m_a = 18$ kg/min;

$\eta_T = 76\%$;

$c_p = 1$ kJ/kg K; $\gamma = 1.4$.

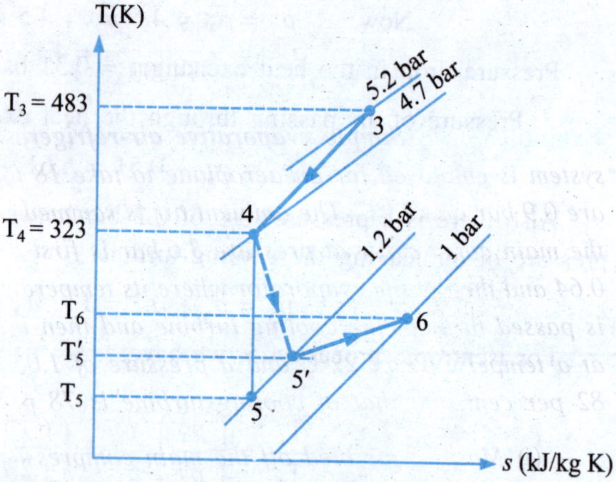

Fig. 2.30. T-*s* diagram

The schematic diagram for the simple air-refrigeration system is shown in Fig. 2.29. The various processes on T-*s* diagram are shown in Fig. 2.30:

- The point 3 represents the air delivered from the comperssor to heat exchanger;
- The point 4 shows the condition of air leaving the heat exchanger;
- The vertical line 4-5 represents the isentropic expansion of air in the cooling turbine;
- The curve 4-5′ shows the actual expansion of air in the cooling turbine due to internal friction;
- The line 5′-6 represents the refrigeration process.

(*i*) **The temperature of air leaving the expander,** $T_{5'}$:

For isentropic expansion 4-5, we have:

$$\frac{T_5}{T_4} = \left(\frac{p_5}{p_4}\right)^{\frac{\gamma-1}{\gamma}} = \left(\frac{1.2}{4.7}\right)^{\frac{1.4.1}{1.4}} = 0.677$$

$$\therefore \quad T_5 = T_4 \times 0.677 = 323 \times 0.677 = 218.7 \text{ K}$$

Now, $$\eta_T = 0.76 = \frac{T_4 - T_{5'}}{T_4 - T_5} = \frac{323 - T_{5'}}{323 - 218.7}$$

$$\therefore \quad T_{5'} = 323 - 0.76\,(323 - 218.7) = \mathbf{243.7\ K} \quad \textbf{(Ans.)}$$

(*ii*) **Power delivered to the ram air which is passed through the heat exchanger:**

Power delivered to the ram air which is passed through the heat exchanger

$$= m_a \times c_p \times (T_4 - T_{5'}) = 18 \times 1 \times (323 - 243.7) = 1427.4 \text{ kJ/min}$$

$$= \frac{1427.4}{60} = \mathbf{23.79\ kW} \quad \textbf{(Ans.)}$$

(*iii*) **Refrigeration load in tonnes:**

Refrigeration load (taken from the cabin)

$$= m_a \times c_p \times (T_6 - T_{5'}) = 18 \times 1 \times (297 - 243.7) = 959.4 \text{ kJ/min}$$

$$= \frac{959.4 \times 60}{14000} = \mathbf{4.11\ tonnes} \quad \textbf{(Ans.)} \qquad (\because 1 \text{ tonne of ref.} = 14000 \text{ kJ/h})$$

Example 2.39. (***Simple evaporative air-refrigeration system***) *A simple evaporative air-refrigeration system is employed for an aeroplane to take 18 tonnes of refrigeration load. The ambient conditions are 0.9 bar and 23°C. The ambient air is rammed isentropically to a pressure of 1 bar. The air leaving the main compressor at pressure 3.6 bar is first cooled in the heat exchanger having effectiveness of 0.64 and then in the evaporator where its temperature is reduced by 6.2°C. The air from the evaporator is passed through the cooling turbine and then it is supplied to the cabin which is to be maintained at a temperature of 22°C and a pressure of 1.05 bar. If the internal efficiency of the compressor is 82 per cent and that of cooling turbine is 78 per cent, determine the following:*

(*i*) *Mass of air bled off the main compressor;*

(*ii*) *Power required for the refrigerating system;*

(*iii*) *C.O.P. of the refrigerating system.* **[Banglore University]**

Solution: Refer to Fig. 2.31.

Given: Q = 18 tonnes; $T_1 = 23 + 273 = 296$ K; $p_1 = 0.92$ bar; $p_2 = 1$ bar; $p_3 = p_{3'} = 3.6$ bar; $\in = 0.62$; $T_6 = 22 + 273 = 295$ K; $p_6 = 1.05$ bar; $\eta_C = 82\%$; $\eta_T = 78\%$.

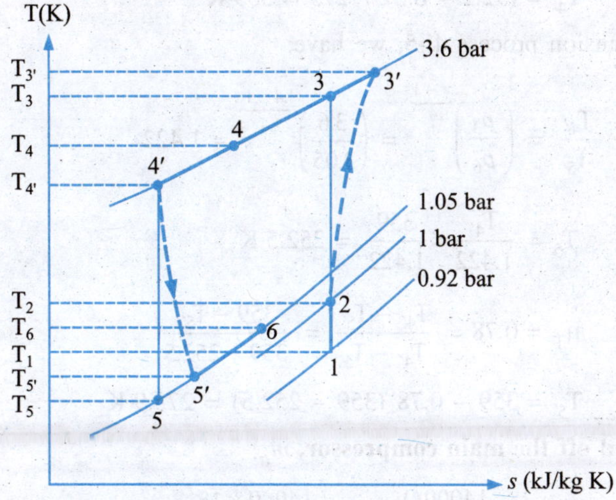

Fig. 2.31. T-*s* diagram.

Let, T_2 = Temperature of air entering the main compressor,

T_3 = Temperature of air after isentropic compression in the main compressor,

$T_{3'}$ = Actual temperature of air leaving the main compressor, and

T_4 = Temperature of air entering the evaporator.

For isentropic ramming process 1-2, we have:

$$\frac{T_2}{T_1} = \left(\frac{p_2}{p_1}\right)^{\frac{\gamma-1}{\gamma}} = \left(\frac{1}{0.92}\right)^{\frac{1.4-1}{1.4}} = 1.024$$

$$T_2 = T_1 \times 1.024 = 296 \times 1.024 = 303.1 \text{ K}$$

For the isentropic compression process 2-3, we have:

$$\frac{T_3}{T_2} = \left(\frac{p_3}{p_2}\right)^{\frac{\gamma-1}{\gamma}} = \left(\frac{3.6}{1}\right)^{\frac{1.4-1}{1.4}} = 1.442$$

$$\therefore \quad T_3 = T_2 \times 1.442 = 303.1 \times 1.442 = 437 \text{ K}$$

Also, $$\eta_C = 0.82 = \frac{T_3 - T_2}{T_{3'} - T_2} = \frac{437 - 303.1}{T_{3'} - 303.1}$$

$$\therefore \quad T_{3'} = 303.1 + \left(\frac{437 - 303.1}{0.82}\right) = 466.4 \text{ K}$$

Effectiveness of beat exchanger,

$$\epsilon = 0.62 = \frac{T_{3'} - T_4}{T_{3'} - T_2} = \frac{466.4 - T_4}{466.4 - 303.1}$$

$$\therefore \quad T_4 = 466.4 - 0.62\,(466.4 - 303.1) = 365.2 \text{ K} = 92.2^\circ\text{C}$$

Since the temperature of air in the evaporator is reduced by 6.2°C, therefore the temperature of air leaving the evaporator is 92.2°C and entering the turbine,

$$T_{4'} = (92.2 - 6.2) + 273 = 359 \text{ K}$$

For isentropic expansion process 4′-5, we have:

$$\frac{T_{4'}}{T_5} = \left(\frac{p_3}{p_6}\right)^{\frac{\gamma-1}{\gamma}} = \left(\frac{3.6}{1.05}\right)^{\frac{1.4-1}{1.4}} = 1.422$$

$$\therefore \quad T_5 = \frac{T_{4'}}{1.422} = \frac{359}{1.422} = 252.5 \text{ K}$$

$$\eta_T = 0.78 = \frac{T_{4'} - T_{5'}}{T_{4'} - T_5} = \frac{359 - T_{5'}}{359 - 252.5}$$

$$\therefore \quad T_{5'} = 359 - 0.78\,(359 - 252.5) = 275.9 \text{ K}$$

(*i*) **Mass of air bled off the main compressor, m_a:**

$$m_a = \frac{14000\,Q}{c_p\,(T_6 - T_{5'})} = \frac{14000 \times 18}{1(295 - 275.9)} = \mathbf{13193.7 \text{ kg/h}} \quad \textbf{(Ans.)}$$

(*ii*) **Power required for the refrigerating system, P:**

$$P = \frac{m_a \times c_p \times (T_{3'} - T_2)}{60 \times 60}$$

$$= \frac{13193.7 \times 1 \times (466.4 - 303.1)}{60 \times 60} = \mathbf{598.5\ kW} \quad \textbf{(Ans.)}$$

(*iii*) **C.O.P. of the refrigerating system:**

$$\text{C.O.P.} = \frac{14000Q}{P \times 3600} = \frac{14000 \times 18}{598.5 \times 3600} = \mathbf{0.117} \quad \textbf{(Ans.)}$$

2.4.5.2 Boot-strap air-refrigeration system

(*a*) ***Without evaporative cooling:***

Refer to Fig. 2.32. In Boot-strap system the ram air is utilised in *two heat exchangers* for cooling purpose. The air from the main compressor is first cooled by the air in first heat exchanger (HE-1). The same air is again compressed in compressor run by the cooling expander turbine and then led to second heat exchanger (HE-2) where again it is cooled by the ram air before passing on to the cooling turbine. The ram air is provided with short path and exit jets and there is no fan to blow in and discharge ram air.

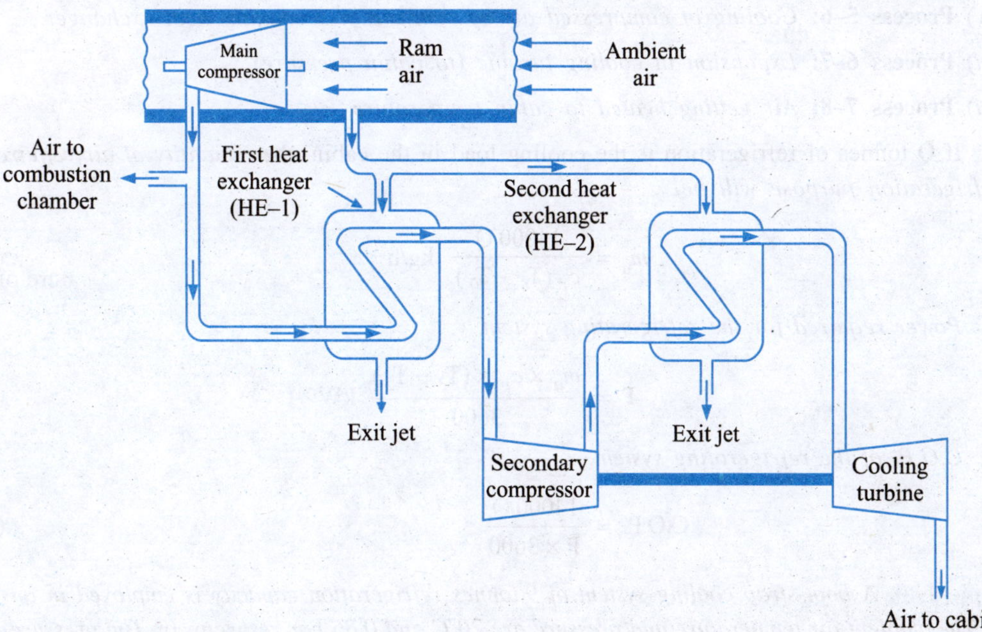

Fig. 2.32. Boot-strap system.

- *This type of cooling system is used in transport aircraft.*

Fig. 2.33 shows the various processes on T-*s* diagram. The processes shown are ideal (without pressure and friction losses).

(*i*) **Process 1–2:** *Ramming of air.*

(*ii*) **Process 2–3:** *Compression of air in main compressor.*

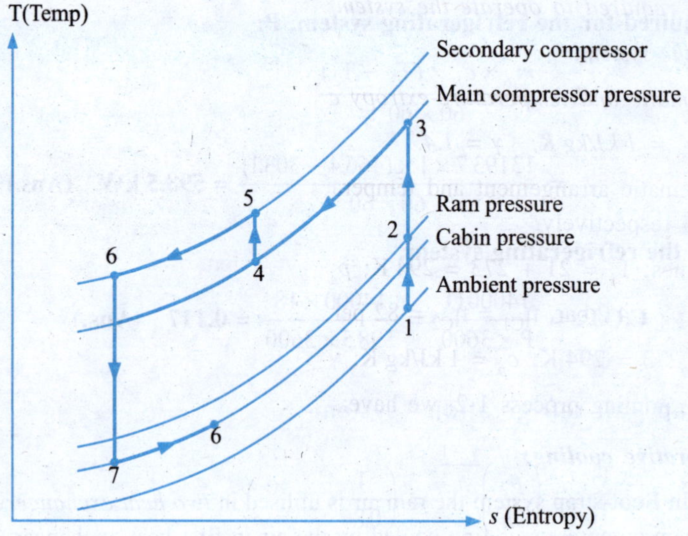

Fig. 2.33. T-*s* diagram.

(*iii*) **Process 3–4:** *Cooling of compressed air by ram air.*

(*iv*) **Process 4–5:** *Compression of air in secondary compressor.*

(*v*) **Process 5–6:** *Cooling of compressed air by ram air in secondary heat exchanger.*

(*vi*) **Process 6–7:** *Expansion in cooling turbine (to cabin pressure).*

(*vii*) **Process 7–8:** *Air getting heated to cabin temperature.*

• If Q tonnes of refrigeration is the cooling load in the cabin, then *quantity of air required for the refrigeration purpose will be,*

$$m_a = \frac{14000\,Q}{c_p\,(T_8 - T_7)} \text{ kg/h} \qquad ...(2.18)$$

• *Power required for the refrigerating system,*

$$P = \frac{m_a \times c_p \times (T_3 - T_2)}{3600} \text{ kW} \qquad ...(2.19)$$

• *C.O.P. of the refrigerating system,*

$$\text{C.O.P.} = \frac{14000\,Q}{P \times 3600} \qquad ...(2.20)$$

Example 2.40. *A boot-strap cooling system of 9 tonnes refrigeration capacity is employed in an aeroplane. The ambient air temperature and pressure are 20°C and 0.86 bar respectively. The pressure of air increases from 0.86 bar to 1 bar due to ramming action of air. The pressure of air discharged from the main compressor in 3.2 bar. The discharge pressure of air from the auxiliary compressor is 4.2 bar. The isentropic efficiency of each of the compressor is 82 per cent, while that of turbine is 86 per cent. 45 per cent of the enthalpy of air discharged from the main compressor is removed in the first heat exchanger and 32 percent of the enthalpy of air discharged from the auxiliary compressor is removed in the second heat exchanger using rammed air. Assuming ramming action to be isentropic, the required cabin pressure of 0.92 bar and temperature of the air leaving the cabin not more than 21°C, find:*

(i) *The power required to operate the system, and*

(ii) *C.O.P. of the system.*

Draw the schematic and temperature-extropy diagrams of the system.

Take for air: $c_p = 1$ *kJ/kg K;* $\gamma = 1.4$.

Solution: The schematic arrangement and temperature-entropy diagram of the system are shown in Figs. 2.32 and 2.34 respectively.

Given: Q = 9 tonnes; $T_1 = 21 + 273 = 294$ K; $p_1 = 0.86$ bar; $p_2 = 1$ bar; $p_3 = p_{3'} = 3.2$ bar;

$p_5 = p_{5'} = p_6 = 4.2$ bar, $\eta_{C1} = \eta_{C2} = 82$ per cent; $\eta_T = 86$ per cent; $p_7 = p_{7'} = p_8 = 0.92$ bar;

$T_8 = 21 + 273 = 294$ K; $c_p = 1$ kJ/kg K; $\gamma = 1.4$.

For isentropic ramming process 1-2, we have:

$$\frac{T_2}{T_1} = \left(\frac{p_2}{p_1}\right)^{\frac{\gamma-1}{\gamma}} = \left(\frac{1}{0.86}\right)^{\frac{1.4-1}{1.4}} = 1.044$$

$$\therefore \quad T_2 = 294 \times 1.044 = 306.9 \text{ K} = 33.9°\text{C}$$

For isentropic process 2-3, we have:

$$\frac{T_3}{T_2} = \left(\frac{p_3}{p_2}\right)^{\frac{\gamma-1}{\gamma}} = \left(\frac{3.2}{1}\right)^{\frac{1.4-1}{1.4}} = 1.394$$

$$\therefore \quad T_3 = T_2 \times 1.394 = 306.9 \times 1.394 = 427.8 \text{ K} = 154.8°\text{C}$$

Isentropic efficiency of the main compressor,

$$\eta_{C_1} = 0.82 = \frac{T_3 - T_2}{T_{3'} - T_2} = \frac{427.8 - 306.9}{T_{3'} - 306.9}$$

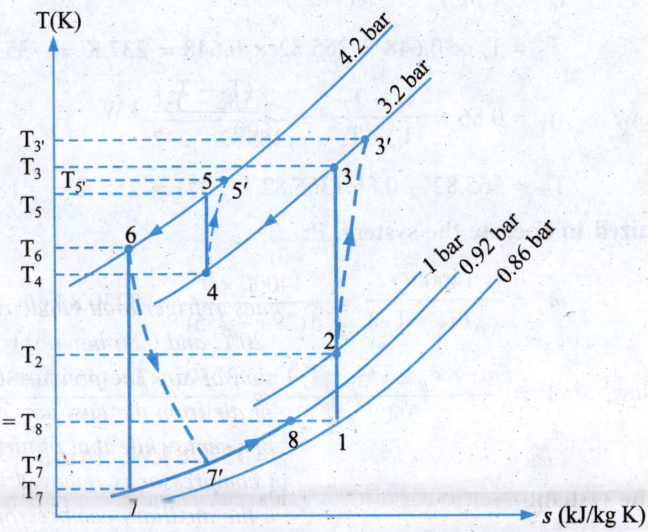

Fig. 2.34. T-*s* diagram.

$$T_{3'} = 306.9 + \left(\frac{427.8 - 306.9}{0.82}\right) = 454.3 \text{ K} = 181.3°\text{C}$$

Since 45% of the enthalpy of air discharged from the main compressor is removed in the first heat exchanger (i.e., during the process 3′-4), therefore temperature of air leaving the first heat exchanger,

$$T_4 = 0.55 \times 181.3 = 99.7°\text{C or } 372.7 \text{ K}$$

For isentropic compression process 4-5, we have:

$$\frac{T_5}{T_4} = \left(\frac{p_5}{p_4}\right)^{\frac{\gamma-1}{\gamma}} = \left(\frac{4.2}{3.2}\right)^{\frac{1.4-1}{1.4}} = 1.081$$

$$\therefore \quad T_5 = T_4 \times 1.081 = 372.7 \times 1.081$$

$$= 402.9 \text{ K} = 129.9°\text{C}$$

Isentropic efficiency of the secondary or auxiliary compressor,

$$\eta_{C_2} = 0.82 = \frac{T_5 - T_4}{T_{5'} - T_4} = \frac{402.9 - 372.7}{T_{5'} - 372.7}$$

$$\therefore \quad T_{5'} = 372.7 + \left(\frac{402.9 - 372.7}{0.82}\right) = 409.5 \text{ K} = 136.5°\text{C}$$

Since 32% of the enthalpy of air discharged from the secondary compressor is removed in the second heat exchanger (i.e. during the process 5′-6), therefore temperature of air leaving the second heat exchanger,

$$T_6 = 0.68 \times 136.5 = 92.82°\text{C} = 365.82 \text{ K}$$

For the isentropic expression process 6-7, we have:

$$\frac{T_7}{T_6} = \left(\frac{p_7}{p_6}\right)^{\frac{\gamma-1}{\gamma}} = \left(\frac{0.92}{4.2}\right)^{\frac{1.4-1}{1.4}} = 0.648$$

$$T_7 = T_6 \times 0.648 = 365.82 \times 0.648 = 237 \text{ K} = -35.9°\text{C}$$

Now, $$\eta_T = 0.86 = \frac{T_6 - T_{7'}}{T_6 - T_7} = \frac{365.82 - T_{7'}}{365.82 - 237}$$

$$\therefore \quad T_{7'} = 365.82 - 0.86\,(365.82 - 237) = 255 \text{ K} = -18°\text{C}$$

(*i*) Power required to operate the system, P:

$$m_a = \frac{14000\text{Q}}{c_p(T_8 - T_{7'})} = \frac{14000 \times 9}{1(294 - 255)} = 3230.8 \text{ kg/h}$$

Now, $$P = \frac{m_a \times c_p \times (T_{3'} - T_2)}{3600} = \frac{3230.8 \times 1 \times (454.3 - 306.9)}{3600}$$

$$= \mathbf{132.28 \text{ kW}} \quad \textbf{(Ans.)}$$

(*ii*) C.O.P. of the system:

$$\text{C.O.P.} = \frac{14000\text{Q}}{P \times 3600} = \frac{14000 \times 9}{132.28 \times 3600} = \mathbf{0.264} \quad \textbf{(Ans.)}$$

Example 2.41. *In an aircraft a bootstrap refrigeration system is used to take 18 tonnes of refrigeration load. The ambient conditions are – 50°C and 0.225 bar. The pressure ratio of the main compressor is 3.5 and required air for refrigeration is bled-off from the main compressor. The air taken from the main compressor is further compressed in secondary compressor which is run by cooling air-turbine. The output of the cooling air-turbine is just sufficient to run the secondary turbine. The refrigerating air coming out from secondary compressor is cooled by ram air to 50°C. The air pressure coming out of cooling turbine and supplied to the cabin is 1 bar. The temperature maintained in the cabin is 25°C. Take the following data:*

$$\eta_{Ram} = \eta_{C1} = \eta_{C2} = 90 \text{ per cent}; \quad \eta_T = 80 \text{ per cent};$$

Effectiveness of both the heat exchangers, ϵ_1, ϵ_2 = 0.4.

When the aircraft is moving with a speed of 1000 km/h, determine the following:

(i) *Delivery pressure of the secondary compressor;*

(ii) *Mass flow rate supplied to the cabin; and*

(iii) *C.O.P. of the system.*

Take for air: c_p = 1 kJ/kg K; γ = 1.4. **[PTU]**

Solution: Fig. 2.35 shows the schematic arrangement and T-*s* diagram for the system.

Given: Q = 18 tonnes refrigeration; $T_1 = -50 + 273 = 223$ K; $p_1 = 0.225$ bar; $p_3 = p_{3'} = 3.5\, p_2$; $W_T = W_{C2}$; $T_6 = 50 + 273 = 323$ K; $p_7 = p_{7'} = 1$ bar; $\eta_{Ram} = \eta_{C1} = \eta_{C2} = 20\%$; $\eta_T = 80\%$;

$$\epsilon_1 = \epsilon_2 = 0.4, \; C = 1000 \text{ km/h} = \frac{1000 \times 1000}{60 \times 60} = 277.8 \text{ m/s}; \; c_p = 1 \text{ kJ/kg K}; \; \gamma = 1.4.$$

Air ramming:
$$T_{2'} (= T_2) = T_1 + \frac{C^2}{2000 c_p} = 223 + \frac{(277.8)^2}{2000 \times 1} = 261.6 \text{ K}$$

$$\frac{p_2}{p_1} = \left(\frac{T_2}{T_1}\right)^{\frac{\gamma}{\gamma - 1}} = \left(\frac{261.6}{223}\right)^{\frac{1.4}{1.4 - 1}} = 1.75$$

$$\therefore \quad p_2 = p_1 \times 1.75 = 0.225 \times 1.75 = 0.394 \text{ bar}$$

Now,
$$\eta_{Ram} = \frac{p_{2'} - p_1}{p_2 - p_1}$$

$$\therefore \quad p_{2'} = p_1 + \eta_{Ram} (p_2 - p_1) = 0.225 + 0.9 (0.394 - 0.225) = 0.377 \text{ bar}$$

Main compressor:

$$\frac{p_3}{p_{2'}} = 3.5 \qquad \text{...(Given)}$$

$$\therefore \quad p_3 = 0.377 \times 3.5 = 1.32 \text{ bar}$$

$$\frac{T_3}{T_{2'}} = \left(\frac{p_3}{p_{2'}}\right)^{\frac{\gamma - 1}{\gamma}}$$

$$= (3.5)^{\frac{1.4 - 1}{1.4}} = 1.43$$

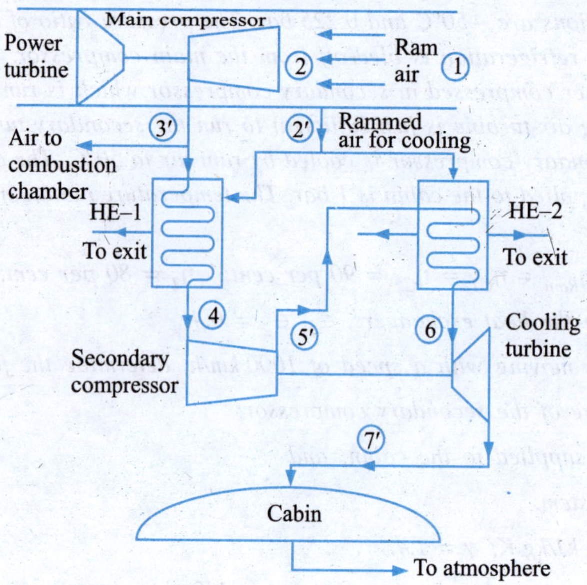

(a) Schematic arrangement of the Boot-strap system.

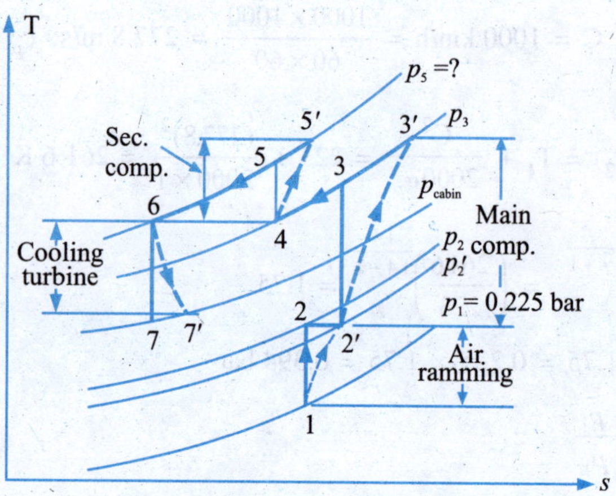

(b) T-s diagram for the system.

Fig. 2.35

$$\therefore \quad T_3 = 261.6 \times 1.43 = 374 \text{ K}$$

$$\eta_{C1} = 0.9 = \frac{T_3 - T_{2'}}{T_{3'} - T_{2'}} = \frac{374 - 261.6}{T_{3'} - 261.6}$$

$$\therefore \quad T_{3'} = 261.6 + \left(\frac{374 - 261.6}{0.9}\right) = 386.5 \text{ K}$$

H.E. (Heat exchanger)-1:

$$\in_1 = 0.4 = \frac{T_{3'} - T_4}{T_{3'} - T_{2'}} = \frac{386.5 - T_4}{386.5 - 261.6}$$

$$\therefore \quad T_4 = 386.5 - 0.4\ (386.5 - 261.6) = 336.5 \text{ K}$$

(*i*) **Delivery pressure of secondary compressor, p_5:**

Secondary compressor:

$$W_{C2} = \frac{c_p (T_5 - T_4)}{\eta_{C2}} = \frac{c_p T_4}{\eta_{C2}}\left[\frac{T_5}{T_4} - 1\right] = \frac{c_p T_4}{\eta_{C2}}\left[\left(\frac{p_5}{p_4}\right)^{\frac{\gamma-1}{\gamma}} - 1\right]$$

$$= \frac{1 \times 336.5}{0.9}\left[\left(\frac{p_5}{1.32}\right)^{\frac{1.4-1}{1.4}} - 1\right] \text{ per kg of air} \qquad (\because p_4 = p_3 = 1.32 \text{ bar})$$

$$= 373.9\left[\left(\frac{p_5}{1.32}\right)^{0.2857} - 1\right] \qquad ...(i)$$

Cooling turbine:

$$W_T = c_p (T_6 - T_7)\ \eta_T = c_p\ \eta_T T_6\left[1 - \left(\frac{p_7}{p_6}\right)^{\frac{\gamma-1}{\gamma}}\right]$$

$$= 1 \times 0.8 \times T_6\left[1 - \left(\frac{p_7}{p_6}\right)^{\frac{\gamma-1}{\gamma}}\right] = 1 \times 0.8 \times (50 + 273)\left[1 - \left(\frac{1}{p_5}\right)^{\frac{1.4-1}{1.4}}\right]$$

$$= 258.4\ [1 - (p_5)^{-0.2857}] \qquad ...(ii)$$

Now, $\quad W_{C2} = W_T \quad$ ...(Given)

$$\therefore \quad 373.9\left[\frac{x}{(1.32)^{0.2857}} - 1\right] = 258.4\left[1 - \frac{1}{x}\right], \qquad \text{where, } x = (p_5)^{0.2857}$$

$$\text{or,} \quad \frac{373.9}{258.4}\ (0.924\,x - 1) = \frac{x-1}{x}$$

$$1.45\ (0.924x - 1) = \frac{x-1}{x}, \quad \text{or,} \quad 1.34x^2 - 1.45x = x - 1, \quad \text{or,} \quad 1.34x^2 - 2.45x + 1 = 0$$

$$x = \frac{2.45 \pm \sqrt{(2.45)^2 - 4 \times 1.34 \times 1}}{2 \times 1.34} = \frac{2.45 \pm 0.802}{2.68} = 1.2, \quad \text{or,} \quad 0.6.$$

$\therefore \quad x = (p_5)^{0.2857} = 1.2$ as 0.86 is not possible

$$\therefore \quad p_5 = (1.2)^{\frac{1}{0.2857}} = \textbf{1.9 bar} \quad \textbf{(Ans.)}$$

(*ii*) **Mass flow rate supplied to the cabin, m_a:**

$$\frac{T_7}{T_6} = \left(\frac{p_7}{p_6}\right)^{\frac{\gamma-1}{\gamma}} = \left(\frac{1}{1.9}\right)^{\frac{1.4-1}{1.4}} = 0.832 \qquad (\because\ p_6 = p_5 = 1.9 \text{ bar})$$

$$\therefore \quad T_7 = T_6 \times 0.832 = (50 + 273) \times 0.832 = 268.7 \text{ K}$$

$$\eta_T = 0.8 = \frac{T_6 - T_{7'}}{T_6 - T_7} = \frac{(50+273) - T_{7'}}{(50+273) - 268.7}$$

$$\therefore \quad T_{7'} = (50 + 273) - 0.8\,[(50 + 273 - 268.7)] = 279.6 \text{ K}$$

Now,
$$m_a = \frac{14000\,Q}{c_p\,(T_{cabin} - T_{7'})} = \frac{14000 \times 18}{1[(25+273) - 279.6]} = \mathbf{13695.6\ kg/h} \quad \textbf{(Ans.)}$$

(*iii*) **C.O.P. of the system:**

$$\text{C.O.P.} = \frac{14000\,Q}{m_a c_p\,(T_{3'} - T_{2'})} = \frac{14000 \times 18}{13695.6 \times 1 \times (386.5 - 261.6)} = \mathbf{0.147} \quad \textbf{(Ans.)}$$

Example 2.42. *For ideal Bootstrap aircraft refrigeration cycle, the compressor and turbine efficiencies are 100%; air temperature after cooling in both the heat exchangers is equal to Ram air temperature. If the power input to auxiliary compressor is equal to turbine power, show that the maximum pressure is given by* $p_2\,[R + \sqrt{R^2 - Rd}\,]^{\gamma/(\gamma-1)}$ *where* $R = (p_3 / p_2)^{(\gamma-1)/\gamma}$ *and* $d = (p_i / p_2)^{(\gamma-1)/\gamma}$. *The Ram pressure and cabin pressure are* p_2 *and* p_i *respectively.* (p_3/p_2) *is pressure ratio across main compressor.*

[Bombay University]

Solution: Fig. 2.36 shows the schematic diagram and T-*s* diagram for the arrangements.

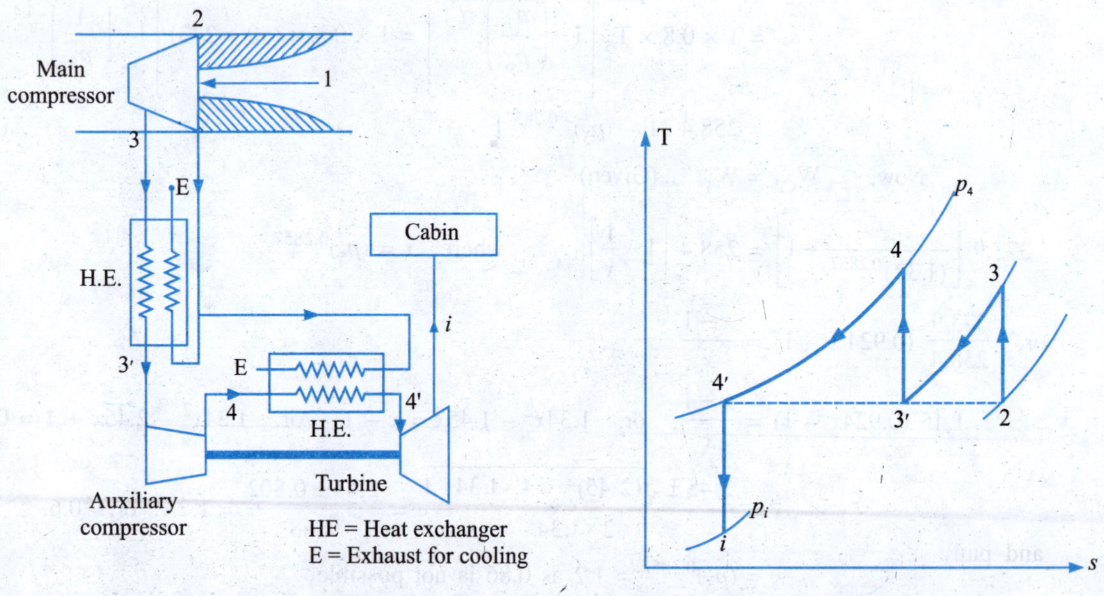

(*a*) Schematic arrangement. (*b*) T-*s* diagram.

Fig. 2.36

$$(W)_{aux.\,comp.} = (W)_{Turbine} \quad \text{...(Given)}$$

or, $$c_p(T_4 - T_3') = c_p(T_4' - T_i)$$

$$c_p(T_4 - T_3') = c_p(T_4' - T_i)$$

or, $$T_3'\left[\frac{T_4}{T_3'} - 1\right] = T_4'\left[1 - \frac{T_i}{T_4'}\right]$$

or, $$\left(\frac{p_4}{p_3}\right)^{\frac{\gamma-1}{\gamma}} - 1 = 1 - \left(\frac{p_i}{p_4}\right)^{\frac{\gamma-1}{\gamma}}, \text{ as } T_3' = T_4' \quad \text{...(1)}$$

Let, $(p_4)^{\frac{\gamma-1}{\gamma}} = x$, $(p_3)^{\frac{\gamma-1}{\gamma}} = a$ and $(p_i)^{\frac{\gamma-1}{\gamma}} = b$

∴ Equation (1) becomes,

$$\frac{x}{a} - 1 = 1 - \frac{b}{x}, \text{ or, } \frac{x}{a} - 2 + \frac{b}{x} = 0$$

or, $x^2 - 2ax + ab = 0$ (Multiplying the equation by ax)

The solution of the quadratic equation is,

$$x = (p_4)^{\frac{\gamma-1}{\gamma}} = \frac{2a + \sqrt{4a^2 - 4ab}}{2} = a + \sqrt{a^2 - ab}$$

$$= (p_3)^{\frac{\gamma-1}{\gamma}} + \sqrt{(p_3)^{\frac{2(\gamma-1)}{\gamma}} - (p_3)^{\frac{\gamma-1}{\gamma}}(p_i)^{\frac{\gamma-1}{\gamma}}}$$

$$p_{max} = p_4 = \left[(p_3)^{\frac{\gamma-1}{\gamma}} + \sqrt{(p_3)^{\frac{2(\gamma-1)}{\gamma}} - (p_3)^{\frac{\gamma-1}{\gamma}}(p_i)^{\frac{\gamma-1}{\gamma}}}\right]^{\frac{\gamma}{\gamma-1}}$$

$$= p_2\left[\left(\frac{1}{p_2}\right)^{\frac{\gamma-1}{\gamma}}\left\{(p_3)^{\frac{\gamma-1}{\gamma}} + \sqrt{(p_3)^{\frac{2(\gamma-1)}{\gamma}} - (p_3)^{\frac{\gamma-1}{\gamma}}(p_i)^{\frac{\gamma-1}{\gamma}}}\right\}\right]^{\frac{\gamma}{\gamma-1}}$$

$$= p_2\left[\left(\frac{p_3}{p_2}\right)^{\frac{\gamma-1}{\gamma}} + \sqrt{\left(\frac{p_3}{p_2}\right)^{\frac{2(\gamma-1)}{\gamma}} - \left(\frac{p_3}{p_2}\right)^{\frac{\gamma-1}{\gamma}}\left(\frac{p_i}{p_2}\right)^{\frac{\gamma-1}{\gamma}}}\right]^{\frac{\gamma}{\gamma-1}}$$

and putting $\left(\frac{p_3}{p_2}\right)^{\frac{\gamma-1}{\gamma}} = R$, and $\left(\frac{p_i}{p_2}\right)^{\frac{\gamma-1}{\gamma}} = d$, we get:

$$p_{max} = p_4 = p_2\left[R + \sqrt{R^2 - Rd}\right]^{\frac{\gamma}{\gamma-1}} \quad \text{...Proved}$$

(*b*) ***With evaporative cooling:***

In Boot strap evaporative cooling system an *evaporative cooler* is added in the high pressure line before the cooling turbine. Water, alcohol or ammonia may be used as a heat sink for evaporative cooler. For the same temperature to be maintained in the cabin, this system gives lower turbine discharge temperature as compared to simple cooling system. *This results in the lower mass flow rate per tonne of refrigeration.* Fig. 2.37 shows the schematic arrangement of the system.

Fig. 2.38 shows the various processes on T-s diagram. The processes shown are ideal (*without pressure and friction losses*).

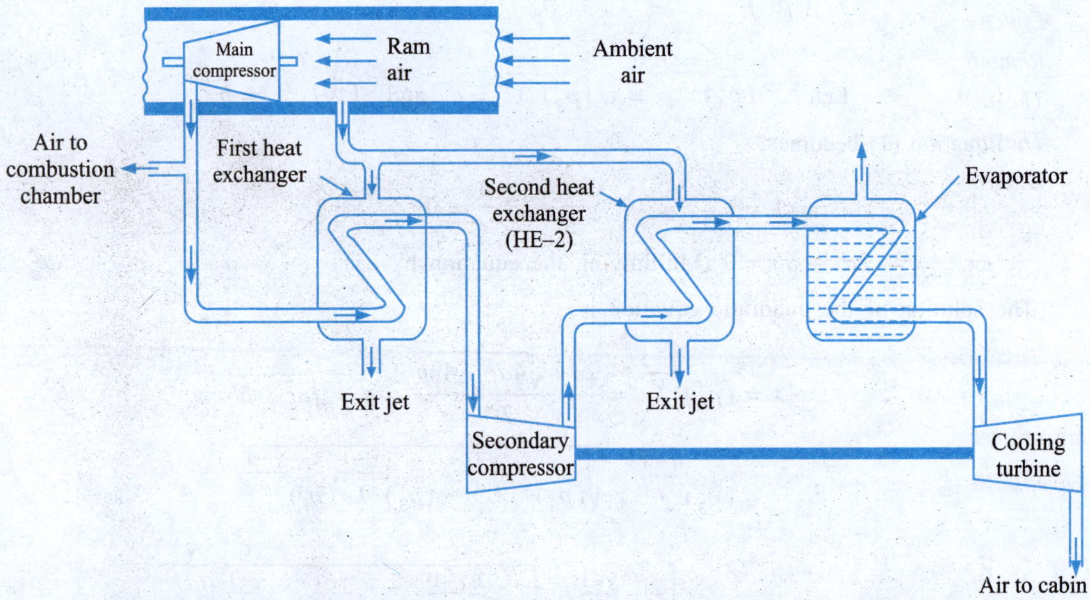

Fig. 2.37. Schematic arrangement of Boot-strap system with evaporative cooling.

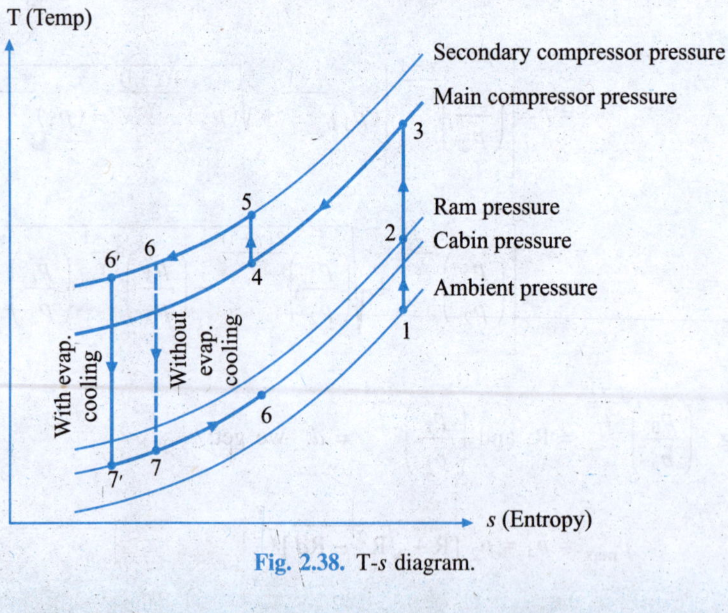

Fig. 2.38. T-*s* diagram.

Example 2.43. *A boot-strap evaporative air-refrigeration system is employed in an aeroplane, moving with a Mach. number of 1.3, to take 18 tonnes of refrigeration load of the cabin. The cooling air is taken from the main compressor at 4.2 bar and further compressed to 5.2 bar which is run by cooling turbine.*

The following data may be used:

The ambient conditions ... 0.52 bar, –12°C

Isentropic efficiency of main compressor ... 88%

Isentropic efficiency of secondary compressor ... 82%

Isentropic efficiency of cooling turbine ... 80%

Effectiveness of both heat exchangers ... 0.42

Ramming efficiency ... 90%

The temperature of air coming out of evaporator cooler ... 100°C

The pressure and temperature maintained in the cabin ... 1 bar, 27°C

Determine the following:

(*i*) *Mass taken from the main compressor for cabin cooling,*

(*ii*) *Power required to run the refrigeration system, and*

(*iii*) *C.O.P. of the system.*

Take for air: $c_p = 1$ *kJ/kg K and* $\gamma = 1.4$.

Solution: The arrangement of the components of the system is shown in Fig. 2.37 and the processes are represented on T-*s* diagram as shown in Fig. 2.39.

Given: M = 1.3; Q = 18 tonnes; $p_1 = 0.52$ bar;

$T_1 = -12 + 273 = 261$ K; $\eta_{C1} = 0.88$, $\eta_{C2} = 0.82$;

$\eta_T = 0.8$; $\in_1 = \in_2 = 0.42$; $\eta_{Ram} = 0.9$;

$T_{6'} = 100 + 273 = 373$ K; $p_7 = p_8 = 1$ bar;

$t_8 = 27 + 273 = 300$ K.

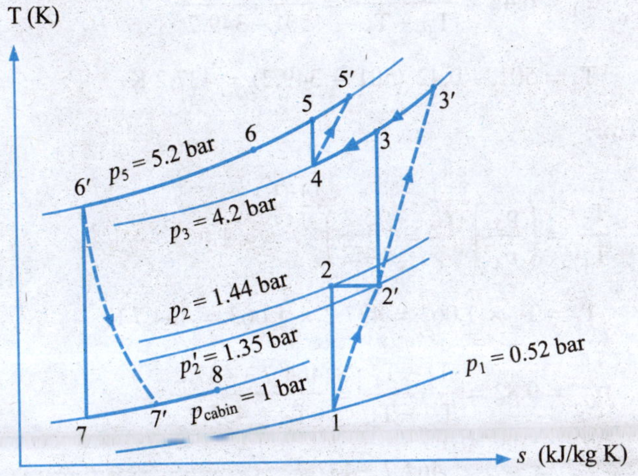

Fig. 2.39. T-*s* diagram.

Ramming process 1-2′:

$$\frac{T_2'}{T_1} = 1 + \frac{\gamma - 1}{2} M^2 = 1 + \frac{1.4 - 1}{2} \times (1.3)^2 = 1.338$$

$$\therefore \quad T_{2'} = 261 \times 1.338 = 349.2 \text{ K}$$

Also,
$$\frac{p_2}{p_1} = \left(\frac{T_2}{T_1}\right)^{\frac{\gamma}{\gamma - 1}} = (1.338)^{\frac{1.4}{1.4-1}} = 2.77 \qquad (\because T_2 = T_{2'})$$

$$\therefore \quad p_2 = 0.52 \times 2.77 = 1.44 \text{ bar}$$

$$\eta_{Ram} = 0.9 = \frac{p_{2'} - p_1}{p_2 - p_1}$$

$$\therefore \quad p_{2'} = p_1 + 0.9\,(p_2 - p_1) = 0.52 + 0.9\,(1.44 - 0.52) = 1.35 \text{ bar}$$

The main compressor:

$$\frac{T_3}{T_{2'}} = \left(\frac{p_3}{p_{2'}}\right)^{\frac{\gamma - 1}{\gamma}} = \left(\frac{4.2}{1.35}\right)^{\frac{1.4-1}{1.4}} = 1.383$$

$$\therefore \quad T_3 = T_{2'} \times 1.383 = 349.2 \times 1.383 = 482.9 \text{ K}$$

$$\eta_{C_1} = 0.88 = \frac{T_3 - T_{2'}}{T_{3'} - T_{2'}} = \frac{482.9 - 349.2}{T_{3'} - 349.2}$$

$$\therefore \quad T_{3'} = 349.2 + \frac{482.9 - 349.2}{0.88} = 501 \text{ K}$$

HE (Heat exchanger)-1:

$$\in_1 = 0.42 = \frac{T_{3'} - T_4}{T_{3'} - T_{2'}} = \frac{501 - T_4}{501 - 349.2}$$

$$\therefore \quad T_4 = 501 - 0.42\,(501 - 349.2) = 437.2 \text{ K}$$

Secondary compressor:

$$\frac{T_5}{T_4} = \left(\frac{p_5}{p_4}\right)^{\frac{\gamma - 1}{\gamma}} = \left(\frac{5.2}{4.2}\right)^{\frac{1.4-1}{1.4}} = 1.063$$

$$\therefore \quad T_5 = T_4 \times 1.063 = 437.2 \times 1.063 = 464.7 \text{ K}$$

$$\eta_{C_2} = 0.82 = \frac{T_5 - T_4}{T_{5'} - T_4} = \frac{464.7 - 437.2}{T_{5'} - 437.2}$$

$$\therefore \quad T_{5'} = 437.2 + \frac{464.7 - 437.2}{0.82} = 470.7 \text{ K}$$

HE (Heat exchanger)-2:

$$\in_2 = 0.42 = \frac{T_{5'} - T_6}{T_{5'} - T_{2'}} = \frac{470.7 - T_6}{470.7 - 349.2}$$

$$\therefore \quad T_6 = 470.7 - 0.42\,(470.7 - 349.2) = 419.7\text{ K}$$

The air is further cooled in evaporator cooler to 100°C = 323 K = $T_{6'}$.

Cooling turbine:

$$\frac{T_7}{T_{6'}} = \left(\frac{p_7}{p_6}\right)^{\frac{\gamma-1}{\gamma}} = \left(\frac{1}{5.2}\right)^{\frac{1.4-1}{1.4}} = 0.624$$

$$\therefore \quad T_7 = T_{6'} \times 0.624 = 373 \times 0.624 = 232.7\text{ K}$$

$$\eta_T = \frac{T_{6'} - T_{7'}}{T_{6'} - T_7}$$

$$0.8 = \frac{373 - T_{7'}}{373 - 232.7}$$

$$\therefore \quad T_{7'} = 373 - 0.8\,(373 - 232.7) = 260.8\text{ K}$$

(*i*) **Mass taken from the main compressor for cabin cooling, m_a:**

$$m_a = \frac{14000\,Q}{c_p\,(T_8 - T_{7'})}\ \text{kg/h}$$

$$= \frac{14000 \times 18}{1(300 - 260.8)} = \mathbf{6428.57\ kg/h}, \quad \text{or,} \quad \mathbf{1.786\ kg/s} \quad \textbf{(Ans.)}$$

(*ii*) **Power required to run the refrigeration system, P:**

$$P = m_a c_p\,(T_{3'} - T_{2'}) = 1.786 \times 1\,(501 - 349.2) = \mathbf{271.1\ kW} \quad \textbf{(Ans.)}$$

(*iii*) **C.O.P. of the system:**

$$\text{C.O.P.} = \frac{14000\,Q}{P \times 3600} = \frac{14000 \times 18}{271.1 \times 3600} = \mathbf{0.258} \quad \textbf{(Ans.)}$$

2.4.5.3 Regenerative air-refrigeration system

The regenerative system is used when the *cooling turbine discharge temperature of a simple system is too high.*

The diagrammatic layout of regenerative cooling system is given in Fig. 2.40. It consists of a *primary air-to-air heat exchanger*, a *regenerative heat exchanger*, and a *cooling turbine*. Higher pressure, high temperature air coming from compressor is cooled first in the primary heat exchanger, then in regenerative heat exchanger and finally in the cooling turbine. Ram air is used as heat sink in the primary heat exchanger and part of the cold air from the turbine discharge is used as the heat sink in the regenerative heat exchanger.

- *This system is used for supersonic air-planes.*

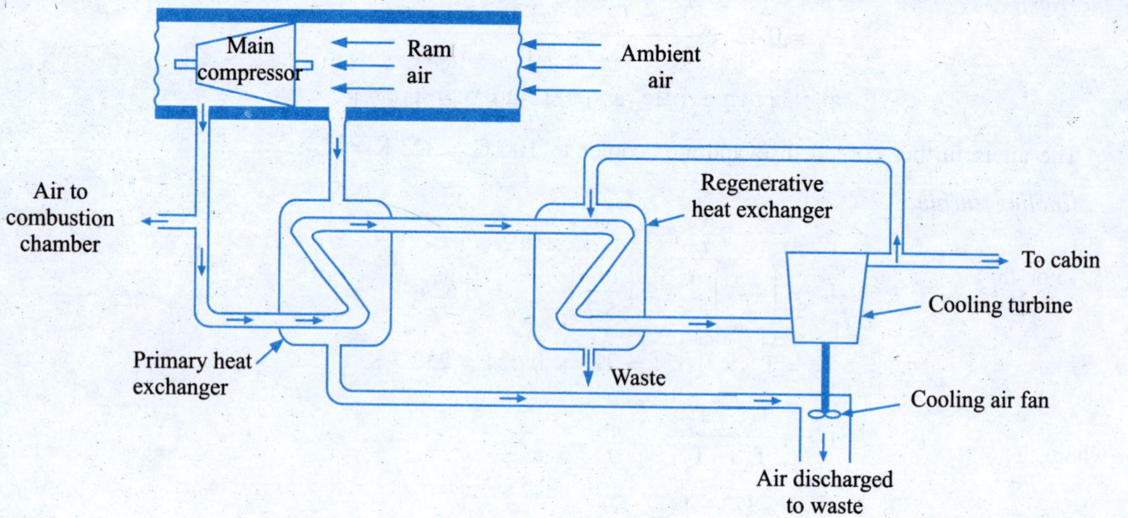

Fig. 2.40. Regenerative air-refrigeration system.

Fig. 2.41 shows the various processes on T-*s* diagram.

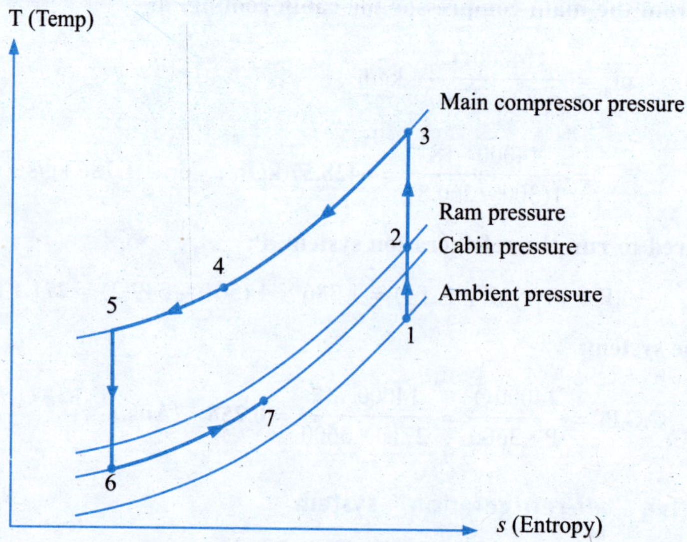

Fig. 2.41. T-*s* diagram.

Process 1–2: *Ramming of air.*

Process 2–3: *Compression of air in main compressor.*

Process 3–4: *Cooling of compressed air by ram air in the primary heat exchanger.*

Process 4–5: *Cooling of air in regenerative heat exchanger.*

Process 5–6: *Expansion in the cooling turbine.*

Process 6–7: *Air getting heated to cabin temperature.*

- If Q tonnes of refrigeration is the cooling load in the cabin, the *quantity of air required for the refrigeration purpose* will be,

$$m_a = \frac{14000\,Q}{c_p\,(T_7 - T_6)}\ \text{kg/h}$$

- *Energy balance of regenerative heat exchanger:*

Let, m_1 = Total mass of air bled from the main compressor, and

m_2 = Mass of cold air bled from the cooling turbine for regenerative heat exchanger.

For the energy balance of regenerative heat exchanger, we have:

$$m_2 c_p\,(T_8 - T_6) = m_1 c_p\,(T_4 - T_5)$$

$$\therefore \quad m_2 = \frac{m_1\,(T_8 - T_6)}{(T_4 - T_5)}$$

where, T_8 = Temperature of air leaving to atmosphere from the regenerative heat exchanger.

Note: $m_a = m_1 - m_2$

- *Power required for refrigeration system,*

$$P = \frac{m_1 c_p\,(T_3 - T_2)}{3600}\ \text{kW}$$

- *C.O.P. of the system,*

$$\text{C.O.P.} = \frac{14000Q}{P \times 3600}$$

Example 2.44. *A regenerative air-refrigeration system is designed to take 18 tonnes of refrigeration load of an air-craft cabin. The ambient air at pressure 0.82 bar and temperature 12°C is rammed isentropically till the pressure rises to 1.22 bar. The air bled off the main compressor at 4.4 bar is cooled by the ram air in the heat exchanger whose effectiveness is 0.62. The air from the heat exchanger is further cooled to 64°C in the regenerative heat exchanger with a portion of the air bled after expansion in the cooling turbine. The cabin is to be maintained at a temperature of 22°C and a pressure of 1 bar. If the isentropic efficiencies of the compressor and turbine are 88 per cent and 82 per cent respectively, determine the following:*

(i) Mass of the air bled from cooling turbine to be used for regenerative cooling,

(ii) Power required for maintaining the cabin at required condition, and

(iii) C.O.P. of the system.

Assume that the temperature of air coming out of regenerative heat exchanger is 100°C.

Solution: The T-*s* diagram for the given system with the given conditions is shown in Fig. 2.42.

Given: Q = 18 tonnes of refrigeration;

$p_1 = 0.82$ bar;

$T_1 = 12 + 273 = 285$ K;

$p_2 = 1.22$ bar;

$p_3 = p_4 = p_5 = 4.4$ bar;

$\in = 0.62$;

$T_5 = 64 + 273 = 337$ K;

$T_7 = 22 + 273 = 295$ K;

$p_7 = p_6 = p_{6'} = 1$ bar;

$\eta_C = 88\%$;

$\eta_T = 82\%$,

$T_8 = 100 + 273 = 373$ K.

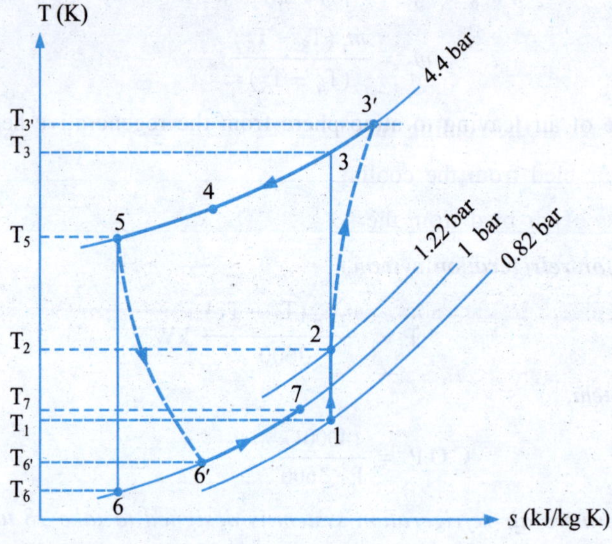

Fig. 2.42. T-*s* diagram

Air ramming:

$$\frac{T_2}{T_1} = \left(\frac{p_2}{p_1}\right)^{\frac{\gamma-1}{\gamma}} = \left(\frac{1.22}{0.8}\right)^{\frac{1.4-1}{1.4}} = 1.12$$

$$\therefore \quad T_2 = T_1 \times 1.12 = 285 \times 1.12 = 319.2 \text{ K}$$

Main compressore:

$$\frac{T_3}{T_2} = \left(\frac{p_3}{p_2}\right)^{\frac{\gamma-1}{\gamma}} = \left(\frac{4.4}{1.2}\right)^{\frac{1.4-1}{1.4}} = 1.45$$

$$\therefore \quad T_3 = T_2 \times 1.45 = 319.2 \times 1.45 = 462.8 \text{ K}$$

$$\eta_C = 0.88 = \frac{T_3 - T_2}{T_{3'} - T_2} = \frac{462.8 - 319.2}{T_{3'} - 319.2}$$

$$\therefore \quad T_{3'} = 319.2 + \left(\frac{462.8 - 319.2}{0.88}\right) = 482.4 \text{ K}$$

Primary heat exchanger:

$$\in = 0.62 = \frac{T_{3'} - T_4}{T_{3'} - T_2} = \frac{482.4 - T_4}{482.4 - 319.2}$$

$$\therefore \quad T_4 = 482.4 - 0.62\ (482.4 - 319.2) = 381.2\ K$$

Cooling turbine:

$$\frac{T_5}{T_6} = \left(\frac{p_5}{p_6}\right)^{\frac{\gamma - 1}{\gamma}} = \left(\frac{4.4}{1}\right)^{\frac{1.4-1}{1.4}} = 1.527$$

$$\therefore \quad T_6 = T_5/1.527 = 337/1.527 = 220.7\ K$$

$$\eta_T = 0.82 = \frac{T_5 - T_{6'}}{T_5 - T_6} = \frac{337 - T_{6'}}{337 - 220.7}$$

$$\therefore \quad T_{6'} = 337 - 0.82\ (337 - 220.7) = 241.6\ K$$

(*i*) **Mass of air bled from the cooling turbine to be used for regenerative cooling, m_2:**

Let, m_a = Mass of air bled from the cooling turbine to be used for cabin cooling,

m_1 = Total mass of air bled from the main compressor, and

m_2 = Mass of cold air bled from the cooling turbine for regenerative heat exchanger.

Now, mass of air supplied to the cabin,

$$m_a = m_1 - m_2 = \frac{14000\ Q}{c_p\,(T_7 - T_{6'})} = \frac{14000 \times 18}{1(295 - 241.6)} = 4719.1\ \text{kg/h} \quad ...(i)$$

and,
$$m_2 = \frac{m_1\,(T_4 - T_5)}{(T_8 - T_{6'})} = \frac{m_1\,(381.2 - 337)}{(373 - 241.6)} = 0.336\, m_1 \quad ...(ii)$$

From eqns. (*i*) and (*ii*), we have:

$$m_1 - m_2 = 4719.1\ , \quad \text{or,} \quad m_1 - 0.336\, m_1 = 4719.1$$

$$\therefore \quad m_1 = \frac{4719.1}{(1 - 0.336)} = 7107.1\ \text{kg/h}$$

and, $m_2 = 0.336 \times 7107.1 =$ **2388 kg/h (Ans.)**

(*ii*) **Power required for maintaining the cabin at the required condition, P:**

$$P = \frac{m_1\, c_p\,(T_{3'} - T_2)}{3600} = \frac{7107.1 \times 1 \times (482.4 - 319.2)}{3600} = \mathbf{322.2\ kW} \quad \textbf{(Ans.)}$$

(*iii*) **C.O.P. of the system:**

$$\textbf{C.O.P.} = \frac{14000\ Q}{P \times 3600} = \frac{14000 \times 18}{322.2 \times 3600} = \mathbf{0.217} \quad \textbf{(Ans.)}$$

2.4.5.4 Reduced ambient air-refrigeration system

The reduced ambient air-refrigeration system shown in Fig. 2.43 includes *two cooling turbines and one heat exchanger.*

— The air reduced for the refrigeration system is bled off from the main compressor. This high pressure high temperature air is cooled initially in the heat exchanger. The air required for cooling is taken from the first cooling turbine (T-1) which lowers the temperature of rammed air which is quite high due to ramming action.

— The cooled air from the heat exchanger is passed through the second cooling turbine (T-2) from where the air is supplied to the cabin.

— The work of the cooling turbines is used to drive the exhaust fan (through reduction gears) which draws the cooling air from the heat exchanger.

• This system shows promise for applications in an *exceptionally high speed aircraft when the ram air temperature is too high.*

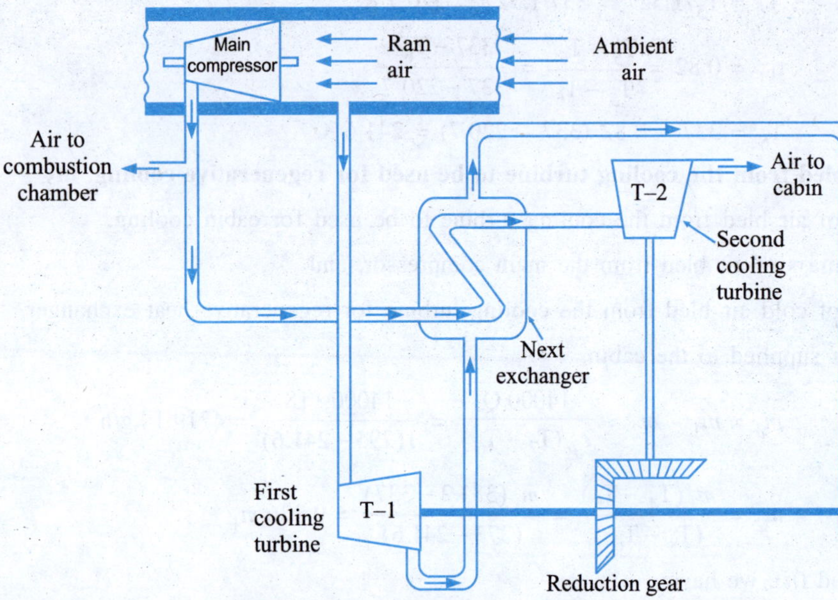

Fig. 2.43. Reduced ambient air-refrigeration system.

Fig. 2.44 shows the various processes (without pressure and friction losses) on T-*s* diagram.

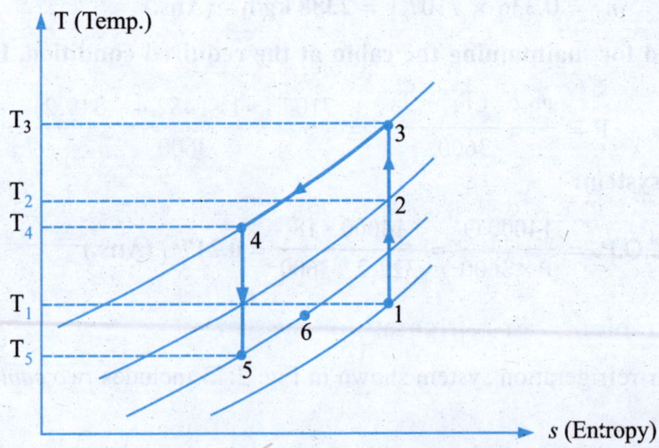

Fig. 2.44. T-*s* diagram

(i) **Process 1–2:** *Ramming of air.*

(ii) **Process 2–3:** *Compression in the main compressor.*

(iii) **Process 3–4:** *Cooling of compressed air by ram air which after passing through the first cooling turbine (T-1) is led to the heat exchanger.*

(iv) **Process 4–5:** *Expansion of air in the second cooling turbine (T-2) upto the cabin pressure.*

(v) **Process 5–6:** *Heating of air upto cabin temperature (T_6).*

- If Q tonnes of refrigeration is the cooling load in the cabin, then the *quantity of air required for the refrigeration purpose will be,*

$$m_a = \frac{14000\,Q}{c_p\,(T_6 - T_5)}\ \text{kg/h}$$

- *Power required for the refrigeration system is given by,*

$$P = \frac{m_a \times c_p \times (T_3 - T_2)}{3600}\ \text{kW}$$

- *C.O.P. of the system,*

$$\text{C.O.P.} = \frac{14000\,Q}{P \times 3600}$$

Example 2.45. *The reduced ambient air-refrigeration system used for an aircraft consists of two cooling turbines with heat exchanger in between. The output of both the turbines is used to run the fan. The speed of the aircraft is 1500 km/h. The ambient air conditions are 0.24 bar and – 38°C. The air passing through the heat exchanger is discharged to the atmosphere with the help of fan. The compressed air is cooled to 48°C in the heat exchanger. The pressure ratio of the main compressor run by a turbine from which air is bled off for the refrigeration purposes is 4.2. There is a pressure loss of 0.06 bar at the supply air nozzle to the cabin. The cabin condition is maintained at 1 bar pressure and 22°C. If ramming efficiency = 88 per cent, isentropic efficiency of compressor = 82 per cent and isentropic efficiency of each turbine = 80 per cent, determine the following:*

(i) *Mass flow rate of air supplied to the cabin if the cooling load in the cabin is 22 tonnes;*

(ii) *Air flow rate of ram air passed over the heat exchanger, if the maximum rise in temperature is limited to 120 K;*

(iii) *Power used to drive the fan;*

(iv) *C.O.P. of the system.*

Take for air: $c_p = 1$ *kJ/kg K;* $\gamma = 1.4$.

Solution: The T-*s* diagram for the reduced ambient air-refrigeration system with the given conditions is shown in Fig. 2.45.

Given: $C = 1500\ \text{km/h} = \frac{1500 \times 1000}{60 \times 60} = 417\ \text{m/s};\ p_1 = p_{1'} = 0.24\ \text{bar};\ T_1 = -38 + 273 = 235\ \text{K};$

$T_4 = 48 + 273 = 321\ \text{K};\ p_3 = (p_{3'} = p_4) = 4.2\ p_{2'};\ p_6 = 1\ \text{bar};\ T_6 = 22 + 273 = 295\ \text{K};$

$\eta_{Ram} = 88\%;\ \eta_C = 82\%;\ \eta_{T-1} = \eta_{T-2} = 80\%;\ c_p = 1\ \text{kJ/kg K};\ \gamma = 1.4;\ Q = 22\ \text{tonnes}.$

$$T_2 = T_{2'} = T_1 + \frac{C^2}{2000\,c_p} = 235 + \frac{(417)^2}{2000 \times 1} = 321.9\ \text{K}$$

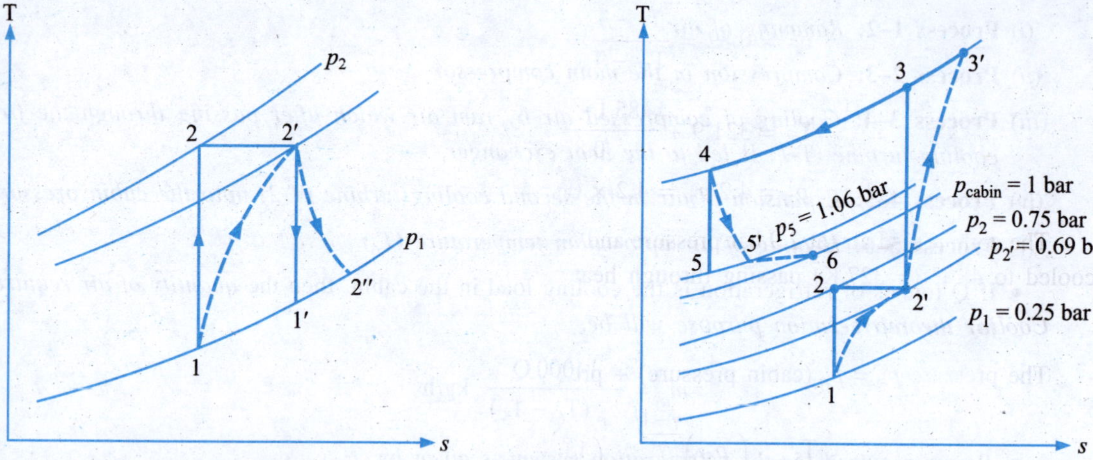

(*a*) First cooling turbine (T-1) (*b*) Complete system.

Fig. 2.45. T-*s* diagram.

Air ramming:

$$\frac{p_2}{p_1} = \left(\frac{T_2}{T_1}\right)^{\frac{\gamma}{\gamma-1}} = \left(\frac{321.9}{235}\right)^{\frac{1.4}{1.4-1}} = 3$$

$$\therefore \quad p_2 = 0.25 \times 3 = 0.75 \text{ bar}$$

$$\eta_{Ram} = 0.88 = \frac{p_{2'} - p_1}{p_2 - p_1}$$

$$\therefore \quad p_{2'} = p_1 + 0.88\ (p_2 - p_1) = 0.24 + 0.88\ (0.75 - 0.24) = 0.69 \text{ bar.}$$

Therefore, stagnation temperature and stagnation pressure of the air entering the main compressor and first turbine (T-1) are 321.9 K and 0.69 bar.

Cooling turbine, T-1:

$$\frac{T_{1'}}{T_{2'}} = \left(\frac{p_{1'}}{p_{2'}}\right)^{\frac{\gamma-1}{\gamma}} = \left(\frac{0.24}{0.69}\right)^{\frac{1.4-1}{1.4}} = 0.739$$

$$\therefore \quad T_{1'} = T_{2'} \times 0.739 = 321.9 \times 0.739 = 237.9 \text{ K}$$

$$\eta_{T\text{-}1} = 0.8 = \frac{T_{2'} - T_{2''}}{T_{2'} - T_{1'}} = \frac{321.9 - T_{2''}}{321.9 - 237.9}$$

$$\therefore \quad T_{2''} = 321.9 - 0.8\ (321.9 - 237.9) = 254.7 \text{ K}$$

Main compressor:

$$\frac{T_3}{T_{2'}} = \left(\frac{p_3}{p_{2'}}\right)^{\frac{\gamma-1}{\gamma}} = (4.2)^{\frac{1.4-1}{1.4}} = 1.507$$

$$\therefore \quad T_3 = T_{2'} \times 1.507 = 321.9 \times 1.507 = 485.1 \text{ K}$$

$$\eta_C = 0.82 = \frac{T_3 - T_{2'}}{T_{3'} - T_{2'}} = \frac{485.1 - 321.9}{T_{3'} - 321.9}$$

$$\therefore \quad T_{3'} = 321.9 + \frac{485.1 - 321.9}{0.82} = 520.9 \text{ K}$$

$$p_3 = 4.2\, p_{2'} = 4.2 \times 0.69 = 2.9 \text{ bar}$$

The compressed air (at 2.9 bar pressure and 520.9 K) from the main compressor is bled off and then cooled to 48°C (= 327 K) passing through heat exchanger.

Cooling turbine T-2:

The pressure $p_5 = p_6$ (cabin pressure) + pressure loss at supply nozzle = 1 + 0.06 = 1.06 bar

$$\frac{T_5}{T_4} = \left(\frac{p_5}{p_4}\right)^{\frac{\gamma - 1}{\gamma}} = \left(\frac{1.06}{2.9}\right)^{\frac{1.4-1}{1.4}} = 0.75$$

$$\therefore \quad T_5 = T_4 \times 0.75 = 321 \times 0.75 = 240.7 \text{ K}$$

$$\eta_{T\text{-}2} = 0.8 = \frac{T_4 - T_{5'}}{T_4 - T_5} = \frac{321 - T_{5'}}{321 - 240.7}$$

$$\therefore \quad T_{5'} = 321 - 0.8\,(321 - 240.7) = 256.8 \text{ K}$$

(*i*) **Mass flow rate of air supplied to the cabin, m_a:**

$$m_a = \frac{14000\,Q}{c_p\,(T_6 - T_{5'})} = \frac{14000 \times 22}{1 \times (295 - 256.8)}$$

$$= \mathbf{8062.8\ kg/h} \quad \text{or} \quad \mathbf{2.24\ kg/s} \quad \textbf{(Ans.)}$$

(*ii*) **Air flow rate of ram air passed over the heat exchanger if the maximum rise in temperature is limited to 120 K:**

Let, m_R = Quantity of ram air passing the heat exchanger.

Considering perfect heat transfer in heat exchanger, we have:

Heat lost by compressed air = Heat gained by cooling air

$$m_a \times c_p \times (T_{3'} - T_4) = m_R \times c_p \times 120, \quad \text{or,} \quad 8062.8 \times 1 \times (520.9 - 321) = m_R \times 1 \times 120$$

$$\therefore \quad m_R = \mathbf{13431.3\ kg/h}, \quad \text{or,} \quad \mathbf{3.73\ kg/s} \quad \textbf{(Ans.)}$$

(*iii*) **Power used to drive fan, P_{fan}:**

$$P_{fan} = W_{T\text{-}1} + W_{T\text{-}2}$$

$$= m_R \times c_p \times (T_{2'} - T_{2''}) + m_a \times c_p \times (T_4 - T_{5'})$$

$$= 3.73 \times 1 \times (321.9 - 254.7) + 2.24 \times 1 \times (321 - 256.8) = \mathbf{394.5\ kW} \quad \textbf{(Ans.)}$$

(*iv*) **C.O.P. of the system:**

$$\text{C.O.P.} = \frac{14000\,Q}{m_a \times c_p \times (T_{3'} - T_{2'})} = \frac{14000 \times 22}{8062.8 \times 1 \times (520.9 - 321.9)} = \mathbf{0.192} \quad \textbf{(Ans.)}$$

Example 2.46. *The reduced ambient system of air-refrigeration for cooling an aircraft cabin consists of the cooling turbines, one heat exchanger and one fan. The first cooling turbine is supplied with the ram air at 1.15 bar and 12°C and delivers after expansion to the heat exchanger at 0.88 bar for cooling the air bled off from the main compressor at 3.6 bar. The cooling air from the heat exchanger is sucked by a fan and discharged to the atmosphere. The cooled air from the heat exchanger is expanded upto 1 bar in the second cooling turbine and discharged into air cabin to be cooled. The air from the cabin is exhausted at 24°C. The refrigeration capacity required is 9 tonnes. If the compression index for the main compressor is 1.45 and the expansion index for both the cooling turbines is 1.3, determine the following:*

(i) *Mass flow rate of the cabin air;*

(ii) *Cooling capacity of heat exchanger and flow rate of ram air when compressed air is to be cooled to 55°C in the heat exchanger and temperature rise in the heat exchanger for the ram air is not to exceed 35 K;*

(iii) *Combined output of both cooling turbines driving the fan with transmission efficiency of 62%;*

(iv) *C.O.P. of the refrigerating system considering only power input to the compressor.*

Solution: Fig. 2.46 shows the T-s diagram of the system with given conditions. The process of cooling the ram air in cooling turbine T-1 is shown by curve 2-2′.

Given: $p_2 = 1.15$ bar; $T_2 = 12 + 273 = 285$ K; $p_{2'} = 0.88$ bar; $p_3 = p_4 = 3.6$ bar; $p_5 = p_6 = 1$ bar;

$T_6 = 24 + 273 = 297$ K; Q = 9 tonnes; $n_c = 1.45$, $n_1 = n_2 = 1.3$; $\eta_{T\text{-}1} = \eta_{T\text{-}2} = 62\%$,

$T_4 = 55 + 273 = 328$ K.

Let $T_{2'}$ = Temperature of ram air after expansion in the cooling turbine T-1, and

T_3 = Temperature of air bled off from the main compressor.

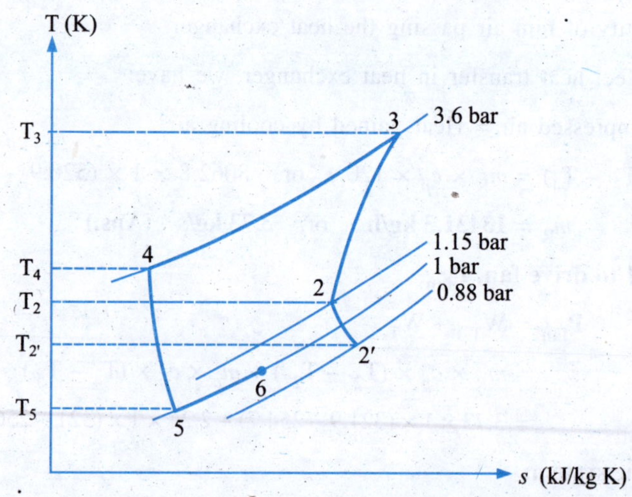

Fig. 2.46. T-s diagram.

For the process 2–2′ (expansion in cooling turbine T-1), we have:

$$\frac{T_2}{T_{2'}} = \left(\frac{p_2}{p_{2'}}\right)^{\frac{n_1-1}{n_1}} = \left(\frac{1.15}{0.88}\right)^{\frac{1.3-1}{1.3}} = \left(\frac{1.15}{0.88}\right)^{0.2308} = 1.064$$

$$\therefore \quad T_{2'} = \frac{T_2}{1.064} = \frac{285}{1.064} = 267.8 \text{ K}$$

For the process 2-3, we have:

$$\frac{T_3}{T_2} = \left(\frac{p_3}{p_2}\right)^{\frac{n_c-1}{n_c}} = \left(\frac{3.6}{1.15}\right)^{\frac{1.45-1}{1.45}} = \left(\frac{3.6}{1.15}\right)^{0.3103} = 1.425$$

$$\therefore \quad T_3 = T_2 \times 1.425 = 285 \times 1.425 = 406 \text{ K}$$

For the process 4–5, we have:

$$\frac{T_4}{T_5} = \left(\frac{p_4}{p_5}\right)^{\frac{n_2-1}{n_2}} = \left(\frac{3.6}{1}\right)^{\frac{1.3-1}{1.3}} = (3.6)^{0.2308} = 1.344$$

$$\therefore \quad T_5 = \frac{T_4}{1.344} = \frac{328}{1.344} = 244 \text{ K}$$

(*i*) Mass flow rate of cabin air, m_a:

$$m_a = \frac{14000\,Q}{c_p(T_6 - T_5)} = \frac{14000 \times 9}{1(297 - 244)} = \mathbf{2377.4 \ kg/h} \ \text{ or } \ \mathbf{0.66 \ kg/s} \quad \textbf{(Ans.)}$$

(*ii*) Cooling capacity of the heat exchanger and flow rate of ram air (m_R):

Cooling capacity of the heat exchanger

$$= m_a \times c_p \times (T_3 - T_4)$$

$$= 0.66 \times 1 \times (406 - 328) = \mathbf{51.48 \ kJ/s} \quad \textbf{(Ans.)}$$

Considering perfect heat transfer in heat exchanger, we have:

Heat lost by compressed air = Heat gained by cooling air

$$0.66 \times c_p \times (T_3 - T_4) = m_R \times c_p \times (\Delta T)$$

$$0.66 \times 1 \times (406 - 328) = m_R \times 1 \times 35$$

$$\therefore \quad m_R = \mathbf{1.47 \ kg/s} \quad \textbf{(Ans.)}$$

(*iii*) Combined output of both cooling turbines:

Output of the cooling turbine T-1

$$= m_R \times c_p \times (T_2 - T_{2'}) = 1.47 \times 1 \times (285 - 267.8) = 25.28 \text{ kW}$$

Output of the cooling turbine T-2

$$= m_a \times c_p \times (T_4 - T_5) = 0.66 \times 1 \times (328 - 244) = 55.44 \text{ kW}$$

Since the transmission efficiency is 62%, therefore combined output of both the cooling turbines

$$= (25.28 + 55.44) \times 0.62 = \mathbf{50.05 \ kW} \quad \textbf{(Ans.)}$$

(*iv*) **C.O.P. of the refrigeration system:**

$$\text{C.O.P.} = \frac{14000Q}{m_a \times c_p \times (T_3 - T_4)} = \frac{14000 \times 9}{2377.4 \times 1 \times (406 - 328)} = \mathbf{0.68} \quad \textbf{(Ans.)}$$

2.4.6 Comparison of Various Air-cooling Systems used for Aircraft

- *The simple air cooling system gives maximum cooling effect on the ground surface and its effect with flight decreases.*
- In case of Boot-strap unit air is required when the plane is to be in flight so that ram air can be used for cooling in the heat exchangers. This *drawback* can be taken care of by using part of work derived from the turbine to drive a fan which pulls air over the secondary heat exchanger, thus combining the features of a simple and boot-strap system.
- With the increase in the speed of the aeroplane, the temperature of the ambient air entering the heat exchanger rises due to isentropic rise in temperature. Thus high speed leads to high discharge temperature from the turbine.

— At speeds much higher than *Mach number of 1.7, the discharge temperature of the simple cooling system is too high for effective cabin cooling. The evaporative Boot-strap and regenerative systems, however, give lower discharge temperatures than the simple system and therefore adaptable to high speed aircraft.*

Fig. 2.47 shows the comparison of DART (Dry air rated temperature) vs. mach number variation for common aircraft refrigeration system. The following brief conclusions may be drawn:

(*i*) The simple system is preferable at low speeds.

(*ii*) Above a speed of 1000 km/h, it is imperative to use regenerative system. The Boot-strap system can also be employed.

(*iii*) Reduced ambient system may be useful for supersonic aircrafts and rockets.

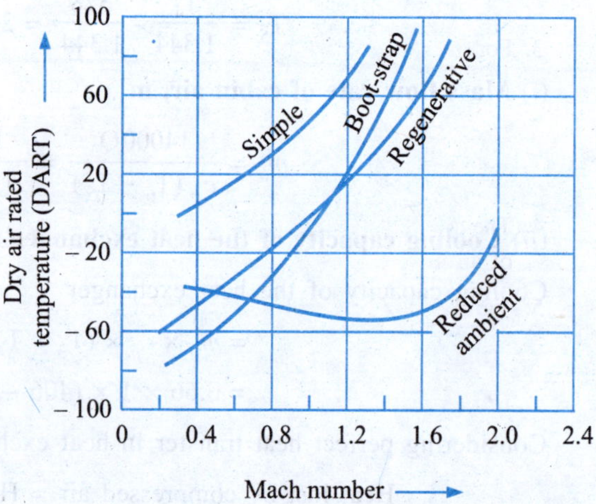

Fig. 2.47. Comparison of DART vs. Mach number variation for common aircraft refrigeration systems.

- In automatic control air-conditioning system, the following three controls are generally incorporated for the efficient working of the air-conditioning system:

(*i*) Temperature control in cabin.

(*ii*) Air-flow (quantity) control in the cabin.

(*iii*) Water separator and anti-ice control.

HIGHLIGHTS

1. The main characteristic feature of air-refrigeration system is that throughout the cycle the refrigerant remains in gaseous state.

   The air refrigeration system may be of two types:

   (*i*) Closed system, and

   (*ii*) Open system,

2. Co-efficient of performance of a '*refrigerator*' working on a reversed Carnot cycle,

$$(\text{C.O.P.})_{\text{ref.}} = \frac{T_2}{T_1 - T_2}$$

   For a Carnot cycle '*heat pump*',

$$(\text{C.O.P.})_{\text{heat pump}} = \frac{T_1}{T_1 - T_2}$$

   For a Carnot cycle '*heat engine*',

$$(\text{C.O.P.})_{\text{heat engine}} = \frac{T_1 - T_2}{T_1}$$

3. For air-refrigeration system working on reversed *Brayton cycle*,

$$\text{C.O.P.} = \frac{(T_3 - T_2)}{\left(\frac{n}{n-1}\right)\left(\frac{\gamma - 1}{\gamma}\right)(T_4 - T_3 + T_2 + T_1)}$$

4. The *dry air rated temperature* (DART) is the temperature in terms of which the comparison of different types of air-refrigeration systems is made.

5. Various types of air-refrigeration systems are:

   (*i*) Basic or simple air-refrigeration system; (*ii*) Boot-strap air-refrigeration system;

   (*iii*) Regenerative air-refrigeration system; (*iv*) Reduced ambient air-refrigeration system.

OBJECTIVE TYPE QUESTIONS

Choose the Correct Answer:

1. The co-efficient of performance of a refrigerator working on Reversed Carnot cycle with T_1 being higher temperature and T_2 being lower temperature is

   (*a*) $\frac{T_1 - T_2}{T_1}$ (*b*) $\frac{T_1}{T_1 - T_2}$ (*c*) $\frac{T_2}{T_1 - T_2}$ (*d*) $\frac{T_1 - T_2}{T_2}$.

2. The relationship between $COP_{\text{heat pump}}$ and $COP_{\text{refrigerator}}$ for the same range of temperature operation is

   (*a*) $COP_{\text{heat pump}} = COP_{\text{refrigerator}}$ (*b*) $COP_{\text{heat pump}} - COP_{\text{refrigerator}} = 1$

   (*c*) $COP_{\text{heat pump}} = 2COP_{\text{refrigerator}}$

   (*d*) not possible to predict without knowing working substance.

3. Co-efficient of performance of a Reversed Carnot cycle refrigerator working between higher temperature T_1 and lower temperature T_2
   (*a*) will increase with increase in T_2 keeping T_1 fixed
   (*b*) will decrease with increase in T_2 keeping T_1 fixed
   (*c*) will first increase with increase in T_2 and then decrease with increase in T_2 keeping T_1 fixed
   (*d*) none of the above.
4. Two Carnot Refrigerators are employed, one for ice making and other for comfort cooling
   (*a*) The COP of refrigerator for ice making is higher than that for the other
   (*b*) The COP of refrigerator for ice making is lower than that for the other
   (*c*) The COP of refrigerator for ice making is same as that for the other
   (*d*) The COP of Carnot refrigerator will depend on refrigerant used.
5. Reversed Carnot cycle comprises
   (a) two isentropic processes and two adiabatic processes
   (*b*) two isentropic processes and two isothermal processes
   (*c*) two isentropic processes and two constant pressure processes
   (*d*) two isentropic processes and two constant volume processes.
6. A reversible refrigerator working between two fixed temperatures
   (*a*) has the same COP whatever the working substance
   (*b*) has its COP increased for working substance with high enthalpy of evaporation
   (*c*) has its COP increased for working substance with high specific heats
   (*d*) none of the above.
7. A carnot refrigerator operates between 300.3 K and 273 K. The fraction of cooling effect required as work input in
   (*a*) 20% (*b*) 10% (*c*) 50%
   (*d*) not possible to find with the data.
8. The ratio of high temperature to low temperature for Reversed Carnot refrigerator is 1.25. The COP will be
   (*a*) 2 (*b*) 3 (*c*) 4 (*d*) 5.
9. In Reversed Carnot cycle working on perfect gas
   (*a*) Isothermal work of compression is equal to isothermal work of expansion
   (*b*) Isentropic work of compression is equal to isentropic work of expansion
   (*c*) Net work of the cycle is zero (*d*) Net heat transfer of the cycle is zero.
10. Reversed Carnot cycle assumes that all processes in the cycle are
    (*a*) Non-flow only (*b*) Steady flow only
    (c) Non-flow or steady flow (*d*) Transient flow.
11. A reverse Carnot cycle has a COP of 4. The ratio of higher temperature to lower temperature will be
    (*a*) 1.5 (*b*) 2 (*c*) 1.25 (*d*) 2.5.
12. In a Reversed Carnot cycle working on vapour
    (*a*) Isentripic work of compression is equal to isentripic work of expansion
    (*b*) Isothermal heat absorption is equal to isothermal heat rejection
    (*c*) There is no work done during isothermal processes
    (*d*) There is no work done during isentropic processes.
13. Air refrigeration cycle is used in
    (*a*) commercial refrigerators (*b*) domestic refrigerators
    (c) air-conditioning (*d*) gas liquefaction.

14. In a refrigerating machine, heat rejected is .... heat absorbed.
(*a*) equal to (*b*) less than (*c*) greater than (*d*) any of these.

15. Air refrigerator works on
(*a*) Carnot cycle (*b*) Rankine cycle
(*c*) reversed Carnot cycle (*d*) Bell-Coleman cycle
(*e*) both (*a*) and (*b*) (*f*) both (*c*) and (*d*).

16. The air cooling system mostly used in transport type aircrafts is
(*a*) simple air cooling system (*b*) simple evaporative air cooling system
(*c*) boot-strap air cooling system (*d*) all of the above.

17. The reduced ambient air cooling system has
(*a*) one cooling turbine and one heat exchanger
(*b*) one cooling turbine and two heat exchangers
(*c*) two cooling turbines and one heat exchanger
(*d*) two cooling turbines and two heat exchangers.

18. The cooling system used for supersonic aircrafts and rockets is
(*a*) simple air cooling system (*b*) boot-strap air cooling system
(*c*) reduced ambient air cooling system (*d*) regenerative air cooling system.

19. A boot-strap air cooling system has
(*a*) one heat exchanger (*b*) two heat exchangers
(*c*) three heat exchangers (*d*) four heat exchangers.

20. In a boot-strap air evaporative cooling system, the evaporator is provided
(*a*) between the combustion chamber and the first heat exchanger
(*b*) between the first heat exchanger and the secondary compressor
(*c*) between the secondary compressor and the second heat exchanger
(*d*) between the second heat exchanger and the cooling turbine.

ANSWERS

1. (c)	2. (b)	3. (a)	4. (b)	5. (b)	6. (a)	7. (b)	8. (c)
9. (b)	10. (c)	11. (c)	12. (c)	13. (d)	14. (c)	15. (f)	16. (c)
17. (c)	18. (d)	19. (b)	20. (d)				

THEORETICAL QUESTIONS

1. What is main characteristic feature of an air-refrigeration system?
2. Differentiate clearly between open and closed air-refrigeration systems.
3. Explain briefly an air-refrigerator working on a reversed Carnot cycle. Derive an expression for its C.O.P.
4. Derive an expression for C.O.P. for an air-refrigeration system working on reversed Brayton cycle.
5. State merits and demerits of an air-refrigeration system.
6. Describe the working of a simple air cycle cooling system used for aircrafts.
7. Explain with a neat sketch a Boot-strap cycle of air-refrigeration system.
8. Describe with a schematic diagram and draw the T-*s* representation of the processes of boot-strap evaporation type aircraft refrigeration system.
9. Describe with a neat sketch a regenerative air-refrigeration system.
10. Explain with a neat sketch the reduced ambient air-refrigeration system.

UNSOLVED EXAMPLES

1. A Carnot refrigerator requires 1.25 kW per tonne of refrigeration to maintain the temperature of –30°C. Find:
   (*i*) Co-efficient of performance, and
   (*ii*) Temperature at which heat is rejected **[Ans.** (*i*) 3.1, (*ii*) 48.4°C]
2. The co-efficient of performance of a Carnot refrigerator, when it extracts 8350kJ/min from a heat source, is 5. Find power required to run the compressor. **[Ans.** 27.8 kW]
3. A reversed cycle has refrigerating C.O.P. of 4,
   (*i*) Determine the ratio T_1/T_2
   (*ii*) If the work done on the cycle is 10 kW, determine the maximum refrigeration effect in tonnes,
   (*iii*) If this cycle is used as heat pump, determine the C.O.P. and heat delivered.
   [Ans (*i*) 1.25, (*ii*) 10.28 tonnes, (*iii*) 5, 50 kW]
4. A reversed carnot cycle delivers 60 000 kJ/min to a conditioned space when used as heat pump. If the space is maintained at 25ºC when the ambient temperature is 10ºC, determine:
   (*i*) Amount of heat drawn from the surroundings;
   (*ii*) Power required to run the heat pump;
   (*iii*) If the required heat is supplied by electric heaters, determine the power consumed.
   [Ans: (*i*) 949.7 kJ/s; (*ii*) 50.3 kW; (*iii*) 1000 kW]
5. A Carnot refrigerator operates between temperatures of –50ºC and 50ºC. Determine C.O.P. of the refrigerator. If the C.O.P. is to be made 4 by changing the temperatures such that increase or decrease in upper temperature is equal to decrease or increase in lower temperature, determine the new temperatures.
   [Ans. 2.23; 203.8 K, 342.2 K]
6. A cold storage is to be maintained at –5ºC while surroundings are at 35ºC. The heat leakage from the surroundings into the cold storage is estimated to be 29 kW. The actual C.O.P. of the refrigeration plant is $\frac{1}{3}$rd of an ideal plant working between the same temperature. Find the power required to drive the plant.
   [Ans. 12.99 kW]
7. An ice plant produces 10 tonnes of ice per day at 0°C, using water at room temperature of 20°C. Estimate the power rating of the compressor motor if the C.O.P. of the plant is 2.5 and overall elector-mechanical efficiency is 0.9. Take latent heat of freezing for water = 335kJ/kg and specific heat of water = 4.18 kJ/kg.
   [Ans. 21.44 kW]
8. An air-refrigeration system operating on Bell-Coleman cycle, takes in air from cold room at 268K and compresses it from 1.0 bar to 5.5 bar. The index of compression being 1.25. The compressed air is cooled to 300 K. The ambient temperature is 20°C. Air expands in an expander where the index of expansion is 1.35.
   Calculate:
   (*i*) C.O.P. of the system,
   (*ii*) Mass of air circulated per minute for production of 1500 kg of ice per day at 0°C from water at 20°C, and
   (*iii*) Capacity of the plant in terms of kJ/s.
   Take c_p = 4.18 kJ/kg K for water, c_p = 1.005 kJ/kg for air. Latent heat of ice = 335 kJ/kg.
   [Ans. 1.974; 5.814 kg/min; 7.27 kJ/s]

9. In an aircraft cooling system, air enters the compressor at 0.1 MPa, 4°C, and is compressed to 0.3 MPa with an isentropic efficiency of 72%. After being cooled to 55°C at constant pressure in a heat exchanger the air then expands in a turbine to 0.1 MPa with an isentropic efficiency of 78%. The low temperature air absorbs a cooling load of 3 tonnes of refrigeration at constant pressure before re-entering the compressor which is driven by the turbine. Assuming air to be an ideal gas, determine:

   (*i*) The C.O.P. of the refrigerator,

   (*ii*) The driving power required, and

   (*iii*) The air mass flow rate. **[Ans.** (*i*) 0.24; (*ii*) 48.6 kW; (*iii*) 0.66 kg/s**]**

10. A simple air-cooled system is used for an aeroplane having a load of 9 tonnes. The atmospheric pressure and temperature are 0.9 bar and 10°C respectively. The pressure increase to 1.013 bar due to ramming. The temperature of the air is reduced by 50°C in the heat exchanger. The pressure in the cabin is 1.01 bar and the temperature of air leaving the cabin is 25°C. Determine:

   (*i*) Power required to take the load of cooling in the cabin, and

   (*ii*) C.O.P. of the system.

   Assume that all the expansions and compressions are isentropic. The pressure of the compressed air is 3.5 bar.

   [Ans. (*i*) 106.4 kW; (*ii*) 0.33**]**

11. An aircraft refrigeration plant has to handle a cabin load of 27 tonnes. The atmospheric temperature is 17°C. The atmospheric air is compressed to a pressure of 0.95 bar and temperature of 30°C due to ram action. This air is then further compressed in a compressor to 4.75 bar, cooled in a heat exchanger to 67°C, expanded in a turbine to 1 bar pressure and supplied to the cabin. The air leaves the cabin at a temperature of 27°C. The isentropic efficiencies of both compressor and turbine are 0.9 each. Calculate the mass of alr circulated per minute and the C.O.P. For air, c_p = 1.004 kJ/kg K and c_p/c_v = 1.4. **[Ans.** 89.64 kg/min; 0.356**]**

12. An aircraft moving with speed of 1000 km/h uses simple gas refrigeration cycle for air-conditioning. The ambient pressure and temperature are 0.35 bar and –10°C respectively. The pressure ratio of compressor is 4.5. The heat exchanger effectiveness is 0.95. The isentropic efficiencies of compressor and expander are 0.8 each. The cabin pressure and temperature are 1.06 bar and 25°C respectively. Determine the volume flow rate through compressor inlet and expander outlet for 90 tonnes of refrigeration. Take c_p = 1.005 kJ/kg K; R = 0.287 kJ/kg K and c_p/c_v = 1.4 for air. **[Ans.** 776 m^3/min, 351.7 m^3/min**]**

13. The cock-pit of a jet plane flying at a speed of 1200 km/h is to be cooled by a simple air cooling system. The cock-pit is to be maintained at 25°C and the pressure in the cock-pit is 1 bar. The ambient air pressure and temperature are 0.85 bar and 30°C respectively. The other data available is as follows:

   Cock-pit cooling load = 9 tonnes of refrigeration; Main compressor pressure ratio = 4; Ram efficiency = 90%; Temperature of air leaving the heat exchanger and entering the cooling turbine = 60°C; Pressure drop in the heat exchanger = 0.5 bar; Pressure loss between the cooler turbine and cock pit = 0.2 bar.

   Assuming the isentropic efficiencies of main compressor and cooler turbine as 80%, find the quantity of air passed through the cooling turbine and C.O.P. of the system. Take g = 1.4 and c_p = 1 kJ/kg K.

   [Ans. 36.6 kg/min; 0.264**]**

14. In an aeroplane a simple air-refrigeration system is used. The main compressor delivers the air at 5 bar and 200°C. The bled air taken from compressor is passed through a heat exchanger, cooled with the help of ram air so that the temperature of air leaving the heat exchanger is 45°C and the pressure is 4.5 bar. The cooling turbine drives the exhaust fan which is used to force the ram air through the heat exchanger. The air leaving the heat exchanger passes through the cooling turbine and then supplied to cabin at 1 bar. The pressure loss between the cooling turbine and cabin is 0.2 bar. If the rate of flow of air through the cooling turbine is 25 kg/min, determine the following:

1. The temperature of the air leaving the expander,
2. The power delivered to the ram air which is passed through the heat exchanger, and
3. The refrigeration load in tonnes when the temperature of the air leaving the cabin is limited to 25°C.

Assume that the isentropic efficiency of the cooling turbine is 75% and no loss of heat from air between the cooling turbine and cabin. Take $\gamma = 1.4$ and $c_p = 1$ kJ/kg K

[Ans. (*i*) 242.85K; (*ii*) 31.31 kW; (*iii*) 5.9 tonnes]

15. A boot-strap air cooling system of 18 tonnes capacity is required for an aircraft cabin. The temperature and pressure conditions of the atmosphere are 20°C and 0.8 bar. The pressure of air from 0.8 bar is increased to 0.96 bar due to ramming action. The pressures of air leaving the main compressor and auxiliary compressor are 3.5 bar and 5.25 bar respectively. The isentropic efficiencies of both compressors are 85% each and that of the expansion turbine 80%. The first heat exchanger removes 60% of the total heat of air leaving the main compressor and 35% of the total heat of air leaving the auxiliary compressor is removed in the second heat exchanger. Assuming the ramming action to be isentropic, the cabin pressure 1.03 bar and temperature of air leaving cabin to be 27°C, find:

    (*i*) Power required to take the load in cabin,

    (*ii*) C.O.P. of the system. **[Ans.** (*i*) 289. 36 kW; (*ii*); 0.301]

16. A boot-strap cooling system of 9 tonnes capacity is used in an aeroplane. The ambient air temperature and pressure are 20°C and 0.85 bar respectively. The pressure of air increases from 0.85 bar to 1 bar due to ramming action of air. The pressure of air discharged from the main compressor is 3 bar. The discharge pressure of air from the auxiliary compressor is 4 bar. The isentropic efficiency of each of the compressor is 80% while that of turbine is 85%. 50% of the enthalpy of air discharged form the main compressor is removed in the first heat exchanger and 30% of the enthalpy of air discharged from the auxiliary compressor is removed in the second heat exchanger using rammed air. Assuming ramming action to be isentropic, the required cabin pressure of 0.9 bar and temperature of the air leaving the cabin not more than 20°C, find:

    (*i*) The power required to operate the system, and

    (*ii*) The C.O.P. of the system.

    Draw the schematic and temperature – entropy diagrams of the system.

    Take $\gamma = 1.4$ and $c_p = 1$ kJ/kg K. **[Ans.** (*i*) 130 kW; (*ii*) 0.27]

17. A boot-strap evaporative air-refrigeration system is used to take 18 tonnes refrigeration load of aeroplane cabin when it is moving with a Mach. number of 1.3. The ambient conditions are 0.5 bar and –10°C. The cooling air is taken from the main compressor at 4 bar and further compressed to 5 bar which is run by cooling turbine. Take the following data:

    Isentropic efficiency of main compressor = 90%

    Isentropic efficiency of secondary compressor = 80%

    Isentropic efficiency of cooling turbine = 80%

    Effectiveness of both the heat exchangers = 0.4 each

    Ramming efficiency = 90%

    The temperature of air coming out of evaporative cooler = 100°C.

    The pressure and temperature maintained in the cabin are 1 bar and 25°C.

    Determine the following:

    (*i*) Mass taken form the main compressor for cabin cooling,

    (*ii*) Power required to run the refrigeration system, and

    (*iii*) C.O.P. of the system.

    Assume $c_p = 1$ kJ/kg.K and $\gamma = 1.4$ for air. **[Ans.** (*i*) 1.994 kg /s; (*ii*) 296 kW; (*iii*) 0.236]

18. A regenerative air cooling system is used for an air plane to take 18 tonnes of refrigeration load. The ambient air at pressure 0.8 bar and temperature 10°C is rammed isentropically till the pressure rises to 1.2 bar. The air bled off the main compressor at 4.5 bar is cooled by the ram air in the heat exchanger whose effectiveness is 60%. The air from the heat exchanger is further cooled to 60°C in the regenerative heat exchanger with a portion of the air bled after expansion in the cooling turbine. The cabin is to be maintained at a temperature of 25°C and a pressure of 1 bar. If the isentropic efficiencies of the compressor and turbine are 90% and 80% respectively, determine the following:
   1. Mass of the air bled from cooling turbine to be used for regenerative cooling;
   2. Power required for maintaining the cabin at the required condition; and
   3. C.O.P. of the system.

   Assume the temperature of air leaving to atmosphere from the regenerative heat exchanger at 100°C.

   [**Ans.** (*i*) 0.703 kg/s (*ii*) 307 kW; (*iii*) 0.23]

19. A regenerative air-refrigeration system for an aeroplane is designed to take a load of 27 tonnes. The temperature and pressure conditions of the atmosphere are 5°C and 0.85 bar. The pressure of the air is increased from 0.85 bar to 1.2 bar due to ramming action. The pressure of air leaving the main compressor is 4.8 bar, 60% of the total heat of the air leaving the main compressor is removed in the heat exchanger and the it is passed through the cooling turbine. The temperature of the rammed air which is used for cooling purposes in the heat exchanger is reduced to 50°C by mixing the air coming out from the cooling turbine. The isentropic efficiencies of the compressor and turbine are 90%C and 80% respectively. The pressure and temperature required in the cabin are 1 bar and 25°C respectively. Assuming isentropic ramming and mass of cooled air passing through the heat exchanger equal to the mass of cooling air, find. (*i*) The ratio of by-passed air to ram air used for cooling purposes; and. (*ii*) The power required for maintaining the cabin at required condition. [**Ans.** 41.2%; 445 kW]

20. The reduced ambient air-refrigeration system used for an aircraft consists of two cooling turbines, one heat exchanger and one air cooling fan. The speed of aircraft is 1500 km/h. The ambient air conditions are 0.8 bar and 10°C. The ram efficiency may be taken as 90%. The rammed air used for cooling is expanded in the first cooling turbine and leaves it at a pressure of 0.8 bar. The air bled from the main compressor at 6 bar is cooled in the heat exchanger and leaves it at 100°C. The cabin is to be maintained at 20°C and 1 bar. The pressure loss between the second cooling turbine and cabin is 0.1 bar. If the isentropic efficiency for the main compressor and both of the cooling turbines are 85% and 80% respectively, determine the following:

   (*i*) Mass flow rate of air supplied to cabin to take a cabin load of 9 tonnes of refrigeration;

   (*ii*) Quantity of air passing through the heat exchanger if the temperature rise of ram air is limited to 80 K;

   (*iii*) Power used to drive the cooling fan; and

   (*iv*) C.O.P. of the system. [**Ans.** (*i*) 1.0167 kg/s; (*ii*) 2.072 kg/s; (*iii*) 253 kW; (*iv*) 0.21]

☐☐☐

Chapter 3

Simple Vapour Compression Refrigeration System

3.1 INTRODUCTION

- Out of all refrigeration system, the vapour compression system is the most important system from the view point of *commercial* and *domestic utility. It is the* ***most practical form of refrigeration***.
- In this system the *working fluid is a vapour*. It readily evaporates and condenses or changes alternately between the vapour and liquid phases without leaving the refrigerating plant. *During evaporation, it absorbs heat from the cold body. This heat is used as its latent heat for converting it from the liquid to vapour. In condensing or cooling or liquifying, it rejects heat to external body, thus creating a cooling effect in the working fluid.* This refrigeration system thus acts as a latent heat pump since it pumps its latent heat from the cold body or brine and rejects it or delivers it to the external hot *body or cooling medium.*
- The principle upon which the vapour compression system works *applies to all the vapours* for which tables of Thermodynamic properties are available.

3.2 SIMPLE VAPOUR COMPRESSION SYSTEM

In a simple vapour compression system *fundamental processes are completed in one cycle*. These are:

1. Compression
2. Condensation
3. Expansion
4. Vaporisation.

The flow diagram of such a cycle is shown in Fig. 3.1.

The *vapour at low temperature and pressure* (state '2') enters the **"compressor"** where it is compressed isentropically and subsequently its temperature and pressure increase considerably (state '3'). The vapour after leaving the compressor enters the **"condenser"** where it is condensed into *high pressure liquid* (state '4') and is collected in a *"receiver tank"*. From receiver tank it passes through the **"expansion valve"**, here it is *throttled down to a lower pressure and has a low temperature* (state '1'). After finding its way through expansion valve it finally passes on to **"evaporator"** where it extracts *heat from the surroundings or circulating fluid being refrigerated and vaporises to low pressure vapour* (state '2').

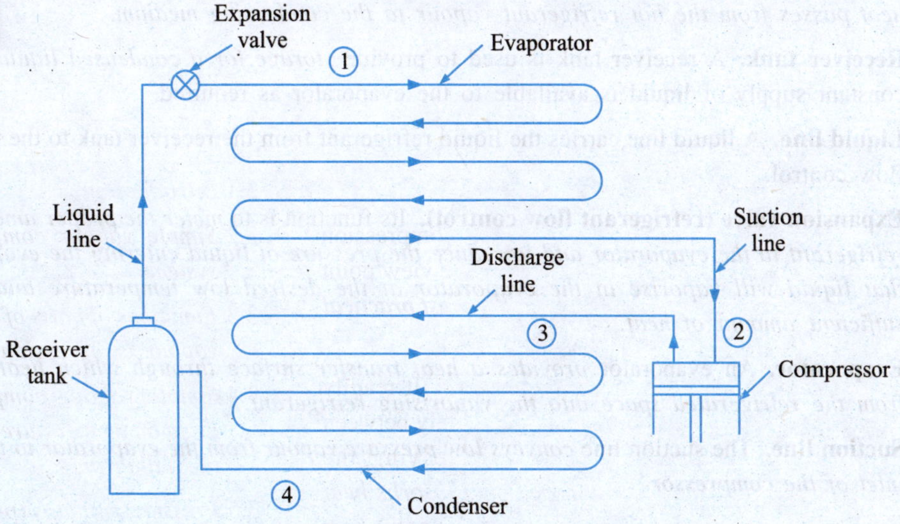

Fig. 3.1. Vapour compression system.

Merits and Demerits of Vapour Compression System over Air Refrigeration System

Merits:

1. C.O.P. is quite high as the working of cycle is very near to that of reversed Carnot cycle.
2. When used on ground level the running cost of vapour-compression refrigeration system is only 1/5th of air refrigeration system.
3. For the same refrigeration effect the size of the evaporator is smaller.
4. The required temperature of evaporator can be achieved simply by adjusting the throttle valve of the same unit.

Demerits:

1. Initial cost is high.
2. The major disadvantages are inflammability, leakage of vapours and toxity. These have been overcome to great extent by improvement in design.

3.3 FUNCTIONS OF PARTS OF A SIMPLE VAPOUR COMPRESSION SYSTEM

Here follows the brief description of various parts of a simple vapour compression system shown in Fig. 3.1.

1. **Compressor.** The function of a compressor is to remove the *vapour from the evaporator and to raise its temperature and pressure to a point such that it (vapour) can be condensed with available condensing media.*
2. **Discharge line (or hot-gas line).** A hot gas or discharge line *delivers the high-pressure, high temperature vapour from the discharge of the compressor to the condenser.*
3. **Condenser.** The function of a condenser is to *provide a heat transfer surface through which heat passes from the hot refrigerant vapour to the condensing medium.*
4. **Receiver tank.** A receiver tank is used to provide *storage for a condensed liquid* so that a constant supply of liquid is available to the evaporator as required.
5. **Liquid line.** A liquid line carries the liquid refrigerant from the receiver tank to the refrigerant flow control.
6. **Expansion valve (refrigerant flow control).** Its function is to *meter the proper amount of the refrigerant to the evaporator and to reduce the pressure of liquid entering the evaporator so that liquid will vaporise in the evaporator at the desired low temperature and take out sufficient amount of heat.*
7. **Evaporator.** An evaporator *provides a heat transfer surface through which heat can pass from the refrigerated space into the vaporising refrigerant.*
8. **Suction line.** The suction line *conveys low pressure vapour from the evaporator to the suction inlet of the compressor.*

3.4 SIMPLE VAPOUR COMPRESSION CYCLE ON TEMPERATURE-ENTROPY (T-s) DIAGRAM

We shall consider the following *three cases:*

1. When the vapour is dry and saturated at the end of compression:

Fig. 3.2. represents the vapour compression cycle, on T-s diagram; the points 1, 2, 3, and 4 correspond to the points 1, 2, 3 and 4 in the Fig. 3.1.

— At point '2' the vapour which is at low temperature (T_2) and low pressure enters the compressor's cylinder and is compressed adiabatically to '3' *when its temperature increases to the temperature* T_3.

— It is then condensed in the condenser (line 3-4) where it gives up its latent heat to the condensing medium.

— It then undergoes *throttling expansion* while passing through the expansion valve and it again reduces to T_2, it is represented by the line 4-1-shown broken; throttling is an *irreversible process*. From the T-*s* diagram it may noted due to this expansion the liquid partially evaporates, as its *dryness fraction* is represented by the ratio $\frac{b1}{b2}$.

— At '1' it enters the evaporator where it is further evaporated at constant pressure and constant temperature to the point '2' and the cycle is completed.

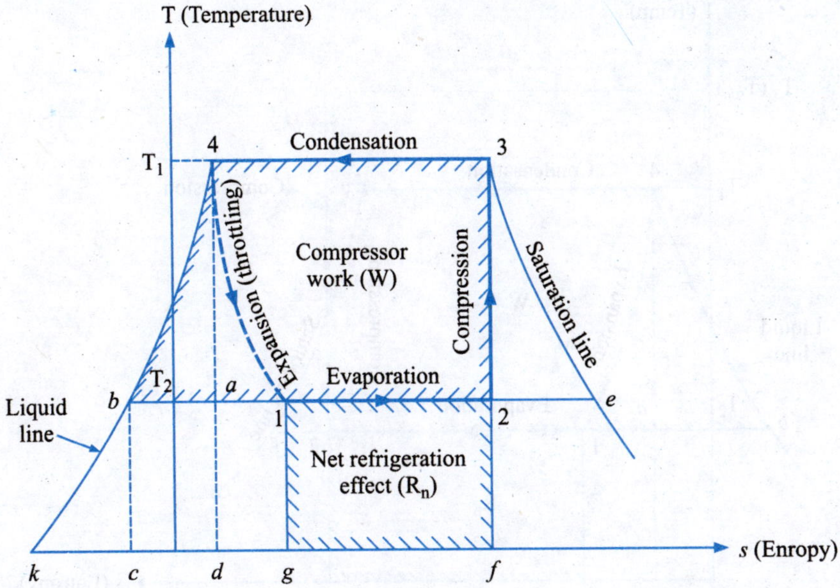

Fig. 3.2. T-*s* diagram.

Work done by the compressor = W = area '2-3-4-*b*-2'

Heat absorbed = area '2-1-*g*-*f*-2'

$$\therefore \quad \text{C.O.P.} = \frac{\text{Heat extracted or refrigerating effect}}{\text{Work done}} = \frac{\text{area '}2-1-g-f-2\text{'}}{\text{area '}2-3-4-b-2\text{'}}$$

$$\text{or,} \quad \text{C.O.P.} = \frac{h_2 - h_1}{h_3 - h_2} \quad \text{...[3.1 (a)]}$$

$$= \frac{h_2 - h_{f4}}{h_3 - h_2} \quad \text{...[3.1 (b)]}$$

($\because$ $h_1 = h_4$ since during the throttling expansion 4-1 the *total heat content remains unchanged.*)

2. When the vapour is superheated after compression :

— If the compression of the vapour is continued after it has become dry, the vapour will be superheated, its effect on T-*s* diagram is shown in Fig. 3.3.

— The vapour enters the compressor at condition '2' and is compressed to '3' where it is superheated to temperature T_{sup}.

— Then it enters the condenser. Here firstly superheated vapour cools to temperature $T_{3'}$ (represented by line 3-3') and then it condenses at constant temperature along the line 3'-4; the remaining of the cycle, however, is the same as before.

Now, work done = area '2-3-3'-4-*b*-2'

and, heat extracted/absorbed = area '2-1-*g*-*f*-2'

$$\text{C.O.P.} = \frac{\text{Heat extracted}}{\text{Work done}} = \frac{\text{area '}2-1-g-f-2\text{'}}{\text{area '}2-3-3'-4-b-2\text{'}} = \frac{h_2 - h_1}{h_3 - h_2} \quad \text{... [3.1.(c)]}$$

In this case $h_3 = h'_3 + c_p\,(T_{sup} - T_{sat'})$ and h'_3 = enthalpy of dry and saturated vapour at the point 3'.

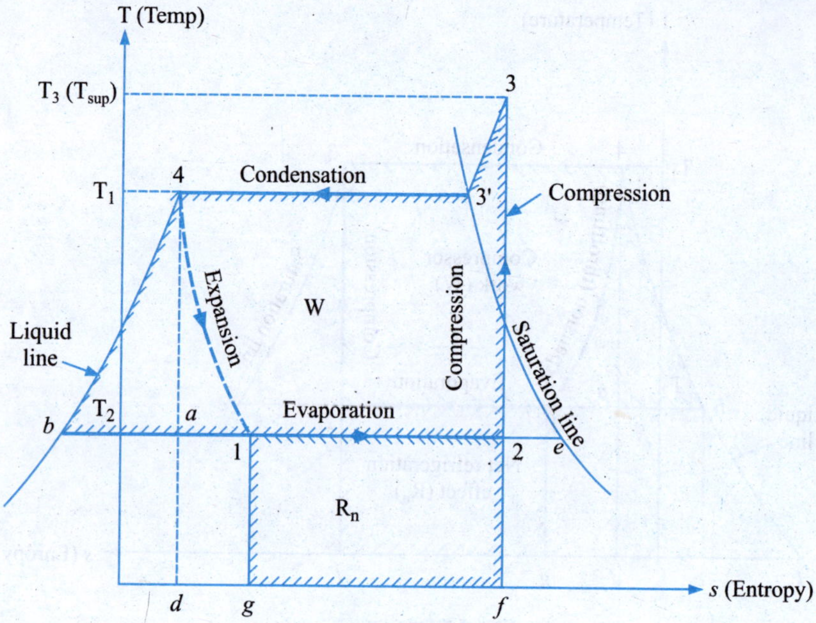

Fig. 3.3. T-*s* diagram.

3. When the vapour is wet after compression: Refer to Fig. 3.4.

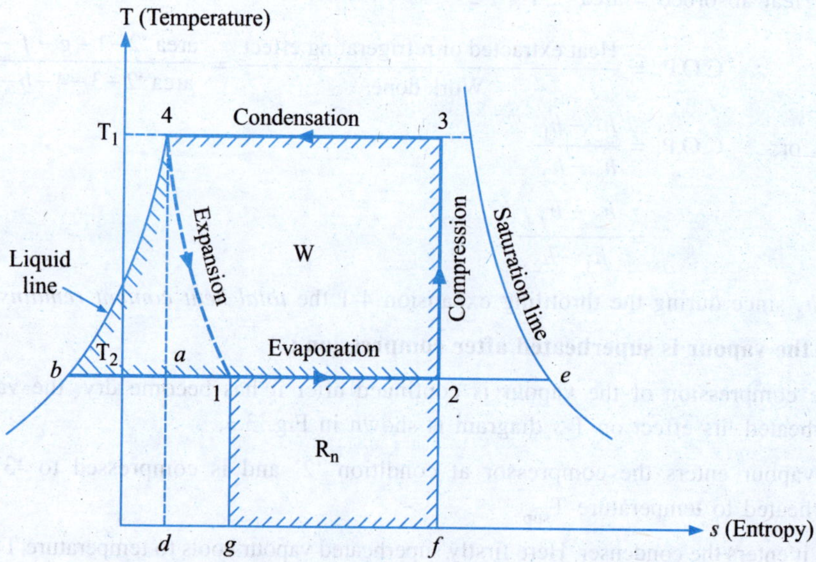

Fig. 3.4. T-*s* diagram.

Work done by the compressor = area '2-3-4-*b*-2'

Heat extracted = area '2-1-*g*-*f*-2'

$$\therefore \quad \text{C.O.P} = \frac{\text{Heat extracted}}{\text{Work done}} = \frac{\text{area '}2-1-g-f-2\text{'}}{\text{area '}2-3-4-b-2\text{'}} = \frac{h_2 - h_1}{h_3 - h_2}$$

Note: If the vapour is not superheated after compression, the operation is called 'WET COMPRESSION' and if the vapour is superheated at the end of compression, it is known as 'DRY COMPRESSION'. *Dry compression*, in actual practice, *is always preferred as it gives higher volumetric efficiency and mechanical efficiency and there are less chances of compressor damage.*

3.5 PRESSURE ENTHALPY (p-h) CHART

The diagrams commonly used in the analysis of the refrigeration cycle are :

(*i*) Pressure-enthalpy (*p-h*) Chart

(*ii*) Temperature-entropy (T-*s*) Chart

Of the two, the *pressure-enthalpy diagram seems to be more useful.*

The condition of the refrigerant in any thermodynamic state can be represented as a point on the *p-h* chart. The point on the *p-h* chart that represents the condition of the refrigerant in any one particular thermodynamic state may be located if any two properties of the refrigerant for that state are known, the other properties of the refrigerant of that state can be determined directly from the chart for studying the performance of the machines.

- Refer to Fig. 3.5. The chart is divided into *three areas* that are separated from the each other by the saturated liquid and saturated vapour lines.
  - — The region on the chart to the *left* of the saturated liquid line is called the **sub-cooled region**. At any point in the sub-cooled region the refrigerant is in the liquid phase and its temperature is below the saturation temperature corresponding to its pressure.
  - — The area to the *right* of the saturated vapour line is **superheated region** and the refrigerant is in the form of a superheated vapour.
  - — The section of the chart between the saturated liquid and saturated vapour lines is the *two phase region* and represents the change in phase of the refrigerant between liquid and vapour phases. At any point between two saturation lines the refrigerant is in the form of a liquid vapour mixture. *The distance between the two lines along any constant pressure line, as read on the enthalpy scale at the bottom of the chart, is the latent heat of vaporisation of the refrigerant at that pressure.*
  - — *The horizontal lines extending across the chart are lines of constant pressure and the vertical lines are lines of constant enthalpy. The lines of 'constant temperature' in the sub-cooled region are almost vertical on the chart and parallel to the lines of constant enthalpy. In the centre section, since the refrigerant changes state at a constant temperature and pressure, the lines of constant temperature are parallel to and coincide with the lines of constant pressure.* At the saturated vapour line the lines of constant temperature change direction again and, in the superheated vapour region, fall off sharply toward the bottom of the chart.
  - — *The straight lines which extend diagonally and almost vertically across the superheated vapour are lines of constant entropy. The curved, nearly horizontal lines crossing the superheated vapour region are lines of constant volume.*
- *p-h chart gives directly the changes in enthalpy and pressure during a process for thermodynamic analysis.*

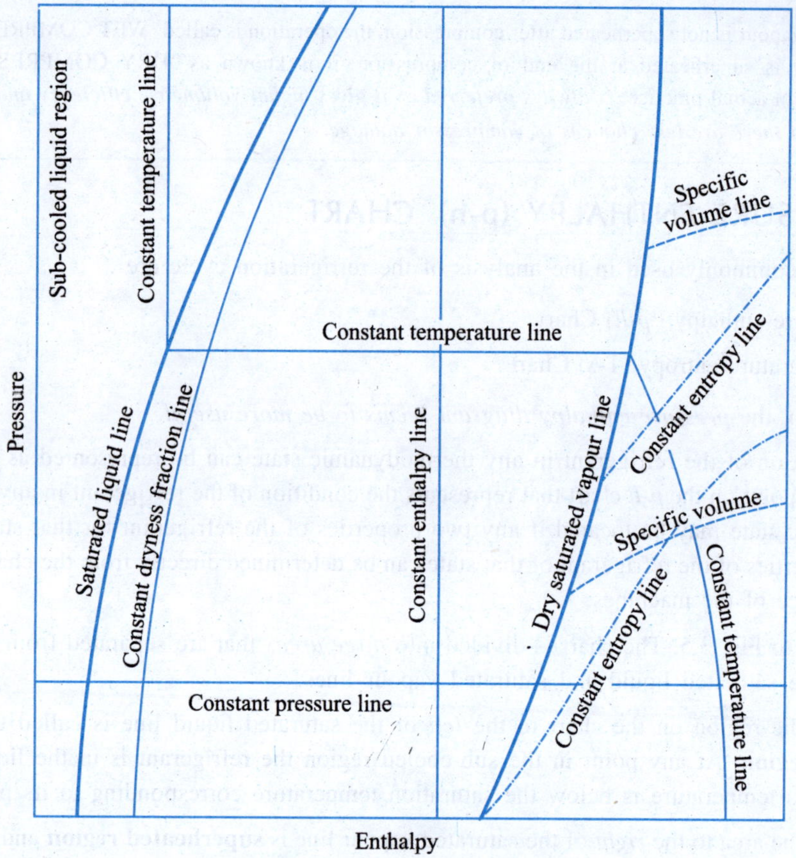

Fig. 3.5. Pressure-enthalphy (*p-h*) chart.

3.6 SIMPLE VAPOUR COMPRESSION CYCLE ON p-h CHART

Fig. 3.6 shows a simple vapour compression cycle on a *p-h* chart. The points 1, 2, 3 and 4 correspond to the points marked in Fig. 3.1.

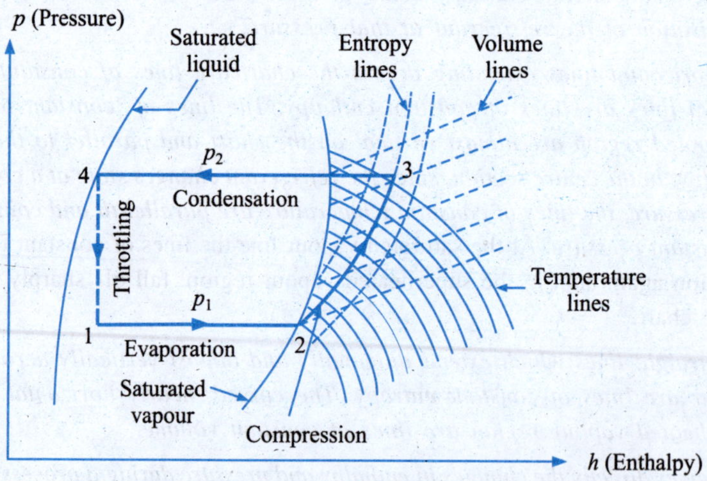

Fig. 3.6. Simple vapour compression cycle on *p-h* chart.

— The dry saturated vapour (at state 2) is drawn by the compressor from evaporator at lower pressure p_1 and then it (vapour) is compressed isentropically to the upper pressure p_2.

— The isentropic compression is shown by the line 2-3. Since the vapour is dry and saturated at the start of compression it becomes superheated at the end of compression as given by point 3.

— The process of *condensation which takes place at constant pressure* is given by the line 3-4.

— The vapour now reduced to saturated liquid is throttled through the expansion valve and the process is shown by the line 4-1.

— At the point 1 a mixture of vapour and liquid enters the evaporator where it gets dry saturated as shown by the point 2. The cycle is thus completed.

Heat extracted (or net refrigerating effect produced),

$$R_n = h_2 - h_1$$

Work done $\quad W = h_3 - h_2$

$$\therefore \quad \text{C.O.P.} = \frac{R_n}{W} = \frac{h_2 - h_1}{h_3 - h_2}$$

The values of h_1, h_2 and h_3 can be directly read for *p-h* chart.

3.7 FACTORS AFFECTING THE PERFORMANCE OF A SIMPLE VAPOUR COMPRESSION SYSTEM

The factors which affect the performance of a vapour compression system are as follows:

1. Effect of suction pressure: The effect of *decrease* in suction pressure is shown in Fig. 3.7.

The C.O.P. of the original cycle,

$$\text{C.O.P.} = \frac{h_2 - h_1}{h_3 - h_2}$$

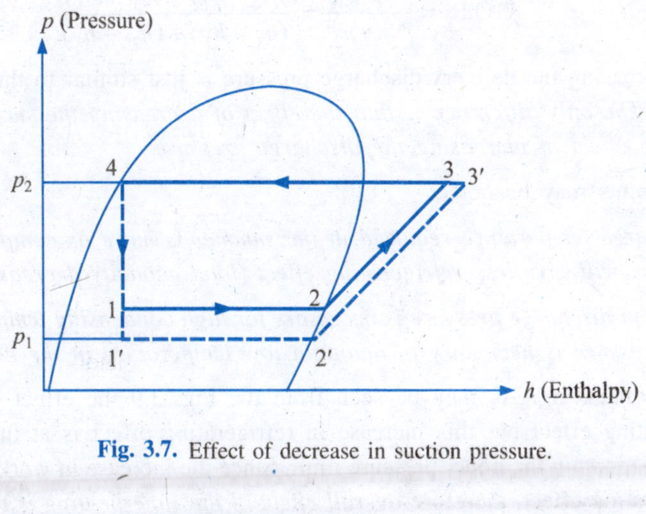

Fig. 3.7. Effect of decrease in suction pressure.

The C.O.P. of the cycle when suction pressure is decreased,

$$\text{C.O.P} = \frac{h_{2'} - h_{1'}}{h_{3'} - h_{2'}} = \frac{(h_2 - h_1) - (h_2 - h_{2'})}{(h_3 - h_2) - (h_{2'} - h_2) + (h_{3'} - h_3)} \qquad (\because\ h_1 = h_{1'})$$

This shows that the *refrigerating effect is decreased and work required is increased. The net effect is to reduce the refrigerating capacity of the system (with the same amount of refrigerant flow) and the C.O.P.*

2. Effect of delivery pressure: Fig. 3.8 shows the effect of *increase in delivery pressure.*

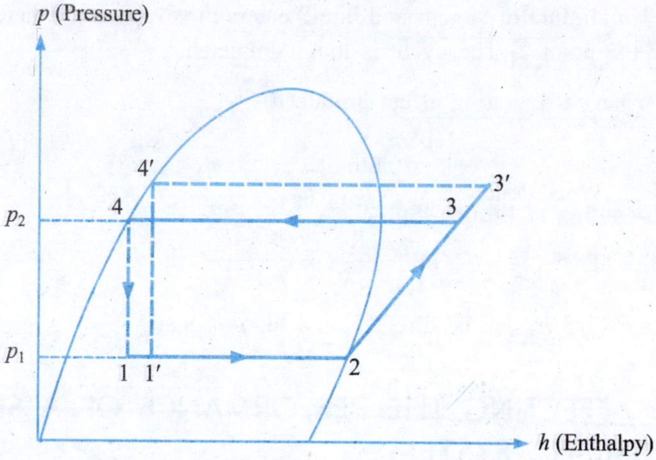

Fig. 3.8. Effect of increase in delivery pressure.

C.O.P. of the original cycle,

$$\text{C.O.P} = \frac{h_2 - h_1}{h_3 - h_2}$$

C.O.P. of the cycle when *delivery pressure is increased,*

$$\text{C.O.P.} = \frac{h_2 - h_{1'}}{h_{3'} - h_2} = \frac{(h_2 - h_1) - (h_{1'} - h_1)}{(h_3 - h_2) + (h_{3'} - h_3)}$$

The effect of increasing the delivery/discharge pressure is just similar to the effect of decreasing the suction pressure. *The only difference is that the effect of decreasing the suction pressure is more predominant than the effect of increasing the discharge pressure.*

The following points may be noted:

(*i*) *As the discharge temperature required in the summer is more as compared with winter, the same machine will give less refrigerating effect (load capacity decreased) at a higher cost.*

(*ii*) *The increase in discharge pressure is necessary for high condensing temperature and decrease in suction pressure is necessary to maintain low temperature in the evaporator.*

3. Effect of superheating. As may be seen from the Fig. 3.9 the effect of superheating is to increase the refrigerating effect but this increase in refrigerating effect is at the cost of increase in amount of work spent to attain the upper pressure limit. Since the increase in work is more as compared to increase in refrigerating effect, therefore *overall effect of the superheating is to give a low value of C.O.P.*

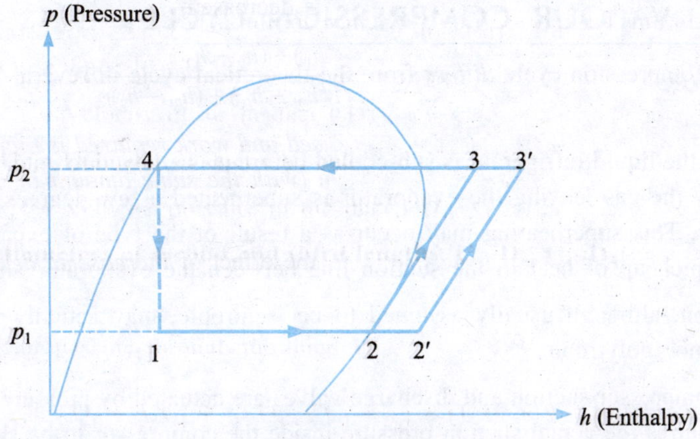

Fig. 3.9. Effect of superheating.

4. Effect of sub-cooling of liquid. 'Sub-cooling' is the process of *cooling the liquid refrigerant below the condensing temperature for a given pressure*. In the Fig. 3.10 the process of sub-cooling is shown by 4-4′. As is evident from the figure the effect of sub-cooling is to increase the refrigerating effect. Thus *sub-cooling results in increase of C.O.P. provided that no further energy has to be spent to obtain the extra cold coolant required.*

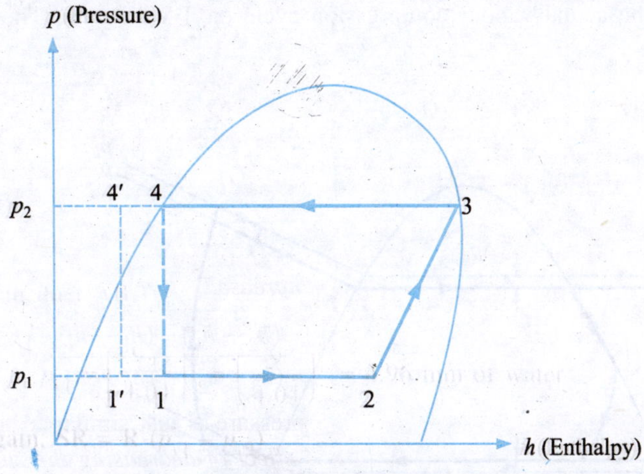

Fig. 3.10. Effect of sub-cooling of liquid.

The sub-cooling or under-cooling may be done by any of the following methods:

(*i*) Inserting a special coil between the condenser and the expansion valve.

(*ii*) Circulating greater quantity of cooling water through the condenser.

(*iii*) Using water cooler than main circulating water.

5. Effect of suction temperature and condenser temperature. The performance of the vapour compression refrigerating cycle varies considerably with both vaporising and condensing temperature. Of the two, the vaporising temperature has far the greater effect. It is seen that the capacity and performance of the refrigerating system improve as the vaporising temperature increases and the condensing temperature decreases. Thus *refrigerating system should always be designed to operate at the highest possible vaporising temperature and lowest possible condensing temperature, of course, keeping in view the requirements of the application.*

3.8 ACTUAL VAPOUR COMPRESSION CYCLE

The actual vapour compression cycle *differs* from the theoretical cycle in several ways because of the following *reasons*:

1. Frequently the liquid refrigerant is sub-cooled before it is allowed to enter the expansion valve and usually the gas leaving the evaporator is superheated a few degrees before it enters the compressor. This superheating may occur as a result of the type of expansion valve used or through a pick up of heat in the suction line between the evaporator and compressor.
2. Compression, although usually assumed to be isentropic, may actually prove to be neither isentropic nor polytropic.
3. Both the compressor suction and discharge valves are actuated by pressure difference, and this process requires the actual suction pressure inside the compressor to be slightly below that of the evaporator and the discharge pressure to be above that of condenser.
4. Although isentropic compression assumes no transfer of heat between the refrigerant and the cylinder walls, yet the cylinder walls are hotter than the incoming gases from the evaporator and colder than the compressed gases discharged to the condenser.
5. Pressure drop in long suction and liquid line piping and any vertical differences in heat created by locating the evaporator and condenser at different elevations.

Fig. 3.11 shows the actual vapour compression cycle on T-*s* diagram. The various processes are discussed below:

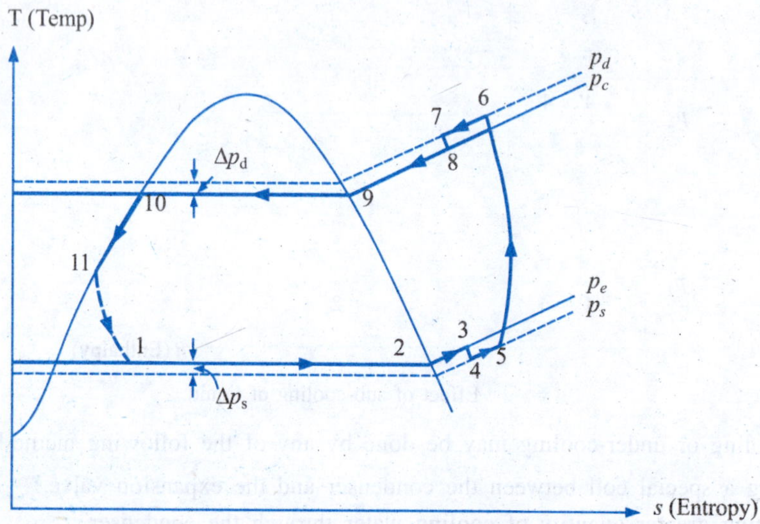

Fig. 3.11. Actual vapour compression cycle (T-*s* diagram).

Process 1-2-3: This process represents passage of refrigerant through the evaporator, with 1-2 indicating gain of latent heat of vaporisation, and 2-3, the gain of superheat before entrance to compressor. Both of these processes approach very closely to the constant pressure conditions (assumed in theory).

Process 3-4-5-6-7-8: This path/process represents the passage of the vapour refrigerant from entrance to the discharge of the compressor. Path 3-4 represents the throttling action that occurs during passage through the suction valves, and path 7-8 represents the throttling during passage through exhaust valves. Both of these actions are accompanied by an entropy increase and a slight drop in temperature.

Compression of the refrigerant occurs along path 5-6, which is actually neither isentropic nor polytropic. The heat transfer indicated by paths 4-5 and 6-7 occurs essentially at constant pressure.

Process 8-9-10-11: This process represents the passage of refrigerant though the condenser with 8-9 indicating removal of superheat, 9-10 the removal of latent heat, 10-11 removal of heat of liquid or sub-cooling.

Process 11-1: This process represents passage of the refrigerant through the expansion valve, both theoretically and practically an *irreversible adiabatic path.*

3.9 VOLUMETRIC EFFICIENCY

A compressor which is theoretically perfect would have neither clearance nor losses of any type and would pump on each stroke a quantity of refrigerant equal to piston displacement. No actual compressor is able to do this, since it is impossible to construct a compressor without clearance or one that will have wire-drawing through the suction and discharge valves, no superheating of the suction gases upon contact with the cylinder walls, or no leakage of gas past the piston or the valve. All these factors affect the volume of gas pumped or the capacity of the compressor, some of them affect requirements per tonne of refrigeration developed.

Volumetric efficiency *is defined as the ratio of actual volume of gas drawn into the compressor (at evaporator temperature and pressure) on each stroke to the piston displacement.* If the effect of clearance alone is considered, the resulting expression may be termed *clearance volumetric efficiency.* The expression used for grouping into one constant all the factors affecting efficiency may be termed *total volumetric efficiency.*

Clearance volumetric efficiency. '*Clearance volume*' is the volume of space between the end of the cylinder and piston when the latter is in dead centre position. The clearance volume is expressed as a percentage of piston displacement. In the Fig. 3.12 the piston displacement is shown as 4′–1.

During the suction stroke 4′–1, the vapour filled in clearance space at a discharge pressure p_d expands along 3–4 and the suction valve opens only when pressure has dropped to suction pressure p_s, therefore actual sucked volume will be $(V_1 - V_4)$ while the swept volume. $(V_1 - V_4')$. *The ratio of actual volume of vapour sucked to the piston displacement is defined as* ***clearance volumetric efficiency***.

Thus, clearance volumetric efficiency,

$$\eta_{cv} = \frac{V_1 - V_4}{V_1 - V_4'} = \frac{V_1 - V_4}{V_1 - V_3} \qquad (\because V_{4'} = V_3)$$

Considering polytropic expansion process 3-4, we have:

$$p_s V_4^n = p_d V_3^n, \quad \text{or,} \quad \frac{p_d}{p_s} = \left(\frac{V_4}{V_3}\right)^n, \quad \text{or,} \quad V_4 = V_3 \left(\frac{p_d}{p_s}\right)^{\frac{1}{n}}$$

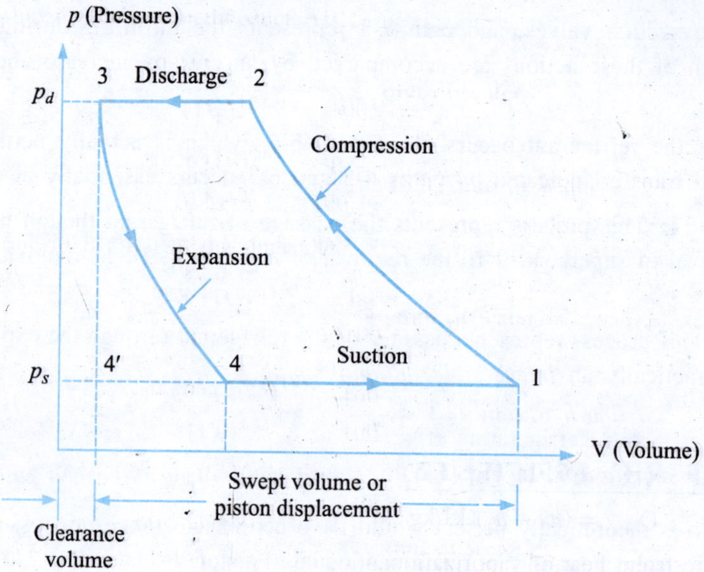

Fig. 3.12

If the clearance ratio,

$$C = \frac{V_3}{V_1 - V_3} = \frac{\text{Clearance volume}}{\text{Swept volume}}$$

Then, $$\eta_{cv} = \frac{V_1 - V_4}{V_1 - V_3} = \frac{(V_1 - V_{4'}) - (V_4 - V_{4'})}{(V_1 - V_3)}$$

$$= \frac{(V_1 - V_3) - (V_4 - V_3)}{(V_1 - V_3)} \qquad (\because V_{4'} = V_3)$$

$$= 1 - \frac{V_4 - V_3}{V_1 - V_3} = 1 - \frac{V_3\left(\frac{p_d}{p_s}\right)^{1/n} - V_3}{V_1 - V_3}$$

$$= 1 + \frac{V_3}{V_1 - V_3}\left[1 - \left(\frac{p_d}{p_s}\right)^{1/n}\right] = 1 + C - C\left(\frac{p_d}{p_s}\right)^{1/n}$$

Hence clearance volumetric efficiency,

$$\eta_{cv} = 1 + C - C\left(\frac{p_d}{p_s}\right)^{\frac{1}{n}} \qquad \text{...(3.2)}$$

Total volumetric efficiency. The total volumetric efficiency (η_{tv}) of a compressor is best obtained by actual laboratory measurements of the amount of refrigerant compressed and delivered to the condenser. It is very difficult to predict the effects of wire-drawing, cylinder wall heating, and piston leakage to allow any degree of accuracy in most cases. The total volumetric efficiency can be approximately calculated if the pressure drop through the suction valve and the temperature of the gases at the end of the suction stroke are known and if it is assumed that there is no leakage past the piston during compression, it can be calculated (by modifying the eqn. 3.2) by using the following equation:

$$\eta_{tv} = \left[1 + C - C\left(\frac{p_d}{P_s}\right)^{1/n}\right] \times \frac{p_c}{p_s} \times \frac{T_s}{T_c} \quad ...(3.3)$$

where the subscript '*c*' refers to compressor cylinder and '*s*' refers to the evaporator or the suction line just adjacent to the compressor.

3.10 MATHEMATICAL ANALYSIS OF VAPOUR COMPRESSION REFRIGERATION

(*i*) **Refrigerating effect.** Refrigerating effect is the *amount of heat absorbed by the refrigerant in its travel through the evaporator*. In Fig. 3.2 this effect is represented by the expression,

$$Q_{evap} = (h_2 - h_1) \text{ kJ/kg} \quad ....(3.4)$$

In addition to the latent heat of vaporization it may include any heat of superheat absorbed in the evaporator.

(*ii*) **Mass of refrigerant.** Mass of refrigerant circulated (per hour per tonne of refrigerant) may be calculated by dividing the amount of heat by the refrigerating effect.

∴ Mass of refrigerant circulated,

$$m = \frac{14000}{(h_2 - h_1) \times 60} \text{ kg/min-tonne} \quad ...(3.5)$$

because one tonne of refrigeration means cooling effect of 14000 kJ/h.

(*iii*) **Theoretical piston displacement.** Theoretical piston displacement (per tonne of refrigeration per minute) may be found by multiplying the mass of refrigerant to be circulated (per tonne of refrigeration per minute) by the specific volume of the refrigerant gas, $(v_g)_2$, at its entrance of compressor. Thus,

Piston displacement (theoretical)

$$= \frac{14000}{(h_2 - h_1) \times 60} (v_g)_2 \text{ m}^3\text{/tonne-min} \quad ...(3.6)$$

(*iv*) **Power (theoretical) required.** Theoretical power per tonne refrigeration is the power theoretically required to compress the refrigerant. Here volumetric and mechanical efficiencies are not taken into consideration. Power required may be calculated as follows :

(*a*) **When compression is isentropic :**

Work of compression = $(h_3 - h_2)$ kJ/kg

$$\therefore \quad \text{Power required} = \frac{m \times (h_3 - h_2)}{60} \text{ kW per tonne} \quad ...(3.7)$$

(where m = mass of refrigerant in kg/min-tonne)

(*b*) **When compression follows the general law pv^n = constant :**

$$\text{Work of compression} = \frac{n}{n-1}(p_3v_3 - p_2v_2) \text{ Nm/kg}$$

$$\therefore \quad \text{Power required} = \frac{n}{n-1}\frac{(p_3v_3 - p_2v_2)}{60} \times m \text{ kW per tonne} \quad ...(3.8)$$

(*v*) **Heat rejected to compressor cooling water.** If the compressor cylinders are jacketed, an appreciable amount of heat may be rejected to the cooling water during compression. If the suction and discharge compression conditions are known, this heat can be determined as follows :

Heat rejected to compressor cooling water

$$= \left[\frac{n}{n-1}(p_3v_3 - p_2v_2) - (h_3 - h_2)\right] \text{kJ / kg} \quad ...[3.9(a)]$$

$$= m\left[\frac{n}{n-1}(p_3v_3 - p_2v_2) - (h_3 - h_2)\right] \text{kJ / min-tonne} \quad ...[3.9(b)]$$

(*vi*) **Heat removed through condenser.** Heat removed through condenser includes all heat removed through the condenser, either as latent heat, heat of superheat, or heat of liquid. This is equivalent to the heat absorbed in the evaporator plus the work of compression.

$\therefore$ Heat removed through condenser $= m\ (h_3 - h_4)$ kJ/min-tonne ...(3.10)

3.11 METHODS OF IMPROVING SIMPLE SATURATION CYCLE

The simple saturation (vapour compression) cycle can be improved by using the following *methods:*

1. By using a *flash chamber* between the expansion valve and the evaporator.
2. By sub-cooling the liquid refrigerant by using vapour refrigerant.
3. By sub-cooling the liquid refrigerant by external cooling source.
4. By sub-cooling the liquid refrigerant with liquid refrigerant.
5. By using the accumulator or pre-cooler.

3.11.1 By Using a Flash Chamber Between the Expansion Valve and the Evaporator

When the high pressure liquid refrigerant from the condenser passes through the expansion valve, it gets evaporated partially. *This partial evaporation of the liquid refrigerant is known as* **Flash** ("Flash" may be defined as the *mass of vaporised refrigerant per kg leaving the expansion/throtlle valve*). This formed vapour, which cannot produce any refrigerating effect, is bypassed around the evaporator and supplied directly to the compressor through a flash chamber. The flash chamber is located between the evaporator and the expansion valve as shown in Fig. 3.13 (*a*). Fig. 3.13 (*b*) shows the *p-h* diagram. It (Flash chamber) is an insulated container and separates the liquid and vapour due to '*centrifugal effect*':

- This arrangement has *no effect on the thermodynamic cycle.*
- The power (P) and C.O.P. will remain as a simple saturation cycle (theoretically).
- *There is reduction in the mass of refrigerant flowing through the evaporator and hence the reduction in size of evaporator.*

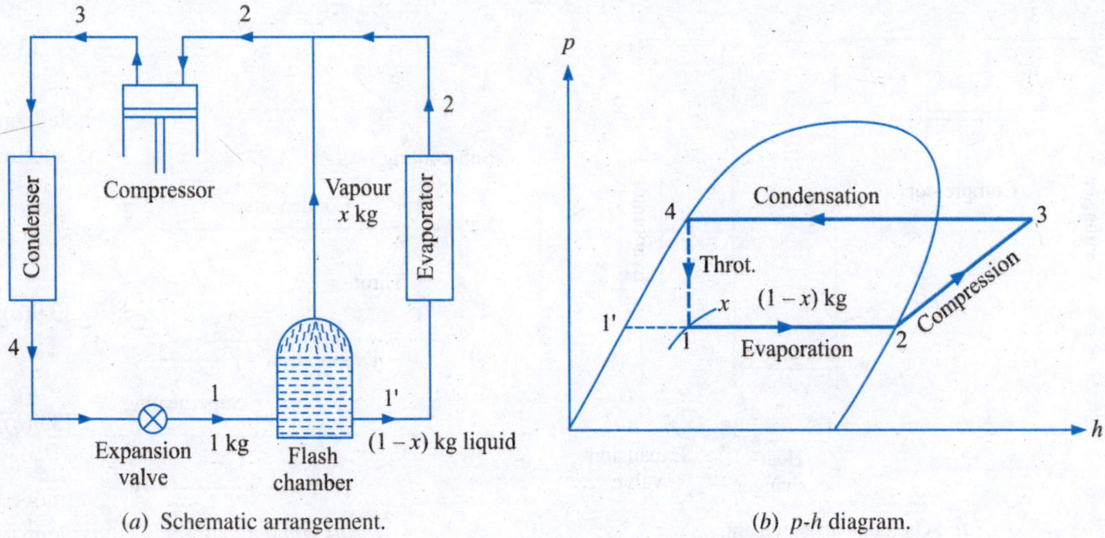

Fig. 3.13. Simple saturation cycle with flash chamber.

3.11.2 By Sub-cooling of Liquid Refrigerant by Using Vapour Refrigerant

The temperature of the liquid refrigerant coming out of the condenser is more than the temperature of the vapour refrigerant leaving the evaporator. The liquid refrigerant can be sub-cooled by passing it through a heat exchanger which is supplied with saturated vapour from the evaporator. The schematic arrangement and *p-h* diagram for this method is shown in Fig. 3.14.

Considering the energy balance in the heat exchanger, we have:

Heat lost by liquid refrigerant = Heat gained by the vapour refrigerant.

$$m_l \times c_{pl} \times (T_{4'} - T_4) = m_v \times c_{pv} \times (T_{2'} - T_2)$$

where, m_l = Mass of the liquid refrigerant,

m_v = Mass of the vapour refrigerant,

T_4 = Temperature of the liquid refrigerant entering the heat exchanger,

$T_{4'}$ = Temperature of the liquid refrigerant leaving the heat exchanger,

T_2 = Temperature of the vapour refrigerant entering the heat exchanger,

$T_{2'}$ = Temperature of vapour refrigerant leaving the heat exchanger,

c_{pl} = Specific heat of the liquid refrigerant, and

c_{pv} = Specific heat of the vapour refrigerant.

As $m_l = m_v$ and $c_{pl} > c_{pv}$

$$\therefore \quad (T_4 - T_{4'}) < (T_{2'} - T_2), \quad \text{or,} \quad (\Delta T)_{liquid} < (\Delta T)_{vapour}$$

By energy balance $h_{f_4} - h_{f_{4'}} = h_{2'} - h_2$ ...(3.11)

Under ideal conditions of heat transfer in heat exchanger

$$T_{4'} = T_{2'} = T_{m\ (say)} \qquad ...(3.12)$$

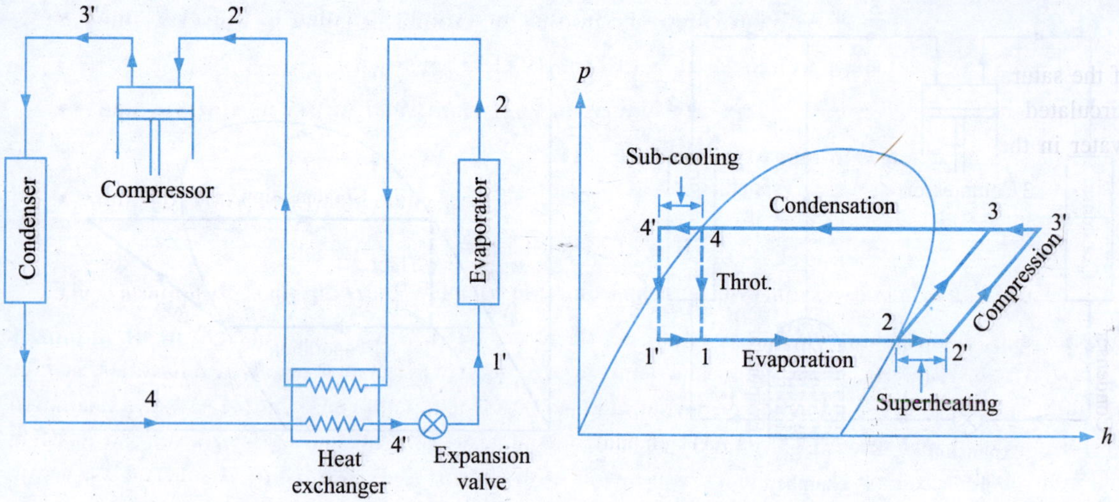

(*a*) Schematic arrangement. (*b*) *p-h* diagram.

Fig. 3.14. Simple saturation cycle with sub-cooling of liquid refrigerant by vapour refrigerant.

Knowing the conditions of points 2 and 4, the conditions of points 2′ and 4′ can be obtained by trial and error method on *p-h* chart.

If Q tonnes of refrigeration is the load on the evaporator, then *mass of the refrigerant required to be circulated through the evaporator for sub-cooled cycle is given by,*

$$m_1 = \frac{14000\,Q}{(h_2 - h_{1'}) \times 60} \text{ kJ/min} \qquad (\text{where } h_{1'} = h_{f_{4'}}) \qquad ...(3.13)$$

Power required to drive the compressor for *sub-cooled cycle,*

$$P_1 = \frac{m_1\,(h_{3'} - h_{2'})}{60} \text{ kW}$$

$$= \frac{14000\,Q}{(h_2 - h_{1'}) \times 60} \times \frac{(h_{3'} - h_{2'})}{60} \quad (\text{Substituting for } m_1 \text{ from eqn. (3.13)})$$

$$= \frac{14000\,Q}{3600}\left[\frac{h_{3'} - h_{2'}}{h_2 - h_{1'}}\right] \qquad ...(3.14)$$

Power required to drive the compressor *without heat exchanger* (i.e., for simple saturation cycle)

$$P_2 = \frac{m_2\,(h_3 - h_2)}{60} \text{ kW} = \frac{14000\,Q}{(h_2 - h_1) \times 60} \times \frac{(h_3 - h_2)}{60}$$

$$= \frac{14000\,Q}{36000}\left[\frac{h_3 - h_2}{h_2 - h_1}\right] \qquad ...(3.15)$$

Extra power required in sub-cooled cycle,

$$P_{excess} = P_1 - P_2 = \frac{14000\,Q}{3600}\left[\frac{h_{3'} - h_{2'}}{h_2 - h_{1'}} - \frac{h_3 - h_2}{h_2 - h_1}\right] \text{ kW} \qquad ...(3.16)$$

3.11.3 By Sub-cooling the Liquid Refrigerant by External Cooling Source

If the saturation temperature of vapour coming out of compressor is *less* than the temperature of water circulated in the condenser then it is possible to *sub-cool* the liquid refrigerant by circulating *more* water in the condenser.

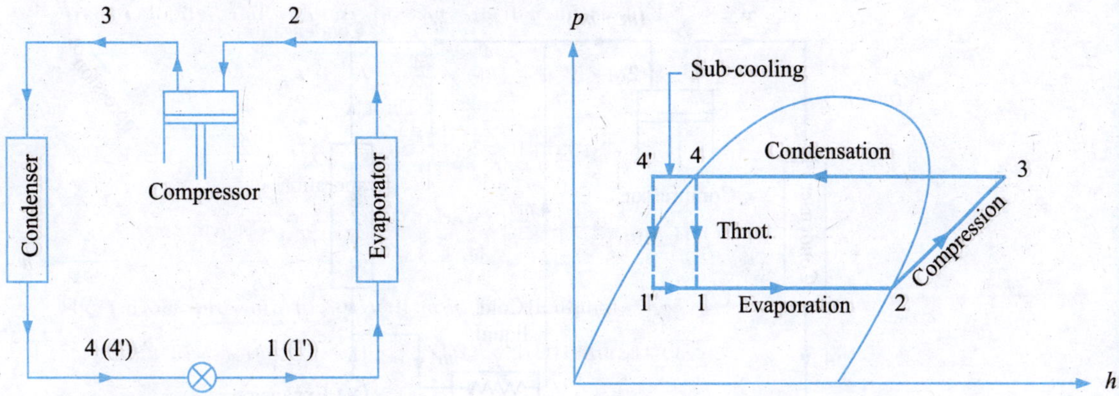

(*a*) Schematic arrangement. (*b*) *p-h* diagram.

Fig. 3.15. Sub-cooling the liquid refrigerant by external cooling source.

Refer to Fig. 3.15. If Q tonnes of refrigeration is the load on the evaporator then mass of refrigerant required to be circulated through the evaporator for *simple cycle* (1-2-3-4) is given by,

$$m_1 = \frac{14000\,\text{Q}}{(h_2 - h_1) \times 60} \text{ kg/min} \qquad ...(3.17)$$

and, power required to drive the compressor,

$$\text{P}_1 = \frac{m_1 (h_3 - h_2)}{60} \text{ kW} = \frac{14000\,\text{Q}}{3600} \left(\frac{h_3 - h_2}{h_2 - h_1} \right) \text{ kW} \qquad ...(3.18)$$

If *sub-cooling* is used, then the mass of refrigerant required is given by,

$$m_2 = \frac{14000\,\text{Q}}{(h_2 - h_{1'}) \times 60} \text{ kg/min.} \qquad ...(3.19)$$

and, Power required to drive the compressor,

$$\text{P}_2 = \frac{m_2 (h_3 - h_2)}{60} \text{ kW} = \frac{14000\,\text{Q}}{3600} \left(\frac{h_3 - h_2}{h_2 - h_{1'}} \right) \text{ kW} \qquad ...(3.20)$$

Net reduction in power

$$= \text{P}_1 - \text{P}_2 = \frac{14000\,\text{Q}}{3600} \left[\frac{h_3 - h_2}{h_2 - h_1} - \frac{h_3 - h_2}{h_2 - h_{1'}} \right] \qquad ...(3.21)$$

$$= \frac{14000\,\text{Q}}{3600} (h_3 - h_2) \left[\frac{1}{h_2 - h_1} - \frac{1}{h_2 - h_{1'}} \right] \qquad ...(3.21)$$

3.11.4 By Sub-cooling the Liquid Refrigerant with Liquid Refrigerant

The liquid refrigerant leaving the expansion value is at a lower temperature than the liquid leaving the condenser; thus, cooling of liquid coming out from condenser is possible by using the liquid coming out of expansion valve. Fig. 3.16 shows the schematic arrangement and the *p-h* diagram.

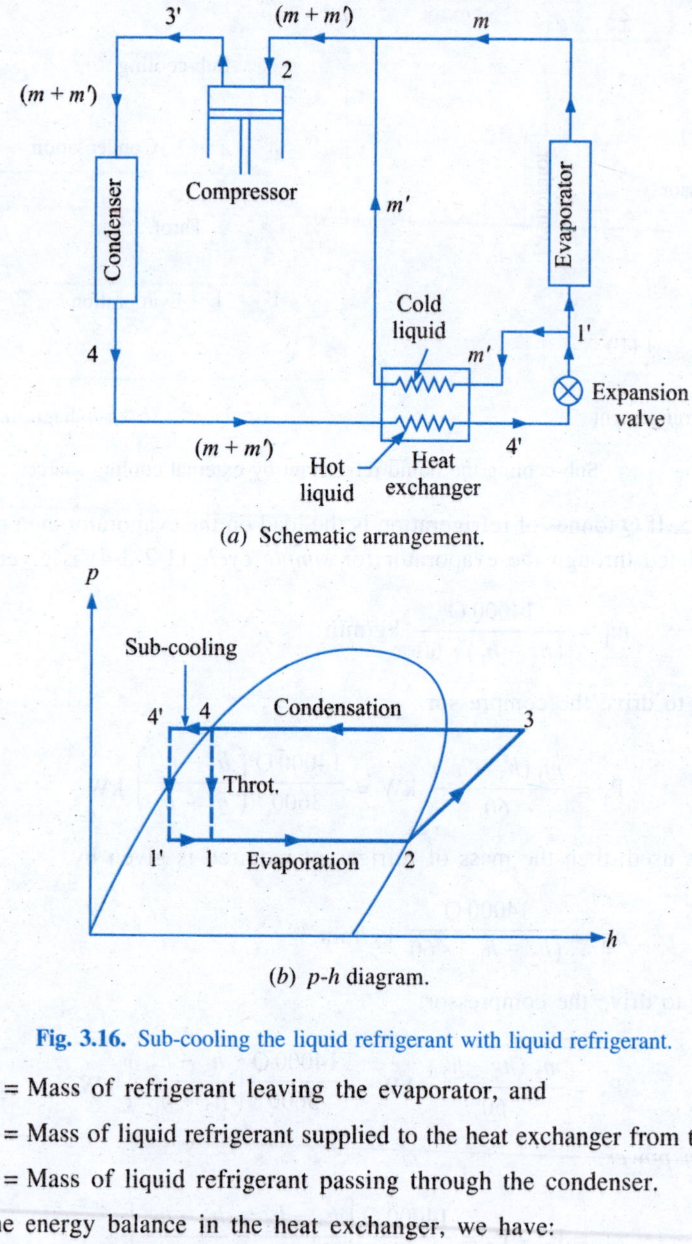

Fig. 3.16. Sub-cooling the liquid refrigerant with liquid refrigerant.

Let, m = Mass of refrigerant leaving the evaporator, and

m' = Mass of liquid refrigerant supplied to the heat exchanger from the expansion valve.

Then, $(m + m')$ = Mass of liquid refrigerant passing through the condenser.

Considering the energy balance in the heat exchanger, we have:

Heat lost by liquid refrigerant from condenser

= Heat gained by liquid refrigerant from expansion valve.

$$(m + m')\,(h_{f_4} - h_{f_{4'}}) = m'\,(h_2 - h_{1'}) \quad \text{...(3.22)}$$

It is assumed that the cold fluid entering into the heat exchanger is saturated liquid and it is coming out as saturated vapour.

From eqn. (3.22) $m' = m\left[\dfrac{h_{f_4} - h_{f_{4'}}}{h_2 - h_{f_4}}\right]$ as $h_{f_{4'}} = h_{1'}$ ...(3.23)

Power required to drive the compressor,

$$P = \frac{(m+m')(h_3 - h_2)}{60} \text{ kW}$$

Substituting the value of m' in the above eqn., we have:

$$P = m\left[1 + \frac{h_{f_4} - h_{f_{4'}}}{h_2 - h_{f_4}}\right]\left[\frac{h_3 - h_2}{60}\right] = m\left[\frac{h_2 - h_{f_{4'}}}{h_2 - h_{f_4}}\right]\left[\frac{h_3 - h_2}{60}\right] \quad ...(3.24)$$

It Q is the load in tonnes of refrigeration on evaporator, then:

$$m = \frac{14000\text{ Q}}{(h_2 - h_{1'}) \times 60} \text{ kg/min}$$

Substituting this in the above equation and using,

$h_{f_{4'}} = h_{1'}$ and $h_{f_4} = h_1$, we get:

$$\mathbf{P} = \frac{14000\text{ Q}}{(h_2 - h_{1'}) \times 60}\left[\frac{h_2 - h_{1'}}{h_2 - h_1}\right]\left[\frac{h_3 - h_2}{60}\right] = \frac{14000\text{ Q}}{3600}\left[\frac{h_3 - h_2}{h_2 - h_1}\right] \quad ...(3.25)$$

Now, $$\textbf{C.O.P.} = \frac{R_n}{W} = \frac{m(h_2 - h_{1'})}{(m+m')(h_3 - h_2)} = \frac{m(h_2 - h_{1'})}{m\left[\dfrac{h_2 - h_{1'}}{h_2 - h_1}\right](h_3 - h_2)}$$

$$= \frac{(h_2 - h_{1'})}{\left[\dfrac{h_2 - h_{1'}}{h_2 - h_1}\right](h_3 - h_2)} = \frac{h_2 - h_1}{h_3 - h_2} \quad ...(3.26) \quad \left[\because h_{f_{4'}} = h_{1'},\ h_{f_4} = h_1\right]$$

- Eqn. 3.23 shows that the mass of refrigerant needed in the heat exchanger in exactly equal to the mass of flash that would occur in a simple saturation cycle operating between the same pressure limits.
- Since the power (P) required to drive the compressor and C.O.P. of the system are the same as that of simple saturation cycle, therefore, this arrangement of sub-cooling the liquid refrigerant is of *no advantage*. This arrangement, in other words, is thermodynamically identical with simple cycle.

3.11.5 By Using the Accumulator or Pre-cooler

It has been found that when an accumulator is used in the system, the refrigerating effect, C.O.P. and power (P) required are same as the simple saturation cycle. *The accumulator is used only to protect the liquid refrigerant to flow into the compressor and thus dry compression is always assured.* When the compressor is supplied with liquid along with vapour refrigerant, then it has to do an additional work of evaporating and raising the temperature of liquid refrigerant, thereby upsetting its normal working of compressing the pure vapour refrigerant.

WORKED EXAMPLES

Example 3.1. *A refrigerating plant works between temperature limits of –5°C and 25°C. The working fluid ammonia has a dryness fraction of 0.62 at entry to compressor. If the machine has a relative efficiency of 55%, calculate the amount of ice formed during a period of 24 hours. The ice is to be formed at 0°C from water at 15°C and 6.4 kg of ammonia is circulated per minute. Specific heat of water is 4.187 kJ/kg and latent heat of ice is 335 kJ/kg.*

Properties of NH_3 (datum –40°C).

Temperature (°C)	*Liquid heat (kJ/kg)*	*Latent heat (kJ/kg)*	*Entropy of liquid (kJ/kg K)*
25	298.9	1167.1	1.124
–5	158.2	1280.8	0.630

Solution: Fig. 3.17 shows the T-*s* diagram of the cycle.

Given: $T_3 = T_4 = 298$ K; $T_1 = T_2 = 268$ K; $(\text{C.O.P.})_{relative} = 0.55$; $m = 6.4$ kg/min; $x_2 = 0.62$;

$c_{pw} = 4.187$ kJ/kg °C; $(h_{fg})_{ice} = 335$ kJ/kg.

Enthalpy at point '2' $\quad h_2 = h_{f2} + x_2\, h_{fg2} = 158.2 + 0.62 \times 1280.8 = 952.3$ kJ/kg

Enthaply at point '1' $\quad h_1 = h_{f_4} = 298.9$ kJ/kg

Also, entropy at point '2' = entropy at point '3'

$$\text{i.e.,}\quad s_2 = s_3$$

$$s_{f2} + x_2\, s_{fg2} = s_{f3} + x_3\, s_{fg3}$$

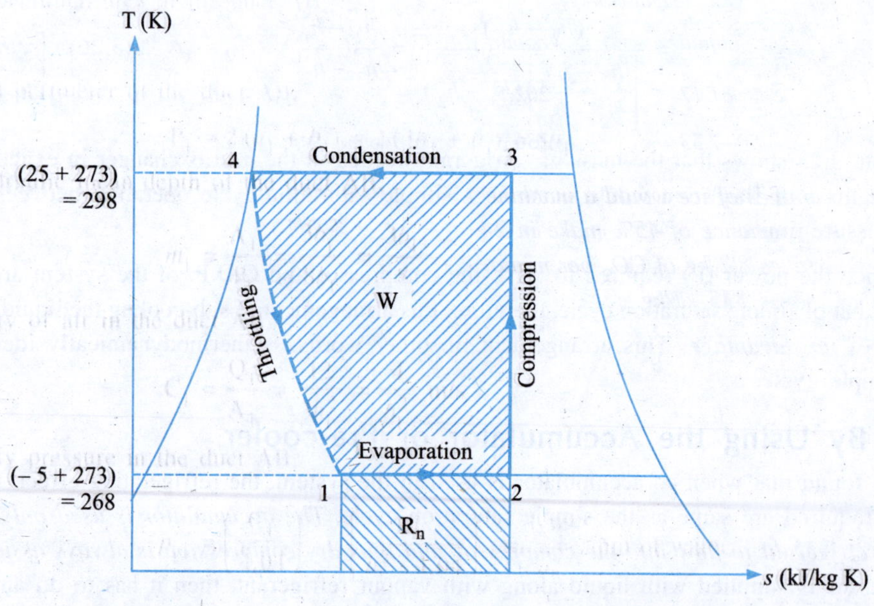

Fig. 3.17

$$0.630 + 0.62 \times \frac{1280.8}{(-5+273)} = 1.124 + x_3 \times \frac{1167.1}{(25+273)}$$

i.e., $x_3 = 0.63$

$\therefore$ Enthalpy at point '3' $h_3 = h_{f3} + x_3\, h_{fg\,3} = 298.9 + 0.63 \times 1167.1 = 1034.17$ kJ/kg

$$\text{C.O.P.}_{\text{(theoretical)}} = \frac{h_2 - h_1}{h_3 - h_2} = \frac{952.3 - 298.9}{1034.17 - 952.3} = \frac{653.4}{81.87} = 7.98$$

$$\text{C.O.P.}_{\text{(actual)}} = (\text{C.O.P.})_{\text{relative}} \times 7.98 = 0.55 \times 7.98 = 4.39$$

Work done per kg of refrigerant $= h_3 - h_2 = 1034.17 - 952.3 = 81.87$ kJ/kg

Refrigerant in circulation, $m = 6.4$ kg/min

$\therefore$ Work done per second $= 81.87 \times \dfrac{6.4}{60} = 8.73$ kJ/s

Heat extracted per second $= (\text{C.O.P.})_{\text{actual}} \times 8.73 = 4.39 \times 8.73 = 38.32$ kJ/s

Heat extracted per kg of ice formed

$$= 1 \times c_{pw}\,(15 - 0) + (h_{fg})_{\text{ice}} = 15 \times 4.187 + 335 = 397.8 \text{ kJ}$$

Amount of ice formed in 24 hours,

$$m_{\text{ice}} = \frac{38.32 \times 3600 \times 24}{397.8} = \mathbf{8322.9\ kg} \quad \textbf{(Ans.)}$$

Example 3.2. *Determine the theoretical C.O.P. for CO_2 machine working between the temperature range of 25°C and –5°C. The dryness fraction of CO_2 gas during the suction stroke is 0.6.*

The following properties are given:

Temperature (°C)	*Heat (kJ/kg)*		*Latent heat (kJ/kg)*	*Entropy (kJ/kg K)*	
	Liquid	*Vapour*		*Liquid*	*Vapour*
25	*81.17*	*202.5*	*121.34*	*0.251*	*0.644*
–5	*–7.53*	*236.8*	*245.2*	*– 0.042*	*0.841*

How many tonnes of ice would a machine working between the same limits and having a relative coefficient of performance of 45% make in 24 hours? The water for the ice is supplied at 15°C and the compressor takes 8.2 kg of CO_2 per minute. Specific heat of water may taken as 4.18 kJ/kg °C and latent heat of ice as 335 kJ/kg.

Solution: Refer to Fig. 3.18.

Given: $T_3 = T_4 = 25 + 273 + 298$ K; $T_1 = T_2 = -5 + 223 = 218$ K; $x_2 = 0.6$;

$(\text{C.O.P.})_{\text{relative}} = 0.45$; $m = 8.2$ kg/min; $c_{pw} = 4.18$ kJ/kg °C; $h_{fg\,(\text{ice})} = 335$ kJ/kg

Let us first find dryness at point '3' (x_3)

Entropy at '2' = Entropy at '3'

$$s_{f2} + \frac{x_2\, h_{fg_2}}{T_2} = s_{f3} + \frac{x_3\, h_{fg_3}}{T_3}$$

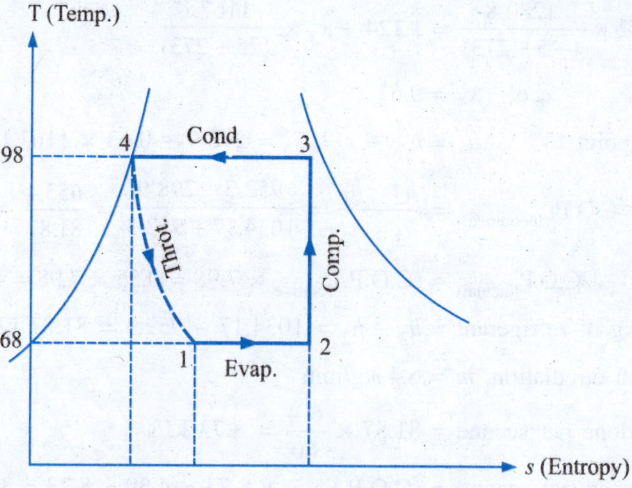

Fig. 3.18

or, $$-0.042+\frac{0.6\times 245.2}{(-5+273)}=0.251+\frac{x_3\times 121.34}{(25+273)}$$

or, $$0.042+0.549=0.251+0.407\,x_3\,,\quad \text{or,}\quad 0.507=0.251+0.407\,x_3$$

or, $$\text{dryness fraction, } x_3=\frac{(0.507-0.251)}{0.407}=0.63$$

$$\text{Work done, W} = h_3 - h_2$$

$$h_3 = h_{f3} + x_3\,h_{fg3} = 81.17 + 0.63 \times 121.34 = 157.6 \text{ kJ/kg}$$

$$h_2 = h_{f2} + x_2\,h_{fg2} = -7.53 + 0.6 \times 245.2 = 139.6 \text{ kJ/kg}$$

$$\therefore \quad \text{Work done, W} = 157.6 - 139.6 = 81 \text{ kJ/kg}$$

$$\text{Refrigerating effect (net), R}_n = h_2 - h_1$$

$$= 139.6 - h_{f4} = 139.6 - 81.17 = 58.43 \text{ kJ/kg} \qquad (\because h_1 = h_{f4})$$

$$\therefore \quad \textbf{(C.O.P.)}_{\text{theoretical}} = \frac{\text{R}_n}{\text{W}} = \frac{58.43}{18} = \mathbf{3.24} \quad \textbf{(Ans.)}$$

$$\text{Relative co-efficient of performance} = \frac{(\text{C.O.P.})_{\text{actual}}}{(\text{C.O.P.})_{\text{theoretical}}}, \quad \text{or,} \quad 0.45 = \frac{(\text{C.O.P.})_{\text{actual}}}{3.24}$$

$$\therefore (\text{C.O.P.})_{\text{actual}} = 0.45 \times 3.24 = 1.458$$

$$\text{Now heat extracted per minute} = (\text{C.O.P.})_{\text{actual}} \times [m \times (h_3 - h_2)]$$

$$= 1.458 \times [8.2\,(157.6 - 139.6)]$$

$$= 215.2 \text{ kJ/min}$$

$$\text{Heat extracted per day} = 215.2 \times 60 \times 24 = 309888 \text{ kJ/day}$$

Heat extracted per kg of water (to produce ice at 0ºC)

$$= 1 \times c_{pw} \times (15 - 0) + (h_{fg})_{\text{ice}}$$

$$= 1 \times 4.18 \times (15 - 0) + 335 = 397.7 \text{ kJ/kg}$$

$\therefore$ **Mass of ice made per day**

$$= \frac{309888}{397.7\times 1000} \text{ tonne} = \mathbf{0.779 \text{ tonne}} \quad \textbf{(Ans.)}$$

Example 3.3. *A refrigeration machine is required to produce ice at 0°C from water at 20°C. The machine has a condenser temperature of 298 K while the evaporator temperature is 268 K. The relative efficiency of the machine is 50% and 6 kg of Freon-12 refrigerant is circulated through the system per minute. The refrigerant enters the compressor with a dryness fraction of 0.6. Specific heat of water is 4.187 kJ/kg K and the latent heat of ice is 335 kJ/kg. Calculate the amount of ice produced in 24 hours. The table of properties of Freon-12 is given below:*

Temperature (K)	*Liquid heat (kJ/kg)*	*Latent heat (kJ/kg)*	*Entropy of liquid (kJ/kgK)*
298	59.7	138.0	0.2232
268	31.4	154.0	0.1251

(U.P.S.C.)

Solution: *Given:* m = 6 kg/min; $\eta_{relative}$ = 50%; x_2 = 0.6; c_{pw} = 4.187 kJ/kg K;

Latent heat of ice = 335 kJ/kg

$$h_{f2} = 31.4 \text{ kJ/kg}; \; h_{fg2} = 154.0 \text{ kJ/kg}; \; h_{f3} = 59.7 \text{ kJ/kg};$$

$$h_{fg3} = 138 \text{ kJ/kg}, \; h_{f4} = 59.7 \text{ kJ/kg}$$

$$h_2 = h_{f2} + x_2 \, h_{fg2} = 31.4 + 0.6 \times 154 = 123.8 \text{ kJ/kg}$$

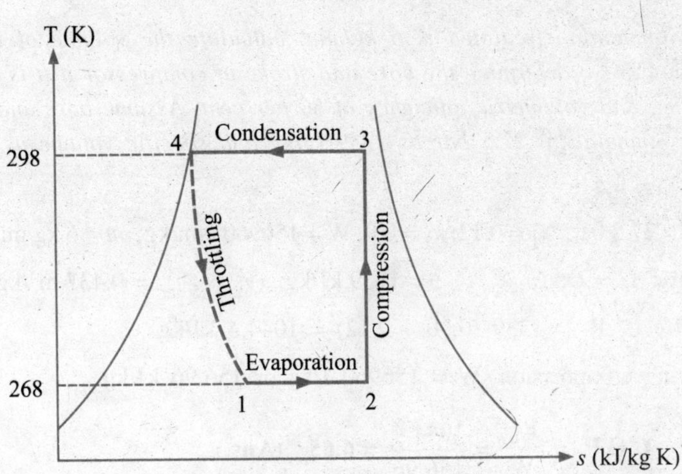

Fig. 3.19

For isentropic compression 2-3,

$$s_3 = s_2$$

$$s_{f3} + x_3 \frac{h_{fg3}}{T_3} = s_{f2} + x_2 \frac{h_{fg2}}{T_2}$$

$$0.2232 + x_3 \times \frac{138}{298} = 0.1251 + 0.6 \times \frac{154}{268} = 0.4698$$

$$\therefore \quad x_3 = (0.4698 - 0.2232) \times \frac{298}{138} = 0.5325$$

Now, $h_3 = h_{f3} + x_3\, h_{fg3} = 59.7 + 0.5325 \times 138 = 133.2$ kJ/kg

Also $h_1 = h_{f4} = 59.7$ kJ/kg

$$\text{Theoretical C.O.P} = \frac{R_n}{W} = \frac{h_2 - h_1}{h_3 - h_2} = \frac{123.8 - 59.7}{133.2 - 123.8} = 6.82$$

$$\text{Actual C.O.P.} = \eta_{relative} \times (\text{C.O.P})_{theoretical} = 0.5 \times 6.82 = 3.41$$

Heat extracted from 1 kg of water at 20°C for the formation of 1 kg of ice at 0°C

$$= 1 \times 4.187 \times (20 - 0) + 335 = 418.74 \text{ kJ/kg.}$$

Let m_{ice} = Mass of ice formed in kg/min.

$$(\text{C.O.P.})_{actual} = 3.41 = \frac{R_{n\,(actual)}}{W} = \frac{m_{ice} \times 418.74}{m(h_3 - h_2)} = \frac{m_{ice} \times 418.74\,(\text{kJ/min})}{6(133.2 - 123.8)\,(\text{kJ/min})}$$

$$\therefore \quad m_{ice} = \frac{6(133.2 - 123.8) \times 3.41}{418.74} = 0.459 \text{ kg/min.}$$

$$= \frac{0.459 \times 60 \times 24}{1000} \text{ tonne (in 24 hours)} = \mathbf{0.661\ tonne} \quad \textbf{(Ans.)}$$

Example 3.4. *In the evaporator of an ammonia refrigerator, the pressure is 27.5 bar and dryness fraction of ammonia at entry and exit are 0.12 and 0.9 respectively. If the work done during compression is 156960 Nm/kg, find the co-efficient of performance.*

If the mass of ammonia circulated is 6 kg/min, calculate the volume of vapour entering the compressor per minute. Also determine the bore and stroke of compressor if it is single-acting having a speed of 100 r.p.m., and volumetric efficiency of 80 per cent. Assume bore and stroke to be equal. Take latent heat of ammonia at 27.5 bar as 1339 kJ/kg and specific volume as 0.437 m³/kg.

Solution: Refer to Fig. 3.4

Given: $p_1 = p_2 = 27.5$ bar; $x_1 = 0.12$; $x_2 = 0.9$; W = 156960 Nm/kg; $m = 6$ kg/min; N = 100 r.p.m.;

$\eta_{vol} = 80\%$; L = D; $(h_{fg})_{27.5\text{ bar}} = 1339$ kJ/kg; $(v_g)_{27.5\text{ bar}} = 0.437$ m³/kg.

Refrigerating effect, $R_n = 1339\ (0.90 - 0.12) = 1044.4$ kJ/kg

Work done during compression, W = 156960 J/kg or 156.96 kJ/kg

$$\textbf{C.O.P.} = \frac{R_n}{W} = \frac{1044.4}{156.96} = \mathbf{6.65} \quad \textbf{(Ans.)}$$

Volume of ammonia at entrance of compressor,

$$v_2 = x_2\,(v_g)_{27.5\text{ bar}} = 0.9 \times 0.437 = 0.3933 \text{ m}^3\text{/kg}$$

Total volume of ammonia vapour entering the compressor per minute

$$= 0.3933 \times 6 = \mathbf{2.36\ m^3/min} \quad \textbf{(Ans.)}$$

$$\text{Volume of compressor} = \frac{\text{Volume of ammonia vapour dealt with}}{\text{Volumetric efficiency}} = \frac{2.36}{0.8} = 2.95 \text{ m}^3\text{/min}$$

Stroke volume of the compressor (single-acting)

$$= \frac{2.95}{100} = 0.0295 \text{ m}^3$$

If D is the diameter of the compressor cylinder in meters, then,

$$\text{Stroke volume} = \frac{\pi}{4} \times D^2 \times L = 0.0295 \qquad (\therefore \text{ Stroke length, } L = D)$$

$$\text{or,} \quad \frac{\pi}{4} \times D^2 \times D = 0.0295, \quad \text{or,} \quad D^3 = \frac{0.0295 \times 4}{\pi} = 0.03756$$

or, **D = 0.335 m or 335 mm (Ans.)**

and, **L = D = 335 mm (Ans.)**

Example 3.5. *A refrigerator works between –7°C and 27°C. The vapour is dry at the end of adiabatic compression. There is no under-cooling and evaporation is by throttle valve. Determine:*

(i) The co-efficient of performance;

(ii) Power of the compressor to remove 180 kJ/min.

The properties of the refrigerant are as under:

Temp. (°C)	*Enthalpy (kJ/kg)*		*Entropy (kJ/kg K)*	
	Liquid (h_f)	*Latent (h_{fg})*	*Liquid (s_f)*	*Vapour (s_g)*
– 7	*– 30*	*1298*	*– 0.108*	*4.75*
27	*115*	*1173*	*427*	*4.33*

Solution: Refer to Fig. 3.20

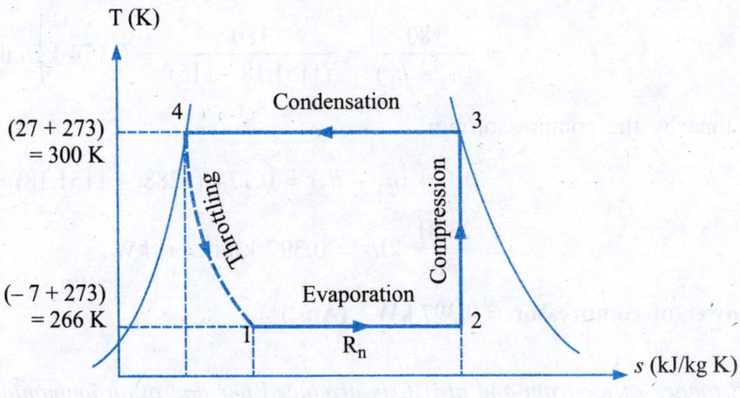

Fig. 3.20

Given: $T_3 = T_4 = 300$ K; $T_1 = T_2 = 266$ K; $x_3 = 1$;

Heat to be removed = 180 kg/min

(i) Co-efficient of performance, C.O.P.:

$$\text{C.O.P.} = \frac{R_n}{W} = \frac{h_2 - h_1}{h_3 - h_2} \qquad ...(1)$$

$$\text{Here,} \quad h_3 = h_{f3} + h_{fg3} = 115 + 1173 = 1288 \text{ kJ/kg};$$

$$h_1 = h_{f4} = 115 \text{ kJ/kg}$$

To find h_2, let us first find dryness fraction at point 2.

Entropy at '2' = Entropy at '3' (process 2-3 being isentropic),

$$\text{i.e.,}\quad s_2 = s_3$$

$$s_{f2} + x_2\, s_{fg2} = s_{g3}$$

$$-0.108 + x_2 \times \frac{h_{fg2}}{T_2} = 4.33$$

$$-0.108 + x_2 \times \frac{1298}{266} = 4.33$$

$$\therefore \quad x_2 = (4.33 + 0.108) \times \frac{266}{1298} = 0.91$$

$$\therefore \quad h_2 = h_{f2} + x_2\, h_{fg2} = -30 + 0.91 \times 1298 = 1151.18 \text{ kJ/kg}$$

Substituting the values in (*i*), we get:

$$\text{C.O.P.} = \frac{1151.18 - 115}{1288 - 1151.18} = \mathbf{7.57} \quad \textbf{(Ans.)}$$

(*ii*) **Power of the compressor:**

Heat required to be removed = 180 kJ/min

Mass of refrigerant circulated/min

$$= \frac{180}{(h_2 - h_1)} = \frac{180}{(1151.18 - 115)} = 0.174 \text{ kg/min}$$

Total work done by the compressor/min

$$= 0.174\,(h_3 - h_2) = 0.174\,(1288 - 1151.18) = 23.81 \text{ kJ/min}$$

$$= \frac{23.81}{60} \text{ kJ/s} = 0.397 \text{ kJ/s , or, kW}$$

$\therefore$ **Power of compressor = 0.397 kW (Ans.)**

Example 3.6. *28 tonnes of ice from and at 0°C is produced per day in an ammonia refrigerator. The temperature range in the compressor is from 25°C to – 15°C. The vapour is dry and saturated at the end of compression and an expansion valve is used. Assuming a co-efficient of performance of 62% of the theoretical, calculate the power required to drive the compressor.*

The following properties are given:

Temperature (°C)	*Enthaply (kJ/kg)*		*Entropy of liquid (kJ/kg K)*	*Entropy of vapour (kJ/kg K)*
	Liquid	*Vapour*		
25	*100.04*	*1319.22*	*0.3473*	*4.4852*
– 15	*– 54.56*	*1304.99*	*– 2.1338*	*5.0585*

Take latent heat of ice = 335 kJ/kg.

Solution: Refer to Fig. 3.21.

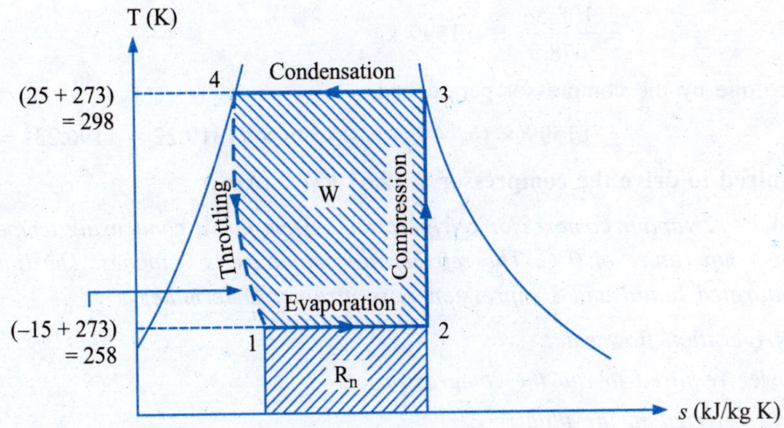

Fig. 3.21

Given: $T_3 = T_4 = 298$ K; $T_1 = T_2 = 258$ K; $(\text{C.O.P.})_{\text{actual}} = 0.62 \times (\text{C.O.P.})_{\text{theoretical}}$; $(h_{fg})_{\text{ice}} = 335$ kJ/kg.

$$\text{Theoretical C.O.P.} = \frac{h_2 - h_1}{h_3 - h_2}$$

Here, $h_3 = 1319.22$ kJ/kg

$h_1 = h_4 = 100.04$ kJ/kg

To find h_2, let us first find dryness at point 2.

Entropy at '2' = Entropy at '3' (Process 2-3 being isentropic)

$$s_{f2} + x_2\, s_{fg2} = s_{g3}$$

$$-\,2.1338 + x_2 \times [5.0585 - (-\,2.1338)] = 4.4852$$

$$\therefore \quad x_2 = \frac{4.4852 + 2.1338}{5.0585 + 2.1338} = 0.92$$

$$\therefore \quad h_2 = h_{f2} + x_2\, h_{fg2} = -\,54.56 + 0.92 \times [1304.99 - (-\,54.56)] = 1196.23 \text{ kJ/kg}$$

$$\therefore \quad \text{C.O.P.}_{\text{(theoretical)}} = \frac{1196.23 - 100.04}{1319.22 - 1196.23} = 8.91$$

$\therefore \quad \text{C.O.P.}_{\text{(actual)}} = 0.62 \times \text{C.O.P.}_{\text{(theoretical)}}$ ...(Given)

i.e. $\quad \text{C.O.P.}_{\text{(actual)}} = 0.62 \times 8.91 = 5.52$

Actual refrigerating effect per kg

$= \text{C.O.P.}_{\text{(actual)}} \times \text{work done} = 5.52 \times (h_3 - h_2)$

$= 5.52 \times (1319.22 - 1196.23) = 678.9$ kJ/kg

Heat to be extracted per hour

$$= \frac{28 \times 1000 \times 335}{24} = 390833.33 \text{ kJ}$$

Heat to be extracted per second

$$= \frac{390833.33}{3600} = 108.56 \text{ kJ/s}$$

∴ Mass of refrigerant circulated per second

$$= \frac{108.56}{678.9} = 0.1599 \text{ kg}$$

Total work done by the compressor per second

$$= 0.1599 \times (h_3 - h_2) = 0.1599 \times (1319.22 - 1196.23) = 19.67 \text{ kJ/s}$$

i.e., Power required to drive the compressor = 19.67 kW (Ans.)

Example 3.7. *A F–12 vapour compressor refrigeration system has a condensing temperature of 50°C and evaporating temperature of 0°C. The refrigeration capacity is 7 tonnes. The liquid leaving the condenser is saturated liquid and compression is isentropic. Determine:*

(*i*) *The refrigeration flow rate;*

(*ii*) *The power required to run the compressor;*

(*ii*) *The heat rejected in the plant;*

(*iv*) *C.O.P. of the system.*

The properties of F–12 are:

Temperature (°C)	*Pressure* (*bar*)	h_f (*kJ/kg*)	h_g (*kJ/kg*)	s_f (*kJ/kg K*)	s_g (*kJ/kg K*)
50	*12.199*	*84.868*	*206.298*	*0.3034*	*0.6792*
0	*3.086*	*36.022*	*187.397*	*0.1418*	*0.6960*

Take enthalpy at the end of isentropic compression = 210 kJ/kg.

Solution: The cycle is shown on T-*s* diagram in Fig. 3.22.

Given: $h_{f4} = h_1 = 84.868$ kJ/kg;

$h_2 = 187.397$ kJ/kg;

$h_3 = 210$ kJ/kg.

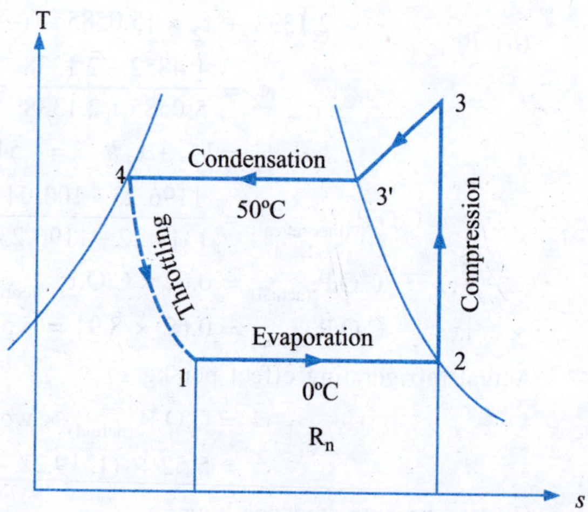

Fig. 3.22

(*i*) **The refrigerant flow rate,**

$$\dot{m} = \frac{7 \times 14000}{(h_2 - h_1) \times 3600} \text{ kg/s}$$

$$= \frac{7 \times 14000}{(187.397 - 84.868) \times 3600}$$

$$= \mathbf{0.266 \text{ kg/s}} \quad \textbf{(Ans.)}$$

(*ii*) **The power required to run the compressor**

$$= \dot{m}\,(h_3 - h_4)$$

$$= 0.266\,(210 - 187.397)$$

$$= \mathbf{6.01 \text{ kW}} \quad \textbf{(Ans.)}$$

(*iii*) **The heat rejected in plant**

$$= \dot{m}\,(h_3 - h_4) = 0.266\,(210 - 84.868) = \mathbf{53.28 \text{ kW}} \quad \textbf{(Ans.)}$$

(*iv*) $$\text{C.O.P.} = \frac{R_n}{W} = \frac{h_2 - h_1}{h_3 - h_2} = \frac{187.397 - 84.868}{210 - 187.397} = \mathbf{4.54} \quad \textbf{(Ans.)}$$

Example 3.8. *In a standard vapour compression refrigeration cycle, operating between an evaporator temperature of –10°C and a condenser temperature of 40°C, the enthalpy of the refrigerant, Feron-12, at the end of compression is 220 kJ/kg. Show the cycle diagram on T-s plane.*

Calculate:

(*i*) *The C.O.P. of the cycle.*

(*ii*) *The refrigerating capacity and the compressor power assuming a refrigerant flow rate of 1 kg/min. You may use the extract of Freon-12 property table given below:*

t (°C)	p (MPa)	h_f (kJ/kg)	h_g (kJ/kg)
– 10	0.2191	26.85	183.1
40	0.9607	74.53	203.1

(GATE)

Solution: The cycle is shown on T-*s* diagram in Fig. 3.23

Given: Evaporator temperature = – 10°C

Condenser temperature = 40°C

Enthalpy at the end of compressor,

$$h_3 = 220 \text{ kJ/kg}$$

From the table (given), we have

$$h_2 = 183.1 \text{ kJ/kg};\ h_1 = h_{f4} = 26.85 \text{ kJ/kg}$$

(*i*) **The C.O.P. of the Cycle:**

$$\text{C.O.P.} = \frac{R_n}{W} = \frac{h_2 - h_1}{h_3 - h_2} = \frac{183.1 - 26.85}{220 - 183.1} = \mathbf{4.23} \quad \textbf{(Ans.)}$$

(*ii*) **Refrigerating capacity:**

Refrigerating capacity = $m\,(h_2 - h_1)$

[where m = mass flow rate of refrigerant = 1 kg/min ... (Given)]

$= 1 \times (183.1 - 26.85) = \mathbf{156.25 \text{ kJ/min}}$ **(Ans.)**

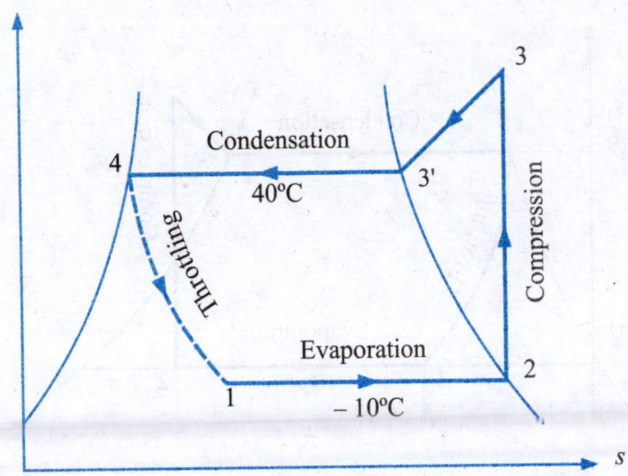

Fig. 3.23

Compressor power:

$$\text{Compressor power} = m\,(h_3 - h_2) = 1 \times (220 - 183.1) = 36.9 \text{ kJ/min. or } 0.615 \text{ kJ/s}$$
$$= \mathbf{0.615\ kW} \quad \textbf{(Ans.)}$$

Example 3.9. *An ammonia vapour compression refrigerator has an effective swept volume of 0.298 m^3 per minute. Condensation and evaporation take place at 28.9°C and –12.2°C respectively. There is no under-cooling and the gas temperature after compression is 51.2°C. Taking c_p for the superheated vapour as 2.89 determine:*

(i) The dryness fraction of the vapour as it enters the compressor;

(ii) The rate of circulation of ammonia in kg/min;

(iii) The rate of extraction of heat in kJ/min;

(iv) The heat rejected in the condenser per minute.

The relevant properties of ammonia are:

Temperature (°C)	*Enthalpy (kJ/kg)*		*Entropy (kJ/kg-K)*		*Specific volume*
	Liquid (h_f)	*Latent (h_{fg})*	*Liquid (s_f)*	*Evap. (s_{fg})*	*$m^3/kg(v_g)$*
28.9	*320.53*	*1148.06*	*1.082*	*3.798*	*0.106*
– 12.2	*125.7*	*1307.28*	*0.506*	*5.00*	*0.406*

Solution: Refer to Fig. 3.24.

(*i*) **The dryness fraction of vapour at entry to compressor, x_2:**

Since 2-3 is an isentropic compression, therefore,

$$\text{Entropy at '2'} = \text{Entropy at '3'}$$

$$\text{i.e.,}\quad s_2 = s_3$$

$$s_{f2} + x_2 \times s_{fg2} = s_{3'} + c_{ps} \ln\left(\frac{51.2 + 273}{28.9 + 273}\right)$$

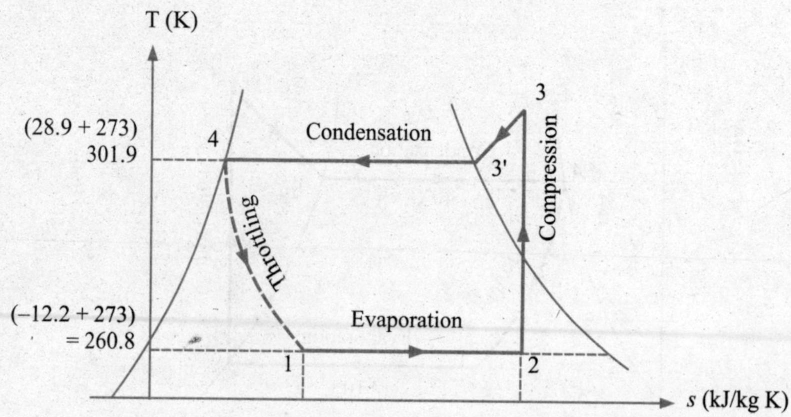

Fig. 3.24

or, $\quad 0.506 + x_2 \times 5 = (1.082 + 3.798) + 2.89 \ln \left(\frac{324.2}{301.9}\right)$

or, $\quad 506 + 5x_2 = 4.88 + 0.206$

$$x_2 = \mathbf{0.916} \quad \textbf{(Ans.)}$$

(*ii*) **The rate of circulation of ammonia in kg/min:**

Total heat at '2', $h_2 = h_{f2} + x_2\, h_{fg2} = 125.7 + 0.916 \times 1307.28 = 1323.17$ kJ/kg

Neglecting the volume of liquid the volume per kg at '2'

$$= x_2 \times v_{g_2} = 0.916 \times 0.406 = 0.372 \text{ m}^3/\text{kg}$$

∴ Mass of refrigerant flowing per minute

$$= \frac{0.298}{0.372} = \mathbf{0.8\ kJ/\ min} \quad \textbf{(Ans.)}$$

(*iii*) **The rate of extraction of heat in kg/min:**

Refrigerating effect per kg $= h_2 - h_1 = 1323.17 - 320.53 = 1002.64$ kJ/kg $(\because h_1 = h_4 = 320.53$ kJ/kg$)$

∴ Refrigerating effect = 1002.64 × 0.8 = 802.1 kJ/min

Hence rate of extraction of heat = **802.1 kJ/min (Ans.)**

(*iv*) **The heat rejected to the condenser per minute:**

$$h_3 = h'_3 + c_{ps}\,(T_3 - T'_3) = (h_4 + 1148.06) + 2.89\,(51.2 - 28.9)$$

$$= h_4 + 1212.5 \quad \text{or} \quad h_3 - h_4 = 1212.5 \text{ kJ/kg}$$

∴ Heat rejected to the condenser per minute = 1212.5 × 0.8 = **970 kJ/min (Ans.)**

Example 3.10. *An ammonia refrigerator produces 20 tonnes of ice per day from and at 0°C. The condensation and evaporation take place at 20°C and –20°C respectively. The temperature of vapour at the end of isentropic compression is 50°C and there is no under-cooling of the liquid. The actual C.O.P. is 70% of the theoretical C.O.P. Determine:*

(*i*) *The rate of NH_3 circulation;*

(*ii*) *The size of single acting-compressor when running at 240 r.p.m., assuming L = D and volumetric efficiency of 80%.*

Take h_{fg} (fusion of ice) = 335 kJ/kg.

Use the properties of NH_3 as listed below:

Temperature (°C)	*Enthalpy (kJ/kg)*		*Entropy (kJ/kg K)*	
	h_f	h_g	s_f	s_g
20	274.98	1461.58	1.04341	5.0919
– 20	89.72	1419.05	0.3682	5.6204

Take (v_{sup}) specific volume of dry vapour at –20°C = 0.624 m³/kg.

Specific heat of superheated vapour = 2.8 kJ/kg °C.

Solution: The cycle is represented on T-*s* diagram as shown in Fig. 3.25.

The refrigerating capacity of the plant

$$= \frac{20 \times 1000 \times 335}{24 \times 3600} = 77.55 \text{ kJ/s}$$

From given table, $h_2 = 1419.05$ kJ/kg,

$$h_{f4} = 274.98 \text{ kJ/kg} = h_1$$

$$h_3 = 1461.58 + 2.8\ (50 - 20)$$

$$= 1545.58 \text{ kJ/kg}$$

$$\text{Actual C.O.P.} = \frac{h_2 - h_1}{h_3 - h_2} \times 0.7$$

$$= \frac{1419.05 - 274.98}{1545.58 - 1419.05} \times 0.7$$

$$= 6.33$$

$$\text{Compressor power} = \frac{\text{Plant capacity}}{\text{Actual C.O.P.}}$$

$$= \frac{77.55}{6.33} = 12.25 \text{ kW}$$

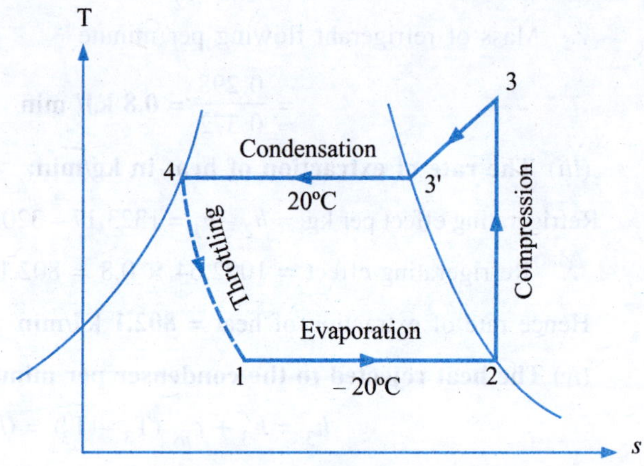

Fig. 3.25

(i) The rate of NH_3 circulation, $\dot{m}$:

$$\text{Mass flow rate, } \dot{m} = \frac{\text{Actual compressor power}}{h_3 - h_2} = \frac{12.25}{1545.58 - 1419.05} = \mathbf{0.097 \text{ kg/s}} \quad \textbf{(Ans.)}$$

(ii) L.D. of compressor:

$$\text{Volume flow rate per second} = \left(\frac{\pi}{4} D^2 L \eta_{vol.}\right) \times \frac{N}{60} = 0.097 \times v_{sup}$$

$$\frac{\pi}{4} D^3 \times 0.8 \times \frac{240}{60} = 0.097 \times 0.624 \qquad (\because L = D)$$

$$\therefore \quad D^3 = \frac{0.097 \times 0.624}{\pi \times 0.8} = 0.0241$$

$$\therefore \quad D = 0.289 \text{ m} = \mathbf{28.9 \text{ cm}} \text{ and } \mathbf{L = D = 28.9 \text{ cm}} \quad \textbf{(Ans.)}$$

Example 3.11. *A Freon-12 refrigerator producing a cooling effect of 20 kJ/s operates on a simple cycle with pressure limits of 1.509 bar and 9.607 bar. The vapour leaves the evaporator dry saturated and there is no undercooling. Determine the power required by the machine.*

If the compressor operates at 300 r.p.m. and has a clearance volume of 3% of stroke volume, determine the piston displacement of the compressor. For compressor assume that the expansion following the law $pv^{1.13}$ = *constant.*

Given:

Temperature (°C)	p_s (*bar*)	v_g (m^3/kg)	*Enthalpy (kJ/kg)* h_f	h_g	*Entropy (kJ/kg K)* s_f	s_g	*Specific heat* (*kJ/kg K*)
– 20	1.509	0.1088	17.8	178.61	0.073	0.7082	—
40	9.607	—	74.53	203.05	0.2716	0.682	0.747

Solution: *Given:* (From the table above):

$$h_2 = 178.61 \text{ kJ/kg};\ h_{3'} = 203.05 \text{ kJ/kg};\ h_{f4} = 74.53 \text{ kJ/kg} = h_1$$

Now, cooling effect $= \dot{m}\ (h_2 - h_1)$

$$20 = \dot{m}\ (178.61 - 74.53)$$

$$\therefore \quad \dot{m} = \frac{20}{(178.61 - 74.53)}$$

$$= 0.192 \text{ kg/s}$$

Also, $s_3 = s_2$

$$s_{3'} + c_{ps} \ln\left(\frac{T_3}{T_{3'}}\right) = 0.7082$$

$$0.682 + 0.747 \ln\left(\frac{T_3}{313}\right) = 0.7082$$

or, $$\ln\left(\frac{T_3}{313}\right) = \frac{0.7082 - 0.682}{0.747}$$

$$= 0.03507$$

or, $$\left(\frac{T_3}{313}\right) = e^{0.03507} = 1.0357$$

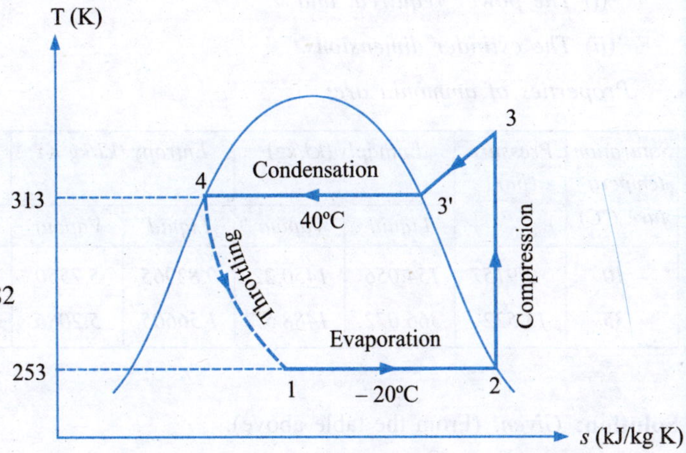

Fig. 3.26

$$\therefore \quad T_3 = 313 \times 1.0357 = 324.2 \text{ K}$$

Now, $$h_3 = h_{3'} + c_{ps}\ (324.2 - 313)$$

$$= 203.05 + 0.747\ (324.2 - 313) = 211.4 \text{ kJ/kg}$$

Power required:

Power required by the machine $= \dot{m}\ (h_3 - h_2)$

$$= 0.192\ (211.4 - 178.61) = \mathbf{6.29\ kW} \quad \textbf{(Ans.)}$$

Piston displacement, V:

Volumetric efficiency, $$\eta_{vol} = 1 + C - C\left(\frac{p_d}{p_s}\right)^{1/n}$$

$$= 1 + 0.03 - 0.03\left(\frac{9.607}{1.509}\right)^{1/1.13} = 0.876 \text{ or } 87.6\%$$

The volume of refrigerant at the intake conditions is

$$\dot{m} \times v_g = 0.192 \times 0.1088 = 0.02089 \text{ m}^3/\text{s}$$

$$\text{Hence the swept volume} = \frac{0.02089}{\eta_{vol.}} = \frac{0.02089}{0.876} = 0.02385 \text{ m}^3/\text{s}$$

$$\therefore \quad V = \frac{0.02385 \times 60}{300} = \mathbf{0.00477 \text{ m}^3} \quad \textbf{(Ans.)}$$

Example 3.12. *A food storage locker requires a refrigeration capacity of 50 kW. It works between a condenser temperature of 35°C and an evaporator temperature of –10°C. The refrigerant is ammonia. It is subcooled by 5°C before entering the expansion valve by the dry saturated vapour leaving the evaporator. Assuming a single-cylinder single-acting compressor operating at 1000 r.p.m. with stroke equal to 1.2 times the bore, determine:*

(i) *The power required, and*

(ii) *The cylinder dimensions.*

Properties of ammonia are:

Saturation temperature (°C)	*Pressure (bar)*	*Enthaply (kJ/kg)*		*Entropy (kJ/kg K)*		*Specific volume, (m^3/kg)*		*Specific heat (kJ/kg K)*	
		Liquid	*Vapour*	*Liquid*	*Vapour*	*Liquid*	*Vapour*	*Liquid*	*Vapour*
– 10	*2.9157*	*154.056*	*1450.22*	*0.82965*	*5.7550*	*—*	*0.417477*	*—*	*2.492*
35	*13.522*	*366.072*	*1488.57*	*1.56605*	*5.2086*	*1.7023*	*0.095629*	*4.556*	*2.903*

(U.P.S.C.)

Solution: *Given:* (From the table above):

$$h_2 = 1450.22 \text{ kJ/kg}; \quad h_{3'} = 1488.57 \text{ kJ/kg}; \quad h_{f4} = 366.072 \text{ kJ/kg};$$

$$h_{f4'} = h_1 = h_{f4} - 4.556\,(308 - 303) = 366.07 - 4.556\,(308 - 303) = 343.29 \text{ kJ/kg}$$

Also, $$s_3 = s_2, \quad \text{or,} \quad s_{3'} + c_{ps} \ln\left(\frac{T_3}{T_{3'}}\right) = 5.755$$

$$5.2086 + 2.903 \ln\left(\frac{T_3}{308}\right) = 5.755$$

or, $$\ln\left(\frac{T_3}{308}\right) = \frac{5.755 - 5.2086}{2.903} = 0.1882$$

$$\frac{T_3}{308} = e^{0.1882}, \quad \text{or,} \quad T_3 = 371.8 \text{ K}$$

Now, $$h_3 = h_{3'} + c_p\,(T_3 - T_{3'}) = 1488.57 + 2.903\,(371.8 - 308) = 1673.8 \text{ kJ/kg}$$

$$\text{Mass of refrigerant, } \dot{m} = \frac{50}{h_2 - h_1} = \frac{50}{1450.22 - 343.29} = 0.04517 \text{ kg/s}$$

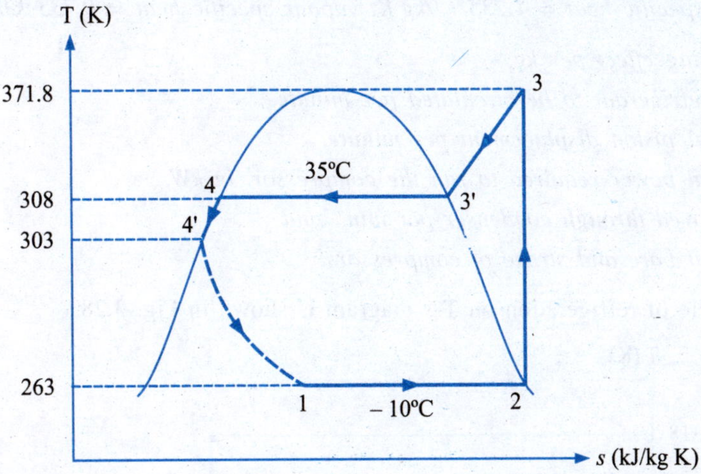

Fig. 3.27

(*i*) **Power Required:**

Power required = $\dot{m}\ (h_3 - h_2)$ = 0.04517 (1673.8 – 1450.22) = **10.1 kW** **(Ans.)**

(*ii*) **Cylinder Dimensions:**

$$\dot{m} = \frac{\pi}{4} D^2 \times L \times \frac{N}{60} \times 0.417477 = 0.04517 \quad \text{(calculated above)}$$

or $$\frac{\pi}{4} D^2 \times 1.2\,D \times \frac{1000}{60} \times 0.417477 = 0.04517$$

or $$D^3 = \frac{0.04517 \times 4 \times 60}{\pi \times 1.2 \times 1000 \times 0.417477} = 0.006888$$

∴ Diameter of cylinder, **D** = $(0.006888)^{1/3}$ = **0.19 m** **(Ans.)**

and, Length of the cylinder, **L** = 1.2 D = 1.2 × 0.19 = **0.228 m** **(Ans.)**

Example 3.13. *A food storage locker requires a refrigeration system of 2400 kJ/min. capacity at an evaporator temperature of 263 K and a condenser temperature of 303 K. The refrigerant used is freon-12 and subcooled by 6ºC before entering the expansion valve and vapour is superheated by 7ºC before leaving the evaporator coil. The compression of refrigerant is reversible adiabatic. The refrigeration compressor is two cylinder single-acting with stroke equal to 1.25 times the bore and operates at 1000 r.p.m.*

Properties of Freon-12:

Saturation temperature (K)	*Absolute pressure (bar)*	*Specific volume vapour (m³/kg)*	*Enthalpy*		*Entropy*	
			Liquid (kJ/kg)	*Vapour (kJ/kg)*	*Liquid (kJ/kg K)*	*Vapour (kJ/kg K)*
263	*2.19*	*0.0767*	*26.9*	*183.2*	*0.1080*	*0.7020*
303	*7.45*	*0.0235*	*64.6*	*199.6*	*0.2399*	*0.6854*

Take: Liquid specific heat = 1.235 kJ/kg K, vapour specific heat = 0.733 kJ/kgK. Determine:

(i) Refrigerating effect per kg,

(ii) Mass of refrigerant to be circulated per minute,

(iii) Theoretical piston displacement per minute,

(iv) Theoretical power required to run the compressor, in kW,

(v) Heat removed through condenser per min., and

(vi) Theoretical bore and stroke of compressors.

Solution: The cycle of refrigeration on T-*s* diagram is shown in Fig. 3.28.

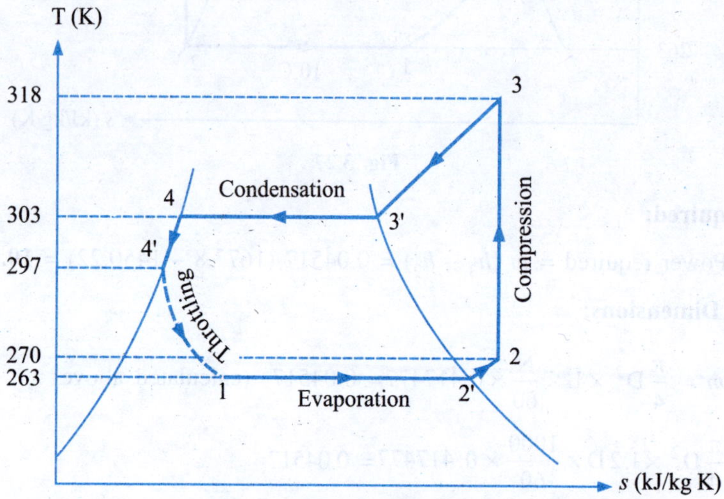

Fig. 3.28

Given: $T_3 = 318$

Enthalpy at '2' $\qquad h_2 = h_{2'} + c_p (T_2 - T_2')$

From the given table, $h_2 = 183.2$ kJ/kg

$(T_2 - T_{2'})$ = degree of superheat as the vapour enters the compressor = 7°C

$$\therefore h_2 = 183.2 + 0.733 \times 7 = 188.33 \text{ kJ/kg}$$

Also, entropy at '2', $s_2 = s_{2'} + c_p \ln \dfrac{T_2}{T_2'} = 0.7020 + 0.733 \ln\left(\dfrac{270}{263}\right) = 0.7212$ kJ/kg K

For isentropic process 2-3, we have:

Entropy at '2' = Entropy at '3'

$$0.7212 = s_{3'} + c_p \ln \frac{T_3}{T_3'} = 0.6854 + 0.733 \ln\left(\frac{T_3}{303}\right)$$

$$\therefore \ln\left(\frac{T_3}{303}\right) = 0.0488$$

i.e., $T_3 = 318$ K

Now, enthalpy at '3' $= h_{3'} + c_p (T_3 - T_{3'}) = 199.6 + 0.733\ (318 - 303) = 210.6$ kJ/kg

Also, enthalpy at 4′, $h_{f4'} = h_{f4} - (c_p)_{\text{liquid}} (T_4 - T_{4'}) = 64.6 - 1.235 \times 6 = 57.19$ kJ/kg

For the process 4′ – 1, we have:

Enthaply at '4' = Enthalpy at 1 = 57.19 kJ/kg

For specific volume at 2,

$$\frac{v_{2'}}{T_{2'}} = \frac{v_2}{T_2}$$

$$\therefore \quad v_2 = \frac{v_{2'}}{T_{2'}} \times T_2 = 0.0767 \times \frac{270}{263} = 0.07874 \text{ m}^3/\text{kg}$$

(*i*) Refrigerating effect per kg

$= h_2 - h_1 = 188.33 - 57.19 =$ **131.14 kJ/kg (Ans.)**

(*ii*) Mass of the refrigerant to be circulated per minute for producing effect of 2400 kJ/min

$= \frac{2400}{131.14} =$ **18.3 kg/min. (Ans.)**

(*iii*) Theoretical piston displacement per minute

= Mass flow/min × specific volume at suction

= 18.3 × 0.7874 = **1.441 m³/min. (Ans.)**

(*iv*) Theoretical power required to run the compressor

= Mass flow of refrigerant per sec. × compressor work/kg

$$= \frac{18.3}{60}(h_3 - h_2) = \frac{18.3}{60}(210.6 - 188.33) \text{kJ/s}$$

= **6.79 kJ/s or 6.79 kW (Ans.)**

(*v*) Heat removed through the condenser per min

= Mass flow of refrigerant × heat removed per kg of refrigerant

$= 18.3\,(h_3 - h_{f4'}) = 18.3\,(210.6 - 57.19) =$ **2807.4 kJ/min. (Ans.)**

(*vi*) Theoretical bore (D) and stroke (L):

Theoretical piston displacement per cylinder

$$= \frac{\text{Total displacement per minute}}{\text{Number of cylinders}} = \frac{1.441}{2} = \mathbf{0.7205\ m^3/min.}$$

Also, length of stroke = 1.25 × diameter of piston

$$\text{Hence,} \quad 0.7205 = \frac{\pi}{4} D^2 \times (1.25D) \times 1000$$

i.e., D = 0.09 m, or, 90 mm **(Ans.)**

and, L = 1.25 D = 1.25 × 90 = **112.5 mm (Ans.)**

Example 3.14. *In an ammonia vapour compression refrigerator condensation and evaporation take place at 11.28 bar and 2.57 bar respectively. The temperature at the end of compression is 50ºC and there is no undercooling. One tonne of ice is to be formed per hour at –5ºC from water at 10ºC. Assuming specific heat of ice as 2.09, the latent heat 335 kJ/kg and c_p of the superheated ammonia vapour as 2.93, calculate the power required to drive the machine; neglect mechanical losses.*

For ammonia:

Pressure (bar)	*Saturation temperature (ºC)*	*Enthalpy (kJ/kg)*		*Entropy per (kJ/kg K)*	
		Liquid	*Latent*	*Liquid*	*Latent*
11.28	*30*	*322.2*	*1146.4*	*1.203*	*3.77*
2.57	*– 12.25*	*124.7*	*1297*	*0.505*	*4.99*

Solution: Refer to Fig. 3.29.

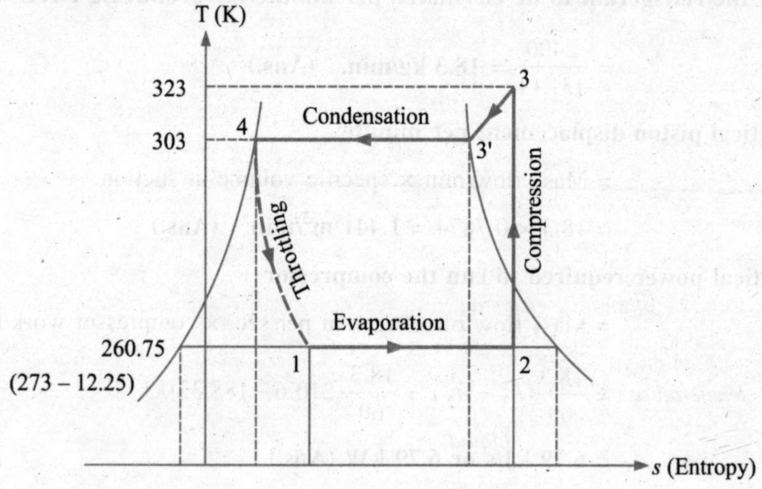

Fig. 3.29

Entropy at point '3'

$$s_3 = \text{entropy at } 3' + c_{ps} \ln \frac{T_3}{T_{3'}} \Rightarrow (1.203 + 3.77) + 2.93 \ln \frac{(50+273)}{(30+273)}$$

$$= 4.973 + 2.93 \ln \frac{323}{303} = 4.973 + 0.187 = 5.16$$

Now, entropy at '3' = entropy at '2'

$$\therefore \quad 5.16 = s_{f2} + x_2\, s_{fg2} = 0.505 + x_2 \times 4.99$$

or, dryness fraction, $x_2 = \dfrac{5.16 - 0.505}{4.99} = 0.932$

Enthalpy at '3', $h_3 = h_{3'} + c_p\,(50 - 30) = (322.2 + 1146.4) + 2.93\,(50 - 30) = 1527.2$ kJ/kg

Enthalpy at '2', $h_2 = h_{f2} + x_2\, h_{fg2} = 124.7 + 0.932 \times 1297 = 1333.5$ kJ/kg

Work done, $W = h_3 - h_2 = 1527.2 - 1333.5 = 193.7$ kJ/kg

Refrigerating effect, $R_n = h_2 - h_1 = h_2 - h_{f4}$ $(\because h_1 = h_{fg2})$

$= 1333.5 - 322.2 = 1011.3$ kJ/kg

Heat actually required to be extracted per hour

$= 1000[1 \times 4.184 \times (10 - 0) + 335 + 1 \times 2.09 \{0 - (-5)\}]$

$= 1000 (41.84 + 335 + 10.45) = 387290$ kJ/h

∴ Mass of refrigerant required

$$= \frac{387290}{1011.3} = 382.96 \text{ kg/h}$$

Total work done per hour $= 193.7 \times 382.96 = 74179.35$ kJ/h

∴ *Power required to drive the machine*

$$= \frac{74179.35}{3600} \text{ kW} = \mathbf{20.6\ kW} \quad \textbf{(Ans.)}$$

Example 3.15. *A single-cylinder, single-acting compressor having bore and stroke of 16 cm and 24 cm respectively runs at a speed of 110 r.p.m. and the indicated mean effective pressure is 2.06 bar. The pressure limits of the refrigerant are 9.66 bar and 2.66 bar and the temperatures at entry to and at exit from condenser are 37.7°C and 16°C, flow of cooling water is 14 kg/min and the inlet and outlet temperatures are 16°C and 24°C. The weight of ice produced per hour from water at 20°C is 56 kg. Assuming the latent heat of ice as 335 kJ/kg and using the following table, find the following:*

(*i*) *The coefficient of performance,*

(*ii*) *The mass of flow of ammonia/min, and*

(*iii*) *The condition of ammonia entering the compressor, neglecting leakage.*

Pressure (bar)	*Saturation temperature (°C)*	*Enthaply (kJ/kg)*		*Specific heat*	
		Liquid	*Vapour*	*Liquid*	*Vapour*
9.66	24	292.88	1462.2	4.60	2.84
2.66	– 12.2	124.68	1426.74	—	—

Solution: Refer to Fig. 3.30.

$$\text{Compressor work} = p_m \text{ LAN} = 2.06 \times 10^5 \times 0.24 \times \frac{\pi}{4} \times 0.16^2 \times 110$$

$$= 109345.5 \text{ Nm/min} = 109.34 \text{ kJ/min}$$

$$\text{Heat abstracted per minute} = \frac{(335 + 20 \times 4.184) \times 56}{60} = 390.77 \text{ kJ/min}$$

(*i*) **C.O.P.:** $\quad \text{C.O.P.} = \frac{\text{Heat abstracted}}{\text{Work done}} = \frac{390.77}{109.34} = \mathbf{3.57}$ **(Ans.)**

(*ii*) **The mass flow of ammonia /min, m:**

Heat carried away by cooling water = Mass of water/min × specific heat × rise of temp.

$= 14 \times 4.184 \times (24 - 16) = 468.6$ kJ/min

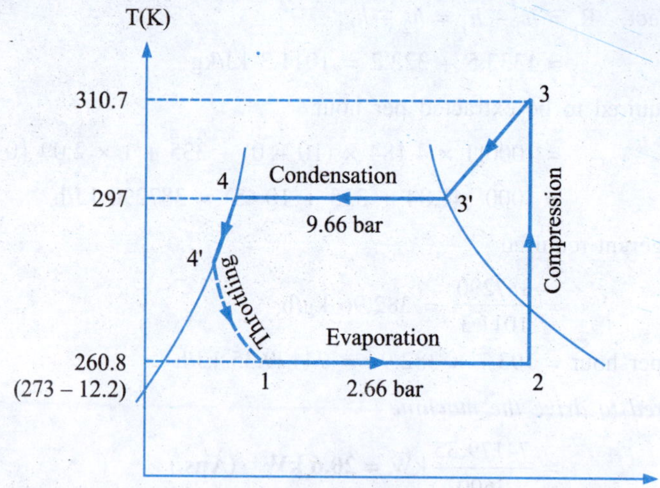

Fig. 3.30

Also, $468.6 = \text{Heat lost by the refrigerant} = \text{Mass of refrigerant} \times (h_3 - h_{4'})$

Now, $h_4 - h_{4'} = 4.60\ (24 - 16) = 36.8\ \text{kJ/kg}$ ...(1)

$h_{3'} - h_4 = 1462.2 - 292.88 = 1169.32\ \text{kJ/kg}$ ...(2)

$h_3 - h_{3'} = 2.84\ (310.7 - 297) = 38.91\ \text{kJ/kg}$ ...(3)

Adding equations (1), (2) and (3), we get:

$$h_3 - h_{4'} = 36.8 + 1169.32 + 38.91 = 1245.03\ \text{kJ/kg}$$

Mass of flow of ammonia,

$$m = \frac{468.6}{1245.03} = \mathbf{0.376\ kg/min} \quad \textbf{(Ans.)}$$

(*iii*) **Condition of ammonia entering the compressor:**

$$\text{Refrigerating effect/kg} = h_2 - h_1 = \frac{390.77}{0.376} = 1039.3\ \text{kJ/kg}$$

But, $h_1 = h_{f4'} = 292.88 - 36.8 = 256.08$

and, $h_2 - h_1 = 1039.3$ (as above)

$\therefore$ $h_2 = h_1 + 1039.3 = 256.08 + 1039.3 = 1295.4\ \text{kJ/kg}$

Also, $h_2 = h_{f2} + x_2 h_{fg2}$

$$1295.4 = 124.68 + x_2 \times (1426.74 - 124.68)$$

or, dryness fraction, $x_2 = \dfrac{(1295.4 - 124.68)}{(1426.74 - 124.68)} = \dfrac{1170.72}{1302.06} = \mathbf{0.899}$ **(Ans.)**

Example 3.16. *Following results were obtained in a test conducted on a vapour compression refrigerator: Evaporator temperature = –28.5°C, condenser pressure = 2.75 bar; Refrigerant entering the condenser is 3°C superheat, refrigerant leaving the condenser is at 12.8°C. Determine the C.O.P. The following properties are given:*

Pressure (bar)	*Saturation temp. (°C)*	*Enthalpy (kJ/kg)*		*Entropy Vapour*	*Specific heat (kJ/kg K) at constant pressure*	
		Liquid	*Vapour*	*(kJ/kg K)*	*Liquid*	*vapour*
2.75	*14*	*438.48*	*802.9*	*5.5287*	*1.381*	*0.669*
0.412	*– 28.5*	*381.58*	*783.24*	*5.6852*	*—*	*—*

Solution: Refer to Fig. 3.31.

$$s_3 = s_{3'} + c_{p_s} \ln\frac{290}{287} = 5.5287 + 0.669 \ln\frac{290}{287} = 5.535$$

Since the compression in isentropic,

∴ Entropy at '3' = Entropy at '2'

∴ Change in entropy,

$$s_c = 5.6852 - 5.535 = 0.15$$

Now, $$s_c = (1-x)\frac{h_{fg}}{T} \text{ at 0.412 bar}$$

∴ $$0.15 = (1-x)\left(\frac{783.24 - 381.58}{244.5}\right) = (1-x) \times 1.643$$

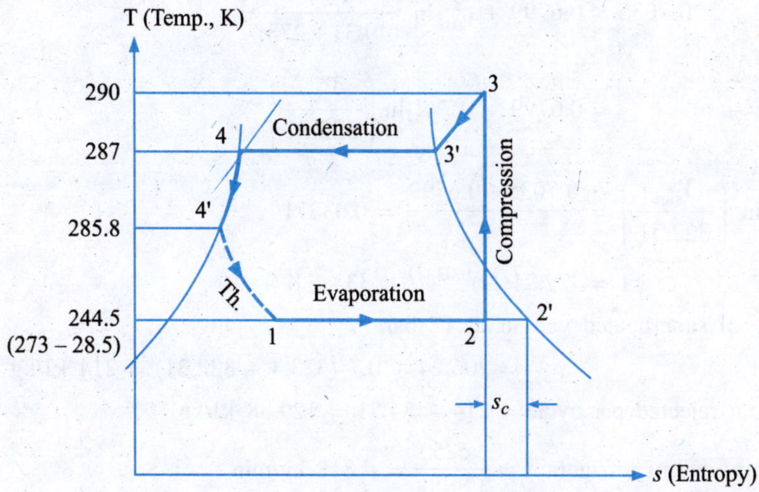

Fig. 3.31

or, dryness fraction, $x = 1 - \dfrac{0.15}{1.643} = 0.909$

∴ Enthalpy at '2', $h_2 = h_{f2} + xh_{fg2} = 381.58 + 0.909 \times (783.24 - 381.58) = 746.69$ kJ/kg

Enthalpy at '3', $h_3 = h_{3'} + c_p (T_3 - T_{3'}) = 802.7 + 0.669 (290 - 287) = 804.9$ kJ/kg

Also, $h_1 = h_{4'} = h_4 - c_p (T_4 - T_{4'}) = 438.48 - 1.381 (287 - 285.8) = 436.8$ kJ/kg

∴ Work done, $W = h_3 - h_2 = 804.9 - 746.69 = 58.21$ kJ/kg

Refrigerating effect, $R_n = h_2 - h_1 = 746.69 - 436.8 = 309.89$

∴ C.O.P. $= \dfrac{R_n}{W} = \dfrac{309.89}{58.21} = \mathbf{5.32}$ **(Ans.)**

Example 3.17. *A vapour compression heat pump is driven by a power cycle having a thermal efficiency of 25%. For the heat pump, refrigerant-12 is compressed from saturated vapour at 2.0 bar to the condenser pressure of 12 bar. The isentropic efficiency of the compressor is 80%. Saturated liquid enters the expansion valve at 12 bar. For the power cycle 80% of the heat rejected by it is transferred to the heated space which has a total heating requirement of 500 kJ/min. Determine the power input to the heat pump compressor. The following date for refrigerant-12 may be used:*

Pressure	*Temperature*	*Enthalpy (kJ/kg)*		*Entropy (kJ/kg K)*	
(bar)	*(°C)*	*Liquid*	*Vapour*	*Liquid*	*Vapour*
20	*– 12.53*	*24.57*	*182.07*	*0.0992*	*0.7035*
12.0	*49.31*	*84.21*	*206.24*	*0.3015*	*0.6799*

Vapour specific heat at constant pressure = 0.7 kJ/kg K. **(U.P.S.C.)**

Solution: Heat rejected by the cycle $= \frac{500}{0.8} = 625$ kJ/min.

Assuming isentropic compression of the refrigerant, we have entropy of dry saturated vapour at 12 bar = entropy of superheated vapour at 12 bar

$$0.7035 = 0.6799 + c_{ps} \ln \frac{T_{sup}}{(49.31 + 273)}$$

$$= 0.6799 + 0.7 \times \ln \left(\frac{T_{sup}}{322.31}\right)$$

or, $$\ln \left(\frac{T_{sup}}{322.31}\right) = \frac{0.7035 - 0.6799}{0.7} = 0.03371$$

or, $$T = 322.21\,(e)^{0.03371} = 333.4 \text{ K}$$

∴ Enthaply of superheated vapour at 12 bar

$$= 206.24 + 0.7\,(333.4 - 322.31) = 214 \text{ kJ/kg}$$

Heat rejected per cycle = 214 – 84.21 = 129.88 kJ/kg

$$\text{Mass flow rate of refrigerant} = \frac{625}{129.88} = 4.812 \text{ kg/min}$$

Work done on compressor = 4.832 (214 – 182.07) = 153.65 kJ/min = 2.56 kW

$$\text{Actual work of compressor} = \frac{2.56}{\eta_{compressor}} = \frac{2.56}{0.8} = 3.2 \text{ kW}$$

Hence, power input to the pump compressor = **3.2 kW** **(Ans.)**

Example 3.18. *A refrigerator operating on standard vapour compression cycle has a co-efficient of performance of 6.5 and is driven by a 50 kW compressor. The enthalpies of saturated liquid and saturated vapour refrigerant at the operating condensing temperature of 35°C are 69.55 kJ.kg and 201.45 kJ/kg respectively. The saturated refrigerant vapour leaving evaporator has an ethalpy of 187.53 kJ/kg. Find the refrigerant temperature at compressor discharge. The c_p of refrigerant vapour may be taken to be 0.6155 kJ/kg°C.* **(GATE)**

Solution: *Given:* C.O.P. = 6.5;

$$W = 50 \text{ kW},$$
$$h_{3'} = 201.45 \text{ kJ/kg};$$
$$h_{f_4} = h_1 = 69.55 \text{ kJ/kg};$$
$$h_2 = 187.53 \text{ kJ/kg};$$
$$c_p = 0.6155 \text{ kJ/kg °C}.$$

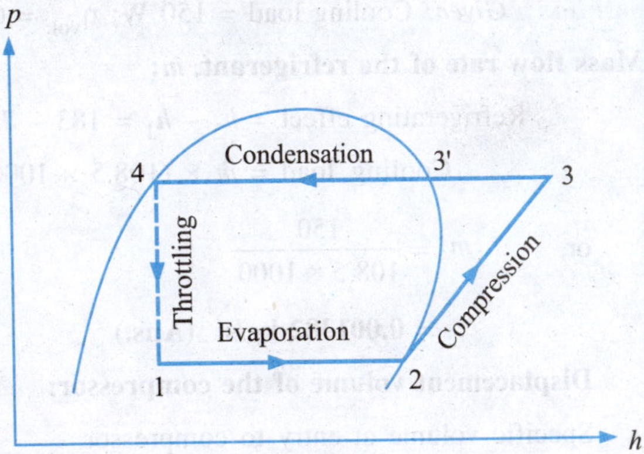

Fig. 3.32

Temperature, t_3:

$$\text{Refrigerating capacity} = 50 \times \text{C.O.P.} = 50 \times 6.5 = 325 \text{ kW}$$

Heat extracted per kg of refrigerant

$$= 187.53 - 69.55 = 117.98 \text{ kJ/kg}$$

$$\text{Refrigerant's flow rate} = \frac{325}{117.98} = 2.755 \text{ kg/s}$$

Compressor power = 50 kW

$$\therefore \quad \text{Heat input per kg} = \frac{50}{2.755} = 18.15 \text{ kJ/kg}$$

Enthalpy of vapour after compression $= h_2 + 18.15$

$$= 187.53 + 18.15 = 205.68 \text{ kJ/kg}$$

$$\text{Superheat} = 205.68 - h_{3'} = 205.68 - 201.45 = 4.23 \text{ kJ/kg}$$

But, $\quad 4.23 = 1 \times c_p\,(t_3 - t_3') = 1 \times 0.6155 \times (t_3 - 35).$

$$\therefore \quad t_3 = \frac{4.23}{0.6155} + 35 = \mathbf{41.87°C} \quad \textbf{(Ans.)}$$

Note: The compressor rating of 50 kW is assumed to be the enthalpy of compression, in the absence of any data on the efficiency of compressor.

Example 3.19. *A refrigeration cycle uses Freon-12 as the working fluid. The temperature of the refrigerant in the evaporator is –10°C. The condensing temperature is 40°C. The cooling load is 150 W and the volumetric efficiency of the compressor is 80%. The speed of the compressor is 720 r.p.m. Calculate the mass flow rate of the refrigerant and the displacement volume of the compressor.*

Properties of Freon-12:

Temperature (°C)	*Saturation Pressure (MPa)*	*Enthalpy (kJ/kg)*		*Specific Volume (m^3/kg)*
		Liquid	*Vapour*	*Saturated vapour*
– 10	0.22	26.8	183.0	0.08
40	0.96	74.5	203.1	0.02

(GATE)

Solution: *Given:* Cooling load = 150 W; $\eta_{vol.}$ = 0.8; N = 720 r.p.m.

Mass flow rate of the refrigerant, *m*:

Refrigerating effect = $h_2 - h_1$ = 183 – 74.5 = 108.5 kJ/kg

Cooling load = $\dot{m} \times (108.5 \times 1000) = 150$

or, $$\dot{m} = \frac{150}{108.5 \times 1000}$$

= 0.001382 kg/s (Ans.)

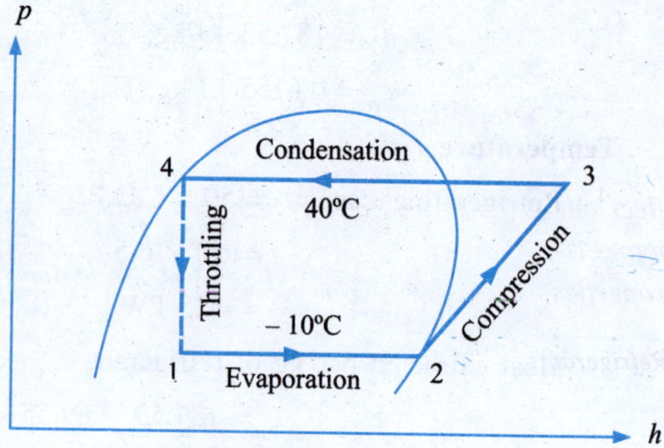

Fig. 3.33

Displacement volume of the compressor:

Specific volume at entry to compressor

v_2 = 0.08 m³/kg (From the table)

∴ Displacement volume of compressor

$$= \frac{\dot{m}v_2}{\eta_{vol.}} = \frac{0.001382 \times 0.08}{0.8}$$

= 0.0001382 m³/s (Ans.)

Example 3.20. *In a simple vapour compression cycle following are the properties of the refrigerant R-12 at various points:*

Compressor inlet	: h_2 = 183.2 kJ/kg	v_2 = 0.0767 m³/kg
Compressor discharge	: h_3 = 222.6 kJ/kg	v_3 = 0.00164 m³/kg
Condenser exit	: h_{f4} = 84.9 kJ/kg	v_4 = 0.00083 m³/kg

The piston displacement volume for compressor is 1.5 litres per stroke and its volumetric efficiency is 80%. The speed of the compressor is 1600 r.p.m.

Find: (i) *Power rating of the compressor (kW);*

(ii) *Refrigerating effect (kW).* **(GATE)**

Solution: Piston displacement volume

$$= \frac{\pi}{4}D^2 \times L = 1.5 \text{ litres}$$

$$= 1.5 \times 1000 \times 10^{-6} \text{ m}^3/\text{stroke}$$

$$= 0.0015 \text{ m}^3/\text{revolution}$$

(*i*) Power rating of the compressor (kW):

Compressor discharge

$$= 0.0015 \times 1600 \times 0.8 \ (\eta_{vol.})$$

$$= 1.92 \text{ m}^3/\text{min.}$$

Mass flow rate of compressor,

$$m = \frac{\text{Compressor discharge}}{v_2}$$

$$= \frac{1.92}{0.0767} = 25.03 \text{ kg/min}$$

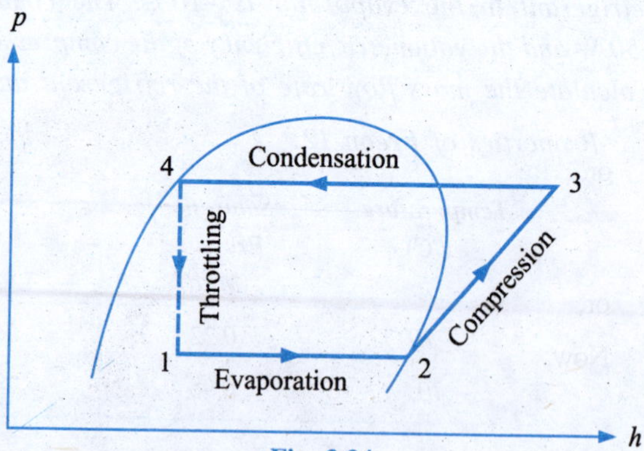

Fig. 3.34

Power rating of the compressor $= m\,(h_3 - h_2)$

$$= \frac{25.03}{60}(222.6 - 183.2) = \mathbf{16.44\ kW} \quad \textbf{(Ans.)}$$

(*ii*) **Refrigerating effect (kW) :**

Refrigerating effect $= m\,(h_2 - h_1) = m\,(h_2 - h_{f4})$ $\qquad (\because h_1 = h_{f4})$

$$= \frac{25.03}{60}(183.2 - 84.9) = \mathbf{41\ kW} \quad \textbf{(Ans.)}$$

Example 3.21. *A refrigerator is to be designed to operate between –45°C and 0°C. You are asked to select one among the three given refrigerants, namely Freon-12, NH_3 and CO_2, on the basis of the factors (i) COP, (ii) Power required per tonne, and (iii) the condenser and evaporator pressures. The properties of the refrigerants are as given below:*

Refrigerants	t_s °C	p_s bar	h_f kJ/kg	h_g kJ/kg	s_f	s_g	c_{pg}
F-12	–45	0.505	–4.4	167.84	0.0190	0.7360	—
	0	3.09	36.25	188.69	0.1420	0.7008	0.62
NH_3	–45	0.535	–22.4	1378.76	0.0961	6.0475	—
	0	4.24	180.88	1443.34	0.7139	5.3368	2.72
CO_2	–45	8.218	–10.68	319.38	0.0456	1.4043	—
	0	34.81	85.58	320.47	0.3257	1.1878	1.26

Determine the values for the above factors and suggest your choice for the most suitable refrigerant.

(UPTU)

Solution: Refer to Fig. 3.35.

Refrigerant "F-12" :

For the compression process 2-3, we have:

$$s_2 = s_3$$

$$s_{g2} = s_{g3} + c_{pg} \ln\left(\frac{t_3 + 273}{0 + 273}\right)$$

$$0.7360 = 0.7008 + 0.62 \ln\left(\frac{t_3 + 273}{273}\right)$$

or, $\ln\left(\frac{t_3 + 273}{273}\right) = \frac{0.7360 - 0.7008}{0.62}$

$$= 0.0568$$

or, $t_3 = 15.96$°C

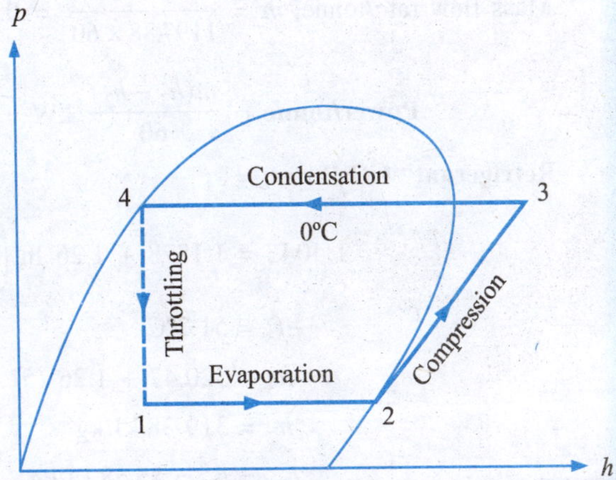

Fig. 3.35

Now, $h_3 = h_{g3} + c_{pg}\,(15.96 - 0)$

$$= 188.69 + 0.62\,(15.96 - 0)$$

$$= 198.58 \text{ kJ/kg}$$

$$h_2 = 167.84 \text{ kJ/kg}$$

$$h_{f4} = h_1 = 36.25 \text{ kJ/kg}$$

$$\therefore \quad \text{Work/kg} = h_3 - h_2 = 198.58 - 167.84 = 30.74 \text{ kJ/kg}$$

$$\text{Ref. effect/kg} = h_2 - h_1 = 167.84 - 36.25 = 131.59 \text{ kJ/kg}$$

$$\therefore \quad \textbf{C.O.P.} = \frac{\text{Ref. effect/kg}}{\text{Work/kg}} = \frac{131.59}{30.74} = \mathbf{4.281}$$

Mass flow rate per tonne,

$$m = \frac{14000}{131.59 \times 60} = 1.773 \text{ kg/min-tonne}$$

$$\therefore \quad \textbf{Power/tonne} = \frac{m(h_3 - h_2)}{60} \text{ kW} = \frac{1.773 \times 30.74}{60} = \mathbf{0.9084\ kW}$$

Refrigerant "NH_3": $s_2 = s_3$

$$6.0475 = 5.3368 + 2.72 \ln\left(\frac{t_3 + 273}{273}\right)$$

$$\therefore \quad t_3 = 81.5°\text{C}$$

$$h_3 = 1443.34 + 2.72\ (81.5 - 0) = 1665.02 \text{ kJ/kg}$$

$$h_2 = 1378.76 \text{ kJ/kg}$$

$$h_{f4} = h_1 = 180.88 \text{ kJ/kg}$$

$$\therefore \quad \text{Work/kg} = h_3 - h_2 = 1665.02 - 1378.76 = 286.26 \text{ kJ/kg}$$

$$\text{Ref. effect/kg} = h_2 - h_1 = 1378.76 - 180.88 = 1197.88 \text{ kJ/kg}$$

$$\therefore \quad \textbf{C.O.P.} = \frac{1197.88}{286.26} = \mathbf{4.184}$$

$$\text{Mass flow rate/tonne, } m = \frac{14000}{1197.88 \times 60} = 0.1948 \text{ kg/min-tonne}$$

$$\therefore \quad \textbf{Power/tonne} = \frac{m(h_3 - h_2)}{60} \text{ kW} = \frac{0.1948 \times 286.26}{60} = \mathbf{0.9294\ kW}$$

Refrigerant "CO_2": $s_2 = s_3$

$$1.4043 = 1.1878 + 1.26 \ln\left(\frac{t_3 + 273}{273}\right)$$

$$\therefore \quad t_3 = 51.2°\text{C}$$

$$h_3 = 320.47 + 1.26\ (51.2 - 0) = 384.98 \text{ kJ/kg}$$

$$h_2 = 319.38 \text{ kJ/kg}$$

$$h_{f4} = h_1 = 85.58 \text{ kJ/kg}$$

$$\therefore \quad \text{Work/kg} = h_3 - h_2 = 384.98 - 319.38 = 65.6 \text{ kJ/kg}$$

$$\text{Ref. effect/kg} = h_2 - h_1 = 319.38 - 85.58 = 233.8 \text{ kJ/kg}$$

$$\therefore \quad \textbf{C.O.P.} = \frac{233.8}{65.6} = \textbf{3.564}$$

$$\text{Mass flow rate/tonne} = \frac{14000}{233.8 \times 60} = 0.998 \text{ kg/min-tonne}$$

$$\therefore \quad \textbf{Power/tonne} = \frac{m(h_3 - h_2)}{60} \text{ kW} = \frac{0.998 \times 65.6}{60} = \textbf{1.091 kW}$$

The comparison for the three refrigerants is summarised in the table below:

S. No.	*Refrigerant*	*C.O.P.*	*Power/tonne (kW)*	*'p' condenser (bar)*	*'p' evaporator (bar)*
1.	*F-12*	4.281	0.9084	3.09	0.505
2.	NH_3	4.184	0.9294	4.24	0.535
3.	CO_2	3.564	1.091	34.81	8.218

The choice for the most suitable refrigerant depends upon the following requirements:

(*i*) If suitable evaporator pressure (positive) and evaporator pressure (not high multiple of *p* evap.) is the requirement, the choice is CO_2.

(*ii*) If safe refrigerant is the requirement, the choice is F-12 or CO_2.

(*iii*) If power consumption (low) or C.O.P. (high) is the requirement, the choice is F-12 or NH_3.

Example 3.22. *For a vapour compression refrigeration system using R-22 as refrigerant, condenser outlet temperature is 40°C and evaporator inlet temperature is –20°C. In order to avoid flashing of refrigerant, a liquid-suction vapour heat-exchanger is provided where liquid is subcooled to 26°C. The refrigerant leaves the evaporator as saturated vapour. The compression process is isentropic. Find the power requirement and coefficient of performance if capacity of the system is 10 kW at –20°C. Show cycle on pressure-enthalpy diagram.* c_p *of vapour is 1.03 kJ/kg K. The thermodynamic properties are given below:*

t (°C)	*p* (bar)	v_f (m^3/kg)	v_g (m^3/kg)	h_f (kJ/kg)	h_g (kJ/kg)	s_f (kJ/kg K)	s_g (kJ/kg K)
– 20	2.448	0.741	0.0928	177.1	397.5	0.9139	1.784
26	10.723	0.840	0.0220	231.6	413.5	1.109	1.717
40	15.335	0.884	0.0151	249.7	416.6	1.167	1.699

[PTU]

Solution: Refer to Fig. 3.36.

From the given table, we have:

$$h_{f4} = 249.7 \text{ kJ/kg};$$

$$h_{2'} = 397.5 \text{ kJ/kg};$$

$$h_{f4'} = 231.6 \text{ kJ/kg}$$

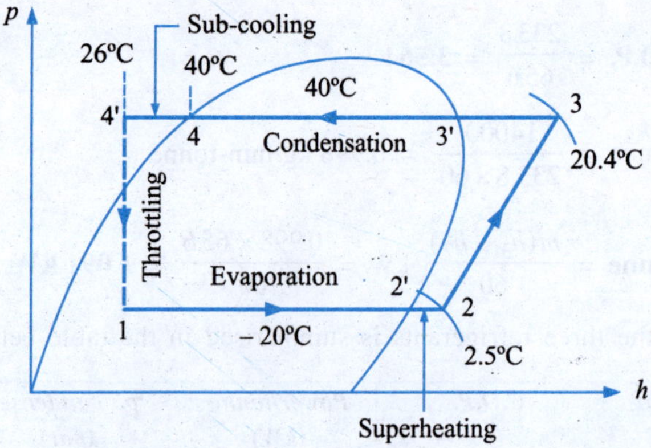

Fig. 3.36

Heat exchanged in the heat exchanger

$$= h_{f4} - h_{f4'} = h_2 - h_{2'}$$

$\therefore$ $h_2 = (h_{f4} - h_{f4'}) + h_{2'} = (249.7 - 231.6) + 397.5 = 415.6$ kJ/kg

Also, $c_{pv}\,(t_2 - t_{2'}) = h_2 - h_{2'} = 415.6 - 397.5 = 18.1$ kJ/kg

or, $1.03\,[t_2 - (-20)] = 18.1$ $\quad [\because t_{2'} = -20$ºC ...Given]

$\therefore$ $t_2 = \dfrac{18.1}{1.03} - 20 = -2.5$ºC

Also, $s_2 = s_3$

or, $s_{g2} + c_{pv}\ \ln\left(\dfrac{t_2 + 273}{-20 + 273}\right) = s_{g3} + c_{pv}\ \ln\left(\dfrac{t_3 + 273}{40 + 273}\right)$

or, $1.784 + 1.03\ \ln\left(\dfrac{-2.5 + 273}{-20 + 273}\right) = 1.699 + 1.03\ \ln\left(\dfrac{t_3 + 273}{40 + 273}\right)$

$$1.853 = 1.699 + 1.03\ \ln\left(\frac{t_3 + 273}{313}\right)$$

or, $\ln\left(\dfrac{t_3 + 273}{313}\right) = \left(\dfrac{1.853 - 1.699}{1.03}\right) = 0.1495$

or, $\dfrac{t_3 + 273}{313} = (e)^{0.1495} = 1.161$, or, $t_3 = 90.4$ºC

$\therefore$ $h_3 = h_{g3'} + c_{pv}\ (90.4 - 40) = 416.6 + 1.03\ (90.4 - 40) = 468.5$ kJ/kg

Refrigerating effect, $= h_{2'} - h_1 = 397.5 - 231.6 = 165.9$ kJ/kg $\quad (\because h_1 = h_{f_{4'}})$

Work/kg, W $= h_3 - h_2 = 468.5 - 415.6 = 52.9$ kJ/kg

Flow rate, $\dot{m} = \dfrac{10}{165.9} = 0.0603$ kg/s

$\therefore$ Power required, **P** $= \dot{m}\ (h_3 - h_2) = 0.0603 \times 52.9 =$ **3.19 kW** **(Ans.)**

C.O.P. $= \dfrac{\text{Refrigerating effect}}{\text{Power}} = \dfrac{10}{3.19} =$ **3.13** **(Ans.)**

Example 3.23. *A compressor having a stroke volume of 500 c.c. runs at 500 r.p.m. and works with CO_2 refrigerant. The evaporator and condenser temperatures are –15°C and 25°C, respectively. The condenser liquid is subcooled to 15°C before passing on to the expansion valve. Assumme wet compression with an initial quality x of CO_2 as 0.9. Assuming isentropic compression and volumetric efficiency of the compressor as 0.85, calculate the following: (i) The COP, (ii) The power required by the compressor (kW), and (iii) The refrigerating capacity of the compressor in tonnes of refrigeration.*

The following data for CO_2 may be used:

t, °C	p, bar	v, m³/kg		h, kJ/kg		s, kJ/kg K	
		Liquid	Vapour	Liquid	Vapour	Liquid	Vapour
– 15	22.88	0.00101	0.0166	49.62	322.86	0.1976	1.2567
15	50.92	0.00130	0.0063	127.75	308.08	0.4697	1.0959
25	64.32	0.00147	0.0042	164.17	283.63	0.5903	0.9912

Take specific heat c_p of CO_2 gas = 2.4 kJ/kg K. **(AMIE)**

Solution: Refer to Fig. 3.37.

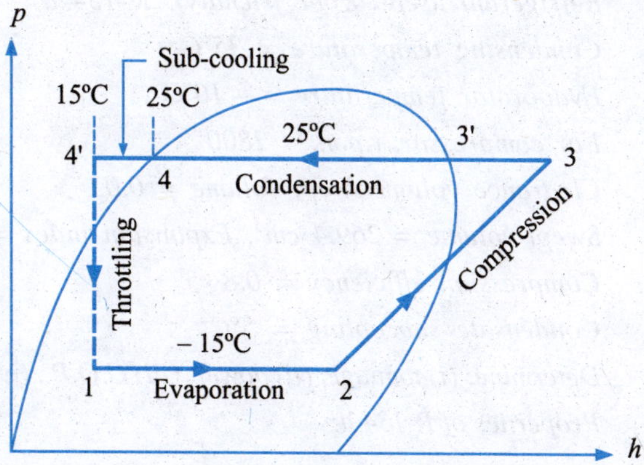

Fig. 3.37

Given: $V_s = 500 \text{ cm}^3 = 500 \times 10^{-6} \text{ m}^3$;

$N = 500$ r.p.m.; $x_2 = 0.9$; $\eta_{vol.} = 0.85$;

$c_{pg} = 2.4$ kJ/kg K.

Assuming single-acting, the compressor displacement $= 500 \times 10^{-6} \times 500 = 0.25 \text{ m}^3/\text{min}$.

Specific volume at compressor inlet,

$$v_2 = x_2 v_{g2} + (1 - x_2)\, v_{f2}$$
$$= 0.9 \times 0.0166 + (1 - 0.9) \times 0.00101$$
$$= 0.01504 \text{ m}^3/\text{kg}$$

Mass flow rate of refrigerant,

$$m = \frac{0.25 \times 0.85}{0.01504} = 14.129 \text{ kg/min}$$

For isentropic compression process 2-3, we have:

$$s_2 = s_3$$

$$s_{f2} + x_2\,(s_{g2} - s_{f2}) = s_{g3} + c_{pg} \ln\left(\frac{t_3 + 273}{25 + 273}\right)$$

$$0.1976 + 0.9\,(1.2567 - 0.1976) = 0.9912 + 2.14 \ln\left(\frac{t_3 + 273}{298}\right)$$

$\therefore \quad t_3 = 45.5°\text{C}$

$\therefore \quad h_3 = h_{g3'} + c_{pg}\,(45.5 - 25) = 283.63 + 2.4 \times 20.5 = 332.83$ kJ/kg

$h_2 = h_{f2} + x_2\,(h_{g2} - h_{f2}) = 49.62 + 0.9\,(322.86 - 49.62) = 295.5$ kJ/kg

$h_{f4'} = h_1 = 127.75$ kJ/kg

$$\text{Ref. effect/kg} = h_2 - h_1 = 295.5 - 127.75 = 167.75 \text{ kJ/kg}$$

$$\text{Work/kg} = h_3 - h_2 = 332.83 - 295.5 = 37.33 \text{ kJ/kg}$$

(*i*) **C.O.P.:** $\text{C.O.P.} = \dfrac{h_2 - h_1}{h_3 - h_2} = \dfrac{167.75}{37.33} = \mathbf{4.49}$ **(Ans.)**

(*ii*) **The power required by the compressor, P:**

$$P = \frac{14.129}{60}(h_3 - h_2) = \frac{14.129}{60} \times 37.33 = \mathbf{8.79\ kW} \quad \textbf{(Ans.)}$$

(*iii*) **Refrigerating capacity:**

$$\text{Refrigerating capacity} = \frac{m(h_2 - h_1) \times 60}{14000} = \frac{14.129 \times 167.75 \times 60}{14000}$$

$$= \mathbf{10.16\ tonnes} \quad \textbf{(Ans.)}$$

Example 3.24. *The following data refer to a single-stage vapour compression system.*

Refrigerant used (ozone friendly): R–134 a

Condensing temperature = 35ºC

Evaporator temperature = – 10ºC

For compressor, r.p.m. = 2800

Clearance volume/swept volume = 0.03

Swept volume = 269.4 cm³, Expansion index = 1.12

Compression efficiency = 0.8

Condensate subcooling = 5ºC

Determine: (i) tonnage, (ii) power, (iii) C.O.P., (iv) heat rejection to condenser and (v) P.I.

Properties of R-134 a:

t,ºC	*p, bar*	$V_g, m^3/kg$	$h_f, kJ/kg$	h_g, *kJ/kg*	s_f, *kJ/kg K*	s_g, *kJ/kg K*
– 10	2.014	0.0994	186.7	392.4	0.9512	1.733
35	8.870	—	249.1	417.6	1.1680	1.715

Assume: Specific heat of vapour at 8.87 bar: 1.1 kJ/kg.K., 1.458 kJ/kg K;

Suction vapour dry saturated; isentropic compression. **(M.U.)**

Solution: Refer to Fig. 3.38.

For isentropic compression process 2-3, we have:

$$s_2 = s_3$$

$$1.733 = 1.715 + 1.1 \ln\left(\frac{t_3 + 273}{35 + 273}\right)$$

$$\therefore \quad t_3 = 40.08ºC$$

$$h_3 = 417.6 + 1.1\ (40.08 - 35) = 423.19 \text{ kJ/kg}$$

$$h_2 = 392.4 \text{ kJ/kg}$$

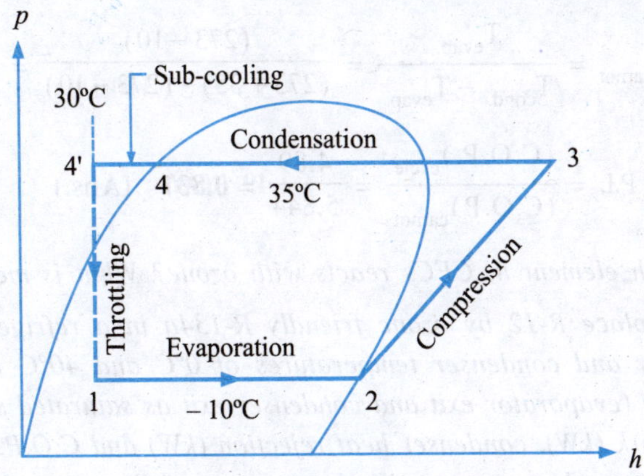

Fig. 3.38

$$h_1 = h_{f4'} = h_{f4} - c_{p_l} \times (35 - 30) = 249.1 - 1.458 \times 5 = 241.81 \text{ kJ/kg}$$

$$v_{g2} = 0.0994 \text{ m}^3\text{/kg}$$

$$\eta_{vol.} = 1 + C - C\left(\frac{p_2}{p_1}\right)^{1/n}$$

$$= 1 + 0.03 - 0.03\left(\frac{8.870}{2.014}\right)^{1/1.12} = 0.9173 \qquad \text{...(Eqn. 3.2)}$$

Volume of refrigerant admitted at state "2"

$$= (269.4 \times 10^{-6}) \times 0.9173 = 247.12 \times 10^{-6} \text{ m}^3\text{/cycle}$$

Assuming the compressor to be singe-acting, mass of air circulated,

$$m = \frac{247.12 \times 10^{-6} \times 2800}{0.0994} = 6.961 \text{ kg/min}$$

(i) **Tonnage:** $\text{Tonnage} = \frac{m(h_2 - h_1) \times 60}{14000} = \frac{6.961(392.4 - 241.81) \times 60}{14000} = \mathbf{4.49 \text{ tonnes}}$ **(Ans.)**

(ii) **Power, P:** $P = \frac{m(h_3 - h_2)}{60} = \frac{6.961(423.19 - 392.4)}{60} = \mathbf{3.572 \text{ kW}}$

(iii) **C.O.P.:** $\text{C.O.P.} = \frac{\text{Ref. effect (kW)}}{\text{Power (kW)}} = \frac{m(h_2 - h_1)/60}{3.572}$

$$= \frac{6.961(392.4 - 241.81)/60}{3.572} = \mathbf{4.89} \quad \textbf{(Ans.)}$$

(iv) **Heat rejected to condenser:**

Heat rejected to condenser = $m\,(h_3 - h_{4'})$

$$= 6.961\ (423.19 - 241.81) = \mathbf{1262.6 \text{ kJ/min}} \quad \textbf{(Ans.)}$$

[It is also equal to : Ref. effect + power = 6.961 (392.4 – 241.8) + 3.572 × 60 = 1262.6 kJ/min.]

(*v*) **P.I.:** $\quad (\text{C.O.P.})_{carnot} = \dfrac{T_{evap.}}{T_{cond.} - T_{evap.}} = \dfrac{(273-10)}{(273+35)-(273-10)} = 5.844$

$$\text{P.I.} = \frac{(\text{C.O.P.})_{cycle}}{(\text{C.O.P})_{carnot}} = \frac{4.89}{5.844} = \mathbf{0.837} \quad \textbf{(Ans.)}$$

Example 3.25. *(a) Which element in CFCs reacts with ozone? What is meant by global warming?*

(b) It is proposed to replace R-12 by ozone friendly R-134a in a refrigeration plant of 9 tonnes capacity with evaporator and condenser temperatures of 0°C and 40°C respectively. Considering standard saturation cycle (evaporator exit and condenser exit as saturated states), compare the mass flow rate, compressor work (kW), condenser heat rejection (kW) and C.O.P. for the two refrigerants. The saturation properties and vapour specific heats are as follows:

t (°C)	h_f (kJ/kg)	h_g (kJ/kg)	s_f (kJ/kg K)	s_g (kJ/kg K)	c_p (kJ/kg K)
R-134a					
0	200.81	398.78	1.0025	1.7261	—
40	255.73	419.63	1.1884	1.7128	1.068
R-12					
0	36.1	187.5	0.0142	0.6966	—
40	74.6	203.2	0.2718	0.6825	0.776

(Anna University)

Solution: (*a*) • The element **chlorine** in CFCs *reacts with ozone.*

• **Global warming** *means the increase in average temperature of the earth.* The causes of global warming are *increase in CO_2 concentration, NO_x emission and use of CFC refrigerants,* all of these causing the depletion and degradation of the protective ozone layer. The ozone layer absorbs ultra violet rays (UVR) from solar radiation and prevents it from reaching the earth's surface. A major fraction of UVR is thermal radiation. *The effect of global warming is that the ice at the two poles is slowly melting, and consequently the sea level is rising* by *about 1 mm/year.* It will affect the cities situated near the sea.

(*b*) Refer to Fig. 3.39:

For R-134*a*: $\quad h_2 = 398.78$ kJ/kg

$\quad h_{f4} = h_1 = 255.73$ kJ/kg

For the isentropic compression process 2-3, we have:

$$s_2 = s_3$$

$$1.7261 = 1.7128 + 1.068 \ln\left(\frac{t_3 + 273}{40 + 273}\right)$$

$$\therefore \quad t_3 = 43.92°\text{C}$$

$$\therefore \quad h_3 = 419.63 + 1.068\,(43.92 - 40)$$

$$= 423.8 \text{ kJ/kg}$$

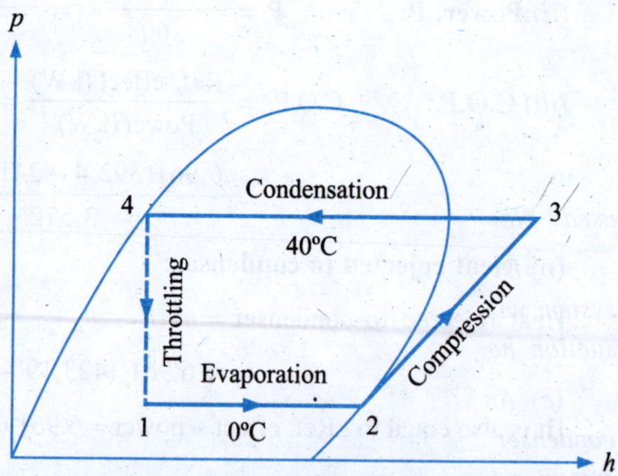

Fig. 3.39

$$\text{Ref. effect/kg} = h_2 - h_1$$
$$= 398.78 - 255.73 = 143.05 \text{ kJ/kg}$$

$$\textit{Mass flow rate, } m = \frac{14000 \times 9}{143.05 \times 60} = \mathbf{14.68 \text{ kg/min}}$$

$$\textit{Compressor work} = \frac{m(h_3 - h_2)}{60} \text{ kW} = \frac{14.68(423.8 - 398.78)}{60} = \mathbf{6.12 \text{ kW}}$$

$$\textit{Condenser heat rejection} = m\,(h_3 - h_{f4}) = 14.68\,(423.8 - 255.73) = \mathbf{2467.3 \text{ kJ/min}}$$

$$\text{C.O.P.} = \frac{\text{Ref. effect}}{\text{Work}} = \frac{(h_2 - h_1)}{(h_3 - h_2)} = \frac{143.05}{(423.8 - 398.78)} = \mathbf{5.717}$$

For R-12: $h_2 = 187.5$ kJ/kg

$$h_{f4} = h_1 = 74.6 \text{ kJ/kg}$$

Now, $s_2 = s_3$

or, $$0.6996 = 0.6825 + 0.776 \ln\left(\frac{t_3 + 273}{40 + 273}\right)$$

$$\therefore \quad t_3 = 46.97°\text{C}$$

$$h_3 = 203.2 + 0.776\,(46.97 - 40) = 208.6 \text{ kJ/kg}$$

$$\text{Ref. effect/kg} = h_2 - h_1 = 187.5 - 74.6 = 112.9 \text{ kJ/kg}$$

$$\textit{Mass flow rate, } m = \frac{14000 \times 9}{112.9 \times 60} = 18.6 \text{ kg/min}$$

$$\textit{Compressor work} = \frac{m(h_3 - h_2)}{60} = \frac{18.6(208.6 - 187.5)}{60} = \mathbf{6.54 \text{ kW}}$$

$$\textit{Condenser heat rejection} = m\,(h_3 - h_{f4}) = 18.6\,(208.6 - 74.6) = \mathbf{2492.4 \text{ kJ/min}}$$

$$\text{C.O.P.} = \frac{h_2 - h_1}{h_3 - h_2} = \frac{112.9}{(208.6 - 187.5)} = \mathbf{5.35}$$

The above results are tabulated here for the purpose of comparison.

Refrigerant	*Mass flow rate (kg/min)*	*Compressor work (kW)*	*Compressor load (kJ/min)*	*C.O.P.*
R-134a	14.68	6.12	2467.3	5.717
R-12	18.6	6.54	2492.4	5.35

Example 3.26. *(a) Compare primary and secondary refrigerants. Why secondary refrigerants are used? Name at least two secondary refrigerants and corrosion inhibitors.*

(b) Discuss the principle of liquid suction heat exchanger in simple vapour compression refrigeration system with a suitable schematic diagram. Show the simple saturated cycle and the cycle using liquid suction heat exchanger on pressure-enthalpy diagram.

(c) An R-12 single-stage saturation cycle operates at temperatures of 35°C and –15°C for the condenser and evaporator respectively. If a liquid vapour regenerative heat exchanger is installed in the system, with the temperature of the vapour leaving the heat exchanger at 15°C, determine the C.O.P.

TABLES FOR R-12

t °C	h_f kJ/kg	h_g kJ/kg	s_g kJ/kg K	Superheated 20 K h kJ/kg	20 K s kJ/kg K	40 K h kJ/kg	40 K s kJ/kg K
35	69.5	201.5	0.6839	216.4	0.731	231.0	0.7741
– 15	—	181.0	0.7052	193.2	0.751	205.7	0.7942

(AMIE)

Solution: ***(a) Comparison of primary and secondary refrigerants.*** *Primary refrigerants directly remove heat from the object/space, to be cooled, by evaporation in a vapour compression system. Secondary refrigerants carry refrigeration from the plant room to the space where it is to be usefully applied.*

There are many applications which forbid the direct use of a refrigerant due to safety considerations. For example the toxic refrigerants cannot be used for the air-conditioning of residential buildings or cinema halls. For non-toxic refrigerants, the amount of refrigerant in a big cold storage or shopping complex would be so large that its cost would be prohibitive or it would cause excessive pressure drop in evaporator coils. Under such circumstances the cheaper grade suitable cooling media such as water, brine solutions etc. are used for indirect cooling. Thus a *comparatively smaller size evaporator containing much less refrigerant is used for chilling the secondary refrigerant for direct use.*

Names of secondary refrigerants. Sodium chloride or calcium chloride brines, ethylene glycol, water, etc.

Corrosion inhibitors. Glycols and sodium dichromate are corrosion inhibitors. Use of paints like lead oxide also inhibits corrosion.

(b) A liquid suction heat exchanger is used to sub-cool the liquid refrigerant from the condenser by exchanging heat with cold suction vapour. The subcooling would increase the refrigerating effect per kg of refrigerant. Also the suction vapour gets superheated, and it ensures that no liquid mist (droplets) should enter the compressor (thereby preventing any damage to the compressor valve). But at the same time the compressor work would increase. However, there may be some improvement in C.O.P. of the cycle.

A schematic arrangement is shown in Fig. 3.40.

The *p-h* diagram for simple saturated cycle is shown as 1-2-3-4-1 in Fig. 3.41, and that using liquid suction heat exchanger is shown as 1-2-3-4′-1 in Fig. 3.42.

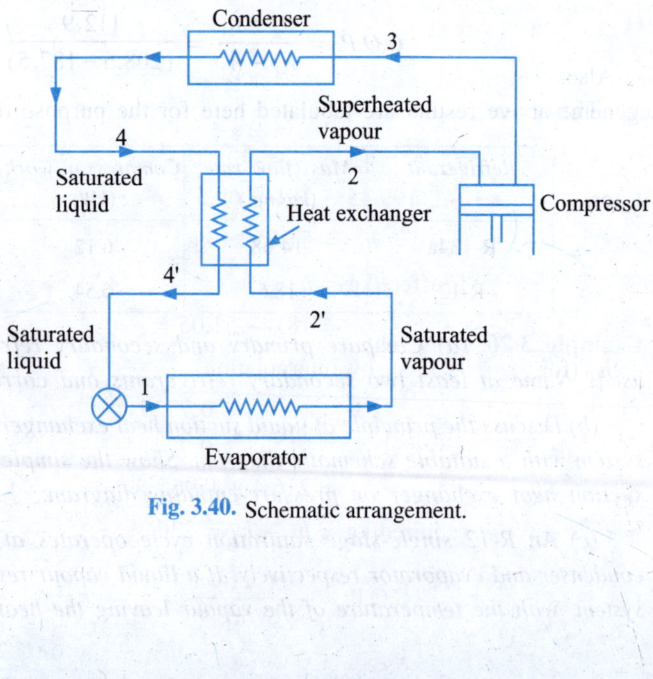

Fig. 3.40. Schematic arrangement.

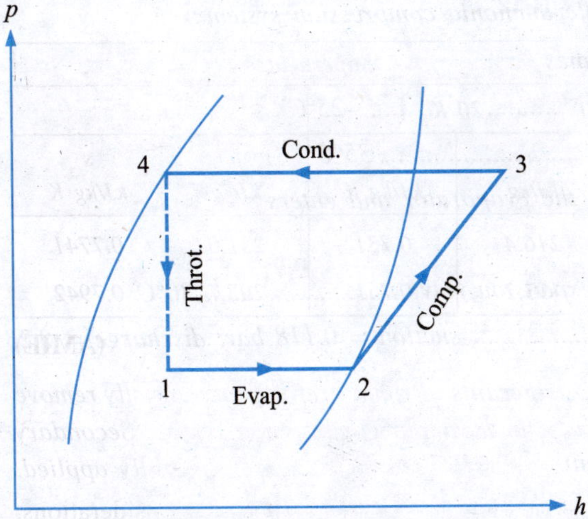

Fig. 3.41. Simple saturated cycle

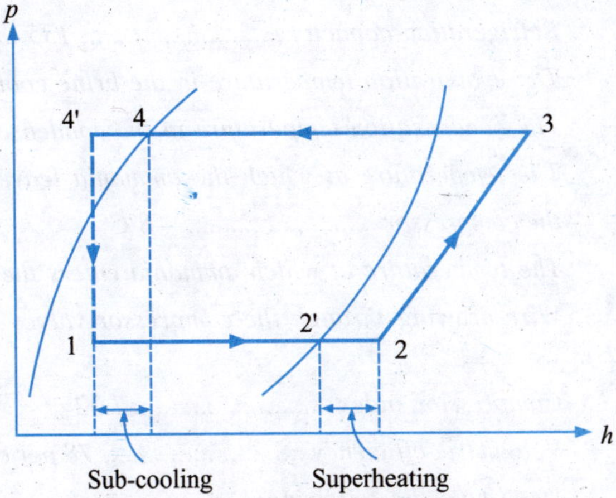

Fig. 3.42. Sub-cooling by using liquid suction heat exchanger

- *This type of heat exchanger is mostly used with capillary tube and hermetic compressor.*

(*c*) Fig. 3.43 shows *p-h* diagram of the cycle. Degree of superheat at suction stage 2

$$= 15 - (-15) = 30°C$$

By interpolation for superheated vapour,

$$h_2 = \frac{193.2 + 205.7}{2}$$

$$= 199.45 \text{ kJ/kg}$$

$$s_2 = \frac{0.751 + 0.7942}{2}$$

$$= 0.7726 \text{ kJ/kg K}$$

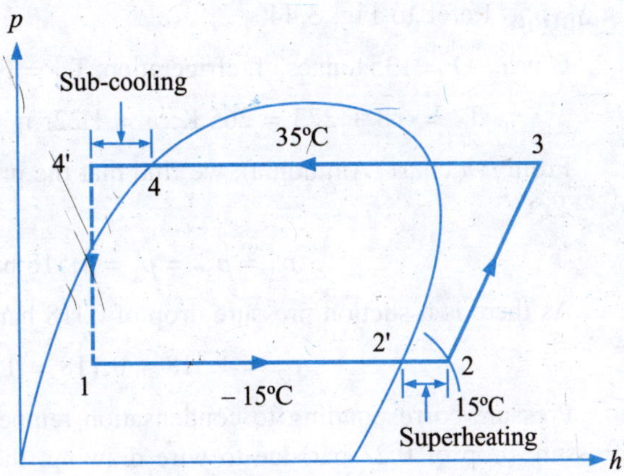

Fig. 3.43

Also, because of installation of a liquid vapour regenerative heat exchanger,

$$h_2 - h_{2'} = h_{f4} - h_{f4'}$$

or, $$199.45 - 181 = 69.5 - h_{f4'}$$

$$\therefore \quad h_{f4'} = 51.05 \text{ kJ/kg}$$

$$\text{Ref. effect/kg} = h_{2'} - h_1$$

$$= 181 - 51.05 = 129.95 \text{ kJ/kg}$$

h_3 (for $s_3 = 0.7726$), by interpolation ($\because \; s_2 = s_3 = 0.7726$)

$$= 216.4 + \frac{0.7726 - 0.731}{0.7741 - 0.731} \times (231 - 216.4)$$

$$= 216.4 + 0.965 \times (231 - 216.4) = 230.49 \text{ kJ/kg}$$

$$\text{Work required/kg} = h_3 - h_2 = 230.49 - 199.45 = 31.04 \text{ kJ/kg}$$

$$\therefore \quad \text{C.O.P.} = \frac{\text{Ref. effect}}{\text{Work}} = \frac{129.95}{31.04} = \mathbf{4.186} \quad \textbf{(Ans.)}$$

Example 3.27. *The following data refer to a simple ammonia compression system :*

Refrigeration capacity .... 135 tonnes

The evaporation temperature in the brine cooler .... –25°C

The condensation temperature in the condenser .... 35°C

The temperature at which the ammonia leaves the evaporator and enters the compressor .... –8°C

The temperature at which ammonia enters the expansion valve .... 30°C

Wire drawing through the compressor valves .... suction = 0.118 bar; discharge = 0.23 bar

Compression index 1.22

Volumetric efficiency .... 78 per cent

Determine the following:

(i) *Power required to drive the compressor,*

(ii) *Heat transferred to cylinder water jacket,*

(iii) *Piston displacement,*

(iv) *Heat transfer in condenser, and*

(v) *Co-efficient of performance.*

Solution: Refer to Fig. 3.44.

Given: Q = 135 tonnes of refrigeration; $T_{3''} = T_4 = 35 + 273 = 308$ K; $T_{2''} = T_1 = -25 + 273 = 248$ K; $T_2 = -8 + 273 = 265$ K; $n = 1.22$, $\eta_{vol.} = 0.78$.

From *p-h* chart (Ammonia), we find that the pressure corresponding to the evaporation temperature of –25°C,

$$p_2 = p_{2''} = p_1 = 1.518 \text{ bar}$$

As there is a suction pressure drop of 0.118 bar due to wire drawing, therefore pressure at point 2′,

$$p_{2'} = 1.518 - 0.118 = 1.4 \text{ bar}$$

Pressure corresponding to condensation temperature of 35°C = 13.5 bar. As there is a discharge pressure drop of 0.23 bar due to wire drawing, therefore, pressure at point 3′,

$$p_{3'} = 13.5 + 0.23 = 13.73 \text{ bar}$$

In the Fig. 3.44 the points 2 and 2′ represent the inlet and outlet of the suction valve respectively, the points 3′ and 3 represent the inlet and outlet of discharge valve respectively.

Again, from *p-h* diagram:

$$h_2 = h_{2'} = 1440 \text{ kJ/kg}; \quad v_{2'} = 0.8 \text{ m}^3\text{/kg}; \ T_{2'} = -9^\circ\text{C} = 264 \text{ K}$$

Let $v_{3'}$ = Specific volume of point 3′.

Then, for the compression process 2′-3′, we have:

$$p_{2'}(v_{2'})^n = (p_{3'}v_{3'})^n, \quad \text{or,} \quad v_{3'} = v_{2'}\left(\frac{p_{2'}}{p_{3'}}\right)^{1/n} = 0.8\left(\frac{1.4}{13.73}\right)^{1/1.22} = 0.1231 \text{ m}^3\text{/kg}$$

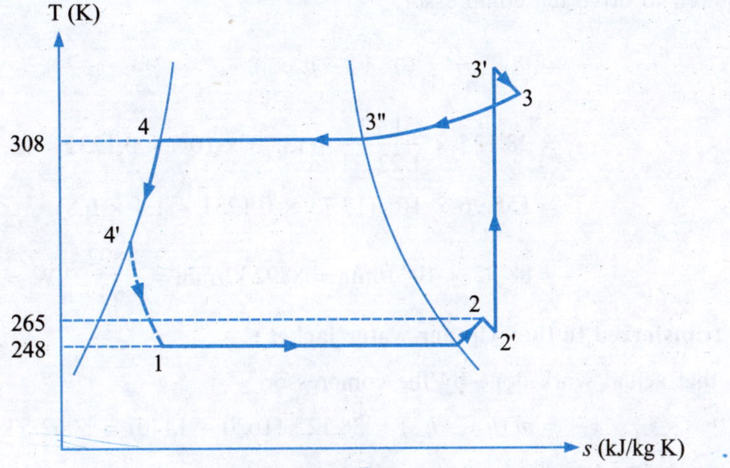

(*a*) T-*s* diagram.

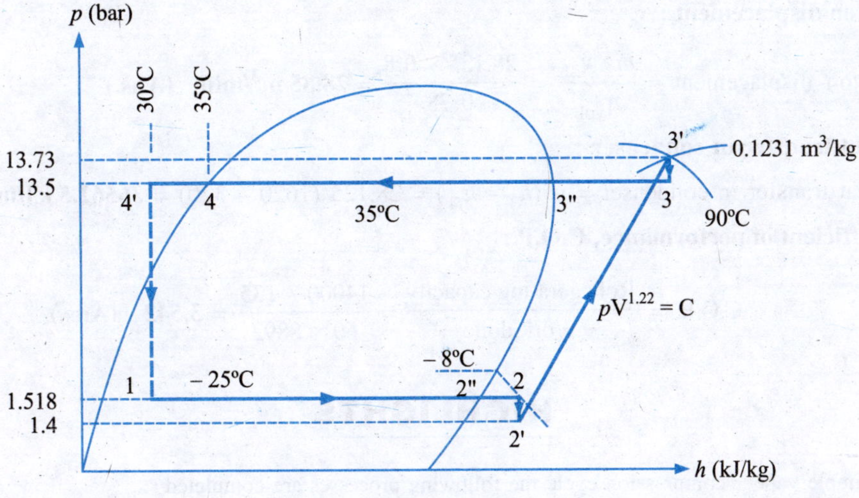

(*b*) *p-h* diagram.

Fig. 3.44

Now plot a point 3′ on the *p-h* diagram corresponding to $p_{3'}$ = 13.73 bar and $v_{3'}$ = 0.1231 m³/kg. From the *p-h* chart, we find:

$$h_3 = h_{3'} = 1620 \text{ kJ/kg}; \quad t_{3'} = 90°\text{C}$$

$$h_{f4'} = h_1 = 320 \text{ kJ/kg}$$

(*i*) **Power required to drive the compressor, P :**

Mass flow of refrigerant,

$$m = \frac{14000 \times 135}{(h_2 - h_1) \times 60} = \frac{14000 \times 135}{(1440 - 320) \times 60} = 28.125 \text{ kg/min}$$

Work required to drive the compressor,

$$= m \times \frac{n}{n-1}\,(p_{3'}v_{3'} - p_{2'}v_{2'})$$

$$= 28.125 \times \frac{1.22}{1.22-1}\,[(13.73 \times 10^5) \times 0.1231 - (1.4 \times 10^5) \times 0.8] \text{ J/min}$$

$$= 155.96 \times 10^5\,(13.73 \times 0.1231 - 1.4 \times 0.8)$$

$$= 88.92 \times 10^5 \text{ J/min} = 8892 \text{ kJ/min} = \frac{8892}{60} \text{ kW} = \mathbf{148.2\ kW} \quad \textbf{(Ans.)}$$

(*iii*) **Heat transferred to the cylinder water jacket :**

We know that actual work done by the compressor

$$= m\,(h_{3'} - h_{2'}) = 28.125\,(1620 - 1440) = 5062.5 \text{ kJ/min}$$

∴ Heat transferred to cylinder water jacket

$$= 8892 - 5062.5 = \mathbf{3829.5\ kJ/min} \quad \textbf{(Ans.)}$$

(*ii*) **Piston displacement :**

$$\text{Piston displacement} = \frac{m \times v_{2'}}{\eta_{vol.}} = \frac{28.125 \times 0.8}{0.78} = \mathbf{28.85\ m^3/min} \quad \textbf{(Ans.)}$$

(*iv*) **Heat transfer in condenser :**

$$\text{Heat transfer in condenser} = m\,(h_3 - h_{f4'}) = 28.125\,(1620 - 320) = \mathbf{36562.5\ kJ/min} \quad \textbf{(Ans.)}$$

(*v*) **Coefficient of performance, C.O.P.:**

$$\text{C.O.P.} = \frac{\text{Refrigerating capacity}}{\text{Work done}} = \frac{14000 \times 135}{60 \times 8892} = \mathbf{3.542} \quad \textbf{(Ans.)}$$

HIGHLIGHTS

1. In a simple vapour compression cycle the following processes are completed :
   (*i*) Compression (*ii*) Condensation
   (*iii*) Expansion, and (*iv*) Vaporisation
2. The various parts of a simple vapour compression cycle are: compressor, discharge line (or hot line), condenser, receiver tank, liquid line, expansion valve, evaporator and suction line.
3. If the vapour is not superheated after compression, the operation is called 'Wet compression' and if the vapour is superheated at the end of compression as 'Dry compression'. *Dry compression, in actual practice is always preferred as it gives higher volumetric efficiency and mechanical efficiency and there are less chances of compressor damage.*
4. *p-h* chart gives directly the changes in enthalpy and pressure during a process for thermodynamic analysis.
5. When suction pressure is decreased, the refrigerating effect is decreased and work required is increased. The net effect is to reduce the refrigerating capacity of the system and the C.O.P.
6. The overall effect of superheating is to give a low value of C.O.P.
7. 'Sub-cooling' results in increase of C.O.P. provided that no further energy has to be spent to obtain the extra cold coolant required.

8. The refrigerating system should always be designed to operate at the higher possible vaporising temperature and lowest possible condensing temperature, of course, keeping in view the requirement of the application.

9. '*Volumetric efficiency*' is defined as the ratio of the actual volume to gas drawn into the compressor (at evaporator temperature and pressure) on each stroke to the piston displacement. If the effect of clearance alone is considered, the resulting expression may be termed' clearance volumetric efficiency'. The expression used for grouping into one constant all the factors efficiency may be termed 'total volumetric efficiency'.

10. Clearance volumetric efficiency,

$$\eta_{cv} = 1 + C - C\left(\frac{p_d}{p_s}\right)^{1/n}$$

where, $$C = \frac{\text{Clearance volume}}{\text{Swept volume}}$$

p_d = Discharge pressure, and p_s = suction pressure.

11. Total volumetric efficiency,

$$\eta_{tv} = \left[1 + C - C\left(\frac{p_d}{p_s}\right)^{1/n}\right] \times \frac{p_c}{p_s} \times \frac{T_s}{T_c}$$

where subscript '*c*' refers to compressor cylinder and '*s*' refers to the evaporator on the suction line just adjacent to the compressor.

12. Methods of improving simple saturation cycle:

(i) *By using a flash chamber.*

(ii) *By sub-cooling the liquid refrigerant by using vapour refrigerant.*

(iii) *By sub-cooling the liquid refrigerant by external cooling source.*

(iv) *By sub-cooling the liquid refrigerant with liquid refrigerant.*

(v) *By using the accumulator or pre-cooler.*

OBJECTIVE TYPE QUESTIONS

A. Choose the correct Answer :

1. The highest temperature during the cycle, in a vapour compression refrigeration system, occurs after

(*a*) compression (*b*) condensation

(*c*) expansion (*c*) evaporation

2. Heat is rejected by the refrigerant, during vapour compression refrigeration cycle in

(*a*) condenser (*b*) evaporator

(*c*) throttle valve (*c*) compressor.

3. The process of undercooling is generally brought about by

(*a*) circulating more quantity of cooling water through the condenser

(*b*) using water colder than the main circulating water

(*c*) employing a heat exchanger

(*d*) any one of the above

4. Increase in condenser temperature keeping evaporator temperature fixed, in vapour compression refrigerating cycle results in

(*a*) decrease in refrigeration capacity.

(*b*) increase in power consumption

(c) decrease in volumetric efficiency of compressor
(d) all of the above.

5. The sub-cooling is a process of cooling the refrigerant in vapour compression refrigeration system
(a) before compression (b) after compression
(c) before throttling (d) after throttling

6. In a Simple saturated vapour compression cycle, the refrigeration is in superheated condition
(a) before entering compressor (b) before entering condenser
(c) before entering throttle valve (d) before entering evaporator.

7. In a vapour compression refrigeration system, the effect of superheating the vapour before suction to compression
(a) increases the work of compression
(b) increases the heat rejection in the condenser
(c) may increase or decrease C.O.P. depending upon the refrigerant used
(d) all of the above.

8. Vapour compression cycle using R-12 gives maximum C.O.P. when
(a) suction state to compressor is in wet region
(b) suction state to compressor is in superheat region
(c) suction state to compressor is dry saturated
(d) suction state to compressor is any.

9. A domestic refrigerator capacity may be approximately
(a) 1 ton (b) 0.1 ton
(b) 1.5 ton (d) 10 ton.

10. The net effect of superheating suction in vapour compression refrigeration systems using R-717 (Ammonia) theoretically results in
(a) reduction in refrigeration capacity of the system
(b) increase in refrigeration capacity of system
(c) keeping refrigeration capacity same of the system
(d) none of the above.

11. Vapour compression refrigeration system is used as Heat pump to heat a room and also direct radiation electrical heater is used to heat similar room. For the same heating effect the former arrangement uses
(a) more electrical energy than the latter (b) less electrical energy than the latter
(c) same electrical energy as the latter (d) more or less as it cannot be predicted.

12. The effect of superheating the vapour before suction to compression in vapour compression refrigeration system
(a) increases the work of compression
(b) increases the heat rejection in condenser
(c) may increase C.O.P. or decrease C.O.P. depending upon the refrigerant used
(d) all of the above.

13. In vapour compression refrigeration system, if expansion cylinder is used in place of throttle valve, the C.O.P. will
(a) increase (b) decrease
(c) will remain same (d) not be predictable unless refrigerant used is known.

14. In a domestic vapour compression refrigerator, the refrigerant commonly used is
(a) CO_2 (b) Ammonia
(c) Freon-12 (c) all of these.

15. In actual large vapour compression refrigeration installations
    (*a*) all the pipes are lagged and insulated
    (*b*) only pipes between evaporator and compressor are lagged and insulated
    (*c*) only pipes between compressor and condenser are lagged and insulated
    (*d*) none of the pipes are lagged and insulated.

16. In a vapour compression refrigeration system, the lowest temperature during the cycle occurs after
    (*a*) compression (*b*) condensation
    (*c*) expansion (*d*) evaporation

17. A Simple saturated refrigeration cycle has the following state points (*i*) Enthalpy after compression = 425kJ/kg. Enthalpy after throttling = 125 kJ/kg, enthalpy before compression = 375 kJ/kg. The C.O.P. refrigeration is given by
    (*a*) 5 (*b*) 3.5
    (*c*) 6 (*d*) not possible to find with this data.

18. In vapour compression refrigeration cycle for ice making the condensing temperature for better COP is desired to be
    (*a*) near critical temperature of refrigerant
    (*b*) above critical temperature of refrigerant
    (*c*) much below critical temperature of refrigerant
    (*d*) any as it is not affected by condensing temperature.

19. The effect of superheating suction vapours in vapour compression refrigeration systems using R-12 theoretically results in
    (*a*) reduction in refrigeration capacity of the system
    (*b*) increase in refrigeration capacity of system
    (*c*) keeping refrigeration capacity same of the system
    (*d*) none of the above.

20. A domestic window type air conditioner capacity may be approximately
    (*a*) 1 ton (*b*) 0.1 ton
    (*c*) 5 ton (*d*) 10 ton.

21. The order in which main components of vapour compression refrigeration system are used is
    (*a*) compressor-evaporator-condenser-throttle valve
    (*b*) compressor-condenser-evaporator-throttle valve
    (*c*) compressor-condenser-throttle valve-evaporator
    (*d*) compressor-throttle valve-evaporator-condenser.

22. Liquid-suction heat exchanger is used in vapour compression refrigeration system to
    (*a*) improve COP of cycle (*b*) avoid useless superheating in pipes
    (*c*) to reduce heat rejection in condenser (*d*) none of the above.

23. Use of liquid-suction heat exchanger or liquid-vapour regenerative heat exchanger in vapour compression system in justified because
    (*a*) COP of the cycle improves irrespective of any refrigerant used
    (*b*) horse power per ton decreases irrespective of any refrigerant used
    (*c*) suction volume per ton decreases irrespective of any refrigerant used
    (*d*) superheating in liquid-suction exchanger is preferable to superheating in evaporator itself.

24. Wet compression vapour compression refrigeration cycle means
    (*a*) vapour compression takes place in wet region
    (*b*) vapour compression in dry region, but evaporation in wet region
    (*c*) vapour compression in wet region, but leaves in superheated region
    (*d*) none of the above.

25. With reciprocating compressor in vapour compression refrigeration system, wet compressor is not desirable because
    (*a*) liquid trapped up in the head of cylinder may damage the compressor valves
    (*b*) COP of the cycle decreases
    (*c*) volumetric efficiency of compressor decreases
    (*d*) mass flow rate per ton of refrigerant increases.
26. The COP of vapour compression refrigerating cycle using liquid suction heat exchanger as compared to simple saturated cycle
    (*a*) is higher definitely
    (*b*) may be higher of lower depending upon refrigerant used
    (*c*) remains same (*d*) is unpredictable
27. In vapour compression refrigeration system, with simple saturated cycle, the isentropic expansion of carnot cycle is replaced by throttling because
    (*a*) positive work of expander is very small to justify cost of expander
    (*b*) throttling device gives better C.O.P.
    (*c*) throttling device is easy to operate
    (*d*) throttling device increases refrigeration capacity.
28. In a Simple saturated vapour compression refrigeration cycle the following results were obtained. Heat rejected in condenser = 160 kJ/kg, Compressor work = 32 kJ/kg. The C.O.P. refrigeration is
    (*a*) 5 (*b*) 4
    (*c*) 6 (*d*) not possible to find with this date.
29. In vapour compression refrigeration system the heat is absorbed from the environments to be cooled by
    (*a*) evaporator (*b*) condenser
    (*c*) compressor (*d*) throttle valve.
30. The effect of liquid sub-cooling in vapour compression refrigeration system is
    (*a*) to reduce the rate of flow of refrigerant per tonne of refrigeration
    (*b*) to reduce the volume of vapour handled by compressor per tonne of refrigeration
    (*c*) to reduce power per tonne of refrigeration
    (*d*) all of the above.
31. C.O.P. of domestic air conditioner as compared to that of domestic refrigerator will be
    (*a*) lower (*b*) higher
    (*c*) same (*d*) un-predictable.
32. The oil separator is incorporated in vapour compression refrigeration system
    (*a*) between evaporator and compressor (*b*) between compressor and condenser
    (*c*) between condenser and throttle valve (*d*) between throttle valve and evaporator.

ANSWERS

1. (*a*)	2. (*a*)	3. (*d*)	4. (*d*)	5. (*c*)	6. (*b*)	7. (*d*)	8. (*b*)
9. (*b*)	10. (*a*)	11. (*b*)	12. (*d*)	13. (*a*)	14. (*c*)	15. (*b*)	16. (*d*)
17. (*a*)	18. (*c*)	19. (*b*)	20. (*a*)	21. (*c*)	22. (*b*)	23. (*d*)	24. (*a*)
25. (*a*)	26. (*b*)	27. (*a*)	28. (*b*)	29. (*a*)	30. (*d*)	31. (*b*)	32. (*b*)

B. Fill in the Blanks:

1. Out of all refrigeration systems, the _______ system is the most important system from the stand point of commercial domestic utility.
2. The function of a _______ is to remove the vapour from the evaporator and to raise its temperature and pressure to a point such that it (vapour) can be condensed with normally available condensing media.
3. The function of a _______ is to provide a heat transfer surface through which heat passes from the hot refrigerant vapour to the condensing medium.
4. The function of _______ is to meter the proper amount of refrigerant to the evaporator and to reduce the pressure of liquid entering the evaporator so that liquid will vaporise in evaporator at the desired low temperature.
5. _______ provides a heat transfer surface through which heat can pass from the refrigerated space or product into the vaporising refrigerant.
6. If the vapour is not superheated after compression, the operation is called _______.
7. If the vapour is superheated at the end of compression, the operation is called _______.
8. When the suction pressure decreases the refrigerating effect and C.O.P. are _______.
9. _______ results in increase of C.O.P. provided that no further energy has to be spent to obtain the extra cold coolant required.
10. _______ efficiency is defined as the ratio of actual volume of gas drawn into the compressor (at evaporator temperature and pressure) on each stroke to the piston displacement.

ANSWERS

1. vapour compression,	2. compressor;	3. condenser;
4. expansion valve;	5. Evaporator;	6. wet compression;
7. dry compression;	8. reduced;	9. Sub-cooling;
10. Volumetric		

THEORETICAL QUESTIONS

1. Describe a simple vapour compression cycle giving clearly its flow diagram.
2. State merits and demerits of 'Vapour compression system' over 'Air-refrigeration system'.
3. State the functions of the following parts of a simple vapour compression system:
   (*i*) Compressor, (*ii*) Condenser,
   (*iii*) Expansion valve, and (*iv*) Evaporator.
4. Show the vapour compression cycle on 'Temperature-Entropy' (T-*s*) diagram for the following cases:
   (*i*) When the vapour is dry and saturated at the end of compression.
   (*ii*) When the vapour is superheated after compression.
   (*iii*) When the vapour is wet after compression.
5. What is the difference between 'Wet compression' 'and Dry compression'?
6. Write a short noted on 'Pressure Enthalpy (*p-h*) chart'.

7. Show the simple vapour compression cycle on a *p-h* chart.
8. Discuss the effect of the following on the performance of a vapour compression system:
   (*i*) Effect of suction pressure
   (*ii*) Effect of delivery pressure
   (*iii*) Effect of superheating
   (*iv*) Effect of sub-cooling of liquid
   (*v*) Effect of suction temperature and condenser temperature.
9. Show, with the help of diagrams the difference between theoretical and actual vapour compression cycle.
10. Define the terms 'Volumetric efficiency' and 'Clearance volumetric efficiency'.
11. Derive an expression for 'Clearance volumetric efficiency'.
12. Explain briefly the term 'Total volumetric efficiency'.
13. Why in practice a throttle valve is used in vapour compression refrigerator rather than an expansion cylinder to reduce pressure between the condenser and the evaporator?

UNSOLVED EXAMPLES

1. The temperature in a refrigerator coil is 267 K and that in the condenser coil is 295K. Assuming that the machine operates on the reversed cannot cycle, calculate :
   (*i*) $\text{C.O.P.}_{\text{ref}}$.
   (*ii*) The refrigerating effect per kW of input work.
   (*iii*) The heat rejected to the condenser. **[Ans.** (i) 9.54, (ii) 9.54 kW, (iii) 10.54 kW**]**
2. A refrigerator using Freon-12 operates between saturation temperature of –10ºC and 60ºC, at which temperature the latent heats are 156.32 kJ/kg and 113.52 kJ/kg respectively. The refrigerant is dry saturated at entry to the compressor and the liquid is not under-cooled in the condenser. The specific heat of liquid freon is 0.970 kJ/kg K and that of the superheated freon vapour is 0.865 kJ/kg K. The vapour is compressed isentropically in the compressor. Using no other information than that given, calculate the temperature at the compressor delivery, and the refrigerating effect per kg of freon. **[Ans.** 69.6ºC; 88.42 kJ/kg**]**
3. A heat pump using ammonia as the refrigerant operates between saturation temperatures of 6ºC and 36ºC. The refrigerant is compressed isentropically from dry saturation and there is 6 K of under-cooling in the condenser. Calculate:
   (*i*) $\text{C.O.P.}_{\text{(heat pump)}}$.
   (*ii*) The mass flow of refrigerant, and
   (*iii*) The heat available per kilowatt input. **[Ans.** 8.8; 25.06 kg/h; 8.8 kW**]**
4. An ammonia vapour compression refrigerator has a single-stage, single-acting reciprocating compressor which has a bore of 127 mm, a stroke of 152 mm and a speed of 240 r.p.m. The pressure in the evaporator is 1.588 bar and that in the condenser is 13.89 bar. The volumetric efficiency of the compressor is 80% and its mechanical efficiency is 90%. The vapour is dry saturated on leaving the evaporator and the liquid leaves the condenser at 32ºC. Calculate the mass flow rate of refrigerant, the refrigerant effect, and the power ideally required to drive the compressor. **[Ans.** 0.502 kg/min; 9.04 kW; 2.73 kW**]**
5. An ammonia refrigerator operates between evaporating and condensing temperatures of – 16ºC and 50ºC respectively. The vapour is dry saturated at the compressor, the compression process is isentropic and there is no under-cooling of the condensate. Calculate:
   (*i*) The refrigerating effect per kg,
   (*ii*) The mass flow and power input per kW of refrigeration, and
   (*iii*) The $\text{C.O.P.}_{\text{ref}}$. **[Ans.** 1003.4 kJ/kg; 3.59 kg/h; 0.338 kW; 2.96**]**

6. 30 tonnes of ice from and at 0ºC is produced in a day of 24 hours by an ammonia refrigerator. The temperature range in the compressor is from 298 K to 258 K. The vapour is dry saturated at the end of compression and expansion valve is used. Assume a co-efficient of performance of 60% of the theoretical and calculate the power in kW required to drive the compressor. Latent heat to ice is 334.72 kJ/kg.

Temperature (K)	*Enthaply (kJ/kg)*		*Entropy of liquid*	*Entropy of Vapour*
	Liquid	*Vapour*	*(kJ/kg K)*	*(kJ/kg K)*
298	100.04	1319.22	0.3473	4.4852
258	– 54.56	1304.99	– 2.1338	5.0585

[Ans. 21.59 kW]

7. A refrigerant plant works between temperature limits of –5ºC (the evaporator) and 25ºC (in the condenser). The working fluid ammonia has a dryness fraction of 0.6 at entry of the compressor. If the machine has a relative efficiency of 50% calculate the amount of ice formed during a period of 24 hours. The ice is to be formed at 0ºC from water at 20ºC and 6 kg of ammonia is circulated per minute. Specific heat of water is 4.187 kJ/kg and latent heat of ice is 335 kJ/kg.

Properties of ammonia (datum –40ºC):

Temperature (K)	*Liquid heat (kJ/kg)*	*Latent heat (kJ/kg)*	*Entropy of liquid (kJ/kgºC)*
298	298.9	1167.1	1.124
268	158.2	1280.8	0.630

[Ans. 1640.5 kg]

8. A food storage locker requires a refrigerant of 2500 kJ/min capacity at an evaporator temperature of –10ºC and condenser temperature of 30ºC. The refrigerant used is freon-12 and sub-cooled by 5ºC before entering the expansion valve and vapour is superheated by 6ºC before leaving the evaporator coil. The compression of refrigerant is reversible adiabatic. The refrigeration compressor is two-cylinder, single-acting with stroke equal to 1.3 times the bore and operates at 975 r.p.m. Determine (using thermodynamic tables of properties for Freon-12).

(*i*) Refrigerating effect per kg.

(*ii*) Mass of refrigerant to be circulated per minute,

(*iii*) Theoretical piston displacement per minute,

(*iv*) Theoretical power required to run the compressor, in kW,

(*v*) Heat removed through the condenser per minute, and

(*vi*) Theoretical bore and stroke of compressor.

Properties of Freon-12:

Saturation temperature (ºC)	*Absolute pressure (bar)*	*Specific volume of vapour, (m^3/kg)*	*Enthalpy*		*Entropy*	
			Liquid (kJ/kg)	*Vapour (kJ/kg)*	*Liquid (kJ/kg K)*	*vapour (kJ/kg K)*
– 10ºC	2.19	0.0767	26.9	183.2	0.1080	0.7020
30ºC	7.45	0.0235	64.6	199.6	0.2399	0.6854

Take : Liquid specific heat = 1.235 kJ/kg K

Vapour specific heat = 0.735 kJ/kg K

[**Ans.** (*i*) 129.17 kJ/kg; (*ii*) 19.355 kg/min; (*iii*) 1.518 m^3/min; (*iv*) 7.2 kW; (*v*) 2931 kJ/min; (*vi*) 91 mm, 118 mm]

9. A vapour compression refrigerator uses methyl chloride and works in the pressure range of 11.9 bar and 5.67 bar. At the beginning of the compression, the refrigerant is 0.96 dry and at the end of isentropic compression, it has temperature of 55ºC. The refrigerant liquid leaving the condenser is saturated. If the mass flow rate of a refrigerant is 1.8 kg/min. Determine :

(*i*) Co-efficient of performance,

(*ii*) The rise in temperature of condenser cooling water if the water flow rate is 16 kg/min., and

(*iii*) The ice produced in the evaporator in kg/hour from water at 15ºC and ice at 0ºC.

Properties of methylchloride:

Saturation temperature	*Pressure (bar)*	*Enthalpy (kJ/kg)*		*Entropy (kJ/kg K)*	
		h_f	h_g	s_f	s_g
– 20	11.9	30.1	455.2	0.124	1.803
25	5.67	100.5	476.8	0.379	1.642

Take: Specific enthaply of fusion of ice = 336 kJ/kg

Specific heat of water = 4.187 kJ/kg [**Ans.** 4.97, 10.9 ºC, 91.3 kg]

10. A vapour compression refrigerator circulates 4.5 kg of NH_3 per hour. Condensation takes place at 30ºC and evaporation at –15ºC. There is no undercooling of the refrigerant. The temperature after isentropic compression is at 75ºC and specific heat of superheated vapour is 2.82 kJ/kgK. Determine:

(*i*) Co-efficient of performance,

(*ii*) Ice product in kg per hour in the evaporator from water at 20ºC and ice at 0ºC. Take enthalpy of fusion of ice = 336 kJ/kg and specific heat of water = 4.187 kJ/kg.

(*iii*) The effective swept volume of the compressor in m^3/min.

Properties of ammonia:

Saturated temperature	*Enthalpy (kJ/kg)*		*Entropy (kJ/kg K)*		*Volume (m^3/kg)*	
(*K*)	h_f	h_g	s_f	s_g	v_f	v_g
303	323.1	1469	1.204	4.984	0.00168	0.111
258	112.3	1426	0.457	5.549	0.00152	0.509

[**Ans.** 4.95; 682 kg/h; 2.2 m^3/min]

❑❑❑

Vapour Compression Refrigeration Systems with Multiple Evaporators and Compressors

Chapter 4

4.1 INTRODUCTION

4.1.1 Multistage or Compound Compression

- The simple vapour compression system is two-pressure system. Systems with more than two pressures may arise either due to *multistage or compound compression* to *minimise work*, or due to feeding of a refrigerant to a *multi-evaporator system.*
  - — ***Cascade systems*** which employ more than one refrigerant also have multipressures although each refrigerant circuit operates on a two-pressure system only.
- *For a given condensation temperature, the lower the evaporator temperature, the higher becomes the compressor pressure ratio.* For a reciprocating compressor, *a high pressure ratio across a single stage means low volumetric efficiency.* Also, with dry compression the higher pressure ratio results in high compressor discharge temperature which may damage the refrigerant. (High head temperatures can cause oil breakdown at the valve plates, hardening of neoprene O-rings used on valve plates, cylinder sleeves etc., thereby rendering the sealing ineffective.) *To reduce the work of compression and improve the C.O.P., multistage compression with intercooling may be adopted.* Since the intercooler temperature may be below the temperature of available cooling water used for the condenser, the refrigerant itself may be used as the intercooling medium.
  - — For low-temperature applications where the compression ratios are high, two or more multistage compression system is adopted. Most of the compressor manufacturers *specify a maximum compression ratio of nine for single-stage operation. Generally where the compression ratio exceeds seven multistage system is adopted.*

4.1.2 Multi-evaporator Systems

Invariably a situation arises when varied types of cooling loads are connected to the same refrigeration system. Each load may require an evaporator working at a different refrigeration temperature. The whole system may, therefore, be operated either at a sunction pressure equal to the lowest evaporator pressure leading to a single-compressor system, or at various suction pressures with individual compressors for each evaporator leading to a multi-compressor system.

> **Note:** A higher C.O.P. indicates the better performance of a refrigeration system. C.O.P. of the system can be *increased* either by decreasing the work of compression or increasing the refrigerating effect or *both*. It is possible to *reduce the compression work* to considerable extent by compressing the refrigerant *very close to the saturation line*. This can be accomplished by "*compressing the refrigerant in more stages with intermediate intercoolers*". *The refrigerating effect can be increased* by maintaining the condition of the refrigerant in *more liquid stage at the entrance of evaporator* which is achieved by expanding the refrigerant very close to the liquid line. *The expansion can be brought close to the liquid line by "sub-cooling" the refrigerant and by "removing the flashed vapour"*. By incorporating the flash-chamber in the working cycle, the *evaporator size is reduced as unwanted vapour formed is removed* before the liquid refrigerant enters the evaporator.

4.2 MULTISTAGE OR COMPOUND COMPRESSION

4.2.1 Advantages of Multistage or Compound Compression with Intercoolers

The following are the main advantages of multistage or compound compression *over single stage compression*:

1. Reduced work done per kg of refrigerant, for the same delivery pressure.
2. Improved volumetric efficiency, for the given pressure ratio.
3. The leakage loss considerably reduced.
4. Small size flywheel needed (Torque being more uniform).
5. Reduced compressor cost.
6. Effective lubrication obtainable (due to lower temperature range).
7. Possible to adjust the sizes of two cylinders (i.e., high pressure and low pressure) to suit the volume and pressure of the refrigerant.

4.2.2 Types of Multistage Vapour Compression with Intercooler(s)

Following cases will be discussed:

1. Two stage compression with intercooling and sub-cooling by external cooling source.
2. The compound compression with water intercooler and single expansion valve.
3. Compound compression with flash intercooling with single expansion valve along the flow line to the evaporator.
4. Single evaporator with multi-expansion valves and flash chambers.
5. Compound compression with multiple expansion valves in series with flash chambers but no intercooling.
6. Compound compression with multiple expansion valves and flash intercooling.

The mentioned cases are discussed below:

4.2.2.1 *Two-stage compression with intercooling and sub-cooling by external cooling source*

The schematic arrangement is shown in Fig. 4.1 (*a*) and its corresponding *p-h* diagram is shown in Fig. 4.1 (*b*).

The mass of the refrigerant passing through the evaporator is given by,

$$m_e = \frac{14000\,Q}{60(h_2 - h_1)} \text{ kg/min} \qquad ...(4.1)$$

where, Q is the load on the evaporator in tonnes of refrigeration.

Mass of the vapour formed in the flash chamber (m_{vf}) is given by,

$$m_{vf} = \frac{m_e}{1}\left(\frac{x_9}{1 - x_9}\right)$$

The vapour refrigerant m_e coming out from compressor C_1 with the condition (3) is mixed with the vapour refrigerant m_{vf} coming out of the flash chamber with the condition (5) at the same pressure before entering into the intercooler. By considering the *energy balance*, the condition (3′) is given by:

$$(m_e + m_{vf})\,h_{3'} = m_e h_3 + m_v h_5$$

$$\therefore \quad h_{3'} = \frac{m_e h_3 + m_{vf} h_5}{(m_e + m_{vf})}$$

The vapour (m_e + m_{vf}) is cooled in the intercooler from condition 3′ to 4.

- *The cooling capacity of the intercooler*

$$= (m_e + m_{vf})\,(h_{3'} - h_4) \qquad ...(4.2)$$

- *Power required to drive the system,*

$$P = \frac{m_e (h_3 - h_2) + (m_e + m_{vf})(h_6 - h_4)}{60} \text{ kW} \qquad ...(4.3)$$

- *The cooling capacity of sub-cooler*

$$= (m_e + m_{vf})\,(h_7 - h_8) \qquad ...(4.4)$$

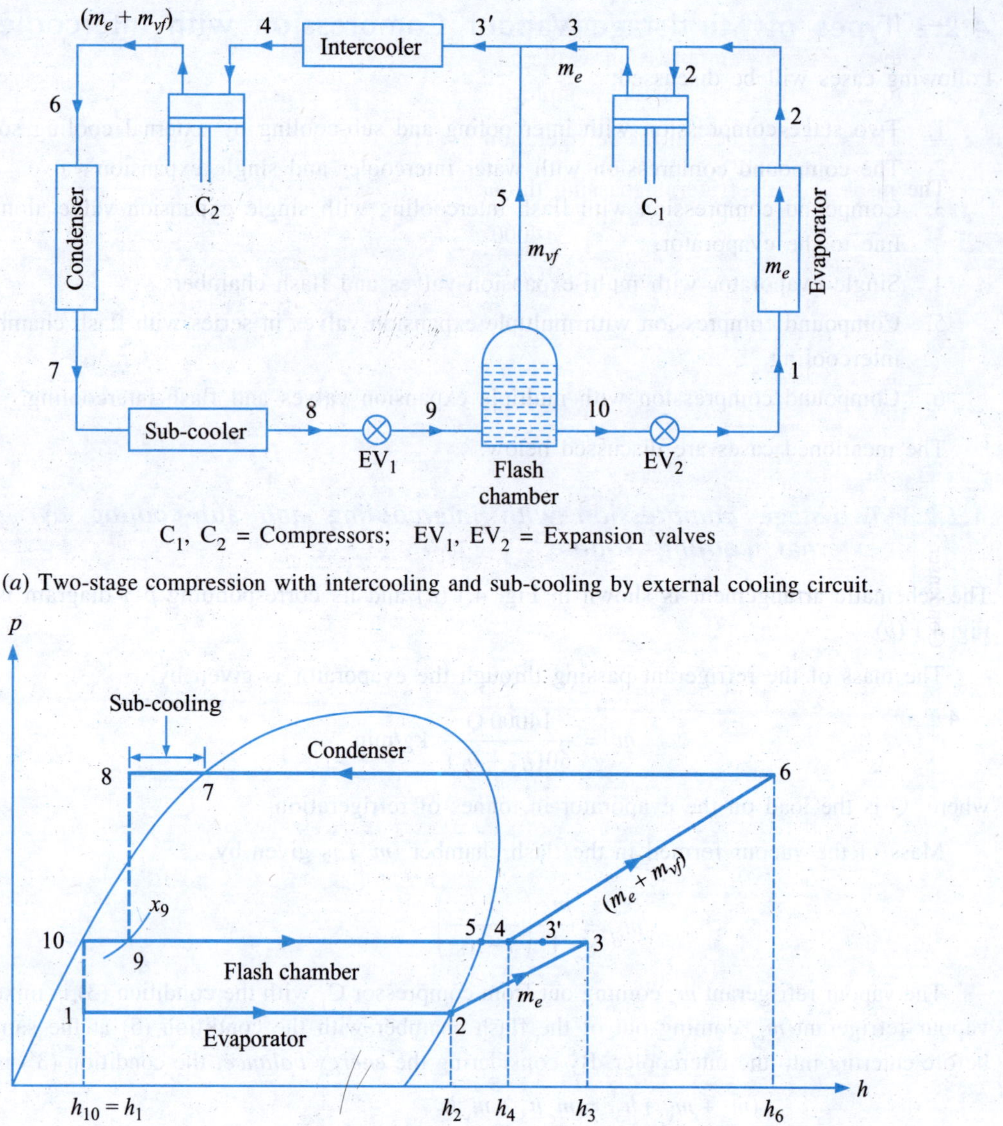

(*a*) Two-stage compression with intercooling and sub-cooling by external cooling circuit.

(*b*) *p-h* diagram.

Fig. 4.1

4.2.2.2 *The compound compression with water intercooler and single expansion valve*

Fig. 4.2 (*a*) and (*b*) shows the schematic arrangement and *p-h* diagram of the system respectively.

— The compression work can be reduced considerably by using intercooling between the stages. The *intermediate cooling leads to reduction in degree of superheat at the suction of each compressor and specific volume* (resulting in less capacity compressor requirement).

— Since the compression process is maintained near the saturation vapour line the total work done/kg will be reduced.

— It is not possible, in actual practice, to have the complete desuperheating of the vapour after every stage of compression as the cooling water temperature is not available sufficiently low to desuperheat the vapour completely.

— There is *no* effect on the refrigeration per kg of refrigerant.

The mass of the refrigerant passing through the *evaporator*,

$$m_e = \frac{14000\,Q}{60(h_2 - h_1)} \text{ kg/min.} \qquad ...(4.5)$$

where, Q is the load on the evaporator in tonnes of refrigeration.

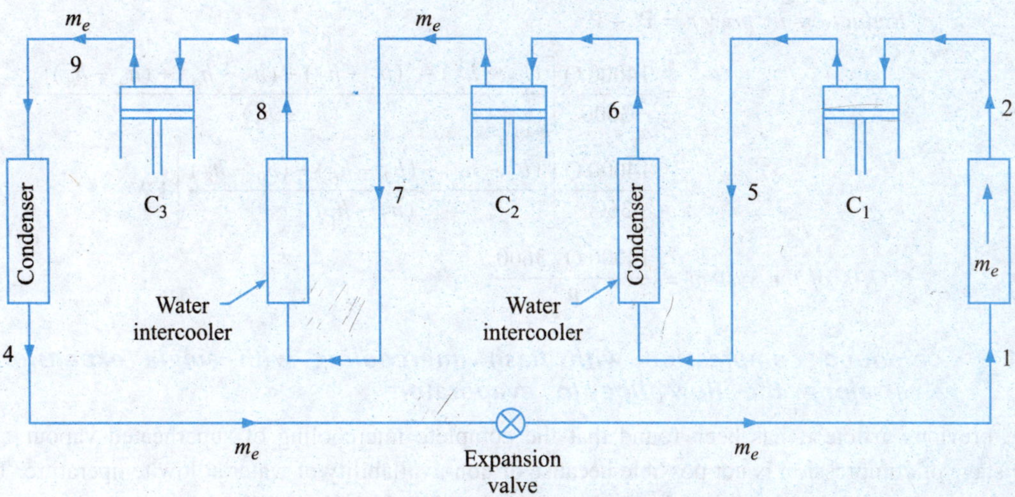

(*a*) Compound compression with water intercooler.

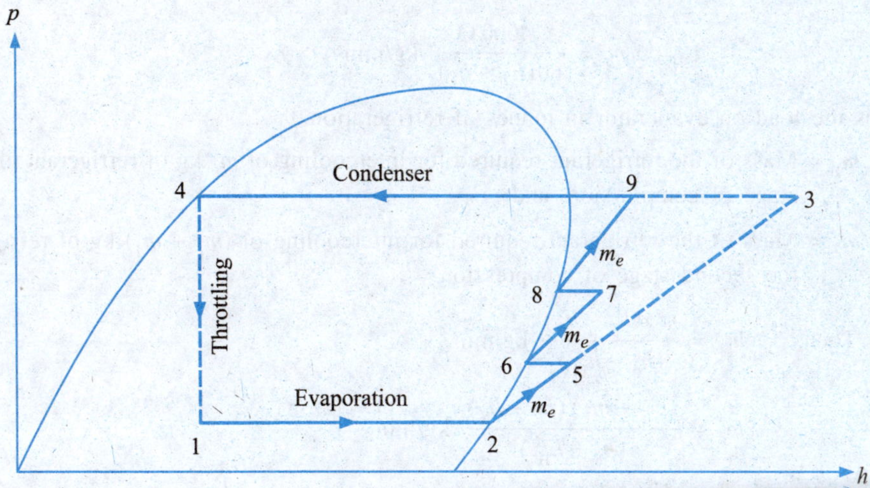

(*b*) *p-h* diagram.

Fig. 4.2

- *Power required for simple cycle,*

$$P' = \frac{m_e (h_3 - h_2)}{60} = \frac{14000\,Q}{3600}\left(\frac{h_3 - h_2}{h_2 - h_1}\right) \text{ kW}$$

- *Actual power required for the given cycle/system* (shown in Fig. 4.2),

$$P = \frac{m_e}{60}\,[(h_3 - h_2) + (h_7 - h_6) + (h_9 - h_8)]$$

$$= \frac{14000\,Q}{3600}\left[\frac{(h_5 - h_2) + (h_7 - h_6) + (h_9 - h_8)}{(h_2 - h_1)}\right] \text{ kW} \quad ...(4.6)$$

$\therefore$ *Reduction in power* = P′ – P

$$= \frac{14000\,Q}{3600}\left[\frac{(h_3 - h_2) - \{(h_5 - h_2) + (h_7 - h_6) + (h_9 - h_8)\}}{(h_2 - h_1)}\right]$$

$$= \frac{14000\,Q}{3600}\left[\frac{(h_3 - h_5) - (h_7 - h_6) - (h_9 - h_8)}{(h_2 - h_1)}\right] \text{ kW} \quad ...(4.7)$$

- *C.O.P. of the system* $= \dfrac{14000\,Q/3600}{P}$ ...(4.8)

4.2.2.3 Compound compression with flash intercooling with single expansion valve along the flow line to evaporator

In the previous article it has been found that the complete intercooling of superheated vapour after every stage of compression is not possible because of non-availability of water at low temperature. This difficulty can be removed by using the flash intercoolers in which the cooling is carried out by using the liquid refrigerant itself as shown in Fig. 4.3 (*a*); Fig. 4.3 (*b*) shows the *p-h* diagram of the cycle/system.

The mass of the refrigerant flowing through the *evaporator*,

$$m_e = \frac{14000\,Q}{60(h_2 - h_1)} \text{ kg/min}$$

where, Q is the load on evaporator in tonnes of refrigeration.

Let m_1 = Mass of the refrigerant required for intercooling of m_e kg of refrigerant after the first stage of compression, and

m_2 = Mass of the refrigerant required for intercooling of ($m_e + m_1$) kg of refrigerant after the second stage of compression.

Then, $$m_1 = \frac{m_e (h_5 - h_6)}{(h_6 - h_{11})} \text{ kg/min.} \quad ...(4.8)$$

and, $$m_2 = \frac{(m_e + m_1)(h_7 - h_8)}{(h_8 - h_{10})} \text{ kg/min.} \quad ...(4.8a)$$

- *Power required to drive the system,*

$$P = \frac{m_e (h_5 - h_2) + (m_e + m_1)(h_7 - h_6) + (m_e + m_1 + m_2)(h_9 - h_8)}{60} \text{ kW} \quad ...(4.9)$$

Power required for simple cycle,

$$P' = \frac{m_e\,(h_3 - h_2)}{60} \qquad ...(4.10)$$

Reduction in power = P′ – P ...(4.11)

- **C.O.P.** *of the system* $= \dfrac{14000\,Q/3600}{P}$ ...(4.12)

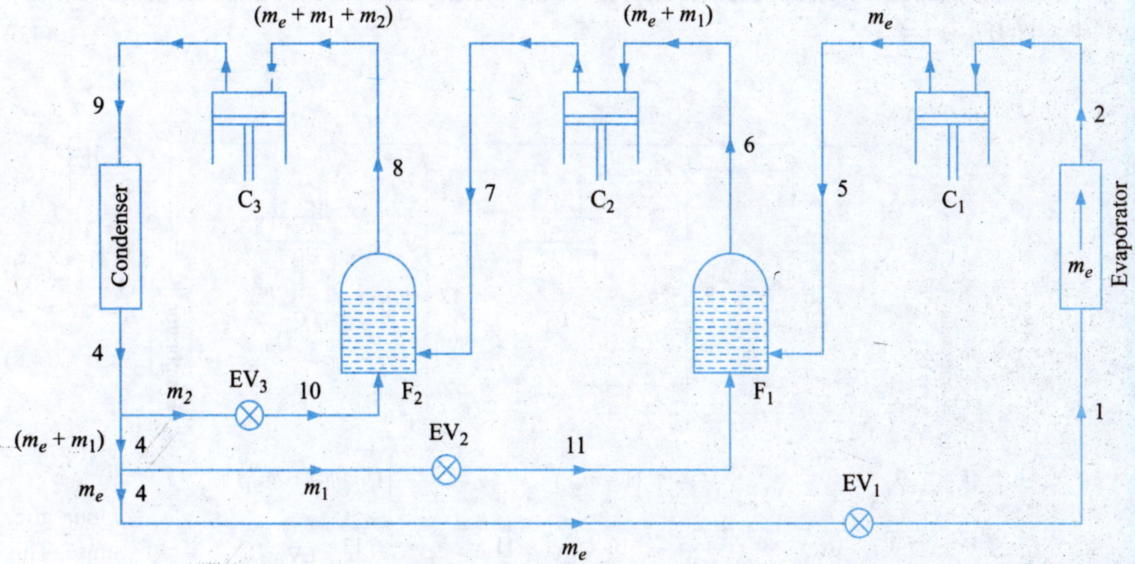

C_1, C_2, C_3 = Compressors; F_1, F_2 = Flash chambers; EV_1, EV_2, EV_3 = Expansion valves

(*a*) Compound compressors with flash intercooling with single expansion valve along the flow line to evaporator.

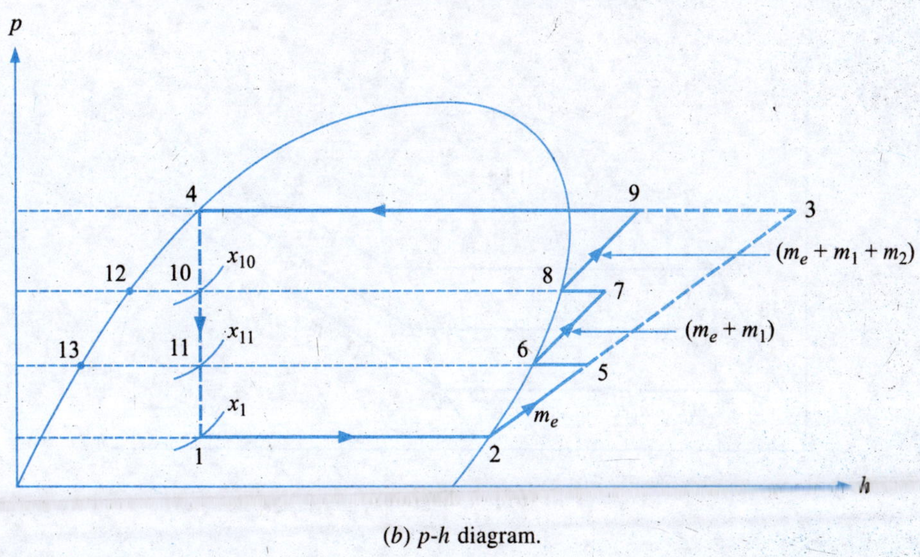

(*b*) *p-h* diagram.

Fig. 4.3

4.2.2.4 Single evaporator with multi-expansion valves and flash chamber

— *With the use of multi-expansion valves, the throttling operation is maintained near liquid line, and by using flash chambers the total work done per kg of refrigerant is reduced.*

— This cycle thermodynamically, will give *more C.O.P.* compared to the saturation cycle for the same operating pressure range, *because of total work done per kg is reduced without affecting the refrigerating effect:*

The schematic arrangement and *p-h* diagram of the cycle/system are shown in Fig. 4.4 (*a*) and (*b*), respectively.

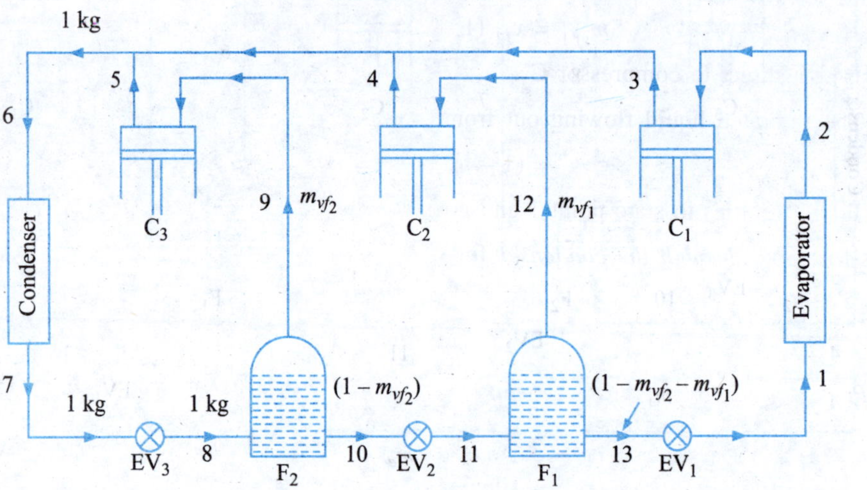

C_1, C_2, C_3 = Compressors; F_1, F_2 = Flash chambers; EV_1, EV_2, EV_3 = Expansion valves.

(*a*) Single evaporator with multi-expansion valves and flash chambers.

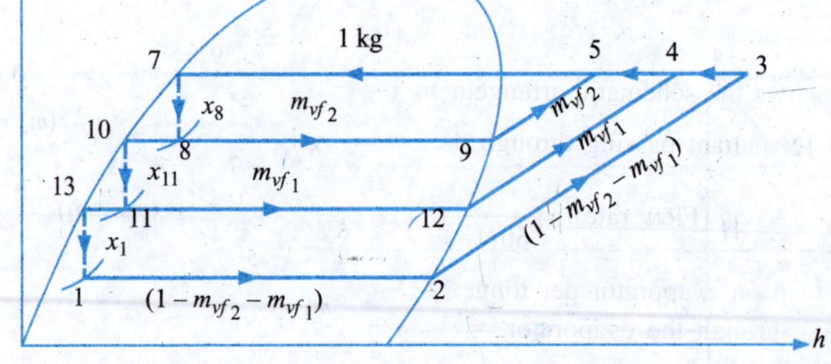

(*b*) *p-h* diagram.

Fig. 4.4

— For 1 kg of refrigerant flow through the condenser, the dry saturated vapour removed in the flash chamber F_2 at state point 9 is given by,

$$m_{vf2} = x_8 \times 1 = x_8 \text{ kg} \quad ...(i)$$

This is sent direct to the compressor C_3 and the saturated liquid flowing out of the flash chamber F_2 at the state point 10

$$= (1 - m_{vf2}) \text{ kg} = (1 - x_8) \text{ kg} \quad ...(ii)$$

The liquid at state point 10 throttles further to state point 11 into the flash chamber F_1.

— Again for 1 kg flow in the condenser, the dry saturated vapour removed in the flash chamber F_1 at stage point 11 in given by,

$$m_{vf1} = x_{11}\,(1 - m_{vf2}) \text{ kg} \quad ...(iii)$$

This is sent direct to compressor C_2.

— The saturated liquid flowing out from flash chamber F_1 at state point 13

$$= (1 - m_{vf2} - m_{vf1}) = (1 - x_{11})\,(1 - x_8) \text{ kg} \quad ...(iv)$$

This liquid throttles to state point 1 and goes to the evaporator for producing refrigerating effect.

Mass flow rate *through the condenser* for Q tonnes of refrigeration

$$= \frac{14000\,Q}{60(1 - x_{11})(1 - x_8)(h_2 - h_1)} \text{ kg/min} \quad ...(4.13)$$

- *Power required to run the system,*

$$\mathbf{P} = P_{C1} + P_{C2} + P_{C3}$$

$$= \frac{\text{Flow rate}}{60}\,[(1 - m_{vf2} - m_{vf1})\,(h_3 - h_2) + m_{vf1}\,(h_4 - h_{12}) + m_{vf2}\,(h_5 - h_9)] \text{ kW}$$

- *Coefficient of performance of the system,*

$$\textbf{C.O.P.} = \frac{14000\,Q/3600}{P} \quad ...(4.14)$$

4.2.2.5 Compound compression with multiple expansion valves in series with flash chambers but no intercooling

Fig. 4.5 (*a*), (*b*) shows the schematic arrangement and *p-h* diagram of the cycle/system, respectively.

The mass of refrigerant passing through the condenser,

$$m \text{ (Flow rate)} = \frac{14000\,Q}{60(1 - x_{13})(1 - x_{11})(h_2 - h_1)} \text{ kg/min} \quad ...(4.15)$$

where, Q is the load on evaporator per tonne of refrigeration, and $m\,(1 - x_2\,(1 - x_3)$ is the mass of refrigerant passing through the evaporator.

- *Power required to drive the system,*

$$\mathbf{P} = \frac{m[(1 - x_{13})(1 - x_{11})(h_3 - h_2) + (1 - x_{11})(h_6 - h_4) + 1\,(h_9 - h_7)]}{60} \text{ kW} \quad ...(4.16)$$

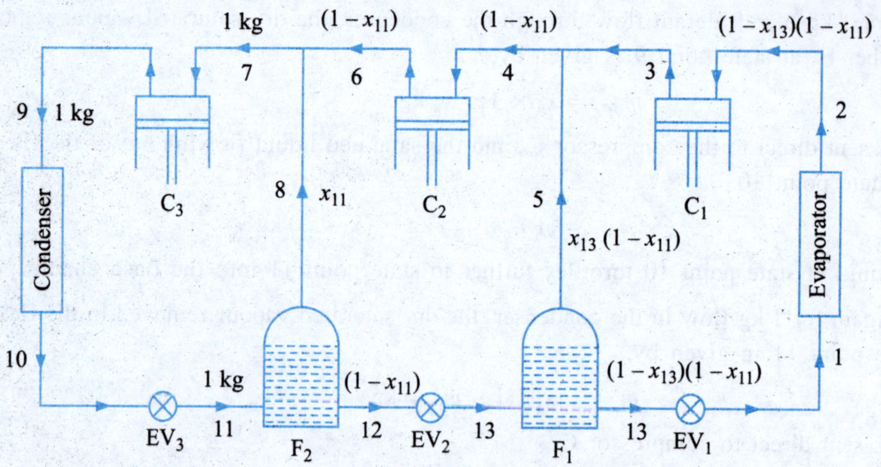

C_1, C_2, C_3 = Compressors; F_1, F_2 = Flash chambers; EV_1, EV_2, EV_3 = Expansion valves.

(*a*) Compound compression with multiple expansion valves in series with flash chambers but no intercooling.

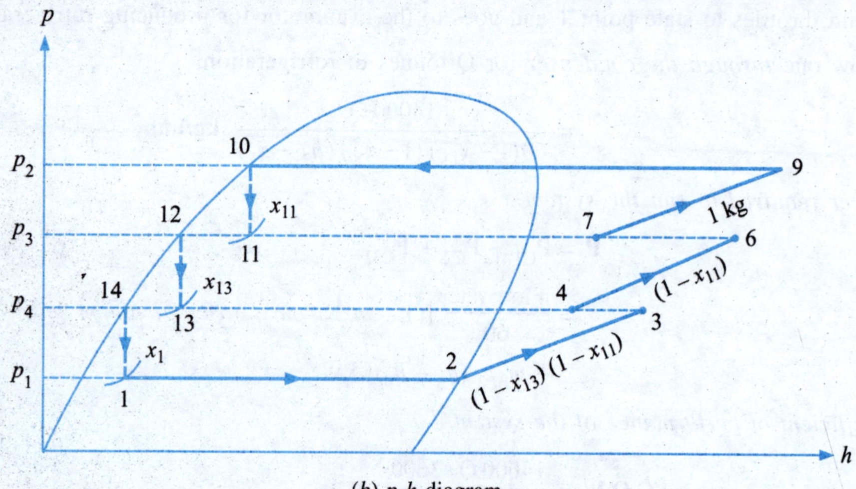

(*b*) *p-h* diagram.

Fig. 4.5

The values of h_4 and h_7 are calculated by using the following equations:

— x_{13} (1 – x_{11}) kg of vapour at condition 5 is mixed with (1 – x_{13}) (1 – x_{11}) kg of vapour at condition 3. Equations the enthalpies, we have:

$$x_{13}(1 - x_{11})\,h_5 + (1 - x_{13})(1 - x_{11})\,h_3 = [x_{13}(1 - x_{11}) + (1 - x_{13})(1 - x_{11})]\,h_4$$

or,
$$x_{13}(1 - x_{11})\,h_5 + (1 - x_{13})(1 - x_{11})\,h_3 = (1 - x_{11})\,h_4$$

$$\therefore \quad h_4 = \frac{x_{13}(1 - x_{11})\,h_5 + (1 - x_{13})(1 - x_{11})\,h_3}{(1 - x_{11})} \qquad \text{...(4.17)}$$

— x_{11} kg of vapour at condition 8 is mixed with (1 – x_{11}) kg of vapour at the condition 6. Equating the enthalpies, we get:

$$x_{11}h_8 + (1 - x_{11})h_6 = [x_{11} + (1 - x_{11})]\,h_7 = h_7$$

$$\therefore \quad h_7 = \frac{x_{11}h_8 + (1 - x_{11})\,h_6}{1} \qquad \text{...(4.18)}$$

After finding out the values of h_4 and h_7, locate these values on enthalpy coordinate and draw the vertical lines where they will cut the p_4 and p_3 lines at the points 4 and 7 as shown in Fig. 4.5 (*b*). After locating the points 4 and 7, the constant entropy lines 4–6 and 7–9 can be drawn (in actual problems).

- **C.O.P.** *of the system* $= \dfrac{14000\,Q/3600}{P}$ ...(4.18a)

4.2.2.6 Compound compression with multiple expansion valves and flash intercooling system

This system, which gives high C.O.P. comparatively is shown in Fig. 4.6 (*a*) alongwith *p-h* diagram in Fig. 4.5 (*b*).

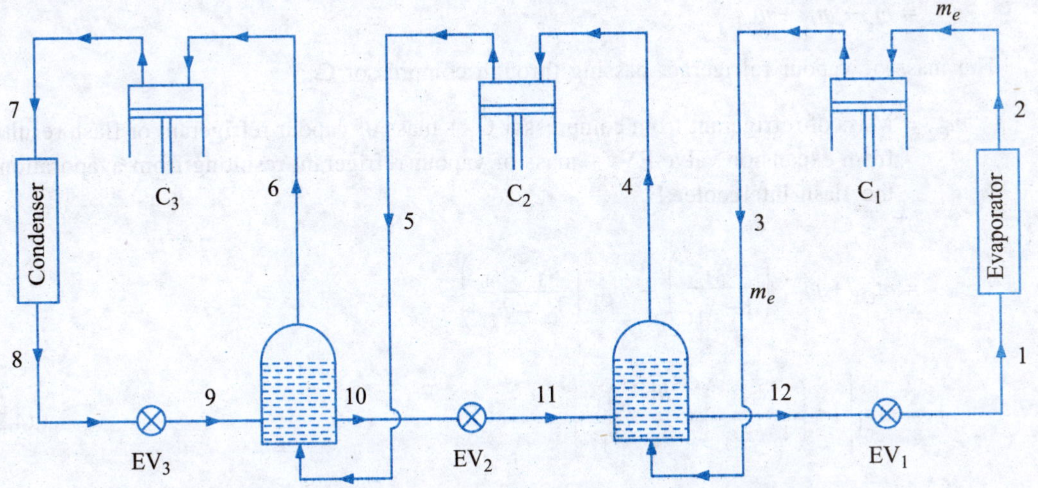

C_1, C_2, C_3 = Compressors; F_1, F_2 = Flash chambers; EV_1, EV_2, EV_3 = Expansion valves.

(*a*) Compound compression with multiple expansion valves and flash intercooling.

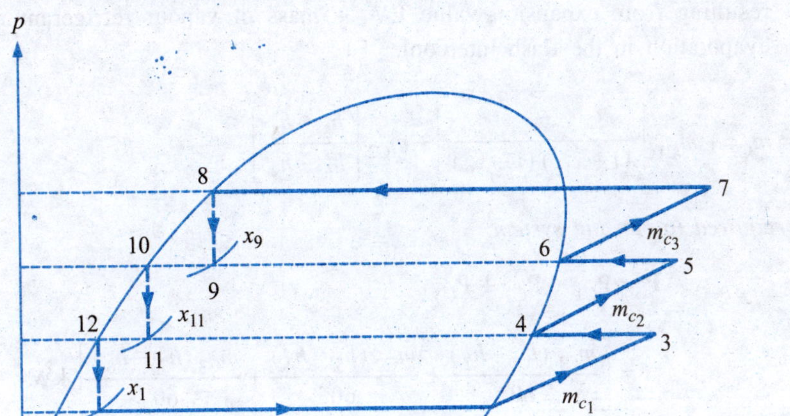

(*b*) *p-h* diagram.

Fig. 4.6

Mass of refrigerant passing through evaporator or compressor C_1,

$$m_e = m_{C1} = \frac{14000\,Q}{60(h_2 - h_1)} \text{ kg/min} \qquad ...(4.19)$$

where, Q is the load on evaporator in tonnes of refrigeration.

Now, the mass of refrigerant vapour passing through the compressor C_2,

m_{C2} = Mass of refrigerant coming from C_1 (m_e) + mass of liquid required to desuperheat the vapour m_e which is m_{l1} + mass of vapour associated from flash with total liquid ($m_e + m_{l1}$) which is m_{v1}

$$= m_e + m_{l1} + m_{v1}$$

∴ The mass of vapour refrigerant passing through compressor C_2,

m_{C2} = Mass of refrigerant from compressor C_1 + mass of vapour refrigerant or flash resulting from expansion valve EV_2 + mass of vapour refrigerant resulting from evaporation in the flash intercooler F_1

$$= m_{C1} + m_{C1}\left(\frac{x_{11}}{1 - x_{11}}\right) + m_{C1}\left(\frac{h_3 - h_4}{h_4 - h_{11}}\right)$$

$$= m_{C1}\left[1 + \left(\frac{x_{11}}{1 - x_{11}}\right) + \frac{h_3 - h_4}{h_4 - h_{11}}\right] \qquad ...(4.20)$$

Similarly, the mass of vapour refrigerant passing through compressor C_3,

m_{C3} = Mass of vapour refrigerant from compressor C_2 + mass of vapour refrigerant or flash resulting from expansion value EV_3 + mass of vapour refrigerant resulting from evaporation in the flash intercooler F_2

$$= m_{C2} + m_{C1}\left[\frac{x_9}{(1 - x_{11})(1 - x_9)}\right] + m_{C2}\left[\frac{h_5 - h_6}{h_6 - h_9}\right] \qquad ...(4.21)$$

- *Power required to run the system,*

$$\mathbf{P} = P_{C1} + P_{C2} + P_{C3}$$

$$= \left[\frac{m_{C1}(h_3 - h_2)}{60} + \frac{m_{C2}(h_5 - h_4)}{60} + \frac{m_{C3}(h_7 - h_6)}{60}\right] \text{kW} \qquad ...(4.22)$$

$$\textbf{C.O.P.} \textit{ of the system} = \frac{(14000\,Q/3600)}{P} \qquad ...(4.23)$$

WORKED EXAMPLES

Note. When load on the evaporator is given in *tons* may be converted to *tonnes* by using the following relation:

1 ton of tefrigeration = 0.9 tonne of refrigeration (S.I. units)

[1 ton = (210 × 60) kJ/h] (1 tonne = 14000 kJ/h)

Example 4.1. *A two-stage vapour compression refrigeration system with a direct contact heat exchanger (flash chamber) operates with ammonia as the refrigerant. The evaporator and condenser temperatures are – 30°C and 40°C respectively. If the capacity of the plant is 25 tonnes of refrigeration, estimate the total work of compression and the C.O.P.*

Had the compression been done in a single stage, what would have been the percentage increase in the work of compression? What is the percentage increase in the C.O.P. owing to the staging of the compression process?

Solution: The schematic arrangement of the system and its corresponding *p-h* diagram are shown in Fig. 4.7 (*a*) and 4.7 (*b*) respectively.

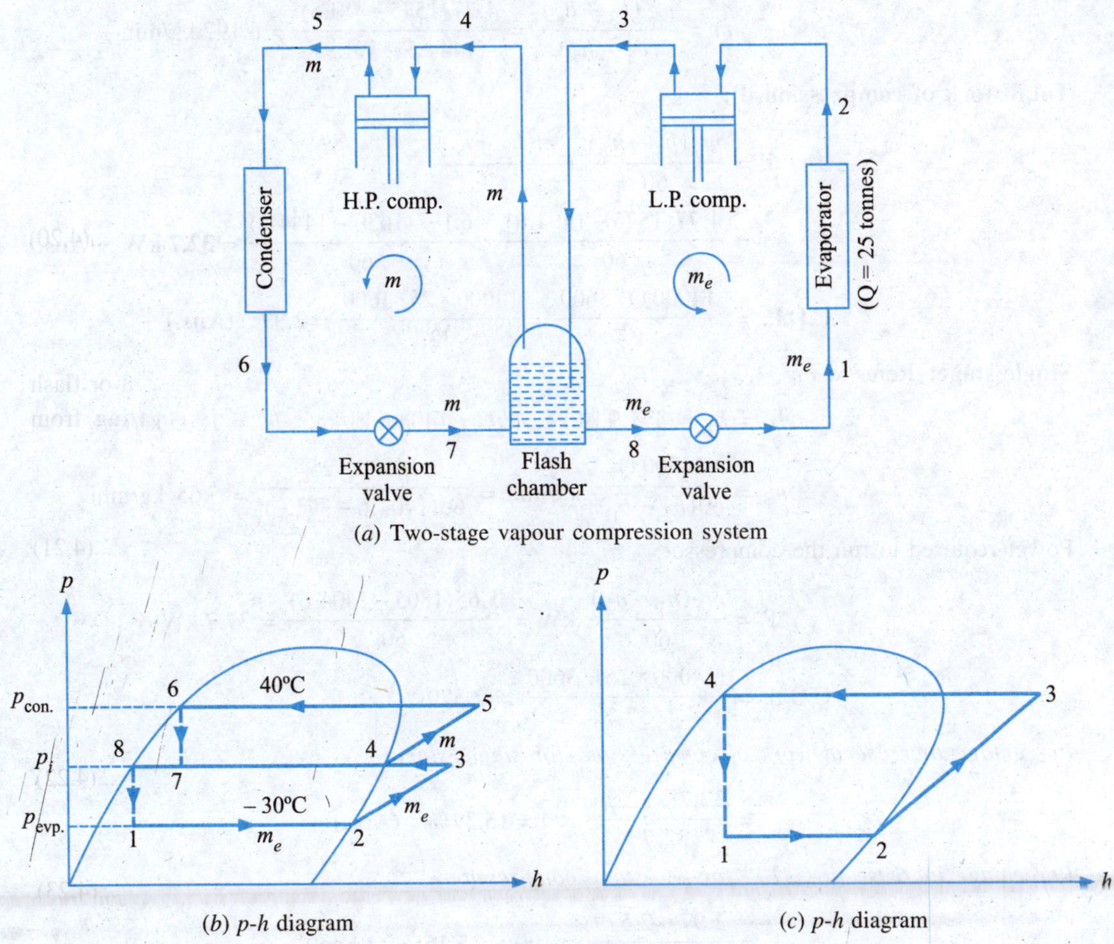

Fig. 4.7

From tables for ammonia, we have:

$$\text{At } 40^{\circ}\text{C}, \; p_{con.} = 15.545 \text{ bar}; \quad \text{At } -30^{\circ}\text{C}, \; p_{evp.} = 1.196 \text{ bar}$$

For minimum work, the intercooler pressure p_i is the geometric mean of the evaporator and condenser pressures, or,

$$p_i = \sqrt{p_{\text{evp}} \times p_{\text{con.}}} = \sqrt{15.545 \times 1.196} = 4.31 \text{ bar}$$

The enthalpies at the various points are:

$$h_8 = h_1 = 181.5 \text{ kJ/kg}; \; h_2 = 1404.6 \text{ kJ/kg}; \; h_4 = 1444.5 \text{ kJ/kg}; \; h_6 = h_7 = 371.9 \text{ kJ/kg};$$
$$h_3 = 1575 \text{ kJ/kg}; \; h_5 = 1630 \text{ kJ/kg}.$$

Mass flux rate through the evaporator,

$$m_e = \frac{14000 \text{ Q}}{60(h_2 - h_1)} \text{ kg/min} = \frac{14000 \times 25}{60(1404.6 - 181.5)} = 4.77 \text{ kg/min}$$

Considering energy balance for direct heat exchanger/flash chamber, we have:

$$m_e \times h_3 + m \times h_7 = m \times h_4 + m_e \times h_8, \quad \text{or,} \quad m_e (h_3 - h_8) = m (h_4 - h_7)$$

$$\therefore \quad m = \frac{m_e (h_3 - h_8)}{(h_4 - h_7)} = \frac{4.77(1575 - 181.5)}{(1444.5 - 371.9)} = 6.197 \text{ kg/min}$$

Total work of compression, P:

$$\text{P} = \frac{m_e (h_3 - h_2)}{60} + \frac{m(h_5 - h_4)}{60}$$

$$= \frac{4.77(1575 - 1404.6)}{60} + \frac{6.197(1630 - 1444.5)}{60} = \mathbf{32.7 \text{ kW}} \quad \textbf{(Ans.)}$$

$$\textbf{C.O.P.} = \frac{14000 \text{ Q}/3600}{\text{P}} = \frac{14000 \times 25/3600}{32.7} = \mathbf{2.97} \quad \textbf{(Ans.)}$$

Single stage: Refer to Fig. 4.7 (*c*)

$$h_4 = h_1 = 371.9 \text{ kJ/kg}; \quad h_2 = 1404.6 \text{ kJ/kg}; \quad h_3 = 1805 \text{ kJ/kg}$$

$$m_e = \frac{14000 \text{ Q}}{60(h_2 - h_1)} \text{ kg/min} = \frac{14000 \times 25}{60(1404.6 - 371.9)} = 5.65 \text{ kg/min}$$

Power required to run the compressor,

$$\text{P} = \frac{m_e (h_3 - h_2)}{60} \text{ kW} = \frac{5.65(1805 - 1404.6)}{60} = 37.7 \text{ kW}$$

$$\text{C.O.P.} = \frac{(14000 \times 25)/3600}{37.7} = 2.579$$

Percentage increase in work of compression (for single stage)

$$= \frac{37.7 - 32.7}{32.7} \times 100 = \mathbf{15.29\%} \quad \textbf{(Ans.)}$$

Percentage increase in C.O.P. for two-stage compression

$$= \frac{2.97 - 2.579}{2.579} \times 100 = \mathbf{15.16\%} \quad \textbf{(Ans.)}$$

Example 4.2. *The following data refer to a two-stage vapour compression system:*

Condensing temperature .... *40°C*

Flash intercooler temperature .... *15°C*

Evaporating temperature .... *–15°C*

Refrigerant used .... *R-717* (NH_3)

Refrigerant flow through H.P. side .... *0.4 kg/s*

Condensate is sub-cooled in condenser .... *by 5°C*

Compression efficiencies .... *80% each*

Determine the following:

(*i*) *Flow rate through evaporator;*

(*ii*) *Tonnes of refrigeration;*

(*iii*) *Total power input;*

(*iv*) *Co-efficient of performance;*

(*v*) *Heat transfer to condenser.* **[N.U.]**

Solution: The schematic arrangement and its corresponding *p-h* diagram are shown in Fig. 4.8 (*a*) and Fig. 4.8 (*b*) respectively.

Assume suction vapour in each compressor as dry saturated.

- The various values read from property tables/*p-h* diagram, for ammonia (R-717), are as follows:

h_2 = 1426.6 kJ/kg; $h_{3'}$ = 1579 kJ/kg; h_4 = 1458.6 kJ/kg; $h_{5'}$ = 1563.2 kJ/kg; $h_8 = h_1$ = 251.4 kJ/kg

Now, $$\eta_{C1} = 0.8 = \frac{h_{3'} - h_2}{h_3 - h_2}, \quad \text{or,} \quad h_3 = h_2 + \frac{h_{3'} - h_2}{0.8}$$

$$\therefore \quad h_3 = 1426.6 + \frac{1579 - 1426.6}{0.8} = 1617.1 \text{ kJ/kg}$$

Again, $$\eta_{C2} = 0.8 = \frac{h_{5'} - h_4}{h_5 - h_4}, \quad \text{or,} \quad h_5 = h_4 + \frac{h_{5'} - h_4}{0.8}$$

$$\therefore \quad h_5 = 1458.6 + \frac{1563.2 - 1458.6}{0.8} = 1589.3 \text{ kJ/kg kJ/kg}$$

$$h_6 = h_7 = 348 \text{ kJ/kg}$$

(*i*) **Flow rate through the evaporator, m_e:**

Considering *energy balance for the flash intercooler*, we have:

$$m_{C1} h_3 + m_{C2} h_7 = m_{C2} h_4 + m_{C1} h_8, \quad \text{or,} \quad m_{C1}(h_3 - h_8) = m_{C2}(h_4 - h_7)$$

$$\therefore \quad m_{C1} (= m_e) = \frac{m_{C_2}(h_4 - h_7)}{(h_3 - h_8)}$$

$$\therefore \quad m_e = \frac{0.4(1458.6 - 348)}{(1617.1 - 251.4)} = \mathbf{0.325 \text{ kg/s}}$$ **(Ans.)** ($\because$ m_{C2} = 0.4 kg/s.....Given)

(*ii*) **Tonnes of refrigeration, Q:**

$$Q = \frac{m_e(h_2 - h_1) \times 3600}{14000}$$

$$= \frac{0.325(1426.6 - 251.4) \times 3600}{14000} = \mathbf{98.2\ tonnes} \quad \textbf{(Ans.)}$$

(*iii*) **Total power input P:**

$$P = P_{C1} + P_{C2} = m_{C1}(h_3 - h_2) + m_{C2}(h_5 - h_4)$$

$$= 0.325\ (1617.1 - 1426.6) + 0.4\ (1589.3 - 1458.6) = \mathbf{114.2\ kW}$$

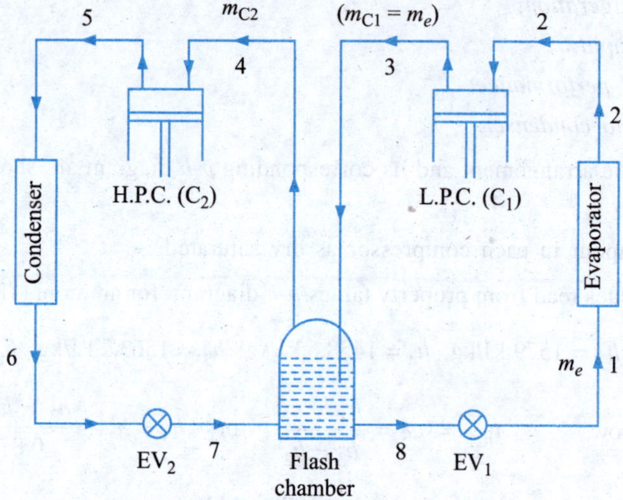

L.P.C. = Low pressure compressor; H.P.C. = High pressure compressor;
EV_1, EV_2 = Expansion valves.

(*a*) Schematic arrangement

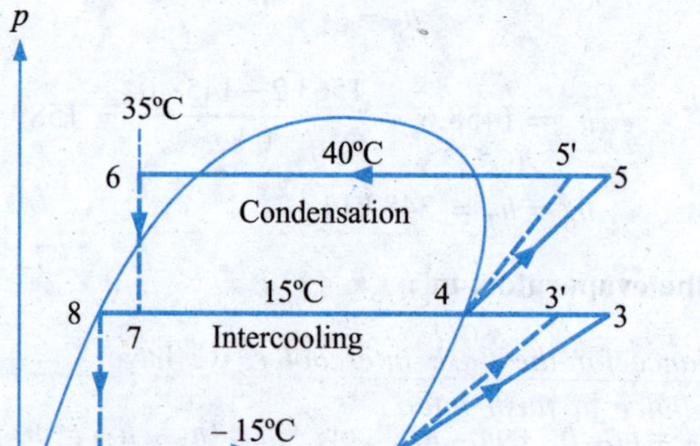

(*b*) *p-h* diagram

Fig. 4.8

(*iv*) **Co-efficient of performance, C.O.P.:**

$$\text{C.O.P.} = \frac{\text{RE (kW)}}{\text{Power input (kW)}} = \frac{14000\,Q/3600}{P}$$

$$= \frac{(14000 \times 98.2/3600)}{114.2} = \mathbf{3.344} \quad \textbf{(Ans.)}$$

Alternatively:

$$\text{C.O.P.} = \frac{R_n(\text{kW})}{W(\text{kW})} = \frac{m_e(h_2 - h_1)}{m_{C1}(h_3 - h_2) + m_{C2}(h_5 - h_4)}$$

$$= \frac{0.325(1426.6 - 251.4)}{114.2} = \mathbf{3.344}$$

(*v*) **Heat transfer to condenser:**

$$\text{Heat transfer to condenser} = m_{C2}(h_5 - h_6)$$

$$= 0.4\,(1589.3 - 348) = \mathbf{496.5\ kJ/s} \quad \textbf{(Ans.)}$$

Example 4.3. *In a 12 tonnes refrigeration ammonia plant, compression is carried out in two stages with water and flash intercooling and water sub-cooling. Condenser pressure, evaporator pressure and flash intercooler pressures are 12 bar, 3 bar and 6 bar respectively. If the limiting temperature for intercooling and sub-cooling is 20°C, determine the following:*

(*i*) *The co-efficient of performance of the plant;*

(*ii*) *The power required for each compressor;*

(*iii*) *The swept volume for each if the volumetric efficiency of each of the compressors is 82%.*

Solution: *Given:* $Q = 12$ *tonnes refrigeration;* $p_{cond.} = 12$ *bar;* $p_{fl} = 6$ *bar;* $p_{evp.} = 3$ *bar;* $t_4 = t_8 = 20°C$; $\eta_{vol.} = 82\%$

The schematic arrangement of the system and the corresponding *p-h* diagram are sown in Fig. 4.9 (*a*) and Fig. 4.9 (*b*) respectively.

The enthalpies at various state points as read from *p-h* chart are as follows:

$h_8 = h_9 = 275.2$ kJ/kg; $h_{10} = h_1 = 224.5$ kJ/kg; $h_2 = 1434$ kJ/kg; $h_3 = 1510$ kJ/kg;

$h_4 = 1468$ kJ/kg; $h_5 = 1442$ kJ/kg; $h_6 = 1535$ kJ/kg; $v_2 = 0.415$ m³/kg; $v_5 = 0.212$ m³/kg.

(*i*) **The coefficient of performance of the plant, C.O.P.:**

The mass of refrigerant passing through the evaporator (or L.P. compressor),

$$m_e = \frac{14000 \times Q}{60(h_2 - h_1)}\ \text{kg/min} = \frac{14000 \times 12}{60(1434 - 224.5)} = 2.315\ \text{kg/min}$$

Considering *energy balance at flash intercooler*, we have:

$$m \cdot h_9 + m_e \cdot h_4 = m \cdot h_5 + m_e \cdot h_{10}$$

or $$m(h_5 - h_9) = m_e(h_4 - h_{10})$$

or $$m = \frac{m_e(h_4 - h_{10})}{(h_5 - h_9)} = \frac{2.315(1468 - 224.5)}{(1442 - 275.2)} = 2.467\ \text{kg/min}$$

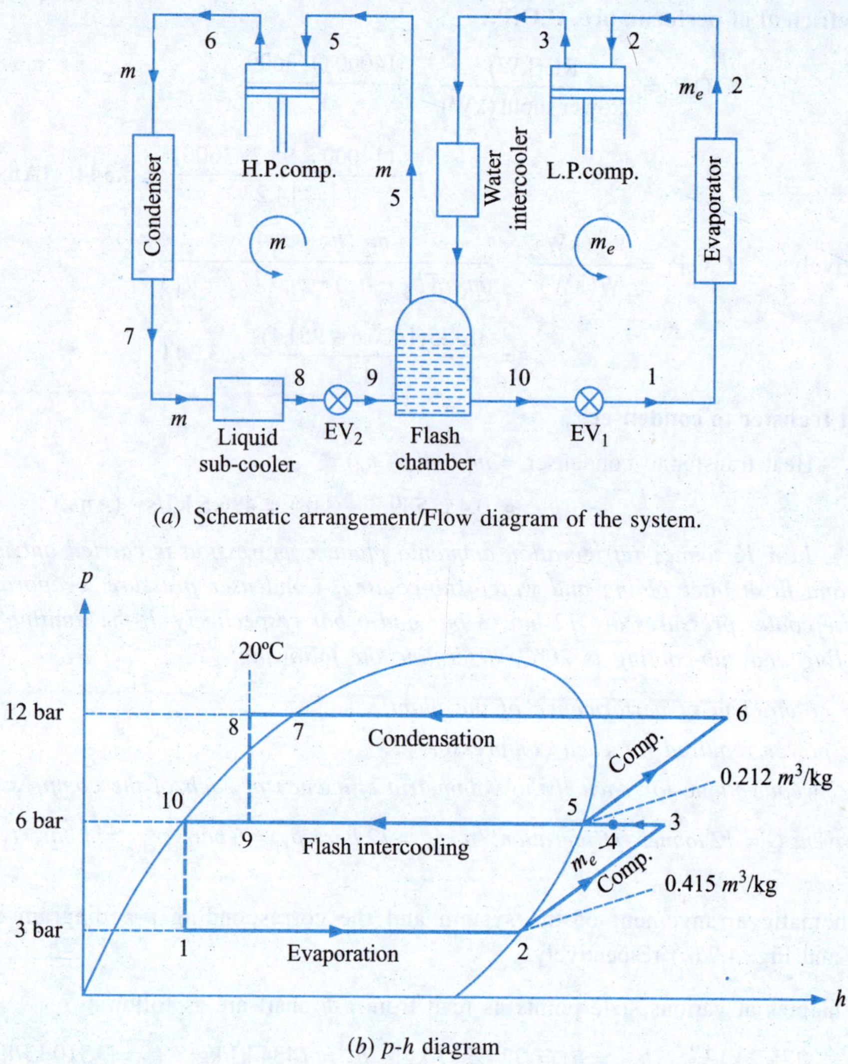

Fig. 4.9

$$\therefore \quad \text{C.O.P.} = \frac{\text{RE}}{\text{Power}} = \frac{m_e(h_2 - h_1)}{m_e(h_3 - h_2) + m(h_6 - h_5)}$$

$$= \frac{2.315(1434 - 224.5)}{2.315(1510 - 1434) + 2.467(1535 - 1442)} = \mathbf{6.9} \quad \textbf{(Ans.)}$$

(*ii*) **The power required for each compressor; $P_{L.P}$, $P_{H.P}$:**

$$P_{L.P.} = \frac{m_e(h_3 - h_2)}{60} \text{ kW} = \frac{2.315(1510 - 1434)}{60} = \mathbf{2.93\ kW} \quad \textbf{(Ans.)}$$

$$P_{H.P.} = \frac{m(h_6 - h_5)}{60} \text{ kW} = \frac{2.467(1535 - 1442)}{60} = \mathbf{3.82\ kW} \quad \textbf{(Ans.)}$$

(*iii*) **The swept volume for each compressor:**

$$\text{Swept volume for L.P. compressor} = \frac{m_e \times v_2}{\eta_{vol.}} = \frac{2.315 \times 0.415}{0.82} = \mathbf{1.172\ m^3/min} \quad \textbf{(Ans.)}$$

$$\text{Swept volume for H.P. compressor} = \frac{m \times v_5}{\eta_{vol.}} = \frac{2.467 \times 0.212}{0.82} = \mathbf{0.638\ m^3/min} \quad \textbf{(Ans.)}$$

Example 4.4. *Calculate the power required to compress 18 kg/min of NH_3 from saturated vapour at 1.4 bar to condensing pressure of 10 bar by a two stage compression with flash intercooling by liquid refrigerant at 4 bar. Assume the refrigerant leaves evaporator as saturated vapour and leaves condenser as saturated liquid. Also find out the load in evaporator in tonnes of refrigeration. Calculate power required without multi-staging.* **(P.U.)**

Solution: The schematic arrangement of the system and its corresponding *p-h* diagram are shown in Fig. 4.10 (*a*) and Fig. 4.10 (*b*) respectively,

- First locate point '8' (pressure 4 bar, and saturated liquid condition is known) and then locate points '1', '2' and '3'.
- Then locate point '4' as the vapour at condition '3' is cooled to condition '4' in the intercooler with the help of liquid refrigerant at condition '8'.
- Then locate the points '5', '6' and '7' on *p-h* chart.

The enthalpies, from the *p-h* chart, at various points are:

$$h_8 = h_1 = 173\ \text{kJ/kg};\quad h_2 = 1410\ \text{kJ/kg};\quad h_3 = 1545\ \text{kJ/kg};\quad h_4 = 1442\ \text{kJ/kg};$$

$$h_5 = 1560\ \text{kJ/kg};\quad h_6 = h_7 = 298\ \text{kJ/kg};\quad h_{10} = 1670\ \text{kJ/kg};\ x = 0.1.$$

- Assume 1 kg of NH_3 is coming out from the condenser and out of that *x* kg is vapour and $(1 - x)$ kg is liquid when it enters the flash chamber. Further assume that *m* kg is the liquid taken from flash chamber into intercooler out of $(1 - x)$ kg in flash chamber. Therefore the mass of NH_3 entering with the evaporator and passing through L.P. compressor is $(1 - x - m)$ kg. Considering the *energy balance at intercooler*, we have:

Heat lost by $(l - x - m)$ kg vapour = Heat gained by *m* kg liquid

$$\therefore \quad (l - x - m)(h_3 - h_4) = m(h_4 - h_8)$$

$$(1 - 0.1 - m)(1545 - 1442) = m(1442 - 173)$$

$$\text{or,} \quad (0.9 - m) \times 103 = 1269\,m, \quad \text{or,} \quad 92.7 - 103\,m = 1269\,m$$

$$\therefore \quad m = 0.067\ \text{kg}$$

∴ Mass of NH_3 passing through evaporator,

$$m_e = (l - x - m) = (1 - 0.1 - 0.067) = 0.833\ \text{kg}$$

(*i*) **Power required to run the system, P:**

$$P = \frac{18}{60}[0.833(h_3 - h_2) + 1(h_5 - h_4)]\ \text{kW}$$

$$= \frac{18}{60}[0.833(1545 - 1410) + 1(1560 - 1442)] = \mathbf{69.14\ kW} \quad \textbf{(Ans.)}$$

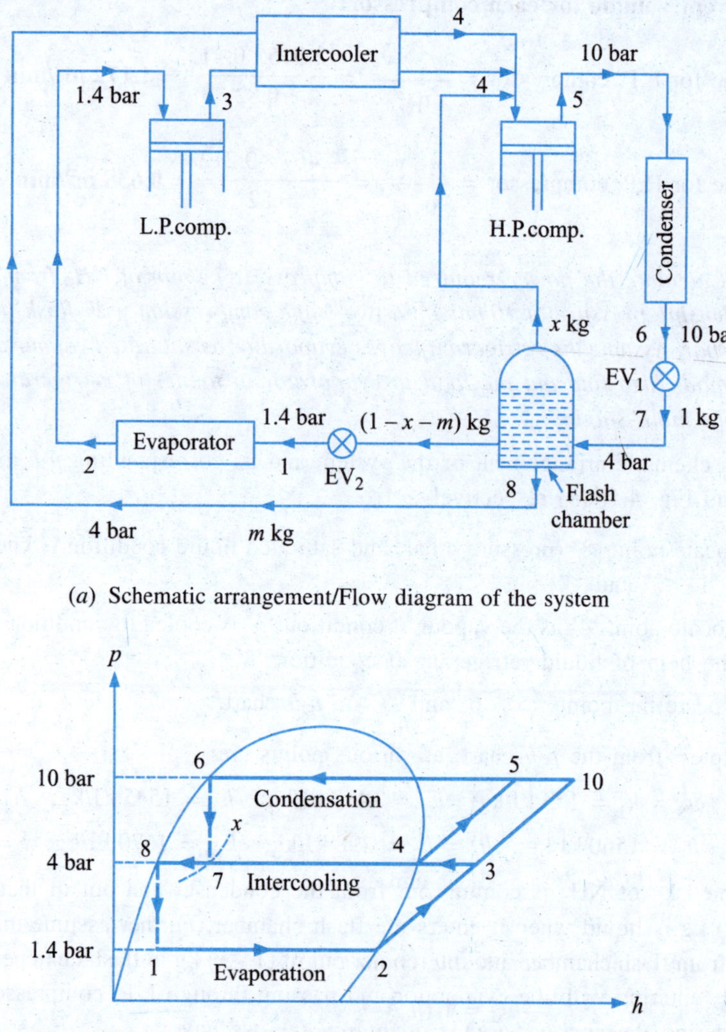

(*a*) Schematic arrangement/Flow diagram of the system

(*b*) *p-h* diagram

Fig. 4.10

(*ii*) Load in the evaporator (in tonnes of refrigeration) Q:

$$Q = \frac{m_e (h_2 - h_1) \times 60}{14000}$$

$$= \frac{18[0.833(1410 - 173)] \times 60}{14000} = \mathbf{79.49\ tonnes} \quad \textbf{(Ans.)}$$

(*iii*) Power required if the compression is carried out in one compressor, P′:

$$P' = \frac{18(h_{10} - h_2)}{60} = \frac{18(1670 - 1410)}{60} = \mathbf{78\ kW} \quad \textbf{(Ans.)}$$

Example 4.5. *A 45 tonnes capacity compound NH_3- compression system with water intercooler after low pressure compressor and liquid flash chamber has a condensing pressure 14 bar. Evaporator pressure is 1.3 bar and intercooler and flash chamber pressure of 5 bar. Volumetric efficiency of both the compressors is 85%. Ammonia is cooled to 32°C in water intercooler and liquid NH_3 is cooled to 30°C in the refrigerant vapour heat exchanger. Assume both the compressions are isentropic and low pressure compressor gases are directly taken to the H.P. compressor after intercooling. Find the following:*

(i) *Ammonia passing through evaporator per minute;*

(ii) *Ammonia passing through H.P. compressor per minute;*

(iii) *Total power required to run the system;*

(iv) *C.O.P. of the system;*

(v) *L.P. compressor displacement per minute.*

Draw flow diagram for the system and represent the processes on p-h chart.

(N.U.)

Solution: *Given:* Q = 45 tonnes of refrigeration.

The schematic arrangement and its corresponding *p-h* diagram are shown in Fig. 4.11 (*a*) and Fig. 4.11 (*b*) respectively.

- From the *p-h* chart/property lables, we have:

$$h_1 = h_{12} = 199 \text{ kJ/kg}; \quad h_2 = 1405 \text{ kJ/kg}; \quad h_6 = 1450 \text{ kJ/kg}; \quad h_9 = 354.5 \text{ kJ/kg};$$

$$h_{10} = h_{11} = 323 \text{ kJ/kg}; \quad x = 0.1 \text{ (at point 11)}; \quad h_5 = 1510 \text{ kJ/kg}.$$

- Now, *Sub-cooling* of 1 kg from 9 to 10 = *Superheating* of (1 – *x*) kg from 2 to 3.

Locate the point 12 at pressure 5 bar and saturated liquid condition is known, as well as locate point '6'. Then locate '1' and '2'.

For locating the point '3', *equate the enthalpies at heat exchanger,*

i.e. $(1 - x)(h_3 - h_2) = 1(h_9 - h_{10})$;

Point 10 can be located on *p-h* chart as pressure (14 bar) and temperature (30°C) are known; $x = 0.1$ at point 11.

$$\therefore \quad (1 - 0.1)(h_3 - 1405) = 1(354.5 - 323)$$

$$\therefore \quad h_3 = 1405 + \left(\frac{354.5 - 323}{0.9}\right) = 1440 \text{ kJ/kg}$$

- Now locate the points '3' and '4' also as 3-4 is isentropic; $v_3 = 0.91 \text{ m}^3/\text{kg}$; $h_4 = 1632$ kJ/kg point (from *p-h* chart).
- Now locate the point '5' as temperature 32°C and pressure are known.
- Now for locating the point '7', use the enthalpy balance of mixing the refrigerant (1 – *x*) kg at h_5 with *x* kg at h_6.

i.e. $(1 - x) h_5 + x h_6 = 1 \times h_7$, or, $(1 - 0.1) \times 1510 + 0.1 \times 1450 = 1 \times h_7$

$$\therefore \quad h_7 = 1504 \text{ kJ/kg}$$

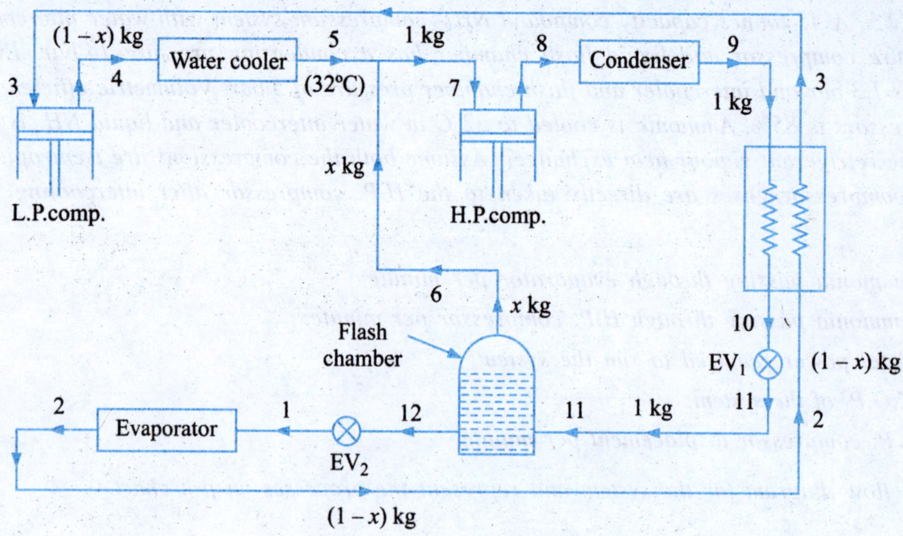

(*a*) Schematic arrangement/Flow diagram

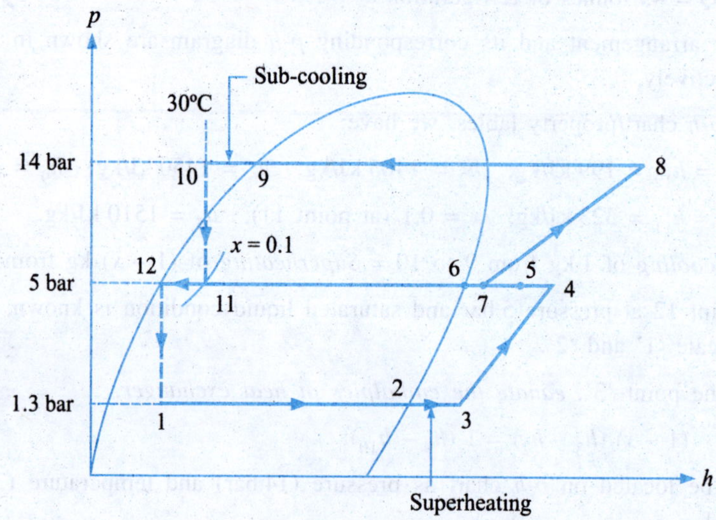

(*b*) *p-h* diagram

Fig. 4.11

- Now locate the point '7' and then locate the point '8' as 7-8 is isentropic compression; h_8 = 1660 kJ/kg (from *p-h* chart).

(*i*) Ammonia passing through the evaporator per minute, m_e ($m_{L.P.}$):

$$m_e\ (= m_{L.P.}) = \frac{14000 \times 45}{60(h_2 - h_1)} = \frac{14000 \times 45}{60(1405 - 199)} = \mathbf{8.706\ kg/min\ (Ans.)}$$

(*ii*) Mass of ammonia passing through H.P. compression, $m_{H.P.}$:

$$m_{H.P.} = \frac{8.706}{1 - x} = \frac{8.706}{1 - 0.1} = \mathbf{9.673\ kg/min\ \ (Ans.)}$$

(*iii*) **Total power required to run the system, P:**

$$P = \frac{m_{L.P.}(h_4 - h_3) + m_{H.P.}(h_8 - h_7)}{60} \text{ kW}$$

$$= \frac{8.706(1632 - 1440) + 9.673(1680 - 1504)}{60} = \mathbf{53\ kW} \quad \textbf{(Ans.)}$$

(*iv*) **C.O.P. of the system:**

$$\text{C.O.P.} = \frac{\text{RE}}{\text{Power}} = \frac{14000 \times 45}{3600 \times 53} = \mathbf{3.3} \quad \textbf{(Ans.)}$$

(*v*) **L.P. compressor displacement per minute:**

$$\text{Piston displacement of L.P. cylinder} = m_{L.P.} \times v_3 = 8.706 \times 0.91$$

$$= \mathbf{7.922\ m^3/min} \quad \textbf{(Ans.)}$$

Example 4.6. *A vapour compression system with NH_3 as the refrigerant works between the pressure limits of 2 bar and 14 bar. It is fitted with expansion valves and flash chambers such that vapours are extracted at 6 bar and 10 bar and sent to the appropriate compressor suction. (Auxiliary compressors to handle only flash gas). For a refrigerating load of 18 tonnes, calculate the power required to drive the three compressors.*

How does the C.O.P. of the system compare with a simple saturated cycle with the same pressure limits?

Solution: The schematic arrangement/flow diagram of the system and its corresponding *p-h* diagram are shown in Fig. 4.12 (*a*) and Fig. 4.12 (*b*) respectively.

From *p-h* chart/properties tables, the enthalpies at various points are:

$h_7 = h_8 = 354.8$ kJ/kg; $h_{10} = h_{11} = 298.4$ kJ/kg; $h_9 = 1465.2$ kJ/kg; $h_{13} = h_1 = 225.4$ kJ/kg; $h_{12} = 1453.5$ kJ/kg; $h_2 = 1421.7$ kJ/kg; $h_3 = 1740$ kJ/kg; $h_4 = 1615$ kJ/kg; $h_5 = 1550$ kJ/kg; $x_1 = 0.102$; $x_{11} = 0.048$; $x_8 = 0.05$.

- For 1 kg of refrigerant flow through the condenser, the dry saturated vapour removed in the flash chamber F_2 at state point 9 is given by,

$$m_{vf2} = x_8 \times 1 = 0.05 \times 1 = 0.05 \text{ kg} \quad ...(i)$$

This is sent direct to the compressor C_3 for compression from 10 bar to 14 bar and the saturated liquid flowing out of the flash chamber F_2 at state point 10

$$= (1 - m_{vf2}) = (1 - 0.05) = 0.95 \text{ kg} = (1 - x_8) \quad ...(ii)$$

The liquid at state point 10 throttles further to stage point 11 into the flash chamber F_1.

- Again for 1 kg flow in the condenser the dry saturated vapour removed in the flash chamber F_1 at state point 11 is given by,

$$m_{vf1} = x_{11}(1 - m_{vf2}) = 0.048 \times 0.95 = 0.0456 \text{ kg} \quad ...(iii)$$

This is sent direct to compressor C_2 for compression from 6 bar to 14 bar.

- The saturated liquid flowing out from flash chamber F_1, at state point 13

$$= (1 - m_{vf2} - m_{vf1}) = (1 - 0.05 - 0.0456) = 0.9044 \text{ kg}$$

$$= (1 - x_{11})(1 - x_8) \quad ...(iv)$$

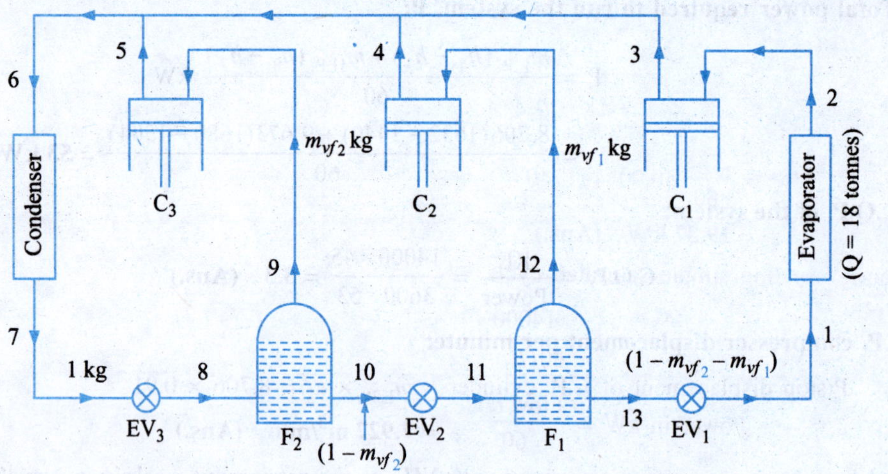

C_1, C_2, C_3 = Compressors; F_1, F_2 = Flash chambers; EV_1, EV_2, EV_3 = Expansion valves.

(*a*) Schematic arrangement/Flow diagram of the system

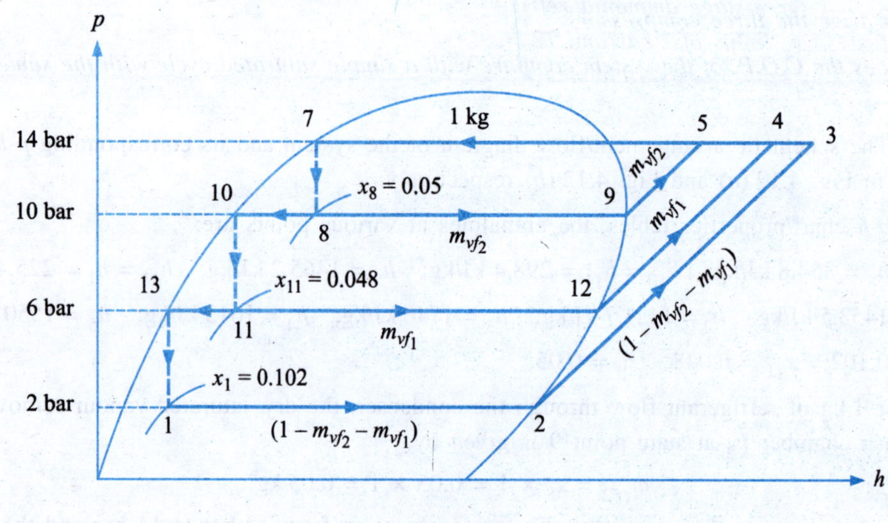

(*b*) *p-h* diagram

Fig. 4.12

This liquid throttles to state point 1 and goes to the evaporator for producing refrigerating effect.

Thus for 1 kg flow through the condenser, the flow through the evaporator is only 0.9044 kg with change of state in the evaporator from 1 to 2. Therefore, *mass flow rate through the condenser* for 18 tonnes of refrigeration

$$= \frac{14000\ Q}{60(1 - x_{11})(1 - x_8)(h_2 - h_1)} \text{ kg/min}$$

$$= \frac{14000 \times 18}{60 \times 0.9044\ (1421.7 - 225.4)} = 3.88 \text{ kg/min}$$

Indicated power of three compressor,

$$P = P_{C_1} + P_{C_2} + P_{C_3}$$

$$= \frac{\text{Flow rate}}{60} [(1 - m_{vf_2} - m_{vf_1})(h_3 - h_2) + m_{vf_1}(h_4 - h_{12}) + m_{v_2}(h_5 - h_9)]$$

$$= \frac{3.88}{60} [0.9044 (1740 - 1421.7) + 0.0456 (1615 - 1453.5) + 0.05 (1550 - 1465.2)]$$

$$= \mathbf{19.37\ kW} \quad \textbf{(Ans.)}$$

For corresponding simple saturated cycle to give 18 tonnes of representation,

$$m_e = \frac{14000 \times 18}{60(h_2 - h_7)} \text{ kg/min} = \frac{14000 \times 18}{60(1421.7 - 354.8)} = 3.94 \text{ kg/min}$$

$$\text{Power in kW} = \frac{\text{Flow rate}}{60}(h_3 - h_2) = \frac{3.94}{60}(1740 - 1421.7) = \mathbf{20.9\ kW} \quad \textbf{(Ans.)}$$

Since the refrigerating effect is same, therefore,

$$\%\text{age increase in C.O.P.} = \frac{20.9 - 19.37}{19.37} \times 100 = \mathbf{7.9\%} \quad \textbf{(Ans.)}$$

Example 4.7. *A three-stage ammonia refrigeration system with flash intercooling operates between the overall pressure limits of 2 bar and 12 bar. The flash intercooling pressures are 4 bar and 8 bar. The load on the evaporator is 10 tonnes of refrigeration.*

(*i*) *Find the power required to run the system;*

(*ii*) *Compare the C.O.P. of the system with that of simple saturation cycle working between the same overall pressure limits.*

Solution: The schematic arrangement/flow diagram of the system and its corresponding *p-h* diagram are shown n Fig. 4.13 (*a*) and Fig. 4.13 (*b*) respectively.

The enthalpies read, from *p-h* chart, at various states points are:

$h_8 = h_9 = h_{10} = h_1 = 328$ kJ/kg, $h_2 = 1420$ kJ/kg; $h_4 = 1442.5$ kJ/kg; $h_6 = 1460$ kJ/kg;
$h_3 = 1500$ kJ/kg; $h_5 = 1520$ kJ/kg; $h_7 = 1500$ kJ/kg; $h_{3'} = 1670$ kJ/kg.

(*i*) **The power required to run the system, P:**

The mass of refrigerant passing through the evaporator or compressor C_1,

$$m_e = m_{C1} = \frac{14000\ Q}{60(h_2 - h_1)} \text{ kg/min} = \frac{14000 \times 10}{60(1420 - 328)} = 2.14 \text{ kg/min}$$

Mass of liquid refrigerant evaporated in the flash intercooler after the first stage of compression,

$$m_1 = \frac{m_{C1}(h_3 - h_4)}{(h_4 - h_{10})} = \frac{2.14(1500 - 1442.5)}{(1442.5 - 328)} = 0.11 \text{ kg/min}$$

∴ Mass of refrigerant passing through compressor C_2,

$$m_{C2} = m_1 + m_{C1} = 0.11 + 2.14 = 2.25 \text{ kg/min}$$

Mass of liquid refrigerant evaporated in the flash intercooler after the second stage of compression,

$$m_2 = \frac{m_{C2}(h_5 - h_6)}{(h_6 - h_9)} = \frac{2.25(1520 - 1460)}{(1460 - 328)} = 0.12 \text{ kg/min.}$$

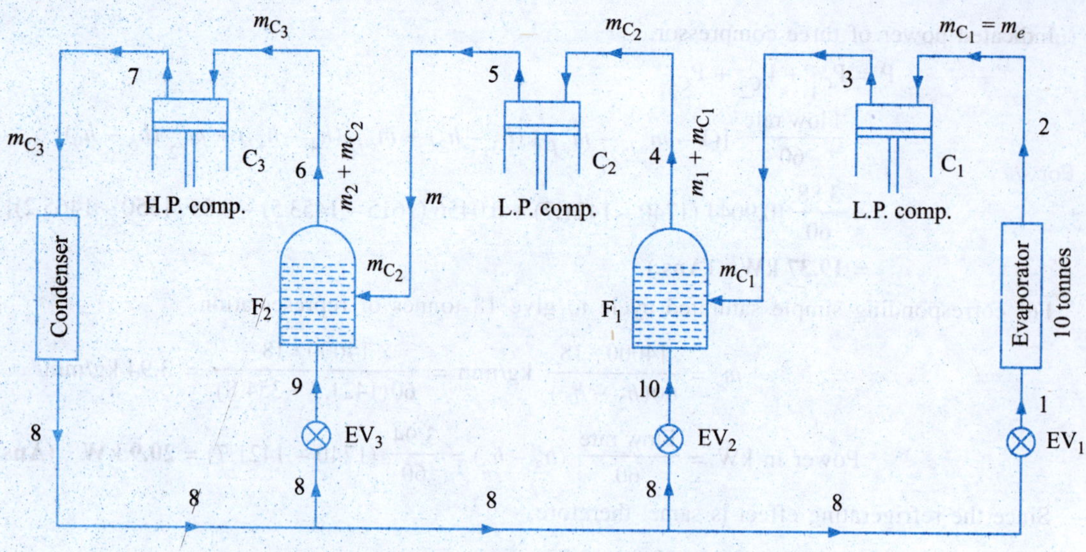

F_1, F_2 = Flash chambers; EV_1, EV_2, EV_3 = Expansion valves.

(*a*) Schematic arrangement/Flow diagram of the system

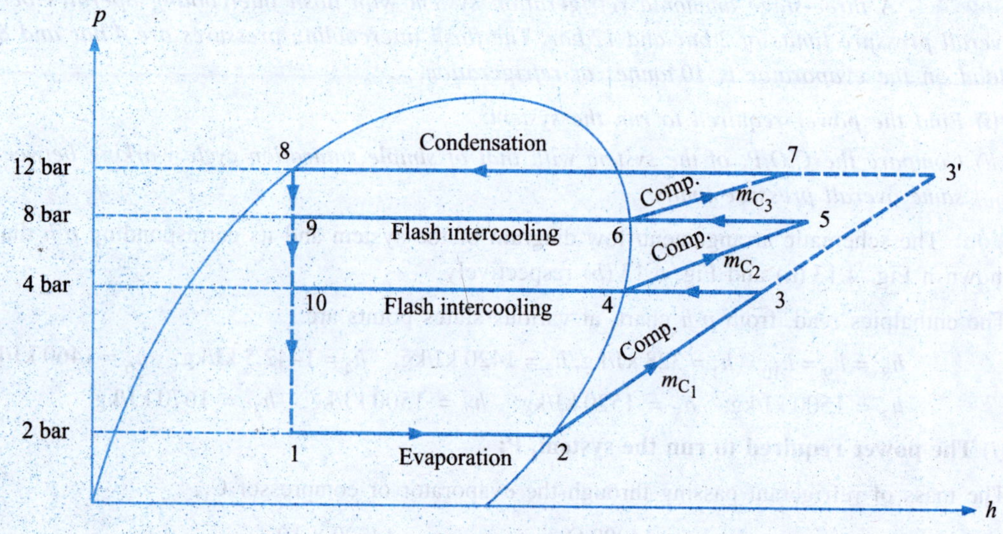

(*b*) *p-h* diagram

Fig. 4.13

∴ Mass of refrigerant passing through compressor C_3,

$$m_{C3} = m_1 + m_{C2} = 0.12 + 2.25 = 2.37 \text{ kg/min}$$

Now, power required to run the system,

$$P = P_{C_1} + P_{C_2} + P_{C_3}$$

$$= \frac{m_{C_1}(h_3 - h_2)}{60} + \frac{m_{C_2}(h_5 - h_4)}{60} + \frac{m_{C_3}(h_7 - h_6)}{60}$$

$$= \frac{1}{60}[2.14\ (1500 - 1420) + 2.25\ (1520 - 1442.5) + 2.37\ (1500 - 1460)]$$

$$= \mathbf{7.34\ kW} \quad \textbf{(Ans.)}$$

(*ii*) **Compression of C.O.P. of the system with that of simple saturation cycle:**

$$\text{C.O.P. of the system} = \frac{14000\,Q/3600}{P} = \frac{14000 \times 10/3600}{7.34} = 5.29$$

For *simple saturation cycle* between the same pressure limits of 2 bar and 12 bar, $h_{3'} = 1670$ kJ/kg.

$$\text{Power required} = \frac{m_e (h_{3'} - h_2)}{60}$$

$$= \frac{2.14(1670 - 1420)}{60} = 8.92 \text{ kW}$$

$\therefore$ C.O.P. of the simple saturation cycle $= \dfrac{14000 \times 10/3600}{8.92} = 4.36$

$\therefore$ *Percentage increase in C.O.P. as compared to simple saturation cycle*

$$= \frac{5.29 - 4.36}{4.36} \times 100 = \mathbf{21.3\%} \quad \textbf{(Ans.)}$$

Example 4.8. *The following data refer to a 3-stage compression ammonia refrigeration system with 3-stage expansion valve and flash intercooling as shown in Fig. 4.14:*

Load on the evaporator .... *10 tonnes of refrigeration*

Evaporator pressure .... *2 bar*

Flash intercooler pressures .... *4 bar and 8 bar*

Condenser pressure .... *12 bar*

(*i*) *Find the power required to drive the system;*

(*ii*) *Compare the C.O.P. of this system with that of simple saturation cycle working between the same overall pressure limits.*

Solution: The schematic arrangement/flow diagram of the system and its corresponding *p-h* diagram are shown in Fig. 4.14 (*a*) and Fig. 4.14 (*b*).

The following are the properties, as read from *p-h* chart, at various state points:

$h_{12} = h_1 = 172$ kJ/kg; $h_2 = 1420$ kJ/kg; $h_3 = 1500$ kJ/kg; $h_4 = 1442.5$ kJ/kg; $h_5 = 1520$ kJ/kg; $h_6 = 1460$ kJ/kg; $h_7 = 1500$ kJ/kg; $h_8 = h_9 = 328$ kJ/kg; $h_{10} = h_{11} = 265$ kJ/kg; $h_{3'} = 1670$ kJ/kg; $x_9 = 0.056$; $x_{11} = 0.072$.

(*i*) **Power required to drive the system, P:**

Mass of refrigerant passing through the *evaporator* or *compressor* C_1,

$$m_e = m_{C1} = \frac{14000\,Q}{60(h_2 - h_1)} \text{ kg/min} = \frac{14000 \times 10}{60(1420 - 172)} = 1.87 \text{ kg/min}$$

The mass of vapour refrigerant passing through *compressor* C_2,

m_{C2} = Mass of vapour refrigerant from compressor C_1 + mass of vapour refrigerant or flash resulting from expansion valve EV_2 + mass of vapour refrigerant resulting from evaporation in the flash intercooler F_1

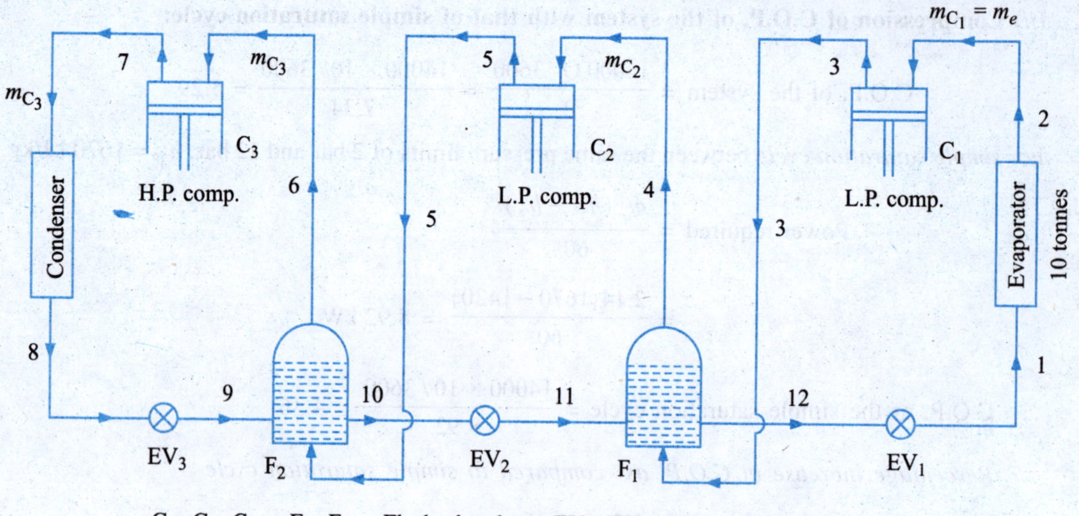

C_1, C_2, C_3 = F_1, F_2 = Flash chambers; EV_1, EV_2, EV_3 = Expansion valves.

(*a*) Schematic arrangement/Flow diagram of the system

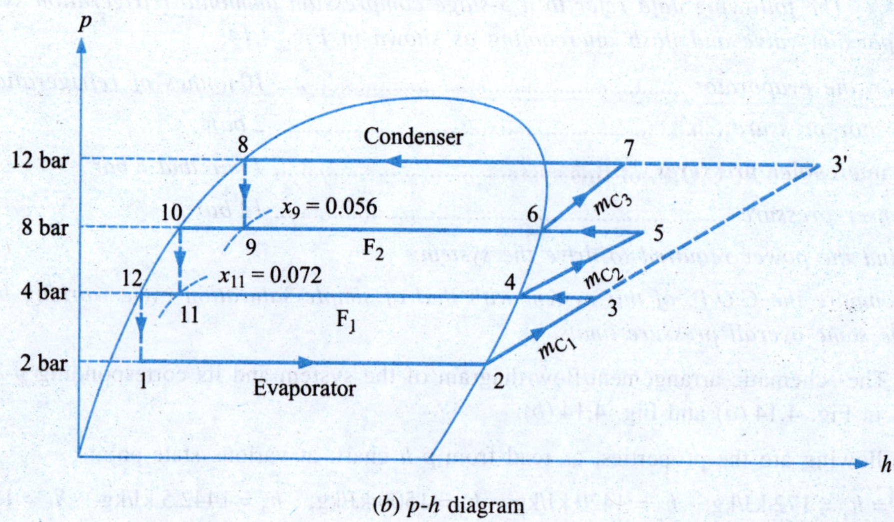

(*b*) *p-h* diagram

Fig. 4.14

$$= m_{C1} + m_{C1}\left(\frac{x_{11}}{1-x_{11}}\right) + m_{C1}\left(\frac{(h_3 - h_4)}{(h_4 - h_{11})}\right) = m_{C1}\left[1 + \frac{x_{11}}{1-x_{11}} + \frac{h_3 - h_4}{h_4 - h_{11}}\right]$$

$$= 1.87\left[1 + \frac{0.072}{1-0.072} + \frac{1500-1442.5}{1442.5-265}\right] = 1.87\ (1 + 0.077 + 0.049) = 2.1 \text{ kg/min}$$

Similarly, the mass of vapour refrigerant passing through compressor C_3,

m_{C3} = Mass of vapour refrigerant from compressor C_2 + mass of vapour refrigerant or flash resulting from expansion valve EV_3 + mass of vapour refrigerant resulting from evaporation in the flash intercooler F_2

$$= m_{C2} + m_{C1}\left[\frac{x_9}{(1-x_{11})(1-x_9)}\right] + m_{C2}\left[\frac{h_5 - h_6}{h_6 - h_9}\right]$$

$$= m_{C2}\left[1 + \left(\frac{h_5 - h_6}{h_6 - h_9}\right)\right] + m_{C1}\left[\frac{x_9}{(1-x_{11})(1-x_9)}\right]$$

$$= 2.1\left[1 + \frac{(1520 - 1460)}{(1460 - 328)}\right] + 1.87\left[\frac{0.056}{(1-0.072)(1-0.056)}\right]$$

$$= 2.21 + 0.12 = 2.33 \text{ kg/min}$$

∴ Power required to drive the system,

$$P = P_{C1} + P_{C2} + P_{C3}$$

$$= \left[\frac{m_{C1}(h_3 - h_2)}{60} + \frac{m_{C2}(h_5 - h_4)}{60} + \frac{m_{C3}(h_7 - h_6)}{60}\right] \text{kW}$$

$$= \frac{1}{60}\,[1.87\,(1500 - 1420) + 2.1\,(1520 - 1442.5) + 2.33\,(1500 - 1460)]$$

$$= \mathbf{6.76\ kW}$$

(*ii*) **Comparison of C.O.P. of this system with that of saturation cycle:**

$$\text{C.O.P. of the system} = \frac{14000\,Q/3600}{P} = \frac{14000 \times 10/3600}{6.76} = 5.75$$

For *simple saturation cycle* between the same pressure limits of 2 bar and 12 bar, $h_{3'}$ = 1670 kJ/kg.

$$\text{Power required} = \frac{m_e\,(h_{3'} - h_2)}{60} = \frac{1.87\,(1670 - 1420)}{60} = 7.79 \text{ kW}$$

$$\therefore \quad \text{C.O.P. of the simple saturation cycle} = \frac{14000 \times 10/3600}{7.79} = 4.99$$

∴ *Percentage increase in C.O.P. as compared to simple saturation cycle*

$$= \frac{5.75 - 4.99}{4.99} \times 100 = \mathbf{15.23\%} \quad \textbf{(Ans.)}$$

4.3 MULTI-EVAPORATOR SYSTEMS

Following cases will be discussed:

1. Multi-evaporators at same temperature.
2. Multi-evaporators at different temperatures with back pressure valves.
3. Multi-evaporators at different temperatures with multiple expansion valves and back pressure valves.

4.3.1 Multi-evaporators at Same Temperature

This arrangement is generally employed when a number of loads at same temperatures are to be handled by the refrigerating plant. The schematic arrangement and *p-h* diagram of the cycle/system are shown in Fig. 4.15 (*a*), and Fig. 4.15 (*b*) respectively.

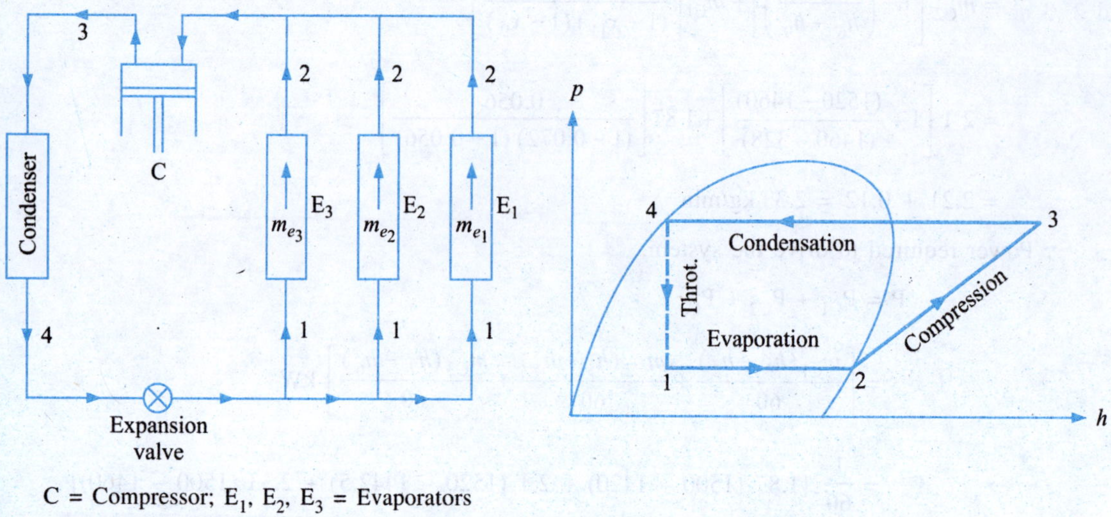

(*a*) Multi-evaporators at same temperature (*b*) *p-h* diagram.

Fig. 4.15

Assuming Q_1, Q_2, and Q_3 are the refrigerating loads (in tonnes) on evaporators E_1, E_2 and E_3, then masses of the refrigerant flowing through E_1, E_2 and E_3 are given by:

$$m_{e1} = \frac{14000 Q_1}{60(h_2 - h_1)} \text{ kg/min} \quad ...(i)$$

$$m_{e2} = \frac{14000 Q_2}{60(h_2 - h_1)} \text{ kg/min} \quad ...(ii)$$

$$m_{e3} = \frac{14000 Q_3}{60(h_2 - h_1)} \text{ kg/min} \quad ...(iii)$$

Since the temperatures in all evaporators are same, the *same thermodynamic cycle will be used for all.*

- *Power required to drive the system,*

$$\mathbf{P} = \frac{(m_{e_1} + m_{e_2} + m_{e_3})(h_3 - h_2)}{60} \text{ kW} \quad ...(4.24)$$

- *Coefficient of performance,*

$$\text{C.O.P.} = \frac{14000(Q_1 + Q_2 + Q_3)/3600}{P} \quad ...(4.25)$$

4.3.2 Multi-evaporators at Different Temperatures with Back Pressure Valves

The arrangement of the system and its *p-h* diagram are shown in Fig. 4.16 (*a*) and Fig. 4.16 (*b*) respectively.

1, 8 and 7 are the conditions of refrigerant entering into the evaporators E_1, E_2 and E_3 respectively and 2, 3 and 4 are corresponding conditions of the refrigerant leaving the respective evaporators.

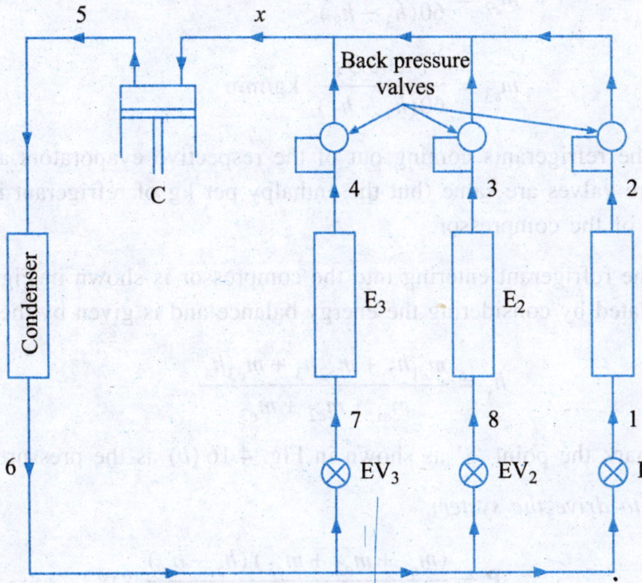

C = Compressor; EV_1, EV_2, EV_3 = Expansion valves.

(*a*) Multi-evaporators at different temperatures with back pressure valves

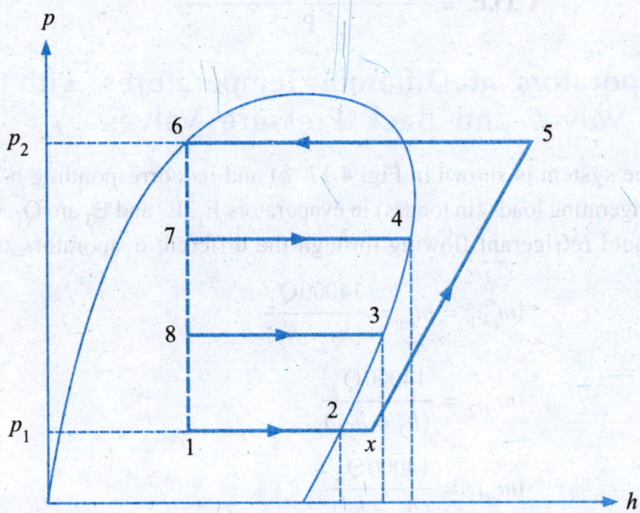

(*b*) *p-h* diagram

Fig. 4.16

If the loads (in tonnes) in the evaporators E_1, E_2 and E_3 are Q_1, Q_2 and Q_3 respectively, then the masses of refrigerant flowing through the evaporators are given by:

$$m_{e1} = \frac{14000\, Q_1}{60(h_2 - h_1)} \text{ kg/min} \quad ...(i)$$

$$m_{e2} = \frac{14000\, Q_2}{60(h_3 - h_8)} \quad ...(ii)$$

$$m_{e3} = \frac{14000\, Q_3}{60(h_4 - h_7)} \text{ kg/min} \quad ...(iii)$$

The pressures of the refrigerants coming out of the respective evaporators and after leaving the respective back pressure valves are same (but the enthalpy per kg of refrigerant is different) and that is the suction pressure of the compressor.

The condition of the refrigerant entering into the compressor is shown in Fig. 4.16 (*b*) by a point '*x*' and it can be calculated by considering the energy balance and is given by the following equation:

$$h_x = \frac{m_{e1}h_2 + m_{e2}h_3 + m_{e3}h_4}{m_{e1} + m_{e2} + m_{e3}} \quad ...(4.26)$$

After getting h_x, mark the point '*x*' as shown in Fig. 4.16 (*b*) as the pressure p_1 is known

- *Power required to drive the system,*

$$\mathbf{P} = \frac{(m_{e1} + m_{e2} + m_{e3})(h_5 - h_x)}{60} \text{ kW} \quad ...(4.27)$$

- *Coefficient of performance of the system,*

$$\textbf{C.O.P.} = \frac{14000(Q_1 + Q_2 + Q_3)}{P} \quad ...(4.28)$$

4.3.3 Multi-evaporators at Different Temperatures with Multiple Expansion Valves and Back Pressure Valves

The arrangement of the system is shown in Fig. 4.17 (*a*) and its corresponding *p-h* diagram is shown in Fig. 4.17 (*b*). If the refrigerating loads (in tonnes) in evaporators E_1, E_2 and E_3 are Q_1, Q_2 and Q_3 respectively, then the masses of liquid refrigerant flowing through the different evaporators are given by:

$$(m_{el})_1 = m_{e1} = \frac{14000\, Q_1}{(h_2 - h_1)} \quad ...[4.29\ (a)]$$

$$(m_{el})_2 = \frac{14000\, Q_2}{(h_3 - h_{10})} \quad ...[4.29\ (b)]$$

$$(m_{el})_3 = \frac{14000\, Q_3}{(h_4 - h_8)} \quad ...[4.29\ (c)]$$

It may be noted that m_{e2} and m_{e3} are the masses of liquid refrigerant passing through E_2 and E_3 to take the refrigerating loads; m_{v2} and m_{v3} are the masses of vapour passing through E_2 and E_3 without taking any part in refrigeration.

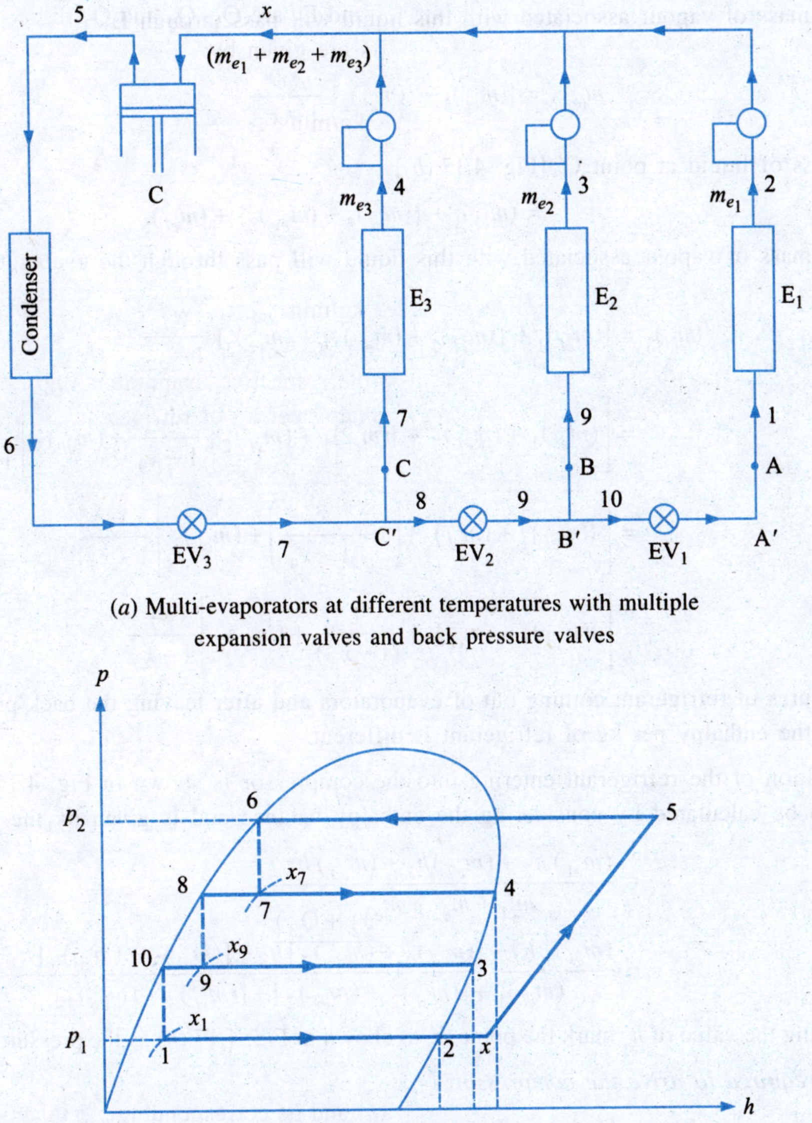

(*a*) Multi-evaporators at different temperatures with multiple expansion valves and back pressure valves

(*b*) *p-h* diagram

Fig. 4.17

The total masses coming out through the evaporators E_1, E_2 and E_3 are:

$$m_{e1} = (m_{el})_1 \quad \text{...[4.30 (a)]}$$

$$m_{e2} = (m_{el})_2 + (m_{ev})_2 \quad \text{...[4.30 (b)]}$$

$$m_{e3} = (m_{el})_3 + (m_{ev})_3 \quad \text{...[4.30 (c)]}$$

Now, let us calculate the masses $(m_{ev})_2$, and $(m_{ev})_3$.

Total mass of liquid at point B′ [as shown in Fig. 4.17 (*b*)]

$$= (m_{el})_1 + (m_{el})_2$$

∴ The mass of vapour associated with this liquid will pass through E_2.

$$\therefore \quad m_{(ev)2} = [(m_{el})_1 + (m_{el})_2] \frac{x_9}{1-x_9} \qquad \text{...(4.31)}$$

Total mass of liquid at point C′ [Fig. 4.17 (*b*)]

$$= (m_{el})_1 + \{(m_{el})_2 + (m_{ev})_2\} + (m_{el})_3$$

∴ The mass of vapour associated with this liquid will pass through the evaporator E_3.

$$\therefore \quad (m_{ev})_3 = [(m_{el})_1 + \{(m_{el})_2 + (m_{ev})_2\} + (m_{el})_3] \frac{x_7}{1-x_7}$$

$$= \left[(m_{el})_1 + (m_{el})_2 + \{(m_{el})_1 + (m_{el})_2\} \frac{x_9}{1-x_9} + (m_{el})_3\right] \frac{x_7}{1-x_7}$$

$$= \left[\{(m_{el})_1 + (m_{el})_2\} \left(1 + \frac{x_9}{1-x_9}\right) + (m_{el})_3\right] \frac{x_7}{1-x_7}$$

$$= \left[\{(m_{el})_1 + (m_{el})_2\} \frac{1}{(1-x_9)} + (m_{el})_3\right] \frac{x_7}{1-x_7} \qquad \text{...(4.32)}$$

The pressures of refrigerant coming out of evaporators and after leaving the back pressure valves are same but the enthalpy per kg of refrigerant is different.

The condition of the refrigerant entering into the compressor is shown in Fig. 4.17 (*b*) by point '*x*' and it can be calculated by considering the enthalpy balance and is given by the equation:

$$h_x = \frac{(m_{e1})\, h_2 + (m_{e2})\, h_3 + (m_{e3})\, h_4}{m_{e1} + m_{e2} + m_{e3}}$$

$$= \frac{(m_{el})_1\, h_2 + [(m_{el})_2 + (m_{ev})_2]\, h_3 + [(m_{el})_3 + (m_{ev})_3]\, h_4}{(m_{el})_1 + [(m_{el})_2 + (m_{ev})_2] + [(m_{el})_3 + (m_{ev})_3]}$$

After getting the value of h_x mark the point '*x*' as shown in Fig. 4.17 (*b*) as the pressure p_1 is known.

- *Power required to drive the compressor*

$$\mathbf{P} = \frac{[(m_{el})_1 + \{(m_{el})_2 + (m_{ev})_2\} + \{(m_{el})_3 + (m_{ev})_3\}]\,(h_5 - h_x)}{60} \text{ kW}$$

- *The C.O.P. of the system,*

$$\text{C.O.P.} = \frac{14000(Q_1 + Q_2 + Q_3)/3600}{P} \qquad \text{...(4.33)}$$

Example 4.9. *A multi-load refrigeration system uses three separate evaporators E_1, E_2 and E_3 at temperatures – 10°C, 0°C and 10°C respectively. While from the condenser the refrigerant Freon-12 liquid is coming to the different evaporators at a temperature 40°C, the liquid from the condenser could be assumed to be saturated. The loads on evaporators E_1 and E_3 are 10 tonnes each and the load on E_2 is 20 tonnes (at 0°C). After each of the above three evaporators, separate pressure reducing valves bring the pressure to the suction pressure of the compressor. Single compressor and individual expansion valves are used in the system.*

Make the neat flow diagram and a cycle diagram on p-h plane for the above system and find the following:

(*i*) *Power required to run the system, and*

(*ii*) *C.O.P. of the system.* **[Benglore University]**

Solution: *Given:* $Q_1 = Q_3 = 10$ tonnes; $Q_2 = 20$ tonnes

The flow diagram and cycle diagram on a *p-h* plane for the system are shown in Fig. 4.18.

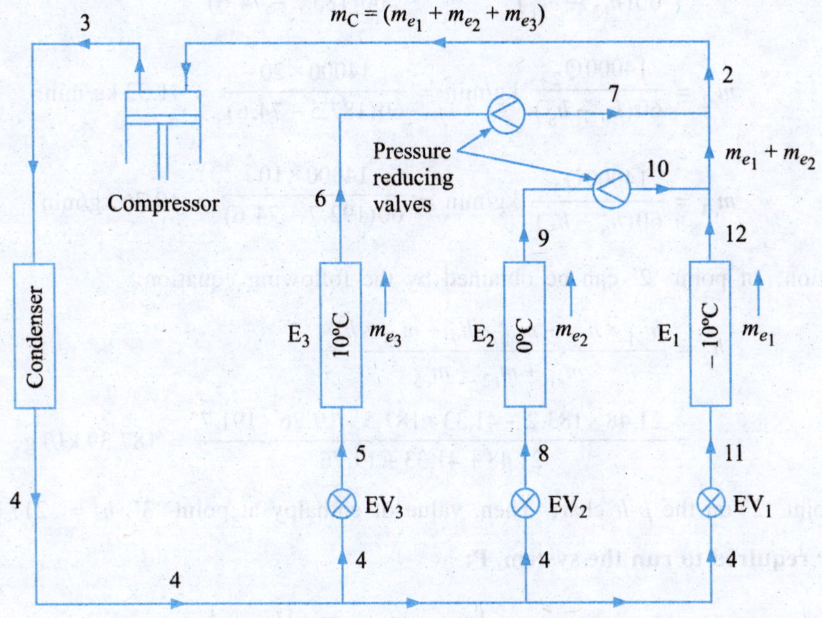

EV_1, EV_2, EV_3 = Expansion valves

(*a*) Flow diagram

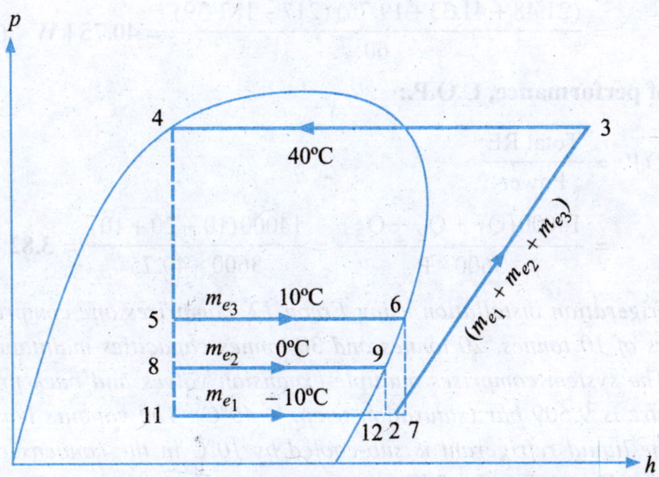

(*b*) *p-h* diagram

Fig. 4.18. Multi-load refrigeration system.

- From property tables/*p-h* diagram, the enthalpies are:

$$h_4 = h_5 = h_8 = h_{11} = 74.6 \text{ kJ/kg}; \quad h_6 = h_7 = 191.7 \text{ kJ/kg};$$

$$h_9 = h_{10} = 187.5 \text{ kJ/kg}; \quad h_{12} = 183.2 \text{ kJ/kg}.$$

The flow rates through the evaporators E_1, E_2, E_3 respectively are:

$$m_{e1} = \frac{14000\,Q_1}{60(h_{12} - h_{11})} \text{ kg/min} = \frac{14000 \times 10}{60(183.2 - 74.6)} = 21.48 \text{ kg/min}$$

$$m_{e2} = \frac{14000\,Q_2}{60(h_9 - h_8)} \text{ kg/min} = \frac{14000 \times 20}{60(187.5 - 74.6)} = 41.33 \text{ kg/min}$$

$$m_{e3} = \frac{14000\,Q_3}{60(h_6 - h_5)} \text{ kg/min} = \frac{14000 \times 10}{60(192.7 - 74.6)} = 19.76 \text{ kg/min}$$

The location of point '2' can be obtained by the following equation:

$$h_2 = \frac{m_{e1} \times h_{12} + m_{e2} \times h_9 + m_{e3} \times h_6}{m_{e1} + m_{e2} + m_{e3}}$$

$$= \frac{21.48 \times 183.2 + 41.33 \times 187.5 + 19.76 \times 191.7}{21.48 + 41.33 + 19.76} = 187.39 \text{ kJ/kg}$$

Locate point '2' on the *p-h* chart. Then, value of enthalpy at point '3', $h_3 =$ 217 kJ/kg

(*i*) **Power required to run the system, P:**

$$P = \frac{m_C(h_3 - h_2)}{60} = \frac{\left(m_{e_1} + m_{e_2} + m_{e_3}\right)(h_3 - h_2)}{60} \text{ kW}$$

$$= \frac{(21.48 + 41.33 + 19.76)(217 - 187.39)}{60} = \mathbf{40.75 \text{ kW}} \quad \textbf{(Ans.)}$$

(*ii*) **Coefficient of performance, C.O.P.:**

$$\text{C.O.P.} = \frac{\text{Total RE}}{\text{Power}}$$

$$= \frac{14000(Q_1 + Q_2 + Q_3)}{3600 \times P} = \frac{14000(10 + 20 + 10)}{3600 \times 40.75} = \mathbf{3.82} \quad \textbf{(Ans.)}$$

Example 4.10. *A refrigeration installation using Freon-12, comprises one compressor, one condenser and three evaporators of 10 tonnes, 20 tonnes and 30 tonnes capacities maintained at 10°C, 5°C and –10°C respectively. The system comprises multiple-expansion valves and back pressure check valves. The condenser pressure is 9.609 bar (saturation temp. = 40°C). The vapours leaving the evaporators are dry saturated. The liquid refrigerant is sub-cooled by 10°C in the condenser sub-cooler. Assume isentropic compression. Determine the following:*

(*i*) *Refrigerating effect/kg in each evaporator;*

(*ii*) *Mass-flow rate in each evaporator;*

(iii) *Compressor power in kW;*

(iv) *C.O.P. of the system.*

Solution: *Given*: Q_1 = 30 tonnes; Q_2 = 20 tonnes, Q_3 = 10 tonnes.

The schematic arrangement and p-h diagram of the system are shown in Fig. 4.19 (*a*) and Fig. 4.19 (*b*) respectively.

From *p-h* diagram/properties tables, we have the following enthalpies:

$$h_4 = h_5 = 64.6 \text{ kJ/kg}; \quad h_8 = h_9 = 45.4 \text{ kJ/kg}; \quad h_{12} = h_{13} = 40.7 \text{ kJ/kg}; \quad h_6 = 191.7 \text{ kJ/kg};$$

$$h_{10} = 189.7 \text{ kJ/kg}; \quad h_{14} = 183.2 \text{ kJ/kg}.$$

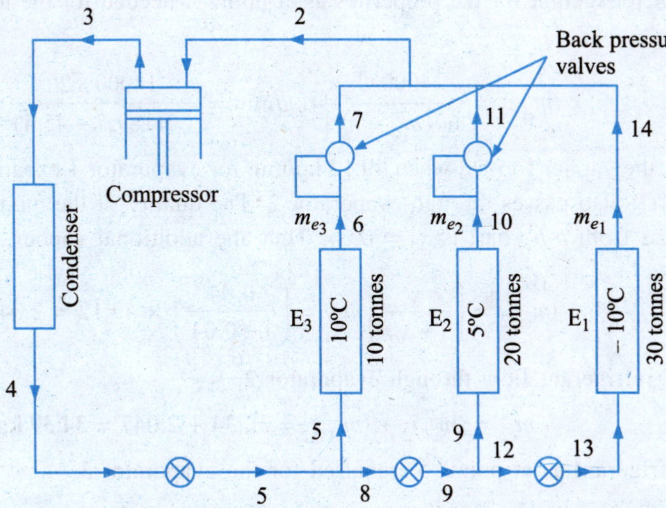

E_1, E_2, E_3 = Evaporators; EV_1, EV_2, EV_3 = Expansion valves.

(*a*) Schematic arrangement/Flow diagram of the system

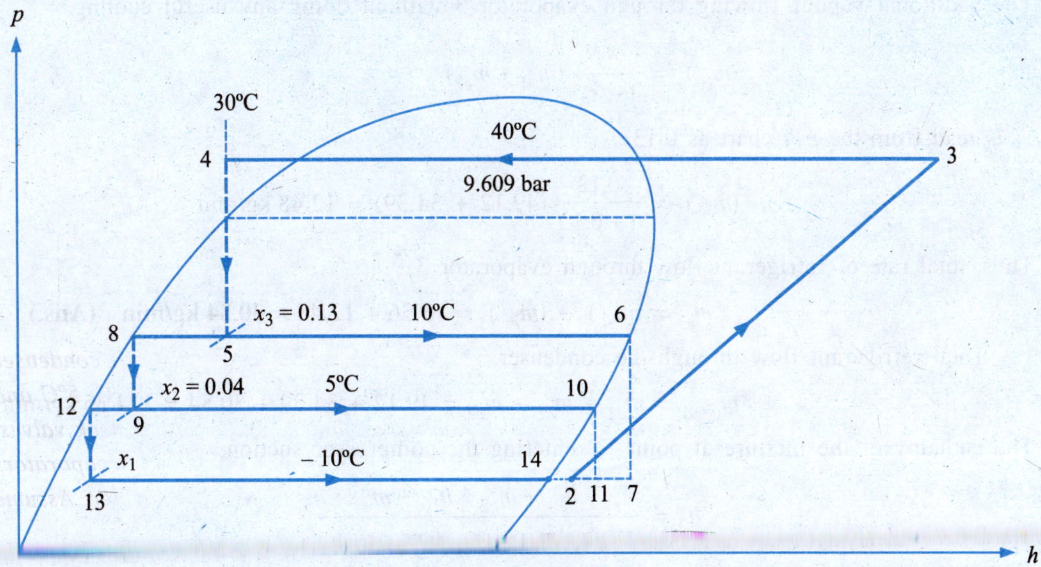

(*b*) *p-h* diagram

Fig. 4.19

(*i*) **Refrigerating effect (RE) in each evaporator:**

RE in evaporator $E_1 = (h_{14} - h_{13}) = 183.2 - 40.7 = \mathbf{142.5\ kJ/kg}$ **(Ans.)**

RE is evaporator $E_2 = (h_{10} - h_9) = (189.7 - 45.4) = \mathbf{144.3\ kJ/kg}$ **(Ans.)**

RE in evaporator $E_3 = (h_6 - h_5) = (191.7 - 64.6) = \mathbf{127.1\ kJ/kg}$ **(Ans.)**

(*ii*) **Mass flow rate in each evaporator; m_{e1}, m_{e2}, m_{e3}:**

$$m_{e1} = \frac{14000\,Q_1}{60\,(h_{14} - h_{13})}\ \text{kg/min} = \frac{14000 \times 30}{60 \times 142.5} = \mathbf{49.12\ kg/min}\quad \textbf{(Ans.)}$$

- The quantity of refrigerant for the properties as at point 9, needed for the load of 20 tonnes of evaporator is given by,

$$(m_{el})_2 = \frac{14000\,Q_2}{60(h_{10} - h_9)}\ \text{kg/min} = \frac{14000 \times 20}{60(189.7 - 45.4)} = 32.34\ \text{kg/min}$$

In addition to this, the vapour formed when 49.12 kg/min for evaporator 1 expanded from 4.234 bar (10°C) to 3.627 bar (5°C) also passes through evaporator 2. The quality of the mixture leaving second expansion valve as read from *p-h* chart is $x_2 = 0.04$. Thus the additional vapour,

$$(m_{ev})_2 = \left(\frac{x}{1 - x_2}\right) m_{el} = \left(\frac{0.04}{1 - 0.04}\right) \times 49.12 = 2.047\ \text{kg/min}$$

∴ Total rate of refrigerant flow through evaporator 2,

$$m_{e2} = (m_{el})_2 + (m_{ev})_2 = 32.34 + 2.047 = \mathbf{34.39\ kg/min}\quad \textbf{(Ans.)}$$

- The rate of refrigerant as at point '5' required for the evaporator 3,

$$(m_{el})_3 = \frac{14000 \times Q_3}{60(h_6 - h_5)}\ \text{kg/min} = \frac{14000 \times 10}{60(191.7 - 64.6)} = 18.36\ \text{kg/min}$$

The additional vapour flowing through evaporator 3 without doing any useful cooling,

$$(m_{ev})_3 = \frac{x_3}{1 - x_3}(m_{e1} + m_{e2})$$

x_3 is read from the *p-h* chart as 0.13.

$$\therefore\quad (m_{ev})_3 = \frac{0.13}{1 - 0.13}(49.12 + 34.39) = 12.48\ \text{kg/min}$$

Thus, total rate of refrigerant flow through evaporator 3,

$$m_{e3} = (m_{el})_3 + (m_{ev})_3 = 18.36 + 12.48 = \mathbf{30.84\ kg/min}\quad \textbf{(Ans.)}$$

∴ Total refrigerant flow through the condenser,

$$m_{cond.} = m_{e1} + m_{e2} + m_{e3} = 49.12 + 34.39 + 30.84 = 114.35\ \text{kg/min}$$

The enthalpy of the mixture at point '2' entering the compressor suction,

$$h_2 = \frac{m_{e1} \times h_{14} + m_{e2} \times h_{10} + m_{e3} \times h_6}{m_{e1} + m_{e2} + m_{e3}}$$

$$= \frac{49.12 \times 183.2 + 34.39 \times 189.7 + 30.84 \times 191.7}{114.35} = 187.45\ \text{kJ/kg}$$

Thus point 2 is located, and following the isentrope passing through till it intersects the line for pressure equal to 9.609 bar gives point '3'. At point '3', from the *p-h* chart, we have,

$$h_3 = 225 \text{ kJ/kg}$$

(*iii*) **Compressor power, P:**

$$P = \frac{(m_{e_1} + m_{e_2} + m_{e_3})(h_3 - h_2)}{60} \text{kW}$$

$$= \frac{114.35(225 - 187.45)}{60} = \mathbf{71.56\ kW} \quad \textbf{(Ans.)}$$

(*iv*) **C.O.P. of the system:**

$$\text{C.O.P.} = \frac{\text{Total RE}}{\text{Power}} = \frac{14000(Q_1 + Q_2 + Q_3)/3600}{71.56}$$

$$= \frac{14000(30 + 20 + 10)/3600}{71.56} = \mathbf{3.26} \quad \textbf{(Ans.)}$$

4.4 MULTIPLE EVAPORATOR AND COMPRESSOR SYSTEMS

Following cases will be discussed:

1. Two evaporators at different temperatures and individual compressor.
2. Two evaporators at different temperatures with compound compression.
3. Two evaporators at different temperatures with compound compression and perfect intercooling.
4. Two evaporators with individual compressors, with multi-expansion valves and flash chamber.
5. Two evaporators with compound compression valves and with flash intercooling.
6. Multi-evaporator with multi-expansion valves with compound compression and flash intercooling.
7. Multi-evaporators with multi-expansion valves and individual compressors.

4.4.1 Two Evaporators at Different Temperatures and Individual Compressor

Fig. 4.20 (*a*) shows the schematic arrangement of the system and Fig. 4.20 (*b*) shows its corresponding *p-h* diagram.

If Q_1 and Q_2 are the refrigeration loads in tonnes, then the masses of the refrigerant flowing through the evaporators E_1 and E_2 are given by:

$$m_{e1} = m_{C1} = \frac{14000 Q_1}{60(h_2 - h_1)} \text{ kg/min} \qquad ...(i)$$

$$m_{e2} = m_{C2} = \frac{14000 Q_2}{60(h_4 - h_8)} \text{ kg/min} \qquad ...(ii)$$

- *Power required to drive the system,*

$$P = \frac{m_{e1}(h_3 - h_2) + m_{e2}(h_5 - h_4)}{60} \text{ kW} \qquad ...(4.34)$$

- *Coefficient of performance of the system,*

$$C.O.P. = \frac{14000 (Q_1 + Q_2)}{60[m_{e1}(h_3 - h_2) + m_{e2}(h_5 - h_4)]} \qquad ...(4.35)$$

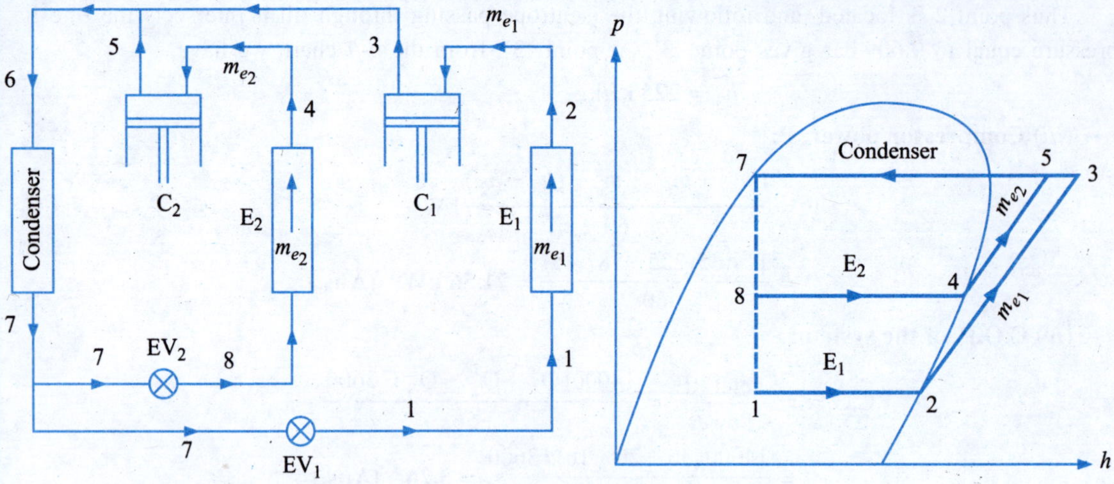

C_1, C_2 = Compressors; E_1, E_2 = Evaporators;
EV_1, EV_2 = Expansion valves.

(*a*) Two evaporators at different temperatures and individual compressor

(*b*) *p-h* diagram

Fig. 4.20

4.4.2 Two Evaporators at Different Temperatures with Compound Compression

The arrangement of the system and its corresponding *p-h* diagram are shown in Fig. 4.21 (*a*) and Fig. 4.21(*b*) respectively.

If Q_1 and Q_2 are the refrigeration loads in tonnes on the evaporators E_1 and E_2 then the masses of refrigerants flowing through the evaporators are given by:

$$m_{e1} = \frac{14000Q_1}{60(h_2 - h_1)} \text{ kg/min} \quad ...(i)$$

$$m_{e2} = \frac{14000Q_2}{60(h_5 - h_4)} = \frac{14000Q_2}{60(h_5 - h_7)} \text{ kg/min} \quad ...(ii) \quad (\because h_4 = h_7)$$

The mass of the refrigerant coming out from compressor C_1 is mixed with refrigerant coming out from evaporator E_2 so the condition of the mixture can be calculated by using enthalpy balance and is given by,

$$h_x = \frac{m_{e1}h_3 + m_{e2}h_5}{m_{e1} + m_{e2}} \quad ...(4.36)$$

This point can be marked along the pressure p_2 as shown in Fig. 4.21 (*b*) (as h_x in known).

- *Power required to drive the system,*

$$P = \frac{m_{e1}(h_3 - h_2) + (m_{e1} + m_{e2})(h_6 - h_x)}{60} \text{ kW} \quad ...(4.37)$$

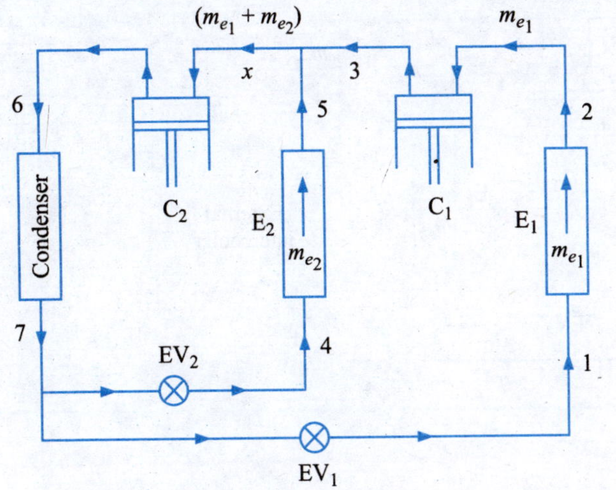

(*a*) Two evaporators at different temperatures with compound compression.

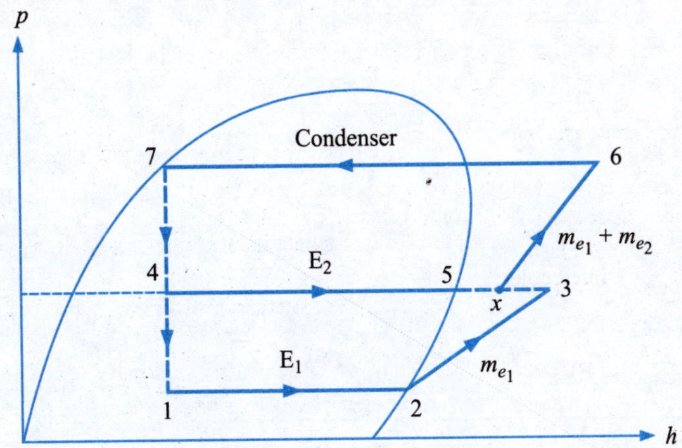

(*b*) *p-h* diagram

Fig. 4.21

- *Coefficient of performance of the system,*

$$C.O.P. = \frac{14000(Q_1 + Q_2)}{60[m_{e1}(h_3 - h_2) + (m_{e1} + m_{e2})(h_6 - h_x)]} \quad ...(4.38)$$

4.4.3 Two Evaporators at Different Temperatures with Compound Compression and Perfect Intercooling

Fig. 4.22 (*a*) and Fig. 4.22 (*b*) show the arrangements of the system and their corresponding *p-h* diagram (same for both the arrangements) is shown in Fig. 4.22 (*c*).

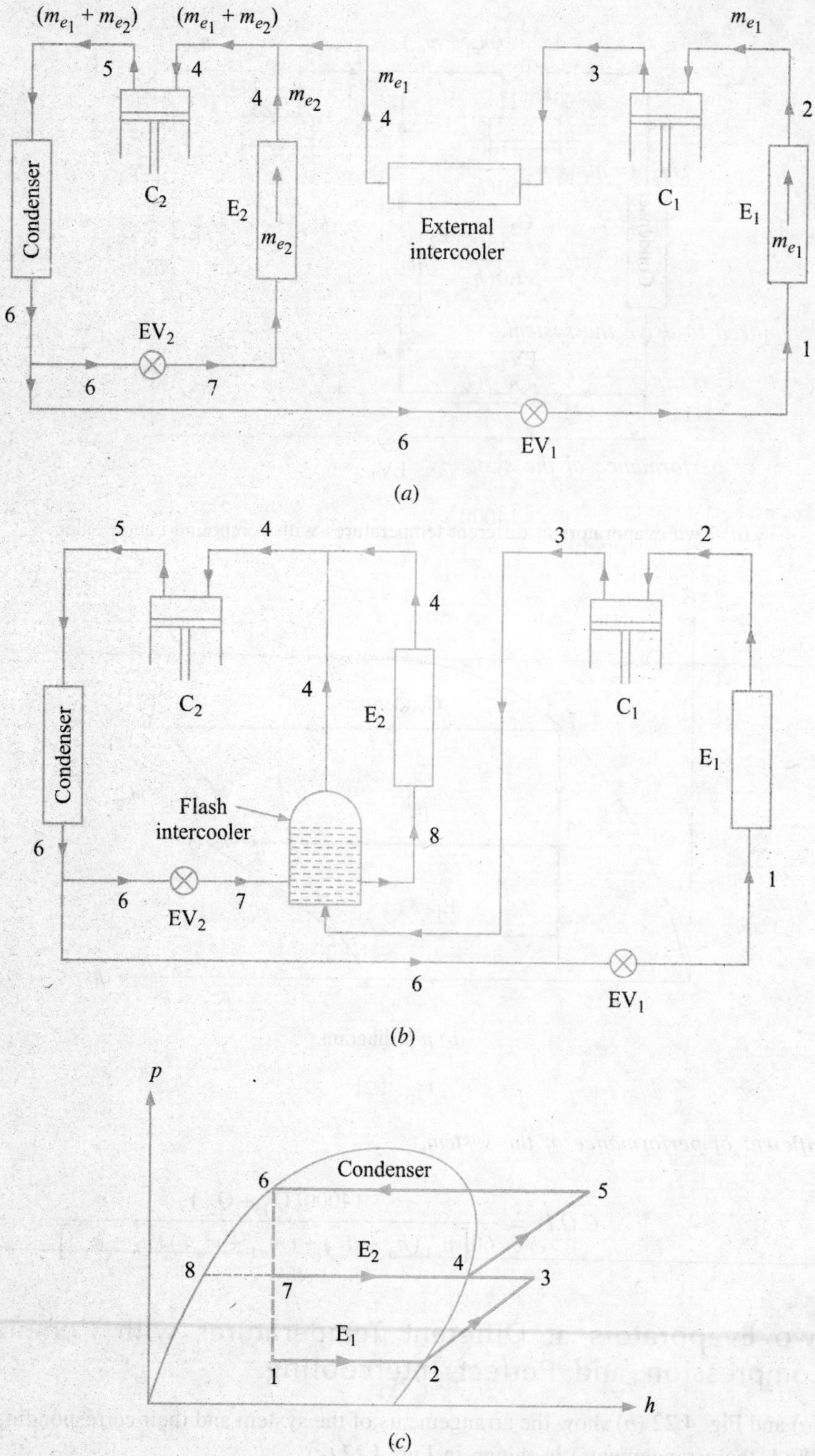

Fig. 4.22. Two evaporators at different temperatures with compound compression and perfect intercooling.

Case I. Calculations for the arrangement (a): Refer to Fig. 4.22 (*a*)

Assuming the loads (in tonnes) on the evaporators E_1 and E_2 as Q_1 and Q_2, the masses of refrigerant flowing through E_1 and E_2 are given by:

$$m_{e1} (= m_{C1}) = \frac{14000\, Q_1}{60(h_2 - h_1)} \text{ kg/min.} \quad ...(i)$$

$$m_{e2} = \frac{14000 Q_2}{60(h_4 - h_7)} \text{ kg/min} \quad ...(ii)$$

• *Power required to drive the system,*

$$\mathbf{P} = \frac{m_{e1}(h_3 - h_2) + (m_{e1} + m_{e2})(h_5 - h_4)}{60} \text{ kW} \quad ...(4.39)$$

• *Coefficient of performance of the system,*

$$\textbf{C.O.P.} = \frac{14000(Q_1 + Q_2)/3600}{P} \quad ...(4.40)$$

Case II. Calculations for the arrangement (*b*): The mass of refrigerant passing through compressor, C_1, $m_{C1} = m_{e1}$ (same as in case I).

The mass of refrigerant passing through compressor C_2,

m_{C2} = Mass of refrigerant passing through E_1 + mass of refrigerant passing through E_2 + mass of liquid refrigerant required in the flash chamber for desuperheating the refrigerating vapour m_{e1} coming out from C_1 (say m_l) + mass of vapour formed with the liquid $(m_{e2} + m_{e1})$ [where $m_{e2} = \frac{14000\, Q_2}{(h_4 - h_8)}$ which is different from eqn. (*ii*) in case I]

$$\therefore \quad m_{C2} = m_{e1} + m_{e2} + m_l + m_v$$

where, $m_{e1} = \dfrac{14000\, Q_1}{(h_2 - h_1)}$, $m_{e2} = \dfrac{14000\, Q_2}{(h_4 - h_8)}$, $m_l = \dfrac{m_{e_1}(h_3 - h_4)}{(h_4 - h_8)}$,

$$m_v = (m_{e1} + m_l)\frac{x_7}{(1 - x_7)} = \left[m_{e1} + \frac{m_{e_1}(h_3 - h_4)}{(h_4 - h_8)}\right]\frac{x_7}{1 - x_7}$$

Power required for driving the system,

P = Power required to drive the compressor C_1 + Power required to drive the compressor C_2.

$$\text{or,} \quad \mathbf{P} = \frac{m_{e1}(h_3 - h_2)}{60} + \frac{m_{C2}(h_5 - h_4)}{60} \quad ...(4.41)$$

Coefficient of performance,

$$\textbf{C.O.P.} = \frac{14000(Q_1 + Q_2)/3600}{P} \quad ...(4.42)$$

4.4.4 Two Evaporators with Individual Compressors, with Multi-expansion Valves and Flash Chamber

The arrangement of the system and its corresponding *p-h* diagram are shown in Fig. 4.23 (*a*) and Fig. 4.23 (*b*) respectively.

If Q_1 and Q_2 are the loads (in tonnes) on the evaporators E_1 and E_2, then the masses of refrigerant flowing through them are given by:

$$m_{e1} = \frac{14000\, Q_1}{(h_2 - h_1)} \text{ kg/min} \quad ...(i)$$

$$m_{e2} = \frac{14000\, Q_2}{60(h_8 - h_7)} \quad ...(ii)$$

The mass of the refrigerant passing through compressor C_2,

$$m_{C2} = m_{e2} + (m_{e1} + m_{e2}) \frac{x_6}{1 - x_6} \quad ...(4.43)$$

where, $(m_{e1} + m_{e2}) \dfrac{x_6}{1 - x_6}$ is the mass of vapour formed in the flash chamber.

Power required to drive the system,

$$\mathbf{P} = \frac{m_{e1}(h_3 - h_2) + m_{C2}(h_4 - h_8)}{60} \text{ kW} \quad ...(4.44)$$

Coefficient of performance of the system,

$$\mathbf{C.O.P.} = \frac{14000(Q_1 + Q_2)/3600}{P} \quad ...(4.45)$$

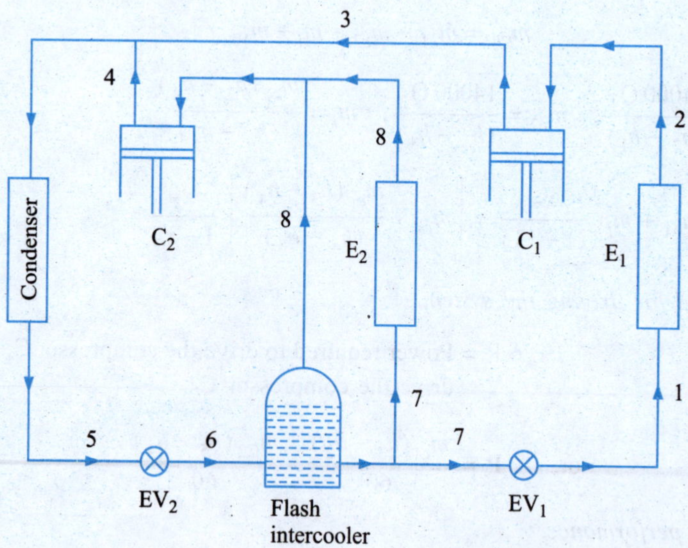

(*a*) Two evaporators with individual compressors, with multi-expansion valves and flash chamber.

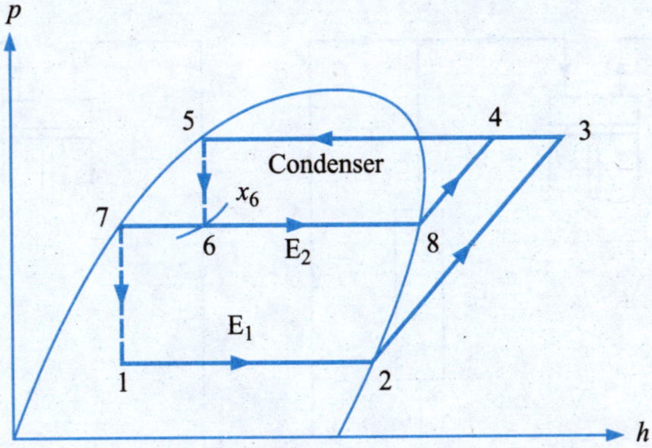

(b) p-h diagram

Fig. 4.23

4.4.5 Two Evaporators with Compound Compression, with Multiple Expansion Valves and with Flash Intercooling

The arrangement of the system and its *p-h* diagram are shown in Fig. 4.24 (*a*) and Fig. 4.24 (*b*) respectively. Assuming Q_1 and Q_2 are the refrigerating loads (in tonnes) on evaporators E_1 and E_2 respectively, then the masses of refrigerant flowing through them are:

$$m_{e1} \,(= m_{C1}) = \frac{14000\,Q_1}{60(h_2 - h_1)} \text{ kg/min} \qquad ...(i)$$

$$m_{e2} = \frac{14000\,Q_2}{60(h_4 - h_7)} \text{ kg/min.} \qquad (ii)$$

The mass of refrigerant passing through compressor C_2 is given by:

$$m_{C2} = m_{e1} + m_{e2} + m_l + m_v$$

m_l is the mass of liquid refrigerant required in flash chamber to desuperheat the vapour m_{e1} from condition 3 to condition 4 coming out of compressor C_1.

$$\therefore \quad m_l = \frac{m_{e1}(h_3 - h_4)}{(h_4 - h_8)} \qquad ...(4.46)$$

The vapour associated with the liquid $(m_{e1} + m_l)$,

$$m_v = (m_{e1} + m_l)\,\frac{x_7}{1 - x_7}$$

Power required to drive the system,

$$\mathbf{P} = \frac{m_{e1}(h_3 - h_2) + m_{C_2}(h_5 - h_4)}{60} \text{ kW} \qquad ...(4.47)$$

Coefficient of performance of the system,

$$\textbf{C.O.P.} = \frac{14000\,(Q_1 + Q_2)/3600}{P} \qquad ...(4.48)$$

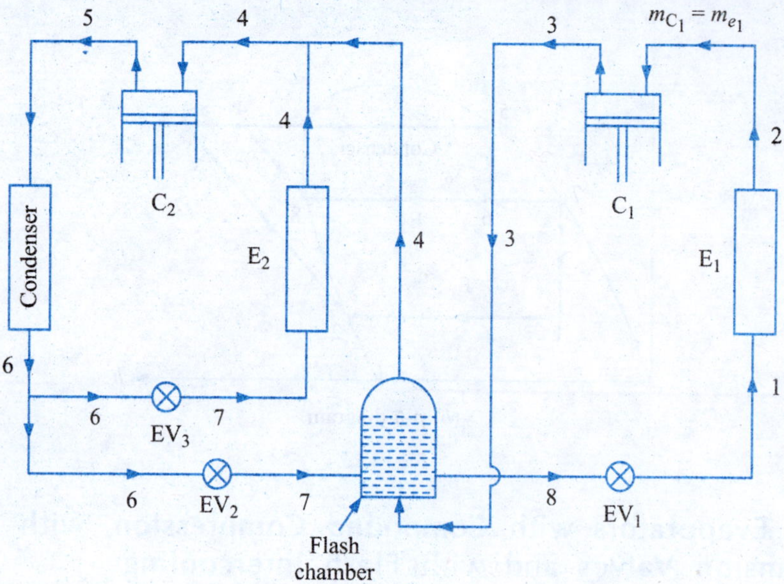

(*a*) Two evaporators with multiple expansion valves, with compound compression and flash intercooling

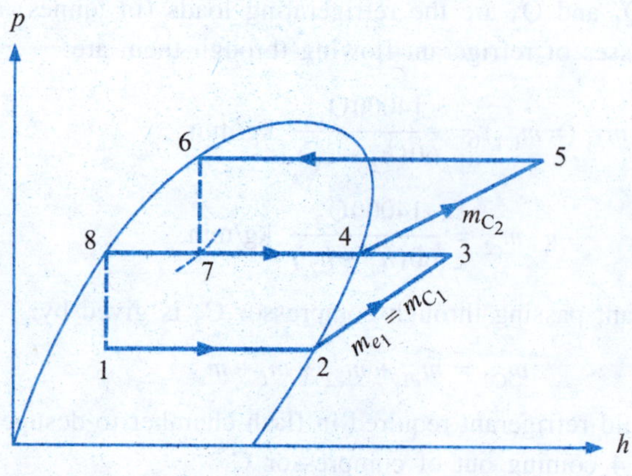

(*b*) *p-h* diagram

Fig. 4.24

4.4.6 Multi-evaporators and Multi-expansion Valves with Compound Compression and Flash Intercooling

The arrangement of the system is shown in Fig. 4.25 (*a*) and its corresponding *p-h* diagram is shown in Fig. 4.25 (*b*).

Assuming Q_1 and Q_2 as the loads (in tonnes) on the evaporators E_1 and E_2 respectively, then the masses of the refrigerant flowing through them are given by:

$$m_{e1} = \frac{14000\, Q_1}{60(h_2 - h_1)} \text{ kg/min} \quad \text{...(i)}$$

$$m_{e2} = \frac{14000\, Q_2}{60(h_4 - h_8)} \quad \text{...(ii)}$$

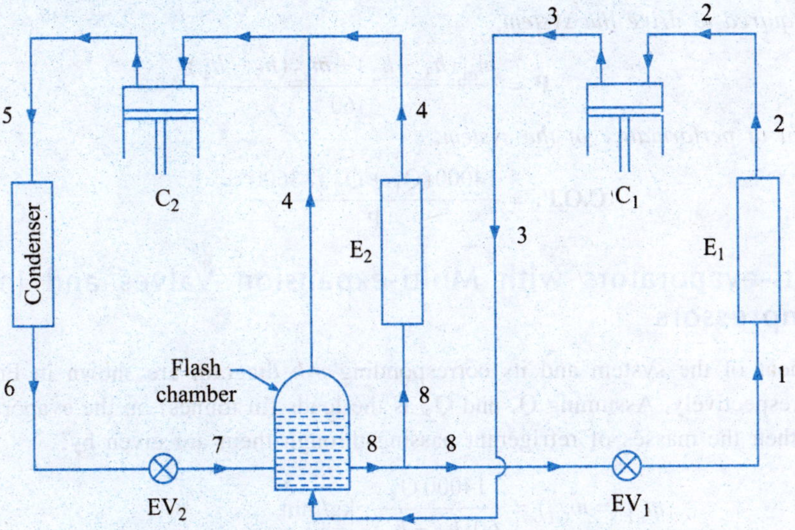

(*a*) Multi-evaporators and multi-expansion valves with compound compression and flash intercooling.

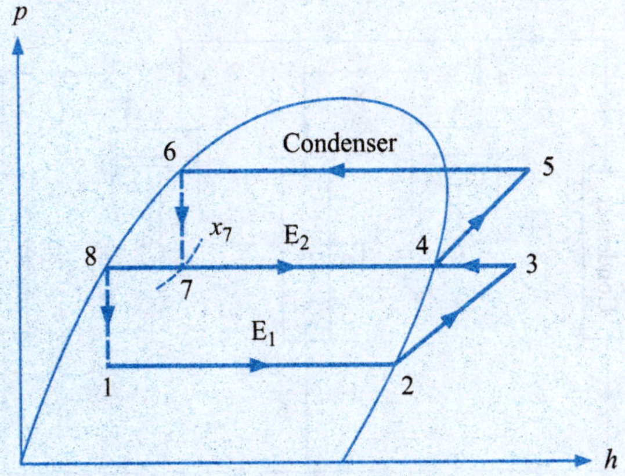

(*b*) *p-h* diagram

Fig. 4.25

The mass of the refrigerant passing through the compressor C_2,

$$m_{C2} = m_{e1} + m_{e2} + m_l + m_v$$

where m_l is the mass of liquid required in the flash chamber to desuperheat the vapour m_{e1} coming out of from the compressor C_1 and it is given by,

$$m_l = \frac{m_{e1}(h_3 - h_4)}{(h_4 - h_8)} \qquad \text{...(4.49)}$$

and m_v is the vapour associated with the liquid refrigerant $(m_{e1} + m_{e2} + m_l)$, and it is given by,

$$m_v = (m_{e1} + m_{e2} + m_l)\frac{x_7}{1 - x_7} \qquad \text{...(4.45)}$$

Power required to drive the system,

$$\mathbf{P} = \frac{m_{e1}(h_3 - h_2) + m_{e2}(h_5 - h_4)}{60} \qquad ...(4.51)$$

Coefficient of performance of the system,

$$\text{C.O.P.} = \frac{14000(Q_1 + Q_2)/3600}{P} \qquad ...(4.52)$$

4.4.7 Multi-evaporators with Multi-expansion Valves and Individual Compressors

The arrangement of the system and its corresponding *p-h* diagram are shown in Fig. 4.26 (*a*) and Fig. 4.26 (*b*) respectively. Assuming Q_1 and Q_2 as the loads (in tonnes) on the evaporators E_1 and E_2 respectively, then the masses of refrigerant passing through them are given by:

$$m_{e1}\ (= m_{C1}) = \frac{14000\ Q_1}{60(h_2 - h_1)}\ \text{kg/min} \qquad ...(i)$$

$$m_{e2} = \frac{14000\ Q_2}{60(h_8 - h_6)}\ \text{kg/min} \qquad ...(ii)$$

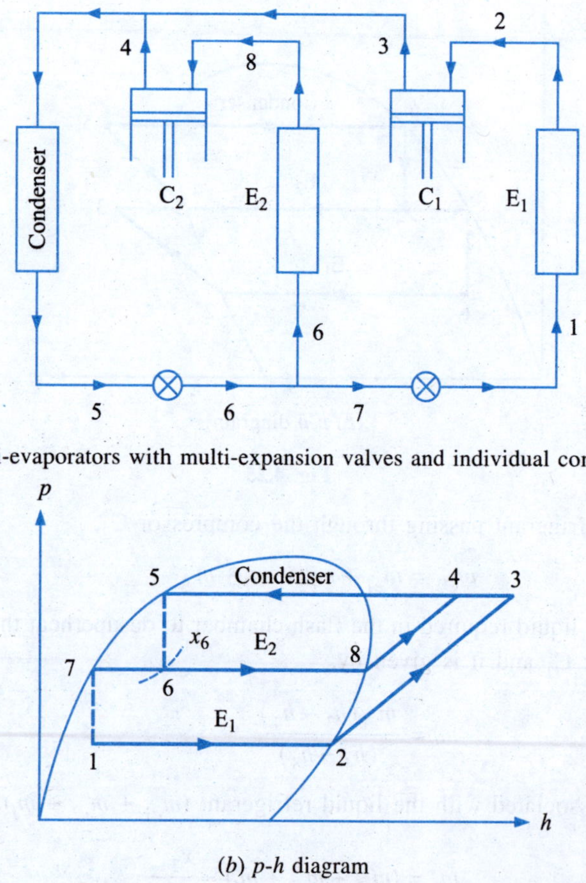

(*a*) Multi-evaporators with multi-expansion valves and individual compressors

(*b*) *p-h* diagram

Fig. 4.26

The mass of the refrigerant vapour passing through compressor C_2 is given by,

$$m_{C2} = m_{e1} + m_{e2} + m_v$$

where m_v, is the mass of refrigerant vapour carried with the liquid refrigerant of $(m_{e1} + m_{e2})$, and is given by:

$$m_v = (m_{e1} + m_{e2}) \frac{x_6}{1 - x_6} \qquad ...(4.52)$$

Power required to drive the system,

$$\mathbf{P} = \frac{m_{C1}(h_3 - h_2) + m_{C2}(h_4 - h_8)}{60} \text{ kW} \qquad ...(4.53)$$

Coefficient of performance of the system,

$$\textbf{C.O.P.} = \frac{14000(Q_1 + Q_2)/3600}{P} \qquad ...(4.54)$$

Example 4.11. *A compound refrigeration system is used for multi-load purposes, as shown in Fig. 4.27. Freon-12 is used as refrigerant. There is no undercooling. Using p-h chart, determine the following:*

(i) The power required to run the system, and

(ii) The coefficient of performance of the combined system.

Solution: The *p-h* diagram of the given system is shown in Fig. 4.28. The various values as read from *p-h* chart are as follows:

$h_7 = h_8 = h_1 = 64.6$ kJ/kg; $h_2 = 183.2$ kJ/kg; $h_4 = 191.7$ kJ/kg; $h_3 = 202$ kJ/kg

The mass of refrigerant passing through the evaporator E_1 or compressor C_1,

$$m_{e1} (= m_{C1}) = \frac{14000 Q_1}{60(h_2 - h_1)} \text{ kg/min} = \frac{14000 \times 9}{60(183.2 - 64.6)} = 17.7 \text{ kg/min}$$

The mass of refrigerant passing through the evaporator E_2,

$$m_{e2} = \frac{14000 Q_2}{60(h_4 - h_8)} \text{ kg/min} = \frac{14000 \times 18}{60(191.7 - 64.6)} = 33.04 \text{ kg/min}$$

The refrigerant leaving the compressor C_1 at point 3 is mixed with the refrigerant coming out the evaporator E_2 at point 4 before entering into compressor C_2. The condition of refrigerant after mixing is shown by point 5 on the *p-h* diagram. The enthalpy at point 5 is given by,

$$h_5 = \frac{m_{e1} h_3 + m_{e2} h_4}{m_{e1} + m_{e2}} = \frac{17.7 \times 202 + 33.04 \times 191.7}{17.7 + 33.04} = 195.3 \text{ kJ/kg}$$

Mark point 5 on *p-h* diagram such that $h_5 = 195.3$ kJ/kg. The enthalpy at point 6 as read from *p-h* diagram is,

$$h_6 = 206 \text{ kJ/kg}$$

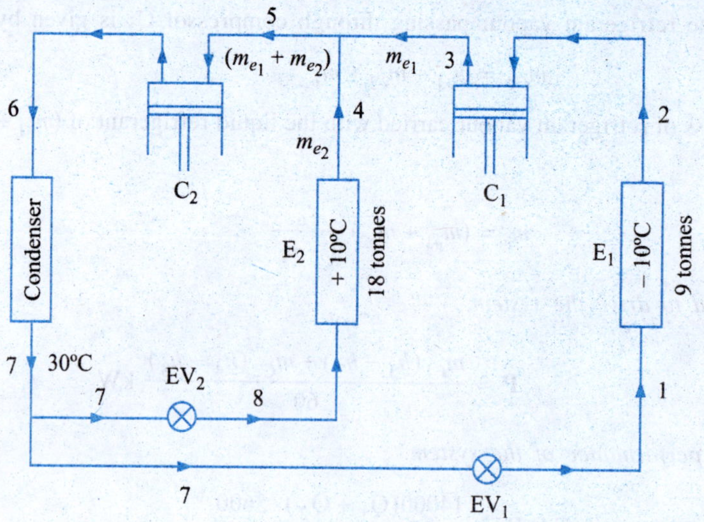

Fig. 4.27

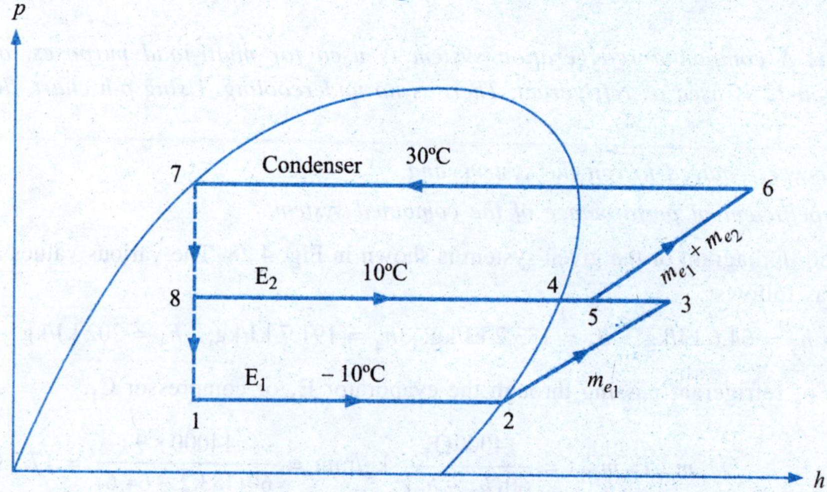

Fig. 4.28. *p-h* diagram

***(i)* The power required to run the system, P:**

$$P = P_{C1} + P_{C2} = \left[\frac{m_{C1}(h_3 - h_2)}{60} + \frac{m_{C2}(h_6 - h_5)}{60}\right] \text{kW}$$

$$= \left[\frac{m_{e1}(h_3 - h_2)}{60} + \frac{(m_{e1} + m_{e2})(h_6 - h_5)}{60}\right] \text{kW}$$

$$= \frac{1}{60}\,[m_{e1}\,(h_3 - h_2) + \left(m_{e1} + m_{e2}\right)(h_6 - h_5)]\ \text{kW}$$

$$= \frac{1}{60}\,[17.7\,(202 - 183.2) + (17.7 + 33.04)\,(206 - 195.3)] = \mathbf{14.59\ kW}$$

***(ii)* The coefficient of performance of the combined system,**

$$\text{C.O.P.} = \frac{\text{RE}}{\text{Power}} = \frac{14000\,(Q_1 + Q_2)/3600}{P} = \frac{14000(9 + 18)/3600}{14.59} = \mathbf{7.19} \quad \textbf{(Ans.)}$$

Example 4.12. *A refrigeration plant comprises three evaporators of capacities 10 tonnes at 10°C, 20 tonnes at 5°C and 30 tonnes at –10°C with individual expansion valves and individual compressors but one condenser operating at 40°C and sub-cooling the liquid to 30°C. All the evaporators discharge dry saturated refrigerant Freon 12 to the compressor. Assume isentropic compression in each compressor. Use p-h diagram for R-12. Determine:*

(*i*) *Refrigeration effect in each evaporator;*

(*ii*) *Mass flow rate in each evaporator;*

(*iii*) *Compressor power required in each compressor;*

(*iv*) *Heat rejected in the condenser;*

(*v*) *Coefficient of performance.* **[Anna University]**

Solution: *Given:* Q_1 = 30 tonnes; Q_2 = 20 tonnes; Q_3 = 10 tonnes.

The arrangement/flow diagram of the plant is shown in Fig. 4.29 (*a*) and the cycle is shown on *p-h* diagram in Fig. 4.29 (*b*).

From *p-h* chart for R-12, we have:

$$h_1 = h_4 = h_7 = h_{11} = 64.6 \text{ kJ/kg}; \quad h_2 = 183.2 \text{ kJ/kg}; \quad h_5 = 189.7 \text{ kJ/kg}; \quad h_8 = 191.7 \text{ kJ/kg};$$

$$h_3 = 210 \text{ kJ/kg}; \quad h_6 = 208 \text{ kJ/kg}; \quad h_9 = 206 \text{ kJ/kg}.$$

(*i*) **Refrigeration effect (RE) in each evaporator:**

RE in evaporator, $E_1 = (h_2 - h_1) = (183.2 - 64.6) =$ **118.6 kJ/kg (Ans.)**

RE in evaporator $E_2 = (h_5 - h_4) = (189.7 - 64.6) =$ **125.1 kJ/kg (Ans.)**

RE in evaporator $E_3 = (h_8 - h_7) = (191.7 - 64.6) =$ **127.1 kJ/kg (Ans.)**

(*ii*) **Mass flow rate in each evaporator; m_{e1}, m_{e2}, m_{e3}:**

$$m_{e1} = \frac{14000\,Q_1}{60(h_2 - h_1)} = \frac{14000 \times 30}{60 \times 118.6} = \mathbf{59.02 \text{ kg/min}} \quad \textbf{(Ans.)}$$

$$m_{e2} = \frac{14000\,Q_2}{60(h_5 - h_4)} = \frac{14000 \times 20}{60 \times 125.1} = \mathbf{37.3 \text{ kg/min}} \quad \textbf{(Ans.)}$$

$$m_{e3} = \frac{14000\,Q_3}{60(h_8 - h_7)} = \frac{14000 \times 10}{60 \times 127.1} = \mathbf{18.36 \text{ kg/min}} \quad \textbf{(Ans.)}$$

(*iii*) **Compressor power required in each compressor; P_{C1}, P_{C2}, P_{C3}:**

$$P_{C1} = \frac{m_{C1}(h_3 - h_2)}{60} = \frac{m_{e1}(h_3 - h_2)}{60} \text{ kW}$$

$$= \frac{59.02(210 - 183.2)}{60} = \mathbf{26.36 \text{ kW}} \quad \textbf{(Ans.)}$$

$$P_{C2} = \frac{m_{C2}(h_6 - h_5)}{60} = \frac{m_{e2}(h_6 - h_5)}{60} = \frac{37.3(208 - 189.7)}{60} = \mathbf{11.37 \text{ kW}} \quad \textbf{(Ans.)}$$

$$P_{C3} = \frac{m_{C3}(h_9 - h_8)}{60} = \frac{m_{e3}(h_9 - h_8)}{60} = \frac{18.36(206 - 191.7)}{60} = \mathbf{4.37 \text{ kW}} \quad \textbf{(Ans.)}$$

Total work, $P = P_{C1} + P_{C2} + P_{C3} = 26.36 + 11.37 + 4.37 = 42.1$ kW

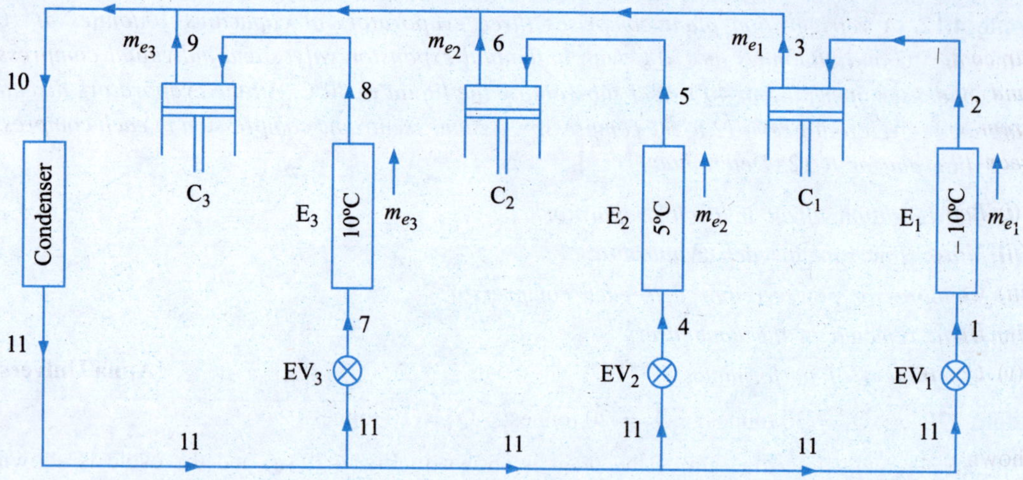

C_1, C_2, C_3 = Compressors; E_1, E_2, E_3 = Evaporators; EV_1, EV_2, EV_3 = Expansion valves.

(*a*) Flow diagram

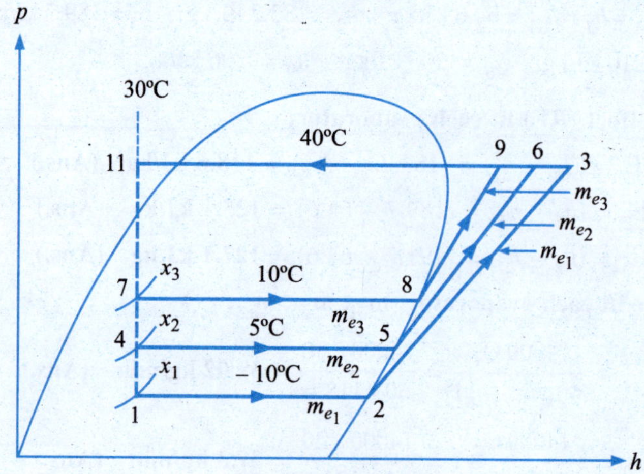

(*b*) *p-h* diagram

Fig. 4.29

(*iv*) **Heat rejected in the condenser:**

$$\text{Heat rejected in the condenser} = m_{e1}(h_3 - h_{11}) + m_{e2}(h_6 - h_{11}) + m_{e3}(h_9 - h_{11})$$

$$= 59.02\ (210 - 64.6) + 37.3\ (208 - 64.6) + 18.36\ (206 - 64.6)$$

$$= \mathbf{16526.43\ kJ/min} \quad \textbf{(Ans.)}$$

(*v*) **Coefficient of performance, C.O.P.:**

$$\text{C.O.P.} = \frac{\text{Total RE}}{\text{Total power}} = \frac{14000\,(Q_1 + Q_2 + Q_3)}{3600 \times P}$$

$$= \frac{14000(30 + 20 + 10)}{3600 \times 42.1} = \mathbf{5.54} \quad \textbf{(Ans.)}$$

Example 4.13. *In a multiple expansion valve and individual compressor and evaporator system of vapour compression plant there are three evaporators of capacities 10 tonnes at 10°C, 20 tonnes at 5°C and 30 tonnes at –10°C. One condenser at 40°C handles the refrigerant flow and sub-cools the exit from it to 30°C. The discharge from each evaporator is dry saturated and the refrigerant used in Freon-12. Compression in each compressor may be assumed isentropic. Determine the following:*

(*i*) *Refrigerating effect/kg in each evaporator;*

(*ii*) *Mass flow rate in each evaporator;*

(*iii*) *Compressor power;*

(*iv*) *Heat rejected in the condenser;*

(*v*) *Co-efficient of performance of the plant.*

Solution: The schematic arrangement/flow diagram of the system and its corresponding *p-h* diagram are shown in Fig. 4.30 (*a*) and Fig. 4.30 (*b*) respectively.

From *p-h* chart/properties tables, we have the following enthalpies:

$h_9 = h_{10} = 64.6$ kJ/kg; $h_{11} = h_{12} = 45.4$ kJ/kg; $h_6 = 191.7$ kJ/kg; $h_{13} = h_1 = 40.7$ kJ/kg;
$h_4 = 189.7$ kJ/kg; $h_2 = 183.2$ kJ/kg; $h_3 = 210$ kJ/kg; $h_5 = 208$ kJ/kg; $h_7 \simeq 206$ kJ/kg.

(*i*) **Refrigerating effect in each evaporator:**

RE in evaporator $E_1 = (h_2 - h_1) = (183.2 - 40.7) =$ **142.5 kJ/kg (Ans.)**

RE in evaporator $E_2 = (h_4 - h_{12}) = (189.7 - 45.4) =$ **144.3 kJ/kg (Ans.)**

RE in evaporator $E_3 = (h_6 - h_{10}) = (191.7 - 64.6) =$ **127.1 kJ/kg (Ans.)**

(*ii*) **Mass flow rate in each evaporator; m_{e1}, m_{e2}, m_{e3}:**

$$m_{e1} = \frac{14000\,Q_1}{60(h_2 - h_1)} = \frac{14000 \times 30}{60 \times 142.5} = \mathbf{49.12\ kg/min} \quad \textbf{(Ans.)}$$

- The quantity of refrigerant of the properties as at point 12 needed for the total load of 20 tonnes for evaporator E_2 is given by,

$$(m_{el})_2 = \frac{14000\,Q_2}{60(h_4 - h_{12})} = \frac{14000 \times 20}{60 \times 144.3} = 32.34 \text{ kg/min}$$

In addition to this, the vapour formed when 49.12 kg/min for evaporator E_1 expanded from 4.234 bar (10°C) to 3.627 bar (5°C) also passes through evaporator 2. The quality of the mixture leaving second expansion valve as read from *p-h* chart is $x_2 = 0.04$. Thus the additional vapour,

$$(m_{ev})_2 = \left(\frac{x_2}{1 - x_2}\right) m_{el} = \left(\frac{0.04}{1 - 0.04}\right) \times 49.12 = 2.047 \text{ kg/min}$$

∴ Total rate of refrigerant flow through evaporator E_2,

$$m_{e2} = (m_{el})_2 + (m_{ev})_2 = 32.34 + 2.047 = 34.39 \text{ kg/min}$$

- The rate of refrigerant as at point '10' required for the evaporator 3,

$$(m_{el})_3 = \frac{14000 \times 10}{60(h_6 - h_{10})} = \frac{14000 \times 10}{60(191.7 - 64.6)} = 18.36 \text{ kg/min}$$

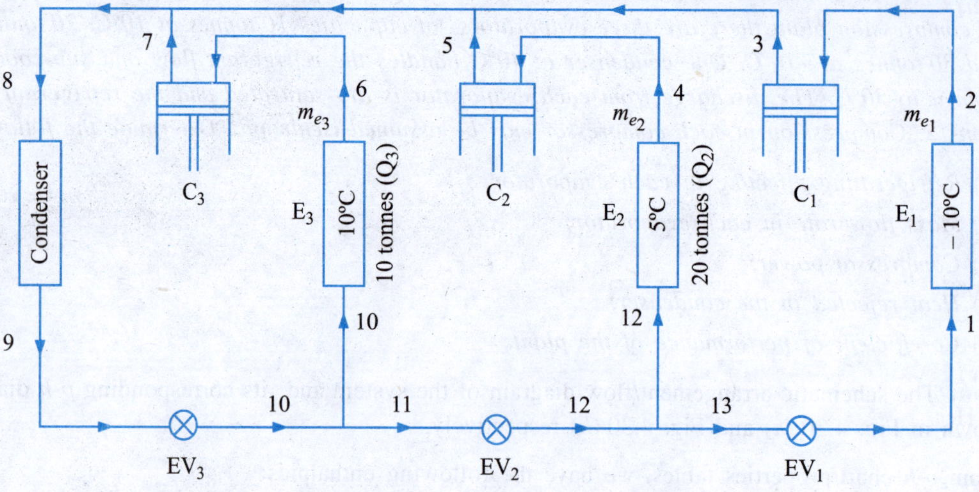

(*a*) Schematic arrangement/Flow diagram of the system

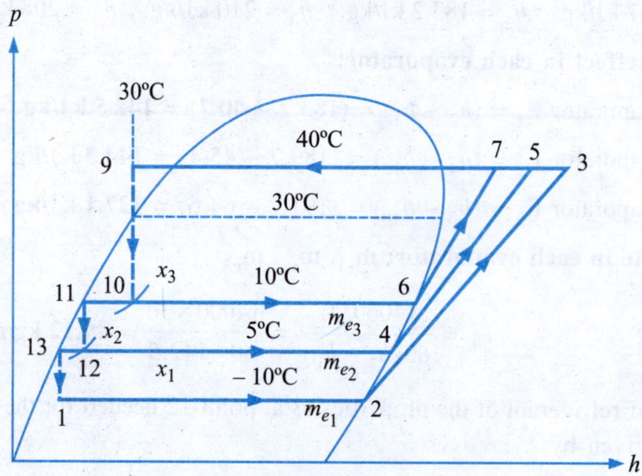

(*b*) *p-h* diagram

Fig. 4.30

The additional vapour flowing through evaporator E_3 without doing any useful cooling,

$$(m_{ev})_3 = \frac{x_3}{1 - x_3}\left(m_{e1} + m_{e2}\right)$$

x_3 read from *p-h* chart is 0.13.

$$\therefore \quad (m_{ev})_3 = \frac{0.13}{1 - 0.13}\,(49.12 + 34.39) = 12.48 \text{ kg/min}$$

Thus, total rate of refrigerant flow through evaporator E_3,

$$m_{e3} = (m_{el})_3 + (m_{ev})_3 = 18.36 + 12.48 = 30.84 \text{ kg/min.}$$

(*iii*) **Compressor power; P:**

$$P_{C1} = \frac{m_{e_1}(h_3 - h_2)}{60} \text{ kW} = \frac{49.12(210 - 183.2)}{60} = 21.94 \text{ kW}$$

$$P_{C2} = \frac{m_{e_2}(h_5 - h_4)}{60} \text{ kW} = \frac{34.39(208 - 189.7)}{60} = 10.49 \text{ kW}$$

$$P_{C3} = \frac{m_{e_3}(h_7 - h_6)}{60} \text{ kW} = \frac{30.84(206 - 191.7)}{60} = 7.35 \text{ kW}$$

Total power input to the plant,

$$P = P_{C1} + P_{C2} + P_{C3} = 21.94 + 10.49 + 7.35 = 39.78 \text{ kW}$$

(*iv*) **Heat rejected in the condenser:**

Heat rejected in the condenser $= m_{e1}(h_3 - h_9) + m_{e2}(h_5 - h_9) + m_{e3}(h_7 - h_9)$

$= 49.12 (210 - 64.6) + 34.39 (208 - 64.6) + 30.84 (206 - 64.6)$

$=$ **16434.35 kJ/min (Ans.)**

(*v*) **Co-efficient of performance of the plant, C.O.P.:**

$$\text{C.O.P.} = \frac{14000(Q_1 + Q_2 + Q_3)/3600}{P}$$

$$= \frac{14000(30 + 20 + 10)/3600}{39.78} = \mathbf{5.86} \quad \textbf{(Ans.)}$$

Example 4.14. *A compound refrigeration system using R-12 as refrigerant consists of three evaporators of capacities 18 tonnes at – 5°C, 27 tonnes at 0°C and 9 tonnes at 5°C. The vapours leaving the evaporators are dry and saturated. The system is provided with individual expansion valves and flash intercoolers, as shown in Fig. 4.31. The condenser temperature is 40°C and the liquid refrigerant leaving the condenser is sub-cooled to 30°C. Assuming isentropic compression at each stage, determine the following:*

(*i*) *The mass of refrigerant passing through each compressor;*

(*ii*) *The power required to drive the system;*

(*iii*) *The coefficient of performance of the system.*

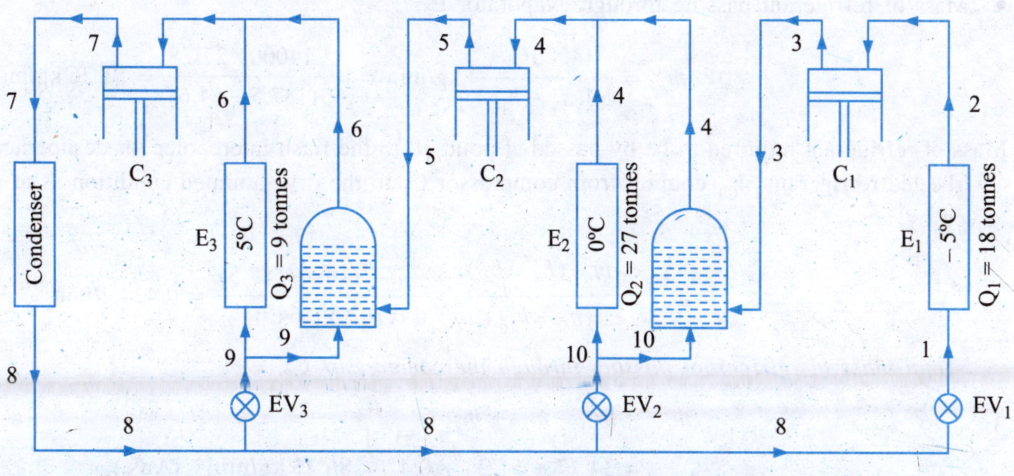

Fig. 4.31. The schematic arrangement of the system.

Solution: Fig. 4.32 shows the *p-h* diagram of the system. The values of enthalpies, as read from *p-h* chart, at various state points, are as follows:

$$h_8 = h_9 = h_{10} = h_1 = 64.6 \text{ kJ/kg}; \quad h_2 = 185.4 \text{ kJ/kg}; \quad h_4 = 187.5 \text{ kJ/kg}; \quad h_6 = 189.6 \text{ kJ/kg};$$

$$h_3 = 189 \text{ kJ/kg}; \quad h_5 = 191 \text{ kJ/kg}; \quad h_7 = 206 \text{ kJ/kg}.$$

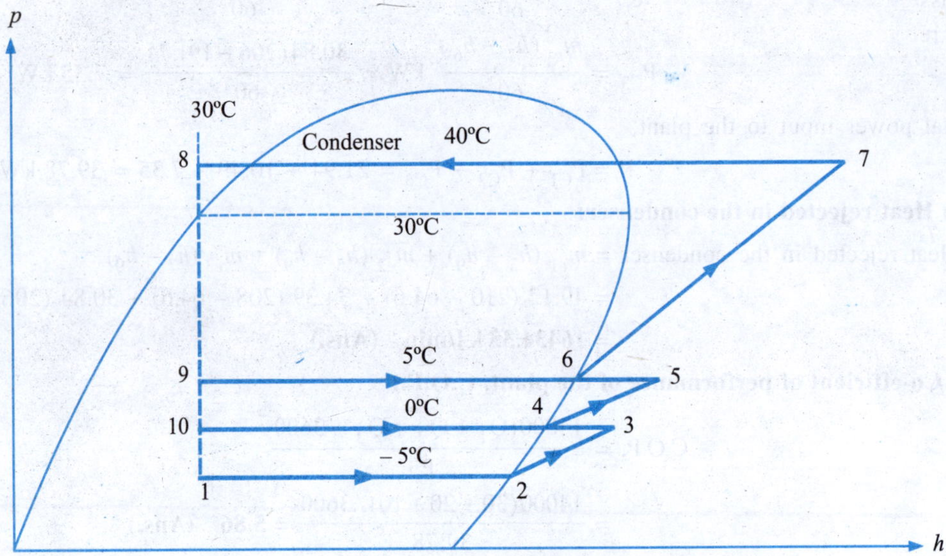

Fig. 4.32. *p-h* diagram.

(*i*) **The mass of refrigerant passing through each compressor:**

- Mass of refrigerant passing through evaporator E_1 or compressor C_1,

$$m_{e1} \ (= m_{C1}) = \frac{14000 Q_1}{60(h_2 - h_1)} \text{ kg/min}$$

$$\text{or} \quad m_{C1} = \frac{14000 \times 18}{60(185.4 - 64.6)} = \textbf{34.77 kg/min} \quad \textbf{(Ans.)}$$

- Mass of refrigerant passing through evaporator E_2,

$$m_{e2} = \frac{14000 Q_2}{60(h_4 - h_{10})} \text{ kg/min} = \frac{14000 \times 27}{60(187.5 - 64.6)} = 51.26 \text{ kg/min}$$

Mass of refrigerant required to be by-passed at point 10 to the flash intercooler for desuperheating the superheated refrigerant m_{C1} coming from compressor C_1 to the dry saturated condition as at point 4 is given by,

$$m'_{e2} = \frac{m_{C_1}(h_3 - h_4)}{(h_4 - h_{10})} = \frac{34.77(189 - 187.5)}{(187.5 - 64.6)} = 0.42 \text{ kg/min}$$

∴ *Total mass of refrigerant passing through the compressor* C_2,

$$m_{C2} = m_{e1} + m_{e2} + m'_{e2}$$

$$= 34.77 + 51.26 + 0.42 = \textbf{86.45 kg/min} \quad \textbf{(Ans.)}$$

- Mass of refrigerant passing through the evaporator E_3,

$$m_{e3} = \frac{14000Q_3}{60(h_6 - h_9)} \text{ kg/min} = \frac{14000 \times 9}{60(189.6 - 64.6)} = 16.8 \text{ kg/min}$$

Mass of refrigerant by-passed at point 9 to the flash intercooler for desuperheating the superheated vapour refrigerant (m_{C2}) coming from compressor C_2 to the dry saturated condition at point 6 is given by,

$$m'_{e3} = \frac{m_{C_2}(h_5 - h_6)}{(h_6 - h_9)} = \frac{86.45(191 - 189.6)}{(189.6 - 64.6)} = 0.97 \text{ kg/min}$$

∴ *Total mass of refrigerant passing through compressor C_3*

$$m_{C3} = m_{C2} + m_{e3} + m'_{e3}$$

$$= 86.45 + 16.8 + 0.97 = \mathbf{104.22\ kg/min} \quad \textbf{(Ans.)}$$

(*iii*) **The power required to drive the system, P:**

$$P = P_{C1} + P_{C2} + P_{C3}$$

$$= \frac{m_{C1}(h_3 - h_2)}{60} + \frac{m_{C2}(h_5 - h_4)}{60} + \frac{m_{C3}(h_7 - h_6)}{60}$$

$$= \frac{1}{60}[m_{C1}(h_3 - h_2) + m_{C2}(h_5 - h_4) + m_{C3}(h_7 - h_6)]$$

$$= \frac{1}{60}[34.77(189 - 185.4) + 86.45(191 - 187.5) + 104.22(206 - 189.6)]$$

$$= \mathbf{35.6} \quad \textbf{(Ans.)}$$

(*ii*) **The coefficient of performance of the system, C.O.P.:**

$$\text{C.O.P.} = \frac{\text{RE}}{\text{Power}} = \frac{14000(Q_1 + Q_2 + Q_3)/3600}{P}$$

$$= \frac{14000(18 + 27 + 9)/3600}{35.6} = \mathbf{5.89} \quad \textbf{(Ans.)}$$

Example 4.15. *A compound compression refrigeration system using R-12 as refrigerant consists of evaporators of capacities 27 tonnes at –10°C, 18 tonnes at 5°C and 9 tonnes at 10°C. The vapours leaving the evaporators are dry and saturated. The system is provided with multiple expansion valves and flash intercoolers as shown in Fig. 4.33. The condenser temperature is 40°C and the liquid refrigerant leaving the condenser is sub-cooled to 30°C. Assuming isentropic at each stage, determine the following:*

(*i*) *The mass of refrigerant passing through each compressor;*

(*ii*) *The power required to drive the system;*

(*iii*) *The coefficient of performance of the system.*

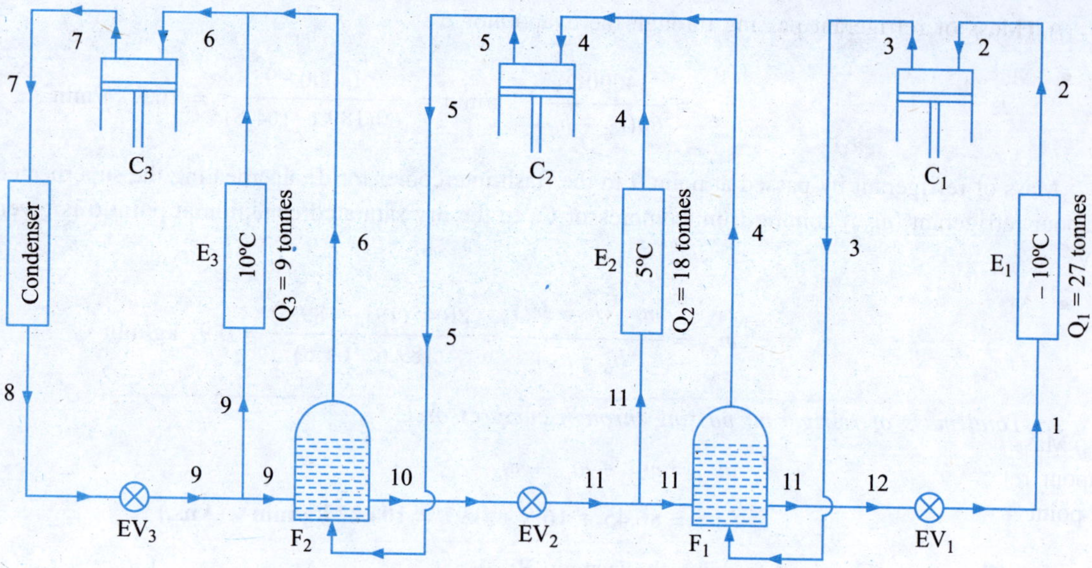

C_1, C_2, C_3 = Compressors; E_1, E_2, E_3 = Evaporators;
EV_1, EV_2, EV_3 = Expansion valves; F_1, F_2 = Flash chambers.

Fig. 4.33. The schematic arrangement/Flow diagram of the system.

Solution: *Given:* Q_1 = 27 tonnes; Q_2 = 18 tonnes; Q_3 = 9 tonnes.

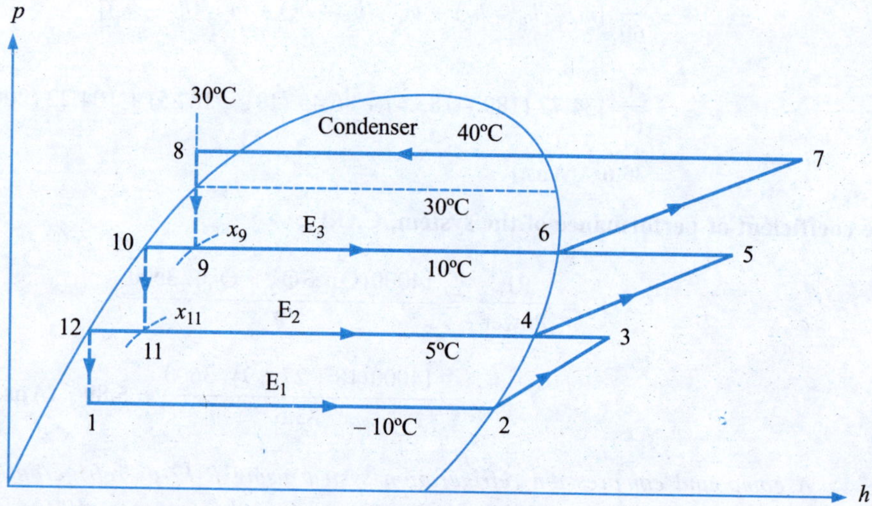

Fig. 4.34. *p-h* diagram.

Fig. 4.34 shows the *p-h* diagram of the system. The values of enthalpies etc. at various state points, as read from *p-h* chart, are as follows:

$h_{12} = h_1$ = 40.7 kJ/kg; h_2 = 183.2 kJ/kg; h_3 = 194.2 kJ/kg; h_4 = 189.7 kJ/kg;

h_5 = 195.4 kJ/kg; h_6 = 191.7 kJ/kg; h_7 = 207.2 kJ/kg; $h_8 = h_9$ = 64.6 kJ/kg;

$h_{10} = h_{11}$ = 45.4 kJ/kg; x_9 = 0.13; x_{11} = 0.04.

(*i*) **The mass of refrigerant passing through each compressor:**

- Mass of refrigerant passing through evaporator E_1 or compressor C_1,

$$\boldsymbol{m_{C1}} = m_{e1} = \frac{14000\,Q_1}{60(h_2 - h_1)} \text{ kg/min}$$

$$= \frac{14000 \times 27}{60(183.2 - 40.7)} = \mathbf{44.2\ kg/min} \quad \textbf{(Ans.)}$$

- Mass of refrigerant required to be circulated through the evaporator E_2,

$$m_{e_2} = \frac{14000\,Q_2}{60(h_4 - h_{11})} \text{ kg/min} = \frac{14000 \times 18}{60(189.7 - 45.4)} = 29.1 \text{ kg/min.}$$

Mass of refrigerant required in the flash intercooler at point 11 for desuperheating the superheated vapour refrigerant m_{C1} = 44.2 kg/min coming from compressor C1, to the dry saturated condition as at point 4 is given by,

$$m'_{e2} = \frac{m_{C_1}(h_3 - h_4)}{(h_4 - h_{11})} = \frac{44.2(194.2 - 189.7)}{(189.7 - 45.4)} = 1.38 \text{ kg/min.}$$

The evaporator 2 is also supplied with vapours formed during expansion of m_{C1} = 44.2 kg/min refrigerant while passing through the expansion valve EV_2.

∴ Mass of vapour formed by m_{C1} = 44.2 kg/min. while passing through EV_2,

$$m''_{e2} = m_{C1}\left(\frac{x_{11}}{1 - x_{11}}\right) = 44.2\left(\frac{0.04}{1 - 0.04}\right) = 1.84 \text{ kg/min}$$

∴ Total mass of refrigerant passing through the compressor C_2,

$$\boldsymbol{m_{C2}} = m_{C1} + m_{e2} + m'_{e2} + m''_{e2}$$

$$= 44.2 + 29.1 + 1.38 + 1.84 = \mathbf{76.52\ kg/min} \quad \textbf{(Ans.)}$$

- Now, mass of refrigerant required to be circulated through the evaporator E_3,

$$m_{e3} = \frac{14000\,Q_3}{60(h_6 - h_9)} \text{ kg/min} = \frac{14000 \times 9}{60(191.7 - 64.6)} = 16.52 \text{ kg/min}$$

Mass of refrigerant required in flash intercooler at point 9 for desuperheating the superheated vapour refrigerant m_{C2} = 76.52 kg/min. coming from the compressor C_2 to the dry saturated condition as at point 6 is given by,

$$m'_{e3} = \frac{m_{C2}(h_5 - h_6)}{(h_6 - h_9)} = \frac{76.52(195.4 - 191.7)}{(191.7 - 64.6)} = 2.23 \text{ kg/min}$$

The evaporator 3 is also supplied with the vapour formed by the expansion of m_{C2} = 76.52 kg/min. refrigerant while passing through the expansion valve EV_3.

∴ Mass of vapours formed by m_{C2} = 76.52 kg/min while passing through EV_3,

$$m''_{e3} = m_{C2}\left(\frac{x_9}{1 - x_9}\right) = 76.52\left(\frac{0.13}{1 - 0.13}\right) = 11.43 \text{ kg/min}$$

∴ Total mass of refrigerant passing through compressor C_3,

$$m_{C3} = m_{C2} + m_{e3} + m'_{e3} + m''_{e3}$$

$$= 76.52 + 16.52 + 2.23 + 11.43 = \mathbf{106.7\ kg/min} \quad \textbf{(Ans.)}$$

(*ii*) **Power required to drive the system, P:**

$$P = P_{C1} + P_{C2} + P_{C3}$$

$$= \left[\frac{m_{C1}(h_3 - h_2)}{60} + \frac{m_{C2}(h_5 - h_4)}{60} + \frac{m_{C3}(h_7 - h_6)}{60}\right] \text{kW}$$

$$= \frac{1}{60}\,[m_{C_1}(h_3 - h_2) + m_{C_2}(h_5 - h_4) + m_{C_3}(h_7 - h_6)] \text{ kW}$$

$$= \frac{1}{60}\,[44.2\,(194.2 - 183.2) + 76.52\,(195.4 - 189.7) + 106.7\,(207.2 - 191.7)]$$

$$= \mathbf{42.94\ kW} \quad \textbf{(Ans.)}$$

OBJECTIVE TYPE QUESTIONS

Choose the correct Answer:

1. A system with multiple evaporators at different temperatures with compound compression will
   (*a*) increase the power requirements
   (*b*) reduce the power requirements
   (*c*) neither increase nor reduce the power requirements
   (*d*) any of the above.
2. When water intercooling is used in a multi-stage compression, it
   (*a*) reduces the work to be done in high pressure compressor
   (*b*) reduces the specific volume of the refrigerant
   (*c*) requires a compressor of less stroke volume
   (*d*) all of the above.
3. For the same refrigeration capacity, some evaporator pressure, same condenser pressure, same intermediate pressure
   (*a*) the mass flow rate in L.P. cylinder and H.P cylinder is same for flash cooler system and flash intercooler system
   (*b*) the mass flow rate in L.P cylinder is same for flash cooler and flash intercooler system but mass flow rate in H.P cylinder is more for flash intercooler system
   (*c*) the mass flow rate in L.P cylinder is same for flash cooler and flash intercooler system but mass flow rate in H.P. cylinder is less for flash intercooler system
   (*d*) the mass flow rate in L.P. cylinder is more for flash intercooler system than for flash cooler system.
4. For the same condenser pressure and the same evaporator pressure the multi-staging as compared to single will increase the refrigeration capacity
   (*a*) more for R-717 than for R-12
   (*b*) more for R-12 than for R-717
   (*c*) same for R-717 and for R-12
   (*d*) may be more or may less for R-12 and R-717 depending upon actual values of condenser.
5. For minimum work input to the compressor the optimum intermediate pressure for two stage compressor is
   (*a*) the geometric mean of the suction and the delivery pressure for R-12
   (*b*) very near to the geometric mean of the suction and delivery pressure for R-12
   (*c*) the geometric mean of the suction and the delivery pressure for R-717
   (*d*) very near to the geometric mean of the suction and the delivery pressure for R-717.

6. In compound compression system with flash cooling the suction vapour to the second stage of compression are
   (*a*) normally superheated (*b*) normally wet
   (*c*) normally dry saturated (*d*) may be dry saturated or superheated.
7. For a given evaporator temperature and given condenser temperature the work input to compressor having two stages as compared to that having single stage for the same refrigerating capacity is
   (*a*) less (*b*) more (*c*) same (*d*) not predictable.
8. Compound compression is adopted for refrigeration system
   (*a*) to improve volumetric efficiency (*b*) to reduce mechanical work input per unit mass
   (*c*) to take care of very large pressure ratios (*d*) to achieve all of the above.
9. For the same range of evaporator pressure and condenser pressure, the total power consumption in case of two stage compression with intercooling by liquid refrigerant as compared to that for single stage compression is
   (*a*) less for R-12 and less for R-717 (*b*) more for R-12 and more for R-717
   (*c*) more for R-12 and less for R-717 (*d*) less for R-12 and more for R-717.
10. For the same clearance factor of compressor and same evaporator and condenser temperatures, single stage compressor as compared to two-stage compressor gives
    (*a*) lower volumetric efficiency (*b*) higher volumetric efficiency
    (*c*) same volumetric efficiency (*d*) none of the above.
11. In compound compression system with intercooling the suction vapours to the second stage of compression are
    (*a*) superheated (*b*) wet
    (*c*) dry saturated (*d*) may be dry saturated or superheated.
12. Multistage compression with intercooling
    (*a*) permits better lubrication as the discharge temperatures are lowered
    (*b*) permits better lubrication as the discharge pressures are lowered
    (*c*) does not affect lubrication as the discharge pressure and temperature are un-affected
    (*d*) does not permit any lubrication in higher stages of compression.
13. For the same refrigerant, same eveporator pressure, same condenser pressure, same intermediate pressure for two-stage vapour compression system
    (*a*) flash intercooling gives better C.O.P. than flash cooling
    (*b*) flash cooling gives better C.O.P. than flash intercooling
    (*c*) both flash cooling and flash intercooling give same C.O.P.
    (*d*) none of the above.
14. In a system with multiple evaporators at different temperatures with compound compression and flash intercooler, the suction vapour to the high pressure stage is
    (*a*) dry saturated (*b*) superheated (*c*) saturated liquid
15. In multiple evaporators at different temperatures and individual expansion valve with compound compression and flash intercoolers system of ideal vapour compression cycle
    (*a*) suction vapour to high pressure stage compressor is dry saturated vapour
    (*b*) inlet to the high pressure throttle valves is wet vapour
    (*c*) inlet to the low pressure throttle valves is wet vapour
    (*d*) all of the above.
16. In multiple evaporators at different temperatures with multiple expansion valves with compound compression and flash intercooler system of ideal vapour compression cycle
    (*a*) suction vapour to the high pressure stage is dry saturated
    (*b*) suction vapour to the high pressure stage is superheated

(*c*) inlet to the high pressure evaporator is saturated liquid
(*d*) inlet to the low pressure evaporator is saturated liquid.

17. In multiple evaporators at different temperature with multiple expansion valves with compound compression and flash intercooler system of ideal vapour compression cycle
(*a*) suction vapour to the high pressure stage is dry saturated
(*b*) inlet to the high pressure throttle valve is liquid
(*c*) inlet to the low pressure throttle valve is liquid
(d) all of the above.

ANSWERS

1. (*b*)	2. (*d*)	3. (*b*)	4. (*a*)	5. (*b*)	6. (*a*)	7. (*a*)	8. (*d*)
9. (*b*)	10. (*a*)	11. (*c*)	12. (*a*)	13. (*a*)	14. (*a*)	15. (*d*)	16. (*a*)
17. (*d*)							

THEORETICAL QUESTIONS

1. What are the advantages of compound compression with intercooler over single stage compression?
2. Why is compound refrigeration system employed when the condenser temperature to evaporator temperature range is very large?
3. Explain the different between compound vapour compression system with flash cooling and flash intercooling.
4. What is the function of flash intercooler provided in a compound vapour compression refrigerating system?
5. Explain two stage compression with intercooling and sub-cooling by external cooling source.
6. Explain compound compression with flash intercooling with single expansion valve along the flow line to evaporator.
7. Explain compound compression with multiple expansion valves in series with flash chambers but no intercooling.
8. Explain compound compression with multiple expansion valves and flash intercooling system.
9. Explain by flow diagram and on *p-h* diagram the two evaporators at different temperatures with compound compression and perfect intercooling, vapour compression refrigeration system.
10. Explain by flow diagram and on *p-h* diagram, the two evaporators with individual compressor, with multi-expansion valves and flash chamber, vapour compression refrigeration system.
11. Explain by flow diagram and on *p-h* diagram the two evaporators with compound compression, with multiple expansion valves and with flash intercooling, vapour compression refrigeration system.

UNSOLVED EXAMPLES

1. A two-stage compression machine with a flash intercooler is to produce 22.5 tonnes of refrigeration while working between –30ºC and 42ºC. The pressure in the flash intercooler is the geometric mean of the upper and lower pressure limits. Sketch the system and cycle on *p-h* diagram giving relative property values. The enthalpy at the end of L.P. compressor is 195 kJ/kg and that of H.P. compressor 211 kJ/kg. The working medium is R-12. Determine the following:

(*i*) Coefficient of performance of the system;

(*ii*) Power required to run the system;

(*iii*) Fluid flows through L.P. and H.P. compressors;

(*iv*) Volume handled by L.P. compressor.

[**Ans.** (*i*) 2.6; (*ii*) 33.73 kW; 0.6346 kg/s, 0.974 kg/s; 0.1015 m^3/s]

2. The following data refer to a two-stage compression ammonia refrigeration system with water intercooler, liquid sub-cooler and a liquid flash chamber:

Condenser pressure = 14 bar; Evaporator pressure = 1.55 bar; Intercooler pressure = 4.92 bar; Temperature of ammonia leaving the intercooler = 32°C; Temperature of ammonia leaving the sub-cooler = 30°C; Volumetric efficiency of L.P. compressor = 85%; Volumetric efficiency of H.P. compressor = 80%.

If the load on the evaporator is 45 tonnes of refrigeration, determine the following:

(*i*) Rate of ammonia circulation per minute.

(*ii*) Power required to drive the compressors;

(*iii*) Piston displacement for both L.P. and H.P. compressors in m^3/min;

(*iv*) Diameters of the cylinders for the single acting L.P. and H.P. compressors, when the speed of compressors is 300 r.p.m. and the stroke is equal to the diameter of L.P. cylinder;

(*v*) C.O.P. of the system.

[**Ans.** (*i*) 8.62 kg/min; (*ii*) 53 kW; (*iii*) 7.9 m^3/min; 3.35 m^3/min; (*iv*) 322 mm, 210 mm; (*v*) 3.32]

3. The following data relate to a 9 tonnes refrigeration three-stage compression system:

Condenser pressure = 15 bar; Evaporator pressure = 2.5 bar; Intermediate pressures = 5 bar and 10 bar.

(*i*) Find the power required to drive the system when it is provided with (*a*) water intercooling; and (*b*) flash intercooling.

(*ii*) Compare the C.O.P. of the above system with that of simple saturation cycle working between the same overall pressure limits.

[**Ans.** (*i*) 8 kW, 8.5 kW; (*ii*) 6.96% increase, 0.87% increase]

4. A single compressor, using Freon-12 as refrigerant, has three evaporators of capacities 9 tonnes, 27 tonnes and 18 tonnes of refrigeration. The temperatures in all the three evaporators is to be maintained at – 5°C. The vapours leaving the evaporators are dry saturated. The condenser temperature is 40°C. The liquid refrigerant leaving the condenser is sub-cooled to 30°C. Assuming isentropic compression, find the following:

(*i*) Power required to drive the compressor, and

(*ii*) Coefficient of performance of the system. [**Ans.** 41.4 kW, (*ii*) 5.07]

5. A single compressor system using R-12 as refrigerant has three evaporators of capacities 18 tonnes, 27 tonnes and 9 tonnes of refrigeration. The temperatures in the three evaporators are to be maintained at –5°C, 0°C and 7°C respectively. The system is provided with multiple expansion valves and back pressure valves. The condenser pressure is 9.61 bar. The liquid refrigerant leaving the condenser is sub-cooled to 30°C. The vapours leaving the evaporators are dry and saturated. Assuming isentropic compression, determine the following:

(*i*) Mass of refrigerant flowing through each evaporator;

(*ii*) Power required to drive the compressor;

(*iii*) Coefficient of performance of the system.

[**Ans.** (*i*) 28.14 kg/min, 44.9 kg/min, 29.6 kg/min; (*ii*) 43.7 kW; (*ii*) 4.78]

6. A refrigeration system using R-12 as refrigerant consists of three evaporators of capacities 18 tonnes at –5ºC, 27 tonnes at 0ºC and 9 tonnes at 5ºC. The vapours leaving the three evaporators are dry and saturated. The system is provided with individual compressors and multiple expansion valves. The condenser temperature is 40ºC and the liquid refrigerant leaving the condenser is saturated. Assuming isentropic compression in each compressor, determine the following:

   (*i*) Mass of refrigerant flowing through each evaporator,

   (*ii*) The power required to drive the system, and

   (*iii*) The coefficient of performance of the system.

[**Ans.** (*i*) 27.5 kg/min, 42.3 kg/min, 34.2 kg/min; (*ii*) 38.46 kW; (*iii*) 5.58]

❑❑❑

Chapter 5

Vapour Absorption Refrigeration System

5.1 INTRODUCTION

- *The vapour absorption refrigeration is a heat operated system.* It is quite similar to the mechanical vapour compression system which employs reciprocating, centrifugal or rotary compressors. In both the systems, we have the evaporator and condenser. The process of evaporation and condensation of the refrigerant takes place at two different pressure levels to achieve refrigeration in both the cases. *The difference between the two systems lies in the method employed to create the two pressure levels in the system for evaporation and condensation of the refrigerant. They also differ in the manner by which circulation of the refrigerant in the system is achieved.*
- The refrigeration by mechanical vapour compression system is an efficient method, however, the energy input in the shaft-work which is high grade energy and therefore *very expensive. The work required is relatively large because of compression of vapour which undergo large changes in specific volumes* (For example, for an ideal vapour compression system having C.O.P. of 4, the shaft-work will be 25 per cent of the cooling effect obtained). Thus, if *means are available for raising pressure of the refrigerant from evaporator pressure to condenser pressure without much changing its volume it is possible to reduce the shaft-work substantially* (Refer Example 5.1). In order to achieve this, in the *vapour absorption system*, the refrigerant vapour is dissolved in an inert liquid at the same pressure as the evaporator and the solution so formed is pumped to a container at condenser pressure. Thus, *liquid which is practically incompressible and undergoes practically no change in specific volume, requires very little work for raising its pressure.* After raising the pressure, the refrigerant is separated from solution by heating. This vapour which is at condenser pressure, goes to the condenser, expansion valve and evaporator just as in vapour compression cycle.

5.2 SIMPLE VAPOUR ABSORPTION SYSTEM

Refer to Fig. 5.1 for a simple vapour absorption system. It consists of an *absorber*, a *pump*, a *generator* and a *pressure reducing valve* to replace the compressor of vapour compression system. The other components of the system and *condenser, expansion valve* and *evaporator* as in the vapour compression system.

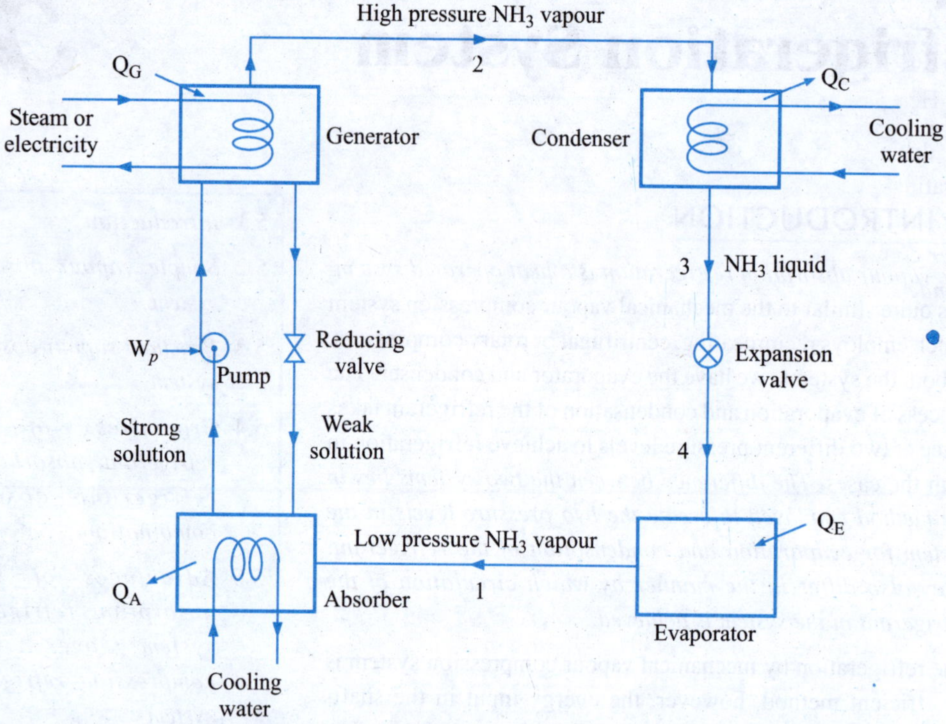

Fig. 5.1. Simple vapour absorption system.

Working of the system:

- The ammonia vapour leaving the **evaporator** at point '1' is readily absorbed in the low temperature hot solution in the **absorber**, releasing the latent heat of condensation. The temperature of the solution tends to rise, while the absorber is cooled by the circulating water, absorbing the heat of solution (Q_A), and maintaining a *constant temperature*.
- *Strong solution*, rich in ammonia, is pumped (pump increases the pressure of the solution upto 10 bar) to the **generator** where heat (Q_G) is supplied from an external source (steam, electricity, gas flame, etc.) Since the boiling point of ammonia is less than that of water, the ammonia vapour is given off from the aqua-ammonia solution at high pressure, and the weak solution returns to the **absorber** through a pressure reducing valve.
- The high pressure ammonia vapour from the generator is condensed in the **condenser** to a high pressure liquid ammonia.
- This liquid ammonia is throttled by the **expansion valve**, and then evaporates, absorbing the heat of evaporation from the surroundings or brine to be chilled. This completes the simple vapour absorption cycle.

Note: The solubility of ammonia in water at low temperatures and pressures is *higher* than it is at high temperatures and pressures.

5.3 PRACTICAL VAPOUR ABSORPTION SYSTEM

Although a simple vapour absorption system can provide refrigeration yet its *operating efficiency is low* (and not very economical).

In order to improve the performance and working of the plant following ***accessories*** are fitted.

1. Analyser
2. Rectifier
3. Heat exchangers

In driving the ammonia vapour out of the solution in the generator, it is impossible to avoid evaporating some of the water. This water vapour going to the condenser along with the ammonia vapour, after condensation may get frozen to ice and block the expansion valve. So an *analyser-rectifier* combination (Fig. 5.2) is *used to eliminate water vapour from the ammonia vapour going into the condenser*.

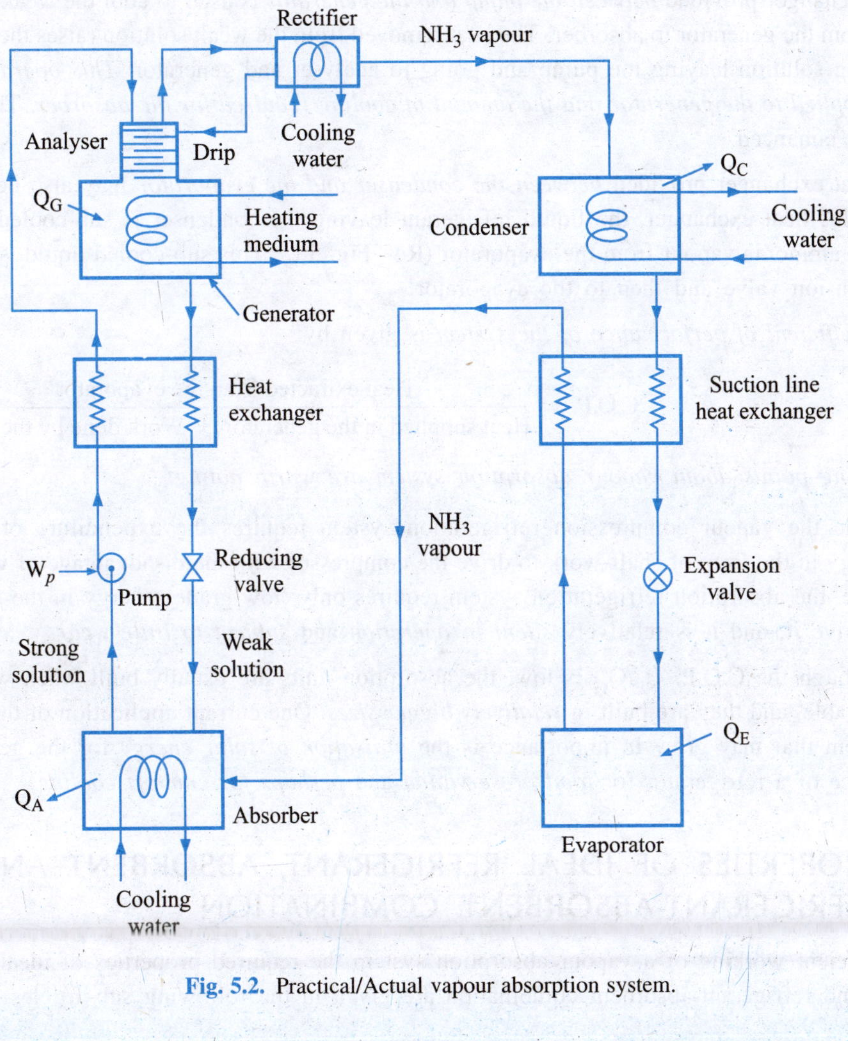

Fig. 5.2. Practical/Actual vapour absorption system.

1. Analyser:

The *analyser* (may be built as an integral part of the generator or made as a separate piece of equipment) is a direct heat exchanger consisting of a series of trays mounted above the generator. The strong solution from the absorber *flows downward* over the trays to cool the outgoing vapours. Since the saturation temperature of water is higher than that of ammonia at a given pressure, it is the water vapour which condenses first. As the vapour passes upward through the analyser, it is cooled and enriched by ammonia, and the liquid is heated. Thus the vapour going to the condenser is lower in temperature and richer in ammonia, and the *heat input to the generator is decreased.*

2. Rectifier:

A *rectifier* is a water-cooled heat exchanger which condenses water vapour (and some ammonia) and sends back (by a drip return pipe) to the generator. Thus *final reduction or elimination of the percentage of water vapour takes place in the rectifier.*

3. Heat exchangers:

The heat exchanger provided *between the pump and the generator* is used to cool the weak hot solution returning from the generator to absorber. The heat removed from the weak solution raises the temperature of the strong solution leaving the pump and going to analyser and generator. *This operation reduces the heat supplied to the generator and the amount of cooling required for the absorber.* Thus the plant economy is enhanced.

The heat exchanger provided *between the condenser and the evaporator* may also be called *sub-cooler*. In this heat exchanger, the liquid refrigerant leaving the condenser is sub-cooled by the low temperature ammonia vapour from the evaporator (Ref. Fig. 5.2). This sub-cooled liquid is now passed to the expansion valve and then to the evaporator.

The *coefficient of performance of the system* is given by:

$$\text{C.O.P.} = \frac{\text{Heat extracted from the evaporator}}{\text{Heat supplied in the generator} + \text{Work done by the pump}}$$

Following points about vapour absorption system are worth noting:

- While the vapour compression refrigeration system requires the expenditure of high grade energy in the form of shaft-work to drive the compressor with the disadvantage of vibration and noise, the absorption refrigeration system requires only 'low grade' energy in the *form of heat* to drive it, and it is relatively *silent in operation* and *subject to little wear*.
- Although the C.O.P. Q_E/Q_G is low, the absorption units are usually built when waste heat is available, and they are built in *relatively bigger sizes*. One current application of the absorption system that may grow in importance is the *utilisation of solar energy* for the generator heat source of a refrigerator for *food preservation* and *perhaps for comfort cooling*.

5.4 PROPERTIES OF IDEAL REFRIGERANT, ABSORBENT AND REFRIGERANT-ABSORBENT COMBINATION

For the efficient working of a vapour absorption system the required properties of ideal refrigerant; absorbent and refrigerant-absorbent combination are given in the following sub-articles.

5.4.1 Properties of Ideal Refrigerant

1. The vapour characteristics should be such that will permit boiling at 2ºC to 10ºC and condensation at 40ºC or above, at pressure not far from atmospheric.
2. Large latent heat of vaporisation.
3. High critical temperature.
4. Low specific heat.
5. Stability in complete cycle.

5.4.2 Properties of Ideal Absorbent

1. Greater affinity for the refrigerant.
2. When refrigerant is absorbed by the absorbent, heat liberated should be minimum.
3. High boiling point.
4. Low specific heat.
5. Chemical stability.

5.4.3 Properties of Ideal Refrigerant-absorbent Combination

The combination should possess the following qualities:

1. The refrigerant should have high affinity for the absorber at low temperature and less affinity at high temperature.
2. The combination should have high degree of negative deviation from Raoult's law.
3. The mixture should have low specific heat and low viscosity.
4. The mixture (solution) should be non-corrosive.
5. The mixture should have a small heat.
6. The mixture should have low freezing point.
7. There should be a large difference in the normal boiling points of the refrigerants and the absorbent.

- Out of the many combinations tried so far (Hoirworth listed 49 refrigerants coupled with various absorbents to give 180 combinations as possibilities) the following combinations remained in common use in air-conditioning applications:

  (*i*) *Ammonia-water combination.*

  (*ii*) *Lithium-bromide water combination.*

  Ammonia-water absorption system finds a significant place in *large tonnage industrial applications.*

- Ammonia-water combination which is largely used possesses most of the desirable qualities which are listed below:

  (*i*) 1 m^3 of water absorbs 800 m^3 of NH_3.

  (*ii*) Latent heat of NH_3 at –15ºC = 1314 kJ/kg.

  (*iii*) Critical temperature of NH_3 = 132.6ºC.

  (*iv*) Boiling point at atmospheric pressure = –33.3ºC.

The only *disadvantage* is that it *generates more heat in absorber when it is absorbed, therefore cooling must be provided and special material should be used as it is corrosive to metals.*

5.4.4 Comparison of Refrigerant-liquid Absorbent Combination (say NH_3–water) with Refrigerant-solid Absorbent Combination (say NH_3–$CaCl_2$)

The ***solid absorbent system*** has the following two main **advantages**:

(1) The amount of refrigerant cycled in relation to the amount of absorbent is larger for the solid absorbent than for the liquid absorbent. Therefore, in solid absorbent system the thermal capacity of the salt contributes little, and hence *heat required in the generator is less as compared to that for water in the aqua-NH_3 system.*

(2) Solid-absorbent system is *extremely robust, not only in mechanical sense, but also with respect to adverse operating conditions.* The *performance is remarkably insensitive* to *changes in both condensing* and *evaporating temperatures.* However the solid-absorption cycle has one **major drawback**. The *heat of reaction is large compared with those found* for *liquid absorbent, approximately twice the latent heat of vaporization.*

5.5 ADVANTAGES OF VAPOUR ABSORPTION REFRIGERATION SYSTEM OVER VAPOUR COMPRESSION REFRIGERATION SYSTEM

The following are the *advantages* of vapour absorption system over a mechanical vapour compression refrigeration system:

1. No moving part in the system except the aqua pump-motor. The pump-motor is quite small as compared to the compressor-motor in vapour compression system.
2. Quiet in operation, very little wear, and low maintenance cost.
3. Any readily source of thermal energy (process steam, hot exhaust from furnaces, solar energy) may be used for the generator. Hence can be used in places where electrical energy is not available, or is costly.
4. Can be built in capacities much larger than that for compressor unit. (The vapour absorption system can be built in capacities well above 1000 tonnes of refrigeration each of which is the largest size for single compressor units).
5. Steam which is used for winter heating, same can be used in summer for heating in the generator.
6. In practice the temperature of vapour leaving the generator is of the order 340 K against about 420 K in case of vapour compression system. Therefore smaller condenser is required in the former than in the latter.
7. The capacity of the absorption system is controlled (and maintained) by adjusting the steam or generator temperature even if the evaporator pressure falls. On the other hand the capacity of the compression system decreases rapidly with the reduced evaporator pressure.
8. The absorption system can be operated at the designed C.O.P. even at part loads or reduced evaporator conditions by the appropriate control of generator temperature.
9. The load variations do not affect the performance of a vapour absorption system. The load variations are met by controlling the quantity of aqua circulated and the quantity of steam supplied to the generator. The performance of vapour compression system is poor at partial loads.

10. The absorption system is much more compact and less bulky than the compression system for large capacity.
11. The space requirements and automatic control requirements favour the absorption system more and more as the desired evaporator temperature drops.

5.6 COMPARISON BETWEEN VAPOUR ABSORPTION AND VAPOUR COMPRESSION REFRIGERATION SYSTEMS

S.No.	*Particulars*	*Vapour absorption system*	*Vapour compression system*
1.	*Type of energy supplied*	Mainly heat—a low grade energy.	Mechanical—a high grade energy.
2.	*Energy supply*	High (about 150% of refrigerating effect).	Low (50 to 25% of refrigerating effect).
3.	*Wear and tear*	Less	More
4.	*Performance at part loads*	System not affected by load variations.	Poor
5.	*Size of condenser*	Small	Large
6.	*Evaporator pressure*	The capacity of the system is controlled by adjusting the steam or generator temperature even if the evaporator pressure falls.	The capacity of the system decreases rapidly with the reduced evaporator pressure.
7.	*Leakage of refrigerant*	No chance as there is no compressor or any reciprocating component to cause leakage.	More chances of leakage of refrigerant.
8.	*Maintenance*	Much less comparatively.	More
9.	*Space*	More bulky for small units and much more compact and less bulky for large capacity comparatively.	Less bulky for small units but more bulky for large capacity units.
10.	*Cost for large tonnage systems*	Less costly.	Much more costly comparatively.
11.	*Charging of refrigerant*	Difficult	Simple
12.	*Damage*	Liquid traces of refrigerant present in piping at the exit of evaporator constitute no danger.	Liquid traces in suction line may damage the compressor.
13.	*Suitability*	Can also be used at remote places as it can work even with a simple kerosene lamp (of course in small capacities).	Used where high grade energy is available.

5.7 COEFFICIENT OF PERFORMANCE OF AN IDEAL VAPOUR ABSORPTION SYSTEM

As earlier discussed, in an ideal vapour absorption system, we have:

Q_G = Heat given to refrigerant in the generator,

Q_C = Heat discharged to the atmosphere or cooling water from the condenser and absorber,

Q_E = Heat absorbed by the refrigerant in the evaporator, and

Q_P = Heat added to the refrigerant due to pump work.

Let, T_G = Temperature at which heat (Q_G) is given to the generator,

T_C = Temperature at which heat Q_C is discharged to atmosphere or cooling water from the condenser and absorber, and

T_E = Temperature at which heat (Q_E) is absorbed in the evaporator.

As per first law of thermodynamics, neglecting Q_P, we have:

$$Q_C = Q_G + Q_E \quad ...(i)$$

As the vapour absorption system can be considered as a perfectly reversible system, therefore,

The initial entropy of the system = Entropy of the system after the change in its condition

$$\text{i.e.}\quad \frac{Q_G}{T_G} + \frac{Q_E}{T_E} = \frac{Q_C}{T_C} \quad ...(ii)$$

$$= \frac{Q_G + Q_E}{T_C} = \frac{Q_G}{T_C} + \frac{Q_E}{T_C} \quad \text{[From eqn. } (i)\text{]}$$

or,
$$\frac{Q_G}{T_G} - \frac{Q_G}{T_C} = \frac{Q_E}{T_C} - \frac{Q_E}{T_E}$$

or,
$$Q_G\left[\frac{T_C - T_G}{T_G \times T_C}\right] = Q_E\left[\frac{T_E - T_C}{T_C \times T_E}\right]$$

or,
$$Q_G = Q_E\left[\frac{T_E - T_C}{T_C \times T_E}\right]\left[\frac{T_G \times T_C}{T_C - T_G}\right] = Q_E\left[\frac{T_C - T_E}{T_C \times T_E}\right]\left[\frac{T_G \times T_C}{T_G - T_C}\right]$$

$$= Q_E\left[\frac{T_C - T_E}{T_E}\right]\left[\frac{T_G}{T_G - T_C}\right] \quad ...(iii)$$

Maximum coefficient of performance of the system,

$$(\text{C.O.P.})_{max} = \frac{Q_E}{Q_G} = \frac{Q_E}{Q_E\left[\frac{T_C - T_E}{T_E}\right]\left[\frac{T_G}{T_G - T_C}\right]}$$

$$= \left[\frac{T_E}{T_C - T_E}\right]\left[\frac{T_G - T_C}{T_G}\right] \quad ...(5.1)$$

Here, $\frac{T_E}{T_C - T_E}$ = C.O.P. of Carnot refrigerator working between the temperature limits of T_E and T_C, and

$\frac{T_G - T_C}{T_G}$ = Efficiency of a Carnot engine working between the temperature limits of T_G and T_C.

Thus an ideal vapour refrigeration system may be regarded as a *combination of a Carnot engine and a Carnot refrigeration,*

$$(\text{C.O.P})_{max} = (\text{C.O.P.})_{Carnot} \times \eta_{Carnot} \quad ...(5.2)$$

• When the heat is discharged at different temperatures in condenser and absorber, then

$$(\text{C.O.P.})_{max} = \left(\frac{T_E}{T_C - T_E}\right)\left(\frac{T_G - T_A}{T_G}\right) \quad ...(5.3)$$

where, T_A is the temperature at which heat is discharged in the absorber.

Example 5.1. *A vapour compression cycle with ammonia as the refrigerant works between the limits of saturated suction temperature of – 20ºC and saturated condensing temperature 30ºC. It is a simple saturated cycle and compression is isentropic, determine the work of compression per kg of ammonia.*

Compare the same, if ammonia vapour leaving the evaporator at – 20ºC is absorbed by water so that the mass concentration in the solution reaches about 40 per cent, and its solution is pumped to the condenser pressure.

The specific volume of the solution may be assumed as 0.001161 m³/kg.

Solution: Refer to Fig. 5.3.

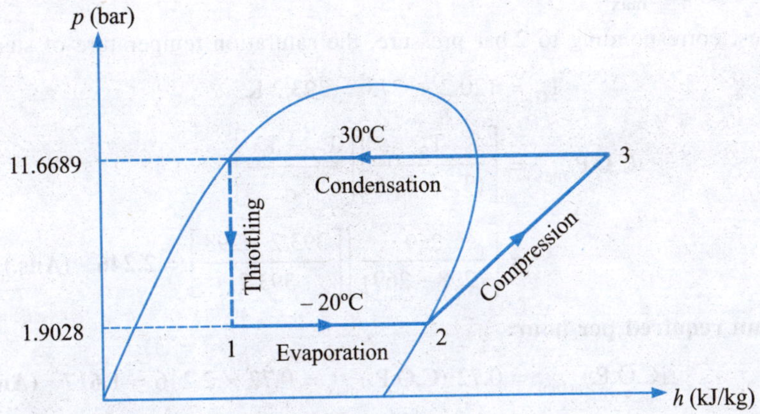

Fig. 5.3

For NH_3: At 30ºC : $p_{cond.}$ = 11.6689 bar

At –20ºC : $p_{evp.}$ = 1.9028 bar

From *p-h* chart, for isentropic compression,

$$h_2 = 1420 \text{ kJ/kg}; \quad h_3 = 1668 \text{ kJ/kg}$$

The *compressor work per kg of* NH_3

$$= h_3 - h_2 = 1668 - 1420 = \mathbf{248\ kJ/kg} \quad \textbf{(Ans.)}$$

For 40 percent concentration, the mass of solution per kg of ammonia dissolved

$$= \frac{1}{0.40} = 2.5 \text{ kg/kg of ammonia dissolved}$$

Thus shaft-work = Mass of solution (kg) × sp. volume (m³/kg) × $(p_{cond.} - p_{evp.}) \times \frac{10^5}{10^3}$ kJ

$$= 2.5 \times 0.001161 \times (11.6689 - 1.9028) \times \frac{10^5}{10^3} \text{ kJ}$$

$$= 2.835 \text{ kJ/kg of dissolved ammonia.}$$

Considering the irreversibilities and efficiencies to be same in both cases for the compression process the shaft-work required by vapour absorption system as compared to vapour compression system

$$= \frac{2.835}{248} = 0.01143 \quad \text{or} \quad \mathbf{1.143\%} \quad \textbf{(Ans.)}$$

Example 5.2. *In an absorption type refrigerator, the heat is supplied to NH_3 generator by condensing steam at 2 bar and 88 per cent dry. The temperature in the refrigerator is to be maintained at –4°C. Find the maximum C.O.P. possible.*

If the refrigeration load is 18 tonnes and actual C.O.P. is 72 per cent of the maximum C.O.P., find the mass of steam required per hour. Take the temperature of the atmosphere as 25°C.

Solution: *Given:* $p = 2$ bar; $x = 88\% = 0.88$; $T_E = -4 + 273 = 269$ K; $Q = 18$ tonnes;

$(C.O.P.)_{actual} = 0.72\ (C.O.P.)_{max}$; $T_C = 25 + 273 = 298$ K.

Maximum C.O.P., $(C.O.P)_{max}$:

From steam tables, corresponding to 2 bar pressure, the saturation temperature of steam,

$$T_G = 120.2 + 273 = 393.2 \text{ K}$$

$$\therefore \quad (C.O.P)_{max} = \left[\frac{T_E}{(T_C - T_E)}\right]\left[\frac{T_G - T_C}{T_G}\right] \quad \text{...[Eqn. (5.1)]}$$

$$= \left[\frac{269}{(298 - 269)}\right]\left[\frac{393.2 - 298}{393.2}\right] = \mathbf{2.246} \quad \textbf{(Ans.)}$$

Mass of steam required per hour:

$$(C.O.P.)_{actual} = 0.72\ (C.O.P.)_{max} = 0.72 \times 2.246 = \mathbf{1.617} \quad \textbf{(Ans.)}$$

$$\therefore \quad \text{Actual heat supplied} = \frac{\text{Refrigeration load}}{(C.O.P.)_{actual}} = \frac{(14000 \times 18)}{1.617} = 155844 \text{ kJ/h}$$

Assuming that only latent heat of steam is used for heating purposes, the latent heat of steam from steam tables, corresponding to 2 bar pressure,

$$h_{fg} = 2201.6 \text{ kJ/kg}$$

$$\therefore \textit{Mass of steam required per hour} = \frac{\text{Actual heat supplied}}{x \times h_{fg}} = \frac{155844}{0.88 \times 2201.6} = \mathbf{80.44\ kg/h} \quad \textbf{(Ans.)}$$

Example 5.3. *In a 18-tonnes absorption refrigeration system the heating in generator is carried out by using steam at 3 bar and 85% dry. The refrigeration temperature is –10°C. The condensation of the refrigerant is carried out at 30°C using cooling water. Determine:*

(i) Maximum possible C.O.P. of the system; and

(ii) Quantity of steam required per hour to run the plant if the steam leaves the generater as saturated water at same pressure. Assume relative C.O.P. = 0.4. **[UPTU]**

Solution: *Given:* $Q = 18$ tonnes; $p = 3$ bar, $x = 0.85$, $T_E = -10 + 273 = 263$ K;

$T_C = 30 + 273 = 303$ K; $(C.O.P.)_{relative} = 0.4$.

(*i*) **Maximum C.O.P., (C.O.P.)$_{max}$:**

From steam tables, corresponding to 3 bar pressure, the saturation temperature of steam,

$$T_G = 133.5 + 273 = 406.5 \text{ K}$$

$$\therefore \quad (\text{C.O.P.})_{max} = \left[\frac{T_E}{(T_C - T_E)}\right]\left[\frac{T_G - T_C}{T_G}\right]$$

$$= \left[\frac{263}{(303 - 263)}\right]\left[\frac{406.5 - 303}{406.5}\right] = \mathbf{1.67} \quad \textbf{(Ans.)}$$

(*ii*) **Quantity of steam required per hour:**

$$(\text{C.O.P.})_{actual} = 0.4 \times 1.67 = 0.668$$

$$\therefore \quad \text{Actual heat supplied} = \frac{\text{Refrigeration load}}{(\text{C.O.P.})_{actual}} = \frac{14000 \times 18}{0.668} = 377245 \text{ kJ/h}$$

h_{fg} at 3 bar (from steam tables) = 2163.2 kJ/kg

∴ *Quantity of steam required per hour*

$$= \frac{\text{Actual heat supplied}}{x \times h_{fg}} = \frac{377245}{0.85 \times 2163.2} = \mathbf{205.17 \ kg} \quad \textbf{(Ans.)}$$

Example 5.4. *In an acqua-ammonia absorption refrigeration system of 9 tonnes refrigeration capacity the vapours leaving the generator are 100% pure NH_3 saturated at 40°C. The evaporator, absorber, condenser and generator temperatures are –20°C, 30°C, 40°C and 170°C respectively. At absorber exit (strong solution) the concentration of ammonia in solution is x = 0.35 and enthalpy h = 22 kJ/kg. At generator exit (weak solution) x = 0.1 and h = 695 kJ/kg.*

(*i*) *Determine the mass flow rate of ammonia in the evaporator;*

(*ii*) *Carry out overall mass conservation and mass conservation of ammonia in absorber to determine mass flow rates of weak and strong solution;*

(*iii*) *Determine the heat rejection in absorber and condenser, heat added in generator and C.O.P.*

Solution: The schematic arrangement of the system is shown in Fig. 5.4.

Given: $x_{1'} = 0.35$; $h_{1'} = 22$ kJ/kg; $x_{2'} = 0.1$; $h_{2'} = 695$ kJ/kg.

Also, $h_2 = 1473.3$ kJ/kg (Saturation vapour at 40°C)

$h_3 = 371.9$ kJ/kg (Saturated liquid at 40°C)

$h_1 = 1420$ kJ/kg (Saturation vapour at – 20°C)

(*i*) **Mass flow rate of ammonia in the evaporator, m:**

$$\text{Refrigerating effect (RE)} = (h_1 - h_4)$$

$$= (h_1 - h_3) = 1420 - 371.9 = 1048.1 \text{ kJ/kg} \qquad (\because h_3 = h_4)$$

$$\therefore \quad m = \frac{(14000 \times 9)/3600}{1048.1} = \mathbf{0.0334 \ kg/s} \quad \textbf{(Ans.)}$$

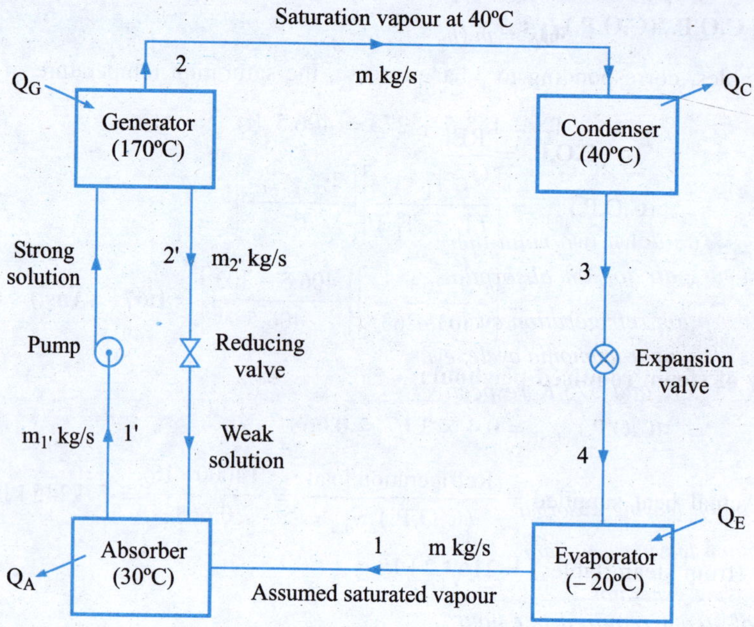

Fig. 5.4

(ii) **Overall mass conservation and mass conservation of ammonia in the absorber:**

Overall mass balance for absorber gives

$$m + m_{2'} = m_{1'}$$

Ammonia balance for the absorber gives,

$$m + 0.1\, m_{2'} = 0.35\, m_{1'} = 0.35\,(m + m_{2'}) \quad \text{or} \quad m + 0.1\, m_{2'} = 0.35\, m + 0.35\, m_{2'}$$

or,

$$m_{2'}\,(0.35 - 0.1) = m\,(1 - 0.35)$$

$$\therefore \quad m_{2'} = \frac{m(1-0.35)}{(0.35-0.1)} = \frac{0.0334(1-0.35)}{(0.35-0.1)} = \mathbf{0.0868\ kg/s} \quad \textbf{(Ans.)}$$

and

$$m_{1'} = m + m_{2'} = 0.0334 + 0.0868 = \mathbf{0.12\ kg/s} \quad \textbf{(Ans.)}$$

(iii) **Heat rejection in absorber and condenser and heat added in generator and C.O.P.:**

Making *energy balance for absorber*, we have:

$$mh_1 + m_{2'}h_{2'} = m_{1'}h_{1'} + Q_A$$

$$0.0334 \times 1420 + 0.0868 \times 695 = 0.12 \times 22 + Q_A$$

$$\therefore \quad Q_A = \mathbf{105.11\ kJ/s \quad (kW) \quad (Ans.)}$$

Energy balance for generator gives,

$$m_{1'}h_{1'} + Q_G = m_{2'}h_{2'} + mh_2$$

$$0.12 \times 22 + Q_G = 0.0868 \times 695 + 0.0334 \times 1473.3$$

$$\therefore \quad Q_G = \mathbf{106.88\ kJ/kg\ (kW) \quad (Ans.)}$$

$$Q_C = m\,(h_2 - h_3) = 0.0334\ (1473.3 - 371.9)$$

$$= \mathbf{36.79\ kJ/s\ (kW)}\quad \textbf{(Ans.)}$$

$$\text{C.O.P.} = \frac{RE}{Q_G} = \frac{(14000 \times 9)/3600}{106.88} = \mathbf{0.327}\quad \textbf{(Ans.)}$$

Example 5.5. *(a) Write down two main thermodynamic requirements and desirable characteristics of refrigerant-absorbent pair for an absorption refrigeration system.*

(b) Simple absorption refrigeration cycle does not consider the rectification column and preheating heat exchanger. In such aqua-ammonia cycle, evaporator, absorber, condenser and generator temperatures are 233 K, 303 K, 313 K and 373 K respectively. The properties of aqua-ammonia are as follows:

	Concentration kg of NH_3/kg solution	*Enthalpy kJ/kg*
Strong solution leaving absorber	*0.421*	*30*
Weak solution leaving generator	*0.375*	*340*
Vapour leaving generator	*0.945*	*1870*
Liquid leaving condenser	*0.945*	*470*
Vapour leaving evaporator	*0.945*	*1387.5*

(*i*) *Draw schematic diagram of the system.*

(*ii*) *For 1 tonne refrigeration capacity determine the mass flow rate of solution in evaporator.*

(*iii*) *Consider overall mass balance and mass balance of NH_3 in absorber. This will give two equations, solve them to determine mass flow rates of strong and weak solutions.*

(*iv*) *Consider energy balance for absorber and generator to find absorber heat rejection and generator heat transfer. Find condenser heat rejection also.*

(*v*) *Check the overall energy balance neglecting pump work, then find C.O.P.*

(*vi*) *Discuss the need of rectification column and preheating heat exchanger.* **[P.U.]**

Solution: (*a*) The two-main *thermodynamic requirements* of the refrigerant absorbent pair are:

(*i*) **Solubility requirement.** The refrigerant should have more than Raoult's law solubility in the absorbent so that a strong solution, highly rich in the refrigerant, is formed in the absorber by the absorption of the refrigerant vapour.

(*ii*) **Boiling points requirement.** There should be a large difference in the normal boiling points of the two substances, at least 200°C, so that the absorbent exerts negligible vapour pressure at the generator temperature. Thus, almost absorbent-free refrigerant is boiled off from the generator and the absorbent alone returns to the absorber.

In addition, the pair should possess the following *characteristics*:

(*i*) Low viscosity to minimize pump work.

(*ii*) Non-corrosive.

(*iii*) Good chemical and thermal stability

(*iv*) Absence of irreversible chemical reactions such as decomposition and polymerization.

(*v*) To yield a solution with as small a heat of solution as is compatible with other properties.

(*b*) (*i*) *Fig. 5.5 shows the* ***schematic diagram of the system.***

Given: $T_E = 233$ K; $T_A = 303$ K; $T_C = 313$ K; $T_G = 373$ K.

The concentration (x) and enthalpy (h, kJ/kg) at various points is as given below:

State point	*x (mass concentration of NH_3 in solution)*	*Enthalpy h, kJ/kg*
1′	0.421	30
2′	0.375	340
2	0.945	1870
3, 4	0.945	470
1	0.945	1388

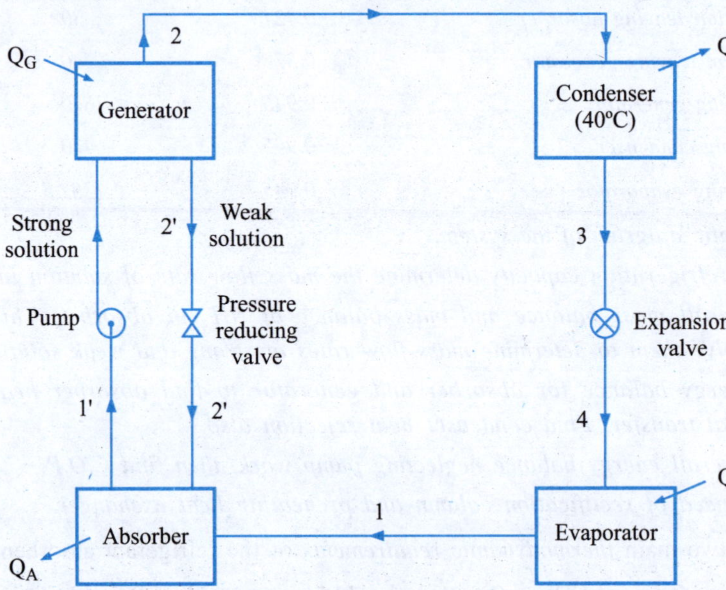

Fig. 5.5

(*ii*) **Mass flow rate of solution in evaporator for 1 tonne refrigeration capacity, *m*:**

$$m\,(= m_1) = \frac{(14000 \times 1)/3600}{(h_1 - h_4)} = \frac{(14000 \times 1)/3600}{(1388 - 470)} = \mathbf{0.00424\ kg/s} \quad \textbf{(Ans.)}$$

(*iii*) **Mass flow rates of strong and weak solutions; $m_{1'}$, $m_{2'}$:**

The two equations are:

$$m_1 + m_{2'} = m_{1'} \quad \text{.......... } \textit{overall mass balance for absorber}$$

$$m_1 x_1 + m_{2'} x_{2'} = m_{1'} x_{1'} = (m_1 + m_{2'})\, x_{1'} \text{ .... } \textit{Ammonia balance for the absorber}$$

$$0.00424 \times 0.945 + 0.375\, m_{2'} = (0.00424 + m_{2'})\ 0.421$$

$$\text{or,}\quad 0.004 + 0.375\, m_{2'} = 0.00178 + 0.421\, m_{2'}$$

or, $$m_{2'} = \frac{(0.004 - 0.00178)}{(0.421 - 0.375)} = \mathbf{0.0483\ kg/s} \quad \textbf{(Ans.)}$$

and, $$m_{1'} = m_1 + m_{2'} = 0.00424 + 0.0483 = \mathbf{0.0525\ kg/s} \quad \textbf{(Ans.)}$$

(*iv*) **Absorber heat rejection (Q_A); Generator heat transfer (Q_G); Condenser heat rejection:**

- Energy balance for *absorber* gives,

$$Q_A + m_{1'}h_{1'} = m_1h_1 + m_{2'}h_{2'}$$

$$\therefore \quad \mathbf{Q_A} = m_1h_1 + m_{2'}h_{2'} - m_{1'}h_{1'}$$

$$= 0.00424 \times 1388 + 0.0483 \times 340 - 0.0525 \times 30$$

$$= \mathbf{20.73\ kJ/s} \quad \textbf{(Ans.)}$$

- Energy balance for *generator* gives,

$$Q_G + m_{1'}h_{1'} = m_{2'}h_{2'} + m_2h_2 = m_{2'}h_{2'} + m_1h_2 \qquad (\because m_2 = m_1 = m)$$

$$\therefore \quad \mathbf{Q_G} = m_{2'}h_{2'} + m_1h_2 - m_{1'}h_{1'}$$

or, $$= 0.0483 \times 340 + 0.00424 \times 1870 - 0.0525 \times 30$$

$$= \mathbf{22.77\ kJ/s} \quad \textbf{(Ans.)}$$

- *Condenser heat rejection*, $\mathbf{Q_C} = m_2(h_2 - h_3) = 0.00424\,(1870 - 470)$

$$= \mathbf{5.94\ kJ/s} \quad \textbf{(Ans.)}$$

(*v*) **Overall energy balance, C.O.P:**

$$\text{Overall energy balance} = Q_G + Q_E - Q_A - Q_C$$

$$= 22.77 + (14000 \times 1)/3600 - 20.73 - 5.94 = 0.$$

This checks the overall energy balance.

$$\textbf{C.O.P} = \frac{Q_{ref.}}{Q_A} = \frac{(14000 \times 1)/3600}{20.73} = \mathbf{0.187} \quad \textbf{(Ans.)}$$

(*vi*) **Need for rectification:** Refer to Fig. 5.2.

During heating of the strong solution in the generator NH_3 vapours are evolved. Along with NH_3, some water also boils off and the NH_3 vapour leaving the generator contains 5 to 10% water vapour also. This is *not desirable*, as *water may freeze in the expansion valve or in the evaporator, thereby choking the flow*. An *analyser* column is used to condense most of the water vapour. However the NH_3 vapour leaving the analyser contains some water vapour. The *rectification column is used to condense this water vapour.*

Need for preheating heat exchanger: Refer to Fig. 5.2.

The weak solution leaving the generator and being sent to absorber is hot, and needs cooling to enable it to absorb NH_3 vapour. The strong solution leaving the absorber has to be heated in the generator to evolve back NH_3 vapour. A **"heat exchanger"** is *provided for transferring heat from strong solution to weak solution. This would save heating required in the generator, and at the same time the cooling required in the absorber.*

5.8 DOMESTIC ELECTROLUX REFRIGERATOR

- The domestic absorption type refrigerator was developed from an invention by Carl Munters and Baltzer Von Platen. This system is often called "*Munters Platen System*".
- This type of refrigerator is also called "*Three-fluids absorption system*". The three fluids used in this system are *ammonia, hydrogen and water*.
  - — The "*ammonia*" is used as a refrigerant because it possesses most of the desirable properties. Though it is toxic, and not otherwise preferred in domestic appliances, it is very safe in this system due to absence of any moving parts in the system and, therefore, there is the *least chance of any leakage*.
  - — The "*hydrogen*" being the lightest gas, is used to increase the rate of evaporation (the lighter the gas, faster is the evaporation) of the liquid ammonia passing through the evaporator. The hydrogen is also non-corrosive and insoluble in water. This is used in the *low-pressure side of the system*.
  - — The "*water*" is used as a *solvent* because it has the ability to *absorb ammonia readily*.

Principle and Working of Electrolux Refrigerator:

Fig. 5.6 shows a schematic diagram of an '*Electrolux refrigerator*'. It is a domestic refrigerator and is the best known absorption type of refrigerator. *Here pump is dispensed with.* The small energy supply is by means of a heater which may be electric or gas.

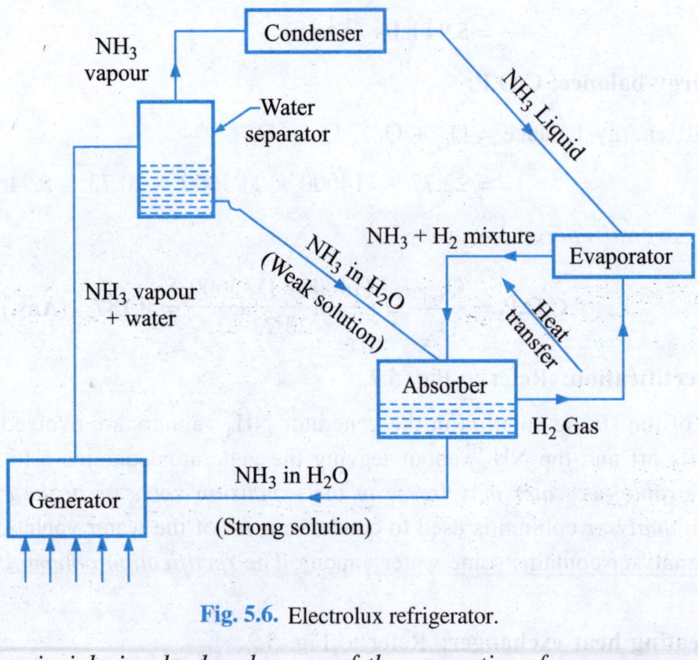

Fig. 5.6. Electrolux refrigerator.

Principle. *The principle involved makes use of the properties of gas-vapour mixtures. If a liquid is exposed to an inert atmosphere, it will evaporate until the atmosphere is saturated with the vapour of the liquid.* This evaporation requires heat which is taken from the surroundings in which the evaporation takes place. A cooling effect is thus produced. The partial pressures of the refrigerant vapour (in this case ammonia) must be low in the evaporator, and higher in condenser. The *total* *pressure throughout the circuit must be constant so that the only movement of the working fluid is by convection currents. The partial pressure of ammonia is kept low in requisite parts of the circuit by concentrating hydrogen in those parts.*

Working. The ammonia liquid leaving the *condenser* enters the *evaporator* and evaporates into the hydrogen at the low temperature corresponding to its low partial pressure. The mixture of ammonia and hydrogen passes to the absorber into which is also admitted water from the *separator*. The water absorbs the ammonia and the hydrogen returns to the evaporator. In the absorber the ammonia therefore passes from the ammonia circuit into water circuit as ammonia in water solution. This strong solution passes to the *generator* where it is heated and the vapour given off rises to the separator. The water with the vapour is separated out and a weak solution of ammonia is passed back to the absorber, thus completing the water circuit. The ammonia vapour rises from the separator to the condenser where it is condensed and then returned to the evaporator.

The actual plant includes refinements and practical modifications (which are not included here).

The following *points are worth noting*:

- The complete cycle is carried out entirely by gravity flow of the refrigerant.
- The hydrogen gas circulates only from the absorber to the evaporator and back.
- With this type of machine efficiency is not important since the energy input is small.
- It has not been used for industrial applications as the C.O.P. of the system is very low.

Role of Hydrogen. By the presence of hydrogen it is possible to maintain uniform total pressure throughout the system and at the same time permit the refrigerant to evaporate at low temperature in the evaporator corresponding to its partial pressure. Thus the condenser and evaporator pressures of the refrigerant are maintained as below:

(*i*) In the condenser only ammonia is present, and the total pressure is the condensing pressure.

(*ii*) In the evaporator hydrogen and ammonia are present, their relative masses are adjusted such that the partial pressure of ammonia is the required evaporator pressure.

These are achieved without the use of pumps or valves.

Advantages and Disadvantages of Electrolux Refrigerator over Conventional Refrigerators:

Advantages:

1. No pump or compressor is required.
2. No mechanical troubles, maintenance cost is low.
3. No lubrication problem; no wear and tear.
4. Completely leak proof.
5. Noiseless.
6. No chance of pressure unbalancing and no need of valves.
7. System may be designed to use any available source of thermal energy-process steam, exhaust from engines or turbines, solar energy etc.
8. Easy control, simply by controlling heat input.

Disadvantages:

1. More complicated in construction and working.
2. C.O.P. very low.
3. The major disadvantage of this type of refrigerator is that if it is spoiled once, it cannot be repaired and has to be replaced fully.

Example 5.6. *The total pressure maintained in an electrolux refrigerator is 14.71 bar. The temperature obtained in the evaporator is –15°C.*

The quantities of heat supplied in the generator are:

(i) 418.7 kJ to dissociate 1 kg of vapour and (ii) 1465.4 kJ/kg for increasing the total enthalpy of NH_3*. The enthalpy of* NH_3 *entering the evaporator is 335 kJ/kg. Take the following properties of* NH_3 *at –15°C: Pressure = 2.45 bar; enthalpy (h), of* NH_3 *vapour = 1666 kJ/kg; specific volume* $(v_s) = 0.5\ m^3/kg$*.*

The hydrogen enters the evaporator at 25°C.

Gas constant for H_2 *= 4.218 kJ/kg-°C*

c_p *(for* H_2*) = 12.77 kJ/kg-°C*

Find the C.O.P. of the system.

Assume NH_3 *leaves the evaporator in saturated condition.*

Solution: Fig. 5.7 shows the schematic arrangement of an electrolux refrigerator.

Consider the flow of 1 kg of NH_3, $Q_G = 418.7 + 1465.4 = 1884.1$ kJ/kg.

In the evaporator: $p_t = 14.71$ bar; $p_{NH_3} = 2.45$ bar

$$\therefore \quad p_{H_2} = 14.71 - 2.45 = 12.26 \text{ bar}$$

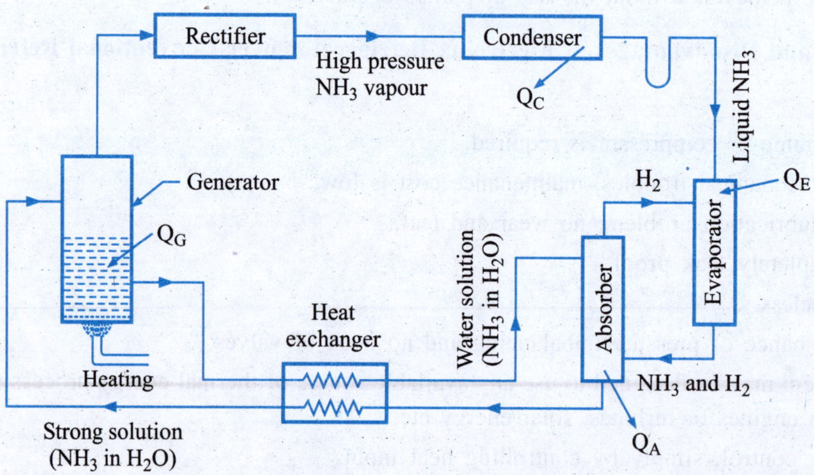

Fig. 5.7

Refer to Fig. 5.8.

For NH_3: $h_1 = 335$ kJ/kg; $h_2 = 1666$ kJ/kg.

Let m kg of H_2 flows through evaporator per kg of NH_3. NH_3 occupies $v = 0.5\ m^3/kg$; same is the volume of m kg of H_2 at $t = -15°C$

$$\therefore \quad m = \frac{p_{H_2} v}{R_{H_2} T} = \frac{12.26 \times 10^5 \times 0.5}{(4.218 \times 10^3)(273 - 15)}$$

$$= 0.563 \text{ kg}$$

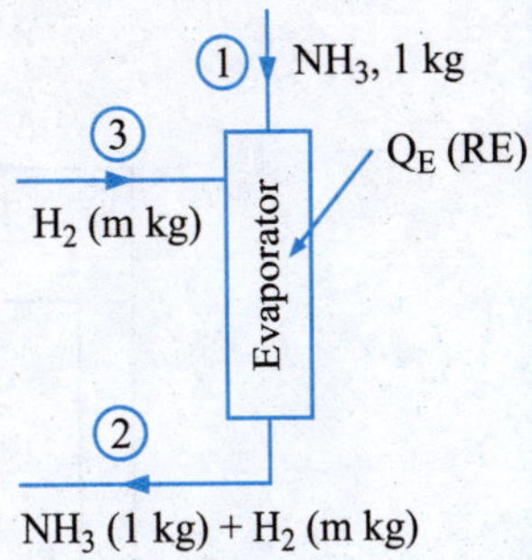

Fig. 5.8

Energy balance for evaporator gives,

$$1 \times (h_2 - h_1) = Q_E + m\, c_{pH_2} (t_3 - t_2)$$

$$1 \times (1666 - 335) = Q_E + 0.563 \times 12.77\ [25 - (-15)]$$

$$\text{or,} \quad Q_E = 1043.4 \text{ kJ/kg of } NH_3$$

$$\therefore \quad \textbf{C.O.P.} = \frac{Q_E}{Q_G} = \frac{1043.4}{1884.1} = \mathbf{0.554} \quad \textbf{(Ans.)}$$

5.9 LITHIUM BROMIDE ABSORPTION REFRIGERATION SYSTEM

In such a system *lithium bromide salt (highly hygroscopic) is employed as the absorbent and water as the refrigerant.* The lithium bromide solution has a *strong affinity for water vapour because of its very low pressure.* Since water is the refrigerant, the refrigerant operating temperature in the evaporator has to be above the freezing point (°C) of water.

Lithium bromide solution is *corrosive*, as such, inhibitor like lithium chromate is often used to protect the metal parts of the system against corrosion.

Fig. 5.9 shows a schematic arrangement of lithium bromide absorption refrigeration system.

- The *absorber and evaporator* are housed in one cylindrical shell as they operate at about the same pressure.
- The *generator and condenser* are enclosed in another cylindrical shell as their operating pressures are near about the same.
- The evaporator contains refrigerant (water) and the absorber has a charge of lithium bromide solution. The solution from the absorber is pumped to the generator by the solution pump.

Chilled water tube bundle passes through the evaporator section of the shell, carrying the water to be cooled.

- The absorber and condenser sections have water tube bundles passing through, in which cooling tower water is circulated. The condenser portion of the generator shell has tube bundle to circulate cooling lower water to cool (and condensate) the refrigerant (water) vapour.
- Refrigerant water is sprayed (through nozzles) over the chilled water tube bundle by the evaporator pump.
- A *heat exchanger* is added to improve the cycle efficiency.

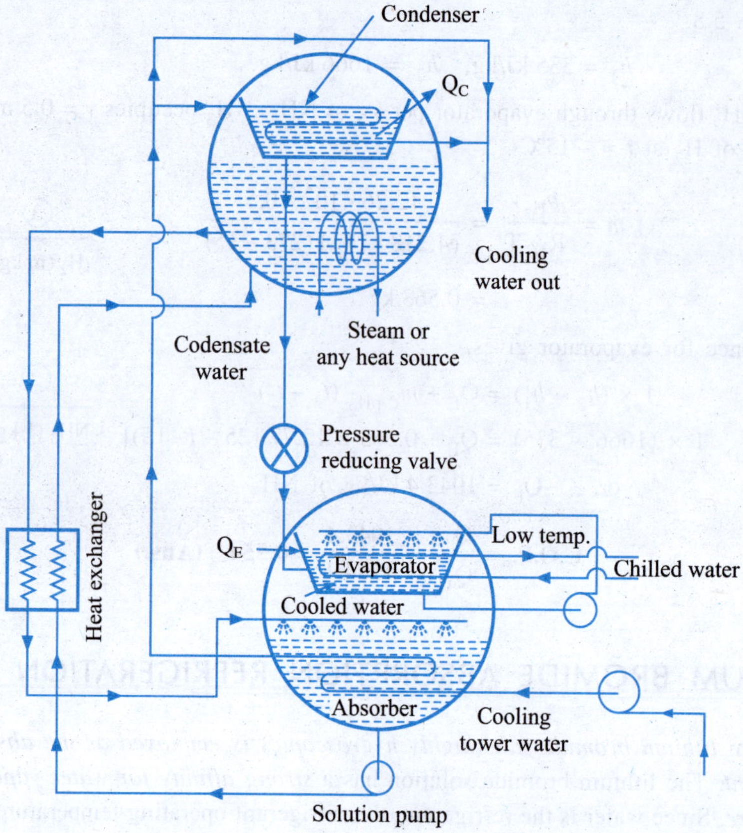

Fig. 5.9. Lithium bromide absorption refrigeration system.

Major fields of application. Since water is used as a refrigerant in lithium bromide water absorption system, it can be used where the temperatures required are above 4°C. Hence major fields of applications are:

1. Air-conditioning;
2. Chilling of drinking water;
3. Chilling of beverages and fruit juices.

Example 5.7. (*a*) *Explain with a neat sketch a water-cooled condensing unit.*

(*b*) *Prepare a list of components used in a vapour compression system from practical consideration (not from thermodynamic cycle consideration).*

(*c*) *The following data refer to a LiBr + H_2O absorption system:*

Generator temperature : 80°C

$T_{condenser} = T_{absorber} = 30°C$

Evaporator temperature : 10°C

Condensate temperature : 25°C

Steam enters the generator heating coil at 120°C (dry-saturated state steam) and leaves it at 100°C as condensate.

Concentration of liquid leaving generator is 0.65 and its enthalpy – 75 kJ/kg, concentration of liquid leaving absorber is 0.51 and its enthalpy – 170 kJ/kg. Enthalpy of vapour leaving generator is 2620 kJ/kg. Flow rate through evaporator is 0.4 kg/s. Determine the following:

(*i*) *Pressures in generator, condenser, evaporator and absorber in mm of murcury head;*

(*ii*) *Tonnage;*

(*iii*) *Heat rejection to condenser and absorber;*

(*iv*) *Rate of mass leaving absorber;*

(*v*) *C.O.P.* **[UPTU]**

Solution: (*a*) **Water-cooled condensing unit:**

The compressor, hot gas line, condenser, and receiver tank, along with the compressor drive (usually an electric motor) are often combined into one compact unit. Such an assembly is called a ***condensing unit***, *because its function in the system is to reclaim the vapour and condense it back into the liquid state.*

A watercooled condensing unit (employing water as the condensing medium) is shown in the Fig. 5.10. The reciprocating compressor is driven by an electric motor through belt and pulley arrangement, and is mounted on the top of the frame. It receives low pressure vapour from the evaporator, and the compressed high temperature high pressure vapour is delivered to the shell-and-tube type watercooled condenser. The two-tube-pass condenser, shown in the figure, is mounted on the frame below the compressor. The liquid refrigerant condensed and collected at the bottom of the condenser is drained into a receiver. *The receiver is located at the lowest level.* The high pressure liquid refrigerant, from the receiver of the condensing unit, is supplied to the evaporator of the fan-coil units through the expansion device. (not shown in the figure).

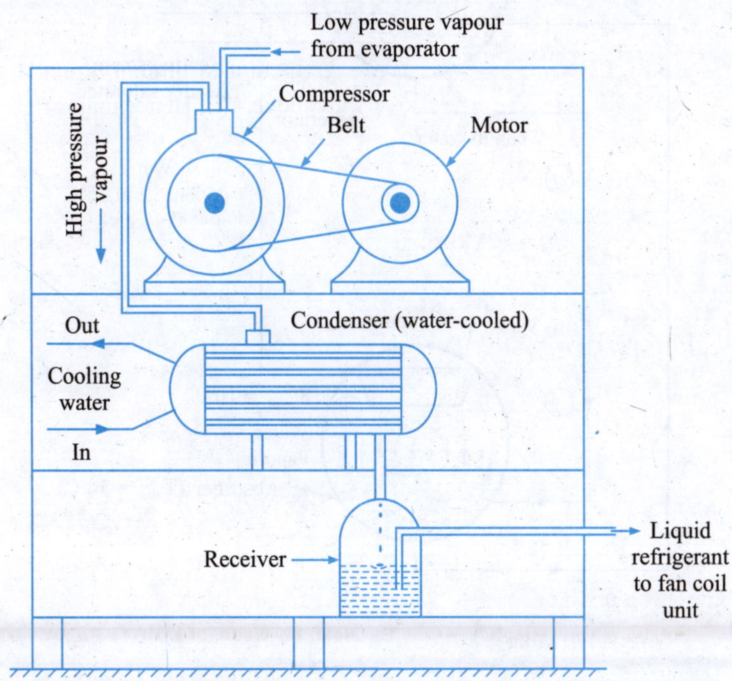

Fig. 5.10. Arrangement of a typical water-cooled condensing unit.

(*b*) The list of components in a vapour compression system, *from practical considerations*, is as given below:

1. Compressor
2. Compressor motor
3. HP and LP cut outs
4. Compressor service valves
5. Condenser (air-cooled or water-cooled)
6. Condenser fan with motor for air-cooled units
7. Cooling tower for water-cooled units
8. Receiver
9. Oil separator
10. Filter-cum-drier
11. Expansion device (thermostatic or capillary tube)
12. Hand shut-off valves
13. Thermostats and other controls.

(*c*) Fig. 5.11 shows the schematic arrangement of the system.

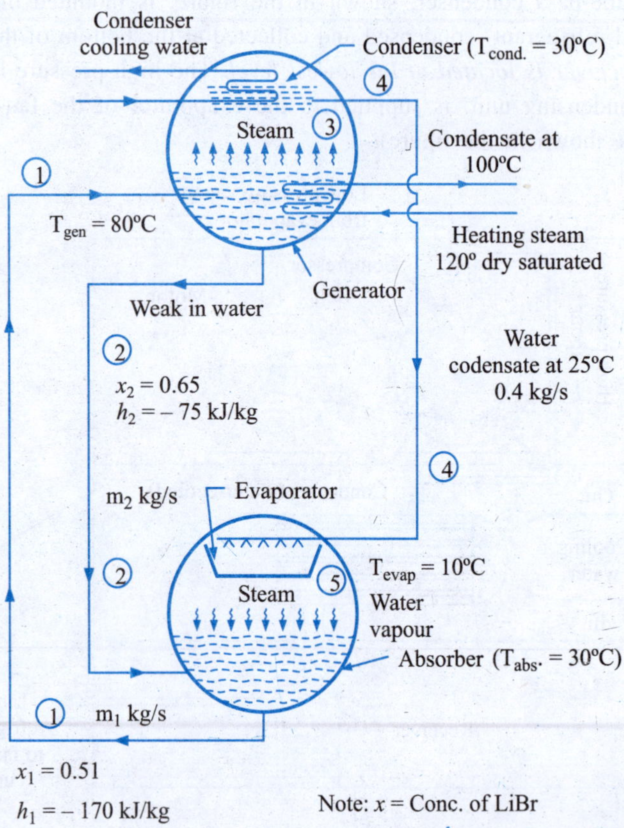

Fig. 5.11

Given: $h_3 = 2620$ kJ/kg, and

$$m_3 = m_4 = m_5 = 0.4 \text{ kg/s}$$

Further, from steam tables,

$h_4 = 104.77$ kJ/kg (liquid water at 25ºC),

and, $h_5 = 2519.9$ kJ/kg (assuming that the water vapour leaving the evaporator is dry saturated at 10ºC)

(*i*) **Pressures in the generator, condenser, evaporator and absorber in *mm* of Hg head:**

Pressure in the generator, condenser, evaporator and absorber will correspond to temperature existing there

Location	*Temp.,ºC*	*Pressure, bar*	*Pressure, mmHg*
Condenser	30	0.04241	31.81
Evaporator	10	0.01228	9.21
Absorber	30	0.04241	31.81
Generator	80	0.4736	355.24

(*ii*) **Cooling capacity (tonnes of refrigeration), Q:**

$$\text{Cooling capacity} = m_4(h_5 - h_4) = 0.4\ (2519.9 - 104.77) = 966.05 \text{ kJ/s (kW)}$$

$$= \frac{966.05 \times 3600}{14000} = 248.4 \quad \textbf{tonnes of refrigeration} \quad \textbf{(Ans.)}$$

(*iii*) **Heat rejection to condenser and absorber, Q_C, Q_A:**

Heat rejected to *condenser*,

$$Q_C = m_4\,(h_3 - h_4) = 0.4\ (2620 - 104.77) = \mathbf{1006.1\ kJ/s\ \ (kW)} \quad \textbf{(Ans.)}$$

For making energy balance for *absorber*, mass rates m_1 and m_2 are required.

Mass balance for Li-Br in absorber gives,

$$m_1x_1 = m_2x_2\,, \quad \text{or,} \quad m_1 = \frac{m_2x_2}{x_1} = \frac{m_2 \times 0.65}{0.51} = 1.2745\ m_2$$

Total mass balance for absorber:

$$m_1 = m_2 + 0.4\,, \quad \text{or,} \quad 1.2745\ m_2 = m_2 + 0.4$$

or, $$m_2 = \frac{0.4}{(1.2745 - 1)} = 1.457 \text{ kg/s}$$

and, $$m_1 = 1.457 + 0.4 = 1.857 \text{ kg/s}$$

Now making enthalpy balance for absorber, we have

$$m_2h_2 + m_5h_5 = m_1h_1 + Q_A$$

or, $$Q_A = m_2h_2 + m_5h_5 - m_1h_1 = 1.457 \times (-75) + 0.4 \times 2519.9 - 1.857 \times (-170)$$

$$= \mathbf{1214.4\ kJ/s\ (kW)} \quad \textbf{(Ans.)}$$

(*iv*) **Rate of mass leaving absorber, m_1:**

Rate of mass leaving absorber, m_1 = **1.857 kg/s (Ans.)** (Calculated above)

(*v*) **C.O.P:**

$$Q_G = Q_C + Q_A - Q_E = 1006.1 + 1214.4 - 966.05 = 1254.45 \text{ kJ/s (kW)}$$

$$\text{C.O.P.} = \frac{Q_{ref.}}{Q_G} = \frac{966.05}{1254.45} = \mathbf{0.77} \text{ (Ans.)}$$

HIGHLIGHTS

1. In a vapour absorption system the refrigerant is absorbed on leaving the evaporator, the absorbing medium being a solid or a liquid.
2. The following accessories are fitted in a simple vapour absorption system to improve its efficiency.

   (*i*) Analyser, (*ii*) Rectifier; and (*iii*) Heat exchangers
3. $$\text{C.O.P.} = \frac{\text{Heat extracted from the evaporator}}{\text{Heat supplied in the generator + Work done by the liquid pump}}$$
4. 'Electrolux refrigerator' is a domestic refrigerator and is best known absorption type of refrigerator. The principle involved makes use of the properties of gas-vapour mixtures.
5. In a lithium bromide absorption refrigeration system lithium bromide (highly hygroscopic) is employed as the absorbent and water as the refrigerant.

OBJECTIVE TYPE QUESTIONS

1. An electrolux refrigerator is called a

   (*a*) single fluid absorption system (*b*) two fluids absorption system

   (*c*) three fluids absorption system (*d*) none of the above.
2. In a lithium bromide absorption refrigeration system

   (*a*) lithium bromide is used as a refrigerant and water as an absorbent

   (*b*) water is used as a refrigerant and lithium bromide as an absorbent

   (*c*) ammonia is used as a refrigerant and lithium bromide as an absorbent

   (*d*) none of these.
3. The refrigerant, commonly used in vapour absorption system, is

   (*a*) water (*b*) ammonia (*c*) freon (*d*) aqua-ammonia

   (*e*) sulphur dioxide.
4. Comparing mechanical vapour compression refrigeration system with absorption refrigeration system, the compressor of the former is replaced in the latter by

   (*a*) an absorber and a liquid pump

   (*b*) an absorber, a generator, a liquid pump and a pressure reduction valve

   (*c*) an absorber, an evaporator, a liquid pump and an expansion valve

   (*d*) a generator, an evaporator, a liquid pump and an expansion valve.
5. A vapour absorption system

   (*a*) gives noisy operation (*b*) gives quiet operation

   (*c*) requires little power consumption (*d*) cools below 0ºC.

6. In electrolux refrigerator
   (*a*) ammonia is absorbed in hydrogen (*b*) ammonia is absorbed in water
   (*c*) ammonia evaporates in hydrogen (*d*) hydrogen evaporates in ammonia.

7. The fluids used in the Electrolux refrigerator are
   (*a*) water and hydrogen (*b*) ammonia and hydrogen
   (*c*) ammonia, water and hydrogen (*d*) none of these.

8. In aqua-ammonia and Li-Br water absorption refrigeration systems, the refrigerants are respectively
   (*a*) water and water (*b*) water and Li-Br (*c*) ammonia and Li-Br (*d*) ammonia and water.

9. Hydrogen is used in electrolux refrigeration system so as to .... the vapour pressure ammonia in evaporator.
   (*a*) equalise (*b*) reduce (*c*) increase.

10. In aqua-ammonia absorption system, incomplete rectification leads to accumulation of water in
   (*a*) condenser (*b*) evaporator (*c*) absorber (*d*) any of the these.

ANSWERS

1. (*c*)	2. (*b*)	3. (*d*)	4. (*b*)	5. (*b*)	6. (*c*)	7. (*c*)	8. (*d*)	9. (*b*)	10. (*a*)

THEORETICAL QUESTIONS

1. What is simple vapour absorption system? State how its efficiency can be improved.
2. What is the basic function of a compressor is vapour compression refrigeration system? How this function is achieved in vapour absorption system?
3. Explain briefly with a neat diagram 'Practical vapour absorption system'.
4. State the functions of the following components in an absorption system:
   (*i*) Absorber, (*ii*) Rectifier, (*iii*) Analyser, and (*iv*) Heat exchangers.
5. State the advantages of vapour absorption refrigeration system over vapour compression refrigeration system.
6. Give the comparison between vapour absorption and vapour compression systems.
7. Derive an expression for C.O.P. of an ideal vapour absorption system in terms of T_G, T_E and T_C,
   where T_G = Temperature at which heat (Q_G) is supplied to generator,
   T_E = Temperature at which heat (Q_E) is absorbed in the evaporator, and
   T_C = Temperature at which heat (Q_C) is discharged from condenser and absorber.
8. Draw a neat diagram of 'Electrolux refrigerator' and explain its working principle. What is the important role of hydrogen in this refrigeration system?
9. Mention the function of each fluid in a three-fluid vapour absorption system.
10. State the advantages and disadvantages of 'Electrolux refrigerator' over conventional refrigerators.
11. Draw a neat compact diagram of Lithium bromide water absorption refrigeration system and explain its working. List-out the major fields of applications of this refrigeration system.

UNSOLVED EXAMPLES

1. Find the maximum C.O.P. of a vapour absorption refrigeration system in which heating, cooling and refrigeration takes place at the temperatures of 100ºC, 20ºC and –5ºC respectively. **[Ans.** 2.3]

2. In a vapour absorption system heating, cooling and refrigeration takes place at temperatures of 197ºC, 17ºC and –3ºC respectively. Find the ideal C.O.P. of the system. **[Ans.** 5.16]

3. In an absorption type refrigerator, the heat is supplied to NH_3 generator by condensing steam at 2 bar and 90 per cent dry. The temperature in the refrigerator is to be maintained at –5ºC. Find the maximum C.O.P. possible.

   If the refrigeration load is 18 tonnes and actual C.O.P. is 70 per cent of the maximum C.O.P., find the mass of steam required per hour.

   Take temperature of the atmosphere as 30ºC. **[Ans.** 1.756; 93.12 kg/h]

❑❑❑

Chapter 6

Steam Jet Refrigeration System

6.1 INTRODUCTION

- Water is a refrigerant which, like air, is *perfectly safe. At low temperatures the saturation pressures are low and the specific volumes high.* The temperatures which can be attained using water as a refrigerant are not low enough for most applications of refrigeration but are in the range which may satisfy the conditions of air-conditioning. *The large volumes can be dealt by means of a steam jet, and so this application is more attractive if a supply of medium pressure steam is available. The steam jet can be applied to high-temperature cooling and can be adjusted quickly to meet variations in load.* In the summer, when steam for heating is not required, a supply of steam is readily available for refrigeration and the equipment of a stem-jet refrigeration can be installed for a low capital expenditure. *Steam jets are used when temperatures below 5ºC are required.*
- The steam jet refrigeration cycle is quite similar to more conventional refrigeration cycle, with an *evaporator, a compression device, a condenser and refrigerant as the basic system components.* This system (steam jet), instead of mechanical compression device characteristically *employs a steam ejector or booster to compress the refrigerant to the condenser pressure level.*
- Here *water is used as the refrigerant and the cooling effect is produced in the steam jet refrigeration continuous cycle by vaporisation of a part of the water in the evaporator at a low absolute pressure level* (When chilling water from 10ºC to 5ºC, about 1 *per cent of water* flowing through the evaporator must be vaporised).

6.2 ADVANTAGES, DISADVANTAGES AND APPLICATIONS OF STEAM JET REFRIGERATION SYSTEM

Advantages:

1. Simple and rugged in construction.
2. Vibration free system (because pumps are the only moving parts).
3. Highly reliable.
4. Low production cost.
5. Low maintenance cost.
6. Relatively less plant mass (kg/tonne of refrigeration).
7. Water is used as a refrigerant which is perfectly safe (non-poisonous and non-inflammable).
8. Low running cost (Above 70 tonnes of refrigeration capacity the running cost is less than 50% compared to vapour compression system).
9. Load variations quickly adjusted.
10. It is most suitable for places where steam is available such as process industry, steel plant, petroleum plants, thermal power plants etc. Moreover, it is a boon to the places where electricity is not available since a pump operating on steam can be incorporated.

Disadvantages:

1. Not suitable for water temperature below 4ºC.
2. The *maintenance of high vacuum in the evaporator* is necessary for proper functioning of the system. This is achieved by direct vaporisation to produce chilled water which is usually limited as a very large volume of vapour needs to be handled.
3. Since the ejected vapour and motive steam condenses in the condenser, the magnitude of heat transfer is about three to four times that needed in the vapour-compression system per tonne per hour. Thus *large size condenser is needed in the former.*
4. C.O.P. is very low of the order of about 0.4 to 0.6.
5. Since the volume of vapour handled is enormous, therefore, either centrifugal or ejector devices have to be used. Centrifugal compressors are rarely used since they are much more costlier than reciprocating ones. On the other hand ejectors are most suitable devices for large volume displacement and thus most commonly used. But the *efficiency of ejector is not good thereby rendering a rather inefficient refrigeration system.*

Applications:

The steam jet refrigeration system, despite having the advantage of simplicity, ruggedness of design, freedom from vibration, high reliability, low maintenance and low cost, has not enjoyed wide application in the field of comfort air-conditioning. However the system is *widely used in*:

1. Food processing plants for precooling of vegetables and concentrating fruit juices;
2. Gas Plants;
3. Breweries;
4. Rubber and vulcanising industries;
5. Paper and pulp industries etc.

6.3 PRINCIPLE OF STEAM JET REFRIGERATION SYSTEM

When external pressure on a liquid changes, its boiling point also changes. Normally, atmospheric pressure acts on the surface of any liquid. However, if by some means this pressure is reduced then liquid will start boiling at lower temperature. This *basic principle of boiling of liquid at low temperature by reducing the pressure on its surface is used in steam jet refrigeration system.*

The pure water boils at 100ºC at standard pressure of 760 mm of Hg (1.013 bar). It has been observed that water boils at 12ºC and 7ºC if the pressures on the surface of water are kept at 0.014 bar and 0.01 bar respectively. The *very low pressure or high vacuum on the surface of the water can be maintained by throttling the steam through the jets or nozzles.*

6.4 WORKING OF STEAM JET REFRIGERATION SYSTEM

Refer to Fig. 6.1. A steam jet refrigeration system consists of an *evaporator* (or flash chamber), a *steam jet ejector* (or booster), a *condenser*, and a *two-stage ejector non-condensable pump*. *"Auxiliary components"* include a *chilled water circulating pump,* a *condenser condensate pump*, and necessary *valves and controls.*

The *working* of steam jet refrigeration system is as follows:

- The steam from the boiler also called *motive steam* expands through the nozzle of an ejector (the pressure normally varies from 1.7 to 8 bar). The high velocity vapour imparts the momentum to the vapour of the flash chamber and thereby the flash vapour moves along with the motive steam through the ejector. The process of momentum exchange between the flash chamber vapour and the motive steam and carry over of the former by the latter is often termed as *entrainment.* The entrainment is *extremely irreversible process* due to which the kinetic energy of the motive steam is considerably decreased. Additional reduction in the momentum of motive steam occurs due to mixing with the flash vapour. The *vapour mixture is then finally compressed in the ejector to the condenser pressure where circulating water causes its condensation.* The condensate is then pumped back to the boiler pressure while extra water is purged into atmosphere.

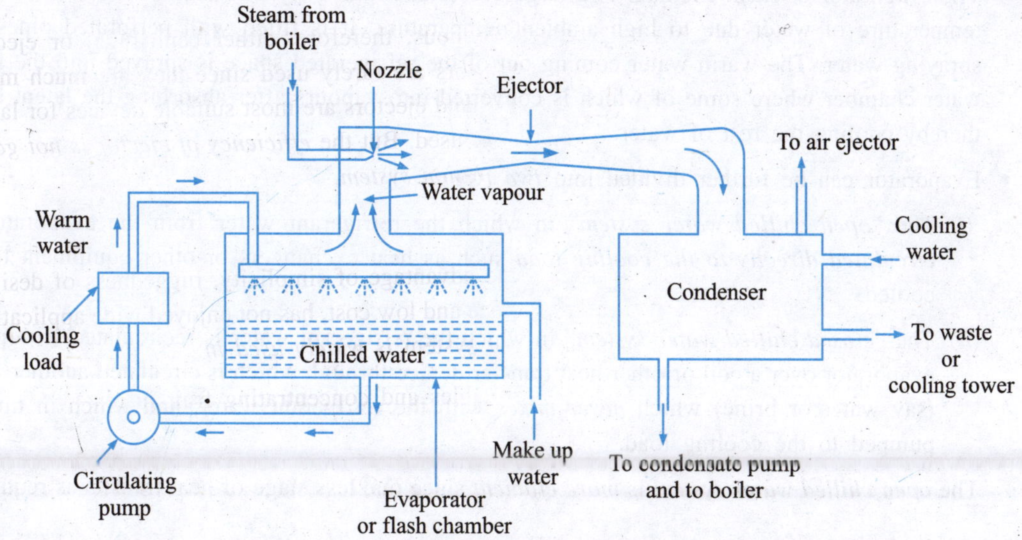

Fig. 6.1. Steam jet refrigeration system.

- The chilled water of the flash chamber is pumped through the load (may be a water cooler, air-conditioning system etc.). The warmed up water due to the heat-load is sprayed into the flash chamber.
- Since the cooling is caused as a result of vaporisation of water, the make-up water is supplied through a *float valve, keeping a constant water level in the flash chamber.* If this water is free from non-condensable gases like air, then the condenser (Fig. 6.1) serves the purpose. Unfortunately water contains non-condensable gases which have to be purged out of the condenser with the help of a pump or an auxiliary ejector. Otherwise the condenser temperature will go up, thereby reduction in the performance of the system takes place.
- In steam jet refrigeration system, the *control of capacity is possible either by throttling the steam inlet pressure*, or *by providing more than one nozzle operating in parallel, each driving a certain amount of vapour from the flash chamber.*

Note. Consider a flash chamber containing 100 kg of water. If 1 kg of water is suddenly removed by boiling as pressure is reduced due to throttling of steam through steam nozzle, approximately 2400 kJ heat (say at 0.085 bar) which is required for its evaporation, will be removed from water. Consequently the remaining water undergoes a falls in temperature which is given by,

$$\Delta t_F = \frac{24000}{(100-1)\times 4.187} = 5.79°C$$

It means that by evaporating one more kg of water, the remaining water will become cooler by another 5.79°C. Thus if process is continued the remaining water can freeze. As a consequence of this *limitation,* steam jet refrigeration is used for high temperature cooling and air-conditioning.

Description of System Components

The components of steam jet refrigeration system are briefly described below.

1. Evaporator:

- The evaporator or flash chamber is a large vessel and is heavily insulated to avoid the rise in temperature of water due to high ambient temperature. It is fitted with perforated pipes for spraying water. The warm water coming out of the refrigerated space is sprayed into the flash water chamber where some of which is converted into vapours after absorbing the latent heat, thereby cooling the rest of water.
- Evaporator can be further divided into *two types*:

  (i) The "*open chilled water system*" in which the refrigerant water from the evaporator is *circulated directly to the cooling load* such as heat exchangers or other equipment to be cooled.

  (ii) The "*closed chilled water system*" in which the refrigerant water is recirculated within the evaporator over a coil or other heat transfer device through which is circulated another fluid (say water or brine) which *never mixes* with the refrigerant water; and which in turn is pumped to the cooling load.

  The *open chilled water system* is *more efficient* since one less stage of heat transfer is required.

2. Steam jet ejector:

- A steam ejector, also known as jet pump or thermo-compressor. is shown in Fig. 6.2. *It is used to compress the water vapour coming out of the flash chamber. It utilizes the energy of fast moving jet steam to entrain the vapour from the flash chamber and then compress it.*

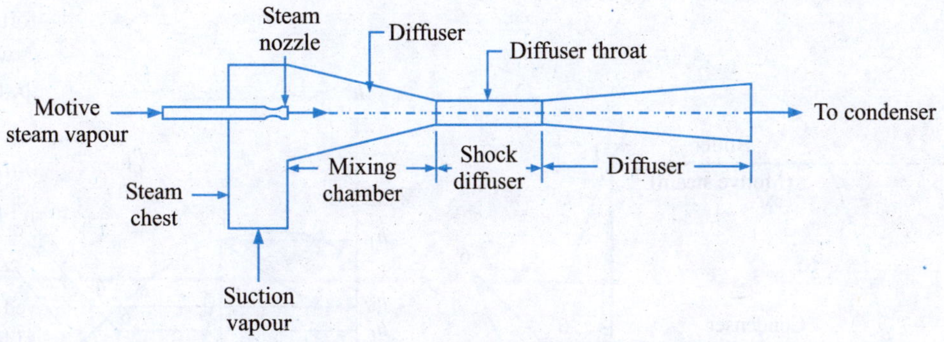

Fig. 6.2. Schematic diagram of steam ejector.

- High pressure steam from the boiler (generally called *primary fluid* or *motive steam/vapour*) is admitted to the steam chest and expanded in a convergent-divergent nozzle to a very low pressure and it attains supersonic velocities in the range of 1000 m/s to 1350 m/s. The flash chamber is connected to the region of low pressure of the ejector. The water vapour from the flash is entrained in the high velocity jet of steam and both are mixed in the mixing section at constant pressure. The mean velocity of the mixture will be supersonic after the completion of the mixing process. This supersonic steam gets *a normal shock* in the constant area throat of diffuser, resulting in the rise of pressure and subsonic flow. In the diverging portion of the diffuser the velocity head is recovered as pressure head and finally high pressure steam is condensed in the condenser.

3. Condenser:

The condenser may be one of *three basic types*:

(*i*) ***A conventional surface condenser.*** It is employed because of its compact size and ability to be located anywhere including inside the building. In this case steam consumption is higher and larger surface condenser and cooling water will be required.

(*ii*) ***An evaporative condenser.*** It operates at a condensing temperature 2ºC to 4ºC below that of a surface condenser with cooling water, and this in turn results in favourable decrease in steam consumption.

(*iii*) ***A barometric condenser.*** It is used to advantage in those installations which do not have space limitations (primary vertical) and where it is economically feasible to treat the evaporator and boiler water on a once through basis. These condensers are not used in majority of comfort air-conditioning applications but they have found acceptance in industrial and other applications.

4. Air ejector:

The air or non-condensable gas ejector typically employs two small steam powered ejectors in *series*. An inter and after condenser is often employed to improve economy. In a tight system, an air ejector unit will have capacity of 0.05 kg of air per tonne of refrigeration.

6.5 ANALYSIS OF STEAM JET REFRIGERATION SYSTEM

The T-s (temperature-entropy) and h-s (enthalpy-entropy) diagrams for a steam jet refrigeration system are shown in Fig. 6.3 (*a*) and Fig. 6.3 (*b*) respectively.

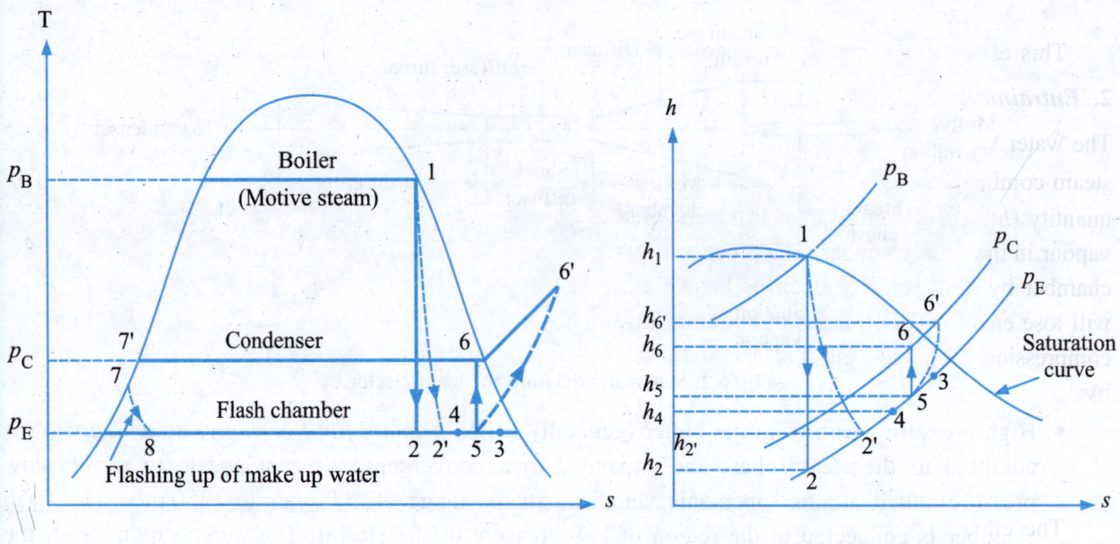

(*a*) T-s diagram. (*b*) h-s diagram.

Fig. 6.3

- The point '1' represents the initial condition of motive steam before passing through the nozzle and the point '2' is the final condition of the steam, *assuming isentropic expansion.*
- The point '3' represents the initial condition of water vapour in the evaporator or flash chamber.
- The point '4' represents the condition of motive steam just before mixing with the water vapour.
- The point '5' is the condition of mixture of high velocity steam from the nozzle and the entrained water vapour before compression.
- The point '6' is the final condition of the mixture discharged to the condenser, *assuming isentropic compression.*
- The make-up water is supplied at point 7 which temperature is slightly lower than the condenser temperature and is throttled to point 8 in the flash chamber.

It may be noted that, due to certain unavoidable losses during expansion and compression, the actual expansion of motive steam is represented by 1-2′ and the actual compression of the mixture of motive steam and water vapour is represented by 5-6′.

Efficiencies: The various efficiencies used in steam jet refrigeration system are:

1. Nozzle efficiency
2. Entrainment efficiency
3. Compression efficiency.

1. *Nozzle efficiency* (η_{nozzle}):

It is defined as the *ratio of actual enthalpy drop to isentropic enthalpy drop of the motive steam passing through the nozzle.*

i.e.,
$$\eta_{nozzle} = \frac{\text{Actual enthalpy drop}}{\text{Isentropic enthalpy drop}} = \frac{h_1 - h_{2'}}{h_1 - h_2} \quad ...(6.1)$$

This efficiency varies from 85 to 90%.

2. *Entrainment efficiency* (η_{ent}):

The water vapour formed in the flash chamber has negligible velocity compared with the velocity of the steam coming out of the nozzle which is equivalent to $\sqrt{2 \times 1000(h_1 - h_{2'})}$ = *$44.72\sqrt{h_1 - h_{2'}}$ m/s. The quantity $(h_1 - h_{2'})$ is equivalent to kinetic energy of motive steam, now available for entrainment of the vapour in the flash chamber. The process of giving the momentum to the water vapour formed in the flash chamber by high velocity steam is known as *entrainment of vapour.* During the entrainment, the steam will lose energy. This process of entrainment is *inefficient* and part of original motive force available for compression is *reduced* and it is taken into account by a factor known as *entrainment efficiency*; it is given by:

$$\eta_{ent.} = \frac{h_1 - h_4}{h_1 - h_{2'}} \quad ...(6.2)$$

The entrainment efficiency may be taken as 65%.

3. *Compression efficiency* ($\eta_{comp.}$):

It is defined as the *ratio of the isentropic enthalpy increase to the actual enthalpy increase required for the compression of the mixture of motive steam and the water vapour, in the diffuser.*

i.e.,
$$\eta_{comp.} = \frac{h_5 - h_6}{h_5 - h_{6'}} \quad ...(6.3)$$

The compression efficiency may be taken as 75 to 80%.

- The compression efficiency is also known as *diffuser efficiency.*

Mass of motive steam required:

Let, m_s = Mass of motive steam supplied in kg/min,

m_v = Mass of water vapour formed from the evaporator or flash chamber in kg/min, and

$m_{mix.}$ = Mass of the mixture for compression in kg/min.

$= m_s + m_v$

Available energy for compression $= m_s (h_1 - h_4)$ ...(i)

Energy required for compression $= m_{mix} (h_{6'} - h_5)$ ...(ii)

$= (m_s + m_v)(h_{6'} - h_5)$

According to the law of conservation of energy, *the available energy for compression must be equal to the energy required for compression.*

* For further details, refer Author's popular book on "Thermal Engineering".

Thus, $$m_s\,(h_1 - h_4) = (m_s + m_v)\,(h_{6'} - h_5) \qquad \text{...}(iii)$$

Now, $$\eta_{nozzle} = \frac{h_1 - h_{2'}}{h_1 - h_2}, \quad \text{or,} \quad h_1 - h_{2'} = \eta_{nozzle}\,(h_1 - h_2) \qquad \text{...}(iv)$$

and, $$\eta_{ent.} = \frac{h_1 - h_4}{h_1 - h_{2'}}, \quad \text{or,} \quad h_1 - h_4 = \eta_{ent.}\,(h_1 - h_{2'}) \qquad \text{...}(v)$$

and, $$\eta_{comp.} = \frac{h_6 - h_5}{h_{6'} - h_5}, \quad \text{or,} \quad h_{6'} - h_5 = \frac{h_6 - h_5}{\eta_{comp.}} \qquad \text{...}(vi)$$

Substituting the values of $(h_1 - h_4)$ and $(h_{6'} - h_5)$ from eqns. (*v*) and (*vi*) in eqn. (*iii*), we get:

$$m_s \times \eta_{ent.}\,(h_1 - h_{2'}) = (m_s + m_v)\left(\frac{h_6 - h_5}{\eta_{comp.}}\right)$$

or, $$m_s \times \eta_{ent.} \times \eta_{nozzle}\,(h_1 - h_2) = (m_s + m_v)\left(\frac{h_6 - h_5}{\eta_{comp.}}\right) \qquad \text{...[From eqn. }(iv)]$$

or, $$m_s \times \eta_{ent.} \times \eta_{nozzle} \times \eta_{comp.}\,(h_1 - h_2) = m_s\,(h_6 - h_5) + m_v\,(h_6 - h_5)$$

or, $$m_s\,[(h_1 - h_2)\,\eta_{ent.} \times \eta_{nozzle} \times \eta_{comp.} - (h_6 - h_5)] = m_v\,(h_6 - h_5)$$

or, $$\frac{m_s}{m_v} = \frac{(h_6 - h_5)}{[(h_1 - h_2)\,\eta_{nozzle} \times \eta_{ent.} \times \eta_{comp.} - (h_6 - h_5)]} \qquad \text{...(6.4)}$$

where $\frac{m_s}{m_v}$ = Mass of motive steam required per kg of water vapour produced in the flash chamber.

As earlier discussed, the make-up water is supplied at point '7' whose temperature is slightly lower than the condenser temperature and is throttled to point '8' in the flash chamber and leaves it corresponding to the condition at point '3'. Since the enthalpy of water at point '7' is equal to the enthalpy of water at point '8', therefore, for *each kg of water vapour formed, heat absorbed is* $(h_3 - h_{f7})$ kJ/kg. In other words, net refrigerating effect, $R_n = m_v\,(h_3 - h_{f7})$ kJ/min.

If the refrigerating load is Q tonnes, then mass of vapour formed,

$$m_v = \frac{14000\,Q}{60(h_3 - h_{f8})} \text{ kg/min} \qquad \text{...(6.5)}$$

Since one kg of water vapour requires m_s kg of motive steam, therefore,

Mass of motive steam required per Q tonnes of refrigerating load

= Mass of water vapour/min. × motive steam required per kg of vapour

$$= m_v \times m_s = \frac{14000\,Q}{60(h_3 - h_{f8})} \times m_s \qquad \text{...(6.6)}$$

The *coefficient of performance* of the system is given by,

$$\text{C.O.P.} = \frac{m_v\,(h_3 - h_{f7'})}{m_s\,(h_1 - h_{f7'})} \qquad \text{...(6.7)}$$

6.6 PERFORMANCE OF STEAM JET REFRIGERATION SYSTEM

The *performance* of steam jet refrigeration system is defined as *the mass of steam required per tonne of refrigeration with the decrease in temperature of chilled water (evaporating temperature) or increase in pressure of condensing vapour (condensing temperature).*

In a steam jet refrigeration in which components have been properly designed, the *steam rate is a function of 'steam pressure', 'condenser temperature', and 'evaporator temperature'.*

- Fig. 6.4 shows the effect of both condenser temperature and evaporator temperature at a constant steam pressure of 7 bar. In industrial practice, the actual steam consumption will vary within 5 percent of the curves shown.
  - — The *amount of steam required to produce one tonne of cooling effects decreases with decreasing condenser temperature and with increasing chilled water temperature.*
- Fig. 6.5 shows a plot between steam consumption and condenser pressure with evaporation temperature as parameter.
- Fig. 6.6 shows a plot between steam consumption and chilled water temperature, with condensing vacuum as parameters.
- Fig. 6.7 shows the effect of steam pressure on the steam requirements for fixed condenser temperature of 30ºC.

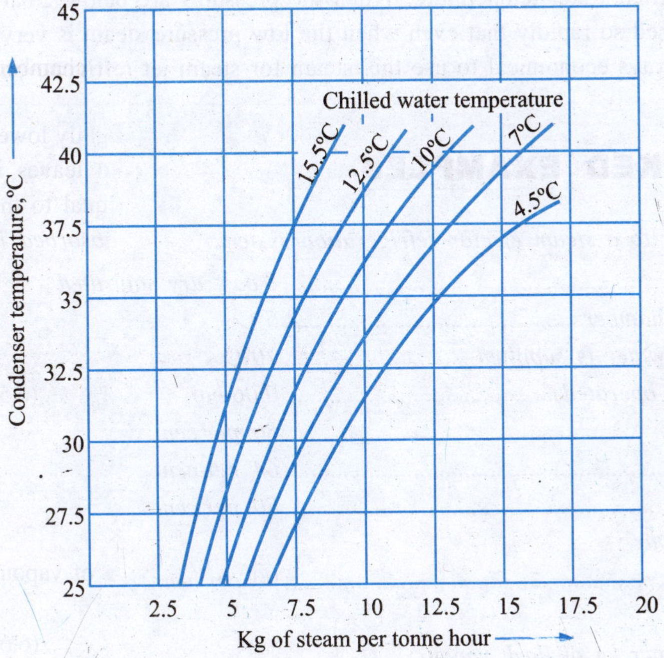

Fig. 6.4. Plot between condensing temperature and steam consumption (Chilled water temperature as parameter).

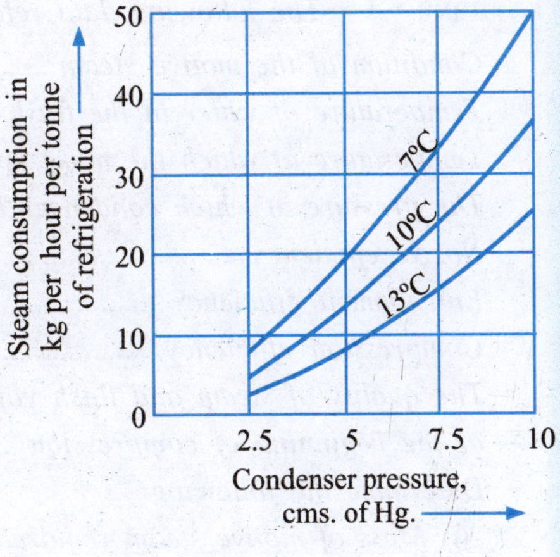

Fig. 6.5. Plot between steam consumption and condenser pressure (Evaporation temperature as parameter).

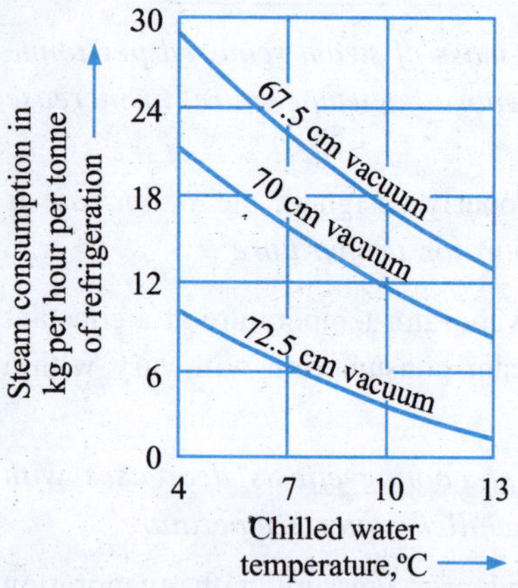

Fig. 6.6. Plot between steam consumption and chilled water temperature (Condenser vacuum as parameter).

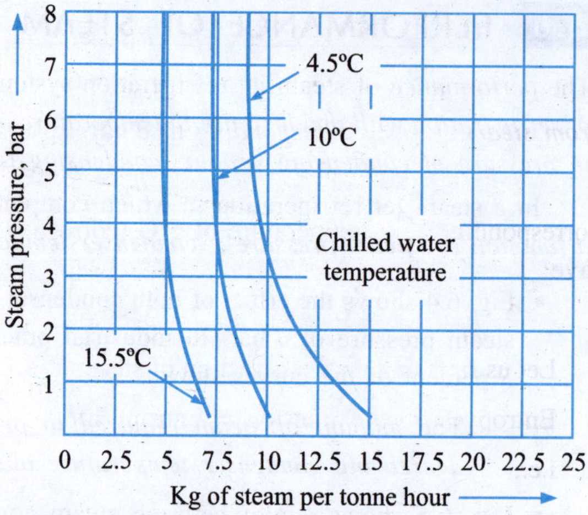

Fig. 6.7. Plot between steam pressure and steam consumption (Chilled water temperature as parameter).

— The curve shows that it is more economical to operate at the highest steam pressure available although pressures above 7 bar have diminishing effect on the steam requirements.

— The influence of steam pressure is significantly more pronounced with chilled water temperature below 10°C than at higher chilled water temperature. When the pressures are below 2 bar steam requirements are increased so rapidly that even when the low pressure steam is very cheaply available, it is not always economical to use this steam for steam jet refrigeration cooling.

WORKED EXAMPLES

Example 6.1. *The following data refer to a steam ejector refrigeration system:*

Condition of the motive steam .... *8 bar, dry saturated*
Temperature of water in the flash chamber .... *5°C*
Temperature at which the make-up water is supplied .... *20°C*
The pressure at which condenser is operated .... *0.06 bar*
Nozzle efficiency .... *86 per cent*
Entrainment efficiency .... *64 per cent*
Compression efficiency .... *80 per cent*
The quality of steam and flash vapour
at the beginning of compression .... *90 per cent.*

Determine the following:

(i) *Mass of motive steam required per kg of flash vapour;*
(ii) *The quality of vapour flashed from the flash chamber;*
(iii) *Refrigerating effect per kg of flash vapour;*
(iv) *Mass of motive steam required per hour per tonne of refrigeration;*
(v) *Volume of vapour removed from the flash chamber per hour per tonne of refrigeration;*
(vi) *The coefficient of performance of the system.*

Solution: *Given:* Refer to Fig. 6.8. p_B = 8 bar; t_{wf} = 5ºC; p_C = 0.06 bar; η_{nozzle} = 86%;

$\eta_{ent.}$ = 64%; $\eta_{comp.}$ = 80%; x_5 = 90%.

From steam tables, corresponding to **8 bar** (dry saturated), we have:

$$h_1 = 2767.5 \text{ kJ/kg}; \quad s_1 = 6.6596 \text{ kJ/kg K}; \quad t_1 = 170.4\text{ºC}.$$

Corresponding to a temperature of **5ºC** (corresponding pressure is 0.0087 bar), *from steam tables*, we have:

$$h_{f2} = 21 \text{ kJ/kg K}; \quad h_{fg2} = 2489.6 \text{ kJ/kg}; \quad s_{f2} = 0.076 \text{ kJ/kg K}; \quad s_{fg\,2} = 8.95 \text{ kJ/kg K}.$$

Let us first find dryness fraction of steam at point 2 (i.e., x_2). For isentropic expansion 1–2, we have:

Entropy before expansion = Entropy after expansion

i.e., $$s_1 = s_2$$

$$6.6596 = s_{f2} + x_2\, s_{fg2} = 0.076 + x_2 \times 8.95$$

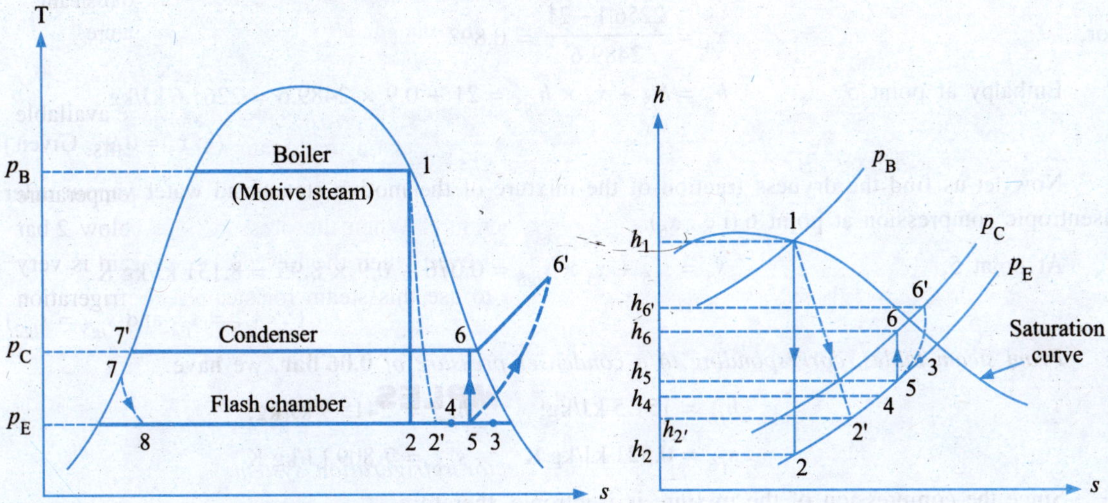

(*a*) T-*s* diagram. (*b*) *h-s* diagram.

Fig. 6.8

$$\therefore \quad x_2 = \frac{6.6596 - 0.076}{8.95} = 0.73$$

Enthalpy at point 2, $$h_2 = h_{f2} + x_2 h_{fg\,2} = 21 + 0.73 \times 2489.6 = \mathbf{1838.4\ kJ/kg} \quad \textbf{(Ans.)}$$

Nozzle efficiency, $$\eta_{nozzle} = \frac{h_1 - h_{2'}}{h_1 - h_2}$$

or, $$0.86 = \frac{2767.5 - h_{2'}}{2767.5 - 1838.4}$$

or, $$h_{2'} = 2767.5 - 0.86\,(2767.5 - 1838.4) = 1968.5 \text{ kJ/kg}$$

Now let us find dryness fraction at point 2′ (i.e. $x_{2'}$). Since the points 2, 2′, 4 and 5 lie on the same pressure line (corresponding to 5ºC), therefore,

$$h_{f2} = h_{f2'} = h_{f4} = h_{f5} = 21 \text{ kJ/kg}$$

and, $$h_{fg2} = h_{fg2'} = h_{fg4} = h_{fg5} = 2489.6 \text{ kJ/kg}$$

Now, $$h_{2'} = h_{f2'} + x_{2'} \, h_{fg2'}$$

or, $$1968.5 = 21 + x_{2'} \times 2489.6$$

or, $$x_{2'} = \frac{1968.5 - 21}{2489.6} = 0.78$$

Also, Entrainment efficiency, $$\eta_{ent.} = \frac{h_1 - h_4}{h_1 - h_{2'}}$$

$\therefore$ $$0.64 = \frac{2767.5 - h_4}{2767.5 - 1968.5}$$

or, $$h_4 = 2767.5 - 0.64\,(2767.5 - 1968.5) = 2256.1 \text{ kJ/kg}$$

Enthalpy at point 4, $$h_4 = h_{f4} + x_4 \times h_{fg4}$$

$\therefore$ $$2256.1 = 21 + x_4 \times 2489.6$$

or, $$x_4 = \frac{2256.1 - 21}{2489.6} = 0.898$$

Enthalpy at point 5, $$h_5 = h_{f5} + x_5 \times h_{fg5} = 21 + 0.9 \times 2489.6 = 2261.6 \text{ kJ/kg}$$

($\because x_5 = 0.9$ ...Given)

Now let us find the dryness fraction of the mixture of the motive steam and water vapour after isentropic compression at point 6 (i.e., x_6).

At point 5, $$s_5 = s_{f5} + x_5 \times s_{fg5} = 0.076 + 0.9 \times 8.95 = 8.131 \text{ kJ/kg K}$$

($\because s_{f5} = s_{f2}$ and $s_{fg5} = s_{fg2}$)

From steam tables, corresponding to a condenser pressure of **0.06 bar**, we have:

$$h_{f6} = 151.5 \text{ kJ/kg}; \qquad h_{fg6} = 2415.9 \text{ kJ/kg};$$
$$s_{f6} = 0.521 \text{ kJ/kg K} \qquad s_{fg6} = 7.809 \text{ kJ/kg K}$$

Since the compression of the mixture is isentropic, therefore,

Entropy before compression (s_5) = Entropy after compression (s_6)

$$8.131 = s_{f6} = x_6 s_{fg6} = 0.521 + x_6 \times 7.809$$

$\therefore$ $$x_6 = \frac{8.131 - 0.521}{7.809} = 0.974$$

Now, $$h_6 = h_{f6} + x_6 h_{fg6} = 151.5 + 0.974 \times 2415.9 = 2504.6 \text{ kJ/kg}$$

Also, compression efficiency,

$$\eta_{comp.} = \frac{h_6 - h_5}{h_{6'} - h_5}$$

or $$0.80 = \frac{2504.6 - 2261.6}{h_{6'} - 2261.6}$$

$\therefore$ $$h_{6'} = 2261.6 + \left(\frac{2504.6 - 2261.6}{0.80}\right) = 2565.3 \text{ kJ/kg}$$

(*i*) **Mass of motive steam required per kg of the flash vapour:**

The mass of motive steam required per kg of flash vapour,

$$\frac{m_s}{m_v} = \frac{(h_6 - h_5)}{\left[(h_1 - h_2)\,\eta_{\text{nozzle}} \times \eta_{\text{ent.}} \times \eta_{\text{comp.}} - (h_6 - h_5)\right]} \quad \text{...[Eqn. (6.4)]}$$

$$= \frac{(2504.6 - 2261.6)}{[(2767.5 - 1838.4) \times 0.86 \times 0.64 \times 0.8 - (2504.6 - 2261.6)]}$$

$$= \frac{243}{(409.1 - 243)} = \mathbf{1.463\ kg/kg\ of\ flash\ vapour} \quad \textbf{(Ans.)}$$

(*ii*) **Quality of vapour flashed from flash chamber, x_3:**

Let us first find the enthalpy at point 3.

We know that, $m_v h_3 + m_s h_4 = (m_s + m_v)\, h_5$

$$h_3 + \frac{m_s}{m_v} \times h_4 = \left(\frac{m_s}{m_v} + 1\right) h_5$$

$$h_3 + 1.463 \times 2256.1 = (1.463 + 1) \times 2261.6$$

$\therefore$ $h_3 = 2269.6$ kJ/kg

Also, $h_3 = h_{f3} + x_3\, h_{fg3}$

or, $2269.6 = 21 + x_3 \times 2489.6$ $\quad (\because h_{f3} = h_{f2}\ ;\ h_{fg3} = h_{fg2})$

$$\therefore \quad x_3 = \frac{2269.6 - 21}{2489.6} = \mathbf{0.903} \quad \textbf{(Ans.)}$$

(*iii*) **Refrigerating effect per kg of flash vapour:**

Refrigerating effect per kg of flash vapour

$$= h_3 - h_{f3} = 2269.6 - 84 = \mathbf{2185.6\ kJ/kg} \quad \textbf{(Ans.)}$$

($\because h_{f7}$ at 20°C, from stream tables = 84 kJ/kg)

(*iv*) **Mass of motive steam required per hour per tonne of refrigeration:**

Mass of motive steam required per hour per tonne of refrigeration $= \dfrac{14000\text{Q}}{(h_3 - h_{f7})} \times \dfrac{m_s}{m_v}$

$$= \frac{14000 \times 1}{2185.6} \times 1.463 = \mathbf{9.37\ kg/h/tonne\ of\ ref.} \quad \textbf{(Ans.)}$$

(*v*) **Volume of vapour removed from the flash chamber per hour per tonne of refrigeration:**

Volume of vapour (per kg) removed from the flash chamber,

v_3 = Volume of liquid at 3 + x_3 (volume of saturated vapour – volume of liquid)

$= 1 + 0.903\ (147.1 - 1) = 132.93\ \text{m}^3/\text{kg}$

... (Volume of saturated vapour corresponding to 5ºC, from steam tables = 147.1 m³/kg)

∴ Volume of vapour removed from the flash chamber per hour per tonne of refrigeration

$$= v_3 \times \frac{14000Q}{(h_3 - h_{f7})} = 132.93 \times \frac{14000 \times 1}{2185.6} = \mathbf{851.5\ m^3/h/tonne\ of\ ref.} \quad \textbf{(Ans.)}$$

(*vi*) **Coefficient of performance of the system, C.O.P.:**

From steam tables, corresponding to a condenser pressure of 0.06 bar,

$$h_{f7'} = 151.5 \text{ kJ/kg}$$

$$\therefore \quad \text{C.O.P.} = \frac{m_v (h_3 - h_{f7})}{m_s (h_1 - h_{f7'})} \quad \text{...[Eqn. (6.7)]}$$

$$= \frac{1(2269.6 - 84)}{1.463(2767.5 - 151.5)} = \mathbf{0.57} \quad \textbf{(Ans.)}$$

Example 6.2. *The following data refer to a steam jet refrigeration system:*

Capacity of thermal power plant .... 2000 kW

Condition of steam supplied from the boiler .... 25 bar, 350°C

Refrigeration load taken up by the steam jet refrigeration system .... 110 tonnes

Pressure at which steam is bled off the steam turbine .... 4 bar

Vacuum maintained in the condenser .... 730 mm of Hg

Pressure and condition at which steam enters in thermo-compressor...... 0.01 bar, 0.92 dry

Temperature at which make-up water enters into flash chamber .... 20°C

Isentropic efficiency of the steam turbine .... 88 per cent

Nozzle efficiency............ 88 percent

Entrainment efficiency .... 64 per cent

Thermo-compressor efficiency .... 64 per cent.

Determine the following, using Mollier diagram:

(*i*) *Dryness fraction of steam leaving the flash chamber,*

(*ii*) *Mass of motive steam bled off from the steam turbine,*

(*iii*) *Mass of steam generated in the boiler,*

(*iv*) *Mass of additional steam to be generated by the boiler due to refrigeration load, and*

(*v*) *Coefficient of performance of the system.*

Solution: *Given:* P = 2000 kW; p_1 = 25 bar; t_1 = 350°C; Q = 110 tonnes; Q = 110 tonnes;

$p_2 = p_{2'} = 4$ bar; $p_7 = p_{7'} = p_8 = p_{8'} = 760 - 730 = 30$ mm Hg (= 30 × 0.00133) = 0.04 bar;

$p_3 = p_{3'} = p_4 = p_5 = p_6 = 0.01$ bar; $x_4 = 0.92$; $\eta_{turbine}$ = 88% or 0.88; η_{nozzle} =88% or 0.88;

$\eta_{ent.}$ = 64% or 0.64; $\eta_{comp.}$ = 64% or 0.64.

The schematic diagram of the thermal power plant and various processes on the Mollier diagram are shown in Fig. 6.9 and Fig. 6.10 respectively.

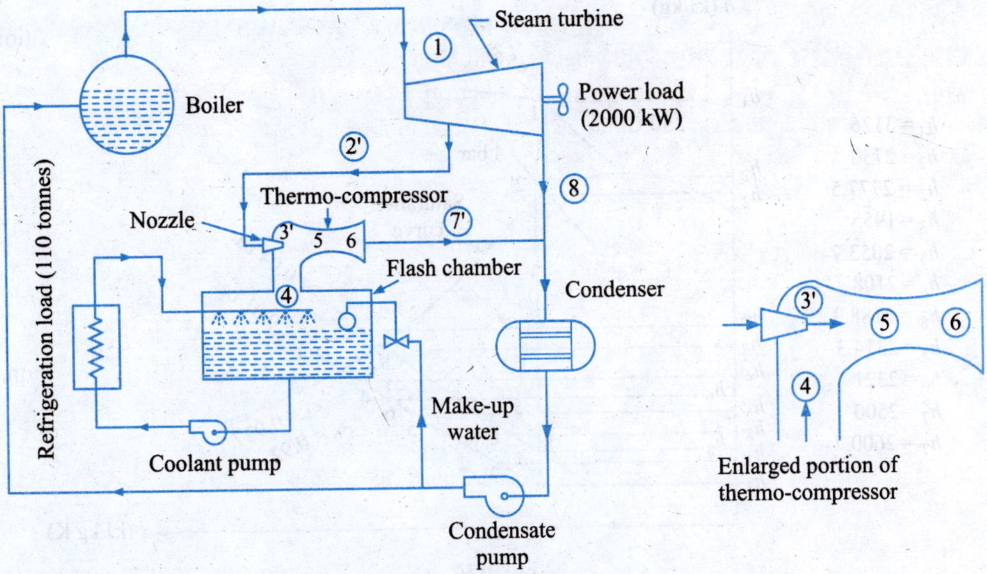

Fig. 6.9

- The point 1 represents the entry of steam at a pressure of 25 bar and 350ºC from the boiler to the steam turbine.
- The process 1-2 represents the *isentropic expansion* whereas the process 1-2′ represents the *actual expansion* of steam in the steam turbine.

From the mollier diagram, h_1 = 3126 kJ/kg; h_2 = 2730 kJ/kg

Now,
$$\eta_{turbine} = \frac{h_1 - h_{2'}}{h_1 - h_2}$$

or,
$$0.88 = \frac{3126 - h_{2'}}{3126 - 2730}$$

$$\therefore \quad h_{2'} = 3126 - 0.88\ (3126 - 2730) = 2777.5 \text{ kJ/kg}$$

The steam required for steam jet refrigeration system is bled off from the turbine at point 2′, i.e., at 4 bar and enthalpy $h_{2'}$ = 2777.5 kJ/kg.

- The process 2′-3 represents the *isentropic expansion* of steam whereas the process 2′-3′ represents the *actual expansion* of steam through the nozzle upto the flash chamber pressure of 0.01 bar.

From Mollier diagram : h_3 = 1955 kJ/kg

Also,
$$\eta_{nozzle} = \frac{h_{2'} - h_{3'}}{h_{2'} - h_3}$$

$$\therefore \quad 0.88 = \frac{2777.5 - h_{3'}}{2777.5 - 1955}$$

or,
$$h_{3'} = 2777.5 - 0.88\ (2777.5 - 1955) = 2053.7 \text{ kJ/kg}$$

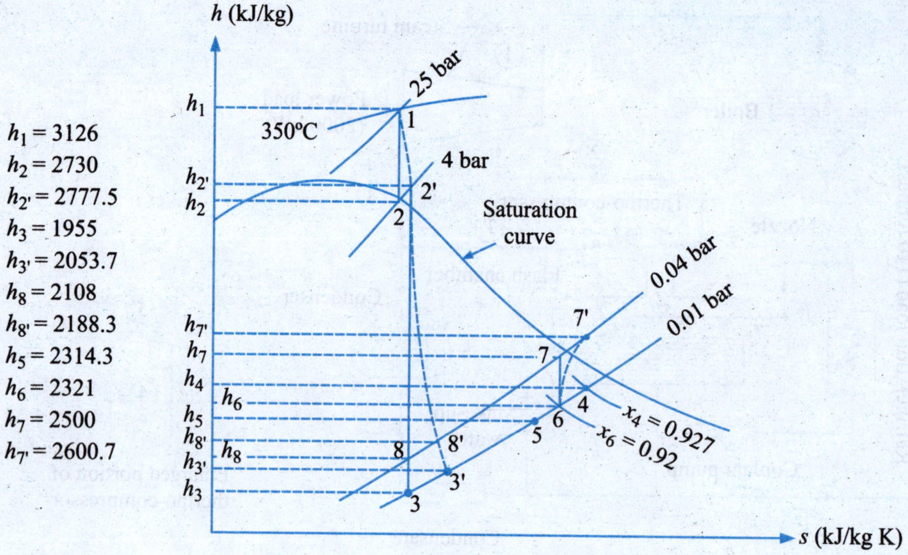

Fig. 6.10

- The remaining steam expands in the turbine from 2′ to a condenser pressure of 0.04 bar.
- The process 2′-8 represents the *isentropic expansion* whereas 2′-8 represents the *actual expansion* of the remaining steam in the turbine.

From Mollier diagram, h_8 = 2108 kJ/kg

Again,
$$\eta_{turbine} = \frac{h_{2'} - h_{8'}}{h_{2'} - h_8}$$

or
$$0.88 = \frac{2777.5 - h_{8'}}{2777.5 - 2108}$$

$$\therefore \quad h_{8'} = 2777.5 - 0.88\ (2777.5 - 2108) = 2188.3 \text{ kJ/kg}$$

- The steam coming out from the nozzle at a very high velocity at point 5 entrains the water vapour from the flash chamber at a low velocity.

Now,
$$\eta_{ent.} = \frac{h_{2'} - h_5}{h_{2'} - h_{3'}}$$

or,
$$0.64 = \frac{2777.5 - h_5}{2777.5 - 2053.7}$$

$$\therefore \quad h_5 = 2777.5 - 0.64\,(2777.5 - 2053.7) = 2314.3 \text{ kJ/kg}$$

- The steam at point 5 mixes with water vapour from the flash chamber at point 4 (which is unknown). The resulting condition of the mixture is represented by point 6 and its condition is given i.e., pressure 0.01 bar and dryness fraction (x_6) 0.92.

From Mollier diagram, h_6 = 2321 kJ/kg.

- The steam at point 6 is now compressed in the thermo-compressor. The process 6-7 represents the *isentropic compression* whereas the process 6-7′ represents the *actual compression.*

From Mollier diagram, $h_7 = 2500$ kJ/kg

Also, $$\eta_{comp.} = \frac{h_7 - h_6}{h_{7'} - h_6}, \quad \text{or,} \quad 0.64 = \frac{2500 - 2321}{h_{7'} - 2321}$$

$$\therefore \quad h_{7'} = 2321 + \frac{(2500 - 2321)}{0.64} = 2600.7 \text{ kJ/kg}$$

(*i*) **Dryness fraction of steam leaving the flash chamber, x_4:**

The mass of motive steam required per kg of flash vapour produced in the flash chamber,

$$\frac{m_s}{m_v} = \frac{(h_7 - h_6)}{(h_{2'} - h_3)\,\eta_{nozzle} \times \eta_{ent.} \times \eta_{comp.} - (h_7 - h_6)}$$

$$= \frac{(2500 - 2321)}{(2777.5 - 1955) \times 0.88 \times 0.64 \times 0.64 - (2500 - 2321)}$$

$$= 1.524 \text{ kg/kg of flash vapour}$$

Considering enthalpy balance, we have:

$$m_v h_4 + m_s h_5 = (m_s + m_v)\, h_6$$

or $$h_4 + \frac{m_s}{m_v} h_5 = \left(\frac{m_s}{m_v} + 1\right) h_6$$

or $$h_4 + 1.524 \times 2314.3 = (1.524 + 1) \times 2321$$

$$\therefore \quad h_4 = 2331.2 \text{ kJ/kg}$$

Now mark point 4 on the Mollier diagram on a pressure line of 0.01 bar and enthalpy of $h_4 = 2331.2$ kJ/kg. From the Mollier diagram, it is found that dryness fraction of steam leaving the flash chamber (i.e., at point 4),

$$x_4 = \mathbf{0.927} \quad \textbf{(Ans.)}$$

(*ii*) **Mass of motive steam bled off from the steam turbine:**

The refrigerating effect per kg of flash vapour

$$= h_4 - \text{enthalpy of make-up water}$$

$$= 2331.2 - 4.187 \times 20 = 2247.5 \text{ kJ/kg of flash vapour}$$

Amount of flashed vapour per hour for 110 tonnes of refrigeration

$$= \frac{14000 \times 110}{2247.5} = 685.2 \text{ kg/h}$$

∴ Mass of motive steam bled off from the steam turbine

$$= \frac{m_s}{m_v} \times 685.2 = 1.524 \times 685.2 = \mathbf{1044.2 \text{ kg/h}} \quad \textbf{(Ans.)}$$

(*iii*) **Mass of steam generated in the boiler:**

Let, m_B = Mass of steam generated in the boiler (or mass of steam supplied to the power turbine)

$$\therefore \quad m_B (h_1 - h_{2'}) + (m_B - 972)(h_{2'} - h_{8'}) = \text{Power in kW} \times 3600 \quad (\because 1\text{kW} = 1\text{kJ/s})$$

or, $m_B (3126 - 2777.5) + (m_B - 1044.2)(2777.5 - 2188.3) = 2000 \times 3600$

or, $348.5\, m_B + 589.2\, m_B - 615242.6 = 7.2 \times 10^6$

$\therefore \quad m_B =$ **8334.5 kg/h (Ans.)**

(*iv*) **Mass of additional steam to be generated by the boiler due to refrigeration load:**

In the absence of refrigeration load, the mass of steam required for 2000 kW is given by,

$$m_{B'} = \frac{2000 \times 3600}{(h_1 - h_{8'})} = \frac{2000 \times 3600}{(3126 - 2188.3)} = 7678.4 \text{ kg/h}$$

$\therefore$ Mass of additional steam to be generated by the boiler due to refrigeration load

$= m_B - m_{B'} = 8334.5 - 7678.4 =$ **656.1 kg/h (Ans.)**

(*v*) **Coefficient of performance of the system, C.O.P.:**

Enthalpy of steam supplied to the refrigeration system at point 2′,

$$h_{2'} = 2777.5 \text{ kJ/kg}$$

The recoverable heat from the steam = Enthalpy of water at 0.04 bar

= 121.5 kJ/kg (From steam tables)

$\therefore$ Heat to be actually supplied per kg of steam for refrigeration purposes

$= h_{2'}$ – enthalpy of water at 0.04 bar

$= 2777.5 - 121.5 = 2656$ kJ/kg

$$\therefore \quad \text{C.O.P.} = \frac{\text{Refrigeration load (kJ / h)}}{\text{Mass of motive steam bled / h} \times \text{heat supplied / kg of steam}}$$

$$= \frac{14000 \times 110}{1044.2 \times 2656} = \mathbf{0.555} \quad \textbf{(Ans.)}$$

HIGHLIGHTS

1. Steam jets are used when temperatures below 5ºC are required.
2. C.O.P. of steam jet refrigeration system is of the order of about 0.4 to 0.6.
3. The basic principle of boiling of liquid at low temperature by reducing the pressure on its surface is used in steam jet refrigeration system.
4. The main components of steam jet refrigeration system are : Evaporator (or Flash chamber); Steam jet ejector; condenser; Air ejector.
5. The *performance* of steam jet refrigeration system is defined as the mass of steam required per tonne of refrigeration with the decrease in temperature of chilled water (evaporating temperature) or increase in pressure of condensing vapour (condensing temperature).

OBJECTIVE TYPE QUESTIONS

Choose the Correct Answer:

1. The velocity of steam at the nozzle exit is
   (*a*) subsonic (*b*) sonic (*c*) supersonic (*d*) any of these.
2. In a steam jet refrigeration system, the expansion of motive steam takes place in
   (*a*) convergent nozzle (*b*) divergent nozzle
   (*c*) convergent-divergent nozzle (*d*) any of these.
3. The compression device employed in a steam jet refrigeration system is a
   (*a*) liquid pump (*b*) steam ejector (*c*) vapour compressor (*d*) diffuser.
4. The C.O.P. of the steam jet refrigeration system varies from
   (*a*) 0.1 to 0.2 (*b*) 0.4 to 0.6 (*c*) 3 to 5 (*d*) 6 to 10.
5. The ratio of isentropic enthalpy increase to the actual enthalpy increase required for the compression of the motive steam and the water vapour, is known as
   (*a*) boiler efficiency (*b*) compression efficiency
   (*c*) nozzle efficiency (*d*) entrainment efficiency.

ANSWERS

1. (*c*) 2. (*c*) 3. (*b*) 4. (*b*) 5. (*b*).

THEORETICAL QUESTIONS

1. State the principle of a steam jet refrigeration system.
2. Describe the working of a steam jet refrigeration system with the help of a neat sketch.
3. What are the advantages and disadvantages of steam jet refrigeration system over other types of refrigeration systems.
4. Derive an expression for finding out the mass of motive steam required per kg of water vapour produced.
5. Draw the T-*s* and *h*-*s* diagrams for a set jet refrigeration system and write the expressions for the following efficiencies:
   (*i*) Nozzle efficiency
   (*ii*) Entrainment efficiency
   (*iii*) Compression efficiency.

UNSOLVED EXAMPLES

1. The following data relate to a steam ejector refrigeration system:
   Condition of the motive steam .... 7 bar, dry saturated
   Temperature of water in the flash chamber 1.5ºC
   Temperature at which the make-up water is supplied 18ºC
   The pressure at which condenser is operated .... 0.058 bar

Nozzle efficiency .... 88 per cent

Entrainment efficiency .... 65 per cent

Compression efficiency .... 80 per cent

The quality of steam and flash vapour at the beginning of compression 0.92

Determine the following:

(*i*) Mass of motive steam required per kg of flash vapour,

(*ii*) The quality of vapour flashed from the flash chamber,

(*iii*) Refrigerating effect per kg of flash vapour,

(*iv*) Mass of motive steam required per hour per tonne of refrigeration,

(*v*) Volume of vapour removed from the flash chamber per hour per tonne of refrigeration, and

(*vi*) The coefficiency of performance of the system.

[**Ans.** (*i*) 1.523 kg/kg of flash vapour; (*ii*) 0.96; (*iii*) 2331.3 kJ/kg; (*iv*) 9.146 kg/h/tonne of refrigeration; (*v*) 868.7 m^3/h/tonne; (*vi*) 0.586]

2. The following data refer to a steam jet refrigeration system:

Capacity of thermal power plant .... 2200 kW

Condition of steam supplied from the boiler .... 30 bar, 400ºC

Refrigeration load taken up by the steam jet refrigeration system .... 135 tonnes

Pressure at which steam is bled off the steam turbine .... 5 bar

Vacuum maintained in the condenser .... 730 mm of Hg

Pressure and condition at which steam enters in thermo-compressor .... 0.01 bar, 0.94 dry

Temperature at which make-up water enters into flash chamber 18ºC

Isentropic efficiency of the steam turbine .... 90 per cent

Nozzle efficiency .... 90 per cent

Entrainment efficiency .... 65 per cent

Thermo-compressor efficiency .... 65 per cent

Determine the following using Mollier diagram:

(*i*) Dryness fraction of steam leaving the flash chamber,

(*ii*) Mass of motive steam bled off from the steam turbine,

(*iii*) Mass of steam generated in the boiler,

(*iv*) Mass of additional steam to be generated by the boiler due to refrigeration load, and

(*v*) Coefficient of performance of the system.

[**Ans.** (*i*) 0.96; (*ii*) 972 kg/h; (*iii*) 8183 kg/h; (*iv*) 604 kg/h; (*v*) 0.716]

3. The data relate to a steam jet refrigeration system:

Temperature of the cold water to be supplied to an industrial plant 10ºC

Temperature of make-up and recirculated water 25ºC

Condition of motive steam .... 8.5 bar, 190ºC

Quality of vapour leaving the flash chamber and entering the ejector 0.95

Condenser pressure .... 60 mm of Hg

Nozzle efficiency .... 93 per cent

Entrainment efficiency .... 65 per cent

Compression efficiency .... 75 per cent

Quality of steam and flash vapour at the beginning of compression .... 0.918.

Determine the following:

(*i*) Mass of motive steam required to produce unit mass of flash vapour at 10ºC;

(*ii*) Total motive steam required to produce 1 tonne of refrigeration per hour; and

(*iii*) Coefficient of performance of the system.

[**Ans.** (*i*) 1.667 kg; (*ii*) 9.167 kg/h/tonne of ref.; (*iii*) 0.58]

❑❑❑

Chapter 7

Low Temperature Refrigeration (Cryogenics)

7.1 INTRODUCTION

- In refrigeration, the temperatures from –100°C to –273°C (or absolute zero) are treated as *low temperatures*.
- The term '*Cryogenic*' is derived from the Greek word Kryos which means *cold or frost*. Cryogenic technology came in the field as a tool for research and development, but its ultimate importance to man lies in its varied industrial, nuclear and electronic applications.

  It is frequently applied to very low temperature refrigeration applications such as in the liquefaction of gases and in study of physical phenomenon at temperatures approaching absolute zero.
- The first low temperature refrigeration system was developed primarily for the *solidification of* CO_2 and the liquifaction of *gases* such *as air,* O_2 *and* N_2. *Oxygen* was liquified in 1877 by Coilletet and Pictet. *Hydrogen* was liquified in 1898 by Dervar using *Joule Thomson effect.* Liquefaction of *helium* was accomplished by Kameslingh Oanes in 1908 in the famous cryogenic laboratories of the University of Leiden.
- In 1933, Debye and Gianque *used the adiabatic demagnetisation of paramagnetic salts producing temperature of 0.7 K.* Now-a-days, a *temperature of 0.001 K is successfully achieved by using cascading of demagnetisation method.*

7.2 LIMITATIONS OF VAPOUR COMPRESSION REFRIGERATION SYSTEMS FOR PRODUCTION OF LOW TEMPERATURE

The vapour compression systems with single compressor for different refrigerants are limited to an evaporator temperature of – 40°C. The adoption of vapour compression system below this temperature (i.e., – 40°C) entails the following *drawbacks*:

1. The refrigerant must have a freezing point much below the required temperature to be attained. The freezing point of some of the refrigerants commonly used for vapour compression systems, simple and compound compression are listed below:

Table 7.1 Freezing temperatures for commonly used refrigerants.

Refrigerant	R-11	R-12	R-13	R-21	R-22	R-113	R-114	R-717 (NH_3)	R-744 (CO_2)	R-764 (SO_2)
Freezing temperature (°C) at atmospheric pressure	– 111	– 157.5	– 181	– 135	– 160	– 35	– 94	– 77.8	– 56.7	– 75.6

Thus R-717 (NH_3), R-744 (CO_2), R-764(SO_2) and R-113 *cannot* be used for cryogenics.

2. When a refrigerant with high boiling temperature is used, the pressure in the evaporator is extremely low (below atmospheric) and the suction volume is very large.
3. The pressure in the condenser is very excessive when a refrigerant with low boiling temperature is used.
4. Owing to very high pressure ratios, the C.O.P. is very low.
5. At very low temperatures the difficulties are encountered in the operation of mechanical equipment.

- The ***"multistage compression"*** *system is employed when low evaporator temperature is required and the pressure ratio is high.*
  — A two-stage compression system using R-12 is used for evaporator temperatures upto –60°C and three-stage compression system upto –68°C.
  — The compression systems with R-22 operate satisfactorily at 5°C to 10°C lower temperatures.
  — R-13 can be used to temperatures well below –100°C, but must be cascaded to prevent extremely high condenser pressure as the critical temperature is 28.8°C.
- If the vapour compression systems are to be employed for the production of low temperature, the common alternative to stage compression is the **cascade system**.

7.3 CASCADE REFRIGERATION SYSTEM

Cascade system is just similar to the Binary-vapour cycle used for the power plants. In a Binary-vapour cycle, a condenser for mercury works as a boiler for water. Similarly in the cascade system, the *condenser for low temperature cycle works as an evaporator for the high temperature cycle.* In cascade system, a series of refrigerants with *progressively lower freezing points* is used in a series of single stage units.

- A two-stage cascade system using two refrigerants is shown in Fig. 7.1 (*a*) and its corresponding *p-h* and T-*s* diagrams are shown in Fig. 7.1 (*b*) and Fig. 7.1 (*c*) respectively.

In this system, a *cascade condenser serves as an evaporator* for the high temperature cascade system and a condenser for the low temperature cascade system. The only *useful refrigerating effect is produced in the evaporator of the low temperature cascade system.* Thus it permits the use of two different refrigerants, with thermodynamic properties favourable for the two temperature ranges.

Further, the lubricating oil from one compressor *cannot* be carried away to another compressor.

— The difference in low temperature cascade condenser temperature and high temperature cascade evaporator temperature is called ***temperature overlap*** (which is necessary for heat transfer). If these temperatures are equal, then it is known as ***intermediate temperature***.

— The low temperature cascade system uses a refrigerant with low boiling temperature (such as R-13 or R = 13BI). These low boiling temperature refrigerants have *extremely high pressure which ensures a smaller compressor displacement in the low temperature cascade system and a higher C.O.P.*

The cascade system was first used by Pictet in 1877 for the liquefaction of oxygen. This installation employed SO_2 and CO_2 as intermediate refrigerants. Since that time, among other combinations used are *methyl chloride, ethylene and oxygen, and ammonia, ethylene, methane and nitrogen.*

A three-stage cascade system is shown in Fig. 7.2 (*a*) and its corresponding *p-h* and T-*s* diagrams are shown in Fig. 7.2 (*b*) and Fig. 7.2 (*c*) respectively.

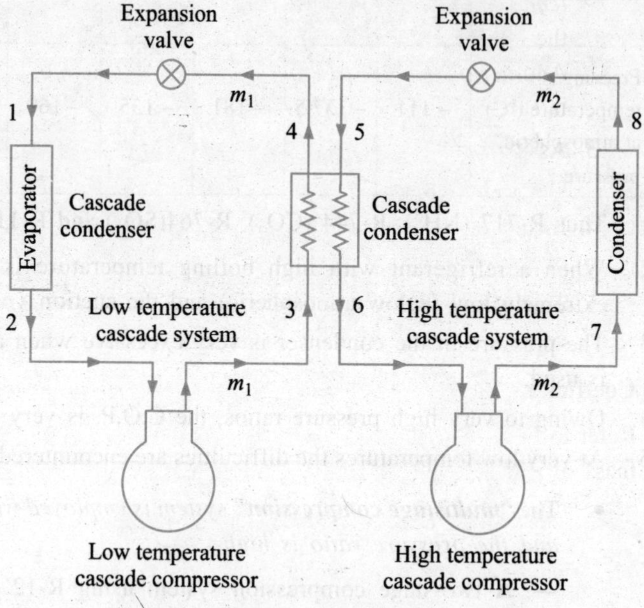

(*a*) Schematic arrangement of a two-stage cascade system.

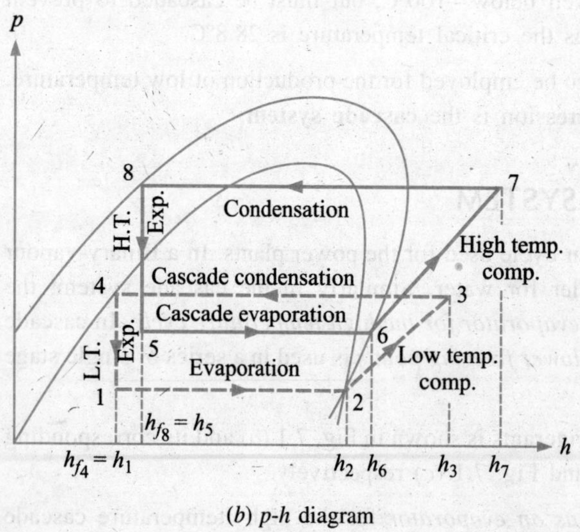

(*b*) *p-h* diagram

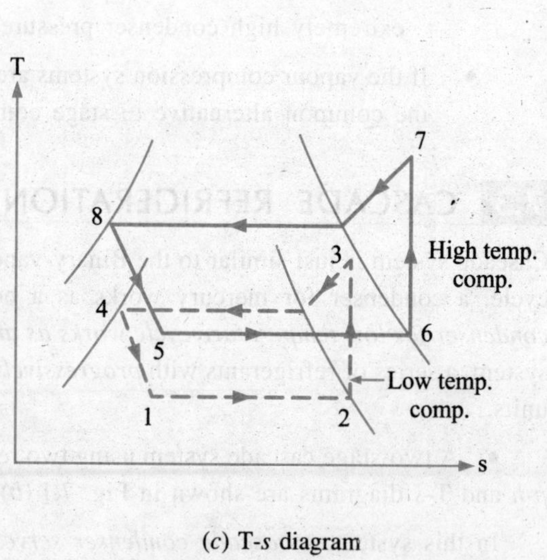

(*c*) T-*s* diagram

Fig. 7.1. Two-stage cascade system

Advantages of Cascading

The following are the *advantages* of cascading:

1. It *permits the use of two or more different refrigerants;* thus each refrigerant circuit is comparatively simple.
2. Each refrigerant can be chosen that *operates best within the required comparatively narrow temperature and pressure range.* For example, a refrigerant which does not require a vacuum in the evaporator to produce a low suction temperature is selected for the low temperature stage. Another refrigerant such as R-22 which does not require an excessive pressure in the condenser is selected for use in the high stage system.
3. As each cycle is separately operating, there is *no difficulty of lubrication.* Oil from one compressor cannot wander to another compressor as it generally happens in multi-stage systems.
4. Inclusion of intercooling with liquid refrigerants in the cascade system improves the performance of the system.
5. The use of a low boiling point and therefore a high pressure refrigerant ensures a smaller compressor displacement in the low temperature cascade, and a higher C.O.P.

Coefficient of Performance of a Two-stage Cascade System

Refer to Fig. 7.1. If low temperature cascade system is under a refrigeration load of Q tonnes, then the mass of refrigerant (m_1) flowing through it is given by,

$$m_1 = \frac{14000Q}{60(h_2 - h_1)} \text{ kg/min.} \qquad ...(7.1)$$

The mass of refrigerant m_2 required in the high temperature cascade system in order to liquefy the refrigerant of low temperature cascade system in the cascade condenser, may be obtained by energy (heat) balance of both the systems, as follows:

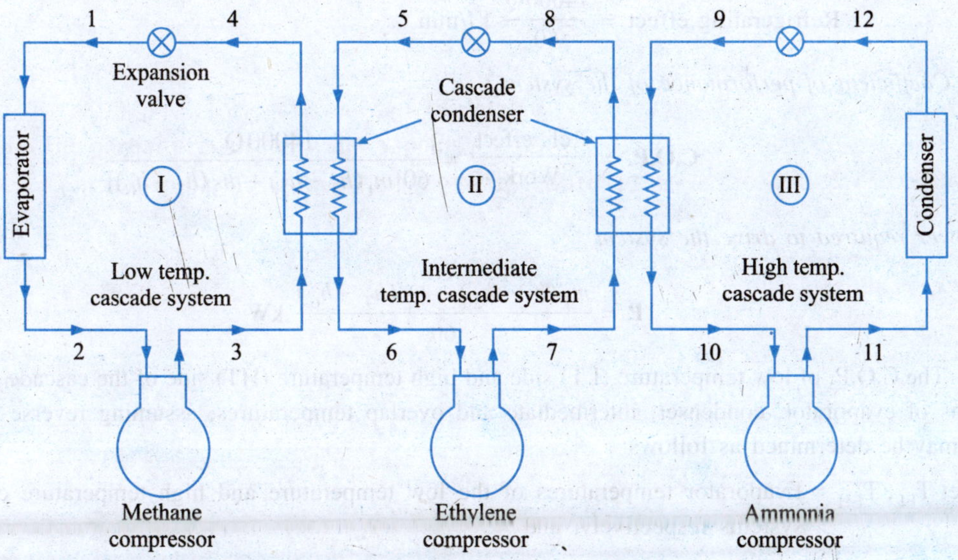

(*a*) Schematic arrangement of a three-stage cascade system

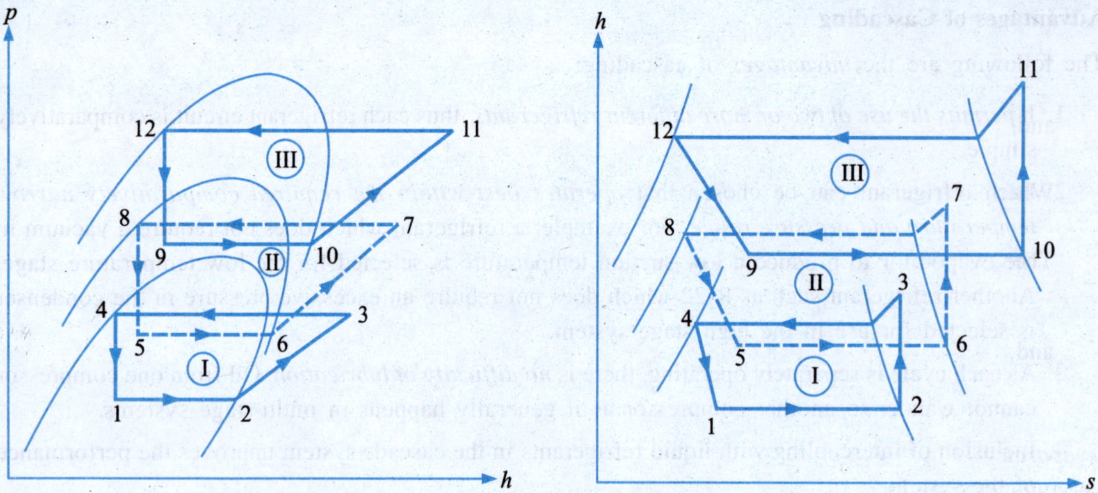

(b) p-h diagram (c) T-s diagram

Fig. 7.2. Three-stage cascade system

Heat absorbed in the high temperature cascade system

= Heat rejected in the low temperature cascade system

i.e., $m_2 (h_6 - h_5) = m_1 (h_3 - h_{f4})$

$$\therefore \quad m_2 = \frac{m_1 (h_3 - h_{f4})}{(h_6 - h_5)} = \frac{m_1 (h_3 - h_1)}{(h_6 - h_5)} \text{ kg/min} \quad ...(7.2)$$

Total work done by the system,

$$W = m_1 (h_3 - h_2) + m_2 (h_7 - h_6) \quad ...(7.3)$$

$$\text{Refrigerating effect} = \frac{14000\,Q}{60} \text{ kJ/min}$$

$\therefore$ *Coefficient of performance of the system,*

$$\textbf{C.O.P.} = \frac{\text{Ref. effect}}{\text{Work}} = \frac{14000\,Q}{60[m_1 (h_3 - h_2) + m_2 (h_7 - h_6)]} \quad ...(7.4)$$

Power required to drive the system,

$$\textbf{P} = \frac{m_1 (h_3 - h_2) + m_2 (h_7 - h_6)}{60} \text{ kW} \quad ...(7.5)$$

- The C.O.P. of low temperature (LT) side and high temperature (HT) side of the cascade system in terms of evaporator, condenser, intermediate and overlap temperatures, assuming reverse Carnot cycle, may be determined as follows:

Let T_{EL}, T_{EH} = Evaporator temperatures of the low temperature and high temperature cascade systems respectively, and

T_{CL}, T_{CH} = Condenser temperatures of the low temperature and high temperature cascade systems respectively

$$\therefore \quad (\text{C.O.P.})_{LT} = \frac{T_{EL}}{T_{CL} - T_{EL}} \quad ...(7.6)$$

and,
$$(\text{C.O.P.})_{HT} = \frac{T_{EH}}{T_{CH} - T_{EH}} \quad ...(7.7)$$

When, $T_{CL} = T_{EH} = T_I$ (Intermediate temperature),

Then,
$$(\text{C.O.P.})_{LT} = \frac{T_{EL}}{T_I - T_{EL}} \quad ...(7.8)$$

and,
$$(\text{C.O.P.})_{HT} = \frac{T_I}{T_{CL} - T_I} \quad ...(7.9)$$

Example 7.1. *A cascade refrigeration system is designed to supply 9 tonnes of refrigeration at an evaporator temperature of –55°C and a condenser temperature of 20°C. The load at –55°C is absorbed by a unit using R-22 as the refrigerant and is rejected to a cascade condenser at –15°C. The cascade condenser is cooled by a unit using R-12 as the refrigerant and operating between –25°C evaporating temperature and 20°C condenser temperature. The refrigerant leaving the R-12 condenser is subcooled to 15°C, but there is no undercooling of R-22 refrigerant. The gas leaving both the evaporators is dry and saturated and the compressions are isentropic. Neglecting losses, determine the following:*

(i) Compression ratio for each unit,

(ii) Quantity of refrigerant circulated per minute for each unit,

(iii) C.O.P. for each unit,

(iv) C.O.P. of the whole system, and

(v) Theoretical power required to run the system.

Solution: *Given:* Q = 9 tonnes of refrigeration; t_E (R-22) = –55°C; t_E (R-22) = –15°C;
t_E (R-12) = – 25°C; t_C (R-12) = 20° C, t_8 = 15° C.

The schematic arrangement of a two-stage cascades refrigerating system is shown in Fig. 7.3 (*a*) and its corresponding *p-h* diagram is shown in Fig. 7.3 (*b*).

— The cycle 1-2-3-4 ... for R-22 unit
— The cycle 5-6-7-8 ... for R-12 unit
} Both these cycles are superimposed in the *p-h* diagram

For **R – 22,** from *p-h* diagram, we have:

$p_1 = p_2 = 0.496$ bar; $p_3 = p_4 = 2.969$ bar

$h_{f4} = h_1 = 27.8$ kJ/kg; $h_2 = 226.2$ kJ/kg; $h_3 = 270$kJ/kg

For **R – 22,** from *p-h* diagram, we have:

$p_5 = p_6 = 1.238$ bar; $p_7 = p_8 = 5.674$ bar

$h_5 = 50.1$kJ/kg; $h_6 = 176.5$ kJ/kg; $h_7 = 215$ kJ/kg

(*i*) **Compression ratio for each unit:**

$$\text{Compression ratio for R-22 unit} = \frac{2.969}{0.469} = \mathbf{6.33\ (Ans.)}$$

$$\text{Compression ratio for R-12 unit} = \frac{5.674}{1.238} = \mathbf{4.58\ (Ans.)}$$

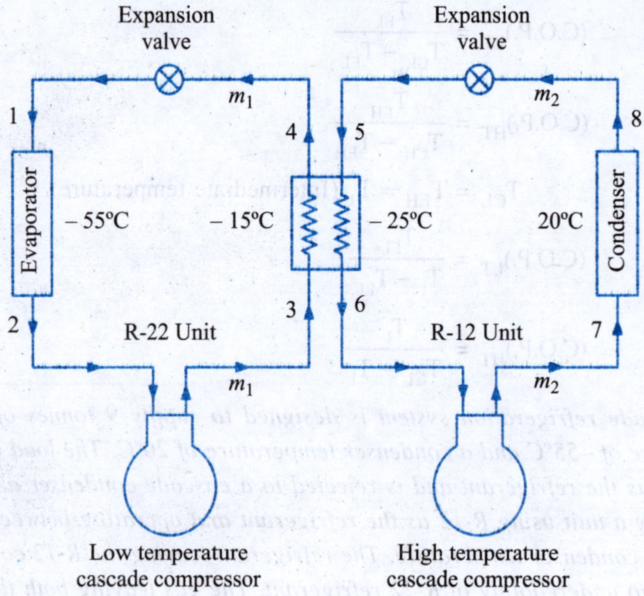

(*a*) Schematic arrangement of a two-stage cascade refrigerating system

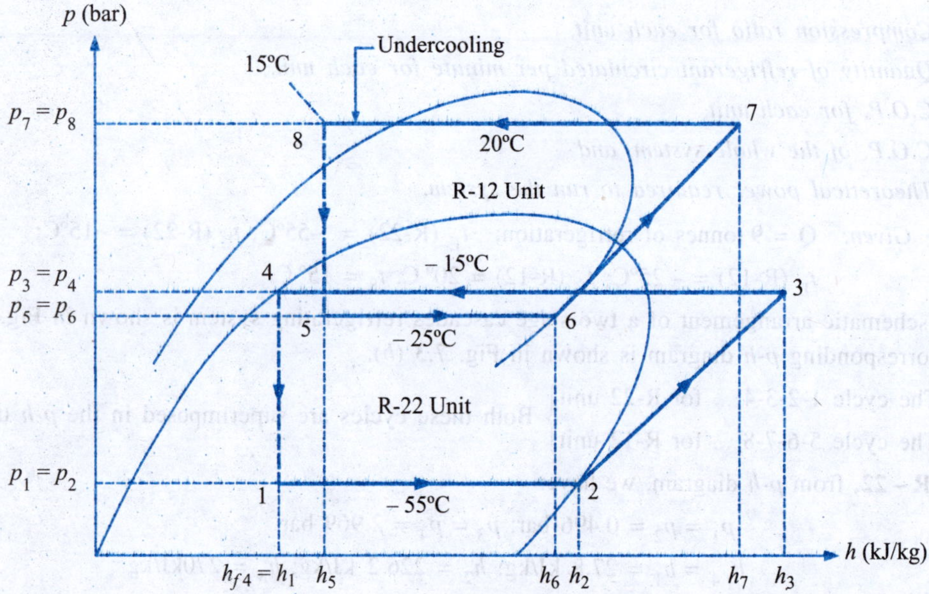

(*b*) *p-h* diagram

Fig. 7.3

(*ii*) **Quantity of refrigerant circulated per minute for each unit:**

Mass of refrigerant circulated for R-22 unit

$$m_1 = \frac{14000Q}{60(h_2 - h_1)} \text{ kg/min}$$

$$= \frac{14000 \times 9}{60(226.2 - 27.8)} = 10.58 \text{ kg/min.}$$

Mass of refrigerant circulated for R-12 unit;

$$m_2 = \frac{m_1 (h_3 - h_1)}{(h_6 - h_5)} \text{ kg/min} \quad \text{...[Eqn. 7.2]}$$

$$= \frac{10.58(270 - 27.8)}{(176.5 - 50.1)} = \mathbf{20.27 \text{ kg/min (Ans.)}}$$

(*iii*) **C.O.P. for each unit:**

$$(\text{C.O.P})_{\text{R-22}} = \frac{\text{Ref. effect / min}}{\text{Work / min}} = \frac{m_1 (h_2 - h_1)}{m_1 (h_3 - h_2)} = \frac{h_2 - h_1}{h_3 - h_2} = \frac{226.2 - 27.8}{270 - 226.2} = \mathbf{4.53 \text{ (Ans.)}}$$

$$(\text{C.O.P})_{\text{R-12}} = \frac{m_2 (h_6 - h_5)}{(h_7 - h_6)} = \frac{h_6 - h_5}{h_7 - h_6} = \frac{176.5 - 50.1}{215 - 176.5} = \mathbf{3.28 \text{ (Ans.)}}$$

(*iv*) **C.O.P. of the whole system, $(C.O.P.)_{system}$:**

$$(\text{C.O.P}) \text{ system} = \frac{14000 \text{ Q}}{60[m_1 (h_3 - h_2) + m_2 (h_7 - h_6)]} \quad \text{...[Eqn. (7.4)]}$$

$$= \frac{14000 \times 9}{60[10.58 (270 - 226.2) + 20.27 (215 - 176.5)]} = \mathbf{1.69 \text{ (Ans.)}}$$

(*v*) **Theoretical power required to run the system, P**

$$\text{P} = \frac{m_1 (h_3 - h_2) + m_2 (h_7 - h_6)}{60} \text{ kW} \quad \text{...[Eqn. (7.5)]}$$

$$= \frac{10.58(270 - 226.2) + 20.27(215 - 176.5)}{60} = \mathbf{20.73 \text{ kW (Ans.)}}$$

7.4 SOLID CARBON DIOXIDE OR DRY ICE

The solid carbon dioxide (or dry ice) is used for refrigeration purposes due to the following **advantages:**

1. It is *non-toxic* and *non-corrosive*.
2. It can be *easily cut* and *handled*.
3. It possesses the advantages over water ice of *lower temperature, the absence of objectionable liquid and smaller space requirement for a required cooling capacity*.
4. At atmospheric pressure, it sublimates directly from solid to gaseous phase at a temperature of –78.3°C and may therefore be used to maintain a wide range of temperatures required for the preservation of frozen perishables. By increasing the pressure, the temperature of sublimation can be raised to triple point of –6.5°C.

The major **disadvantage** of using solid carbon dioxide as a refrigerant is its *high cost of production and careful control due to its very low temperature.*

Solid carbon dioxide, despite few disadvantages is generally *used for transportation and distribution of foods and perishable materials* as *it can be easily handled.*

> **Note.** The *triple point* is the point of intersection of sublimation and vaporisation curves. Here solid, liquid and vapour phases are in equilibrium.

– *Below* the triple point temperature, the heat required to change the substance from a solid directly to a gas is termed as *latent heat of sublimation.*

– *At the triple point temperature,* the heat required to change from a solid to a liquid is called latent heat of fusion.

– *Above* the triple point temperature, the heat required to change from a liquid to a vapour is termed as *latent heat of vaporisation.*

Manufacture of Dry ice

The carbon dioxide can be liquefied by compressor to pressure of 60 to 70 bar and then condensed by using water. Solidification of this liquid CO_2 can be done by using the following *methods*:

(*i*) By removing the latent heat of solidification using auxiliary refrigeration – this method is *uneconomical* because of very low temperature requirement of the auxiliary refrigeration system.

(*ii*) By removing the heat through evaporation of a portion of liquid carbon dioxide – this method is *not* a practical means of production.

(*iii*) By *throttling the liquid* CO_2 *to pressure below that of triple point.* – This method is *widely used* for production of dry ice.

The basic cycle for producing dry ice is almost identical to the vapour compression cycle with the exception that means must the provided for supplying the make-up CO_2 gas and for removal of the CO_2 snow. The schematic arrangement of producing dry ice is shown in Fig. 7.4. The corresponding *p-h* diagram is shown in Fig. 7.5 and details are as follows:

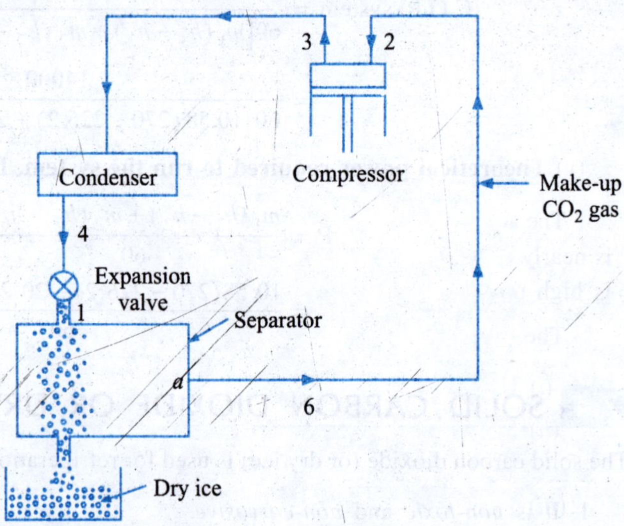

Fig. 7.4. Schematic arrangement of producing dry ice.

- **Process 2-3 :** Isentropic compression of CO_2 in a compressor to a sufficiently high pressure.
- **Process 3-4 :** Condensation of CO_2 in the *condenser,* by cooling water.
- **Process 4-1 :** Expansion of high pressure CO_2 in an *expansion valve* to pressure below that of triple point pressure. The CO_2 at point 1 is a mixture of solid CO_2 and gaseous CO_2.
- The solid CO_2 or dry ice can be removed at point '*a*' and pressed mechanically into blocks.
- The CO_2 gas removed at point '*b*' is mixed with make-up CO_2 gas (equal to the weight of solid CO_2 removed) at point '*c*' before being compressed at point 2.

CO_2 gas may be obtained from flue gas, oil wells, by product of chemical processes, by burning the carbonaceous substances or acid treatment of carbonates. All these sources have other gases as well, therefore a separator plant to remove them is a must.

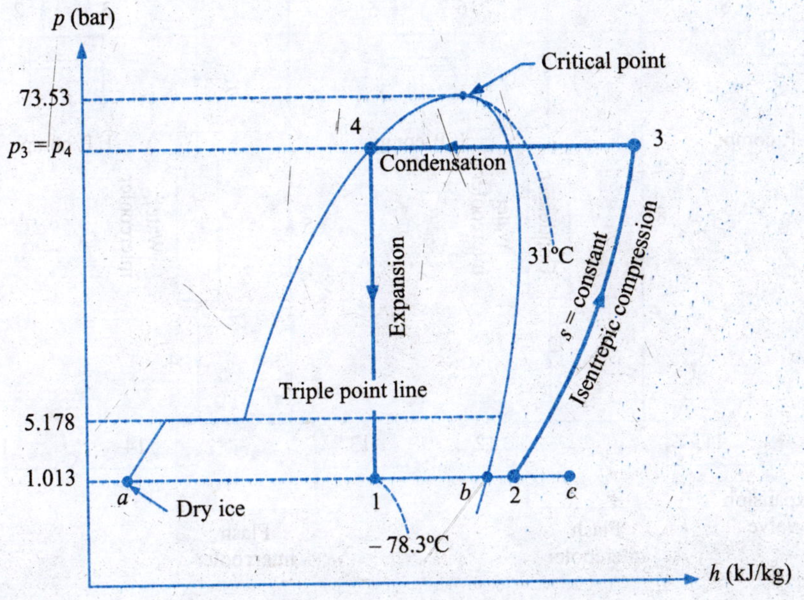

Fig. 7.5. *p-h* diagram for production of dry ice.

The above system is very uneconomical due to high consumption of power. The rate of consumption is nearly 360 kW hour per tonne of solid CO_2 produced. One of the main reasons for high consumption is high pressure ratio which is about 70.

The production cost of solid CO_2 can be reduced by using the following *modifications*:

(*i*) Use of multistage cost with intercoolers to reduce cost of *compression*.

(*ii*) Snow-chamber may be operated at a pressure only slightly below that corresponding to the triple point instead of atmospheric pressure.

(*iii*) Flash chambers also can be used for sub-cooling the liquid and de superheating the vapour.

With the above modifications, the cost of the system can be reduced from 360 kW hour to 160 kW hour per tonne of solid CO_2 produced.

Fig 7.6 shows a diagrammatic layout of three-stage compression system with water and flash intercooler for production of solid CO_2 or dry ice. Its corresponding *p-h* diagram is shown in Fig. 7.7.

- One of the problems which is encountered during the manufacture of dry ice is the formation of solid snow in the expansion valve which will block the flow to the snow chamber. This may be eliminated by first producing liquid CO_2 in low pressure flash intercooler as pressure (say p_{FL}) above the triple point pressure (i.e., 5.178 bar), and then reducing its pressure to one atmosphere (i.e., 1.013 bar) in the snow chamber, as shown in Fig. 7.7. Thus the pressure ratio in the low pressure stage has to be increased; this increase in pressure ratio is compensated by the larger mass flow rate in the higher pressure stages at lower pressure ratios.

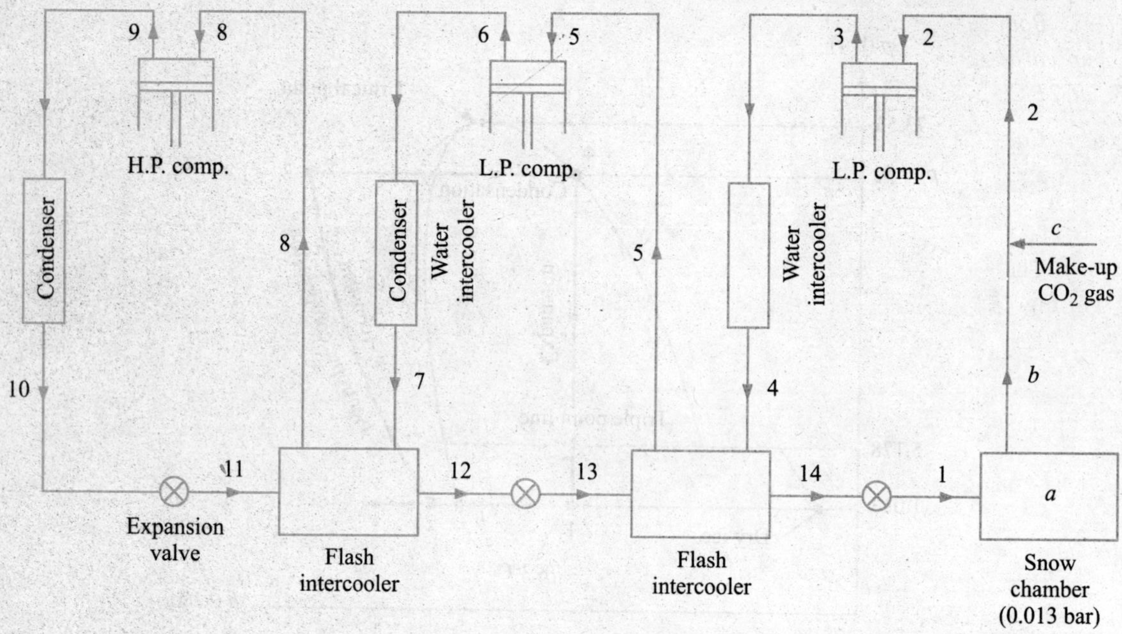

Fig. 7.6. Diagrammatic layout of three-stage compression system with water and flash intercoolers for production of solid CO_2 or dry ice.

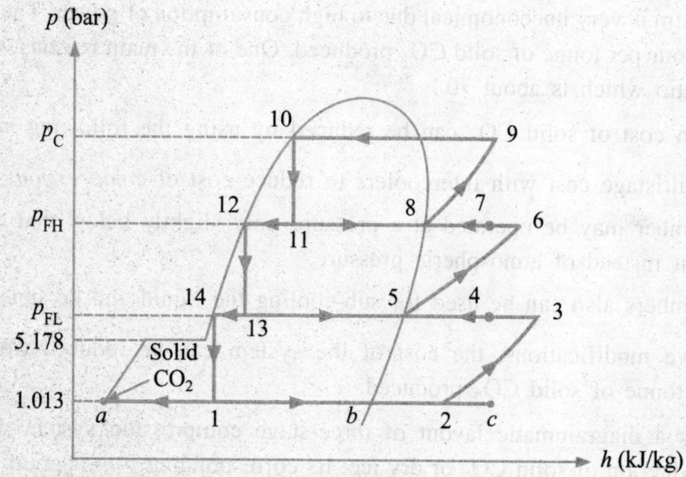

Fig. 7.7. *p-h* diagram of three-stage compression system with water and flash intercoolers for production of solid CO_2 or dry ice.

Example 7.2. *The following data refer to a three-stages compression system for the manufacture of one tonne of solid carbon dioxide per hour:*

Condenser pressure	...65 bar
Pressure in high pressure flash intercooler	...27.5 bar
Pressure in low pressure flash intercooler	...5.8 bar
Pressure in snow chamber	...1.013 bar

Both the flash and water intercoolers are used between the stages as shown in Fig. 7.6. The make-up carbon dioxide gas enters at 1.013 bar and 21°C. The gas leaving both the water intercoolers is at 21°C.

Calculate the theoretical power required to run the system.

Solution: *Given.* m_a = 1 tonne/h = 1000kg/h; p_C = 65 bar; p_{FH} = 27.5 bar; p_{FL} = 5.8 bar;

p_{SC} (pressure in snow chamber) = 1.013 bar; $t_C = t_4 = t_7$ = 20°C

The *p-h* diagram of a three-stage compression system with flash and water intercoolers is shown in Fig. 7.8.

— At point 1, the CO_2 is a *mixture of solid CO_2 and gaseous CO_2.*

— The solid CO_2 is removed at point '*a*' and the gaseous CO_2 at point '*b*'.

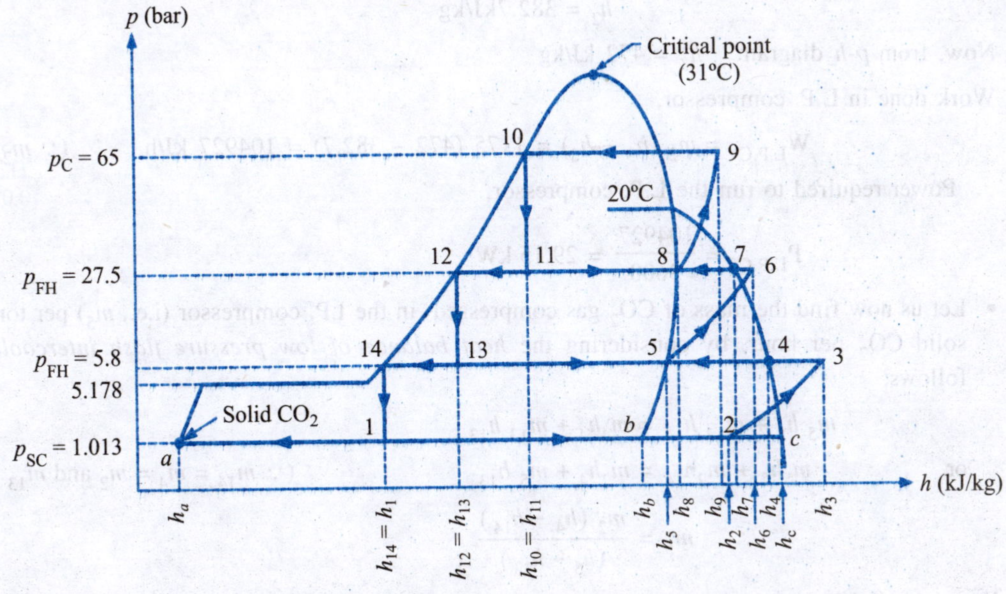

Fig. 7.8. *p-h* diagram of a three-stage system with flash and water intercoolers.

Let, m_1 = Mass of solid CO_2 and gaseous CO_2

$= m_a + m_b$,

m_a = Mass of solid CO_2 removed at point '*a*' = 1000 kg/h ....(Given)

m_b = Mass of gaseous CO_2 removed at point '*b*'

Considering the heat balance of snow chamber, we get:

$$m_1 h_1 = m_a h_a + m_b h_b \quad ...(i)$$

From *p-h* diagram: $h_1 = -24.2$ kJ/kg , $h_a = -82.2$ kJ/kg, $h_b = 307$ kJ/kg

Substituting the valves in eqn. (*i*), we get:

$$m_1 \times (-24.2) = 1000 \times (-82.2) + (m_1 - 1000) \times 307$$

or, $\quad -24.2\, m_1 = -82200 + 307\, m_1 - 307000 \quad (\because m_b = m_1 - m_a)$

or, $\quad m_1 = 1175$ kg/h

- In order to obtain enthalpy at point 2 let us consider heat balance at points 2, *b* and *c*.

$$m_2 h_2 = m_b h_b + m_c h_c \quad \text{...}(ii)$$

The make-up CO_2 gas supplied at a temperature 20° C at point '*c*' is equal to the mass of solid CO_2 removed at point '*a*' i.e.,

$$m_c = m_a = 1000 \text{ kg/h}$$

and

$$m_2 = m_1 = 1175 \text{ kg/h}$$

From *p-h* diagram; h_c = 396 kg/kg

Substitutions the values in eqn. (*ii*), we have:

$$1175 \times h_2 = (1175 - 1000) \times 307 + 1000 \times 396$$

$$\therefore \quad h_2 = 382.7 \text{kJ/kg}$$

Now, from *p-h* diagram: h_3 = 472 kJ/kg

Work done in L.P. compressor,

$$W_{L.P.C.} = m_2 (h_3 - h_2) = 1175 (472 - 382.7) = 104927 \text{ kJ/h} \qquad (\because m_2 = m_1)$$

∴ Power required to run the L.P. compressor,

$$P_{L.P.C.} = \frac{104927}{3600} = 29.15 \text{ kW}$$

- Let us now find the mass of CO_2 gas compressed in the I.P. compressor (i.e., m_5) per tonne of solid CO_2 per hour, by considering the *heat balance of low pressure flash intercooler*, as follows:

$$m_5 h_5 + m_{14} h_{14} = m_4 h_4 + m_{13} h_{13}$$

or,

$$m_5 h_5 + m_2 h_{14} = m_2 h_4 + m_5 h_{13} \qquad (\because m_{14} = m_4 = m_2 \text{ and } m_{13} = m_5)$$

$$\therefore \quad m_5 = \frac{m_2 (h_4 - h_{14})}{(h_5 - h_{13})} \quad \text{...}(iii)$$

From *p-h* diagram:

$$h_5 = 312 \text{ kJ/kg}; \; h_{14} = -24.2 \text{ kJ/kg}; \; h_4 = 387 \text{ kJ/kg};$$

$$h_{13} = 63.5 \text{ kJ/kg}; \; h_6 = 382 \text{ kJ/kg}$$

Substituting the values in eqn. (*iii*), we get:

$$m_5 = \frac{1175[387 - (-24.2)}{(312 - 63.5)} = 1944.3 \text{ kg/h}$$

Work done in I.P. compressor,

$$W_{I.P.C} = m_5 (h_6 - h_5) = 1944.3 (382 - 312) = 136101 \text{ kJ/h}$$

∴ Power required to run the I.P. compressor,

$$P_{I.P.C.} = \frac{136101}{3600} = 37.8 \text{ kW}$$

- Let us now find the mass of CO_2 gas compressed in the H.P. compressor (i.e., m_8) per tonne of solid CO_2 per hour, by considering the *heat balance of high pressure flash intercooler*, as follows:

$$m_8h_8 + m_{12}h_{12} = m_7h_7 + m_{11}h_{11}$$

or, $$m_8h_8 + m_5h_{12} = m_5h_7 + m_8h_{11} \quad (\because m_{12} = m_7 = m_5 \text{ and } m_{11} = m_8)$$

$$\therefore \quad m_8 = \frac{m_5(h_7 - h_{12})}{(h_8 - h_{11})} \quad ...(iv)$$

From *p-h* diagram:

$$h_7 = 360 \text{ kJ/kg}; \; h_{12} = 63 \text{ kJ/kg};$$
$$h_8 = 318 \text{ kJ/kg}; \; h_{11} = 158 \text{ kJ/kg};$$
$$h_9 = 354 \text{ kJ/kg}$$

Substituting the values in eqn. (*iv*), we have:

$$m_8 = \frac{1944.3(360-63)}{(318-158)} = 3609.1 \text{ kg/h}$$

Work done in H.P. compressor,

$$W_{H.P.C.} = m_8(h_9 - h_8) = 3609.1\,(354 - 318) = 129928 \text{ kJ/h}$$

$\therefore$ Power required to run the H.P. compressor,

$$P_{H.P.C.} = \frac{129928}{3600} = 36.09 \text{ kW}$$

Hence, total power required to run the system,

$$P = P_{L.P.C} + P_{I.P.C} + P_{H.P.C}$$
$$= 29.15 + 37.8 + 36.09 = \mathbf{103.04\ kW} \quad \textbf{(Ans.)}$$

7.5 LIQUEFACTION OF GASES

For producing low temperature for refrigeration purposes the gas must be liquefied. The gases may be expanded by the following two methods:

1. Isentropic expansion.
2. Free, irreversible expansion.

1. Isentropic expansion. This method is employed in air refrigeration cycle which is utilised in aircraft refrigeration system. Since by isentropic expanssion very low temperature of the gas may be obtained, therefore, it is used for comparatively high temperatures only.

2. Free, irreversible expansion. It is known that *free, irreversible expansion of a "perfect gas" results in no change of temperature*, but no gas is really perfect. Thus the actual drop in temperature experienced by most gases upon free expansion between two pressures may be *used as a measure of the degree of imperfection of that gas and also as a means of refrigeration.*

The change in temperature with drop in pressure at constant enthalpy is termed as *Joule-Thomson coefficient* (μ).

Mathematically,

$$\mu = \left(\frac{dT}{dp}\right)_h$$

It varies with both the temperature and pressure of the gas.

The magnitude of μ is a measure of the imperfection of a gas a its deviation from perfect gas behaviour. Depending upon the thermodynamic state of the gas, μ for real gases may be either positive or negative.

– The temperature at which $\mu = 0$, is called the *inversion temperature* for a given pressure.
– If μ is *greater than zero*, the *temperature of the gas decreases with throttling*.
– When μ *is less than zero*, the *temperature of the gas increases with throttling*.

Thus, in *cooling* of a gas by throttling, gas should show a *large positive value of μ*.

- For all gases under high temperatures and pressures the values of μ are negative. The throttling, under these conditions, results in a warming of the gas. For most common gases the inversion temperatures are above those of temperatures normally encountered. Therefore, if throttling occurs at ordinary temperatures and from high pressures that are not too high, then a cooling effect results. However, hydrogen and helium behave at ordinary temperatures as most gases do at high temperatures, since their temperatures are –77.8° C and –250° C respectively. Therefore, *if throttling of these gases is to be used to accomplish refrigeration, it is first necessary to cool them, by some other means, below their inversion temperatures.*

7.6 LINDE SYSTEM FOR LIQUEFACTION OF AIR

This system (also known as *Hampson system*) was developed by Linde in Germany in 1895.

The schematic arrangement of Linde system and its corresponding T-*s* diagram are shown in Fig. 7.9 (*a*) and Fig. 7.9 (*b*) respectively. The equipment includes a *compressor,* a *heat exchanger* and a *separator*. The working of the system is as follows:

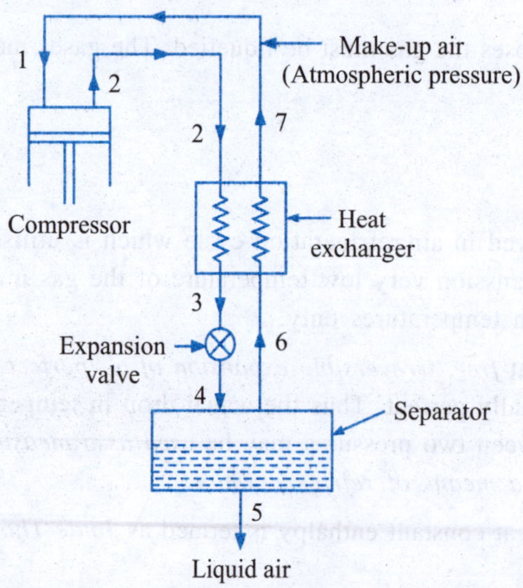

(*a*) Linde system

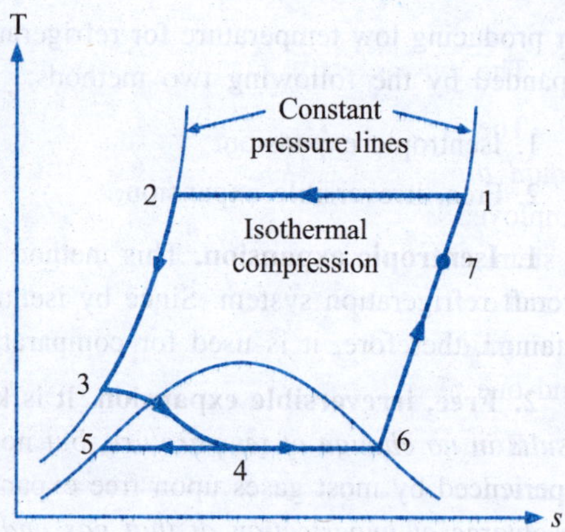

(*b*) T-*s* diagram

Fig. 7.9. Linde system for liquefaction of air

- **Process 1-2.** The atmospheric air is compressed isothermally to pressure of 100 to 200 atmosphere.
- **Process 2-3.** The high pressure air is cooled to about –106.7°C in the heat exchanger.
- **Process 3-4.** The cooled air from the heat exchanger is throttled to atmospheric pressure and a temperature of –190°C.
- **At point 5.** A portion of the air which is liquefied is removed from the separator. The remainder of the cold air leaves the separator at **point 6** and returns to the compressor through the heat-exchanger where it cools the incoming high pressure air.
- **At point 7** The make-up air from the atmosphere (equal to the amount of air liquefied) is supplied to the compressor along with the air removed from the heat exchanger.

Let m_2 = Mass of air compressed in the compressor (or mass of high pressure air passing through the heat exchanger, between points 2 and 3),

m_5 = Mass of air liquefied in the separator, and

m_6 = Mass of low pressure air passing through the heat exchanger between points 6 and 1 $= (m_2 - m_5)$

Considering *heat balance of heat exchanger*, we have:

$$m_2 (h_2 - h_3) = m_6 (h_1 - h_6) = (m_2 - m_5) (h_1 - h_6) \quad ...(i) \quad (\because m_6 = m_2 - m_5)$$

Considering heat balance of the *separator*, we get:

$$m_2h_4 = m_5h_5 + m_6h_6 = m_5h_5 + (m_2 - m_5) h_6 \quad ...(ii)$$

For throttling process 3-4, we have:

$$h_3 = h_4 \quad ...(iii)$$

The mass of air liquefied by the system may be obtained from the eqns. (*i*) and (*ii*).

The simple Linke system is comparatively *inefficient* and is employed only when *quantity of liquid air need is small.* In order to get an increased efficiency, a dual pressure Linde system is employed as shown in Fig. 7.10 (*a*). Its corresponding T-*s* diagram is shown in Fig. 7.10 (*b*). In this system, a portion of air after the first throttling process from point 5 to point 6 is *bled back* through a heat exchanger into the discharge of the first (low pressure) stage of compression. This system is more economical since the majority of the air undergoes only one expansion to the intermediate pressure and one compression from this intermediate pressure to the high pressure.

Both the systems described above may be improved if the *high pressure compressed air is precooled by an evaporator of a vapour compression system* as shown in Fig. 7.11. The best performance is obtained when the compressed air is precooled to a relatively low temperature.

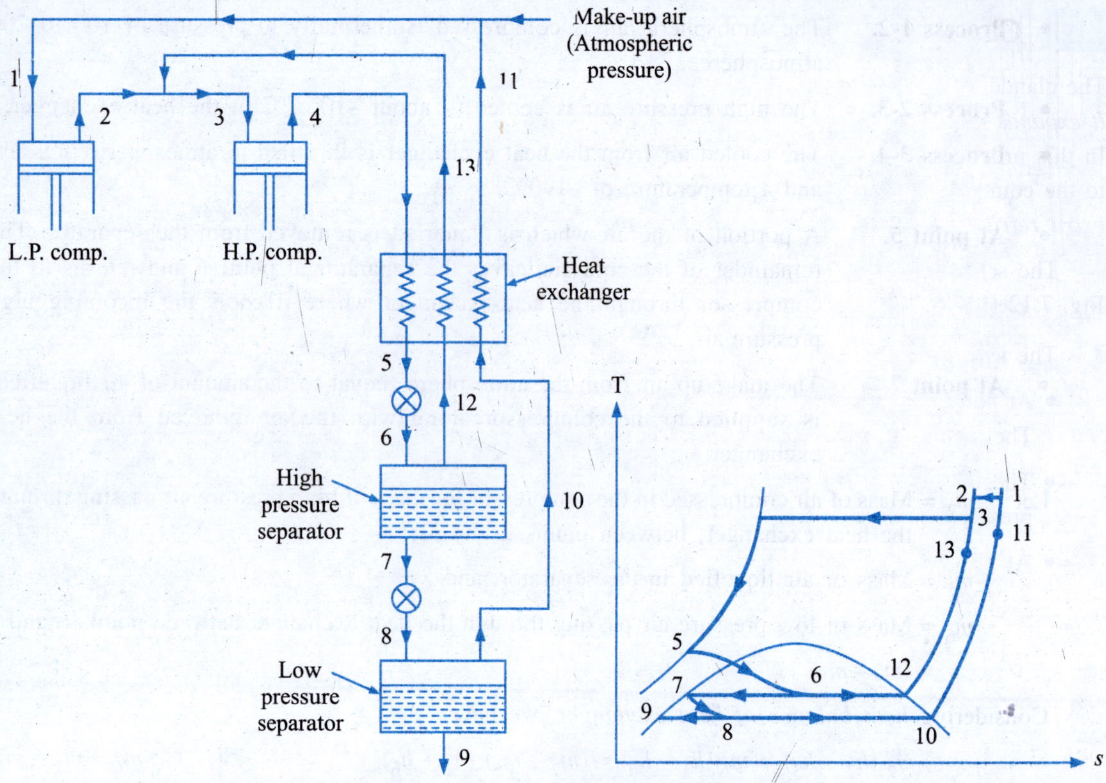

(*a*) Schematic arrangement for dual pressure Linde system (*b*) T-*s* diagram for dual pressure Linde system

Fig. 7.10. Dual pressure Linde system.

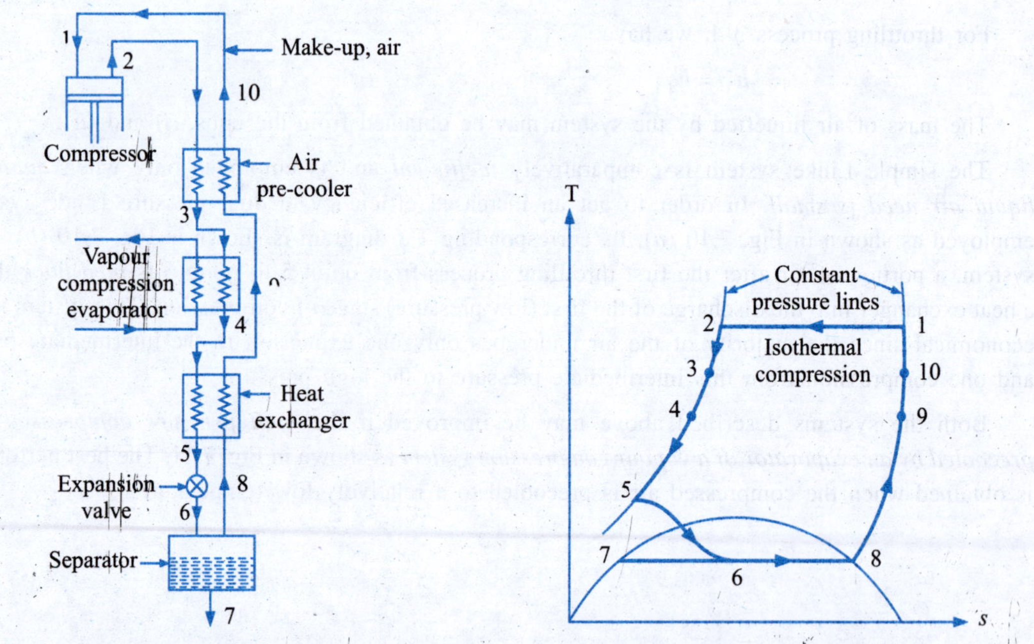

(*a*) Linde system with pre-cooling by vapour compression system evaporator (*b*) T-*s* diagram

Fig. 7.11

7.7 CLAUDE SYSTEM FOR LIQUEFACTION OF AIR

The clande system of liquefying air *utilizes the principle that a gas which performs external work as it expands experiences a greater drop in temperature than when it expands through throttling valve.* In this process, a large part of the air is expanded in an engine which transfers the energy produced to the compressor to compress the incoming air. Thus energy removed from expanding air *represents extra refrigeration produced in the cycle.*

The schematic diagram of the system is shown in Fig. 7.12 (*a*) and its T-*s* diagram is shown in Fig. 7.12 (*b*)

The ***working*** *of the system* is as follows:

- At point 3 about 80% of air is *bled* and cooled by expansion in an expander between 3 and 8. The remaining 20% portion of air passes through second heat exchanger between point 3 and 4.
- The air from the second heat exchanger is throttled between points 4 and 5 at atmospheric pressure.
- At point 6, the liquid air is removed from the separator.

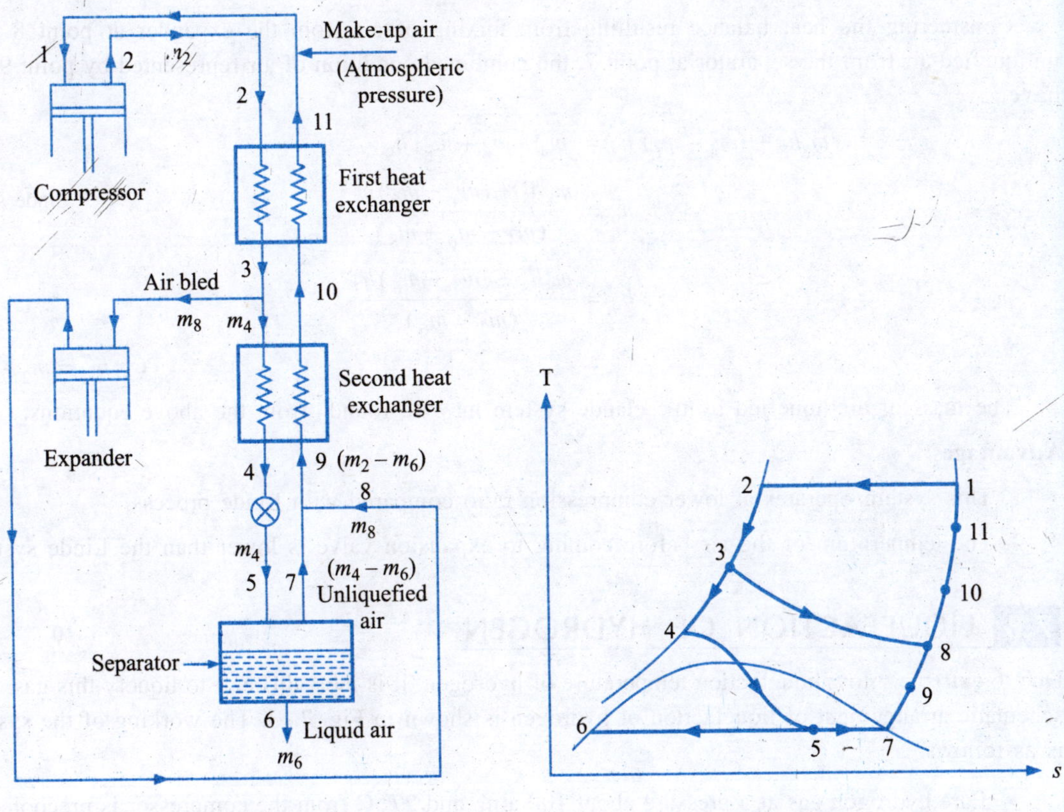

(*a*) Schematic arrangement for Claude system for air liquefaction

(*b*) T-*s* diagram

Fig. 7.12. Claude system for liquefaction of air.

- The unliquefied air at point 7 is mixed with the low temperature air from the expander at point 8, giving increased mass flow of air at point 9. This air passes back through the two heat exchangers to the compressor.

Let m_2 = Mass of compressed air,

m_8 = Mass of air bled/passed to expander,

m_4 = Mass of air passing through second heat exchanger, and

$= (m_2 - m_8)$, and

m_6 = Mass of air liquified in the separator.

Considering the heat balance of *first heat exchanger*, we have:

$$m_2 = (h_2 - h_3) = (m_2 - m_6)\,(h_1 - h_{10}) \quad ...(i)$$

Considering the heat balance of the *second heat exchanger*, we get:

$$m_4\,(h_3 - h_4) = (m_2 - m_6)\,(h_{10} - h_9) \quad ...(ii)$$

Considering the heat balance of the *separator*, we have:

$$m_4 h_5 = m_6 h_6 + (m_4 - m_6)\,h_7 \quad ...(iii)$$

Considering the heat balance resulting from mixing of air from the expander at point 8 with unliquefied air from the separator at point 7, the combined condition of air represented by point 9, we have:

$$m_8 h_8 + (m_4 - m_6)\,h_7 = (m_4 - m_6 + m_8)\,h_9$$

$$\therefore \quad h_9 = \frac{m_8 h_8 + (m_4 - m_6) h_7}{(m_4 - m_6 + m_8)}$$

or

$$h_9 = \frac{m_8 h_8 + (m_4 - m_6) h_7}{(m_2 - m_6)} \quad ...(iv)$$

$$(\because m_4 = m_2 - m_8)$$

The mass of air liquefied by the claude system may be found from the above equations.

Advantages:

1. This system operates at lower compression ratio compared with Linde process.
2. The temperature of the air before coming to expansion valve is lower than the Linde system.

7.8 LIQUEFACTION OF HYDROGEN

Due to extremely low liquefaction temperature of hydrogen, it is most difficult to liquefy this gas. The schematic arrangement of liquefaction of hydrogen is shown in Fig. 7.13. The working of the system is as follows:

- Pure hydrogen gas at a pressure about 100 atm. and 27° C from the compressor is precooled in two heat exchangers (1) and (2). In the heat exchanger (1), the incoming high pressure hydrogen is cooled by outgoing low pressure hydrogen while in heat exchanger (2), it is cooled by nitrogen.

- The high pressure hydrogen gas from both the heat exchangers is passed through heat exchanger (3) where the hydrogen is further cooled to about –209°C *by nitrogen boiling under reduced pressure.*
- The hydrogen gas is further cooled to about –230°C in the heat exchanger (4) by the low pressure hydrogen gas returning from the separator.
- The liquid hydrogen is produced by throttling the hydrogen gas from the heat exchanger (4) to atmospheric pressure.

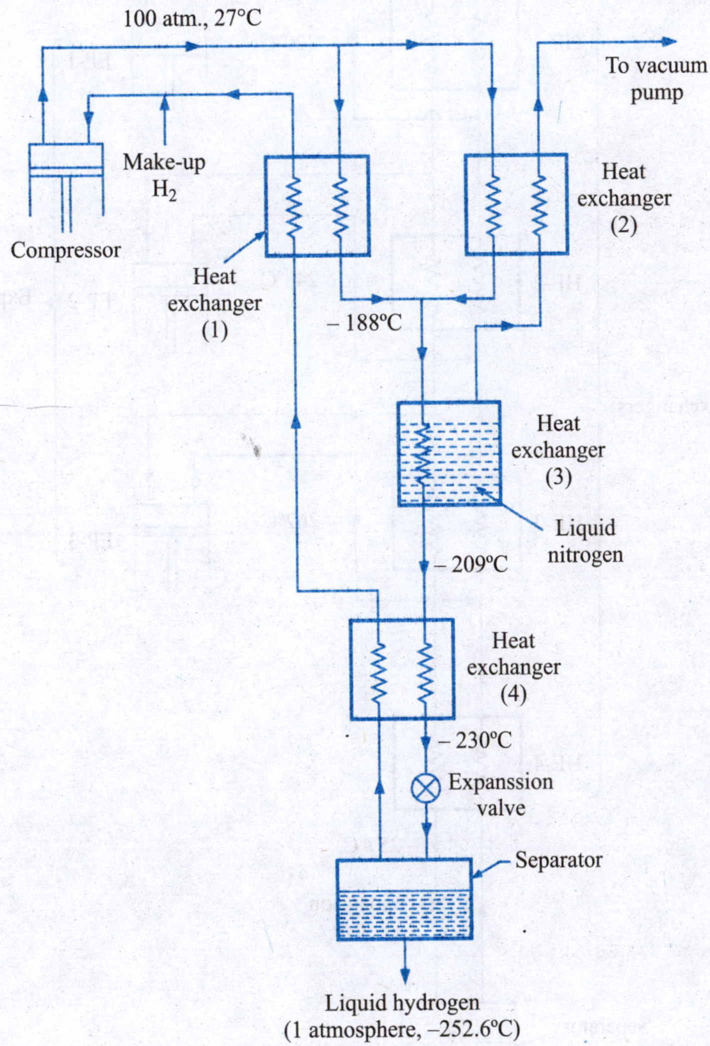

Fig. 7.13. Hydrogen liquefier.

7.9 LIQUIFACTION OF HELIUM

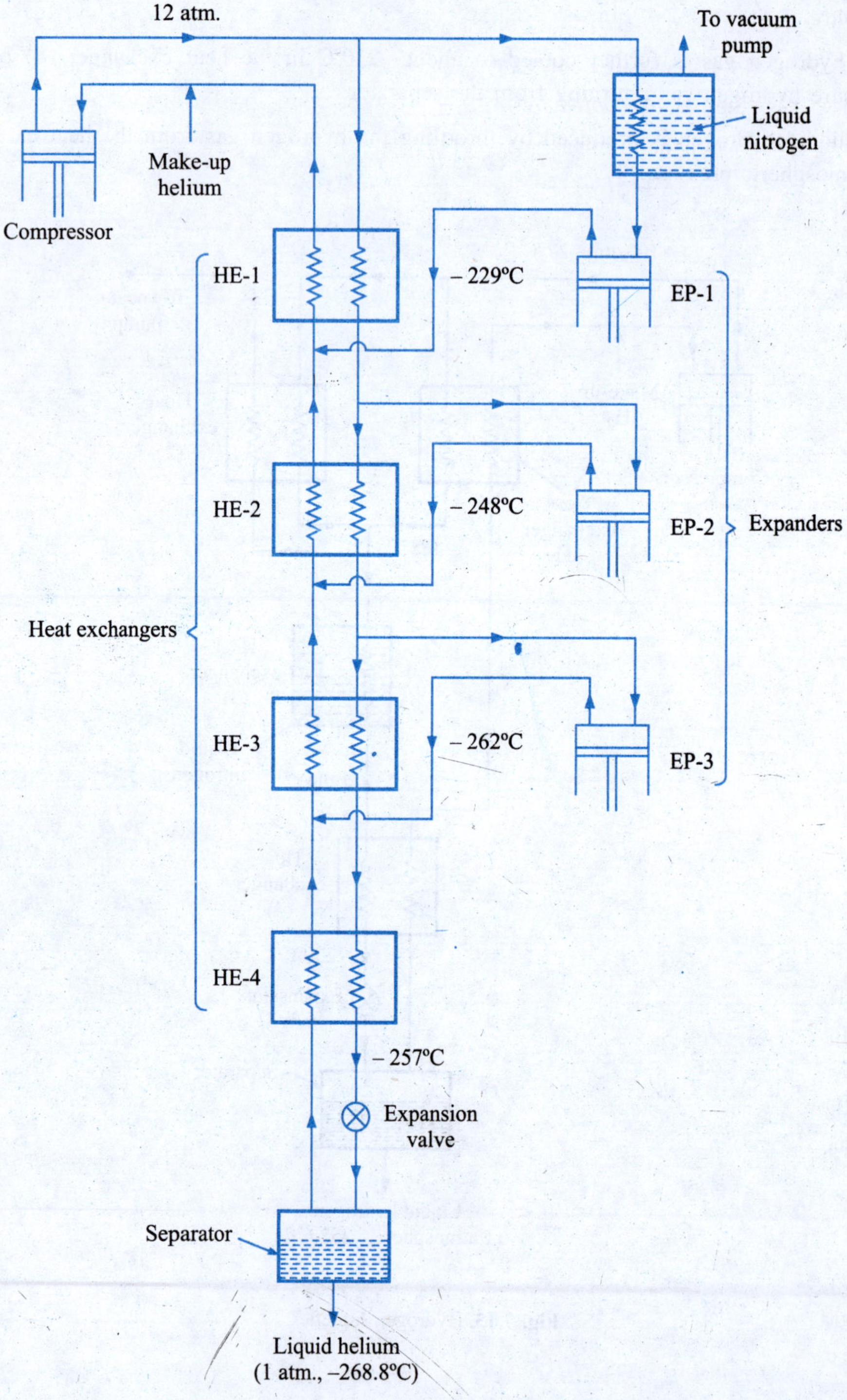

Fig. 7.14. S.C. Collins' helium liquefier.

Helium is most difficult among all gases to liquefy. It boils at –268.8° C at atmospheric pressure and its maximum inversion temperature is –233° C. Fig. 7.14 shows S.C. Collins helium liquifier. The working of this system is as follows:

- Helium gas at a pressure of 12 atm. (approx) is supplied to the liquifier by a four-stage compressor.
- A part of helium is *precooled by liquid nitrogen.*
- The high pressure helium gas may be cooled to –257° C by the use of *heat exchangers.* (HE-1, HE-2, HE-3, HE-4) and expanders (EP-1, EP-2, EP-3 and EP-4) as per arrangement shown in Fig. 7.14).
- The helium gas obtained above is then *throttled* to atmospheric pressure to produce liquid helium at –268.8°C.

It is found that one such unit may provide a liquefaction rate of 22 to 28 litres per hour with 40 kW power.

7.10 APPLICATIONS OF LOW TEMPERATURE

Following are a *few applications of low temperatures*:

1. Expansion fitting;
2. *Cryobiology* (This term relates to the effects of low temperatures on materials of biological origin);
3. *Semen preservation*;
4. *Cryosurgery*;
5. *Space research*;
6. *Computers*;
7. *Underground power lines*;
8. *Miscellaneous uses:*
   – *Infrared detectors;*
   – *Frictionless superconductive magnetic bearings etc.*

Notes. The following types of low temperature insulations are used in the various applications:

(*i*) High vacuum-reflective insulation;
(*ii*) Evacuated powders;
(*iii*) Rigid Foams;
(*iv*) Super-insulation.

HIGHLIGHTS

1. In *cascades system,* a series of refrigerants with progressively lower freezing points is used in a series of single stage units.
2. The carbon dioxide can be liquefied by compression to pressure of 60 to 70 bar and then condensed by using water.
3. Air may be liquefied by Linde system and Claude system.
4. S.C. Collins' helium liquefier is comparatively efficient and safer for liquefaction of helium.

OBJECTIVE TYPE QUESTIONS

Fill in the Blanks or Say "Yes" or "No":

1. In refrigeration, the temperatures from – 100° C to – 273° C (absolute zero) are treated as .... temperatures.
2. Oxygen was liquefied in 1877 by ....
3. Cascade system is just similar to the .... cycle used for the power plants.
4. The solid CO_2 is toxic and corrosive.
5. The .... point is the point of intersection of sublimation and vaporisation curves.
6. The liquid oxygen boils at 90.2 K.
7. The triple point for CO_2 at 5.18 bar is ....
8. The solidification of liquid CO_2 is done by expanding it to a pressure .... that of its triple point pressure.
9. The Claude air liquefaction system is .... efficient than the Linde system.
10. In the simple Linde system, the high pressure air from the compressor is cooled in a heat exchanger to a temperature of –106, 7°C.

ANSWERS

1. low	2. Coilletet	3. Primary-vapour	4. No.	5. Triple
6. Yes	7. – 56.6°C	8. below	9. more	10. Yes.

THEORETICAL QUESTIONS

1. What do you understand by 'Cryogenics'?
2. State the limitations of vapour compression refrigeration system for preduction of low temperatures.
3. Explain with a neat sketch a cascades refrigeration system. Show the cycle on p-h and T-s diagrams.
4. What are the advantages of cascading?
5. State the difference between overlap temperature and intermediate temperature.
6. Explain with a neat sketch the working of a system used for production of dry ice.
7. Explain briefly with help of a neat diagram the working of Claude system for liquefaction of air.
8. Describe briefly the methods used for the liquefaction of hydrogen and helium.
9. List a few applications of low temperature refrigeration.

UNSOLVED EXAMPLES

1. A cascade refrigeration system is designed to supply 9 tonnes of refrigeration at an evaporator temperature of –60°C and a condenser temperature of 25°C. The load at –60°C is absorbed by a unit using R-22 as the refrigerant and rejected to a cascade condenser at –20°C. The cascade condenser is cooled by a unit using R-12 as the refrigerant and operating between –30°C evaporating temperature and 25°C condenser temperature. The refrigerant leaving the R-12 condenser is subcooled to 20°C, but there is no sub-cooling of R-22 refrigerant. The gas leaving both the evaporators is dry and saturated and the compressions are isentropic. Neglecting losses, determine:

   (*i*) Compression ratio for each unit,

   (*ii*) Quantity of refrigerant circulated per minute for each unit,

   (*iii*) C.O.P for each unit,

   (*iv*) C.O.P. of the whole system, and

   (*v*) Theoretical power required to run the system.

   [**Ans.** (*i*) 6.56, 6.24; (*ii*) 10.4 kg/min., 22.04 kg/min (*iii*) 3.93, 3.64; (*iv*) 1.67; (*v*) 20.94 kW]

2. A cascade refrigeration system using R-22 and R-13 is required to produce 18 tonnes of refrigeration at – 70°C. Heat is rejected in cascade condenser by R-13 at –15°C to R-22 at –15°C. The condensation in the water-cooled condenser is at 40°C. Assume simple saturated cycles for both the circuits. Determine:

   (*i*) Pressure ratio and mass flow rate of each cascade,

   (*ii*) C.O.P. and piston displacement of each cascade, and

   (*iii*) C.O.P. of the cascade system.

   [**Ans.** (*i*) 5.25, 9.14; (*ii*) 2.545, 1.622, 0.0646 m^3/s, 0.086 m^3/s; (*iii*) 0.8]

3. The following data refer to a three-stage compression system for the manufacture of one tonne of solid CO_2 per hour:

   Condenser pressure = 65.3 bar;

   Pressure in the high pressure flash intercooler = 28 bar;

   Pressure in the low pressure flash intercooler = 6 bar; Pressure in snow chamber = 1.013 bar

   Both the flash and water intercoolers are used between the stages as shown in Fig. 7.6. The gas leaving both the water intercoolers is at 21°C.

   Calculate the theoretical power required to run the system. [**Ans.** 102.8 kW]

❑❑❑

Non-Conventional Refrigeration Systems

Chapter **8**

8.1 THERMOELECTRIC REFRIGERATION SYSTEM

8.1.1. Introduction

- In 1821 *Seebeck*, a German physicist observed that if a closed circuit was made of two *dissimilar metals*, an electric current flowed (e.m.f. produced) in the circuit when the two junctions were maintained at *different temperatures*. However, he was unable to make use of this invention.
- In 1834, *Jean Peltier*, a French scientist discovered a reversed phenomenon to that of Seebeck. He found that *there is heating or cooling of a junction of a pair of dissimilar substances, if direct current is passed through them.* However, he was not only unable to realise the utility of the same but also failed to correlate between his invention and that of seebeck.
- In 1838 Lenz, a German scientist clearly showed the importance of both seebeck's and Peltier's inventions by placing a drop of water on a junction of two dissimilar metals and passing a direct current through the circuit. When the current flowed in one direction, the water froze; and when the current was reversed the ice melted. This led to the *concept of thermoelectric refrigeration.* However, the idea did not materialise into commercial products for about a century because of unavailability of suitable materials until 1930's when semiconductors were developed.
- In 1851 Thomson pointed out the *third effect.* He *related the heat absorbed or evolved in a single conductor to the temperature gradient along it and current flowing through it.* This effect takes place in addition to the joule (I^2R) heating.

8.1.2 Advantages, Disadvantages and Applications

Advantages:

1. No moving parts; noiseless.
2. Simple and fewer parts required.
3. Can operate in any position.
4. These units are much more flexible than conventional units.
5. Easy control.
6. No leakage problem.
7. Compact in size.
8. Suitable for low capacity.
9. Very long life.
10. It can take over-load simply by increasing power input.
11. More reliable.
12. The weight per unit refrigeration is considerably lower than conventional refrigeration system.
13. An interchange of heating and cooling process can be exercised by reversing the polarity.
14. These units being static are more reliable than rotating or reciprocating equipments.
15. An important advantage of this refrigeration system is the *independence of C.O.P. on the size of thermo-electric refrigerator and this makes it particularly attractive to use Peltier cooling when the cooling capacity required is high.*

Disadvantages:

1. Low C.O.P.
2. More power is needed to run the system.
3. Advantageous only for units of smaller capacity.
4. Unavailability of suitable materials of high figure of merit.

Applications:

1. Peltier refrigerators (widely used in several western countries).
2. Peltier cooling is also used in air-conditioning in air-conditioning of rooms where large cooling capacities are required but the temperature differences need not be so large.

   Such a system has a great advantage that it can be used for heating the room in winter merely by reversing the direction of current.
3. The *thermoelectric cooling can be effectively used* in the following:

   (*i*) Constant low temperature bath and chambers;

   (*ii*) Cooled baffles for oil diffusion pump in vacuums system;

   (*iii*) Dew point hygrometer for determining absolute humidity;

   (*iv*) Photo multiplier cooler;

   (*v*) Serum coolers for preservation of blood plasma and serums etc.

8.1.3 Thermoelectric Effects

When direct current passes through a pair of thermocouples with junctions maintained at different temperatures, the following *five* effects are observed:

1. Seebeck effect.
2. Peltier effect.
3. Thomson effect.
4. Joulean effect.
5. Conduction effect.

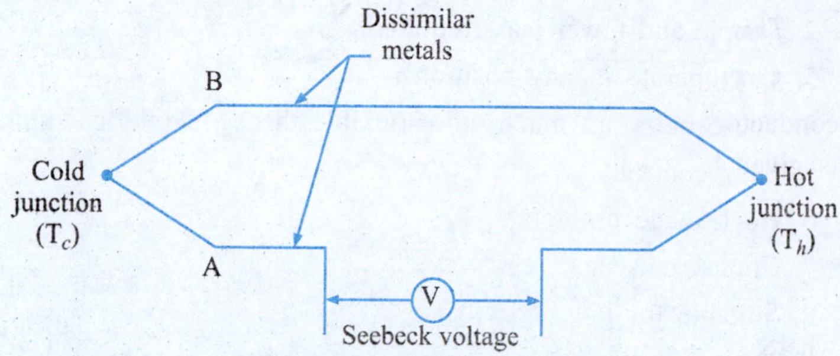

Fig. 8.1. Seebeck effect.

1. Seebeck effect. *When the two junctions of a pair of dissimilar metals are maintained at different temperatures, e.m.f. (electromotive force) is generated.* The phenomenon of generation of e.m.f. is called *seebeck effect.*

Refer to Fig. 8.1. For small temperature difference between the two junctions of materials A and B, the *open circuit voltage developed is proportional to the temperature difference and is given by:*

$$\Delta V = \alpha_{ab} \cdot \Delta T \quad \text{...(8.1)}$$

where, ΔV = Open circuit voltage developed, and

$\alpha_{ab} = (\alpha_a - \alpha_b)$ = Seebeck coefficient (the thermoelectric power) for materials A and B (or P and N or *p* and *n*) of materials A and B, and

ΔT = Temperature difference between the junctions of materials A and B.

— The absolute Seebeck coefficient (α_{abs}) of material is determined with respect to a material such as *lead* in which Seebeck coefficient is negligible.

— Whereas α_{abs} in *semiconductors* varies from 0.0002 to 0.00025 volt/°C, it does not exceed 0.0005 volt/°C in *metals.*

This effects is reversible is nature.

2. Peltier effect. If direct current is passed through a pair of dissimilar metals Fig. (8.2), there is a heating at one junction, *cooling* at the other depending upon the material combinations.

At the function between the two dissimilar metals the *heat evolved or absorbed in unit time is proportional to the current* and is given by,

$$Q_p = \pi_{ab}\, I \quad \text{...(8.2)}$$

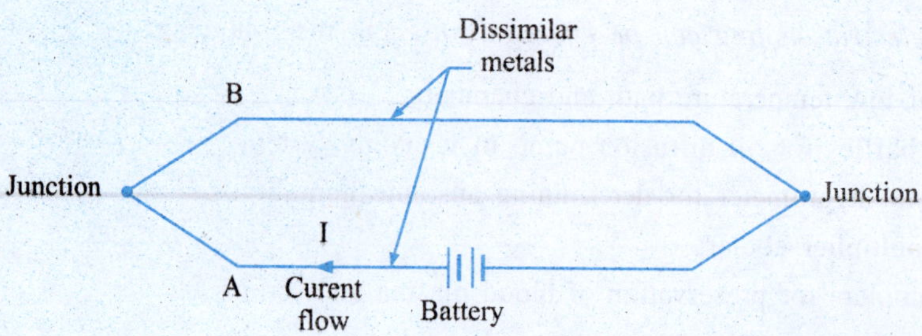

Fig. 8.2. Peltier effect.

where, Q_p = Heat evolved or absorbed in unit time, watts,

$\pi_{ab} = (\pi_a - \pi_b)$ = Peltier coefficient, and

I = Direct current, amperes.

This effect is reversible in nature.

3. Thomson effect. When a current passes through a single conductor having a temperature gradient (Fig. 8.3), heat transfer is given by,

$$\frac{\delta Q}{dx} = \tau I \left(\frac{dT}{dx}\right) \quad ...(8.3)$$

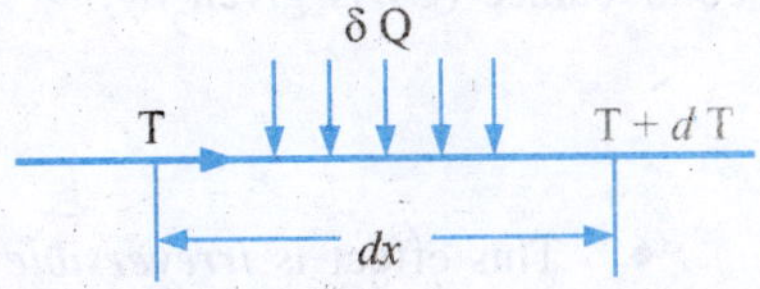

Fig. 8.3. Thomson effect.

where, $\frac{\delta Q}{dx}$ = Thomson heat transfer,

τ = Thomson coefficient,

I = Flow of current through the conductor, and

$\frac{dT}{dx}$ = Temperature gradient along the conductor.

- Thomson effect is *a reversible* thermoelectric phenomenon.
- Zemansky obtained the following *two relations* connecting these three coefficients (α, π, τ) by applying the first and second laws of thermodynamics to a simple thermoelectric circuit:

$$\pi_{ab} = \alpha_{ab} \cdot T \quad ...(8.4)$$

$$\tau_a - \tau_b = T \times \frac{d\alpha_{ab}}{dT} \quad ...(8.5)$$

— From the above equations it may be noted that the thomson coefficient has been defined for a single conductor, the Seebeck and Peltier coefficients refer to a junction between two materials

— Using eqn: (Fig. 8.4) into eqn. (8.2), we have:

$$Q_p = \alpha_{ab} \cdot I \cdot T \quad ...(8.6)$$

This equation indicates that to get the high value of cooling or heating, α_{ab} should be high, otherwise large current would be required.

4. Joulean effect. When the electric current flows through a conductor, there is a dissipation of electrical energy. According to *Joule* it is related as:

$$Q_J = I^2 R$$

where, I and R are the current and electrical resistance respectively.

- The Joulean effect is *irreversible in nature.*

5. Conduction effect. When the ends of any element are maintained at different temperatures, there is heat transfer from the heat end to the cold end and is related by:

$$Q_{cond.} = U(T_h - T_c) \quad ...(8.7)$$

where, U = Overall conductance,

T_h = Temperature at hot end, and

T_c = Temperature at cold end.

If there is only one conductor of cross-sectional area A, conductivity k and length L, the overall conductance (U) is given by:

$$U = \frac{kA}{L} \quad \text{...(8.8)}$$

- This effect is *irreversible in nature.*

8.1.4. Desirable Properties of a Thermoelectric Material

A thermoelectric material should posses the following ***properties***:

1. It must be *excellent conductor of electricity* so that ohmic losses are minimum.
2. It must be a *very poor conductor of heat* because the heat must be absorbed at one end, and rejected at the other.
3. It must have *high thermoelectric power.* This means it must have a high rate of change in voltage with temperature.
   - A *good thermoelectric material* should have a *high electrical conductivity, low thermal conductivity, and a high Seeback coefficient.* **Semiconductor** is the proper material for thermoelectric refrigeration (since metals are good conductors of heat and insulators are poor conductors).

8.1.5. Analysis of Thermoelectric Refrigeration

The following **assumptions** are made in analysing the magnitude of refrigeration effect and work done:

1. Heat absorption and heat rejection take place only at the junctions.
2. Thermal conductivity and electrical resistance do not change over the temperature range under considerations.
3. No energy exchange between the conductors through space separating them.
4. Thomson coefficient in negligible.

Now cooling and heating due to the thermoelectric effect is given by (Peltier effect):

$$Q_C = \alpha_{ab} I T_c \quad \text{...(8.9)}$$

and,

$$Q_h = \alpha_{ab} I T_h \quad \text{...(8.10)}$$

Refer to Fig. 8.4 (*b*). The heat generation in absence of other effects will lead to heat transfer from both ends equal to $\frac{1}{2} I^2 R$. From Fig. 8.4 (*a*), considering the control volume, we have:

For the *cold junction*:

$$Q_c + \frac{1}{2} I^2 R + U(T_h - T_c) = \alpha_{ab} I T_c \quad \text{...(8.11)}$$

For the *hot junction*:

$$Q_h + U(T_h - T_c) = \alpha_{ab} I T_h + \frac{1}{2} I^2 R \quad \text{...(8.12)}$$

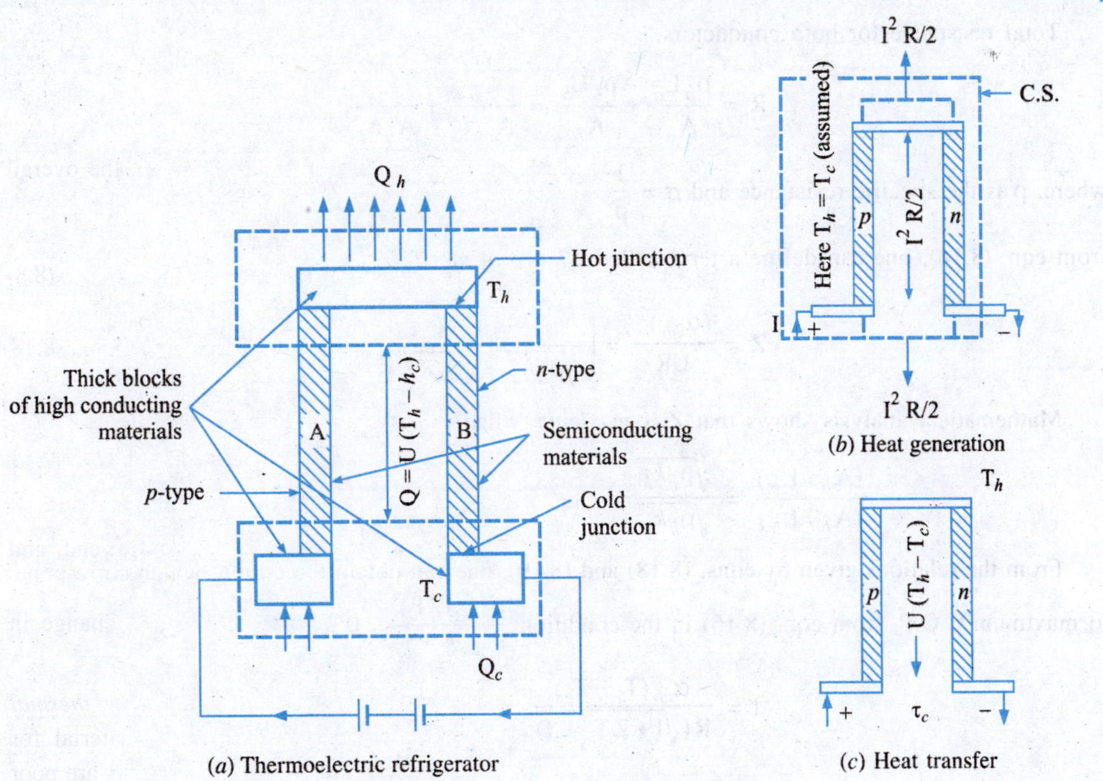

Fig. 8.4. Energy transfer in thermoelectric system.

Thus, the "*thermoelectric cooling*" is:

$$Q_C = \alpha_{ab}\, IT_C - \frac{1}{2} I^2R - U(T_h - T_c) \qquad ...(8.13)$$

and, *heating* is :

$$Q_h = \alpha_{ab}\, I\, T_h + \frac{1}{2} I^2R - U(T_h - T_c) \qquad ...(8.14)$$

Now energy input to the system from outside, as per first law of thermodynamics, is given as:

$$\oint \delta Q = Q_{net} = -Q_h + Q_c$$

$$= -[\alpha_{ab}\, I\,(T_h - T_c) + I^2R] \qquad ...(8.15)$$

Negative sign indicates that energy has to be supplied to the system.

Now, $$\textbf{C.O.P.} = \frac{Q_c}{\text{Energy supplied}}$$

$$= \frac{\alpha_{ab}\, IT_c - \frac{1}{2} I^2R - U(T_h - T_c)}{[\alpha_{ab}\, I\,(T_h - T_c) + I^2R]} \qquad ...(8.16)$$

where overall conductance (U) for both the conductors is given as:

$$U = \frac{A_a k_a}{L_a} + \frac{A_b k_b}{L_b} \qquad ...(8.17)$$

Total resistance for both conductors,

$$R = \frac{\rho_a L_a}{A_a} + \frac{\rho_b L_b}{A_b} = \frac{L_a}{A_a \sigma_a} + \frac{L_a}{A_b \sigma_b}$$

where, ρ is the specific resistance and $\sigma = \frac{1}{\rho}$

from eqn. (8.16), one can define a term called *"figure of merit"* Z as

$$Z = \frac{(\alpha_{ab})^2}{UR} = \left[\frac{\alpha_{ab}}{\sqrt{\rho_a k_a} + \sqrt{\rho_b k_b}}\right]^2 \quad ...(8.18)$$

Mathematical analysis shows that *Z is maximum* when,

$$\frac{(A_a / L_a)}{(A_b / L_b)} = \frac{\sqrt{\rho_a k_b}}{\sqrt{\rho_b k_a}} \quad ...(8.19)$$

From the relations given by eqns. (8.18) and (8.19), one can obtain the current which corresponds to maximum C.O.P. from eqn. (8.16) in the condition $\frac{d(\text{C.O.P})}{dI} = 0$ as,

$$I = \frac{\alpha_{ab}(T_h - T_c)}{R(\sqrt{1 + ZT_m} - 1)} \quad ...(8.20)$$

where, T_m = mean temperature = $\frac{(T_h + T_c)}{2}$

The maximum C.O.P, is given by

$$(\text{C.O.P})_{max} = \left(\frac{T_c}{T_h - T_c}\right)\left[\frac{\sqrt{1 + ZT_m} - (T_h / T_c)}{\sqrt{1 + ZT_m} + 1}\right] \quad ...(8.21)$$

$$= (\text{C.O.P})_{Carnot} \times (\text{C.O.P.}_{ref.})_{rel.} \quad ...[8.21(a)]$$

where, $(\text{C.O.P})_{Carnot}$ = It is the Carnot value for refrigeration system operating between the upper and lower temperature limits, and

$$(\text{C.O.P}_{ref})_{rel.} = \left[\frac{\sqrt{1 + ZT_m} - (T_h / T_c)}{\sqrt{1 + ZT_m} + 1}\right] \quad ...(8.22)$$

Form the eqn. (8.22) it is evident that $(\text{C.O.P.}_{ref.})_{rel} \rightarrow 1$ when $Z \rightarrow \infty$. Then, from eqn. (8.21 a),

$$(\text{C.O.P})_{max} \rightarrow (\text{C.O.P})_{Carnot}$$

$$i.e., \ (\text{C.O.P.})_{max} = \frac{T_c}{T_h - T_c} \quad ...(8.23)$$

The maximum cooling is obtained from the condition $\frac{dQ_c}{dI} = 0$, which gives,

$$I_{opt} = \frac{\alpha_{ab} T_c}{R} = ZT_c \sqrt{U / R} \quad ...(8.24)$$

The corresponding cooling is given by,

$$(Q_c)_{max} = U\left[\frac{1}{2}ZT_c^2 - (T_h - T_c)\right] \quad ...(8.25)$$

and,

$$(C.O.P)_{opt} = \frac{\frac{1}{2}ZT_c^2 - (T_h - T_c)}{ZT_h T_c} \quad ...(8.26)$$

If the heat source is removed the $(C.O.P)_{opt}$ becomes zero *trivially*. Thus the maximum temperature difference is then,

$$(T_h - T_c)_{max} = \frac{1}{2}ZT_c^2 \quad ...(8.27)$$

From all the above derivations it is evident that is order to get reasonable amount of refrigeration effect and C.O.P. Z *should be as high as possible.* The value of Z is about 0.006/K.

- Fig. 8.5 shows the variation of C.O.P. with temperature difference $(T_h - T_c)$, with figure of merit (Z) as parameters where T_h has been kept 313 K (40ºC).

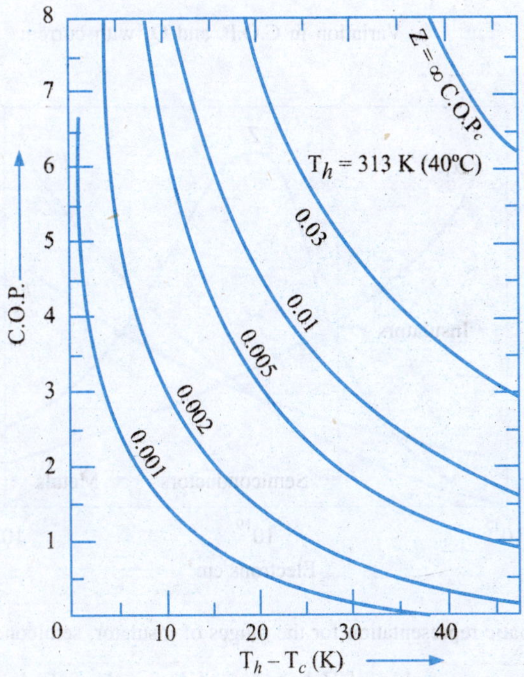

Fig. 8.5. Variation in C.O.P. with temperature difference.

- Fig. 8.6 shows the variation in C.O.P. and Q_c with current for given value of Z and other parameters.
- Fig. 8.7 exhibits the schematic representation for the ranges of insulator, semiconductor and metal. It is seen that Z is small for both the insulators and conductors while the *same is maximum for semiconductors for a given electron density.*

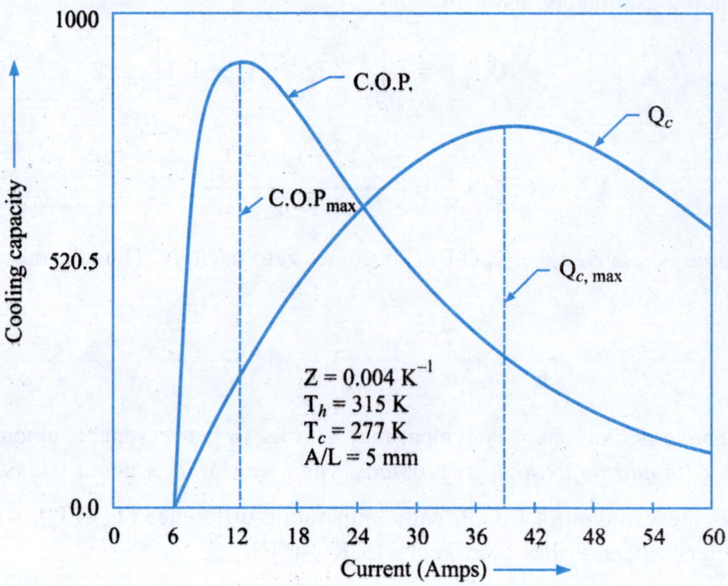

Fig. 8.6. Variation in C.O.P. and Q_c with current.

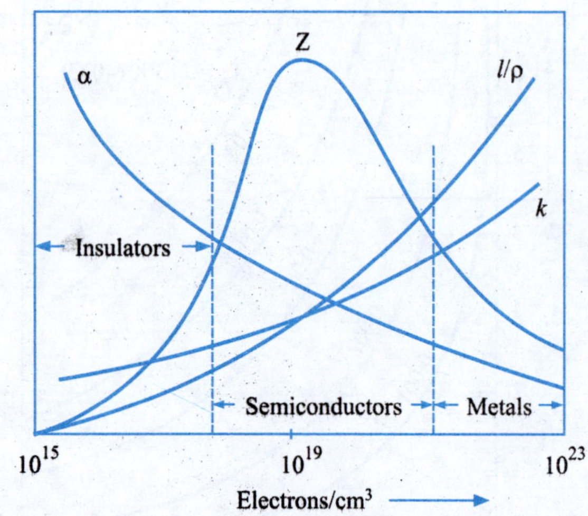

Fig. 8.7. Schematic representation for the ranges of insulator, semiconductor and metal.

— Insulators show the poor value of Z because of low value of electrical conductivity whereas conductors have low value of Z because of high thermal conductivity and low thermoelectric power.

Materials used in Thermoelectric refrigeration system

There are various combinations of thermoelectric materials which are used in thermoelectric refrigeration system. These are listed in the Table 8.1.

Table 8.1 Some Thermoelectric Elements

p-type(Positive)		*n-type(Negative)*	
Material	$Z_p (K^{-1})$	*Material*	$Z_n(k^{-1})$
Pb Te	0.0012	Pb Te	0.0015
Pb Se	0.0012	$Bi_2 Te_3$	0.0022
$Sb_2 Te_3$	0.0012	$Bi_2 (Se\ Te)_3$	0.003
Bi_2Te_3	0.0018		
$(Bi\ Sb)_2 Te_3$	0.0033		

In the above list, the symbols for various elements stand for:

Pb = lead; Te = Tellurium; Se = Selenium; Sb = Antimony.

— Z, an *index used in rating thermoelectric converters*, depends on the properties of thermoelectric materials used. A high value of Z is obtained by using materials of:

(*i*) *Large Seebeck co-efficient*; (*ii*) *Small thermal conductivity*; (*iii*) *Small electrical resistivity.*

— In recent times, the most commonly used material for thermoelectric converters is *lead telluride* [a compound of lead and tellurium, containing small amounts of either bismuth (*n*-type) or sodium (*p*-type)]. The efficiency of such a thermoelectric converter is, however, only about 5 to 7 per cent.

— Taking into account mechanical characteristics, stability under operating conditions and ease of fabrication, Bismuth telluride appears to be amply suitable material. It can be alloyed with such materials as Bismuth selenide, Antimony telluride, Lead selenide and tin telluride to give improved properties.

— Research is being made to find more efficient thermocouple materials. For high temperature applications, semiconductors based on silicon-germanium and compounds of selenium appear to be promising.

— To achieve *higher efficiency, thermoelectric material should have a high value of Z and be able to operate upto very high temperature.* The following points are worth noting in this regard:

1. The component thermal conductivity of semiconductor should be as low as possible.
2. The mobility of current carriers (electrons or holes, should be as high as is compatible with condition 1).
3. One of the arms should consist of a purely hole type and the other of a purely electronic type semiconductor.
4. In the low temperature zone the impurity concentration should be lower than in the higher temperature zone.
5. So that the thermoelectric material may not crack under the effect of stresses it should possess the following properties:

    (*i*) It should be able to resist chemical influences such as oxidation etc.

    (*ii*) It should have good mechanical strength.

    (*iii*) It should be amply elastic.

8.1.6. Comparison between Thermoelectric Refrigeration and Vapour Compression System

Fig. 8.8. shows the two systems in which the direction of flow of electrons for the thermoelectric refrigeration system and refrigerant vapour flow for the vapour compression system are exhibited.

- In the thermoelectric refrigeration system the electrons are pumped by the *battery* whereas in case of vapour compression system the vapour from the evaporator (to a condenser) is delivered by the *compressor* at a high pressure.
- When the electrons reach the junction of the dissimilar reconductors, their energy decreases due to heat transfer to the surroundings. This corresponds to the cooling or condensation of compressed vapour.
- After this the electrons are reduced to lower potential which resembles the process of *throttling* as in vapour compression system).
- The heat transfer to the cold junction imparts energy to electrons which again move to battery similar to transfer of vapour from evaporator to the compressor.

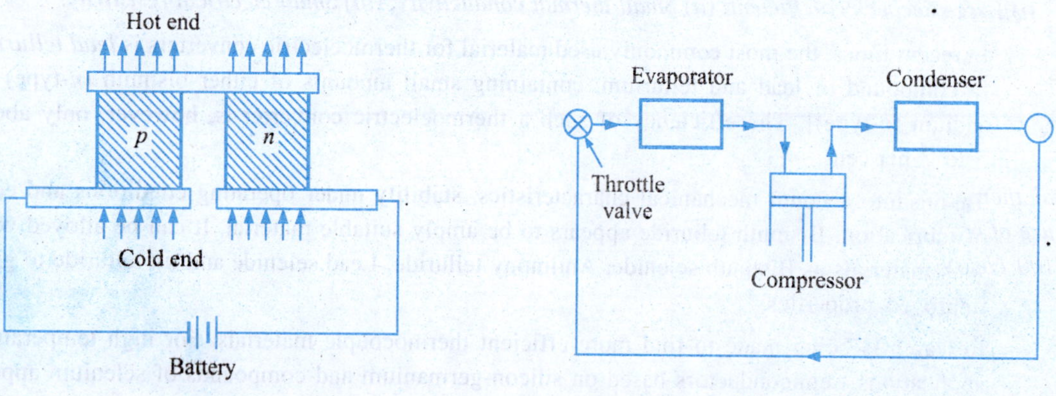

(*a*) Thermoelectric refrigeration system (*b*) Vapour compression system

Fig. 8.8. Analogy between thermoelectric and vapour compression systems

Example 8.1. *The Seebeck coefficient of a material is given by*

$$\alpha = [14 \times 10^{-5} + 350\,T - T^2] \times 10^{-7} \text{ volts/K}$$

Calculate Peltier and Thomson coefficients.

Solution: *Peltier coefficient,*

$$\pi = \alpha \cdot T = [14 \times 10^{-5} + 350\,T - T^2]\,T \times 10^{-7} \text{ Volts} \quad \textbf{(Ans.)}$$

Thomson coefficient,

$$\tau = T \times \frac{d\alpha}{dT} = \frac{d}{dT}[14 \times 10^{-5} + 350\,T - T^2] \times 10^{-7}$$

$$= (350 - 2T) \times 10^{-7} \text{ volts/K} \quad \textbf{(Ans.)}$$

Example 8.2. *Two junctions of a thermoelectric pair are maintained at 25°C and –25 C°. Find the Peltier amount of heat absorbed and heat rejected at the junctions. Take the following properties of the thermoelectric materials A and B which are used for the pair:*

$$\alpha_a = [10^5 + 80\,T - T^2] \times 10^{-7} \text{ volts/K, where T is in °C and}$$

$$\alpha_b = -\alpha_a$$

The current flow is limited to 8 amperes.

Solution: *Given:* $T_h = 25 + 273 = 298$ K; $T_c = -25 + 273 = 248$ K;

$$\alpha_a = [10^5 + 80T - T^2] \times 10^{-7} \text{ volts/K}; \; \alpha_b = -\alpha_a \; ; \quad I = 8 \text{ A.}$$

Peltier heat absorbed at cold junction,

$$Q_c = \pi_{ab} \cdot I = \alpha_{ab} \cdot I \cdot T_c$$

Now,

$$\alpha_{ab} = \alpha_a - (-\alpha_a) = 2\alpha_a$$

$\therefore$

$$Q_c = 2\alpha_a \times I \cdot T_c$$

$$= 2\,[10^5 + 80\,(-25) - (-25)^2] \times 10^{-7} \times 8 \times 248 = \mathbf{38.64\ (Ans.)}$$

Peltier heat rejected at heat junction,

$$Q_h = a_{ab} \cdot I \cdot T_h$$

$$= 2[10^5 + 80\,(25) - (25)^2] \times 10^{-7} \times 8 \times 298 = \mathbf{48.33\ W\ (Ans.)}$$

Example 8.3. *In a thermoelectric cooling system each couple has total electric resistance 0.003 Ω, total thermal conductance 0.025 W/K and thermoelectric power of 420 × 10^{-6} V/K. Calculate the figure of merit. T_h = 310 K, T_c = 270 K, I = 20 A. Find the cooling capacity in tonnes of refrigeration if 100 couples are used.* **(Roorkee University)**

Solution: *Given.* R = 0.003 Ω; U = 0.025 W/K; $\alpha_{ab} = 420 \times 10^{-6}$ volts/K;

$$I = 20 \text{ A} \; ; \; T_h = 310 \text{ k}; \; T_c = 270 \text{ K}$$

Figure of merit,

$$\mathbf{Z} = \frac{(\alpha_{ab})^2}{UR} \qquad \text{...[Eqn. (8.18)]}$$

$$= \frac{(420 \times 10^{-6})^2}{0.025 \times 0.003} = \mathbf{0.002352 \quad (Ans.)}$$

Heat absorbed at the cooling junction per couple,

$$Q_c = \alpha_{ab} \cdot I \cdot T_c - \frac{1}{2} I^2 R - U\,(T_h - T_c) \qquad \text{...[Eqn. (8.13)]}$$

$$= 420 \times 10^{-6} \times 20 \times 270 - \frac{1}{2} \times 20^2 \times 0.003 - 0.025\,(310 - 270)$$

$$= 0.668 \text{ W}$$

$\therefore$ Cooling capacity for 100 couples

$$= \frac{(0.668 \times 3600 / 1000) \times 100}{14000} = \mathbf{0.01718\ tonne \quad (Ans.)}$$

($\because$ 1 tonne = 14000 kJ/h)

Example 8.4. *The following data refer to a thermoelectric refrigeration system consisting of N pairs, connected in series and working under a solar battery:*

Voltage of solar battery .... *6 V*

Temperature of hot junction .... *38°C*

Temperature of cold junction .... *9°C*

Properties of thermoelectric materials: $\alpha_p = 0.00016$ V/K;

$\alpha_n = -\ 0.0002$ VK

$R_{pair} = 0.003\ \Omega$

$U_{pair} = 0.02$ W/K.

Determine for the cases (a) for maximum C.O.P. and (b) for maximum cooling effect the following:

(i) *Number of thermoelectric couples,*

(ii) *Tonnage (cooling) of the unit,*

(iii) *Power consumed, and*

(iv) *Compare the C.O.P with Carnot value.*

Solution: *Given.* Voltage of solar battery = 6 V; $T_h = 38 + 273 = 311$ K;

$T_c = -\ 9 + 273 = 264$ K; $\alpha_p = 0.00016$ V/K; $\alpha_n = -\ 0.0002$ V/K;

$R_{pair} = 0.003\ \Omega$; $U_{pair} = 0.02$ W/K.

Let us first find figure of merit, Z.

$$\alpha_{pn} = \alpha_p - \alpha_n = 0.00016 - (-\ 0.0002) = 0.00036 \text{ V/K}$$

$$\therefore \quad Z = \frac{(\alpha_{pn})^2}{UR} \qquad \text{...[Eqn. (8.18)]}$$

$$= \frac{(0.00036)^2}{(0.02)(0.003)} = 0.00216 \text{ K}^{-1}$$

Case (a) Maximum C.O.P.:

(i) Number of thermoelectric couples, N:

For maximum C.O.P. the current is given by,

$$I = \frac{\alpha_{pn}(T_h - T_c)}{R(\sqrt{1 + ZT_m} - 1} = \frac{0.00036(311 - 264)}{0.003\left[\sqrt{(1 + 0.00216 \times 287.5)} - 1\right]} = 20.64 \text{ A}$$

where, $$T_m = \frac{T_h + T_c}{2} = \frac{311 + 264}{2} = 287.5 \text{ K}$$

and, $$\sqrt{1 + ZT_m} = \sqrt{1 + 0.00216 \times 287.5} = 1.273$$

The overall voltage drop for N couples in series is given by:

$$V = N\,[\alpha_{pn}(T_h - T_c) + IR]$$

or, $$6 = N\,[0.00036\,(311 - 264) + 20.64 \times 0.003]$$

$$N = \textbf{76 couples} \quad \textbf{(Ans.)}$$

(*ii*) **Tonnage (cooling) of the unit:**

Cooling effect is given by:

$$Q_c = N\left[\alpha_{pn}\cdot I\cdot T_c - \frac{1}{2}I^2R - U(T_h - T_c)\right] \quad \text{...[Eqn. (8.13)]}$$

$$= 76\left[0.00036 \times 20.64 \times 264 - \frac{1}{2} \times (20.64)^2 \times 0.003 - 0.02(311-264)\right] = 29\text{W}$$

$$\text{Tonnage} = \frac{(29 \times 3600/1000)}{14000} = \mathbf{0.007457 \ tonne \ (Ans.)}$$

(*iii*) **Power consumed:**

Power consumed by the system = VI = 6 × 20.64 = **123.84 W (Ans.)**

(*iv*) **Comparison of C.O.P. with Carnot value, (C.O.P.)/(C.O.P)$_{\textbf{carnot}}$:**

$$\text{C.O.P.} = \left(\frac{T_c}{T_h - T_c}\right)\left[\frac{\sqrt{1+ZT_m} - (T_h/T_c)}{\sqrt{1+ZT_m}+1}\right] \quad \text{...[Eqn. (8.21)]}$$

$$= \left(\frac{264}{311-264}\right)\left[\frac{1.273 - (311/264)}{1.273+1}\right]$$

$$= 5.617 \times 0.0418 = 0.2348$$

$$(\text{C.O.P})_{\text{Carnot}} = \frac{T_c}{T_h - T_c} = \frac{264}{311-264} = 5.617$$

$$\therefore \quad \frac{\text{C.O.P.}}{(\text{C.O.P.})_{\text{Carnot}}} = \frac{0.2348}{5.617} = \mathbf{0.0418 \ (Ans.)}$$

Case (b) Maximum Cooling:

(*i*) **Number of thermoelectric couples N:**

When the cooling is to be maximum, the current is given by:

$$I = \frac{\alpha_{pn}\,T_c}{R} \quad \text{...[Eqn. (18.24)]}$$

$$= \frac{(0.00036)(264)}{0.003} = 31.68 \text{ A}$$

∴ Number of couples to be connected in series,

$$N = \frac{V}{[\alpha_{pn}(T_h - T_c) + IR]}$$

$$= \frac{6}{[0.00036(311-264) + 31.68 \times 0.003]} = 53.6 \simeq \mathbf{54 \ couples} \quad \textbf{(Ans.)}$$

(*ii*) **Tonnage (cooling) of the unit:**

The maximum cooling is given by:

$$(Q_c)_{\max} = N\left[U\left\{\frac{1}{2}ZT_c^2 - (T_h - T_c)\right\}\right] \quad \text{...[Eqn. (8.25)]}$$

$$= 54\left[0.02\left\{\frac{1}{2}\times 0.00216\times(264)^2-(311-264)\right\}\right] = 30.53 \text{ W}$$

$$\therefore \quad \text{Tonnage} = \frac{(30.53\times 3600)/1000}{14000} = \textbf{0.00785 tonne (Ans.)}$$

(*iii*) **Power consumed:**

Power consumed by the system

$$= \text{VI} = 6\times 31.68 = \textbf{190.08 W (Ans.)}$$

(*iv*) **Comparison of C.O.P with Carnot value, (C.O.P)/(C.O.P)$_{\text{Carnot}}$**

$$\text{C.O.P.} = \frac{\frac{1}{2}\text{Z}\,\text{T}_c^2-(\text{T}_h-\text{T}_c)}{\text{Z}\cdot\text{T}_h\cdot\text{T}_c} \quad \text{...[Eqn. 8.26)]}$$

$$= \frac{\frac{1}{2}\times 0.00216\times(264)^2-(311-264)}{0.00216\times 311\times 264} = 0.1594$$

$$\therefore \quad \frac{\text{C.O.P.}}{(\text{C.O.P.})_{\text{Carnot}}} = \frac{0.1594}{5.617} = \textbf{0.02838 (Ans.)}$$

This example shows that the C.O.P. of the refrigeration system decreases to $\frac{0.1594}{0.2348}$ i.e., 0.6788, but the refrigeration is increased to $\frac{0.00785}{0.007457}$ i.e., 1.053.

8.2 VORTEX TUBE

8.2.1 Construction and Working

- The *vortex tube* is a simple device for *producing cold* and was invented by Ranque and further developed by Hilsch. The vortex tube which shows a good promise is due to Pamlekar and is called *short Vortex Tube.*
- Fig. 8.9 shows the cross-section of a 16 mm vortex tube of the type experimented on by Timothy and Hingue. It consists of the following *parts*:

1. Nozzle,
2. Diaphragm,
3. Cold air side,
4. Hot air side, and
5. Throttling valve.

Air at room temperature and about 5 bar is passed through the nozzle (which is located between warm and cold ends). After expansion through the nozzle it (air) flows tangentially into the vortex chamber. The diaphragm forces the vortex to deviate towards the warm side. When the throttle valve is adjusted, let only a fraction of the air leave at warm side; it is found that two streams of air emerge, one at the warm end and another at cold end. The warm side temperature is higher than that of air at inlet and cold side temperature lower. *By controlling the opening of valve, the quantity of cold air and its temperature can be varied.*

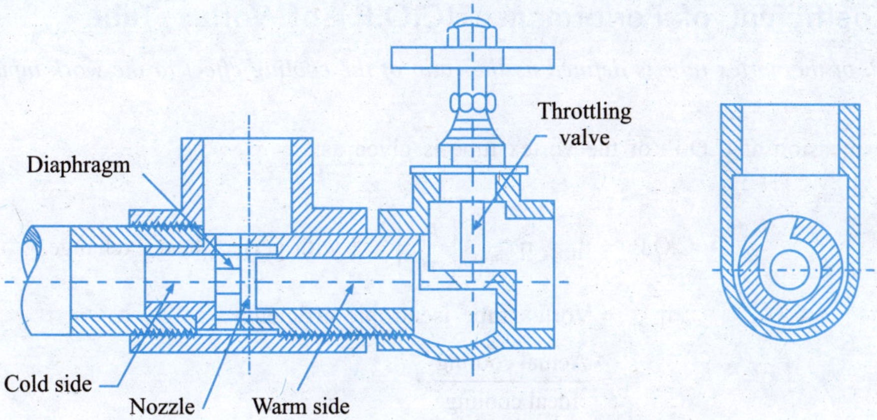

Fig. 8.9. Vortex tube.

8.2.2 Advantages, Disadvantages and Applications of Vortex Tube

Advantages:

(*i*) There is no leakage problem as it uses only air as the refrigerant.

(*ii*) It is simple in design, and needs only control of valves for appropriate functioning.

(*iii*) No moving parts.

(*iv*) Light in weight.

(*v*) Requires less space.

(*vi*) Initial cost is low. At places where compressed air is readily available, its working expenses are also low.

(*vii*) No expertise attendance is required.

Disadvantages:

(*i*) Low C.O.P.

(*ii*) Limited capacity.

Applications:

The vortex tube finds use in the following fields:

1. Air suits.
2. Aviation.
3. Cooling of gas turbine rotor blades.
4. Spot cooling.
5. Cutting tools.
6. Shrink fitting.
7. Laboratory sample cooler.
8. Heating and cooling requirements, simultaneously.
9. Cooling enclosures.

8.2.3 Coefficient of Performance (C.O.P.) of Vortex Tube

*The **C.O.P.** of the vortex tube is defined as the ratio of the cooling effect to the work input to the air compressor.*

The expression of C.O.P. of the vortex tube is given as:

$$\text{C.O.P.} = \eta_{isen}\,\eta_{comp.}\left(\frac{p_a}{p_i}\right)^{\frac{\gamma-1}{\gamma}}, \text{ with perfect heat exchange,} \quad \text{...(8.28)}$$

where, η_{isen} = Vortex tube isentropic efficiency.

$= \dfrac{\text{Actual cooling}}{\text{Ideal cooling}}$,

$\eta_{comp.}$ = Compressor efficiency,

p_i = Pressure of air at inlet to the nozzle, and

p_a = Ambient pressure.

If isothermal compression is considered,

$$(\text{C.O.P.})_{isothermal} = \frac{\gamma}{\gamma-1}\eta_{isen}\cdot\eta_{comp.}\frac{\left[1-\left(\frac{p_a}{p_i}\right)^{\frac{\gamma-1}{\gamma}}\right]}{\ln\left(\frac{p_i}{p_a}\right)} \quad \text{...(8.29)}$$

The general expression for C.O.P. is given as;

$$\text{C.O.P.} = \frac{m_c\,c_p\,\Delta T_c}{\frac{m_i\,c_p}{\eta_{comp.}}\left[\left(\frac{p_i}{p_a}\right)^{\frac{\gamma-1}{1}}-1\right]} \quad \text{...(8.30)}$$

where, suffices *i and c* stand for inlet to nozzle, and cold end respectively and $\Delta T_c = T_i - T_c$.

- It has been observed that $\eta_{comp.}$ is always considerably small, therefore, C.O.P. of the vortex tube will also be very small of the order of 0.15 to 0.2 under normal operating condition.
- To obtain high C.O.P., the pressure ratio should be small, which will give very low refrigeration capacity. Hence it is not *applicable to large capacity plants.*

8.3 PULSE TUBE REFRIGERATION SYSTEM

- Pulse tube refrigeration is a *phenomenon of nature* observed in 1961 but reported in 1964 by Gifford and Langsworth.
- This is a *simple but fascinating method of obtaining low temperatures* and the problem of lubrication at low temperature is basically absent as there are *no movings parts.*

- *In this system sudden expansion and release of gas are employed to get the refrigeration effect.*
- *This method of obtaining low temperatures with moderate pressure and without the use of moving parts can be used to construct* ***compact and reliable miniature refrigerators.***

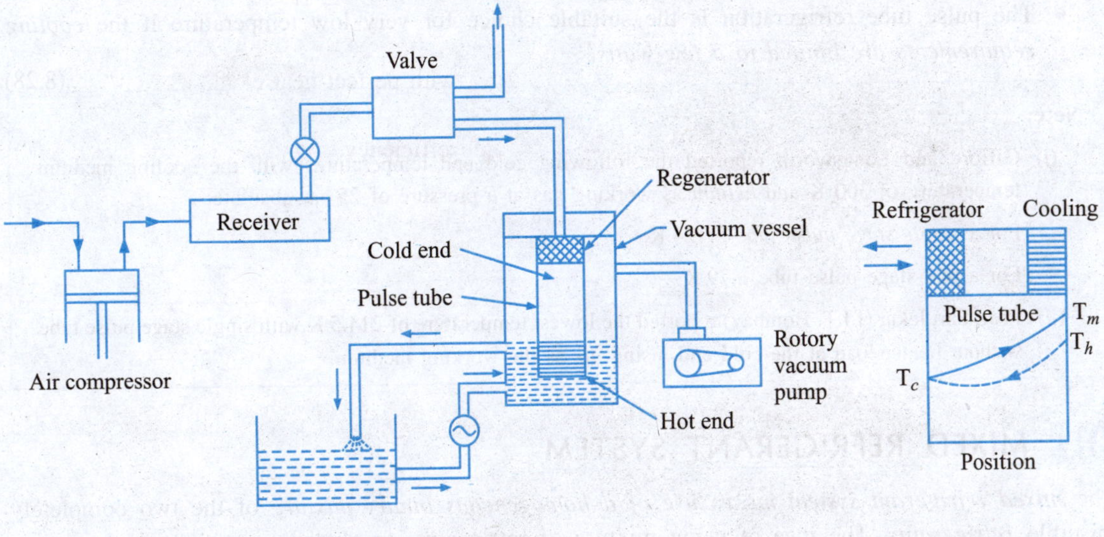

(*a*) Schematic arrangement of pulse tube refrigerator (*b*) Temperature variation with position

Fig. 8.10. Pulse tube refrigeration system.

Construction and Working:

Fig. 8.10 (*a*) shows a simple circuit of the pulse tube refrigeration system. The major components of the system are:

(*i*) Pulse tube (*ii*) Regenerator
(*iii*) Valve mechanism with drive (*iv*) Vacuum system
(*v*) Cooling system.

The *working of the system* is as follows:

- The compressed gas is admitted by valve mechanism to the pulse tube, through a regenerator, where it is cooled to refrigeration (cold end) temperature. This gas entering the tube acts as a fictitious (gas) piston. Subsequently the gas already present inside the pulse tube gets compressed resulting in the increase in the temperature varying from minimum at left end to the maximum at the right end (Fig. 8.10 *b*). Thereafter the heat transfer to the cooling medium reduces the temperature to some value T_h, the temperature close to the cooling medium. The supply of the high pressure gas stops after the inlet valve is closed.
- Thereafter, the gas in the pulse tube is connected to the exhaust with the help of valve mechanism thus releasing the gas that entered the pulse tube during pressure built-up process. This results in the expansion of the tube gas leading to decrease in temperature at the cold end (much lower than that of ambient value) which takes care of the refrigeration load. The cycle is completed and repeated.

Fields of application of Pulse tube refrigeration system

- *This system is most suitable for cryogenics where the requirement of very low temperature cannot be met with even after employing a multi-stage or cascade vapour compression system of best C.O.P.*
- The pulse tube refrigeration is the suitable choice for very low temperature if the *cooling requirements are limited to a few watts.*

Note:

(*i*) Gifford and Longsworth reported the following cold end temperatures with the cooling medium temperature of 300 K and *helium* as working gas at a pressure of 25 bar absolute:

For a *single-stage pulse tube* ...124 K;

For a two-stage pulse tube ...79 K

(*ii*) Prof. Parulekar (I.I.T. Bombay) reported the lowest temperature of 214.5 K with single stage pulse tube without heater load at the cold end, using *air* as the working medium.

8.4 MIXED REFRIGERANT SYSTEM

The *mixed refrigerant system makes use of a homogeneous binary mixture* of the two completely miscible *refrigerants.* The idea of using mixture of refrigerants to obtain properties which are not available with a single refrigerant is not a new one but serious thought has been given recently. The use of mixtures *offers the air-conditioning engineer a means of optimising system performance at any desire temperature.*

The mixture should have the following *desirable properties*:

(*i*) It should be *completely miscible* with the main components in the range of evaporator and condenser temperatures and pressures.

(*ii*) The addition of a small percentage of the additive component should *result in correct boiling and condensing temperature ranges.*

(*iii*) It should have *large latent heat of vaporisation* to counteract the loss in capacity due to increase in volume.

(*iv*) It must be *stable* and *non-reactive.*

(*v*) For the required duty *pressure ratio should not increase.*

Advantages:

(*i*) The capacity of the same compressor can be increased or decreased by adding different percentages of different refrigerants.

(*ii*) Evaporation temperature variation at a given refrigerating effect or refrigerating effect variations at a given evaporation temperature is feasible.

(*iii*) It is possible to reduce condenser pressure by addition of higher boiling point refrigerant.

(*iv*) Oil miscibility is improved in case of mixtures compared with single refrigerant.

(*v*) If one component lacks stability this can be increased by adding a component which is more stable.

Disadvantages:

(*i*) It is difficult to design machines working on this system as the thermodynamic properties of all mixtures are not available in detail.

(*ii*) The normal operation of the machines is upset if the leakage of one component is in greater proportion to the other.

(*iii*) From heat transfer point of view, the non-isothermal condensation may prove disadvantageous.

(*iv*) It is imperative to know heat transfer co-efficients of various mixtures at different compositions if suitable refrigerants in proportions to be chosen.

Applications:

The applications of mixture refrigerant include the following:

1. Process refrigeration– Refineries, natural gasoline and petrochemical industries.
2. Multi-evaporator system.
3. Low temperature refrigeration.

8.5 COOLING BY ADIABATIC DEMAGNETISATION

8.5.1. Introduction

- By evaporating liquefied gases in insulated containers, the following extremely low temperatures can be obtained:

  Liquid air .... 90 K

  Liquid oxygen .... 54.3 K

  Liquid nitrogen .... 35.6 K

  Liquid hydrogen .... 14 K

  Liquid helium.. 1 K [The lowest temperature reached by cooling liquid helium under smallest pressure obtainable (0.2 mm Hg pressure) is 0.71 K.]

- The temperatures below 0.7 K may be obtained by *abiabatic demagnetisation of certain "paramagnetic salts" previously cooled by liquid helium and subjected to strong magnetic field.* This entirely different process is known as **"magnetic cooling"**.

8.5.2 Principle of Magnetic Cooling

All materials/substances possess magnetic properties to a greater or lesser degree and fall within one of these following classes:

— Those materials which are *strongly attracted* by a magnetic called **"ferromagnetic"**.

— Those materials which are repelled by a magnetic pole called **"diamagnetic"**

— Those materials that are weakly attracted by a magnet-called **"paramagnetic"**.

When *paramagnetic salts* (such as gadolinium sulphate) are precooled to a very low temperature, the motion of the molecules is reduced to a minimum, the molecules may than the considered as elementary magnets that will be aligned when subjected to a strong magnetic field. The alignment of molecules requires work which is converted into heat causing a rise of temperature unless the heat is removed by some form of cooling. If the salt/substance is then demagnetised, work is performed without external heat exchange. In the absence of external heat exchange, the *internal energy of the salt decreases* and consequently the salt cools (by adiabatic cooling process).

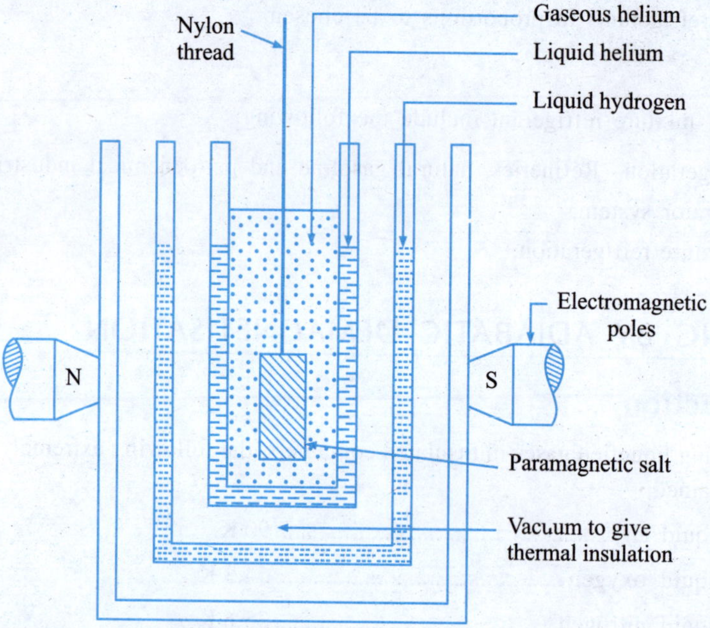

Fig. 8.11. Adiabatic demagnetisation apparatus.

Fig. 8.11 shows a schematic arrangement for adiabatic demagnetisation of a paramagnetic salt. For producing low temperature the paramagnetic salt is suspended by a thread in the tube containing a low pressure *gaseous helium.* This tube is surrounded by a bath of *liquid helium,* which, in turn, is surrounded by a bath of *liquid hydrogen.*

The *magnetic cooling*, in actual practice, *is accomplished in the following* **steps**:

1. The paramagnetic *salt is cooled to about 0.8 K* by surrounding it with liquid helium boiling under reduced pressure.
2. The *salt is exposed to a strong magnetic field.* The heat produced by magnetisation of the salt is transferred to the liquid helium without change in temperature of the salt.
3. The tube containing the salt is evacuated of gaseous helium and the substance/salt is *isolated* at temperature below 1 K under the stress of a strong magnetic field.
4. On the release of magnetic field, the salt temperature *decreases* in an almost *perfectly isentropic way.*

The lowest temperature thus produced by adiabatic demagnetisation of paramagnetic salt is 0.001 K approx.

— The commonly used paramagnetic salts are : *Ferricalum* and *godolinium sulphate.*

— The strength of the magnetic field generally used is of the order of 80 kilogauses.

Very low temperature (i.e., close to absolute zero) can not be measured by ordinary means but *can be calculated by measuring the magnetic susceptibility.* According to Curie's law, the magnetic susceptibility (χ) is inversely proportional to the absolute temperature (T),

$$\text{i.e.}\quad \chi \propto \frac{1}{T}, \quad \text{or,} \quad \chi = \frac{C}{T}.$$

8.6 ADSORPTION REFRIGERATION SYSTEM

In this system an absorbing solid (silica gel) is placed in an equipment to which refrigerant vapour from the evaporator flows, the vapour being condensed in capillary spaces within the adsorbent. The cycle is of the intermittent flow type in contrast to the steady flow that characterises the absorption cycle; the absorbent must be cooled to maintain an isothermal process during adsorption. When the concentration of the refrigerant in the adsorbent reaches the design value, the equipment in question is disconnected (by closing a valve) from the evaporator and connected (by opening a valve) to the condenser, heat is supplied and the same equipment which initially acted as the equivalent of the absorber now acts as a equivalent of a generator.

The **important features** of this system are:

(*i*) Unlike the absorbent of an absorption system, the *adsorbent* is solid and neither mixes with refrigerant, nor contaminates it, nor flows through any part of the cycle.

(*ii*) It does not have pump, hence no moving parts and no shaft work is required. Only little work is required for opening and closing the valve.

(*iii*) *A single equipment acts intermittently through a cycle in which it first absorbs refrigerant vapour at low temperature and evaporator pressure and subsequently generates refrigerant vapour at high temperature and condenser pressure.*

(*iv*) Owing to intermittency of operation, duplicate adsorbent charges (within identical equipment) must be provided; one operates in the adsorption phase while the other is generating refrigerant vapour.

HIGHLIGHTS

1. When the two junctions of a pair of dissimilar metals are maintained at different temperatures e.m.f. is generated. The phenomenon of generation of e.m.f. is called "*Seebeck effect*".
2. According to "*Peltier effect*", if direct current is passed through a pair of dissimilar metals, there is a heating at one junction, cooling at the other depending upon the material combinations.
3. Some important formulae for thermoelectric refrigeration system:

(*i*) Figure of merit, $Z = \dfrac{(\alpha_{ab})^2}{UR}$

(*ii*) $(\text{C.O.P.})_{max} = \dfrac{T_c}{(T_h - T_c)}\left[\dfrac{\sqrt{1 + ZT_m} - (T_h / T_c)}{\sqrt{1 + ZT_m} + 1}\right]$ where, $T_m = \dfrac{T_h + T_c}{2}$

(*iii*) $(\text{C.O.P.})_{opt} = \dfrac{\frac{1}{2} Z T_c^2 - (T_h - T_c)}{Z T_h T_c}$, and $(T_h - T_c)_{max} = \dfrac{1}{2} Z T_c^2$

4. The vortex tube is a simple device for producing cold. The general expression for C.O.P. is given as:

$$\text{C.O.P.} = \frac{\dot{m}_c\, c_p\, \Delta T_c}{\dfrac{M_i\, c_p\, T_a}{\eta_{comp.}}\left[\left(\dfrac{p_i}{p_a}\right)^{\frac{\gamma-1}{\gamma}} - 1\right]}$$

where suffices i and c stand for inlet to nozzle and cold end respectively and $\Delta T_c = T_i - T_c$.

5. In pulse tube refrigeration system sudden expansion and release of gas are employed to get refrigeration effect.
6. The mixed refrigerant system makes use of a homogeneous binary mixture of two completely miscible refrigerants.
7. The temperature below 0.7 K may be obtained by adiabatic demagnetisation of certain "paramagnetic salts" previously cooled by liquid helium and subjected to a strong magnetic field. This process is called *"magnetic cooling."*

OBJECTIVE TYPE QUESTIONS

Fill in the blanks or Say "Yes" or "No":

1. In 1821 .... observed that if a closed circuit was made of two dissimilar metals, an electric currents flowed in the circuit when the two junctions were maintained at different temperatures.
2. In 1834, Jean Peltier observed that there is heating or cooling of a junction of a pair of dissimilar substances, if .... current in passed through them.
3. .... related the heat absorbed or evolved in a single conductor to the temperature gradient along it and current flowing through it.
4. Thermoelectric refrigeration units are much more flexible than conventional units.
5. Seebeck effect is irreversible in nature.
6. Peltier effect is reversible in nature.
7. Joulean effect is reversible in nature.
8. Semiconductor is the proper material for thermoelectric refrigeration.
9. The vertex tube was invented by Ranque.
10. In .... refrigeration system sudden expansion and release of gas are employed to get the refrigerating effect.
11. The .... system makes use of a homogeneous binary mixture of the two completely miscible refrigerants.
12. Those materials which are weakly attracted by a magnet are called ....
13. The lowest temperature produced by adiabatic demagnetisation of paramagnetic salt is 0.001 K approximately.

ANSWERS

1. Seebeck	2. direct	3. Thomson	4. Yes	5. No	6. Yes
7. No	8. Yes	9. Yes	10. pulse tube	11. mixed refrigerant	
12. paramagnetic	13. Yes.				

THEORETICAL QUESTIONS

1. Explain briefly the following thermoelectric effects:
   (*i*) Seebeck effect; (*ii*) Peltier effect; (*iii*) Thomson effect;
   (*iv*) Joulean effect; (*v*) Conduction effect
2. What do you mean by "Peltier cooling"?
3. Explain the working principle of thermoelectric refrigeration system.
4. State the advantages and disadvantages of thermoelectric refrigeration system.
5. List the applications of thermoelectric refrigeration system.
6. Define Seebeck effect; Peltier effect and Thomson effects and prove that their inter-relation is given by

$$\pi_{ab} = \tau_{ab} \times T, \quad \text{and} \quad \tau_a - \tau_b = T.\frac{d\alpha_{ab}}{dT}$$

7. Define the figure of merit related to Thermoelectric refrigeration system and explain its effect on C.O.P of the system.
8. Name two pairs of thermoelectric elements which are used for the commercial purposes.
9. Explain the meaning of *p* and *n* type thermoelectric elements.
10. How can you use a thermoelectric refrigerator as a heat pump?
11. Obtain an expression for the current corresponding to maximum C.O.P. in terms of figure of merit; temperature of the cold body, resistance and thermal conductance.
12. Who invented vortex tube?
13. Explain the working principle of vortex tube.
14. What are the advantages of vortex-tube over other refrigeration systems?
15. List the disadvantages of vortex tube.
16. State the applications of vortex tube.
17. Give the general expression for C.O.P. of the vortex tube.
18. Describe briefly the working principle of a pulse tube system.
19. What are the advantages of a pulse tube cooling?
20. State the fields of application of pulse tube cooling.
21. What is mixed refrigerant system?
22. List out the advantages and disadvantages of mixture refrigeration systems.
23. What are the applications of mixture refrigeration systems?
24. Discuss the arrangement used for producing low temperatures by adiabatic demagnetisation of a paramagnetic salt.

UNSOLVED EXAMPLES

1. Two junctions of a thermoelectric pair are maintained at 30ºC and –30ºC. Find the Peltier amount of heat absorbed and heat rejected at the junctions. Take the following properties of the thermoelectric materials A and B which are used for the pair:

$$\alpha_a = [10^5 + 100T - T^2] \times 10^{-7} \text{ Volts/K, where T is in ºC}$$

and, $$\alpha_b = -\alpha_a$$

The current flow is limited to 10 A. **[Ans.** 46.7 W; 61.87 W]

2. In a thermoelectric cooling system each couple has total electric resistance R_e = 0.003 ohm, total thermal conductance U = 0.025 W/K and thermoelectric power of 450 × 10^{-6} Volts/K. Calculate the figure of merit. The temperature of cold junction is 260 K and hot junction is 310 K. The current flow is 20 A. Find the cooling capacity in tonnes of refrigeration if 120 couples are used.

[Ans. 0.02074 tonne]

3. A thermoelectric refrigeration system consists of N pairs, connected in series. It works under a 6 volt solar battery. The hot and cold junctions are maintained at 40ºC and – 10º C respectively. The properties of thermoelectric materials are α_p = 0.00016 V/K and α_n = – 0.0002 V/K, the resistance and conductance for each pair 0.003 ohm and 0.02 W/K respectively.

Determine for the cases (*a*) for maximum C.O.P and (*b*) for maximum cooling effect the following:

(*i*) Number of thermoelectric couples, (*ii*) Tonnage (cooling) of the unit,

(*iii*) Power consumed, and (*iv*) Compare the C.O.P. with Carnot value.

[Ans. (*a*) 72 couples, 0.006369 tonne; 131.39 W; 0.03688,

(*b*) 54 couples, 0.00686 tonne; 189.36 W; 0.0264]

❑❑❑

Chapter 9

Refrigerants, Driers and Lubricants

9.1 INTRODUCTION TO REFRIGERANTS

- A **'refrigerant'** *is defined as any substance that absorbs heat through expansion or vaporisation* and loses it through *condensation in a refrigeration system.*
- The term '*refrigerant*' in the broadest sense is also applied to such *secondary cooling mediums* as *cold water or brine, solutions.* Usually refrigerants include only those working mediums which pass through the cycle of *evaporation, recovery, compression, condensation and liquefaction. These substances absorb heat at one place at low temperature level and reject the same at some other place having higher temperature and pressure. The rejection of heat takes place at the cost of some mechanical work.* Thus circulating cold mediums and cooling mediums (such as ice and solid carbondioxide) are not primary refrigerants. In the early days only four refrigerants, *Air, ammonia* (NH_3), *Carbondioxide* (CO_2), *Sulphur Dioxide* (SO_2), possessing chemical, physical and thermodynamic properties permitting their efficient application and service in the practical design of refrigeration equipment were used. All the refrigerants change from liquid state to vapour state during the process.
- Refrigerants have to be physiologically non-toxic and non-flammable substances. Thermodynamically, they are essentially characterised by their respective normal boiling points. The *normal boiling point of a refrigerant should be little lower than the evaporator temperature.* This will *ensure positive pressure both in the condenser and the evaporator*, and as such in whole refrigeration system. Positive pressure in the system is desirable *so that atmospheric air may not enter into the system* through gap across compressor shaft seal.

The *choice* of a particular refrigerant depends upon the application in *terms of*:

(*i*) The refrigerating capacity (very small, small, medium or large);

(*ii*) The type of compressor used generally reciprocating or centifugal or screw);

(*iii*) The refrigeration temperature required (whether for air-conditioning, cold storage or food freezing).

9.2 CLASSIFICATION OF REFRIGERANTS

The refrigerants are classified as follows:

1. Primary refrigerants;
2. Secondary refrigerants.

- ***Primary refrigerants*** are those working mediums or heat carries which *directly take part in the refrigeration system and cool the substance by the absorption of latent heat* e.g., *Ammonia, Carbondioxide, Sulphur dioxide, Methyl chloride, Methylene chloride, Ethyl chloride* and *Freon group* etc.
- ***Secondary refrigerants*** are those circulating substances *which are first cooled with the help of the primary refrigerants and are then employed for cooling purposes, e.g., ice, carbondioxide etc*. These refrigerants cool substances by *absorption of their sensible heat.*

9.3 PRIMARY REFRIGERANTS

The primary refrigerants are further classified into the following groups:

1. Halocarbon compounds
2. Azeotropes
3. Hydrocarbons
4. Inorganic compounds
5. Unsaturated organic compounds.

9.3.1 Halocarbon Compounds

- In 1928, Charles Kettening and Dr. Thomas Mighey invented and developed this group of refrigerants. In this group are included refrigerants which contain one or more of three halogens, chlorine and bromine and they are sold in the market under the names as *Freon, Genetron, Isotron* and *Areton*. Since the refrigerants belonging to this group have *outstanding merits over the group's refrigerants*, therefore they find wide field of application in *domestic, commercial and industrial purposes.*
- ASHARAE (American Society of Heating and Air-conditioning Engineers) identified, 42 halocarbon compounds as refrigerants, but only a few of them are commonly used. Some of the commonly used halocarbon compounds are given in the Table 9.1.

Table 9.1 Commonly used Halocarbon Refrigerants

Refrigerant number	*Chemical name*	*Chemical formula*
R-11	Trichloromonofluoromethane	CCl_3F
R-12	Dichlorodifluoromethane	CCl_2F_2
R-13	Monochlorotrifluoromethane	$CClF_3$
R-14	Carbontetrafluoride	CF_4
R-21	Dichloromonofluoromethane	$CHCl_2F$
R-22	Monochlorodifluoromethane	$CHClF_2$
R-30	Methylene chloride	CH_2Cl_2
R-40	Methyl chloride	CH_3Cl
R-100	Ethyl chloride	C_2H_5Cl
R-113	Trichlorotrifluorothane	CCl_2FCClF_2
R-114	Dichlorotetrafluoroethane	$CClF_2CClF_2$
R-115	Monochloropentafluoroethane	$CClF_2CF_3$
R-152	Difluoroethane	$C_2F_4Cl_2$

9.3.2 Azeotropes

The refrigerants belonging to this group consist of mixtures of different substances. These substances *cannot be separated into components by distillation*. They possess *fixed thermodynamic properties and do not undergo any separation with changes in temperature and pressure*. An azeotrope behaves like a simple substance.

Some of the Azeotropes are given in the Table 9.2.

Table 9.2 Azeotropes

Refrigerant number	*Azeotropic mixing refrigerants*	*Chemical formula*
R-500	73.8% R-12 and 26.2% R-152	CCl_2F_2/CH_3CHF_2
R-502	48.8% R-22 and 51.2% R-115	$CHClF_2/CClF_2CF_3$
R-503	40.1% R-23 and 59.9% R-13	$CHF_3/CClF_3$
R-504	48.2% R-32 and 51.8% R-115	$CH_2F_2/CClF_2CF_3$

9.3.3 Hydrocarbons

Most of the refrigerants of this group are organic compounds. Several hydrocarbons are used successfully in commercial and industrial installations. Most of them possess satisfactory thermodynamic properties but are highly inflammable. Some of the important refrigerants of this group are given in the Table 9.3.

Table 9.3 Hydrocarbons

Refrigerant number	*Chemical name*	*Chemical formula*
R-50	Methane	CH_4
R-170	Ethane	C_2H_6
R-290	Propane	C_2H_8
R-600	Butane	C_4H_{10}
R-601	Isobutane	$[CH(CH_3)_3]$

9.3.4 Inorganic Compounds

Before the introduction of hydrocarbon group these refrigerants were most commonly used for all purposes.

The important refrigerants of this group are:

Table 9.4 Inorganic Compounds

Refrigerant number	*Chemical name*	*Chemical formula*
R-717	Ammonia	NH_3
R-718	Water	H_2O
R-729	Air	Mixturer of O_2, N_2,CO_2 etc.
R-744	Carbon dioxide	CO_2
R-764	Sulphur dioxide	SO_2

9.3.5 Unsaturated Organic Compounds

The refrigerants belonging to this group possess ethylene or propylene as their constituents. They are given in the Table 9.5.

Table 9.5 Unsaturated Compounds

Refrigerant name	*Chemical name*	*Chemical formula*
R-1120	Trichloroethylene	$C_3H_4Cl_3$
R-1130	Dichloroethylene	$C_2H_4Cl_2$
R-1150	Ethylene	C_2H_4
R-1270	Propylene	C_3H_6

9.4 DESIGNATION OF REFRIGERANTS

The international designation committee of refrigerants uses Refrigerant or R as the designation followed by certain numerals (e.g., R-11, R-12, R-22 etc.)

– A refrigerant followed by a *'two digit number'* indicates that a refrigerant is derived from ***'methane base'*** while *'three digit number'* represents ***'ethane base'***.

• *The general chemical formula for a compound derived from a "saturated hydrocarbon" is given as:*

$$C_a H_b F_c Cl_d$$

where, $b + c + d = 2a + 2$,

and, a = Number of carbon atoms,

b = Number of hydrogen atoms,

c = Number of fluorine atoms, and

d = Number of chlorine atoms.

The complete designation/number of the refrigerant is given by:

$$R\,(a - 1)\,(b + 1)\,c$$

Examples: 1. *Dichloro-tetrafluoro-ethane:*

In this refrigerant there are,

(*i*) Two chlorine atoms, i.e., $d = 2$

(*ii*) Four flourine atoms, i.e., $c = 4$

(*iii*) No, hydrogen atom, i.e., $b = 0$

We know that,

$$b + c + d = 2a + 2$$

$$0 + 4 + 2 = 2a + 2$$

or, $a = 2$ i.e., number of carbon atoms = 2

Thus the chemical formula for dichloro-tetrafluoro-ethane becomes $\mathbf{C_2Cl_2F_4}$ and the designation/number of the refrigerant becomes:

$R\,(a - 1)\,(b + 1)\,c = R\,(2 - 1)\,(0 + 1)\,4$ or **R-114**

2. **In case of CCl_3F (Trichloromonofluoromethane):**

$$a = 1, b = 0, c = 1, d = 3$$

∴ Designation/number of the refrigerant is: R (1 – 1) (0 + 1) (1) i.e., **R-11**

3. **In case of CCl_2F_2 (Dichlorodifluoromethane):**

$$a = 1, b = 0, c = 2, d = 2$$

∴ Designation is: R (1 – 1) (0 + 1) (2) i.e., **R-12**

4. **In case of $CHClF_2$ (Monochlorodifluoromethane):**

$$a = 1, b = 1, c = 2, d = 1$$

∴ Designation is: R (1 – 1) (1 + 1) (2) i.e., **R-22.**

5. **In case of $C_2\,Cl_3\,F_3$ (Trichlorotrifluoroethane):**

$$a = 2, b = 0, c = 3, d = 3$$

∴ Designation is: R (2 – 1) (0 + 1) (3) = **R-113**

6. **In case of CH_4 (Methane):**

$$a = 1, b = 4, c = 0$$

∴ Designation is: R (1 – 1) (4 + 1) (0) i.e., **R-50**

7. **In case of C_2H_6 (Ethane):**

$$a = 2, b = 6, c = 0$$

∴ Designation is: R (2 – 1) (6 + 1) (0) i.e., **R-170**

- The **brominated refrigerants** are denoted by putting an additional B and a number to denote as to how many chlorine atoms are replaced by bromine atoms.

Example: R13 B1 is derived from R13 with the replacement of one chlorine atom by a bromine atom. Its chemical formula is therefore CF_3Br.

- In the case of **isomers**, i.e., compounds with the same chemical formula but different molecular structure subscripts, *l, m,* etc., are used after the designation.
- **Unsaturated compounds** for which $(b + c + d) = 2a$, are distinguished by putting the digit 1 before $(a – 1)$. Thus ethylene is 1150.
- In case of **common inorganic refrigerants**, numerical designations have been given according to their molecular weight added to 700. Thus ammonia whose weight is 17 is designated as R-717. Similarly, carbon dioxides designated as 744 and water as 718.

Cryogenic refrigerants

- The temperature range of – 157° C to – 273° C is known as *cryogenic range,* and the refrigerants for producing is these temperatures are known as *"Cryogenic refrigerants"*. These temperatures may be produced by evaporating cryogenic liquids which have *very low boiling points* at *atmospheric pressures.* The *common cryogenic refrigerants are:*

Air, O_2, N_2, H_2 and He

- The containers used for storing these liquids are made of *special materials capable of withstanding extremely low temperatures.* Containers are *heavily insulated to reduce the loss of cryogenic refrigerant by boiling.*

9.5 DESIRABLE PROPERTIES OF AN IDEAL REFRIGERANT

An ideal refrigerant should possess the following *properties*:

A. Thermodynamic Properties:

(*i*) Low boiling point

(*ii*) Low freezing point

(*iii*) Positive pressure (but nor very high) in evaporator and condenser

(*iv*) High saturation temperature

(*v*) High latent heat of vapourisation.

B. Chemical Properties:

(*i*) Non-toxicity

(*ii*) Non-flammable and non-explosive
(*iii*) Non-corrosiveness
(*iv*) Chemical stability in reacting
(*v*) No effect on the quality of stored (food and other) products like flowers, with other materials, i.e., furs and fabrics.
(*vi*) Non-irritating and odourless.

C. Physical Properties:

(*i*) Low specific volume of vapour
(*ii*) Low specific heat
(*iii*) High thermal conductivity
(*iv*) Low viscosity
(*v*) High electrical insulation.

D. Other Properties:

(*i*) Ease of leakage location
(*ii*) Availability and low cost
(*iii*) Ease of handling
(*iv*) High C.O.P.
(*v*) Low power consumption per tonne of refrigeration
(*vi*) Low pressure ratio and pressure difference.

*Some **important properties** (mentioned above) are discussed below*:

1. Boiling temperature. The boiling temperature of the refrigerant at atmospheric pressure should be *low*. If the boiling temperature of the refrigerant is high at atmospheric pressure, the compressor should be operated at high vacuum. The *high boiling temperature reduces the capacity and operating cost of the system.* The Table (9.6) below shows the boiling temperatures at atmospheric pressure of some commonly used refrigerants.

Table 9.6 Boiling Temperatures

Refrigerant	*Boiling temp. (°C) at atmospheric pressure*	*Refrigerant*	*Boiling temp. (°C) at atmospheric pressure*
R-11	+ 23.77	R-40	– 23.7
R-12	– 29.80	R-113	+ 3.6
R-21	+ 8.92	R-717	– 33.3
R-22	– 40.80	R-744	– 73.6
R-30	39.80	R-764	– 10.0

2. Freezing point. As the refrigerant must operate in the cycle above its freezing point, it is evident that the same for the refrigerant *must be lower than system temperatures*. It is found that except in the case of water for which the freezing point is 0° C, other refrigerants have reasonably low values. Water, therefore, can be used only in air-conditioning applications which are above 0°C. The Table 9.7 shows the freezing temperature of some common refrigerants.

Table 9.7 Freezing Temperatures

Refrigerant	*Freezing temp. (°C)*	*Refrigerant*	*Freezing temp. (°C)*
R-11	– 111	R-40	– 97.5
R-12	– 157.5	R-113	– 35
R-21	– 135	R-717	– 77.8
R-22	– 160	R-744	– 56.7
R-30	– 96.6	R-764	– 75.6

3. Evaporator and condenser pressures. The evaporating pressure should be *positive* and should be as near atmospheric as possible. If it is *too low*, it would result in a *large volume of the suction vapour*. If it is *too high, overall high pressures including condenser pressure would result necessitating stronger equipment and consequently higher cost. A positive pressure is required in order to eliminate the possibility of the entry of air and moisture into the system. The normal boiling point of the refrigerant should, therefore, be lower than the refrigerant temperature.* The Table 9.8 shows the evaporator and condensers pressures for the commonly used refrigerants.

Table 9.8 Evaporator and Condenser Pressures

Refrigerant	*Evaporator pressure (p_E) at – 15°C, in bar*	*Condenser pressure (p_C) at + 30°C in bar*	*Compression ratio (p_C/p_E)*
R-11	0.2021	1.2607	6.24
R-12	1.8262	7.4510	4.08
R-21	0.3618	2.1540	5.95
R-22	2.9670	12.0340	4.05
R-30	0.0807	0.7310	9.06
R-40	1.4586	6.5310	4.47
R-113	0.0676	0.5421	8.02
R-717	2.3634	11.67	4.94
R-744	22.90	71.930	3.14
R-764	0.8145	4.5830	5.63

— The *'reciprocating compressors'* are used with refrigerants having *low specific volumes, high operating pressures and high pressure ratios.*

— The *'centrifugal compressors'* are used with refrigerants having *high specific volumes, low operating pressures and low pressure ratios.*

4. Critical temperature and pressure. Generally, for high C.O.P. the *critical temperature should be very high so that the condenser temperature line up on p-h diagram is far removed from the critical point.* This ensures reasonable refrigerating effect as it is very small with the state of liquid before expansion near the critical point. The *critical pressure should be low so as to give low condensing pressure.* The Table 9.9 shows the critical temperatures and pressures for the commonly used refrigerants.

Table 9.9 Critical Temperatures and Pressures

Refrigerant	*Critical temperature (ºC)*	*Critical pressure (bar)*
R-11	198	43.8
R-12	112	41.2
R-21	178.5	51.65
R-22	96	49.38
R-30	216	44.14
R-40	143	66.83
R-113	214	34.14
R-717	133	113.86
R-744	31	73.8
R-764	157	78.7

— The critical temperature for most of the common refrigerants is well above normal condensing temperature with exception of carbon dioxide (R-744) whose critical temperature in 31ºC.

5. Latent heat of vaporisation. *It should be as large as possible to reduce the weight of the refrigerant to be circulated in the system. This reduces initial cost of the refrigerant. The size of the system will also be small and hence low initial cost.* The Table (9.10) shows the refrigerating effect for common refrigerants operating between –15ºC evaporator temperature and 30ºC condenser temperature. It also shows the latent heat, mass of refrigerant circulated for *ton* of refrigeration and the volume of the liquid refrigerant per *ton* of refrigeration.

Table 9.10 Refrigerating Effect, Latent Heat of Vaporisation, Mass of Refrigerant and Volume of Liquid Refrigerant

Refrigerant	*Refrigerating effect for standard cycle of –15ºC to +30ºC, in kJ/kg*	*Latent heat of vaporisation at –15ºC, in kJ/kg*	*Mass of refrigerant circulated per standard ton, in kg/kmin*	*Volume of liquid refrigerant circulating per standard ton at 30ºC in litres/min*
R-11	157.3	195.7	1.34	0.918
R-12	116.5	159.0	1.81	1.404
R-22	161.5	218.1	1.31	1.115
R-30	313.6	377.7	0.676	0.507
R-40	350.0	421.0	0.603	0.67
R-113	125.1	164.5	1.7	1.1
R-717	1105.4	1316.5	0.19	0.32
R-744	129.3	274.0	1.63	2.73
R-764	329.5	394.7	0.64	0.474

6. Toxicity. Taking into consideration comparative hazard to life due to gases and vapours underwriters Laboratories have divided the compounds into *six groups. Group six contains compounds*

with a very low degree of toxicity. It includes R_{12}, R_{114}, R_{13}, B_1 *etc. Group one, at the other end of the scale, includes the most toxic substances such as* SO_2.

Ammonia is not used in comfort air-conditioning and in domestic refrigeration because of inflammability and toxicity.

7. Inflammability. Hydrocarbons (e.g., methane, ethane etc.) are highly explosive and inflammable. Fluorocarbons are neither explosive nor inflammable. *Ammonia is explosive in* a mixture with air in concentration of 16 to 25% by volume of ammonia.

8. Volume of suction vapour. *The size of the compressor depends on the volume of suction vapour per unit (say per tonne) of refrigeration. Reciprocating compressors are used with refrigerants with high pressures and small volumes of the suction vapour. Centrifugal or turbo-compressors are used with refrigerants with low pressure and large volumes of the suction vapour.* A high volume flow rate for a given capacity is required for centrifugal compressors to permit flow passage of sufficient width to minimise drag and obtain high efficiency.

Table 9.11 Specific Volume and Theoretical Piston Displacements

Refrigerant	*Specific volume of the refrigerant vapour at – 15° C* (m^3/kg)	*Vapour displacement for standard ton of refrigeration* (m^3/min)
R-11	0.77	1.016
R-12	0.091	0.163
R-22	0.078	0.101
R-30	3.12	2.08
R-40	0.28	0.167
R-113	1.7	2.08
R-717	0.51	0.096
R-744	0.0167	0.027
R-764	0.401	0.254

9. Thermal conductivity. *For a high transfer co-efficient a high thermal conductivity is desirable.* R-22 has better heat transfer characteristics than R-12 ; R-21 is still better, R-13 has poor heat transfer characteristics.

10. Viscosity. For a *high heat transfer co-efficient a low viscosity is desirable.*

11. Leak tendency. *The refrigerants should have low leak tendency.* The greatest drawback of *fluorocarbons* is the fact that they are *odourless*. This at times, results in a complete loss of costly gas from leaks without being detected. An ammonia leak can be very easily detected by pungent odour.

12. Refrigerant cost. The cost factor is only relevant to the extent of the price of the initial charge of the refrigerant which is very small compared to the total cost of the plant and its installation. The cost of losses due to leakage is also important. In small capacity units requiring only a charge of the refrigerant, the cost of refrigerant is immaterial.

The *cheapest refrigerant is Ammonia*. R-12 is *slightly cheaper than* R-22, R-12 and R-22 have replaced ammonia in the dairy and frozen food industry (and even in cold storage) *because of the tendency of ammonia to attack some food products.*

13. Co-efficient of performance and power per ton of refrigeration. Practically all common refrigerants have approximately same C.O.P. and power requirement.

Table 9.12 gives the values of C.O.P. and power per ton of refrigeration for some important refrigerants.

Table 9.12 Coefficient of Performance and Power Per Ton of Refrigeration

Refrigerant	*Coefficient of performance*	*kW/ton*
R-11	5.09	0.694
R-12	4.70	0.746
R-22	4.66	0.753
R-30	4.90	0.716
R-40	4.90	0.716
R-113	4.92	0.716
R-717	4.76	0.738
R-729	5.74	0.619
R-744	2.56	1.372
R-764	4.87	0.724

— Practically, all common refrigerants have approximately the same coefficient of performance and power requirement.

14. Action with oil. No chemical reaction between refrigerant and lubricating oil of the compression should take place. Miscibility of the oil is quite important as some oil should be carried out of the compressor crankcase with the hot refrigerant vapour to lubricate the pistons and discharge valves properly.

15. Reaction with materials of construction. While selecting a material to contain the refrigerant, this material should be given a due consideration. Some metals are attacked by the refrigerants; e.g., **ammonia** *reacts with copper, brass or other cuprous alloys in the presence of water*, therefore, in *ammonia systems the common metals used are iron and steel.* **Freon group** does not react with steel, copper, brass, zinc, tin and aluminium but is corrosive to Magnesium and Aluminium having magnesium more than 2%. Freon group refrigerants tend to dissolve natural rubber in packing and gaskets but synthetic rubber such as neoprene are entirely suitable. The **hydrogenated hydrocarbons** may react with zinc but not with copper, aluminium, iron and steel.

16. Dielectric strength. The dielectric strength of a refrigerant is important in *hermetically sealed units in which the electric motor is exposed to the refrigerant.* The relative dielectric strength of the refrigerant is the *ratio of the dielectric strength of nitrogen and the refrigerant vapour mixture to the dielectric strength of nitrogen at atmospheric pressure and room temperature.*

The Table 9.13 gives the refrigerants with the chemical formulae and a few of their more important thermal and physical properties.

Table 9.13 Refrigerants with Chemical Formulae and Some Important Properties

Refrigerant	*Chemical formula*	*Molecular weight*	*Boiling point, °C*	*Freezing point, °C*	*Critical point*		*Specific gravity at atmospheric pressure*	*Liquid specific heat from – 15°C to30°C, kJ/kg K*	*Specific heat of vapour – 15°C to 30°C, kJ/kg K*	$\frac{c_p}{c_v} = \gamma$	*Gas constant R, kJ/KgK*
					Temp, °C	*Pressure, bar*					
Air (R-729)	—	—	—	—	—	—	—	—	1.0048	1.4 (– 30°C to 10°C)	0.287
Ammonia (R-717)	NH_3	17.031	– 33.3	– 77.8	133	114.24	– 0.684	4.689	2.19	1.31 (21° C)	—
R-11	CCl_3F	137.38	+ 23.77	– 111	198	43.78	1.468	0.879	0.578	1.136 (48°C)	0.0605
R-12	CCl_2F_2	120.93	– 29.8	157.5	112	41.16	1.480	0.963	0.615	1.16 (10°C)	0.0686
R-13	$CClF_3$	104.47	– 81.5	– 181	28.8	38.6	—	—	0.578	1.172 (–15°C)	—
R-21	$CHCl_2F$	102.93	8.92	– 135	178.5	51.68	—	—	0.586	1.175	—
R-22	$CHClF_2$	86.48	– 40.8	– 160	96	49.37	1.411	1.4	0.657	1.184 (48°C)	0.096
R-113	$C_2Cl_3F_3$	187.39	3.6	– 35	214	34.13	1.559	0.913	—	1.08 (60°C)	0.044
R-114	$C_2Cl_2F_4$	170.93	3.55	– 94	145.7	32.58	—	0.996	0.653	1.088	0.0486
Methyle chloride (R-40)	CH_3Cl	—	– 22.8	– 97.5	152	66.78	1.002	1.59	1.0048	1.28 (25° C)	—
R-500 (Azeotrope)	73.8% CCl_2F_2 CH_2CHF_2	99.29	– 33.3	– 159	105	43.5	—	1.26	0.716	0.13	0.0837

9.6 PROPERTIES AND USES OF COMMONLY USED REFRIGERANTS

1. Air (R-729)

Properties:

1. No cost involved; easily available.
2. Completely non-toxic.
3. Completely safe.
4. The C.O.P. of air cycle operating between temperatures of 80°C and –15°C is 1.67.

Uses:

1. Air was one of the earliest refrigerants and was widely used even as late as World War I wherever a completely non-toxic medium was needed.
2. Because of low C.O.P., it is used only where *operating efficiency is secondary as in aircraft refrigeration.*

2. Ammonia (R-717)

Properties:

1. It is highly toxic, flammable, irritating and food destroying.
2. It has excellent thermal properties.
3. It has the *highest refrigerating effect per kg of refrigerant.*
4. Low volumetric displacement.
5. Low cost.
6. Low weight of liquid circulated per tonne of refrigeration.
7. High efficiency.
8. Anhydrous ammonia has no effect on lubricating oil but in presence of moisture, ammonia forms an emulsion with oil that causes operating difficulties. The formed emulsion becomes effective when the percentage of water exceeds 0.01%. It is not oil miscible therefore it will not dilute the oil in the crankcase of the compressor.
9. It is highly volatile and becomes explosive when mixed with air and compressed, therefore air leaks must be avoided in ammonia refrigeration systems.
10. Ammonia attacks on non-ferrous metals in the presence of water therefore *copper and brass are never used with ammonia refrigeration systems.*
11. Ammonia can be used economically for –70ºC evaporator temperature and its application for further low temperature becomes highly uneconomical and difficult to maintain low vacuum required in the evaporator.
12. The evaporator and condenser pressure are 3.4 bar and 12.9 bar respectively at standard conditions of –15ºC and 30ºC.

Uses:

1. It is widely used in large industrial and commercial reciprocating compression systems where high toxicity is secondary.
2. It is widely used as the refrigerant in *absorption systems.*
3. It is extensively used in *ice plants, packing plants, large cold storages and skating rinks etc.*

Ammonia is commonly used in ice factories and breweries without exception. The **reasons** for this are:

(*i*) Cheapest refrigerant.

(*ii*) Lowest running cost.

(*iii*) Smallest pipelines.

(*iv*) Leaks easy to detect.

(*v*) Low side equipment can be readily added to or taken away with no operating problems.

(*vi*) Different forms of refrigerant-circuiting can be incorporated with the main system, e.g., dry expansion, flooded, overfeed or recirculation.

(*vii*) The possibility of fire hazard with NH_3 is minimum.

(*viii*) Easier maintenance with no evacuation and drying.

(*ix*) Largest latent heat per unit mass combined with the smallest liquid density and low viscosity allows smaller liquid pumps and liquid supply lines in overfeed system.

Disadvantage of NH_3:

The major engineering disadvantage of NH_3 is its *high temperature which requires more efficient separators and can cause oil carbonisation problems when it is single staged with high compression ratios*. However, this is not a problem with oil injected screw compressors which can operate successfully with compression ratio of 25 down to suction temperature of –45°C and condensing at 35°C. The toxic nature of NH_3 is the another disadvantage as pigs were killed when exposed to concentration 1/2% (5000 ppm) for half an hour. This can be avoided by making the system leak-high.

> **Note:**
>
> (*i*) *Ammonia should never be used with copper, brass and other copper alloys; iron and steel should be used in ammonia systems instead.*
>
> (*ii*) *In ammonia systems, to detect the leakage a sulphur candle is used which gives off a dense white smoke when ammonia vapour is present.*

3. SO_2 (R-764)

Properties:

1. It is a colourless gas or liquid.
2. It is extremely toxic and has a pungent irritating odour.
3. It is non-explosive and non-flammable.
4. It has a liquid specific gravity of 1.36.
5. Works at low pressures.
6. Possesses small latent heat of vaporisation.
7. It forms sulphurous acid in the presence of water and sulphuric acid in the presence of water and air which are highly corrosive to most of the metals.
8. It has *low refrigerating effect* and high specific volume per kg of refrigerant compared with other refrigerants, therefore, *large compressors with higher speed are required* with this refrigerant for same refrigerating capacity. Its piston displacement is 2.3 times greater than NH_3 and 8.5 times greater than CO_2.
9. Sulphur dioxide is not oil miscible. Liquid SO_2 is heavier than oil therefore oil floats on the top surface of liquid sulphur dioxide which simplifies the problems of oil removal and return.

Uses: It finds little use these days. However its use was made in small machines in early days.

- The leakage of sulphur dioxide may be detected by bringing *aqueous ammonia near the leak; this gives off white smoke.*

4. CO_2 (R-744)

Properties:

1. It is a colourless and odourless gas, and is heavier than air.
2. It has liquid specific gravity of 1.56.

3. It is non-toxic, non-flammable, non-poisonous but causes death due to suffocation when present in large quantities.
4. It is non-explosive and non-corrosive.
5. It has *extremely high operating pressures.*
6. It gives very *low refrigerating effect.*
7. It is chemically stable under all pressure and temperature conditions occuring in the system. It remains unaffected either with metals or oil in the presence of air and water, therefore, any metal can be used with this refrigerant.
8. It is immiscible in oil and therefore will not dilute the oil in the crankcase of the compressor.

Disadvantages:

1. The main disadvantages of CO_2 is its *high operating pressure under standard temperature conditions.* This *requires heavy piping and robust condenser and evaporator.*
2. Another disadvantage of CO_2 is *high power requirement.* The H.P. required per tonne of refrigeration using CO_2 as refrigerant is *nearly twice* than any other commonly used refrigerant.
3. Another major *disadvantage* of this refrigerant is the *requirement of low temperature coolant in the condenser because of its low critical temperature (31ºC).*

Uses: This refrigerant has received only limited use because of the high power requirements per tonne of refrigeration and the high operating pressures. In *former years* it was selected for *marine refrigeration, for theatre air-conditioning systems* and for *hotel and institutional refrigeration* instead of ammonia because it is non-toxic.

- The *leak detection* of CO_2 is done by *soap solution.*
- *All hydrocarbon refrigerants are made from two elements, carbon and hydrogen.* The refrigerants of this group are *highly explosives and flammable in the presence of air. Most of them do not absorb moisture and they do not attack the metals used.* All of them are *extremely miscible with oil.* The main refrigerants of this group which are in used for different purposes are methane, ethane, butane, propane, ethylene and isobutane. *Ethane, methane and ethylene are used in low temperature applications. Isobutane was used for domestic purposes and now it is widely used in water desalting plants.*

5. Isobutane [$(CH_3)_3$ CH]:

Butane and isobutane are made from same number of hydrogen and carbon atoms but they have different chemical structure [Butane C_4H_{10} and *Isobutane* $(CH_3)_3$ CH]. Isobutane has more advantages over butane as refrigerant. *It has a low boiling point and high vapour density at atmospheric pressure compared with butane. The compressor ratio is also low compared with butane for the required temperature range.*

Properties:

1. It is inflammable in the presence of air.
2. It is highly miscible with oil.
3. It has slight sweetish odour.
4. It does not affect metals but does affect the rubber.
5. It is popular with trade name ***Freezol.***

The refrigerants of *chlorine of halogenated hydrocarbon are better* in thermodynamic properties than the straight hydrocarbons.

6. Methyl chloride (CH_3Cl) (R-40)

Properties:

1. It is a colourless liquid with a faint, sweet, non-irritating odour.
2. It has liquid specific gravity of 1.002.
3. It is flammable and explosive when mixed with air in concentrations between 8 to 17.5%.
4. It is corrosive to zinc, aluminium and magnesium and it forms the explosive compounds with the action on these metals, therefore these metals should not be used with this refrigerant.
5. It forms weak HCI in the presence of moisture which is corrosive to both ferrous and non-ferrous metals. Natural and synthetic rubbers are soluble in liquid methyl chloride, so these cannot be used as gasket materials with methyl chloride system.
6. It is oil miscible so oil return in methyl chloride system is simplified.

Uses: It has been used in the past in both domestic and commercial applications. *It should never be used with Aluminium.*

7. Methylene chloride (CH_2Cl_2) (R-30):

Properties:

1. It is non-flammable and non-toxic.
2. It is non-corrosive to metals in the presence of moisture.
3. The volume of vapour handled is considerably large as evaporator and condenser pressures are both below atmospheric pressure. The centrifugal compressors are adopted to handle large volume of vapour at low pressure. Oil miscibility of methylene chloride carries less importance. Oil and refrigerants do not come in contact with each other due to the use of centrifugal compressor.

Uses: The refrigerant was extensively used for air-conditioning theatres, audiotoriums and office buildings because of its safety properties.

- The refrigerants from ***fluorinated hydrocarbon group*** *meet all demands of an ideal refrigerant. These are used for ultra low to high temperature ranges.* The refrigerants of this group are normally *non-toxic, non-irritating, non-flammable and non-corrosive in the absence of moisture.* They do not act chemically with lubricating oil and maintain the required properties. The properties of few refrigerants of this group are described below.

8. R-11 (Trichloromonofluromethane) Freon-11 (CCl_3F):

It is a fluorocarbon methane series. It is a synthetic chemical product especially made for refrigeration.

Properties:

1. It is composed of one carbon, three chlorine and one fluorine atoms (or parts by weight) and *non-corrosive, non-toxic and non-flammable*.
2. It dissolves natural rubber.
3. It has a boiling point of +23.77°C at atmospheric pressure.
4. It mixes completely with mineral lubricating oil under all conditions.

5. This refrigerant is considered a low pressure refrigerant. It has *high specific volume* of suction vapour at any evaporator temperature rendering it *unsuitable for reciprocating compressors*. It has a low side pressure at –15ºC of 0.2025 bar and high side pressure of 1.2630 bar at +30ºC.
6. The latent heat at –15ºC is 195.66 kJ/kg.

Uses: It is employed for 50 tonnes capacity and over in small office buildings and factories. A centrifugal compressor is used in the plants employing this refrigerant.

- *Its leakage may be detected by an electronic detector or a halide torch.*

9. R-12 (Dichlodifluoromethane) or Freon-12 (CCl_2F_2):

This refrigerant is most widely used and must popular refrigerant of this group.

Properties:

1. It is *non-toxic, non-flammable, non-explosive and non-corrosive therefore, it is most suitable refrigerant.*
2. It is fully oil miscible therefore it simplifies the problem of oil return.
3. Its latent heat at –15ºC is 161.3 kJ/kg.
4. C.O.P. = 4.61.
5. It does not break even under the extreme operating conditions.
6. It condenses at moderate pressure and under atmospheric conditions, and boils at –29.8ºC at atmospheric pressure.
7. The operating pressures of R-12 in evaporator and condenser are 1.83 bar abs. at –15ºC and 7.44 bar abs. at +30ºC.
8. R-12 is more critical as to its water content as compared to R-22. It is soluble in oil down to –70ºC. Oil will begin to separate at this temperature and because it is lighter than refrigerant, will collect on the surface of the liquid.
9. This refrigerant is commercially available in different cylinder sizes. Generally 0.7 kg of refrigerant is required in refrigeration system per cubic metre of air-conditioned space.
10. This *refrigerant can be used with all types of condensing units, water-cooled, air-cooled and evaporative type.*

Uses:

1. It is suitable for high, medium and low temperature applications.
2. It is used for domestic applications.
3. It is excellent electric insulator, therefore, it is universally used in sealed type compressors.

- *Leaks may be detected by halide torch or electronic detector*. The cylinder colour code is *white.*

10. R-22 (Monochloro difluoro methane) or Freon-22 ($CHClF_2$):

It is a synthetic refrigerant specially developed for refrigeration installations that *need a low evaporating temperature.*

R-22 refrigerant is superior to R-12 in many respects. It has following *properties* and *uses*:

Properties:

1. The compressor displacement per tonne of refrigeration with R-22 is 60% less than the compressor displacement with R-22 as refrigerant.
2. R-22 is miscible with oil at condenser temperature but tries to separate at evaporator temperature when the system is used for very low temperature applications (–90ºC). Oil separators must be incorporated to return the oil from the evaporator when the system is used for such low temperature applications.
3. The pressures in the evaporator and condenser at standard ton of refrigeration are 2.9 bar abs. and 11.9 bar abs. (app.)
4. The latent heat at –15ºC is low and is 218.1 kJ/kg.
5. The pressures in the evaporator are above atmosphere for the evaporator temperatures between –30ºC to 40ºC with R-22 whereas the evaporator pressures with R-12 are below atmosphere for this temperature range.
6. Toxicity of R-22 is about same as that of CO_2 which we breathe all the time. The safe concentration of R-22 is 10.0 ppm, whereas the highest number for any substance except CO_2 is 5000 ppm.
7. Water is more soluble in R-22 than R-22 by a ratio 3 : 1. Therefore, water must be kept a minimum in these refrigerants, and consequently driers and desiccants should be used to remove moisture.
8. It has good solubility down to –10ºC. The oil will begin to separate at this point.
9. R-22 can be used with all types of condensing units, water-cooled, air-cooled and evaporative type. Normally the water-cooled and evaporative type condensers are designed for a condensing temperature of 38ºC corresponding to the pressure of 13.5 bar whereas air-cooled installations are designed for a condensing temperature of 49ºC corresponding to the pressure of 18 bar.

- The *major* **disadvantage** of R-22 *compared with R-12 is the high discharge temperature which requires water cooling of the compressor head and cylinder.*

Uses: R-22 is *universally used in commercially and industrial low temperature systems.*

- The *leaks may be detected with a soap solution,* a *halide torch or with an electronic detector.*
- The cylinder colour code for R-22 is green.

11. R-113 (Trichlorotrifluoroethane) or Freon-13 (CCl_2FCClF_2 or $C_2Cl_3F_3$):

Properties:

1. It has a boiling point of 47.6 ºC at atmospheric pressure.
2. It remains liquid at room temperatures and pressures and as such it can be carried in sealed tins rather than cylinders.

Uses: It is used in *commercial and industrial air-conditioning with centrifugal compressor systems.*

12. R-114 (Dichlorotetrafluoroethane) or Freon -14 ($CClF_2CClF_2$ or $C_2Cl_2F_4$):

Properties:

1. It is *non-toxic, non-explosive* and *non-corrosive even in the presence of water.*
2. It has a boiling point of 3.6º C at atmospheric pressures.

3. It evaporates at a pressure of 0.54 bar at –15ºC, and it condenses at a pressure of 1.5 bar at +30ºC.
4. Its latent heat of vaporisation at –15ºC is 143 kJ/kg.

Uses: It is used in *fractional power household refrigerating systems and drinking water-coolers* employing rotary-wave type compressors.

13. R-500

It is an azeotropic mixture of 73.8% R-12 (CCl_2F_2) and 26.2% of R-152 (CH_3CHF_2).

Properties:

1. It is *non-flammable, low in toxicity and non-corrosive.*
2. It has a fairly constant vapour pressure temperature curve which is different from the vaporising curves for either R-152 or R-12.
3. If offers about 20% greater refrigerating capacity than R-12 for the same size of motor when used of the same purpose.
4. The evaporator pressure is 1.37 bar at –15ºC and its condensing pressure is 7.78 bar at 30ºC.
5. It has a boiling point of –33ºC at atmospheric pressure.
6. Its latent heat at –15ºC is 192 kJ/kg.
7. The solubility of water in this refrigerant is highly critical. It has fairly high solubility with oil.

Uses: It is used in *both industrial and commercial applications but only in systems with reciprocating compressors.*

- *The leakage may be detected by using soup solution, halide torch an electronic leak detector or a coloured tracing agent.*
- The cylinder colour code for this refrigerant is *yellow*.

14. R-502

It is an azeotropic mixture of 48.8% R-22 ($CHClF_2$) and 51.2% of R-15 ($CClF_2CF_3$). It has replaced R-22 because of difficulties of high discharge temperature and poor oil return experienced with R-22.

Properties:

1. It is a *non-flammable, non-corrosive*, and *practically non-toxic liquid*. It is a good refrigerant for obtaining medium and low temperatures. It is suitable where temperatures from –18ºC to –51ºC are needed.
2. Its boiling point at atmospheric pressure is –46ºC.
3. Its evaporating pressure at –15ºC is 2.48 bar and the condensing pressure at 30ºC is 12.06 bar.
4. It latent heat at –29ºC is 168.6 kJ/kg.

The refrigerant R-502 has replaced R-22 because of difficulties of high discharge temperature and poor oil return experienced with R-22.

Two major advantages of R-502 compared with R-22 are:

(*i*) *Less compressor displacement* (which *reduces size and cost*).

(*ii*) *Less power input* (which *reduces running cost*).

— The added advantages of positive evaporator pressure at low temperature and simple compressor cooling have superseded R-22 and all new commercial low temperature cold stores use R-502 exclusively.

Uses: It is often used in *frozen food lockers, frozen food processing plants, frozen food display cases* and in *storage units for frozen foods and icecream.*

— It is *only used with reciprocating compressors.*

- R-502 has replaced many refrigerants because of the following *reasons*:
  (*i*) The compressor discharge temperatures and motor winding temperatures are lower compared with other refrigerants and lower temperatures in compressor mean generator reliability, longer life and less maintenance.
  (*ii*) Lower compression ratios.
  (*iii*) Lower oil temperatures.
  (*iv*) Greater capacity in low-temperature applications.
  (*v*) The capacity of R-502 is same as that of R-22 at air-conditioning temperatures and 10 to 20% greater than that of R-22 at low evaporating temperatures.
  (*vi*) The discharge temperature of R-502 is always sufficiently lower than R-22.
  (*vii*) The flow rate of R-502 is nearly 60% greater than R-22 so that the compressor gives 60% greater capacity where R-502 is used as refrigerant instead of R-22.
- The cylinder colour code for this refrigerant is *orchid.*

15. R-503

It is an azeotropic mixture of 40.1% R-23 (CHF_3) and 59.9% of R-13 ($CClF_3$).

Properties:

1. It is a *non-flammable, non-corrosive,* and *practically non-toxic liquid.*
2. Its boiling temperature at atmospheric pressure is –88ºC (which is lower than R-23 or R-13).
3. Its evaporating pressure at –15ºC is 17.15 bar.
4. Its critical temperature is 20º C and its critical pressure is 41.15 bar.
5. Its latent heat of vaporisation at atmospheric pressure is 173 kJ/kg.

Uses: It is good for use in the low state of cascade systems which require temperatures in the range of –73ºC to –87ºC.

- The leaks may be detected with the use of *soap solution,* a *halide torch or electronic leak detector.*
- The cylinder code for this refrigerant is *aquamarine.*

16. R-504

It is an azeotropic mixture of 48.2% R-32 (CH_2F_2) and 51.8% R-115 ($CClF_2CF_3$).

Properties:

1. It is *non-flammable, non-corrosive* and *non-toxic.*
2. The boiling temperature at atmospheric pressure is –57ºC.
3. Its evaporating pressure at –15ºC is 5.88 bar and its critical pressure is 48 bar.

Uses: It is used in *industrial processes where a low temperature range of –40°C to –62°C is desired.*

- The leakage of the refrigerant may be detected by using scap *solution, halide torch or an electronic leak detector.*
- The cylinder colour code for this refrigerant is *tan*.

Substitutes for Chlorofluorocarbons (CFCs)

- Presently the most commonly used refrigerants, besides ammonia, are chlorofluorocarbons, often referred to as CFCs, also called Freons under the trade name of E.I. du Point de Nemours and Co. Internationally designated as R-11, R-12, R-22 etc., they are some of the chorofluoro derivatives of methane and ethane. Thus we have a series of refrigerants with widely different normal boiling points to suit the varied requirements of different applications.
- The *ozone layer* of earth in the upper atmosphere is needed for the absorption of harmful ultra violet rays from the sun which are responsible for causing skin cancer. The CFCs have been linked to the depletion of this ozone layer. They have varying degree of ozone depletion potential (ODP). In addition they act as *greenhouse gases*. Hence, they have global warning potential (GWP) also. It is established that *it is the chlorine atom in the molecule which is responsible for the ODP.*
- Hydroflurocarbons (HFCs) provide an alternative to fully hydrogenated CFC refrigerants. They contain *no chlorine atom* at all and, therefore have zero ODP. Even hydrochlorofluoro carbons (HCFCs) which contain some Cl-atoms, but in association with H-atoms, have *reduced* ODP. However, because of the H-content of HFCs, these *may be slightly inflammable* (Pure hydrocarbons are of course, highly inflammable.
- The common refrigerants for which substitutes have to be found are: R-11, R-12, R-113. Of these, R-11 and R-12 have equally high ODP. (R-22 which is an HCFC has one twentieth ODP of R-11 and R-12. Hence it can continue to be used for some time). However, all these have the GWP. Because of this even R-22 will have to be ultimately phased out.

— At present, the following substitutes are available.

(*i*) HCFC **R-123** ($CHCl_2-CF_3$) in place of CFC *R-11* (CCl_3F)

(*ii*) HFC *134a* (CH_2F-CF_3) and isobutance in place of CFC *R-12* (CCl_2F_2)

(*iii*) R-69S in place of R-22 and R-502.

9.7 SECONDARY REFRIGERANTS

- There are several applications which forbid the direct use of a refrigerant due to safety consideration.
  — As an *example*, the toxic refrigerants cannot be used for the air-conditioning of residential buildings. Similarly the amount of refrigerant for the circulation in a big cold storage would be so large that its cost may be comparable with the refrigeration systems. Under such circumstances the cheaper grade suitable cooling media such as water, brine solution of sodium chloride or calcium chloride etc., is selected to have indirect cooling.
- The ***advantages*** of *using secondary refrigerants* are as follows:

  (*i*) Different temperatures can be maintained in different parts of the building by controlling the amount of brine flowing to that part.

(*ii*) Easy to handle and easy to control.

(*iii*) The size of pipeline is considerably smaller.

(*iv*) It keeps the toxic refrigerant away from the load place.

- The commonly used, secondary refrigerants are:

— Water;

— Sodium chloride brine;

— Calcium chloride brine;

— Ethylene glycol, Propylene glycol etc.

1. Water:

- Water is universally used as secondary refrigerant when the required *temperature to be maintained is above the freezing point of water.*
- It is used in *air-conditioning plants* and *industrial cooling installations.* In air-conditioning plants, chilled water is used for cooling and dehumidifying the air either with the help of cooling or with the help of water spray unit.

2. Brines:

Brine is a *solution of a salt in water.* It may be noted that when a salt is mixed in water, then the freezing *temperature* of the solution becomes *lower than that of the water.* This is due to the fact that the salt while dissolving in water takes off its latent heat from the solution and cools it below the freezing point of water. The mass of the salt in the solution expressed as the percentage of the mass of the solution is known as *'concentration'* of the solution. *As the concentration of the solution increases, its freezing point decreases.* But if the concentration of the salt is increased beyond a certain point, the freezing point increases instead of decreasing. The point at which the freezing temperature is minimum, is known as *'eutectic temperature'* and the concentration at this point is known as *'eutectic concentration'*. The brine used in a particular application should have a concentration for which the freezing point of the brine is at least *5°C to 8°C lower than the brine temperature required.*

The brines commonly used are: *Sodium chloride (NaCl), calcium chloride* and *glycols* such as ethylene glycol, propylene glycol etc.

— *Sodium chloride* brine solutions are most common upto about –15°C while *calcium chloride* brine can be used upto –50°C. Unfortunately these solutions are very *corrosive* to metals such as brass, copper, aluminium etc. That is why mild steel pipings and mild steel sheets are used for such systems. However, mild steels sheets and also corroded to a certain extent. In order to increase the life of the brine solution containers, usually *lead oxide or similar corrosion resisting paints are coated over the tank surfaces time to time once or twice a year on regular basis.*

- In some cases instead of salt solution some liquid chemicals often called *antifreeze are mixed with water in varying proportions to suppress the freezing temperature, 0°C. Ethylene and propylene glycol* have a number of good properties. Since they are *non-corrosive and non-electrolytic even in the presence of water*, therefore, these brines are *most extensively used as antifreeze elements.*

9.8 COMPARISON OF REFRIGERANTS

A. Table 9.14 gives the Comparison of Some Common Refrigerants.

Table 9.14 Comparative Study of Common Refrigerants

S. No.	Aspects	Refrigerants				
		Ammonia	*Carbon dioxide*	*Freon-12 or R-12*	*Sulphur dioxide*	*Methyl chloride*
1.	Chemical formula	NH_3	CO_2	CCl_2F_2	SO_2	CH_3Cl
2.	Molecular weight	17	44	120.9	64	50.5
3.	Boiling point at atmospheric pressure (ºC)	– 33	– 78.5	– 29.8	– 10.0	– 23.7
4.	Evaporator pressure at – 15ºC (bar)	2.36	22.9	1.826	0.807	1.46
5.	Condenser pressure at 30ºC (bar)	11.67	71.95	7.44	4.62	6.53
6.	Cost	Very cheap	Cheap	Very costly	Cheap	Costly
7.	Flammability	Yes	Nil	Nil	Yes	Yes
8.	Explosion risk	Yes	Nil	Nil	Nil	Yes
9.	Toxicity	High	Medium	Low	High	Medium
10.	Odour	Pungent	Nil	Sweat	Pungent	Sweat
11.	Corrosion	On non-ferrous metal	Nil	Nil	Rubber ferrous metals	Rubber, Al.
12.	Chemical stability	Good	Excellent	Excellent	Excellent	Excellent
13.	Specific volume at –15ºC saturated vapour (m^3/kg)	0.509	0.017	0.093	0.406	0.279
14.	Latent heat at –15ºC (kJ/kg)	1310.6	272.8	161.3	393.55	418.12
15.	C.O.P.	4.76	2.56	4.61	4.73	4.85
16.	H.P. per ton of refrigeration	0.899	1.84	1.0	0.995	0.962
17.	Miscibility with oil	Small	Nil	Large	Small	Great
18.	Electrical insulation	Poor	Good	Excellent	Poor	Poor
19.	Critical temperature (ºC)	132.6	31.0	111.7	142.5	157.2
20.	Order of preference for small plants	—	—	First	Third	Second
21.	Order of preference for large plants	First	Third	—	Second	—

B. Comparison of Physical Properties of Conventional and Future Refrigerants: Refer to Table 9.15

Table 9.15

	R-11	*R-123*	*R-12*	*R-134*	*R-22*	*R-502*	*R-69S*	*R-717*
Formula	CCl_3F	$CHCl_2CF_3$	CCl_2F_2	CH_2FCF_3	$CHClF_2$	$CHClF_2$ + CF_3 $CClF_2$	$CHClF_2$ $+C_3F_8$ C_3H_8	NH_3
Molecular weight	137	153	121	102	86.5	112	86	17
Boiling point (°C)	23.8	27.1	– 29.8	– 26.22	– 40.8	– 45.6	– 43.3	– 33.3
$\gamma = \frac{c_p}{c_v}$	1.136	1.11	1.139	1.116	1.18	1.135	1.15	1.29
Flammability (vol % in air)	Nil	Nil	Nil	Nil	Nil	Nil	Nil	16 to 25
Toxicity	No		No	No	No	No	No	Yes
(TLV)	(1000)	(5 – 10)	(1000)	(1000)	(1000)	—	—	(25)
Latent heat (kJ/kg) at 15°C	194	175	159	205	217	156	195	1314
ODP	1.00	0.02	0.98	0.00	0.05	0.23	0.042	0.00
HGWP	1.00	0.02	3.05	0.285	0.365	5.1	1.26	0.00
Availability	Yes	Yes	Yes	Yes	Yes	Yes	Yes	Yes
Ideal C.O.P.	7.78	7.63	6.91	6.77	7.06	—	—	7.28

C. Comparison of R-69S with R-22 and 502: Refer to Table 9.16.

Table 9.16

S.No.	*Aspects*	*R-69S*	*R-22*	*R-502*
1.	*Suction pressure (bar)*	3.00	2.8	3.1
2.	*Discharge pressure (bar)*	11.25	10.5	10.6
3.	*Pressure ratio*	3.74	3.75	3.4
4.	*Discharge temperature (°C)*	73.0	77.6	61.6
5.	*Motor winding temp (°C)*	67.60	81.6	74.6
6.	*Power consumption (watts)*	652	630	676.0
7.	*Refrigerating effect (watts)*	1578	1270	1580.0
8.	*C.O.P.*	2.42	2.02	2.34
9.	*Ice production (kg/24-hrs)*	420	400	419.0

9.9 APPLICATIONS OF REFRIGERANTS

Table 9.17 Freon Group Applications

S.No.	*Refrigerant*	*Applications*	*Compressor type*
1.	*Freon-11 (R-11)*	Air-conditioning system ranging from 200 to 2000 tonnes in capacity. It is used where low freezing point and non-corrosive properties are important.	Centrifugal
2.	*Freon-12 (R-12)*	Employed for air-conditioning plants, refrigerator frozen food and ice cream cabinets, food locker plants, window air-conditioners, automobile air-conditioning.	Reciprocating, centrifugal, rotary
3.	*Freon-13 (R-13)*	For low temperature refrigeration up –90ºC in cascades systems.	Reciprocating
4.	*Freon-14 (R-14)*	For low temperature upto –130ºC in triple cascade system.	Reciprocating
5.	*Freon-22 (R-22)*	Mainly used in commercial air-conditioning. Also used in frozen food storages, frozen food plants, display cases and low temperature applications.	Reciprocating, centrifugal
6.	*Freon-113 (R-113)*	For small to medium air-conditioning systems and industrial cooling.	Centrifugal
7.	*Freon-114 (R-114)*	For household refrigerators and rotary compressors used in large industrial process cooling.	Centrifugal, rotary
8.	*Freon-500 (R-500)*	For commercial air-conditioning and house-hold refrigeration mostly where 50 cycle current is common.	Reciprocating
9.	*Freon-502 (R-502)*	For frozen food and ice cream display cases and ware houses, food freezing plants, truck refrigeration and heat pumps.	Reciprocating
10.	*Freon-503 (R-503)*	For low temperature application upto –90ºC.	Reciprocating

Table 9.18 Applications of Common Refrigerants

S.No.	*Refrigerants*	*Uses*
1.	R-11, R-113, CH_2Cl_2 (*low pressure and high temperature refrigerants*)	For air-conditioning theatres, offices, factories, auditoriums. Also used for industrial cold water supply used for equipment cooling and distillation.
2.	R-12, SO_2, R-500, CH_3Cl and Isobutane (*Intermediate pressure and intermediate temperature refrigerants*)	For domestic purposes as house-hold refrigerators, water coolers and small air-conditioners. Also used for commercial purposes as display cases, refrigerated carriers etc.
3.	NH_3, R-12, R-22 and CO_2 (*High pressure and low temperature refrigerants*)	For domestic and commercial freezers, industrial purposes as cold storage plants and ice cream factories.
4	R-13, R-14, C_2H_6 and Ethylene (*High pressure and ultra low temperature refrigerants*).	For liquefaction of gases, metallurgical operations and in aerodynamic wind tunnels.

9.10 LEAK DETECTION

One of the major problems in refrigeration equipment is the leak detection. Depending upon the pressure maintained in the system the leak may be inwards or outwards. Due to 'leak' the *cooling capacity of the system is reduced*, therefore it becomes necessary to detect the leaks immediately and repair the leaking spots or joints at once.

Leak detection methods for some important refrigerants are given below:

Ammonia. Leak in ammonia system can be located in the following *three ways*:

(*i*) **Sulphur test.** Melt some sulphur flowers with slow heat and dip short pieces or heavy strings into sulphur and dry it. A sulphur fumigating candle can be used. In order to locate the leak, light the candle and hold it near the suspected leak. *A white vapour will appear when the sulphur fumes come in contact with ammonia.*

(*ii*) **Litmus paper.** *Litmus paper will turn blue if ammonia is present or leaking* when it is brought to the suspected leak point or condenser water containing ammonia.

(*iii*) **Test paper.** Since ammonia is alkaline, a soft paper can be dipped in a solution consisting of about half a gram of phenolphthalene in a pint of alcohol. This paper can be allowed to dry and then cut up ir small strips. To use the test paper, wet it with water and hold it near suspected leak. If there is any ammonia present, the *paper will turn pink*. This is the *best method of checking leaks into condenser water.*

Freon-12 (R-12). '*Halide torch*' may be used to detect the *leak of freon vapour*. This halide torch detector (acetylene) produces nearly colourless flame. Air for combustion is drawn through a tube at the base of the burner. The flame burns through an element of metallic copper. The air tube is held near the suspected leak and if any Freon is present, the flame changes to brilliant green. This type of torch is able to detect *Freon concentration as low as 0.01 per cent*. Another type of leak detector burns alcohol, but operates on the same principle. The detector which is of pump type similar to blow torch is not satisfactory because, if there is any Freon vapour in the air round the torch, the flame will burn green continuously. The leak detection described for Freon-12 is applicable for all Freon group refrigerants namely Freon-11 (R-11), Freon 22 (R-22) etc.

In case *Freon has escaped into the room in such a large quantity that the detector burns with green flame continuously then a soap method must be used.*

- **Electronic leak detectors** are *available which produce buzz when brought near the point of leakage. The intensity of buzz is proportional to the leakage rate.*

9.11 DRIER OR DEHYDRATOR

- **Drier or dehydrator** *is a device employed to remove moisture from a refrigerant.* It is generally placed in a liquid line. The drying material used is called the *desiccant*. A desiccant (a solid substance) is capable of removing moisture from a gas, liquid or solid.
- A drier is in the shape of shell made of steel of brass. It is provided with inlet and outlet openings. A strainer is provided at the outlet of the shell. It is charged with moisture absorbing chemicals namely: (*i*) Silica gel, (*ii*) Activated alumina, (*iii*) Calcium sulphate and (*iv*) Calcium chloride.

There are two types of driers:

(*i*) Sealed type

(*ii*) Refillable type.

In ***sealed type*** of drier the shell charged with moisture absorbing chemical is permanently sealed. Here the *moisture absorbing chemical cannot be replaced if it loses its effectiveness*. This type of drier is used on *small sealed units where the complete* drier will be replaced if found ineffective.

In ***refillable type*** of drier there is a provision for *filling the moisture absorbing chemical as and when required*.

> **Note.** The following liquids are injected in a refrigerating system for absorbing moisture:
>
> (*i*) *Alcohol,* (*ii*) *Thaizone* and (*iii*) *Maizone*.

Effects and causes of moisture in refrigerating systems:

The presence of moisture in a refrigerating system can produce the following detrimental *effects*:

(*i*) A single small drop of water present in the system may *freeze and clog the expansion valve or capillary tube and stop the circulation of refrigerant*.

(*ii*) The moisture also *causes corrosion* which damages the tubing, valves, seals, etc. and thus reduces the life of equipment.

(*iii*) The *presence of moisture in hermetic units may causes short-circuiting of motor winding.*

(*iv*) Moisture *deteriorates the quality of lubricating oil* in the refrigeration system.

The probable causes of moisture in a refrigeration system are:

(*i*) The equipment may be *incompletely* dried before charging.

(*ii*) Moisture entering into the system in dissolved form in liquid refrigerant during its charging.

(*iii*) Air leakage on low pressure side. Air always contains some quantity of water vapour in it.

(*iv*) Water leakage in system in the watercooled condensers.

(*v*) Formation of water due to oxidation of certain hydrocarbons in lubricating oil.

Removal of moisture from the system:

When moisture is present in the system it can be *removed* by using the *following methods*:

(*i*) By heating the entire assembly.

(*ii*) By circulating the dehumidified air through the system.

(*iii*) By creating vacuum within the system using mechanical vacuum pump and removing moisture by the combined use of alcohol and carbon-dioxide.

(*iv*) *Small quantity of water can be removed satisfactorily by installing a suitable drier and strainer which carries a drying agent. A drier is placed in liquid line between condenser and expansion valve.*

Two types of *drying agents/driers* commonly used are:

(*i*) **Chemical driers.** These are very *quick in action*, so these should however never be left in system longer than 24 hours after their installation. Some common chemical driers are: *Anhydrous calcium sulphate* (Drierite), *calcium oxide or quick lime and calcium chloride*.

(*ii*) **Absorbent type driers.** These chemicals remove moisture *slowly than chemical driers*. Therefore, these are left in system for long periods and sometimes permanently. *Examples: Granular aluminium oxide* (activated alumina) and *silicon dioxide* (silica gel).

9.12 LUBRICANTS

9.12.1 Characteristics of a Good Lubricant

A good lubricant to be used in a refrigeration system should possess the following *properties*:

(*i*) It should have *minimum oil-foaming tendency*.

(*ii*) It should have *a very low acidity*.

(*iii*) It should have *high miscibility* with refrigerant used in the system.

(*iv*) is should have *high flash point*.

(*v*) It should have *low pour point*.

(*vi*) It must be free *from dissolved air*.

(*vii*) It must have *low solubility in water*.

(*viii*) It should have *low sulphur content*.

(*ix*) It should be *free from wax separation* upon a lower temperature level.

9.12.2 Lubrication Methods

In a refrigeration system the only parts which need lubrication are: (*i*) *Compressors* and (*ii*) *Bearings of motors*. The two methods used for lubrication of these parts are: (*i*) *Splash lubrication* and (*ii*) *Pressure lubrication*.

1. Splash lubrication. This method is commonly employed for compressors of small capacity. Here crankcase is filled with the correct oil upto the bottom of the main bearings or to the middle of the crankshaft main bearings. At each crankshaft revolution, the crank-throw or the eccentric dips into the oil and splashes it around the inside of the compressor. Oil is thrown on to the cylinder walls, piston pin bearings and into small openings where it can drain into the main bearings. Some compressor connecting rods have little dips or scoops attached to the lower ends which scoop up small amounts of oil and sling it around other parts to be lubricated.

2. Pressure lubrication. In pressure (or forced feed) lubrication system a *small oil pump is used to force oil to main bearings, lower connecting rod bearings* and in some *piston pins*. In this system, besides providing pump, the crankshaft and connecting rod must have oil passages drilled in them.

- *Mineral oil is a commonly used lubricant in a refrigeration system for lubrication purposes.*

HIGHLIGHTS

1. A *refrigerant* is defined as any substance that absorbs heat through expansion or vaporisation.
2. *Primary refrigerants* are those working mediums or heat carriers which directly take part in the refrigeration system e.g., Ammonia, Carbon-dioxide, Sulphur-dioxide, Methyl chloride, Methylene chloride, Ethyl chloride and Freon group etc.

Secondary refrigerants are those circulating substances which are first cooled with the help of the primary refrigerants and are then employed for cooling purposes e.g., Ice, Solid carbon-dioxide etc.

3. *Desirable properties of an ideal refrigerants are*:

(*i*) It should be non-toxic non-injurious to human being and should have no harmful effect on taste and colour of the food stuffs.

(*ii*) The freezing point must be lower than any temperature likely to encounter in the system.

(*iii*) The latent heat of vapourisation must be high.

(*iv*) The volume of the refrigerant should be as small as possible.

(*v*) It should not attack any substance in the system.

(*vi*) It should be miscible with oil so as to carry it in a sufficient quantity for lubrication.

(*vii*) It should provide a high rate performance.

(*viii*) It should be cheap and easily available.

(*ix*) It should make possible to leak detection by simple means.

4. *Drier or dehydrator* is a device employed to remove moisture from a refrigerant. The drying material used is called '*Desiccant*'.

Two types of driers commonly used are:

(*i*) Absorbent type driers: Granular aluminium oxide and silica gel (Silicon dioxide).

(*ii*) Chemical driers: Anhydrous calcium sulphate, calcium oxide or quick lime and calcium chloride.

5. The following methods are used to lubricate the various parts in a refrigeration system.

(*i*) Splash lubrication

(*ii*) Pressure lubrication.

OBJECTIVE TYPE QUESTIONS

A. Choose the Correct Answer:

1. The refrigerant R-744 stand for

(*a*) Ammonia (*b*) Carbon dioxide (*c*) Sulphur dioxide (*d*) Methyl chloride.

2. Among the refrigerants, namely R-11, R-22, R-12 and R-717, the latent heat of vaporisation at a particular temperature is the highest for

(*a*) R-12 (*b*) R-22 (*c*) R-11 (*d*) R-717.

3. A good refrigerant should have

(*a*) low specific heat ratio (γ) (*b*) high latent heat

(*b*) high thermal conductivity (*d*) all of the above.

4. Low boiling point refrigerants are preferably suited for

(*a*) reciprocating compressors (*b*) large centrifugal compressors

(*c*) large axial flow rotary compressors (*d*) all of the above.

5. Among the refrigerants R-717, R-22, R-12 and R-11, the ratio of condenser to evaporator pressures –15°C and 30°C range is highest for

(*a*) R-717 (*b*) R-22 (*c*) R-12 (*d*) R-11.

6. Freon-12 is used with

(*a*) reciprocating refrigerant compressors in small units like domestic refrigerators

(*b*) reciprocating refrigerant compressors in large units for central air-conditioning plants

(c) centrifugal refrigerant compressors for larger units for central air-conditioning plants.
(d) all of the above.

7. The refrigerant which gives evaporator pressure
(a) as high as possible is desirable (b) as low as possible is desirable
(c) near about atmospheric pressure but below atmospheric pressure is desirable
(d) near about atmospheric pressure but above atmospheric pressure is desirable.

8. C.O.P. of refrigeration cycle for a fixed evaporator temperature and fixed condenser temperature
(a) is higher for refrigerant having high latent heat of vaporisation and high ratio of specific heats (γ)
(b) is higher for refrigerants having small latent heat of vaporisation and small ratio of specific heats (γ)
(c) is higher for refrigerants having high latent heat of vaporisation and low ratio of specific heats (γ)
(d) is independent of the ratio of specific heats (γ) and latent heat of vaporisation.

9. The refrigerant R-764 stands for
(a) Ammonia (b) Carbon dioxide (c) Sulphur dioxide (d) Methyl chloride.

10. Ammonia is used with
(a) reciprocating refrigerant compressors (b) centrifugal refrigerant compressors
(c) axial flow refrigerant compressors (d) all of the above.

11. Freon-22 is used with
(a) reciprocating refrigerant compressors in large units such as package units or central air-conditioning plants
(b) centrifugal refrigerant compressors in large units and central air-conditioning plants.
(c) only small domestic refrigerators or water-coolers
(d) all of the above.

12. Freezing point temperature for R-717 (NH_3) is
(a) –94ºC (b) –77.8ºC (c) –157.5ºC (d) –160ºC.

13. For high C.O.P.
(a) critical temperature in general should be low
(b) critical temperature in general should be high
(c) critical temperature is insignificant
(d) critical pressure is important.

14. It is desirable for refrigerant to have
(a) high thermal conductivity for high heat transfer co-efficient.
(b) low thermal conductivity for low heat transfer co-efficient
(c) any thermal conductivity as it does not affect performance
(d) all of the above depending upon components of plant.

15. The refrigerant R-717 stands for
(a) Ammonia (b) Carbon dioxide (c) Sulphur dioxide (d) Methyl chloride.

16. The following refrigerant in not miscible with oil.
(a) R-717 (b) R-11 (c) R-22 (d) R-12.

17. For fixed pressure ratio the isentropic compression work is the lowest for
(a) R-11 (b) R-12 (c) R-22 (d) R-717.

18. The refrigerant R-22 stands for
(a) $CFCl_3$ (b) CF_2Cl_2 (c) CF_3Cl (d) CHF_3Cl.

19. High boiling point refrigerants are preferably suited for
(a) Reciprocating compressors (b) Centrifugal compressors
(c) Small screw type compressors (d) For all the above.

20. For a fixed pressure ratio, the isentropic compressor work is the highest for
(*a*) R-717 (*b*) R-11 (*c*) R-22 (*d*) R-12.

21. Refrigerant R-12 as compared to R-717 is
(*a*) costly (*b*) cheap (*c*) same cost (*d*) none of these.

22. Freon-11 is used with
(*a*) reciprocating refrigerant compressors for central plants
(*b*) centrifugal refrigerant compressors for large capacity central plants.
(*c*) window type air-conditioning units
(*d*) all of the above.

23. For vapour compression refrigeration system using R-717 (NH_3) the materials of pipes are
(*a*) Iron and steel (*b*) Copper (*c*) Brass (*d*) Aluminium.

24. The refrigerant R-290 stands for
(*a*) C_2H_4 (*b*) C_2H_8 (*c*) C_4H_{10} (*d*) $(CH_2)_2$ CH.

25. The following refrigerants is most miscible with oil
(*a*) R-717 (*b*) R-11 (*c*) R-22 (*d*) R-12.

ANSWERS

1. (*b*)	2. (*d*)	3. (*d*)	4. (*a*)	5. (*d*)	6. (*a*)	7. (*d*)	8. (*c*)
9. (*c*)	10. (*a*)	11. (*a*)	12. (*b*)	13. (*b*)	14. (*a*)	15. (*a*)	16. (*a*)
17. (*a*)	18. (*c*)	19. (*b*)	20. (*a*)	21. (*a*)	22. (*b*)	23. (*a*)	24. (*b*)
25. (*a*)							

B. Fill in the Blanks:

1. .... is any substance that absorbs heat through expansion or vaporisation.
2. .... refrigerants are those working mediums or heat carries which directly take part in the refrigeration system.
3. .... refrigerants are those circulating substances which are first cooled with the help of the primary refrigerants and are then employed for cooling purposes.
4. .... should be as large as possible to reduce the weight of the refrigerant to be circulated in the system.
5. .... pressure of a refrigerant should be low so as to give low condensing pressure.
6. .... is not used in comfort air-conditioning and in domestic refrigeration because of inflammability and toxicity.
7. .... compressors are used with low pressure and large volumes of suction vapour.
8. For a high heat transfer co-efficient a .... viscosity is desirable.
9. .... is the cheapest refrigerant.
10. .... may be used to detect leak of Freon vapour.

ANSWERS

1. Refrigerant;	2. Primary;	3. Secondary	
4. Latent heat of vaporisation;	5. Critical;	6. Ammonia;	7. Centrifugal;
8. low;	9. Ammonia;	10. Halide torch.	

THEORETICAL QUESTIONS

1. Define a 'refrigerant'. How are refrigerants classified?
2. Differentiate between 'Primary' and 'Secondary' refrigerants.
3. How are refrigerants designated?
4. Can water be used as a refrigerant?
5. Discuss why refrigerators are so selected that evaporator pressures and condenser pressures are greater than atmospheric pressure?
6. Write down the chemical formulae for the following refrigerants:

   R-11, R-12, R-21, R-170, R-764, R-717.
7. Discuss the physical and thermodynamic properties of the following refrigerants and how the leaks are detected in each case:

   R-11, R-12, R-22, R-717.
8. Enumerate the desirable properties of an ideal refrigerant.
9. State the properties and uses of any three of the following refrigerants:
   (*i*) Ammonia
   (*ii*) Sulphur dioxide
   (*iii*) Carbon dioxide
   (*iv*) R-11 (Trichloromonofluoromethane)
   (*v*) R-12 (Dichlorodifluoro methane)
   (*vi*) R-22 (Monochlorodifluoro methane).
10. State how will you carry out 'leak detection' in the following refrigerants?
    (*i*) Ammonia
    (*ii*) R-12 (Freon-12).
11. Name three refrigerants that are commonly used in commercial refrigerators. Discuss their relative merits and demerits.
12. State the factors which are considered in selecting a refrigerant in the following systems and mention the name of refrigerant used in each system.
    (*i*) Air-conditioning plant for picture house;
    (*ii*) Household refrigerator;
    (*iii*) Ice-producing plant.
13. Which refrigerants require centrifugal compressors? Where these refrigerants are in practice? Give four examples.
14. What is a secondary refrigerant? Where is it used?
15. What is the critical temperature of a refrigerant?
16. What is a drier/dehydrator? Explain briefly refillable type of drier.
17. State the effects and causes of moisture in refrigeration systems.
18. How can moisture be removed from a refrigeration system?
19. State the characteristics of a good lubricant.

20. Discuss briefly the following systems of lubrication:

    (*i*) Splash lubrication system

    (*ii*) Pressure lubrication system.

21. (*a*) Give the chemical formulae for the following refrigerants:

    (*i*) R-12 (*ii*) R-22 (*iii*) R-113 (*iv*) R-717.

    (*b*) Why are R-12 and R-22 used in hermetic compressors?

    (*c*) Why is R-12 system susceptible to expansion valve freeze up?

    (*d*) Among R-12, R-22 and R-21 systems, which one requires an oil separator and why?

    (*e*) Amongst R-11, R-12, R-21 and R-22 which are best suited for centrifugal compressors and why?

    (*f*) Why is evaporator pressure kept above atmospheric in most refrigeration systems?

Solution: (*a*) The chemical formula for the refrigerants are:

R-12 : CCl_2F_2

R-22 : $CHClF_2$

R-113 : CCl_2FCClF_2 or $C_2Cl_3F_3$

R-717 : NH_3

(*b*) R-12 and R-22 have favourable properties with regards to evaporator and condenser pressures for use in domestic applications. Further these refrigerants satisfy all safety requirements, and hence are ideally suited for window air-conditioners and home refrigerators. Hermetic compressors are suitable only for small units (due to cooling problem of compressor motors) like domestic refrigerators and window air-conditioners. Therefore *R-12 and R-22 are used in hermetic compressors.*

(*c*) Water is only slightly soluble in R-12. At –18°C it will hold 6 parts per million of water. Hence if any water is present it will freeze up in expansion valve. To prevent this dryer is used before the expansion valve in R-12 system.

(*d*) R-22 requires an oil separator. It is miscible with lubricating oil at condenser temperature, but tries to separate at evaporator temperature. Oil separator is used to ensure the return of oil from the evaporator to compressor.

(*e*) R-11 and R-21 are best suited for centrifugal compressor due to low evaporation pressures and large volumes to be handled. Also big size air-conditioning plants using R-12 where large volume is to be handled employ centrifugal compressors.

(*f*) The evaporator pressure is kept above atmospheric (i.e., positive pressure) to prevent the leakage of air and moisture into the system.

❑❑❑

Refrigeration Components and Controls

Chapter 10

10.1 INTRODUCTION

In a vapour compression system the following are the *major refrigeration components*:

1. Compressors
2. Condensers
3. Evaporators

Refrigeration controls *are the devices used to control the flow of refrigerant at various points throughout the refrigeration cycle.* The following are the *six* basic types of *refrigerant flow controls*:

(*i*) Hand expansion valve

(*ii*) Automatic expansion valve

(*iii*) Thermostatic expansion valve

(*iv*) Capillary tube

(*v*) Low-side float

(*vi*) High-side float.

10.2 REFRIGERANT COMPRESSORS

A **compressor** is *the heart of vapour compression system.* It pumps and circulates refrigerant through the system just as the heart pumps and circulates blood through the body. In general, a compressor performs the following *functions*:

(*i*) It removes low temperature and low pressure vapours from the cooling coil through the line called ***suction line***.

(*ii*) It compresses these vapours by increasing the pressure and temperature resulting in an increase of boiling point of the refrigerant.

(*iii*) It discharges the vapours in high temperature and pressure form to condenser through the line called ***discharge line***.

The refrigeration compressor is different from those used in other applications, such as air or gas compressors. The function of the air or gas compressors is totally unrelated to the use to which the compressed air or gas is put to. So their capacity, susceptibility to breakdown, etc., are not governed by what happens to the compressed fluid once it is discharged to the receiver. On the other hand, the function of the refrigeration compressor, being an integral part of the system, is related to other components. Hence its capacity, life, breakdown, etc. are very much influenced by the performance of the other components like condenser, evaporators, throttling device etc.

Types of Refrigeration Compressors:

Types of refrigeration compressors used are:

1. ***Reciprocating compressors :*** These are very suitable for refrigerants (e.g., R-12, R-22 R-502, R-717 etc.) which require *relatively small displacement volumes and high condensing pressures*.
2. ***Centrifugal compressors.*** These are very suitable for refrigerants (e.g., R-11, R-12, R-113 etc.) which require *relatively large displacement volumes and high condensing pressures*.
3. ***Rotary compressors.*** These are very suitable for refrigerants (e.g., R-21, R-114) having *moderate or low condensing pressure*. These are mainly used for domestic refrigerators.
4. ***Screw compressors.***

The general arrangement of a compressor set is shown diagrammatically in Fig. 10.1.

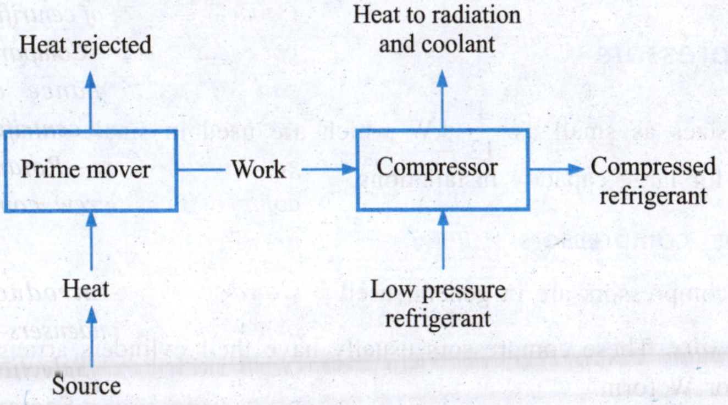

Fig. 10.1. General arrangement of a compressor set.

Important terms:

Important terms relating compressors are given below:

1. **Suction pressure (p_s).** The absolute pressure of refrigerant at the inlet of a compressor is termed as *suction pressure*.
2. **Discharge pressure (p_d).** The absolute pressure of refrigerant at the outlet of a compressor is called *discharge pressure*.
3. **Compression ratio (or pressure ratio).** The ratio of the absolute discharge pressure (p_d) to the absolute suction pressure (p_s) is called *compression ratio* (*or pressure ratio*). It value is more then unity (since $p_d > p_s$). The compression ratio may also be defined as the total cylinder volume to the clearance volume.
4. **Suction Volume ($V_{suc.}$).** It is the volume of refrigerant sucked by the compressor during its suction stroke.
5. **Swept volume or stroke volume or piston displacement volume (V_s).** It is the volume swept by the piston when it moves from its top or inner dead position to bottom or outer dead centre position.

$$V_s = \frac{\pi}{4} \times D^2 \times L$$

where, D = Diameter of the cylinder, and

L = Length of piston stroke.

6. **Clearance factor.** It is the ratio of clearance volume (V_c) to the piston displacement/swept volume (V_s).

i.e., $$C = \frac{V_c}{V_s}$$

7. **Compression capacity.** It is the volume of the actual amount of refrigerant passing though the compressor in a unit time. It is equal to suction volume ($V_{suc.}$).
8. **Volumetric efficiency ($\eta_{vol.}$).** It is the ratio of compressor capacity or the suction volume ($V_{suc.}$) to the swept volume (V_s).

i.e., $$\eta_{vol.} = \frac{V_{suc.}}{V_s}$$

The volumetric efficiency of a good compressor is 70% to 80%.

10.2.1 Reciprocating Compressors

These compressors are available in sizes as small as $\frac{1}{12}$ kW which are used in small domestic refrigerators and upto about 150 kW for large capacity installations.

10.2.1.1 Types of reciprocating compressors

The following types of reciprocating compressors are in general use:

(*i*) *Single acting vertical compressors.* These compressors usually have their cylinders arranged vertically, radially or in a V or W form.

(*ii*) *Double acting horizontal compressors.* They have their cylinders arranged horizontally.

Reciprocating compressors may also be *subdivided* as follows:

(*i*) Open type;

(*ii*) Semi-sealed type;

(*iii*) Hermetically sealed type.

(*i*) **Open type vertical reciprocating compressor:**

Fig. 10.2 shows the working of an open type vertical reciprocating compressor. It is *flexible* in the sense that by varying speed of the compressor, different capacities can be obtained. It can be operated by any type of prime mover, such as an electric motor (of different voltage and frequency), petrol/diesel, steam engines etc. *In the field, the motor can be easily changed in case of a motor burn-out.* The refrigeration system is *not* affected by burn-out.

- A **disadvantage** in this type is the *shaft seal which is most vulnerable point for the leakage of the refrigerant.*

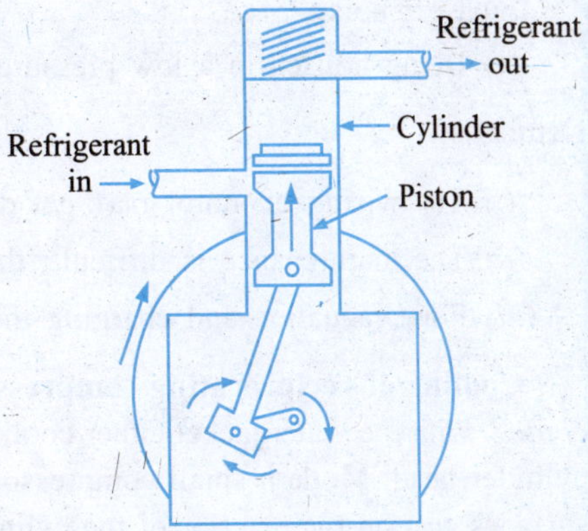

Fig. 10.2. Open type vertical reciprocating compressor.

(*ii*) **Semi-sealed type:**

- In case of semi-sealed compressor the motor and compressor are enclosed in common housing.
- *Normally the cylinder head, end plates and base plate can be removed for servicing the internal mechanism.*
- *The compressor and motor are coupled together by a common shaft inside the housing.*

(*iii*) **Hermetically sealed type:**

- In a hermetically sealed compressor the motor and compressor are inside a welded steel dome which *cannot be removed easily for internal servicing* (Fig. 10.3).
- The basic difference between this compressor and serviceable hermetic compressor is in the *means of access to the compressor and motor assembly.* In this case too the compressor and motor are coupled together by a common shaft inside welded steel dome. Here losses due to drive package and shaft seal friction are eliminated, *i.e., the power required per tonne of refrigeration is less than that of the open type.*

Fig. 10.3. Hermetically sealed compressor.

- This type of compressor can be *operated only by A.C. electric supply* and that too of the voltage and frequency the particular assembly is designed for. In the event of a motor burn out, highly corrosive hydrochloric and hydrofluoric acids are formed due to disintegration of the fluorocarbon refrigerants (R-12, R-22, R-11 etc.) at the high temperature associated with a motor burn out. The system, therefore, gets contaminated and before a repaired or new 'motor compressors' assembly is fixed, the system has to be *thoroughly flushed and cleaned.*

Hermetically sealed type compressors entail the following *merits and demerits* :

Merits:

(*i*) Complete prevention of leakage of refrigerant.

(*ii*) More compact.

(*iii*) Requires less space.

(*iv*) Less noisy.

(*v*) Being housed in a low pressure area, the motor is cooled more effectively.

Demerits:

(*i*) The wiring of motor may get damaged/harmed in the presence of moisture.

(*ii*) The maintenance is difficult; the welded joint has to be broken open for repairs.

(*iii*) For evacuation and charging the refrigerant a separate pump is needed.

Cooling of reciprocating compressors. Large sized reciprocating compressors are usually *water-cooled*, with the water jacket either cooling the cylinder walls or both the cylinder walls and compressor cylinder head. Modern small compressors are either *water-cooled or air-cooled*, with extended finned surfaces cast on the exterior of the cylinder. In a few cases small compressors may be found in which there is no attempt at any purposive cooling other than through transfer from non-finned surfaces to the lower temperature ambient air. A water-cooled compressor usually run lower in temperature than an air-cooled unit, but even under the best conditions cylinder cooling removes only a portion of the super-heat of the refrigerant gas. However, *whatever superheat is removed from the gas results in some decrease in the work of compression as well reduction in the load on the condenser.*

10.2.1.2 Construction and working of a single-stage reciprocating compressor

Fig. 10.4 (*a*) shows a sectional view of a single-stage reciprocating compressor. It consists of a piston which reciprocates is a cylinder, driven through a connecting rod and crank mounted in a crankcase. There are inlet and delivery valves mounted in the head of the cylinder. These valves are usually of the pressure differential type, meaning that they will operate as the result of the difference of pressures across the valve. The *working* of this type of compressor is as follows:

As shown in Fig. 10.4 (*b*), the piston is moving down the cylinder and any residual compressed refrigerant left in the cylinder after the previous compression will expand and will eventually reach a pressure slightly below intake pressure early or in the stroke. This means that the pressure outside the inlet valve is now higher than on the inside and hence the inlet valve will lift off its seat. A stop is provided to limit its lift and to retain it in its valve seating. Thus a fresh charge of refrigerant will be aspirated into the cylinder for the remainder of the *induction stroke*, as it is called. During this stoke the delivery valve will remain closed, since the compressed refrigerant on the outside of this valve is at a much higher pressure than the induction stroke.

As shown in Fig. 10.4 (*c*) the piston is now moving upwards. At the beginning of this upward stroke, a slight increase in cylinder pressure will have closed the inlet valve. Since both the inlet and delivery valves are now closed, the pressure of refrigerant will rapidly rise because it is now locked up in the cylinder. Eventually a pressure will be reached which is slightly in excess of the compressed refrigerant pressure on the outside of the delivery valve and hence the delivery valve will lift. The compressed refrigerant is now delivered from the cylinder in the remainder of the stroke. Once again there is a stop on the delivery valve to limit its lift and to retain it in its seating. At the end of compression stroke piston once again begins to move down the cylinder, the delivery valve closes ; the inlet valve eventually opens and the cycle is repeated.

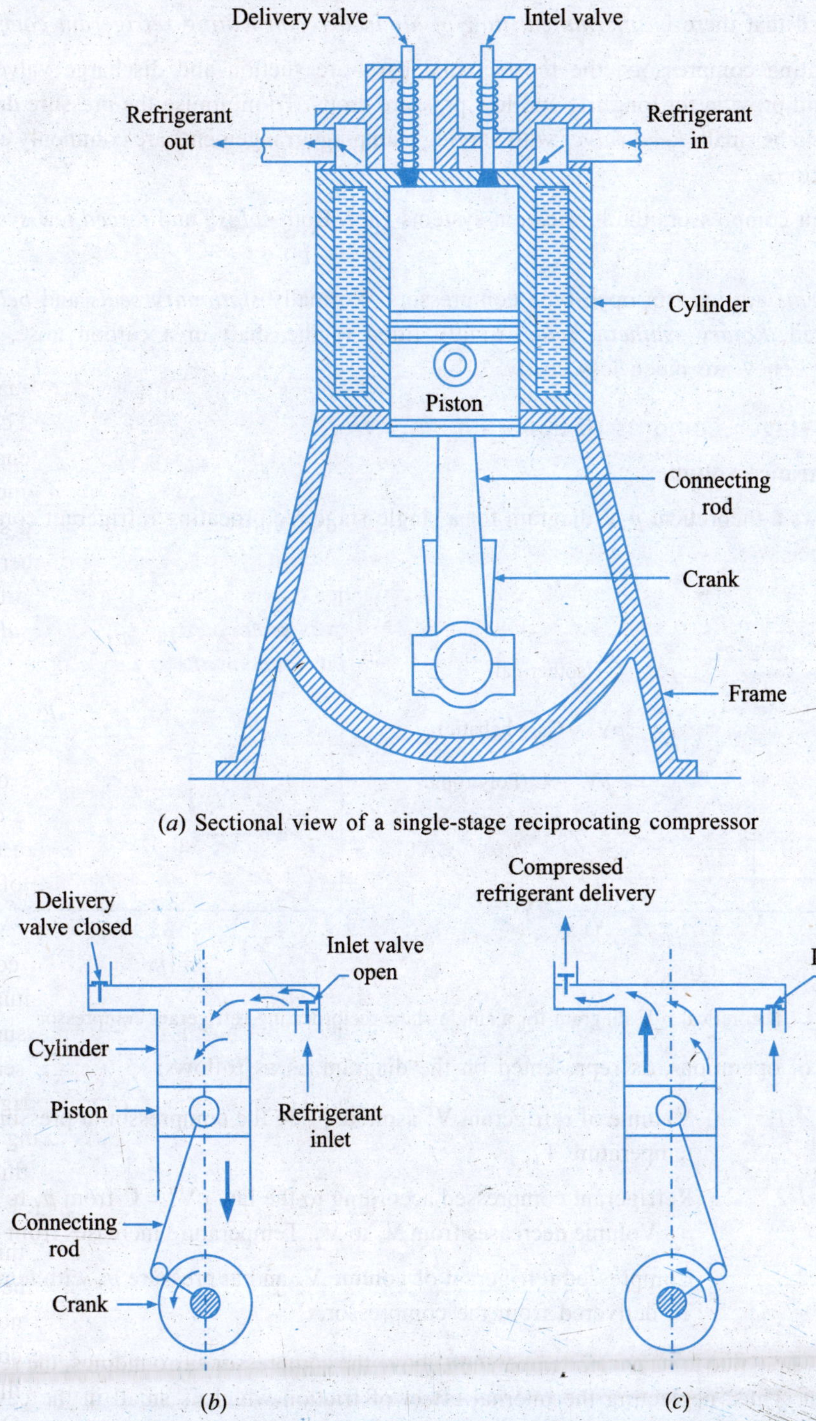

Fig. 10.4

As refrigerant is locked up in the cylinder of a reciprocating compressor then the *compression pressure* for this type of compressor *can be very high. It is limited by the strength of the various parts of the compressor and the power of the driving motor.*

It may be noted that there is *intermittent flow of air in a reciprocating refrigerant compressors.*

- In reciprocating compressors, the most crucial parts are suction and discharge valves. Their design should provide for long life and low pressure drops. To minimise the pressure drop, their weight should be small. *Reed valves* with varying clamping arrangements are commonly employed in refrigeration.
- In refrigerant compressors the lubrication systems vary. Both *splash* and *forced few systems* are employed.
  - — *Shaft seals* are used in open-type compressors. Normally *stationary seals* and *bellows* are employed. *Rotary synthetic seals,* tightly fitted to the shaft in a carbon nose, are also popular. They are much less costly.

10.2.1.3 Single-stage compressor-equation for work

(*i*) **Neglecting clearance volume:**

In Fig. 10.5 is shows a theoretical *p*-V diagram for a single-stage reciprocating refrigerant compressor, *neglecting clearance.*

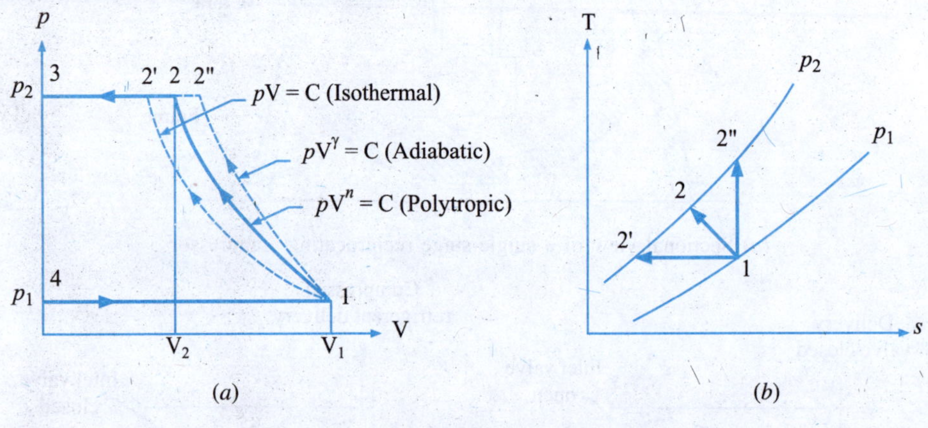

Fig. 10.5. Theoretical *p*-V diagram for a single-stage reciprocating refrigerant compressor

The sequence of operations, as represented on the diagram, is as follows:

(*i*) *Operation 4-1* : Volume of refrigerant V_1 aspirated into the compressor at pressure p_1 and temperature T_1.

(*ii*) *Operation 1-2* : Refrigerant compressed according to the law $pV^n = C$ from p_1 to pressure p_2. Volume decreases from V_1 to V_2. Temperature increases from T_1 to T_2.

(*iii*) *Operation 2-3* : Compressed refrigerant of volume V_2 and at pressure p_2 with temperature T_2 delivered from the compressor.

During compression, due to its excess temperature above the compressor surroundings, the refrigerant will lose some heat. Thus, neglecting the internal effect of friction which is small in the case of the reciprocating compressors, the index *n is less than* γ, the adiabatic index. Since work must be put into the refrigerant compressor to run it, every effort is made to reduce this amount of work input. Inspection of p-V diagram shows the frictionless adiabatic as 1–2″ and that if compression were along the isothermal 1-2′ instead of polytropic 1-2 then the work done, given by the area of the diagram, would be reduced and, infact, would then be '*minimum*'. *Isothermal compression cannot be achieved in practice but an attempt is made to approach the isothermal case by cooling the compressor either by addition of cooling fins or a water jacket to the compressor cylinder.* For a reciprocating compressor, a comparison between the actual work done during compression and the ideal isothermal work done is made by means of the **isothermal efficiency**.

This is defined as,

$$\text{Isothermal efficiency} = \frac{\text{Isothermal work done}}{\text{Actual work done}}$$

Thus, the higher the isothermal efficiency, the more nearly has the actual compression approached the ideal isothermal compression.

Total shaft work done/cycle, W = Area 41234

or W = Area under 4-1 – area under 1-2 – area under 2-3

$$= p_1V_1 - \frac{p_2V_2 - p_1V_1}{n-1} - p_2V_2$$

$$= (p_1V_1 - p_2V_2) - \left(\frac{p_2V_2 - p_1V_1}{n-1}\right) = (p_1V_1 - p_2V_2) + \left(\frac{p_1V_1 - p_2V_2}{n-1}\right)$$

$$= \left(1 + \frac{1}{n-1}\right)(p_1V_1 - p_2V_2)$$

$$\therefore \quad W = \left(\frac{n}{n-1}\right)(p_1V_1 - p_2V_2) \qquad ...(10.1)$$

This equation can be modified as follows:

$$W = \frac{n}{n-1}(p_1V_1 - p_2V_2) = \frac{n}{n-1} \cdot p_1V_1\left(1 - \frac{p_2V_2}{p_1V_1}\right) \qquad ...(10.2)$$

Now $p_1V_1^n = p_2V_2^n$

$$\therefore \quad \frac{V_2}{V_1} = \left(\frac{p_1}{p_2}\right)^{1/n}$$

and substituting this into eqn. (10.2), we have:

$$W = \frac{n}{n-1}p_1V_1\left\{1 - \frac{p_2}{p_1}\left(\frac{p_1}{p_2}\right)^{1/n}\right\} = \frac{n}{n-1}p_1V_1\left\{1 - \frac{p_2}{p_1}\left(\frac{p_1}{p_2}\right)^{-1/n}\right\}$$

$$= \frac{n}{n-1}p_1V_1\left\{1 - \left(\frac{p_2}{p_1}\right)^{1-\frac{1}{n}}\right\} = \frac{n}{n-1}p_1V_1\left\{1 - \left(\frac{p_2}{p_1}\right)^{\frac{n-1}{n}}\right\} \qquad ...(10.3)$$

The solution to this equation will always come out *negative* showing that work must be done *on* the compressor. Since only the *magnitude* of the work done is required from the expression then it is often written,

$$W = \frac{n}{n-1} p_1 V_1 \left\{ \left(\frac{p_2}{p_1} \right)^{\frac{n-1}{n}} - 1 \right\} \quad ...(10.4)$$

$$= \frac{n}{n-1} mRT_1 \left\{ \left(\frac{p_2}{p_1} \right)^{\frac{n-1}{n}} - 1 \right\} \quad ...(10.4\text{ (a)})$$

If the air delivery temperature T_2 is required then this can be obtained by using this equation :

$$\frac{T_2}{T_1} = \left(\frac{p_2}{p_1} \right)^{\frac{n-1}{n}}, \quad \text{or,} \quad T_2 = T_1 \left(\frac{p_2}{p_1} \right)^{\frac{n-1}{n}} \quad ...(10.5)$$

Power required to drive a single-stage reciprocating compressor,

$$P = \frac{W\,(Nm) \times N_{ws}}{60} \text{ watts} \quad (10.6)$$

where, N_{ws} = number of working strokes per min.

= N ... For single-acting compressor

= 2N ... For double-acting compressor.

(*ii*) **With clearance volume:**

In practice, all reciprocating compressors will have a *clearance volume.* The clearance volume is *that volume which remains in the cylinder after the piston has reached the end of its inward stroke.*

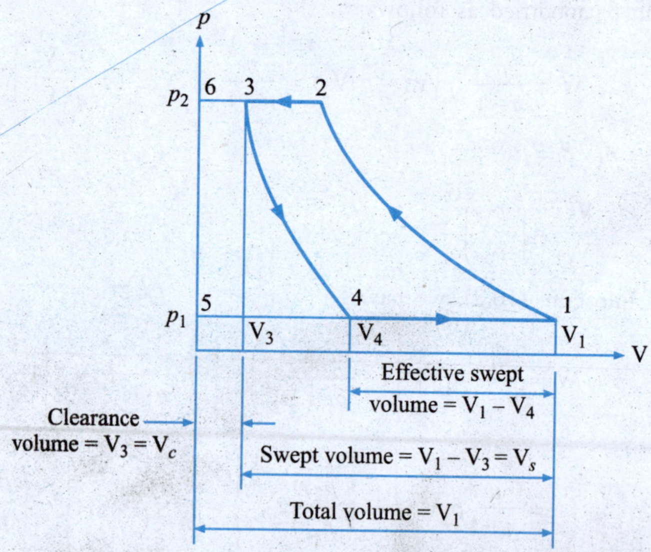

Fig. 10.6

Refer to Fig. 10.6. At point 1, the cylinder is full of intake refrigerant, volume V_1 and the piston is about to commence its compression stroke. The refrigerant is compressed polytropically according to some law $pV^n = C$ to delivery pressure p_2 and volume V_2. At 2 the delivery valve theoretically opens and for the remainder of the stroke, 2 to 3, the compressed refrigerant is delivered from the cylinder. At 3 the piston has reached the end of its inward stroke and so on, delivery of compressed refrigerant ceases at 3. V_3 is the clearance volume and is filled at this stage with compressed refrigerant. As the piston begins the intake stroke *this residual refrigerant will expand,* according to some polytropic law $pV^n = C$, and it is not *until the pressure has reduced to intake pressure* at 4 that the inlet valve will begin to open thus permitting the intake of a fresh charge of refrigerant. For the remainder of the intake stroke a fresh charge is taken into the cylinder. This volume refrigerant $(V_1 - V_4)$ is *effective swept volume*.

Work done/cycle, W = Net area 12341 = Area 51265 – area 54365

Assuming the polytropic index to be same for both compression and clearance expansion, then,

$$W = \frac{n}{n-1} p_1 V_1 \left\{ \left(\frac{p_2}{p_1} \right)^{\frac{n-1}{n}} - 1 \right\} - \frac{n}{n-1} p_4 V_4 \left\{ \left(\frac{p_3}{p_4} \right)^{\frac{n-1}{n}} - 1 \right\} \quad \text{...(10.7)}$$

But, $p_4 = p_1$ and $p_3 = p_2$, then eqn. (10.7) becomes:

$$W = \frac{n}{n-1} p_1 V_1 \left\{ \left(\frac{p_2}{p_1} \right)^{\frac{n-1}{n}} - 1 \right\} - \frac{n}{n-1} p_1 V_4 \left\{ \left(\frac{p_2}{p_1} \right)^{\frac{n-1}{n}} - 1 \right\}$$

$$= \frac{n}{n-1} p_1 (V_1 - V_4) \left\{ \left(\frac{p_2}{p_1} \right)^{\frac{n-1}{n}} - 1 \right\} \quad \text{...(10.8)}$$

Volumetric efficiency:

Refer to Fig. 10.6. *The* **volumetric efficiency** *of a compressor is the ratio of refrigerant delivered to the displacement of the compressor.* It is also the *ratio of effective swept volume to the swept volume.*

$$\text{i.e.,} \quad \text{Volumetric efficiency} = \frac{\text{Effective swept volume}}{\text{Swept volume}} = \frac{V_1 - V_4}{V_1 - V_3} \quad \text{...(10.9)}$$

Because of presence of clearance volume, volumetric efficiency is always *less than unity*. As a percentage, it usually varies from 60 to 80.

$$\text{The ratio,} \quad \frac{\text{Clearance volume}}{\text{Swept volume}} = \frac{V_3}{V_1 - V_3} = \frac{V_c}{V_s} = C \quad (10.10)$$

is the *clearance ratio.*

As a percentage, this ratio will have a value, in general, of between 4% and 10%. The greater the pressure ratio through a reciprocating compressor, then the greater will be the effect of the clearance volume since the clearance air will now expand through a greater volume before intake conditions are reached. The cylinder size and stroke being fixed, however will mean that $(V_1 - V_4)$, the effective swept volume, will reduce as the pressure ratio increases and thus the volumetric efficiency reduces.

$$\text{Volumetric efficiency, } \eta_{vol.} = \frac{V_1 - V_4}{V_1 - V_3} = \frac{(V_1 - V_3) + (V_3 - V_4)}{(V_1 - V_3)} = 1 + \frac{V_3}{V_1 - V_3} - \frac{V_4}{V_1 - V_3}$$

$$= 1 + \frac{V_3}{V_1 - V_3} - \frac{V_4}{V_1 - V_3} \cdot \frac{V_3}{V_3} = 1 + \frac{V_3}{V_1 - V_3} - \frac{V_3}{V_1 - V_3} \cdot \frac{V_4}{V_3}$$

$$= 1 + C - C \cdot \frac{V_4}{V_3} \qquad \left| \begin{array}{l} p_3 V_3^n = p_4 V_4^n \\ \frac{V_4}{V_3} = \left(\frac{p_3}{p_4}\right)^{1/n} \end{array} \right|$$

$$= 1 + C - C \left(\frac{p_3}{p_4}\right)^{1/n}$$

or, $$\eta_{vol.} = 1 + C - C \left(\frac{p_2}{p_1}\right)^{1/n} \qquad (\because p_3 = p_2, p_4 = p_1), \qquad ...(10.11)$$

or, $$\eta_{vol.} = 1 + C - C \left(\frac{V_1}{V_2}\right) \qquad ...(10.12)$$

The *volumetric efficiency is lowered* by any of the following conditions :

(*i*) Very high speed,

(*ii*) Leakage past the piston,

(*iii*) Too large a clearance volume,

(*iv*) Obstruction at inlet valves,

(*v*) Overheating of air by contact with hot cylinder walls, and

(*vi*) Inertia effect of air in suction pipe.

By paying careful attention in the design of the compressor to these causes of loss, an improvement in volumetric efficiency can be obtained.

Factors affecting total volumetric efficiency:

The various factors which affect the total volumetric efficiency are enumerated and discussed as under :

1. Clearance factor.
2. Wire drawing of throttling at suction valve and delivery valve.
3. Cylinder heating and re-expansion index.
4. Valve and piston leakage.

1. ***Clearance factor*:**

We know that, $$\eta_{vol.} = 1 + C - C \left(\frac{p_2}{p_1}\right)^{1/n} = 1 + C \left[1 - \left(\frac{p_2}{p_1}\right)^{1/n}\right]$$

Since the ratio $\left(\frac{p_2}{p_1}\right)^{1/n}$ will always be greater than unity the valve of bracket $\left[1-\left(\frac{p_2}{p_1}\right)^{1/n}\right]$ will always be *negative*. Thus, any increase in clearance factor C will increase the quantity to be subtracted from 1 to give $\eta_{vol.}$. Hence the $\eta_{vol.}$ *decreases as the clearance factor increases.*

2. ***Wire drawing or throttling at suction valve and delivery valve:***

The p-V diagram of a reciprocating compressor considering wire drawing at the suction and discharge valves is shown in Fig. 10.7. The compression actually will start from pressure p_s, and the expansion from pressure p_d at point 3′. The pressure $p_1 = p_4$ is the pressure at the flange to the compressor at intake and pressure $p_2 = p_3$ is the pressure at the flange to the compressor at discharge.

Applying pV^n = C to points 1 and 1′, we get:

$$p_1 V_1^{n_c} = p_s (V_c + V_s)^{n_c}$$

$$\therefore \quad V_1 = (V_c + V_s)\left(\frac{p_s}{p_1}\right)^{\frac{1}{n_c}} \quad ...(10.13)$$

Similarly, applying $pV_c^{n_e}$ = C to points 3′ and 4, we have:

$$p_d V_c^{n_e} = p_1 (V_4)^{n_e}$$

$$\therefore \quad V_4 = V_c\left(\frac{p_d}{p_1}\right)^{\frac{1}{n_e}} \quad ...(10.14)$$

Now,
$$\eta_{vol.} = \frac{V_1 - V_4}{V_s}$$

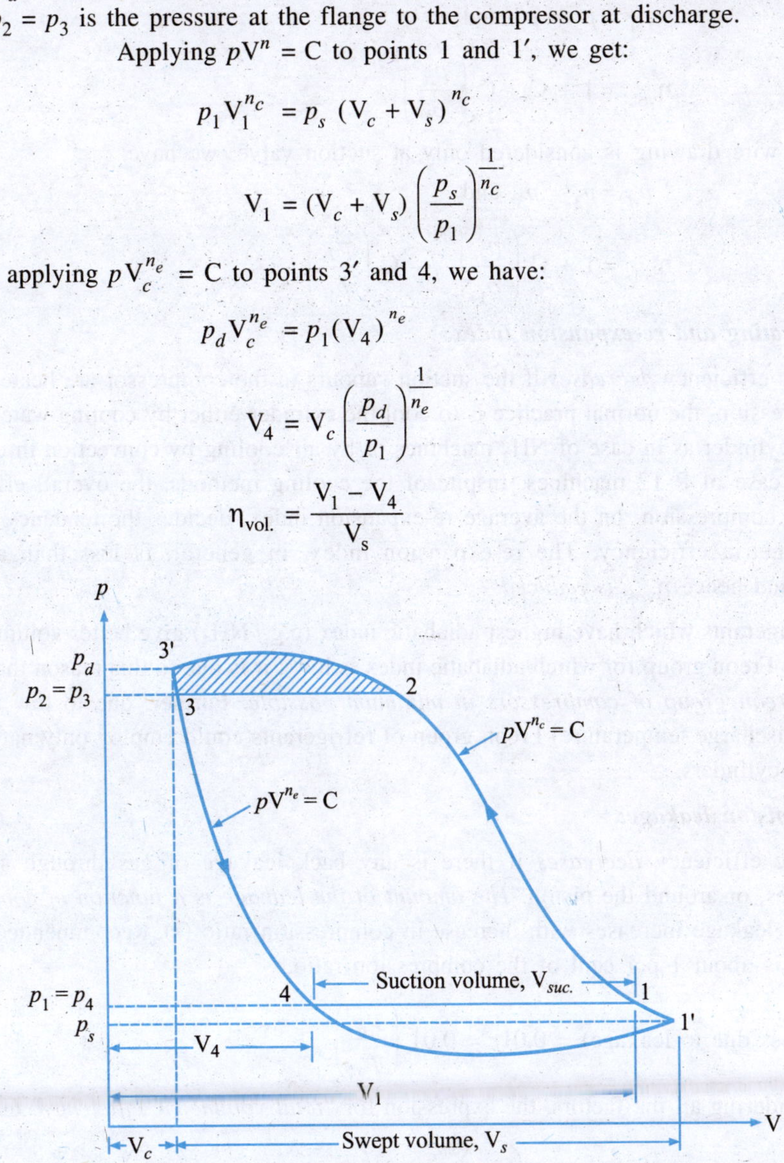

Fig. 10.7. p-V diagram considering wire drawing at the suction and discharge valves.

Substituting the valves of V_1 and V_4 from eqns. (10.13) and (10.14), we get:

$$\eta_{vol.} = \frac{(V_c + V_s)(p_s/p_1)^{1/n_c} - V_c(p_d/p_1)^{1/n_e}}{V_s}$$

$$= (C+1)\left(\frac{p_s}{p_1}\right)^{1/n_c} - C\left(\frac{p_d}{p_1}\right)^{1/n_e} \qquad \left(\because \frac{V_c}{V_s} = C\right)$$

$$= (1+C)\left(\frac{p_s}{p_1}\right)^{1/n_c} - C\left(\frac{p_d}{p_1}\right)^{1/n_e} \qquad ...(10.15)$$

If wire drawing is considered only at delivery valve, we have:

$$p_1 = p_4 = p_s, \text{ and}$$

$$\eta_{vol.} = 1 + C - C\left(\frac{p_d}{p_1}\right)^{1/n_e} \qquad ...(10.16)$$

When the wire drawing is considered only at suction valve, we have:

$$p_d = p_2 = p_3, \text{ and}$$

$$\eta_{vol.} = 1 + C\left(\frac{p_s}{p_1}\right)^{1/n_c} - C\left(\frac{p_2}{p_1}\right)^{1/n_e} \qquad ...(10.17)$$

3. ***Cylinder heating and re-expansion index:***

The volumetric efficiency *decreases* if the suction vapours in the compressor are heated. For reducing work of compression, the normal practice is to cool the cylinder either by cooling water through water jackets in the cylinder as in case of NH_3 machines or by air cooling by convection through fins of the cylinder as in case of R-12 machines. Inspite of the cooling methods, the overall effect of cylinder heating due to compression, on the average re-expansion index, decides the tendency to decrease, or increase volumetric efficiency. The re-expansion index, in general, is less than adiabatic index (i.e., $n_e < \gamma$) and hence $\eta_{vol.}$ *is reduced.*

Those refrigerants which have highest adiabatic index (e.g., NH_3) give better volumetric efficiency as compared to Freon group for which adiabatic index is low ; it is due to this reason that the *clearance volumes for Freon group of compressors in minimum possible.* Further, due to low adiabatic index (causing low discharge temperature) Freon group of refrigerants could employ only natural convection air cooling of cylinders.

4. ***Valve and piston leakage:***

The volumetric efficiency *decreases* if there is any back leakage of gas through suction and the discharge valves, or around the piston. *The amount of the leakage is a function of compression ratio.* The amount of leakage increases with increase in compression ratio (r). Recommended value of such a loss of $\eta_{vol.}$ is about 1 per cent of the compression ratio,

i.e., $\eta_{vol.}$ (Loss due to leakage) $= 0.01r = 0.01\left(\frac{p_d}{p_s}\right)$

After considering all the factors, the expression for "*total volumetric efficiency*" may be given as:

$$\eta_{vol.\,(total)} = 1 + C\left(\frac{p_d}{p_1}\right)^{1/n_c} - C\left(\frac{p_d}{p_1}\right)^{1/n_e} - 0.01\left(\frac{p_d}{p_1}\right) \qquad ...(10.18)$$

10.2.1.4 Multistage compression

In a single-stage reciprocating compressor if the delivery is restricted the delivery pressure will increase. If the delivery pressure is increased too far, however, certain disadvantages appear. Referring to Fig. 10.8 assume that the single-stage compressor is compressing to pressure p_2, the complete cycle to 1234. Clearance refrigerant expansion will be 3-4 and mass flow through the compressor will be controlled by the effective swept volume ($V_1 - V_4$). Assume now that a restriction is now placed on delivery. The delivery pressure becomes p_5 say, and the cycle is then 1567, clearance expansion being 6-7. The mass flow through the compressor is now controlled by effective swept volume ($V_1 - V_7$), which is less than ($V_1 - V_4$). In the limit, assuming the compressor to be strong enough, the compression would take place 1-8, where V_s = clearance volume, in which case there would be no delivery. It is seen, therefore, that as the *delivery pressure for a single-stage, reciprocating compressor is increased so the mass flow through the compressor decreases.* Note, also that as the *delivery pressure is increased, so also will the delivery temperature increase.* Referring to Fig. 10.8, $T_8 > T_5 > T_2$. *If high temperature refrigerant is not a requirement of the compressed refrigerant delivered, then, any increase in temperature represents an energy loss.*

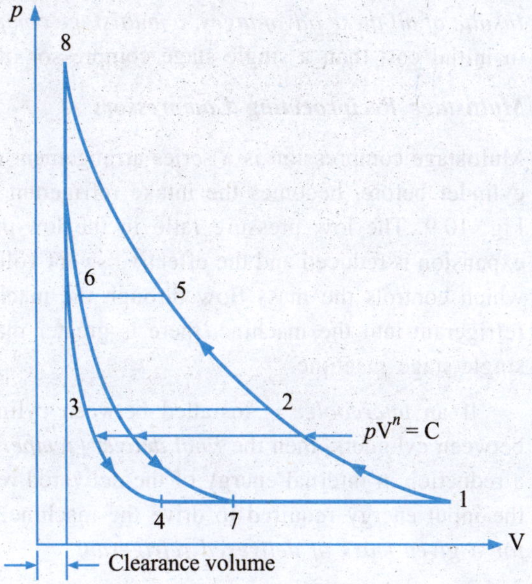

Fig. 10.8

If high pressure is to be delivered by a single-stage machine then it will require heavy working parts in order to accommodate the high pressure ratio through the machine. This will increase the balancing problem and the high torque fluctuation will require a heavier flywheel installation. Such disadvantages can be overcome by *multistage compression.*

Multistage compression is *very efficient* and is now a days almost universally adopted except for compressors where the overall pressure rise required is small. The method is not only advantageous from a thermodynamic point of view, but also has mechanical advantages over single-stage compression.

Advantages:

The important *advantages of multistage compression* can be summed up as follows:

1. The *power* required to drive a multistage machine is *less* than would be required by a single-stage machine delivering the same quantity of refrigerant at the same delivery pressure.
2. Multistage machines have *better mechanical balance.*
3. The pressure range (and hence also the temperature range) may be kept within desirable limits. This results in *(i) reduced losses due to refrigerant leakage (ii) improved lubrication, due to lower temperatures,* and *(iii) improved volumetric efficiency.*
4. The cylinder, in a single-stage machine, must be robust enough to withstand the delivery pressure. The down pressure cylinders of a multistage machine may be *lighter in construction* since the maximum pressure there in is low.

5. The sizes of the two cylinders (i.e., high pressure and low pressure cylinders) may be adjusted to suit the volume and pressure of refrigerant.
6. It gives more uniform torque, and hence a smaller size flywheel is required.
7. The cost of the compressor is reduced.

Disadvantages:

Inspite of all these advantages, a multistage compressor with intercoolers is likely to be more expensive in initial cost than a single-stage compressor of the same capacity.

Multistage Reciprocating Compressors

Multistage compression is a series arrangement of cylinders in which compressed refrigerant from the cylinder before, becomes the intake refrigerant for the cylinder which follows. This is illustrated in Fig. 10.9. The low pressure ratio in the low-pressure cylinder means that the clearance refrigerant expansion is reduced and the effective swept volume of this cylinder is increased. Since in this cylinder which controls the mass flow through the machine, because it is this cylinder which introduces the refrigerant into the machine, there is greater mass flow through the multistage arrangement than the single-stage machine.

If an *intercooler* is installed between cylinders, in which the compressed refrigerant is cooled between cylinders, then the *final delivery temperature is reduced.* This reduction in temperature means a reduction in internal energy of the delivered refrigerant, and since this energy must have come from the input energy required to drive the machine, this results in a *decrease in input work requirement for a given mass of delivered refrigerant.*

It is common to find machines with either two or three stages of compression. *The complexity of the machinery limits the number of stages.*

Refer to Fig. 10.9. The cylinders are shown with diameters which decrease as the pressure increases. This is because, as the pressure increases, so the volume of a given mass of gas decreases. There is continuity of mass flow through a compressor and hence each following cylinder will require a smaller volume due to its increased pressure range. This reduction in volume is usually accomplished by reducing the cylinder diameter.

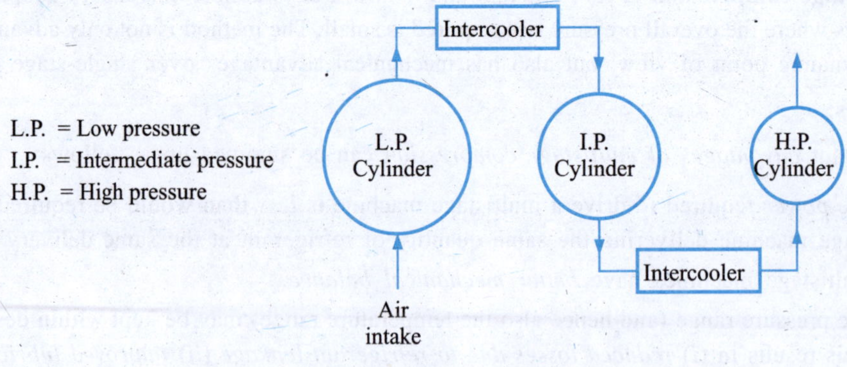

Fig. 10.9. 3-stage compressor.

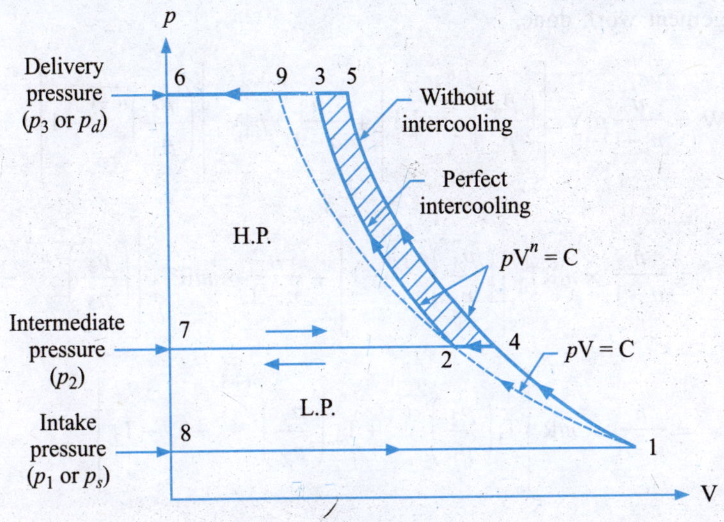

Fig. 10.10

Fig. 10.10 shows cycle arrangements in the development of the *ideal conditions* required for multistage compression. For simplicity, clearance is neglected.

Referring to Fig. 10.10, the overall pressure range is p_1 to p_3. Cycle 8156 is that of single-stage compressor. Cycles 8147 and 7456 are that of a two-stage compressor without intercooling between cylinders. Cycles 8147 and 7236 are that of a two-stage compressor with *perfect intercooling* between cylinders.

'*Perfect intercooling' means that after the initial compression in the L.P. cylinder, with its consequent temperature rise, the air is cooled in an intercooler back to its original temperature.* This means, referring to Fig. 10.10, $T_2 = T_1$, in which case point 2 lies on isothermal through point 1. This shows that multistage compression, with perfect intercooling, approaches more closely the ideal isothermal compression than in the case with single-stage compression.

Ideal conditions for multistage compressors:

Case 1. *Single-stage compressors*:

As earlier stated cycle 8156 is that of a single-stage compressor, neglecting clearance. For this cycle,

$$W = \frac{n}{n-1} p_1 V_1 \left[\left(\frac{p_5}{p_1} \right)^{\frac{n-1}{n}} - 1 \right] \quad ...(10.19)$$

Delivery temperature,

$$T_5 = T_1 \left(\frac{p_5}{p_1} \right)^{\frac{n}{n-1}} \quad ...(10.20)$$

Case 2. *Two-stage compressor*:

(*i*) Without intercooling

8147 ...... Low pressure cycle

7456 ...... High pressure cycle.

For this arrangement work done,

$$W = \frac{n}{n-1} p_1 V_1 \left[\left(\frac{p_4}{p_1} \right)^{\frac{n}{n-1}} - 1 \right] + \frac{n}{n-1} \cdot p_4 V_4 \left[\left(\frac{p_5}{p_4} \right)^{\frac{n}{n-1}} - 1 \right] \quad ...(10.21)$$

$$= \frac{n}{n-1} \times mRT_1 \left[\left(\frac{p_4}{p_1} \right)^{\frac{n}{n-1}} - 1 \right] + \frac{n}{n-1} \times mRT_2 \left[\left(\frac{p_5}{p_4} \right)^{\frac{n-1}{n}} - 1 \right]$$

$$= \frac{n}{n-1} \times mR \left[T_1 \left(\frac{p_2}{p_1} \right)^{\frac{n-1}{n}} + T_4 \left(\frac{p_3}{p_2} \right)^{\frac{n-1}{n}} - T_1 - T_2 \right] \quad ...(10.21\ (a))$$

$(\because p_4 = p_2, P_5 = P_3)$

This will give the same result as that of eqn. (10.19). The final delivery temperature will also be given by eqn. (10.20), because there is no intercooling.

(*i*) **With perfect intercooling:**

8147 ...... Low pressure cycle

7236 ...... High pressure cycle.

For this arrangement,

$$W = \frac{n}{n-1} p_1 V_1 \left[\left(\frac{p_4}{p_1} \right)^{\frac{n}{n-1}} - 1 \right] + \frac{n}{n-1} p_2 V_2 \left[\left(\frac{p_3}{p_2} \right)^{\frac{n}{n-1}} - 1 \right] \quad ...(10.22)$$

Delivery temperature is given by:

$$T_3 = T_2 \left(\frac{p_3}{p_2} \right)^{\frac{n}{n-1}} = T_1 \left(\frac{p_3}{p_2} \right)^{\frac{n}{n-1}}, \text{ Since } T_2 = T_1, \quad ...(10.23)$$

Now, since $T_2 = T_1$, then:

$$p_2 V_2 = p_1 V_1 \quad ...(10.24)$$

Also

$$p_4 = p_2 \quad ...(10.25)$$

Inserting eqns. (10.24) and (10.25) in eqn. (10.22)

$$W = \frac{n}{n-1} p_1 V_1 \left[\left(\frac{p_2}{p_1} \right)^{\frac{n}{n-1}} + \left(\frac{p_3}{p_2} \right)^{\frac{n}{n-1}} - 2 \right] \quad ...(10.26)$$

Now, inspection of Fig. 10.10 *shows the shaded area 2453 which is the work saving and occurs as a result of using an intercooler.*

Minimum work required for a two-stage reciprocating compressor:

As discussed above, the maximum work is saved in a two-stage reciprocating compressor with complete or perfect intercooling. But in actual practice, *complete or perfect intercooling in refrigerant compressors cannot be achieved.* We know that the total work done (W) by the compressor per cycle, with incomplete or imperfect cooling is given by;

$$W = \frac{n}{n-1} mR \left[T_1 \left(\frac{p_2}{p_1}\right)^{\frac{n}{n-1}} + T_2 \left(\frac{p_3}{p_2}\right)^{\frac{n}{n-1}} - T_1 - T_2 \right] \quad \text{...(i)}$$

The work done by the compressor is minimum, when:

$$\frac{dW}{dp_2} = 0$$

In order to find p_2 for minimum work let us consider following *three cases*:

1. ***When the temperature at the end of cooling in intercooler is fixed:***

When adequate quantity of cooling water supply is available, it is possible to cool the refrigerant to a fixed temperature T_2 for any temperature T_4 of the refrigerant entering the intercooler.

For minimum work, $\frac{dW}{dp_2} = 0$

$$\frac{d}{dp_2}\left[\frac{n}{n-1} \times mR \left\{ T_1 \left(\frac{p_2}{p_1}\right)^{\frac{n}{n-1}} + T_2 \left(\frac{p_3}{p_2}\right)^{\frac{n}{n-1}} - T_1 - T_2 \right\}\right] = 0$$

Now substituting $\frac{n}{n-1} = k$ (a constant) in the above equation, we have:

$$\frac{d}{dp_2}\left[\frac{mR}{k}\left\{ T_1 \left(\frac{p_2}{p_1}\right)^{k} + T_2 \left(\frac{p_3}{p_2}\right)^{k} - T_1 - T_2 \right\}\right] = 0$$

$$T_1 k p_2^{k-1} p_1^{-k} + T_2 p_3^k (-k)(p_2)^{-k-1} - 0 - 0 = 0$$

$$\frac{T_1 p_2^{k-1}}{p_1^k} - \frac{T_2 p_3^k}{p_2^{k+1}} = 0$$

$$\frac{T_1 p_2^{k-1}}{p_1^k} = \frac{T_2 p_3^k}{p_2^{k+1}}$$

$$p_2^{k-1} \times p_2^{k+1} = \frac{T_2}{T_1}(p_1 p_3)^k$$

$$p_2^{2k} = \frac{T_2}{T_1}(p_1 p_3)^k$$

$$p_2 = \left[\frac{T_2}{T_1}(p_1 p_3)^k\right]^{\frac{1}{2k}} = \left(\frac{T_2}{T_1}\right)^{\frac{1}{2k}} (p_1 p_3)^{\frac{1}{2}} = \left(\frac{T_2}{T_1}\right)^{\frac{n}{2(n-1)}} (p_1 p_3)^{\frac{1}{2}} \quad \text{...(ii)}$$

or, $$p_2 = \sqrt{p_1 p_3 \left(\frac{T_2}{T_1}\right)^{\frac{n}{n-1}}} \qquad \text{...(iii)}$$

Substituting the value of p_2 from eqn. (*ii*) in eqn. (*i*),

$$W = \frac{n}{n-1} \times mR \left[\frac{T_1}{(p_1)^{\frac{n-1}{n}}} \left\{ \left(\frac{T_2}{T_1}\right)^{\frac{n}{2(n-1)}} (p_1 p_3)^{\frac{1}{2}} \right\}^{\frac{n-1}{n}} \right.$$

$$\left. + T_2 \, (p_3)^{\frac{n-1}{n}} \left\{ \left(\frac{T_2}{T_1}\right)^{\frac{n}{2(n-1)}} (p_1 p_3)^{\frac{1}{2}} \right\}^{\frac{n-1}{n}} - T_1 - T_2 \right]$$

$$= \frac{n}{n-1} \times mR \left[\frac{T_1}{(p_1)^{\frac{n-1}{n}}} \left\{ \left(\frac{T_2}{T_1}\right)^{\frac{1}{2}} (p_1 p_3)^{\frac{n-1}{2n}} \right\} \right.$$

$$\left. + T_2 \, (p_3)^{\frac{n-1}{n}} \left\{ \left(\frac{T_2}{T_1}\right)^{-\frac{1}{2}} (p_1 p_3)^{\frac{1-n}{2n}} \right\} - T_1 - T_2 \right]$$

$$= \frac{n}{n-1} \times mR \left[\sqrt{T_1 T_2} \left(\frac{p_3}{p_1}\right)^{\frac{n-1}{2n}} + \sqrt{T_1 T_2} \left(\frac{p_3}{p_1}\right)^{\frac{n-1}{2n}} - T_1 - T_2 \right] \qquad \text{...(10.27)}$$

For *complete or perfect* intercooling,

$$T_2 = T_1$$

$$p_2 = \sqrt{p_1 p_3},$$

and, $$W = 2 \times \frac{n}{n-1} \times mRT_1 \left[\left(\frac{p_3}{p_1}\right)^{\frac{n-1}{2n}} - 1 \right]$$

2. ***When the cooling ratio is fixed:***

The "*cooling ratio*" is defined as the *ratio of the heat abstracted by the cooling system to the heat abstracted to bring the refrigerant to the initial temperature.*

Cooling ratio, $$q = \frac{T_4 - T_2}{T_4 - T_1}$$

If the cooling method used in such that cooling ratio is *constant,* then:

$$T_2 = T_4 - q\,(T_4 - T_1)$$

$$= T_1\left[q + (1-q)\left(\frac{T_4}{T_1}\right)\right]$$

$$= T_1\left[q + (1-q)\left(\frac{p_2}{p_1}\right)^{\frac{n-1}{n}}\right] \qquad \text{...}(iv)$$

Differentiating this expression with respect to p_2, we get:

$$\frac{dT_2}{dp_2} = \frac{T_1\left[0 + (1-q)\left(\frac{n-1}{n}\right)(p_2)^{-\frac{1}{n}}\right]}{(p_1)^{\frac{n-1}{n}}}$$

$$= \frac{T_1\,(1-q)\left(\frac{n-1}{n}\right)(p_2)^{-\frac{1}{n}}}{(p_1)^{\frac{n-1}{n}}} \qquad \text{...}(v)$$

We know that *for minimum work,*

$$\frac{dW}{dp_2} = 0$$

$$\frac{d}{dp_2}\left[\frac{n}{n-1}\times mR\left\{T_1\left(\frac{p_2}{p_1}\right)^{\frac{n-1}{n}} - T_2\left(\frac{p_3}{p_2}\right)^{\frac{n-1}{n}} - T_1 - T_2\right\}\right] = 0 \qquad \text{...[From eqn. }(i)]$$

$$\frac{T_1\left(\frac{n-1}{n}\right)(p_2)^{-\frac{1}{n}}}{(p_1)^{\frac{n-1}{n}}} + \left[T_2\,(p_3)^{\frac{n-1}{n}}\left(-\frac{n-1}{n}\right)(p_2)^{\frac{1-2n}{n}} + \left(\frac{p_3}{p_2}\right)^{\frac{n-1}{n}}\frac{dT_2}{dp_2} - 0 - \frac{dT_2}{dp_2}\right] = 0 \qquad \text{...}(vi)$$

Substituting the value of $\frac{dT_2}{dp_2}$ from eqn. (*iv*), we get:

$$\frac{T_1\left(\frac{n-1}{n}\right)(p_2)^{-\frac{1}{n}}}{(p_1)^{\frac{n-1}{n}}} + \left[T_2\,(p_3)^{\frac{n-1}{n}}\left(\frac{1-n}{n}\right)(p_2)^{\frac{1-2n}{n}} + \left(\frac{p_3}{p_2}\right)^{\frac{n-1}{n}}\left\{T_1\,(1-q)\left(\frac{n-1}{n}\right)(p_2)^{-\frac{1}{n}}\right\}\right.$$

$$\left. - \frac{T_1\,(1-q)\left(\frac{n-1}{n}\right)(p_2)^{-1}}{(p_1)^{\frac{n-1}{n}}}\right] = 0$$

Dividing throughout by $T_1\left(\frac{n-1}{n}\right)$, we get:

$$\frac{(p_2)^{-\frac{1}{n}}}{(p_1)^{\frac{n-1}{n}}}+\frac{T_3}{T_1}(p_3)^{\frac{n-1}{n}}(-1)(p_2)^{\frac{1-2n}{n}}+\left(\frac{p_3}{p_2}\right)^{\frac{n-1}{n}}\frac{(1-q)(p_2)^{-\frac{1}{n}}}{(p_1)^{\frac{n-1}{n}}}-\frac{(1-q)(p_3)^{-\frac{1}{n}}}{(p_1)^{\frac{n-1}{n}}}=0$$

$$(p_2)^{-\frac{1}{n}}(p_1)^{\frac{n-1}{n}}-\frac{T_3}{T_1}(p_3)^{\frac{n-1}{n}}(p_2)^{\frac{1-2n}{n}}+(p_3)^{\frac{n-1}{n}}(p_2)^{-1}(1-q)(p_1)^{\frac{1-n}{n}}$$

$$-(1-q)(p_2)^{-\frac{1}{n}}(p_1)^{\frac{n-1}{n}}=0$$

Now substituting the value of $\frac{T_2}{T_1}$ from equation (*iv*),

$$(p_2)^{-\frac{1}{n}}(p_1)^{\frac{n-1}{n}}\left\{q+(1-q)\left(\frac{p_2}{p_1}\right)^{\frac{n-1}{n}}\right\}\left\{(p_3)^{\frac{n-1}{n}}(p_3)^{\frac{1-2n}{n}}\right\}$$

$$+(p_3)^{\frac{n-1}{n}}(p_2)^{-1}(1-q)(p_1)^{\frac{1-n}{n}}-(1-q)(p_2)^{-\frac{1}{n}}(p_1)^{\frac{1-n}{n}}=0$$

$$(p_2)^{-\frac{1}{n}}(p_1)^{\frac{1-n}{n}}-q(p_3)^{\frac{n-1}{n}}(p_2)^{\frac{1-2n}{n}}-(1-q)(p_2)^{\frac{n-1}{n}}(p_1)^{\frac{1-n}{n}}(p_3)^{\frac{n-1}{n}}(p_2)^{\frac{1-2n}{n}}$$

$$+(p_3)^{\frac{n-1}{n}}(p_2)^{-1}(1-q)(p_1)^{\frac{1-n}{n}}-(1-q)(p_2)^{-\frac{1}{n}}(p_1)^{\frac{1-n}{n}}=0$$

$$(p_2)^{-\frac{1}{n}}(p_1)^{\frac{1-n}{n}}-q(p_3)^{\frac{n-1}{n}}(p_2)^{\frac{1-2n}{n}}-(p_2)^{-1}(p_1)^{\frac{1-n}{n}}(p_3)^{\frac{n-1}{n}}$$

$$+q(p_2)^{-1}(p_1)^{\frac{1-n}{n}}(p_3)^{\frac{n-1}{n}}+(p_3)^{\frac{n-1}{n}}(p_2)^{-1}(p_1)^{\frac{1-n}{n}}$$

$$-q(p_3)^{\frac{n-1}{n}}(p_2)^{-1}(p_1)^{\frac{1-n}{n}}-(p_2)^{-\frac{1}{n}}(p_1)^{\frac{1-n}{n}}+q(p_2)^{-\frac{1}{n}}(p_1)^{\frac{1-n}{n}}=0$$

$$(p_2)^{-\frac{1}{n}}(p_1)^{\frac{1-n}{n}}-q(p_3)^{\frac{n-1}{n}}(p_2)^{\frac{1-2n}{n}}-(p_2)^{-\frac{1}{n}}(p_1)^{\frac{1-n}{n}}+q(p_2)^{-\frac{1}{n}}(p_1)^{\frac{1-n}{n}}=0$$

$$-q(p_3)^{\frac{n-1}{n}}(p_2)^{\frac{1-2n}{n}}+q(p_2)^{-\frac{1}{n}}(p_1)^{\frac{1-n}{n}}=0$$

$$(p_3)^{\frac{n-1}{n}}(p_2)^{\frac{1-2n}{n}}=(p_2)^{-\frac{1}{n}}(p_1)^{\frac{1-n}{n}}$$

$$\frac{(p_2)^{\frac{1-2n}{n}}}{(p_2)^{-\frac{1}{n}}}=\frac{(p_1)^{\frac{1-n}{n}}}{(p_3)^{\frac{n-1}{n}}}$$

$$(p_2)^{\frac{1-2n}{n}}(p_2)^{\frac{1}{n}}=(p_1)^{\frac{1-n}{n}}(p_3)^{\frac{1-n}{n}}$$

$$(p_2)^{\frac{2(1-n)}{n}}=(p_1p_3)^{\frac{1-n}{n}}$$

$$p_2^2=p_1\,p_3$$

$$\therefore \quad p_2=\sqrt{p_1p_3} \qquad \text{...(10.28)}$$

Thus the *intercooler or intermediate pressure is the geometric mean of the suction or evaporator pressure* (p_1) *and condenser pressure* (p_3).

Substituting the value of p_2 in equation (*i*).

$$W = \frac{n}{n-1} \times mR \left[T_1 \left(\frac{\sqrt{p_1 p_3}}{p_1} \right)^{\frac{n-1}{n}} + T_2 \left(\frac{p_3}{\sqrt{p_1 p_3}} \right)^{\frac{n-1}{n}} - T_1 - T_2 \right]$$

$$= \frac{n}{n-1} \times mR \left[T_1 \left(\frac{p_3}{p_1} \right)^{\frac{n-1}{2n}} + T_2 \left(\frac{p_3}{p_1} \right)^{\frac{n-1}{2n}} - T_1 - T_2 \right]$$

where, $$T_2 = T_1 \left[q + (1-q) \left(\frac{p_2}{p_1} \right)^{\frac{n-1}{2n}} \right] \qquad ...(10.29)$$

[When the intercooling is perfect, then $T_2 = T_1$, and

$$W = 2 \times \frac{n}{n-1} \times mRT_1 \left[\left(\frac{p_3}{p_1} \right)^{\frac{n-1}{2n}} - 1 \right]$$
]

3. ***When the temperature after intercooling (T_2) is equal to the saturation temperature at intermediate or intercooler pressure:***

In case of refrigeration compressors, it is *always desirable to cool the refrigerant to a temperature T_2 which is also the saturation temperature at the intermediate pressure selected for minimum work.* For such a case, we can write,

$$T_2 = f(p_2)$$

where f is a function which correlates the saturation temperature of the refrigerant with the pressure.

Hence to find the optimum pressure p_2, we have:

$$\frac{dT_2}{dp_2} = f'(p_2) = \phi(p_2)$$

Substituting the values of T_2 and $\frac{dT_2}{dp_2}$ in eqn. (*vi*) and multiplying by $\left(\frac{n}{n-1} \times \frac{1}{T_1} \right)$, we get:

$$(p_2)^{-\frac{1}{n}} (p_1)^{\frac{1-n}{n}} + \left[\frac{f(p_2)(p_3)^{\frac{n-1}{n}} (-1)(p_2)^{\frac{1-2n}{n}}}{T_1} + \frac{\left(\frac{p_3}{p_2} \right)^{\frac{n-1}{n}} \phi(p_2) \times n}{(n-1)\, T_1} - \phi(p_2) \left(\frac{n}{n-1} \right) \left(\frac{1}{T_1} \right) \right] = 0$$

$$(p_2)^{-\frac{1}{n}}(p_1)^{\frac{1-n}{n}} - \frac{f(p_2)}{T_1}(p_3)^{\frac{n-1}{n}}(p_2)^{\frac{1-2n}{n}} + \frac{n}{n-1} \times \frac{\phi(p_2)}{T_1}\left(\frac{p_3}{p_2}\right)^{\frac{n-1}{n}}$$

$$- \frac{n}{n-1} \times \frac{\phi(p_2)}{T_1} = 0 \quad \text{...(10.30)}$$

By solving the above expression the value of optimum pressure p_2 can be found out ; the complexity depends upon the function f.

10.2.1.5 Capacity control of reciprocating compressors

Capacity control of reciprocating compressors running at constant speed consists in controlling the quantity of gas delivered to match the fluctuating load. *In an efficient control system, the power consumption of the compressor should be proportional to the amount of gas delivered.*

Following are the various *methods used for controlling the capacity of reciprocating compressors*:

1. Using multi-speed motor,
2. On and off control,
3. Holding the valves open,
4. Hot-gas bypass, and
5. Using multiple units.

1. ***Using multi-speed motor:***

- It is a simple and satisfactory method of controlling capacity of a reciprocating compressor.
- *The capacity of a compressor is proportional to speed of the driving motor.* When the suction pressure of the refrigeration system is high, the motor speed and compressor capacity must be *increased.* Because of this reason, electric motors with two or more speeds are used. The variable speed motors find limited application since the motors and their controls are expensive to use on large installations.

2. ***On and off control:***

- In small unitary equipment (e.g., refrigerators, air-conditioners, water-coolers (etc.) simple on-off or start-stop control in conjunction with a thermostat is used to advantage.
- This method is particularly suitable *where there is a sudden large demand followed by periods of small or no demand on capacity.*
- Switching off is preferable to running at partial load because of the low part-load efficiency of squirrel cage induction motors employed in refrigerant compressors.

3. ***Holding the valves open:***

- This is the most commonly used method of unloading in multicylinder V/W compressors. It is the best method when the condition is of intermittent unloading that requires idling for short periods only.
- It is accomplished by lifting of suction valves, usually of two cylinders together, by means of depressors or fingers. The depressor rods are operated by a cam which is turned either manually or by means of a thermostat or a pressure–stat (for suction pressure).

4. ***Hot-gas bypass:***

- In this system/method a part of the hot compressed gas is by-passed back to the suction line, through a constant-pressure throttle valve. The valve admits hot gas in the suction line as the evaporator pressure tends to drop, thus *maintaining constant suction pressure.*
- This method is *inefficient* as the suction temperature tends to increase as more and more hot gas is by-passed. The compressor work also tends to increase, thus resulting in the loading of the compressor.
- The method, however, is *simple* and *can be used in conjunction with other methods.*

5. ***Using multiple units:***

- The multiple units of the same capacity can be employed to provide capacity control. The operation of all units will provide the maximum desired capacity and operation of various combinations of the units will permit efficient capacity reduction.
- Some units may serve the purpose of stand-by equipment as well, thus allowing partial load operation in the case of breakdown of other units.
- In this method, the units are usually installed with common suction and discharge headers.

10.2.1.6. Performance characteristics of refrigerant reciprocating compressor

The performance of a refrigerant reciprocating compressor is measured in terms of its:

— Refrigerating capacity;

— Brake power (B.P) per tonne of refrigeration;

— Total brake power.

The two parameters which are significant in the design of a refrigerant reciprocating compressor are:

(*i*) Evaporator temperature or saturated suction temperature.

(*ii*) Condenser temperature or saturated condensation temperature.

The effects of these two parameters on the performance of the compressor are discussed below:

1. **Effect of suction temperature:**

The effect of suction temperature on refrigerating capacity (in tonnes) and compressor B.P per tonne of refrigeration is shown in Fig. 10.11.

— The refrigerating capacity of a reciprocating compressor *decreases with the decrease in evaporator or suction temperature.*

— The compressor B.P. per tonne of refrigeration *decreases with increase in suction temperature,* when the condenser temperature remains unaltered. The *total B.P. of the compressor decreases with the decrease in suction temperature.*

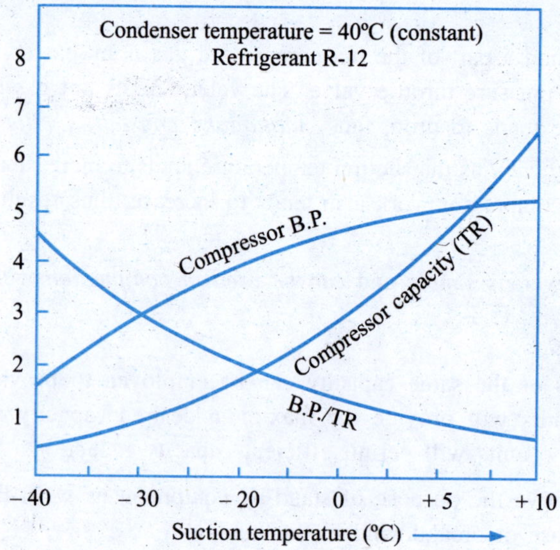

Fig. 10.11. Effect of suction temperature.

2. Effect of condenser temperature:

Fig. 10.12 exhibits the effect of condenser temperature on the compressor refrigerating capacity, B.P. per tonne of refrigeration and the total B.P., the suction temperature being kept constant in all the cases. It may be observed that the *increase in condenser temperature results in :*

— *decrease* of compressor refrigerating capacity;

— *Increase* in B.P. per tonne of refrigeration and the total B.P.

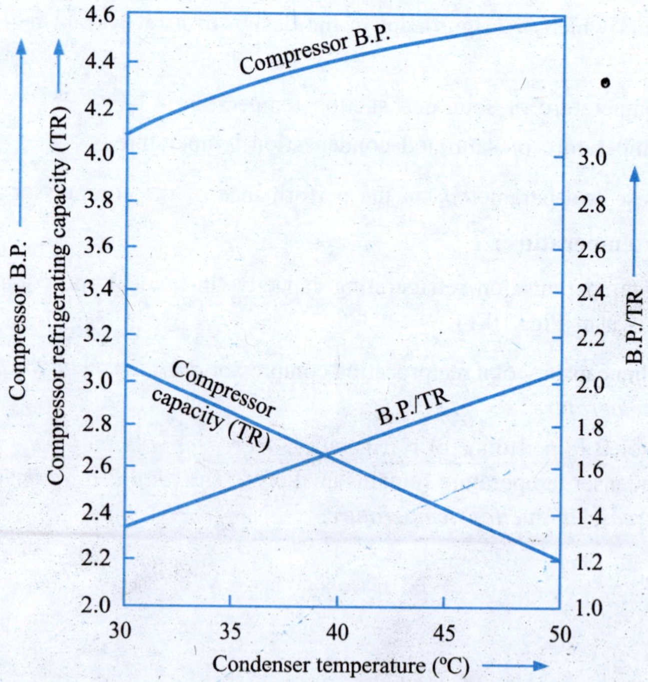

Fig. 10.12. Effect of condenser temperature.

10.2.2 Centrifugal Compressors

- Dr. Willis H. Carrier in 1922, developed the centrifugal compressor for refrigeration systems.
- In this compressor, centrifugal force changes low pressure vapour with high pressure vapour.
- These compressors are generally used for refrigerants that require *large displacement and low condensing pressure* such as R-11 and R-113. However, R-12 refrigerant is also used for large capacity and low temperature applications.
- Centrifugal refrigeration compressors find a wide application because they are *characterised by* the following :

  – Large capacity;

  – At very low temperatures they have the ability to carry varying loads with efficiency at all loads within their operating range.

10.2.2.1 Construction and working

The centrifugal compressors are similar in construction to centrifugal pumps, where the incoming gas enters the eye of the spinning impeller (Fig. 10.13) and is thrown by centrifugal force to the periphery of the impeller. That causes the blades of the impeller to impart high velocity to the gas, also to build up the pressure. From the impeller the gas then flows into diffuser blades, or into a volute, where some of the kinetic energy is converted into pressure. Essentially in high speed units, centrifugal compressors may be directly driven by steam turbine, electric motor, or internal combustion engine.

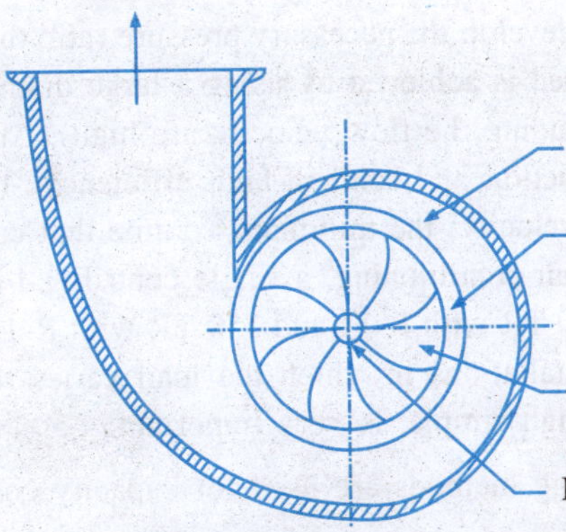

Fig. 10.13. Centrifugal compressor.

— In case of a single-stage centrifugal compressor, the compression ratio that an impeller can develop is limited to about 4.5. *When high compression ratio is required, multistage centrifugal compressors with intercoolers are employed.*

— The centrifugal compressors have no valves, pistons and cylinders. The main bearings are the only wearing parts.

10.2.2.2 Advantages and disadvantages of centrifugal compressors over reciprocating compressors

Advantages:

1. Due to the absence of unbalanced masses, these compressors operate with little or no vibration.
2. As compared to reciprocating compressors, these compressors due to high speed can handle large volume of vapour refrigerant.
3. Working life is more owing to the absence of valves, pistons, cylinders, connecting rod etc.
4. Operation is quite and calm.
5. These compressors can be directly connected to electric motors or steam turbines as they run at high speeds (3000 r.p.m. and above).

6. Considerably high efficiency.
7. As compared to reciprocating compressors, the large size centrifugal compressors require less floor area.
8. Especially adapted for systems ranging from 50 to 5000 tonnes. Also employed for temperature ranges between –90ºC and +10°C.

Disadvantages:

1. "*Surging*" (the reversal of flow of refrigerant from compressor to evaporator when refrigeration load decreases) occurs when the refrigeration load decreases to about 35% of the rated capacity and causes severe conditions in the compressor.
2. Pressure increase per stage in less as compared to reciprocating compressors.
3. Practicable below 50 tonnes capacity load.
4. Require refrigerants which should have high specific volume.

10.2.2.3 Capacity control of centrifugal compressors

In order to develop the necessary pressure ratio the centrifugal compressors require high tip speeds. The high tip speed is achieved by using a large diameter impeller or high r.p.m. or both (the velocities in general including the flow velocity are high). Also, there must be reasonable width of the shrouds to minimise friction and achieve high efficiency. Thus, because of the sufficiently large flow area and large flow velocity; the minimum volume that can be handled is about 50 m^3/min. Therefore, for the purpose of air-conditioning, a single centrifugal compressor can be designed for a minimum capacity of about 250 TR with R-11 and 150 TR with R-13. Thus, the centrifugal compressors are used in large capacity installations in which the load varies through a wide range and hence, capacity control in them, although simple, is very important.

Following methods are used for capacity control of centrifugal compressors:

1. ***Speed control system:*** One of the common methods for capacity control is by varying the compressor speed through a *speed-reduction gear.* The decrease in speed results in an operation on a lower head-flow characteristic giving a lower volume flow rate corresponding to the same pressure ratio.

2. ***Inlet vane control system:*** The use of variable inlet whirl vanes is another method of capacity control with a constant-speed drive. *The capacity is varied by changing the angle at which the gas enters the impeller.* The gas then enters with prerotation and this results in a decrease in flow.

Both the methods, (discussed above) are *efficient and work done remains proportional to the volume handled.*

10.2.2.4 Comparison of performance of reciprocating and centrifugal compressors

As compared to the reciprocating compressor, a centrifugal compressor has a *high efficiency* outer a large range of load and a large volume of suction vapour and hence a *large capacity for it size.* Furthermore, one of the most advantageous features of the centrifugal compressor is the *flat head-capacity characteristic* as compared to a reciprocating compressor.

- **Fig. 10.14** shows that for a centrifugal compressor, evaporator temperature changes only from 2°C to 7.5°C when the load changes from 100 to 240 tonnes of refrigeration, whereas the evaporation temperature changes from –11°C to 6°C for the same load change for a reciprocating compressor.

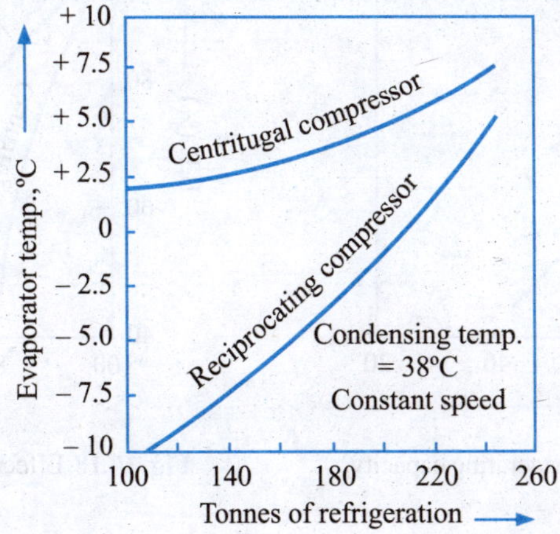

Fig. 10.14. Capacity variation with evaporator temperature for centrifugal and reciprocating compressors.

- **Fig. 10.15** shows the *non-overloading characteristic* of centrifugal compressor as compared to that of reciprocating compressor. It may be seen that for a centrifugal compressor, there is a reduction in power requirement with increase in condensing temperature. Thus there is no overloading of the motor with the increasing condensing temperature. However, for reciprocating compressors, there is a small increase in power requirement with the increase in condensing temperature. Thus there will be overloading of this motor.

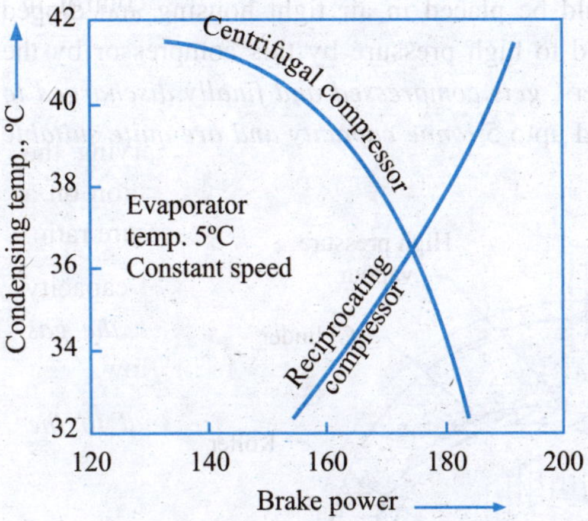

Fig. 10.15. Non-overloading characteristic of centrifugal compressor as compared to that of reciprocating compressor.

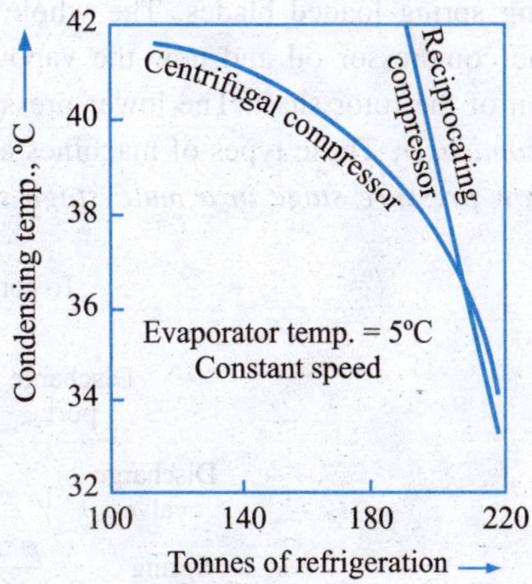

Fig. 10.16. Capacity variation with condenser temperature for centrifugal and reciprocating compressors.

- **Fig. 10.16** shows a reduction in refrigerating capacity of the centrifugal compressor with the increase in condensing temperature. There is a small decrease in refrigerating capacity of reciprocating compressor with increase in condensing temperature.
- **Fig 10.17** and **Fig. 10.18** show the effect of speed on the refrigerating capacity and power requirements respectively for both the centrifugal and reciprocating compressors.

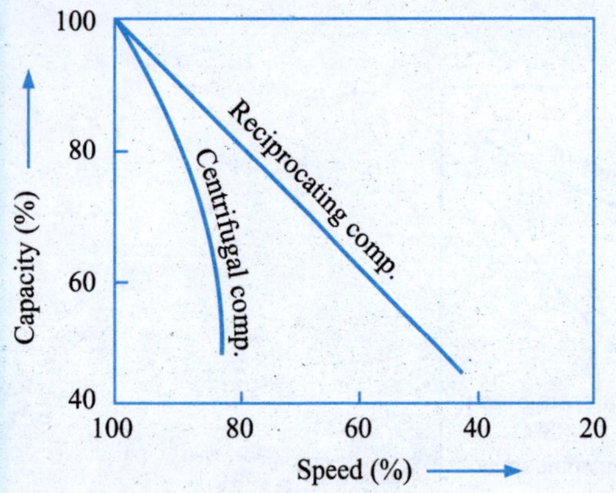

Fig. 10.17. Effect of speed on the refrigerating capacity.

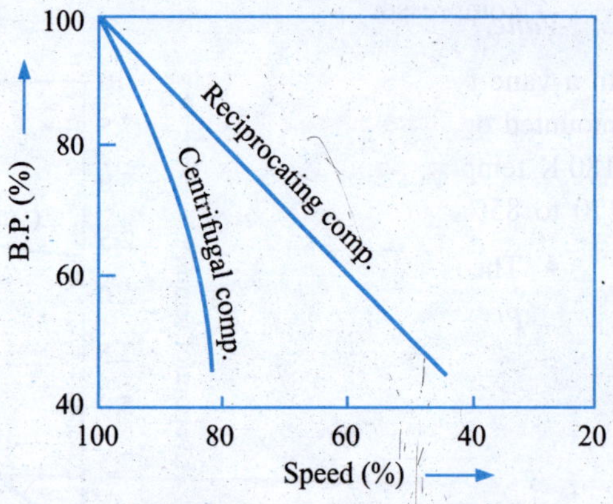

Fig. 10.18. Effect of speed on power requirements.

10.2.3 Rotary Compressors

The *'rotary compressors'* are *suitable for fraction tonnage capacity machines.*

These compressors are of the following *two types:*

1. Roller type rotary compressors.
2. Vane type rotary compressors.

1. ***Roller type rotary compressor:***

Refer Fig. 10.19. In this type of compressor the roller is eccentrically mounted on a motor shaft space having spring loaded blades. The whole assembly should be placed in air tight housing and dipped in the compressor oil and now the vapour is compressed to high pressure by this compressor by the action of the rotor shaft. The lower pressure vapour enters, *gets compressed and finally discharged to the condenser.* These types of machines are manufactured upto 5 *tonne capacity and are quite suitable for low-pressure stage in a multi-stage system.*

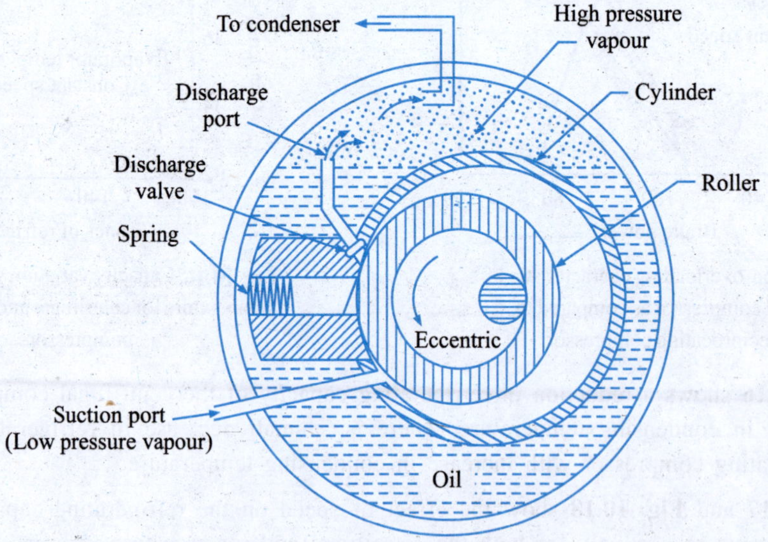

Fig. 10.19. Roller type rotary compressor.

2. ***Vane type rotary compressor:***

In a vane type rotary compressor two or more number of vanes are mounted in the slots of the rotor mounted on the eccentric. The vanes position is controlled as the rotor rotates. They can be used upto 180 K temperatures. The *compression ratio is limited to 7 : 1 and has very high volumetric efficiency* (70 to 85%) since no residual gas expands in this case.

- The rotary compressor is *most suited to pumping refrigerants having moderate or low condensing pressures* (*such as R-21 and R-114*). *This is mainly used for domestic refrigerators.*

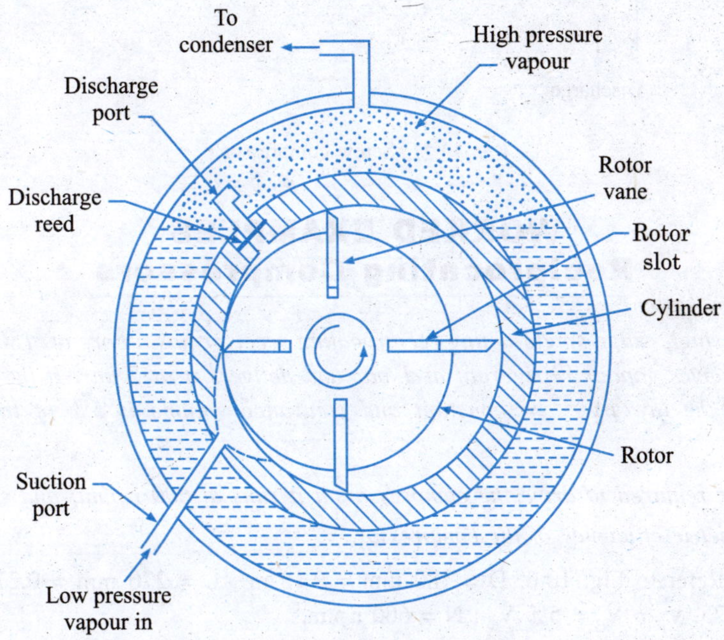

Fig. 10.20. Vane type rotary compressor.

10.2.4 Screw Compressors

Fig. 10.21 shows a screw compressor. The helical rotary or screw compressor is a *positive displacement compressor* in which compression is accomplished by the enmeshing of two mating helically groved rotors suitably housed in a cylinder equipped with appropriate inlet and discharge ports (Fig. 10.21). The male rotor is normally the driving rotor and consist of series of lobes (usually four) along the length of the rotor that mesh with similarly formed corresponding helical flutes (usually six) on the female (driven) rotor. As the rotors turn, gas is drawn through the inlet opening to fill the space between the male lobe and female flute. As the rotors continue to rotate, gas is moved past the suction port and sealed in the interlobe space. The gas so trapped in the interlobe space is moved both axially and radially and is compressed by direct volume reduction as the enmeshing of the lobes of the compressor progressively reduces the space occupied by the gas. Compression of the gas continues until the interlobe space communicates with the discharge ports in the cylinder and the compressed gas leaves the cylinder through these ports.

- A *screw compressor is available from 15 H.P. to 800 H.P. It applies to both halocarbon and ammonia refrigerants.*

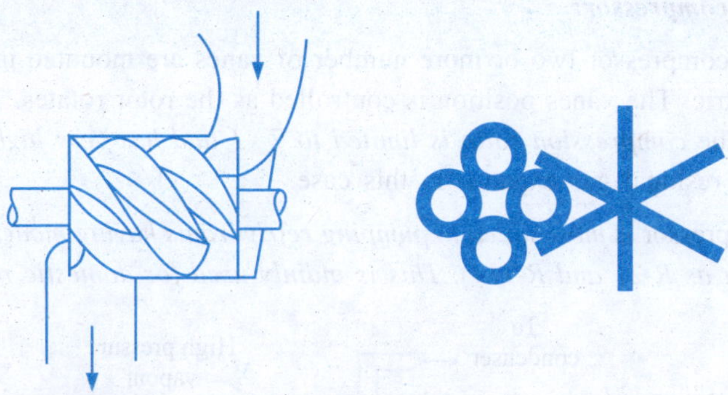

Fig. 10.21. Screw compressor.

WORKED EXAMPLES
Reciprocating Compressors

Example 10.1. *A single-stage single-acting reciprocating receiver has a bore of 180 mm and a stroke of 270 mm. It receives vapour refrigerant at 1 bar and delivers it at 5 bar. If the compression and expansion follow the law $pV^{1.25}$ = constant and clearance volume is 5% of the stroke volume, determine :*

(i) *The power required to drive the compressor, if it runs at 600 r.p.m.; and*

(ii) *The volumetric efficiency of the compressor.*

Solution: *Given*: Refer to Fig. 10.6. D = 180 mm = 0.18 m ; L = 270 mm = 0.27 m ; p_1 = 1 bar, p_2 = 5 bar; n = 1.25, $V_c = V_3 = 5\%\ V_s$; N = 600 r.p.m.

The stroke volume, $V_s = \dfrac{\pi}{4} \times D^2 \times L = \dfrac{\pi}{4} \times (0.18)^2 \times 0.27 = 0.00687 \text{ m}^3$

Clearance volume, $V_c = 5\%\ V_s = \dfrac{5}{100} \times 0.00687 = 0.0003435 \text{ m}^3$

$\therefore$ Total cylinder volume, $V_1 = V_c + V_s = 0.0003435 + 0.00687 = 0.007213 \text{ m}^3$

Now, $p_4 V_4^n = p_3 V_3^n$

or,
$$V_4 = V_3\left(\frac{p_3}{p_4}\right)^{\frac{1}{n}} = V_c\left(\frac{p_2}{p_1}\right)^{\frac{1}{n}} = 0.0003435\left(\frac{5}{1}\right)^{\frac{1}{1.25}} = 0.001245 \text{ m}^3$$

(i) **The power required to drive the compressor, P:**

The work due by the compressor,

$$W = \frac{n}{n-1} \times p_1 (V_1 - V_4)\left[\left(\frac{p_2}{p_1}\right)^{\frac{n-1}{n}} - 1\right] \qquad \text{...[Eqn. (10.8)]}$$

$$= \frac{1.25}{1.25-1} \times (1 \times 10^5) \times (0.007213 - 0.001245) \left[\left(\frac{5}{1}\right)^{\frac{1.25-1}{1.25}} - 1 \right]$$

$$= 2984\ (1.38 - 1) = 1133.9 \text{ Nm}$$

$$\therefore \quad P = \frac{W \times N_{ws}}{60} = \frac{1133.9 \times 600}{60} = 11339 \text{ W}, \quad \text{or,} \quad \textbf{11.339 kW (Ans.)}$$

(N_{ws} = N for single-acting compressor.)

(*ii*) **Volumetric efficiency of the compressor, $\eta_{vol.}$:**

$$\eta_{vol.} = \frac{V_1 - V_4}{V_1 - V_3} \quad \text{...[Eqn. (10.9)]}$$

$$= \frac{0.007213 - 0.001245}{0.007213 - 0.0003435} = 0.8687, \quad \text{or,} \quad \textbf{87.87\% (Ans.)}$$

Example 10.2. *The bore and stroke of a single-cylinder, single-acting reciprocating compressor using R-12 refrigerant are 100 mm and 80 mm respectively. The compressor runs at 1500 r.p.m. If the condensing temperature is 40°C, and evaporator temperature (a) 10°C and (b) –10°C, find the following :*

(*i*) *Mass of refrigerant circulated per minute;*

(*ii*) *Refrigerating capacity of the compressor;*

(*iii*) *Power per tonne of refrigeration;*

(*iv*) *Total power required to drive the compressor.*

Determine the change in results when the compression index is 1.25 and clearance factor is 5 per cent:

Solution: *Given:* D = 100 mm = 0.10 m; L = 80 mm = 0.08 m; N = 1500 r.p.m ; t_4 = 40°C ; t_2 = 10°C; $t_{2'}$ = –10°C ; C = 5% = 0.05; γ = 1.25.

$$V_2 = \text{Piston displacement per min.}$$

$$= \frac{\pi}{4} \times D^2 \times L \times N = \frac{\pi}{4} \times (0.10)^2 \times 0.08 \times 1500 = 0.9425 \text{ m}^3/\text{min}$$

Case I. When there is no clearance:

(*a*) ***Evaporation or suction temperature t_2 = 10°C***

From p-h diagram, for R-12 refrigerant (Fig. 10.22) at point 2 (i.e., at 10°C) we have:

Specific volume, $v_2 = 0.041 \text{ m}^3/\text{kg}$; h_2 = 191.7 kJ/kg ;

$$h_{f4} = h_1 = h_{1'} = 74.6 \text{ kJ/kg}, \; h_3 = 206.5 \text{ kJ/kg}$$

(*i*) **Mass of refrigerant circulated per minute, m:**

$$m = \frac{V_2}{v_2} = \frac{0.9425}{0.041} = \textbf{22.99 kg/min (Ans.)}$$

(*ii*) **Refrigerating capacity of the compressor, $R_{capacity}$:**

$$R_{capacity} = \frac{m(h_2 - h_1) \times 60}{14000} \text{ tonnes of refrigeration}$$

$$= \frac{22.99\,(191.7 - 74.6) \times 60}{14000} = \mathbf{11.45\ tonnes\ (Ans.)}$$

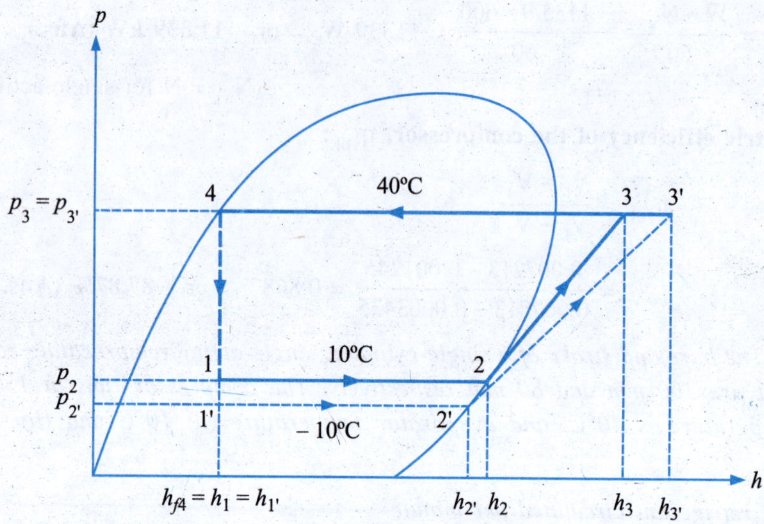

Fig. 10.22. *p-h* diagram for the cycle.

(*iii*) **Power per tonne of refrigeration P:**

$$P = \frac{m(h_3 - h_2)}{60 \times 11.54} = \frac{22.99\,(206.5 - 191.7)}{60 \times 11.54} = \mathbf{0.49\ kW/tonne\ (Ans.)}$$

(*iv*) **Total power required to drive the compressor, P_{total} :**

$$P_{total} = \frac{m(h_3 - h_2)}{60} \text{ kW}$$

$$= \frac{22.99\,(206.5 - 191.7)}{60} = \mathbf{5.67\ kW\ (Ans.)}$$

(*b*) ***Evaporator or suction temperature*** $t_{2'} = -10$°C

From *p-h* diagram, for R-12 refrigerant at point 2′ (i.e., at –10°C), we have :

Specific volume, $v_{2'} = 0.077$ m^3/kg ; $h_{2'} = 183.2$ kJ/kg ; $h_{3'} = 220$ kJ/kg

(*i*) **Mass of refrigerant circulated per minute *m′*:**

$$m' = \frac{V_2}{v_{2'}} = \frac{0.9425}{0.077} = 12.24 \text{ kg/min.}$$

(*ii*) **Refrigerating capacity of the compressor, $R'_{capacity}$:**

$$R'_{capacity} = \frac{m'(h_{2'} - h_{1'}) \times 60}{14000} \text{ tonnes of refrigeration}$$

$$= \frac{12.24\,(183.2 - 74.6) \times 60}{14000} = \mathbf{5.7\ tonnes\ (Ans.)}$$

(*iii*) **Power per tonne of refrigeration, P' :**

$$P' = \frac{m'(h_{3'} - h_{2'})}{60 \times 5.7} = \frac{12.24\,(210 - 183.2)}{60 \times 5.7} = \mathbf{0.959\ kW/tonne\ (Ans.)}$$

(*iv*) **Total power required to drive the compressor, P'_{total} :**

$$P'_{total} = \frac{m'(h_{3'} - h_{2'})}{60}\ \text{kW} = \frac{12.24\,(210 - 183.2)}{60} = \mathbf{5.47\ kW\quad (Ans.)}$$

Case II. When clearance is considered:

From *p-h* chart, corresponding to:

40°C saturation temp. : $p_3 = p_{3'} = 9.61$ bar

10°C saturation temp. : $P_2 = 4.23$ bar

– 10°C saturation temp. : $p_{2'} = 2.2.$ bar

Volumetric efficiency when suction temperature is 10°C,

$$\eta_{vol.} = 1 + C - C\left(\frac{p_3}{p_2}\right)^{\frac{1}{\gamma}}$$

$$= 1 + 0.05 - 0.05\left(\frac{9.61}{4.23}\right)^{\frac{1}{1.25}} = 0.9536 \text{ or } 95.36\%$$

Volumetric efficiency when suction temperature is – 10°C,

$$\eta_{vol.} = 1 + C - C\left(\frac{p_{3'}}{p_{2'}}\right)^{\frac{1}{\gamma}}$$

$$= 1 + 0.05 - 0.05\left(\frac{9.61}{2.2}\right)^{\frac{1}{1.25}} = 0.8874 \text{ or } 88.74\%.$$

(*a*) ***When evaporator or suction temperature t_2 = 10°C:***

(*i*) *Mass of refrigerant circulated per minute*

$$= m \times \eta_{vol.} = 22.99 \times 0.9536 = \mathbf{21.92\ kg/min\ (Ans.)}$$

(*ii*) *Refrigerating capacity of the compressor*

$$= R_{capacity} \times \eta_{vol.} = 11.54 \times 0.9536 = \mathbf{11\ tonnes\ (Ans.)}$$

(*iii*) *Power required per tonne of refrigeration*

$$= P \times \eta_{vol.} = 0.49 \times 0.9536 = \mathbf{0.467\ kW/tonne\ (Ans.)}$$

(*iv*) *Total power required to drive the compressor*

$$= P_{total} \times \eta_{vol.} = 5.67 \times 0.9536 = \mathbf{5.4\ kW\ (Ans.)}$$

(*b*) ***When evaporator or suction temperature, t_2 = – 10°C:***

(*i*) *Mass of refrigerant circulated per minute*

$$= m' \times \eta_{vol.} = 12.24 \times 0.8874 = \mathbf{10.86\ kg/min\ (Ans.)}$$

(ii) *Refrigerating capacity of the compressor*

$$= R'_{capacity} \times \eta_{vol.} = 5.7 \times 0.8874 = \mathbf{5.06\ tonnes\ (Ans.)}$$

(iii) Power required per tonne of refrigeration

$$= P' \times \eta_{vol.} = 0.959 \times 0.8874 = \mathbf{0.85\ kW/tone\ (Ans.)}$$

(iv) *Total power required to drive the compressor*

$$= P'_{total} \times \eta_{vol.} = 5.47 \times 0.8874 = \mathbf{4.85\ kW\ (Ans.)}$$

10.3 CONDENSERS

10.3.1. Introduction

A **condenser** *is a heat exchanger is which desuperheating of high temperature vapour changes the phase from vapour to liquid and sub-cooling of condensate occurs.* However, in some cases condensation may *not* be accompanied with desuperheating of vapour or subcooling of condensate.

The heat exchanger should have the following *features*:

(i) It should be *inexpensive, compact, strong and light*;

(ii) It should be *easily cleaned and adaptable for replacement* of *worn out tubes*; etc.

(iii) The *manufacture* should be *quick*;

(iv) The *pressure drop should be minimum.*

10.3.2 Types of Condensers

10.3.2.1 General aspects

Condensers are of *three* general types:

1. Air-cooled condensers
2. Water-cooled condensers
3. Evaporative condensers.

As their names imply, "***Air cooled condensers***" use *air* as the cooling medium and "***Water-cooled***" condensers use *water* as the medium. The "***Evaporative condenser***" is a combination of the above, i.e., uses both water and air. Although, in this condenser, there is some increase in the temperature of the air passing through the condenser, condensation of the refrigerant in the condenser results primarily from the evaporation of the water sprayed over the condenser. The *function of air is to increase the rate of evaporation by carrying water vapour which results from the evaporating process.*

- Since the heat can flow only from a higher-temperature medium to a lower-temperature medium, the temperature of the condensing medium determines the condensing temperature. "*Condensing temperature*" is the *temperature at which the refrigerant gas turns into its liquid state in the condenser.*
  - *The* air-cooled condenser is cooled by the ambient air. The ambient air temperature will be higher than the temperature of water. Hence the *condensing temperature will be higher in an air-cooled condenser than in water-cooled condenser.*

— Since the heat-transfer efficiency of a heat exchanger increases when the heat-transfer medium is liquid, therefore, the heat-transfer efficiency in a water-cooled condenser will be *more* than in an air-cooled condenser. The comparatively poor heat-transfer efficiency of an air-cooled condenser is to come extent offset by maintaining a higher temperature difference between the refrigerant and air.

10.3.2.2 Selection of a condenser

The *selection of a condenser* depends upon the following *factors*:

1. The capacity of the refrigerating system.
2. The type of refrigerant used.
3. The type of cooling medium available.

10.3.2.3 Factors affecting the condenser capacity

The "**condenser capacity**" *is the ability of the condenser to transfer heat from the hot water vapour refrigerant to the condensing medium*. Following are the factors on which the heat-transfer capacity of a condenser depends :

1. ***Material.*** *Higher the ability* of a material to transfer heat, the *smaller* will be the size of condenser.

2. ***Amount of contact.*** The condenser capacity may be varied by controlling the amount of contact between the condenser surface and the condensing medium. This can be achieved by varying the surface area of the condenser and flow rate of the condensing medium over the surface of condenser.

3. ***Temperature difference.*** With the increase in temperature difference the condenser capacity increases (due to increased heat-transfer rate). Normally, this temperature difference cannot be controlled. However, under difficult conditions, devices are available that control condenser capacity by changing the amount of condensing surface and air-flow rate. Most air-cooled condensers are designed to operate with a temperature difference of 14°C.

10.3.2.4 Heat rejection factor (HRF)

The heat rejected by the condenser per tonne of refrigeration is a *function of the evaporator and condenser temperatures*. The condenser has to reject heat absorbed at the evaporator (R_n) and also the heat of compression added to the system by the compressor (W). Thus the lead of the condenser (Q_C) is given by:

$$Q_C = R_n + W$$

The ***heat rejection factor*** for a condenser is *defined as the ratio of heat rejected at the condenser to the refrigerating effect.*

i.e., $$\text{H.R.F.} = \frac{Q_C}{R_n} = \frac{R_n + W}{R_n} = 1 + \frac{1}{\text{C.O.P.}} \quad \ldots(10.31) \quad \left(\because \text{C.O.P.} = \frac{R_n}{W}\right)$$

Thus, HRF depends upon the C.O.P. which in turn depends on the condenser and evaporator temperatures.

- The HRF is about 1.25 in actual air-conditioning equipment for R-12 and R-22, operating at a condenser temperature of 40°C and an evaporator temperature of 5°C.

Effect of compression ratio on HRF

For the same load on the evaporator, the heat load on the condenser will increase if the heat of compression increases. Since the heat of compression is the heat developed from the work done on the suction gas in compressing it and raising its pressure to the head pressure, the heat of compression will increase if the compression ratio increases. So *the load on the condenser (HRF) will increase with an increase in compression ratio.*

The HRF of a plant varies with the operating conditions of the plant, i.e., the evaporator temperature and the temperature of the condensing medium. *So to obtain optimum performance from a plant, the head pressure should be kept reasonably low; by maintaining the condenser tubes and condenser water system (or air system) clean.*

10.3.3 Air-cooled Condensers

In these condensers, heat is removed by air using either *natural convection* or *forced convection circulation.* These condensers are made of steel, copper and aluminium tubing provided with *fins* [Fig. 10.23 (*a*)] to improve air-aided transfer.

Natural convection condenser: Refer Fig. 10.23 (*a*) and (*b*).

In such a condenser, air movement over the surface of condenser tubes is by *natural convection.* As air comes in contact with the warm-condenser tubes, it absorbs, heat from the refrigerant and thus the temperature of air increases. Warm air being lighter, rises up and in its place cooler air from below rises to take away the heat from the condenser. This cycle goes on. Since the *air moves very slowly by natural connection,* the *rate of flow of heat from the refrigerant to air will be small.* Thus a natural convection condenser is not capable of rejecting heat rapidly. Therefore a *relatively large surface area of the condenser is required.*

- The use of this type of condenser is limited to small units such as *domestic refrigerators* and *small water-coolers* which use vertical wire and tube or plate and tube [Fig. 10.23 (*b*)].
- It requires very little maintenance.

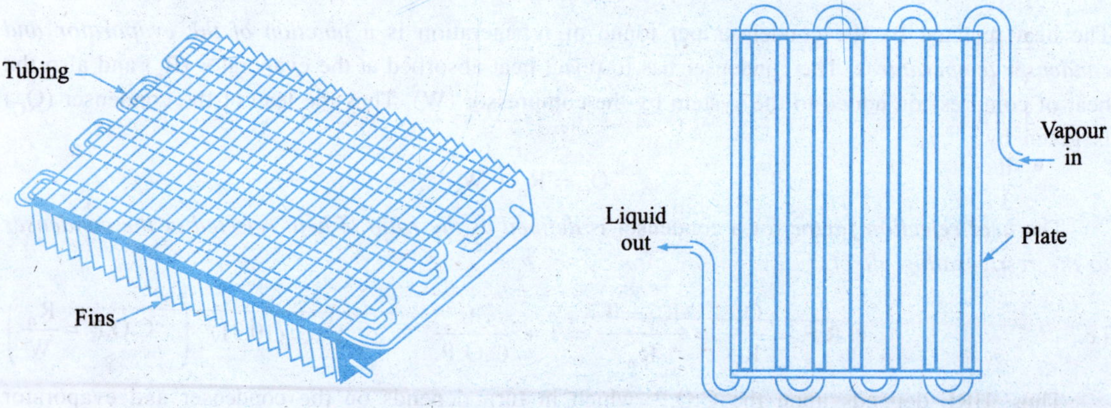

(*a*) Air condenser having tubing provided with fins. (*b*) Plate and tube air condenser.

Fig. 10.23. Air-cooled condensers (Natural convection)

Forced-air circulation condenser. Refer to Fig. 10.24.

In this type of air-cooled condenser a fan or blower is employed to move air over the condenser coil at a certain velocity. The condenser coil is of the finned type. Fins in such coils are closely spaced (ranging between 5 and 7 fins per cm). The space between the fins gets choked with dirt and lint. Therefore *to obtain optimum capacity, the fins should be kept clean.*

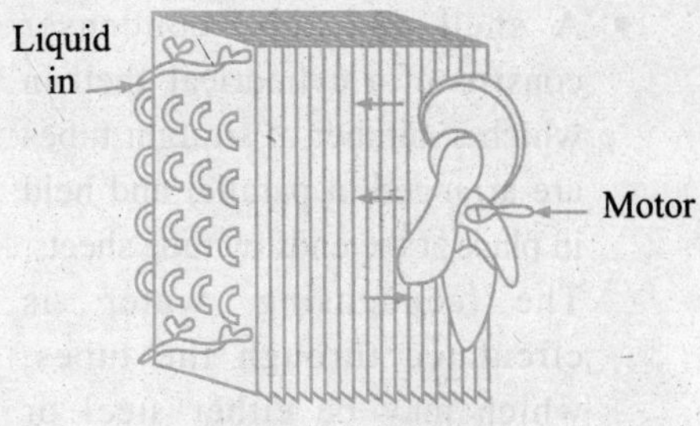

Fig. 10.24. Air-cooled condenser (Forced circulation type)

The forced-air circulation condensers may be divided into the following *two categories* :

(*i*) Base mounted air-cooled condensers;

(*ii*) Remote air-cooled condensers.

(*i*) ***Base mounted air-cooled condensers:***

- These condensers use propeller fans and are mounted on the same base of compressor, motor receiver and other controls. The whole arrangement is called a "*condensing unit.*"
- These condensing units are used on packaged refrigeration systems of 10 tonnes or less.

(*ii*) ***Remote air-cooled condensers***:

- These condensers may be horizontal or vertical. They can be located either outside or inside the building and usually use propeller fans (because they have low resistance to air flow and free air discharge).
- They are used on systems above 10 tonnes and are available upto 125 tonnes. The systems above 125 tonnes usually have two or more condensers.

Advantages of air-cooled condensers:

Air-cooled condenser claims the following ***advantages:***

1. It is easy to clean dust or gummy matter that collects over fins and tubes (just by blowing).
2. It is quite compact and easy to mount.
3. As surfaces are visible its cleaning can be easily checked without special care.
4. There is no problem about the disposal of hot air flowing through fins.

- It case of air-cooled system the temperature rise varies between 8 K to 15 K as compared to about 3 K to 6 K for the water-cooled system. Thus the *increased head pressure in case of air-cooled system would result in reduced rating because of reduced volumetric efficiency of compressor, decreased refrigerating effect and increased mechanical energy input.*

10.3.4 Water-cooled Condensers

10.3.4.1 Types of water-cooled condensers

Water-cooled condensers are of *three* types:

1. Shell and tube condenser
2. Shell and coil condenser
3. Double tube condenser.

1. ***Shell and tube condenser.*** Refer to Fig. 10.25.

 - A shell and tube condenser consists of a cylindrical shell in which a number of straight tubes are arranged in parallel and held in place at the ends by tube sheets. The condensing water is circulated through the tubes, which may be either steel or copper, bare or extended surface. The refrigerant is contained in the steel shell between the tube sheets. Water circulates in the annular spaces between the tube sheets and the end plates, the end plates being baffled to act as manifolds to guide water flow through the tubes. The arrangement of the end-plate baffling determines the number of passes the water makes through the condensers from one end to the other before leaving the condenser. The number of passes may vary from 2 to 20.

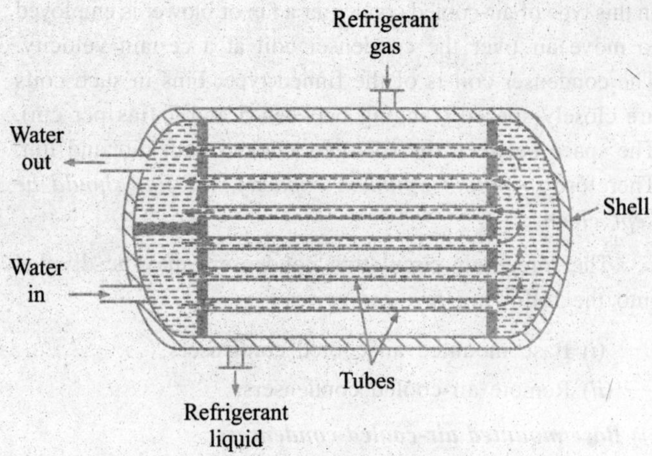

Fig. 10.25. Shell and tube condenser.

 - In such condensers, the upper tubes act as *desuperheaters* while the tubes at the bottom yield *subcooled condensate*. If there is any oil associated with refrigerant, it flows with the condensate.

2. ***Shell and coil condenser.*** Refer to Fig. 10.26.

 - In a shell and coil condenser vapour enters the shell whereas the cooling water flowing through the coil causes condensation.
 - The main *advantages* of such condensers are:

   (*i*) *Simplicity in construction.*

   (*ii*) *High overall heat transfer coefficient due to curvature.*

   But there is a *severe cleaning problem* of inside tube surface due to curvature.

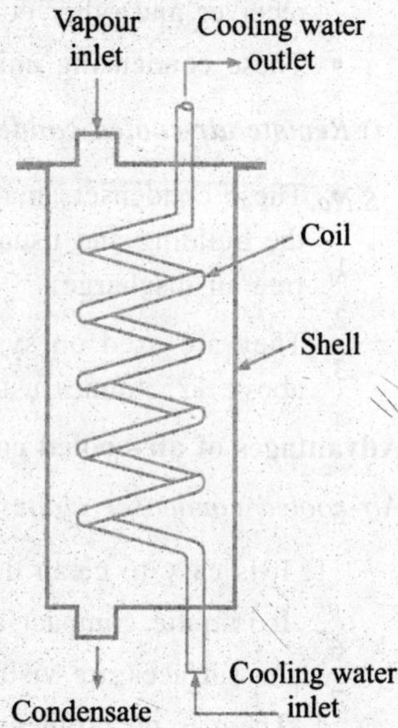

Fig. 10.26. Shell and coil condenser.

 - As a general rule, shell and coil condensers are used for *small installations* (upto 35 kW).

3. ***Double tube condenser.*** Refer to Fig. 10.27.

 - A double tube condenser has the condensing water tube inside the refrigerant tube. The refrigerant flows in the space between the tubes while the water is pumped through the inner tube. *Water flows in the opposite direction to the refrigerant, with the coolest water in contact with the coolest refrigerant and the warmest water with the warmest refrigerant.* Since copper tubing cannot be used with ammonia, this type of condenser is made up of *steel pipe for ammonia.*
 - Though the counter-flow principle with double tube condensers *makes good use of water available,* the large number of points and connections needed in large installations increase chances of leakage.

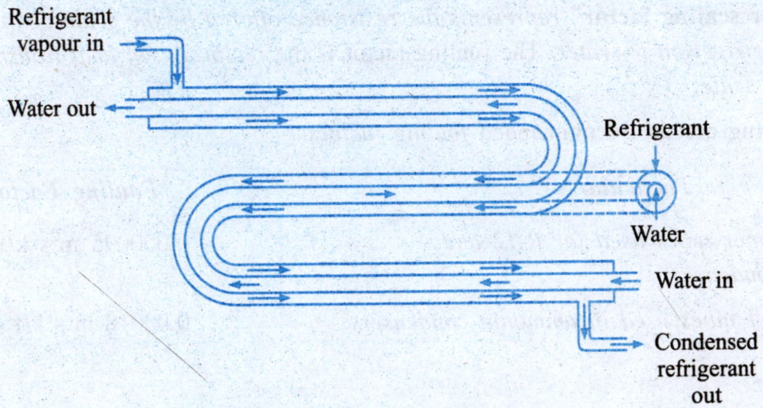

Fig. 10.27. Double tube condenser.

- These condensers are *difficult to clean and do not provide enough space for separation of gas and liquid.* For these reasons they are not *much used* in larger modern installations. Some small units are used in new installations, but they must be cleaned chemically. In case of leak entire unit must be replaced.

10.3.4.2 Comparison between water-cooled and air-cooled condensers

S.No.	Particulars	Water-cooled condensers	Air-cooled condensers
1.	*Construction*	Complicated	Simple
2.	*Maintenance cost*	High	Low
3.	*Handling of the unit*	Difficult	No handling problem
4.	*Carrying of water*	Pipes are needed to carry water to and from the condenser.	No pipe arrangement is required for carrying water.
5.	*Disposal of used water*	There is a no problem of disposing of used water unless a recirculation system is provided.	No problem in disposing of air.
6.	*Corrosion*	Corrosion occurs.	Corrosion does not occur.
7.	*Fouling effect*	More	Less
8.	*Heat transferring capacity*	High	Low
9.	*Fan noise*	No fan noise	In case of forced convection type of air condensers fan noise is there which is unwanted/objectionable.
10.	*Flexibility*	Low	High
11.	*Applications*	Used for large capacity plants.	Used for low capacity plants.

10.3.4.3 Fouling factor

In case of water cooled condensers the water used, depending upon it source, contains a certain amount of minerals and other foreign materials. These materials form deposits inside the condenser tubes, called "*water fouling*".

"Fouling or scaling factor" *represents the resistance offered by the scale or fouling to the heat flow from the refrigerant to water.* The fouling factor is the *reciprocal of heat transfer coefficient for the material of scale.*

The following are the recommended *fouling factors*:

Particulars	***Fouling Factors***
(*i*) For *copper tubes used for R-12 and R-22 condensers*	0.000095 m^2s k/J
(*ii*) For *steel tubes used in ammonia condensers*	0.00018 m^2s k/J

10.3.4.4 Heat transfer in condenser

In a water-cooled condenser, the heat transfer (Q) is given by:

$$Q = UA\,\Delta t = \frac{\Delta t}{R} \qquad ...(10.32)$$

where, U = Overall heat-transfer coefficient,

A = Surface area of the condenser,

Δt = Overall temperature difference, and

R = Overall thermal resistance of the condenser

$$= \frac{1}{UA}$$

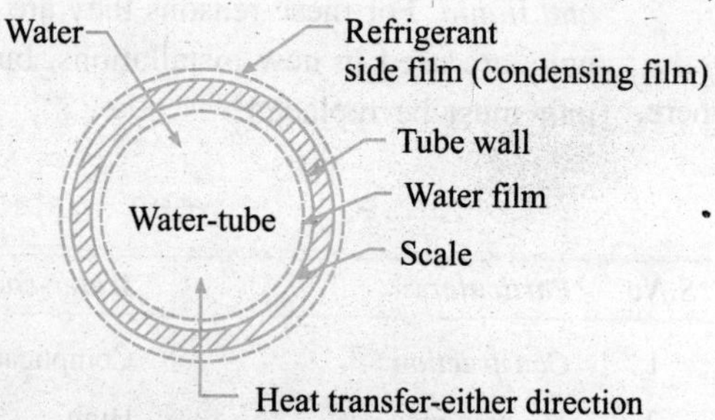

Fig. 10.28. Thermal resistances in a water-cooled condenser.

Fig. 10.28 shows the *components* of the heat-transfer resistance in a water-cooled condenser viz., the *outside refrigerant film, metal wall, scaling on water-side surface* and *inside water film.* All these resistances are in series and the overall resistance (R) is obtained by adding them as given below:

$$R = \frac{1}{U_0A_0} = \frac{1}{h_0A_0} + \frac{x}{kA_m} + \frac{1}{h_fA_i} + \frac{1}{h_iA_i} \qquad ...(10.33)$$

$$= \frac{1}{U_0} = \frac{1}{h_0} + \frac{xA_0}{kA_m} + \frac{A_0}{h_fA_i} + \frac{A_0}{h_iA_i} \qquad ...(10.33(a))$$

where, U_0 = Overall heat-transfer coefficient based on the outside surface area,

A_0 = Outside or refrigerant-side area,

h_0 = Condensing film coefficient of heat-transfer,

x = Thickness of the metal-tube wall,

k = Thermal conductivity of the tube material,

A_m = Mean surface area of the tube,

h_f = Coefficient of heat-transfer through the scale,

A_i = Inside or water-side area, and

h_i = Water-side coefficient of heat transfer.

After finding out the individual resistances, the overall heat-transfer coefficient can be obtained from eqn. (10.33).

10.3.4.5 Condensing heat transfer coefficient

In condensation, refrigerants which are usually good wetting agents follow a film-wise pattern. Nusselt, using his laminar liquid film theory, suggested the following expression for condensing heat transfer coefficients:

1. The overall coefficient of heat transfer for condensation on a vertical surface of height x,

$$h = 0.943 \left[\frac{k_f^3\, \rho_f\, (\rho_f - \rho_g)\, g h_{fg}}{\mu_f \times x \times \Delta t}\right]^{1/4} \quad \text{...(10.34)}$$

2. Average coefficient of heat transfer for vapour condensing on the outside of horizontal tubes of diameter D,

$$h_o = 0.725 \left[\frac{k_f^3\, \rho_f^2\, g\, h_{fg}}{\mathrm{N\,D}\,\mu_f\, \Delta t}\right]^{1/4} \quad \text{...(10.35)}$$

where, k_f = Thermal conductivity of condensate film,

ρ_f = Density of condensate film,

ρ_g = Density of vapour (very small compared to ρ_f),

g = Acceleration due to gravity,

h_{fg} = Latent heat of condensation,

μ_f = Viscosity of the condensate film,

Δt = Difference of temperatures between condensing vapour and outside surface,

N = Number of tubes in a vertical row, and

D = Diameter of tube.

10.3.4.6 Water-side coefficient

In case of forced connection for turbulent flow inside tubes, the water-side coefficient (h_i) is given by Dittus-Boelter equation as follows:

$$h_i = 0.023 \times \frac{k}{\mathrm{D}} \left(\frac{\mathrm{D V}\rho}{\mu}\right)^{0.8} \left(\frac{c_p \mu}{k}\right)^{0.4} \quad \text{...(10.36)}$$

where, k = Thermal conductivity of water,

D = Inside diameter of tube,

V = Average velocity of water,

ρ = Density of water,

μ = Viscosity of water, and

c_p = Specific heat of water at constant pressure.

The expressions $\frac{\mathrm{DV}\rho}{\mu}$ and $\frac{c_p \mu}{k}$ are called *Reynolds number* and *Prandtl number* respectively.

10.3.4.7 Air-side coefficient

In refrigerant condensers with forced convection, air flows outside the tubes in cross-flow to the refrigerant. Average heat-transfer coefficients for the forced convection of air across a tube may be calculated from the Grimson's equation which is given as :

$$\frac{hD}{k} = C\left(\frac{Du_{\infty}\rho}{\mu}\right)(Pr)^{1/3} \qquad ...(10.37)$$

where constants C and n are tabulated as functions of Reynolds number in Table 10.1. Pr stands for Prandtl number.

Table 10.1 Values of constants C and n

Reynolds number (Re)	*C*	*n*
0.4-4	0.989	0.33
4-40	0.911	0.385
40-4000	0.683	0.466
4000-40000	0.193	0.618
40000-400000	0.0266	0.805

In several heat exchanger arrangements multiple rows of tubes are involved, which may be *staggered* or *in-line* as shown in Fig. 10.29.

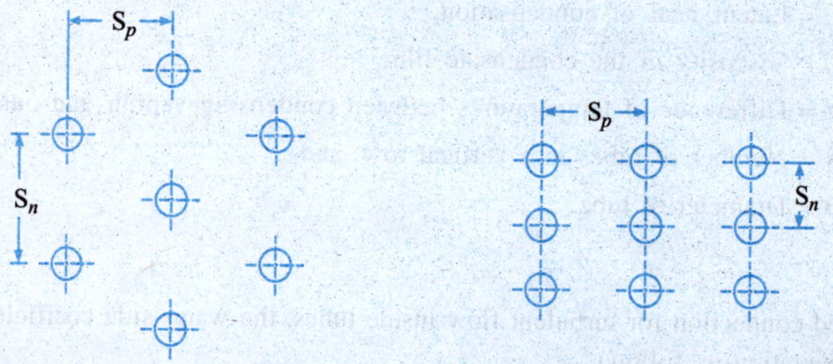

(*a*) Staggered tube rows (*b*) In-line tube rows

Fig. 10.29. Tube arrangements.

The Reynolds number is based on velocity u_{∞} at a distance from the tube. In case of tube banks, u_{∞} is the *face velocity*, given by:

$$u_{\infty} = \frac{\text{Volume flow rate}}{\text{Face area}} = \frac{Q_v}{A_F}$$

The Reynolds number, however, is based on the maximum velocity (u_{max}), the value of which in case of tube bank is given as:

$$u_{max} = u_{\infty} \times \frac{S_n}{S_n - D} \qquad ...(10.38)$$

where S_n in the vertical pitch of tubes which may be staggered or in line type as shown in Fig. 10.29.

- In case of *natural connection* of air on horizontal tubes, the following simplified relations may be used :

For *laminar flow,* $10^4 < Gr.Pr < 10^9$

$$h = 1.32\ (\Delta T/D)^{1/4} \quad ....(10.39)$$

For *turbulent flow,* $Gr \cdot Pr > 10^9$

$$h = 1.24\ (\Delta T)^{1/3} \quad ...(10.40)$$

where, Gr and Pr are the Grashaff's and Prandtl's numbers respectively.

10.3.4.8 Finned tubes

The eqn. 10.33 (*a*) may be written on the basis of inside area as follows:

$$\frac{1}{U_i} = \frac{A_i}{A_0 h_0} + \frac{xA_i}{kA_m} + \frac{1}{h_f} + \frac{1}{h_i} \quad ...(10.43)$$

From the above equation we find that U_i can be increased by increasing $\frac{A_0}{A_i}$, the ratio of outside to inside surface areas. Thus if *the outside of the tubes in finned, the ratio* $\frac{A_0}{A_i}$ *is increased.*

- For R-12 and R-22 condensers the finned tubes (fins to be used on the refrigerant side) are popular.
- For ammonia condensers the *finned tubes should not be used* because the condensing coefficients for ammonia are very high, approximately 8000 W/m^2 K, as compared with approximately 1600 W/m^2 K for R-12.

10.3.5 Evaporative Condensers

- In an evaporative condenser *both air and water are employed as a condensing medium.*
- Fig. 10.30 shows a schematic diagram of such a condenser. Water is pumped from the sump of the evaporative condenser to a spray header and sprayed over the condenser coil. At the same time a ***fan*** draws air from the bottom side of the condenser and dischargers out at the top of the condenser. An ***eliminator*** *is provided above the spray header to stop particles of water escaping along with the discharge air.* The spray water coming in contact with the condenser-tube surface, evaporates into the air stream, the amount of heat for vaporising the water being taken from the refrigerant, thereby condensing the gas.

  The cold water that drops down into a sump is recirculated. In order to make up the deficiency caused by the evaporated water, additional water in supplied to the sump. The make-up supply is controlled by a float valve (not shown in the figure).

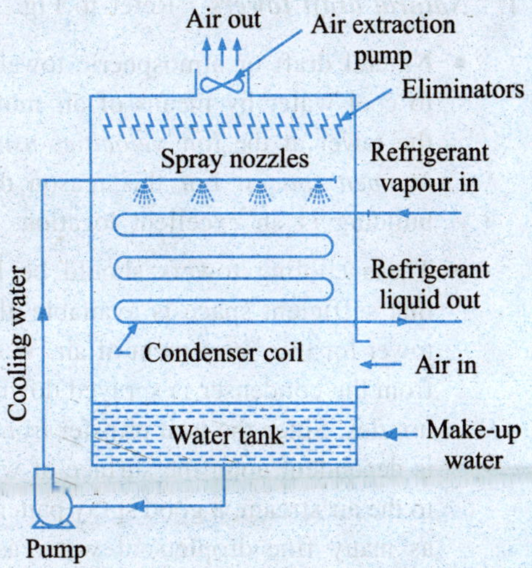

Fig. 10.30. Evaporative condenser.

- An *evaporative condenser occupies less space and needs less power than a water-cooled condenser.*
- The most troublesome point in this condenser is the difficulty of keeping the surface of the condenser coil clean. Because of the maintenance problem, evaporative condensers are not much in favour.

10.3.6 Cooling Towers and Spray Ponds

The use of cooling towers and spray ponds is made when *water-cooled condensers are desired and water is scarce or expensive.*

— A *cooling tower* is an enclosed tower like structure through which atmospheric air circulates to cool large quantities of warm water by direct contact.

— A *spray-pond* consists of a piping and spray nozzle arrangement suspended over an outdoor open reservoir or pond. It can also cool large quantity of warm water.

The capacity of cooling towers and spray ponds depends upon the amount of evaporation of water that takes place, which in turn depends upon the following factors:

(*i*) Amount of surface of water exposed to air,

(*ii*) Length of exposure time,

(*iii*) Velocity of air passing over the water droplets formed in the cooling powers, and

(*iv*) Wet bulb temperature of the atmospheric air.

Cooling towers:

There are used *for large installation, absorption refrigeration units, steam-jet systems, and centrifugal refrigeration units.*

On the basis of air circulation, cooling towers are *classified* as:

1. Natural draft towers
2. Mechanical draft towers.

1. ***Natural draft towers :*** Refer to Fig. 10.31.

- Natural draft or atmospheric tower is designed to cool water by means of air moving through the tower at the *low velocities usually existing in open spaces.* For this reason the roof of a building is an excellent location.
- These cooling towers should be located such that sufficient space is available all around the tower for free movement of air. Water returning from the condenser is sprayed down through the nozzles. Since the heat transfer from water to air is dependent upon the surface of water exposed to the air stream, a good spray pattern to produce as many fine droplets of water as possible is essential for good performance of the cooling tower. So *clear nozzles are a must.*

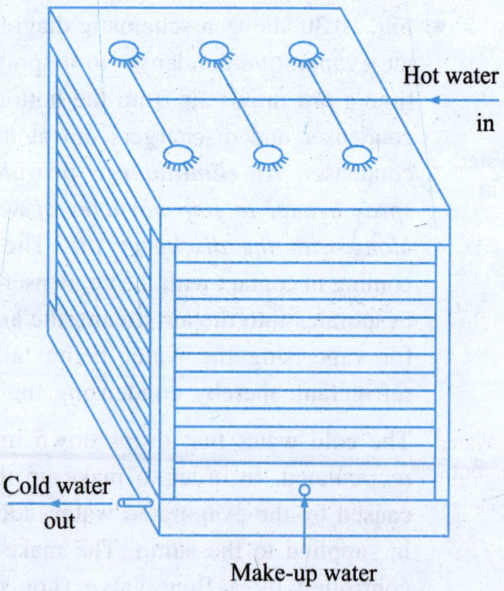

Fig. 10.31. Natural draft tower.

2. ***Mechanical draft towers:***

- In a mechanical draft tower the draft (air movement) is *mechanical, created by fan.*
- It is a vertical shell of wood, metal or masonry etc. Water is distributed near the top of the shell and falls to a collecting basin. As it falls, it passes through air that is circulated from the bottom to the top of the shell by either *forced or induced-draft fans.* Because it passes against the flow of water, a given quantity of air picks up more heat than an equal quantity of air on natural-draft equipment. So less air is needed to cool the same amount of water. Since the air is supplied by fans, the air quantity must be held to a minimum for low operating cost. To prevent excessive water loss and nuisance caused by draft *baffles* are usually installed at the top of the tower. *Loss of water by evaporation and drift is very small.*
- These towers may be *forced-draft* or *induced-draft.*

(*i*) ***Forced-draft cooling tower.*** Refer Fig. 10.32.

- It works well with *corrosive waters, since the fan can be near the ground,* the parts most likely to corrode are easily serviced. However, *the fan maintenance and depreciation costs of these units are high.*
- The *main objective in these units is to have heated air leave the top of the lower at a low velocity.* But at times the air recirculates to the inlet of the fan, and with unfavourable winds this cuts efficiency as much as 20 per cent. During cold weather recirculation can cause ice formation on heavy equipment, buildings, or in the fan ring of the tower, sometimes breaking the fan.

The forced draft cooling towers entail the following *advantages*:

(*i*) Neat appearance.

(*ii*) More adaptable to architectural surroundings.

(*iii*) By locating the fan at the top of the tower, noise is minimized.

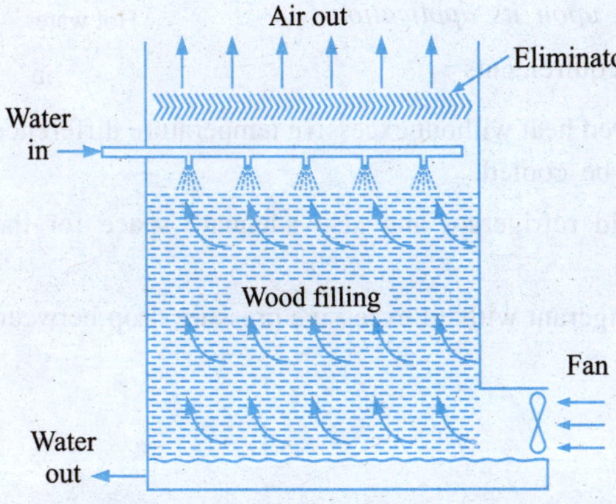

Fig. 10.32. Forced-draft cooling tower.

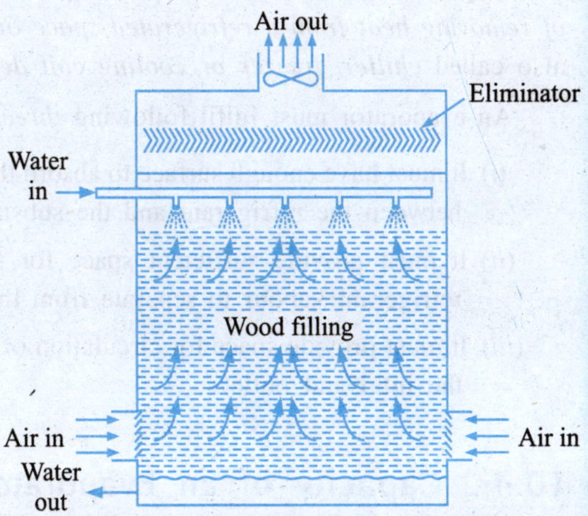

Fig. 10.33. Induced-draft cooling tower.

(*ii*) ***Induced-draft cooling tower:*** Refer to Fig. 10.33.

- In this type of cooling tower the *fan is mounted at the top*. The air movement is vertical, across the filling, and upward at a high velocity to prevent recirculation. If the unit is to handle small loads, the fan is mounted at one end to give a horizontal air flow.
- This type of tower is *normally installed outdoors* but, because the draft is created mechanically, many other installations including indoors with ductwork are also possible.

Spray ponds:

Natural ponds were first used with water discharged near the shore and the intake water taken from the centre and as far below the surface as possible. In order to provide a great area of exposed water, ***spray ponds*** (Fig. 10.34) were developed. For these, nozzles are located so as to spray the discharge water into the air. They should break up the water into droplets but not into a mist, which would easily drift off. In addition to providing more water-surface area, the water on a pool, the air velocity is greater at a distance above the pool than at the pool surface, and greater volume of air contacts the water. Ponds are located on the ground or on treated roofs. In many instances sufficient horizontal area is not available, and cooling towers are used.

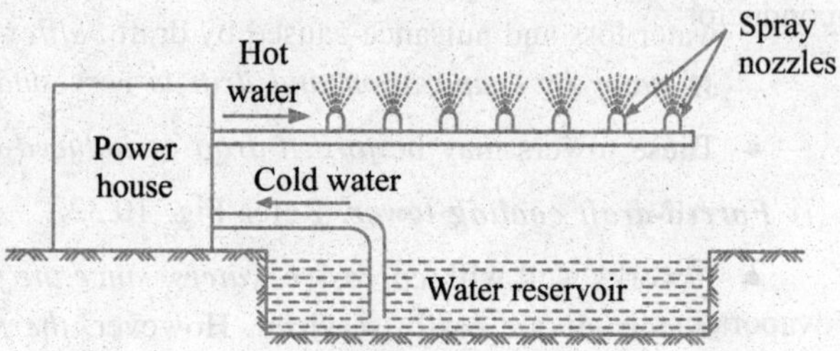

Fig. 10.34. Spray pond.

10.4 EVAPORATORS

10.4.1 Introduction

An **evaporator** *is any heat transfer surface in which a volatile liquid is vapourised for the purpose of removing heat from a refrigerated space or product and also cooling its own coils.* Evaporator is also called *chiller, freezer or cooling coil depending upon its application.*

An evaporator must fulfil following *three* main requirements :

(*i*) It must have enough surface to absorb the required heat without excessive temperature difference between the refrigerant and the substance to be cooled.

(*ii*) It must provide sufficient space for the liquid refrigerant and also adequate space for the refrigerant vapour to separate from the liquid.

(*iii*) It must provide space for circulation of the refrigerant without excessive pressure drop between the intel and outlet.

10.4.2 Capacity of an Evaporator

The **capacity of an evaporator** is *defined as the amount of heat absorbed by it over a given period of time.* The heat transfer capacity (i.e., heat absorbed) of an evaporator is given as :

$$Q = UA\,(t_2 - t_1) \text{ watts.}$$

where, Q = Overall heat transfer coefficient, W/m^2 °C,

A = Area of evaporator surface, m^2,

t_2 = Temperature of the medium to be cooled (or temperature outside the evaporator), °C, and

t_1 = Saturation temperature of refrigerant at evaporator pressure (or temperature inside the evaporator), °C.

Following are some important *factors which* affect the heat transfer capacity of an evaporator:

1. **Material.** The material to be used for the construction of an evaporator should be good conductor of heat and not be affected by the refrigerant.

— Iron and steel can be used with all common refrigerants.

— Brass and copper are used with all refrigerants *except ammonia.*

— Aluminium should *not* be used with freon.

2. **Temperature difference.** The difference of temperature between the refrigerant within the evaporator and the product to be cooled plays a significant role in the heat transfer capacity of an evaporator.

- A *too low temperature difference* (below 8°C) may *cause slime on certain products such as meat or poultry.*
- A *too high temperature difference causes excessive dehydration.*

3. **Velocity of refrigerant.** The heat transfer capacity of an evaporator is also affected by the velocity of refrigerant. With the increase in velocity of the refrigerant the overall heat transfer coefficient also increases, but the increased velocity leads to greater loss of pressure in the evaporator. Therefore, *the only recommended velocities for different refrigerants which give high heat transfer rates and allowable pressure loss should be used.*

4. **Wall thickness of the evaporator coil.** The wall thickness of the evaporator coil has a slight effect on total heat transfer capacity because the evaporators are usually made from highly conductive materials.

5. **Contact surface area.** The contact surface available between the walls of evaporator coil and the medium being cooled is an important factor which affects the heat transfer capacity of an evaporator. It depends on the physical size and shape of the coil of the evaporator.

10.4.3 Heat Transfer in Evaporators

In evaporators there are three heat-transfer resistances; these are:

(*i*) *Refrigerant side* for the *transfer of heat from solid surface to the liquid refrigerant.*

(*ii*) *Metal wall.*

(*iii*) *Cooling medium side* which could be due to air, water, brine or any other fluid or a wetted surface on a cooling and dehumidifying coil.

Here, the transfer of heat from a solid surface of the evaporating refrigerant is significant. However, the mechanism of boiling is so complex that it is difficult to predict the heat transfer coefficient because of the influence of such factors as surface tension, saturation temperature, latent heat and nature of solid surface.

The boiling process, in commercial equipment, occurs in the following two types of situations:

(*i*) **Pool boiling** .... as in *flooded evaporators*;

(*ii*) **Flow or forced-convection boiling** .... as in *direct expansion evaporators.*

10.4.3.1 Pool boiling

As illustrated in Fig. 10.35, pool boiling exhibits the following three distinct regimes:

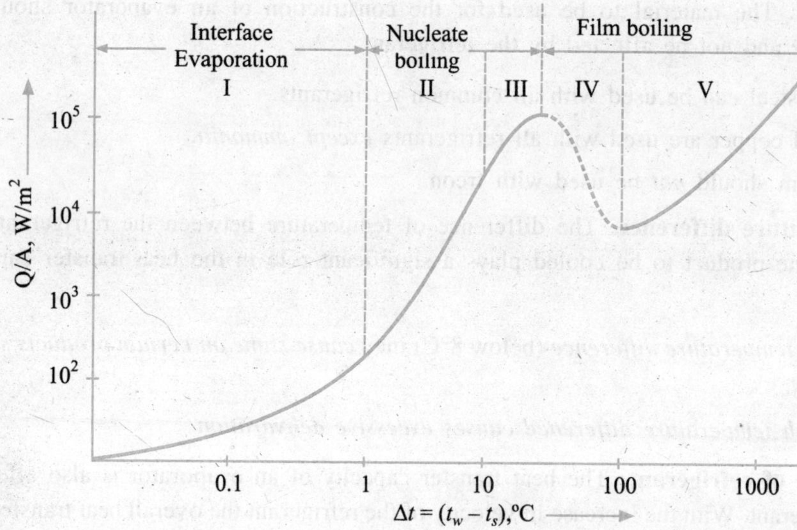

Fig. 10.35. Pool boiling—Heat flux vs temperature difference.

(*i*) ***Interface evaporation.*** Interface evaporation (evaporation process with no bubble formation) exists in region I, called the *free convection zone*. Here the excessive temperature $\Delta t = t_w - t_s$ (t_w is the solid wall temperature and t_s is the saturation temperature of evaporating substance) is very small and ≈5°C. In this region the liquid near the surface is superheated slightly, the convection currents circulate the liquid and evaporation takes place at the liquid surface.

(*ii*) ***Nucleate boiling.*** This type of boiling exists in regions II and III. With increase in excess temperature (Δt) the formation of bubbles on the metal surface at certain localised spots commences. The bubbles condense in the liquid without reaching the liquid surface. In fact it is the region II where nucleate boiling starts. With further increase in Δt the bubbles are forced more rapidly and rise to the surface of the liquid resulting in rapid evaporation, as indicated in region III. The *nucleate boiling* is thus characterised by *formation of bubbles at the nucleation sites and the resulting liquid agitation. The bubble agitation induces considerable mixing that promotes substantial increase in the heat flux and the boiling heat transfer coefficient.*

(*iii*) ***Film boiling.*** When excessive temperature (Δt) is increased further, the heating surface gets covered with vapour bubbles, resulting in a process called *film boiling*.

- Heat transfer in refrigerant evaporators corresponds to regime II of the nucleate boiling in which bubbles formed condense in the liquid before reaching the surface.

Heat transfer coefficient for nucleate boiling:

Rohsenow suggested the following relation for nucleate pool boiling:

$$\frac{c_f\,(t_w - t_s)}{h_{fg}} = C_{st}\left[\frac{Q/A}{\mu_f\,/h_{fg}}\left\{\frac{\sigma}{g\,(\rho_f - \rho_g)}\right\}^{1/2}\right]^r\left(\frac{c\mu}{k}\right)_f^s \qquad ...(10.44)$$

where, c_f = Specific heat of saturated liquid,

$(t_w - t_s)$ = Excess temperature,

h_{fg} = Enthalpy of vaporisation,

C_{sf} = A function of the particular fluid-heating surface combination (Its value for CCl_4 – copper combination is 0.013, the same value can also be used for copper and R-12, R-22 etc.),

Q/A = Heat flux in W/m^2,

μ_f = Viscosity of saturated liquid,

σ = Vapour-liquid surface tension of the fluid,

g = Acceleration due to gravity,

ρ_f = Density of saturated liquid, and

ρ_g = Density of saturated vapour.

Rohsenow obtained the value of the exponent r as 0.33 by plotting the experimental data of fluid (Pr = constant) for the first term on the R.H.S. versus the L.H.S. Similarly, a cross plot of $\frac{c_f\,\Delta t}{h_{fg}}$ vs Pr for constant values of the first term on the R.H.S. gives s equal to 1.0 for water and 1.7 for freons.

10.4.3.2 Flow or forced-convection boiling

- *"Flow or forced-conection boiling" takes place when a liquid flows by forced connection over a surface maintained at a higher temperature than the saturation temperature of the fluid.* In this type of boiling heat transfer rates are *substantially higher* than those of pool boiling.
- The forced-convection boiling processes occurring in *dry evaporators* are *very complex.* The flow is of a low quality at the inlet end. As the evaporation proceeds, the quality increases, a critical condition (tube-wall dry out) appears and then the remainder of the tube may have superheated vapour with liquid droplets.

10.4.3.3 Fluid-side heat transfer

The general *expression for fluid-side heat transfer coefficient* is given by:

$$h_f = C\,\frac{k}{D}\left(\frac{VD\rho}{\mu}\right)^{0.8}\left(\frac{c_p\,\mu}{k}\right)^{0.4} \qquad ...(10.45)$$

where, h_f = Fluid-side heat transfer coefficient,

C = An experimentally determined constant,

k = Conductivity,

D = Inside diameter of the tube,

V = Velocity of fluid,

ρ = Density of fluid,

μ = Viscosity of fluid, and

c_p = Specific heat at constant pressure.

The following situations are common :

(i) *Liquid flows through the evaporation tubes* : For h_f refer to eqn. (10.45).

(ii) *Liquid flows through the evaporation shell* :

$$h_f = C\,(m)^{0.5} \quad \text{...(10.46)}$$

where, m = Mass flow rate of liquid, and

C = Constant for a given evaporator and fluid

(iii) *Air flows over the finned coil by forced convection* :

$$h_f = C\,(m)^{0.4} \quad \text{...(10.47)}$$

where, m = Rate of air flow, and

C = Experimental constant for a given evaporator.

(iv) *Air flows over the cold pipes by natural convection* :

$$h_f = 0.2\left(\frac{t_{db} - t_m}{D}\right)^{0.25} \quad \text{...(10.48)}$$

where, t_{db} = Dry bulb temperature of air, and

t_m = Temperature of pipe.

10.4.4 Classification of Evaporators

Evaporators may be classified as follows:

(a) **On the basis of operating conditions:**

1. Flooded type evaporators.
2. Dry or direct expansion type evaporators.

(b) **On the basis of construction of the surfaces:**

1. Bare-tube evaporator
2. Plate-surface evaporator
3. Finned-tube evaporator.

The above mentioned evaporators are discussed in the following articles.

10.4.5 Flooded Type Evaporators

Fig. 10.36 shows a *flooded type evaporator.*

A flooded or overfeed evaporator is one wherein *the amount of liquid refrigerant circulated through the evaporator is considerably in excess of that which can be vaporised.* Here a *constant refrigerant liquid level is maintained.* A ***float valve*** is *used as the throttling device which maintains a constant liquid level in the evaporator.* Due to the heat supplied by the substance to be cooled, the liquid refrigerant is vaporised and so the liquid level falls down. The float valve opens to admit more liquid and thus maintains a constant liquid level . As a result the evaporator is always filled with liquid to the level determined by the float adjustment and the inside surface is wetted with the liquid. Thus this type is called *flooded evaporator.*

To prevent liquid carry over to compressor, ***accumulator*** is generally used with a flooded evaporator. Accumulator also serves as the chamber for the liquid level float valve. The evaporator coil is connected to the accumulator and the liquid flow from the accumulator to the evaporator coil is generally by gravity. The vapour formed by the vaporising of the liquid in the coil being lighter, rises up and passes on to the top of the accumulator from where it enters the suction line. Sometime *liquid eliminators* are provided in the accumulator top to prevent the possible carry over of the liquid particles from the accumulator to the suction line.

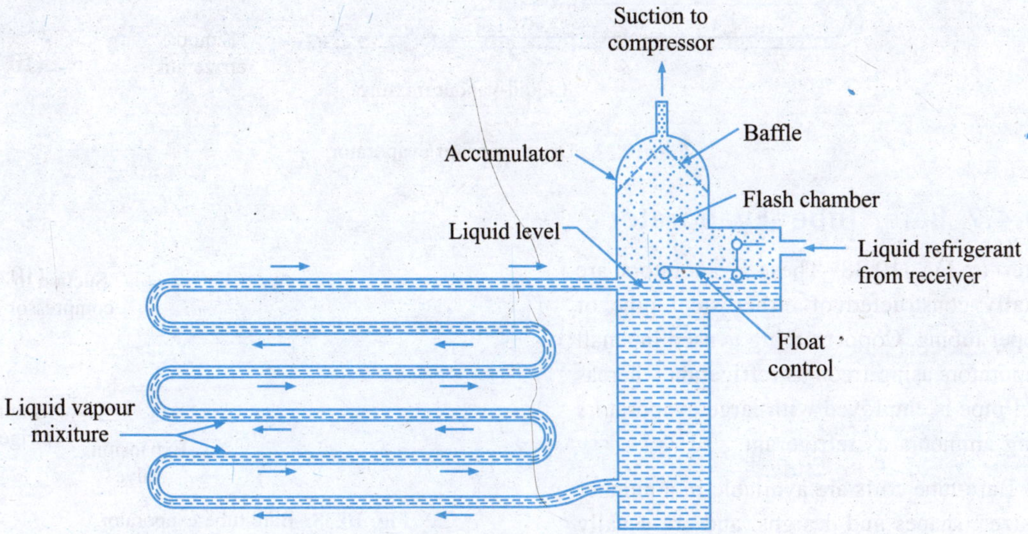

Fig. 10.36. Flooded type evaporator.

- These evaporators *give high rate of heat transfer as the whole surface of the evaporator coil remains in contact with liquid refrigerant.* These *are bulky in construction and require large amount of refrigerant for their working.*
- The flooded evaporators are used in *large installations, especially in chemical and food processing industries.*

10.4.6 Dry Expansion Evaporator

Refer to Fig. 10.37. In a dry expansion evaporator the *amount of liquid refrigerant fed into the evaporator is limited to that which can be completely vaporised by the time it reaches the end of the evaporator.* The expansion device feeds the evaporator at such a rate that whole of the refrigerant gets gradually vaporised by the time it reaches the end of the coil. *Feeler bulb of the expansion valve controls the rate of flow through the orifice of the flow control.*

- Such an evaporator is one of the *most widely used device* for producing refrigeration and is used on systems having capacity below 150 tonnes of refrigeration.

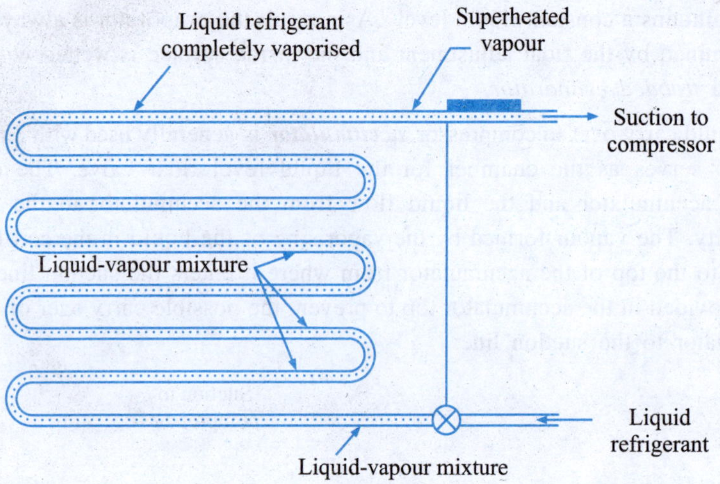

Fig. 10.37. Dry expansion evaporator.

10.4.7 Bare Tube Evaporator

Refer to Fig. 10.38. These evaporators are usually constructed of either steel pipe or copper tubing. Copper tubing is used in small evaporators using freon as refrigerant whereas steel pipe is employed with large evaporators using ammonia as refrigerant.

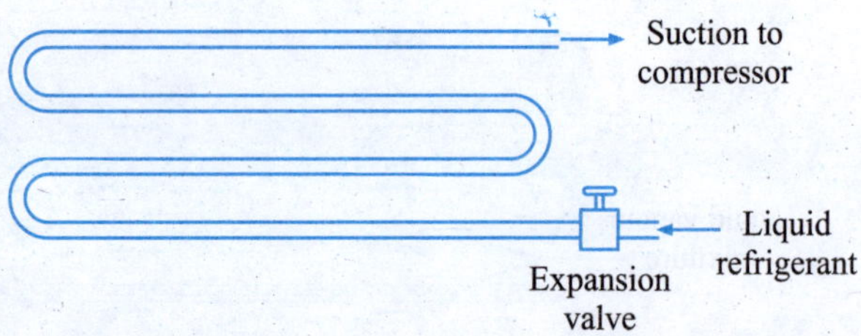

Fig. 10.38. Bare tube evaporator.

Bare-tube coils are available in a number of sizes, shapes and designs, and are usually custom made to the individual application.

- Its use is limited to applications where the box temperatures are under 0°C and in liquid cooling, because the accumulation of ice or frost on these evaporators has less effect on the heat transfer than on those equipped with fins.
- Since these evaporators are easier to clean. They are also extensively used in *household refrigerators.*

10.4.8 Plate Surface Evaporator

Refer to Fig. 10.39. These evaporators are made in different designs. In some cases tubing is attached to the plates whereas in other cases two sheets may be pressed together to provide a path for refrigerant flow between them.

- Such evaporators are used in *domestic refrigerators or frozen food industry.*

Fig. 10.39. Plate surface evaporator.

10.4.9 Finned Tube Evaporator

Refer Fig. 10.40. Finned coils are bare-tube coils upon which metal plates or fins have been installed. The fins increase the surface area of evaporator which *means increased heat transfer and capacity*. The size and spacing of fins depends upon the type of application. Such evaporators have *always a direct expansion feed of liquid refrigerant*.

- Finned tube evaporators are *best suited for window type air-conditioners and other air cooling application where temperature is maintained above 1°C*. For such an application the capacity of evaporator is further increased by adding a fan to the finned evaporator.

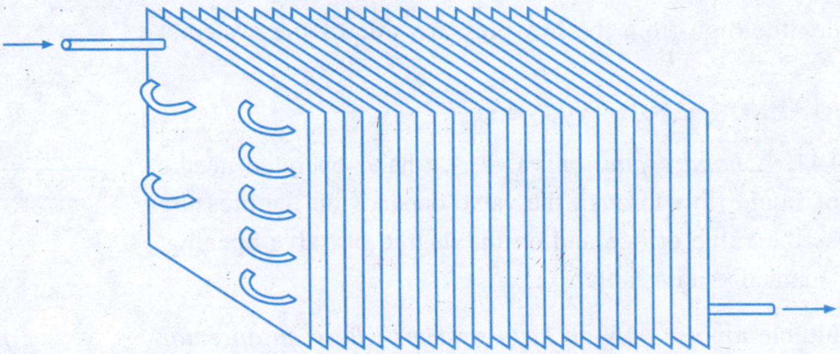

Fig. 10.40. Finned tube evaporator.

10.5 REFRIGERATION CONTROLS

10.5.1 General Aspects

Refrigerant controls *are devices used to control the flow of refrigerant at various points throughout the refrigeration cycle*. Below are the *seven points* where refrigerant controls are used in a system.

(*i*) *"Expansion valves," used to regulate the flow of refrigerant liquid to the evaporator.*

(*ii*) *"Suction-line regulators," used to control the flow of refrigerant gas from the evaporator coil.*

(*iii*) *"Hold back valves,"* used to *limit the flow of gas to the compressor to prevent surge or excessive loads from overloading the compressor motor.*

(*iv*) *"Solenoid valves," used in liquid, suction or discharge lines to interrupt the flow on demand from any one of several types of temperature or pressure-sensing devices.*

(v) "*Check valves,*" *used to prevent the flow of gas from the condenser back to the compressor during off cycles.*

(vi) "*Reversing valves,*" used in heat-pump applications *to defrost cycles to change the flow of refrigerant.*

(vii) "*Hot-gas defrost*" *for modern multiplex systems.*

There are *six basic types of* **refrigerant flow controls:**

1. Hand expansion valve
2. Automatic expansion valve
3. Thermostatic expansion valve
4. Capillary tube
5. Low-side float
6. High-side float.

"**Expansion valve**", is the term usually used in industry, for any *device that meters or regulates the flow of liquid refrigerant to an evaporator.* It has two *purposes*:

1. To reduce the pressure of the liquid refrigerant.
2. To regulate the flow of refrigerant to the evaporator.

It thus divides the high-from the low-pressure side of the system.

10.5.2 Hand Expansion Valve

Refer to Fig. 10.41. A *hand expansion* valve is a hand operated needle valve. The rate of liquid flow through the valve depends on the pressure differential across the valve orifice and on the degree of valve opening, the latter being manually adjustable.

- This is suitable *for use only on large systems where an operator is on duty and where the load on the system is relatively constant* (e.g., ice making plants and cold storages).
- The main **advantages** of a hand expansion valve are its *simplicity* and *low initial cost*. Also, because of its simple construction there is very little that can get out of order. The principle **disadvantage** is that an *operator must be available at all times to make the necessary adjustment to meet the changing load conditions.*

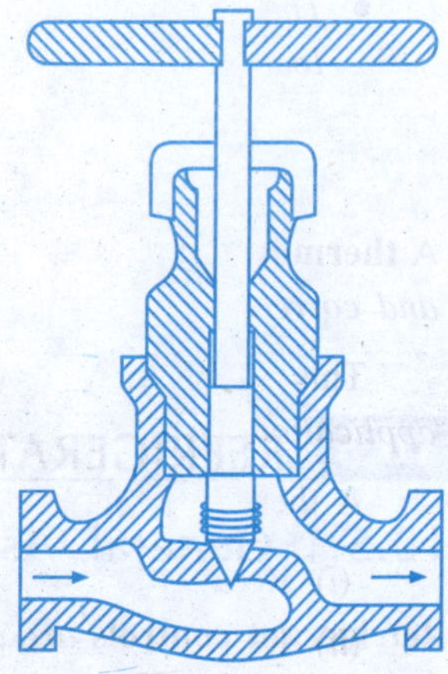

Fig. 10.41. Hand expansion valve.

10.5.3 Automatic Expansion Valve

- The ***automatic expansion valve*** *is a pressure-reducing device*. It is activated by the evaporator pressure which it keeps constant since the pressure of the refrigerant in the evaporator determines evaporator temperature.
- Fig. 10.42 shows a schematic sketch of an automatic expansion valve. *It is a diaphragm-or bellows operated valve with the evaporator pressure acting on the lower side of the diaphragm and atmospheric pressure plus adjustable spring pressure acting on the upper side.* As the

compressor operates to remove the gas from the evaporator, reducing the pressure in the evaporator and under the diaphragm, the adjusting spring pressure pushes the diaphragm down. This motion is transmitted through push rods (or by needle valve steam) to the valve needle, opening it enough to allow more refrigerant to flow to the evaporator. As more liquid enters the evaporator, the pressure increases forcing the diaphragm upward and allowing the valve to close. A properly sized valve will pass enough liquid refrigerant to maintain constant temperature and pressure conditions.

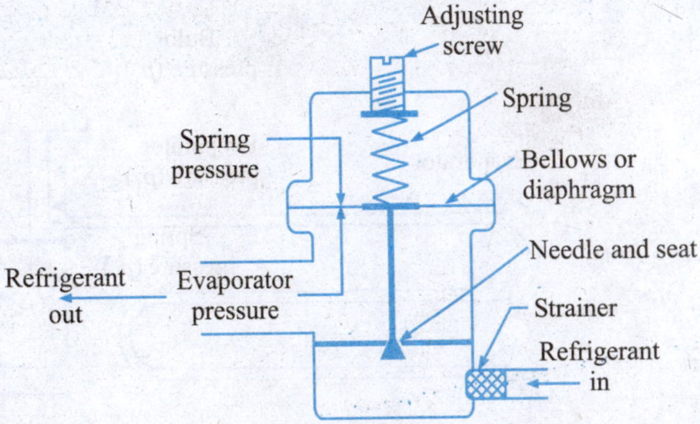

Fig. 10.42. Automatic expansion valve.

- An automatic expansion valve must be set to prevent over feeding on low-load conditions and therefore cannot feed enough on high-load conditions. If *the heat load drops off quickly, the evaporator pressure drops suddenly, opening the valve wide again in trying to raise evaporator pressure to the pressure setting of the valve. As a result the liquid refrigerant can flood back to the compressor and cause much damages. If the heat load increases suddenly, the evaporator pressure will increase rapidly, forcing the diaphragm up and allowing the valve to close.* This '*starves*' the evaporator until the compressor can reduce the pressure and allow more refrigerant to pass into the evaporator.
- *The automatic expansion valves find their greatest use in the systems with relatively constant loads and in systems with only one evaporator coil.*

10.5.4 Thermostatic Expansion Valve

A **thermostatic expansion valve** *is a throttling device which works automatically, maintaining proper and correct liquid flow as per the requirements of the load on the evaporator.*

This valve finds a wide application because of *its adaptability to any type of dry expansion application, automatic operation, high efficiency and ability to prevent liquid flood backs.*

A thermostatic expansion valve performs the following **functions:**

(*i*) Reduces the pressure of liquid from the condenser pressure to evaporator pressure.

(*ii*) Keeps the evaporator fully active.

(*iii*) Modulates the flow of liquid to the evaporator according to the load requirement of the evaporator so as to prevent flood back of liquid refrigerant to the compressor.

Fig. 10.43 shows a *thermostatic expansion valve.*

The following are the important *parts* of the valve:

1. Power element with a feeler bulb
2. Valve seat and needle
3. Adjustment spring
4. Bellows or diaphragm.

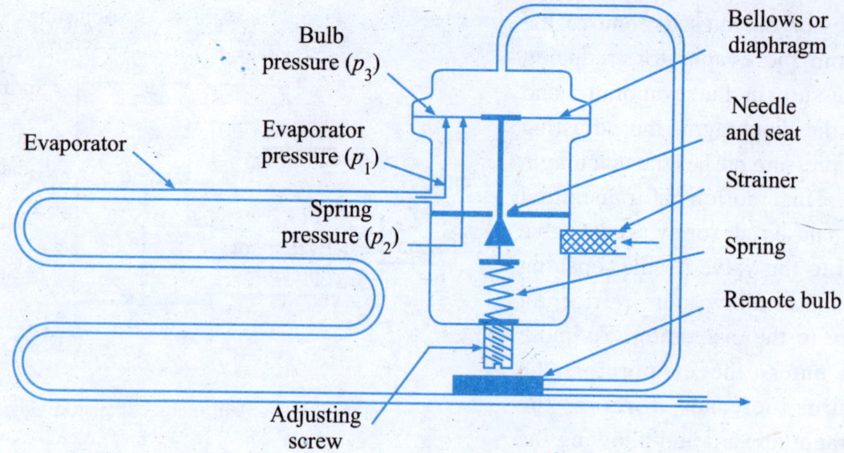

Fig. 10.43. Thermostatic expansion valve.

The remote bulb charged with fluid which is open on one side of the diaphragm through a capacity tube is clamped firmly to the evaporator outlet. The temperature of the saturated liquid in vapour mixture is the same as the temperature of the superheat gas leaving the evaporator at the location. The *pressure of the liquid in the bulb* (p_3) *tends to open the valve.* This *pressure is balanced by pressure due to spring* (p_2) *plus pressure in the evaporator* (p_1).

The *performance characteristics of thermostatic-expansion valves are most suitable for application in air-conditioning and refrigerant plants.* When the cooling load '*increases*', the refrigerant evaporates at a *faster rate* in the evaporator than the compressor can suck. As a result the pressure and degree of superheat in the evaporator increase. The increase in *superheat causes the valve to open more* and to *allow more refrigerant to enter the evaporator.* At the same time, the increase in suction pressure also enables the compressor to delivery increased refrigerating capacity. When the *cooling load 'decreases'* the refrigerant evaporates at a *slower rate* than the compressor can suck. As a result, the evaporator pressure drops and the degree of superheat decreases. The valve tends to close and the compressor delivers less refrigerating capacity at a decreased suction pressure. Thus the thermostatic-expansion valve, as opposed to the automatic expansion valve, is *capable of meeting varying load requirement.*

- *Most thermostatic expansion valves are set for 5°C of superheat and are usually rated in tonnes of refrigeration.*

10.5.5 Capillary Tube

A capillary tube is a *fixed restriction-type device.* It is the simplest of the refrigerant flow controls, consisting merely of a fixed length (from 0.5 m to 5 m) of small diameter (0.5 mm to 2.25 mm) tubing installed between the condenser and the evaporator, usually in place of the conventional liquid line (Fig. 10.44).

The pressure drop through the capillary tube is due to the following *two factors:*

(*i*) *Friction,* due to fluid viscosity, resulting in *'frictional pressure drop'*.

(*ii*) *Acceleration, due* to the flashing, of the liquid refrigerant into vapour *resulting 'momentum pressure drop'*.

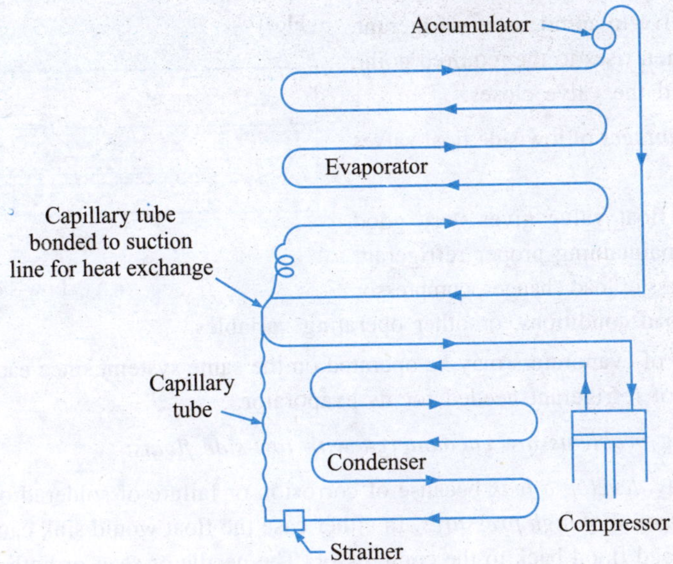

Fig. 10.44. Capillary tube.

The cumulative pressure drop must be equal to the difference in pressures at the two ends of the tube. The flow through the capillary tube will, therefore, adjust so that the pressure drop through the tube just equals the difference in pressures between the condenser and evaporator. For a *given state of the refrigerant, the pressure drop is directly proportional to the length and inversely proportional to the bore diameter of the tube.*

To obtain the desired flow and pressure drop a number of combinations of length and bore are possible for a capillary tube. However, once a capillary tube has been selected, it will be suitable only for design pressure drop. It *cannot satisfy the flow requirements with changing condenser and evaporator pressures.* Even then the *capillary tube is most commonly used in small refrigerators, window type air-conditioners, water-coolers, etc.*

Advantages:

Capillary tube claims the following ***advantages:***

1. Simplicity.
2. Low cost.
3. Absence of moving parts.
4. It is found *most advantageous with on-off control because of its unloading characteristics. Thus when the compressor stops, it allows high and low pressures to equalize, thereby enabling the compressor motor to re-start on no load. Accordingly, smaller low starting torque motors can be used.*

10.5.6 Low-side Float

Refer to Fig. 10.45. A low-side float is essentially a hollow ball, pan or inverted bucket connected through linkages and pivots to open or close needle valve. It *maintains a predetermined liquid level in an evaporator*. As refrigerant is evaporated, the liquid level falls, lowering the float. The connecting linkage opens the valve to admit more refrigerant. As the liquid level then rises to the required point, the float is lifted and the valve closes.

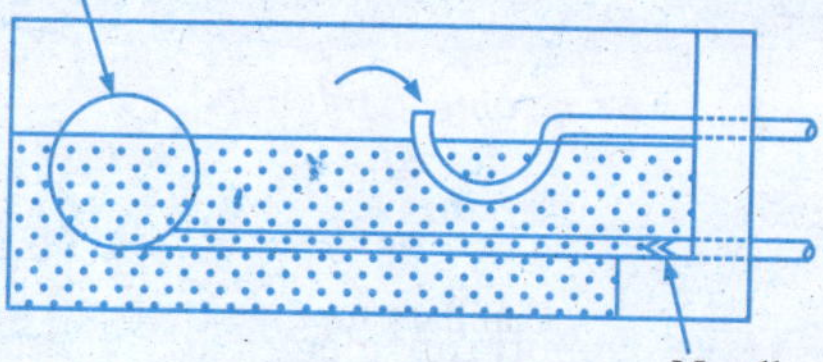

Fig. 10.45. Low-side float.

Some of the ***advantages*** of low-side float valves are:

(*i*) A low-side float valve gives very good control by maintaining proper refrigerant level regardless of load changes, compressor off cycles load conditions, or other operating variables.

(*ii*) Any number of evaporators may be operated on the same system, since each valve drains only the amount of refrigerant needed for its evaporator.

- The following *problems are encountered with low-side floats*:

The float ball may *develop a leak* because of corrosion or failure of soldered or welded joint. The float may *collapse because of high pressures*. In either case the float would sink causing the refrigerant to feed continuously and flood back to the compressor. The needle or seat, or both, may wear, causing it to leak continuously. The float ball may operate erratically because of the boiling action of refrigerant. In such cases the float assembly is placed in a separate chamber.

10.5.7 High-side Float

Fig. 10.46 shows a schematic sketch of a high-side float. It has the same elements as a low-side float: a hollow ball, linkages, and a needle valve. It *differs from the low-side float in that it is on the high-pressure side of the system and that a rising liquid level opens the valve*. It is mounted below the condenser and passes the liquid refrigerant to the evaporator as rapidly as it is condensed, but it does not allow passage of uncondensed vapour. This requires that the major portion of the refrigerant charge be located in the evaporator.

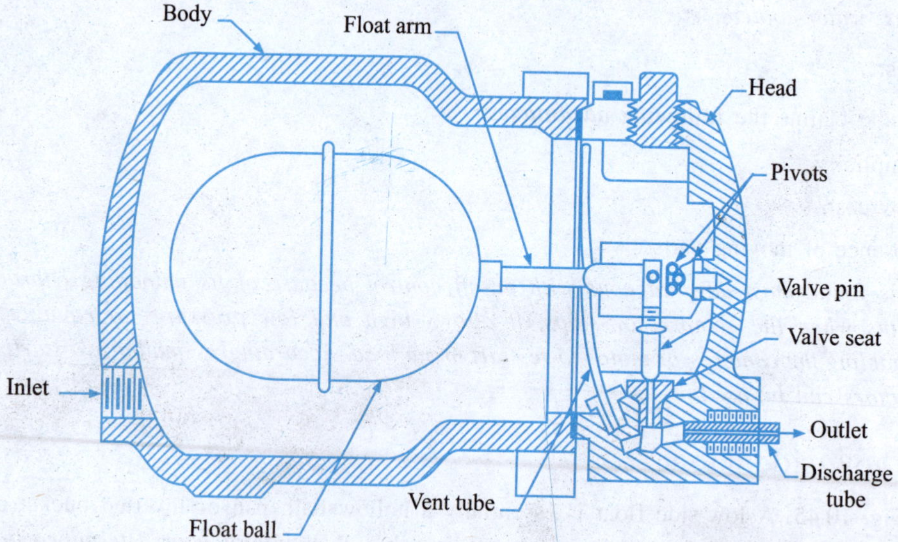

Fig. 10.46. High-side float.

- High-side floats find wide applications in *domestic and small commercial systems* with a single compressor, evaporator, and condenser. They are also used in *discharge oil traps to return oil to the system and some types of automatic purgers.*

10.5.8 Solenoid Valves

A *solenoid valve* is simply an electrically operated valve which consists essentially of a coil of insulated copper wire and iron core or armature (sometimes called a *plunger*) which is drawn into the centre of the coil magnetic field when the coil is energized. By attaching a valve stem and pin to the coil armature, a valve port can be opened and closed as the coil is energized and de-energized respectively.

Solenoid valves are of two types:

(*i*) Direct acting.

(*ii*) Pilot operated.

A direct acting solenoid valve (Fig. 10.47) is opened when the plunger pulls the valve pin away from the valve seat. To keep the solenoid coil to a reasonable size, the direct-acting valve is limited to smaller sizes.

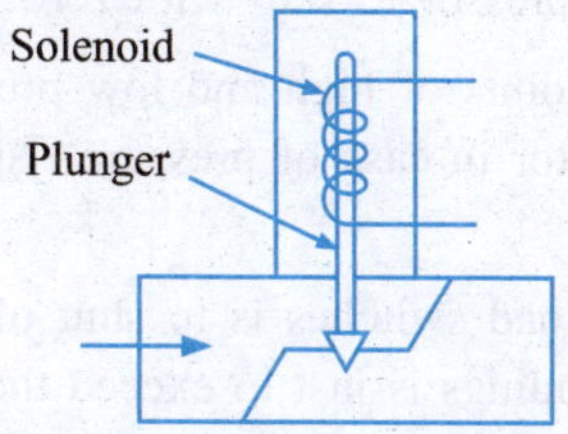

Fig. 10.47. Direct-acting solenoid valve.

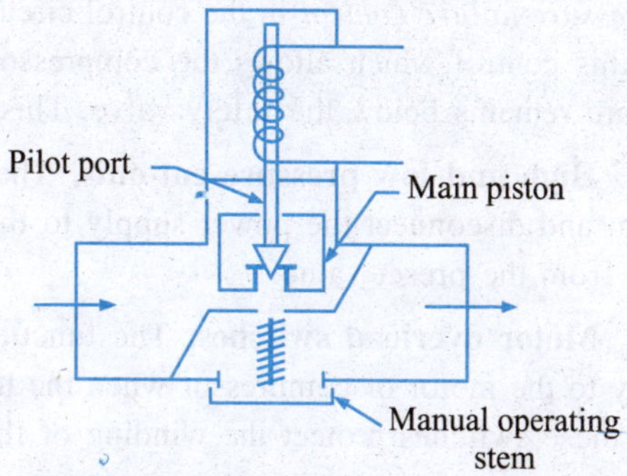

Fig. 10.48. Pilot-operated solenoid valve.

The **pilot-operated solenoid valve** (Fig. 10.48) does not open directly but first opens a pilot port which releases the pressure on top of valve seat. The higher upstream pressure then acts on the valve to open it.

- *Solenoid valves are tight shut off valves if kept in good repair*, but because of being held closed by the pressure across the valve, they stop the float in one direction only. If the pressure conditions reverse, that is, if pressure on the down stream side valve builds up to a higher pressure than the upstream side, the valve will open, permitting flow.
- Solenoid valves are widely *used in refrigerant, water and brine lines in place of manually operated stop valves to provide automatic operation.*
- The common places of use of a solenoid valve are:
  - (*i*) It is installed just ahead of the expansion valve to prevent flow of refrigerant, liquid evaporator when the compressor is not working, and to provide individual temperature control in each refrigerated space of a multiple system.
  - (*ii*) Solenoid valves are frequently used to control the flow of condensing water. A solenoid valve reduces the water cost by stopping the flow of water as soon as the compressor stops. It will open the supply only when the compressor starts again.

10.6 SAFETY DEVICES

To safeguard and protect various components of a refrigeration system a large number of safety devices are used ; a few important of these are discussed below:

1. **Pressure relief valves.** These are safety valves designed to *relieve the pressure in the system to the atmosphere, or to the outdoors through a vent line, in the event that the pressure in the system rises to unsafe level for any reason.* Most refrigerating systems have at least one pressure relief valve mounted on the receiver tank or water-cooled condenser. In many instances, additional relief valves are required at other points in the system.

2. **Fusible plug.** A fusible plug is sometimes substituted for the pressure relief valve. A fusible plug is simply a pipe plug which has been drilled and filled with a metal alloy *designed to melt at some pre-determined fixed temperature;* the melting temperature depends on the pressure temperature relationship of the refrigerants used in the system.

3. **Oil safety switches.** In large compressors forced lubrication is done for which an oil pump is employed. If the pump does not work due to some reason the parts will not be lubricated and these parts will get damaged. Therefore some means must be provided to stop the compressor in case the pressure of lubricating oil falls below a pre-determined minimum value. This is achieved by using an *oil pressure failure control* in the control circuit of refrigeration equipment. A time delay realy is used with this control which allows the compressor to run 90 seconds to 120 seconds even when the oil pressure remains below the safety valve. This enables the compressor to start with zero oil pressure.

4. **High and low pressure cut-outs.** These are used on points of high and low pressure of the system and disconnect the power supply to the compressor motor in case of pressure rising or going down from the preset value.

5. **Motor overload switches.** The function of motor overload switches is to shut off the power supply to the motor of compressor when the temperature of windings is just to exceed the safe value. Thus these switches protect the winding of the motor from burn-outs.

HIGHLIGHTS

1. A compressor is the heart of a vapour compression system.
2. Reciprocating compressors are very suitable for refrigerant (e.g., R-12, R-22, R-502, R-717 etc.) which require relatively small displacement volumes and high condensing pressures.
3. Centrifugal compressors are very suitable for refrigerants (e.g., R-11, R-12, R-113 etc.) which require relatively large displacement volumes and high condensing pressures.
4. Rotary compressors are very suitable for refrigerants (e.g., R-21, R-114) having moderate or low condensing pressures.
5. Multistage compression is very efficient and is now-a-days almost universally adapted except for compressors where the overall pressure rise required is small.
6. **Capacity control of compressors :**

   Reciprocating : (*i*) Using multi-speed motor,

   (*ii*) On and off control,

   (*iii*) Holding the valves open,

(iv) Hot gas by-pass. and

(v) Using multiple units.

Centrifugal : (i) Speed control system, and

(ii) Inlet vane control system.

7. Rotary compressors are suitable for fraction tonnage capacity machines.
8. The helical rotary or screw compressor is a positive displacement compressor in which compression is accomplished by the enmeshing of two mating helically groved rotors suitably housed in a cylinder equipped with appropriate inlet and discharge ports.

Condensers

9. A *condenser* is a heat exchanger in which desuperheating of high temperature vapour changes the phase from vapour to liquid and sub-cooling of condensate occurs.
10. *Air-cooled condensers* use air as the cooling medium, water-cooled condensers use water as the medium and the evaporative condenser is a combination of the above, i.e., uses both water and air.
11. "*Fouling or scaling factor*" represents the resistance offered by the scale or fouling to the heat flow from the refrigerant to water.
12. An *evaporative condenser* occupies less space and needs less power than a water-cooled condenser.
13. A *cooling tower* is an enclosed tower like structure through which atmospheric air circulates to cool large quantities of warm water by direct contact. Cooling towers are used for large installations, absorption refrigeration units, steam-jet systems and centrifugal refrigeration units.
14. A *spray pond* consists of a piping and spray nozzle arrangement suspended over an outdoor open reservoir or pond.

Evaporators

15. An *evaporator* is any heat transfer surface in which a volatile liquid is vaporised for the purpose of removing heat from a refrigerated space or product and also cooling its own coils.
16. *Capacity of an evaporator* is defined as the amount of heat absorbed by it over a given period of time.
17. The boiling process, in commercial equipment, occurs in the following two types of situations:

    (i) *Pool boiling* .... as in flooded evaporators.

    (ii) *Flow or forced-convection boiling* .... as in direct expansion evaporators.
18. *Flooded evaporators* give high rate of heat transfer as the whole surface of the evaporator coil remains in contact with liquid refrigerant. These are bulky in construction and require large amount of refrigerant for their working.
19. *Dry expansion* evaporator is one of the most widely used device for producing refrigeration and is used on systems having capacity below 150 tonnes of refrigeration.
20. *Finned tube evaporators* always have a direct expansion feed of liquid refrigerant.

Refrigeration Controls

21. *Refrigerant controls* are devices used to control the flow of refrigerant at various points throughout the refrigeration cycle. Six basic types of refrigerant flow controls are:

    (i) Hand expansion valve,

    (ii) Automatic expansion valve,

    (iii) Thermostatic expansion valve,

(*iv*) Capillary tube,

(*v*) Low-side float, and

(*vi*) High-side float.

22. A *solenoid valve* is an electrically operated valve which consists essentially of a coil of insulated copper wire and iron core or armature which is drawn into the centre of the coil magnetic field when the coil is energized. Solenoid valves are of two types:

(*i*) Direct acting;

(*ii*) Pilot operated.

OBJECTIVE TYPE QUESTIONS

Compressors

Choose the Correct Answer:

1. The clearance factor is the ratio of
   (*a*) swept volume of the cylinder to the clearance volume
   (*b*) total volume of the cylinder to the clearance volume
   (*c*) clearance volume to the swept volume of the cylinder
   (*d*) clearance volume to the total volume of the cylinder.

2. For a centrifugal refrigerant compressor the pressure ratio
   (*a*) increases as the adiabatic index of compression increases
   (*b*) increases as the adiabatic index of compression decreases
   (*c*) remain same with change in adiabatic index of compression
   (*d*) first increases and then decreases as adiabatic index of compression increases.

3. With suction pressure being fixed, increase in delivery pressure with fixed clearance volume
   (*a*) increases volumetric efficiency
   (*b*) decreases volumetric efficiency
   (*c*) does not change volumetric efficiency
   (*d*) first increases volumetric efficiency and then decreases it.

4. The reciprocating refrigerant compressors are very suitable for
   (*a*) small displacements and low condensing pressures
   (*b*) large displacements and high condensing pressures
   (*c*) small displacements and high condensing pressures
   (*d*) large displacements and low condensing pressures.

5. A refrigerant compressors is used to
   (*a*) raise the pressure of the refrigerant
   (*b*) raise the temperature of the refrigerant
   (*c*) circulate the refrigerant through the refrigerating system
   (*d*) all of the above.

6. For operation with centrifugal refrigerant compressors the least suitable refrigerant is
   (*a*) R-12 (*b*) R-11 (*c*) R-22 (*d*) R-717.

7. The pressure at the inlet of a refrigerant compressor is called
   (*a*) suction pressure (*b*) discharge pressure (*c*) critical pressure (*d*) back pressure.

8. For a centrifugal refrigerant compressor the pressure ratio increases when
   (*a*) density of fluid increases (*b*) adiabatic index of compression decreases
   (*c*) as the blade tip velocity increases (*d*) all of the above happen.

9. Work input to the refrigerant compressor with n as the index of compression
   (*a*) increases with increase in value of n (*b*) decreases with increase in value of n
   (*c*) remains same whatever the value of n
   (*d*) first increases and then decreases with increase in value of n.
10. The clearance volume in reciprocating refrigerant compressors ...... the workdone and the power required for compressing the refrigerant.
    (*a*) does not effect (*b*) increases (*c*) decreases.
11. For centrifugal refrigerant compressor to operate for suction temperature of 10°C and condensing temperature of 40°C with minimum blade tip speed and positive suction pressure the choice of refrigerant would be
    (*a*) R-12 (*b*) R-717 (*c*) R-11 (*d*) R-22.
12. The pressure at the outlet of a refrigerant compressor is called
    (*a*) suction pressure (*b*) discharge pressure (*c*) critical pressure (*d*) back pressure.
13. A thrust bearing is generally necessary in
    (*a*) reciprocating refrigerant compressors (*b*) centrifugal refrigerant compressors
    (*c*) rotary vane type refrigerant compressors (*d*) none of these.
14. Equation of work for single-stage, single-acting refrigerant compressor without clearance on 1 kg of refrigerant delivered when compression index is n is given by (refrigerant is superheated and treated as perfect-gas)
    (*a*) $\frac{n}{n-1} R (T_2 - T_1)$ (*b*) $\frac{n}{n-1} (p_2v_2 - p_1v_1)$
    (*c*) $\frac{n}{n-1} p_1v_1 \left[\left(\frac{p_2}{p_1}\right)^{(n-1)/n} - 1\right]$ (*d*) all of these.
15. For centrifugal refrigerant compressor to operate for suction temperature of 10°C and condensing temperature of 40°C with minimum condensing pressure, refrigerant used is
    (*a*) R-12 (*b*) R-717 (*c*) R-11 (*d*) R-22.
16. The function of compressor in vapour compression refrigeration system is
    (*a*) to continuously circulate the refrigerant through the system
    (*b*) to compress the vapour from low pressure to high pressure
    (*c*) to compress the vapour from low temperature to high temperature
    (*d*) to do all of the above.
17. The work requirement for a reciprocating compressor is minimum when the compression process is
    (*a*) isothermal (*b*) isentropic (*c*) polytropic (*d*) adiabatic
18. Function of providing clearance volume in reciprocating refrigerant compressors is
    (*a*) to reduce the work done per kg of refrigerant delivered
    (*b*) to increase the volumetric efficiency of compressor
    (*c*) to accommodate valves in the head of the compressor
    (*d*) to create turbulence in the air to be delivered.
19. The reciprocating refrigerant compressors are very suitable for
    (*a*) small displacements and high condensing pressure
    (*b*) large displacements and high condensing pressures
    (*c*) small displacements and low condensing pressure
    (*d*) large displacement and low condensing pressures.
20. The refrigerant R 11 is very suitable for
    (*a*) large reciprocating refrigerant compressors
    (*b*) large centrifugal refrigerant compressors

(*c*) small rotary positive displacement refrigerant compressors
(*d*) small reciprocating refrigerant compressors.

21. The refrigerant supplied to a compressor most be
(*a*) superheated vapour refrigerant (*b*) dry saturated liquid refrigerant
(*c*) a mixture of liquid and vapour refrigerant (*d*) any of these.

22. The work input to air compressor is minimum if the compression law followed is
(*a*) Isentropic $pV^{\gamma} = C$ (*b*) Isothermal $pV = C$ (*c*) $pV^{1.2} = C$ (*d*) $pV^{1.35} = C$.

23. Centrifugal refrigerant compressors are employed for the following refrigerants
(*a*) R-113 (*b*) R-22 (*c*) R-717 (*d*) R-21.

24. With increase in clearance volume, the ideal work of compressing 1 kg of refrigerant
(*a*) increases (*b*) decreases
(*c*) remains same (*d*) first increases and then decreases.

ANSWERS

1. (*c*)	2. (*b*)	3. (*b*)	4. (*c*)	5. (*d*)	6. (*d*)	7. (*a*)	8. (*d*)
9. (*a*)	10. (*a*)	11. (*a*)	12. (*b*)	13. (*b*)	14. (*d*)	15. (*c*)	16. (*d*)
17. (*a*)	18. (*c*)	19. (*a*)	20. (*c*)	21. (*a*)	22. (*b*)	23. (*a*)	24. (*c*).

Condensers

Choose the Correct Answer:

1. In a shell and coil condenser
(*a*) water flows in the shell and the refrigerant in the coil
(*b*) water flows in the coil and the refrigerant in the shell
(*c*) only water flows through the shell as well as coil
(*d*) only refrigerant flows through the shell as well as coil.

2. For the suction vapour cooled hermetic compressor refrigeration units
(*a*) the electric motor heat is also load on condenser
(*b*) the electric motor heat is not a load on condenser
(*c*) may be any one (*a*) or (*b*)
(*d*) none of the above.

3. For one ton refrigeration plant, if the heat rejection ratio of condenser is 1.2, the C.O.P. of the plant is
(*a*) 3 (*b*) 4 (*c*) 5 (*d*) 6.

4. For water-cooled refrigerant condensers, the typical values of temperature difference between refrigerant and entering water is
(*a*) 3°C to 4°C (*b*) 14°C to 17°C (*c*) 8°C to 10°C (*d*) 20°C to 30°C.

5. A window air-conditioner uses
(*a*) natural convection air-cooled condenser (*b*) shell and coil type condenser
(*c*) forced convection air-cooled condenser (*d*) double pipe condenser.

6. For one ton refrigeration plant, if the C.O.P. of the plant is 5, the heat rejection ratio of the condenser is
(*a*) 2.2 (*b*) 1.2 (*c*) 2.333 (*d*) 1.333.

7. In a double pipe condenser
   (*a*) water flows through inner pipe and refrigerant through annular outer pipe
   (*b*) refrigerant flows through inner pipe and water through annular outer pipe
   (*c*) both (*a*) and (*b*) are used (*d*) none of the above.

8. For air-cooled refrigeration condensers the typical values of the temperature difference between refrigerant and air is
   (*a*) 3°C to 4°C (*b*) 14°C to 17°C (*c*) 8°C to 10°C (*d*) 20°C to 30°C.

9. For one ton refrigeration plant, if the heat rejected at the condenser is 280 kJ, heat rejection ratio is
   (*a*) 2.333 (*b*) 1.333 (*c*) 1.2 (*d*) 2.2

10. The heat rejection ratio is defined as
   (*a*) $\frac{\text{Heat absorbed at evaporator}}{\text{Heat rejected at condenser}}$ (*b*) $\frac{\text{Heat rejected at condenser}}{\text{Heat absorbed at evaporater}}$
   (*c*) $\frac{\text{Heat rejected at condenser}}{\text{Work done on the compressor}}$ (*d*) $\frac{\text{Heat absorbed at evaporator}}{\text{Work done on compressor}}$

11. For a fixed condensing temperature, the heat rejection ratio
   (*a*) increases with decrease in suction temperature
   (*b*) decreases with decrease in suction temperature
   (*c*) remains unaffected with decrease in suction temperature
   (*d*) is not predictable with decrease in suction temperature.

12. In shell and tube condenser
   (*a*) water flows in the shell and refrigerant flows in the tubes
   (*b*) water flow in the tubes and refrigerant flows in the shell
   (*c*) any one of (*a*) and (*b*) may be there
   (*d*) none of the above.

13. A domestic refrigerator uses
   (*a*) natural convection air-cooled condenser (*b*) shell and coil type condenser
   (*c*) forced convection air-cooled condenser (*d*) double pipe condenser.

14. In shell and coil condenser
   (*a*) water flows in the coil and refrigerant in shell
   (*b*) water flows in the shell and refrigerant in coil
   (*c*) any one (*a*) or (*b*) could be there (*d*) none of the above.

15. In a refrigeration system having water cooled condenser, in a hot humid day
   (*a*) the head pressure will be lower (*b*) the head pressure will be higher
   (*c*) the head pressure of water-cooled condenser is not affected by high humidity.
   (*d*) the heat pressure of water-cooled condenser is not affected by dry bulb temperature.

16. The heat rejection factor (HRF) is given by
   (*a*) 1 + C.O.P. (*b*) 1 – C.O.P. (*c*) $1 + \frac{1}{\text{C.O.P.}}$ (*d*) $1 - \frac{1}{\text{C.O.P.}}$

17. For ammonia refrigerating systems, the tubes of a shell and tube condenser are made of
   (*a*) copper (*b*) aluminium (*c*) steel (*d*) brass.

18. Evaporative condenser uses
   (*a*) only air as the cooling medium (*b*) water and air both as cooling media
   (*c*) only water as the cooling medium (*d*) none of these.

19. The natural convection air-cooled condensers are used in
   (*a*) domestic refrigerators (*b*) water-coolers
   (*c*) room air-conditioners (*d*) all the these

20. In actual air-conditioning application for R-12 and R-22 and operating at a condenser temperature of 40°C and an evaporator temperature of 5°C, the heat rejection factor is about
   (*a*) 1 (*b*) 1.25 (*c*) 2.15 (*d*) 5.12

21. Heat rejection ratio for hermetic units as compared to open units for same evaporator pressure and same condenser pressure is
   (*a*) less (*b*) same (*c*) more (*d*) not predictable.

22. A water-cooled condenser is fitted with
   (*a*) water pressure regulator connection (*b*) liquid refrigerant shut off value connection
   (*c*) purge cock connection (*d*) all of these.

23. The condensing medium used in evaporative condensers is
   (*a*) air only (*b*) water only (*c*) both air and water (*d*) any of these.

24. Most air-cooled condensers are designed to operate with a temperature difference of
   (*a*) 5°C (*b*) 8°C (*c*) 14°C (*d*) 22°C

ANSWERS

1. (*b*)	2. (*a*)	3. (*c*)	4. (*c*)	5. (*b*)	6. (*b*)	7. (*a*)	8. (*b*)
9. (*b*)	10. (*b*)	11. (*a*)	12. (*b*)	13. (*a*)	14. (*a*)	15. (*b*)	16. (*c*)
17. (*c*)	18. (*c*)	19. (*a*)	20. (*b*)	21. (*b*)	22. (*d*)	23. (*c*)	24. (*c*).

Evaporators

Choose the Correct Answer:

1. Dry evaporator is one
   (*a*) in which the exit from evaporator is dry saturated refrigerant
   (*b*) in which the exit from evaporator is superheated refrigerant
   (*c*) in which the exit from evaporator has refrigerant with high dryness fraction
   (*d*) in which the inlet to the evaporator has refrigerant with some dryness fraction.

2. The evaporator changes the low pressure liquid refrigerant from the expansion valve into
   (*a*) high pressure liquid refrigerant
   (*b*) low pressure liquid and vapour refrigerant
   (*c*) low pressure vapour refrigerant
   (*d*) none of the above.

3. Air side heat transfer co-efficient for evaporator when air flows over the finned coil by forced connection is given by
   (*a*) $h_a = C\ (m)^{0.4}$ (*b*) $h_a = C\ (m)^{0.5}$ (*c*) $h_a = C\ (m)^{0.3}$ (*d*) $h_a = C\ (m)^{0.25}$

4. The evaporator generally used for wine cooling and in petroleum industry for chilling oil is
   (*a*) plate evaporator (*b*) finned evaporator
   (*c*) tube-in-tube evaporator (*d*) shell and tube evaporator.

5. The evaporator generally used in home freezers, ice cream cabinets etc. is
   (*a*) plate evaporator (*b*) finned evaporator
   (*c*) shell and tube evaporator (*d*) shell and coil evaporator.

6. Direct Expansion coil evaporator is
   (*a*) flooded type evaporator (*b*) dry type evaporator
   (*c*) wet type evaporator (*d*) none of these.
7. Flooded evaporator has to be fitted with
   (*a*) accumulator (*b*) float valve (*c*) Liquid eliminator (*d*) all of thses.
8. In a liquid chiller when the liquid to be chilled is in the shell and refrigerant flows in the tubes it is fitted with
   (*a*) float valve (*b*) thermostatic expansion valve
   (*c*) capillary tube (*d*) none of these.

ANSWERS

1. (*b*)	2. (*c*)	3. (*a*)	4. (*c*)	5. (*a*)	6. (*b*)	7. (*d*)	8. (*b*).

Refrigeration Controls

Choose the Correct Answer:

1. With thermostatic expansion valve used as expansion device in vapour compression refrigeration system, if there is high suction pressure and high degree of superheat, the cause is
   (*a*) compressor undersized (*b*) evaporator too large
   (*c*) compressor discharge valve leaking (*d*) any or all of these.
2. Thermostatic expansion valve operates on
   (*a*) the changes in pressure in the evaporator
   (*b*) the changes in the temperature of the evaporator
   (*c*) the changes in the degree of superheat at exit from evaporator
   (*d*) none of the above.
3. For varying suction pressures and temperatures of the evaporator
   (*a*) same charge thermostatic expansion valve is recommended
   (*b*) cross-charge thermostatic expansion valve is recommended
   (*c*) any one of the above is equally satisfactory.
   (*d*) none of the above as varying pressure is not related to thermostatic expansion valve charge.
4. The capillary tube used as expansion device in vapour compression system works on the principle of
   (*a*) isothermal expansion causing pressure drop
   (*b*) adiabatic expansion causing pressure drop
   (*c*) throttle expansion causing pressure drop
   (*d*) flow through pipe with friction causing pressure drop.
5. With thermostatic expansion valve used as expansion device in vapour compression refrigeration system, if the discharge pressure is very high, the cause is
   (*a*) insufficient cooling water
   (*b*) cooling water above design temperature
   (*c*) air or non-condensable gases in condenser
   (*d*) any or all of the above.
6. The thermostatic expansion valve is also called
   (*a*) constant pressure valve (*b*) constant temperature valve
   (*c*) constant superheat valve (*d*) none of these.

7. With the capillary as expansion device in vapour compression refrigeration system, at fixed condensing temperature
   (*a*) mass flow rate of compressor increases with increase in suction pressure, and mass flow rate through capillary increases with suction pressure.
   (*b*) mass flow rate of compressor decreases with increase in suction pressure, and mass flow rate through capillary decreases with increase in suction pressure.
   (*c*) mass flow rate of compressor increases with increase in suction pressure and mass flow rate through capillary decreases with increase in suction pressure.
   (*d*) mass flow rate of compressor decreases with increase in suction pressure and mass flow rate through capillary increases with suction pressure.
8. Externally equalised thermostatic expansion valve has equalising tapping taken from
   (*a*) inlet to the evaporator (*b*) middle point in the evaporator
   (*c*) tail end of the evaporator (*d*) anywhere in the evaporator.
9. Hunting of the thermostatic expansion valve is
   (*a*) variation of the evaporator load with the degree of superheat
   (*b*) variation of the pressure of the evaporator with variation of load
   (*c*) alternate overfeeding and starving of the refrigerant flow to the evaporator
   (*d*) not there at all. The term is not used at all.
10. In a refrigeration system, the expansion device is connected between the
   (*a*) compressor and condenser (*b*) condenser and receiver
   (*c*) receiver and evaporator (*d*) evaporator and compressor
11. With thermostatic expansion valve used as expansion device in vapour compression refrigeration system, if the suction pressure is fluctuating, the case is
   (*a*) Improper superheat adjustment (*b*) Trapped suction line
   (*c*) Improper remote bulb location or application (*d*) any of these.
12. Thermostatic expansion valve is used in .... type of evaporators.
   (*a*) flooded (*d*) DX coil (*c*) dry (*d*) any of these.
13. Most thermostatic expansion valves are set for a superheat of
   (*a*) 5°C (*b*) 10°C (*c*) 15°C (*d*) 20°C.
14. With capillary used as expansion device in vapour compression refrigeration system, if the refrigerant charge is excessive
   (*a*) it will lead to rise of condenser pressure
   (*b*) it will lead to warming up of evaporator quickly during shut down period
   (*c*) it will lead to defrosting and short-cycling of the evaporator
   (*d*) it will cause all of the above.
15. The pressure in a capillary tube decreases due to
   (*a*) frictional resistance offered by the tube wall (*b*) acceleration of refrigerant in the tube
   (*c*) heat transfer from the tube (*d*) both (*a*) and (*b*)
   (*e*) both (*b*) and (*c*).
16. The capillary tube, as an expansion device, is used in
   (*a*) domestic refrigerators (*b*) water-coolers
   (*c*) room air-conditioners (*d*) all of these.
17. Capillary tube is not used in large capacity refrigeration systems because
   (*a*) cost is too high (*b*) capacity control is not possible
   (*c*) it is made of copper (*d*) required pressure drop cannot be achieved.

ANSWERS

1. (*d*)	2. (*c*)	3. (*b*)	4. (*d*)	5. (*d*)	6. (*c*)	7. (*c*)	8. (*c*)
9. (*c*)	10. (*c*)	11. (*d*)	12. (*c*)	13. (*a*)	14. (*d*)	15. (*d*)	16. (*d*)
17. (*b*).							

THEORETICAL QUESTIONS

Compressors

1. Enumerate the major refrigeration components of a vapour compression system.
2. What are refrigeration controls?
3. What are the functions of a refrigeration compressor?
4. List the types of refrigerating compressors.
5. How are reciprocating compressor classified?
6. Explain briefly the following types of reciprocating compressors:
   (*i*) Open type vertical reciprocating compressor
   (*ii*) Semi-sealed type
   (*iii*) Hermetically sealed type.
7. With the help of neat sketches give the construction and working of a single-stage reciprocating compressor.
8. What is the effect of clearance volume in a reciprocating compressor?
9. Define "volumetric efficiency" of a reciprocating compressor.
10. What is a clearance ratio or factor?
11. Derive an expression for clearance volumetric efficiency of a reciprocating compressor.
12. Explain the factors which affect the total volumetric efficiency.
13. State the effects of the following on the total volumetric efficiency:
    (*i*) Clearance factor
    (*ii*) Wire drawing or throttling at suction valve and delivery valve
    (*iii*) Cylinder heating and re-expansion index (*iv*) Valve and piston leakage.
14. What is multistage compression?
15. State the advantages and disadvantages of multistage compression ?
16. Derive an expression for the work done of a two-stage reciprocating compressor with intercooler.
17. Obtain the conditions for the minimum work required for a two-stage reciprocating compressor.
18. Discuss briefly the various methods used for capacity control of reciprocating compressors.
19. Describe with a neat sketch, a centrifugal compressor.
20. Where are centrifugal compressors preferred over reciprocating compressors in refrigeration system?
21. What are the advantages and disadvantages of centrifugal compressors over reciprocating compressors?
22. Compare the performance of reciprocating and centrifugal compressors.
23. Describe briefly the following:
    (*i*) Roller type rotary compressor (*ii*) Vane type rotary compressor
24. Explain briefly a screw compressor.

Condensers

25. What is a condenser?
26. State the function of a condenser in a refrigeration system.
27. What factors should be considered while selecting a condenser?
28. Explain briefly the factors which affect the condenser capacity.
29. What is "heat rejection factor" for a condenser? On which factor does it depend?
30. Explain briefly the effect of compression ratio on "Heat rejection factor."
31. Give the main types of condensers in use with specific application of each type.
32. Explain briefly the following air-cooled condensers:

    (*i*) Natural convection condenser (*ii*) Forced-air circulation condenser.
33. Describe briefly the following air-cooled condensers:

    (*i*) Base mounted air-cooled condensers (*ii*) Remote air-cooled condensers.
34. What are the advantages of air-cooled condensers ?
35. Describe with neat sketches any two of the following water-cooled condensers:

    (*i*) Shell and tube condenser (*ii*) Shell and coil condenser

    (*iii*) Double tube condenser.
36. Give the comparison between water-cooled and air-cooled condensers.
37. What is "fouling factor"?
38. Explain with a neat sketch working of an evaporative condenser.
39. What is a cooling tower?
40. Explain briefly the following cooling towers:

    (*i*) Natural draft cooling towers (*ii*) Mechanical draft cooling tower.
41. Describe briefly with neat sketches the following:

    (*i*) Forced draft cooling tower (*ii*) Induced-draft cooling tower.
42. Write a short note on "Spray pond".

Evaporators

43. What is an evaporator?
44. Which are the three main requirements which an evaporator must fulfil?
45. How is "capacity of an evaporator" defined?
46. Explain briefly the factors which affect the heat transfer capacity of an evaporator.
47. Describe briefly the following:

    (*i*) Pool boiling (*ii*) Flow or forced-convection boiling.
48. Describe the three distinct regimes of boiling process.
49. Give the classification of evaporators.
50. Explain with neat sketches the following evaporators:

    (*i*) Flooded type evaporator (*ii*) Dry expansion evaporator.
51. Describe briefly any *two* of the following evaporators:

    (*i*) Bare-tube evaporator (*ii*) Plate-surface evaporator

    (*iii*) Finned-tube evaporator.

Refrigeration Controls

52. What do you mean by 'Refrigerant controls'? State the seven points where refrigerant controls are used in a system.
53. Name six basic types of refrigerant flow controls and explain any one of them in detail.
54. Explain briefly any two of the following:
    (*i*) Hand expansion valve (*ii*) Automatic expansion valve
    (*iii*) Thermostatic expansion valve (*iv*) Capillary tube.
55. Discuss the operation of a capillary tube in a refrigeration system.
56. Explain briefly as to why capillary tube is preferred to other throttling devices in household refrigerator.
57. Differentiate between low-side float valve and high-side float-valve.
58. Write short notes on any two of the following:
    (*i*) Low-side float (*ii*) High-side float (*iii*) Solenoid valves.

UNSOLVED EXAMPLES

1. A single-stage single-acting reciprocating compressor has a bore of 200 mm and a stroke of 300 mm. It receives vapour refrigerant at 1 bar and delivers it at 5.5 bar. If the compression and expansion follow the low $pV^{1.3}$ = constant and clearance volume is 5% of the stroke volume, determine:
    (*i*) The power required to drive the compressor, if it runs at 500 r.p.m., and
    (*ii*) The volumetric efficiency of the compressor. **[Ans.** (*i*) 14.12 kW; (*ii*) 86.5%]
2. The bore and stroke of a single-cylinder, single-acting reciprocating compressor using R-12 refrigerant are 80 mm and 60 mm respectively. The compressor runs at 1450 r.p.m. If the condensing temperature is 40°C and evaporator temperature (*a*) 10°C and (*b*) –10°C, find the following:
    (*i*) Mass of refrigerant circulated per minute, (*ii*) Refrigerating capacity of the compressor,
    (*iii*) Power per tonne of refrigeration, and (*iv*) Total power required to drive the compressor.

    Determine the change in results when the compression index is 1.25 and clearance factor is 5 per cent:

    [Ans. (*i*) 10.67 kg/min; (*ii*) 5.355 tonnes; (*iii*) 0.49 kW/tonne; (*iv*) 2.62 kW;
    (*i*) 5.68 kg/min; (*ii*) 2.64 tonnes; (*iii*) 0.96 kW/tonne; (*iv*) 2.53 kW
    (*i*) 10.1 kg/min; (*ii*) 5.07 tonnes; (*iii*) 0.464 kW/tonne; (*iv*) 2.48 kW
    (*i*) 4.92 kg/min; (*ii*) 2.286 tonnes; (*iii*) 0.83 kW/tonne; (*iv*) 1.897 kW]

Chapter 11

Defrosting Methods

11.1 FROST FORMATION AND DEFROSTING—INTRODUCTION

Frost formation

- For cold storage and low temperature applications, the "*evaporators*" necessarily must be at temperatures *lower than* 0°C, the *freezing point of water*. Therefore, the moisture from the air will not only condense but also freeze (or frost, as it is called) on the coil surface. Hence for such applications, frost formation on the evaporator coil is inevitable. Also it is obvious that *lower the evaporator temperature, greater will be the formation of frost.*
- When the finned coils are employed, frost formation between the fins will obstruct the air passages and thus affect the quantity of air over the coil. This again will reduce the evaporator temperature, leading to thicker frost formation, i.e., a vicious cycle starts. *The frost on the evaporator also acts as an insulation and retards the heat transfer. With decreased heat-transfer, the evaporator temperature will tend to go down, cousing thicker frost and a further decrease in the evaporator temperature and capacity.* As the frost formation increases between the fins, and at times, the fins can get moved and thus loosen the bond of the fin on the evaporator tube. Once this happens the evaporator capacity will come down substantially necessating the coil to be changed.

Defrosting:

- In view of the above reasons, the frost formation on evaporator coils of low temperature installation should be melted out periodically. *The process of removing the frost by melting it is called defrosting the evaporators.*

- *The lower the evaporator temperature, more frequent should be the defrost* (e.g., for deep-freeze coils, the defrosting will have to be done every 3 or 4 hours).
- The evaporator fan should be stopped during the defrost cycle, otherwise the water of the melted frost will be sprayed into the room. The system will have to be pumped down to remove the liquid refrigerant from the evaporator during the defrost cycle.
- *Time switches* (or a timer) are used for accomplishing *automatic defrosting.*

11.2 METHODS OF DEFROSTING AN EVAPORATOR

The following *methods/systems* are used for defrosting the evaporators:

1. Manual defrosting method.
2. Automatic periodic defrosting method.
3. Solid absorbent system.
4. Water defrosting method.
5. Reverse cycle defrosting method.
6. Simple hot gas defrosting method.
7. Automatic hot gas defrosting method.
8. Thermobank defrosting method.
9. Electric defrosting method.

All these methods are discussed in details in the following articles.

11.3 MANUAL DEFROSTING METHOD

- Manual defrosting method is the earliest and simplest method of defrosting. Here, the system is shut down manually until the evaporator warms up enough to melt the accumulated frost.
- The major *drawback* of this method is long defrosting period.
- This method is *still employed on large installations.*

11.4 AUTOMATIC PERIODIC DEFROSTING METHOD

In this case, the stopping of the system after objectionable accumulation of frost and starting of the system after *natural defrosting* is done ***automatically***. The starting and stopping of the system is carried out automatically as per the ***change in pressure*** or ***change in temperature*** in the evaporator as shown in Figs. 11.1 and 11.2. respectively.

(*i*) **Pressure control.** Refer to Fig. 11.1.

One account of accumulation of the frost on the surface of the evaporator the heat transfer rate between air and evaporator coil is reduced. Consequently the compressor runs at a lower suction pressure. When the pressure falls below predetermined value due to objectionable frosting, the pressure operated control comes into action and stops the compressor. The control switch is set to permit the coil to worm upto 0°C and defrost before it starts again. There is an automatic stopping and starting of the system according to the change in pressure.

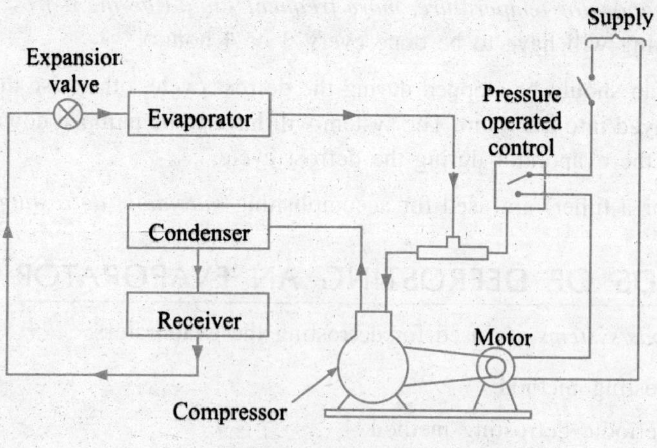

Fig. 11.1. Pressure control defrost method.

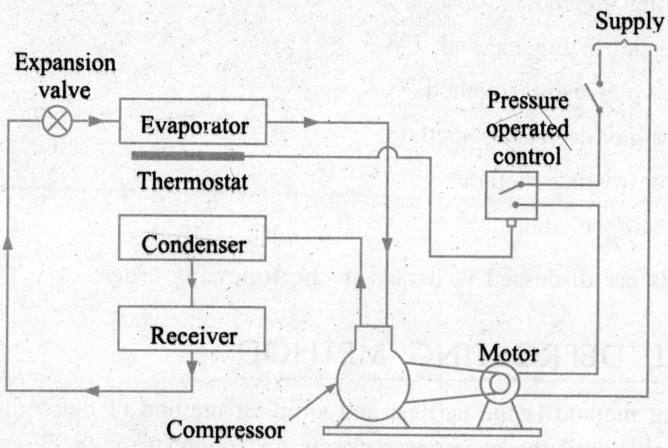

Fig. 11.2. Temperature control defrost method.

(*i*) **Temperature control.** Refer to Fig. 11.2.

This method is just similar to pressure control method except the *control is operated by the increase in temperature of air.* The temperature of air passing over the coil increases due to the reduction in heat transfer caused by frosting the evaporator coil. The stopping and starting of the system is done automatically as per change in temperature.

- This method *makes it possible to maintain temperatures regardless of variations in load.*
- It is *preferred* to the pressure control defrost method for cold locations because the operation of the compressor is controlled directly by the temperature.
- The *disadvantage* of this method is that it *does not assure complete defrost especially in hot weather or other high load conditions.* (This disadvantage can be overcome by an occasional manual defrost method).

11.5 SOLID ABSORBENT SYSTEM

Fig. 11.3 shows the solid absorbent system *in which most of the moisture in the air is removed before passing the air over the evaporator coil*; this avoids the frost formation on the coil. Here, *silica get, activated alumina are used for dehumidifying the air*.

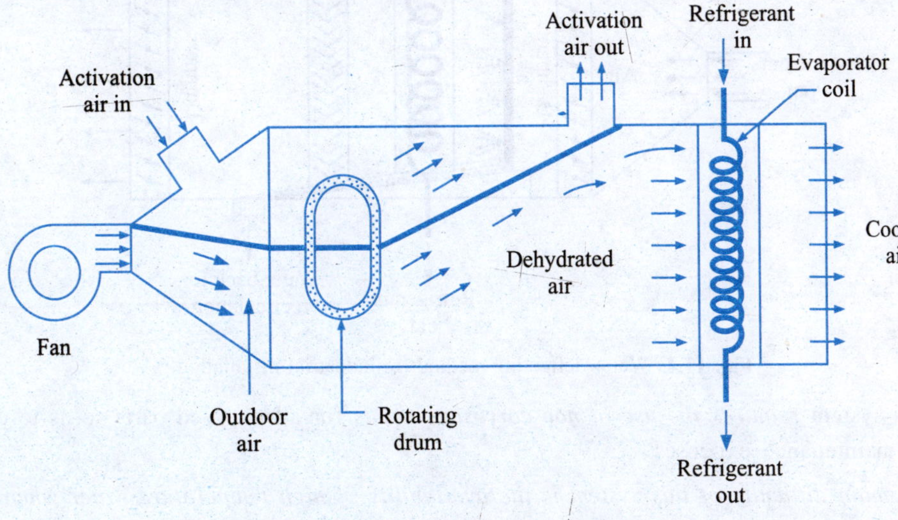

Fig. 11.3. Solid absorbent system

As shown in the figure, the solid drying agent is carried in a continuously rotating perforated drum ahead of the evaporator coil. The drum rotation is so adjusted that the dehydration of the air and the reactivation of the absorbent will be completed efficiently. The lower portion and upper portion of the drum are used for dehydration of air and reactivation of absorbent respectively. For reactivation, hot air (heated by electric heater or by steam coil) at 100°C to 130°C is used.

- This type of system to avoid the frosting of the evaporator coil is *only used when the humidity of the air is not of much importance.*
- The main **disadvantage** of the system is that its *initial cost is high.*

11.6 WATER DEFROSTING METHOD

When the temperature required in the refrigerator is below 0°C it is not desirable to shut down the system for defrosting, since the time required for defrosting in considerably large.

Where ample quantity of water at considerable high temperature (20°C) is easily available *water spray is a simple and economical method of defrosting* the evaporator coil.

The schematic arrangement of the system is shown in Fig. 11.4. During the defrosting cycle, the water is sprayed in ample quantity with sufficient velocity to wash out the ice formed on the surface of the coil. During the defrosting operation the *fan should be kept off to avoid the reduction of water temperature* which is possible due to evaporation of water into air. After defrosting operation, before starting the refrigeration system, it is necessary to *start the fan*. This is necessary to *remove the water from the coil surface* before the temperature of the coil is reduced below freezing. When the outdoor temperatures are below 0°C water from deep wells, lakes and manufacturing processes can be used.

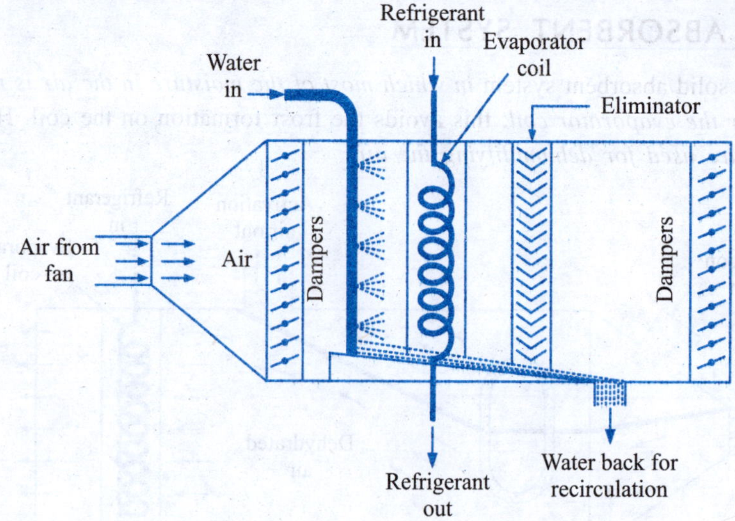

Fig. 11.4. Water defrosting method for industrial installation.

- This system *requires the use of non-corrosive metals* for all exposed parts so as to eliminate the maintenance expenses.
- The *main difficulty of this system is the availability of high temperature water when outdoor temperature is below* 0°C.
- This system is *used on many big installations.*

11.7 REVERSE CYCLE DEFROSTING METHOD

Fig. 11.5 shows the normal cycle of operation in a refrigerating system. The evaporator may be defrosted by reversing the cycle of operation as shown in Fig. 11.6. When the cycle is reversed, the hot gas from the compressor is passed to the evaporator. The hot gas condenses and heats the evaporator coil and thus the frost accumulated in the evaporator surface is melted out. The condensed gas from the evaporator bypasses the expansion valve by means of a check valve. The refrigerant evaporates in the condenser and is returned to the compressor in the vapour state.

— During the defrosting operation, the air is heated instead of cooled as it flows over the evaporator coil. If this air is supplied to an air-conditioned space occupied by the people, it causes discomfort. To overcome this difficulty, the defrosting cycle can be timed to occur at night or during periods when the space is not occupied.

- This method is *never used for industrial installations because these systems run continuously.*

11.8 AUTOMATIC HOT GAS DEFROSTING METHOD

Fig. 11.7 shows the schematic diagram of the automatic hot gas defrosting method.

During the **normal cycle**, the solenoid valve is closed and the hot gas discharged from the compressor is passed through the condenser, receiver, heat exchanger, thermostatic expansion valve, evaporator and back to the compressor. During this normal operation *fan A remains in working condition* while the fan *B* remains in non-working condition.

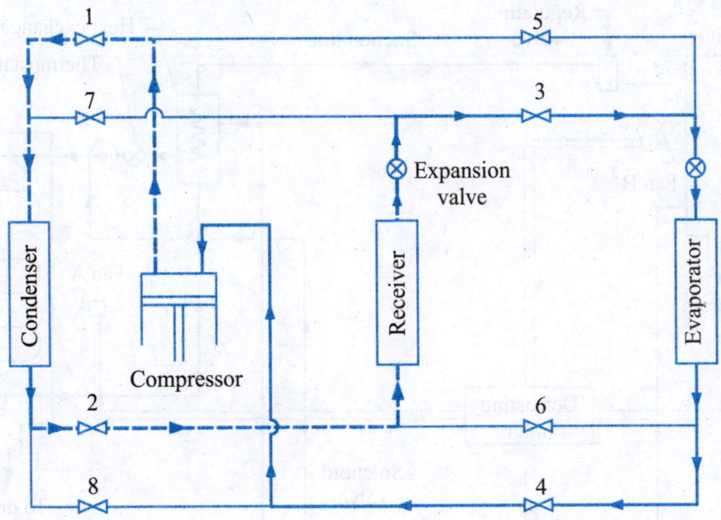

Fig. 11.5. Normal refrigeration system (Valves 1, 2, 3, 4 are *open* and valves 5, 6, 7, 8 are *closed*).

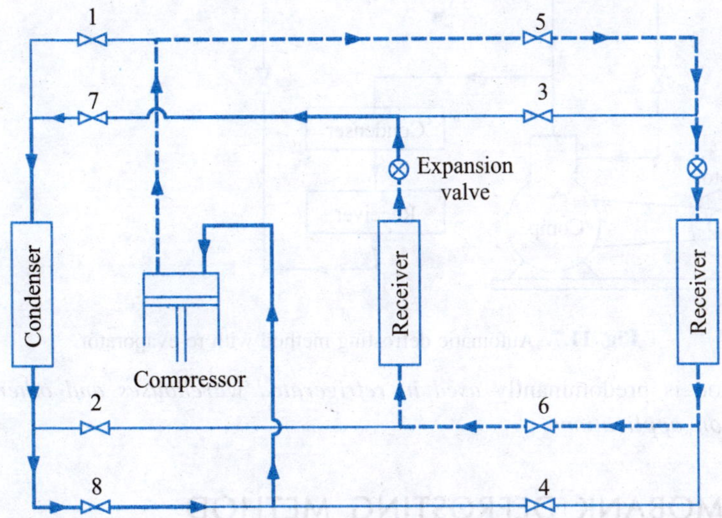

Fig. 11.6. Reverse cycle defrosting system (Valves 5, 6, 7, 8 are *open* and valves 1, 2, 3, 4 are *closed*).

The **defrosting cycle** is started (at regular intervals, usually 3 to 6 hours depending upon setting of the valve) by a defrosting timer by opening the solenoid valve in the hot gas line. During this operation, the hot gas (vapour refrigerant) discharged from the compressor enters the evaporator between the expansion valve and the evaporator. The hot gas gives up heat to the cold evaporator and subsequently the frost accumulated on the surface of the evaporator melts. The hot gas coming in contact with the cold surface of the evaporator gets condensed with a liquid refrigerant. This liquid refrigerant is now passed to a re-evaporator where it is again evaporated before being fed to the compressor. During this cycle, the *fan B remains in working condition* and fan A in non-working condition The opening of solenoid valve, stopping of fan A and starting of fan B is done simultaneously by the automatic defrosting timer. When the defrosting is completed, the defrosting timer shuts off the solenoid valve to stop the hot gas flow to the evaporator and puts the system on normal operation.

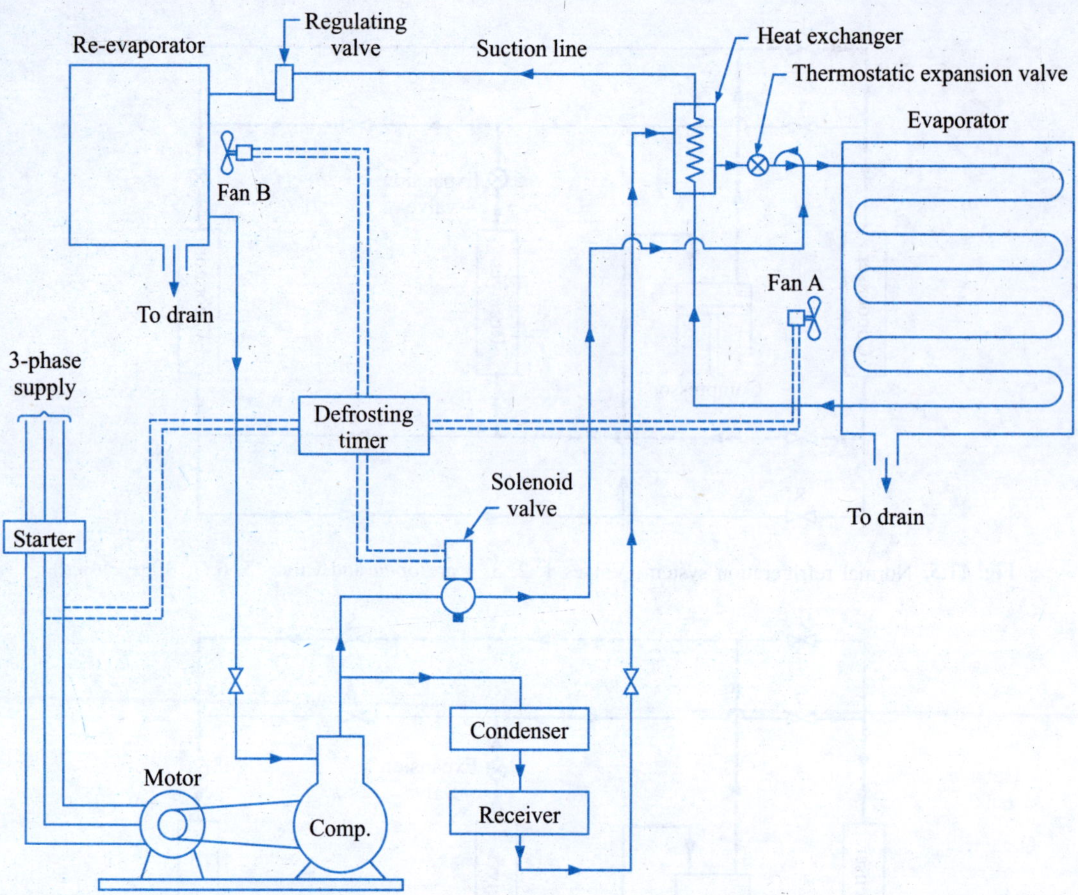

Fig. 11.7. Automatic defrosting method with re-evaporator.

- This method is predominantly *used in refrigerated warehouses and other low temperature refrigeration applications.*

11.9 THERMOBANK DEFROSTING METHOD

Thermobank frosting system (developed by Kramer Trenton & Company) *uses hot gas as a defrosting medium and provides a heat accumulator in the suction line.*

Thermobank, as the name suggests, is a *bank for storing heat during normal operation of the cycle*. Heat from the hot gas coming out of the compressor is stored in a small liquid tank. This stored heat during defrosting cycle goes to the suction line instead of directly to the hot gas line, this prevents *liquid slugging of the compressor.*

The schematic arrangement of the system is shown in Fig. 11.8. The thermobank consists of a tank within another tank, the inner tank being longer than the outer tank. The suction connection is made at the lot of the inner tank and the outer tank is filled with an *antifreeze liquid* in which there is a submerged coil. The coil carries the hot gas discharged from the compressor. Both the tanks are heavily insulated to prevent the heat loos from the thermobank. In the hot gas line a solenoid valve is placed which is electrically connected to the self starting timer.

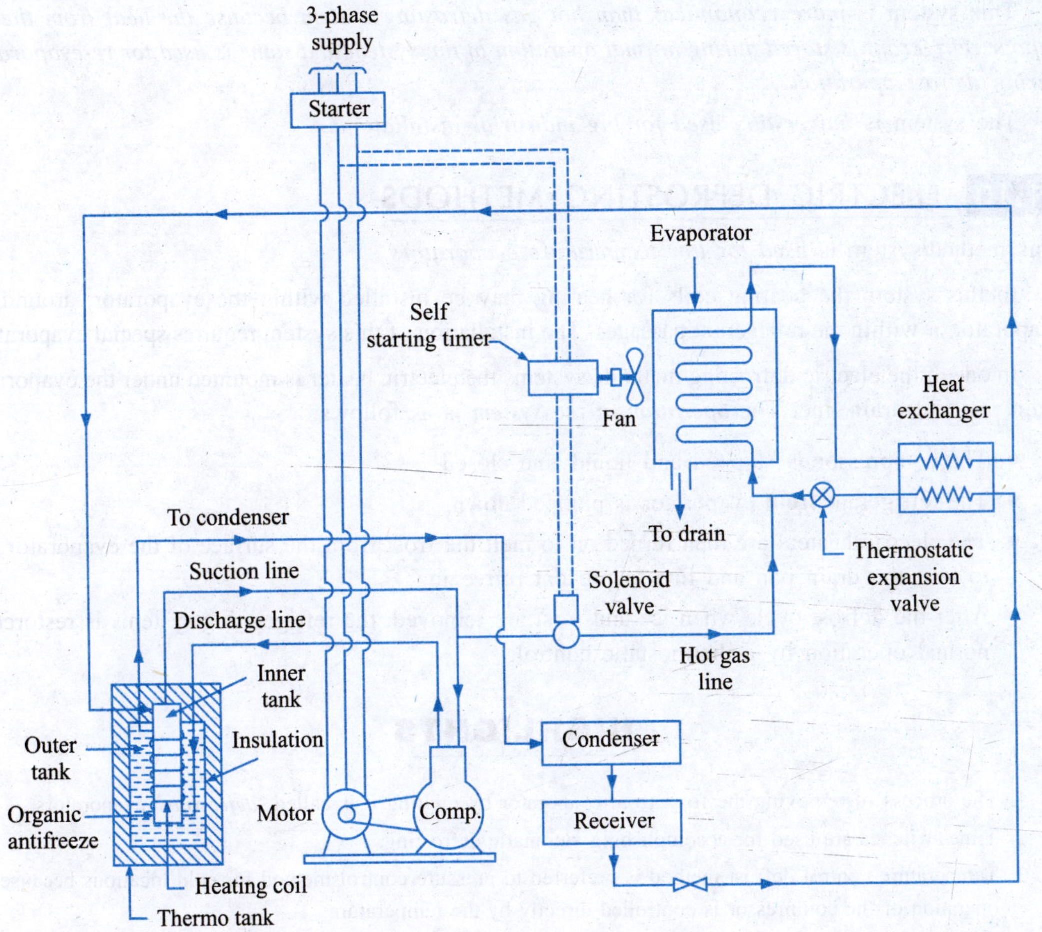

Fig. 11.8. Thermobank defrosting method.

The working of the system is as follows:

- During the **normal operation**, the *solenoid valve is closed* and the hot gas discharged from the compressor passes through the submerged coil in the outer tank, heats the antifreeze liquid and then is condensed in the condenser. The liquid refrigerant from receiver through heat exchanger, expansion valve and evaporator comes to the inner tank of thermobank since the *vapour refrigerant* (i.e., suction gas) *enters and leaves the tank at the upper end, therefore, it does not absorb any heat*. When a sufficient frost in accumulated on the surface of the evaporator coil, the timer automatically opens the solenoid valve and *shuts off the fan motor* and compressor continues to operate.
- During the **defrosting cycle**, the hot gas goes through the solenoid valve to the evaporator where it melts the frost. When the hot gas comes in contact with the cold surface of evaporator, it condenses into a liquid refrigerant, which then flows through the suction line to the bottom of the inner tank of thermobank. It then gets evaporated by absorbing the heat from the antifreeze liquid which was accumulated during normal operation, and goes to compressor. The defrosting cycle continues for approximately six minutes and after that, the solenoid valve is closed automatically and *starts the fan motor*. The system then operates on normal cycle.

This system is ***more economical*** *than hot gas defrosting system because the heat from the hot vapour refrigerant is stored during normal operation of the system and same is used for re-evaporation during defrost operation.*

The system is *universally used for big industrial installations.*

11.10 ELECTRIC DEFROSTING METHODS

This method/system is *used for low temperature evaporators.*

In this system the heating coils for heating may be installed within the evaporator, around the evaporator or within the refrigerant passages. The installation of this system requires special evaporators.

In one of the electric defrosting methods/systems the electric heater is mounted under the evaporator, drain pan and drain line. The *operation of the system* is as follows:

- The compressor is stopped and liquid line closed.
- The refrigerant from evaporator is pumped down.
- The electric heaters are then turned on to melt the frost from the surface of the evaporator and to heat the drain pan and line to prevent refreezing.
- After the defrost cycle when ice and frost are removed, the refrigerating systems is restored to normal operation by a thermostatic control.

HIGHLIGHTS

1. The process of removing the frost from evaporator by melting it is called '*defrosting*' evaporators.
2. Time switches are used for accomplishing automatic defrosting.
3. Temperature control defrost method is preferred to pressure control method for cold locations because the operation of the compressor is controlled directly by the temperature.
4. Automatic hot gas defrosting method is predominantly used in refrigerated warehouses and other low temperature refrigeration applications.
5. Thermobank defrosting method is universally used for big industrial installations.

OBJECTIVE TYPE QUESTIONS

Fill in the Banks or Say "Yes" or "No":

1. The .... on the evaporator acts as an insulation and retards the heat transfer.
2. The lower the evaporator temperature, the lesser will be the formation of frost.
3. The process of removing the frost from evaporator by melting it is called .... the evaporator.
4. The lower the evaporator temperature, more frequent should be the defrost.
5. The evaporator fan should not be stopped during the defrost cycle.
6. .... switches are used for accomplishing automatic defrosting.
7. Temperature control method makes it possible to maintain temperatures regardless of variations in load.
8. The main disadvantage of solid absorbent system is that its initial cost is high.
9. .... defrosting method requires the use of non-corrosive metals for all exposed parts so as to eliminate the maintenance expenses.
10. Thermobank defrosting method is universally used for big industrial installations.

ANSWERS

1. frost	2. No	3. defrosting	4. Yes	5. No	6. Time	7. Yes
8. Yes	9. Water	10. Yes.				

THEORETICAL QUESTIONS

1. How is frost formed?
2. What is the effect of frost on the cooling coil?
3. Enumerate three common methods of defrosting evaporator coils of commercial refrigerators.
4. How is defrosting carried out manually?
5. Explain with a neat diagram the working of an automatic periodic defrosting system.
6. Describe briefly solid absorbent system.
7. Explain briefly with neat diagram(*s*) reverse cycle defrosting method.
8. Describe briefly with a neat sketch an automatic hot gas defrosting method.
9. What is the principle of 'thermobank defrosting'?
10. Explain briefly with a neat diagram the working of a thermal defrosting system.
11. Write a short note on methods of defrosting.
12. State the factors which should be considered while selecting a particular type of defrosting system for a particular type of application.

❑❑❑

PART II: AIR-CONDITIONING

Chapters:

Chapter 12

Psychrometrics

12.1 CONCEPT OF PSYCHROMETRY AND PSYCHROMETRICS

- Air consists of fixed gases principally, *nitrogen and oxygen with an admixture of water vapour in varying amounts.* In atmospheric air water is always present and its relative weight averages less than 1% of the weight of atmospheric air in temperate climates and less than 3% by weight under the most extreme natural climatic conditions, it is nevertheless one of most important factors in human comfort and has significant effects on many materials. Its effect on human activities is in fact altogether disproportionate to its relative weights.
- *The art of measuring the moisture content of air is termed* **"psychrometry"**.
- The *science which investigates the thermal properties of moist air, considers the measurement and control of the moisture content of air, and studies the effect of atmospheric moisture on material and human comfort may properly be termed* **"psychrometrics"**.

12.2 DEFINITIONS

Some of the more important definitions are given below:

1. Dry air:

- The international joint committee on psychrometric Data has adopted the following exact compositions of air expressed in mole fractions (volumetric): *Oxygen 0.2095, Nitrogen 0.7809, Argon 0.0093, Carbon dioxide 0.0003.* Traces of rare gases are neglected.
- Molecular weight of air for all air-conditioning calculations will be taken as 28.97. Hence the gas constant,

$$R_{air} = \frac{8.3143}{28.97} = 0.287 \text{ kJ/kg K}$$

- Dry air is never found in practice. Air always contains some moisture. Hence the common designation "air" *usually means moist air.*
- The term '*dry air*' is used to indicated the *water free contents* of air having any degree of moisture.

2. Saturated air:

- Moist air is said to be saturated when its condition is such that it can co-exist in natural equilibrium with an associated condensed moisture phase presenting a flat surface to it.
- For a given *temperature, a given quantity of air can be saturated with a fixed quantity of moisture.*
- *At saturation, vapour pressure of moisture in air corresponds to the saturation pressure given in steam tables corresponding to the given temperature of air.*

3. Dry-bulb temperature (DBT). *It is the temperature of air as registered by an ordinary thermometer* (t_{db}).

4. Wet-bulb temperature (WBT). *It is the temperature registered by a thermometer when the bulb is covered by a wetted wick* and is exposed to a current of rapidly moving air (t_{wb}).

5. Adiabatic saturation temperature. *It is the temperature at which the water or ice can saturate air by evaporating adiabatically into it.* It is numerically equivalent to the measured wet bulb temperature (as corrected, if necessary for radiation and conduction) (t_{wb}).

6. Wet bulb depression. *It is the difference between dry-bulb and wet bulb temperature* ($t_{db} - t_{wb}$).

7. Dew point temperature (DPT). *It is the temperature to which air must be cooled at constant pressure in order to cause condensation of any of its water vapour.* It is equal to steam table saturation temperature corresponding to the actual partial pressure of water in the air (t_{dp}).

8. Dew point depression. *It is the difference between the dry bulb and dew point temperature* ($t_{db} - t_{dp}$).

9. Specific humidity (Humidity ratio)

- *It is the ratio of the mass of water vapour per unit mass of dry air in the mixture of vapour and air. It is generally expressed as grams of water per kg of dry air.*
- For given barometric pressure it is a function of *dew point temperature alone.*

10. Relative humidity (RH) (ϕ). *It is the ratio of the partial pressure of water vapour in the mixture to the saturated partial pressure at the dry bulb temperature, expressed as percentage.*

11. Sensible heat. *It is the heat that changes the temperature of a substance when added to or abstracted from it.*

12. Latent heat. *It is the heat that does not affect the temperature but changes the state of a substance when added to or abstracted from it.*

13. Enthalpy. *It is the combination energy which represents the sum of internal and flow energy in a steady flow process.* It is determined from an arbitrary datum point for the air mixture and is expressed as kJ per kg of dry air (h).

Note: *When air is saturated DBT, WBT, DPT are equal.*

12.3 PSYCHROMETRIC RELATIONS

1. Pressure:

- **Dalton's law of partial pressure** *is* employed to determine the pressure of a mixture of gases. This law *states that the total pressure of a mixture of gases is equal to the sum of partial pressures which the component gases would exert if each existed alone in the mixture volume at the mixture temperature.*
- Precise measurements have shown that this law as well as Boyle's and Charle's laws are only approximately correct. Modern tables of atmospheric air properties are based on the correct versions.
- For calculating partial pressure of water vapour in the air many equations have been proposed, probably Dr. Carrier's equation is most widely used.

$$p_v = (p_{vs})_{wb} - \frac{[p_t - (p_{vs})_{wb}](t_{db} - t_{wb})}{1527.4 - 1.3\, t_{wb}} \quad ...(12.1)$$

where, p_v = Partial pressure of water vapour,

p_{vs} = Partial pressure of water vapour when *air is fully saturated,*

p_t = Total pressure of moist air,

t_{db} = Dry bulb temperature (°C), and

t_{wb} = Wet bulb temperature (°C).

2. Specific humidity W:

$$\text{Specific humidity} = \frac{\text{Mass of water vapour}}{\text{Mass of dry air}}$$

or,
$$W = \frac{m_v}{m_a}$$

Also,
$$m_a = \frac{p_a V}{R_a T} \quad ...(12.2)$$

and,
$$m_v = \frac{p_a \times V}{R_v \times T} \quad ...(12.3)$$

where, p_a = Partial pressure of dry air,

p_v = Partial pressure of water vapour,

V = Volume of mixture,

R_a = Characteristic gas constant for dry air, and

R_v = Characteristic gas constant for water vapour.

From eqns. (12.2) and (12.3)

$$W = \frac{p_v \times V}{R_v \times T} \times \frac{R_a T}{p_a V} = \frac{R_a}{R_v} \times \frac{p_v}{p_a}$$

But,
$$R_a = \frac{R_0}{M_a}\left(= \frac{8.3143}{28.97} = 0.287 \text{ kJ/kg K in SI units}\right)$$

$$\therefore \qquad R_v = \frac{R_0}{M_v}\left(= \frac{8.3143}{18} = 0.462 \text{ kJ/kg K in SI units}\right)$$

where R_0 = Universal gas constant,

M_a = Molecular weight of air, and

M_v = Molecular weight of water vapour.

$$\therefore \qquad W = \frac{0.287}{0.462} \times \frac{p_v}{p_a} = 0.622 \frac{p_v}{p_t - p_v}$$

i.e.,

$$W = 0.622 \frac{p_v}{p_t - p_v} \qquad \text{...(12.4)}$$

The masses of air and water vapour in terms of specific volumes are given by expression as

$$m_a = \frac{V}{v_a} \text{ and } m_v = \frac{V}{v_v}$$

where, v_a = Specific volume of dry air, and

v_v = Specific volume of water vapour.

$$W = \frac{v_a}{v_v} \qquad \text{...(12.5)}$$

3. Degree of saturation (μ) :

$$\text{Degree of saturation} = \frac{\text{Mass of water vapour associated with unit mass of dry air}}{\text{Mass of water vapour associated with unit mass of dry saturated air}}$$

$$\mu = \frac{W}{W_s} \qquad \text{...(12.6)}$$

where, W_s = Specific humidity of air when air is fully saturated.

$$\mu = \frac{0.622\left(\dfrac{p_v}{p_t - p_v}\right)}{0.622\left(\dfrac{p_{vs}}{p_t - p_{vs}}\right)} = \frac{p_v\,(p_t - p_{vs})}{p_{vs}\,(p_t - p_v)}$$

$$= \frac{p_v}{p_s}\left[\frac{\left(1 - \dfrac{p_{vs}}{p_t}\right)}{\left(1 - \dfrac{p_v}{p_t}\right)}\right] \qquad \text{...(12.7)}$$

where, p_{vs} = partial pressure of water vapour when air is fully saturated (p_{vs} can be calculated from steam tables corresponding to the dry bulb temperature of the air).

4. Relative humidity (RH) (ϕ) :

$$\text{Relative humidity,} \quad \phi = \frac{\text{Mass of water vapour in a given volume}}{\text{Mass of water vapour in the same volume if saturated at the same temp.}}$$

$$= \frac{m_v}{m_{vs}} = \frac{\dfrac{p_v T}{R_v T}}{\dfrac{p_{vs} T}{R_v T}} = \frac{p_v}{p_{vs}} \qquad ...(12.8)$$

Inserting the value of eqn. (12.8) into eqn. (12.7), we get:

$$\mu = \phi \left[\frac{1 - \dfrac{p_{vs}}{p_v}}{1 - \dfrac{p_v}{p_t}}\right] = \phi \left[\frac{1 - \dfrac{p_{vs}}{p_t}}{1 - \dfrac{\phi p_{vs}}{p_t}}\right] = \phi \left(\frac{p_t - p_{vs}}{p_t - \phi\, p_{vs}}\right)$$

$\therefore \qquad \phi\,(p_t - p_{vs}) = \mu\,(p_t - \phi\, p_{vs})$

or, $\qquad \phi\,(p_t - p_{vs} + \mu\, p_{vs}) = \mu\, p_t$

$$\therefore \qquad \phi = \frac{\mu\, p_t}{p_t - p_{vs} + \mu\, p_{vs}} = \frac{\mu}{1 - (1 - \mu)\dfrac{p_{vs}}{p_t}} \qquad ...(12.9)$$

Since, $\quad p_{vs} << p_t$

$\therefore \qquad \phi \approx \mu$

Inserting the value of p_v from eqn. (12.4) into eqn. (12.8), we get:

$$\phi = \frac{p_a W}{0.622} \times \frac{1}{p_{vs}} = 1.6\ W\ \frac{p_a}{p_{vs}} \qquad ...(12.10)$$

Note:

1. Relative humidity as compared to specific humidity plays a vital role in comfort air-conditioning and industrial air-conditioning. *Relative humidity signifies the absorption capacity of air. If initial relative humidity of air is less it will absorb more moisture.*
2. W and ϕ cannot be conveniently measured as they require measurement of p_v and p_{vs}. *The value of p_v can be obtained from the measurement of the wet bulb temperature and the value of p_{vs} can be calculated from steam corresponding to given air temperature.*

5. Enthalpy of moist air(h):

It is the sum of enthalpy of dry air and enthalpy of water vapour associated with dry air. It is expressed in kJ/kg of dry air.

$$h = h_{air} + W \cdot h_{vapour}$$
$$= c_p\, t_{db} + W \cdot h_{vapour}$$

where, h = Enthalpy of mixture/kg of dry air,

h_{air} = Enthalpy of 1 kg of dry air,

h_{vapour} = Enthalpy of 1 kg of vapour obtained from steam tables,

W = Specific humidity in kg/kg of dry air, and

c_p = Specific heat of dry air normally assumed as 1.005 kJ/kg K.

Also, $h_{vapour} = h_g + c_{ps}(t_{db} - t_{dp})$

where, h_g = Enthalpy of saturated steam at dew point temperature, and

c_{ps} = 1.88 kJ/kg K

$$\therefore \quad h = c_p t_{db} + W[h_g + c_{ps}(t_{db} - t_{dp})] \quad ...(12.11)$$

$$= (c_p + c_{ps} W) t_{db} + W(h_g - c_p t_{dp})$$

$$= c_{pm} t_{db} + W(h_g - c_{ps} t_{dp}) \quad ...[12.11 (a)]$$

where $c_{pm} = (c_p + c_{ps} W)$ is the specific heat of humid air or humid specific heat.

The value of c_{pm} is taken as 1.022 kJ/kg dry air per K. It is the heat capacity of (1 + W) kg of moisture per kg of dry air.

$h_{vapour} = h_g$ at dry bulb temperature. So,

$$h = c_p t_{db} + W h_g \quad ...(12.12)$$

However, a better approximation is given by the following relationship:

$$h_{vapour} = 2500 + 1.88 t_{db} \text{ kJ/kg of water vapour} \quad ...[12.12 (a)]$$

where, t_{db} is dry bulb temperature in °C, and the datum state is liquid water at 0°C.

$$\therefore \quad h = 1.005 t_{db} + W(2500 + 1.88 t_{db}) \text{ kJ/kg dry air} \quad ...[12.12 (b)]$$

12.4 ADIABATIC SATURATION PROCESS

In an insulated chamber when unsaturated air flows over a long sheet of water (Fig. 12.1), the water evaporates, and the specific humidity of the air increases. As the evaporation takes place both the air and water are cooled. *The process continues until the energy transferred from the air to the water is equal to the energy required to vapourise the water.* When this point is reached, thermal equilibrium exists with respect to water, air and water vapour, and consequently the air is saturated. The equilibrium temperature is called the *adiabatic saturation temperature* or the *thermodynamic wet bulb temperature.* The make-up water is introduced at *this temperature* to make the water level constant. The '*adiabatic*' cooling process is shown in Fig. 12.2 for the vapour in the air-vapour mixture. Although the total pressure of the mixture is constant, the partial pressure of the vapour increases, and in the saturated state corresponds to the adiabatic saturation temperature. The vapour is initially at DBT t_{db_1} and is cooled adiabatically to DBT t_{db_2} *which is equal to the adiabatic saturation* t_{wb_2}. The adiabatic saturation temperature and wet bulb temperature are taken to be equal for all practical purposes. The wet bulb temperature lies between the dry bulb temperature and dew point temperature.

Let us now apply the first law to the entire process. Considering the process to be steady state steady flow (neglecting changes in kinetic and potential energies), we have:

$$h_1 + (W_{2s} - W_1) h_{f2} = h_{2s} \quad ...(12.13)$$

The quantities W_{2s}, h_{2s} and h_{f2} are the functions of temperature t_{db_2}. The term $(W_{2s} - W_1) h_{f2}$ is quite small and if this term is neglected, it can be seen that enthalpy remains constant in adiabatic saturation.

Equation (12.13) may be rewritten as:

$$h_1 - W_1 h_{f2} = h_{2s} - W_{2s} h_{f2}$$

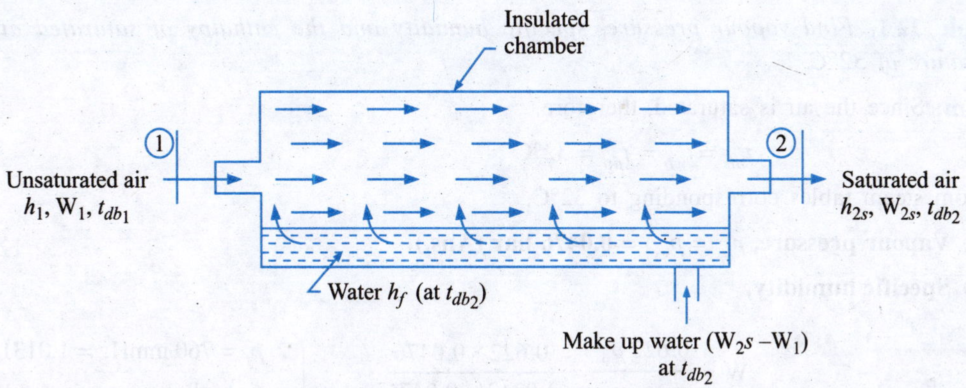

Fig. 12.1. Adiabatic saturation process.

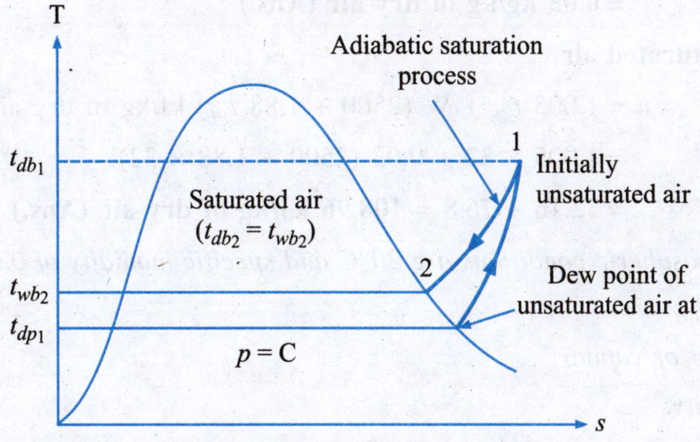

Fig. 12.2. Adiabatic saturation process on T-s chart.

The inlet term can be generalized and the expression can be written as follows:

$$\Sigma = h_{2s} - W_{2s}\, h_{f2} = h_1 - W_1\, h_{f2} = h_x - W_x\, h_{f2} \quad ...(12.14)$$

This means that sigma function (Σ) as defined by the equation, is constant for any wet bulb temperature.

Also,
$$h_1 = h_{2s} - (W_{2s} - W_1)\, h_{f2} \quad ...(12.15)$$

Equation (12.15) indicates that the enthalpy of air-water vapour mixture is equal to the enthalpy of saturated air at the same wet bulb temperature, *less* small correction term $(W_{2s} - W_1)\, h_{f2}$. This correction term is called *enthalpy deviation.*

$$h_1 = h_{air(1)} + W_1\, h_{vapour(1)} \quad ...(12.16)$$

$$h_{2s} = h_{air(2)} + W_{2s}\, h_{vapour(2)} \quad ...(12.17)$$

or,
$$h_{air\,(1)} + W_1 h_{g1} + (W_{2s} - W_1)\, h_{f2} = h_{air\,(2)} + W_{2s}\, h_{g2}$$

$$c_p\, t_{db_1} + W_1 h_{g1} + (W_{2s} - W_1)\, h_{f2} = c_p\, t_{db_2} + W_{2s}\, h_{g2} \quad ...(12.18)$$

After arranging, wet get:

$$W_1 = \frac{c_p\,(t_{db_2} - t_{db_1}) + W_{2s}\,(h_{g2} - h_{f2})}{h_{g1} \quad h_{f2}} \quad ...[12.18)\ (a)]$$

Note: The wet bulb temperature is not a property of moist air as it is influenced by heat and mass transfer rates. Thus in psychrometric equations and psychrometric charts where the wet bulb temperature appears, *it is always the thermodynamic wet bulb temperature that is considered.*

Example 12.1. *Find vapour pressure, specific humidity and the enthalpy of saturated air at a temperature of 32°C.*

Solution: Since the air is saturated, therefore,

$$t_{db} = t_{wb} = t_{dp} = 32°C$$

From steam tables corresponding to 32°C,

(*i*) **Vapour pressure,** p_v $(= p_{vs})$ = **0.0476 bar (Ans.)**

(*ii*) **Specific humidity,**

$$W = \frac{0.622\, p_v}{p_t - p_v} = \frac{0.622 \times 0.0476}{1.0312 - 0.0476} \qquad \left(\because p_t = 760 \text{ mmHg} = 1.01312 \text{ bar} \; \text{...atmospheric pressure}\right)$$

$$= \mathbf{0.03 \text{ kg/kg of dry air (Ans.)}}$$

(*iii*) **Enthalpy of saturated air,**

$$h = 1.005\, t_{db} + W\,(2500 + 1.88\, t_{db}) \text{ kJ/kg of dry air} \quad \text{...[Eqn. 12.12 (b)]}$$

$$= 1.005 \times 32 + 0.03\,(2500 + 1.88 \times 32)$$

$$= 32.16 + 76.8 = \mathbf{108.96 \text{ kJ/kg of dry air (Ans.)}}$$

Example 12.2. *The atmospheric conditions are 20°C and specific humidity of 0.0095 kg/kg of dry air. Calculate the following:*

(*i*) *Partial pressure of vapour;*

(*ii*) *Relative humidity;*

(*iii*) *Dew point temperature.*

Solution: Dry bulb temperature, t_{db} = 20°C

Specific humidity, W = 0.0095 kg/kg of dry air

(*i*) **Partial pressure of vapour,** p_v **:**

The specific humidity is given by

$$W = \frac{0.622\, p_v}{p_t - p_v}$$

$$0.0095 = \frac{0.622\, p_v}{1.0132 - p_v}$$

$$0.0095\,(1.0132 - p_v) = 0.622\, p_v$$

$$0.009625 - 0.0095\, p_v = 0.622\, p_v$$

$$p_v = \mathbf{0.01524 \text{ bar (Ans.)}}$$

(*ii*) **Relative humidity,** ϕ **:**

Corresponding to 20°C, from steam tables,

$$p_{vs} = 0.0234 \text{ bar}$$

$$\therefore \quad \textbf{Relative humidity, } \phi = \frac{p_v}{p_{vs}} = \frac{0.01524}{0.0234} = \mathbf{0.65 \text{ or } 65\% \text{ (Ans.)}}$$

(*iii*) **Dew point temperature, t_{dp} :**

The dew point temperature is the saturation temperature of water vapour at a pressure of 0.01524 bar,

t_{dp} [from steam tables by interpolation]

$$= 13 + \frac{(14 - 13)}{(0.01598 - 0.0150)} \times [0.01524 - 0.0150]$$

$$13 + \frac{0.00024}{0.00098} = \mathbf{13.24°C\ (Ans.)}$$

Example 12.3. *The atmospheric conditions are 32°C and specific humidity of 13.4 g/kg of air. Determine:*

(*i*) *Partial pressure of vapour;*

(*ii*) *Relative humidity;*

(*iii*) *Dew point temperature.*

Atmospheric pressure = 758 mm Hg.

Solution: *Given:* t_{db} = 32°C; W = 13.4 g/kg of air; p_t = 758 mm Hg

(*i*) **Partial pressure of vapour p_v :**

$$W = \frac{0.622}{p_t - p_v}, \quad \text{or,} \quad \frac{13.4}{1000} = \frac{0.622\, p_v}{758 - p_v}$$

or, $\quad 0.134\,(758 - p_v) = 0.622\, p_v, \quad$ or, $\quad 10.157 - 0.0134\, p_v = 0.622\, p_v$

$\therefore \quad p_v = \mathbf{15.98\ mm\ Hg\ (Ans.)}$

(*ii*) **Relative humidity; ϕ:**

$$\phi = \frac{p_v}{p_{vs}}$$

where, p_{vs} = Saturation pressure of dry bulb temperature of 32°C

$$= 0.0476 \text{ bar} = 0.0476 \times \frac{758}{1.0132} = 35.61 \text{ mm Hg}$$

$$\therefore \quad \phi = \frac{15.98}{35.61} = 0.4487, \quad \text{or,} \quad \mathbf{44.87\%\ (Ans.)}$$

(*iii*) **Dew point temperature, t_{dp} :**

The dew point temperature is the saturation temperature of water vapour at a pressure of 15.98 mm Hg (0.0214 bar). From steam tables this temperature is found as follows:

t_{dp} [from steam tables by interpretation]

$$= 18 + \frac{(19 - 18)}{(0.022 \quad 0.0206)} \times [0.0214 - 0.0206]$$

$$= \mathbf{18.57°C\ (Ans.)}$$

Example 12.4. *The air supplied to a room of a building in winter is to be at 17°C and have a relative humidity of 60%. If the barometric pressure is 1.01325 bar, find : (i) The specific humidity, and (ii) The dew point under these conditions.*

Solution: Dry bulb temperature, t_{db} = 17°C

Relative humidity, ϕ = 60%

Barometric or total pressure, p_t = 1.01325 bar

(*i*) **Specific humidity, W:** Corresponding to 17°C, from steam tables,

$$p_{vs} = 0.0194 \text{ bar}$$

Also,
$$\phi = \frac{p_v}{p_{vs}}$$

i.e.,
$$0.6 = \frac{p_v}{0.0194}$$

$\therefore$
$$p_v = 0.6 \times 0.0194 = 0.01164 \text{ bar}$$

Specific humidity,
$$W = \frac{0.622\, p_v}{p_t - p_v} = \frac{0.622 \times 0.01164}{1.01325 - 0.01164}$$

$$= \mathbf{0.00723\ kg/kg\ of\ dry\ air\ (Ans.)}$$

(*ii*) **Dew point temperature, t_{dp}:** If the air is cooled at constant pressure the vapour will begin to condense at the saturation temperature corresponding to 0.01164 bar. By interpolation from steam tables, the dew point temperature, t_{dp} is then:

$$t_{dp} = 9 + (10 - 9) \times \frac{0.01164 - 0.01150}{0.01230 - 0.01150} = \mathbf{9.18°C\ (Ans.)}$$

Example 12.5. *The pressure and temperature of air in a room are 1.0132 bar and 30°C respectively. If the relative humidity is found to be 40 per cent, estimate :*

(*i*) *The partial pressure of water vapour;*

(*ii*) *The dew point temperature;*

(*iii*) *The specific volume of each component;*

(*iv*) *The specific humidity.*

Solution : *Given:* p_v = 1.0132 bar; t_{db} = 30°C; ϕ = 40%.

(*i*) **The partial pressure of water vapour, p_v:**

From steam tables, corresponding to 30°C,

$$p_{vs} = 0.0425 \text{ bar}$$

Also,
$$\phi = \frac{p_v}{p_{vs}}$$

or
$$0.4 = \frac{p_v}{0.0425}$$

$\therefore$
$$p_v = 0.4 \times 0.0425 = \mathbf{0.017\ bar\ (Ans.)}$$

(*ii*) **The dew point, t_{dp} :** The dew point corresponding to p_v = 0.017 bar,

$$t_{dp} = \mathbf{15°C\ (Ans.)}$$

(*iii*) **The specific volume of each component, v_v, v_a ;**

For vapour,
$$v_v = \frac{R_v T}{p_v} = \frac{462 \times (30 + 273)}{0.017 \times 10^5}$$
$$= \mathbf{82.345\ m^3/kg\ (Ans.)}$$

For air,
$$v_a = \frac{R_a T}{p_a} = \frac{287 \times (30 + 273)}{(1.0132 - 0.017) \times 10^5}$$
$$= \mathbf{0.873\ m^3/kg\ (Ans.)}$$

(*iv*) **The specific humidity, W :**

$$W = \frac{m_v}{m_a} = \frac{v_a}{v_v} \quad \text{...[Eqn. (12.5)]}$$
$$= \frac{0.873}{82.345} = \mathbf{0.0106\ kg/kg\ of\ dry\ air\ (Ans.)}$$

Example 12.6. *0.004 kg of water vapour per kg of atmospheric air is removed and temperature of air after removing the water vapour becomes 20°C. Determine:*

(*i*) *Relative humidity;*

(*ii*) *Dew point temperature.*

Assume that condition of atmospheric air is 30°C and 55% R.H. and pressure is 1.0132 bar.

Solution: Corresponding to 30°C, from steam tables, p_{vs} = 0.0425 bar.

$\therefore$ Relative humidity (RH), $\phi = \frac{p_v}{p_{vs}}$

i.e.,
$$0.55 = \frac{p_v}{0.0425}$$

$\therefore$
$$p_v = 0.02337 \text{ bar.}$$

Also the specific humidity,
$$W = \frac{0.622\,p_v}{p_t - p_v} = \frac{0.622 \times 0.02337}{1.0132 - 0.02337}$$
$$= 0.01468 \text{ kg/kg of dry air.}$$

The specific humidity after removing 0.004 kg of water vapour becomes,

0.01468 – 0.004 = 0.01068 kg/kg of dry air and the temperature t_{db} is given as 20°C.

The partial pressure of water vapour, p_v, at this condition can be calculated as follows :

$$W = \frac{0.622\,p_v}{p_t - p_v}$$
$$0.01068 = \frac{0.622\,p_v}{1.0132 - p_v}$$

or,
$$0.01068\,(1.0132 - p_v) = 0.622\,p_v$$

or,
$$0.01082 - 0.01068\,p_v = 0.622\,p_v$$

or,
$$0.6327\,p_v = 0.01082$$

$\therefore$
$$p_v = 0.0171 \text{ bar}$$

Corresponding to 20°C, from steam tables p_{vs} = 0.0234 bar.

(*i*) **Relative humidity,** $\phi = \dfrac{p_v}{p_{vs}} = \dfrac{0.0171}{0.0234}$ = **0.73 or 73% (Ans.)**

(*ii*) **Dew point temperature,** t_{dp} **:**

Corresponding to 0.0171 bar, from steam tables t_{dp} = **15°C (Ans.)**

Example 12.7. *A mixture of dry air and water vapour is at a temperature of 22°C under a total pressure of 730 mm Hg. The dew point temperature is 15°C. Find :*

(*i*) *Partial pressure of water vapour;*

(*ii*) *Relative humidity;*

(*iii*) *Specific humidity;*

(*iv*) *Enthalpy of air per kg of dry air;*

(*v*) *Specific volume of air per kg of dry air.*

Solution: *Given:* t_{db} = 22°C; t_{dp} = 15°C;

$$p_t = 1.01325 \times \frac{730}{760} = 0.9732 \text{ bar}$$

(*i*) From steam tables partial pressure of water vapour at 15°C dew point temperature,

$$p_v = \mathbf{0.017\ bar\ (Ans.)}$$

(*ii*) Saturation pressure of water vapour at 22°C DBT,

$$p_{vs} = 0.0264 \text{ bar}$$

∴ **Relative humidity,**

$$\phi = \frac{p_v}{p_{vs}} = \frac{0.017}{0.0264} = \mathbf{0.644\ (Ans.)}$$

(*iii*) **Specific humidity,**

$$W = \frac{0.622\, p_v}{p_t - p_v}$$

$$= \frac{0.622 \times 0.017}{0.9732 - 0.017} = \mathbf{0.011\ kg/kg} \text{ of dry air } \mathbf{(Ans.)}$$

(*iv*) Enthalpy of air, $h = 1.005\, t_{db} + W\,(2500 + 1.88\, t_{db})$ ...[Eqn. 12.12 (b)]

$$= 1.005 \times 22 + 0.011\,(2500 + 1.88 \times 22)$$

$$= 22.11 + 27.95 = \mathbf{50.06\ kJ/kg\ of\ dry\ air\ (Ans.)}$$

(*v*) Specific volume of air is equal to the volume of 1 kg of dry air or 0.011 kg of water vapour. Based on dry air part

$$v = v_a = \frac{R_a \times T_a}{p_a} = \frac{R_a \times T_a}{(p_t - p_v)}$$

$$= \frac{287 \times (22 + 273)}{(0.9732 - 0.017) \times 10^5} = \mathbf{0.885\ m^3/kg\ of\ dry\ air\ (Ans.)}$$

Example 12.8. *The sling psychrometer in a laboratory test recorded the following readings:*

Dry bulb temperature = 35°C;

Wet bulb temperature = 25°C.

Calculate the following:

(i) Specific humidity;

(ii) Relative humidity;

(iii) Vapour density in air;

(iv) Dew point temperature;

(v) Enthalpy of mixture per kg of dry air.

Take atmospheric pressure = 1.0132 bar.

Solution: For finding the partial pressure of vapour, using the equation:

$$p_v = (p_{vs})_{wb} - \frac{[p_t - (p_{vs})_{wb}](t_{db} - t_{wb})}{1527.4 - 1.3\, t_{wb}}$$

Corresponding to 25°C (from steam tables)

$$(p_{vs})_{wb} = 0.0317 \text{ bar}$$

Substituting the values in the above equation, we get:

$$p_v = 0.0317 - \frac{[1.0132 - 0.0317](35 - 25)}{1527.4 - 1.3 \times 25} = 0.0317 - 0.0065 = 0.0252 \text{ bar}$$

(*i*) **Specific humidity,**

$$W = \frac{0.622\, p_v}{p_t - p_v} = \frac{0.622 \times 0.0252}{1.0312 - 0.0252} = \mathbf{0.01586\ kg/kg\ of\ dry\ air\ (Ans.)}$$

(*ii*) **Relative humidity,** $\phi = \dfrac{p_v}{p_{vs}} = \dfrac{0.0252}{0.0563}$

[p_{vs} = 0.0563 bar corresponding to 35°C, from steam tables]

$$= \mathbf{0.447\ or\ 44.7\%\ (Ans.)}$$

(*iii*) **Vapour density:**

Using characteristic gas equation, we have

$$p_v V_v = m_v R_v T_v$$

$$p_v = \frac{m_v}{V_v} R_v T_v = \rho_v R_v T_v$$

$$\left[\text{where, } \rho_v = \text{Vapour density, } R_v = \frac{\text{Universal gas constant}}{\text{Molecular weight of } H_2O} = \frac{8314.3}{18}\right]$$

$$\therefore \quad 0.0252 \times 10^5 = \rho_v \times \frac{8314.3}{18} \times (273 + 35)$$

$$\therefore \quad \rho_v = \frac{0.0252 \times 10^5 \times 18}{8314.3 \times 308} = \mathbf{0.0177\ kg/m^3\ (Ans.)}$$

(*iv*) **Dew point temperature; t_{dp}:** Corresponding to 0.0252 bar, from steam tables (by interpolation)

$$t_{dp} = 21 + (22 - 21) \times \frac{(0.0252 - 0.0249)}{(0.0264 - 0.0249)}$$

$$= \mathbf{21.2°C \text{ (Ans.)}}$$

(*v*) **Enthalpy of mixture per kg of dry air, *h*:**

$$h = 1.005\, t_{db} + W\,(2500 + 1.88\, t_{db}) \quad \text{...[Eqn. 12.12 (b)]}$$

$$= 1.005 \times 35 + 0.01586\,[2500 + 1.88 \times 35]$$

$$= 35.175 + 40.69$$

$$= \mathbf{75.86 \text{ kJ/kg of dry air (Ans.)}}$$

Example 12.9. *In a laboratory test, a psychrometer recorded dry bulb temperature as 35°C and wet bulb temperature 28°C. Calculate:*

(*i*) *Vapour pressure;*

(*ii*) *Relative humidity;*

(*iii*) *Specific humidity;*

(*iv*) *Degree of saturation;*

(*v*) *Dew point temperature;*

(*vi*) *Enthalpy of mixture.*

Barometric pressure = 1.0132 bar

Solution: *Given:* t_{db} = 35°C; t_{wb} = 28°C; p_t = 1.0132 bar

From steam tables, corresponding to 28°C,

$$(p_{vs})_{wb} = 0.0378 \text{ bar}$$

(*i*) **Vapour pressure** is given by the relation:

$$p_v = (p_{vs})_{wb} - \frac{[p_t - (p_{vs})_{wb}](t_{db} - t_{wb})}{1527.4 - 1.3\, t_{wb}}$$

$$= 0.0378 - \frac{[1.0132 - 0.0378](35 - 28)}{1527.4 - 1.3 \times 28}$$

$$= 0.0378 - \frac{6.8278}{1491} = \mathbf{0.0332 \text{ bar (Ans.)}}$$

(*ii*) **Relative humidity,** $\phi = \dfrac{p_v}{p_{vs}}$

From steam tables corresponding to 35°C, p_{vs} = 0.0563 bar

$$\therefore \quad \phi = \frac{0.0332}{0.0563} = 0.5897 \text{ or } \mathbf{58.97\% \text{ (Ans.)}}$$

(*iii*) **Specific humidity,** $W = \dfrac{0.622\, p_v}{p_t - p_v} = \dfrac{0.622 \times 0.0332}{1.0132 - 0.0332}$

$$= \mathbf{0.021 \text{ kg/kg of dry air (Ans.)}}$$

(*iv*) **Degree of saturation,** $\mu = \dfrac{p_v\,(p_t - p_{vs})}{p_{vs}\,(p_t - p_v)} = \dfrac{0.0332\,(1.0132 - 0.0563)}{0.0563\,(1.0132 - 0.0332)} = \mathbf{0.576}$ **(Ans.)**

(*v*) **Dew point temperature, t_{db},** corresponding to

p_v (i.e., 0.0332 bar) ≈ **26°C (Ans.)** ...From steam tables

(*vi*) **Enthalpy of mixture,** $h = 1.005\,t_{db} + W\,(2500 + 1.88\,t_{db})$ ...[Eqn. 12.12 (b)]

$= 1.005 \times 35 + 0.021\,(2500 + 1.88 \times 35)$

$= 35.175 + 53.882 =$ **89.06 kJ/kg of dry air (Ans.)**

Example 12.10. *Atmospheric air at 0.967 bar has 11°C wet bulb depression from 31°C dry bulb temperature during adiabatic saturation process. Determine :*

(*i*) *Specific humidity from the adiabatic saturation equation;*

(*ii*) *Vapour pressure and RH at 31°C;*

(*iii*) *Dew point temperature.*

Check the result of vapour pressure by equation,

$$p_v = (p_{vs})_{wb} - \frac{[p_t - (p_{vs})_{wb}\,(t_{db} - t_{wb})}{1527.4 - 1.3\,t_{wb}} \qquad \textbf{(UPTU)}$$

Solution: *Given:* p_t = 0.967 bar; t_{db_1} = 31°C, Wet bulb depression = 11°C

$t_{wb_2} = t_{db_1}$ – wet bulb depression = 31 – 11 = 20°C.

At t_{wb_2} (= t_{db_2}) = 20°C: p_{vs} = 0.0234 bar ; h_{f2} = 83.86 kJ/kg

At t_{db_1} = 31°C: p_{v1} (= p_{vs}) = 0.045 bar

At 20°C, for saturated air, $W_{2s} = \dfrac{0.622\,p_{vs}}{p_t - p_{vs}} = \dfrac{0.622 \times 0.0234}{(0.967 - 0.0234)} = 0.0154$ kg/kg of dry air

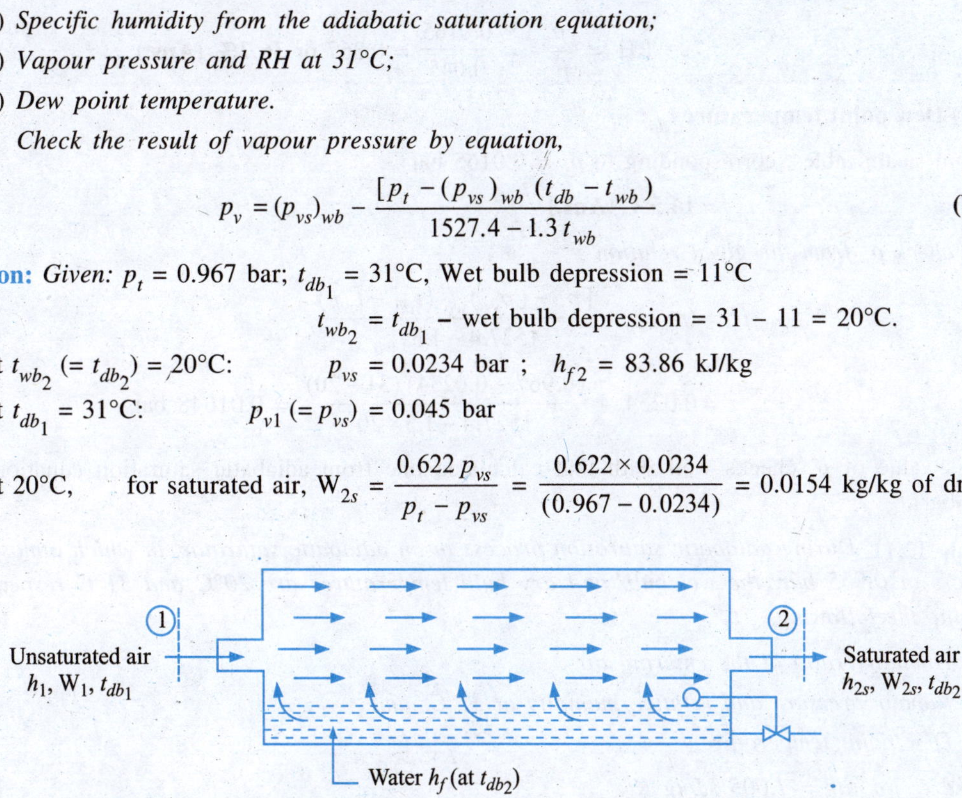

Fig. 12.3

(*i*) **Specific humidity from the adiabatic saturation equation, W_1:** Energy balance at 1 and 2 (Fig. 12.3) gives:

$h_1 + (W_{2s} - W_1)\,h_{f2} = h_{2s}$

$[1.005\,t_{db_1} + W_1\,(2500 + 1.88\,t_{db_1})] + (W_{2s} - W_1)\,h_{f2} = [1.005\,t_{db_2} + W_{2s}\,(2500 + 1.88\,t_{db_2})]$

$1.005\,(t_{db_1} - t_{db_2}) + W_1\,(2500 + 1.88\,t_{db_1} - h_{f2}) = W_{2s}\,(2500 + 1.88\,t_{db2} - h_{f2})$

$1.005\,(31 - 20) + W_1\,(2500 + 1.88 \times 31 - 83.86) = 0.0154\,(2500 + 1.88 \times 20 - 83.86)$

$11.055 + 2474.4\,W_1 = 37.787$

or, W_1 = **0.0108 kg/kg of dry air (Ans.)**

(*ii*) **Vapour pressure and RH at 31°C:**

$$W_1 = \frac{0.622\,p_v}{p_t - p_v}$$

$$0.0108 = \frac{0.622\,p_v}{0.967 - p_v}$$

or, $$0.0108\,(0.967 - p_v) = 0.622\,p_v$$

or, $$0.01044 - 0.0108\,p_v = 0.622\,p_v$$

or, $$p_v = \frac{0.01044}{(0.0108 + 0.622)} = \mathbf{0.0165\ bar\ (Ans.)}$$

$$RH = \frac{p_v}{p_{vs}} = \frac{0.0165}{0.045} = 0.367 \text{ or } \mathbf{36.7\%\ (Ans.)}$$

(*iii*) **Dew point temperature t_{dp}:**

From steam tables, corresponding to $p_v = 0.0165$ bar

$$t_{dp} = \mathbf{14.5°C\ (Ans.)}$$

To check p_v from the given relation :

$$p_v = (p_{vs})_{wb} - \frac{[p_t - (p_{vs})_{wb}\,(t_{db} - t_{wb})}{1527.4 - 1.3\,t_{wb}}$$

$$= 0.0234 - \frac{(0.967 - 0.0234)\,(31 - 20)}{1527.4 - 1.3 \times 20} = 0.01648 \text{ bar}$$

This value of p_v checks well with that calculated one from adiabatic saturation equation (i.e., 0.0165 bar).

Example 12.11. *During adiabatic saturation process in an adiabatic saturator, in which atmospheric air enters at 0.965 bar, the wet bulb and dry bulb temperatures are 20°C and 31°C respectively. Determine the following:*

(*i*) *Humidity ratio of the entering air;*

(*ii*) *Vapour pressure and relative humidity at 31°C; and*

(*iii*) *Dew point temperature.*

Take c_p for air = 1.005 kJ/kg K. **(M.U.)**

Solution: *Given:* $p_t = 0.965$ bar; $t_{wb} = 20°C$; $t_{db} = 31°C$

(*i*) **Humidity ratio of the entering air, W_1 :**

Let W_1 = Humidity ratio of the entering air, and

W_{2s} = Humidity ratio of saturated air.

Let us first find W_{2s} from psychrometric or steam tables.

At 20°C : $p_{v2} = 0.02337$ bar

$h_{2s} = h_{g2} = 2538.2$ kJ/kg

$h_{f2} = 83.86$ kJ/kg

At 31°C : $h_{1s} = h_{g1} = 2558.2$ kJ/kg

We know that, $$W_2 = \frac{0.622\, p_{v2}}{p_t - p_{v2}} = \frac{0.622 \times 0.02337}{0.965 - 0.02337} = 0.0154 \text{ kg/kg of dry air}$$

$$\therefore \quad W_1 = \frac{c_p\,(t_{db_2} - t_{db_1}) + W_{2s}\,(h_{g2} - h_{f2})}{h_{g1} - h_{f2}} \quad \text{...[Eqn. 12.18 (a)]} \quad (\because t_{db_2} = t_{wb_2})$$

$$= \frac{1.005\,(20 - 31) + 0.0154\,(2538.2 - 83.86)}{(2558.2 - 83.86)}$$

$$= \mathbf{0.0108 \text{ kg/kg of dry air (Ans.)}}$$

(*ii*) **Vapour pressure and relative humidity at 31°C, p_{v1}:**

Let p_{v1} = Vapour pressure at 31°C

Now, $$W_1 = 0.0108 = \frac{0.622\, p_{v1}}{p_t - p_{v1}} = \frac{0.622\, p_{v1}}{0.965 - p_{v1}}$$

or, $0.0108\,(0.965 - p_{v1}) = 0.622\, p_{v1}$

or, $0.0104 - 0.0108\, p_{v1} = 0.622\, p_{v1}$

or, $$p_{v1} = \frac{0.0104}{(0.0108 + 0.622)} = \mathbf{0.0164 \text{ bar (Ans.)}}$$

The saturation pressure corresponding to 31°C, from psychrometric or steam tables,

$$p_{vs} = 0.04497 \text{ bar}$$

$$\therefore \text{ Relative humidity, } \phi = \frac{p_{v1}}{p_{vs}} = \frac{0.0164}{0.04497} \sim 0.365 \text{ or } \mathbf{36.5\% \text{ (Ans.)}}$$

(*iii*) **Dew point temperature, t_{dp}:**

Since t_{dp} is the saturation temperature corresponding to the partial pressure of water vapour (p_{v1}), therefore from psychrometric or steam tables, we have:

$$t_{dp} = \mathbf{14.5°C} \text{ (corresponding to 0.0164 bar). } \mathbf{(Ans.)}$$

Example 12.12. *The pressure and temperature of mixture of dry air and water vapour are 736 mm of Hg and 21°C. The dew point temperature of the mixture is 15°C. Determine the following using steam tables :*

(*i*) *Partial pressure of water vapour in the mixture;*

(*ii*) *Relative humidity;*

(*iii*) *Specific humidity;*

(*iv*) *Enthalpy of mixture per kg of dry air;*

(*v*) *Specific volume of mixture per kg of dry air.* **(N.U.)**

Solution: *Given:* p_t = 736 mm of Hg = $\dfrac{(1000 \times 13.6) \times 9.81 \times (736/1000)}{10^5} = 0.98194$ bar ;

$$t_{db} = 21°C;\ t_{dp} = 15°C.$$

(*i*) **Partial pressure of water vapour in the mixture, p_v:**

p_v from steam tables at 15°C (dew point temperature)

$$= \mathbf{0.017039 \text{ bar (Ans.)}}$$

(*ii*) **Relative humidity, ϕ :**

p_{vs} from steam tables at 21°C DBT = 0.024853 bar

$$\therefore \quad \phi = \frac{p_v}{p_{vs}} = \frac{0.017039}{0.024853} = 0.6856 \text{ or } \mathbf{68.56\% \text{ (Ans.)}}$$

(*iii*) **Specific humidity, W:**

$$W = \frac{0.622\, p_v}{p_t - p_v}$$

$$= \frac{0.622 \times 0.017039}{0.98194 - 0.017039} = \mathbf{0.011 \text{ kg/kg of dry air (Ans.)}}$$

(*iv*) **Enthalpy of mixture per kg of dry air, *h* :**

$$h = 1.005\, t_{ab} + W\,(2500 + 1.88\, t_{db}) \text{ kJ/kg of dry air} \quad \text{...[Eqn. 12.12 (b)]}$$

$$= 1.005 \times 21 + 0.011\,(2500 + 1.88 \times 21) = \mathbf{49.04 \text{ kJ/kg (Ans.)}}$$

(*v*) **Specific volume of mixture per kg of dry air *v*:**

$$v = v_a = \frac{R_a \times T_a}{p_a} = \frac{R_a \times T_a}{(p_t - p_v)}$$

$$= \frac{287 \times (21 + 273)}{(0.98194 - 0.017039) \times 10^5} = \mathbf{0.874 \text{ m}^3\text{/kg (Ans.)}}$$

12.5 PSYCHROMETERS

A **psychrometer** *is a device which is used for measuring dry bulb and wet bulb temperatures simultaneously.*

The psychrometers may be classified as follows:

1. Laboratory psychrometer.
2. Sling psychrometer.
3. Aspirating psychrometer.
4. Continuous recording psychrometer.

The description of a *sling psychrometer* is given below:

Refer to Fig. 12.4. The sling psychrometer consists of two thermometers mounted on a plate. The one with the 'sock' is wet-bulb thermometer; the other is dry-bulb. The wet bulb exists below the dry-bulb. This is done purposely so that 'sock' can be dipped in water without wetting the dry-bulb. The handle of the frame helps for rotating the psychrometer to produce necessary air motion. As the psychrometer is rotated it provides necessary air velocity over the thermometer. Fast movement of air past the 'sock' is necessary to bring the air at temperature t_{db} always in immediate contact with the wet 'sock'. The temperature spread between dry bulb and wet bulb readings depends upon the *amount of moisture in the air. Dry air,* or air that has low moisture content has a low wet bulb temperature; humid air that has a high moisture content, has a high wet-bulb temperature.

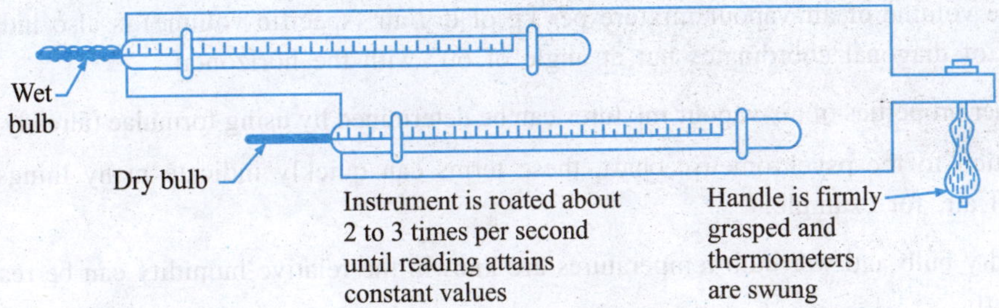

Fig. 12.4. Sling psychrometer.

- *When dry and wet bulb temperatures are known the other psychrometric properties like relative humidity, dew point temperature, degree of saturation, humidity ratio, and volume of air-vapour mixture per kg of dry air are determined by calculations.*

12.6 PSYCHROMETRIC CHARTS

The psychrometric charts are prepared to represent graphically all the necessary moist air properties used for air-conditioning calculations. The values are based on actual measurements verified for thermodynamic consistency.

For psychrometric charts the most convenient co-ordinates are dry bulb temperature of air vapour mixture as the abscissa and moisture content (kg/kg of dry air) or water vapour pressure as the ordinate. Depending upon whether the humidity content is abscissa or ordinate with temperature co-ordinate, the charts are generally classified as Mollier chart and Carrier chart. Carrier chart having t_{db} as the abscissa and W as the ordinate finds a wide applications. The chart is constructed as under :

1. *The dry bulb temperature,* (°C) of unit mass of dry air for different humidity contents or humidity ratios are indicated by *vertical lines drawn parallel to the ordinate.*
2. *The mass of water vapour in* kg (or grams) per kg of dry air is drawn parallel to the abscissa for different values of dry bulb temperature. *It is the major vertical scale of the chart.*
3. *Pressure of water vapour in mm of mercury* is shown in the scale at left and is the absolute pressure of steam.
4. *Dew point temperatures* are temperatures corresponding to the boiling points of water at low pressures of water vapour and are shown in the scale on the *upper curved line*. The dew points for different low pressures are read on *diagonal coordinates*.
5. Constant relative humidity lines in per cent are indicated by marking off vertical distances between the saturation line or the upper curved line and the base of the chart. The relative humidity curve depicts quantity (kg) of moisture actually present in the air as a percentage of the total amount possible at various dry bulb temperatures and masses of vapour.
6. *Enthalpy or total heat* at saturation temperature in kJ/kg of dry air is shown by a diagonal system of co-ordinates. The scale on the diagonal line is separate from the body of the chart and is indicated above the saturation line.
7. *Wet bulb temperatures* are shown on the diagonal co-ordinate coinciding with heat co-ordinates. The scale of wet bulb temperatures is shown on the saturation curve. The diagonals run downwards to the right at an angle of 30°C to the horizontal.

8. The volume of air-vapour mixture per kg of dry air (specific volume) is also indicated by a set of diagonal coordinates but at angle of 60° with the horizontal.

The other properties of air-vapour mixtures can be determined by using formulae (already discussed).

In relation to the psychrometric chart, these terms can quickly indicate many things about the condition of air, for example:

1. If dry bulb and wet bulb temperatures are known, the relative humidity can be read from the chart.
2. If the dry bulb and relative humidity are known, the wet bulb temperature can be determined.
3. If wet bulb temperature and relative humidity are known, the dry bulb temperature can be found.
4. If wet bulb and dry bulb temperatures are known, the dew point can be found.
5. If wet bulb and relative humidity are known, dew point can be read from the chart.
6. If dry-bulb and relative humidity are known, dew point can be found.
7. The quantity (kg) of moisture in air can be determined from any of the following combinations:
   (*i*) Dry bulb temperature and relative humidity;
   (*ii*) Dry bulb temperature and dew point;
   (*iii*) Wet bulb temperature and relative humidity;
   (*iv*) Wet bulb temperature and dew point temperature;
   (*v*) Dry bulb temperature and wet bulb temperature; and
   (*vi*) Dew point temperature alone.

Fig. 12.5 shows the skeleton psychrometric chart.

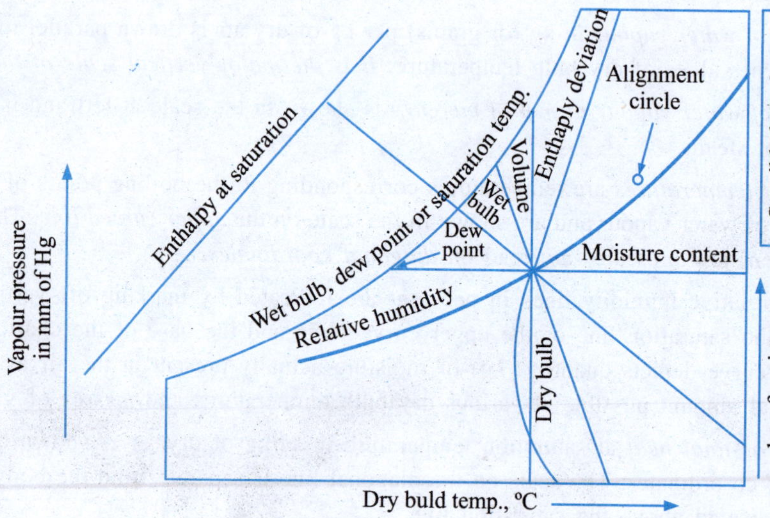

Fig. 12.5. Skelton psychrometric chart.

12.7 PSYCHROMETRIC PROCESSES

In order to condition air to the conditions of human comfort or of the optimum control of an industrial process required, certain processes are to be carried out on the outside air available. *The processes affecting the psychrometric properties of air are called* ***Psychrometric processes***. These processes involve mixing of air streams, heating, cooling, humidifying, dehumidifying, adiabatic saturation and mostly the combinations of these.

The *important psychrometric processes* are enumerated below and explained in the following Articles:

1. Mixing of air streams
2. Sensible heating
3. Sensible cooling
4. Cooling and dehumidification
5. Cooling and humidification
6. Heating and dehumidification
7. Heating and humidification.

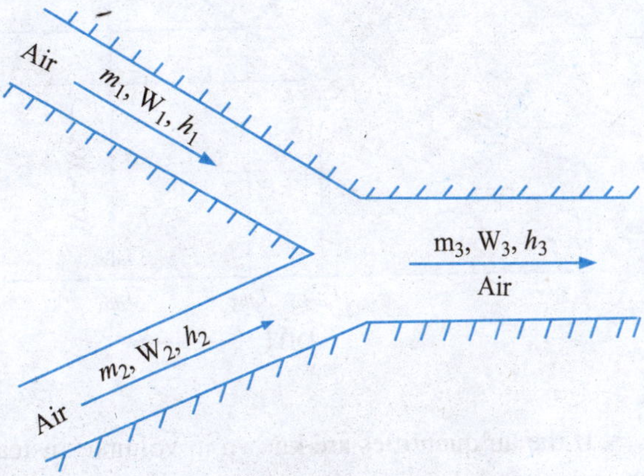

Fig. 12.6. Mixing of air streams.

12.7.1 Mixing of Air Streams

Refer to Figs. 12.6 and 12.7. Mixing of several air streams is the process which is very frequently used in air-conditioning. This mixing normally takes place without the addition or rejection of either heat or moisture, i.e., *adiabatically and at constant total moisture content. Thus we can write the following equations:*

$$m_1 + m_2 = m_3 \quad ...(12.19)$$

$$m_1 W_1 + m_2 W_2 = m_3 W_3 \quad ...(12.20)$$

$$m_1 h_1 + m_2 h_2 = m_3 h_3 \quad ...(12.21)$$

Rearranging of last two equations gives the following:

$$m_1 (W_1 - W_3) = m_2 (W_3 - W_2)$$

$$m_1 (h_1 - h_3) = m_2 (h_3 - h_2)$$

or,

$$\frac{m_1}{m_2} = \frac{W_3 - W_2}{W_1 - W_3} = \frac{h_3 - h_2}{h_1 - h_3} \quad ...[12.21 (a)]$$

where, m = Mass of dry air, W = Specific humidity, h = Enthalpy — at particular state points.

On the psychrometric chart, the specific humidity and enthalpy scales are linear, ignoring enthalpy deviations. Therefore, the final state 3 lies on a straight line connecting the initial states of the two streams mixing, and the final state 3 divides this line into two parts that are in the same ratio as were the two masses of air before mixing.

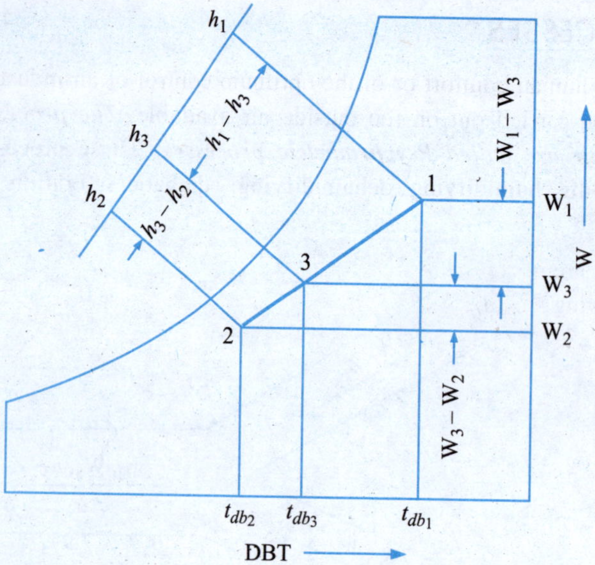

$$\frac{m_1}{m_2} = \frac{W_3 - W_2}{W_1 - W_3} = \frac{h_3 - h_2}{h_1 - h_3}$$

Fig. 12.7

If the air quantities are known in volumes instead of mass units, it is generally sufficiently accurate to units of m^3/min. in the mixing equations. The inaccuracy introduced is due to the difference in specific volumes at two initial states. This difference is small for most of the comfort air-conditioning problems.

12.7.2 Sensible Heating

- *When air passes over a dry surface which is at a temperature greater than its (air) dry bulb temperature, it undergoes* ***sensible heating***. Thus the heating can be achieved by passing the air over heating coil like electric resistance heating coils or steam coils.
- During such a process, the specific *humidity remains constant but the dry bulb temperature rises and approaches that of the surface.* The extent to which it approaches at mean effective surface temperature of the coil is conveniently expressed in terms of the equivalent ***"by-pass factor."***

The ***"by-pass factor (BF)"*** *for the process is defined as the ratio of the difference between the mean surface temperature of the coil and leaving air temperature to the difference between the mean surface and the entering air temperature.* Thus on Fig. 12.8, air at temperature t_{db1}, passes over a heating coil with an aveerage surface temperature t_{db_3} and leaves at temperature t_{db_2}.

The *"by-pass factor" is expressed as follows:*

$$BF = \frac{t_{db_3} - t_{db_2}}{t_{db_3} - t_{db_1}} \qquad ...(12.22)$$

Or, in terms of lengths on the chart (Fig. 12.9) it is $\frac{\text{length } 2\text{-}3}{\text{length } 1\text{-}3}$. The *value of the by-pass factor is a function of coil design and velocity.* The heat added to the air can be obtained directly from the entering and the leaving enthalpies $(h_2 - h_1)$ or it can be obtained from the humid specific heat multiplied by the temperature difference $(t_{db_2} - t_{db_1})$.

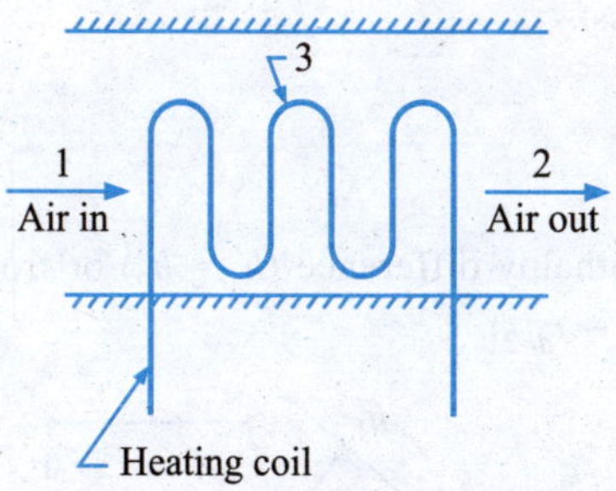

Fig. 12.8. Sensible heating.

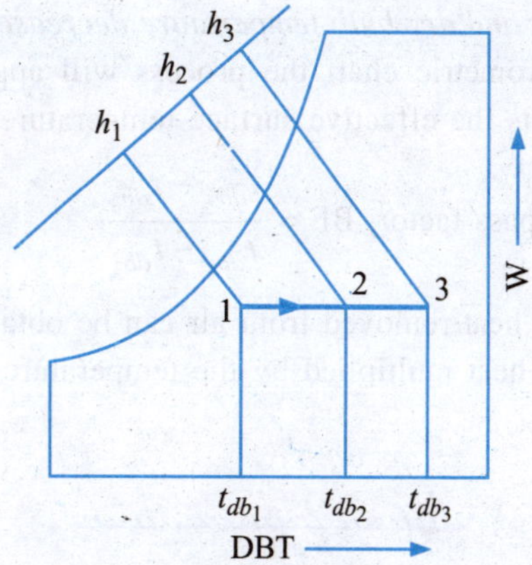

Fig. 12.9

In other words,

Heat added to the air $= c_{pm} \, (t_{db_2} - t_{db_1})$ kJ/kg

$$\left[\begin{aligned} \text{where } c_{pm} &= 1.022 \text{ kJ/kg K} = \text{humidity specific heats} \\ &= c_{pa} + W \times c_{ps} \end{aligned}\right]$$

For air-conditioning purposes, the sensible heat (SH) per minute is given as:

$$\text{SH} = m_a \, c_{pm} \, \Delta t = v_a \rho_a \, c_{pm} \, \Delta t \text{ kJ/min} \qquad (\because m_a = v_a \rho_a)$$

where, v_a = Rate of dry air flowing, m³/min,

ρ_a = Density of moist air at 20°C and 50% RH

= 1.2 kg/m³ of dry air

c_{pm} = Humid specific heat = 1.022 kJ/kg K, and

$\Delta t = t_{db_1} - t_{db_2}$ = Difference of DBTs between the entering and leaving conditions of air, °C.

Substituting the values of ρ and c_{pm} in the above expression, we get:

$$\text{SH} = v_a \times 1.2 \times 1.022 \times \Delta t \text{ kJ/min}$$

$$= \frac{v_a \times 1.2 \times 1.022 \times \Delta t}{60} = 0.02044 \, v_a \, \Delta t \text{ kJ/s, or, kW} \qquad \text{...(Eqn. 12.23)}$$

- In a complete air-conditioning system the *preheating* and *reheating* of air are among the familiar examples of sensible heating.

Note: 'By-pass factor' can be considered to represent the *fraction of air which does not come into contact with coil surface.*

12.7.3 Sensible Cooling

Refer to Fig. 12.10. *Air undergoes* **sensible cooling** *whenever it passes over a surface that is at a temperature less than the dry bulb temperature of the air but greater than the dew point temperature.* Thus sensible cooling can be achieved by passing the air over cooling coil like *evaporating coil of the refrigeration cycle or secondary brine coil*. During the process, the *specific humidity remains* *constant and dry bulb temperature decreases, approaching the mean effective surface temperature.* On a psychrometric chart the process will appear as a horizontal line 1-2 (Fig. 12.11), where point 3 represents the effective surface temperature. For this process:

$$\text{By-pass factor, BF} = \frac{t_{db_2} - t_{db_3}}{t_{db_1} - t_{db_3}} \qquad ...(12.24)$$

The heat removed from air can be obtained from the enthalpy difference $(h_1 - h_2)$ or from humid specific heat multiplied by the temperature difference $(t_{db1} - t_{db2})$.

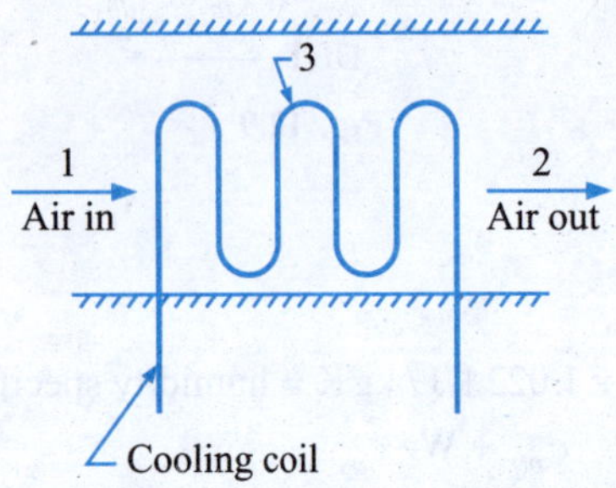

Fig. 12.10. Sensible cooling.

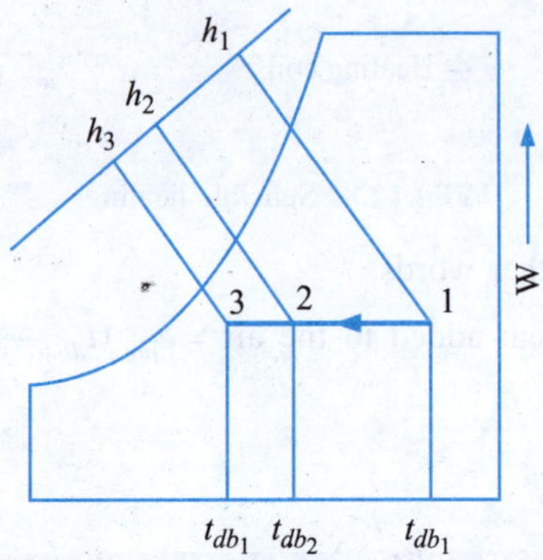

Fig. 12.11

12.7.4 Cooling and Dehumidification

Refer to fig. 12.12. Whenever air is made to pass over a surface or through a spray of water that is at a *temperature less than the dew point temperature of the air, condensation of some of the water vapour in air will occur simultaneously with the sensible cooling process.* Any air that comes into sufficient contact with the cooling surface will be reduced in temperature to the mean surface temperature along a path such as 1-2-3 in Fig. 12.12. with condensation and therefore dehumidification occurring between points 2 and 3. The air that does not come in contact with the surface will be finally cooled by mixing with the portion that does, and the final state point will be somewhere on the straight line connecting points 1 and 3. The actual path of air during the path will not be straight line shown but will be something similarly to the curved dashed line 1-4. It will result from a continuous mixing of air which is connecting a particular part of the coil and air which is by passing it. It is convenient, however to analyse the problem with the straight line shown, and to assume that the final air state results from the mixing of air that has completely by passed the coil with air that has been cooled to the mean effective surface temperature. If there is enough contact between air and surface for all the air to come to the mean surface temperature, the process is one of zero by pass. In any practical system, complete saturation is not obtained and final state will be a point such as 4 in Fig. 12.12. with an equivalent by pass factor equal to $\frac{\text{length 3-4}}{\text{length 3-1}}$. For process involving condensation, the effective surface temperature, e.g., t_{db_3} in Fig. 12.12 is called **'apparatus dew point'** (ADP). The final state point of air passing through a cooling and dehumidifying apparatus is in effect a mixture condition that results from mixing the fraction of the air, which is equal to the equivalent by pass factor (BF) and is at initial state point and the remaining fraction which is equal to one minus by pass factor (1 – BF) and is saturated at the apparatus dew point (ADP).

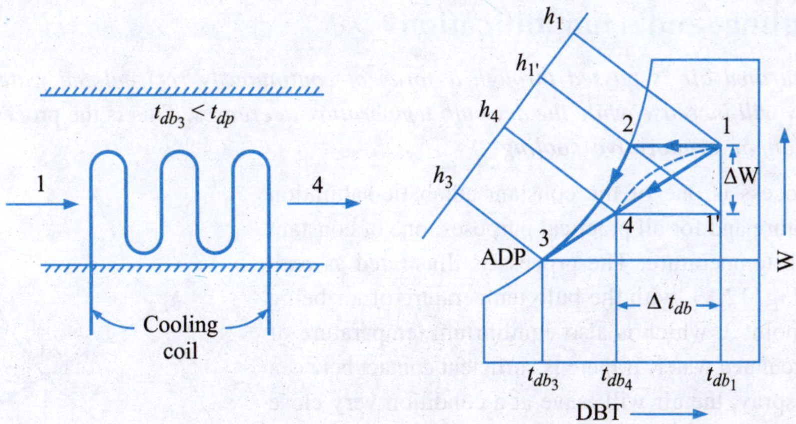

Fig. 12.12. Cooling and dehumilidification.

Total heat (TH) removed from the air is given by

$$\text{TH} = h_1 - h_4 = (h_1 - h_1') + (h_1' - h_4)$$
$$= \text{SH} + \text{LH}$$

where LH = Latent heat removed $(h_1 - h_1')$

SH = Sensible heat removed $(h_1' - h_4)$

The ratio $\frac{\text{SH}}{\text{LH}}$ *is called* **sensible heat factor** (SHF) or *sensible heat ratio* (SHR).

$$\therefore \quad \text{SHF} = \frac{\text{SH}}{\text{SH} + \text{LH}} \qquad \text{...(12.25)}$$

The ratio fixes the slope of the line 1-4 on the psychrometric chart. Sensible heat factor slope lines are given on the psychrometric chart. If the initial conditions and SHF are known for the given process, then the process line can be drawn though the given initial condition at a slope given by SHF on the psychrometric chart.

The capacity of the coiling in *tonnes of refrigeration* is given by,

$$\text{Capacity in tonnes} = \frac{m_a\,(h_1 - h_4) \times 60}{14000} \qquad \text{...(12.26)}$$

where, m_a is the mass of air in kg/min. passing over the coil.

Latent heat load:

In air-conditioning, the latent heat load per minute is given as:

$$\text{LH} = m_a\,\Delta h = m_a\,h_{fg}\,\Delta W = v_a \rho_a\,h_{fg}\,\Delta W \qquad (\because m_a = v_a \rho_a)$$

where, v_a = Rate of dry air flowing, m^3/min,

ρ_a = Density of moist air = 1.2 kg/m^3 of dry air,

h_{fg} = Latent heat of vaporisation = 2500 kJ/kg, and

ΔW = Difference of specific humidity between entering and leaving conditions of air = $(W_2 - W_1)$ for *humidification* and $(W_1 - W_2)$ for *dehumidification*.

Substituting these valves in the above expression, we get:

$$\text{LH} = v_a \times 1.2 \times 2500 \times \Delta W = 3000\, v\, \Delta W \text{ kJ/min}$$

$$= \frac{3000\, v_a\, \Delta W}{60} = 50\, v_a\, \Delta W \text{ kJ/}s\text{, or, kW} \qquad \text{...(12.27)}$$

12.7.5 Cooling and Humidification

- *If unsaturated air is passed through a spray of continuously recirculated water, the specific humidity will increase while the dry bulb temperature decreases.* This is the process of ***adiabatic saturation or evaporative cooling***.
- This process is one of the constant adiabatic-saturation temperature and for all practical purposes, one of constant wet bulb temperature. The process is illustrated as path 1-2 on Fig. 12.13, with the bulb temperature of air being that of point 3, which is also equilibrium temperature of the recirculated water. If there is sufficient contact between air and spray, the air will leave at a condition very close to that of point 3.
- The concept of equivalent by-pass can be applied to this process but another term is more used to describe the performance of a humidifying apparatus. It is the '***saturating***' or '***humidifying efficiency***' which is *defined as the ratio of dry-bulb temperature decrease to the entering wet bulb depression usually expressed as percentage*. Thus, from Fig. 12.13 the *saturating efficiency* is:

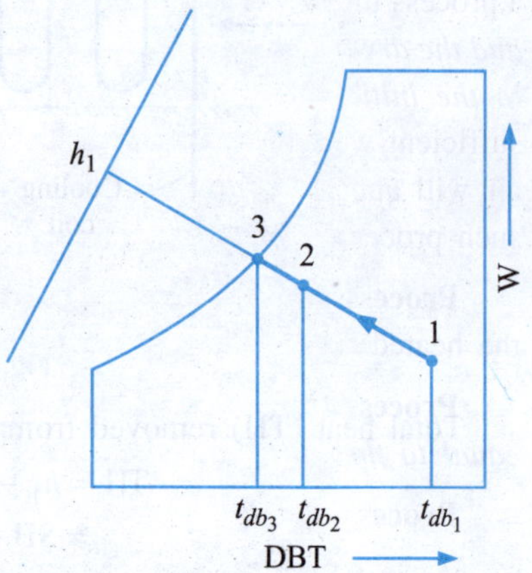

Fig. 12.13. Cooling and humidification.

$$\% \ \eta_{sat} = \left(\frac{t_{db_1} - t_{db_2}}{t_{db_1} - t_{db_3}} \right) \times 100 \quad ...(12.26)$$

As a fraction, it is *equal to one minus the by-pass factor* for the process.

The adiabatic process, for all practical purposes, is line of constant *enthalpy*. The moisture added can be obtained from the increase in specific humidity.

12.7.6 Heating and Dehumidification

If air is passed over a solid adsorbent surface or through a liquid absorbent spray simultaneous heating and dehumidification is accompanied. In either case the dehumidification results from adsorbent or absorbent having a lower water vapour pressure than air. Moisture is condensed out of the air, and consequently the latent heat of condensation is liberated, causing sensible heating of air. If these were the only energies involved, the process would be the inverse of the adiabatic saturation process. There is, however, an additional energy absorbed or liberated by the active material, termed the *heat of adsorption or absorption*. For the solid adsorbent used commercially, such as silica get or activated alumina, and for the more common liquid absorbents, such as solutions of organic salts or inorganic compounds like ethylene, glycol, heat is evolved and results in additional sensible heating. Thus the path lies above a constant wet bulb line on the psychrometric chart such as path 1-2 in Fig. 12.14.

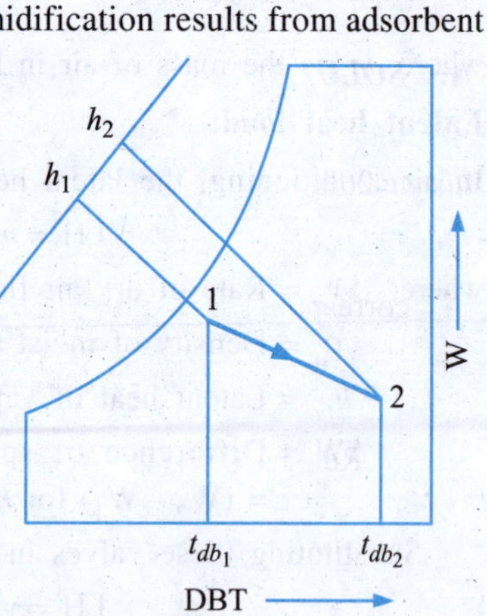

Fig. 12.14. Heating and dehumidication.

12.7.7 Heating and Humidification

If air is passed through a humidifier which has heated water sprays instead of simply recirculated spray, the air is humidified and may be heated, cooled or unchanged in temperature. In such a process the *air increases in specific humidity and the enthalpy, and the dry bulb temperature will increase or decrease according to the initial temperature of the air and that of the spray.* If sufficient water is supplied relative to the mass flow of air, the air will approach saturation at water temperature. Examples of such processes are shown in Fig. 12.15.

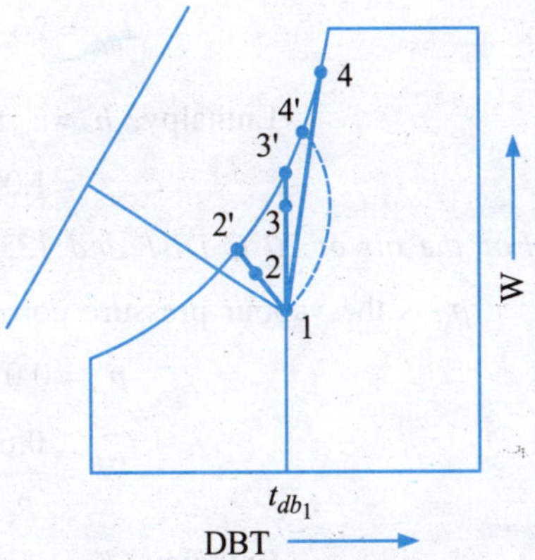

Fig. 12.15. Heating and humidification.

Process 1-2 : Denotes the cases in which the temperature of the heated spray water is *less than the air DBT.*

Process 1-3 : Denotes the cases in which the temperature is *equal to the air DBT.*

Process 1-4 : Denotes the cases in which a spray temperature is *greater than air DBT.*

As in the case of adiabatic saturation, the degree to which the process approaches saturation can be expressed in terms of the by-pass factor or a saturating efficiency.

If the water rate relative to the air quantity is smaller, the water temperature will drop significantly during the process. The resultant process will be a curved line such as the dashed 1-4 where 4 represents the leaving water temperature.

Note : It is possible to accomplish heating and humidification by evaporating from an open pan of heated water, or by direct injection of heated water or steam. The latter is more common. The process line for it is of little value because the process is essentially an instantaneous mixing of steam and the air. The final state point of the air can be found, however, by making a humidity and enthalpy balance for the process. The solution of such a problem usually involves *cut-and-try procedure.*

WORKED EXAMPLES—PSYCHROMETRIC PROCESSES

A. *SOLUTIONS "WITHOUT" USING PSYCHROMETRIC CHARTS*

Example 12.13. (***Adiabatic mixing***) *One kg of air at 35°C DBT and 60% RH is mixed with 2 kg of air at 20°C DBT and 13°C dew-point temperature. Calculate the specific humidity of the mixture.*

Solution: For the air at 35°C DBT and 60% RH,

corresponding to 35°C, from steam tables,

$$p_{vs} = 0.0563 \text{ bar}$$

$$\text{Relative humidity, } \phi = \frac{p_v}{p_{vs}}$$

$$\therefore \quad p_v = \phi\, p_{vs} = 0.6 \times 0.0563 = 0.0338 \text{ bar}$$

$$W = \frac{0.622\, p_v}{p_t - p_v} = \frac{0.622 \times 0.0338}{1.0132 - 0.0338} = 0.0214 \text{ kg/kg of dry air}$$

Corresponding to 0.0338 bar, from steam tables,

$$t_{dp} = 26 + (27 - 26)\frac{(0.0338 - 0.0336)}{(0.0356 - 0.0336)} = 21.6°C$$

Enthalpy, $h = c_p\, t_{db} + W\, h_{vapour} = 1.005\, t_{db} + W\,[h_g + 1.88\,(t_{db} - t_{dp})]$

$= 1.005 \times 35 + 0.0214\,[2565.3 + 1.88\,(35 - 26.1)] = 90.43$ kJ/kg of dry air.

For the air at 20°C DBT and 13°C dew-point temperature:

p_v is the vapour pressure corresponding to the saturation pressure of steam at 13°C.

$\therefore \quad p_v = 0.0150$ bar

$$W = \frac{0.622\, p_v}{p_t - p_v} = \frac{0.622 \times 0.015}{1.0132 - 0.015} = 0.00935 \text{ kg/kg of dry air}$$

Enthalpy, $h = c_p\, t_{db} + W\, h_{vapour} = 1.005 \times 20 + 0.00935\,[h_g + 1.88\,(t_{db} - t_{dp})]$

$= 20.1 + 0.00935\,[2538.1 + 1.88\,(20 - 13)] = 43.95$ kJ/kg of dry air.

Now enthalpy per kg of moist air

$$= \frac{1}{3}\left[\frac{90.43}{1.0214} + \frac{43.95 \times 2}{1.00935}\right] = 58.54 \text{ kJ/kg of moist air}$$

Mass of vapour/kg of moist air

$$= \frac{1}{3}\left[\frac{0.0214}{1.0214} + \frac{0.00935 \times 2}{1.00935}\right] = 0.01316 \text{ kg/kg of moist air}$$

Specific humidity of mixture $= \dfrac{0.01316}{1 - 0.01316} = 0.01333$ **kg/kg of dry air. (Ans.)**

Example 12.14. **(*a*).** *Using Dalton's law of partial pressure show that:*

(*i*) *Humidity ratio* $= \dfrac{0.622\, p_v}{(p_t - p_v)} = W$, *and*

(*ii*) *Relative humidity,* $\phi = \mu\left(\dfrac{0.622 + W_s}{0.622 + W}\right) = \dfrac{\mu}{1 - (1 - \mu)(p_{vs}/p_t)}$

where, p_v = *Partial pressure of water vapour,*

p_{vs} = *Saturation vapour pressure,*

p_t = *Total pressure of moist air,*

W_s = *Humidity ratio at saturation, and*

$\mu = \dfrac{W}{W_s}$ = Degree of saturation.

(*b*) If 100 kg/s of moist air at 30°C DBT and 20°C DPT mixes adiabatically with 30 kg/s of moist air at 40°C DBT and 30% RH, determine:

(*i*) *Humidity ratio of the mixed stream;*

(*ii*) *Temperature of the mixed stream;*

(*iii*) *Enthalpy of the mixed stream.*

Use only given data:

At 20°C : $p_{vs} = 0.0234$ *bar;*

At 30°C : $p_{vs} = 0.0424$ *bar;*

At 40°C : $p_{vs} = 0.0738$ *bar;*

Atmospheric pressure = 1.01325 bar

$h = 1.005\, t_{db} + W\,(2500 + 1.88\, t_{db})$ *kJ/kg of air; t is in °C.* **(GATE)**

Solution: (*a*) For derivation of part (*i*), refer to article 12.3

(*ii*) We now that, $W = \dfrac{0.622\, p_v}{p_t - p_v}$

or, $W\,(p_t - p_v) = 0.622\, p_v$

or, $Wp_t - Wp_v = 0.622\, p_v$

or, $Wp_t = Wp_v + 0.622\, p_v = p_v\,(W + 0.622)$

or, $p_v = \dfrac{Wp_t}{W + 0.622}$, and

$$p_{vs} = \frac{W_s\, p_t}{W_s + 0.622}$$

But, relative humidity, $\phi = \dfrac{p_v}{p_{vs}} = \dfrac{W}{W_s} \cdot \left(\dfrac{W_s + 0.622}{W + 0.622}\right)$

or, $\phi = \mu \left(\dfrac{0.622 + W_s}{0.622 + W}\right)$ ...**Proved.**

and, $\phi = \dfrac{\mu}{1 - (1-\mu)\left(p_{vs}/p_t\right)}$ ... For derivation, refer to article 12.3.

(*b*) *Given:* $m_1 = 100$ kg/s; $t_{db_1} = 30°C$; $t_{dp_1} = 20°C$;

$m_2 = 300$ kg/s;

$t_{db_2} = 40°C$; $\phi_2 = 30\%$

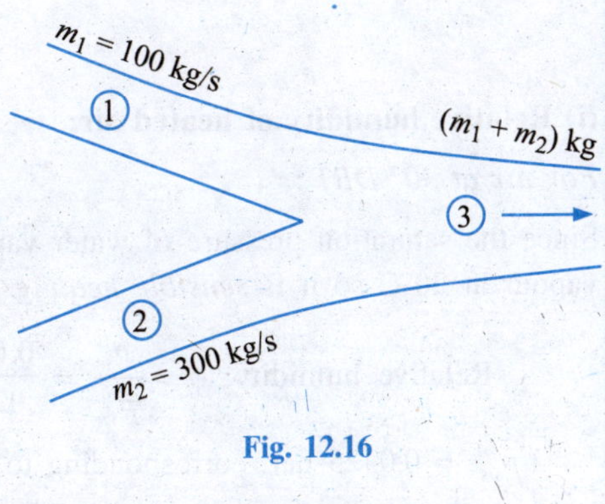

Fig. 12.16

$p_{v1} = p_{vs}$ at 20°C = 0.0234 bar

$W_1 = 0.622\, \dfrac{p_{v1}}{p_t - p_{v1}} = \dfrac{0.622 \times 0.0234}{1.01325 - 0.0234}$

$= 0.0147$ kg/kg of dry air

$p_{v2} = 0.3 \times p_{vs}$ (at 40°C)

$= 0.3 \times 0.738 = 0.02214$ bar

$W_2 = \dfrac{0.622\, p_{v2}}{p_t - p_{v2}} = \dfrac{0.622 \times 0.02214}{1.01325 - 0.02214} = 0.01389$ kg/kg of dry air

$h_1 = 1.005 \times 30 + 0.0147\,(2500 + 1.88 \times 30) = 67.73$ kJ/kg

$h_2 = 1.005 \times 40 + 0.01389\,(2500 + 1.88 \times 40) = 75.97$ kJ/kg (*i*) **Humidity ratio of the mixed stream, W_3:**

For the *mixed air*, $$W_3 = \frac{m_1 W_1 + m_2 W_2}{m_1 + m_2} = \frac{100 \times 0.0147 + 300 \times 0.01389}{100 + 300}$$

$$= \mathbf{0.01409\ kg/kg\ of\ dry\ air\ (Ans.)}$$

(*ii*) **Temperature of the mixed stream, t_{db_3}:**

$$t_{db_3} = \frac{m_1 t_{db_1} + m_2 t_{db_2}}{m_1 + m_2} = \frac{100 \times 30 + 300 \times 40}{100 + 300} = \mathbf{37.5°C\ (Ans.)}$$

(*iii*) **Enthalpy of the mixed stream, h_3:**

$$h_3 = \frac{m_1 h_1 + m_2 h_2}{m_1 + m_2} = \frac{100 \times 67.73 + 300 \times 75.97}{100 + 300}$$

$$= \mathbf{73.91\ kJ/kg\ of\ dry\ air\ (Ans.)}$$

Example 12.15. (***Sensible heating***) *90 m^3 of air per minute at 20°C and 75% RH is heated until its temperature becomes 30°C. Calculate:*

(*i*) *RH of the heated air;*

(*ii*) *Heat added to air per minute.*

Solution: (*i*) *For air at 20°C and 75% RH*:

$$p_{vs} = 0.0234 \text{ bar (from steam tables, at 20°C)}$$

$$p_v = \phi \times p_{vs} = 0.75 \times 0.0234 = 0.01755 \text{ bar}$$

$$t_{dp} = 15 + (16 - 15)\frac{(0.01755 - 0.017)}{(0.0182 - 0.017)} \approx 15.5°C$$

$$W_1 = \frac{0.622\, p_v}{p_t - p_v} = \frac{0.622 \times 0.01755}{1.0132 - 0.01755} = 0.0109 \text{ kg/kg of dry air}$$

Enthalpy, $$h_1 = c_p\, t_{db} + W\, h_{vapour} = 1.005 \times 20 + 0.0109\,[h_g + 1.88\,(t_{db} - t_{dp})]$$

$$= 1.005 \times 20 + 0.0109\,[2538.1 + 1.88\,(20 - 15.5)]$$

$$= 47.85 \text{ kJ/kg of dry air}$$

(*i*) Relative humidity of heated air:

For air at 30° DBT:

Since the saturation pressure of water vapour at 30°C is *higher* than the saturation pressure of water vapour at 20°C so it is *sensible heating where p_v* is same after heating.

Relative humidity, $$\phi = \frac{p_v}{p_{vs}} = \frac{0.01755}{0.0425} = 0.412 \text{ or } 41.2\%$$

(p_{vs} = 0.0425 bar, corresponding to 30°C)

i.e., *Relative humidity of heated air* = **41.2% (Ans.)**

(*ii*) **Heat added to air per minute :**

Enthalpy, $$h_2 = c_p\, t_{db} + W\, h_{vapour}$$

$$= 1.005 \times 30 + 0.0109 \times [h_g + 1.88\,(t_{db} - t_{dp})]$$

$$= 1.005 \times 30 + 0.0109 \, [2556.3 + 1.88 \, (30 - 15.5)]$$

$$= 58.31 \text{ kJ/kg of dry air}$$

Mass of dry air in 90 m^3 of air supplied

$$m_a = \frac{pV}{RT} = \frac{(p_t - p_v)\,V}{RT}$$

$$= \frac{(1.0132 - 0.01755) \times 10^5 \times 90}{287 \times (273 + 20)} = 106.5 \text{ kg/min.}$$

Amount of heat added per minute

$$= 106.5 \, (h_2 - h_1) = 106.5 \, (58.31 - 47.85)$$

$$\approx \mathbf{1114 \text{ kJ (Ans.)}}$$

Example 12.16. **(*Sensible cooling*)** *40 m^3 of air at 35°C DBT and 50% RH is cooled to 25°C DBT maintaining its specific humidity constant. Determine :*

(*i*) *Relative humidity (RH) of cooled air;*

(*ii*) *Heat removed from air.*

Solution: *For air at 35°C DBT and 50% RH:*

$$p_{vs} = 0.0563 \text{ bar (AT 35°C, from steam tables)}$$

$$\phi = \frac{p_v}{p_{vs}}$$

$$\therefore \quad p_v = \phi \times p_{vs} = 0.5 \times 0.0563 = 0.02815 \text{ bar}$$

$$W_1 = \frac{0.622\, p_v}{p_t - p_v} = \frac{0.622 \times 0.02815}{1.0132 - 0.02815} = 0.0177 \text{ kg/kg of dry air}$$

$$h_1 = c_p \, t_{db_1} + W_1 \, [h_{g1} + 1.88 \, (t_{db_1} - t_{dp_1})]$$

$$t_{dp_1} \approx 23°\text{C (corresponding to 0.02815 bar)}$$

$$\therefore \quad h_1 = 1.005 \times 35 + 0.0177 \, [2565.3 + 1.88 \, (35 - 23)] = 89.98 \text{ kJ/kg of dry air}$$

For air at 25°C DBT :

Since the specific humidity remains constant the vapour pressure in the air remains constant.

(*i*) **Relative humidity of cooler air,** $\phi = \dfrac{p_v}{p_{vs}} = \dfrac{0.02815}{0.0317} = 0.888$ or **88.8% (Ans.)**

$$h_2 = c_p \, t_{db_2} + W_2 \, [h_{g2} + 1.88 \, (t_{db_2} - t_{dp_2})]$$

$$= 1.005 \times 25 + 0.0177 \, [2547.2 + 1.88 \, (25 - 23)]$$

$$= 70.27 \text{ kJ/kg of dry air} \quad \left[\begin{array}{l} \because W_1 = W_2 = 0.0177 \text{ kg/kg of dry air} \\ t_{dp_2} = t_{dp_1} = 23°\text{C since } p_v \text{ does not change} \end{array}\right]$$

To find mass of dry air (m_a), using the relation,

$$p_a v_a = m_a R_a T_a, \text{ we get:}$$

$$m_a = \frac{p_a v_a}{R_a T_a} = \frac{(1.0132 - 0.02815) \times 10^5 \times 40}{287 \times (273 + 35)} = 44.57 \text{ kg}$$

(ii) **Heat removed from 40 m³ of air**

$$= m_a (h_1 - h_2) = 44.57 (80.98 - 70.27) = \textbf{477.3 kJ (Ans.)}$$

Example 12.17. **(Cooling and dehumidification)** *120 m³ of air per minute at 35°C DBT and 50% relative humidity is cooled to 20°C DBT by passing through a cooling coil. Determine the following :*

(i) Relative humidity of out coming air and its wet bulb temperature;

(ii) Capacity of cooling coil in tonnes of refrigeration;

(iii) Amount of water vapour removed per hour.

Solution: *For the air at 35°C DBT and 50% RH:*

$$p_{vs} = 0.0563 \text{ bar (At 35°C, from stem tables)}$$

$$p_v = \phi \times p_{vs} = 0.5 \times 0.0563 = 0.02815 \text{ bar}$$

$$W_1 = \frac{0.622\, p_v}{p_t - p_v} = \frac{0.622 \times 0.02815}{1.0132 - 0.02815} = 0.0177 \text{ kg/kg of dry air.}$$

$$h_1 = c_p\, t_{db_1} + W_1 [h_{g1} + 1.88 (t_{db_1} - t_{dp_1})]$$

$$t_{dp_1} \approx 23°\text{C (Corresponding to 0.02815 bar)}$$

$$\therefore \quad h_1 = 1.005 \times 35 + 0.0177 [2565.3 + 1.88 (35 - 23)] = 80.98 \text{ kJ/kg of dry air.}$$

For the air at 20°C: As the saturation vapour pressure at 20°C is 0.0234 bar and is less than the vapour pressure 0.02815 bar at 35°C, so that *condensation takes place* and air will be *saturated at 20°C.*

(i) ∴ **Relative humidity of exit air is 100 per cent (Ans.)**

Since the air is saturated, wet bulb temperature is equal to dry bulb temperature = **20°C. (Ans.)**

$$p_v = p_{vs} = 0.0234 \text{ bar}$$

$$W_2 = \frac{0.622\, p_v}{p_t - p_v} = \frac{0.622 \times 0.0234}{(1.0132 - 0.0234)} = 0.0147 \text{ kg/kg of dry air}$$

$$h_2 = c_p\, t_{db_2} + W_2 [h_{g2} + 1.88 (t_{db_2} - t_{dp_2})]$$

$$= 1.005 \times 20 + 0.0147 [2538.1 + 1.88 (20 - 20)]$$

[∵ When air is saturated $t_{db} = t_{dp}$]

$$= 57.41 \text{ kJ/kg of dry air.}$$

The weight of water vapour removed per kg of dry air

$$= 0.0177 - 0.0147 = 0.003 \text{ kg/kg of dry air}$$

Heat removed per kg of dry air

$$= h_1 - h_2 = 80.98 - 57.41 = 23.57 \text{ kJ/kg of dry air}$$

Mass of dry air passing per minute

$$m_a = \frac{p_a V_a}{R_a T_a} = \frac{(1.0132 - 0.02815) \times 10^5 \times 120}{287 \times (35 + 273)} = 133.7 \text{ kg/min}$$

(*ii*) **Capacity of the cooling coil in tonnes of refrigeration**

$$= \frac{m_a (h_1 - h_2) \times 60}{14000} = \frac{133.7 \times 23.57 \times 60}{14000} = \textbf{13.5 tonnes (Ans.)}$$

(*iii*) **Amount of water vapour removed per hour**

$$= m_a (W_1 - W_2) \times 60 = 133.7\ (0.0177 - 0.0147) \times 60$$

$$= \textbf{245.066 kg/hour (Ans.)}$$

Example 12.18. **(*Adiabatic humidification*)** *150 m³ of air per minute is passed through the adiabatic humidifier. The condition of air at inlet is 35°C DBT and 20 per cent relative humidity and the outlet condition is 20°C DBT and 15°C WBT. Determine the following :*

(*i*) *Dew point temperature.*

(*ii*) *Relative humidity of the exit air.*

(*iii*) *Amount of water vapour added to the air per minute.*

Solution: *For air at 35°C DBT and 20% relative humidity:*

$$p_{vs} = 0.0563 \text{ bar (At 35°C from steam tables)}$$

$$p_v = \varphi \times p_{vs} = 0.2 \times 0.0563 = 0.01126 \text{ bar}$$

$$W_1 = \frac{0.622\, p_v}{p_t - p_v} = \frac{0.622 \times 0.01126}{1.0132 - 0.01126} = 0.00699 \text{ kg/kg of dry air}$$

(*i*) The dew point temperature of air which is the saturation temperature of steam corresponding to the pressure 0.01126 bar is

$$8 + (9 - 8) \times \frac{(0.01126 - 0.01072)}{(0.01150 - 0.01072)} = 8.7°\text{C}$$

i.e., **Dew point temperature = 8.7°C (Ans.)**

(*ii*) **Relative humidity of the exit air:**

For air at 20°C DBT and 15°C WBT,

$$p_v = (p_{vs})_{wb} - \frac{[p_t - (p_{vs})_{wb}](t_{db} - t_{wb})}{1527.4 - 1.3\, t_{wb}}$$

$$= 0.0170 - \frac{[1.0132 - 0.0170](20 - 15)}{1527.4 - 1.3 \times 15} = 0.0137 \text{ bar}$$

$$W_2 = \frac{0.622\, p_v}{p_t - p_v} = \frac{0.622 \times 0.0137}{(1.0132 - 0.0137)} = 0.00852 \text{ kg/kg of dry air}$$

$$\textbf{Relative humidity} = \frac{p_v}{p_{vs}} = \frac{0.0137}{0.0234} = \textbf{0.585 or 58.5\% (Ans.)}$$

[∵ p_{vs} = 0.234 bar, corresponding to 20°C, from steam tables]

The *dew point temperature of air* which is the saturation temperature of steam corresponding to 0.0137 bar is **11°C** (from steam tables). **(Ans.)**

The amount of water vapour per kg of dry air

$$= W_2 - W_1 = 0.00852 - 0.00699 = 0.00153 \text{ kg}$$

The mass of dry air in 150 m^3 of mixture

$$m_a = \frac{p_a V_a}{R_a T_a} = \frac{(1.0132 - 0.01126) \times 10^5 \times 150}{287 \times (35 + 273)} = 170 \text{ kg}$$

(*iii*) **The amount of water vapour added to air per minute**

$$= m_a (W_2 - W_1) = 170 \times 0.00153 = \mathbf{0.26 \text{ kg/min (Ans.)}}$$

Example 12.19. (***Adiabatic saturation process***) *An air-water vapour mixture enters an adiabatic saturator at 28°C and leaves at 18°C, which is the adiabatic saturation temperature. The pressure remains constant at 1.0 bar. Determine the relative humidity and humidity ratio of the inlet mixture.*

Solution: The specific humidity at the exit,

$$W_{2s} = \frac{0.622\, p_s}{p_t - p_s} = \frac{0.622 \times 0.0206}{(1.00 - 0.0206)} = 0.01308 \text{ kg/kg of dry air}$$

The specific humidity at the inlet [equation 12.18(a)]

$$W_1 = \frac{c_p(t_{db_2} - t_{db_1}) + W_{2s}(h_{g2} - h_{f2})}{h_{g1} - h_{f2}}$$

$$= \frac{1.005\,(18 - 28) + 0.01308\,(2534.4 - 75.6)}{2552.6 - 75.6}$$

$$= \frac{42.211}{2477} = 0.01704 \text{ kg/kg of dry air}$$

$$W_1 = \frac{0.622\, p_{v1}}{p_t - p_{v1}}$$

$$\therefore \quad 0.01704 = \frac{0.622\, p_{v1}}{1.0 - p_{v1}}$$

or, $\quad 0.01704\,(1.0 - p_{v1}) = 0.622\, p_{v1}$

or, $\quad 0.01704 - 0.01704\, p_{v1} = 0.622\, p_{v1}$

or, $\quad 0.0639\, p_{v1} = 0.01704$

$\therefore \quad p_{v1} = 0.02666$ bar

Also, Pressure corresponding to 28°C, $p_{s1} = 0.03782$

$$\therefore \quad \text{Relative humidity} = \frac{p_{v1}}{p_{s1}} = \frac{0.02666}{0.03782} = \mathbf{0.70 \text{ or } 70\% \text{ (Ans.)}}$$

Example 12.20. *An air-water vapour mixture enters an air-conditioning unit at a pressure of 1.0 bar, 38°C DBT, and a relative humidity of 75%. The mass of dry air entering is 1 kg/s. The air-vapour mixture leaves the air-conditioning unit at 1.0 bar, 18°C, 85% relative humidity. The moisture condensed leaves at 18°C. Determine the heat transfer rate for the process.*

Solution: $t_{db_1} = 38°C$, RH, $\phi_1 = 75\%$

$t_{db_2} = 18°C$, RH, $\phi_2 = 85\%$

The flow diagram and the process are shown in Fig. 12.17 (*a*) and (*b*) respectively.

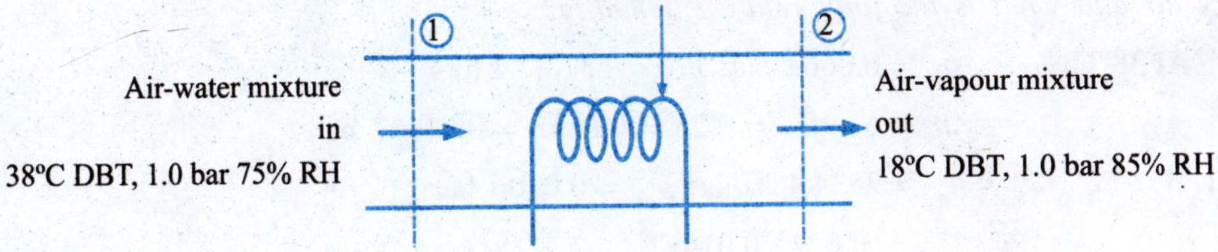

(*a*)

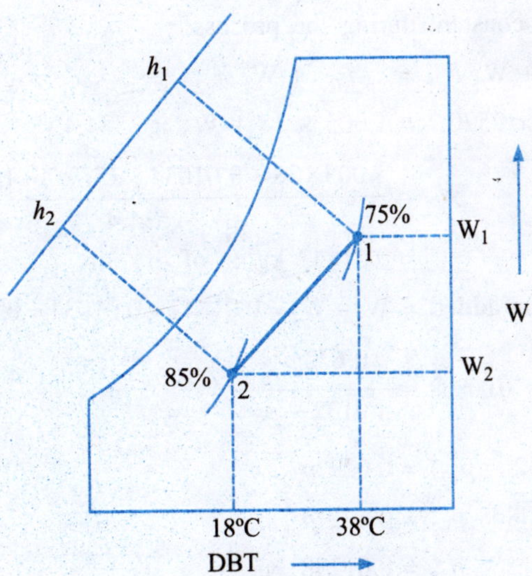

(*b*)

Fig. 12.17

At 38°C:

From steam tables: $p_{vs} = 0.0663$ bar,

$$h_{g1} = 2570.7 \text{ kJ/kg}$$

$$\therefore \quad p_v = \phi \times p_{vs} = 0.75 \times 0.0663 = 0.0497 \text{ bar}$$

$$W_1 = \frac{0.622 \times 0.0497}{1.0 - 0.0497} = 0.0325 \text{ kg/kg of dry air}$$

At 18°C :

From steam tables: $p_{vs} = 0.0206$ bar, $h_{g2} = 2534.4$ kJ/kg

$$h_{f2} = 75.6 \text{ kJ/kg}$$

$$p_v = 0.85 \times 0.0206 = 0.01751 \text{ bar}$$

$$W_2 = \frac{0.622 \times 0.01751}{1 - 0.01751} = 0.01108 \text{ kg/kg of dry air}$$

Heat transfer rate, $q = (W_2 h_{g2} - W_1 h_{g1}) + c_p (t_{db_2} - t_{db_1}) + (W_1 - W_2) h_{f2}$

$$= (0.01108 \times 2534.4 - 0.0325 \times 2570.7) + 1.005\,(18 - 38) + (0.0325 - 0.01108) \times 75.6$$

$$= -55.46 - 20.1 + 1.62$$

$$= \mathbf{-73.94 \text{ kJ/kg of dry air}} \text{ (Ans.)}$$

Example 12.21. **(*Evaporative cooler*).** *Atmospheric air at 38°C and 25% relative humidity passes through an evaporator cooler. If the final temperature of air is 18°C, how much water is added per kg of dry air and what is the final relative humidity?*

Solution: At 38°C: $p_{vs} = 0.0663$ bar, $h_{g1} = 2570.7$ kJ/kg

and $p_v = \phi \times p_{vs} = 0.25 \times 0.0663 = 0.01657$ bar

At 18°C: $h_{g2} = 2534.4$ kJ/kg; $p_{vs} = 0.0206$ bar

$$\therefore \quad W_1 = \frac{0.622 \times 0.01657}{1.0132 - 0.01657} = 0.01034 \text{ kg/kg of dry air}$$

Since enthalpy remains constant during the process,

$$\therefore \quad c_p\, t_{db_1} + W_1 h_{g1} = c_p\, t_{db_2} + W_2 h_{g2}$$

$$1.005 \times 38 + 0.01034 \times 2570.7 = 1.005 \times 18 + W_2 \times 2534.4 \quad (\because \text{At } 18°\text{C}, h_{g2} = 2534.4 \text{ kJ/kg})$$

or,
$$W_2 = \frac{1.005 \times 38 + 0.01034 \times 2570.7 - 1.005 \times 18}{2534.4}$$
$$= 0.01842 \text{ kg/kg of dry air}$$

$\therefore$ **Amount of water added** $= W_2 - W_1 = 0.01842 - 0.01034 =$ **0.00808 kg/kg of dry air (Ans.)**

Also,
$$0.00808 = \frac{0.622\, p_{v_2}}{1.0132 - p_{v_2}}$$

or, $0.00808\ (1.0132 - p_{v2}) = 0.622\, p_{v2}$

or, $0.00818 - 0.00808\, p_{v2} = 0.622\, p_{v2}$

$\therefore \quad p_{v2} = 0.01298$ bar

$$\therefore \quad \textbf{Final relative humidity} = \frac{0.01298}{0.0206} = \textbf{0.63 or 63\% (Ans.)}$$

Example 12.22. *Saturated air at 3°C is required to be supplied to a room where the temperature must be held at 22°C with a relative humidity of 55%. The air is heated and then water at 10°C is sprayed to give the required humidity. Determine :*

(*i*) *The mass of spray water required per m³ of air at room conditions;*

(*ii*) *The temperature to which the air must be heated.*

Neglect the fan power. Assume that the total pressure is constant at 1.0132 bar.

Solution: (*i*) The flow diagram is shown in Fig. 12.18 (*a*) and the process is shown in Fig. 12.18 (*b*).

At 22°C: From steam tables:

$$p_{vs} = 0.0264 \text{ bar}$$

$$\phi_3 = \frac{p_{v3}}{p_{vs3}} = \frac{p_{v3}}{0.0264} = 0.55$$

$$\therefore \quad p_{v_3} = 0.55 \times 0.0264 = 0.01452 \text{ bar}$$

$$\therefore \quad W_3 = \frac{0.622 \times 0.01452}{(1.0132 - 0.01452)} = 0.00904 \text{ kg/kg of dry air}$$

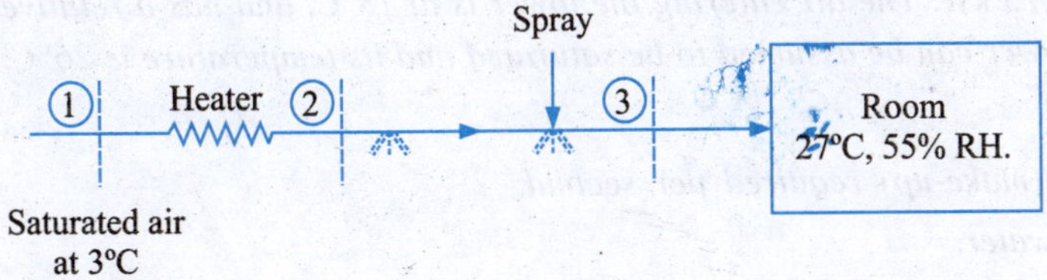

(a)

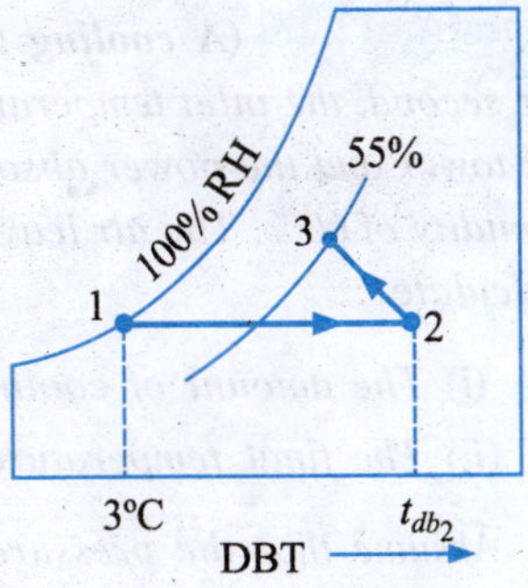

(b)

Fig. 12.18

At 3°C: From steam tables:

$$p_{vs} = 0.0076 \text{ bar}$$

$$\therefore \quad \phi_1 = \frac{p_{v1}}{p_{vs1}} = 1.0$$

$$\therefore \quad p_{v1} = p_{vs1} = 0.0076 \text{ bar}$$

$$W_1 = \frac{0.622 \times 0.0076}{1.0132 - 0.0076} = 0.0047 \text{ kg/kg of dry air}$$

$$W_3 - W_1 = 0.00904 - 0.0047 = 0.00434 \text{ kg/kg of dry air}$$

$$v_{a3} = \frac{R_a T_a}{p_{a3}} = \frac{287 \times (273 + 22)}{(1.0132 - p_{v3}) \times 10^5} = \frac{287 \times 295}{0.9987 \times 10^5}$$

$$= 0.847 \text{ m}^3\text{/kg of dry air}$$

$$\textbf{Spray water} = \frac{0.00434}{0.847} = \textbf{0.005124 kg moisture/m}^3 \textbf{ (Ans.)}$$

(*ii*) Temperature to which the air must be heated, t_{db_2} :

Now, $\quad h_2 + (W_3 - W_2)\, h_4 = h_3$

$$[C_p\, t_{db_2} + W_2\, h_{vapour\,(2)}] + (W_3 - W_2)\, h_2 = c_p\, t_{db_3} + W_3\, h_{vapour\,(3)}$$

$$\therefore \quad c_p\,(t_{db_3} - t_{db_2}) + W_3\, h_{vapour\,(3)} - W_2\, h_{vapour(2)} - (W_3 - W_2)\, h_4 = 0$$

From the steam tables at $p_v = 0.01452$ bar : $h_g = 2524$ kJ/kg

and, $\quad t_{dp} = t_{sat} = 12.5°C$

$$\therefore \quad 1.005\,(22 - t_{db_2}) + 0.00904\,[2524 + 1.88\,(22 - 12.5)] - 0.0047\,[2524 + 1.88\,(t_{db_2} - 12.5)] - (0.00434 \times 4.187 \times 10) = 0$$

$$22.11 - 1.005\, t_{db_2} + 22.97 - 11.86 - 0.0088\, t_{db_2} + 0.11 - 0.1817 = 0$$

$$1.014\, t_{db_2} = 33.148$$

$$t_{db\,2} = \textbf{32.7°C (Ans.)}$$

Example 12.23. **(*A cooling tower*)** *A small-size cooling tower is designed to cool 5.5 litres of water per second, the inlet temperature of which is 44°C. The motor-driven fan induces 9 m³/s of air through the tower and the power absorbed is 4.75 kW. The air entering the tower is at 18°C, and has a relative humidity of 60%. The air leaving the tower can be assumed to be saturated and its temperature is 26°C. Calculate :*

(*i*) *The amount of cooling water (make-up) required per second;*

(*ii*) *The final temperature of the water.*

Assume that the pressure remains constant throughout the tower at 1.013 bar.

Solution: The cooling water is shown diagrammatically in Fig. 12.19.

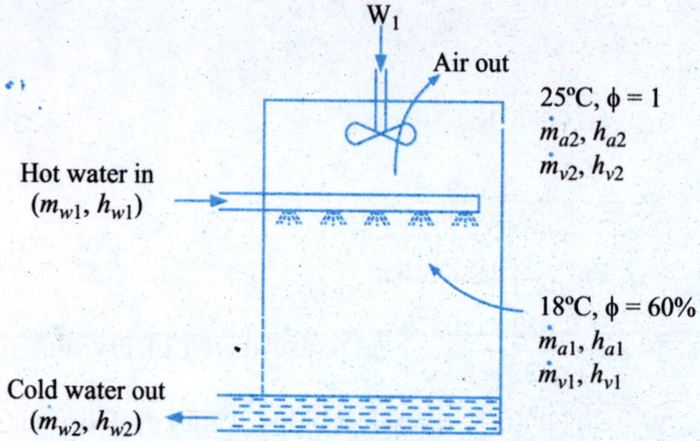

Fig. 12.19

(*i*) **Make-up water required:**

At 18°C : $p_{vs} = 0.0206$ bar,

$\therefore \quad p_v = \phi \times p_{vs} = 0.6 \times 0.0206 = 0.01236$ bar

$\therefore \quad p_{a1} = 1.013 - 0.01236 = 1.00064$ bar

Then,
$$\dot{m}_{a1} = \frac{10^5 \times 1.00064 \times 9}{(0.287 \times 10^3) \times (18 + 273)} = 10.78 \text{ kg/s}$$

and,
$$\dot{m}_{v1} = \frac{10^5 \times 0.01236 \times 9}{(0.4618 \times 10^3) \times (18 + 273)} = 0.0828 \text{ kg/s}$$

(Script '*v*' denotes vapour and the script '*a*' denotes the air).

At exit at 26°C: $p_{vs} = 0.0336$ bar and $\phi = 1$

$\therefore \quad p_v = p_{vs} = 0.0336$ bar

$$W_2 = \frac{0.622\, p_v}{p_t - p_v} = \frac{0.622 \times 0.0336}{1.013 - 0.0336} = 0.02133 \text{ kg}$$

But,
$$W = \frac{m_v}{m_a}$$

$\therefore \quad m_{v2} = 0.02133 \times 10.78 = 0.23$ kg/s

Hence, **make up water required**

$$= 0.23 - 0.0828 = \mathbf{0.1472\ kg/s\ (Ans.)}$$

(*ii*) **Final temperature of the water:**

Also, $\dot{m}_{w1} = 5.05 \times 1 = 5.5$ kg/s

and, $\dot{m}_{w2} = m_{w1}$ – (make-up water)

$= 5.5 - 0.1472 = 5.353$ kg/s

Applying the steady flow energy equation and neglecting changes in kinetic energy and potential energy, we have:

$$W_{in} + \dot{m}_{w1}\, h_{w1} + \dot{m}_{a1}\, h_{a1} + \dot{m}_{v1}\, h_{v1} = \dot{m}_{a2}\, h_{a2} + \dot{m}_{v2}\, h_{v2} + \dot{m}_{w2}\, h_{w2}$$

Now, W_{in} (i.e., work input) = 4.75 kW = 4.75 kJ/s

Evaluating the enthalpies from a datum of 0°C, we have:

$h_{w1} = h_f$ at 44°C = 184.3 kJ/kg,

$h_{a1} = 1.005\ (18 - 0) = 18.09$ kJ/kg,

$h_{v1} = 2519.7 + 1.88\ (18 - 10)$

$= 2534.74$ kJ/kg

[Corresponding to $p_v = 0.01236$ bar, $t_s = t_{dp} \approx 10°C$ i.e., the vapour is superheated]

$h_{v1} = h_g$ at 26°C = 2549 kJ/kg

$h_{a1} = 1.005\ (26 - 0) = 26.13$ kJ/kg

Then, substituting, we get:

$$4.75 + 5.5 \times 184.3 + 10.78 \times 18.09 + 0.0828 \times 2534.74$$

$$= 10.78 \times 26.13 + 0.23 \times 2549 + 5.353 \times h_{w2}$$

or, $5.353\ h_{w2} = 1423.28 - 867.95 = 555.33$

or, $h_{w2} = 103.74$ kJ/kg

By interpolation, $h_f = 103.74$ kJ/kg at 26.7°C

Hence, **final temperature of water = 26.7°C (Ans.)**

B. *SOLUTIONS USING PSYCHROMETRIC CHART*

Example 12.24. *The following data pertains to an air-conditioning system : Unconditioned space DBT = 30°C; Unconditioned space WBT = 22°C. Cold air duct supply surface temperature = 14°C.*

Determine:

(*i*) *Dew point temperature;*

(*ii*) *Whether or not condensation will form on the duct.*

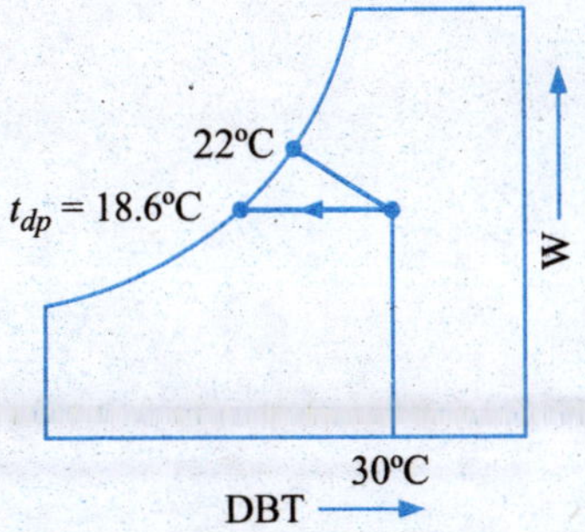

Fig. 12.20

Solution: Refer to Fig. 12.20.

(*i*) To determine the dew point temperature for the given conditions, find the intersection of 30°C DBT and 22°C WBT and move horizontally (as shown by the arrow) to the dew point temperature scale. The dew **point (t_{dp}) is 18.6°C. (Ans.)**

(*ii*) Since the duct temperature (14°C) is less than t_{dp} (18.6°C) therefore *moisture will condense on the duct surface* **(Ans.)**

Example 12.25. ***(Mixing of air streams)*** *400 m³/min of recirculated at 20°C DBT and 10°C DPT is to be mixed with 150 m³/min. of fresh air at 35°C DBT and 45% RH. Determine the enthalpy, specific volume, humidity ratio and dew point temperature of the mixture.*

Solution: *Given:* $V_2 = 400$ m³/min.; $t_{db_2} = 20$°C; $t_{dp_2} = 10$°C; $V_1 = 150$ m³/min.; $t_{db_1} = 35$°C; $\phi_1 = 45\%$.

Refer to Fig. 12.21.

- Locate point 1 at the intersection of 35°C DBT and 45% RH lines.
- Locate point 2 at the intersection of 20°C DBT and 10°C DPT lines.

Join points 1 and 2.

From *psychrometric chart*, we find:

$$v_{s1} = 0.8945 \text{ m}^3\text{/kg of dry air;}$$

$$h_1 = 76.1 \text{ kJ/kg of dry air;}$$

$$W_1 = 0.0158 \text{ kg/kg of dry air;}$$

$$v_{s2} = 0.8414 \text{ m}^3\text{/kg of dry air;}$$

$$h_2 = 39.1 \text{ kJ/kg of dry air;}$$

$$W_2 = 0.0076 \text{ kJ/kg of dry air.}$$

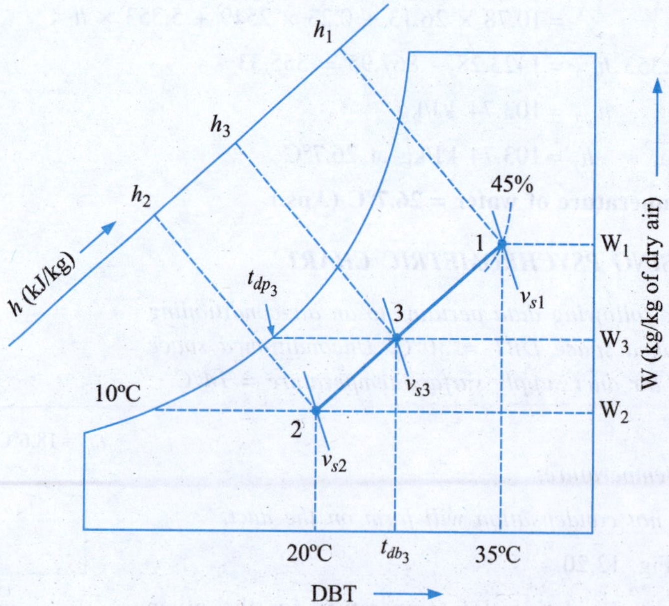

Fig. 12.21

Mass of fresh air at point 1,

$$m_1 = \frac{V_1}{v_{s1}} = \frac{150}{0.8945} = 167.7 \text{ kg/min.}$$

Mass of air recirculated at point 2,

$$m_1 = \frac{V_2}{v_{s2}} = \frac{400}{0.8414} = 475.4 \text{ kg/min.}$$

Enthalpy of mixture h_3:

We known that,

$$\frac{m_1}{m_2} = \frac{h_3 - h_2}{h_1 - h_3} \quad \text{...[Eqn. 12.21 (a)]}$$

or, $$\frac{167.7}{475.4} = \frac{h_3 - 39.1}{76.1 - h_3}$$

or, $$0.3527 = \frac{h_3 - 39.1}{76.1 - h_3}$$

or, $$0.3527\ (76.1 - h_3) = h_3 - 39.1$$

or, $$26.84 - 0.3527\ h_3 = h_3 - 39.1$$

or, $$h_3 = \frac{(26.84 + 39.1)}{(1 + 0.3527)} = \mathbf{48.75 \text{ kJ/kg of dry air (Ans.)}}$$

Specific volume, humidity ratio and dew point temperature of the mixture:

- Locate point 3 on the line joining 1 and 2 corresponding to enthalpy h_3 = 48.75 kJ/kg of dry air, as shown in Fig. 12.21.

From *psychrometric* chart, corresponding to point 3 we find :

Specific volume, v_{s3} = **0.8543 m³/kg of dry air (Ans.)**

Humidity ratio, W_3 = **0.0097 kJ/kg of dry air (Ans.)**

Dew point temp., t_{dp_3} = **13.7°C (Ans.)**

Example 12.26. ***(Sensible heating)*** *200 m³ of air per minute at 15°C DBT and 75% RH is heated until its temperature is 25°C.*

Find: (i) RH of heated air;

(ii) Wet bulb temperature of heated air;

(iii) Heat added to air per minute.

Solution: Refer to Fig. 12.22.

- Locate point 1 on the psychrometric chart at the intersection of 15°C DBT and 75% RH lines.
- Through point 1 draw a horizontal line to cut 25°C DBT line and get point 2.

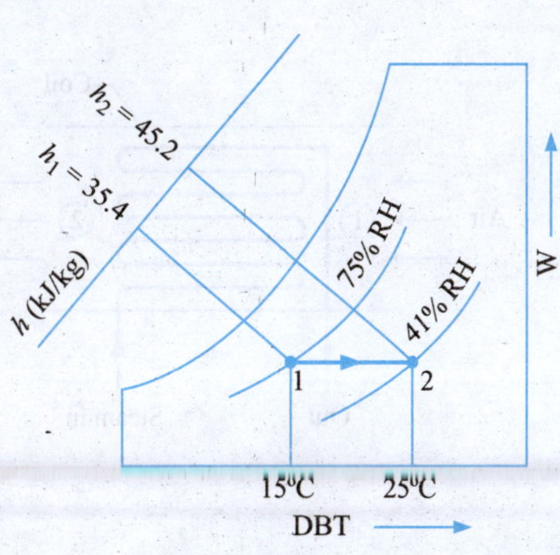

Fig. 12.22

- Read the following values from the psychrometric chart:

h_1 = 35.4 kJ/kg; h_2 = 45.2 kJ/kg; v_{s_1} = 0.8267 m^3/kg

(*i*) **RH of heated air** (read from chart) = **41% (Ans.)**

(*ii*) **WBT of heated air** (read from chart) = **16.1°C (Ans.)**

(*iii*) Mass of air circulated per min., $m_a = \frac{200}{0.8267}$ = 241.9 kg.

∴ **Heat added to air/min.**

$$= m_a (h_2 - h_1)$$
$$= 241.9\ (45.2 - 35.4)$$
$$= \mathbf{2370.6\ kJ\ (Ans.)}$$

Example 12.27. **(*Sensible heating*)** *In a heating application, moist air enters a steam heating coil at 10°C, 50% RH and leaves at 30°C. Determine :*

(*i*) *The sensible heat transfer if the mass flow rate of air is 100 kg/s;*

(*ii*) *The steam flow rate if steam enters saturated at 100°C and condensate leaves at 80°C.*

Solution: *Given:* t_{db_1} = 10°C; ϕ_1 = 50%; t_{db_2} = 30°C; m_a = 100 kg/s

(*i*) **Sensible heat transfer:**

Sensible heat transfer $= m_a\, c_p\, (t_{db_2} - t_{db_1})$ = 100 × 1.005 (30 – 10) = 2010 kJ/s or **2.01 kW (Ans.)**

(*ii*) **Steam flow rate:**

- Locate point 1 at the intersection of 10°C and 50% RH lines.
- Draw a horizontal line from point 1 till it cuts 30°C DBT vertical line at point 2.

From *Psychrometric chart*, we have:

h_1 = 19.6 kJ/kg ; h_2 = 39.8 kJ/kg

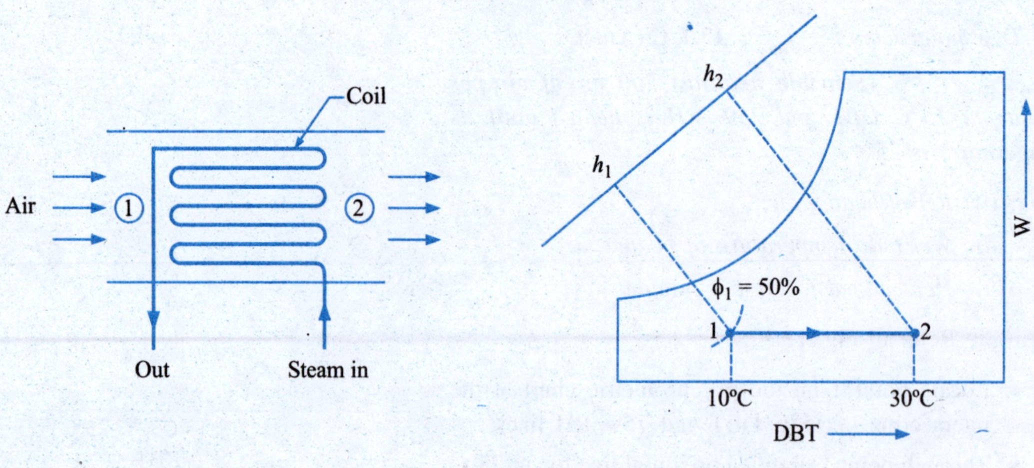

Fig. 12.23

Total heat transferred $= m_a (h_2 - h_1) = 100 (39.8 - 19.6) = 2020$ kJ/s

$$\text{Steam flow rate} = \frac{\text{Total heat transferred}}{h\,(100°\text{C}) - h_f\ (80°\text{C})}$$

From steam stables: h (100°C) = 2676 kJ/kg, and

h_f (80°C) = 334.9 kJ/kg

$$\therefore \quad \text{Steam flow rate} = \frac{2020}{(2676 - 334.9)} = 0.8629 \text{ kg/s or } \mathbf{3106.4\ kg/h\ (Ans.)}$$

Example 12.28. **(*Cooling and dehumidification*)** *In a cooling application, moist air enters a refrigeration coil at the rate of 100 kg/min. at 35°C and 50% RH. The apparatus dew point of coil is 5°C and by-pass factor is 0.15. Determine:*

(*i*) *Outlet state of moist air;*

(*ii*) *Cooling capacity of coil in tonnes of refrigeration.*

Solution: *Given:* t_{db_1} = 35°C; ϕ_1 = 50% ; t_{db_3} (ADP) = 5°C; BF = 0.15

Refer to Fig. 12.24.

- Locate point 1 at the intersection of 35°C DBT and 50% RH lines.
- Locate point 3 on the saturation curve where 5°C DBT line meets the curve.

Join points 2 and 3.

(*i*) **Outlet state of moist air, t_{db_2}, ϕ_2:**

$$\text{BF} = \frac{t_{db_2} - t_{db_3}}{t_{db_1} - t_{db_3}}$$

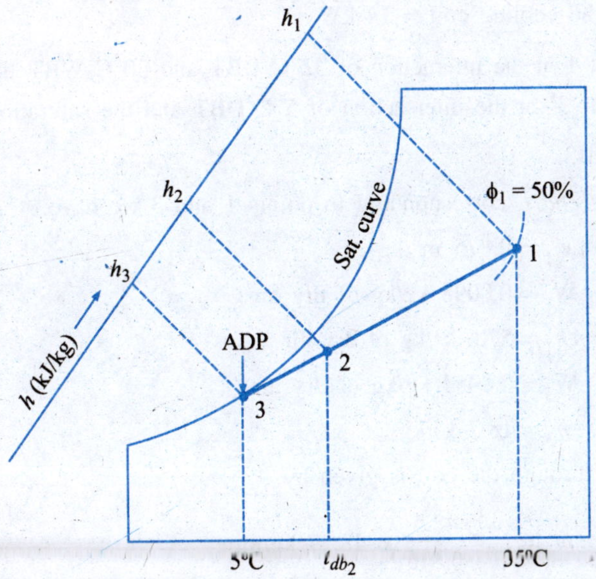

Fig. 12.24

$$0.15 = \frac{t_{db_2} - 5}{35 - 5}$$

$$\therefore \quad t_{db_2} = 9.5°C$$

- Locate point 2 on the line 1-3 where 9.5°C DBT vertical line meets this line.

From the *psychrometric chart*, the outlet state (at point 2) is:

9.5°C DBT; 98% RH. (Ans.)

(*ii*) **Cooling capacity of coil in tonnes of refrigeration:**

From *psychrometric chart*;

$$h_1 = 81 \text{ kJ/kg} \; ; \; h_2 = 28 \text{ kJ/kg}$$

∴ Cooling capacity of coil

$$= \frac{m_a(h_1 - h_2) \times 60}{14000} \text{ tonnes}$$

$$= \frac{100\,(81 - 28) \times 60}{14000} = \textbf{22.7 tonnes (Ans.)}$$

Example 12.29. **(*Cooling and dehumidification*)** *Air at 32°C DBT and 20°C WBT is passed through a cooling coil maintained at 5°C. The heat extracted by the cooling coil from air is 14 kW and air-flow rate is 42.5 m³/min. Determine :*

(*i*) *DBT and WBT of the air leaving the coil; and*

(*ii*) *Coil by-pass factor.*

Solution: *Given:* $t_{db_1} = 32°C$, $t_{wb_1} = 20°C$;

$t_{db_3} = 5°C$, $v = 42.5 \text{ m}^3/\text{min}$.

Heat extracted by the cooling coil = 14 kW

- Locate the point 1 at the interaction of 32°C DBT and 20°C WBT lines.
- Locate the point '3' at the intersection of 5°C DBT and the saturation curve.

Joint points 1 and 3.

From *psychrometric chart*, corresponding to points 1 and 3 we have:

$$v_{s1} = 0.875 \text{ m}^3/\text{kg}$$

$$W_1 = 0.0098 \text{ kg/kg of dry air}$$

$$h_1 = 57.6 \text{ kJ/kg of dry air}$$

$$W_3 = 0.0054 \text{ kg/kg of dry air}$$

$$h_3 = 18.7 \text{ kJ/kg}$$

Mass of air passed through the coil is given by,

$$m_a = \frac{v}{v_1} = \frac{42.5}{0.875} = 48.57 \text{ kg/min}$$

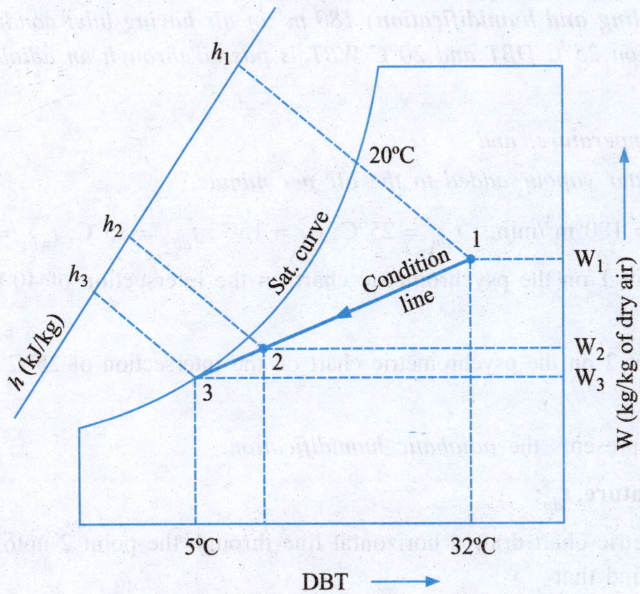

Fig. 12.25

Heat removed per kg of air by the cooling coil,

$$h_1 - h_2 = \frac{14 \times 60}{48.57} = 17.29 \text{ kJ/kg}$$

$$\therefore \quad h_2 = h_1 - 17.29 = 57.6 - 17.29 = 40.31 \text{ kJ/kg}$$

(*i*) DBT and WBT of the air leaving the coil:

The equation for the *condition line* can be written as:

$$\frac{h_1 - h_2}{h_1 - h_3} = \frac{W_1 - W_2}{W_1 - W_3}$$

$$\therefore \quad \frac{57.6 - 40.31}{57.6 - 18.7} = \frac{0.0098 - W_2}{0.0098 - 0.0054}$$

or, $$0.444 = \frac{0.0098 - W_2}{0.0044}$$

or, $$W_2 = 0.00785 \text{ kg}$$

- Corresponding to $h_2 = 40.31$ kJ/kg and $W_2 = 0.00785$, locate the point 2 on the pychrometric chart; and find that:

$$t_{db_2} = 20.3°\text{C} \qquad t_{wb_2} = 14.4°\text{C}$$

(*ii*) Coil by-pass factor, BF:

$$\text{BF} = \frac{h_2 - h_3}{h_1 - h_3} = \frac{40.31 - 18.7}{57.6 - 18.7} = \mathbf{0.555 \text{ (Ans.)}}$$

Example 12.30. **(*Cooling and humidification*)** *180 m³ of air having inlet condition 40°C DBT, 15% RH and outlet condition 25°C DBT and 20°C WBT is passed through an adiabatic humidifier. Find the following:*

(*i*) *Dew point temperature; and*

(*ii*) *Amount of water vapour added to the air per minute.*

Solution: *Given:* V_1 = 180 m³/min. ; t_{db_1} = 25°C; ϕ_1 = 15%; t_{db_2} = 25°C; t_{wb_1} = 20°C.

- Locate the point 1 on the psychrometric chart on the intersection of 40°C DBT and 15% RH lines.
- Locate the point 2 on the psychrometric chart on the intersection of 25°C DBT and 20°C WBT lines.

Then, line 1-2 represents the *adiabatic humidification.*

(*i*) **Dew point temperature, t_{dp}:**

On the psychrometric chart draw a horizontal line through the point 2 upto the saturation curve. From the chart ; we find that

$$t_{dp} = 17.7°\text{C} \textbf{ (Ans.)}$$

(*ii*) **Amount of water added to air per minute:**

From the *psychrometric chart*, specific volume of air at point 1,

$$v_{s1} = 0.895 \text{ m}^3\text{/kg of dry air}$$

Specific humidity at point 1,

$$W_1 = 0.007 \text{ kg/kg of dry air,}$$

and specific humidity at point 2,

$$W_2 = 0.0128 \text{ kg/kg of dry air}$$

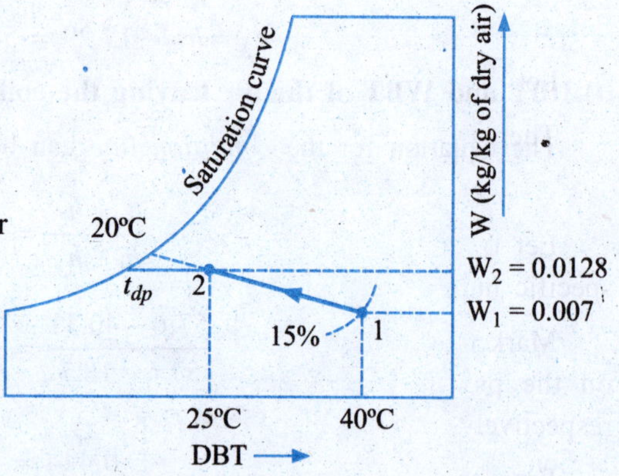

Fig. 12.26

Mass of air supplied, $m_a = \dfrac{V_1}{v_{s1}} = \dfrac{180}{0.895}$

$= 201.1$ kg/min.

∴ Amount of water vapour added to the air

$= m_a (W_2 - W_1)$

$= 201.1\ (0.0128 - 0.007) =$ **1.166 kg/min. (Ans.)**

Example 12.31. **(*Cooling and humidification*)** *Air at 30°C and 50% RH enters an air washer which has humidifying efficiency of 80 per cent. Determine the final DBT and RH of air washed with recirculated spray water.*

Solution: *Given:* t_{db_1} = 30°C; ϕ_1 = 50%; humidifying efficiency (η_H) = 80% = 0.8

Locate the point 1 (30°C DBT and 50% RH) on the psychrometric chart as shown in Fig. 12.27. The wet bulb temperature of the entering air as read from the psychrometric chart is,

$$t_{wb_3} = 22°\text{C} = t_{db_3}$$

Final DBT:

Let, t_{db2} = Final DBT of the air leaving the air washer.

Then, $\eta_H = 0.8$

$$= \frac{t_{db_1} - t_{db_2}}{t_{db_1} - t_{db_3}}$$

$$= \frac{30 - t_{db_2}}{30 - 22} = \frac{30 - t_{db2}}{8}$$

$$\therefore \quad t_{db_2} = 30 - 0.8 \times 8 = 23.6°C$$

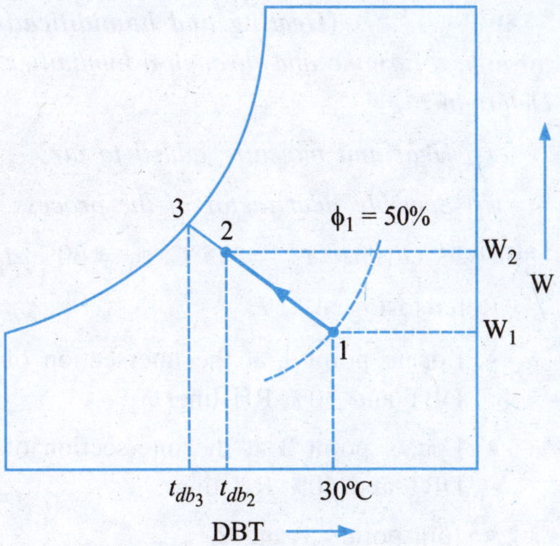

Fig. 12.27

Final relative humidity, ϕ_2:

Mark point 2 on the constant wet bulb temperature line 1-3 such that t_{db_2} = 23.6°C. Now the relative humidity of the air leaving the air washer (corresponding to point 2) as read from the pychrometric chart is,

$$\phi_2 = \mathbf{86\%} \text{ (Ans.)}.$$

Example 12.32. **(*Heating and humidification*)** *It is required to maintain a drying room at 33°C DBT and 25% RH. The sensible heat gain to the room is 165000 kJ/h. The moisture to be evaporated from the objects during drying is 19 kg/h. If there is no direct heat source to provide for evaporation in the room, calculate the state and rate of supply air at 16°C DBT.*

Solution: *Given:* t_{db_2} = 33°C ; ϕ_2 = 25% ; RSH = 165000 kJ/*h* ; m_W = 19 kg/*h* ; t_{db_1} = 16°C

Let, m_a = Mass of supply air.

Then, $165000 = m_a \, c_p \, (t_{db_2} - t_{db_1}) = m_a \times 1.005 \, (33 - 16)$

$\therefore$ m_a = **9657.6 kg/*h* (Ans.)**

Let ϕ_1 and W_1 be the relative humidity and specific humidity at point 1 respectively.

Mark the initial and final conditions of the air on the psychrometric chart as points 1 and 2 respectively, as shown in Fig. 12.28.

From the *psychrometric chart*, corresponding to point 2,

$$W_2 = 0.008 \text{ kg/kg of dry air}$$

Also, $$W_2 = W_1 + \frac{m_w}{m_a}$$

or, $$0.008 = W_1 + \frac{19}{9657.6}$$

$$\therefore \quad W_1 = 0.008 - \frac{19}{9657.6} = 0.006 \text{ kg/kg of dry air}$$

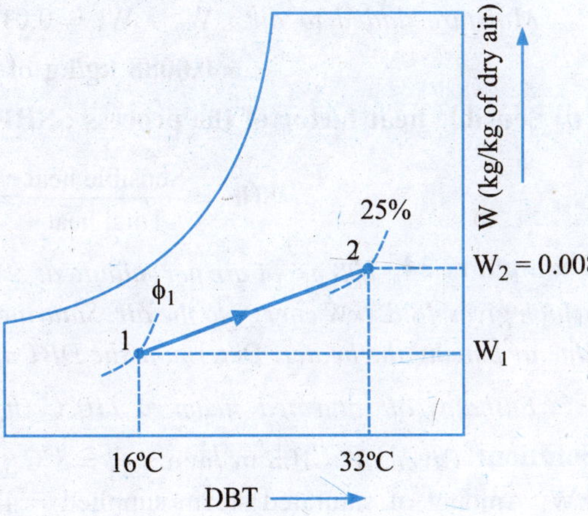

Fig. 12.28

From *psychrometric chart*, corresponding to 16°C and 0.006 kg/kg of dry air,

$$\phi_1 = \mathbf{52.5\%} \text{ (Ans.)}$$

Example 12.33. **(*Heating and humidification*).** *Atmospheric air at 15°C DBT and 30% RH passes through a furnace and through a humidifier, in such a way that the final DBT is 32°C and 40% RH. Determine:*

(*i*) *Heat and moisture added to air;*

(*ii*) *Sensible heat factor of the process.*

Solution: *Given:* t_{db_1} = 15°C; ϕ_1 = 30%; t_{db_2} = 32°C; ϕ_2 = 40%

Refer to Fig. 12.29.

- Locate point 1 at the intersection of 16°C DBT and 30% RH lines.
- Locate point 2 at the intersection of 32°C DBT and 40% RH lines.
- Join points 1 and 2.
- Locate point *A* by drawing horizontal line through point 1 and vertical line through point 2.

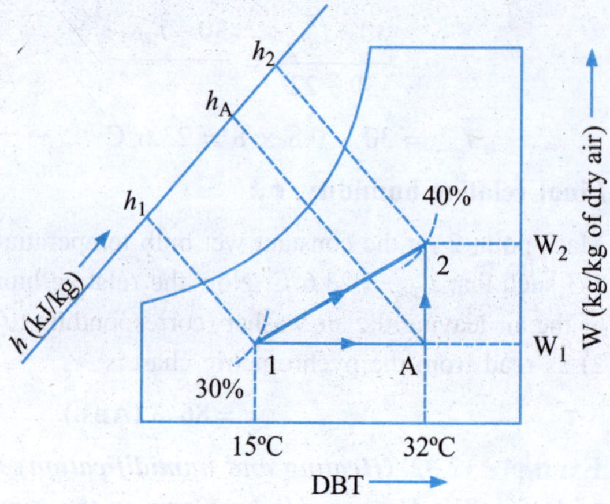

Fig. 12.29

From *psychrometric chart*, we have:

h_1 = 22.7 kJ/kg of dry air;

h_2 = 62.3 kJ/kg of dry air;

h_A = 40 kJ/kg of dry air;

W_1 = 0.003 kg/kg of dry air;

W_2 = 0.0171 kg/kg of dry air.

(*i*) **Heat and moisture added to air:**

Heat added to the air = $h_2 - h_1$ = 62.3 − 22.7 = **39.6 kJ/kg of dry air (Ans.)**

Moisture added to air = $W_2 - W_1$ = 0.0118 − 0.003

= **0.0088 kg/kg of dry air (Ans.)**

(*ii*) **Sensible heat factor of the process ; SHF :**

$$\text{SHF} = \frac{\text{Sensible heat}}{\text{Total heat}} = \frac{(h_A - h_1)}{(h_2 - h_1)} = \frac{40 - 22.7}{62.3 - 22.7} = \mathbf{0.437\ (Ans.)}$$

Example 12.34. *108 m³ of air per minute at 5°C DBT and 2.5°C WBT is passed through a heating coil which gives 48.85 kW energy to the air. Saturated steam at 110°C and with a rate of 48 kg/h is mixed with the air leaving the heater. Determine the DBT and WBT of the air after mixing the steam.*

Enthalpy of saturated steam at 110°C (h_s) = *2691 kJ/kg*

Solution: *Given:* V = 108 m³/min. t_{db1} = 5°C ; t_{wb_1} = 2.5°C ; Energy given by the heating coil = 48.85 kW; Amount of saturated steam supplied = 48 kg/*h*; h_s = 2691 kJ/kg

Refer to Fig. 12.30.

- Locate point 1 at the intersection of 5°C DBT and 2.5°C WBT. Corresponding to this point, from the *psychrometric chart*, we have:

$$v_{s1} = 0.792 \text{ m}^3/\text{kg} \; ; \; h_1 = 14 \text{ kJ/kg} \; ; \; W_1 = 0.0034 \text{ kg/kg of dry air}$$

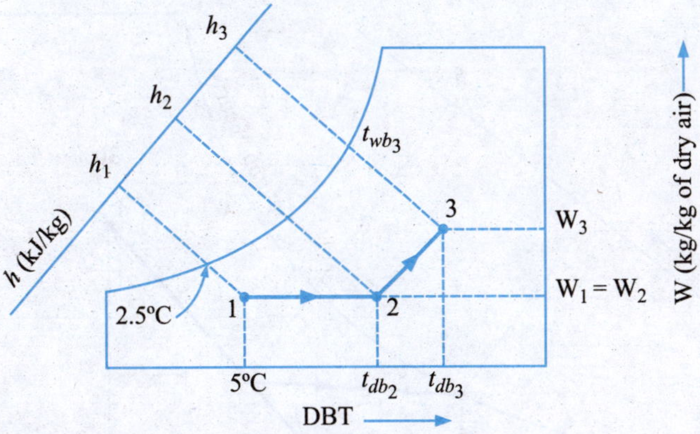

Fig. 12.30

The mass of air passing through the system is given by,

$$m_a = \frac{V}{v_{s1}} = \frac{108}{0.792} = 136.4 \text{ kg/min}$$

Also, $$h_2 = h_1 + \frac{48.85 \times 60}{136.4} = 14 + \frac{48.85 \times 60}{136.4} = 35.5 \text{ kJ/kg}$$

- Draw a horizontal line through 1 till it into the line h_2 (i.e. 35.5 kJ/kg) and then mark the point 2.

$$W_3 = W_2 + \frac{(48/60)}{136.4}$$

$$= W_1 + \frac{(48/60)}{136.4} \quad (\because W_2 = W_1)$$

$$= 0.0034 + \frac{(48/60)}{136.4} = 0.00927 \text{ kg/kg of dry air}$$

$$h_3 = h_2 + \frac{(48/60)}{136.4} \times 2691$$

$$= 35.5 + 15.78 = 51.28 \text{ kJ/kg}$$

- Corresponding to h_3 = 51.28 kJ/kg and W_3 = 0.00927 kg/kg of dry air, locate the point 3 on the psychrometric chart and find that;

$$t_{db_3} = \mathbf{26.8°C}, \text{ and } t_{wb_3} = \mathbf{18°C} \textbf{ (Ans.)}$$

Example 12.35. *Room air at 20°C DBT and 50% RH is mixed with outdoor air at 40°C DBT and 30% RH in the ratio of 4:1. The mixture is passed through a cooling coil whose temperature is maintained constant at 10°C whose by-pass factor is 0.2. Determine the following:*

(i) Condition of air before entering the coil,

(ii) Condition of air leaving the coil, and

(iii) Refrigeration load on the cooling coil when 250 m³/min of air is supplied to the room.

[N.U.]

Solution: Refer Fig. 12.31.

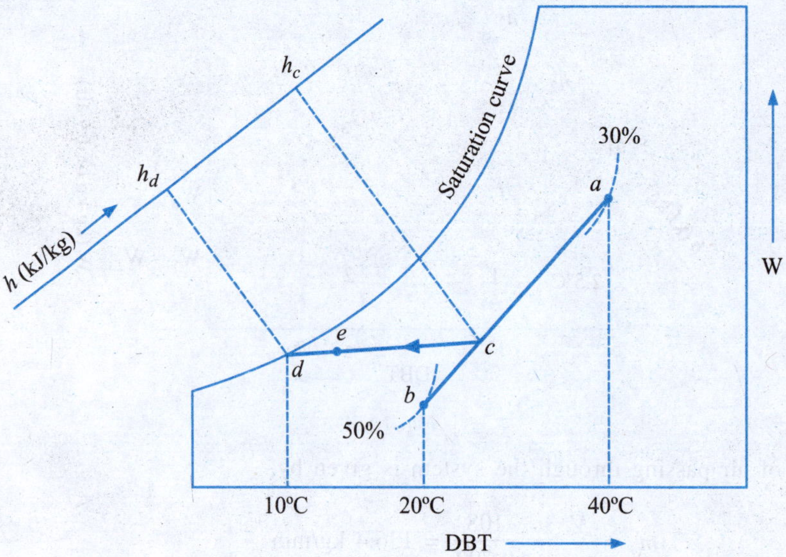

Fig. 12.31

- Locate point '*a*' on the psychrometric chart on the intersection of 40°C DBT and 30% RH lines.
- Locate point '*b*' at the intersection of 20°C DBT and 50% RH lines:
  Join points '*a*' and '*b*'.
- Locate point '*d*' on at the intersection of 10°C and saturation curve.
- Locate the point '*c*' on the line *ab* such that,

$$\frac{ac}{bc} = \frac{4}{1}$$

  Join '*c*' and '*d*'.
- Locate the point '*e*' in such a way that

$$\frac{de}{dc} = 0.2$$

(*i*) **Condition of air before entering the coil:**

From *psychrometric chart*,

Condition of air at point '*c*' = **24°C DBT, 46% RH (Ans.)**

(*ii*) **Condition of air leaving the coil:**

From *psychrometric chart*,

Condition of air at point '*e*' = **13°C, 83% RH (Ans.)**

(*iii*) **Refrigeration load on the cooling coil :**

From *psychrometric chart*: $v_1 = 0.84$ m³/kg; $h_c = 46$ kJ/kg of dry air;

$h_e = 33$ kJ/kg of dry air;

Mass of air passing through room,

$$m_a = \frac{250}{0.84} = 297.6 \text{ kg/min.}$$

$$\therefore \quad \text{Cooling coil load} = \frac{m_a(h_c - h_e) \times 60}{14000} \text{ tonnes}$$

$$= \frac{297.6(46 - 33) \times 60}{14000} = \textbf{16.58 tonnes (Ans.)}$$

Example 12.36. *Air at 12°C DBT and 70% RH is to be heated and humidified to 36.5°C DBT and 21°C WBT. The air is preheated sensibly before passing to the air washer in which water is recirculated. The relative humidity of the air coming out of the air washer is 70%. This air is again reheated sensibly to obtain the final desired condition. Determine:*

(i) Temperature to which the air should be preheated;

(ii) Total heating required;

(iii) Make up water required in the air washer;

(iv) Humidifying efficiency of the air washer.

Solution: *Given:* $t_{db1} = 12°C$; $\phi_1 = 70\%$; $t_{db_4} = 36.5°C$; $t_{wb_4} = 21°C$.

- Locate point 1 at the intersection of 12°C DBT and 70% RH lines.
- Locate point 4 at the intersection of 36.5°C DBT and 21°C WBT lines.
- From point 1 draw a horizontal line to represent sensible heating and from point 4 draw horizontal line to intersect 70% RH curve at point 3. Now from point 3, draw a constant wet bulb temperature line which intersects the horizontal line drawn through point 1 at point 2. *The line 1-2 represents preheating of air, line 2-3 represents humidification and line 3-4 represents reheating to final condition.*

1. **Temperature to which the air should be preheated, t_{db_2}:**

From *psychrometric chart*, corresponding to point 2,

$$t_{db_2} = \textbf{26.6°C (Ans.)}$$

2. **Total heating required:**

From *psychrometric chart*, we find:

$$h_1 = 26.9 \text{ kJ/kg}; \; h_2 = h_3 = 42.1 \text{ kJ/kg}; \; h_4 = 61 \text{ kJ/kg}$$

Total heat required = Heat required for preheating of air + Heat required for reheating of air

$$= (h_2 - h_1) + (h_4 - h_3) = (h_2 - h_1) + (h_4 - h_2) \qquad (\because h_3 = h_2)$$

$$= h_4 - h_1 = 61 - 26.9 = \textbf{34.1 kJ/kg of dry air (Ans.)}$$

(iii) **Make-up water required in the air washer, $(W_4 - W_1)$:**

From *psychrometric chart*, we find :

$$W_1 (= W_2) = 0.006 \text{ kg/kg of dry air;}$$

$$W_4 (= W_3) = 0.0092 \text{ kg/kg of dry air.}$$

$\therefore$ Make-up water required in the air washer

$$= (W_3 - W_2) = (W_4 - W_1) = 0.0092 - 0.006 = \textbf{0.0032 kg/kg of dry air (Ans.)}$$

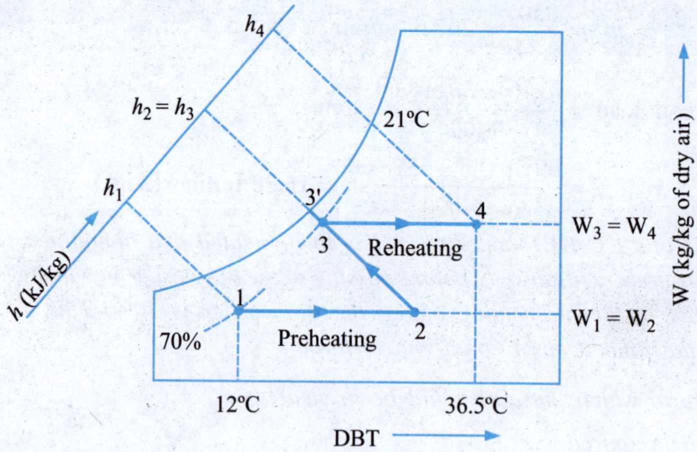

Fig. 12.32

(*iv*) **Humidifying efficiency of the air washer, η_H:**

From *psychrometric chart,* we find:

$$t_{db_3} = 18.5°C;\ t_{db_{3'}} = 15°C$$

$$\eta_H = \frac{\text{Actual drop in DBT}}{\text{Ideal drop in WBT}} = \frac{t_{db_2} - t_{db_3}}{t_{db_2} - t_{db_{3'}}}$$

$$= \frac{26.6 - 18.5}{26.6 - 15} = 0.698 \text{ or } \mathbf{69.8\%\ (Ans.)}$$

Example 12.37. *Saturated air at 19°C is passed through a drier so that its final relative humidity is 25%. The drier uses silica gel adsorbent. The air is then passed through a cooler until its final temperature is 19°C DBT without change in specific humidity. Determine the following:*

(*i*) *Temperature of air at the end of the drying process;*

(*ii*) *Heat rejected during the cooling process;*

(*iii*) *Relative humidity at the end of cooling process;*

(*iv*) *Dew point temperature at the end of the drying process;*

(*v*) *Moisture removed during the drying process.*

Solution: *Given:* $t_{db_1} = t_{db_3} = 19°C;\ \phi_2 = 25\%$.

- Locate point 1 (19°C) on the saturation line as shown in Fig. 12.33.
- Since the drying process is a *chemical dehumidification process,* therefore, it *follows a path along the constant wet bulb temperature or the constant enthalpy line as* shown by the line 1-2. Locate point 2 at relative humidity of 25%.
- The cooling process is represented by the line 2-3, the temperature at point 3 being 19°C DBT.

(*i*) **Temperature of air at the end of the drying process, t_{db_2}:**

From the *psychrometric chart,* corresponding to point 2, we find,

$$t_{db_2} = \mathbf{33.7°C\ (Ans.)}$$

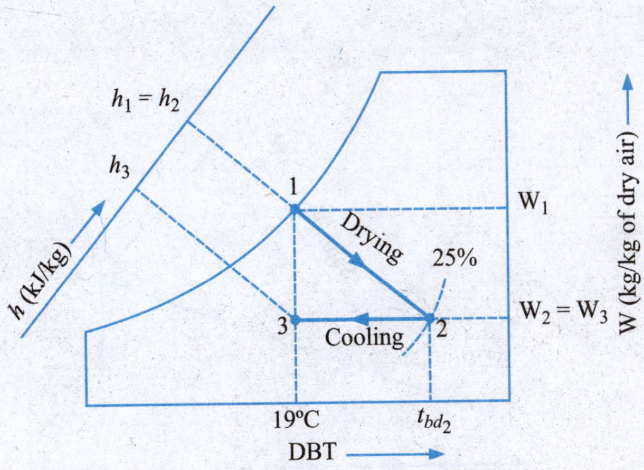

Fig. 12.33

(*ii*) **Heat rejected during the cooling process, ($h_2 - h_3$):**

From *psychrometric chart*, we find:

$$h_2 = 54.5 \text{ kJ/kg}; \quad h_3 = 38.9 \text{ kJ/kg}$$

∴ Heat rejected during the cooling process

$$= h_2 - h_3 = 54.5 - 38.9 = \mathbf{15.6 \text{ kJ/kg (Ans.)}}$$

(*iii*) **Relative humidity at the end of cooling process, ϕ_3:**

From the *psychrometric chart*, we find :

$$\phi_3 = \mathbf{58\% \text{ (Ans.)}}$$

(*iv*) **Dew point temperature at the end of drying process, t_{dp2}:**

From the psychrometric chart, we find:

$$t_{dp_2} = \mathbf{10.3°C \text{ (Ans.)}}$$

(*v*) **Moisture removed during the drying process ($W_1 - W_2$):**

From the *psychrometric chart*, we find:

$$W_1 = 0.0138 \text{ kg/kg of dry air};$$

$$W_2 = 0.0078 \text{ kg/kg of dry air}.$$

∴ Moisture removed during the process

$$= W_1 - W_2 = 0.0138 - 0.0078 = \mathbf{0.006 \text{ kg/kg of dry air (Ans.)}}$$

Example 12. 38. *250 m^3 of air is supplied per minute from outdoor conditions of 38°C DBT and 25°C WBT to an air-conditioned room. The air is dehumidified by a cooling coil having a by-pass factor 0.35 and dew point temperature 13°C and then by a chemical dehumidifier. Air leaves the chemical dehumidifier at 32°C DBT. Air then is passed over a cooling coil whose surface temperature is 13°C and by-pass factor 0.25.*

Calculate the capacities of the two cooling coils and dehumidifier.

Solution: *Given:* $V_1 = 250$ m³/min.; $t_{db_1} = 38°C$; $t_{wb_1} = 25°C$; ADP = 13ºC; $t_{db_3} = 32°C$.

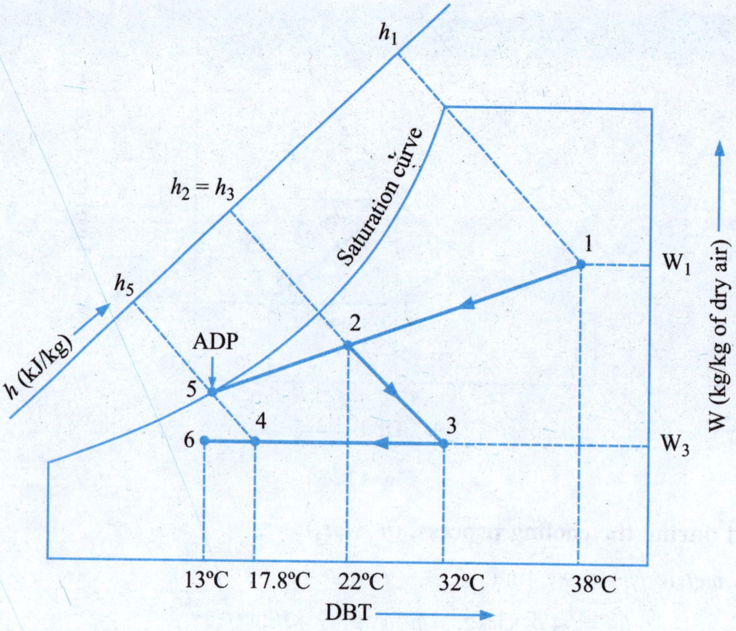

Fig. 12.34

- Locate point 1 at the intersection of 38°C DBT and 25°C WBT lines.
- Locate point 5 on the saturation curve representing the apparatus dew point i.e., ADP = 13°C of the cooling coil.
- Since the by-pass factor of cooling coil is 0.35, therefore locate point 2 on the line 1-5 such that,

  Length of line 5-2 = 0.35 × length of line 5-1.

  The point 2 represents the *condition of air leaving the cooling coil*. This air now passes through a chemical dehumidifier.
- From point 2 draw a constant enthalpy line to intersect the vertical line drawn through 32°C DBT, at point 3.

  The air after leaving the chemical humidifier, passes over another cooling coil having surface temperature of 13°C.
- Through point 3, draw a horizontal line intersecting the vertical line drawn through 13°C DBT, at point 6.
- Since the by-pass factor of this coil is 0.25, therefore mark point 4 on the line 6-3 such that

Length of line 6-4 = 0.25 × length of line 6-3.

The point 4 represents the *condition of air leaving the second cooling coil*.

Capacities of the two cooling coils:

From *psychrometric chart*, we find:

$$v_{s1} = 0.902 \text{ m}^3\text{/kg of dry air;}$$

$$h_1 = 76.8 \text{ kJ/kg of dry air;}$$

$$W_1 = 0.01464 \text{ kg/kg of dry air;}$$
$$h_2 (= h_3) = 50.2 \text{ kJ/kg of dry air;}$$
$$W_3 = 0.007 \text{ kg/kg of dry air;}$$
$$h_4 = 35.9 \text{ kJ/kg of dry air;}$$
$$h_5 = 36.6 \text{ kJ/kg of dry air.}$$

Mass of air supplied,

$$m_a = \frac{V_1}{v_{s1}} = \frac{250}{0.902} = 277.16 \text{ kg/min}$$

Capacity of first cooling coil

$$= m_a (h_1 - h_2) = 277.16 \ (76.8 - 50.2) = \mathbf{7372.5 \ kJ/min \ (Ans.)}$$

Capacity of second cooling coil

$$= m_a (h_3 - h_4) = 277.16 \ (50.2 - 35.9) = \mathbf{3963.4 \ kJ/min \ (Ans.)}$$

Capacity of the dehumidifier:

Capacity of the humidifier $= m_a (W_1 - W_3) = 277.16 \ (0.01464 - 0.007) =$ **2.18 kg/min (Ans.)**

HIGHLIGHTS

1. '*Air-conditioning*' is the simultaneous control of temperature, humidity, motion and purity of the atmosphere in a confined space.
2. '*Psychrometry*' is an art of measuring moisture content of air.

   The science which investigates the thermal properties of moist air, considers the measurement and control of the moisture content of air, and studies the effects of atmospheric moisture on material and human comfort may property be termed '*psychrometrics*'.
3. When air is saturated DBT, WBT, DPT are equal.
4. *Dalton's law of partial pressure* states, that the total pressure of a mixture of gases is equal to the sum of partial pressures which the component gases would exert if each existed alone in the mixture volume at the mixture temperature.
5. Specific humidity $W = \dfrac{0.622 \, p_v}{p_t - p_v}$
6. Degree of saturation

$$(\mu) = \frac{\text{Mass of water vapour associated with unit mass of dry air}}{\text{Mass of vapour associated with saturated unit mass of dry air}}$$

7. Relative humidity, $\phi = \dfrac{p_v}{p_{vs}}$.
8. A '*Psychrometer*' is a device which is used for measuring dry bulb and wet-bulb temperatures simultaneously.
9. The processes which affect the psychrometric properties of air are called '*Psychrometric processes*'.

   Important *Psychrometric processes* are :

   (*i*) Mixing of air streams (*ii*) Sensible heating

   (*iii*) Sensible cooling (*iv*) Cooling and dehumidification (*v*) Cooling and humidification (*vi*) Heating and dehumidification

(*vii*) Heating and humidification.

10. '*Room sensible heat factor (RSHF)* is the ratio of the room sensible heat to the sum of room sensible heat and the room latent heat.

11. *Grand sensible heat factor (GSHF)* is the ratio of total sensible heat to the grand total heat the cooling coil or the conditioning apparatus should handle.

OBJECTIVE TYPE QUESTIONS

A. Choose the Correct Answer:

1. In an unsaturated air the state of a vapour is
   (*a*) wet (*b*) superheated (*c*) saturated (*d*) unsaturated.

2. For saturated air,
   (*a*) wet bulb depression is zero (*b*) wet bulb depression is positive
   (*c*) wet bulb depression is negative
   (*d*) wet bulb depression can be either positive or negative.

3. Which one of the following statements is *correct*?
   (*a*) Dew point temperature can be measured with the help of thermometer.
   (*b*) Dew point temperature is the saturation temperature corresponding to the partial pressure of the water vapour in moist air.
   (*c*) Dew point temperature is the same as the thermodynamic wet bulb temperature.
   (*d*) For saturated air, dew point temperature is less than the wet bulb temperature.

4. During sensible heating of moist air, enthalpy
   (*a*) increases (*b*) decreases (*c*) remains constant (*d*) none of these.

5. During sensible cooling, wet bulb temperature
   (*a*) decreases (*b*) increases
   (*c*) remains constant (*d*) can decreases or increase.

6. Which of the following statements is *correct*?
   (*a*) Evaporative cooling and sensible cooling is the same.
   (*b*) Evaporative cooling is a cooling and humidification process.
   (*c*) Evaporative cooling is a cooling and dehumidification process.
   (*d*) Evaporative cooling is not effective for hot and dry climates.

7. An air washer can work as a
   (*a*) filter only (*b*) humidifier only (*c*) dehumidifier only (*d*) all of these.

8. The relative humidity, during sensible heating
   (*a*) can increase or decrease (*b*) increases
   (*c*) decreases (*d*) remains constant.

9. The vapour pressure, during sensible heating of moist air
   (*a*) increases (*b*) decreases
   (*c*) can increase or decrease (*d*) remains constant.

10. The relative humidity, during heating and humidification
    (*a*) increases (*b*) decreases (*c*) may increase (*d*) remains constant.

11. The relative humidity, during cooling and dehumidification of moist air
(*a*) increases (*b*) decreases
(*c*) can increase or decrease (*d*) remains constant.

12. The minimum temperature to which moist air can be cooled under ideal conditions in a spray washer is
(*a*) dew point temperature of inlet air (*b*) wet bulb temperature of inlet air
(*c*) water inlet temperature (*d*) water outlet temperature.

13. A mixture of dry air and water vapour, when the air has diffused the maximum amount of water vapour into it, is called
(*a*) dry air (*b*) moist air (*c*) saturated air (*d*) specific humidity.

14. The by-pass factor of a cooling coil decreases with
(*a*) decreases in fin spacing and increase in number of rows
(*b*) increase in fin spacing and increase in number of rows
(*c*) increase in fin spacing and decrease in number of rows
(*d*) decrease in fin spacing and decrease in number of rows

15. In order to cool and dehumidify a stream of moist air, it must be passed over the coil at a temperature
(*a*) which lies between the dry bulb and wet bulb temperature of the incoming stream
(*b*) which lies between the wet bulb and dew point temperature of the incoming stream
(c) which is lower than the dew point temperature of the incoming stream
(*d*) of adiabatic saturation of incoming stream.

16. The temperature of air recorded by a thermometer, when it is not effected by the moisture present in it, is called
(*a*) wet bulb temperature (*b*) dry bulb temperature
(*c*) dew point temperature (*d*) none of these.

17. The relative humidity of air is defined as the ratio of
(*a*) mass of water vapour in a given volume to the total mass of the mixture of air and water vapour.
(*b*) mass of water vapour in a given volume to the mass of water vapour, if air is saturated at the same temperature.
(*c*) mass of water vapour in a given volume to the mass of air
(*d*) mass of air to the mass of water vapour in the mixture of air and water vapour.

18. The process generally used in winter air-conditioning to warm and humidify the air, is called
(*a*) humidification (*b*) dehumidification
(*c*) heating and humidification (*d*) cooling and dehumidification.

19. The difference between dry bulb temperature and wet bulb temperature, is called
(*a*) dry bulb depression (*b*) wet bulb depression
(*c*) dew point depression (*d*) degree of saturation

20. The wet bulb depression is zero, when relative humidity is equal to
(*a*) zero (*b*) 50% (*c*) 75% (*d*) 100%.

21. The vertical and uniformly spaced lines on a psychrometric chart indicate
(*a*) dry bulb temperature (*b*) wet bulb temperature
(*c*) dew point temperature (*d*) specific humidity.

22. The curved lines on a psychrometric chart indicates
(*a*) dry bulb temperature (*b*) wet bulb temperature
(*c*) specific humidity (*d*) relative humidity.

23. Air at dry bulb temperature of 33°C and dew point-temperature of 18°C passes through a cooling coil which is maintained at 25°C.

The process would be

(*a*) cooling and dehumidification (*b*) sensible cooling
(*c*) cooling and humidification (*d*) unpredictable.

24. Specific humidity is defined as

(*a*) mass of water vapour contained in air-vapour mixture per kg of dry air
(*b*) mass of water vapour contained per kg of air-vapour mixture
(*c*) mass of dry air contained per kg of air-vapour mixture
(*d*) none of the above.

25. Humidity ratio W may be expressed as (m_a = mass of dry air, m_v = mass of vapour in air vapour mixture)

(*a*) $W = \frac{m_v}{m_a}$ (*b*) $W = \frac{m_v}{m_v + m_a}$ (*c*) $W = \frac{m_a}{m_a + m_v}$ (*d*) $W = \frac{m_v}{m_a - m_v}$.

26. The ambient air temperature as recorded by ordinary thermometer is called

(*a*) wet bulb temperature (*b*) dew point temperature
(*c*) dry bulb temperature (*d*) saturation temperature.

27. Degree of Saturation μ is defined as the ratio of

(*a*) specific humidity of moist air to specific humidity of saturated air at same dry bulb temperature
(*b*) moisture content contained per 1 kg of dry air to moisture content required to saturate 1 kg of dry air at same dry bulb temperature
(*c*) both (*a*) and *b*) (*d*) none of these.

28. Room Sensible Heat Factor is defined as : (RSH = Room sensible heat, RLH = Room latent heat)

(*a*) $= \frac{RSH}{RSH + RLH}$ (*b*) $\frac{RLH}{RSH + RLH}$ (*c*) $\frac{RSH}{RLH}$ (*d*) none of the above.

29. Humidity ratio W is expressed as (p_v = partial pressure of vapour, p_a = partial pressure of air, p_b = barometric pressure of air):

(*a*) $W = 0.622 \frac{p_v}{p_b}$ (*b*) $W = 0.622 \frac{p_v}{p_a}$

(*c*) $W = 0.622 \frac{p_v}{(p_b - p_a)}$ (*d*) $W = 0.622 \frac{p_v}{p_b - p_v}$.

30. Specific humidity is also called

(*a*) relative humidity (*b*) humidity ratio
(*c*) absolute humidity (*d*) saturation humidity

31. Wet-bulb temperature of air whose relative humidity is 100% is

(*a*) less than dew point temperature (*b*) equal to dew point temperature
(*c*) greater than dry point temperature
(*d*) may be more or less depending upon dry bulb temperature

32. Wet-bulb temperature of air whose relative humidity is 100% is

(*a*) less than dry bulb temperature
(*b*) equal to dry bulb temperature
(*c*) greater than dry bulb temperature
(*d*) may be more or less depending upon humidity ratio.

33. If the specific humidity of moist air remains same but its dry bulb temperature increases
   (*a*) its dew point temperature increases
   (*b*) its dew point temperature decreases
   (*c*) its dew point temperature remains same
   (*d*) its dew point temperature may increase or decrease depending upon increase or decrease of relative humidity.

B. Fill in the Blanks:

1. The simultaneous control of temperature, humidity, motion and purity of the atmosphere in a confined space is called ....
2. The art of measuring the moisture content of air is termed ....
3. .... is the temperature of air as registered by an ordinary thermometer.
4. The difference between the dry bulb and wet bulb temperature is called ....
5. .... is the ratio of the partial pressure of water vapour in the mixture to the saturated partial pressure at the dry bulb temperature, expressed as percentage.
6. The heat that does not affect the temperature but changes the state of a substance when added to or subtracted from it is called ....
7. The ratio of mass of water vapour associated with unit mass of water vapour associated with saturated unit mass of dry air is called ....
8. Air undergoes sensible .... whenever it passes over a surface that is at a temperature less than the dry bulb temperature of the air but greater than the dew point temperature.
9. The ratio of the room sensible heat to the sum of room sensible heat and room latent heat is called ....
10. .... is the ratio of total sensible heat to the grand total heat that the cooling coil or the conditioning apparatus should handle.

ANSWERS

A. Choose the Correct Answers:

1. (*b*)	2. (*a*)	3. (*b*)	4. (*a*)	5. (*a*)	6. (*b*)	7. (*d*)	8. (*b*)
9. (*d*)	10. (*a*)	11. (*c*)	12. (*b*)	13. (*c*)	14. (*a*)	15. (*c*)	16. (*b*)
17. (*b*)	18. (*c*)	19. (*b*)	20. (*d*)	21. (*a*)	22. (*d*)	23. (*b*)	24. (*a*)
25. (*b*)	26. (*c*)	27. (*c*)	28. (*a*)	29. (*b*)	30. (*b*)	31. (*b*)	32. (*b*)
33. (*c*).							

B. Fill in the Blanks:

1. Air-conditioning;	2. Psychrometry;	3. DBT;	4. Wet-bulb depression;
5. Relative humidity;	6. Latent heat;	7. Degree of saturation;	
8. Cooling;	9. RSHF;	10. GSHF.	

THEORETICAL QUESTIONS

1. Define the following terms:

   (*i*) Saturated air (*ii*) Dry bulb temperature (*iii*) Dew point temperature

   (*iv*) Relative humidity (*v*) Specific humidity.

2. State 'Dalton's law of partial pressure.

3. Derive an expression for specific humidity and show that it is function of vapour pressure and barometric pressure of air.

4. Derive the following relations:

   (*i*) Specific humidity, $W = 0.622 \dfrac{p_v}{p_t - p_v}$

   (*ii*) Degree of saturation, $\mu = \dfrac{p_v}{p_s}\left[\dfrac{\left(1-\dfrac{p_{vs}}{p_t}\right)}{\left(1-\dfrac{p_v}{p_t}\right)}\right]$

5. Prove that relative humidity, ϕ is given by :

$$\phi = \frac{\mu}{1-(1-\mu)(p_{vs}/p_t)}$$

   where, μ = Degree of saturation,

   p_{vs} = Saturation pressure of vapour in moist air, and

   p_t = Total pressure of moist air.

6. Explain briefly with a neat sketch a 'sling psychrometer'.

7. Describe briefly any two of the following processes:

   (*i*) Sensible heating (*ii*) Cooling and dehumidification

   (*iii*) Heating and humidification (*iv*) Heating and dehumidification

8. Write a short note on 'by-pass factor'.

UNSOLVED EXAMPLES

1. The atmospheric conditions are 30°C and specific humidity of 0.0125 kg/kg of air, Determine:

   (*i*) Partial pressure of air; (*ii*) Relative humidity;

   (*iii*) Dew point temperature.

   Atmospheric pressure = 756 mm hg.

   [**Ans.** (*i*) 14.89 mm of Hg. (*ii*) 46.8%, (*iii*) 17°C]

2. A mixture of air and water vapour at 1 bar and 25°C has a dew point temperature of 15°C. Determine the relative humidity and specific humidity. [**Ans.** 53.8%, 0.01078 kg/kg of dry air]

3. An air-water vapour mixture at 1.24 bar has a temperature of 38°C and relative humidity of 60%. Calculate the kg-mass of water vapour per kg of air and per kg of mixture. Also find the dew point.

   [**Ans.** 0.0203 kg/kg of dry air, 0.0199 kg/kg of mixture, 28.5°C]

4. In a house, the temperature of the windows on a day in winter is 5°C. When the temperature in the room is 23°C, and the barometric pressure is 748.8 mm Hg, what would be maximum relative humidity that could be maintained in the room without condensation on the window panes ? Under these conditions find the partial pressure of the water vapour and air, the specific humidity, and the density of the mixture.

[**Ans.** 30.7%, 0.00872 bar, 0.989 bar, 0.00548 kg/kg of dry air, 1.164 kg/m^3]

5. Atmospheric air enters a heater at 4.5°C and 60% relative humidity and leaves at a temperature of 21°C. Calculate:

(*i*) The heat supplied to the air;

(*ii*) The final relative humidity. [**Ans.** 16.8 kJ/kg, 29.2%]

6. The air supplied to a room of building in winter is to be at 17°C and have a relative humidity of 60%. If the barometric pressure is 1.01325 bar, calculate the specific humidity. What would be the dew point under these conditions ? [**Ans.** 0.00723 kg/kg of dry air, 9.18°C]

7. If air at the condition of example 6, is passed at the rate of 0.5 m^3/s over a cooling coil which is at a temperature of 6°C, calculate the amount of vapour which will be condensed. Assume that the barometric pressure is the same as in example 6, and that the air leaving the coil is saturated.

[**Ans.** 3.082 kg/h]

8. An air and water vapour mixture at 1 bar and 26.7°C has a specific humidity of 0.0085. Determine the percentage saturation. [**Ans.** 37.7%]

9. A mixture of air and water vapour at 1.013 bar and 16°C has dew point of 5°C. Determine the relative and specific humidities. [**Ans.** 48%, 0.0054 kg/kg of dry air]

10. Atmospheric air at a pressure of 760 mm Hg has a temperature of 32°C and a percentage saturation as determined from a psychrometric chart of 52%.

Calculate :

(*i*) The partial pressure of the vapour and the dry air;

(*ii*) The specific humidity;

(*iii*) The dew point;

(*iv*) The density of the mixture.

[**Ans.** (*i*) 0.0247 bar, 0.988 bar, (*ii*) 0.01556, (*iii*) 20.9°C, (*iv*) 1.147 kg/m^3]

11. In a laboratory test, a psychrometer recorded 36°C DBT and 30°C WBT.

Calculate :

(*i*) Vapour pressure; (*ii*) Relative humidity;

(*iii*) Specific humidity; (*iv*) Degree of saturation;

(*v*) Dew point temperature; (*vi*) Enthalpy of the mixture.

[**Ans.** (*i*) 0.0385 bar, (*ii*) 64.5%, (*iii*) 0.025 kg/kg of dry air, (*iv*) 0.63, (*v*) 28°C, (*vi*) 99.2 kJ]

12. The pressure and temperature of the air in a room is 1 bar and 28°C. If the relative humidity is found to be 30 per cent, determine:

(*i*) The partial pressure of the water vapour and dew point,

(*ii*) The specific volume of each constituent, and

(*iii*) The specific humidity.

[**Ans** (*i*) 0.0378 bar, 8.8°C, (*ii*) v_{air} = 0.874 m^3/kg, v_{vap} = 122.7 m^3/kg (*iii*) 0.00712 kg/kg dry air]

13. 100 m^3 of air per minute at 35°C DBT and 60% relative humidity is cooled to 20°C DBT by passing through a cooling coil. Find the following :

(*i*) Capacity of cooling coil in kJ/h;

(*ii*) Amount of water vapour removed per hour;

(*iii*) Relative humidity of air coming out and its wet-bulb temperature.

[**Ans.** (*i*) 1037088 kJ/h, (*ii*) 465.36 kg/h, (*iii*) 100%, 20°C]

14. Atmospheric air at 38°C and 40 per cent relative humidity is to be cooled and dehumidified to a state of saturated air at 10°C. The mass rate of flow of atmospheric air entering the dehumidifier is 45.5 kg/h. Neglecting any pressure drop, determine :

(*i*) The mass of water removed, and (*ii*) The quantity of heat removed.

[**Ans.** (*i*) 0.397 kg/h, (*ii*) 2287 kJ/h]

15. 1 kg of air at 24°C and a relative humidity of 70% is to be mixed adiabatically in a steady state, steady flow device with 1 kg of air at 16°C and a relative humidity of 10%. Assuming that the mixing is to be carried out at a constant pressure of 1.0 atmosphere, determine the temperature and relative humidity of the steam leaving the device. [**Ans.** 19.5°C, 50%]

16. An air-water vapour mixture enters the adiabatic saturator at 30°C and leaves at 20°C, which is the adiabatic saturation temperature. The pressure remains constant at 1 bar. Determine the relative humidity and the humidity ratio of the inlet mixture. [**Ans.** 39.8%, 0.0107 kg/kg of dry air]

Solve The Following Problems by Using Psychrometric Charts:

17. 200 m^3 of air per minute is passed through the adiabatic humidifier. The condition of air at inlet is 40°C DBT and 15% RH and the outlet condition is 25°C DBT and 20°C WBT, Find the following:

(*i*) Dew point temperature, and

(*ii*) Amount of water vapour added to the air per minute. [**Ans.** (*i*) 17.6°C; (*ii*) 1.25 kg/min.]

18. It is required to maintain a drying room at 32°C DBT and 30% RH. The sensible heat gain to the room is 150000 kJ/h. The moisture to be evaporated from the objects during drying in 18 kg/h. If there is no direct heat source to provide for evaporation in the room, calculate the state and rate of supply air at 15°C DBT.

[**Ans.** 8780 kg/h; 65%]

19. Air at 35°C DBT and 50% RH enters an air washer which has humidifying efficiency of 85 per cent. Determine the final DBT and RH of air washed with recirculated spray water. [**Ans.** 27.4°C; $\phi_2 = 90\%$]

20. Moist air enters a chamber at 5°C DBT and 2.5°C WBT at a rate of 90m^3/min. The barometric pressure is 1.01325 bar. While passing through the chamber, the air absorbs sensible heat at the rate of 40.7 kW and picks up 40 kg/h of saturated steam at 110°C. Determine the DBT and WBT of the leaving air.

[**Ans.** 26°C, 18.5°C]

21. 39.6 m^3/min of a mixture of recirculated room air and outdoor air enters a cooling coil at 31°C DBT and 18.5°C WBT. The effective surface temperature of the coil is 4.4°C. The surface area of the coil is such as would give 12.5 kW of refrigeration with the given entering air state. Determine the dry and wet bulb temperatures of the air leaving the coil and the coil by-pass factor. [**Ans.** 18.6°C; 12.5°C; 0.534]

22. Atmospheric air at a 16°C DBT and 25% RH passes through a furnace and then through a humidifier, in such a way that the final DBT is 30°C and 50% RH. Determine:

(*i*) Heat and moisture added to air;

(*ii*) Sensible heat factor of the process.

[**Ans.** (*i*) 41 kJ/kg of dry air, 0.0106 kg/kg of dry air; (*ii*) 0.366]

23. Air at 10°C DBT and 90% RH is to be heated and humidified to 35°C DBT and 22.5°C WBT. The air is preheated sensibly before passing to the air washer in which water is recirculated. The relative humidity of the air coming out of the air-washer is 90%. The air is again reheated sensibly to obtain the final desired condition. Determine:

 (*i*) The temperature to which the air should be preheated;

 (*ii*) The total heating required;

 (*iii*) The make-up water required in the air washer;

 (*iv*) The humidifying efficiency of the air washer.

[**Ans.** (*i*) 32.6°C; (*ii*) 40.8 kJ/kg of dry air; (*iii*) 0.0054 kg/kg of dry air; (*iv*) 92.4%]

24. Saturated air at 21°C is passed through a drier so that its final relative humidity is 20%. The drier uses silica gel adsorbent. The air is then passed through a cooler until its final temperature is 21°C without a change in specific humidity. Determine the following:

 (*i*) Temperature of air at the end of the drying process;

 (*ii*) Heat rejected during the cooling process;

 (*iii*) Relative humidity at the end of cooling process;

 (*iv*) Dew point temperature at the end of the drying process;

 (*v*) Moisture removed during the drying process.

[**Ans.** (*i*) 38.5°C; (*ii*) 18 kJ/kg of dry air; (*iii*) 55%; (*iv*) 11.6°C; (*v*) 0.0073 kg/kg of dry air]

25. 400 m^3/min. of recirculated air at 22°C DBT and 10°C DPT is to be mixed with 150 m^3/min. of fresh air at 30°C DBT and 50% RH. Determine the enthalpy, specific volume, humidity ratio and dew point temperature of the mixture.

[**Ans.** 47.86 kJ/kg of dry air; 0.855 m^3/kg of dry air, 0.0092 kg/kg of dry air; t_{dp} = 13°C]

❑❑❑

Chapter 13

Human Comfort

13.1 THERMODYNAMICS OF HUMAN BODY

- A human body feels comfortable thermodynamically when the *heat produced by the metabolism of human body is equal to the sum of the heat dissipated to the surroundings and the heat stored in human body by raising the temperature of body tissues.* These phenomena can be represented by the following equation.

$$Q_M = \pm Q_S + Q_E \pm Q_R \pm Q_C \quad ...(13.1)$$

where Q_M = *Metabolic heat* produced within the body (kJ/h.)

$\pm Q_S$ = *Stored energy* in temperature rise of body tissues; +ve when the tissue temperature rises and –ve when it falls.

Q_E = *Evaporative* heat loss due to the water evaporating at skin surface and the lungs expressed in kJ/h.

$\pm Q_R$ = Heat loss and gain by *radiation*; +ve when heat is lost to the surrounding, and –ve, when it is gained from the surroundings kJ/h.

Q_C = Heat loss or gain by *conduction and convection*; *+ve* when heat is lost, and –ve, when heat is gained from the surroundings in kJ/h.

- *The rate at which the body produces heat is termed as* the **metabolic rate**, the value of which depends upon a number of factors such as *individual's health, his physical activity* and *his environments.* The metabolic heat produced depends on the rate of food energy consumption in body.

  — A fasting man, the weak or sick, will have less metabolic heat production. If Q_E, Q_R and Q_C are high and +ve, Q_S will become –ve when Q_M is low and hence sick, old weak or fasting man feels more cold when exposed to low temperature, high draft air.

— A man gets fever, when internal body activities increase Q_M to such an extent that Q_S becomes +ve for given Q_E, Q_R and Q_C.

— The stored energy Q_S has the maximum and minimum limits which when exceeded brings death.

— Body temperature when exceeds 40.5°C and falls allow 36.7°C is dangerous. The usual body temperature when $Q_S = 0$ is 37°C for a normal man. There is some kind of *thermo-static control* in human body, which tries to maintain temperature of human body at the normal level of 37°C.

- Whenever the environment temperature changes, the human body reacts by trying many physiological adjustments to keep the deep temperature of the body constant. There are two types of adjustments generally adopted by the human control system.

  (*i*) **Vasomotor control.** This control *regulates the blood supply to the skin.* It acts by causing vasodilation of the peripheral blood vessels, if environment temperature is increased. The increased circulation of the blood increases the convective transport of heat from the interior of the body to the surface. In other words, the thermal conductance of the skin is increased, thus minimising body heating. The vasomotor mechanism alone is sufficient to maintain the heat balance at a low level of heat load, and is therefore known as *"First Line of Defence"*.

  If environment temperature is decreased, the control causes vasocontraction of the peripheral blood vessels. This decreases blood circulation and reduces body cooling.

  (*ii*) **Sudomotor Control.** This control *regulates sweat production.* Whenever the heat loss by convection and radiation becomes negative due to high temperature of atmosphere compared to body temperature, the only mode of heat transfer to dissipate the heat is by evaporation.

  The sudomotor control *acts by initiating sweat gland activity.* The sweating capacity differs according to persons degree of acclimatization to heat and work.

> **Note:** ***Effect of activities on the heat load.*** The rate at which body produces heat is called the *metabolic rate.* The metabolic rate depends upon the type of activity of the human body. The heat produced (the heat load) by a normal healthy person while sleeping is of the order of 60 W. It increases when the human body is active, the value of heat load being dependent on the severity of activity. In extreme case it can be as high as 600 W, say for a person performing very heavy exercise.
>
> The metabolic heat load is both, sensible and latent. For different types of activities the heat loads have been found experimentally, and are available in databook.

13.2 FACTORS AFFECTING COMFORT

Air-conditioning of buildings mainly concerns the comfort of people and not the maintenance of exact conditions as required for products. The conditions conductive to comfort depend upon the following factors:

1. Temperature
2. Humidity
3. Air motion
4. Air purity.

Temperature. Proper control of temperature of the air medium surrounding the body removes a physiological stress of accommodation, thereby making for greater comfort and improved physical well-being and health.

Humidity. It is through evaporation from skin that a large proportion of body heat is lost. Since evaporation is promoted by a low relative humidity of the air and is related by a high humidity control has an important effect on the comfort. Extremes of humidity not only results undesirable physiological relations but also affect (usually adversely) the properties of many substances in the treated space, clothing and furniture in particular.

Air motion. Air movement over the body increases rate of heat and moisture dissipation above the still-air rate, thereby modifying the feeling of warmth or coldness. Depending on the amounts of motion there are pleasing and displeasing effects.

Air Purity. People feel uncomfortable when breathing contaminated air even if it is within acceptable temperature and humidity ranges. Therefore proper filtration, cleaning and purification of air is necessary to keep it away from dust, dirt and other properties.

13.3 EFFECTIVE TEMPERATURE

In addition to major four factors of *temperature, humidity, air movement, purity of air* there are several other factors such as *clothing, economical status of persons, climate* (a person living in the cold climates takes about 7 to 9 days to become acclimatized to the hot environments) to which a person is accustomed to, etc., which influence human comfort and deserve due consideration. Unfortunately it is not practically feasible to study the combined effects of all the factors on human comfort. Therefore, it is usual practice to consider the aforesaid four factors for most applications which are included in a term called the '*effective temperature*'.

- **Effective Temperature is** *defined as a sensory index that combines into a single factor the effects of temperature, humidity and air movement on human comfort in a noise free pure air environment.* The effective temperature corresponds to the *dry-bulb temperature of the saturated air at which a given percentage of people feel comfortable.* For *example* at 21°C, RH = 100%, and air movement 8m/min. most people feel comfortable.

  Although a particular effective temperature may have a humidity ranging all the way from 0 to 100% and air motion from still air to high velocities, it does not follow that all these combinations are equally comfortable. *Each combination produces the same sensation of warmth or coolness but other affects may contribute a feeling of discomfort.*

  — Too low a humidity produces a parched condition of skin, mouth, and nose.

  — *Too high humidity* causes an accumulation of moisture in clothing and promotes the production of body odours.

  — *High air motion* will usually cause annoyance because of its directional effects such as blowing of hair.

  — When the ranges of humidity and air motion are limited to avoid these physiological or psychological regions of discomfort the effective temperature index may be said to measure comfort fairly closely. This is on the assumption that air purity and noise level is maintained at an acceptable standard.

- Refer to Fig. 13.1. This chart shows as to how the human comfort remains invariant with temperature of confinement when the air motion and humidity are changed. For example, if a room is maintained at 25°C, a man feels the same comfort provided the moisture and air movement are increased. Similarly the effects of other parameters can also be explained. The amount of moisture is varied by controlling the wet bulb temperature. The above effects can be better understood with the help of thermal analysis for human body.

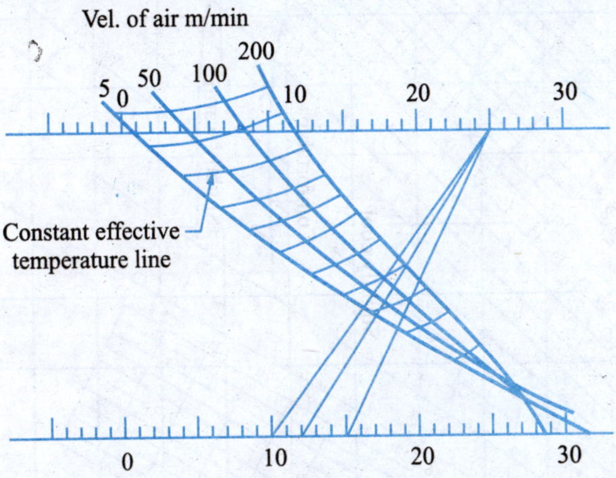

Fig. 13.1. Effective temperature chart.

- *The empirically limited effective temperature chart is known as* **'Comfort chart'**. This chart is prepared on the two scales (Fig. 13.2) having constant relative humidity lines. The effective temperature is represented by *broken lines*. The per cent of people feeling comfortable is shown for summer as well as winter season. Effective temperature for summer air-conditioning is 21°C when 100% people feel comfortable. On the other hand for winter air-conditioning the temperature has to be kept at about 21.7°C for 100% comfort. However, the statistical study on the comfort air-conditioning shows that in the slightly cool condition the comfort of human beings is not influenced by moisture from RH = 30% to 90%. But as the *temperature increases, the comfort is influenced by moisture content.* At t_{db} = 25.3°C, 70% people feel comfortable from RH = 25% to 70%. But any moisture increase beyond this value requires reduction in dry bulb temperature in order to maintain the level of comfort. At around t_{db} = 28°C slightly warm condition and t_{db} = 30.5°C, 40% people feel comfortable from RH = 25% to 60%. Thereafter the increase in moisture requires decrease in DBT.

13.4 FACTOR GOVERNING OPTIMUM EFFECTIVE TEMPERATURE

The optimum effective temperature is affected/governed by the following important factors:

1. **Climatic and seasonal difference.** It is generally agreed that people living in the colder climates are comfortable in effective temperatures lower than for the persons living in warmer regions. The optimum indoor effective temperature appears to bear some relation to the average outdoor effective temperature. This changes slowly with the season of the year, and day to day variations seem to have relatively little effect. The total shift in the optimum from summer to winter is shown in the comfort chart.

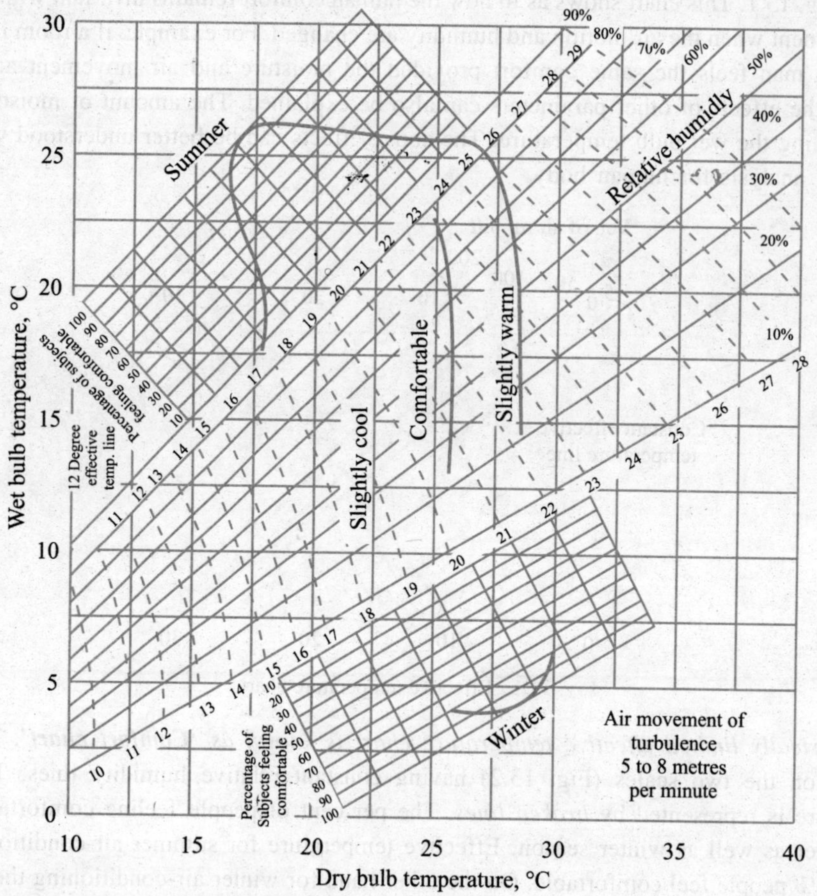

Fig. 13.2. Comfort chart.

2. **Duration of occupancy.** By experience it has been observed that the shorter the occupancy, the more the effective temperature line tends to shift from the continuous-occupancy optimum in the direction of the outdoor effective temperature. This shift, while not great, is large enough to be taken into account.
3. **Clothing.** In winter, much of the effect due to climate and variable occupancy is compensated, because people can dress comfortable to suit both outdoor and indoor conditions by merely adding or removing garments. In considering any application, therefore, allowance should be made for the amount of clothing worn. For example, a formal dining room with women present in evening dress will require a high temperature than the average restaurant.
4. **Age and sex.** As compared to men, women require 0.5°C higher effective temperature. Furthermore, both men and women over forty on the average require an effective temperature about 0.5°C higher than do those below forty.
5. **Activity.** The desirable room effective temperature is influenced by the rate of bodily activity. The men performing heavy manual labour in factories need a lower temperature than occupants seated at rest in auditoriums.

6. **Radiant heat.** The effect of body radiant heat from person to person particularly in a crowded place like theatre or an auditorium is large enough to *require a slight lowering in effective temperature* that could otherwise be maintained.

7. **Latitude.** With every 5° reduction in latitude 0.5°C increase in effective temperature is desired.

8. **Shock effect.** The shock effect is due to sudden entrance from or exit to outdoors, and is likewise mostly a summer problem. This effect is due to two factors: (*i*) relatively inflexibility in the amount of clothing so that evaporation of perspiration makes for a chilling effect and (*ii*) outdoor exertion gives beneficial warmth in winter and objectionable overheating of the body in summer. Careful test have proved, however, that the shock effect is not physiologically harmful.

HIGHLIGHTS

1. Thermodynamically a human body feels comfortable when the heat produced by the metabolism of human body is equal to the sum of heat dissipated to the surroundings and the heat stored in the human body by raising the temperature of body tissues.
2. The following factors affect human comfort:
   (*i*) Temperature (*ii*) Humidity (*iii*) Air motion (*iv*) Air purity.
3. *Effective temperature* is defined as a sensory index that combines into single factor the effects of temperature, humidity and air movement on human comfort in a noise free pure air environment.
4. Effective temperature chart shows as to how the human comfort remain invariant with temperature of confinement when the air motion and humidity are changed.
5. The empirically limited effective temperature chart is known as '*comfort chart*'. It is prepared on the t_{db} and t_{wb} scales, having constant relative humidity lines.

OBJECTIVE TYPE QUESTIONS

A. Choose the Correct Answer:

1. For same degree of comfort with same type of clothes worn
   (*a*) men need higher effective temperature than women
   (*b*) men need lower effective temperature than women
   (*c*) men need same effective temperature as women
   (*d*) any of the above is correct.
2. The index which correlates the combined effects of air temperature, relative humidity and air velocity on the human body is known as
   (*a*) mean radiant temperature (*b*) effective temperature
   (*c*) dew point temperature (*d*) none of these.
3. Dry bulb temperature is 20°C, wet bulb temperature is 20°C, then effective temperature is
   (*a*) 20°C (*b*) less than 20°C (*c*) more than 20°C (*d*) not determinable.
4. The optimum effective temperature for human comfort is
   (*a*) higher in winter than in summer
   (*b*) lower in winter than in summer
   (*c*) same in winter and summer
   (*d*) not dependent on season.

5. The optimum effective temperature for human thermal comfort in the conditioned space
   (*a*) is lower in winter than in summer
   (*b*) is higher in winter than in summer
   (*c*) is same for winter and summer
   (*d*) is not dependent on season.
6. Dry bulb temperature is 25°C and wet bulb temperature is 25°C and the velocity of air passing human body is 6 m/min. If the velocity increases to 20m/min
   (*a*) the effective temperature decreases
   (*b*) the effective temperature increases
   (*c*) the effective temperature remains same
   (*d*) the effective temperature may increase or decrease depending upon relative humidity.
7. When the temperature of the surroundings is higher than the temperature of the body, then the heat loss by convection from the body to the surroundings will be
   (*a*) positive (*b*) negative (*c*) zero (*d*) any of these
8. The heat production from a normal healthy man when asleep is about
   (*a*) 20 watts (*b*) 40 watts (*c*) 60 watts (*d*) 80 watts.
9. If dry bulb temperature is 20°C and wet bulb temperature is 17°C, the effective temperature is 19°C. If wet bulb temperature is 17°C and dry bulb temperature is 21°C, the effective temperature:
   (*a*) will be less than 19°C (*b*) will be more than 19°C
   (*c*) will be equal to 19°C (*d*) cannot be predicted with above data.
10. Effective temperature depends on
    (*a*) only dry bulb temperature
    (*b*) dry bulb temperature and wet bulb temperature
    (*c*) dry bulb temperature and relative humidity
    (*d*) dry bulb temperature, wet bulb temperature and air motion.
11. Relative humidity of 50% is desired in the air-conditioned space during winter and summer. The optimum effective temperature for summer and winter should respectively be
    (*a*) 20°C and 22°C (*b*) 22°C and 20°C (*c*) 22°C and 22°C (*d*) 20°C and 20°C.
12. In summer comfort cooling, the air of the occupied space should not have a relative humidity above
    (*a*) 30% (*b*) 40% (*c*) 50% (*d*) 60%.
13. If dry bulb temperature is 20°C and wet bulb temperature is 17°C the effective temperature is 19°C. If dry bulb temperature is 20°C and wet bulb temperature is 18°C, the effective temperature
    (*a*) will be lass than 19ºC (*b*) will be more than 19°C
    (*c*) either (*a*) or (*b*) (*d*) none of these.
14. At the same dry bulb temperature
    (*a*) effective temperature increases with decrease in relative humidity
    (*b*) effective temperature decreases with decrease in relative humidity
    (*c*) effective temperature remains constant with decrease in relative humidity
    (*d*) effective temperature has no dependence on relative humidity.

B. Fill in the Blanks:

1. The rate at which the body produces heat is termed as the ....
2. There is some kind of .... control in a human body, which tries to maintain temperature of human body at the normal of 37°C.
3. The conditions conductive to comfort depend upon the following four factors:
   (*a*) .... (*b*) .... (*c*) .... (*d*) ....

4. A sensory index that combines into a single factor the effects of temperature, humidity and air movement on human comfort in a noise free pure environment is called .... temperature.
5. The empirically limited effective temperature chart, is known as .... chart.
6. In 'comfort chart' the effective temperature is represented by .... lines.
7. As compared to men, women require .... higher effective temperature.
8. Both men and women over forty on average require an effective temperature about .... higher than do those below forty.
9. The men performing heavy manual labour in factories need a .... effective temperature than occupants seated at rest in auditoriums.
10. With every 5° reduction in latitude .... increase in effective temperature is desired.

ANSWERS

A. Choose the correct Answer:

1. (*a*) 2. (*b*) 3. (*a*) 4 (*b*) 5. (*a*) 6. (*a*) 7. (*b*) 8. (*c*)
9. (*a*) 10. (*d*) 11. (*b*) 12. (*d*) 13. (*b*) 14. (*b*).

B. Fill in the Blanks:

1. Metabolic rate; 2. Thermostatic;
3. Temperature, humidity, air motion and air purity; 4. Effective;
5. Comfort; 6. Broken; 7. 0.5°C;
8. 0.5°C; 9. lower; 10. 0.5°C.

THEORETICAL QUESTIONS

1. Write a short note on 'thermodynamics of a human body'.
2. State and explain the factors which affect comfort.
3. What is an 'Effective temperature'? Explain briefly 'Effective temperature chart' and 'Comfort chart'.
4. State and explain factors which govern optimum effective temperature.
5. Sketch 'comfort chart' and show on it the 'comfort zone'.
6. Explain briefly as to how the human body reacts to changes in temperature of human body. Also explain the effect of activities on the heat load calculation for 'comfort' calculation.

❑❑❑

Chapter 14

Air-Conditioning Systems

14.1 INTRODUCTION

A **air-conditioning system** *is defined as an assembly of different parts of the system used to produce a specified condition of air within a required space of building.*

The *basic elements of air-conditioning systems* (of whatever form) are:

1. **Fans:** For moving air.
2. **Filters:** For cleaning air, either fresh, recirculated or both.
3. **Refrigerating plant:** Connected to heat exchange surface, such as finned coils or chilled water sprays.
4. **Means for warming:** The air, such as hot water or steam heated coils or electrical elements.
5. **Means for humidification; and/or dehumidification.**
6. **Control system:** To regulate automatically the amount of cooling or warming.

14.2 AIR-CONDITIONING CYCLE

Refer to Fig. 14.1. An air-conditioning cycle comprises the *following steps*:

1. The fan forces air into duct-work which is connected to the openings in the room. These openings are commonly called *outlets or terminals.*
2. The duct-work direct the air to the room through the outlets.
3. The air enters the room and either heats or cools as required. Dust particles from the room enter the air stream and are carried along with it.
4. Air then flows from the room through a second outlet (sometimes called the return outlet) and enters the return duct-work, where dust particles are removed by a *filter.*

5. After the air is cleaned, it is either heated or cooled depending upon the condition in the room. If cool air is required, the air is passed over the surface of a *cooling coil*; If warm air required, the air is passed through a *combustion chamber* or over the surface of a *heating coil*.
6. Finally the air flows back to the fan, and the cycle is completed.

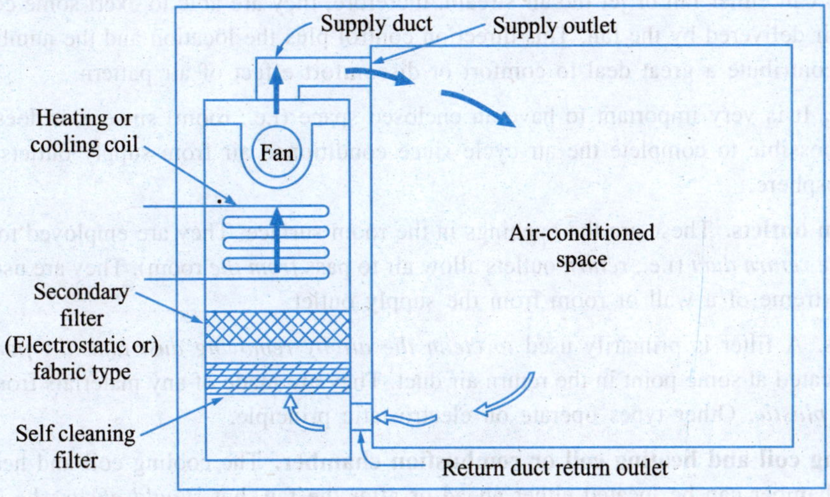

Fig. 14.1. Air-conditioning cycle.

The *main parts* of the equipment in the air-conditioning cycle are:

1. Fan;
2. Supply ducts;
3. Supply outlets;
4. Space to be conditioned;
5. Return outlets;
6. Return ducts;
7. Filter;
8. Heating chamber or cooling coil.

1. **Fan:** The primary function of a fan is to move air *to and from the room*. The air that a fan moves in air-conditioning system is made up of :

(*i*) All outdoor air

(*ii*) All indoor room air (called recirculated air)

(*iii*) A combination of outdoor and indoor air.

The fan *pulls* air from outdoors or from the room but in most systems it *pulls air* from both sources at the same time. The amount of air supplied by the fan must be regulated since drafts in the room cause discomfort, and poor air movement slows the heat rejection processes. This can be achieved by (*i*) choosing a fan that can deliver the correct amount of air, and (*ii*) by controlling the speed of the fan so that air stream in the room provides good circulation but does not cause drafts. Of course, the fan is only one of the pieces of equipment that contributes body comfort.

2. **Supply duct.** The function of a supply duct is to direct the air from fan to the room. In order that air may flow freely it should be *as short as possible* and have *minimum number of turns.*

3. **Supply outlets.** The function of supply outlets is to *distribute the air evenly in a room.* These outlets may (*i*) fan the air, (*ii*) direct air in a jet stream and (*iii*) may do a combination of both. Since supply outlets can either fan or jet the air stream, therefore, they are able to exert some control on the direction of air delivered by the fan. This direction control plus the location and the number of outlets in the room contribute a great deal to comfort or discomfort effect of air pattern.

4. **Space.** It is very important to have an enclosed space (i.e., room) since if it does not exist it would be impossible to complete the air cycle since conditioned air from supply outlets would flow into the atmosphere.

5. **Return outlets.** These are the openings in the room surface. They are employed to *allow room air to enter the return duct* (i.e., *return* outlets allow air to pass *from the* room). They are usually located at opposite extreme of a wall or room from the supply outlet.

6. **Filters.** A filter is primarily used *to clean the air by removing dust and dirt particles.* They are usually located at some point in the return air duct. They are made of any materials from *spun glass to composite plastic.* Other types operate on electrostatic principle.

7. **Cooling coil and heating coil or combustion chamber.** The cooling coil and heating coil or combustion chamber can be located either ahead or after the fan, but *should always be located after the filter.* A filter ahead of the coil is necessary to prevent the excessive dirt, dust and dirt particles from covering the coil surface.

Summer operation. The air-conditioning cycle *cools the air* during summer operation. Return air from the room passes over the surface of *cooling coil,* and the air cooled to the required temperature. If there is too much moisture present, it is removed automatically as the air is cooled by the coil.

Winter operation. The air-conditioning cycle *adds heat to the air during winter operation.* This is achieved by passing the return air from the room over the surface of a *heating coil,* etc.

14.3 CLASSIFICATION OF AIR-CONDITIONING SYSTEMS

The air-conditioning systems may be classified as follows:

1. **According to the arrangement of equipment**

   (*i*) Central systems;

   (*ii*) Zoned systems;

   (*iii*) Unitary systems;

   (*iv*) Unitary-central systems.

2. **According to the purpose:**

   (*i*) Comfort air-conditioning system;

   (*ii*) Industrial air-conditioning system.

3. **According to season of the year:**

   (*i*) Winter air-conditioning system;

   (*ii*) Summer air-conditioning system;

   (*iii*) Year-round air-conditioning system.

Another method of classification of air-conditioning system is as follows:

(*i*) Single-air systems

(*ii*) Dual-air systems

(*iii*) Primary-air systems

(*iv*) Unit systems

(*v*) Panel systems.

14.4 CENTRAL SYSTEM

This type of system is suitable for air-conditioning large space such as *theatres, cinemas, restaurants, exhibition halls, or big factory spaces where no sub-division exists.* The central systems are generally employed for the loads above 25 tonnes of refrigeration and 2500 m^3/min of conditioning air. The unitary systems can be more economically employed for low capacity (below 25 tonnes) units.

In central system, the equipment such as fans, coils, filters and their encasement are designed for assembly in the field.

- A *central system serves differing rooms, requires individual control of each room.* The condenser, compressor, dampers, heating, cooling and humidifying coils and fan are *located at one place say basement.*
- The conditioned air is carried to the different rooms by means of supply ducts and returned back to the control plant through return ducts. Part of the supply air to the rooms may be exhausted outdoors.
- Outdoor air enters from an intake which should be situated on that side of the building least exposed to solar heat. If should not be close to the ground or to dust collected roof.
- The air after passing through damper passes through *filters*. The filters may be of a mechanical cleaned type, replaceable-cell type or may be electrostatic.
- The cleaned air then passes to the conditioning equipment in the following order: *Temperature* (*or preheater*) *coil, cooling coil, humidifier* (*air washer*), *heating coils and finally fan.*

Fig. 14.2 shows a schematic diagram of complete (year round) air-conditioning system.

Advantages of Central System:

1. Low investment cost as compared to total cost of separate unit.
2. Space occupied is unimportant as compared to a room unit conditioner which must be placed in the room.
3. Better accessibility for maintenance.
4. The running cost is less per unit of refrigeration.
5. Noise and vibration troubles are less to the people in air-conditioned places as the air-conditioning plant is far away from the air-conditioned places.
6. The exhaust air can be returned and partly reused with obvious saving in heating and refrigeration.

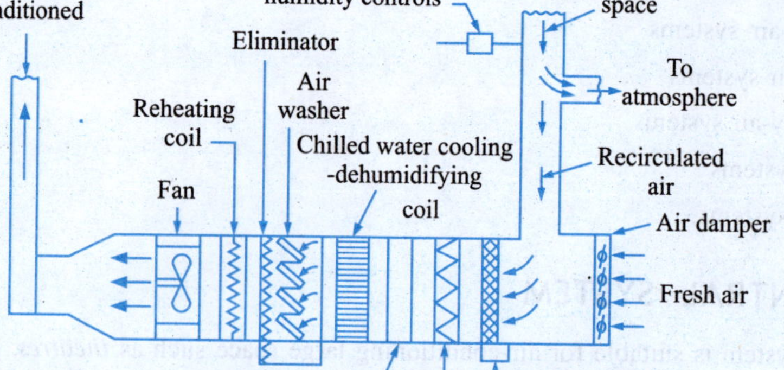

Fig. 14.2. Central system.

Disadvantages:

The main disadvantage of the central system is that it results in *large size ducts which are costly and occupy large space.*

14.5 ZONED SYSTEM

When in a building several rooms or floors are to be served, it is necessary to consider means by which the varying heat gains in the different departments can be dealt with. Some rooms may have some, and other not; some may be crowded and other empty; again some may contain heat-producing equipment. Variations in requirements of this kind are the most common case with which air-conditioning has to deal, and for this central system is unsuitable. *Zoned system is one approach.*

In *zoned system* the building is divided into zones such that as nearly as possible conditions may be expected to exist. *Each zone is provided with its own local recirculating fan and booster cooler or heater, and this unit receives fresh air supply conditioned to some average temperature and corrected for humidity by means of what is in effect a central plant.* Such an arrangement is shown in Fig. 14.3. In this case the central plant is on the roof. The recirculating units are fixed overhead adjacent to the corridor, taking then return air there-from, and distributing ducts are run above the corridor false ceiling delivering into the various rooms, floor by floor each floor, constituting a separate zone. The return air from the room passes through grilles into the corridor which acts as the returning air collecting duct. The cooling or heating booster coils could be served from circulating water mains, each coil being controlled locally according to the requirements of the zone served.

14.6 UNITARY SYSTEMS

Unitary systems are packaged units in which all the equipments and controls, for fulfilling the requirements of air-conditioning, are *assembled in a single casing with decorative covering.* The *components are factory assembled, tested and balanced.* They can be directly installed in or near the vicinity to be air-conditioned. The only utility required is a source of electricity, and for water-cooled units, a supply of cooling water.

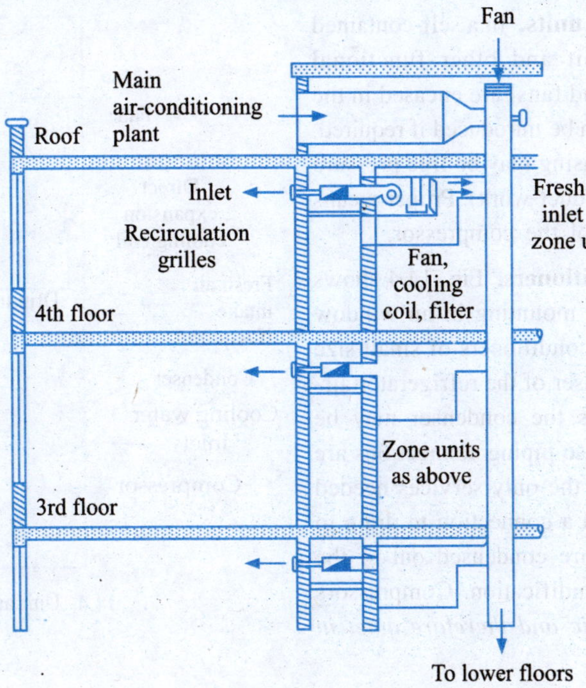

Fig. 14.3. Zoned system.

They are **used** for cooling capacities upto 100 tons (350 kW), with 10 to 50 tons (3.5 kW to 175 kW) installations a common present day practice. *Multiple units are used for zone control due to modular flexibility.*

Various **factory assembled units** available are:

(*i*) Attic (or exhaust) fans.

(*ii*) Remote units.

(*iii*) Self contained units.

(*iv*) Room air-conditioners.

(*v*) Unit air coolers.

(*i*) **Attic fans.** An attic or exhaust fan is a cooling unit without any heat transfer element such as a cooling coil. When the sun sets the temperature of outdoor air reduces to cool levels whereas the indoor temperatures are high. To reduce the inside temperature an attic fan is placed in the attic. It draws outdoor air into several rooms of the building through various doors and windows and finally discharges from attic to outdoors. Consequently a circulation of cool outdoor air is set up in the building. A propeller type fan is usually recommended as it can handle large volume of air at low pressure efficiently.

(*ii*) **Remote units.** A system in which air handling unit is separated from the condensing unit is called a *remote system*. The conditioning or air-handling unit is called *remote unit*. *It* consists of a fan (either propeller or centrifugal type) with its driving motor, cooling coil, heating coil, filters, drip pan, louvers etc. with or without the duct connections at the outlet. Remote units are available in capacities ranging from 2 to 100 tonnes. These units are available for floor mounting or for suspension from ceiling. Some remote units have air-washing and coil wetting features.

(*iii*) **Self contained units.** In a self-contained unit the condensing unit and other functional elements (such as coils and fans) are encased in the same cabinet. Fresh air can be introduced if required. The discharge from the casing may be free pressure type (i.e., with or without duct-work). Proper means should be adopted to cool the compressor.

(*iv*) **Room air-conditioners.** Fig. 14.4 shows a unit air-conditioner for mounting in an window or wall bracket. Unit air-conditioners of small size generally have the condenser of the refrigerator air-cooled but in large sizes the condenser may be water-cooled, in which case piping connections are required. Apart from this the only services needed are an electric supply and a connection to drain to conduct away any moisture condensed out of the atmosphere during dehumidification. Compressors in most units are *hermetic and therefore quiet in running*.

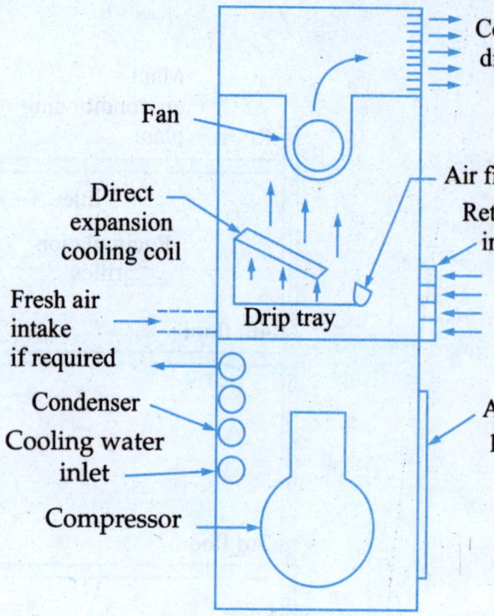

Fig. 14.4. Unit air-conditioner.

Units of considerable size are suitable for industrial applications, in which case ducting may be connected for distribution.

(*v*) **Unit air-coolers.** A unit air-cooler is a special application of remote units. It primarily reduces the temperature in insulated and sealed storage rooms. The rating of these coolers is the basic rating expressed in kJ per hour per C-degree temperature differential between the refrigerant and air. A defrosting coil is necessary when the temperature is below 2°C. It may be mounted on the floor or wall or suspended from the ceiling.

Advantages of unitary system. The unitary system commands following **advantages** over the central system.

1. There is saving in the installation and assembly labour charges.
2. Zoning and duct-work eliminated.
3. In unitary system exact requirement of each separate room is met where as in central system the individual needs of separate rooms cannot be met.
4. Failure of the unit puts off conditioning in only one room whereas the failure of the central plant off-set all the rooms to be served.
5. Only those rooms which need cooling will have their units running, whereas the central plant will have to run all the time for the sake of only a few rooms.
6. The specific feature of a unitary system is that there is *individual room-temperature control.*

14.7 UNITARY–CENTRAL SYSTEMS

In a unitary-central system *each room is provided with a room unit which gets a supply of conditioned air from a central system. The main aim of such systems is to either decrease the size of the ducts or to eliminate them completely.*

The following three unitary-central systems are in common use :

(*i*) Induction units.

(*ii*) All-air high velocity system.

(*iii*) Fan coil units.

(*i*) **Induction units.** An induction units *uses the principle of induction as a means of recirculation.* Each induction unit receives primary conditioned air under pressure from a central plant.

Fig. 14.5 shows an induction unit (*generally placed below the window*). The primary air, in summer is treated in a surface or spray dehumidifier and dehumidified to sufficiently low dew point temperature to compensate for the latent heat gain. The primary air, in winter is treated in a humidifier to a sufficiently high dew point to take care of latent heat losses. The water to the cooling coil in the induction room unit is supplied from a central plant by means of a centrifugal pump. In summer this water is chilled in a central water cooler and in winter it is heated by means of a water heater centrally located. The quantity of water supplied to the induction coil can be controlled by means of a room thermostat for controlling the room temperature.

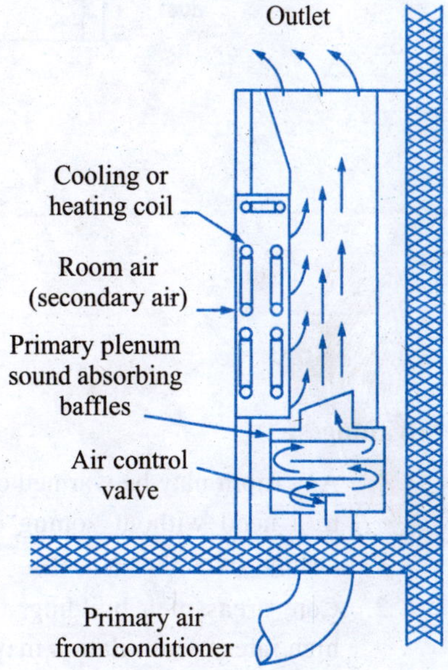

Fig. 14.5. Induction unit.

The high velocity of primary air jets induces a secondary air stream to flow from the room through the coil of the induction unit. The room air is therefore treated to proper temperature while it flows through the coil. The mixture of primary (20%) and secondary (80%) air is then delivered freely at the outlet of the unit. The quantity of primary air required is greatly reduced since the secondary room air takes up part of the room load. The primary air is usually sufficient for ventilation purposes and consists wholly of outdoor air. Consequently there is no need to have return air ducts and zoning of primary air and water circuits.

(*ii*) **All-air high velocity system.** In these systems special control and acoustic equipments are employed since they use high static pressures and high duct velocities. The ducts used are round and are designed carefully in order to reduce friction and noise.

All-air high velocity systems are of the following two types:

1. Single-duct system
2. Dual-duct system.

1. **Single duct system.** Fig. 14.6 shows a schematic diagram of a single-duct system. Each room in a zone has a volume control on the room outlet box which varies the volume of air supplied to each individual room. From the outlet box, the high velocity air is fed into a diffuser. The box itself contains some sound control equipment.

2. **Dual-duct system.** In this system use is made of two ducts, one *conveying warm air and one conveying cool air and* each room contains a blender so arranged with air valves or dampers that all warm, all cool or some mixture of both is delivered into the room. Fig. 14.7 illustrates such a unit, incorporated there in being a means of automatically recirculating the total air-delivery, such that,

regardless of variations in pressure in the system, each unit delivers it correct air quantity. Referred to as 'constant volume control', it is an essential part of this system.

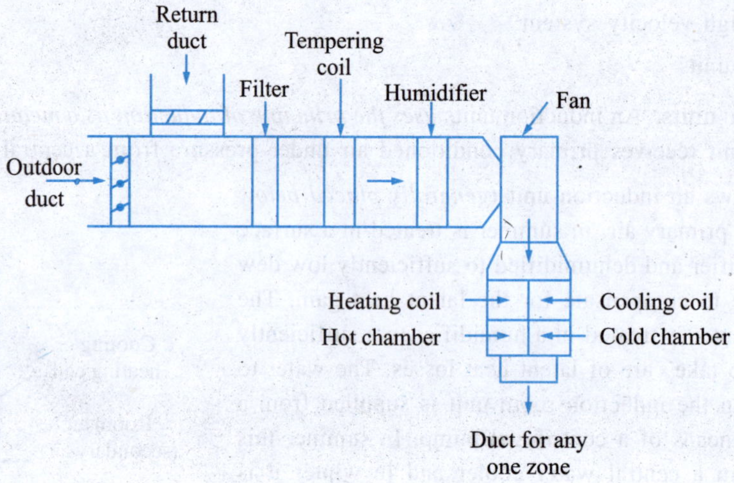

Fig. 14.6. Single-duct system.

Advantages:

1. Any room may be warmed or cooled according to a need without zoning or any problem of change-over thermostats.
2. Core areas of a building, or rooms requiring high rates of ventilation may equally be served off the same systems, no separate plants being necessary.

Disadvantages:

1. Ducts are larger and fan blowers greater.
2. Owing to the greater quantity of air in circulation throughout the building, dual-duct system usually incorporates means for circulation back to the main plant, and this involves return air ducts and shafts in some form.

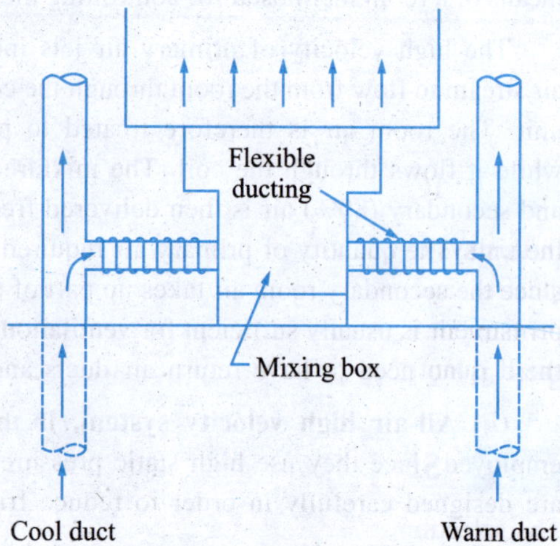

Fig. 14.7. Dual-duct system.

3. Ducts both warm and cool, need to be installed thermally whereas with some forms of induction systems this can be dispensed with.

(*iii*) **Fan-coil units.** The fan-coil units which are installed in individual room are supplied with primary air. The return room air mixes with then primary air and the mixture is then blown over coil by a fan located in the unit. The operation is just similar to induction coil units. The room thermostat is heating-cooling type. It controls the quantity of hot water in winter and chilled water is supplied to the room coil.

These units, like induction units, are located around the periphery of a building, usually under windows.

Difference Between Central and Packaged Air-conditioning Systems. The differences between the central and packaged air-conditioning systems are given below:

S.No.	*Aspects*	*Central air-conditioning system*	*Packaged air-conditioning system*
1.	*Installation*	The primary components are grouped together in a central machine room, and the conditioned air is sent from here to the spaces to be conditioned through extensive duct-work.	They are directly installed in, or immediately adjacent to, the space to be conditioned.
2.	*Assembly*	They are erected in the field following the carefully prepared plans and specification of a licensed engineer.	They are factory assembled in a decorative casing of sheet metal or fibre glass
3.	*Testing and balancing*	They are tested and balanced on site.	They are tested and balanced in the factory.
4.	*Cooling of condenser*	The condenser is invariably water-cooled, and a boiler is used for supplying hot water/steam for heating.	The condenser is usually air-cooled. In bigger sizes (more than 20 tonnes capacity) it may be evaporative type or even water-cooled.
5.	*Applications*	They are used for large capacity plants exceeding 180 tonnes cooling capacity and 2500 m^3/min of conditioned air.	They are used for smaller capacity (generally upto 45 tonnes cooling)

14.8 ICE SYSTEM OF AIR-CONDITIONING

In this system cooling is done by melting ice. This is a practical air-conditioning method for churches, theatres and public halls that have short operating hours and relatively high peak loads. Since the investment in mechanical refrigeration equipment is expensive for short period, ice fills the need. A small quantity of ice in a water-cooling tank can release refrigeration at a rapid rate.

14.9 SELECTION OF SYSTEM

The selection of system type is determined by a great variety of design requisites, and the final choice may vary from a minimum-cost, summer-relief cooling installation to a closely controlled, year-round air-conditioning system. The more important factors to be considered in the selection of a system may be classified as follows:

1. **General:**

 (*i*) New or existing building.

 (*ii*) Nature of building occupancy.

 (*iii*) Type and construction of building.

 (*iv*) Relative magnitudes of cooling air heating loads, particularly including the highly variable effect of solar load.

 (*v*) Ratio of interior to peripheral zones.

2. **Functional:**
   (*i*) Independent control of temperature and humidity.
   (*ii*) Zone or individual room control.
   (*iii*) Simultaneous heating and cooling capacity for reversal of intermediate season loads on exterior zones.
   (*iv*) Air-distribution in space, including effect of control method.
   (*v*) Ventilation-air supply, including flushing action for concentrated loads or contamination sources.
   (*vi*) Mixing of air between spaces.
   (*vii*) Cleanliness of supply air.
   (*viii*) Noise and vibration.
   (*ix*) Adaptability to partial-load operation.
   (*x*) Local code requirements.
3. **Initial cost:**
   (*i*) Required refrigeration capacity.
   (*ii*) Apparatus and equipment requirements.
   (*iii*) Cost of duct-work, piping and miscellaneous services.
   (*iv*) Possible use of existing heating system, ventilation ducts, piping etc.
   (*v*) Costs of required alterations to existing building.
   (*vi*) Cost of design engineering.
4. **Operating and owning cost:**
   (*i*) Relative costs of power, steam, fuel and water.
   (*ii*) Possible use of 100% outside air for cooling without refrigeration during intermediate seasons.
   (*iii*) Amount of cancellation of cooling effect by heating for control purposes.
   (*iv*) Service and maintenance requirements.
   (*v*) Rental value of space occupied by air-conditioning equipment, ducts and pipes.
   (*vi*) Rate of depreciation and obsolescence of equipment.
   (*vii*) Insurance, taxes and financing costs.
5. **Space requirements:**
   (*i*) Equipment and apparatus space size and locations.
   (*ii*) Duct and pipe sizes and locations.
   (*iii*) Possible use of marginal spaces, such as abandoned elevator shafts, closets etc.
6. **Installation problems:**
   (*i*) Disturbance of occupants of existing building.
   (*ii*) Adaptability to progressive or partial installation.
   (*iii*) Adaptability to installation of equipment in isolated or remote areas.
   (*iv*) Skills required by installation mechanics.

7. Miscellaneous:

(*i*) Flexibility for architectural and partition changes.

(*ii*) Simplicity of operation and maintenance.

(*iii*) Appearance.

14.10 ROOM (OR CONDITIONED SPACE) SENSIBLE HEAT FACTOR (RSHF)

"Room sensible heat factor (RSHF)" *is the ratio of room sensible heat to the sum of room sensible heat and the room latent heat.* In other words,

$$\text{RSHF} = \frac{\text{RSH}}{\text{RSH} + \text{RLH}} = \frac{\text{RSH}}{\text{RTH}} \quad \text{...(14.1)}$$

where, RSH = Room sensible heat,

RLH = Room latent heat, and

RTH = Room total heat.

The air which is to be supplied to the conditioned space must have the *capacity to offset simultaneously both the room sensible and room latent heat loads.*

The room and supply air conditions to the space may be plotted on the psychrometric chart as shown in Fig. 14.8. and the points connected by a straight line. This represents the psychrometric process of the supply air within the conditioned space and is called "*Room sensible heat factor line*". The slope of the line gives the ratio of sensible heat load to latent heat load. Thus supply air having dry bulb and wet bulb temperatures, such that the condition point falls on this line, will satisfy the requirements of the room with adequate supply of such air. Thus the supply air with properties as at

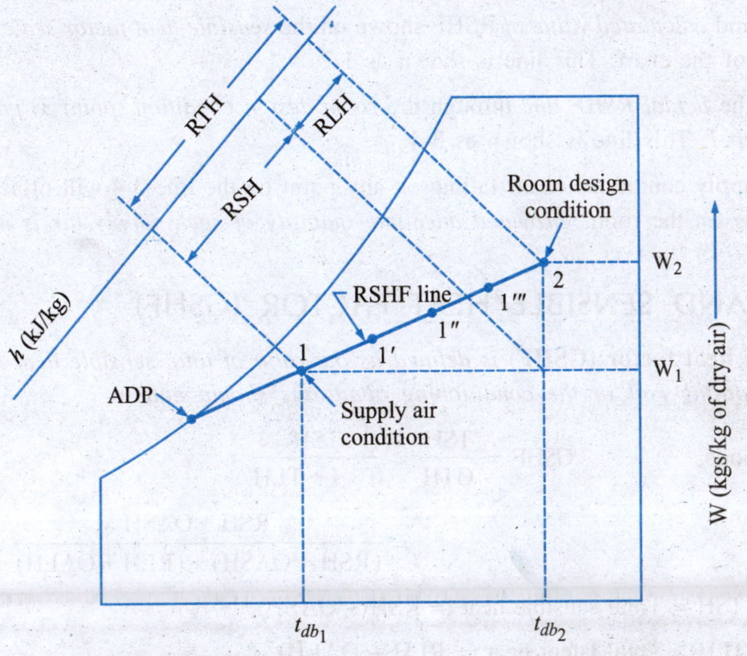

Fig. 14.8

1, 1′, 1″ and 1‴ will satisfy the requirement, but only the quantities of air will be different for different supply of point. Supply condition at 1 requires less air than at 1′ and at 1‴ it is highest of all four supply points.

When the supply air conditions are not known (which is generally required to be found out) the RSHF line can be drawn from the calculated value of RHSF as follows: Refer Fig. 14.9.

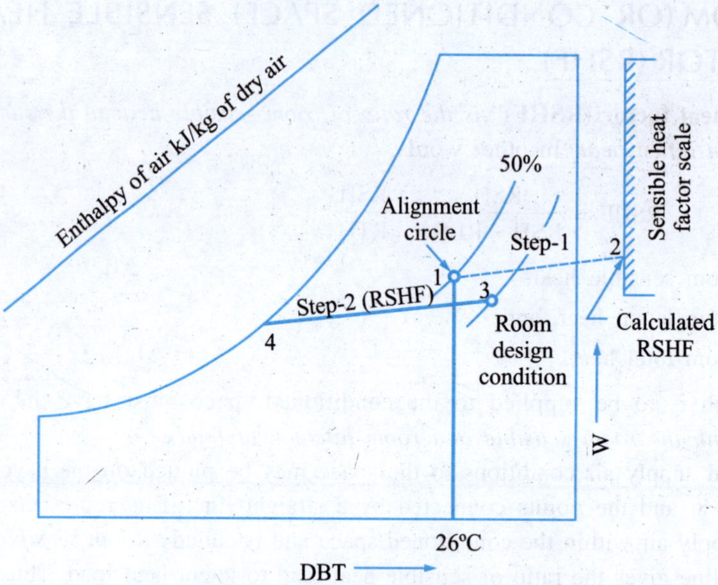

Fig. 14.9. Room sensible heat factor (RSHF).

Step-1: Draw a base line through the alignment circle (28°C DBT and 50% RH-shown on the psychrometric chart) and *calculated value of* RSHF shown on the *sensible heat factor scale* in the upper right corner of the chart. This line is shown as 1-2.

Step-2: Draw the *actual RSHF line* through the *room design condition (point 3) parallel to the base line step-1.* This line is shown as 3-4.

Thus the supply conditions of air falling on any point on the line 3-4 will offset the sensible and latent heat loads on the room, *provided adequate quantity of such supply air is made available.*

14.11 GRAND SENSIBLE HEAT FACTOR (GSHF)

Grand sensible heat factor (GSHF) *is defined as the ratio of total sensible heat to the grand total heat that the cooling coil or the conditioning apparatus should handle.*

In other woods,

$$\text{GSHF} = \frac{\text{TSH}}{\text{GTH}} = \frac{\text{TSH}}{\text{TSH} + \text{TLH}}$$

$$= \frac{\text{RSH} + \text{OASH}}{(\text{RSH} + \text{OASH}) + (\text{RLH} + \text{OALH})}$$

where, TSH = Total sensible heat (= RSH + OASH),

TLH = Total latent heat (= RLH + OALH),

GTH = Grand total heat (=TSH + TLH),

OASH = Outside air sensible heat, and

OALH = Outside air latent heat.

OATH = Outside air total heat (= OASH + OALH)

Also, $$\text{OASH} = 0.02044\ V_1\ (t_{db_1} - t_{db_2})\ \text{kW},$$

$$\text{OALH} = 50\ V_1\ (W_1 - W_2)\ \text{kW}$$

$$\text{OATH} = \text{OASH} + \text{OALH}$$

OATH may also be calculated from the following relation:

$$\text{OATH} = 0.02\ V_1\ (h_1 - h_2)\ \text{kW}$$

Refer to Fig. 14.10.

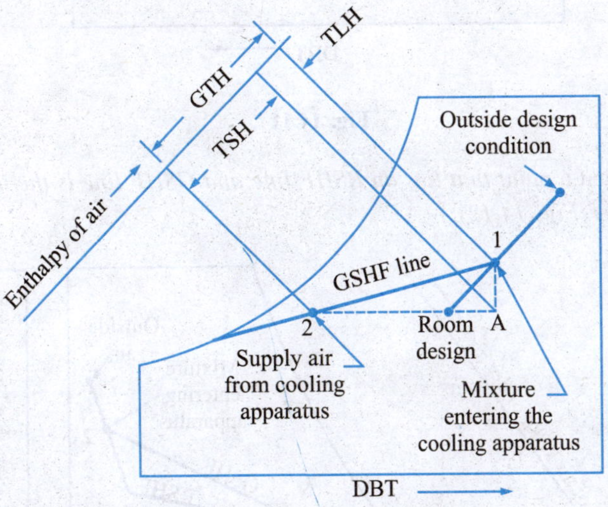

Fig. 14.10. Grand sensible heat factor.

- Under *actual conditions, some outside air for ventilation and some infiltrated air from outside mixes with the room air and subsequently this outside air also adds sensible and latent heat load to the room which is having its internal sensible and latent load.* Thus these mixture conditions reach the cooling coil or conditioning apparatus, get cooled and dehumidified and leave the apparatus (cooling coil) with reduced temperatures and moisture content. The fall in temperature and moisture content is determined by the total sensible heat and latent heat required to be handled by the apparatus.
- If the condition of the air entering the apparatus (cooling coil) and the condition of air leaving the apparatus is joined by a straight line 1–2, this gives the GSHF lines. This line 1–2 gives the psychrometric process undergone by air as it passes over the apparatus. Thus any point on this line will give sensible heat, and the latent heat removed in a fixed proportion as shown in Fig. 14.10.
- If GSHF and the mixture condition entering the apparatus are given, the GSHF line can be drawn with the help of sensible heat factor scale in the same way as explained for RSHF line. This is illustrated in Fig. 14.11.

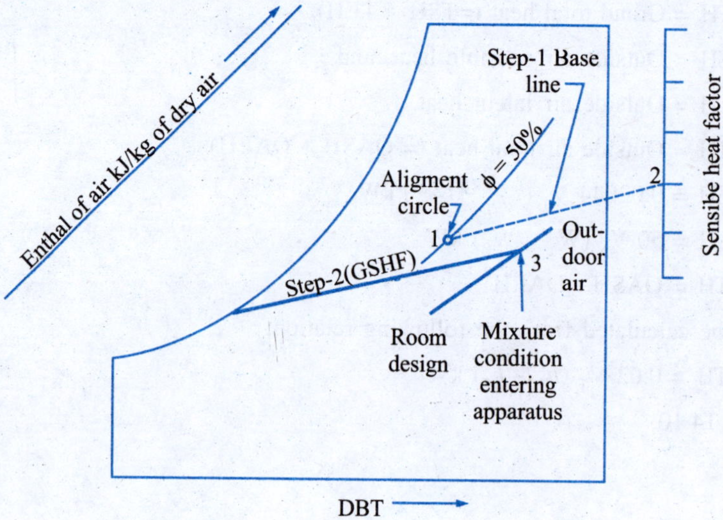

Fig. 14.11

- It may be noted that a *point that lies on RSHF line and GSHF line is the ideal supply conditions to the room* (*Refer Fig. 14.12*).

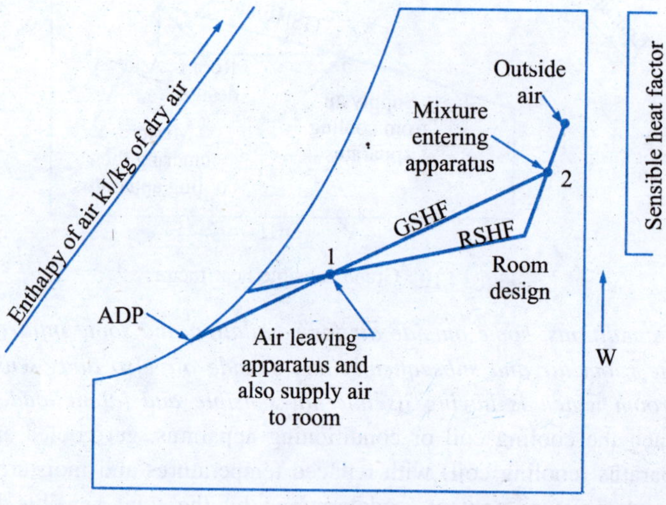

Fig. 14.12

Effective Room Sensible Heat Factor (ERSHF):

Effective room sensible heat factor (ERSHF) *is defined as the ratio of the effective room sensible heat to the effective room total heat.* In other words,

$$\text{ERSHF} = \frac{\text{ERSH}}{\text{ERTH}} = \frac{\text{ERSH}}{\text{ERSH} + \text{ERLH}}$$

where, ERSH = Effective room sensible heat

$= RSH + OASH \times BF$

$= RSH + 0.02044\ V_1\ (t_{db_1} - t_{db_2})\ BF$

ERLH = Effective room latent heat

$= RLH + OALH \times BF$

$= RLH + 50\ V_1\ (W_1 - W_2)\ BF$

ERTH = Effective room total heat

$= ERSH + ERLH$

BF = By-pass factor.

14.12 APPLICATIONS OF AIR-CONDITIONING

Following the are the applications of air-conditioning:

1. Industrial applications:
   (*i*) Food industry.
   (*ii*) Photographic industry.
   (*iii*) Textile industry.
   (*iv*) Printing industry.
   (*v*) Machine tool industry.
2. Theatres air-conditioning.
3. Departmental-store air-conditioning.
4. Transport air-conditioning:
   (*i*) Automobile air-conditioning.
   (*ii*) Train air-conditioning
   (*iii*) Air craft air-conditioning.
   (*iv*) Ship air-conditioning.
5. Air-conditioning for television centre.
6. Air-conditioning for computer centre.
7. Air-conditioning of automatic telephone exchange building.
8. Museum air-conditioning.
9. Hospital air-conditioning.

WORKED EXAMPLES

Example 14.1. *Air at 12°C DBT and 85% RH is to be brought to 36°C DBT and 23.2°C WBT with the help of winter air-conditioner. If the humidified air comes out of the humidifier at 85% RH, draw the various processes involved on a skelton psychrometric chart and determine:*

(*i*) *Temperature to which air should be heated;*

(*ii*) *Efficiency of the air washer.*

Solution: *Given:* $t_{db_1} = 12°C$; $\phi_1 = 85\%$; $t_{db_4} = 36°C$, $t_{wb_4} = 23.2°C$

- Locate point 1 at the interection of 12°C DBT and 85% RH lines.
- Locate point 4 at the interection of 36°C DBT and 23.2°C WBT lines.

Since the final condition of air is obtained with the help of a winter air-conditioner, therefore the processes involved are as follows:

1. *Preheating* of entering air in a preheater;
2. *Humidification* of preheated air in a humidifier or air washer; and

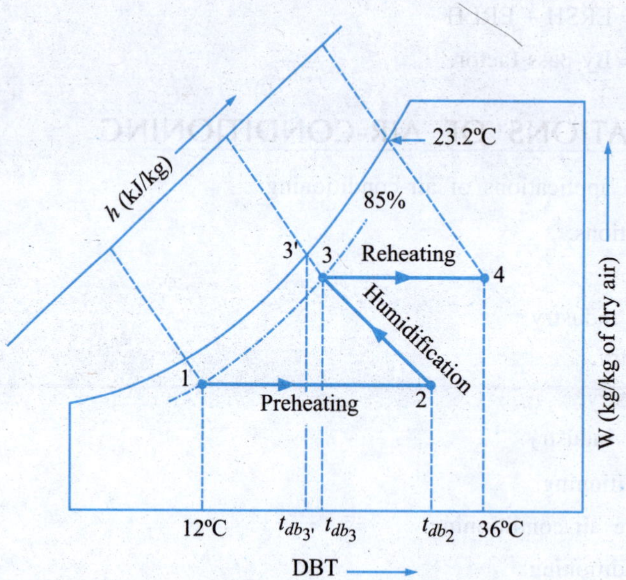

Fig. 14.13

3. *Reheating* of humidified air in a reheater. These processes are shown by 1-2, 2-3 and 3-4 respectively on the psychrometric chart as shown in Fig. 14.13. These processes are obtained on the psychrometric chart as follows:

- Draw horizontal line through point 1 to represent sensible heating of air and from point 4 draw a horizontal line to intersect 85% RH curve at 3.
- From point 3, draw a constant wet bulb temperature line which intersects the horizontal line drawn through point 1 at point 2.

Now line 1-2 represents preheating of air, line 2-3 represents humidification and line 3-4 represents reheating of air to final condition.

(*i*) **Temperature to which air should be preheated, t_{db_2}:**

From the psychrometric chart, the temperature to which the air should be preheated (corresponding to point 2) is

$$t_{db_2} = 34°C \text{ (Ans.)}$$

(*ii*) **Efficiency of air washer:**

From psychrometric chart, we find:

$$t_{db_3} = 20.6°\text{C} \;;\quad t_{db_{3'}} = 18.7°\text{C}$$

$$\therefore \quad \text{Efficiency of air washer} = \frac{\text{Actual drop in DBT}}{\text{Ideal drop in DBT}} = \frac{t_{db_2} - t_{db_3}}{t_{db_2} - t_{db_{3'}}}$$

$$= \frac{34 - 20.6}{34 - 18.7} = 0.876 \text{ or } \mathbf{87.6\%\ (Ans.)}$$

Example 14.2. *250 m³/min of air at atmospheric conditions 12°C DBT and 50% RH is supplied to an air-conditioned hall. The required conditions are 18°C DBT and 60% RH. Determine:*

(*i*) *Sensible heat and latent heat removed from the air per minute, and*

(*ii*) *Sensible heat factor for the system.*

Solution: *Given:* $V_1 = 250$ m³/min; $t_{db_1} = 32°$C; $\phi_1 = 50\%$; $t_{db_2} = 18°$C; $\phi_2 = 60\%$

- Locate point 1 at the intersection of 32°C DBT and 50% RH.
- Locate point 2 at the intersection of 18°C DBT and 60% RH.
- Locate point 3 on the chart by drawing horizontal line through point 2 and vertical line through point 1.

From *psychrometric chart* we find:

$$h_1 = 70.6 \text{ kJ/kg of dry air;}$$

$$h_2 = 37.5 \text{ kJ/kg of dry air;}$$

$$h_3 = 52 \text{ kJ/kg of dry air;}$$

$$v_1 = 0.8848 \text{ m}^3\text{/kg of dry air.}$$

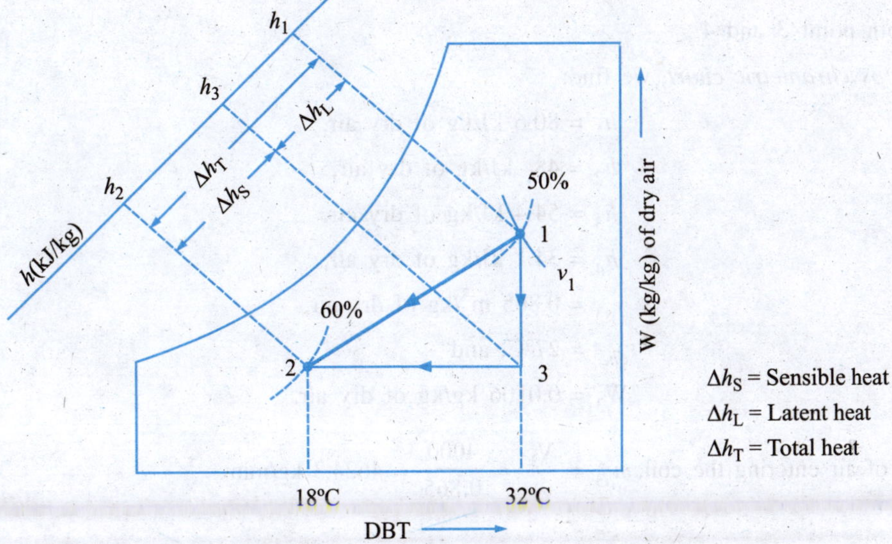

Fig. 14.14

$$\text{Mass or air supplied, } m_2 = \frac{V_1}{v_1} = \frac{250}{0.8848} = 282.5 \text{ kg/min}$$

(*i*) **Sensible heat and latent heat removed from the air per minute, Δh_S, Δh_L:**

$$\Delta h_S = m_a (h_3 - h_2) = 282.5\ (52 - 37.5) = \mathbf{4096.2\ kJ/min\ (Ans.)}$$

$$\Delta h_L = m_a (h_1 - h_3) = 282.5\ (70.6 - 52) = \mathbf{5254.5\ kJ/min\ (Ans.)}$$

(*ii*) **Sensible heat factor for the system, SHF**

$$\text{SHF} = \frac{\Delta h_S}{\Delta h_S + \Delta h_L} = \frac{4096.2}{4096.2 + 5254.5} = \mathbf{0.438\ (Ans.)}$$

Example 14.3. *In an air-conditioning plant, an air handling unit supplies a total of 4000 m³/min of dry air which comprises by mass 20 per cent of fresh air at 39°C DBT and 26°C WBT and 80 percent recirculated air at 24°C DBT and 50% RH. The air leaves the cooling coil at 12°C saturated.*

Calculate the following:

(*i*) *Total cooling load; and* (*ii*) *Room heat gain.*

Solution: *Given:* $V_3 = 4000$ m³/min; $t_{db_1} = 39°C$; $t_{wb_1} = 26°C$; $t_{db_2} = 24°C$; $\phi_2 = 50\%$ RH; t_{db_4} = ADP = 12°C.

- Locate point 1 at the intersection of 39°C DBT and 26°C WBT lines.
- Locate point 2 at the intersection of 24°C DBT and 50% RH lines.

  Join points 1 and 2.
- Locate point 3, which represents 20% of fresh air and 80% of recirculated air, on the line 1-2, such that

  Length 2-3 = 0.2 × length 2-1.
- Locate point 4 on the saturation curve where it meets vertical line drawn at 12°C DBT.

  Join point 3 and 4.

From *psychrometric chart*, we find:

h_1 = 80.6 kJ/kg of dry air,

h_2 = 48 kJ/kg of dry air,

h_3 = 54.4 kJ/kg of dry air,

h_4 = 34.1 kJ/kg of dry air,

v_{s3} = 0.865 m³/kg of dry air,

t_{db_3} = 27°C, and

W_3 = 0.0106 kg/kg of dry air.

$$\text{Mass of air entering the coil, } m_{a3} = \frac{V_3}{v_3} = \frac{4000}{0.865} = 4624.3 \text{ kg/min}$$

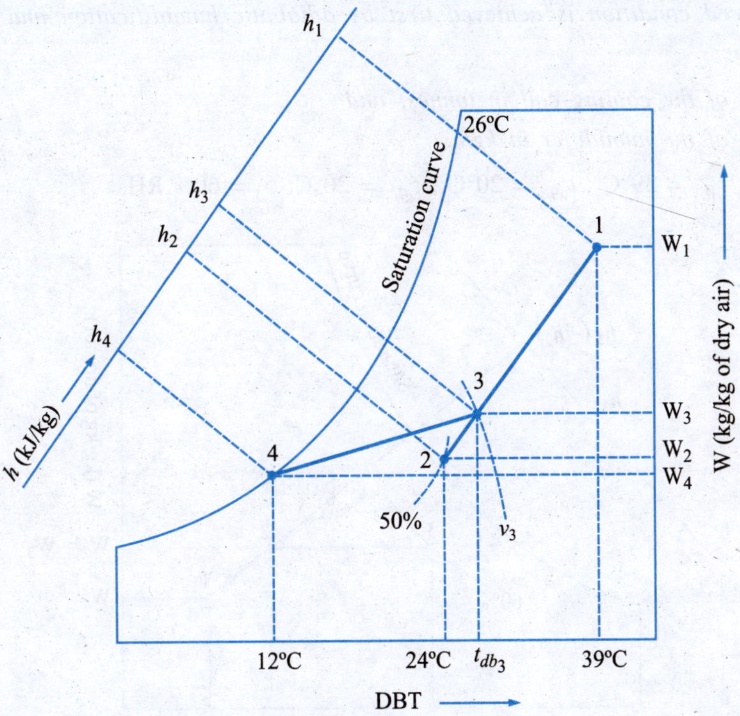

Fig. 14.15

(*i*) **Total cooling load:**

$$\text{Total cooling load} = m_{a3}\,(h_3 - h_4)$$
$$= 4624.3\ (54.4 - 34.1) = 93873.3 \text{ kJ/min}$$
$$= \frac{93873.3 \times 60}{14000} = \mathbf{402.3 \text{ tonnes (Ans.)}}$$

(*ii*) **Room heat gain:**

Since the total mass of air (m_{a3} = 4624.3 kg/min) comprises 20% of fresh air, therefore, mass of fresh air supplied at point 1,

$$m_{a1} = 0.2 \times 4624.3 = 924.86 \text{ kg/min, and}$$

$$\text{Fresh air load} = m_{a1}\,(h_1 - h_2)$$
$$= 924.86\ (80.6 - 48) = 30150.4 \text{ kJ/min}$$
$$= \frac{30150.4 \times 60}{14000} = 129.2 \text{ tonnes}$$

$\therefore$ Room heat gain = 402.3 – 129.2 = **273.1 tonnes (Ans.)**

Example 14.4. *For a hall to be air-conditioned, the following conditions are given:*

Outdoor condition .... *39°C DBT, 20°C WBT*

Required comfort condition .... *20°C DBT, 60% RH*

Seating capacity of hall *1200*

Amount of outdoor air supplied .... *0.3 m³/min per person.*

If the required condition is achieved first by adiabatic humidification and then by cooling determine:

(*i*) *Capacity of the cooling coil in tonnes; and*

(*ii*) *Capacity of the humidifier in kg/h.*

Solution: *Given:* t_{db_1} = 39°C ; t_{wb_1} = 20°C ; t_{db_3} = 20°C; ϕ_3 = 60% RH

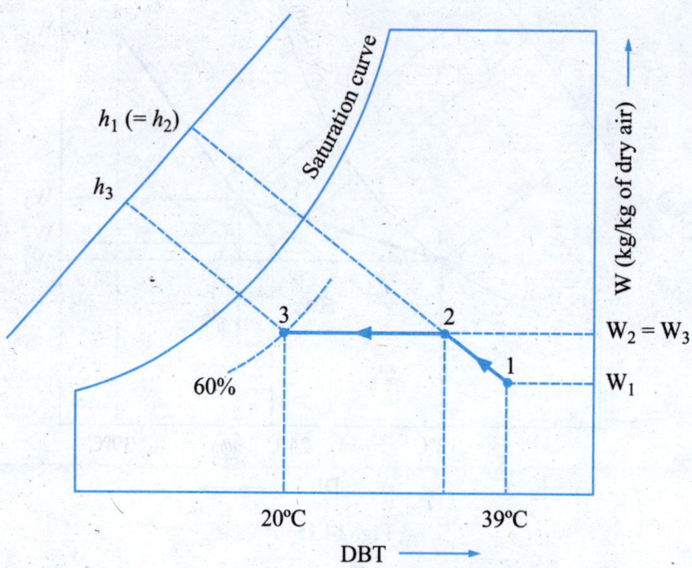

Fig. 14.16

- Locate point 1 at the intersection of 39°C DBT and 20°C WBT.
- Locate point 3 at the intersection of 20°C DBT and 60% RH.
- Locate point 2 at the intersection of horizontal line from point 3 and constant enthalpy line from point 1.

From the psychrometric chart, we find:

$$h_1 = (= h_2) = 57.6 \text{ kJ/kg of dry air;}$$

$$v_1 = 0.893 \text{ m}^3\text{/kg of dry air.}$$

$$h_3 = 42.1 \text{ kJ/kg of dry air;}$$

$$W_1 = 0.007 \text{ kg/kg of dry air;}$$

$$W_2 = W_3 = 0.0086 \text{ kg/kg of dry air.}$$

$$\text{Amount of air supplied, } m_a = \frac{1200 \times 0.3}{0.893} = 403.1 \text{ kg/min}$$

(*i*) **Capacity of the cooling coil in tonnes:**

$$\text{Capacity of the cooling coil} = m_a (h_2 - h_3)$$

$$= 403.1 (57.6 - 42.1) = 6248.05 \text{ kJ/min}$$

$$= \frac{6248.05 \times 60}{14000} = \mathbf{26.78 \text{ tonnes (Ans.)}}$$

(*ii*) **Capacity of the humidifier in kg/h:**

Capacity of the humidifier $= m_a\ (W_2 - W_1) = 403.1\ (0.0086 - 0.007)$ kg/min

$= 403.1\ (0.0086 - 0.007) \times 60 =$ **38.7 kg/h (Ans.)**

Example 14.5. *It is required to design an air-conditioning plant for a small office room for following winter conditions:*

Outdoor conditions .... *12°C DBT and 10°C WBT*

Required conditions .... *20°C DBT and 60% R.H.*

Amount of air circulation .... *0.30 m³/min/person*

Seating capacity of office .... *60.*

The required condition is achieved first by heating and then by adiabatic humidification

Determine the following :

(*i*) *Heating capacity of the coil in kW and the surface temperature required if the by pass factor of coil is 0.4.*

(*ii*) *The capacity of the humidifier.*

Solve the problem by using psychrometric chart.

Solution: Refer to Fig. 14.17.

- Locate the points '1' (12ºC DBT, 10ºC WBT) and '3' (20ºC DBT, 60% RH) on the psychrometric chart.
- Draw a constant enthalpy line through '3' and constant specific humidity line through '1'.
- Locate the point '2' where the above two lines intersect.

From the psychrometric chart:

$$h_1 = 29.3 \text{ kJ/kg}; \; h_2 = h_3 = 42.3 \text{ kJ/kg}$$

$$t_{db_2} = 24.5°\text{C}, \; v_{s1} = 0.817 \text{ m}^3\text{/kg}$$

The mass of air circulated per minute,

$$m_a = \frac{0.30 \times 60}{0.817} = 22.03 \text{ kg/min}$$

(*i*) **Heating capacity of the heating coil**

$$= m_a\ (h_2 - h_1) = 22.03\ (42.3 - 29.3)$$

$$= \mathbf{286.4\ kJ/min\ (Ans.)}$$

The by pass factor (BF) of heating coil is given by:

$$\text{BF} = \frac{t_{db_4} - t_{db_2}}{t_{db_4} - t_{db_1}}$$

$$0.4 = \frac{t_{db_4} - 24.5}{t_{db_4} - 12}$$

$$\therefore \quad 0.4\ (t_{db_4} - 12) = t_{db_4} - 24.5$$

i.e., t_{db_4} (Coil surface temperature) = **32.8°C (Ans.)**

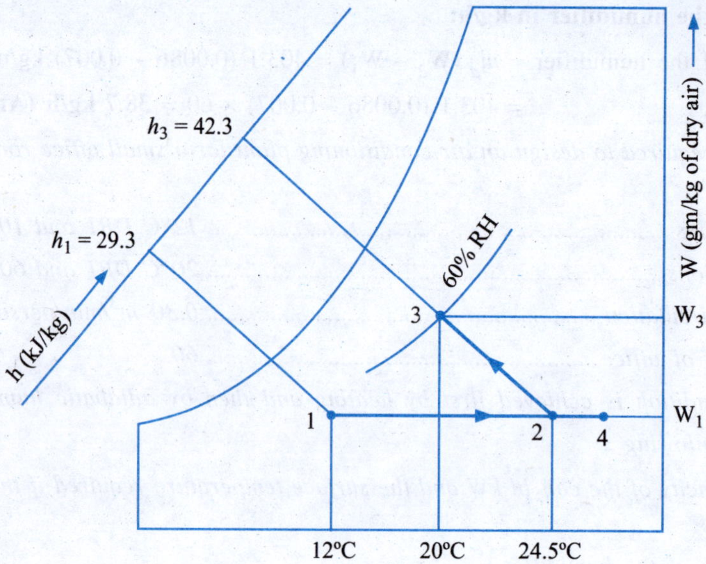

Fig. 14.17

(*ii*) **The capacity of the humidifier** $= \dfrac{m_a (W_3 - W_1)}{1000} \times 60$ kg/h $= \dfrac{22.03\,(8.6 - 6.8)}{1000} \times 60$

$= \mathbf{2.379}$ **kg/h (Ans.)**

Example 14.6. *It is required to design an air-conditioning system for an industrial process for the following hot and wet summer conditions :*

Outdoor conditions .... 32°C DBT and 65% RH

Required air inlet conditions .... 25°C DBT and 60% RH

Amount of free air circulated .... 250 m^3/min

Coil dew temperature .... 13°C.

The required condition is achieved by first cooling and dehumidifying and then by heating. Calculate the following:

(i) The cooling capacity of the cooling coil and its by-pass factor;

(ii) Heating capacity of the heating coil in kW and surface temperature of the heating coil if the by pass factor is 0.3.

(iii) The mass of water vapour removed per hour.

Solution this problem with the use of psychrometric chart.

Solution: Refer to Fig. 14.18.

- Locate the points '1' (32ºC DBT, 65% RH), '5' (13ºC) and '3' (25ºC DBT, 60% RH) as shown on psychrometric chart.
- Join the line 1-5.
- Draw constant specific humidity line through '3' which cuts the line 1-5 at point '2'. The point '2' is located in this way.

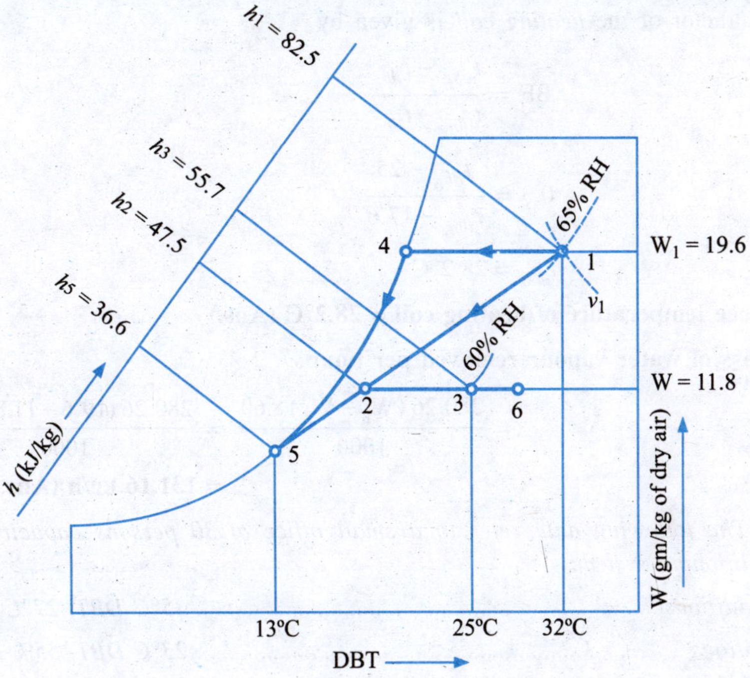

Fig. 14.18

From *psychrometric chart*:

$h_1 = 82.5$ kJ/kg; $h_2 = 47.5$ kJ/kg; $h_3 = 55.7$ kJ/kg; $h_5 = 36.6$ kJ/kg;

$W_1 = 19.6$ gm/kg; $W_3 = 11.8$ gm/kg; $t_{db_2} = 17.6$°C; $v_1 = 0.892$ m^3/kg.

The mass of air supplied per minute,

$$m_a = \frac{250}{0.892} = 280.26 \text{ kg/min}$$

(*i*) **The capacity of the cooling coil**

$$= \frac{m_a(h_1 - h_2) \times 60}{14000} = \frac{280.26\,(82.5 - 47.5) \times 60}{14000}$$

$$= \mathbf{42.04\ TR\ (Ans.)}$$

The **by-pass factor** *of the cooling coil* is given by:

$$\text{BF} = \frac{h_2 - h_5}{h_1 - h_5} = \frac{47.5 - 36.6}{82.5 - 36.6} = \mathbf{0.237\ (Ans.)}$$

(*ii*) **The heating capacity of the heating coil**

$$= m_a\,(h_3 - h_2) = 280.26\,(55.7 - 47.5)$$

$$= 2298.13 \text{ kJ/min} = \frac{2298.13}{60} \text{ kJ/s}$$

$$= \mathbf{38.3\ kW\ (Ans.)}$$

The by-pass factor of the *heating coil* is given by

$$\text{BF} = \frac{t_{db_6} - t_{db_3}}{t_{db_6} - t_{db_2}}$$

$$0.3 = \frac{t_{db_6} - 25}{t_{db_6} - 17.6}$$

$$\therefore \quad t_{db6} = 28.2°\text{C}$$

Hence, **surface temperature of heating coil = 28.2°C (Ans.)**

(*iii*) **The mass of water vapour removed per hour**

$$= \frac{280.26\,(\text{W}_1 - \text{W}_3) \times 60}{1000} = \frac{280.26\,(19.6 - 11.8) \times 60}{1000}$$

$$= \mathbf{131.16\ kg/h\ (Ans.)}$$

Example 14.7. *The following date refer to a small office of 30 persons capacity, provided with summer air-conditioning system:*

Outdoor conditions .... *35°C DBT, 27°C WBT*

Inside conditions .... *23°C DBT, 55% RH*

Volume of air required .... *0.35 m³/min/person*

Sensible heat load in the room .... *121500 kJ/h*

Latent heat load in the room .... *41300 kJ/h.*

Find the sensible heat factor of the plant.

Solution: *Given:* Seating capacity = 30 persons; $t_{db_1} = 35°\text{C}$; $t_{wb_1} = 27°\text{C}$; $t_{db_2} = 23°\text{C}$; $\phi_2 = 55\%$; $\text{V}_1 = 0.35\ \text{m}^3/\text{min/person} = 0.35 \times 30 = 10.5\ \text{m}^3/\text{min}$; Sensible heat load = 121500 kJ/h;

Latent heat load = 41300 kJ/*h*.

- Locate point 1 at the intersection of 35°C DBT and 27°C WBT lines.
- Locate point 2 at the intersection of 23°C DBT and 55% RH lines.
- Locate point 3 by drawing horizontal line through point 2 and vertical line through point 1.

From *psychrometric chart*, we find:

$$h_1 = 85.2\ \text{kJ/kg of dry air};$$

$$h_2 = 48\ \text{kJ/kg of dry air};$$

$$h_3 = 60.2\ \text{kJ/kg of dry air};$$

$$v_1 = 0.9\ \text{m}^3/\text{kg of dry air}$$

$$\text{Mass of air supplied per min.} = \frac{\text{V}_1}{v_1} = \frac{10.5}{0.9} = 11.67\ \text{kg/min}$$

$$\text{Sensible heat removed from air} = m_a\,(h_3 - h_2)$$

$$= 11.67\ (60.2 - 48) = 142.37\ \text{kJ/min or } 8542\ \text{kJ/h}$$

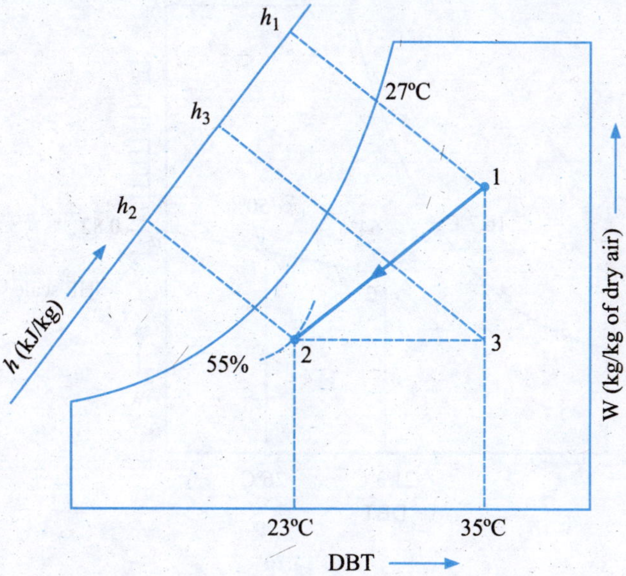

Fig. 14.19

Total sensible heat load of the room

$$SH = 8542 + 121500 = 130042 \text{ kJ/h}$$

Latent heat removed from the air $= m_a (h_1 - h_3)$

$$= 10.67 (85.2 - 60.2) = 291.75 \text{ kJ/min or } 17505 \text{ kJ/h}$$

Total latent heat load of the room,

$$LH = 17505 + 41300 = 58805 \text{ kJ/h}$$

∴ Sensible heat factor of the plant

$$SHF = \frac{SH}{SH + LH} = \frac{130042}{130042 + 58805} = \mathbf{0.689 \text{ (Ans.)}}$$

Example 14.8. *A room has to be maintained at 26°C DBT, 50% RH. It has a sensible heat gain of 22 kW and latent heat gain of 4.8 kW. 190 m³/min of air is supplied to the room. Estimate state of supply air.*

Solution: *Given:* $t_{db_1} = 26°C$; $\phi = 50\%$; **RSH = 22 kW**; RLH = 4.8 kW; $v = 190 m^3/min$.

Let t_{db_2} = Dry bulb temperature of the supply air.

Now, $RSH = 0.02044 \text{ V } (t_{db_1} - t_{db_2})$

$$22 = 0.02044 \times 190 \; (27 - t_{db_2})$$

∴ $t_{db_2} = 21.3°C$

Also,

$$RSHF = \frac{RSH}{RSH + RLH} = \frac{22}{22 + 4.8} = 0.82$$

- Locate point 1 at the intersection of 26°C DBT and 50% RH. This point also represents the alignment circle (i.e., 26°C, 50% RH).
- Mark the calculated value of RSHF = 0.82 on the sensible heat factor (SHF) scale as point *a* and join the point 1 (i.e., alignment circle).

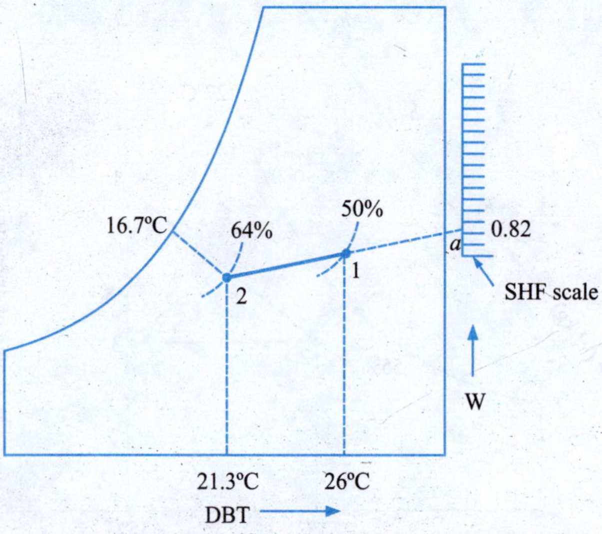

Fig. 14.20

Produce this line upto point 2 such that t_{db_2} = 21.3°C. The point 2 *represents the supply condition of air*. From the *psychrometric chart*, we find that wet bulb temperature of supply air,

$$t_{db_2} = \mathbf{16.7°C}\ ; \quad \phi_2 = \mathbf{64\%}\ \textbf{(Ans.)}$$

Example 14.9. *An air-conditioned auditorium is to be maintained at 27°C DBT and 55% RH. The ambient condition is 39°C DBT and 28°C WBT. The total sensible heat load is 120000 kJ/h and the total latent heat load is 45000 kJ/h. 60 per cent of the return air is recirculated and mixed with 40 per cent of make-up air after the cooling coil. The condition of air leaving the coil is 17°C. Determine:*

(*i*) *Room sensible heat factor*;

(*ii*) *Condition of air entering the auditorium*;

(*iii*) *Amount of make-up air*;

(*iv*) *Apparatus dew point*;

(*v*) *By-pass factor of the cooling coil.*

Solution: *Given:* t_{db_4} = 27°C; ϕ_4 = 55%; t_{db_1} = 39°C; t_{wb_1} = 28°C;
RSH = 120000 kJ/h; RLH = 45000 kJ/h; t_{db_2} = 17°C.

(*i*) **Room sensible heat factor, RSHF:**

$$\text{RSHF} = \frac{\text{RSH}}{\text{RSH} + \text{RLH}} = \frac{120000}{120000 + 45000} = \mathbf{0.727}\ \textbf{(Ans.)}$$

(*ii*) **Condition of air entering the auditorium:**

The line diagram for processes involved in the air-conditioning of the auditorium is shown in Fig. 14.21. The various processes drawn on the psychrometric chart (shown in Fig. 14.22) are discussed below:

- Locate point 1 at the intersection of 39°C DBT and 28°C WBT.
- Locate point 4 at the intersection of 27°C DBT and 55% RH.

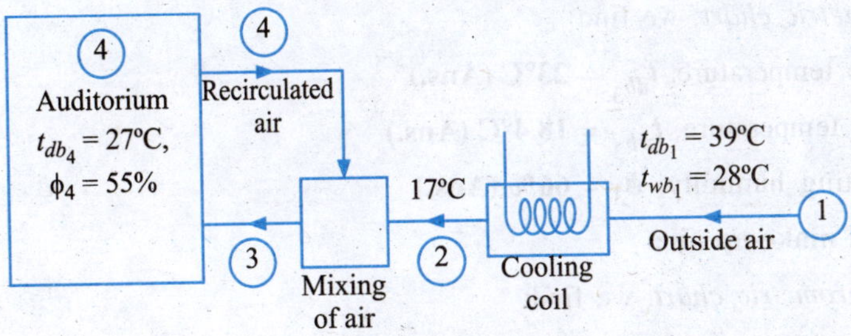

Fig. 14.21

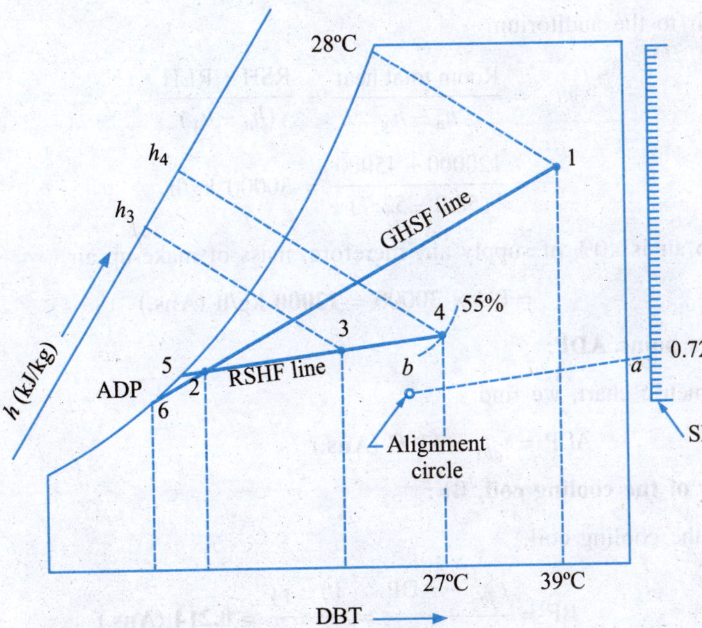

Fig. 14.22

- Mark the calculated value of RSHF = 0.727 on the SHF scale as point *a* and join with the alignment circle (i.e., 26°C DBT and 50% RH) i.e., point *b* as shown in Fig. 14.22.
- From point 4, draw a line 4-5 (known as RSHF line) parallel to the line *a-b*.
- Since the condition of air leaving the cooling coil is 17°C, therefore, locate point 2 such that t_{db_2} = 17°C.
- Joint points 1 and 2 and produce upto point 6 on the saturation curve. The line 1-2-6 is the GSHF line.
- It is given that 60% of the air from the auditorium is recirculated and mixed with 40% of the make-up air after the cooling coil. The mixing condition of air is shown at point 3 such that

$$\text{Length 2-3} = 0.6 \times \text{length 2-4}$$

The point 3 gives the condition of air entering the auditorium.

From *psychrometric chart*; we find:

Dry bulb temperature, t_{db_3} = **23°C (Ans.)**

Wet bulb temperature, t_{wb_3} = **18.4°C (Ans.)**

Relating humidity, ϕ_3 = **66% (Ans.)**

(*iii*) **Amount of make-up air:**

From the *psychrometric chart*, we find:

$$h_4 = 58.2 \text{ kJ/kg of dry air; and}$$

$$h_3 = 52.7 \text{ kJ/kg of dry air.}$$

Mass of supply air to the auditorium

$$m_{sa} = \frac{\text{Room total heat}}{h_4 - h_3} = \frac{\text{RSH + RLH}}{(h_4 - h_3)}$$

$$= \frac{120000 + 45000}{(58.2 - 52.7)} = 30000 \text{ kg/h}$$

Since the make-up air is 40% of supply air, therefore, mass of make-up air

$$= 0.4 \times 30000 = \mathbf{12000 \text{ kg/h (Ans.)}}$$

(*iv*) **Apparatus dew point, ADP:**

From the psychrometric chart, we find:

$$\text{ADP} = t_{db1} = \mathbf{11°C \text{ (Ans.)}}$$

(*v*) **By-pass factor of the cooling coil, BF:**

By-pass factor of the cooling coil,

$$\text{BF} = \frac{t_{db_2} - \text{ADP}}{t_{db_1} - \text{ADP}} = \frac{17 - 11}{39 - 11} = \mathbf{0.214 \text{ (Ans.)}}$$

Example 14.10. *It is required to maintain an air-conditioned hall at 28°C DBT and 20°C WBT. It has sensible heat load of 45.4 kW and latent heat load of 16.5 kW. The air supplied from outside atmosphere at 39°C DBT and 28°C WBT is 22 m^3/min, directly into the room through vertilation and infiltration. Outside air to be conditioned is passed through the cooling coil whose apparatus dew point is 14°C, 60 per cent of the recirculated air from the hall is mixed with the conditioned air after the cooling coil. Determine:*

(*i*) *Condition of air after the coil and before the recirculated air mixes with it;*

(*ii*) *Condition of air entering the hall i.e., after mixing with recirculated air;*

(*iii*) *Mass of fresh air entering the cooler;*

(*iv*) *By-pass factor of the cooling coil;*

(*v*) *Refrigerating load on the cooling coil.*

Solution: *Given:* t_{db_4} = 28°C; t_{wb_4} = 20°C; Q_{S4} = 45.4 kW; Q_{L4} = 16.5 kW; t_{db_1} = 39°C; t_{wb_1} = 28°C; V_1 = 22 m^3/min ; ADP = 14°C

The line diagram for the processes involved in the air-conditioning of the hall is shown in Fig. 14.23. The various processes drawn on the psychrometric chart (shown in Fig. 14.24) are discussed below:

- Locate point 1 at the intersection of 39°C DBT and 28°C WBT.
- Locate point 4 at the intersection of 28°C DBT and 20°C WBT.
- Locate point A by drawing horizontal and vertical lines from point 4 and 1 respectively.
- Since 22 m³/min of outside air (at 39°C DBT and 28°C WBT) is supplied directly into the room through vertilation and infiltration, therefore the sensible heat and latent heat of 22 m³/min of *infiltrated air* are added into the hall in addition to the sensible and latent heat loads of 45.4 kW and 16.5 kW respectively.

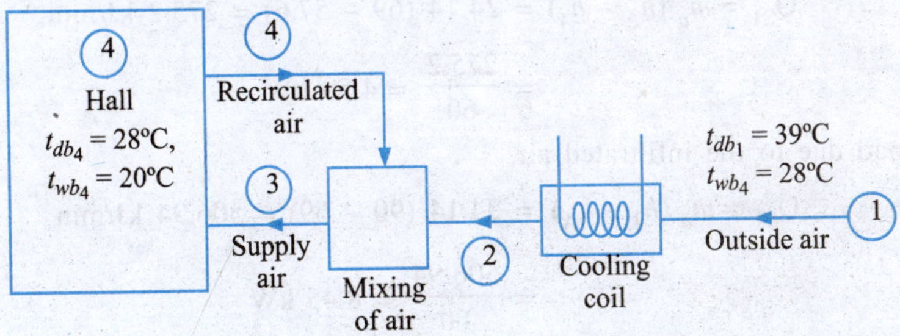

Fig. 14.23

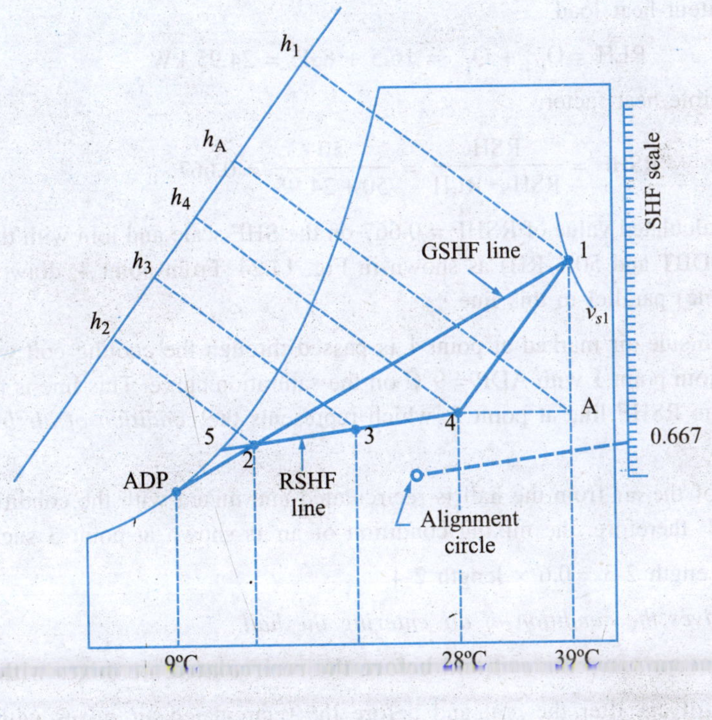

Fig. 14.24

From the *psychrometric chart*, we find:

$$h_1 = 90 \text{ kJ/kg of dry air;}$$

$$h_4 = 57.6 \text{ kJ/kg of dry air;}$$

$$h_A = 69 \text{ kJ/kg of dry air;}$$

$$v_{s1} = 0.9115 \text{ m}^3\text{/kg of dry air;}$$

∴ Mass of air infiltrated into the hall,

$$m_a = \frac{V_1}{v_1} = \frac{22}{0.9115} = 24.14 \text{ kg/min}$$

Sensible heat load due to the infiltrated air,

$$Q_{S1} = m_a (h_A - h_4) = 24.14 (69 - 57.6) = 275.2 \text{ kJ/min}$$

$$= \frac{275.2}{60} = 4.59 \text{ kW}$$

Latent heat load due to the infiltrated air,

$$Q_{L1} = m_a (h_1 - h_A) = 24.14 (90 - 69) = 506.94 \text{ kJ/min}$$

$$= \frac{506.94}{60} = 8.45 \text{ kW}$$

Total room sensible heat load

$$RSH = Q_{S4} + Q_{S1} = 45.4 + 4.59 \simeq 50 \text{ kW}$$

Total room latent heat load

$$RLH = Q_{L4} + Q_{L1} = 16.5 + 8.45 = 24.95 \text{ kW}$$

∴ Room sensible heat factor,

$$RSHF = \frac{RSH}{RSH + RLH} = \frac{50}{50 + 24.95} = 0.667$$

- Make the calculated value of RSHF = 0.667 on the SHF scale and join with the alignment circle (i.e., 26°C DBT and 50% RH) as shown in Fig. 14.24. From point 4, draw a line 4-5 (known as RSHF line) parallel to this line.
- Since the outside air marked at point 1 is passed through the cooling coil whose ADP = 9°C, Therefore, join point 1 with ADP = 9°C on the saturation curve. This line is the GHSF line and intersects the RSHF line at point 2, which represents the *condition of air leaving the cooling coil.*
- State 60% of the air from the hall is recirculated and mixed with the conditioned air after the cooling coil, therefore, the mixing condition of air is shown at point 3 such that,

$$\text{Length 2-3} = 0.6 \times \text{length 2-4}$$

The *point 3 gives the condition of air entering the hall.*

(*i*) **Condition of air after the coil and before the recirculated air mixes with it:**

The condition of air after the coil and before the recirculated air mixes with it, is shown by point 2 on the psychrometric chart (Fig. 14.24).

From the *psychrometric chart*, we find:

Dry bulb temperature, t_{db_2} = **12.3°C (Ans.)**

Wet bulb temperature, t_{wb_2} = **11.8°C (Ans.)**

Relative humidity ϕ_2 = **95% (Ans.)**

(*ii*) **Condition of air entering the hall, i.e., after mixing with recirculated air:**

The condition of air entering hall, i.e., after mixing with recirculated air is shown by point 3 on the psychrometric chart (Fig. 14.24).

From the *psychrometric chart*, we find:

Dry bulb temperature, t_{db_3} = **21.9°C (Ans)**

Wet bulb temperature, t_{wb_3} = **17°C (Ans.)**

Relative humidity, ϕ_3 = **63% (Ans.)**

(*iii*) **Mass of fresh air entering the cooler; m_F:**

$$m_F = \frac{\text{Total heat removed}}{(h_4 - h_2)} = \frac{\text{RSH} + \text{RLH}}{(h_4 - h_2^*)}$$

$$= \frac{(50 + 24.95)}{(57.6 - 33.9)} = 3.162 \text{ kg/s or } \mathbf{189.75 \text{ kg/min (Ans.)}}$$

(h_2^* = 33.9 kJ/kg of dry air ... from the chart).

(*iv*) **By-pass factor of the cooling coil, BF:**

$$\text{BF} = \frac{t_{db_2} - \text{ADP}}{t_{db_1} - \text{ADP}} = \frac{12.3 - 9}{39 - 9} = \mathbf{0.11 \text{ (Ans.)}}$$

(*v*) **Refrigeration load on the cooling coil:**

Refrigeration load on the cooling coil

$$= m_F (h_1 - h_2)$$

$$= \frac{189.75 (90 - 33.9) \times 60}{14000} = \mathbf{45.62 \text{ tonnes (Ans.)}}$$

Example 14.11. *The following data relate to an air-conditioned space :*

Outdoor condition .... *38°C DBT, 50% RH*

Room condition .... *24°C DBT, 50% RH*

Sensible heat load .... *24 kW*

Latent heat load .... *6 kW*

By-pass factor of the cooling coil *0.16*

If the ventilation requirement is such that on mass flow rate basis 20 per cent of fresh air is introduced and 80 per cent of supply air is recirculated, determine :

(*i*) *Supply air flow rate;*

(*ii*) *Outside air sensible heat;*

(*iii*) *Outside air latent heat;*

(*iv*) *Grand total heat;*

(*v*) *Effective room sensible heat factor.*

Solution: *Given:* $t_{db_1} = 38°C$; $\phi_1 = 50\%$; $t_{db_2} = 24°C$; $\phi_2 = 50\%$; RSH = 24 kW ; RLH = 6 kW ; BF = 0.16

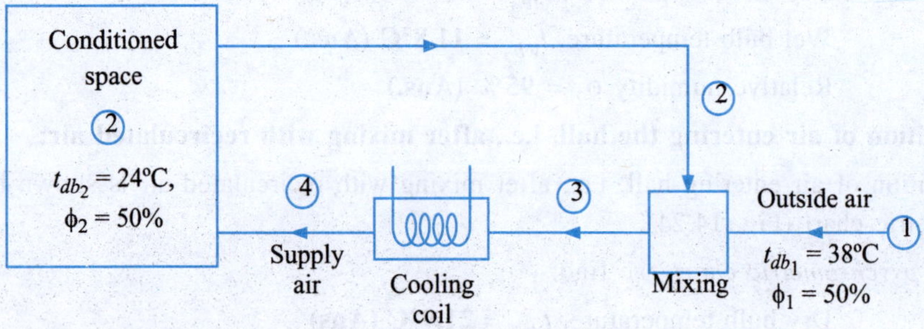

Fig. 14.25

The flow diagram for the air conditioned space in sown in Fig. 14.25. The various processes drawn on the psychrometric chart (Fig. 14.26) are discussed below:

- Locate point 1 at the intersection of 38°C DBT and 50% RH lines.
- Locate point 2 at the intersection of 24°C DBT and 50% RH lines.

Now, $$\text{RSHF} = \frac{\text{RSH}}{\text{RSH} + \text{RLH}} = \frac{24}{24 + 6} = 0.8$$

- Mark the calculated value of RSHF = 0.8 on the SHF scale and join with alignment circle (i.e., 26°C DBT, 50% RH) as shown in Fig. 14.26. From point 2 draw a line 2-5 parallel to this line. The line 2-5 is called *RSHF line.*

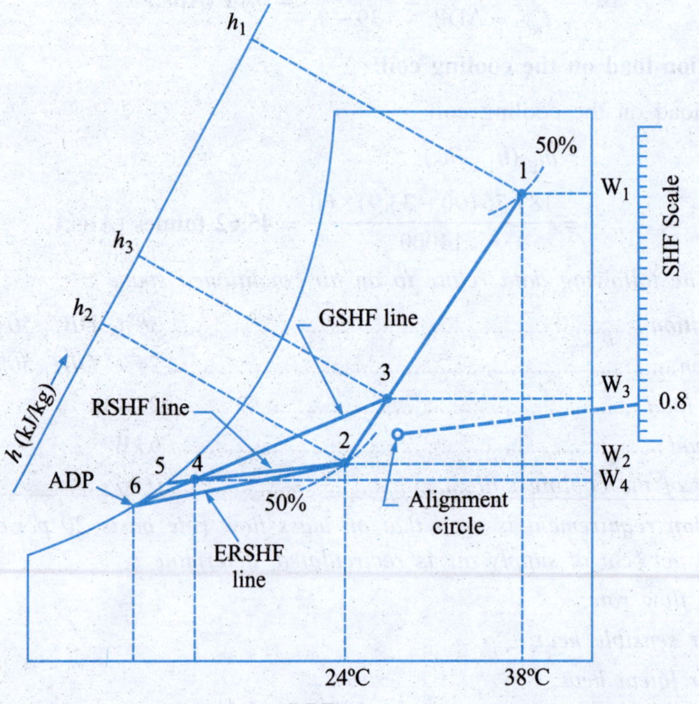

Fig. 14.26

- Since 20% of fresh or outside air is mixed with 80% of supply air, therefore, the condition of air entering the cooling coil after mixing process is marked on the line 1-2 by point 3, such that

$$\text{Length 2-3} = 0.2 \times \text{length 1-2}$$

- Through point 3, draw a line 3-6 (known as *GSHF line*) intersecting the RSHF line at point 4 and saturation curve at point 6, such that,

$$\frac{\text{Length 4-6}}{\text{Length 2-6}} = \text{BF} = 0.16$$

(*i*) **Supply air flow rate:**

Let, V = Supply air flows rate, m^3/min,

t_{db_4} = DBT of air leaving the cooling coil, and

t_{wb_6} = Apparatus dew point (ADP)

From the *psychrometric chart*; we find:

$$t_{db_3} = 26.8°\text{C}$$

Now,
$$\text{BF} = 0.16 = \frac{t_{db_4} - \text{ADP}}{t_{db_3} - \text{ADP}} = \frac{t_{db_4} - t_{db_6}}{26.8 - t_{db_6}}$$

By trial and error, we find that:

$$t_{db_4} = 12.7°\text{C and } t_{db_6} = 10°\text{C}$$

Room sensible heat load,

$$\text{RSH} = 0.02044 \text{ V } (t_{db_2} - t_{db_4})$$

or,
$$24 = 0.02044 \text{ V } (24 - 12.7)$$

$$\therefore \quad \text{V} = \mathbf{103.9\ m^3/min\ (Ans.)}$$

(*ii*) **Outside air sensible heat (OASH):**

Since the outside air is 20% of the supply air, therefore outside air flow rate,

$$\text{V}_0 = 0.2 \text{ V} = 0.2 \times 103.9 = 20.78 \text{ m}^3/\text{min}$$

∴ Outside air sensible heat;

$$\text{OASH} = 0.02044 \text{ V}_0 \ (t_{db_1} - t_{db_2}) \quad \text{...(Eqn. 12.23)}$$

$$= 0.02044 \times 20.78 \ (38 - 24) = \mathbf{6.37\ kW\ (Ans.)}$$

(*iii*) **Outside air latent heat (OALH):**

From the *psychrometric chart*, we find:

$$\text{W}_1 = 0.0212 \text{ kg/kg of dry air, and}$$

$$\text{W}_2 = 0.0092 \text{ kg/kg of dry air.}$$

$$\therefore \quad \text{OALH} = 50 \text{ V}_0 \ (\text{W}_1 - \text{W}_2) \quad \text{...(Eqn. 12.26)}$$

$$= 50 \times 20.78 \ (0.0212 - 0.0092) = \mathbf{12.47\ kW\ (Ans.)}$$

(*iv*) **Grand total heat (GTH):**

$$\text{GTH} = \text{TSH} + \text{TLH}$$
$$= (\text{RSH} + \text{OASH}) + (\text{RLH} + \text{OALH})$$
$$= (24 + 6.37) + (6 + 12.47) = \mathbf{48.84\ kW\ (Ans.)}$$

> **Note:** TSH and TLH may also be calculated as follows:
>
> From *psychrometric chart*, we find:
>
> $$W_3 = 0.0117 \text{ kg/kg of dry air,}$$
> $$W_4 = 0.0082 \text{ kg/kg of dry air}$$
> $$\text{TSH} = 0.02044\ V\ (t_{db_3} - t_{db_4})$$
> $$= 0.02044 \times 103.9\ (26.8 - 12.7) = 29.94 \text{ kW}$$
> $$\text{TLH} = 50\ V\ (W_3 - W_4)$$
> $$= 50 \times 103.9\ (0.0117 - 0.0082) = 18.2 \text{ kW}$$

(*v*) **Effective room sensible heat factor (ERSHF):**

$$\text{ERSH} = \text{RSH} + \text{OASH} \times \text{BF} = 24 + 6.37 \times 0.16 = 25.02 \text{ kW}$$
$$\text{ERLH} = \text{RLH} + \text{OALH} \times \text{BF} = 6 + 12.47 \times 0.16 = 7.99 \text{ kW}$$

$$\therefore \quad \text{ERSHF} = \frac{\text{ERSH}}{\text{ERSH} + \text{ERLH}} = \frac{25.02}{25.02 + 7.99} = \mathbf{0.758\ (Ans.)}$$

Example 14.12. *The following data relate to summer air-conditioning of a building:*

Outside design conditions .... *42°C DBT, 28°C WBT*

Inside design conditions .... *24°C DBT, 50% RH*

Room sensible heat gain .... *82000 kJ/h*

Room latent heat gain .... *18000 kJ/h*

By-pass factor of the cooling coil *0.2*

The return air from the room is mixed with outside air before entry to cooling coil in the ratio of 4 : 1 by mass. Determine:

(*i*) *Apparatus dew point (ADP) of the coil;*

(*ii*) *Entry and exit conditions of air for cooling coil;*

(*iii*) *Fresh air mass flow rate;*

(*iv*) *Refrigeration load on the cooling coil.*

Solution: *Given:* $t_{db_1} = 42°\text{C}$; $t_{wb_1} = 28°\text{C WBT}$; $t_{db_2} = 24°\text{C}$; $\phi_2 = 50\%$; RSH = 82000 kJ/h; RLH = 18000 kJ/h; BF = 0.2

The flow diagram for the conditioned space is shown in Fig. 14.27. The various process shown on the psychrometric chart (Fig 14.28) are discussed below:

- Locate point 1 at the intersection of 42°C DBT and 28°C WBT.
- Locate point 2 at the intersection of 24°C DBT and 50% RH.

$$\text{RSHF} = \frac{\text{RSH}}{\text{RSH} + \text{RLH}} = \frac{82000}{82000 + 18000} = 0.82$$

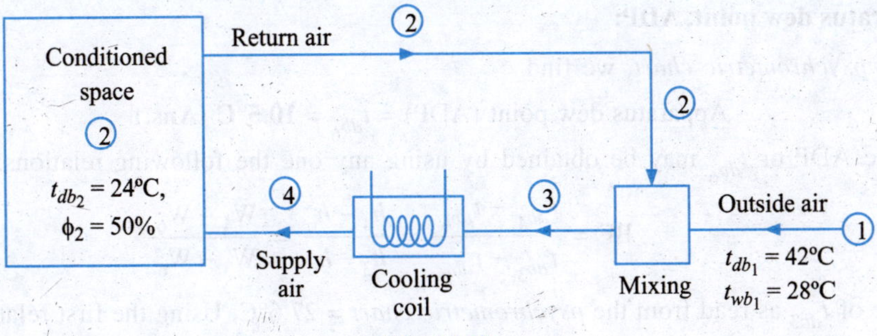

Fig. 14.27

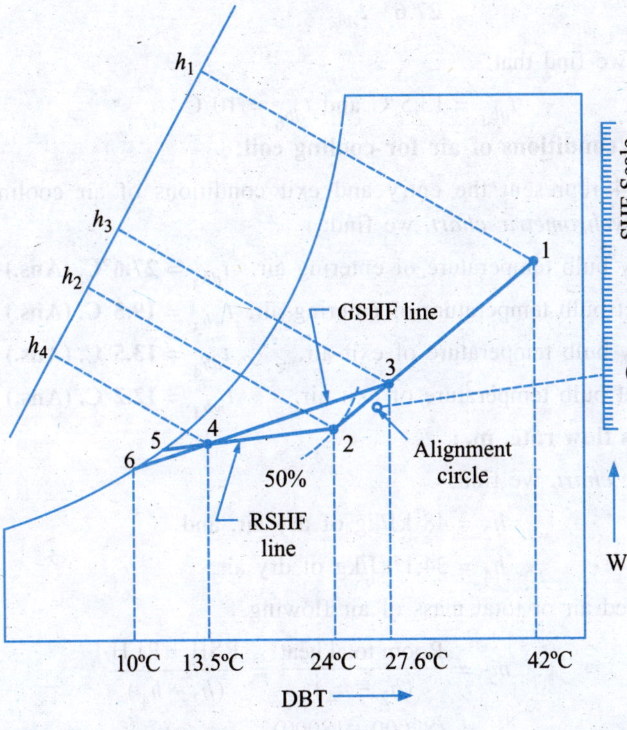

Fig. 14.28

- Mark the calculated value of RSHF = 0.82 on the SHF scale and join with the alignment circle (i.e., 26°C DBT and 50% RH) as, shown in Fig. 14.28. From point 2, draw a line 2-5 parallel to this line. This line 2-5 is called *RSHF line*.
- Since the return air from the conditioned space is mixed with outside air before entry to the cooling coil in the ratio of 4 : 1, therefore, the condition of air entering the cooling coil after mixing process is marked on the line 1-2 by point 3, such that,

$$\text{Length 2-3} = \frac{1}{5} \times \text{length 1-2}$$

- Through point 3, draw a line 3 6 (known as GSHF line) intersecting the RSHF line at point 4 and the saturation curve at point 6, such that,

$$\frac{\text{Length 4-6}}{\text{Length 3-6}} = \text{BF} = 0.2$$

(*i*) **Apparatus dew point, ADP:**

From the *psychrometric chart*, we find

$$\text{Apparatus dew point (ADP)} = t_{db_6} = \mathbf{10.5°C\ (Ans.)}$$

Note: The ADP or t_{db_6} may be obtained by using any one the following relations:

$$BF = \frac{t_{db_4} - t_{db_6}}{t_{db_3} - t_{db_6}} = \frac{h_4 - h_6}{h_3 - h_6} = \frac{W_4 - W_6}{W_3 - W_6}$$

The value of t_{db3} as read from the *psychrometric chart* = 27.6°C. Using the first relation, we have:

$$0.2 = \frac{t_{db_4} - t_{db_6}}{27.6 - t_{db_6}}$$

By trial and error, we find that:

$$t_{db_4} = 13.5°C \text{ and } t_{db_6} = 10°C$$

(*ii*) **Entry and exit conditions of air for cooling coil:**

The points 3 and 4 represent the entry and exit conditions of air cooling coil as shown in Fig. 14.28. From the *psychrometric chart*, we find:

Dry bulb temperature of entering air, t_{db_3} = **27.6°C. (Ans.)**

Wet bulb temperature of entering air, t_{wb_3} = **19.5°C. (Ans.)**

Dry bulb temperature of exit air, t_{db_4} = **13.5°C. (Ans.)**

Wet bulb temperature of exit air, t_{wb_4} = **12.2°C. (Ans.)**

(*iii*) **Fresh air mass flow rate, m_F:**

From *psychrometric chart*, we find:

$$h_2 = 48 \text{ kJ/kg of dry air, and}$$

$$h_4 = 34.1 \text{ kJ/kg of dry air.}$$

Mass of dehumidified air or total mass of air flowing,

$$m_a = \frac{\text{Room total heat}}{(h_2 - h_4)} = \frac{\text{RSH} + \text{RLH}}{(h_2 - h_4)}$$

$$= \frac{(82000 + 18000)}{(48 - 34.1)} = 7194.2 \text{ kg/h}$$

Since this mass of air contains return air and fresh air in the ratio 4 : 1 therefore, fresh air mass flow rate,

$$m_F = \frac{1}{5} \times 7194.2 = \mathbf{1438.84\ kg/h\ (Ans.)}$$

(*iv*) **Refrigeration load on the cooling coil:**

From the *psychrometric chart*, we find:

$$h_3 = 56 \text{ kJ/kg of dry air.}$$

∴ Refrigeration load on the cooling coil

$$= m_a (h_3 - h_4) \text{ kJ/h}$$

$$= \frac{7194.2\,(56 - 34.1)}{14000} = \mathbf{11.25\ tonnes\ (Ans.)}$$

Example 14.13. *The following data relate to an air-conditioned room:*

Outside conditions .... *36°C DBT, 27°C WBT*

Inside conditions .... *24°C DBT, 50% RH*

Sensible heat load .... *12 kW*

Latent heat load .... *7.3 kW*

Apparatus dew point .... *7°C*

By-pass factor of the cooling coil used *0.1*

Return air from the room is mixed with the outside air before entering the cooling coil in the ratio of 4 : 1 and return air from room is also mixed after the cooling coil in the ratio 1 : 4. The air may be reheated, if necessary, before supplying to the conditioned room.

Determine the following:

(*i*) *Supply air condition to the room;*

(*ii*) Refrigeration load;

(*iii*) Quantity of fresh air supplied.

Solution: *Given:* t_{db_1} = 36°C ; t_{wb_1} = 27°C; t_{db_2} = 24°C; ϕ_2 = 50%.

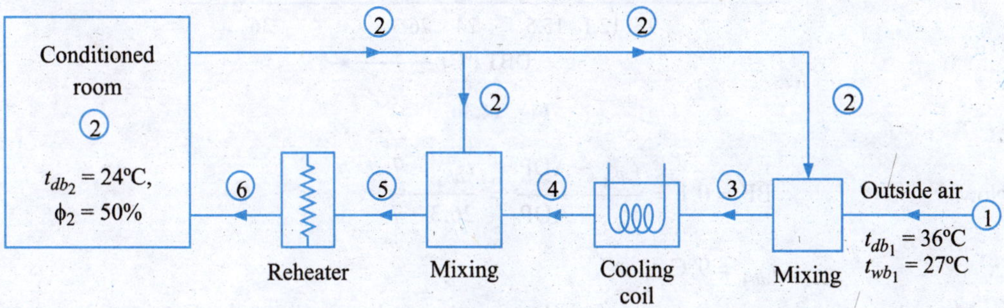

Fig. 14.29

The flow diagram for the processes involved in the air-conditioning of a room is shown in Fig. 14.29. These processes shown on the psychrometric chart (Fig. 14.30) are discussed below:

- Locate point 1 at the intersection of 36°C DBT and 27°C WBT.
- Locate point 2 at the intersection of 24°C DBT and 50% RH.
- Since return air from the room and outside air is mixed before entering the cooling coil in the ratio 4 : 1, therefore, mark the condition of air after mixing and before entering to the cooling coil (i.e., point 3) on the line 1-2 such that,

$$\text{Length 2-3} = \frac{1}{5} \times \text{length 2-1.}$$

From the *psychrometric chart,* we find:

$$t_{db_3} = 26.3°\text{C};\ h_3 = 55.1 \text{ kJ/kg of dry air.}$$

Let t_{db_4} be the DBT of air leaving the coil.

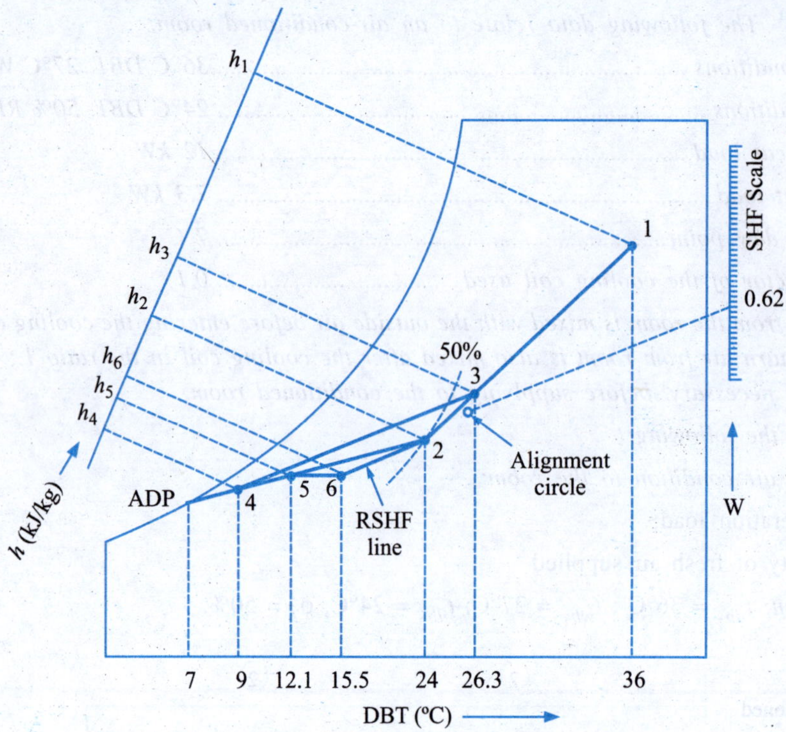

Fig. 14.30

Now, $$BF = 0.1 = \frac{t_{db_4} - ADP}{t_{db_3} - ADP} = \frac{t_{db_4} - 7}{26.3 - 7}$$

$\therefore$ $t_{db4} \simeq 9°C$

- Mark the point 4 (i.e., condition of air leaving the cooling coil) on the line joining point 3 and ADF = 7°C on the saturation curve, such that its DBT is equal to 9°C. From the *psychrometric chart*, we find:

$$h_4 = 26 \text{ kJ/kg of dry air}$$

- Since the return air from the room is also mixed with the air leaving the cooling coil in the ratio 1 : 4, therefore, mark the condition of air after mixing (i.e., point 5) on the line 4-2, such that,

$$\text{Length 4-5} = \frac{1}{5} \times \text{length 4-2}$$

From the *psychrometric chart*, we find:

$$t_{db_5} = 12.1°C \; ; \; h_5 = 30.5 \text{ kJ/kg of dry air.}$$

Now, $$RSHF = \frac{RSH}{RSH + RLH} = \frac{12}{12 + 7.3} = 0.62$$

The reheating of air is necessary to take care of the RSHF which is equal to 0.62.

Draw a horizontal line from point 5 to intersect the RSHF line at *point 6*, which represents the *condition of supply air to the room.*

(*i*) **Supply air condition to the room:**

From the *psychrometric chart* (corresponding to point-6), we find:

Dry bulb temperature, $\mathbf{t_{db_6}}$ = **15.5°C (Ans.)**

Wet bulb temperature, $\mathbf{t_{wb_6}}$ = **12°C (Ans.)**

(*ii*) **Refrigeration load:**

From the *psychrometric chart*, we find:

$h_6 = 34.1$ kJ/kg of dry air,

$h_2 = 48$ kJ/kg of dry air.

Total amount of supply air

$$= \frac{\text{Room total heat}}{(h_2 - h_6)} = \frac{\text{RSH + RLH}}{(h_2 - h_6)}$$

$$= \frac{(12 + 7.3)}{(48 - 34.1)} = 1.388 \text{ kg/s or } 83.28 \text{ kg/min}$$

The amount of dehumidified air,

$$m_D = \frac{4}{5} \times 83.28 = 66.6 \text{ kg/min}$$

∴ Refrigeration load (total)

$$= m_D\,(h_3 - h_4) = 66.6\,(55.1 - 26) = 1938.06 \text{ kJ/min}$$

$$= \frac{1938.06 \times 60}{14000} = \mathbf{8.3 \text{ tonnes (Ans.)}}$$

(*iii*) **Quantity of fresh air supplied $\mathbf{m_F}$:**

$$m_F = \frac{m_D}{5} = \frac{66.6}{5} = \mathbf{13.32 \text{ kg/min (Ans.)}}$$

Example 14.14. *In an air-conditioning system the inside and outside conditions are dry bulb temperature 25°C, relative humidity 50% and dry bulb temperature 40°C, wet bulb temperature 27°C respectively. Room sensible heat factor is 0.8.50% of room air is rejected to atmosphere and an equal quantity of fresh air added before air enters the Air-conditioning apparatus. If the fresh air added is 100 m³/minute, determine the following assuming the bypass factor as zero, density of air = 1.2 kg/m³ at a total pressure of 1.01325 bar. Draw the process diagram and determine the following:*

(*i*) *Room sensible and latent heat load.*

(*ii*) *Sensible and latent heat load due to fresh air.*

(*iii*) *Apparatus dew point.*

(*iv*) *Humidity ratio and dry bulb temperature of air entering Air-conditioning apparatus.*

Solution: *Given:* $t_{db_1} = 40°C$, $t_{wb_1} = 27°C$; $t_{db_2} = 25°C$; $\phi_2 = 50\%$; $v_1 = 100$ m³/min; SHF = 0.8; BF = zero; $\rho_a = 1.2$ kg/m³.

Fig. 14.31 shows the arrangement of the air-conditioned system. The various processes shown on the psychrometric chart (Fig. 14.32) are discussed below:

- Locate point 1 at the intersection of 40°C DBT and 27°C WBT.
- Locate point 2 at the intersection of 25°C DBT and 50% RH. Join points 1 and 2.
- Locate point 3 on the line 1-2 at the intersection of 32.5°C ($t_{db_3} = \frac{t_{db_1} + t_{db_2}}{2} = \frac{40 + 25}{2}$ = 32.5°C) DBT line; this points divides line 1-2 in the ratio 50 : 50.
- Mark value of SHF = 0.8 on SHF scale and join with the alignment circle (i.e., 26°C, 50% RH) as shown in Fig. 14.32.
- From point 2 draw a line parallel to this line and extend upto saturation curve at point 4.

From *psychrometric chart*, we find:

$$h_1 = 85.2 \text{ kJ/kg of dry air,}$$
$$h_2 = 50 \text{ kJ/kg of dry air, and}$$
$$h_4 = 32.9 \text{ kJ/kg of dry air.}$$

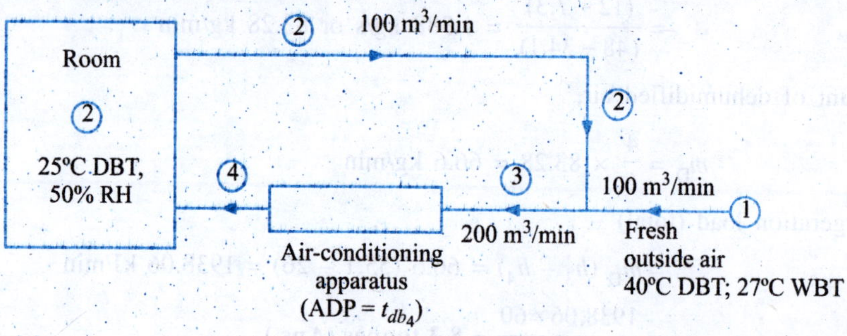

Fig. 14.31

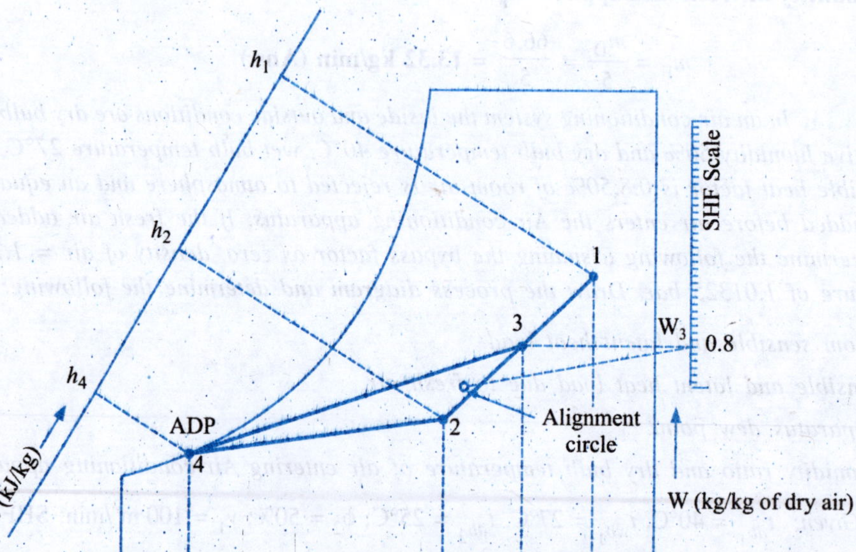

Fig. 14.32

(*i*) **Room sensible and latent heat loads, RSH, RLH:**

$$\text{RSH} = m_a \times c_{pm}\,(t_{db_2} - t_{db_4}) \qquad \left[\begin{array}{r}\text{where, } c_{pm} = \text{humid spacific} \\ \text{heat} = 1.022 \text{ kJ/kg K} \\ \rho_a = 1.2 \text{ kg/m}^3\end{array}\right]$$

$$= \frac{200 \times 1.2}{60} \times 1.022\,(25 - 11.5) = \mathbf{55.2\ kW\ (Ans.)}$$

$$\text{Room total load, RTH} = \frac{200 \times 1.2}{60}\,(h_2 - h_4) = \frac{200 \times 1.2}{60}\,(50 - 32.9) = 68.4 \text{ kW}$$

$$\therefore \qquad \text{RLH} = \text{RTH} - \text{RSH} = 68.4 - 55.2 = \mathbf{13.2\ kW\ (Ans.)}$$

(*ii*) **Sensible and latent heat loads due to fresh air:**

$$\text{Sensible heat load} = m_F \times c_{pm} \times (t_{db1} - t_{db2})$$

$$= \frac{100 \times 1.2}{60} \times 1.022\,(40 - 25) = \mathbf{30.66\ kW\ (Ans.)}$$

Total heat load due to fresh air

$$= \frac{100 \times 1.2}{60}\,(h_1 - h_2) = \frac{100 \times 1.2}{60}\,(85.2 - 50) = 70.4 \text{ kW}$$

$\therefore$ Latent heat load due to fresh air

$$= \text{Total heat load} - \text{sensible heat load}$$

$$= 70.4 - 30.66 = \mathbf{39.74\ kW\ (Ans)}$$

(*iii*) **Apparatus dew point, ADP:**

From *psychrometric chart*, corresponding to point 4, we find:

$$\text{ADP } (= t_{db_4}) = 11.5°\text{C}.$$

(*iv*) **Humidity ratio and DBT of air entering air-conditioning apparatus:**

From *psychrometric chart*, corresponding to point 3, we find:

$$W_3 = \mathbf{0.0135\ kg/kg\ of\ dry\ air\ (Ans.)}$$

$$t_{db3} = \frac{40 + 25}{2} = \mathbf{32.5°C\ (Ans.)}$$

Example 14.15. *Air flowing at the rate of 100 m³/min at 40°C DBT and 50% RH is mixed with another stream flowing at the rate of 20 m³/min at 26°C DBT and 50% RH. The mixture flows over a cooling coil whose ADP temperature is 10°C and by-pass factor is 0.2. Find DBT and RH of air leaving the coil. If this air is supplied to an air-conditioned room where DBT of 26°C and RH of 50% are maintained, estimate room sensible heat factor and cooling load capacity of the coil in tonnes of refrigeration.* **(Anna University)**

Solution: *Given:* $t_{db_1} = 40°\text{C}$; $\phi_1 = 50\%$; $V_1 = 100 \text{ m}^3/\text{min}$; $t_{db_2} = 26°\text{C}$, $\phi_2 = 50\%$; $V_2 = 20 \text{ m}^3/\text{min}$; ADP = 10°C; BF = 0.2

The flow diagram is shown in Fig. 14.33. The various process shown on the psychrometric chart (Fig. 14.34) are discussed below:

- Locate point 1 at the intersection of 40°C DBT and 50% RH lines.
- Locate point 2 at the intersection of 26°C DBT and 50% RH lines.

  Join points 1 and 2.

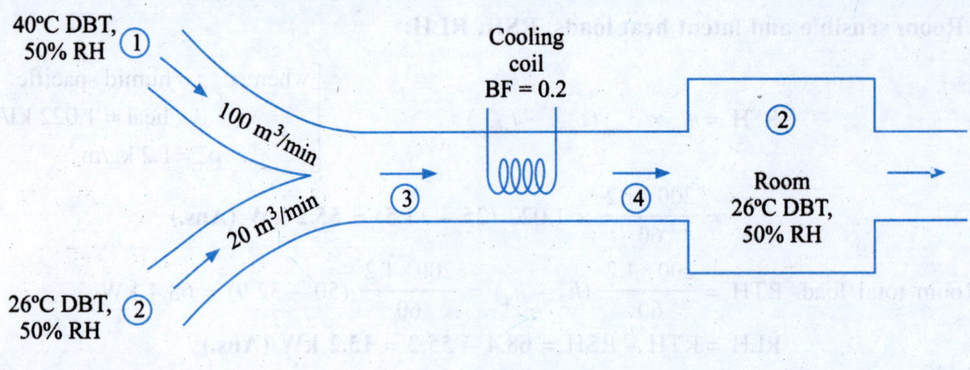

Fig. 14.33

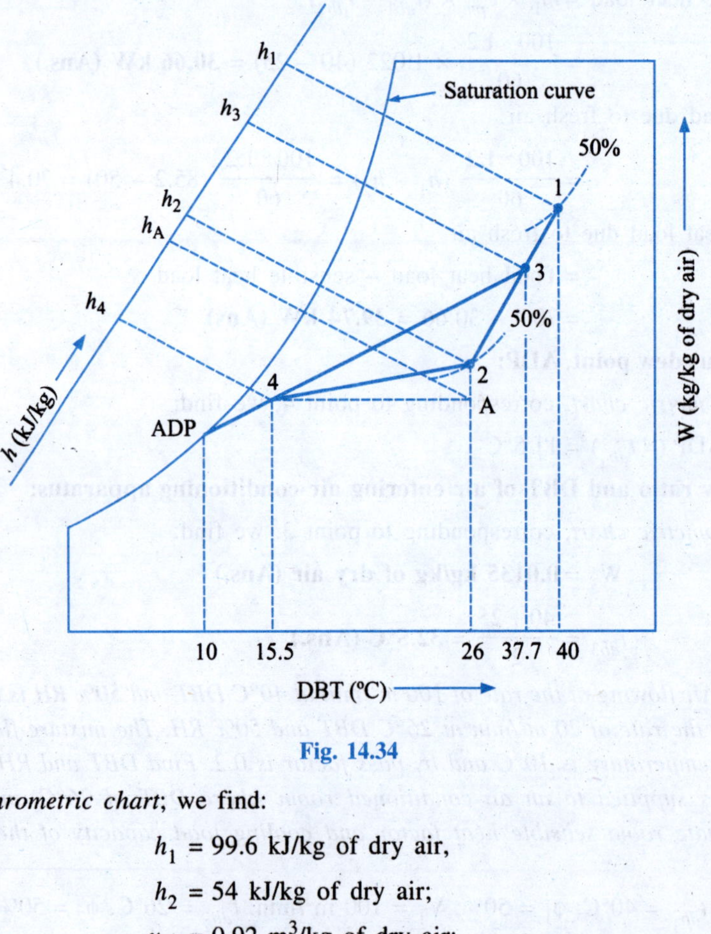

Fig. 14.34

- From *psychrometric chart*; we find:

$$h_1 = 99.6 \text{ kJ/kg of dry air,}$$

$$h_2 = 54 \text{ kJ/kg of dry air;}$$

$$v_{s1} = 0.92 \text{ m}^3\text{/kg of dry air;}$$

$$v_{s2} = 0.862 \text{ m}^3\text{/kg of dry air}$$

Now,

$$m_1 = \frac{V_1}{v_1} = \frac{100}{0.92} = 108.7 \text{ kg/min}$$

$$m_2 = \frac{V_2}{v_2} = \frac{20}{0.862} = 23.2 \text{ kg/min}$$

$$m_3 = \text{Mass of air flowing through the coil}$$
$$= m_1 + m_2 = 108.7 + 23.2 = 131.9 \text{ kg/min}$$

Also, $m_1 h_1 + m_2 h_2 = m_3 h_3$

$$\therefore \quad h_3 = \frac{m_1 h_1 + m_2 h_2}{m_3} = \frac{108.7 \times 99.6 + 23.2 \times 54}{131.9} = 91.6 \text{ kJ/kg}$$

- Locate point 3 on the line 1-2 such that h_3 = 91.6 kJ/kg

  From *psychrometric chart*, we find:

$$t_{db_3} = 37.7°\text{C}$$

- Locate ADP = 10°C on the saturation curve as shown in Fig. 14.34.
- Also, $$\text{BF} = 0.2 = \frac{t_{db_4} - \text{ADP}}{t_{db_3} - \text{ADP}} = \frac{t_{db_4} - 10}{37.7 - 10}$$

or, $$t_{db_4} = 10 + 0.2\,(32.7 - 10) = 15.5°\text{C}$$

- Locate point 4 (t_{db_4} = 15.5°C) on the line joining point 3 and ADP = 10°C.

From the psychrometric chart we find : ϕ_4 = 92%

State of air leaving the coil : 15.5°C DBT, 92% RH (Ans.)

From the psychrometric chart, we find :

$$h_2 = 54 \text{ kJ/kg of dry air}$$
$$h_4 = 42 \text{ kJ/kg of dry air;}$$
$$h_A = 52.5 \text{ kJ/kg of dry air;}$$

Room sensible heat factor, RSHF:

$$\text{RSHF} = \frac{h_A - h_4}{h_2 - h_4} = \frac{52.5 - 42}{54 - 42} = \mathbf{0.87 \ (Ans.)}$$

Cooling load of the coil:

$$\text{Cooling load of the coil} = \frac{m_3 (h_3 - h_4) \times 60}{14000}$$
$$= \frac{131.9\,(91.6 - 42) \times 60}{14000} = \mathbf{28.04 \ tonnes \ (Ans.)}$$

Example 14.16. *The air-handling unit of an air-conditioning plant supplies a total of 4500 m³/min of dry air which comprises by weight 20 per cent fresh air at 40°C DBT and 27°C WBT and 80 per cent recirculated air at 25°C DBT and 50 per cent RH. The air leaves the cooling coil at 13°C saturated state.*

Calculate the total cooling and room heat gain.

Solution: *Given:* t_{db_1} = 40°C; t_{wb_1} = 27°C; t_{db_2} = 25°C; ϕ_2 = 50%; t_{db_4} = 13°C; V_3 = 4500 m³/min.

The schematic arrangement, and the involved psychrometric processes are shown in Fig. 14.35 and Fig. 14.36 respectively; the various process are discussed below:

- Locate point 1 at the intersection of 40°C DBT and 27°C WBT lines.
- Locate point 2 at the intersection of 25°C DBT and 50% RH lines.
- Since 80% air is recirculated and 20% fresh air is used, locate point 3 on the line joining points 1 and 2 such that,

$$\text{Length 2-3} = \frac{1}{5} \times \text{length 2-1}$$

- Locate point 4 on the saturation curve at 13°C as shown in the Fig. 14.36 Join points 3 and 4.

From the *psychrometric chart*, we find:

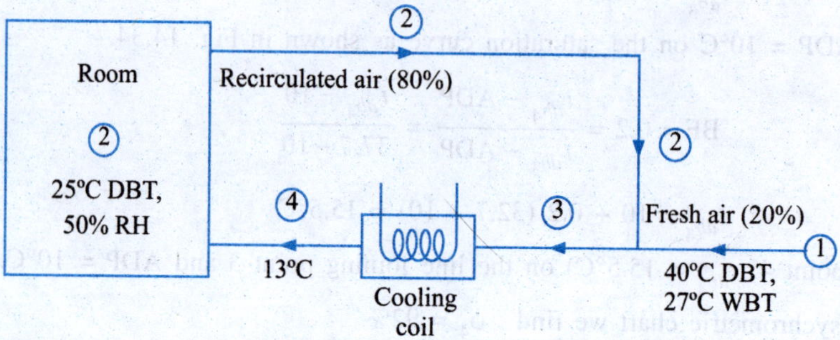

Fig. 14.35

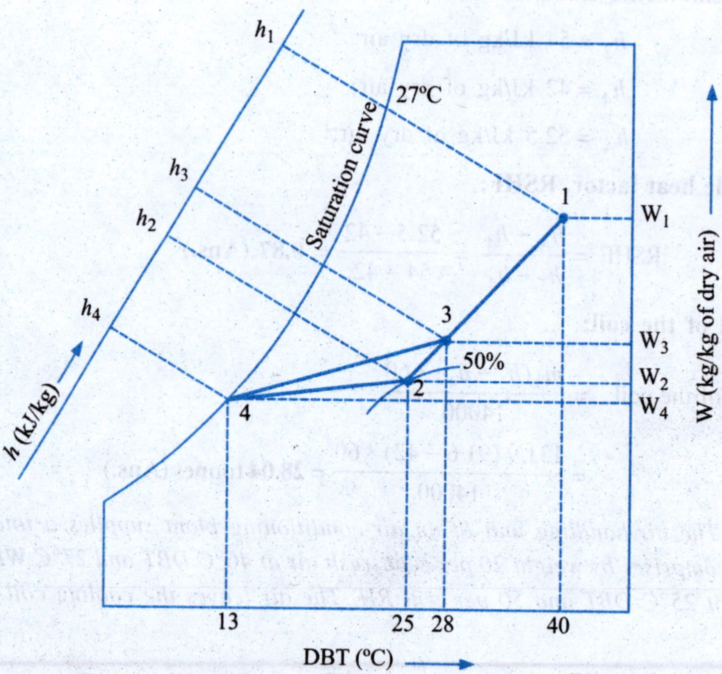

Fig. 14.36

$$h_1 = 89.2 \text{ kJ/kg of dry air};$$
$$h_2 = 51 \text{ kJ/kg of dry air};$$
$$h_3 = 57.3 \text{ kJ/kg of dry air};$$
$$h_4 = 36.6 \text{ kJ/kg of dry air};$$
$$W_1 = 0.0172 \text{ kg/kg of dry air};$$
$$W_2 = 0.01 \text{ kg/kg of dry air};$$
$$W_3 = 0.0113 \text{ kg/kg of dry air};$$
$$W_4 = 0.0093 \text{ kg/kg of dry air};$$
$$v_3 = 0.868 \text{ m}^3\text{/kg of dry air}$$

Mass of air entering the cooling coil,

$$m_a = \frac{V_3}{v_3} = \frac{4500}{0.868} = 5184.3 \text{ kg/min}$$

Total cooling load:

$$\therefore \quad \text{Total cooling load} = \frac{m_a(h_3 - h_4)}{60} \text{ kW}$$

$$= \frac{5184.3\,(57.3 - 36.6)}{60} = \mathbf{1788.6 \ kW \ (Ans.)}$$

Room heat gain:

$$\text{Fresh air load} = \frac{0.2 \times 5184.3 \times (h_1 - h_2)}{60}$$

$$= \frac{0.2 \times 5184.3 \times (85.2 - 51)}{60} = 591 \text{ kW}$$

$$\therefore \quad \text{Room heat gain} = \text{Total cooling load} - \text{Fresh air load}$$

$$= 1788.6 - 591 = \mathbf{1197.6 \ kW \ (Ans.)}$$

Example 14.17. *A retail shop located in a city at 30°N latitude has the following loads:*

Room sensible = 58.15 kW

Room latent heat = 14.54 kW

The summer outside and inside design conditions are:

Outside: 40°C DBT, 27°C WB

Inside: 25°C DBT, 50% RH

70 m³/min. of ventilation air is used. Determine the following if the bypass factor of cooling coil is 0.15:

(i) Ventilation load.

(ii) Grand total heat.

(iii) Effective sensible heat factor.

(iv) Apparatus dew-point.

(v) Dehumidified air quantity.

(vi) Condition of air entering and leaving apparatus.

Assume a bypass factor of 0.15 for the cooling coil. **(Punjabi University)**

Solution: *Given:* t_{db_1} = 40°C; t_{wb_1} = 27°C; t_{db_1} = 25°C; ϕ_2 = 50%, V = 70 m³/min; BF = 0.15

From *psychrometric chart*, we find:

$$h_1 = 85.2 \text{ kJ/kg of dry air;}$$

$$h_2 = 50.5 \text{ kJ/kg of dry air.}$$

Assume standard air density; ρ = 1.2 kg/m³

(*i*) **Ventilation Load, OATH:**

Outside air sensible heat, OASH $= m_0 \times c_{pm} \times (t_{db_2} - t_{db_1})$

$$= \frac{70 \times 1.2}{60} \times 1.022\ (40 - 25) = 21.46 \text{ kW}$$

Outside air total heat OATH,

$$\text{OATH} = m_0\ (h_1 - h_2) = \frac{70 \times 1.2}{60}\ (85.2 - 50.5) = 48.58 \text{ kW}$$

∴ Ventilation load = OATH = **48.58 kW (Ans.)**

(*ii*) **Grant total heat, GTH:**

Outside air latent heat, OALH = OATH – OASH = 48.58 – 21.46 = 27.12 kW

Total sensible heat, TSH = RSH + OASH = 58.15 + 21.46 = 79.61 kW

Total latent heat, TLH = RLH + OALH = 14.54 + 27.12 = 41.66 kW

Grand total heat, GTH = TSH + TLH = 79.61 + 41.66 = **121.27 kW (Ans.)**

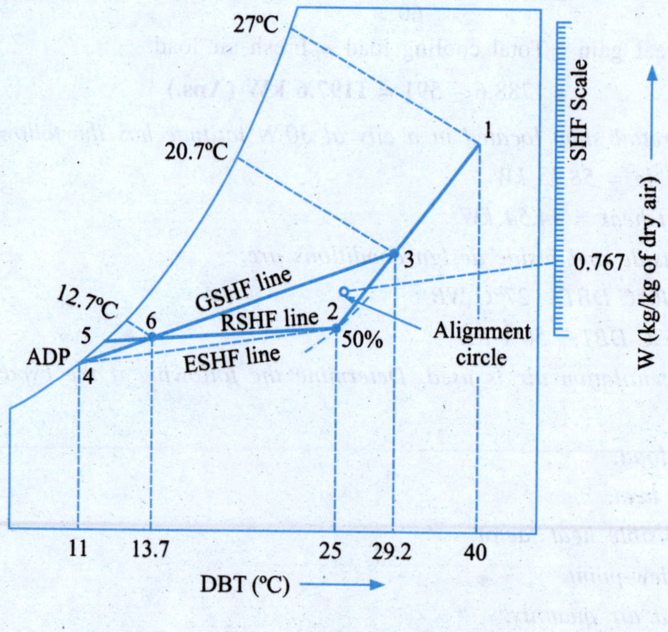

Fig. 14.37

(*iii*) **Effective sensible heat factor, ESHF:**

Effective room sensible heat,

$$\text{ERSH} = 58.15 + 0.15 \times 21.46 = 61.36 \text{ kW}$$

Effective room latent heat,

$$\text{ERLH} = 14.54 + 0.15 \times 27.12 = 18.61 \text{ kW}$$

$$\text{ESHF} = \frac{\text{ERSH}}{\text{ERSH} + \text{ERLH}} = \frac{61.36}{61.36 + 18.61} = 0.767$$

(*iv*) **Apparatus dew point, ADP:**

Draw a line from point 2, parallel to the line joining alignment circle and 0.767 SHF (Fig 14.37), to intersect the saturation curve at point 4. From *psychrometric chart*, we find:

$$\text{ADP } (= t_{db_4}) = \mathbf{11°C\ (Ans.)}$$

(*v*) **Dehumidified air quantity:**

Let, V = Supply air quantity, m^3/min,

Then, $$\text{ERSH} = m \times c_{pm}\,(t_{db_2} - \text{ADP})\,(1 - 0.15)$$

or, $$61.26 = \frac{V \times 1.2}{60} \times 1.022\,(25 - 11) \times 0.85$$

∴ $$V = \mathbf{251.8\ m^3/min\ (Ans.)}$$

Temperature of air entering the coil,

$$t_{db_3} = \frac{181.8 \times 25 + 70 \times 40}{(181.8 + 70)} \simeq 29.2°C$$

Entering air WBT, on the line joining 2-1, and at temperature t_{db_3} is read on the psychrometric chart as 20.7°C.

Hence, *entering air condition* is **29.2°C DBT, 20.7°C WBT (Ans.)**

Supply air temperature (leaving the coil), t_{db_6}, is given by:

$$\text{RSH} = 58.15 = \frac{V \times 1.2}{60} \times c_{pm}\,(t_{db_2} - t_{db_6})$$

or, $$58.15 = \frac{251.8 \times 1.2}{60} \times 1.022\,(25 - t_{db_6})$$

or, $$t_{db_6} = 25 - \frac{58.15 \times 60}{251.8 \times 1.2 \times 1.022} = 13.7°C$$

Leaving air WBT on the line joining 3 and 4 is 12.7°C. Hence *leaving air condition* is **13.7°C DBT, 12.7°C WBT (Ans.)**

Example 14.18. *An air-conditioned space is maintained at 27°C DBT and 50 per cent RH. The ambient conditions are 40°C DBT and 27°C WBT. The space has a sensible heat gain of 14 kW. Air is supplied to the space at 7°C saturated. Calculate:*

(*i*) *Mass of moist air supplied to the space in kg/h;*

(*ii*) *Latent heat gain of space in kW;*

(*iii*) *Cooling load of air washer in kW if 30 per cent of the air supplied to the space is fresh, the remainder being recirculated.* **(PTU)**

Solution: *Given:* $t_{db_1} = 40°C$; $t_{wb_2} = 27°C$; $t_{db_2} = 27°C$; $\phi_2 = 50\%$; SH gain = 14 kW; $t_{db_4} = 7°C$.

The states and psychrometric processes shown in Fig. 14.38 are discussed below:

- Locate point 1 at the intersection of 40°C DBT and 27°C WBT lines.
- Locate point 2 at the intersection of 27°C DBT and 50% RH lines.
  Joint points 1 and 2.
- Locate point 3 on the line 1-2 such that,

  Length 2-3 = 0.3 × length 2-1.
- Locate point 4 (7°C) on the saturation curve as shown in Fig. 14.38 Joint points 3 and 4 and points 2 and 4.

From the *psychrometric chart,* we find:

$$h_1 = 85.2 \text{ kJ/kg of dry air;}$$
$$h_2 = 55.9 \text{ kJ/kg of dry air;}$$
$$h_4 = 22.8 \text{ kJ/kg of dry air;}$$
$$W_1 = 0.0172 \text{ kg/kg of dry air;}$$
$$W_2 = 0.0112 \text{ kg/kg of dry air;}$$
$$W_4 = 0.0062 \text{ kg/kg of dry air;}$$

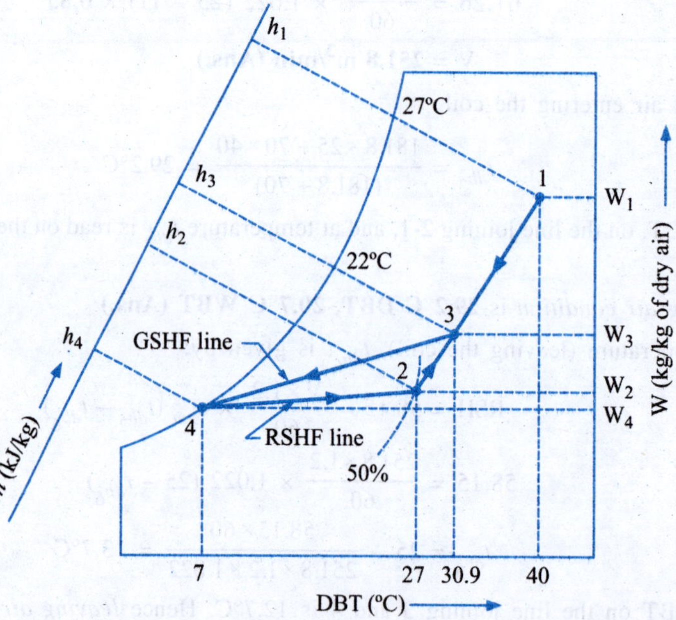

Fig. 14.38

(*i*) **Mass of moist air supplied to the space in kg/h:**

Mass of dry air supplied to space,

$$m_a = \frac{\text{Sensible heat (SH)}}{c_{pm}\,(t_{db_2} - t_{db_4})}$$

$$= \frac{14 \times 3600}{1.022\,(27 - 7)} = 2465.75 \text{ kg/h}$$

Mass of moist air supplied to space

$$= m_a (1 + W_4) = 2465.75 (1 + 0.0062) = \textbf{2481 kg/h (Ans.)}$$

(*ii*) **Latent heat gain of space in kW, LH:**

$$\text{LH gain} = m_a (W_2 - W_4) \times 2500$$

$$= \frac{2465.75 (0.0112 - 0.0062) \times 2500}{3600} = \textbf{8.56 kW (Ans.)}$$

(*iii*) **Cooling load of air washer in kW**

$$t_{db_3} = 0.7 \times 27 + 0.3 \times 40 = 30.9°\text{C}$$

On the line joining 1-2 locate point 3 at 30.9°C. Then from the *psychrometric chart*, we find:

$$h_3 = 64.6 \text{ kJ/kg of dry air}$$

∴ Cooling load of air washer

$$= m_a (h_3 - h_4)$$

$$= \frac{2465.75 (64.6 - 22.8)}{3600} = \textbf{28.63 kW (Ans.)}$$

HIGHLIGHTS

1. An *air-conditioning system* is defined as an assembly of different parts of the system used to produce specified condition of air within a required space on building.
2. The *basic elements of* an air-conditioning system are: Fans, filters, refrigerating plant, means for warming the air, means for humidification and/or dehumidification, and control system.
3. The *main parts* of the equipment in the air-conditioning cycle are: Fan, supply ducts, supply outlets, space to be conditioned, return outlets, return duct, filter and heating chamber or cooling coil.
4. The air-conditioning systems are mainly *classified* as follows:
   (*i*) Central systems (*ii*) Zoned systems
   (*iii*) Unitary systems (*iv*) Unitary-central systems.
5. *Room sensible heat factor* (*RSHF*) is the ratio of the room sensible heat to the sum of room sensible heat and room latent heat.
6. '*Grand sensible heat factor*' (*GSHF*) is defined as the ratio of total sensible heat to the grand total heat that the cooling coil on the conditioning apparatus should handle.

OBJECTIVE TYPES QUESTIONS

A. Choose the correct answer:

1. For winter air-conditioning, the relative humidity should not be more than
   (*a*) 40% (*b*) 60% (*c*) 75% (*d*) 90%.
2. In all-water system of central air-conditioning system
   (*a*) cooling medium is chilled water or brine water and not direct refrigerant
   (*b*) the cooling medium may be supplied from a source away from the conditioned space
   (*c*) the cooling medium may be circulated through the coils of an air terminal
   (*d*) all the above are true.
3. The supply air state of cooling coil with a by-pass factor (*BF*) lies at
   (*a*) intersection of *RSHF* line with saturation curve
   (*b*) intersection of GSHF line with saturation curve

(*c*) point dividing *RSHF* line in proportions of *BF* and (1-*BF*)
(*d*) intersection of RSHF line and *GSHF* line.

4. The air-conditioning system selection
(*a*) should provide correct temperature, humidity, air movement, air cleanliness, ventilation and noise level, otherwise it is rejected
(*b*) should be most economical otherwise it is rejected
(*c*) should be able to take up the peak load of the year, otherwise it is rejected
(*d*) should satisfy the customer's concept of environment desired, otherwise it is rejected.

5. In summer air-conditioning, the air is
(*a*) cooled and humidified (*b*) cooled and dehumidified
(*b*) heated and humidified (*d*) heated and dehumidified.

6. In an All-Air System of central air conditioning
(*a*) the refrigeration plant and air treatment plants may be remotely located in central station apparatus
(*b*) cooling medium or heating medium is air and is sent through the ducts and distributed into conditioned space through outlet or mixing terminals
(*c*) both (*a*) and (*b*) are true (*d*) none of the above is true.

7. Factors for selection of air-conditioning system are
(*a*) purpose, location, orientation, and shape of the space or building
(*b*) temperature, humidity, wind, exposure to sun and shade of environments of the space or building
(*c*) occupancy, lighting, and other heat exchanges and heat storage capability
(*d*) all of the above.

8. The conditioned air supplied to the room must have the capacity to take up
(*a*) room sensible heat load only (*b*) room latent heat load only
(*c*) both room sensible heat and latent heat loads (*d*) any of these

9. The volume control method of taking care of part loads is applicable
(*a*) for load changes of about 50% (*b*) for load changes of about 20%
(*c*) for load changes of about 60% (*d*) for load changes of about 80%.

10. The sensible heat factor for auditorium or cinema hall is generally kept as
(*a*) 0.6 (*b*) 0.7 (*c*) 0.8 (*d*) 0.9.

11. The alignment circle is marked on the psychrometric chart at
(*a*) 20°C *DBT* and 50% *RH* (*b*) 26°C *DBT* and 50% *RH*
(*c*) 20°C *DBT* and 60% *RH* (*d*) 26°C *DBT* and 60% *RH*.

12. In winter air-conditioning, the air is
(*a*) cooled and humidified (*b*) cooled and dehumidified
(*c*) heated and humidified (*d*) heated and dehumidified.

13. In Dual-conduit system of All-Air system of air-conditioning
(*a*) there are two separate air streams
(*b*) one of the two conduits is permanent by supplying cold and variable air volume for fluctuating internal and solar loads
(*c*) one of the two conduits supplies constant volume but variable temperature air for varying building transmission losses
(*d*) all of the above statements are true.

14. For summer air-conditioning, the relative humidity should not be less than
(*a*) 40% (*b*) 60% (*c*) 75% (*d*) 90%

B. Fill in the Blanks:

1. An assembly of different parts of the system used to produce specified condition of air with a required space building is called ....
2. .... are used for moving air.
3. .... are employed for cleaning air.
4. The function of a .... is to direct the air from fan to the room.
5. .... distribute the air evenly in a room.
6. .... is a cooling unit without any heat transfer element such as a cooling coil.
7. A system in which air handing unit is separated from condensing unit is called a .... system.
8. .... unit uses the principle of induction as a means of recirculation.
9. In .... system zoning and duct-work are eliminated.
10. In .... system each room is provided with a room unit which gets a supply of conditioned air from a central system.

ANSWERS

A. Choose the correct answer:

1. (*a*) 2. (*d*) 3. (*d*) 4. (*d*) 5. (*b*) 6. (*d*) 7. (*d*) 8. (*c*)
9. (*b*) 10. (*b*) 11. (*b*) 12. (*c*) 13. (*c*) 14. (*b*).

B. Fill in the blanks:

1. Air-conditioning system; 2. Fans; 3. Filter; 4. Supply duct;
5. Supply outlets; 6. Attic fan; 7. Remote; 8. Induction;
9. Unitary; 10. Unitary-central system.

THEORETICAL QUESTIONS

1. Define an 'air-conditioning system'. Name its basic elements.
2. Explain with a neat sketch an 'air-conditioning cycle'.
3. Enumerate the main parts of the equipment in the air-conditioning cycle.
4. How are air-conditioning system classified?
5. Explain with a heat diagram the working of central system of air-conditioning.
6. With the help of a neat sketch explain the main components of central air-conditioning system.
7. Explain the difference between summer air-conditioning and winter air-conditioning.
8. Write a short note on zone system of air-conditioning.
9. Define a 'Unitary system'. Where is it commonly preferred ? Explain a room air conditioner with a neat sketch.
10. State the advantages of central system over unitary system of air-conditioning.
11. Discuss briefly any two of the following:
    (*i*) Attic fans (*ii*) Remote units (*iii*) Self contained units (*iv*) Unit air coolers.
12. Explain any two of the following unitary central system:
    (*i*) Induction units. (*ii*) All-air high velocity systems. (*iii*) Fan coil units.

13. Write a short note on 'Ice system of air-conditioning.'
14. Define room sensible heat factor. How room sensible heat factor line is drawn on the psychrometric chart?
15. Explain the procedure to draw a ground sensible heat factor line on a psychrometric chart.
16. What do you understand by effective room sensible heat factor?
17. List the applications of air-conditioning.
18. Define room sensible heat factor.
19. Explain the procedure to draw a *GSHF* line on a psychrometric chart.
20. What is "Effective room sensible heat factor"?
21. State the factors which should be taken into consideration while selecting a system of air-conditioning.

UNSOLVED EXAMPLES

1. Following data refer to an air-conditioning system to be designed for an industrial process for hot and wet climate:

   Outside conditions = 30°C DBT and 75% RH; Required inside conditions = 20°C DBT and 60% RH.

   The required condition is to be achieved first by cooling and dehumidifying and then by heating. If 20 m^3 of air is absorbed by the plant every minute, determine:

   (*i*) Capacity of the cooling coil in tonnes of refrigeration;

   (*ii*) Capacity of the heating coil in kW; (*iii*) Amount of water removed per hour.

   [Ans. (*i*) 4.59 tonnes; (*ii*) 3.16 kW; (*iii*) 15.48 kg/h]

2. Air at 10°C DBT and 90% RH is to be brought to 35°C DBT and 22.5°C WBT with the help of winter air-conditioner. If the humidified air comes out of the humidifier at 90% RH, draw the various processes involved on a skelton psychrometric chart. Determine:

   (*i*) Temperature to which the air should be preheated;

   (*ii*) Efficiency of the air washer. **[Ans.** (*i*) 31.2°C; (*ii*) 92.7%]

3. 300 m^3/min of air at atmospheric conditions 35°C DBT and 55% RH is supplied to an air-conditioned hall. The required conditions are 20°C DBT and 60% RH. Determine:

   (*i*) Sensible heat and latent heat removed from the air per minute, and

   (*ii*) Sensible heat factor for the system. **[Ans.** (*i*) 5066.2 kJ/min; 9465.7 kJ/min; (*ii*) 0.348]

4. In an air-conditioning plant, an air-handling unit supplies a total of 4500 m^3/min of air which comprises by mass 20 per cent of fresh air at 40°C DBT and 27°C WBT and 80 per cent recirculated air at 25°C DBT and 50% RH. The air leaves the cooling coil at 13°C saturated. Calculate:

   (*i*) Total cooling load, and

   (*ii*) Room heat gain. **[Ans.** (*i*) 466 tonnes, (*ii*) 314.8 tonnes]

5. An air-conditioning system is designed under the following conditions:

   Outdoor conditions = 30°C DBT, 75% RH; Required indoor conditions = 22°C DBT, 70% RH;

   Amount of free air circulated = 3.33 m^3/s; Coil dew point temperature = 14°C.

   The required condition is achieved first by cooling and dehumidification, and then by heating. Determine:

   (*i*) Capacity of the cooling coil in tonnes, (*ii*) Capacity of the heating coil in kW, and

   (*iii*) Amount of water vapoured removed in kg/s.

   [Ans. (*i*) 33.79 tonnes; (*ii*) 18.77 kW; (*iii*) 0.0319 kg/s]

6. For a hall to be conditioned, the following conditions are given :

Outdoor condition .... 40°C DBT, 20°C WBT

Required comfort condition .... 20°C DBT, 60% RH

Seating capacity of hall 1500

Amount of outdoor air supplied .... 0.3 m^3/min per person

If the required condition is achieved first by adiabatic humidification and then by cooling, determine:

(*i*) Capacity of the cooling coil in tonnes, and

(*ii*) Capacity of the humidifier in kg/h **[Ans.** (*i*) 32.27 tonnes, (*ii*) 69.3 kg/h]

7. The following data refer to air-conditioning of a public hall:

Outdoor conditions = 40°C DBT, 20°C WBT; Required comfort conditions = 20°C DBT, 50% RH;

Seating capacity of hall = 1000; Amount of outdoor air supplied = 0.3 m^3/min/person.

If the required condition is achieved first by adiabatic humidifying and then cooling, find:

(*i*) Capacity of cooling coil and surface temperature of the coil if the by-pass factor is 0.25, and

(*ii*) Capacity of the humidifier and its efficiency. **[Ans.** (*i*) 26.64 tonnes, 14°C; (*ii*) 20.1 kg/h, 10%]

8. An air-conditioning plant is to be designed for a small office for winter conditions:

Outdoor conditions = 10°C DBT and 8°C WBT; Required indoor conditions = 20°C DBT and 60% RH;

Amount of air circulation = 0.3 m^3/min/person; Seating capacity of the office = 50 persons.

The required condition is achieved first by heating and then by adiabatic humidifying. Find:

(*i*) Heating capacity of coil in kW and surface temperature, if the by-pass factor of the coil is 0.32, and

(*ii*) Capacity of the humidifier. **[Ans.** (*i*) 5.5 kW, 35.7°C; (*ii*) 3.3 kg/h]

9. An air-conditioned auditorium is to be maintained at 27°C DBT and 60% RH. The ambient condition is 40°C DBT and 30°C WBT. The total sensible heat load is 10^5 kJ/h and total latent heat load is 4×10^4 kJ/h. 60 per cent of return air is recirculated and mixed with 40 per cent of make-up air after the cooling coil. The condition of air leaving the cooling coil is at 17°C. Determine:

(*i*) Room sensible heat factor.

(*ii*) Condition of air entering the auditorium.

(*iii*) Amount of make-up air.

(*iv*) Apparatus dew point.

(*v*) By-pass factor of the cooling coil.

[Ans. (*i*) 0.714; (*ii*) t_{db} = 23°C, t_{wb} = 19.5°C, ϕ = 72% (*iii*) 11200 kg/h; (*iv*) 13°C; (*v*) 0.185]

10. It is required to maintain an air-conditioned hall at 27°C DBT and 21°C WBT. It has a sensible heat load of 46.5 kW and latent heat load of 17.5 kW. The air supplied from outside atmosphere at 38°C DBT and 27°C WBT is 25m^3/min, directly into the room through ventilation and infiltration. Outside air to be conditioned is passed through the cooling coil whose ADP is 15°C. The quantity of recirculated air from the hall is 60 per cent. This quantity is mixed with the conditioned air after the cooling coil. Determine:

(*i*) Condition of air after the coil and before the recirculated air mixes with it.

(*ii*) Condition of air entering the hall, i.e., after mixing with recirculated air.

(*iii*) Mass of fresh air entering the cooler.

(*iv*) By-pass factor of the cooling coil.

(*v*) Refrigerating load on the cooling coil.

[Ans. (*i*) 19°C DBT, 17.5°C WBT; (*ii*) 24°C DBT, 19.8°C WBT; (*iii*) 375 kg/min; (*iv*) 0.174; (*v*) 57.87 tonnes]

11. The following data refer to an air-conditioned space:

Outdoor condition .... 40°C DBT, 50% RH

Room condition .... 25°C DBT, 50% RH

Sensible heat load .... 25 kW

Latent heat load .... 5 kW

By-pass factor of the cooling coil .... 0.15.

If the ventilation requirement is such that on mass flow rate basis 20 per cent of fresh air is introduced and 80 per cent of supply air is recirculated, determine:

(*i*) Supply air flow rate. (*ii*) Outsider air sensible heat.

(*iii*) Outside air latent heat. (*iv*) Grand total heat.

(*v*) Effective room sensible heat factor.

[**Ans.** (*i*) 108.7 m^3/min, (*ii*) 6.66 kW; (*iii*) 15 kW, (*iv*) 51.66 kW; (*v*) 0.782]

12. The following data refer to summer air-conditioning of a building.

Outside design conditions = 43°C DBT, 27°C WBT; Inside design conditions = 25°C DBT, 50% RH; Room sensible heat gain = 84000 kJ/h; Room latent heat gain = 21000 kJ/h; By-pass factor of the cooling coil used = 0.2

The return air from the room is mixed with the outside air before entry to cooling coil in the ratio of 4 : 1 by mass. Determine:

(*i*) Apparatus dew point of the cooling coil. (*ii*) Entry and exit conditions of air for cooling coil.

(*iii*) Fresh air mass flow rate. (*iv*) Refrigeration load on the cooling coil.

[**Ans.** (*i*) 11°C; (*ii*) 28.8°C DBT; 19.9°C WBT; 14.5°C DBT; 13°C WBT; (*ii*) 1591 kg/h; (*iv*) 11.47 WBT tonnes.]

13. The following data relate to an air-conditioned room:

Outside conditions .... 35°C DBT, 28°C WBT

Inside conditions .... 25°C DBT, 50% RH

Sensible heat load .... 11.5 kW

Latent heat load .... 7.5 kW

Apparatus dew point .... 8°C

By-pass factor of the cooling coil used 0.1

Return air from the room is mixed with the outside air before entering the cooling coil in the ratio of 4 : 1 and return air from the room is also mixed after the cooling coil in the ratio 1 : 4. The air may be reheated, if necessary, before supplying to the conditioned room.

Determine the following:

(*i*) Supply air condition to the room. (*ii*) Refrigeration load.

(*iii*) Quantity of fresh air supplied.

[**Ans.** (*i*) 17°C DBT, 13.2°C WBT; (*ii*) 8.64 tonnes; (*iii*) 13.44 kg/min]

❑❑❑

Air-Conditioning Equipments, Components and Controls

Chapter 15

15.1 AIR-CONDITIONING EQUIPMENT

The air-conditioning equipment can be broadly *classified* in two groups:

1. Package units
2. Central units.

Package units *include window room air-conditioners and all such installations where functional components are packed all together.*

Central units *refer to the installation of the different components at different points of the building.*

As for functions-wise distribution is concerned both types will fall in the *following categories:*

(*i*) Only cooling type system.
(*ii*) Combination cooling and heating type system.
(*iii*) Heat pump system.

15.1.1 Package Units

Package units are of two types:

1. Window type.
2. Console type.

1. Window type package units

- A *package unit is a self-contained unit* because the complete unit including evaporator and condensing unit is *all incorporated in a common enclosure.*
- The normal capacity of such a unit is 1 and 1.5 tonnes.
- These are window mounting models which are normally capable of cooling, heating, cleaning and circulating the air. The air distribution is met by a grill arrangement which also allows fresh air through dampers. The dampers are also provided for exhaust purposes.

- A *window type air-conditioner* is basically designed for cooling of room where it is installed. The entire system consists of following sub-assemblies:

Subassembly	**Parts**
1. System assembly	(*i*) Evaporator
	(*ii*) Capillary
	(*iii*) Condenser
	(*iv*) Strainer
	(*v*) Compressor.
2. Motor, fan and blower assembly	(*i*) Fan
	(*ii*) Blower motor
	(*iii*) Motor mounting brackets.
3. Cabinet and grill assembly	(*i*) Cabinet
	(*ii*) Grill.
4. Switch board panel	(*i*) Selector switch
	(*ii*) Relay
	(*iii*) Thermostat
	(*iv*) Fan motor capacitor
	(*v*) Running capacitor for compressor motor and starting capacitor for compressor motor.

Fig. 15.1 shows a schematic diagram of a *window type air-conditioner*.

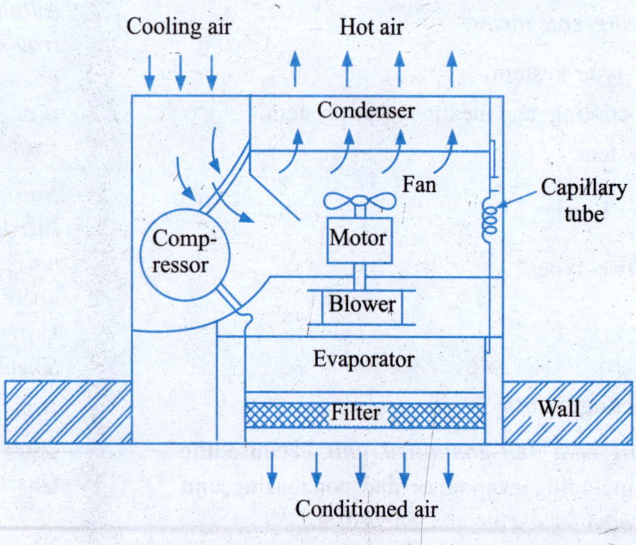

Fig. 15.1. Window type air-conditioner.

2. **Console type package units**

- These units are *usually packaged* in decorative cabinets and are mounted on the floor directly.
- Normally water-cooled condensers are employed with such units.

- The capacities are normally 3 to 20 tonnes.
- For proper delivery of conditioned air to the space a short length of duct-work is attached or built-in grilles are placed on the top. For heating cycle, if desired, electric strip heaters can be installed. These heaters are inter-locked with the blower motor using in addition a heating thermostat for safety and temperature control purposes.

15.1.2 Central Units

- In **central units** (central air-conditioning system) *the air handling unit is generally separated from the condensing unit.*
- Such units are available in two types : Horizontal and vertical depending upon the position of filter and the drain pan.
- The cooling coil in these units may be either direct expansion (DX) or *chilled water type.*
- The *functional elements* of a central unit are:

  (*i*) Equipment for air flow like blower and a motor to drive it.

  (*ii*) Equipment for cooling and/or dehumidification like direct expansion (DX) coil or chilled water coil.

  (*iii*) Equipment for cleaning of air like filters.

  (*iv*) Equipment for dehydration.

  (*v*) Equipment for operational and safety controls.

Fig. 15.2 shows a typical control air-conditioning unit.

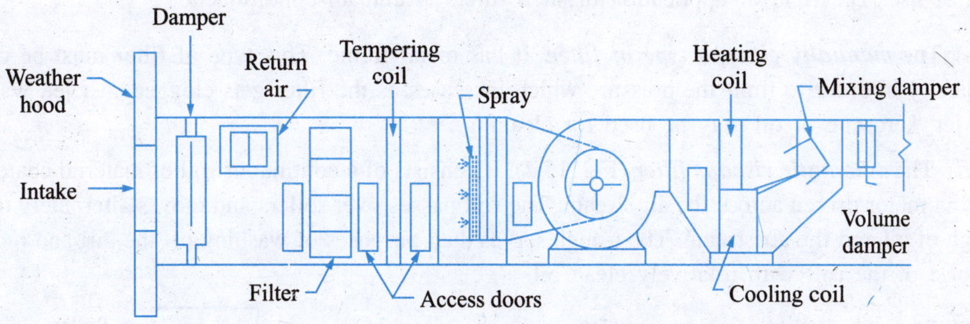

Fig. 15.2. Typical control air-conditioning unit.

15.2 AIR-CONDITIONING COMPONENTS

Some important components used in air-conditioning equipment (systems) are discussed in the following Articles.

15.2.1 Filters

- The function of filters is to *arrest* the solid impurities such as soot, ash, lint, smoke and fumes and even living organisms such as virus, bacteria and fungus spores.
- These are placed *ahead of heating or cooling coils.*

The filters are broadly *classified* as follows:

1. Dry air filters.
2. Viscous impingement filters.
3. Electrostatic filters.

1. Dry air filters. In these filters the filtering medium consists of fabrics made out of wool felt, cotton batting, cellulose fibre etc. The medium is supported in pockets of wire frame or V-shaped plates in order to increase the area of medium exposed to air. Such filters have high dust arrestance including lint. The cleanable filtering medium in permanent frames must be cleaned by application of a vacuum cleaner on the windward side. The *throw away* filters have frame of card board of fibre; these must be replaced by a new one, after collection of full dust load.

2. Viscous filters. In these filters the filtering medium consists of relatively coarse fibres packed between two panels of expanded metal screen enclosed in a frame. The coarse fibres may be of glass, steel wool, wire screen, animal hair, hemp fibres, metal stamping or shavings. The *fibres are dipped in oil or grease called adhesive intended to retain dust particles which come in contact with it.* There are three types of such filters:

(*i*) Throw away or replaceable type.

(*ii*) Manually cleaned type.

(*iii*) Automatic or sell cleaning filter.

(*i*) ***A throw-away or replaceable type*** (Fig. 15.3). This type of filter contains fibrous material such as *glass wool having card-board frames and retaining grilles*. It is designed to be discarded after one period of use. The field of application of such filters is unit air-conditioners.

(*ii*) The ***manually cleaned type of filter.*** It has metal frame. This type of filter must be cleaned periodically in order to limit the pressure which increases as the filter gets clogged. Air jet, water jet, steam jet, kerosene or oil may be used for cleaning.

(*iii*) The ***automatic viscous filter*** (Fig. 15.4). It consists of a continuous roll of material coated with oil and is motor driven across the air stream. The roll passes over rollers and moves alternately through a trough of oil and the air stream. The trough serves dual purposes of washing off the dirt and recoating the fabric of the roll with relatively clean oil.

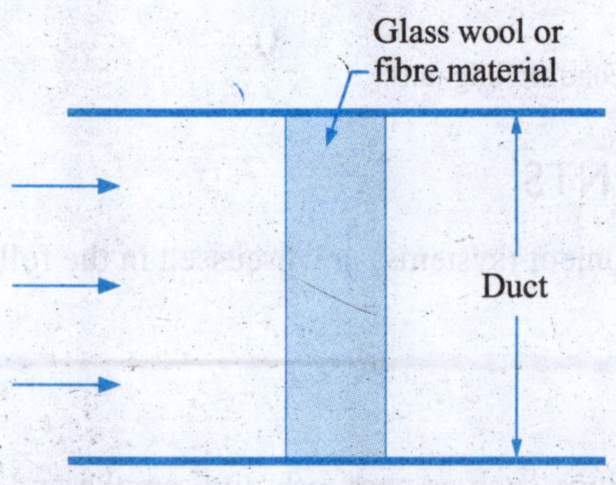

Fig. 15.3. Throw-away or replaceable type filter.

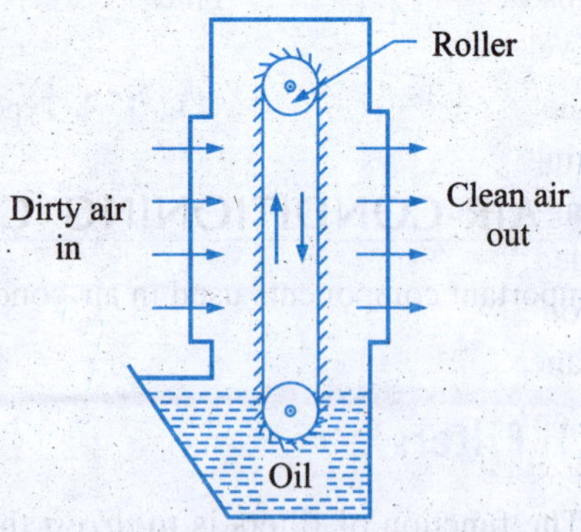

Fig. 15.4. Automatic viscous filter.

3. Electrostatic filters. These filters make use of the *principle of electrostatics*. The dirty stream is made to pass through a series of charging plates where dirt particles get electrically charged. There this air stream passes through a bank of collector plates which are given an opposite charge. Consequently the charged dirt particles adhere to the collector plates of opposite polarity which are removable for cleaning.

The tobacco smoke can be cleaned by electrostatic method where very high D.C. voltage (about 15000 V) is employed. Fig. 15.5. shows a typical electrostatic filter.

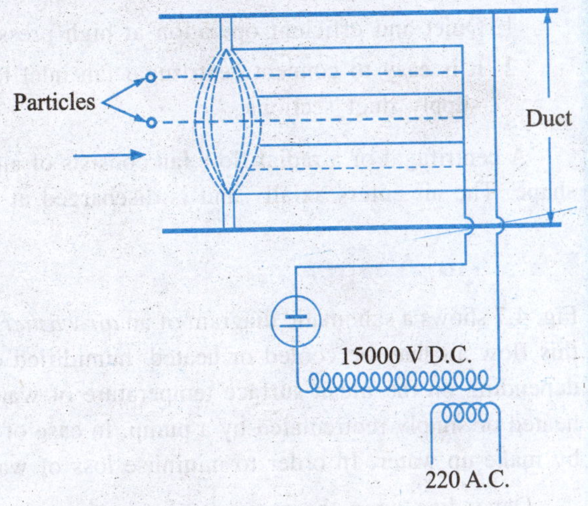

Fig. 15.5. Electrostatic filter.

15.2.2 Fans

The primary function of a fan is to produce air movement through heating, ventilating and air-conditioning apparatus. It essentially consists of a rotating wheel which is surrounded by a stationary member known as a *housing*. The fans can be *classified* in two general groups.

1. Axial flow fans.
2. Centrifugal or radial-flow fans.

Axial flow fans. These fans *produce a flow of air in a direction parallel to the axis of rotation*. These are *suitable for handling large air volumes* and can be used where noise-level considerations are not important. They are therefore, used in industrial ventilation and air-conditioning systems. The majority of fans are centrifugal or radial-flow type. Fig. 15.6. shows a schematic representation of an axial-flow fan. The flow or air is parallel to the axis of the impeller. Blades are of aerofoil section. The tips run with as fine a clearance as practicable in a cylindrical casing. Air enters in the axial direction and leaves with a rotational component due to the work done. The absolute velocity of the leaving air is higher than the axial velocity.

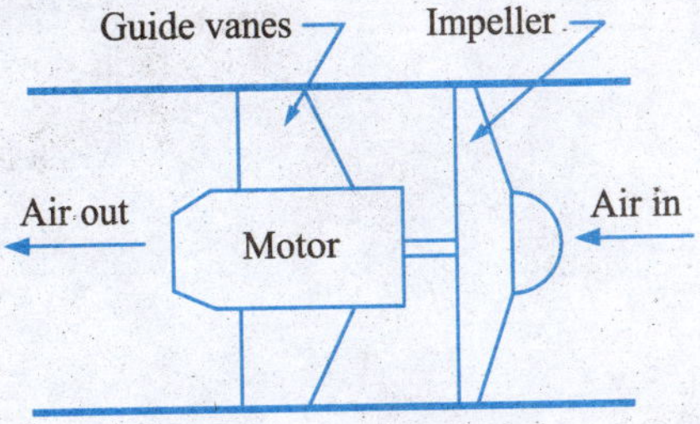

Fig. 15.6. Axial flow fan.

The axial flow fan most commonly used, the *disc or propeller fan* is best adapted to applications requiring flow against low resistances. When an axial-flow fan is placed in a cylindrical drum-type housing equipped with satisfactory directional vanes, it is called *vane-axial-flow* fan. If vanes are not included, the term *tube-axial* is used. These fans can be applied in air-conditioning work and operate at good efficiencies and satisfactory noise levels if properly selected for air delivery and system resistance.

Centrifugal or radial-flow fans. In case a *system has duct-work, centrifugal fans have to be used as the static pressure drop is considerable*. But when there is no duct-work, propellers or axial flow fans can be used. Nevertheless, *in window type and package units, drum-type centrifugal fans only are used, whereas most exhaust fans are of axial type*. A centrifugal fan possesses the following ***advantages:***

1. Quiet and efficient operation at high pressures
2. It is easy to connect centrifugal fan inlet to large apparatus sections and its outlet to smaller supply duct sections.

A centrifugal or a radial flow fan consists of an impeller running in a casing, normally of a volute shape. The air enters axially and is discharged at the periphery.

5.2.3 Air-Washer

Fig. 4.7 shows a schematic diagram of an *air washer*. Here air flows through a spray of water and during this flow air may be cooled or heated, humidified or dehumidified, or simply adiabatically saturated, depending on the mean surface temperature of water. The water is accordingly, externally cooled or heated or simply recirculated by a pump. In case of humidification of air loss of water is compensated by make-up water. In order to minimise loss of water droplets eliminator plates are used.

Depending upon the spray water temperature, the processes which may take place are discussed in detail below:

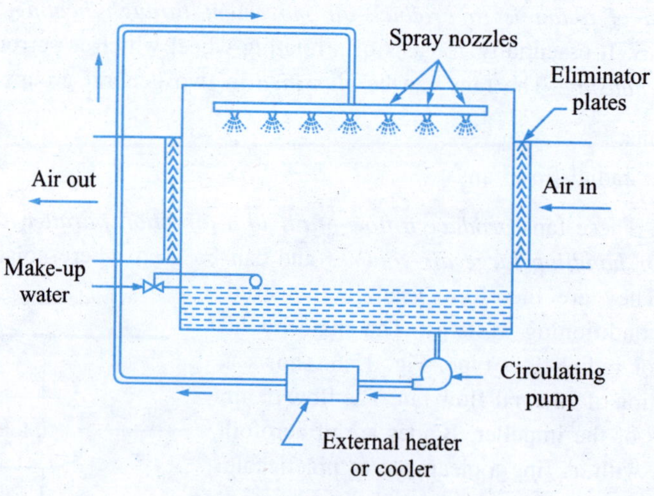

Fig. 15.7. Air washer.

Case I. No heat is supplied to or rejected from the spray water, and the same water is circulated again and again. If there is intimate contact between air and water *adiabatic saturation* will be reached, the process followed is *a-b* as shown on psychrometric chart in Fig. 15.8. The temperature of spray water will reach the thermodynamic wet-bulb temperature entering air. This is a process *along a constant wet-bulb temperature line. The air is cooled and humidified at constant enthalpy.*

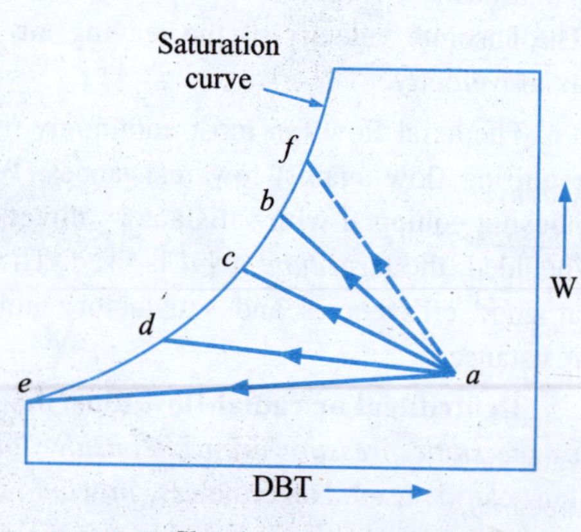

Fig. 15.8. Sketch on psychrometric chart, showing various air-washer processes.

Case II. If the temperature of spray water is maintained at a value *less than the wet-bulb temperature* of the entering air but *higher than the entering dew point temperature* by cooling the spray water by coolers before pumping it back to the spray nozzles, then the process followed is shown as *a-c* on the skeleton psychrometric chart, Fig. 15.8. It is still *cooling and humidification process but the enthalpy of air is reduced.*

Case III. If the temperature of spray water is maintained *equal to the dew point temperature* of entering air, the process is *sensible cooling* process *a-d* in Fig. 15.8. The *water vapour contents of air remains constant during the process.*

Case IV. If the temperature of spray water is *further lowered* and *maintained lower than entering dew point temperature*, the process is *cooling and dehumidification, a-e* in Fig. 15.8.

Case V. If the temperature of spray water is *maintained constant at a value higher than the entering well-bull temperature but lower than the inlet dry-bulb temperature,* by heating of spray water by heaters before pumping back to spray nozzles, the process followed is along *a-f* as shown in Fig. 15.8. There is cooling and humidification of air, but its *enthalpy is increased.*

15.2.4 Radiator

- A radiator is a heating unit/heat transfer element. It is an important component of winter heating equipment. It is exposed to view within the room to be heated. All objects which fall within its visible range are heated by *radiation*. It also imparts heat by conduction to surrounding air which in turn circulates the heat by *natural convection currents*. Thus heat is transferred to room by the radiator both by radiation and convection. When a radiator is placed or covered in such a way that it is not visible in the room, it is said to be a *concealed radiator*. This type of radiator heats the room by *natural convection of air current*. '*Base board radiators*' are installed along the bottom of the walls in places of the conventional board. These are supplied with steam, hot water or electric energy. They operate with gravity circulated (natural convection) room air and have a substantial portion of their surface directly exposed to the room. The **'finned tube units'** consist of metal tubing with metallic fins bonded to the tube. The tubing is supplied with hot water or steam. They may be completely exposed to room or enclosed with top, front or inclined outlets. They operate with gravity or natural circulation of room air. **Pipe coils** are sometimes used in factory buildings and are usually placed under windows or along exposed wall. The output of these units is expressed in kJ/h. or in sq. metre of equivalent direct radiation (EDR).
- Preferably a radiator should be installed near floor and should deliver a horizontal natural current of heated air so that it mixes readily with room air and there by reduce stratification i.e., difference of temperature of air at floor and ceiling.

15.2.5 Convector

A *convector* is also a heating unit. It may or may not be concealed and is placed in the room to be heated. It transfers heat to the room mainly by convection. The convection currents may be natural or forced. It may be supplied (like radiator) with steam, hot water or electrical energy. Fig. 15.9 shows a schematic representation of a convector. Cold room air enters at the bottom while the heated air is discharged at the outlet grill by natural currents set up due to heating. Convectors with electrical resistors usually have ratings from 1 kW to 8 kW. Base board type units are usually supplied with hot water but sometimes with steam. They do not interfere with furniture placement. They give a pleasant appearance and can be easily installed. *Finned tube units (convectors)* are placed in recesses of exposed wall and can be used with either water or steam.

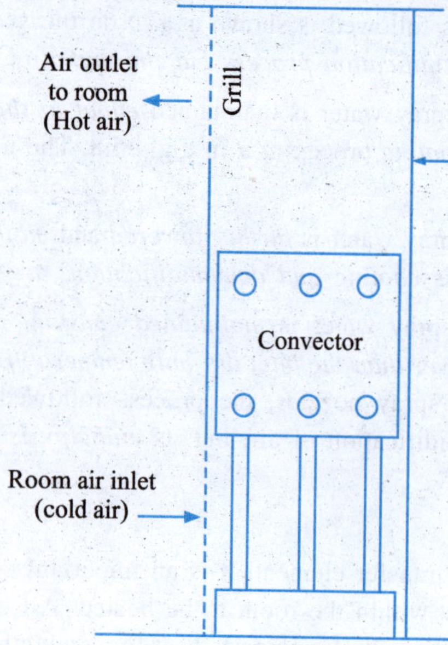

Fig. 15.9. Convector.

15.3 AIR-CONDITIONING CONTROLS

The field of application of air-conditioning control devices depends upon the size of area under consideration, the number and size of the units involved etc. Depending upon particular requirements the following *three* types of controls are used for air-conditioning:

1. Manual controls,
2. Automatic controls, and
3. Semi-automatic controls.

15.3.1 Manual Controlled System

- The manual control can be an *ordinary knife switch which must be operated by hand when the actuating mechanism is started or stopped.*
- This method requires personal attention and regulation at frequent intervals, *especially when the load varies.* Even if the plant is in operation only at certain times of the day, inside and outside conditions may vary so greatly that constant changes in adjustment are necessary.
- In this type of control *constant attention is required* by the operator to maintain the predetermined condition, and the *cost of operation is usually found to be excessive.*
- If a manual control is provided, the use may permit the unit to operate until the room is too cold, or else he may forget to start the machine unit after the room has reached an uncomfortable warm temperature. In either case, the user will not be satisfied, and therefore, *it is usually wise to install an automatic control.*

15.3.2 Automatic System

- To meet the *varying load* almost constantly completely automatic system may be employed.
- The various instruments used in this type of control are:
  - — *Time clocks,*
  - — *Thermostats,*
  - — *Humidity controllers, and*
  - — *Automatic dampers.*
- Automatically operated controls *provide all the features that are considered most desirable for temperature and humidity regulation as well as economy of operation.* In this *system operator or owner is relieved of all responsibility for maintaining the predetermined conditions.*

15.3.3 Semi-automatic Control System

This system makes use of a *close-differential thermostat* and in most cases it is entirely satisfactory. *Time clocks are also very desirable on installations where definite shut-down periods are required.* The fresh air intake from outdoors should be equipped with adjustable louvers, so that maximum amount of air can be regulated for both summer cooling and winter heating. On any unit below 25 tonnes capacity it is desirable to keep the control system as simple and free from service liabilities as possible.

The control devices that may be encountered are briefly outlined as follows:

Thermostats. A *thermostat is a device that operates by changes in room temperature.* In some thermostat a bimetal strip is used to make and break a set of contacts. In another design, the bimetal strip is coiled into a loose helix, with outer coil rigidity fixed to the case. The inner coil is free to move in accordance with temperature changes, and to this free end is secured a glass capsule containing an ample charge of mercury. Two electric wires are seated in one end of the capsule, their base ends projecting into the capsule. When the capsule is tilted in one direction, the mercury flows to the lower end; if this happens to the plain end, the circuit is broken. The electrical circuit is complete only when the capsule is titled in the other direction, so that mercury flows to the end containing the two wire ends. The mercury bridges the space between the two wires and acts as a metallic conductor.

15.3.4 Automatic Humidity Control

Humidistat *is an instrument which controls the humidity automatically.* In the instrument the element is some type of hygroscopic material, such as wood, human hair, paper and similar materials. When the material absorbs water from the air, expansion takes places. It is this action that is used to make and break the circuit controlling humidifying or dehumifying apparatus.

15.3.5 Air Movement Control

Movement of air is usually determined by factors viz., *size of the blowers*, the *size of the ducts* and *supply grilles*, and the *speed of the blower motor*. These factors are established when the system is installed and can only be varied by changing the motor speeds or by the manipulation of suitable dampers.

15.3.6 Automatic Temperature Control

The automatic device that controls the temperature of the unit is the *thermostat*. It contains an element that is sensitive to changes in temperature. This element is usually a bimetal strip or a metal bellows filled with a volatile liquid. An automatic temperature control system is generally operated by making and breaking an electric circuit or by opening and closing a compressed air line. Electrically operated systems are used by manufacturers for practically all installations whereas compressed air operated systems are employed in extremely large central and multiple installations.

15.3.7 Limit Switches

The use of limit switches is made in heating installations to prevent the fan motor from circulating cold air when the heat is off. These are also employed to prevent the humidifier from operating when the heat is off, or to cut off refrigeration when heat is on.

15.3.8 Time Clocks

Timers are merely clock-operated switches that can be placed either with the thermostat or manual switch, thus automatically preventing operating during certain periods.

15.4 NOISE AND NOISE CONTROL

15.4.1 Introduction

- **Sound** (a by-product of many industrial operations) is defined *as a vibration in an elastic medium, generally in the frequency ranges that produce the sensation of hearing.* **Noise** is defined as *unwanted sound and it is often an unwelcome by-product of industrial operations.*

  Sound is *classified* into two basic categories.

  (*i*) ***Airborne sound.*** It *moves from one point to another through air.*

  (*ii*) ***Structure borne sound.*** *It passes through solid material, such as concrete slab or steel beam.*

  Most sound within a plant is airborne. It can be effectively reduced by applying accoustical materials at the following three possible points:

  — At the source;

  — Along the path; or

  — At the receiver.

- The modern air-conditioning systems, in addition to the control of temperature, humidity and air velocity of excessive noise which causes uncomfortable feeling, require control of noise below a particular level. The **object** *of controlling the noise in air-conditioned system is that the occupier should be provided comfortable noise free environment.*

  The noise is contributed to the air-conditioned spaces, by the following:

  — Blowers,

  — Motors,

  — Humidifies etc.

— Noise is also created by air passing through the ducts and grills (and is transmitted to the conditioned space).

- Nowadays the successful design of air-conditioning system is dependent on the successful design and installation of noise control equipments in addition to air-conditioning equipments.

15.4.2 Effects of Noise Pollution (Physiological)

The following are some important *effects of noise pollution*:

1. Liver and kidney may get affected (due to production of ulcers and pathological changes) owing to prolonged exposure of high intensity sound.
2. High level noise exposure may lead to abnormal heart rhythms.
3. Slower and deeper breathings and slowing of digestive processes.
4. High intensity sound can lower electrical skin resistance.
5. Loss of hearing.

15.4.3 Sources of Noise in Air-conditioning Systems

The noises encountered in heating, ventilating, air-conditioning, refrigeration and process industries are caused by the following:

1. Rotating, vibrating and oscillating machines,
2. Moving fluids,
3. Impacts, and
4. Combustion processes.

15.4.4 Noise Control

The following methods are commonly used to *control the noise in refrigeration and air-conditioning systems*:

1. **Resonance.** Resonance is one of the main causes of amplification of noise dependent on how it was originally produced. It can be reduced by: (*i*) Damping ; (*ii*) Isolation and (*iii*) De-tuning
2. **Radiation.**
   - If a fan is bolted rigidly to a roof, it will use the roof to radiate energy to the surroundings. It is due to this reason that fans are usually mounted on anti-vibration *mounts, with flexible connection connecting the fan to ducting.*
   - The noise radiated from ducts or panel can be effectively reduced by lagging them with rockwool and further putting some kind of impervious surface over the rockwool.
   - The noise transmitted in air-flows are reduced by using silencers.
3. **Impending noise transfer.** In this method of noise control source of noise is enclosed in some way so that noise getting to the surrounding area is stopped.

15.4.5 Sound Absorbing Material(s)

Fibre glass crown wool is the most effective material in combating reverberation. It entails the following ***advantages***:

1. Non-flammable.
2. Rot and vermin proof.
3. Non-hygroscopic.
4. Light in weight and easy to handle, cut and install.
5. When applied to ceilings and walls, it serves dual purpose of providing thermal as well as sound insulation.
6. It is particularly suitable as a sound absorbent because the energy of sound waves falling on it is transformed into other forms of energy.

Prefabricated attenuators:

Prefabricated attenuators (duct silencers) offer the following *advantages* in reducing the duct generated noise over the other systems:

(*i*) Permit maximum system design flexibility.

(*ii*) Economical.

(*iii*) Easily installed at the points where noise control is required.

(*iv*) Eliminate the erosion problem and its possible health hazard.

(*v*) Can be accurately selected to control an accurately calculated noise control requirement.

Measures for sound attenuation:

(*i*) Avoid the use of high velocity systems, if possible.

(*ii*) Locate fans away from the people and the work area, possibly on the roof, utilising the attenuation properties of distance.

(*iii*) If quiet operation is necessary, select speeds under 12000 r.p.m.

(*iv*) Add sound attenuation devices such as silencers, vibration isolators, duct liners and sound absorbent coatings to the equipment.

(*v*) In order to block direct-path sound effectively, select materials with a high sound transmission loss (STL) rating.

HIGHLIGHTS

1. Air-conditioning equipment primarily include package units and central units. Package unit may be of window or console type.
2. A *package unit* is self contained unit because the complete unit including evaporator and condensing unit is all incorporated in a common enclosure.
3. *Central units* are available in two types : Horizontal and vertical depending upon position of filter and the drain pan.
4. *Air-conditioning components*: Filters, fans, air washer, radiator, convector etc.

5. Air-conditioning controls may be manual, automatic, semi-automatic.
6. Some important control devices are : Thermostats, automatic humidity control, air movement control, automatic temperature control, limit switches and time clocks.
7. *Sound* is defined as a vibration in an elastic medium, generally in the frequency ranges that produce the sensation of hearing. *Noise* is defined as unwanted sound and it is often an unwelcome by-product of industrial operations.
8. Noise may be controlled by the following methods.
   (*i*) Resonance; (*ii*) Radiation; (*iii*) Impending noise transfer.
9. Fibre glass crown wool is the most effective material in combating reverberation.

OBJECTIVE TYPE QUESTIONS

Fill in the Blanks:

1. Package units may be of .... or .... type.
2. The cooling coil in central units may be either .... or .... type.
3. The function of a .... is to arrest the solid impurities, such as soot, ash, lint, smoke and fumes and even living organisms such a virus, bacteria and fungus spores.
4. The manually cleaned type of filter has .... frame.
5. The primary function of a .... is to produce air movements through heating, ventilating and air-conditioning apparatus.
6. .... fans produce a flow of air in a direction parallel to the axis of rotation.
7. A .... fan consists of an impeller running in a casing, normally of a volute shape.
8. Heat is transferred to the room by the radiator both by .... and ....
9. A convector transfers heat to the room mainly by ....
10. Three types of air-conditioning controls are ...., ... and ....
11. .... is defined as a vibration in an elastic medium, generally in the frequency ranges that produce the sensation of hearing.
12. .... is defined as unwanted sound.
13. .... is the most effective material in combating reverberation.
14. High intensity sound can .... electrical skin resistance.
15. In .... method of noise control source of noise is enclosed in some way so that noise getting to the surrounding area is stopped.

ANSWERS

1. Window, console; 2. Direct expansion, chilled water; 3. Filters;
4. Metal; 5. Fan; 6. Axial flow;
7. Centrifugal or a radial flow; 8. Radiation and convection;
9. Convection; 10. Manual, automatic and semi-automatic. 11. Sound.
12. Noise 13. Fibre glass crown wool. 14. Lower.
15. Impending noise transfer.

THEORETICAL QUESTIONS

1. How are air-conditioning equipments classified? In what ways package units differ from central units?
2. Describe briefly with a neat sketch a window type air-conditioner.
3. Write a short note on 'console type' package units.
4. Enumerate the functional elements of a central unit.
5. What is the function of a filter? How are filters classified? Explain briefly an automatic or self cleaning filter.
6. What purpose is served by a fan? How are fans classified?
7. State the fields of application of 'Axial flow' and 'Centrifugal' fans.
8. Write short notes on any two of the following:
   (*i*) Air washer (*ii*) Radiator (*iii*) Convector.
9. Enumerate air-conditioning controls. Describe briefly any two of them.
10. Explain briefly any two of the following control devices:
    (*i*) Thermostats (*ii*) Automatic humidity control
    (*iii*) Air movement control (*iv*) Automatic temperature control.
11. What is the difference between 'sound' and 'noise'?
12. What are the effects of noise pollution?
13. Enumerate the sources of noise in air-conditioning systems.
14. Explain briefly the commonly used methods for noise control.
15. Write a short note on 'Sound absorbing materials."
16. What are the advantages of prefabricated attenuators?
17. List the measures which should be taken for sound attenuation.

❑❑❑

Chapter 16

Duct and Air-Distribution Systems

16.1 **Duct systems**—*General aspects—Duct size, shape, material and construction—Pressure in ducts—Continuity and Bernoulli's equations for ducts—Pressure losses in ducts—Pressure losses due to friction in ducts—equivalent diameter of a circular duct for a rectangular duct—friction chart for circular ducts—Dynamic losses in ducts—pressure losses due to enlargement in area and static regain—pressure loss due to contraction in area—pressure loss at suction and discharge of a duct—pressure loss due to an obstruction in a duct—Duct design—general aspects—determination of duct size—System resistance, systems in series and in parallel—system resistance—systems in series—systems in parallel—Leakage of air and maintenance of ducts*

16.1 DUCT SYSTEMS

16.1.1 General Aspects

- The *function of a "duct" is to convey the air between two points,* such as between the air-handling unit or air-washer and the room to be conditioned. It also carries the room air back to the air-conditioning apparatus.
- For air-conditioning applications, the following two types of air transmission systems are adopted.

  (*i*) **Low velocity system** .... Air velocity in the main supply-air duct is within 760 m/min.

  (*ii*) **High velocity system** .... Air velocities above 760 m/min.

  "Return-air ducts" wherever used normally have low velocities whether for low velocity or high velocity systems.
- The successful operation of any air-conditioning system is dependent upon the sufficient circulation of air in the air-conditioned space. For selection of proper duct size, it is necessary to have the careful estimation of the pressure losses. *The duct cost in air-conditioning system is about 20 to 30 per cent of the total cost of the equipments required and power required by the fan contributes the substantial part of the running cost.* It is therefore necessary to *design an air-duct system for least capital cost and lowest running cost of the fan.*

16.1.2 Duct Size, Shape, Material and Construction

- The duct has to be so sized that it is accommodated within the available space. Air (like other fluids) while passing through a duct suffers a *pressure drop due to friction.* Larger the quantity of air passing through a given cross-sectional area of the duct (i.e., larger air velocity), greater will be the frictional loss and pressure drop. The fan selected has to deliver the required quantity of air overcoming the resistance offered by the various components (e.g., cooling coil/air-washer, filters, supply and returns-air outlets, dampers etc.) in the air-distribution system, i.e., the fan has to work against a head. As *this head increases the fan will need higher power and furthermore as velocity increases, noise level due to the air motion also increases.* Thus on these two counts the velocity of air in the duct *has to be kept at a reasonably low level. With lower air velocities the size of the duct (and material content of the duct) increases.* Thus the *duct system should be a balance between the initial cost (material, fabrication and erection cost) and the operating cost (fan power consumption) for a given rate of flow of air.*

- The *initial cost depends on the size of the duct*. A smaller sized duct will cost less. But with a smaller duct the air velocity will increase and so pressure loss due to friction also will increase. The fan will then have to use more power to overcome the head, thereby increasing the operating cost. Instead of evaluating the balance of initial cost and operating costs in every case, *designers depend upon past experience and generally accepted recommended air velocities in the ducts are employed.*
- In the duct, pressure of the air falls due to the following *two main factors*:

  (*i*) *Frictional losses,* primarily due to surface friction;

  (*ii*) *Dynamic losses* from sudden changes in the direction of air flow, in velocity and eddy sources.

  — *The pressure loss increases with duct length.*

  — *Pressure losses increase when the direction of air flow is changed because of the change in the direction of the duct route.*

  — Pressure loss occurs when air is abruptly made to pass from a *large duct to a smaller duct*. Eddys will be formed at the abrupt point of contraction, leading to pressure drop. Likewise when there is an *abrupt expansion* in the duct there will be *much higher pressure loss than in the case of an abrupt contraction,* as more of eddys are created.

— ***The pressure** losses can be substantially reduced by making the **change** gradual instead of being abrupt.*

16.2 ***Air-distribution Systems****—Definitions—Principles of air-distribution—Air-handling system—Room air distribution—requirements of good room air-distribution—draft—types of supply air outlets—Arrangement of ducts—Types of air-distribution systems*

Highlights—Objective Type Questions—Theoretical Questions—Unsolved Examples.

Aspect ratio:

- "***Aspect ratio***" *is the ratio of long side to the short side of the duct.*
- As the aspect ratio increases, more metal surface is required for the duct, for the same cross-sectional area. Thus *higher aspect ratios increase not only the initial cost but also the operating cost, by way of increased heat gain or loss*. Thus this ratio is an important factor to be considered in the design of the ducts.
- For the same cross-sectional area a circular duct requires less surface area (i.e., saving in material, fabrication and erection costs) than a rectangular duct. Also the circular duct offers less resistance to flow due to absence of corners. However, it usually becomes *difficult to accommodate the round ducts in the available space. Hence rectangular ducts are generally adopted.*
- Rectangular ducts selected should be *as nearly square as possible*. This is because the *square requires less surface area than a rectangular one for the some cross-sectional area*. The rectangular section should be so sized as to have *low aspect ratio*—the aspect ratio in any case should not exceed 4 : 1.

Since the aspect ratio of a "square section" is one, a square duct or duct very near a square section is best suited for minimising the initial and operating cost.

Duct material:

The ducts and usually made from the following *materials*:

(*i*) Galvanised iron sheet metal,

(*ii*) Aluminium sheet metal, or

(*iii*) Black steel.

— In air-conditioning systems, *galvanised sheet metal is the most commonly used duct material* since the zinc coating of this metal prevents rusting and avoids the painting cost. The sheet thickness of galvanised iron (G.I) duct varies from 26 gauge (0.55 mm) to 16 gauge (1.6 mm).

— The *aluminium* is used because of its *lighter weight and resistance to moisture*.

— The black sheet metal is always painted unless it withstands high temperature.

— *Stainless steel* is used in special applications for maximum resistance to moisture.

The ***non-metal ducts*** are also finding increased application nowadays.

— The *"resin bonded glass fibre" ducts* are used because *they are quite strong and easy to manufacture according to the desired shape and size*. They find use in low velocity applications less than 600 m/min and for static pressures below 5 mm of water gauge.

— The "*cement asbestos*" ducts may be employed for underground air distribution and for exhausting corrosive materials.

— The "*wooden*" ducts are suitable for places where *moisture content* in air is *not large in quantity*.

- Duct used for air-conditioning are generally of the rectangular low pressure category as this calls for only simple joining methods. For medium and high pressure ducts, sheet of steels are exclusively used for strength and rigidity to withstand the higher static pressures.

Duct construction:

The ducts made of sheet metal expand and contract when heated and cooled; the movement is absorbed by fabric joints oftenly. The fabric joints should also be used where ducts fasten to a furnace or an air-conditioner. But in fact most duct joints are made of sheet metal. Fig. 16.1 shows the various types of sheet metal joints used in the construction of ducts. The insulation is fastened to the duct with adhesives or metal clips. For longer ducts and ducts under high pressures, the duct reinforcement as shown in Fig. 16.1 is recommended.

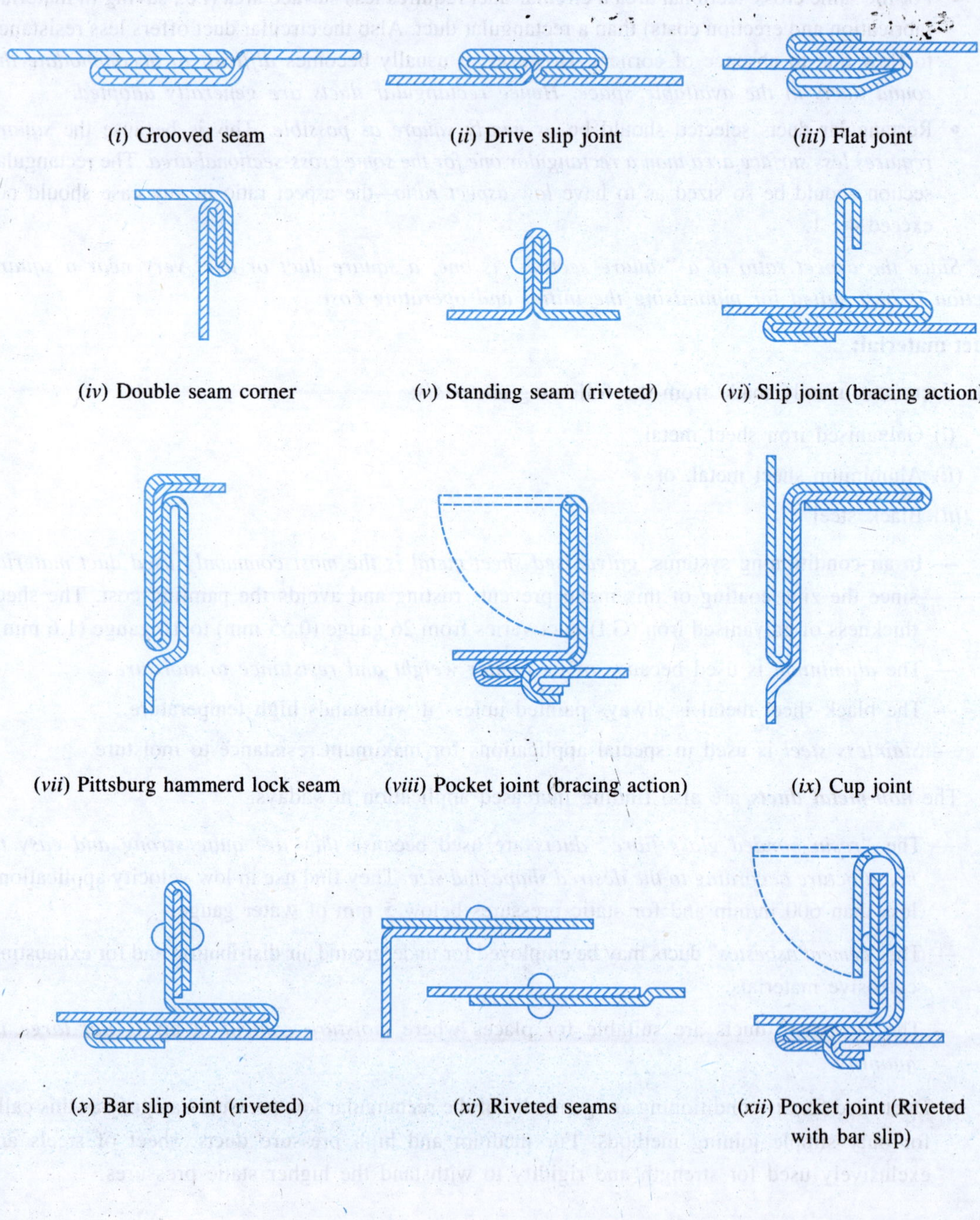

(*i*) Grooved seam

(*ii*) Drive slip joint

(*iii*) Flat joint

(*iv*) Double seam corner

(*v*) Standing seam (riveted)

(*vi*) Slip joint (bracing action)

(*vii*) Pittsburg hammerd lock seam

(*viii*) Pocket joint (bracing action)

(*ix*) Cup joint

(*x*) Bar slip joint (riveted)

(*xi*) Riveted seams

(*xii*) Pocket joint (Riveted with bar slip)

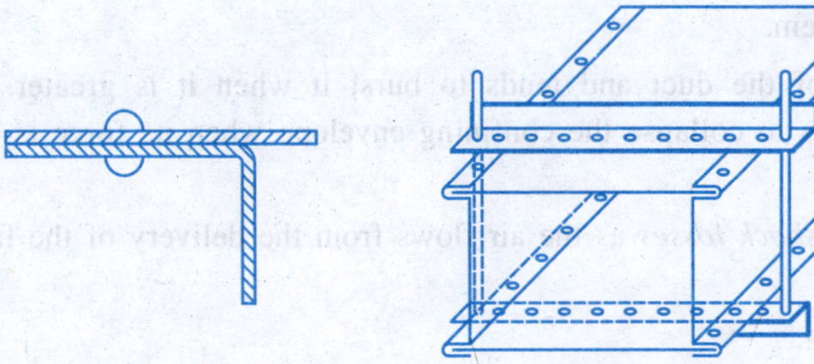

(*xiii*) Channel type seam (*xiv*) Reinforced duct

Fig. 16.1. Sheet metal duct joints.

Summarily, the following *points about duct systems are worth noting*:

(*i*) The velocity of air in the duct to be kept reasonably low so that the frictional losses and noise level are minimised.

(*ii*) The route of the duct should be as direct as possible for limiting the length of the duct and to minimize the frictional losses.

(*iii*) For changing the direction long radius smooth bends should be used.

(*iv*) When a sharp right angle turn is inevitable, turning vanes may be provided.

(*v*) Properly fabricated transformation pieces should be used for contraction and expansion in the duct sizes.

(*vi*) As a smooth surface offers less resistance to the air flow than a rough surface, galvanised iron or aluminium sheet metal should be used for the fabrication of ducts. If other materials are used, pressure losses can increase due to the roughness of the surface and so allowance should be made for this effect when determining the size of the duct.

(*vii*) In rectangular duct, aspect ratio should be kept within 4 : 1. Higher aspect ratios can also create turbulence.

16.1.3. Pressure in Ducts

The ducts may be classified as follows:

(*i*) Supply air duct
(*ii*) Return air duct
(*iii*) Fresh air duct
(*iv*) Low pressure duct
(*v*) Medium pressure duct
(*vi*) High pressure duct
(*vii*) Low velocity duct
(*viii*) High velocity duct.

The following types of *pressures* are involved in a duct system :

1. Static pressure (p_s)
2. Dynamic or velocity pressure (p_v)
3. Total pressure (p_T).

1. Static pressure (p_s):

- The static (or stationary) pressure is so called since it does *not* depend upon the air movement.

- It *always exists* in a duct system.
- It pushes against the walls of the duct and tends to burst it when it is greater than the atmospheric pressure and tends to collapse the confining envelope when its force is less than that of the atmosphere.
- It *overcomes the friction and shock losses* as the air flows from the delivery of the fan to the outlet of the duct.
- It may be *positive or negative*.

2. Dynamic or velocity pressure (p_v) :

- The pressure is *equal to the drop in static pressure necessary to produce a given velocity of flow*. Conversely, it is *equal to the increase of static pressure possible when the velocity is reduced to zero.*
- It is *always positive.*

3. Total pressure (p_T):

- The total pressure is the sum of the static pressure (p_s) and dynamic or velocity pressure (p_v).

i.e., $$p_T = p_s + p_v$$

- It may be *positive* or *negative*.

16.1.4. Continuity and Bernoulli's Equations for Ducts

16.1.4.1. Continuity equation

The continuity equation is based on the *principle of conservation of mass.* It states as follows :

"If no fluid is added or removed from the pipe in any length then the mass passing across different sections shall be same."

- Consider the flow of air through a duct between the two sections 1–1 and 2–2 as shown in Fig. 16.2 (*a*)

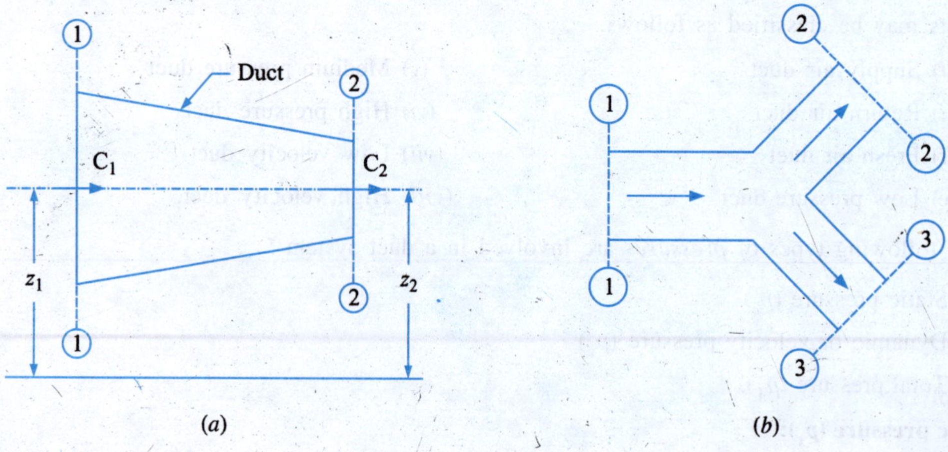

Fig. 16.2. Flow of air through a duct.

Let, A_1 = Cross-sectional area of the duct at section 1–1,

C_1 = Velocity of air at section 1–1, and

ρ_1 = Density of air at section 1–1.

and, A_2, C_2 and ρ_2 are the corresponding values at section 2–2.

Mass flow rate of air at section 1–1,

$$m_1 = \rho_1 A_1 C_1$$

and, mass flow rate of air at section 2–2,

$$m_2 = \rho_2 A_2 C_2$$

As per continuity theorem, we have:

$$\rho_1 A_1 C_1 = \rho_2 A_2 C_2$$

For air-conditioning purposes, normally the density of air (1.2 kg/m^3) is assumed constant, therefore,

$$A_1 C_1 = A_2 C_2 \quad \text{[or, } Q_1 = Q_2\text{) where Q stands for discharge]} \qquad ...(16.1)$$

This is called *continuity equation* for flowing air through ducts.

- When the duct 1 branches into two ducts 2 and 3 as shown in Fig. 16.2 (*b*), then:

$$m_1 = m_2 + m_3$$

or, $$\rho_1 Q_1 = \rho_2 Q_2 + \rho_3 Q_3$$

or, $$Q_1 = Q_2 + Q_3$$

where, $\rho_1 = \rho_2 = \rho_3$, assumed in a duct system ...(16.2)

16.1.4.2 Bernoulli's equation

Bernoulli's equation states as follows:

"*In an ideal, incompressible fluid when the flow is steady and continuous, the sum of pressure energy, kinetic energy and potential (or datum) energy is constant along a streamline*". Mathematically,

$$\frac{p}{w} + \frac{C^2}{2g} + z = \text{constant} \qquad ...(16.3)$$

where, $\frac{p}{w}$ = Pressure energy; $\frac{V^2}{2g}$ = Kinetic energy; z = Datum or elevation energy.

or, $$\frac{p}{\rho g} + \frac{C^2}{2g} + z = \text{constant} \qquad (\because w = \rho g)$$

or, $$\frac{p}{\rho} + \frac{C^2}{2} + gz = \text{constant}$$

or, $$p + \frac{\rho C^2}{2} + \rho g z = \text{constant}$$

Applying this equation to sections 1–1 and 2–2 of a duct, we have :

$$p_{s1} + \frac{\rho_1 C_1^2}{2} + \rho_1 g z_1 = p_{s2} + \frac{\rho_2 C_2^2}{2} + \rho_2 g z_2$$

Since $\rho_1 = \rho_2$ and $z_1 = z_2$, therefore, the above expression reduces to:

$$p_{s1} + \frac{\rho_1 C_1^2}{2} = p_{s2} + \frac{\rho_2 C_2^2}{2} \qquad \text{(where } \rho_1 = \rho_2 = \rho\text{)}$$

$$p_{s1} + p_{v1} = p_{s2} + p_{v2} \qquad ...(16.4)$$

where p_{s1}, p_{s2} and p_{v1}, p_{v2} are static gauge pressures and velocity pressures at sections 1–1 and 2–2 respectively.

When the flow is frictionless and there is no pressure drop between the two sections, then the total pressure in the two sections will be equal. Then the total pressure (p_T),

$$p_T = p_{s1} + p_{v1} = p_{s2} + p_{v2} \qquad ...[16.4 (a)]$$

In actual practice, since *there is always a pressure drop or loss in a duct due to friction and other causes such as sudden changes in cross-section and direction,* the above equation is written as :

$$p_{s1} + p_{v1} = p_{s2} + p_{v2} + p_L \qquad ...(16.5)$$

where, p_L = Total pressure drop between sections 1–1 and 2–2.

If a fan or blower is introduced between the two sections of a duct, then the above equation may be written as:

$$p_{s1} + p_{v1} + p_{tf} = p_{s2} + p_{v2} + p_L \qquad ...(16.6)$$

where, p_{tf} = Pressure rise due to fan work and is known as *fan total pressure.*

Following *points are worth noting:*

1. The pressures in the ducts are *usually expressed in "mm of water".*
2. The air flowing through the duct is taken as standard air unless stated otherwise. The standard air is one which *corresponds to 20°C, an atmospheric pressure of 1.013 bar and 62% RH.* Its *mass density is equal to 1.2 kg/m³* (or 11.72 N/m³).
3. The velocity pressure (p_v) in the duct, when the velocity of air (V) flowing through the duct is in *m/s,* is given by:

$$p_v = \frac{\rho_a C^2}{2} = \frac{1.2\, C^2}{2} = 0.6\, C^2 \text{ N/m}^2$$

$$= \frac{0.6\, C^2}{9.81} = \frac{C^2}{16.35} = \left(\frac{C}{4.04}\right)^2 \text{ mm of water} \qquad ...(16.7)$$

$$\left(\because 1 \text{ N/m}^2 = \frac{1}{9.81} \text{ mm of water}\right)$$

When the velocity of air is in *m/min,* then:

$$p_v = \left(\frac{C}{60 \times 4.04}\right)^2 = \left(\frac{C}{242.4}\right)^2 \text{ mm of water.} \qquad (16.8)$$

16.1.5 Pressure Losses in Ducts

When air flows through a duct, there is a loss of pressure *due to friction* between the moving particles of air (fluid) and the interior surfaces of a duct. When the loss of pressure occurs in a straight duct, it is usually termed as **Friction loss**". *When the pressure is lost dynamically at the changes of direction such as in bends, elbows etc., and at the changes of cross-section of the duct,* the pressure loss is usually termed as **"Dynamic loss"**.

16.1.5.1 Pressure loss due to friction in ducts

The loss of pressure due to friction in ducts may be obtained by using the D'Arcy's, formula or the Fanning's equation, as follows:

$$p_f = \frac{f\,L\rho_a C^2}{2m} \quad ...(16.9)$$

where, p_f = Pressure loss due to friction, N/m^2,

f = Friction factor (dimensionless); it depends upon the surface of the duct,

L = Length of the duct, m,

C = Mean velocity of the air flowing through the duct, m/s, and

m = Hydraulic mean depth, m,

$$= \frac{\text{A (Cross-sectional area of the duct)}}{\text{P (Wetted perimeter of the duct)}}$$

The pressure loss to friction, in air-conditioning, is generally expressed in mm of water.

Equation 16.9 may be written as;

$$p_f = \frac{f\,L}{m}\left(\frac{\rho_a C^2}{2}\right) = \frac{f\,L}{m} \times p_v \text{ mm of water} \quad ...(16.10)$$

Also,

$$p_v = \frac{\rho_a C^2}{2} = \left(\frac{C}{4.04}\right)^2 \text{ mm of water} \quad \text{(Refer eqn. 16.7)}$$

The eqn. (16.10) may now be written as:

$$p_f = \frac{f\,L}{m}\left(\frac{C}{4.04}\right)^2 \text{ mm of water} \quad ...(16.11)$$

The *pressure loss* due to friction in the duct when air is at a temperature t°C, is given by:

$$p_f = \frac{f\,L}{m}\left(\frac{C^2}{4.04}\right)^2\left(\frac{273+20}{273+t}\right)$$

$$= \frac{f\,L}{m}\left(\frac{C^2}{4.04}\right)^2\left(\frac{293}{273+t}\right) \text{ mm of water} \quad ...(16.12)$$

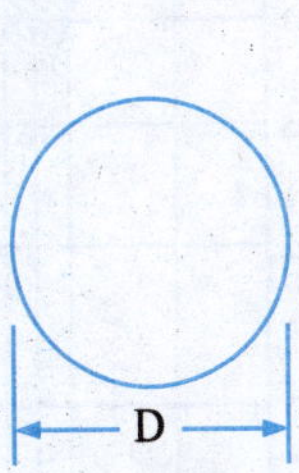

Fig. 16.3. Circular duct

Hydraulic mean depth (m) for circular and rectangular ducts:

Circular duct of diameter D:

$$m = \frac{A}{P} = \frac{\frac{\pi}{4}D^2}{\pi D} = \frac{D}{4}$$

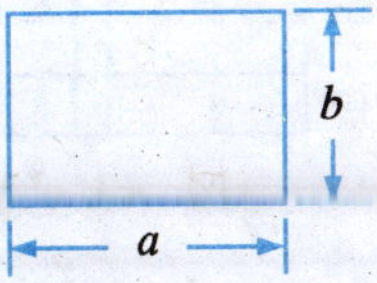

Fig. 16.4. Rectangular duct

Rectangular duct of sides a and b:

$$m = \frac{A}{P} = \frac{a \times b}{2(a+b)}$$

Friction factor for ducts (f):

The following expressions are used for obtaining the value of friction factor (f):

Smooth ducts:

1. *For laminar flow:* $$f = \frac{64}{\text{Re}} \quad \text{...(16.13)}$$
2. *For turbulent flow:* $$f = \frac{0.3164}{\text{Re}} \quad \text{...(16.14)}$$

where, Re = Reynolds number.

"*Reynolds number*" is defined as the *ratio of the inertia forces to the viscous forces*. It is a dimensionless quantity.

Mathematically, $$\text{Re} = \frac{\text{Inertia forces}}{\text{Viscous forces}} = \frac{\rho_a C^2}{\mu C/D} = \frac{\rho_a DC}{\mu} = \frac{DC}{\mu/\rho_a} = \frac{DC}{\nu} \quad \text{...(16.15)}$$

where, ρ_a = Mass density of air, kg/m^3,

D = Diameter of the duct, m,

C = Velocity of air, m/s,

μ = Dynamic (or absolute) viscosity, N-s/m^2, and

ν = Kinematic viscosity $\left(= \frac{\mu}{\rho_a}\right)$, m^2/s.

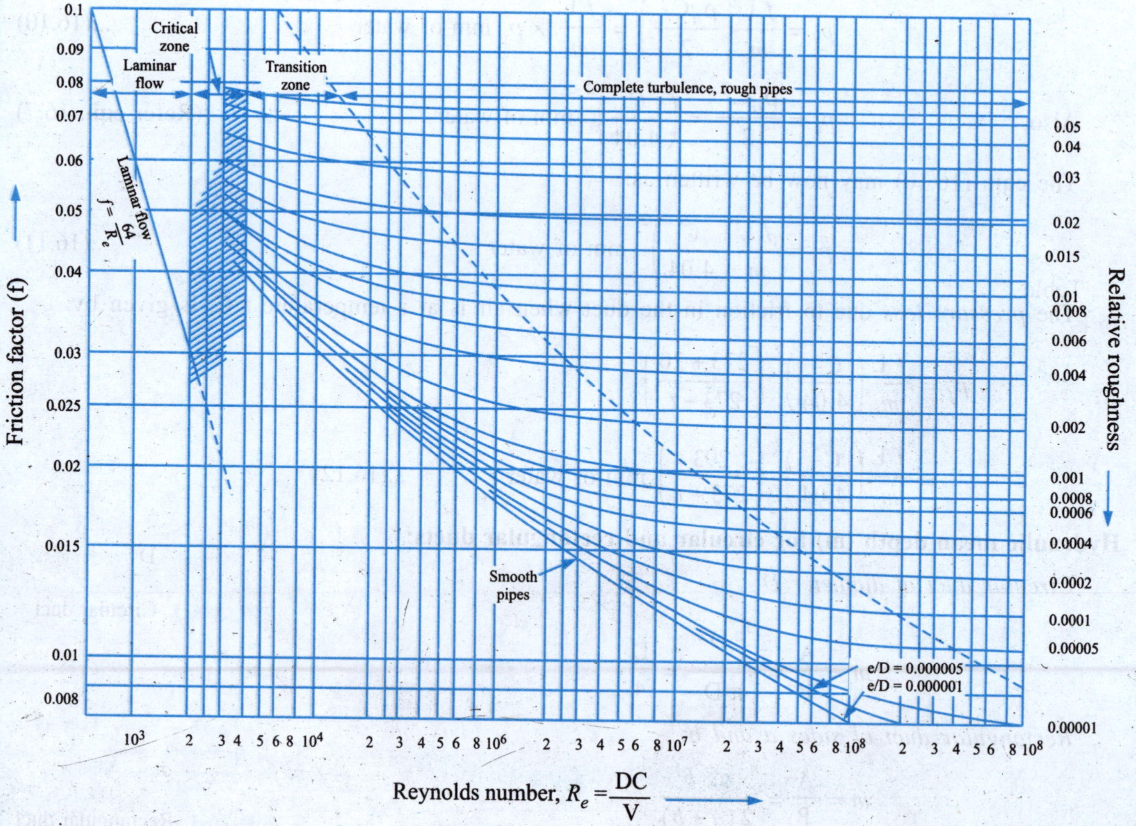

Fig. 16.5. Moody chart showing friction for fluid flow in circular pipes or ducts.

When, Re < 2000 .... Flow is *laminar* (or *viscous*)

Re > 4000 .... Flow is *turbulent*

Re between 2000 and 4000 .... Flow is *unpredictable.*

Rough pipes or ducts:

$$f = \frac{1}{\left[1.74 - 2\log\left(\frac{2e}{D}\right)\right]^2} \quad ...(16.16)$$

The values of '*f*' for different Reynolds numbers and different *roughness factors* $\left(\frac{e}{D}\right)$ *may be read directly from the 'Moody chart' as shown in Fig. 16.5.*

When '*f*' is obtained from the moody chart, then the pressure loss due to fraction in a circular duct of diameter 'D' is given by : $p_f = \frac{f\,L}{D} \times p_v = \frac{f\,L}{D}\left(\frac{C}{4.04}\right)^2$ mm of water.

Table 16.1 shows the values of mass density of air and dynamic viscosity at different temperatures and at atmospheric pressure of 1.013 bar.

Table 16.1 Values of mass density and dynamic viscosity of air at different temperatures

Temperature (°C)	*Dynamic viscosity (μ), N-s/m²*	*Density, kg/m³*
0	17.24×10^{-6}	1.29
10	17.71×10^{-6}	1.25
20	18.18×10^{-6}	1.20
30	18.65×10^{-6}	1.16
40	19.12×10^{-6}	1.13
50	19.60×10^{-6}	1.09

Table 16.2 shows the values of surface roughness *(e)* for different types of commercial pipes or duct work.

Table 16.2 Recommended surface roughness (*e*) for different materials pipes or ducts.

Types of pipe or duct	*Absolute surface roughness (e) in mm*
Smooth-drawn tubes (glass, brass and lead)	0.0015
Commercial steel or wrought iron pipe	0.045
Galvanised iron or steel air ducts	0.15
Cast iron	0.255
Smooth concrete	0.3
Riveted steel (light weight, small rivets)	0.9
Wood-stave conduits	0.9
Average concrete	1.2
Very rough concrete	3.0
Brick conduits	9.0
Riveted steel (heavy weight, large rivets)	9.0

16.1.5.2. Equivalent diameter of a circular duct for a rectangular duct

For finding equivalent diameter of a circular duct for a rectangular duct, the following *two cases* will be considered.

1. When the *quantity of air* passing through the rectangular and circular ducts is *same*.
2. When the *velocity of air* passing through the rectangular and circular ducts is *same*.

1. ***When the quantity of air passing through the rectangular and circular ducts is same:***

Let, a = Longer side of the rectangular duct,

b = Shorter side of the rectangular duct,

D_{eq} = Equivalent diameter of the circular duct.

$A_{rect.}$ = Cross-sectional area of the rectangular duct = $a \times b$

$P_{rect.}$ = Wetted perimeter of the rectangular duct = $2\,(a + b)$

$A_{cir.}$ = Cross-sectional area of the equivalent circular duct

$= \frac{\pi}{4} D_e^2,$

P_{cir} = Wetted perimeter of the equivalent circular duct = $\pi\, D_e,$

ρ_a = Mass density of air, and

Q = Quantity of air passing through the rectangular and circular ducts.

Velocity of air passing through the circular duct, $C_{cir.} = \dfrac{Q}{A_{cir.}}$

Velocity of air passing through the rectangular duct, $C_{rect.} = \dfrac{Q}{A_{rect.}}$

Pressure loss due to friction (f) is given as,

$$p_f = \frac{f\,L\rho_a C^2}{2m} = \frac{f\,L\rho_a}{2m}\left(\frac{Q}{A}\right)^2 \qquad \left(\because C = \frac{Q}{A}\right)$$

and, Hydraulic mean depth,

$$m = \frac{\text{A (Cross-sectinal area of duct)}}{\text{P (Wetted perimeter of duct)}}$$

$\therefore$ Pressure loss due to friction for the *circular duct*,

$$(p_f)_{cir} = \frac{f\,L\rho_a}{2}\left(\frac{P_{cir}}{A_{cir}}\right)\left(\frac{Q}{A_{cir}}\right)^2 = \frac{f\,L\rho_a Q^2}{2}\left[\frac{P_{cir.}}{(A_{cir.})^3}\right] \qquad ...(16.17)$$

and, Pressure loss due to friction for the *rectangular duct*,

$$(p_f)_{rect} = \frac{f\,L\rho_a}{2}\left(\frac{P_{rect}}{A_{rect}}\right)\left(\frac{Q}{A_{rect}}\right)^2 = \frac{f\,L\rho_a Q^2}{2}\left[\frac{P_{rect.}}{(A_{rect.})^3}\right] \qquad ...(16.18)$$

As the pressure loss, friction factor, length, density and quantity of air for the circular and rectangular ducts are same, therefore, from eqns. 16.17 and 16.18, we have:

$$(p_f)_{cir.} = (p_f)_{rect.}$$

or, $$\frac{f L\rho_a Q^2}{2}\left[\frac{P_{cir}}{(A_{cir})^3}\right] = \frac{f L\rho_a Q^2}{2}\left[\frac{P_{rect.}}{(A_{rect.})^3}\right]$$

or, $$\frac{P_{cir.}}{(A_{cir.})^3} = \frac{P_{rect.}}{(A_{rect.})^3}$$

or, $$\frac{\pi D_{eq.}}{\left(\frac{\pi}{4} D_{eq.}^2\right)^3} = \frac{2(a+b)}{(a \times b)^3}$$

or, $$\frac{32}{\pi^2 D_{eq.}^5} = \frac{a+b}{a^3 b^3}$$

or, $$D_{eq.} = \left[\frac{32 a^3 b^3}{\pi^2 (a+b)}\right]^{1/5} = 1.265\left(\frac{a^3 b^3}{a+b}\right)^{1/5} \quad ...(16.19)$$

2. ***When the velocity of air passing through the rectangular and circular ducts is same:***

Let, C = Velocity of air passing through the rectangular and circular ducts.

Pressure loss due to friction for a *circular duct,*

$$(p_f)_{cir.} = \frac{f L\rho_a C^2}{2}\left(\frac{P_{cir}}{A_{cir}}\right) \quad ...(16.20)$$

and, Pressure loss due to friction for a *rectangular duct,*

$$(p_f)_{rect.} = \frac{f L\rho_a C^2}{2}\left(\frac{P_{rect}}{A_{rect}}\right) \quad ...(16.21)$$

As the pressure loss, velocity of air, friction factor, density and length for the circular and rectangular ducts are same, therefore, from eqns. 16.20 and 16.21, we have:

$$(p_f)_{cir.} = (p_f)_{rect.}$$

$$\frac{f L\rho_a C^2}{2}\left(\frac{P_{cir.}}{A_{cir.}}\right) = \frac{f L\rho_a C^2}{2}\left(\frac{P_{rect.}}{A_{rect.}}\right)$$

or, $$\frac{P_{cir.}}{A_{cir.}} = \frac{P_{rect.}}{A_{rect.}}$$

or, $$\frac{\pi D_e}{\frac{\pi}{4} \times D_{eq}^2} = \frac{2(a+b)}{a \times b}$$

or, $$D_{eq} = \frac{2(a \times b)}{a+b} = \frac{2a}{\frac{a}{b}+1} \quad ...(16.22)$$

where $\frac{a}{b}$ is known as "*Aspect ratio*".

- In case of *rectangular ducts the aspect ratio should not be greater than 8 in any case.*

16.1.5.3 Friction chart for circular ducts

The frictional pressure loss in ducts, (p_f) is given as:

$$p_f = \frac{f\,L\rho_a C^2}{2m}$$

But, for circular ducts, $m = \frac{A}{P} = \frac{\frac{\pi}{4}\times D^2}{\pi D} = \frac{D}{4}$

$$\therefore \quad p_f = \frac{f\,L\rho_a C^2}{2}\left(\frac{4}{D}\right) = \frac{4fL}{D}\left(\frac{\rho_a C^2}{2}\right)$$

or,
$$p_f = \frac{4fL}{D} \times p_v \;(\text{N/m}^2) \qquad \ldots(16.23)$$

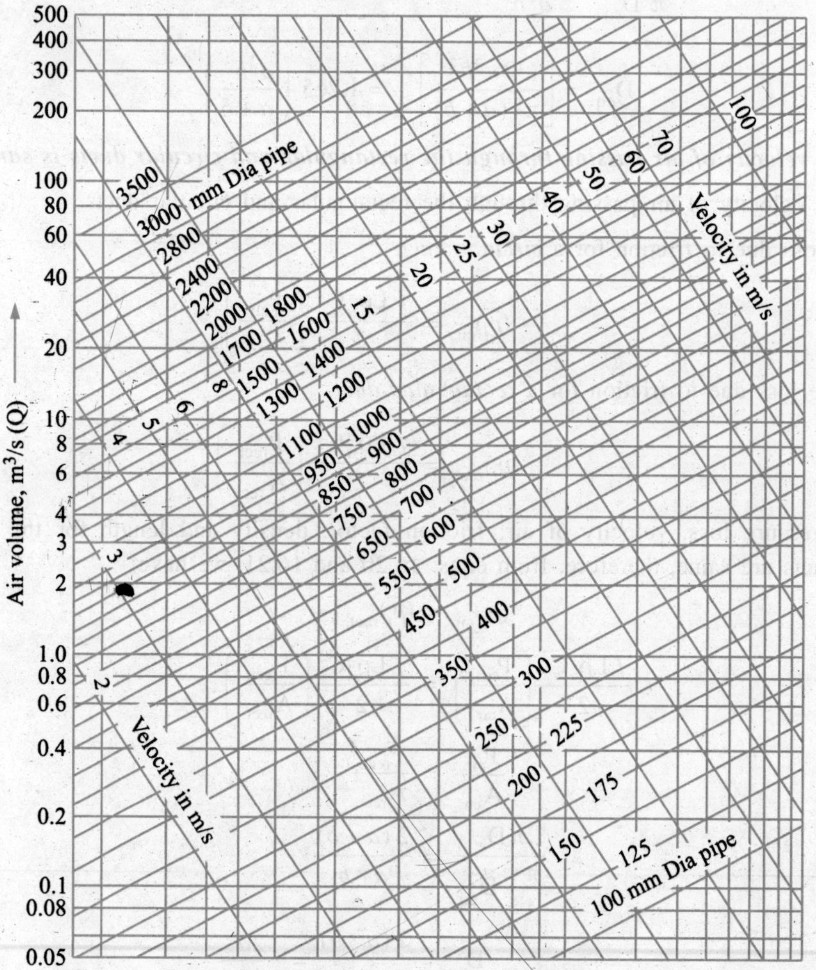

Fig. 16.6. Friction chart for circular ducts.

The frictional pressure loss (p_f) is circular ducts may be obtained by using the following relation, suggested by Fritzsche :

$$p_f = \frac{0.01422\,L\,(C)^{1.852}}{(D)^{1.269}}\ \text{N/m}^2 \qquad ...(16.24)$$

where, C is in m/s and L and D are in metres.

Mean velocity of air flowing through the duct,

$$C = \frac{Q\ (\text{Volume flow vate})}{A\ (\text{Cross-sectional area})} = \frac{Q}{\frac{\pi}{4} \times D^2} = \frac{4Q}{\pi D^2}$$

Substituting the value of C in eqn. (16.24), we get:

$$p_f = \frac{0.22243\,L\,(Q)^{1.852}}{(D)^{4.973}}\ \text{N/m}^2 \qquad ...[16.25\ (a)]$$

$$= \frac{0.002268\,L\,(Q)^{1.852}}{(D)^{4.973}}\ \text{mm of water} \qquad ...[16.25\ (b)]$$

$$= \frac{0.012199\,L\,(C)^{2.4865}}{(Q)^{0.6343}}\ \text{N/m}^2 \qquad ...(\text{interns of C and Q}) \qquad ...[16.25\ (c)]$$

The results obtained from the above Fritzsche relations are quite accurate.

For circular ducts, the frictional pressure loss (in mm of water) for various velocities may be obtained directly from the **"*friction chart*"** as shown in Fig. 16.6. *These charts are valid for* 20°C *and* 1.013 *bar and clean galvanised iron ducts with joints and seams having good commercial practice.*

The friction charts apply to *air only*. These are based on *standard air density*.

- In case the material of the duct is other than galvanised iron (e.g. concrete, wood, fibre glass, plastics etc.) then correction factor should be applied. The correction shall be made according to $p_f \propto \rho_a$ for *small* differences in density of air.
  - — When the air is at other temperatures, the correction shall be made according to $p_f \propto [1/(T)^{0.857}]$
  - — The friction chart may be used to determine the pressure loss in rectangular ducts if the equivalent diameter of a circular duct for the rectangular duct is first calculated (as discussed in Art. 16.1.5.2).

16.1.6 Dynamic Losses in Ducts

The dynamic losses in ducts are caused by the following:

(*i*) Change in direction, i.e., due to bends, elbows etc.

(*ii*) Change in area or velocity, i.e., due to enlargement, contraction, suction and discharge openings, dampers etc.

The dynamic pressure loss,

$$p_d = Kp_v = K\left(\frac{C}{4.04}\right)^2 \ \text{in mm of water} \qquad ...(16.26)$$

where, C = Velocity (normally downstream), in m/s, and

K = Dynamic loss coefficient (found experimentally)

The dynamic loss expressed in terms of an additional equivalent length (L_e) of the duct is given by:

$$p_d = \frac{f\,L_{eq}}{m} \times p_v = \frac{f\,L_e}{m}\left(\frac{C}{4.04}\right)^2 \text{ in mm of water} \quad ...(16.27)$$

The relationship between the dynamic loss coefficient and the equivalent length, from eqns. 16.26 and 16.27, may be given as :

$$K = \frac{f\,L_e}{m}, \quad \text{or,} \quad L_{eq} = \frac{Km}{f} \quad ...(16.28)$$

The method of equivalent additional length is more convenient to use. The *total pressure loss as friction loss only* can be found by simply adding additional length to the actual total length. *Since the additional equivalent length accounts for only the dynamic losses and not the actual frictional losses in lengths of bends or elbows, therefore while considering the actual total length, the length of bends and elbows should also be taken in addition to the straight lengths of the ducts.*

16.1.6.1 Pressure loss due to enlargement in area and static regain

Wherever there is an enlargement in a duct-run, the velocity of fluid decreases and there is a conversion of velocity pressure into static pressure. The increase in static pressure due to the conversion of velocity pressure is termed as **"Static regain"** (SR). If the enlargement is not accompanied with pressure loss, the velocity pressure is completely converted into static pressure. In case the enlargement is accompanied with pressure loss, the increase in the static pressure or static regain is *reduced by the amount of the pressure loss.*

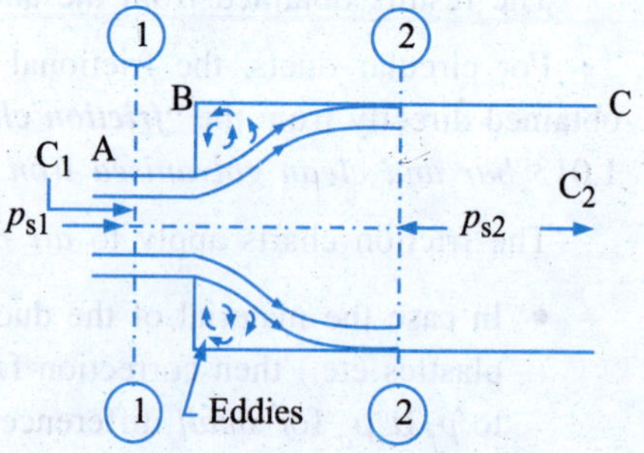

Fig. 16.7. Sudden enlargement

Consider a duct ABC through which air is flowing and having a sudden (or abrupt) enlargement at B. Due to this enlargement, eddies will be formed (Fig. 16.7) leading to loss of pressure at the suddenly enlarged section.

Let, A_1 = Cross-sectional area of the duct at section 1–1,

C_1 = Velocity of air at section 1–1,

p_{s1} = Static pressure at section 1–1,

A_2, C_2, p_{s2} = Corresponding values at section 2–2, and

$(p_L)_{enl.}$ = Pressure loss due to sudden enlargement.

Applying Bernoulli's equation to section 1–1 and 2–2, we get:

$$p_{s1} + p_{v1} = p_{s2} + p_{v2} + (p_L)_{enl.}$$

$$p_{s1} + \frac{\rho_a C_1^2}{2} = p_{s2} + \frac{\rho_a (C_2)^2}{2} + (p_L)_{enl.}$$

$$\therefore \quad (p_L)_{enl.} = (p_{s1} - p_{s2}) + \left[\frac{\rho_a C_1^2}{2} - \frac{\rho_a C_2^2}{2}\right] \quad ...(i)$$

Also, momentum (mass × velocity) of air per second at section 1–1,

$$m_1 C_1 = (\rho_a A_1 C_1) \times C_1 = \rho_a A_1 C_1^2$$

Similarly, momentum of air per second at section 2–2 = $\rho_a A_2 C_2^2$

∴ Change of momentum per second

$$= \rho_a A_1 C_1^2 - \rho_a A_2 C_2^2 \quad ...(ii)$$

As per continuity equation, we have:

$$A_1 C_1 = A_2 C_2 \quad (\because \rho_a \text{ is constant})$$

or,

$$A_1 = \frac{A_2 C_2}{C_1}$$

Substituting this value of A_1 in eqn. *(ii)* we have:

$$\frac{\rho_a A_2 C_2 C_1^2}{C_1} - \rho_a A_2 C_2^2 = \rho_a A_2 C_2 C_1 - \rho_a A_2 C_2^2 \quad ...(iii)$$

The force causing this change of momentum

$$= (p_{s2} - p_{s1}) A_2 \quad ...(iv)$$

Equating eqns. *(iii)* and *(iv)*, we get:

$$(p_{s2} - p_{s1}) A_2 = \rho_a A_2 C_2 C_1 - \rho_a A_2 C_2^2$$

$$p_{s2} - p_{s1} = \rho_a (C_1 C_2 - C_2^2)$$

$$p_{s1} - p_{s2} = \rho_a (C_2^2 - C_1 C_2)$$

Substituting this value of $(p_{s1} - p_{s2})$ in eqn. *(i)* we have:

$$(p_L)_{enl.} = \rho_a (C_2^2 - C_1 C_2) + \left[\frac{\rho_a C_1^2}{2} - \frac{\rho_a C_2^2}{2}\right]$$

$$= \rho_a \left[\frac{2C_2^2 - 2C_1 C_2 + C_1^2 - C_2^2}{2}\right] = \rho_a \left[\frac{(C_1 - C_2)^2}{2}\right] \text{N/m}^2$$

$$\left[\text{or, } (p_L)_{enl.} = \frac{(C_1 - C_2)^2}{2} \; m \text{ of air}\right]$$

Taking $\rho_a = 1.2$, kg/m^3, we have:

$$(p_L)_{enl.} = 1.2 \left[\frac{(C_1 - C_2)^2}{2}\right] = 0.6 (C_1 - C_2)^2 \text{ N/m}^2$$

$$= \frac{0.6}{9.81} (C_1 - C_2)^2 = \left(\frac{C_1 - C_2}{4.04}\right)^2 \text{ mm of water} \quad ...(v)$$

$$\left(\because 1 \text{ N/m}^2 = \frac{1}{9.81} \text{ mm of water}\right)$$

Again, $A_1C_1 = A_2C_2$

$$C_2 = \frac{A_1C_1}{A_2}, \quad \text{or,} \quad C_1 = \frac{A_2C_2}{A_1}$$

Substituting the value of C_2 in eqn. (*v*), we have:

$$(p_L)_{enl.} = \left[\frac{C_1 - \frac{A_1C_1}{A_2}}{4.04}\right]^2 = \left(1 - \frac{A_1}{A_2}\right)^2\left(\frac{C_1}{4.04}\right)^2 \quad ...(16.29)$$

$$= K_1\left(\frac{C_1}{4.04}\right)^2 = K_1 p_{v1} \quad ...[16.29\ (a)]$$

where, $K_1 = \left(1 - \frac{A_1}{A_2}\right)^2$, known as *"loss coefficient"*, and $p_{v1} = \left(\frac{C_1}{4.04}\right)^2$ mm of water

Now substituting the value of C_1 in eqn. (*v*), we have:

$$(p_L)_{enl.} = \left[\frac{\left(\frac{A_2C_2}{A_1} - C_2\right)}{4.04}\right]^2 = \left(\frac{A_2}{A_1} - 1\right)^2\left(\frac{C_2}{4.04}\right)^2 \quad ...(16.30)$$

$$= K_2\left(\frac{C_2}{4.04}\right)^2 = K_2 p_{v2} \quad ...[16.30\ (a)]$$

where, $K_2 = \left[\left(\frac{A_2}{A_1}\right)^2 - 1\right]$, known as *"loss coefficient"*, and

$$p_{v2} = \left(\frac{C_2}{4.04}\right)^2 \text{ mm of water}$$

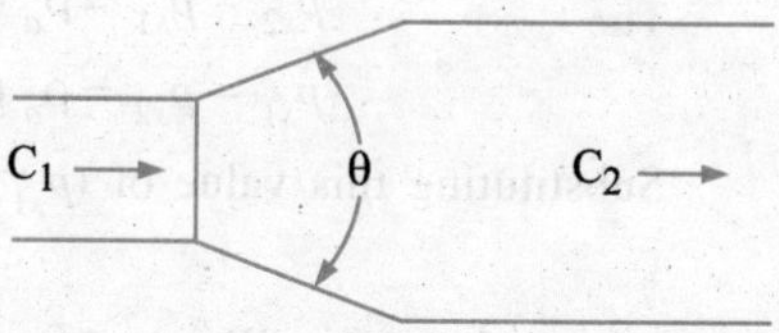

Fig. 16.8. Gradual enlargement.

When there is a *gradual enlargement in area of the duct* as shown in Fig. 16.8, then the pressure loss due to gradual enlargement is given by:

$$(p_L)_{enl.} = K_r K_1\left(\frac{C_1}{4.04}\right)^2 = K_r C_2\left(\frac{C_2}{4.04}\right)^2 \quad ...(16.31)$$

where, K_r = Loss coefficient giving the ratio of the actual loss to the loss for sudden enlargement.

The values of K_r as a function of included angle θ of the sides is given in the Table 16.1.

Table 16.1. Values of Loss Coefficient (K_r)

Included angle (θ)	5°	7°	10°	20°	30°	40°
Loss coefficient (K_r)	0.17	0.22	0.28	0.45	0.59	0.73

16.1.6.2 Pressure loss due to contraction in area

Consider a duct ABC having sudden contraction at B as shown in Fig. 16.9. It may be noted that turbulence (formation of eddies) occurs at two places, viz., at the shoulders of the contraction in the large section, and at a section shortly after the vena contracta. It may be observed that the major source of pressure loss is at the neck. Thus, the *loss corresponds to that of sudden expansion from the velocity* C_1 *at the vena contracta to the down stream velocity* C_2. If A_1 and A_2 are the cross-sectional areas at sections and 2–2 respectively, then pressure loss due to sudden contraction,

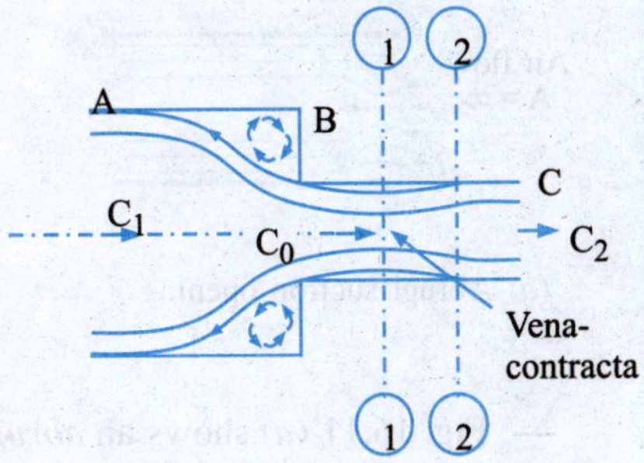

Fig. 16.9. Sudden contraction.

$$(p_L)_{cont.} = \left(1 - \frac{A_1}{A_2}\right)^2 \left(\frac{C_1}{4.04}\right)^2 \quad ...(16.32)$$

$$= K_1 \left(\frac{C_1}{4.04}\right)^2 = K_2 \left(\frac{C_2}{4.04}\right)^2 \quad ...[16.32\ (a)]$$

When there is a gradual contraction in area of the duct as shown in Fig. 16.10, then the loss due to gradual contraction is given by:

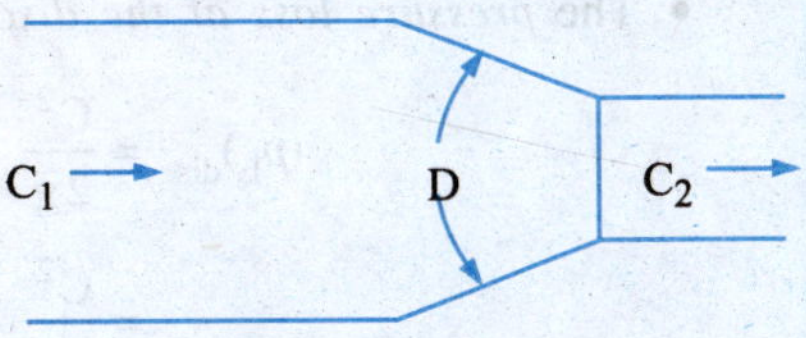

Fig. 16.10. Gradual contraction.

$$(p_L)_{cont} = K_r K_1 \left(\frac{C_1}{4.04}\right)^2 = K_r K_2 \left(\frac{C_2}{4.04}\right)^2 \quad ...(16.33)$$

where, K_r is the loss coefficient.

The values of K_r as a function of included angle θ of the sides is given in the Table 16.2.

Table 16.2. Values of Loss Coefficient (K_r)

Included angle (θ)	30°	45°	60°
Loss coefficient (K_r)	0.02	0.04	0.07

16.1.6.3 Pressure loss at suction and discharge of a duct

- The *pressure loss at suction to the duct* is given by:

$$(p_L)_{suc.} = \frac{KC^2}{2g} \text{ m of air} \quad ...[16.34\ (a)]$$

$$= \frac{KC^2}{2} \rho_a \text{ N/m}^2 \quad ...[16.34\ (b)]$$

$$= K \left(\frac{C}{4.04}\right)^2 \text{ mm of water} \quad ......[16.34\ (c)]$$

where, C – Velocity of air in the duct, m/s, and

K = Loss coefficient.

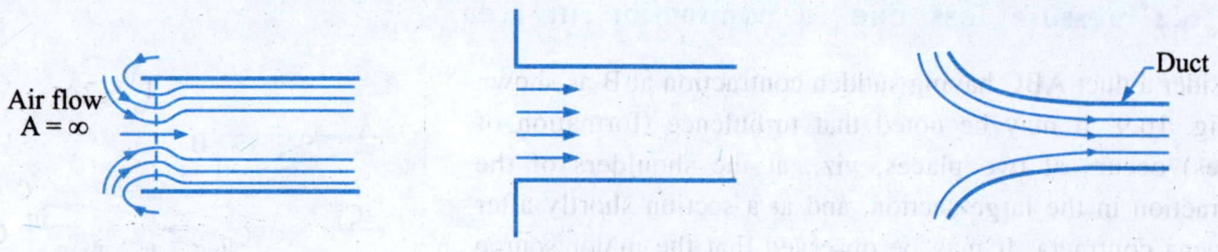

(*a*) Abrupt suction opening (*b*) Flanged entrance (*c*) Formed entrance

Fig. 16.11

— Fig. 16.11 (*a*) shows an *abrupt suction opening*. In this case as the air approaches the opening, it gets accelerated, forming a vena contracta inside the duct. The change in area is from *infinity to the duct area*.

In such a case, the loss coefficient K is taken as 0.85.

— Fig. 16.11 (*b*), shows *flanged entrance*. In this case, the loss coefficient is reduced to 0.34.

— By making *formed entrance* of bell-mouth shape, the loss coefficient can be further reduced to 0.03.

- The *pressure loss at the discharge* or exit of *a duct* is given by,

$$(p_L)_{dis.} = \frac{C^2}{2g} \text{ m of air} \quad \text{...16.35 (a)}$$

$$= \frac{C^2}{2} \times \rho_a \text{ N/m}^2 \quad \text{...16.35 (b)}$$

$$= \left(\frac{C}{4.04}\right)^2 \text{ mm of water.} \quad \text{16.35 (c)}$$

16.1.6.4 Pressure loss due to an obstruction in a duct

Besides frictional and dynamic losses in ducts, the pressure losses also occur due to various obstructions in the path of air flow from the fan to outlet i.e., air-conditioned space. The pressure losses for different obstructions are given in the Table 16.3.

Table 16.3 Pressure Losses for Different Obstructions

Type of obstruction	*Possible pressure loss in mm of water*
Screen grills	2.5 to 5
Air heaters or cooler with several rows	5 to 10
Air filters	5 to 10
Air washers	6.25 to 10

16.1.7 Duct Design

16.1.7.1 General aspects

- The major consideration in duct design is that ducts should carry the necessary volume of conditioned air space *with minimum frictional and dynamic pressure losses and still be economical in size.*

- The duct layout must be made so as to *reach the outlet with least number of bends, obstructions and changes in area.* The area changes must be *gradual* where possible and limited to not more than *20° for diverging area* and *60° for converging area.*
- With rectangular ducts the *aspect ratio of 4 : 1 and less is desirable* but in no case be made greater than 8 : 1.
- *The velocities in the ducts must be high enough to reduce the size of the ducts but low enough to reduce the noise and the pressure losses, to economise power requirement.*
- Ducts should be made of *smooth materials such as galvanized iron (GI) or aluminium sheet metal.* Whenever other materials are used, allowance should be made for the roughness of the material.
- *Dampers should be provided* in each branch outlet for balancing the system.
- Obstructions in the ducts should be avoided.

'Recommended G.I. gauges' and 'Recommended velocities in m/min.' are given in Table 16.4 and Table 16.5 respectively.

Table 16.4. Recommended G.I. Gauges

Longest side in cm			*GI sheet metal gauge*
Low pressure	*Medium pressure*	*High pressure*	
Up to 30	—	—	26
30–75	Up to 45	—	24
75–135	45–120	Up to 120	22
135–210	120–180	120–180	20
Above 210	Above 180	180–240	18
—	—	Above 240	16

Table 16.5. Recommended Velocities in m/min

Designation	*Residences*	*Schools, Theatres and Public Buildings*	*Industrial Buildings*
Out-door air Intakes	150	150	150
Filters	75	90	105
Heating or cooling coils	135	150	180
Air-washers	150	150	150
Fan out-lets	300 to 480	400 to 600	480 to 725
Grills	40 to 60	60 to 80	80 to 100
Main duct	210 to 300	300 to 400	360 to 540
Branch Duct	180	180 to 270	240 to 300
Branch Riser	150	180 to 210	240

16.1.7.2 Determination of duct size

After deciding the layout and calculating the requirements of air quantities at various outlets the size of the ducts may be determined by using one of the following *three methods*:

1. Velocity reduction method,
2. Equal friction (or constant pressure loss) method,
3. Static regain method.

1. ***Velocity reduction method***

In this method of duct design the ducts are designed in such a way that the velocity decreases as the flow proceeds. The pressure drops are calculated for these velocities for respective branches and main duct. The fan is designed to overcome the pressure losses along any single run including the losses of main duct, branch duct, elbows, valves etc. *Balancing is obtained by dampering.*

- It is *usually adopted for very simple systems* and is not recommended for vigorous design of any less simple systems.

Advantages:

(*i*) This method is the *easiest* in sizing ducts.

(*ii*) The *velocities can be adjusted to avoid noise.*

Disadvantage: Considerable experience and judgement is required in selecting velocities such as *to make the system optimum is economy and power.*

2. ***Equal friction (or constant pressure loss) method***

- In this method, the *frictional pressure drop per unit length of the duct is maintained constant throughout the system.* This is *superior* to the velocity reduction method since it *prevents one section of the duct having an excessive resistance compared with another in a system having symmetrical branch and outlet runs.*
- Its use in the design of complicated systems, particularly those with several units of widely different lengths and resistances, or with long runs with many grilles, is *not recommended.* Such a system *will be difficult to balance,* since the method makes *no provision for equatising pressure drops in branches or for providing the same static pressure behind each outlet.* In sizing a rectangular duct with a number of outlets in branches, it is desirable to maintain an aspect ratio as close to 1 as practical, and to change only on of the dimensions at a time. *When there is an appreciable difference in equivalent length between branch ducts, the longest would be sized by equal friction and the total pressure drop of the run should be computed. The pressure drop between the fan and point of take-off for each of the other runs should then be subtracted from this total to give the desired total pressure drop for each branch.* If the same friction loss per unit length is used for all runs, *dampering will be required at the take-offs to shorter runs.* With package equipment where the fan pressure is set, it may be necessary to estimate the total equivalent length of the longest run and divide it into the available pressure to determine the unit pressure drop.
- The *modification in this system to avoid the dampering is to design the duct for consuming the pressure available at the take-off from the main.*

Advantage: In the case of symmetrical duct layout with the same length in each run, no dampers are required to balance the system since this method gives equal pressure loss in various branches.

Disadvantages:

(*i*) When the lengths of the runs are different, then the shortest run will have the minimum drop and air will come out with higher pressure compared with long run ducts.

(*ii*) If the lengths of the runs are different the method does not balance the pressures at the outlets, therefore, the *dampers are required for balancing the pressure drops in various runs.*

3. ***Static regain method***

For achieving *perfect balancing, the pressure losses in various branches should be made equal, so that the pressure at all outlets is the same*. This is possible if the *friction loss in each branch is made equal to the gain in pressure due to reduction in velocity*. The gain in pressure is taken as ***static pressure regain.***

For *'complete regain,'* it may not be possible to design economically very long branches very near to the fan.

In such cases, *it may be sufficient to design the main duct for complete regain and provide same pressure at all outlets from the main duct for branches.*

'Partial regain' may be considered a good practice for a first few outlets from the main duct so that same pressure loss is allowed in the beginning. At the rest of the outlets all frictional losses may be regained by reducing the velocity.

Advantages:

(*i*) Long runs as well as short runs (near the fan) can be designed for complete regain.

(*ii*) It is sufficient to design the main duct for complete regain and use the same pressure at outlets of the branches.

- This method is inferior to equal friction method in that the duct size for static regain method increases thus increasing the cost. The cost of sheet metal increases, but reduced balancing problems give option in favour of this method.

16.1.8 System Resistance, Systems in Series and in Parallel

16.1.8.1 System resistance

When a single duct-line consists of different cross-sections, bends, expanders, reducers, registers, dampers etc., it constitutes a system of resistances in series. In such a system all the pressure losses occurring at a given volumetric flow rate are added together to give the *system resistance or total pressure loss* (at a given volumetric flow rate).

Since the frictional or dynamic pressure loss is directly proportional to the velocity pressure, therefore, total pressure loss in a duct system

$$p_L = K_1 p_{v1} + K_2 p_{v2} + ... \quad ...(i)$$

$$= K_1 C_1^2 + K_2 C_2^2 + K_3 C_3^2 + ... \quad ...(ii)$$

(Since, $p_v \propto C^2$)

Since Q (volumetric flow rate) is same for system in series, therefore,

$$Q = A_1C_1 = A_2C_2 = A_3C_3$$

i.e., $C_1 \alpha Q, C_2 \alpha Q, C_3 \alpha Q$

$\therefore$ Eqn. (*ii*) reduces to

$$p_L = z_1Q^2 + z_2Q^2 + z_3Q^2 + ...$$
$$= (z_1 + z_2 + z_3 + ...) Q^2 = ZQ^2$$

Thus, p_L = R (System resistance) = ZQ^2 ...(16.36)

where, Z is a constant of the system.

16.1.8.2 System in series

Fig. 16.12 (*a*) shows systems R_1, R_2 and R_3 having constants Z_1, Z_2 and Z_3 respectively connected in series and a single equivalent system is depicted in Fig. 16.12 (*b*). The resistance of a single equivalent system or the overall system resistance for the *given flow rate* is obtained by adding the individual system resistances,

i.e., $R_{eq} = R_1 + R_2 + R_3$

or, $ZQ^2 = Z_1Q^2 + Z_2Q^2 + ZQ_3^2$ ...(16.37)

and, $Z = Z_1 + Z_2 + Z_3$

16.1.8.3 Systems in parallel

Fig 16.13 (*a*) shows systems R_1, R_2 and R_3 having constants Z_1, Z_2 and Z_3 respectively connected in parallel and a single equivalent system is shown in Fig. 16.13 (*b*). The resistance of a single equivalent system or the overall system resistance may be obtained as follows:

$$Q = Q_1 + Q_2 + Q_3$$

$$\sqrt{\frac{R_{eq}}{Z}} = \sqrt{\frac{R_1}{Z_1}} + \sqrt{\frac{R_2}{Z_2}} + \sqrt{\frac{R_3}{Z_3}} \quad \left[\because R = ZQ^2 \text{ or, } Q = \sqrt{\frac{R}{Z}}\right]$$

When the *systems are in parallel the pressure loss in same*, i.e.,

$$R_{eq} = R_1 = R_2 = R_3$$

$$\therefore \sqrt{\frac{1}{Z}} = \sqrt{\frac{1}{Z_1}} + \sqrt{\frac{1}{Z_2}} + \sqrt{\frac{1}{Z_2}} \quad ...(16.38)$$

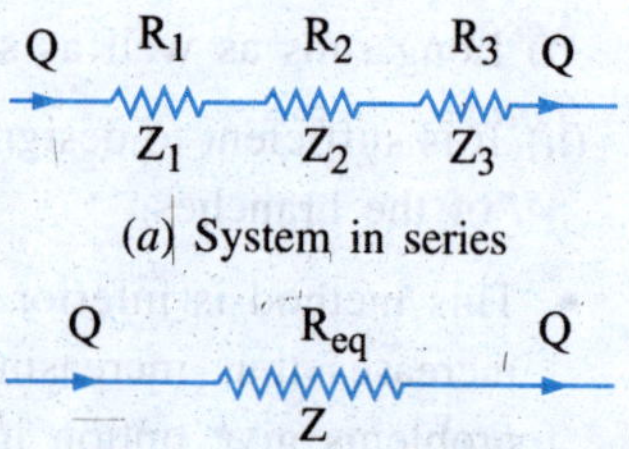

(*a*) System in series

(*b*) Single equivalent system

Fig. 16.12

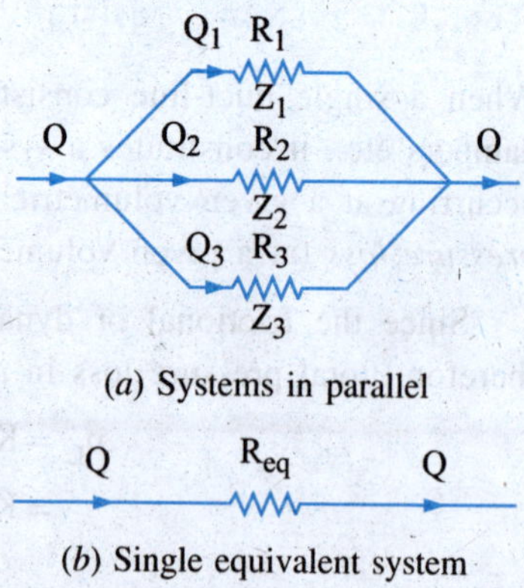

(*a*) Systems in parallel

(*b*) Single equivalent system

Fig. 16.13

16.1.9 Leakage of Air and Maintenance of Ducts

- Depending on air pressure, type of construction, and workmanship air leakage varies over wide limits. *Leakages from 5 to 30% have been observed during actual tests on typical supply systems.* Corner holes normally account for only a small portion. The largest usual source is at transverse seams located against the wall or ceiling in such manner that light joints are almost impossible.

- Allowance should be made for leakage, depending on job conditions. *For supply systems with static pressures in excess of 2.5 cm, calking, felting or soldering is recommended.* Very little maintenance is required for ventilating and air-conditioning ducts. When they are dry, the deterioration by corrosion is usually negligible.
- *Periodic cleaning is important* because even with comparatively efficient air-cleaning devices dirt accumulates over a period of time. A shift in dampers will frequently blow a cloud of dirt into the room. To allow accessibility for cleaning ducts should be provided with *access doors.*

16.2 AIR-DISTRIBUTION SYSTEMS

16.2.1 Definitions

Intake. An opening through which air is *returned* or *exhausted* from the space.

Outlet. An opening through which air is *supplied* to the treated space.

Grille. A functional or decorative covering for an outlet or intake.

Resister. A grille provided with a damper.

Diffuser. An outlet grille (or appurtenance) designed to guide the direction of the air.

Throw. The horizontal or vertical axial distance that an air stream travels on leaving the outlet.

Drop. The vertical distance the lower edge of the air stream drops between the time it leaves the outlet and the time it reaches the end of its throw.

Primary air. The air coming out of the outlet.

Secondary air. The room air picked up by the primary air by entrainment.

Total air. The mixture of primary and secondary air.

Aspect ratio. Grille-dimension ratio, length to width.

16.2.2 Principles of Air-distribution

The following are the *principles of air distribution* :

1. Air should be distributed in the room so that the required temperature, humidity and air velocity are maintained in the occupied zone of about 1.8 m above the floor.
2. Air stratification, temperature difference, dead pockets, high drafts, stragnation layers, convection currents and spot ceiling or heating must be avoided.
3. There should be thorough mixing of the conditioned air discharged into the room through outlets with the air inside the room.
4. The temperature differential in the room should not be larger than 1°C.
5. The velocity should be in the range of 7 m/min to 17 m/min to avoid low or high drafts. Down flow and flow directed to the faces of people is preferred to the upward flow and flow directed to backs or sides of people.
6. The exhaust and inlet points must be so arranged that fresh air is available in all parts of the room.

Location of return air-openings:

For locating return air-openings following *points* should be considered :

(*i*) There should not be short-circuiting of supply air and return air. The return air-opening should be complementary to the flow pattern established by the supply air.

(*ii*) The return air containing dirt, gases or odours should be removed without causing stratification in the conditioned space.

(*iii*) The recommended velocity of air at the inlet of the return air-opening should not exceed 80 m/min as they are nearer to the ear level to avoid undesirable noise. Whereas the velocity of air at the outlet of the supply grilles lies between 180 to 240 m/min.

- The *wall return* near the floor is *considered best.*
- A *ceiling return* should be *avoided* as there is possibility of short-circuiting.
- *Floor returns* should also be avoided as they act as duct collector. If used, a settling chamber should be provided.

16.2.3 Air-handling System

The air handling system consists of:

1. **Air-distribution system.** It comprises various *inlets* for recirculated air and ducts for the supply air.
2. **Duct system.** It includes the *return duct, supply* duct and *air-conditioning apparatus* comprising *dampers, filters, coil or air washer.*
3. **Fan.** It provides necessary energy to move the air.

Fig. 16.14 shows schematic air-flow diagram for an air-conditioning system. It may be seen that in circulation of air a closed loop is formed. In this loop the point is room itself which can be considered at atmospheric pressure. Through the inlets the air enters and it continues to drop in pressure until it reaches the fan. The fan raises the pressure, and this pressure thereafter starts dropping again until the air enters the space. Therefore, the pressure on the discharge side of the fan is positive and on suction side negative.

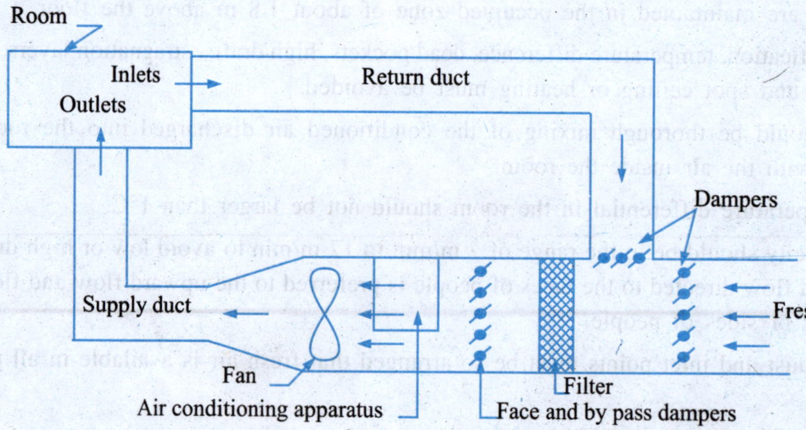

Fig. 16.14. Air-flow diagram.

16.2.4 Room Air-distribution

16.2.4.1 Requirements of good room air-distribution

The primary requirement of good room air distribution is to create a proper combination of temperature, humidity and air motion in the occupied zone which is normally at 1.8 m above the floor level. The maximum variation in temperature in a single room should not be more than 1°C, and within rooms, 2°C. The *desirable air velocity* is 7.5 to 9 m/min.

16.2.4.2 Draft

It is defined as *any localized feeling of coolness or warmth of any portion of the body due to both air movement and air temperature with humidity and radiation, considered constant.* The warmth or coolness of a draft is measured above or below the controlled room condition of 24.4°C DBT at the centre of the room and air moving at approximately 9 m/min.

16.2.4.3 Types of supply air outlets

Number of outlets depends upon the total quantity of air to be distributed, size and type of outlets selected. The *location* of the outlet is determined by height, shape, structural features, window and door openings of the room. Several types of outlets are available and the *selection* of the type of the outlet depends on the required throw and the air quantity to be thrown by each outlet.

Outlets may be divided into *four types:*

1. Grille outlets.
2. Slot diffuser outlets.
3. Ceiling diffuser outlets.
4. Perforated ceiling panels.

1. ***Grille outlets.*** Grilles with *fixed bars* can give direction at the outlet and gives a *constant capacity* at the outlet. This can be used where air capacity can be determined accurately and is not likely to change. *Adjustable bar grilles* provide for *varying the air quantity* if required. Horizontal and vertical, both adjustments, if provided, make control easier.
2. ***Slot diffuser outlets.*** Slotted outlets have performance similar to bar grilles. Long narrow slots give better induction of room air and mixing is perfected in a short throw. They suit mostly to window outlets and linear design of architecture. These should not be installed in ceilings.
3. ***Ceiling diffuser outlets.*** These outlets supply air in multiple layers and induction is rapid at a short distance after the air leaves the outlets. They may also be fitted with dampers.
4. ***Perforated ceiling panels.*** These are particularly suited to low ceilings. It offers unobstructive appearance and sound absorbing device.

16.2.5 Arrangement of Ducts

Ducts are employed to supply conditioned air (from air-conditioning plant) to the outlets which further distribute air in the room or occupied zone of space. The ducts which supply air to the room/space are called "*supply ducts*" and those (ducts) which extract the air from conditioned space and send back to the air conditioning plant are known as "*return ducts*".

The "*Supply ducts*" may be arranged in the following ways :

1. Loop perimeter duct system.
2. Radial perimeter duct system.
3. Extended plenum duct system.

1. ***Loop perimeter duct system.*** In "*loop perimeter duct system*" (Fig. 16.15) the conditioned air is carried in several feeder ducts to a common continuous closed loop duct around the perimeter of the building. The outlets then supply air to the room. This system is well adopted to *concrete slab construction* on ground, for offsetting down draft and cold floors. To avoid waste of heat, edge insulation of the slab is essential.

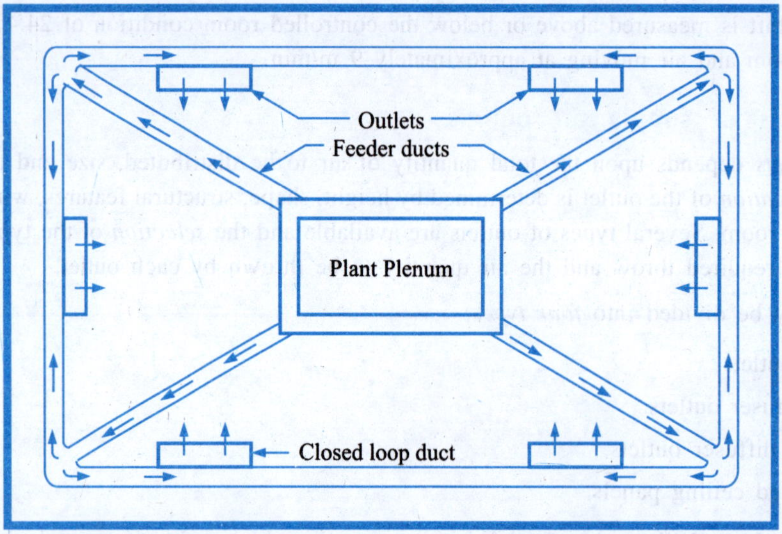

Fig. 16.15. Loop perimeter duct system.

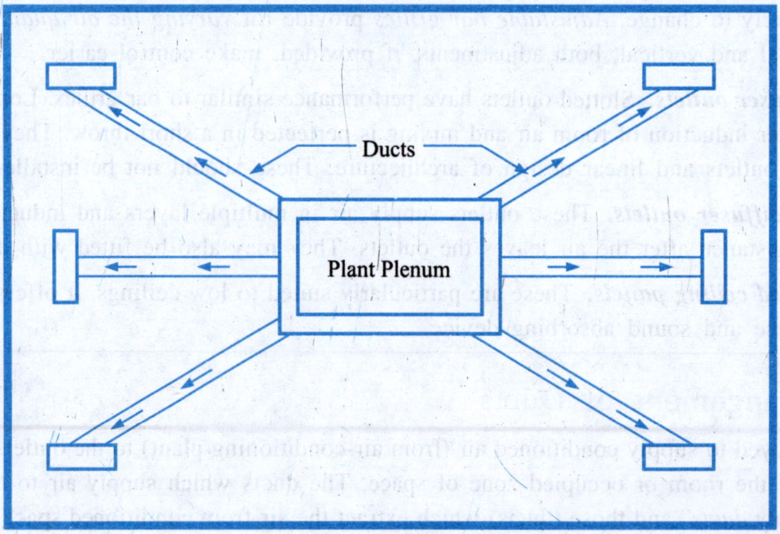

Fig. 16.16. Radial perimeter duct system.

2. ***Radial perimeter duct system.*** In "*radial perimeter duct system*" separate ducts supply conditioned air to the respective supply outlets from the plant plenum as shown in Fig. 16.16. This system is recommended for *crawl-space houses*. It can also be used in *slab floors*, but will not be effective as the loop system in warming the perimeter of the floor.

3. ***Extended plenum duct system.*** In "*Extended plenum duct system*" the plenum is usually extended to one or both sides of the air-conditioning plant. Each supply outlet is fed with conditioned-air from separate duct as shown in Fig. 16.17. This system is recommended for *houses with basement or crawl space*.

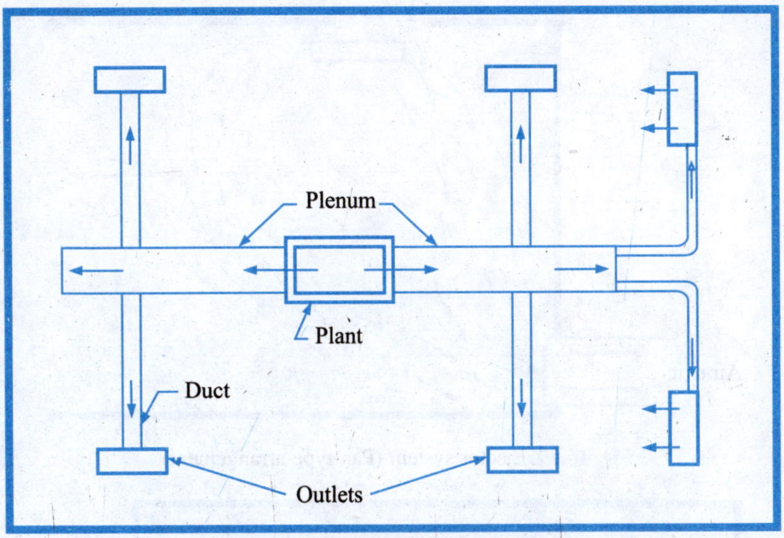

Fig. 16.17. Extended plenum duct system.

16.2.6 Types of Air-distribution Systems

Conditioned air from air-conditioning plant may be fed to a room through supply by any of the following *air-distribution* systems:

1. Ejector system.
2. Downward system.
3. Upward system.

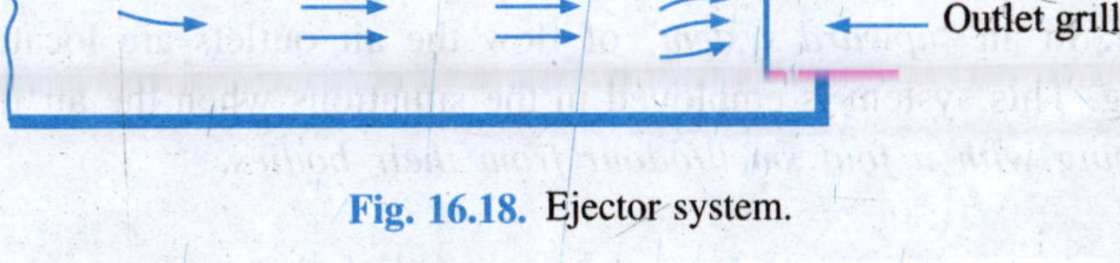

Fig. 16.18. Ejector system.

1. ***Ejector system.*** In an "*ejector system*" the inlet grille ejects the air into the room and inducts sufficient velocity for circulation. Figs. 16.18. and 16.19 show important systems of this type. Former is cheaper and very simple in construction. Letter system is known as "*pan-type*" arrangement. It provides more uniform distribution of air. Fig. 16.20 shows an arrangement which is recommended for place such as recreation places and gymnasiums where the persons occupying the room are moving or doing some exercises.

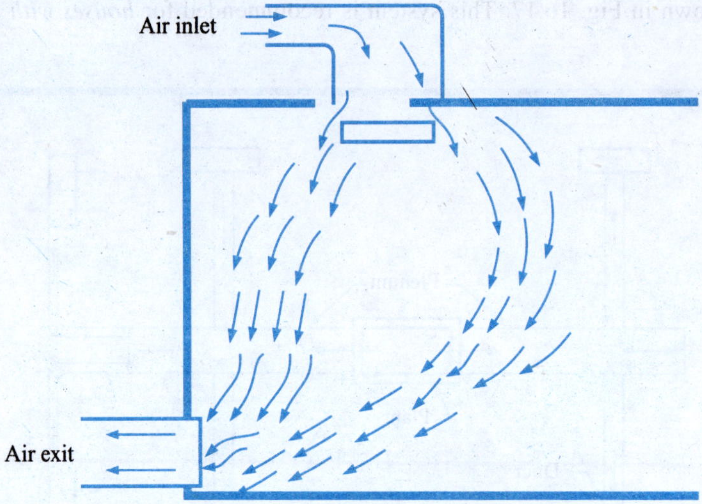

Fig. 16.19. Ejector system (Pan-type arrangement).

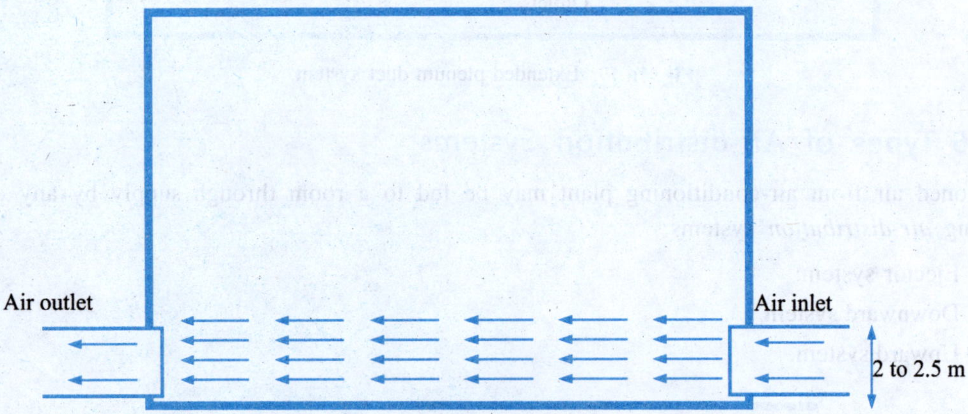

Fig. 16.20

2. ***Downward system.*** In "*downward system*" of air flow (Fig. 16.21) air is introduced through openings located in the ceiling and removed through the openings made in floor or in walls near the floor. This system is costly because of the use of perforated ceiling. It may be employed in *schools, offices, theatres and auditoriums*.

3. ***Upward system.*** In an "*upward system*" of flow the air outlets are located in walls near the ceiling or in the ceiling. This system is employed in the situations when the air in the room occupied by persons *rises carrying with it foul smell/odour from their bodies.*

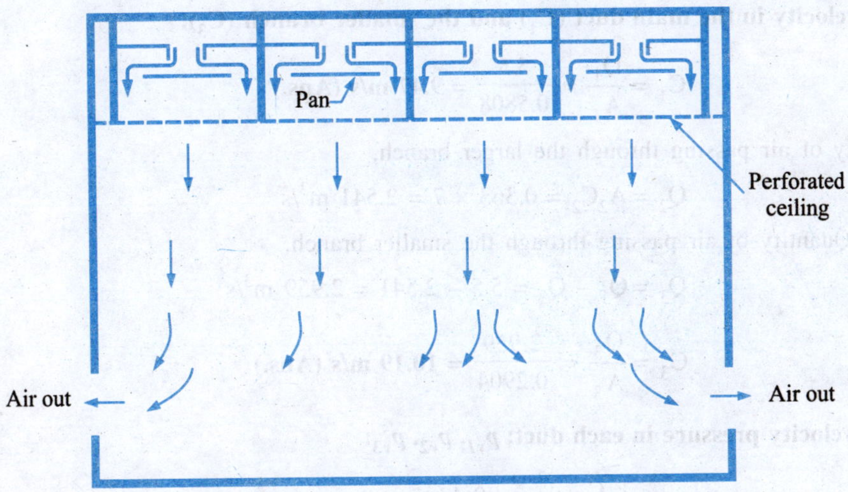

Fig. 16.21. Downward systen..

WORKED EXAMPLES

"DUCT SYSTEMS"

Example 16.1. *In an air-conditioning system, the size of the main air supply duct is 880 mm × 660 mm in cross-section, and carries 330 m^3/min of standard air. The main duct branches into two ducts of cross-section 660 mm × 550 mm and 660 mm × 440 mm. If the mean velocity in the larger branch is 420 m/min, determine :*

(i) Mean velocity in the main duct and the smaller branch.

(ii) Mean velocity pressure in each duct.

Solution: Refer to Fig. 16.22.

Given: a_1 = 880 mm = 0.88 m;

b_1 = 660 mm = 0.66 m;

Q_1 = 330 m^3/min = 5.5 m^3/s;

a_2 = 660 mm = 0.66 m; b_2 = 550 mm = 0.55 m;

a_3 = 660 mm = 0.66 m; b_3 = 440 mm = 0.44 m;

C_2 = 420 m/min = 7 m/s.

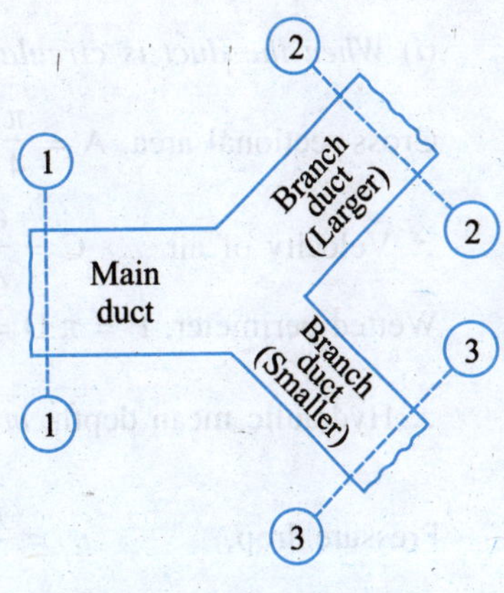

Fig. 16.22

Cross-sectional area of the main duct,

$$A_1 = a_1 \times b_1 = 0.88 \times 0.66 = 0.5808 \text{ m}^2$$

Cross-sectional area of the larger branch duct,

$$A_2 = a_2 \times b_2 = 0.66 \times 0.55 = 0.363 \text{ m}^2$$

Cross-sectional area of the smaller branch duct,

$$A_3 = a_3 \times b_3 = 0.66 \times 0.44 = 0.2904 \text{ m}^2$$

(*i*) **Mean velocity in the main duct (C_1) and the smaller branch (C_3):**

Now, $$C_1 = \frac{Q_1}{A_1} = \frac{5.5}{0.5808} = \mathbf{9.47\ m/s\ (Ans.)}$$

Quantity of air passing through the larger branch,

$$Q_2 = A_2C_2 = 0.363 \times 7 = 2.541\ m^3/s$$

∴ Quantity of air passing through the smaller branch,

$$Q_3 = Q_1 - Q_2 = 5.5 - 2.541 = 2.959\ m^3/s$$

∴ $$C_3 = \frac{Q_3}{A_3} = \frac{2.959}{0.2904} = \mathbf{10.19\ m/s\ (Ans.)}$$

(*ii*) **Mean velocity pressure in each duct; p_{v1}, p_{v2}, p_{v3}:**

$$p_{v1} = \left(\frac{C_2}{4.04}\right)^2 = \left(\frac{9.47}{4.04}\right)^2 = \mathbf{5.49\ mm\ of\ water\ (Ans.)}$$

$$p_{v2} = \left(\frac{C_2}{4.04}\right)^2 = \left(\frac{7}{4.04}\right)^2 = \mathbf{3.0\ mm\ of\ water\ (Ans.)}$$

$$p_{v3} = \left(\frac{C_3}{4.04}\right)^2 = \left(\frac{10.19}{4.04}\right)^2 = \mathbf{6.36\ mm\ of\ water\ (Ans.)}$$

Example 16.2. *A 12 cm long duct passes air at the rate of 1.2 m³/s. If the friction factor is 0.0048, calculate the pressure drop in the following cases:*

(*i*) *When the duct is circular of diameter 280 mm;*

(*ii*) *When the duct is 280 mm square section.*

Solution: *Given:* L = 12 m; Q = 1.2 m³/s; f = 0.0048; D = 280 mm = 0.28 m; $a = b$ = 280 mm = 0.28 m.

Pressure drop :

(*i*) *When the duct is circular :*

Cross-sectional area, $$A = \frac{\pi}{4} D^2 = \frac{\pi}{4} \times (0.28)^2 = 0.06157\ m^2$$

∴ Velocity of air $$C = \frac{Q}{A} = \frac{1.2}{0.06157} = 19.49\ m/s$$

Wetted perimeter, $P = \pi D = \pi \times 0.28 = 0.8796$ m

∴ Hydraulic mean depth, $$m = \frac{A}{P} = \frac{0.06157}{0.8796} = 0.07\ m$$

Pressure drop, $$p_f = \frac{fL}{m}\left(\frac{C}{4.04}\right)^2 \text{ mm of water} \qquad \text{...[Eqn. 16.11]}$$

$$= \frac{0.0048 \times 12}{0.07}\left(\frac{19.49}{4.04}\right)^2 = \mathbf{19.15\ mm\ of\ water\ (Ans.)}$$

(ii) When the duct is of square cross-section:

Cross-sectional area, $A = a \times b = 0.28 \times 0.28 = 0.0784\ m^2$

$\therefore$ Velocity of air, $C = \dfrac{Q}{A} = \dfrac{1.2}{0.0784} = 15.31$ m/s

Wetted perimeter, $P = 2\,(a + b) = 2\,(0.28 + 0.28) = 1.12$ m

$\therefore$ Hydraulic mean depth, $m = \dfrac{A}{P} = \dfrac{0.0784}{1.12} = 0.07$ m

Pressure drop,
$$p_f = \frac{f\,L}{m}\left(\frac{C}{4.04}\right)^2 \text{ mm of water}$$
$$= \frac{0.0048 \times 12}{0.07}\left(\frac{15.31}{4.04}\right)^2 = \mathbf{11.82\ mm\ of\ water\ (Ans.)}$$

Example 16.3. *A duct of rectangular cross-section 600 mm × 400 mm carries 90 m^3/min of air having density of 1.2 kg/m^3. Determine the equivalent diameter of a circular duct:*

(i) When the quantity of air carried in both the cases is same; and

(ii) When the velocity of air in both the cases is the same.

If friction factor (f) = 0.011, find the pressure loss per 100 m length of duct.

Solution; *Given:* $a = 600$ mm = 0.6 m; $b = 400$ mm = 0.4 m; $Q = 90\ m^3$/min;

$\rho_a = 1.2\ kg/m^3$; $f = 0.011$; $L = 100$ m.

Equivalent diameter, D_{eq}:

(i) When the quantity of air carried in both the cases is the same:

In this case,
$$D_{eq} = 1.265\left(\frac{a^3 b^3}{a+b}\right)^{1/5} \quad \text{...[Eqn. (16.19)]}$$
$$= 1.265\left[\frac{(0.6)^3\,(04)^3}{0.6+0.4}\right]^{1/5} = 0.537 \text{ m or } \mathbf{537\ mm\ (Ans.)}$$

(ii) When the velocity of air in both cases in the same:

In this case,
$$D_{eq} = \frac{2ab}{a+b} \quad \text{...[Eqn. (16.22)]}$$
$$= \frac{2 \times 0.6 \times 0.4}{0.6 + 0.4} = 0.48 \text{ m or } \mathbf{480\ mm\ (Ans.)}$$

Pressure loss, p_L:

Velocity of air passing through the duct,
$$C = \frac{Q}{A} = \frac{90}{a \times b} = \frac{90}{0.6 \times 0.4} = 375 \text{ m/min or } 6.25 \text{ m/s}$$

Hydraulic mean depth, $m = \dfrac{A}{P} = \dfrac{ab}{2\,(a+b)} = \dfrac{0.6 \times 0.4}{2\,(0.6+0.4)} = 0.12$ m

$\therefore$ Pressure loss, $p_L = \dfrac{f L \rho_a C^2}{2m}$ ...[Eqn. (16.9)]

$$= \frac{0.011 \times 100 \times 1.2 \times (6.25)^2}{2 \times 0.12}$$

$$= 214.84 \text{ N/m}^2 = \frac{214.84}{9.81} = \textbf{21.9 mm of water (Ans.)}$$

$$\left(\because 1 \text{ N/m}^2 = \frac{1}{9.81} \text{ mm of water} \right)$$

Example 16.4. *An air duct system is provided as shown in Fig. 16.23.*

If the quantity of air discharged at D is 1500 m³/min, Calculate:

(i) The quantity of air discharged at C;

(ii) The static pressure at fan outlet A.

Take value of friction factor as 0.0048.

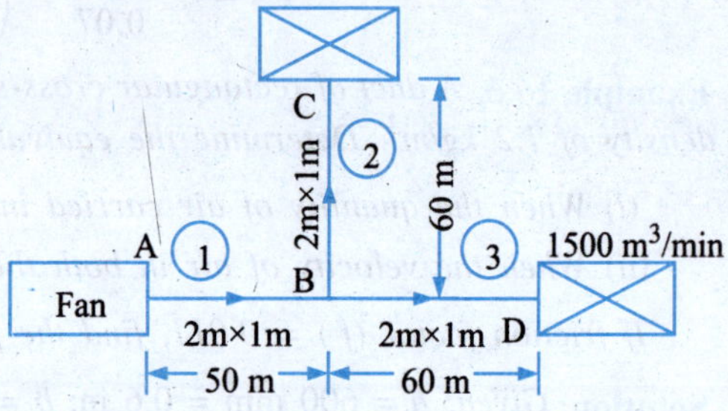

Fig. 16.23

Solution: *Given:* a = 2m; b = 1 m; L_1 = 50 m; L_2 = 60 m; L_3 = 60 m; Q = 1500 m³/min; f = 0.0048.

(*i*) Quantity of air discharged at C, Q_2:

Cross-sectional areas of ducts AB, BC and BD,

$$A_1 = A_2 = A_3 = a \times b = 2 \times 1 = 2 \text{ m}^2$$

Velocity of air in duct BD,

$$C_3 = \frac{Q_3}{A_3} = \frac{1500}{2} = 750 \text{ m/min or } 12.5 \text{ m/s}$$

Velocity pressure, $p_v = \dfrac{\rho_a C^2}{2} = \dfrac{1.2 \times C^2}{2} = 0.6 \, C^2 \text{ N/m}^2$ [Taking ρ_a = 1.2 kg/m³]

$\therefore$ Velocity pressure in duct BD,

$$p_{v3} = 0.6 \, C_3^2 = 0.6 \, (12.5)^2 = 93.75 \text{ N/m}^2$$

Since the size of each duct is 2 m × 1 m, therefore, hydraulic mean depth for each of the ducts AB, BC and BD is,

$$m_1 = m_2 = m_3 = \frac{ab}{2(a+b)} = \frac{2 \times 1}{2(2+1)} = \frac{1}{3}$$

$\therefore$ Pressure loss due to friction in duct BD,

$$p_{f3} = \frac{f L_3}{m_3} (p_{v3}) = \frac{0.0048 \times 60}{(1/3)} \times 93.75 = 81 \text{ N/m}^2$$

Total pressure at B, $p_{TB} = p_{f3} + p_{v3}$

$$= 81 + 93.75 = 174.75 \text{ N/m}^2 \qquad ...(i)$$

Let, p_{v2} = Velocity pressure in duct BC, and

C_2 = Velocity in duct BC.

Then, the pressure loss due to friction in the duct BC,

$$p_{f2} = \frac{f\,L_2}{m_2}\,(p_{v2}) = \frac{0.0048 \times 60}{(1/3)} \times p_{v2} = 0.864\,p_{v2} \quad ...(ii)$$

From eqns. (*i*) and (*ii*), we have:

$$0.864\,p_{v2} = 174.75$$

$$\therefore \quad p_{v2} = 202.26 \text{ N/m}^2$$

Also, $p_{v2} = 0.6\,C_2^2$, or, $202.26 = 0.6\,C_2^2$

or, $$C_2 = \left(\frac{202.26}{0.6}\right)^{1/2} = 18.36 \text{ m/s}$$

∴ Quantity of air discharged at C,

$$Q = A_2C_2 = 2 \times 18.36 = 36.72 \text{ m}^3\text{/sec} = \mathbf{2203.2\ m^3/min\ (Ans.)}$$

(*ii*) **The static pressure at fan outlet A, p_{sA}:**

Quantity of air flowing through the duct AB,

$$Q_1 = Q_2 + Q_3 = 2203.2 + 1500 = 3703.2 \text{ m}^3\text{/min.}$$

∴ Velocity of air in the duct AB,

$$C_1 = \frac{Q_1}{A_1} = \frac{3703.2}{2} = 1851.6 \text{ m/min or } 30.86 \text{ m/s}$$

Velocity pressure in the duct AB,

$$p_{v1} = 0.6\,C_1^2 = 0.6 \times (30.86)^2 = 571.4 \text{ N/m}^2$$

Pressure loss due to friction in duct AB,

$$p_{f1} = \frac{f\,L_1}{m_1} \times p_{v1} = \frac{0.0048 \times 50}{(1/3)} \times 571.4 = 411.4 \text{ N/m}^2$$

Total pressure at the fan outlet (i.e., at A),

$$p_{TA} = p_{TB} + p_{f1} = 174.75 + 411.4 = 586.15 \text{ N/m}^2$$

$$\therefore \quad p_{sA} = T_A - p_{v1} = 586.15 - 571.4 = \mathbf{14.75\ N/m^2\ (Ans.)}$$

Example 16.5. *A 30 m long rectangular duct, 200 mm × 160 mm in size, carries standard air at the rate of 24 m³/min. Assuming the friction factor f = 0.0048, determine:*

(*i*) *Total pressure required at the inlet to the duct to maintain this flow; and*

(*ii*) *Air power required.*

Solution: *Given:* L = 30 m; a = 200 mm = 0.2 m; b = 160 mm = 0.16 m; Q = 2.4 m³/min; f = 0.0048.

(*i*) **Total pressure required at the inlet to the duct, p_T:**

Cross-sectional area, $A = a \times b = 0.2 \times 0.16 = 0.032 \text{ m}^2$

∴ Velocity of air in the duct, $C = \frac{Q}{A} = \frac{24}{0.032} = 750$ m/min or 12.5 m/s

Velocity pressure, $p_v = 0.6\ C^2 = 0.6\ (12.5)^2 = 93.75\ N/m^2$

Wetted perimeter of the duct,

$$P = 2\ (a + b) = 2\ (0.2 + 0.16) = 0.72\ m$$

Hydraulic mean depth of the duct,

$$m = \frac{A}{P} = \frac{0.032}{0.72} = 0.0444\ m$$

Pressure loss due to friction in the duct,

$$p_f = \frac{f\,L}{m} \times p_v = \frac{0.0048 \times 30}{0.0444} \times 93.75 = 304.05\ N/m^2$$

$$\therefore \quad p_T = p_f + p_v = 304.05 + 93.75 = \mathbf{397.8\ N/m^2\ (Ans.)}$$

(*ii*) **Air power required:**

$$\text{Air power required} = Q \times p_T$$

$$= \frac{24}{60} \times 397.8 = \mathbf{159.12\ W\ (Ans.)} \qquad (\because 1\ Nm/s = 1\ W)$$

Example 16.6. *A 700 mm diameter duct is 20 m long and carries 168 m³/min of air at 20°C. Find the pressure loss in the duct by using the friction chart.*

Solution: *Given:* D = 700 mm; L = 20 m; Q = 168 m³/min = 2.8 m³/s

- Draw a horizontal line corresponding to 2.8 m³/s intersecting the diagonal line for 700 mm duct at point A, as shown in Fig. 16.24.

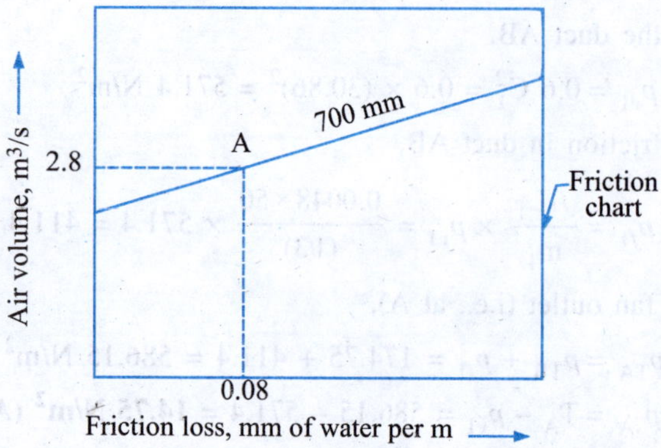

Fig. 16.24

- Read the pressure loss from the bottom of the chart, which is read as:

0.08 mm per meter length of the duct

$\therefore$ Total pressure loss of 20 m long duct = 20 × 0.08 = **1.6 mm of water (Ans.)**

Example 16.7. *A main circular duct consists of three branch ducts taking equal air volumes at equal intervals. Each interval duct has a friction loss of 1.2 mm of water and a static pressure of 4 mm of water is necessary at each branch to cope with its friction loss. If the initial velocity in the main duct of 1.3 m diameter is 9 m/s, calculate the velocities and diameters of the second and third lengths, whereby the static pressure regain is sufficient to overcome the friction loss in the succeeding length of main duct upto the next branch. Assume static pressure regain factor as 0.58.*

Also identify static, velocity and total pressures at the appropriate points of change.

Solution: *Given:* $p_{f1} = p_{f2} = p_{f3} = 1.2$ mm of water; $p_{sB} = p_{sD} = p_{sF} = 4$ mm of water; $D_1 = 1.3$ m; $C_1 = 9$m/s; R = 0.58.

Let, C_2 = Velocity of air in duct CD,

D_2 = Diameter of duct CD, and

p_{v2} = Velocity pressure in the duct CD.

Velocities and diameters of second and third lengths, C_2, D_2, C_3, D_3:

C_2 and D_2:

The quantity of air passing through the duct AB,

$$Q_1 = A_1C_1 = \frac{\pi}{4} D_1^2 \times C_1 = \frac{\pi}{4} \times (1.3)^2 \times 9 = 11.94 \text{ m}^3/\text{s}$$

Quantity of air passing through each branch,

$$Q_{1'} = Q_{2'} = Q_{3'} = \frac{11.94}{3} = 3.98 \text{ m}^3/\text{s}$$

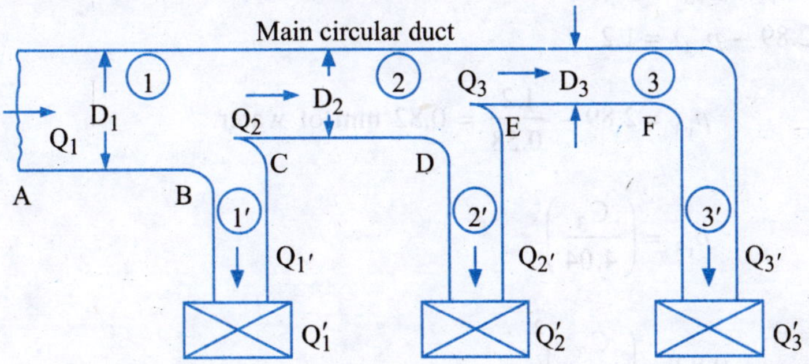

Fig. 16.25

Velocity pressure in duct AB,

$$p_{v1} = \left(\frac{C_1}{4.04}\right)^2 = \left(\frac{9}{4.04}\right)^2 = 4.96 \text{ mm of water}$$

Static pressure regain, SR = R $(p_{v1} - p_{v2})$

Since the static pressure regain is equal to the friction loss in the duct CD, therefore,

$$R\,(p_{v1} - p_{v2}) = p_{f2}$$

or, $$0.58\,(4.96 - p_{v2}) = 1.2$$

or, $$p_{v2} = 4.96 - \frac{1.2}{0.58} = 2.89 \text{ mm of water.}$$

Also, $$p_{v2} = \left(\frac{C_2}{4.04}\right)^2$$

or, $$2.89 = \left(\frac{C_2}{4.04}\right)^2$$

or, $$C_2 = 4.04\sqrt{2.89} = \mathbf{6.87 \text{ m/s (Ans.)}}$$

Quantity of air passing through the duct CD,

$$Q_2 = Q_1 - Q_1' = 11.94 - 3.98 = 7.96 \text{ m}^3/\text{s}$$

Also,
$$Q_2 = 7.96 = \frac{\pi}{4} \times D_2^2 \times C_2 = \frac{\pi}{4} D_2^2 \times 6.87$$

$$\therefore \quad D_2 = \left(\frac{7.96 \times 4}{\pi \times 6.87}\right)^{1/2} = \mathbf{1.214 \text{ m (Ans.)}}$$

C_3 and D_3:

Let, C_3 = Velocity in the duct EF,

D_3 = Diameter of the duct EF, and

p_{v3} = Velocity pressure in the duct EF.

Static pressure regain, SR = R $(p_{v2} - p_{v3})$

Again, $R(p_{v2} - p_{v3}) = p_{f3}$

or, $0.58(2.89 - p_{v3}) = 1.2$

$$\therefore \quad p_{v3} = 2.89 - \frac{1.2}{0.58} = 0.82 \text{ mm of water.}$$

Also,
$$p_{v3} = \left(\frac{C_3}{4.04}\right)^2$$

or,
$$0.82 = \left(\frac{C_3}{4.04}\right)^2$$

or,
$$C_3 = 4.04\sqrt{0.82} = \mathbf{3.66 \text{ m/s (Ans.)}}$$

Quantity of air passing through the duct EF,

$$Q_3 = Q_2 - Q_2' = 7.96 - 3.98 = 3.98 \text{ m}^3/\text{s}$$

Also,
$$Q_3 = 3.98 = \frac{\pi}{4} \times D_3^2 \times C_3 = \frac{\pi}{4} \times D_3^2 \times 3.66$$

$$\therefore \quad D_3 = \left(\frac{3.98 \times 4}{\pi \times 3.66}\right)^{1/2} = \mathbf{1.177 \text{ m (Ans.)}}$$

Static, velocity and total pressures at the appropriate points of change:

Pressure	*Appropriate point of change*				
mm of water	*B*	*C*	*D*	*E*	*F*
Static pressure (p_s)	4	5.2	4	5.2	4
Velocity pressure (p_v)	4.96	2.89	2.89	0.82	0.82
Total pressure (p_T)	8.96	8.09	6.89	6.02	4.82

Example 16.8. *In the Fig. 16.26 is shown a duct ABCDEF, in which air is supplied at A and discharged freely to atmosphere at F. The velocity of air in the duct is 540 m/min. The loss in the expander BC is 0.5 times the velocity pressure in duct AB and loss in the reducer DE is 0.2 times the velocity pressure in the duct EF. Using the expression,* $p_f = \dfrac{0.263\, C^{1.85}}{D^{1.27}}$, *where* p_f *is the pressure loss due to friction in mm of water per 100 m length of duct, C is the duct velocity in m/s and D is the diameter of the duct in m, calculate the static pressure at point A.*

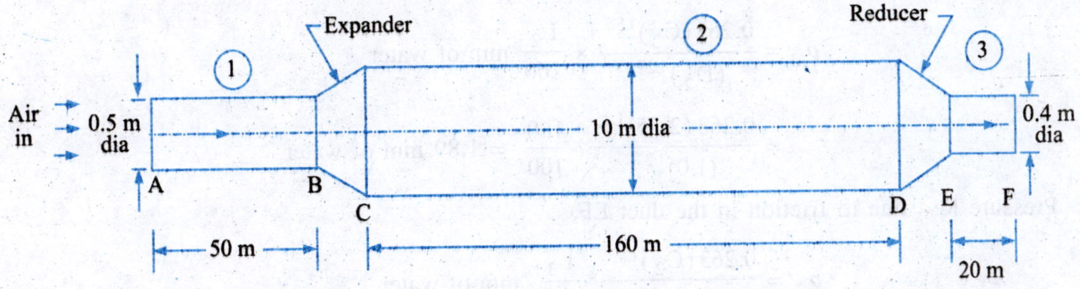

Fig. 16.26

Solution: *Given:* $D_1 = 0.5$ m; $L_1 = 50$ m; $D_2 = 1.0$ m; $L_2 = 160$ m; $D_3 = 0.4$ m; $L_3 = 20$ m; $C_1 = 540$ m/min $= 9$ m/s:

Static pressure of point A:

Let C_2 and C_3 be the velocities of air in duct DC and duct EF respectively.

Also, $$A_1C_1 = A_2C_2 = A_3C_3$$

$$\therefore \quad C_2 = \frac{A_1C_1}{A_2} = C_1\left(\frac{D_1}{D_2}\right)^2 = 9\left(\frac{0.5}{1.0}\right)^2 = 2.25 \text{ m/s}$$

$$C_3 = \frac{A_1C_1}{A_3} = C_1\left(\frac{D_1}{D_3}\right)^2 = 9\left(\frac{0.5}{0.4}\right)^2 = 14.06 \text{ m/s}$$

Assuming standard air, velocity pressure in duct AB,

$$p_{v1} = \left(\frac{C_1}{4.04}\right)^2 = \left(\frac{9}{4.04}\right)^2 = 4.96 \text{ mm of water,}$$

and, velocity pressure in the duct EF,

$$p_{v3} = \left(\frac{C_3}{4.04}\right)^2 = \left(\frac{14.06}{4.04}\right)^2 = 12.11 \text{ mm of water}$$

Pressure loss in expander BC,

$$(p_L)_{exp} = 0.5 \times p_{v1} = 0.5 \times 4.96 = 2.48 \text{ mm of water}$$

Pressure loss in reducer DE,

$$(p_L)_{red} = 0.2 \times p_{v3} = 0.2 \times 12.11 = 2.42 \text{ mm of water}$$

Pressure loss due to friction in the duct AB,

$$p_{f1} = \frac{0.263\,(C_1)^{1.85}}{(D_1)^{1.27}} \times \frac{L_1}{100} \text{ mm of water}$$

$$= \frac{0.263\,(9)^{1.85}}{(0.5)^{1.27}} \times \frac{50}{100} = 18.47 \text{ mm of water}$$

Pressure loss due to friction in the duct CD,

$$p_{f2} = \frac{0.263\,(C_2)^{1.85}}{(D_2)^{1.27}} \times \frac{L_2}{100} \text{ mm of water}$$

$$= \frac{0.263\,(2.25)^{1.85}}{(1.0)^{1.27}} \times \frac{160}{100} = 1.89 \text{ mm of water}$$

Pressure loss due to friction in the duct EF,

$$p_{f3} = \frac{0.263\,(C_2)^{1.85}}{(D_3)^{1.27}} \times \frac{L_3}{100} \text{ mm of water}$$

$$= \frac{0.263\,(14.06)^{1.85}}{(0.4)^{1.27}} \times \frac{20}{100} = 22.39 \text{ mm of water}$$

∴ Total pressure loss in the duct between A and F,

$$\Sigma_{pL} = p_{f1} + (p_L)_{exp} + p_{f2} + (p_L)_{red} + p_{f3}$$

$$= 18.47 + 2.48 + 1.89 + 2.42 + 22.39 = 47.65 \text{ mm of water}$$

Let, p_{s1} = Static pressure at point A.

Now applying Bernoulli's equation between A and F, we have :

$$p_{s1} + p_{v1} = p_{s3} + p_{v3} + \Sigma p_L$$

or, $p_{s1} + 4.96 = 0 + 12.11 + 47.65$

(p_{s3} = 0, since the air is discharged at F to atmosphere)

∴ p_{s1} = **54.8 mm of water (Ans.)**

Example 16.9. *In a system consisting of inlet and outlet ducts, a centrifugal fan of outlet 0.8 m × 0.6 m is moving standard air at a rate of 10 m³/s. The inlet duct is 0.8 m diameter and 12 m long and the discharge duct is 0.9 m diameter and 50 m long. There is a diffuser between the fan discharge and the 0.9 m diameter duct for which the loss of pressure may be taken as 0.3 times the difference between the velocity pressures. The loss at entry to the inlet duct is 0.5 times the velocity pressure there. The friction factor 'f' for the inlet duct is 0.0038 and for the outlet duct is 0.0032. Determine :*

(i) Fan total pressure;

(ii) Static pressures at the fan inlet and outlet;

(iii) Plot the variation of the static pressure, velocity pressure and total pressure along the system.

Assume that the air is sucked in by the inlet duct and delivered by the outlet duct at atmospheric pressure.

Solution: *Given:* $a = 0.8$ m; $b = 0.6$ m; Q = 10 m^3/s; $D_1 = 0.8$ m; $L_{1-2} = 12$ m; $D_4 = 0.9$ m; $L_{4-5} = 50$ m; $f_{1-2} = 0.0038$; $f_{4-5} = 0.0032$.

The duct system is shown in Fig. 16.27 (*a*)

(*i*) **Fan total pressure, $(p_T)_{fan}$:**

Cross-sectional area of the duct 1–2,

$$A_1 = A_2 = \frac{\pi}{4}(D_1)^2 = \frac{\pi}{4}(0.8)^2 = 0.5026 \text{ m}^2$$

∴ Velocity of air in the duct 1–2

$$C_1 = C_2 = \frac{Q}{A_1} = \frac{10}{0.5026} = 19.9 \text{ m/s};$$

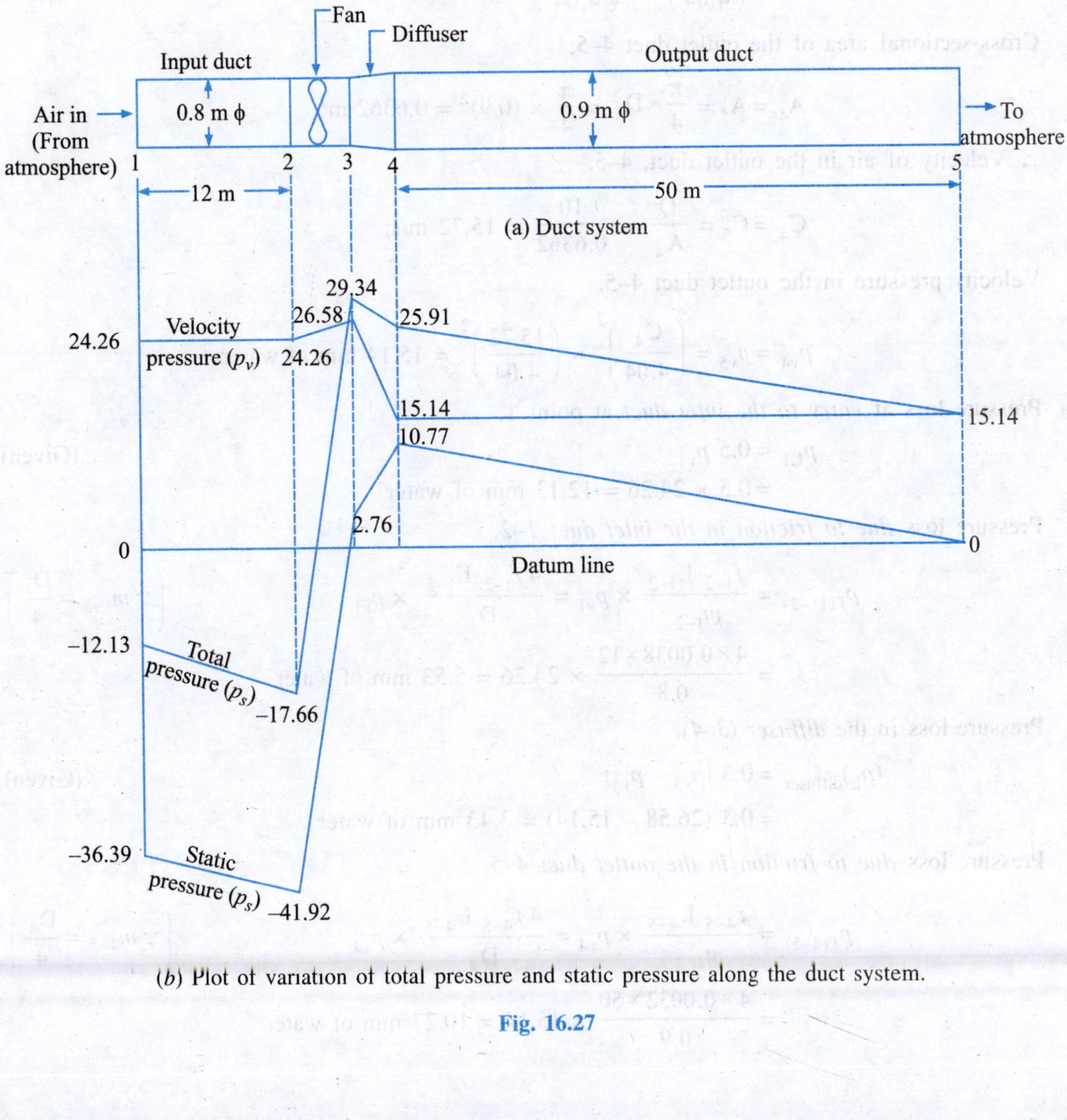

(*b*) Plot of variation of total pressure and static pressure along the duct system.

Fig. 16.27

Velocity pressure in the inlet duct 1–2,

$$p_{v1} = p_{v2} = \left(\frac{C_1}{4.04}\right)^2 = \left(\frac{19.9}{4.04}\right)^2 = 24.26 \text{ mm of water}$$

Cross-sectional area at the fan outlet,

$$A_3 = a \times b = 0.8 \times 0.6 = 0.48 \text{ m}^2$$

∴ Velocity of air at the fan outlet,

$$C_3 = \frac{Q}{A_3} = \frac{10}{0.48} = 20.83 \text{ m/s;}$$

Velocity pressure at the fan outlet,

$$p_{v3} = \left(\frac{C_3}{4.04}\right)^2 = \left(\frac{20.83}{4.04}\right)^2 = 26.58 \text{ mm of water}$$

Cross-sectional area of the outlet duct 4–5,

$$A_4 = A_5 = \frac{\pi}{4} \times D_4^2 = \frac{\pi}{4} \times (0.9)^2 = 0.6362 \text{ m}^2$$

∴ Velocity of air in the outlet duct, 4–5,

$$C_4 = C_5 = \frac{Q}{A_4} = \frac{10}{0.6362} = 15.72 \text{ m/s;}$$

Velocity pressure in the outlet duct 4–5,

$$p_{v4} = p_{v5} = \left(\frac{C_4}{4.04}\right)^2 = \left(\frac{15.72}{4.04}\right)^2 = 15.14 \text{ mm of water}$$

Pressure loss at *entry to the inlet duct* at point 1,

$$p_{L1} = 0.5\, p_{v1} \qquad \text{...(Given)}$$
$$= 0.5 \times 24.26 = 12.13 \text{ mm of water}$$

Pressure loss *due to friction in the inlet duct 1–2*,

$$p_{f(1-2)} = \frac{f_{1-2}\, L_{1-2}}{m_{1-2}} \times p_{v1} = \frac{4 f_{1-2}\, L_{1-2}}{D_1} \times p_{v1} \qquad \left(\because m_{1-2} = \frac{D_1}{4}\right)$$
$$= \frac{4 \times 0.0038 \times 12}{0.8} \times 24.26 = 5.53 \text{ mm of water}$$

Pressure loss in the *diffuser (3–4)*,

$$(p_L)_{diffuser} = 0.3\,(p_{v3} - p_{v4}) \qquad \text{...(Given)}$$
$$= 0.3\,(26.58 - 15.14) = 3.43 \text{ mm of water}$$

Pressure loss *due to friction in the outlet duct 4–5*,

$$p_{f(4-5)} = \frac{f_{4-5}\, L_{4-5}}{m_{4-5}} \times p_{v4} = \frac{4 f_{4-5}\, L_{4-5}}{D_4} \times p_{v4} \qquad \left(\because m_{4-5} = \frac{D_4}{4}\right)$$
$$= \frac{4 \times 0.0032 \times 50}{0.9} \times 15.14 = 10.77 \text{ mm of water}$$

Total pressure loss in the duct system,

$$(p_T)_{duct} = p_{L1} + p_{f(1-2)} + (p_L)_{diffuser} + p_{f(4-5)}$$

$$= 12.13 + 5.53 + 3.43 + 10.77 = 31.86 \text{ mm of water}$$

∴ *Fan total pressure,*

$$(p_T)_{fan} = (p_T)_{duct} + p_{v4} = 31.86 + 15.14 = \mathbf{47 \text{ mm of water (Ans.)}}$$

(*ii*) **Static pressure at fan inlet and outlet:**

Let p_{s2} and p_{s3} be the static pressures at fan inlet and outlet respectively.

Applying Bernoulli's equation to a section at O in the atmosphere upstream of the entrance to the inlet duct and to a point at fan inlet (i.e., at section 2), we have:

Total pressure at O = Total pressure at 2

$$p_{TO} = p_{T2} = p_{s2} + p_{v2} + p_{L1} + p_{f(1-2)}$$

Since the total pressure in the atmosphere (p_{TO}) is zero, therefore:

$$O = p_{s2} + 24.26 + 12.13 + 5.53$$

$$\therefore \quad p_{s2} = \mathbf{-41.92 \text{ mm of water (Ans.)}}$$

Again applying Bernoulli's equation to fan inlet (i.e., section 2) and fan outlet (i.e., section 3), we have:

$$p_{T2} = p_{T3}$$

$$p_{s2} + p_{v2} + (p_T)_{fan} = p_{s3} + p_{v3}$$

$$-41.92 + 24.26 + 47 = p_{s3} + 26.58$$

$$\therefore \quad p_{s3} = \mathbf{2.76 \text{ mm of water (Ans.)}}$$

(*iii*) **Variation of total pressure and static pressure along the system:**

- *Total pressure at entrance to inlet duct* (i.e., at section 1),

$$p_{T1} = p_{TO} - p_{L1} = 0 - 12.13 \text{ mm} = -12.13 \text{ mm of water}$$

- *Static pressure at entrance to inlet duct,*

$$p_{s1} = p_{T1} - p_{v1} = -12.13 - 24.26 = -36.39 \text{ mm of water}$$

- *Total pressure at fan inlet* (i.e., at section 2),

$$p_{T2} = p_{T1} - p_{f(1-2)} = -12.13 - 5.53 = -17.66 \text{ mm of water}$$

- *Static pressure at fan inlet,*

$$p_{s2} = p_{T2} - p_{v2} = -17.66 - 24.26 = -41.92 \text{ mm of water}$$

- *Total pressure at fan outlet or inlet to diffuser* (i.e., *at section 3*)

$$p_{T3} = p_{T2} + (p_T)_{fan} = -17.66 + 47 + 29.34 \text{ mm of water}$$

- *Static pressure at fan outlet,*

$$p_{s3} = p_{T3} - p_{v3} = 29.34 - 26.58 = 2.76 \text{ mm of water} \quad \text{...(same as before)}$$

- *Total pressure at outlet of diffuser or entrance to outlet duct* (i.e., *at section 4*),

$$p_{T4} = p_{T3} - (p_L)_{diffuser} = 29.34 - 3.43 = 25.91 \text{ mm of water}$$

- *Static pressure at outlet of diffuser or entrance to outlet duct,*

$$p_{s4} = p_{T4} - p_{v4} = 25.91 - 15.14 = 10.77 \text{ mm of water}$$

- *Total pressure at exit of outlet duct (i.e., at section 5),*

$$p_{T5} = p_{T4} - p_{f(4-5)} = 25.91 - 10.77 = 15.14 \text{ mm of water}$$

- *Static pressure at exit of outlet duct,*

$$p_{s5} = p_{T5} - p_{v5} = 15.14 - 15.14 = 0$$

The variation of static pressure, velocity pressure and total pressure is shown in Fig. 16.27 (*b*).

Example 16.10. *A centrifugal fan with 90 cm × 70 cm outlet is moving standard air at a rate of 11.5 m³/s through a system which consists of straight inlet and outlet ducts. The inlet duct is 90 cm in diameter and 15 m long. The outlet duct is 100 cm in diameter and 60 m long. There is a fan diffuser between the fan discharge and the 100 cm diameter duct for which the pressure loss is one-third the difference in velocity pressures. The pressure drop at the filter, damper and cooling coil in the inlet duct is 15 mm H_2O. The loss at entry to the inlet is half the velocity pressure. The friction factors for the inlet and outlet ducts are 0.004 and 0.0035 respectively.*

(i) *Determine the fan total pressure.*

(ii) *Determine the static pressures at the fan inlet and outlet.*

(iii) *Calculate and plot the variation of the total pressure and static pressure along the duct system.*

The air is sucked in by the inlet duct and delivered by the outlet at atmospheric pressure.

Solution: *Given:* $Q = 11.5 \text{ m}^3/\text{s}$; $D_2 = 90 \text{ cm} = 0.9 \text{ m}$; A_3 (Fan discharge) $= 0.9 \text{ m} \times 0.7 \text{ m}$;

$D_{3'} = 100 \text{ cm} = 0.1 \text{ m}$; $L_1 = 15 \text{ m}$; $L_3 = 60 \text{ m}$;

Loss at filter, damper and cooling coil = 15 mm H_2O;

$$\text{Loss at entry to the inlet} = \frac{1}{2} \times p_v;$$

$$\text{Loss in fan diffuser} = \frac{1}{3} \times \text{difference in velocity pressures;}$$

$f_1 = 0.004$; $f_3 = 0.0035$.

The duct system is shown in Fig. 16.28 (*a*).

(*i*) **Fan total pressure, $(p_T)_{fan}$:**

Area, velocity and velocity pressure in the inlet duct are:

$$A_1 = A_2 = \frac{\pi}{4} \times (0.9)^2 = 0.636 \text{ m}^2$$

$$C_1 = C_2 = \frac{Q}{A} = \frac{11.5}{0.636} = 18.08 \text{ m/s}$$

$$p_{v1} = p_{v2} = \left(\frac{C_1}{4.04}\right)^2 = \left(\frac{18.08}{4.04}\right)^2 = 20.03 \text{ mm } H_2O$$

Area, velocity and velocity pressure at the fan discharge are:

$$A_3 = 0.9 \times 0.7 = 0.63 \text{ m}^2$$

$$C_3 = \frac{11.5}{0.63} = 18.25 \text{ m/s}$$

$$p_{v3} = \left(\frac{C_3}{4.04}\right)^2 = \left(\frac{18.25}{4.04}\right)^2 = 20.41 \text{ mm } H_2O$$

Area, velocity, and velocity pressure in the outlet duct,

$$A_{3'} = A_4 = \frac{\pi}{4} \times (1)^2 = 0.785 \text{ m}^2$$

$$C_{3'} = C_4 = \frac{11.5}{0.785} = 14.65 \text{ m/s}$$

$$p_{v3'} = p_{v4} = \left(\frac{14.65}{4.04}\right)^2 = 13.15 \text{ mm } H_2O$$

The various pressure losses will now be calculated.

$$\text{Inlet loss} = 0.5\, p_{v1} = 0.5 \times 20.03 = 10.01 \text{ mm } H_2O$$

$$\text{Inlet duct friction} = \frac{4 f_1 L_1}{D_1} \times p_{v1}$$

$$= \frac{4 \times 0.004 \times 15}{0.9} \times 20.03 = 5.34 \text{ mm } H_2O$$

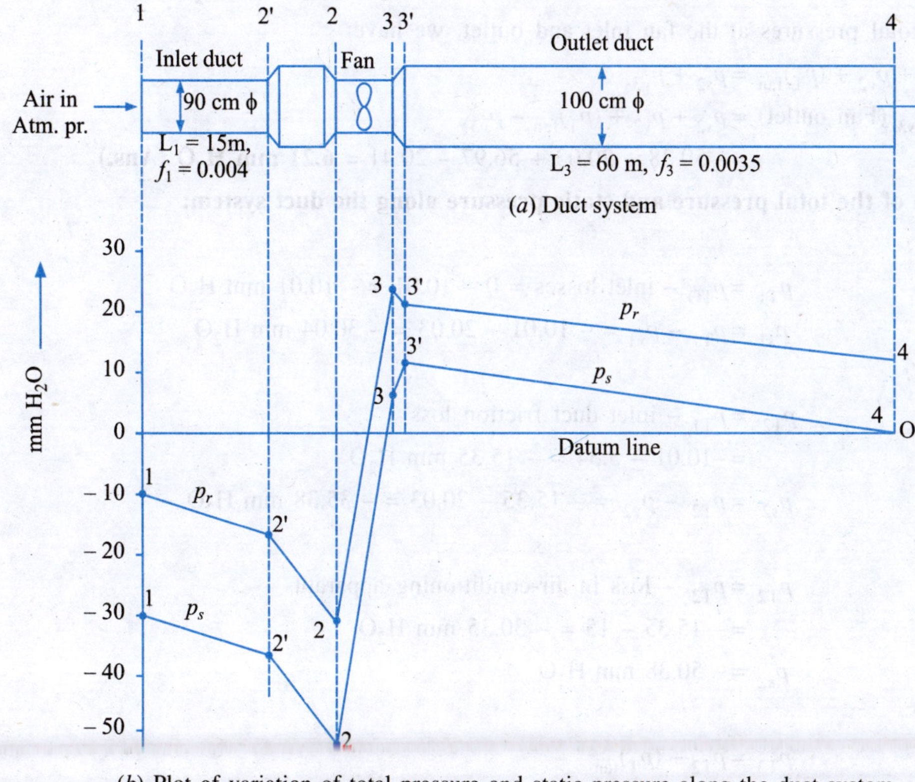

(*b*) Plot of variation of total pressure and static pressure along the duct system.

Fig. 16.28

Loss at the filter, damper and cooling coil = 15 mm H_2O ...(Given)

$$\text{Loss in fan diffuser} = \frac{1}{3}(p_{v3} - p_{v3'}) \quad \text{...(Given)}$$

$$= \frac{1}{3}(20.41 - 13.15) = 2.42 \text{ mm } H_2O$$

$$\text{Outlet duct frictional loss} = \frac{4 f_3 L_3}{D_{3'}} \times p_{v3'}$$

$$= \frac{4 \times 0.0035 \times 60}{1} \times 13.15 = 11.05 \text{ mm } H_2O$$

∴ Total duct loss = Sum of all the pressure drops

$$= 10.01 + 5.34 + 15 + 2.42 + 11.05 = 43.82 \text{ mm } H_2O$$

Fan total pressure, $(p_T)_{fan}$ = Pressure loss in duct system + Outlet velocity pressure

$$= 43.82 + 13.15 = \mathbf{56.97 \text{ mm } H_2O \text{ (Ans.)}}$$

(*ii*) **Static pressure at the fan inlet and outlet; p_{s2}, p_{s3} :**

Equating total pressures in the atmosphere and fan inlet, we have:

$$0 = \Delta p_{O2} + p_{v2} + p_{s2}$$

$$\therefore \; p_{s2} \text{ (Fan inlet)} = -\Delta p_{O2} - p_{v2}$$

$$= -(10.01 + 5.34 + 15) - 20.03 = \mathbf{-50.38 \text{ mm } H_2O \text{ (Ans.)}}$$

Equating total pressures at the fan inlet and outlet, we have:

$$p_{s2} + p_{v2} + (p_T)_{fan} = p_{s2} + p_{v3}$$

$$\therefore \; p_{s3} \text{ (Fan outlet)} = p_{s2} + p_{v2} + (p_T)_{fan} - p_{v3}$$

$$= -50.38 + 20.03 + 56.97 - 20.41 = \mathbf{6.21 \text{ mm } H_2O \text{ (Ans.)}}$$

(*iii*) **Variation of the total pressure and static pressure along the duct system:**

Section 1 :

$$p_{T1} = p_{TO} - \text{inlet losses} = 0 - 10.01 = -10.01 \text{ mm } H_2O$$

$$p_{s1} = p_{T1} - p_{v1} = -10.01 - 20.03 = -30.04 \text{ mm } H_2O$$

Section 2':

$$p_{T2'} = p_{T1} - \text{inlet duct friction loss}$$

$$= -10.01 - 5.34 = -15.35 \text{ mm } H_2O$$

$$p_{s2'} = p_{T2'} - p_{v2'} = -15.35 - 20.03 = -35.38 \text{ mm } H_2O$$

Section 2:

$$p_{T2} = p_{T2'} - \text{loss in air-conditioning apparatus}$$

$$= -15.35 - 15 = -30.35 \text{ mm } H_2O$$

$$p_{s2} = -50.38 \text{ mm } H_2O$$

Section 3:

$$p_{T3} = p_{T2} + (p_T)_{fan}$$

$$= -30.35 + 56.97 = 26.62 \text{ mm } H_2O$$

$$p_{s3} = 6.21 \text{ mm } H_2O$$

Section 3':

$$p_{T3'} = p_{T3} - \text{loss in fan diffuser}$$
$$= 26.62 - 2.42 = 24.2 \text{ mm } H_2O$$
$$p_{s3'} = p_{T3'} - p_{v3'} = 24.2 - 13.15 = 10.05 \text{ mm } H_2O$$

Section 4:

$$p_{T4} = p_{T3'} - \text{outlet duct friction loss}$$
$$= 24.2 - 11.05 = 13.15 \text{ mm } H_2O$$
$$p_{s4} = p_{T4} - p_{v4} = 13.15 - 13.15 = 0$$

The variations of the total pressure and the static pressure along the duct system are plotted as shown in Fig. 16.28 (*b*).

Example 16.11. *Fig. 16.29 shows a duct system in which the cross-section of the main duct is such that the width is always thrice the depth. Air flow and static pressure at A are 720 m³/min and 22 mm of water respectively. At B, a branch 0.8 m × 0.6 m in cross-section and 60 m long is led off. It discharges at atmospheric pressure and there is a loss of 0.5 velocity head at the conversion piece. At C, the branch is so designed that it discharges 150 m³/min.*

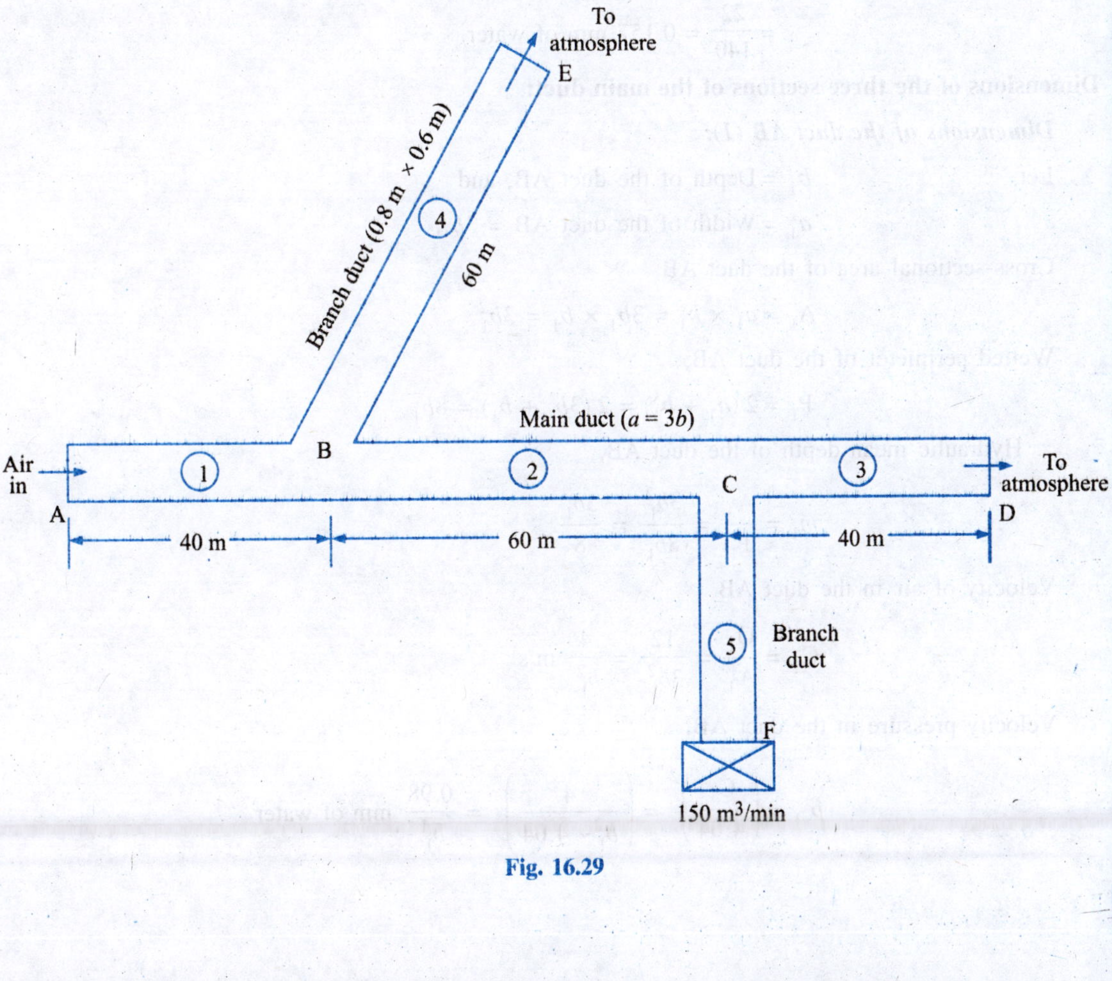

Fig. 16.29

Calculate the dimensions of the three sections of the main duct, assuming the basis of design as uniform pressure drop.

The value of friction factor may be taken as 0.0052. Assume that the discharge pressure at D is atmospheric and velocity changes in the main duct are neglected.

Solution: *Given:* $Q_1 = 720\ m^3/min = 12\ m^3/s$; $p_{sA} = 22$ mm of water; $a_4 = 0.8$ m; $b_4 = 0.6$ m; $L_4 = 60$ m; $p_{L4} = 0.5\, p_{v4}$; $Q_5 = 150\ m^3/min = 2.5\ m^3/s$; $f = 0.0052$; $L_1 = 40$ m; $L_2 = 60$ m; $L_3 = 40$ m.

When the velocity changes in the main duct AD are neglected, then static pressure at A must be equal to the pressure loss due to friction between A and D.

$$\therefore \quad p_{f1} + p_{f2} + p_{f3} = p_{sA} = 22 \text{ mm of water.}$$

(suffices 1, 2, 3, 4, and 5 correspond to the AB, BC, CD, BE and CF duct portions respectively.)

Total length of the duct AD,

$$L = L_1 + L_2 + L_3 = 40 + 60 + 40 = 140 \text{ m}$$

∴ Pressure drop per m length of the duct AD

$$= \frac{22}{140} = 0.157 \text{ mm of water.}$$

Dimensions of the three sections of the main duct:

Dimensions of the duct AB (1):

Let, b_1 = Depth of the duct AB, and

a_1 = Width of the duct AB = $3b_1$.

Cross-sectional area of the duct AB,

$$A_1 = a_1 \times b_1 = 3b_1 \times b_1 = 3b_1^2$$

Wetted perimeter of the duct AB,

$$P_1 = 2(a_1 + b_1) = 2(3b_1 + b_1) = 8b_1$$

∴ Hydraulic mean depth of the duct AB,

$$m_1 = \frac{A_1}{P_1} = \frac{3b_1^2}{8b_1} = \frac{3b_1}{8}$$

Velocity of air in the duct AB,

$$C_1 = \frac{Q_1}{A_1} = \frac{12}{3b_1^2} = \frac{4}{b_1^2} \text{ m/s}$$

Velocity pressure in the duct AB,

$$p_{v1} = \left(\frac{C_1}{4.04}\right)^2 = \left(\frac{4}{b_1^2 \times 4.04}\right)^2 = \frac{0.98}{b_1^4} \text{ mm of water}$$

$\therefore$ Pressure loss due to friction in the duct AB,

$$p_{f1} = \frac{f\,L_1}{m_1} \times (p_{v1}) = \frac{0.0052 \times 40}{(3b_1/8)} \times \frac{0.98}{b_1^4} = \frac{0.544}{b_1^5} \text{ mm of water} \quad ...(i)$$

Also, as the pressure drop per metre length is 0.157 mm of water, therefore, total pressure loss due to friction in the duct AB,

$$p_{f1} = 0.157 \times 40 = 6.28 \text{ mm of water} \quad ...(ii)$$

Equating eqns. (*i*) and (*ii*), we get

$$\frac{0.544}{b_1^5} = 6.28$$

$$\therefore \quad b_1 = \left(\frac{0.544}{6.28}\right)^{1/5} = \mathbf{0.613\ m\ (Ans.)}$$

and, $$a_1 = 3b_1 = 3 \times 0.613 = \mathbf{1.839\ m\ (Ans.)}$$

Dimensions of duct BC (2):

Let, b_2 = Depth of the duct BC, and

a_2 = Width of the duct BC = $3b_1$.

Let us first find out the quantity of air flowing in duct BE (4).

Cross-sectional area, $A_4 = a_4 \times b_4 = 0.8 \times 0.6 = 0.48\ m^2$

Wetted perimeter, $P_4 = 2\,(a_4 + b_4) = (0.8 + 0.6) = 2.8$ m

Hydraulic mean depth, $m_4 = \frac{A_4}{P_4} = \frac{0.48}{2.8} = 0.1714$ m

$\therefore$ Pressure loss due to friction in the duct BE,

$$p_{f4} = \frac{f\,L_4}{m_4} \times (p_{v4}) = \frac{0.0052 \times 60}{0.1714} \times p_{v4} = 1.82\, p_{v4}$$

Pressure loss at the conversion piece = $0.5\, p_{v4}$ ...(Given)

$\therefore$ Pressure at B = $p_{f4} + 0.5\, p_{v4} = 1.82\, p_{v4} + 0.5\, p_{v4} = 2.32\, p_{v4}$ ...(*iii*)

Also pressure at B = $p_{sA} - p_{f1} = 22 - 6.28 = 15.72$ mm of water ...(*iv*)

Equating eqns. (*iii*) and (iv), we have:

$$2.32\, p_{v4} = 15.72, \quad \text{or,} \quad p_{v4} = 6.77 \text{ mm of water}$$

Velocity of air in duct BE,

$$C_4 = 4.04\sqrt{p_{v4}} = 4.04\sqrt{6.77} = 10.51 \text{ m/s} \qquad \left[\because p_v = \left(\frac{C}{4.04}\right)^2\right]$$

$\therefore$ Quantity of air flowing through the duct BE (4),

$$Q_4 = A_4 \times C_4 = 0.48 \times 10.51 = 5.04\ m^3/s$$

Now, quantity of air flowing through duct BC (2),

$$Q_2 = Q_1 - Q_4 = 12 - 5.04 = 6.96\ m^3/s$$

Cross-sectional area of the duct BC (2),

$$A_2 = a_2 \times b_2 = 3b_2 \times b_2 = 3b_2^2$$

Wetted perimeter of the duct,

$$P_2 = 2(a_2 + b_2) = 2(3b_2 + b_2) = 8b_2$$

∴ Hydraulic mean depth of duct BC,

$$m_2 = \frac{A_2}{P_2} = \frac{3b_2^2}{8b_2} = \frac{3b_2}{8}$$

Velocity of air in the duct BC,

$$C_2 = \frac{Q_2}{A_2} = \frac{6.96}{3b_2^2} = \frac{2.32}{b_2^2} \text{ m/s}$$

Velocity pressure in the duct BC,

$$p_{v2} = \left(\frac{C_2}{4.04}\right)^2 = \left(\frac{2.32}{b_2^2 \times 4.04}\right)^2 = \frac{0.33}{b_2^4} \text{ mm of water}$$

Pressure loss due to friction in duct BC,

$$p_{f2} = \frac{f L_2}{m_2}(p_{v2}) = \frac{0.0052 \times 60}{(3b_2/8)} \times \frac{0.33}{b_2^4} = \frac{0.274}{b_2^5} \text{ mm of water} \quad ...(v)$$

Also, as the pressure drop per metre length is 0.157 mm of water, therefore, total pressure loss due to friction in the duct BC,

$$p_{f2} = 0.157 \times 60 = 9.42 \text{ mm of water} \quad ...(vi)$$

Equating eqns. (*v*) and (*vi*), we have:

$$\frac{0.274}{b_2^5} = 9.42, \quad \text{or,} \quad b_2^5 = 0.029$$

∴ $\quad \boldsymbol{b_2 = (0.091)^{1/5} = 0.493 \text{ m (Ans.)}}$

and, $\quad \boldsymbol{a_2 = 3b_2 = 3 \times 0.493 = 1.479 \text{ m (Ans.)}}$

Dimensions of the duct CD (3):

Let, $\quad b_3$ = Depth of the duct CD, and

$\quad a_3$ = Width of the duct CD = $3b_3$.

Cross-sectional area of the duct CD,

$$A_3 = a_3 \times b_3 = 3b_3 \times b_3 = 3b_3^2$$

Wetted perimeter of the duct CD,

$$P_3 = 2(a_3 + b_3) = 2(3b_3 + b_3) = 8b_3$$

∴ Hydraulic mean depth of the duct CD,

$$m_3 = \frac{A_3}{P_3} = \frac{3b_3^2}{8b_3} = \frac{3b_3}{8}$$

Quantity of air flowing through the duct CD,

$$Q_3 = Q_2 - Q_5 = 6.96 - 2.5 = 4.46 \text{ m}^3/\text{s}$$

Velocity of air in the duct CD,

$$C_3 = \frac{Q_3}{A_3} = \frac{4.46}{3b_3^2} = \frac{1.487}{b_3^2} \text{ m/s}$$

Velocity pressure in the duct CD,

$$p_{v3} = \left(\frac{C_3}{4.04}\right)^2 = \left(\frac{1.487}{b_3^2 \times 4.04}\right)^2 = \frac{0.135}{b_3^4} \text{ mm of water}$$

Pressure loss due to friction in the duct CD,

$$p_{f3} = \frac{f\,L_3}{m_3} \times p_{v3} = \frac{0.0052 \times 40}{(3b_3/8)} \times \frac{0.135}{b_3^4} = \frac{0.075}{b_3^5} \text{ mm of water} \quad ...(vii)$$

Also, pressure loss due to friction in the duct CD,

$$p_{f3} = 0.157 \times 40 = 6.28 \text{ mm of water} \quad ...(viii)$$

Equating eqns. (*vii*) and (*viii*), we have:

$$\frac{0.075}{b_3^5} = 6.28, \quad \text{or,} \quad b_3^5 = 0.01194$$

$$\therefore \quad \boldsymbol{b_3 = 0.412 \text{ m (Ans.)}}$$

and, $$\boldsymbol{a_3} = 3b_3 = 3 \times 0.412 = \mathbf{1.236 \text{ m (Ans.)}}$$

Example 16.12. *Fig. 16.30 shows a duct system in which air enters at A with a static pressure of 8 mm of water.*

Considering friction losses only and using friction chart and equal pressure drop method, determine the correct diameters and velocity pressures in lengths AB, BC, CD, branch BE and branch CF.

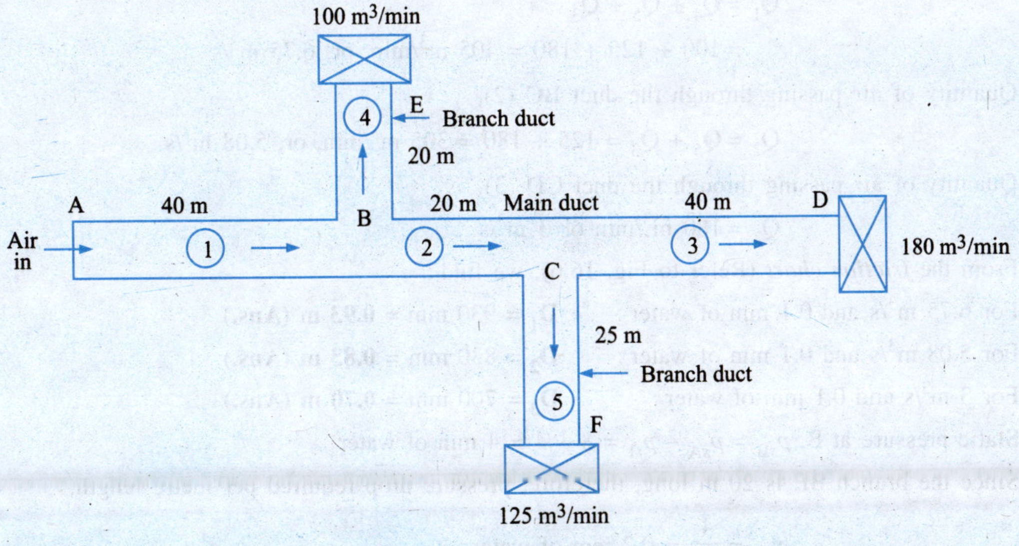

Fig. 16.30

Solution: *Given:* p_{SA} = 8 mm of water; L_1 = 40 m; L_2 = 20 m; L_3 = 40 m; L_4 = 20 m; L_5 = 25 m; Q_4 = 100 m^3/min; Q_3 = 180 m^3/min; Q_5 = 125 m^3/min.

Diameters and velocity pressures is various length:

- ***Diameters of lengths AB, BC, CD, branch BF and branch CF***

As per Bernoulli's equations, we have :

Total pressure at A = The pressure losses between the main duct AD + velocity pressure at D

i.e.,
$$p_{sA} + p_{v1} = p_{f1} + p_{f2} + p_{f3} + p_{v3}$$

Neglecting velocity pressure in the main duct, total pressure losses between the main duct AD,

$$8 = p_{f1} + p_{f2} + p_{f3}$$

Total length of the main duct AD,

$$L = L_1 + L_2 + L_3 = 40 + 20 + 40 = 80 \text{ m}$$

Pressure drop per metre length of the duct

$$= \frac{8}{80} = 0.1 \text{ mm of water}$$

∴ Pressure loss due to friction in the duct AB (1)

$$p_{f1} = 0.1 \times 40 = 4 \text{ mm of water}$$

Pressure loss due to friction in duct BC (2)

$$p_{f2} = 0.1 \times 20 = 2 \text{ mm of water}$$

Pressure loss due to friction in duct CD (3),

$$p_{f3} = 0.1 \times 40 = 4 \text{ mm of water}$$

Quantity of air passing through the duct AB (1)

$$Q_1 = Q_4 + Q_5 + Q_3$$
$$= 100 + 125 + 180 = 405 \text{ m}^3\text{/min, or, } 6.75 \text{ m}^3\text{/s}$$

Quantity of air passing through the duct BC (2),

$$Q_2 = Q_5 + Q_3 = 125 + 180 = 305 \text{ m}^3\text{/min, or, } 5.08 \text{ m}^3\text{/s}.$$

Quantity of air passing through the duct CD (3),

$$Q_3 = 180 \text{ m}^3\text{/min or } 3 \text{ m}^3\text{/s}$$

From the *friction chart* (Refer to Fig. 16.6), we find :

For 6.75 m^3/s and 0.1 mm of water : D_1 = 930 mm = **0.93 m (Ans.)**

For 5.08 m^3/s and 0.1 mm of water : D_2 = 830 mm = **0.83 m (Ans.)**

For 3 m^3/s and 0.1 mm of water : D_3 = 700 mm = **0.70 m (Ans.)**

Static pressure at B, $p_{sB} = p_{sA} - p_{f1} = 8 - 4 = 4$ mm of water

Since the branch BE is 20 m long, therefore, pressure drop required per metre length,

$$p_{f4} = \frac{4}{20} = 0.2 \text{ mm of water}$$

Now, from the *friction chart*, we find :

For 100 m^3/min or 1.67 m^3/s and 0.2 mm of water : $\mathbf{D_4}$ = 480 mm = **0.48 m (Ans.)**

Static pressure at C, $p_{sC} = p_{sB} - p_{f2} = 4 - 2 = 2$ mm of water

Since the branch CF is 25 m long, therefore, pressure drop required per metre length,

$$p_{f5} = \frac{2}{25} = 0.08 \text{ mm}$$

Now, from the *friction chart*, we find :

For 125 m^3/min or 2.08 m^3/s and 0.08 mm of water : $\mathbf{D_5}$ = 640 mm = **0.64 m (Ans.)**

- ***Velocity pressure in lengths AB, BC, CD, branch BE and branch CF:***

The velocity pressures in lengths AB, BC, CD, branch BE and branch CF are calculated in the table given below :

Lengths	*Diameter, D* (*m*)	*Cross-section area* $\left(A = \frac{\pi}{4} D^2\right)$ (*m^2*)	*Quantity of air* (*Q*), *m^3/s*	*Velocity* $C = \frac{Q}{A}$ *m/s*	*Velocity pressure* $p_v = \left(\frac{C}{4.04}\right)^2$ *mm of water*
AB (1)	0.93 m	0.679	6.75	9.94	6.05
BC (2)	0.83 m	0.541	5.08	9.39	5.40
CD (3)	0.70 m	0.385	3	7.79	3.72
BE (4)	0.48 m	0.181	1.67	9.23	5.22
CF (5)	0.64 m	0.322	2.08	6.46	2.56

Example 16.13. *An air duct system is provided as shown in Fig. 16.31.*

(*i*) *Determine the dimensions of AB, BC, CD using the equal friction method. Choose a friction rate of 0.08 mm H_2O per metre length of duct. Use the following formula for friction rate :*

$$\frac{\Delta p_f}{L} = \frac{2.268 \times 10^{-3}\, Q^{1.852}}{D^{4.973}}$$

where, Q is in m^3/s and D and L are in metres.

(*ii*) *Determine the total and static pressures at point A. Assume free exit at each outlet.*

Losses are given by :

For elbow : 0.25 p_v, branch : 0.2 p_v + elbow loss

For straight-through section : 0.25 × difference of velocity pressures.

(*iii*) *Find the diameter of BE so that no dampening is required in the section.*

Solution: (*i*) **Dimensions of AB, BC and CD:**

$$\frac{\Delta p_f}{L} = 0.08 \text{ mm of water per m length of duct}$$

$$= \frac{2.268 \times 10^{-3}\, Q^{1.852}}{D^{4.973}}$$

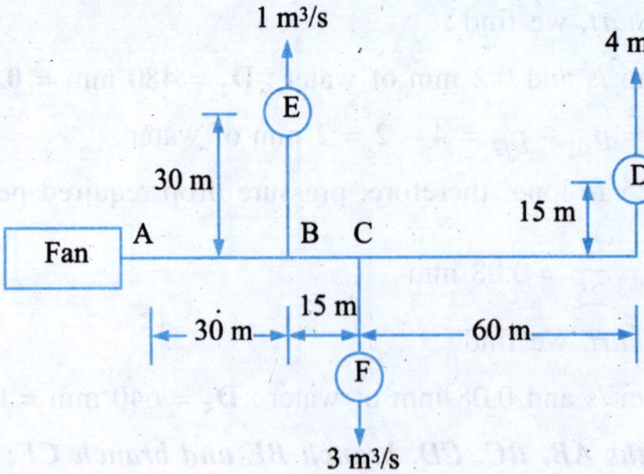

Fig. 16.31

or,
$$D = \left[\frac{2.268 \times 10^{-3}\,Q^{1.852}}{0.08}\right]^{(1/4.973)} = 0.48846\ Q^{0.3724}$$

Hence, the following table can be completed :

Section	*Length* (*L*)	*Q*	*D*	*Area* (*A*)	*Velocity* $\left(C = \dfrac{Q}{A}\right)$	$p_v = \left(\dfrac{C}{4.04}\right)^2$	$\Delta p_f = 0.08\ L$
	m	m^3/s	m	m^2	m/s	mm H_2O	mm H_2O
AB	30	8	1.0596	0.8818	9.072	5.042	2.4
BC	15	7	1.0082	0.7983	8.768	4.710	1.2
CD	75	4	0.8185	0.5262	7.602	3.541	6.0
				Frictional pressure drop from A to D			9.6

Hence, the duct dimensions (diameter) are:

$$\mathbf{D_{AB} = 1.0596\ m;\quad D_{BC} = 1.0082\ m;\quad D_{CD} = 0.8185\ m\ (Ans.)}$$

(*ii*) **Total and static pressures at point A:**

Dynamic losses between A and D, Δp_{dyn}:

(*a*) Loss in discharge opening = p_v = 3.541 mm H_2O

(*b*) Elbow loss = 0.25 p_v = 0.25 × 3.541 = 0.885 mm H_2O

(*c*) Fitting loss = 0.25 $(p_{v1} - p_{v2})$

$$\text{At B}: = 0.25\ (5.042 - 4.710) = 0.083 \text{ mm } H_2O$$

$$\text{At C}: = 0.25\ (4.710 - 3.541) = 0.2922 \text{ mm } H_2O$$

Total dynamic loss, Δp_{dyn}

$$= 3.541 + 0.885 + 0.083 + 0.2922 = 4.8012 \text{ mm } H_2O$$

Total friction and dynamic losses between A and D,

$$\Delta p\ (= p_{TA}) = \Delta p_f + \Delta p_{dyn}$$

$$= 9.6 + 4.8012 = \mathbf{14.4012\ mm\ H_2O\ (Ans.)}$$

This is the total pressure at fan exit A (p_{TA}).

Static pressure at point A,

$$p_{sA} = p_T - p_{vA} = 14.4012 - 5.042 = \mathbf{9.3592\ mm\ H_2O\ (Ans.)}$$

(*iii*) **Diameter of BE so that no dampening is required:**

Total pressure at B,

$$p_{TB} = p_{TA} - (\Delta p_T)_{AB}$$

$$= 14.4012 - 2.4 = 12.0\ mm\ H_2O$$

Pressure loss in BE

$$(\Delta p_T)_{BE} = (0.25 + 0.2 + 1)\ p_v + \Delta p_f = 1.45\ p_v + \Delta p_f.$$

The first term on the right hand side represents the *sum of the dynamic losses in the elbow at B, the area change in the branch and the velocity pressure at discharge.* Equating it to the total pressure at B for complete balancing, we get:

$$1.45\ p_v + \Delta p_f = 12.0$$

The above equation can be solved by *trial and error.*

Assuming, $D = 0.4$ m for BE,

$$\frac{\Delta p_f}{L} = \frac{2.268 \times 10^{-3} (1)^{1.852}}{(0.4)^{4.973}} = 0.216\ mm\ H_2O$$

$$\therefore \quad \Delta p_f = 0.216 \times 30 = 6.48\ mm\ H_2O$$

Velocity,

$$C = \frac{Q}{A} = \frac{1}{\frac{\pi}{4}(0.4)^2} = 7.958\ m/s$$

Velocity pressure,

$$p_v = \left(\frac{C}{4.04}\right)^2 = \left(\frac{7.958}{4.04}\right)^2 = 3.88\ mm\ H_2O$$

∴ Total pressure loss in BE,

$$(\Delta p)_{BE} = 1.45 \times 3.88 + 6.48 = 12.1\ mm\ H_2O$$

which is approximately equal to the total pressure at B. Hence an assumed diameter of **0.4 m** for the section BE is satisfactory. **(Ans.)**

Example 16.14. *A fan delivers air at 8 m/s to the main line of air-conditioning duct. After a straight run of 10 m a branch of 4 m length delivers air to a room at 2 m³/s. Another branch after a 5 m further run is taken from main branch. This branch is 5 m in length and delivers 3 m³/s of air. The main branch runs straight for another 5 m, then turns by 90° and runs for another 5 m length and delivers 3 m³/s of air. Losses are given by:*

For elbow : Loss = 0.25 p_v*, where* $p_v = \left(\frac{C}{4.04}\right)^2$ *mm* H_2O

For branch : 0.2 p_v + Elbow loss

For straight-through section : 0.25 × difference of velocity pressures

Fitting losses where change in area occurs.

Find the size of all the ducts and determine the static pressure at A. Use the following formula for friction rate. Find friction rate in main line and then assume same friction ratio for whole of the duct work.

$$\frac{\Delta p_f}{L} = \frac{0.002268\,(Q)^{1.852}}{(D)^{4.973}}$$

where, Q is in m^3/s and D is in m.

Solution: *Given:* $C_{AB} = 8$ m/s; $Q_{BE} = 2$ m^3/s; $Q_{CF} = 3$ m^3/s; $Q_{CD} = 3$ m^3/s $L_{AB} = 10$ m; $L_{BC} = 5$ m; $L_{CD} = 10$ m; $L_{BE} = 4$ m; $L_{CF} = 5$m

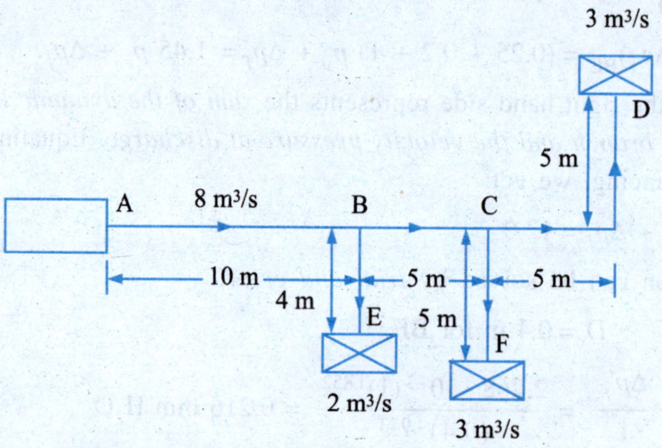

Fig. 16.32

Size of all the ducts and static pressure at A (p_{sA}):

Size of all the ducts:

Volume rate, $$Q_{AB} = Q_{BE} + Q_{CF} + Q_{CD} = 2 + 3 + 3 = 8 \text{ m}^3/\text{s}$$

$$C_{AB} = \frac{Q_{AB}}{A_{AB}} \quad \text{or} \quad 8 = \frac{8}{A_{AB}} \quad \text{or} \quad A_{AB} = 1 \text{ m}^2$$

or, $$\frac{\pi}{4} D_{AB}^2 = 1 \quad \text{or} \quad D_{AB} = \sqrt{\frac{4}{\pi}} = 1.128 \text{ m}$$

$$\frac{\Delta p_f}{L} = \frac{0.002268 \times (Q)^{1.852}}{(D)^{4.973}}$$

$$= \frac{0.002268 \times (8)^{1.852}}{(1.128)^{4.973}} = 0.0586 \text{ mm } H_2O \text{ per m length.}$$

This is *constant for all sections.*

$$\therefore \quad D = \left(\frac{0.002268 \times (Q)^{1.852}}{0.0586}\right)^{1/4.973} = 0.52\,(Q)^{0.372}$$

Hence the following table can be completed :

Sections	Length, L (m)	Flow rate Q (m^3/s)	Diameter, D (m)	Area, A (m^2)	Velocity, C (m/s)	$p_v = \left(\frac{C}{4.04}\right)^2$ mm H_2O	$\Delta p_f = 0.0586$ L (mm H_2O)
AB	10	8	1.128	1	8	3.92	0.586
BC	5	6	1.0127	0.805	7.453	3.40	0.293
CD	10	3	0.7825	0.481	6.237	2.38	0.586
Frictional pressure drop from A to D (Δp_f)							1.465

Hence duct sizes are:

$$\mathbf{D_{AB} = 1.128\ m;\ D_{BC} = 1.0127\ m;\quad D_{CD} = 0.7825\ m\ (Ans.)}$$

Static pressure at A, p_{sA}

(*a*) Loss in discharge opening,

$$\Delta p = p_v = 2.38 \text{ mm } H_2O$$

(*b*) Elbow loss = $0.25 \times p_v = 0.25 \times 2.38 = 0.595$ mm H_2O

(*c*) Fitting losses: $\Delta p = 0.25 \times$ (difference of velocity pressure)

$$\therefore \quad \Delta p_B = 0.25\ (3.92 - 3.40) = 0.13 \text{ mm } H_2O$$

$$\Delta p_C = 0.25\ (3.40 - 2.38) = 0.255 \text{ mm } H_2O$$

Total dynamic losses, $\Delta p_{dyn} = 2.38 + 0.595 + 0.13 + 0.255 = 3.30$ mm H_2O

Total frictional and dynamic losses between A and D

$$= \Delta p_f + \Delta p_{dyn} = 1.465 + 3.36 = 4.825 \text{ mm } H_2O$$

This is the total pressure at fan exit, at A $(p_T)_{fan}$.

$\therefore$ Static pressure at A,

$$p_{sA} = (p_T)_{fan} - p_{vA} = 4.825 - 3.92 = \mathbf{0.905\ mm\ H_2O\ (Ans.)}$$

Example 16.15. *An air-distribution system for an industrial complex is shown in Fig. 16.33. having flow rates as indicated. The velocity in the main duct should not exceed 400 m/min. Using equal pressure drop ($\Delta p/\Delta L$) in the main duct and the balanced pressure drop in the branch line and taking the depth of duct as 2000 mm throughout, determine:*

(*i*) *The duct dimensions;*

(*ii*) *Total pressure drop.*

(*iii*) *Power for the blower if the overall efficiency of blower and motor is 0.7.*

Solution: *Given:* $L_{AB} = 40$ m; $L_{BC} = 50$ m; $L_{CD} = 60$ m; $L_{BE} = 20$ m; $L_{CF} = 30$ m; $Q_{BE} = 400\ m^3/min$; $Q_{CF} = 500\ m^3/min$; $Q_{CD} = 100\ m^3/min$; $C_{AB} = 400$ m/min; $\eta = 0.7$.

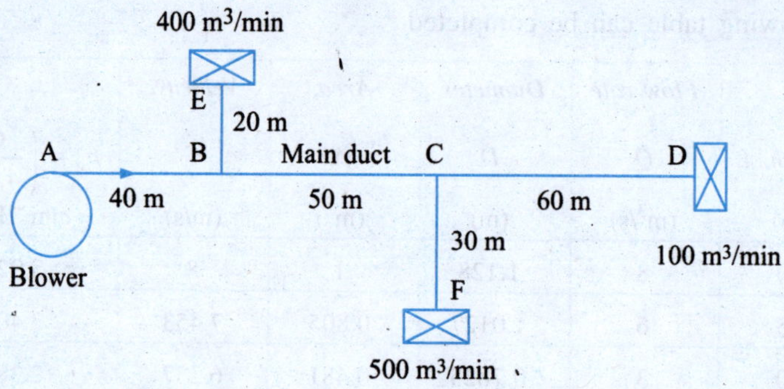

Fig. 16.33

(*i*) **The duct dimensions:**

Duct AB:

The total volume flow rate in the main duct AB,

$$Q_{AB} = Q_{BE} + Q_{CF} + Q_{CD}$$

$$= 400 + 500 + 100 = 1000 \text{ m}^3/\text{min}$$

Let, D_{eq} = Equivalent diameter of the duct,

Then, $$Q_{AB} = 1000 = \frac{\pi}{4}(D_{eq})^2 \times C_{AB}$$

$$= \frac{\pi}{4}(D_{eq})^2 \times 400$$

$$\therefore \quad D_{eq} = \left(\frac{1000 \times 4}{\pi \times 400}\right)^{1/2} = 1.784 \text{ m or } 1784 \text{ mm}$$

Also, $$D_{eq} = 1.265\left(\frac{a^3b^3}{a+b}\right)^{1/5} \quad \text{...[Eqn. (16.19)]}$$

Here, $a = 2000$ mm, b = ?

$$\therefore \quad 1784 = 1.265\left[\frac{(2000)^3 b^3}{2000 + b}\right]^{1/5}$$

or, $$b^3 = \frac{(2000+b)}{(2000)^3} \times \left(\frac{1784}{1.265}\right)^5 = 697319.22\,(2000 + b)$$

or, $$b = 88.677\,(2000 + b)^{1/3}$$

This can be solved by *iteration,* starting with an initial value of b = 1250 mm.

Solving by iteration, b = 1323 mm

∴ *Dimensions* of duct AB ; **2000 mm × 1323 mm (Ans.)**

Also, $$\left(\frac{\Delta p}{L}\right)_{AB} = \frac{2.268 \times 10^{-3}(Q)^{1.852}}{(D_{eq})^{4.973}}$$

$$= \frac{2.268 \times 10^{-3}(1000/60)^{1.852}}{(1.784)^{4.973}} = 0.02335 \text{ mm } H_2O/\text{m length}$$

Duct BC:

Flow through BC is 600 m^3/min, and $\frac{\Delta p}{L}$ = 0.02335 mm H_2O,

$$0.02335 = \frac{2.268 \times 10^{-3}(600/60)^{1.852}}{(D_{eq})^{4.973}}$$

$$\therefore \quad D_{eq} = \left[\frac{2.268 \times 10^{-3}(600/60)^{1.852}}{0.02335}\right]^{1/4.973} = 1.475 \text{ m, or, } 1475 \text{ mm}$$

Here, $a = 2000$ mm ; b = ?

$$\therefore \quad 1475 = 1.265\left[\frac{(2000)^3 b^3}{2000+b}\right]^{1/5}$$

or, $$b^3 = \frac{(2000+b)}{(2000)^3} \times \left(\frac{1475}{1.265}\right)^5 = 269412.3\ (2000 + b)$$

or, $$b = 64.586\ (2000 + b)^{1/3}$$

Again to be solved by iteration, starting with b = 930 mm, the iteration gives, b = 919 m

$\therefore$ *Dimensions* of the duct BC: **2000 mm × 919 mm (Ans.)**

Duct CD:

Flow rate is 100 m^3/min.

Again for $$\frac{\Delta p}{L} = 0.02335 \text{ mm } H_2O$$

$$0.02335 = \frac{2.268 \times 10^{-3}(100/60)^{1.852}}{(D_{eq})^{4.973}}$$

$$\therefore \quad D_{eq} = \left[\frac{2.268 \times 10^{-3}(100/60)^{1.852}}{0.02335}\right]^{1/4.973} = 0.757 \text{ m or } 757 \text{ mm}$$

Here, $a = 2000$ mm; b = ?

$$\therefore \quad 757 = 1.265\left[\frac{(2000)^3 b^3}{(2000+b)}\right]^{1/5}$$

or, $$b^3 = \frac{(2000+b)}{(2000)^3} \times \left(\frac{757}{1.265}\right)^5 = 9592.61\ (2000 + b)$$

or, $$b = 21.25\ (2000 + b)^{1/3}$$

Again, this is to be solved by iteration, starting with b = 220 mm, we get b = 278 mm.

$\therefore$ *Dimensions* of the duct CD : **2000 mm × 278 mm (Ans.)**

Duct CF:

Here, $$(\Delta p)_{CF} = (\Delta p)_{CD} \quad \text{...(Given)}$$

$$= 0.02335 \times 60 = 1.401$$

$$\therefore \quad \frac{(\Delta p)_{CF}}{L_{CF}} = \frac{1.401}{30} = 0.0467 \text{ mm } H_2O \text{ per m.}$$

$$\therefore \quad 0.0467 = \frac{2.268 \times 10^{-3} (500/60)^{1.852}}{(D_{eq})^{4.973}}$$

or,
$$D_{eq} = \left[\frac{2.268 \times 10^{-3} (500/60)^{1.852}}{0.0467}\right]^{1/4.973} = 1.199 \text{ m or } 1199 \text{ mm}$$

Here $\quad a = 2000 \text{ mm} ; b = ?$

$$\therefore \quad 1199 = 1.265 \left[\frac{(2000)^3 b^3}{(2000+b)}\right]^{1/5}$$

or,
$$b^3 = \frac{(2000+b)}{(2000)^3} \times \left(\frac{1199}{1.265}\right)^5 = 95621.0\ (2000 + b)$$

or,
$$b = 45.73\ (2000 + b)^{1/3}$$

Starting with initial value of 910 mm, this is again solved by iteration to give b = 629 mm

∴ *Dimensions* of the duct CF = **2000 mm × 629 mm (Ans.)**

Duct BE:

Here, $\quad (\Delta p)_{BE} = (\Delta p)_{BD}$ ...(Given)

$$= 0.02335 \times 110 = 2.5685$$

$$\therefore \quad \frac{(\Delta p)_{BE}}{L_{BE}} = \frac{2.5685}{20} = 0.128425 \text{ mm } H_2O \text{ per m length}$$

$$\therefore \quad 0.128425 = \frac{2.268 \times 10^{-3} (400/60)^{1.852}}{(D_{eq})^{4.973}}$$

or
$$D_{eq} = \left[\frac{2.268 \times 10^{-3} (400/60)^{1.852}}{0.128425}\right]^{1/4.973} = 0.9 \text{ m or } 900 \text{ mm}$$

Here $\quad a = 2000 \text{ mm} ; b = ?$

$$\therefore \quad 900 = 1.265 \left[\frac{(2000)^3 b^3}{(2000+b)}\right]^{1/5}$$

or,
$$b^3 = \frac{(2000+b)}{(2000)^3} \times \left(\frac{9000}{1.265}\right)^5 = 22786\ (2000 + b)$$

or,
$$b = 28.35\ (2000 + b)^{1/3}$$

Starting with b = 390 mm, by iteration, b = 377 mm.

∴ *Dimensions* of the duct BE = **2000 mm × 377 mm (Ans.)**

(*ii*) **Total pressure drop:**

Total pressure drop, $(\Delta p)_{AD} = (40 + 50 + 60) \times 0.02335$

$$= \mathbf{3.5025 \text{ mm } H_2O \text{ (Ans.)}}$$

(*iii*) **Power for the blower, P_{blower} :**

$$P_{blower} = \frac{Q \times \text{pressure drop (N/m}^2)}{\eta}$$

$$= \frac{(1000/60) \times (1000 \times 9.81 \times 3.5025/1000)}{0.7}$$

$$= 818.1 \text{ W, or, } \mathbf{0.8181 \text{ kW (Ans.)}}$$

Example 16.16. *An air-conditioning duct runs straight from fan over 60 m length. It has four equally spaced outlet diffusers mounted on duct, the last one being at the end of duct. The volume flow rate through each diffuser is 1 m³/s. The velocity at the duct inlet is 15 m/s. Carry out the duct design by static regain method if the static regain factor is 0.75 at each transition and frictional pressure drop is given by,*

$$\frac{\Delta p_f}{L} = \frac{0.002268\,(Q)^{1.852}}{(D)^{4.973}} \quad \text{and} \quad \Delta p_v = \left(\frac{C}{4.04}\right)^2$$

where Q (flow rate) is in m³/s and C (velocity) is in m/s.

Solution: *Given:* $L_{AB} = L_{BC} = L_{CD} = L_{DE} = 15$ m ; $Q_{AB} = 4$ m³/s; $Q_{BC} = 3$ m³/s; $Q_{CD} = 2$ m³/s; $Q_{DE} = 1$ m³/s; $C_A = 15$ m/s; $R = 0.75$.

Duct design by static regain method:

$$Q = AC = \frac{\pi}{4} D^2 \times C$$

or, $$D^2 = \frac{4Q}{\pi C} \quad \text{or} \quad D = \sqrt{\frac{4Q}{\pi C}} \quad ...(i)$$

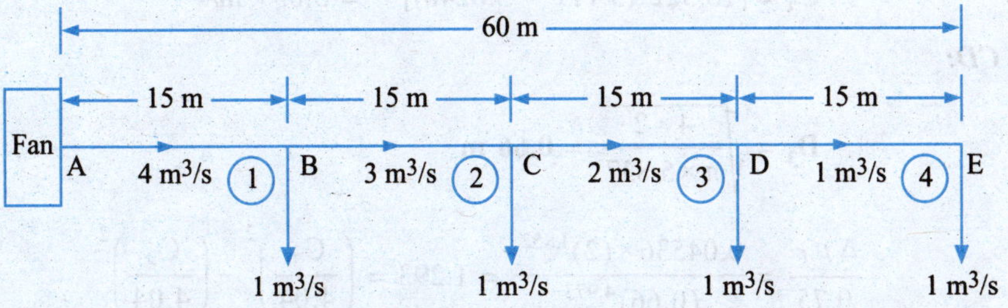

Fig. 16.34

$$\frac{\Delta p_f}{L} = \frac{0.00\,268 \times (Q)^{1.852}}{(D)^{4.973}}$$

The pressure loss in each section (AB = BC = CD = DE = 15 m) is recovered by static regain due to velocity reduction.

$$\therefore \quad \Delta p_f = \frac{0.002268\,(Q)^{1.852} \times 15}{(D)^{4.973}} = 0.75\left[\left(\frac{C_1}{4.04}\right)^2 - \left(\frac{C_2}{4.04}\right)^2\right]$$

or, $$\left(\frac{C_1}{4.04}\right)^2 - \left(\frac{C_2}{4.04}\right)^2 = \frac{\Delta p_f}{0.75} = \frac{0.002268 \times 15 \times (Q)^{1.852}}{0.75\,(D)^{4.973}}.$$

or, $$\frac{\Delta p_f}{0.75} = \left(\frac{C_1}{4.04}\right)^2 - \left(\frac{C_2}{4.04}\right)^2 = \frac{0.04536 \times (Q)^{1.852}}{(D)^{4.973}} \qquad ...(ii)$$

Section AB:

$$D_1 = \sqrt{\frac{4 \times 4}{\pi \times 15}} = \mathbf{0.58269\ m} \qquad ...\text{[From eqn. } (i)]$$

$$\frac{\Delta p_f}{0.75} = \frac{0.04536 \times (4)^{1.852}}{(0.58269)^{4.973}} = 8.673 \qquad ...\text{[From eqn. } (ii)]$$

$$= \left(\frac{C_1}{4.04}\right)^2 - \left(\frac{C_2}{4.04}\right)^2 = \left(\frac{15}{4.04}\right)^2 - \left(\frac{C_2}{4.04}\right)^2 = 13.785 - \frac{C_2^2}{16.322}$$

$\therefore$ $$C_2 = [16.322\ (13.785 - 8.673)]^{1/2} = 9.134 \text{ m/s}$$

Section BC:

$$\mathbf{D_2} = \sqrt{\frac{4 \times 3}{\pi \times 9.134}} = \mathbf{0.647\ m}$$

$$\frac{\Delta p_f}{0.75} = \frac{0.04536 \times (3)^{1.852}}{(0.647)^{4.973}} = 3.0246 = \left(\frac{C_2}{4.04}\right)^2 - \left(\frac{C_3}{4.04}\right)^2$$

$$= \left(\frac{9.134}{4.04}\right)^2 - \left(\frac{C_3}{4.04}\right)^2 = 5.112 - \frac{C_3^2}{16.322}$$

$\therefore$ $$C_3 = [16.322\ (5.112 - 3.0246)]^{1/2} = 5.837 \text{ m/s}$$

Section CD:

$$\mathbf{D_3} = \sqrt{\frac{4 \times 2}{\pi \times 5.837}} = \mathbf{0.66\ m}$$

$$\frac{\Delta p_f}{0.75} = \frac{0.04536 \times (2)^{1.852}}{(0.66)^{4.973}} = 1.293 = \left(\frac{C_3}{4.04}\right)^2 - \left(\frac{C_4}{4.04}\right)^2$$

$$= \left(\frac{5.837}{4.04}\right)^2 - \left(\frac{C_4}{4.04}\right)^2 = 2.0874 - \frac{C_4^2}{16.322}$$

$\therefore$ $$C_4 = [16.322\ (2.0874 - 1.293)]^{1/2} = 3.6 \text{ m/s}$$

Section DE:

$$\mathbf{D_4} = \sqrt{\frac{4 \times 1}{\pi \times 3.6}} = \mathbf{0.5947\ m}$$

$$\frac{\Delta p_f}{0.75} = \frac{0.04536 \times (1)^{1.852}}{(0.5947)^{4.973}} = 0.6013$$

The calculated values are tabulated below:

S. No.	Section	Flow rate, Q (m^3/s)	Velocity, C (m/s)	Diameter, D (m) [Eqn. (i)]	$\left(\frac{C_2}{4.04}\right)^2$	$\frac{\Delta p_f}{0.75}$ [Eqn (ii)]
1	AB	4	15	0.58269	13.785	8.673
2	BC	3	9.134	0.647	5.112	3.0246
3	CD	2	5.837	0.66	2.0874	1.293
4	DE	1	3.6	0.5947	0.794	0.6013

It is surprising to note (from the table) that, though the velocities in successive ducts are decreasing (and it should), the equivalent diameters are *not* decreasing.

Example 16.17 *Fig. 16.35 shows a duct system containing a dehumidifying coil. Air enters at t_{db} = 26°C and ϕ = 60% and leaves at t_{db} = 14°C and ϕ = 90%. The chilled water enters the cooling coil at a rate of 15 kg/s and its temperature rise is 4.03°C. The velocity in the main duct is 10 m/s. The aspect ratio for each duct is 2 : 1. The main duct line is A–B–C with pressure drop per unit length being same and there is no damper in the system. The pressure drop due to 90° bend is equivalent to 3 times the diameter of the duct at the point considered.*

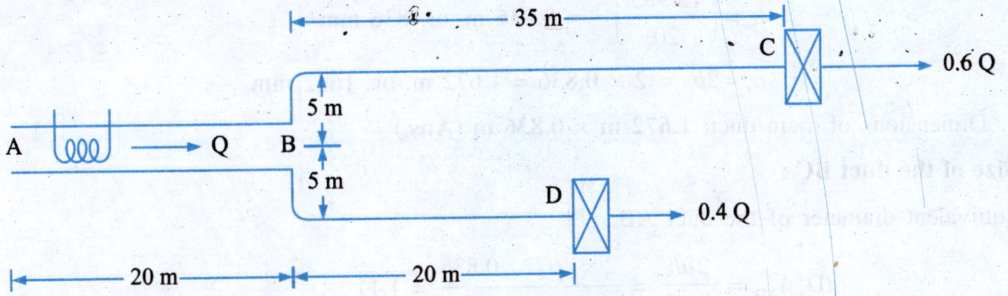

Fig. 16.35

Determine the following:

(*i*) *Air flow rate* (m^3/s)*;*

(*ii*) *Main duct dimensions;*

(*iii*) *Size of duct BC;*

(*iv*) *Total pressure;*

(*v*) *Dimensions of duct BD.*

Solution: *Given:* t_{db1} = 26°C; ϕ_1 = 60%; t_{db2} = 14°C; ϕ_2 = 40%; m_{cw} = 15 kg/s, Δt_{cw} = 4.03°C;

$$C_{AB} = 10 \text{ m/s; Aspect ratio, } \frac{a}{b} = 2:1; \quad (\Delta p)_{bend} = 3 \times D_{duct}$$

(*i*) **Air flow rate, Q_{AB}:**

The enthalpy of air entering and leaving the dehumidifying coils from *psychrometric chart* are, respectively,

$$h_1 = 58.4 \text{ kJ/kg of dry air;}$$

$$h_2 = 36.5 \text{ kJ/kg of dry air;}$$

$$v_{s2} = 0.825 \text{ m}^3\text{/kg of dry air.}$$

Let, m = Mass rate of air, kg/s.

Then the decrease in enthalpy of air is equal to heat carried away by chilled water (cw)

or, $$m\,(h_1 - h_2) = m_{c\omega} \times 4.18 \times (\Delta t)_{c\omega}$$

or, $$m\,(58.4 - 36.5) = 15 \times 4.18 \times 4.03$$

$$\therefore \quad m = \frac{15 \times 4.18 \times 4.03}{(58.4 - 36.5)} = 11.538 \text{ kg/s}$$

Hence, air flow rate, $\mathbf{Q_{AB}} = \dfrac{m}{v_{s2}} = \dfrac{11.538}{0.825}$ = **13.985 m³/s (Ans.)**

(*ii*) **Main duct dimensions:**

If a (width) and b (depth) are the dimensions of the main duct AB, then;

$$a = 2b \qquad \text{...(Given)}$$

Now, $$Q_{AB} = A_B \times C_{AB}$$

or, $$13.985 = (a \times b) \times 10$$

$$= (2b \times b) \times 10 = 20b^2$$

$$\therefore \quad b = \left(\frac{13.985}{20}\right)^{1/2} = 0.836 \text{ m, or, } 836 \text{ mm}$$

and, $$a = 2b = 2 \times 0.836 = 1.672 \text{ m, or, } 1672 \text{ mm}$$

∴ Dimensions of main duct: **1.672 m × 0.836 m (Ans.)**

(*iii*) **Size of the duct BC :**

Equivalent diameter of the duct AB,

$$(D_{eq})_{AB} = \frac{2ab}{a+b} = \frac{2 \times 1.672 \times 0.836}{1.672 + 0.836} = 1.11$$

Pressure loss per m length is given by the relation:

$$\frac{\Delta p_f}{L} = \frac{0.002268\, Q^{1.852}}{(D_{eq})^{4.973}}$$

$$= \frac{0.002268 \times (13.985)^{1.852}}{(1.11)^{4.973}} = 0.1786 \text{ mm } H_2O \text{ per m length}$$

This is *constant for all sections.*

$$\therefore \quad D_{eq} = \left[\frac{0.002268 \times (Q)^{1.852}}{0.1786}\right]^{1/4.973} = 0.4156 \times (Q)^{0.3724}$$

Also, $$D_{eq} = \frac{2ab}{a+b} = \frac{2 \times 2b \times b}{2b + b} = \frac{4b^2}{3b} = \frac{4b}{3}$$

$$\therefore \quad b = \frac{3}{4} D_{eq}, \quad \text{and,} \quad a = 2b$$

Hence, the following table can be completed:

Section	Length, L (m)	Q (m^3/s)	D_{eq} (m)	Dimensions a × b (m × m)	Velocity, C(m/s)	$p_v = \left(\frac{C}{4.04}\right)^2$ (mm H_2O)	Δp_f = 0.1786 L (mm H_2O)
AB	20	13.985	1.11	1.672 × 0.836	10.00	6.127	3.572
BC	35	8.391	0.918	1.377 × 0.6885	8.85	4.799	7.144
BD	20	5.594	0.789	1.1835 × 0.5917	7.99	3.911	4.465

∴ *Size of duct* BC is **1.377 m × 0.6885 m (Ans.)**

(*iv*) **Total pressure:**

Dynamic losses between A and C :

(*a*) Loss in discharge opening C = p_v = 4.799 mm H_2O

(*b*) Elbow loss = 0.25 p_v = 0.25 × 4.799 = 1.1997 mm H_2O

(*c*) Fitting loss Δp = 0.25 × difference of velocity pressures

∴ $(\Delta p)_B$ = 0.25 (6.127 – 4.799) = 0.332 mm H_2O

(*d*) Pressure loss in dehumidifier coil,

(usually 2 to 10 mm H_2O), say = 5.0000 mm H_2O

Total dynamic loss = 4.799 +1.1997 + 0.332 + 5.000 = 11.331 mm

Total pressure (at fan exit)= Friction loss + Dynamic loss

= $(\Delta f)_{AC}$ + Dynamic loss

= (3.572 + 7.144) + 11.331 = **22.047 mm H_2O (Ans.)**

(*v*) **Dimensions of duct BD :**

Dimension of duct BD (from the table above) : **1.1835 m × 0.5917 m (Ans.)**

Example 16.18. *An air-conditioning system has volume flow rate of 7.5 m^3/s and fan outlet velocity is 10 m/s. The duct has four branches with 90° elbows. The first branch is 10 m from fan; distance between branches is 10* m *each and main duct has 90° elbow 10 m after the fourth branch. Volume flow rate in each branch is 1.5 m^3/s. The main duct runs 10 m after the 90° bend.*

(*i*) *Using equal friction method determine the equivalent diameter of duct and dimensions of rectangular duct if one side of the duct is 0.5 m.*

(*ii*) *Determine the total pressure drop, given,*

$$\frac{\Delta p_f}{L} = \frac{0.002268\,(Q)^{1.852}}{(D)^{4.973}} \text{ mm } H_2O$$

$$p_v = \left(\frac{C_2}{4.04}\right)^2 \text{ mm } H_2O$$

$$D = \frac{1.3\,(a\,b)^{0.625}}{(a+b)^{0.25}}$$

Elbow loss = 0.25 p_v;

Straight through section loss = 0.25 × difference of velocity pressures

(Fitting losses where changes in area occur)

Dynamic loss in branch = 0.2 p_v + elbow loss.

Solution: *Given:* $C_A = 10$ m/s; $Q_{AB} = 7.5$ m³/s; $Q_{BC} = 6$ m³/s; $Q_{CD} = 4.5$ m³/s; $Q_{DE} = 3$ m³/s;
$L_{AB} = L_{BC} = L_{CD} = L_{DE} = L_{EF} = 10$ m.

(*i*) **Equivalent diameter of the duct and dimensions of the rectangular duct:**

Duct AB:

Area, $$A = \frac{Q}{C} = \frac{7.5}{10} = 0.75 \text{ m}^2 = a \times b$$

where, the duct sides are *a* and *b*, both in m.

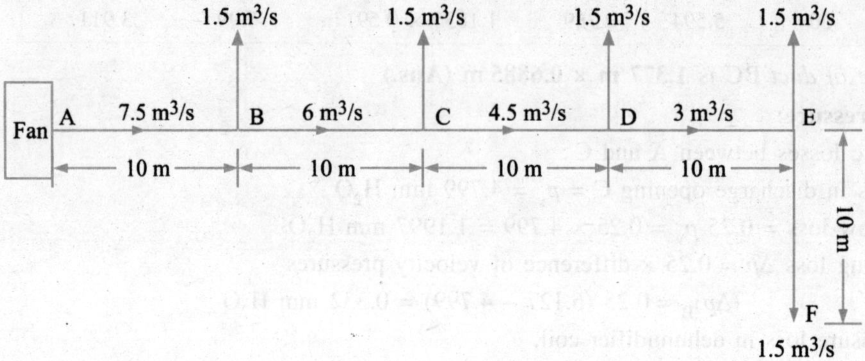

Fig. 16.36

One side, *b* is constant, equal to 0.5 m (given),

$$\therefore \quad a = \frac{0.75}{0.5} = 1.5 \text{ m.}$$

$$\therefore \quad (D_{eq})_{AB} = \frac{1.3\,(1.5 \times 0.5)^{0.625}}{(1.5 + 0.5)^{0.25}} = 0.9133 \text{ m}$$

$$\frac{\Delta p_f}{L} = \frac{0.002268\,(Q)^{1.852}}{(D)^{4.973}}$$

$$= \frac{0.002268\,(7.5)^{1.852}}{(0.9133)^{4.973}} = 0.1486 \text{ mm } H_2O$$

This is *constant for all sections.*

$$\therefore \quad D\,(= D_{eq}) = \left(\frac{0.002268\,(Q)^{1.852}}{0.1486}\right)^{1/4.973} = 0.4313\,(Q)^{0.3724} \quad ...(i)$$

Also, $$D = \frac{1.3\,(0.5)^{0.625}\,(a)^{0.625}}{(a + 0.5)^{0.25}} = \frac{0.843\,(a)^{0.625}}{(a + 0.5)^{0.25}}$$

or, $$a = \left[\frac{D\,(a + 0.5)^{0.25}}{0.843}\right]^{1/0.625} = \left(\frac{D\,(a + 0.5)^{0.25}}{0.843}\right)^{1.6} \quad ...(ii)$$

Hence, the following table can be completed:

Section	Length, L (m)	Flowrate, Q (m^3/s)	Diameter, D (= D_{eq}) (m) [From eqn. (i)]	*Side a (m)	Area, A = a×b (m^2)	Velocity, $C = \frac{Q}{A}$ m/s	$p_v = \left(\frac{C}{4.04}\right)^2$ mm H_2O	Δp_f = 0.1486 L (mm H_2O)
AB	10	7.5	0.9133	1.5	0.75	10	6.1268	1.486
BC	10	6	0.8405	1.2424	0.6212	9.6587	5.7158	1.486
CD	10	4.5	0.7551	0.981	0.4905	9.1743	5.1568	1.486
DE	10	3	0.6493	0.711	0.3555	8.4388	4.3631	1.486
EF	10	1.5	0.5016	0.4217	0.2108	7.116	3.1025	1.486
					Frictional pressure drop from A to F (Δp_f) : 7.43			

*Side *a* is obtained by iteration from eqn. (*ii*)

Hence, equivalent diameters of various sections of the duct and dimensions of rectangular duct are:

Equivalent diameters:

D_{AB} = **0.9133** m; D_{BC} = **0.8405** m; D_{CD} = **0.7551** m; D_{DE} = **0.6493** m; D_{EF} = **0.5016** m **(Ans.)**

Dimensions of the rectangular duct:

AB : 1.5 m × 0.5 m; BC : 1.24242 m × 0.5 m; CD = 0.981 m × 0.5 m ; DE : 0.711 m × 0.5 m; EF : 0.4217 m × 0.5 m (Ans.)

(*ii*) **Total pressure drop (p_T):**

Dynamic losses between A and F, Δp_{dyn}:

(*a*) Loss in discharge opening = 3.1025 mm H_2O

(*b*) Elbow loss = 0.25 × p_v = 0.25 × 3.1025 = 0.7756 mm H_2O

(*c*) Fitting loss, Δp = 0.25 × difference in velocity pressures.

$$\Delta p_B = 0.25\ (p_{v1} - p_{v2})$$
$$\Delta p_C = 0.25\ (p_{v2} - p_{v3})$$
$$\Delta p_D = 0.25\ (p_{v3} - p_{v4})$$
$$\Delta p_E = 0.25\ (p_{v4} - p_{v5})$$
$$\text{Total fitting loss} = 0.25\ (p_{v1} - p_{v5})$$
$$= 0.25\ (6.1268 - 3.1025) = 0.756 \text{ mm } H_2O$$

Total dynamic loss,

$$\Delta p_{dyn} = 3.1025 + 0.7756 + 0.756 = 4.6341 \text{ mm } H_2O$$

Total frictional and dynamic losses between A and F (including loss in discharge opening)

$$p_T = \Delta p_f + \Delta p_{dyn}$$
$$= 7.43 + 4.6341 = \mathbf{12.0641\ mm\ H_2O\ (Ans.)}$$

HIGHLIGHTS

1. The function of a "*duct*" is to convey the air between two points, such as between the air-handling unit or air-washer and the room to be conditioned.
2. "*Aspect ratio*" is the ratio of long side to the short side of the duct.
3. Pressure loss due to friction in ducts,

$$p_f = \frac{f\,L}{m}\left(\frac{C}{4.04}\right)^2 \text{ mm of water}$$

where f = Friction factor,
L = Length of the duct,
m = Hydraulic mean depth, and
C = Velocity of air.

4. Equivalent diameter of a circular duct for a rectangular duct :

(*i*) $D_{eq} = 1.265\left(\frac{a^3b^3}{a+b}\right)$ ... when the *quantity of air passing through the rectangular and circular ducts is same.*

(*ii*) $D_{eq} = \frac{2\,ab}{a+b}$ ... when the *velocity of air passing through the rectangular and circular ducts is* same.

5. The dynamic pressure loss,

$$p_d = Kp_v = K\left(\frac{C}{4.04}\right)^2 \text{ mm of water}$$

where, C = Velocity (normally downstream), m/s, and
K = Dynamic loss coefficient (found experimentally)

Also, $L_{eq} = \frac{Km}{f}$ (m = hydraulic mean depth, m)

6. Pressure loss due to *sudden enlargement* in area,

$$(p_L)_{enl.} = \left(\frac{A_2}{A_1} - 1\right)^2\left(\frac{C_2}{4.04}\right)^2$$

7. Pressure loss due to contraction in area,

$$(p_L)_{cont.} = \left(1 - \frac{A_1}{A_2}\right)^2\left(\frac{C_1}{4.04}\right)^2$$

8. The duct size may be determined by the following methods:
   (*i*) Velocity reduction method.
   (*ii*) Equal friction (or constant pressure loss) method.
   (*iii*) Static regain method.
9. When systems are in series:

$$Z = Z_1 + Z_2 + Z_3$$

When systems are in parallel.

$$\sqrt{\frac{1}{Z}} = \sqrt{\frac{1}{Z_1}} + \sqrt{\frac{1}{Z_2}} + \sqrt{\frac{1}{Z_3}}$$

where Z_1, Z_2 and Z_3 are constants of the systems.

10. *Draft* is defined as any localized feeling of coolness or warmth of any portion of the body due to both air movement and air temperature with humidity and radiation, considered constant.

11. *Outlets* may be of the following types :

(*i*) Grille outlets (*ii*) Slot diffuser outlets

(*iii*) Ceiling diffuser outlets (*iv*) Perforated ceiling panels.

12. Air-handling system comprises of :

(*i*) Air distribution system. (*ii*) Duct system. (*iii*) Fan.

13. The supply ducts may be arranged in the following ways :

(*i*) Loop perimeter duct system. (*ii*) Radial perimeter duct system.

(*iii*) Extended plenum duct system.

14. Air-distribution systems may be divided into three types :

(*i*) Ejector system. (*ii*) Downward system.

(*iii*) Upward system.

OBJECTIVE TYPE QUESTIONS

A. Choose the Correct Answer:

1. Equal friction method of designing air conditioning duct
   (*a*) ensures same velocity in the duct all throughout in all branches
   (*b*) automatically reduces the air velocity in the duct in the direction of flow
   (*c*) ensures constant static pressure at all terminals in the duct
   (*d*) does none of the above.

2. For rectangular ducts, the aspect ratio is equal to
   (*a*) sum of longer and shorter sides
   (*b*) difference of longer and shorter sides
   (*c*) product of longer and shorter sides
   (*d*) ratio of longer and shorter sides.

3. The velocity pressure in air-conditioning ducts is given by (V in m/min, p_v in mm of water column)

   (*a*) $p_v = \dfrac{V^2}{240.6}$ (*b*) $p_v = \dfrac{V^2}{5788.36}$

   (*c*) $p_v = \dfrac{V^2}{16.08}$ (*d*) $p_v = \left(\dfrac{V}{16.08}\right)^2$.

4. When the quantity of air passing through the rectangular and circular ducts in same, then the equivalent diameter (D_{eq}) of a circular duct for a rectangular duct for the same pressure loss per unit length, is given by

   (*a*) $1.265\left(\dfrac{a^3b^3}{a+b}\right)^{1/5}$ (*b*) $1.265\left(\dfrac{a^4b^4}{a+b}\right)^{1/5}$

   (*b*) $1.265\left(\dfrac{a+b}{a^3b^3}\right)^{1/5}$ (*d*) $1.265\left(\dfrac{a+b}{a^4b^4}\right)^{1/5}$

   where, a and b = Longer and shorter sides of the rectangular duct respectively.

5. The Equal Friction method of duct design
(*a*) satisfies the requirements of uniform static pressure at all air terminals
(*b*) does not satisfy the requirements of uniform static pressure at all air terminals
(*c*) does not require splitter dampers to be provided to regulate the flow to each branch
(*d*) none of the above.

6. When the velocity of air passing through the rectangular and circular ducts is same, then the equivalent diameter (D_{eq}) of a circular duct for a rectangular duct for the same pressure loss per unit length, is given by
(*a*) $\frac{a+b}{ab}$ (*b*) $\frac{2ab}{a+b}$ (*c*) $\frac{2a}{a-b}$ (*d*) $\frac{2b}{a+b}$.

7. A duct is said to be a low velocity duct if the velocity of air in the duct is upto
(*a*) 600 m/min (*b*) 800 m/min
(*c*) 1200 m/min (*d*) 1600 m/min.

8. Static Regain Method of designing ducts for air-conditioning as compared to equal friction method
(*a*) reduces the cost of sheet metal for the duct
(*b*) increases the cost of sheet metal for the duct
(*c*) increases balancing problems.
(*d*) increases problems of maintaining pressure at the terminals.

9. The duct is made of
(*a*) galvanised iron (*b*) aluminium
(*c*) fibre glass (*d*) cement asbestos
(*e*) any one of the above.

10. Equal friction method of designing air conditioning ducts
(*a*) is ideal when the system is balanced (*b*) is ideal when the system is not balanced
(*c*) is ideal only for return ducts (*d*) is ideal for none of the above.

11. The aspect ratio, for rectangular ducts, should not be greater than .... in any case.
(*a*) 8 (*b*) 10 (*c*) 12 (*d*) 16

B. Fill in the Blanks

1. An opening through which air is supplied to the treated space is called ....
2. .... is a grille provided with a damper.
3. .... is an outlet grille designed to guide the direction of the air.
4. .... air is the air coming out of the outlet.
5. .... air is the air picked up by the primary air by entrainment.
6. Grille-dimensions ratio is called .... ratio.
7. An opening through which air is returned or exhausted from the space is called ....
8. .... provides necessary energy to move the air.
9. The various duct systems are ...., ... and ....
10. Three methods of sizing the ducts are ...., ... and ....

ANSWERS

A. Choose the Correct Answer.

1. (*b*) **2.** (*d*) **3.** (*b*) **4.** (*a*) **5.** (*b*)
6. (*b*) **7.** (*a*) **8.** (*b*) **9.** (*e*) **10.** (*a*) **11.** (*a*)

B. Fill in the Blanks

1. Outlet; **2.** Register; **3.** Diffuser; **4.** Primary; **5.** Secondary;
6. Aspect; **7.** Intake; **8.** Fan;
9. Loop perimeter, radial perimeter and extended Plenum;
10. Velocity reduction, equal friction and static regain.

THEORETICAL QUESTIONS

1. What is the function of a 'duct'?

2. Why are the ducts used in air-conditioning systems?

3. Name the factors due to which pressure of air falls in a duct.

4. What is "Aspect ratio"?

5. Which material is most commonly used for making ducts in the air-conditioning systems?

6. How are ducts classified?

7. Explain briefly the following types of pressures involved in a duct system:

(*i*) Static pressure (p_s)

(*ii*) Dynamic or velocity pressure (p_v)

(*iii*) Total pressure (p_T).

8. Derive an expression for the equivalent diameter of circular duct corresponding to a rectangular duct of sides *a* and *b*, for the same pressure loss per unit length, when:

(*i*) the quantity of air passing through both the ducts is same;

(*ii*) the velocity of air flowing through both the ducts is the same.

9. Enumerate various methods of air-conditioning duct design and explain in detail the 'Equal friction method'.

10. Explain 'Static regain method' of duct design for air-conditioning. Under what situations is this recommended. What are its main disadvantages?

11. Define the following as applied to 'air distribution'.

Intake, Outlet, Grille, Register, Diffuser, Throw, Drop, Primary air.

12. State the principles of air distribution.

13. What is an air-handling system? Give flow diagram for an air-conditioning system.

14. What are the requirements of good room air-distribution?

15. Describe briefly the following:

(*i*) Grille outlets (*ii*) Slot diffuser outlets

(*iii*) Ceiling diffuser outlets (*iv*) Perforated ceiling panels.

16. What is the functions of ducts ? Enumerate duct systems and explain any one of them with a neat sketch.

17. Explain any two of the following air-distribution systems :

(*i*) Ejector system (*ii*) Downward system

(*iii*) Upward system.

18. Write an short note on 'Air leakage and duct maintenance'.

UNSOLVED EXAMPLES

1. In an air-conditioning system, the size of the main air supply duct is 0.8 m × 0.6 m in cross-section, and carries 5 m^3/s of standard air. The main duct branches into two ducts of cross-section 0.6 m × 0.5 m and 0.6 m × 0.4 m. If the mean velocity in the larger branch is 8 m/s, determine:

(*i*) Mean velocity in the main duct and the smaller branch; and

(*ii*) Mean velocity pressure in each duct. [**Ans.** (*i*) 10.4 m/s; 10.8 m/s; (*ii*) 6.62, 3.92, 7.14 mm of water]

2. A 15 m long duct passes air at the rate of 1.5 m^3/s. If the friction factor is 0.005, calculate the pressure drop in the following cases:

(*i*) When the duct is circular of diameter 300 mm; and

(*ii*) When the duct is of 300 mm square cross-section.

[**Ans.** (*i*) 28.47 mm of water; (*ii*) 17.02 mm of water]

3. A duct of rectangular cross-section 0.5 m × 0.35 m carries 1.25 m^3/s of air having density of 1.2 kg/m^3. Determine the equivalent diameter of a circular duct if the quantity of air carried in both the cases is same. Also find the pressure loss per 100 m length of duct if friction factor (f) = 0.01.

[**Ans.** 0.46 m; 29 mm of water]

4. A 20 m long rectangular duct, 150 mm × 120 mm in size, carries standard air at the rate of 18 m^3/min. Assuming friction factor f = 0.005, determine :

(*i*) Total pressure required at the inlet to the duct to maintain this flow; and

(*ii*) Air power required. [**Ans.** (*i*) 671.85 N/m^2; (*ii*) 201.6 W)

5. A main circular duct consists of three branch ducts taking equal air volumes at equal intervals. Each interval duct has a friction loss of 1.3 mm of water and a static pressure of 5 mm of water is necessary at each branch to cope with friction loss. If the initial velocity in the main duct of 1.2 m diameter is 10 m/s, calculate the velocities and diameters of the second and third lengths, whereby the static pressure regain is sufficient to overcome the friction loss in the succeeding length of main duct upto the next branch.

Assume static pressure regain factor as 0.6.

[**Ans.** C_2 = 8.04 m/s; D_2 = 1.09 m; C_3 = 5.4 m/s; D_3 = 0.94 m]

6. In a duct ABCDEF, air is supplied at A and discharged to atmosphere at F. The portion AB is 0.6 m in diameter and 45 m long, BC is the expander, the portion CD is 1.2 m in diameter and 150 m long, DE is the reducer and the portion EF is 0.45 m in diameter and 15 m long. The velocity of air in duct AB is 10 m/s. The loss in the expander BC is 0.5 times the velocity pressure in duct AB and the loss in the reducer DE is 0.2 times the velocity pressure in the duct EF.

Using the expression, $p_f = \dfrac{0.263\,(C)^{1.85}}{(D)^{1.27}}$ where p_f is the pressure loss due to friction in mm of water per 100 mm length of duct, C is the duct velocity in m/s and D is the diameter of the duct in m, calculate the static pressure at point A.

[**Ans.** 60.28 mm of water]

7. In a system consisting of inlet and outlet ducts, a centrifugal fan of outlet 0.9 m × 0.7 m is moving standard air at 11.33 m^3/s. The inlet duct is 0.9 m diameter and 15 m long and discharge duct is 1 m diameter and 60 m long. There is a diffuser between the fan discharge and 1 m diameter duct for which the loss of pressure may be taken as 0.3 times the difference between velocity pressures. The loss at entry to the inlet duct is 0.5 times the velocity pressure there and the friction factor 'f' for the inlet duct is 0.004 and for the outlet duct is 0.0035. Determine :

(*i*) Fan total pressure; and

(*ii*) Static pressure at the fan inlet and outlet.

[**Ans.** 40.51 mm of water; –34.35 mm of water; 5.76 mm of water)

8. Fig. 16.37 shows a duct system in which air enters at A with a static pressure of 7.5 mm of water. Considering friction losses only and using friction chart and equal pressure drop method, determine the correct diameters and velocity pressures in lengths AB, BC, CD, branch BE and branch CF.

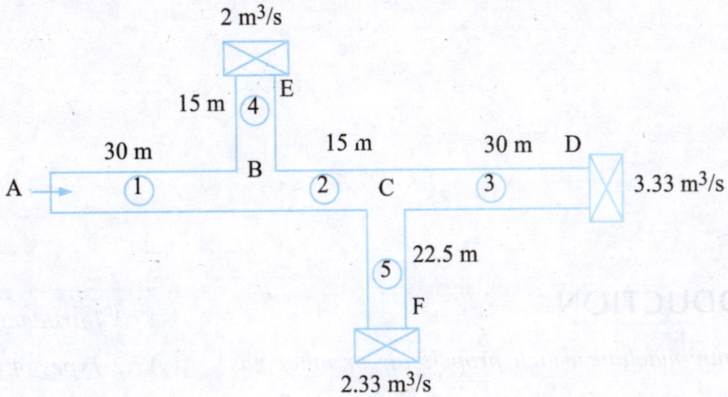

Fig. 16.37

[**Ans.** D_1 = 0.95 m; D_2 = 0.85 m; D_3 = 0.72 m; D_4 = 0.46 m; D_5 = 0.6 m; 7.16, 6.12, 4.11, 8.89, 4.16 respectively in mm of water]

❑❑❑

Chapter 17

Fans

17.1 INTRODUCTION

A **fan** *is rotodynamic machine which propels air or other gas continuously.*

Fans find extensive use in air-conditioning and refrigeration plants for *moving or circulating air over air-cooled condensers and evaporator coils.*

While a pump or a compressor develops higher fluid pressures, the *fans can develop comparatively low pressures.* The pressure developed by the fan *being very low,* is expressed in "*mm of water*".

According to the maximum total pressures they can safely develop, fans may be classified as follows:

(*i*) Maximum pressure upto 95 mm water gauge .... *Class I*

(*ii*) Maximum pressure upto 171 mm water guage .... *Class II*

(*iii*) Maximum pressured upto 311 mm water guage .... *Class III*

(*iv*) Maximum pressure over 311 mm water gauge .... *Class IV*

— For standard systems, *Class I* fans are generally sufficient. Higher-class fans are needed where higher total pressures are required.

Similar to the resistance offered by a pipe and fittings to the flow of water or gas in it, air ducts, outlets and other components offer resistance to air (e.g., the condenser coil in an air-cooled condenser, cooling coil in an air-conditioning or a cold-storage plant etc. offer resistance to the flow of air). Further the resistance increases : (*i*) inversely as the cross-sectional area of the duct and air openings (such as return-air openings in air-conditioned plants), (*ii*) with the increase in the number of rows of pipes in the direction of air flow (as in air-cooled condenser coil or cooling coil), (*iii*) with increase in number of fins, (*iv*) with the wetness of the surface, and (*v*) with increase in the rate of flow of air.

Thus, the *fan has to develop enough pressure to overcome the resistances offered by the various components to the air flow. It is these resistances that create static pressure or static head when air moves in the system.*

- The air at inlet to the fan is below atmospheric pressure while at the exhaust or outlet of fan is above atmospheric pressure. The air feed into the fan is called *induced drafts* while the air axhaust from a fan is called *forced draft.*
- Irrespective of the type of construction of the fans, they may function as either blowers or exhausters. *The blowers discharge their air against a pressure at their outlet whereas exhausters remove gases from a space by suction.*

17.2 TYPES OF FANS

For transmission of air, the following two types of fans are used:

1. Centrifugal or radial flow fans, and
2. Axial-flow fans.

17.2.1 Centrifugal Fans

A **Centrifugal or radial flow fan** *is one in which air enters the impeller axially and is discharged radially from the impeller.*

Centrifugal fans are used for the *air-conditioning and ventilation purposes.*

Depending upon the inclination (inclination determines the performance characteristics of the fan) of the fan blades, the centrifugal fans are further classified as:

(*i*) Forward-curved fans,

(*ii*) Radial blade fans, and

(*iii*) Backward-curved fans.

(*i*) ***Forward-curved fans.*** Refer to Fig. 17.1.

- In this type of fan, the tip of the blade is *curved forward*, i.e., in the direction of rotation of the fan. The forward-covered impellers usually have 24 to 64 blades.
- At any given r.p.m., forward-curved blade fans *deliver more air quantity and produce relatively lower static* pressure than the radial or back word-curved blade fans.

(*ii*) ***Radial blade fans.*** Refer to Fig. 17.2.

- Radial blade has no curvature and the blade is *straight* and is *fixed radially.*
- These fans are generally used in electric motors for internal cooling. The blades are built into the rotor itself in case of small-sized motors. For bigger-sized motors, a separate fan is fixed on the motor shaft (within the body).

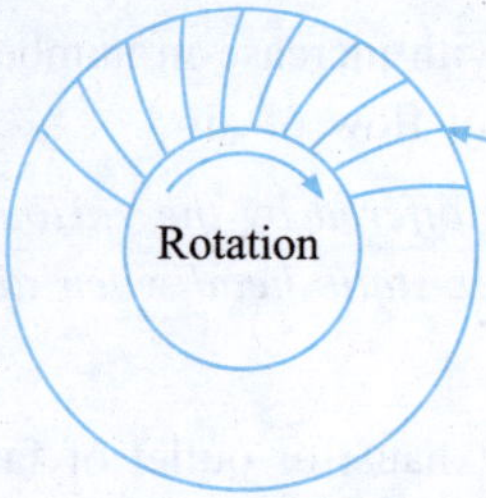

Fig. 17.1. Forward-curved blade fan.

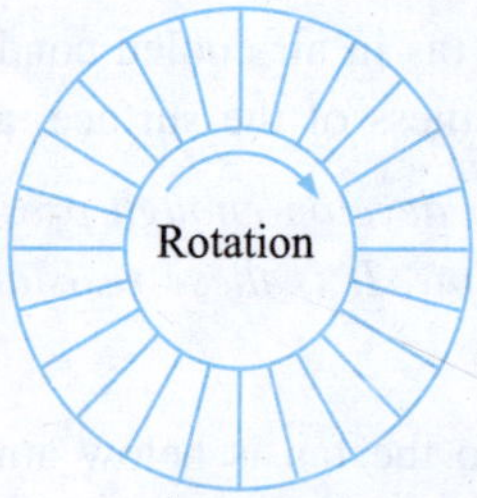

Fig. 17.2. Radial blade fan.

(*iii*) ***Backward-curved fans.*** Refer Figs. 17.3 and 17.4.

- In this type of fan, the tip of the blade is *curved backward* or away from the direction of rotation of the fan wheel.
- These fans are require higher speed and therefore have to be more sturdy in construction. However, *because of their characteristic higher efficiency over a broader range of system resistance,* these types of fans are employed for *bigger installations.*

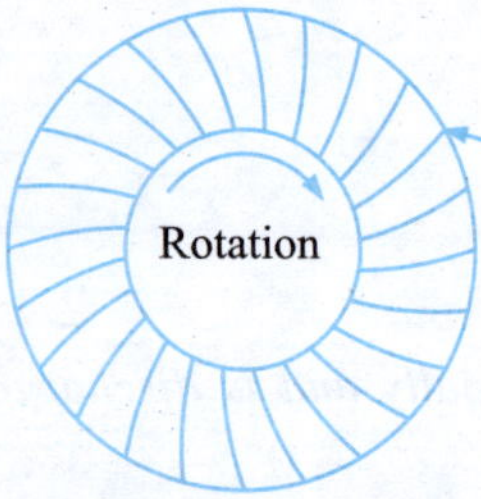

Fig. 17.3. Backward-curved fan.

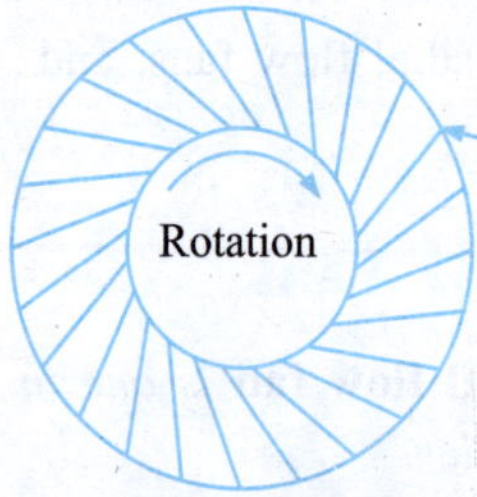

Fig. 17.4. Backward-inclined blade fan.

17.2.2 Axial-flow Fans

When the air flows parallel to the axis of impeller, it is called an **"axial-flow fan".**

Following are the three types of axial-flow fans :

(*i*) Propeller fans;

(*ii*) Tube axial fans; and

(*iii*) Vane axial fans.

(*i*) **Propeller fans.** Refer to Fig. 17.5.

- These fan are *low pressure high-capacity fans.* The static pressure developed by these fans in *rather limited.*
- They deliver large volumes of air with less power than the centrifugal fans but can develop low static pressure. With the increase in the system resistance, the power required also increases; this is in contrast to the centrifugal fan in which the power required decreases with an increase in the system resistance. Thus a propeller fan used for an air-cooled condenser can draw more electric current when the condenser fins get choked with dust. The quantity of air delivered also decreases.

Fig. 17.5. Propeller fan.

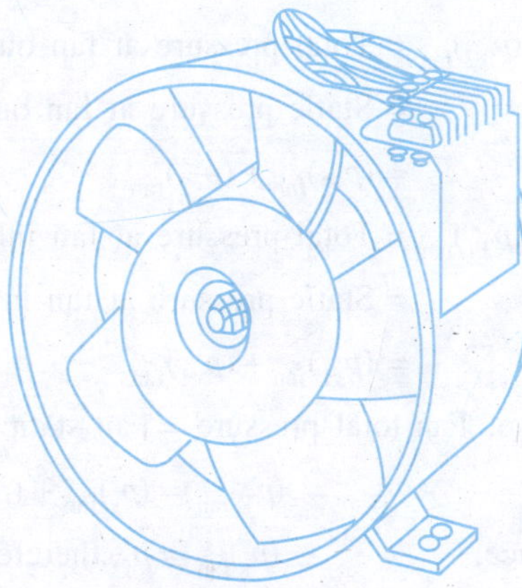

Fig. 17.6. Tube axial fan.

(*ii*) **Tube axial fans.** Refer to Fig. 17.6.

- These fans may be considered as heavy-duty propeller fans mounted in a cylindrical housing.
- They are duct-mounted whereas the propeller fans are wall or diaphragm mounted.
- Since these fans produce high noise level, therefore, their use is limited to applications where the noise level is not of much importance.

(*iii*) **Vane axial fans.** Refer to Fig. 17.7.

- These are *tube axial fans with vanes*. The vanes located at the leaving side of the wheel help to straighten out the spiral flow of air coming out of an axial fan. The straight line flow leaving the fan assures quiet operation.
- The efficiency of operation and the pressure characteristics are *better than those of tube axial fan.*

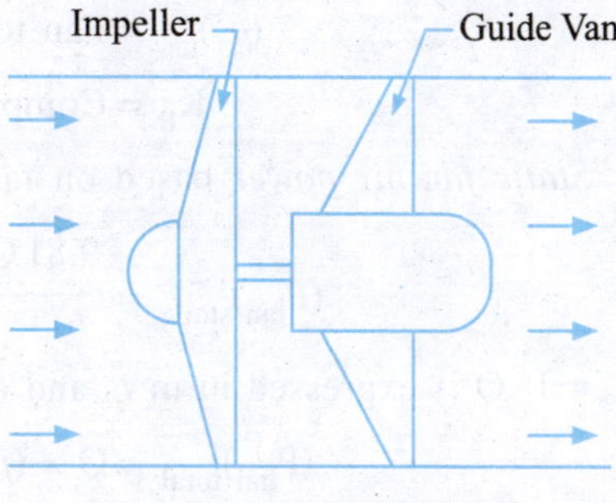

Fig. 17.7. Vane axial fan.

The axial-flow fans are particularly suitable for handling large volumes of air at relatively low pressures. They are *never employed for duct air-conditioning system because they are incapable of developing high pressures.*

17.3 TOTAL PRESSURE DEVELOPED BY A FAN

In case of a fan;

- The ***fan static pressure*** $[(p_s)_{fan}]$ is the *pressure increase produced by a fan.*
- The ***fan velocity pressure*** $[(p_v)_{fan}]$ is the *velocity pressure corresponding to the mean velocity of air at the fan outlet based on the total outlet area without any deductions for motors etc.*
- The ***total fan pressure*** $[(p_T)_{fan}]$ is the *algebraic difference between the total pressure at the fan outlet and the total pressure at the fan inlet.*

In other words,

$$(p_T)_{fan} = (p_{T2})_{fan} - (p_{T1})_{fan}$$

where, $(p_{T2})_{fan}$ = Total pressure at fan outlet

= Static pressure at fan outlet + Velocity pressure at fan outlet

$= (p_{s2})_{fan} + (p_{v2})_{fan}$

$(p_{T1})_{fan}$ = Total pressure at fan inlet

= Static pressure at fan inlet + Velocity pressure at fan inlet

$= (p_{s1})_{fan} + (p_{v1})_{fan}$

Also, Fan total pressure = Fan static pressure + Fan velocity pressure

$$(p_T)_{fan} = (p_s)_{fan} + (p_v)_{fan} \qquad ...(17.1)$$

Since, $(p_v)_{fan} = p_v$, therefore,

$$(p_T)_{fan} = (p_s)_{fan} + p_v \qquad ...[17.1\ (a)]$$

17.4 FAN AIR POWER

The *power outlet of a fan* is expressed in terms of *air power* and represents the work done by the fan.

Total fan air power, based on fan total pressure $[(p_T)_{fan}]$, is given by:

$$(P_{fan})_{total} = \frac{9.81\, Q\, (p_T)_{fan} \times K_P}{60}, \text{ watts} \qquad ...(17.2)$$

where, Q = Total quantity of air flowing at the fan inlet, m^3/min,

$(p_T)_{fan}$ = Fan total pressure, mm of water, and

K_P = Compressibility coefficient.

Static fan air power based on fan static pressure $[(p_s)_{fan}]$ in given by:

$$(P_{fan})_{static} = \frac{9.81\, Q \times (p_s)_{fan} \times K_p}{60} \text{ watts.} \qquad ...(17.3)$$

- If Q is expressed in m^3/s and (p_T)fan and $(p_s)_{fan}$ are in N/m^2, then:

$$(P_{fan})_{total} = Q \times (p_T)_{fan} \times K_P \qquad ...(17.4)$$

$$(P_{fan})_{static} = Q \times (p_s)_{fan} \times K_P \qquad ...(17.5)$$

17.5 FAN EFFICIENCIES

Fan efficiencies are discussed below:

1. **Total fan efficiency $(\eta_{fan})_{total}$:**

It is defined as the *ratio of the total fan air power to the driving power* (*or brake power*) *required at the fan shaft.* It is also called "*mechanical efficiency*" of the fan.

In other words,

$$(\eta_{fan})_{total} = \frac{\text{Total fan air power } [(P_{fan})_{total}]}{\text{Input or brake power (B.P.)}}$$

2. **Static fan efficiency $(\eta_{fan})_{static}$:**

It is defined as the *ratio of the static fan air power to the input or brake power.*

In other words,

$$(\eta_{fan})_{static} = \frac{\text{Static fan air power } [(p_{fan})_{static}]}{\text{Input or brake power (B.P.)}}.$$

17.6 VELOCITY TRIANGLES FOR MOVING BLADES OF A CENTRIFUGAL FAN

Fig. 17.8 shows a portion of the impeller of a centrifugal fan with one backward curved blade and the velocity triangles at the inlet and outlet tips of the blade (air stream entering at C and leaving at D).

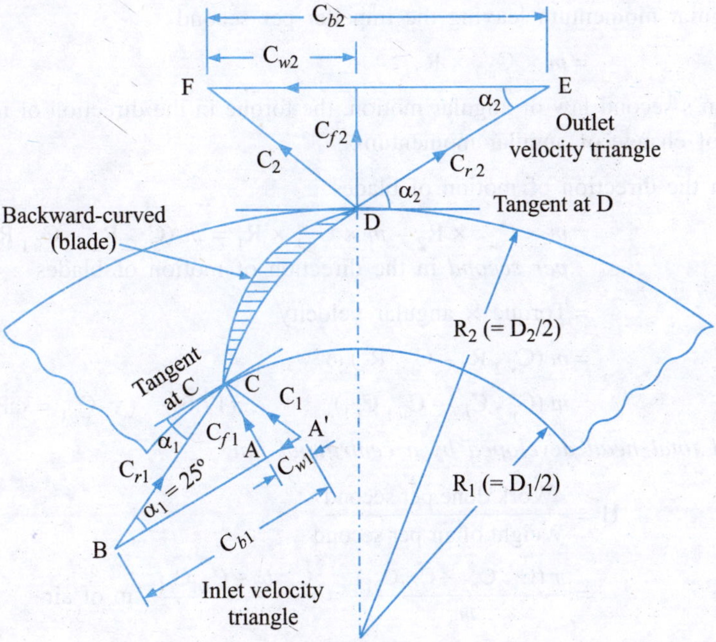

Fig. 17.8

Let, C_{b1} = Linear or tangential velocity of the moving blade at inlet (BA′),

C_1 = Absolute velocity of air entering the blade (A′C),

C_{f1} = Velocity of flow at inlet (AC). It is radial component of C_1

C_{r1} = Relative velocity of air to the moving blade at inlet (BC). It is vectorial difference between C_{b1} and C_1.

C_{w1} = Velocity of whirl at inlet. It is the tangential component of C_1.

α_1 = Blade angle at inlet. It is the angle which the relative velocity (C_{r1}) makes with the tangent at the blade inlet. It is equal to the angle between C_{r1} and C_{b1} (i.e., ∠ CBA).

C_{b2}, C_2, C_{f2}, C_{r2}, C_{w2}, α_2 = Corresponding values at outlet of the blade tip.

Since the air enters and leaves the blades *without any shock* (i.e., tangentially), therefore the shape of the blades will be such that C_{r1} and C_{r2} are *along the tangents* to the blades at inlet and outlet respectively.

Let, R_1 (= $D_1/2$) = Internal radius of the impeller,

R_2 (= $D_2/2$) = External radius of the impeller,

ω = Angular velocity of the impeller in rad./s

$= \dfrac{C_{b1}}{R_1}$ or $\dfrac{C_{b2}}{R_2}$, and

m = Mass of air flowing through the impeller, kg/s.

Angular momentum entering the impeller per second

= Mass of air flowing per second × Velocity of whirl × Radius of impeller

$= m \times C_{w1} \times R_1$

Similarly, angular momentum leaving the impeller per second

$= m \times C_{w2} \times R_2$

As per Newton's second law of angular motion, the torque in the direction of motion of blades is equal to the rate of change of angular momentum.

∴ Torque in the direction of motion of blades

$= m \times C_{w2} \times R_2 - m \times C_{w1} \times R_1 = m\,(C_{w2}\,R_2 - C_{w1}\,R_1)$, and *work done per second* in the direction of motion of blades

= Torque × angular velocity

$= m\,(C_{w2}\,R_2 - C_{w1}\,R_1)\,\omega$

$= m\,(C_{w2}\,C_{b2} - C_{w1}\,C_{b1})$ ...(17.6) $(\because C_{b1} = \omega R_1 \text{ and } C_{b2} = \omega R_2)$

∴ *Theoretical total head developed by a centrifugal fan,*

$$H = \frac{\text{Work done per second}}{\text{Weight of air per second}}$$

$$= \frac{m\,(C_{w2}C_{b2} - C_{w1}C_{b1})}{mg} = \frac{C_{w2}b_2 - C_{w1}C_{b1}}{g}, \text{ m of air} \quad \text{...(17.7)}$$

This expression is called **Euler's equation.**

Total pressure developed by a fan,

$$(p_T)_{fan} = \rho_a \times H = \frac{\rho_a\,(C_{w2}C_{b2} - C_{w1}C_{b1})}{g}, \text{ mm of water} \quad \text{...(17.8)}$$

$$= \rho_a\,(C_{w2}\,C_{b2} - C_{w1}\,C_{b1}), \text{ N/m}^2 \quad \text{...(17.9)}$$

where, ρ_a = Mass density of air, kg/m^3

Further, from *inlet velocity triangle*, we have:

$$C_{b1} = AA' + AB = C_{w1} + C_{r1}\cos\alpha_1$$

From *outlet velocity triangle*,

$$C_{b2} = FG + GE = C_{w2} + C_{r2}\cos\alpha_2$$

or, $$C_{w2} = C_{b2} - C_{r2}\cos\alpha_2 \quad \text{...}(i)$$

Also, $$(C_2)^2 = (C_{b2})^2 + (C_{r2})^2 - 2\,(C_{b2})\,(C_{r2})\cos\alpha_2$$

$$\therefore \quad \cos\alpha_2 = \frac{(C_{b2})^2 + (C_{r2})^2 - (C_2)^2}{2\,(C_{b2})(C_{r2})}$$

Substituting the value of cos α_2 in eqn. (*i*), we get:

$$C_{w2} = C_{b2} - C_{r2}\left[\frac{(C_{b2})^2 + (C_{r2})^2 - (C_2)^2}{2\,(C_{b2})(C_{r2})}\right]$$

$$= C_{b2} - \left[\frac{(C_{b2})^2 + (C_{r2})^2 - (C_2)^2}{2\,C_{b2}}\right]$$

$$= \frac{2\,(C_{b2})^2 - (C_{b2})^2 - (C_{r2})^2 + (C_2)^2}{2\,C_{b2}} = \frac{(C_{b2})^2 - (C_{r2})^2 + (C_2)^2}{2\,C_{b2}}$$

Similarly, for inlet triangle,

$$C_{w1} = \frac{(C_{b1})^2 - (C_{r1})^2 + (C_1)^2}{2\,C_{b1}}$$

Substituting the values of C_{w1} and C_{w2} in eqn. (17.6), we get:

Work done per second

$$= m\left[\frac{(C_{b2})^2 - (C_{r2})^2 + (C_2)^2}{2\,C_{b2}} \times C_{b2} - \frac{(C_{b1})^2 - (C_{r1})^2 + (C_1)^2}{2\,C_{b1}} \times C_{b1}\right]$$

$$= m\left[\frac{(C_{b2})^2 - (C_{b1})^2}{2} + \frac{(C_{r1})^2 - (C_{r2})^2}{2} + \frac{C_2^2 - C_1^2}{2}\right] \text{N-m or J} \qquad ...(17.10)$$

and, **theoretical total head developed,**

$$H = \frac{\text{Work done per second}}{mg}$$

or,
$$H = \frac{(C_{b2})^2 - (C_{b1})^2}{2g} + \frac{(C_{r2})^2 - (C_{r1})^2}{2g} + \frac{(C_2)^2 - (C_1)^2}{2g}, \text{ m of air} \qquad ..(17.11)$$

— The *first term* of this expression represents **"Increase in static head"** *which is due to the forced vortex formed by the rotating impeller.*

— The *second term* represents that **part of the static head increase** *which, according to Bernoulli's equation, results from the change in relative velocity of air passing through the impeller.*

— The *third term* is the **"Change in velocity head"** *of the air flowing through the impeller.*

($C_{b1} = \frac{\pi D_1 N}{60}$; $C_{b2} = \frac{\pi D_2 N}{60}$, where D_1 and D_2 are the internal and external diameters of the impeller respectively, and N is the speed of the impeller in r.p.m.)

Work Done and Theoretical Total Head Developed by a Centrifugal Fan for Radial Entry of Air.

Refer to Fig., 17.7. When the air enters the blades at right angles (i.e., radially) to the direction of motion of the blade, then:

$$C_1 = C_{f1} \quad \text{and} \quad C_{w1} = 0$$

∴ For *radial entry of air*,

Work done per second $= m\,(C_{w2} \times C_{b2})$, N-m or J ...(*ii*) ....[From eqn. (17.6)]

Theoretical head developed,

$$H = \frac{1}{g}\,(C_{\omega 2} \times C_{b2}), \text{ m of air} \quad ...(iii) \quad ...[\text{From eqn. (17.7)}]$$

Let, Q = Quantity of air flowing through the fan, m³/s,

D_1 = Internal diameter of impeller, m

D_2 = External diameter of impeller, m

b_1 = Width of impeller at inlet, m, and

b_2 = Width of impeller at outlet, m.

Since the quantity of air flowing through the impeller is constant, therefore,

$$Q = \pi D_1 b_1 C_{f1} = \pi D_2 b_2 C_{f2}$$

or,

$$C_{f1} = \frac{Q}{\pi D_1 b_1} \quad \text{and} \quad C_{f2} = \frac{Q}{\pi D_2 b_2}$$

From the outlet velocity triangle (Fig. 17.8), we have:

$$\cot \alpha_2 = \frac{GE}{GD} = \frac{C_{b2} - C_{w2}}{C_{f2}} \quad ...(iv)$$

∴

$$C_{b2} - C_{w2} = C_{f2} \cot \alpha_2 = \frac{Q}{\pi D_2 b_2} \times \cot \alpha_2$$

and,

$$C_{w2} = C_{b2} - \frac{Q \cot \alpha_2}{\pi D_2 b_2}$$

Substituting the value of C_{w2} in eqns. (*ii*) and (*iii*), we get:

$$\textbf{Work done per second} = m\left[C_{b2} - \frac{Q \cot \alpha_2}{\pi D_2 b_2}\right] C_{b2}$$

$$= m\left[(C_{b2})^2 - \frac{Q \cot \alpha_2 \times C_{b2}}{\pi D_2 b_2}\right] \text{N-m or J} \quad ...(17.8)$$

Theoretical total head developed,

$$H = \frac{1}{g}\left[(C_{b2})^2 - \frac{Q \cot \alpha_2 \times C_{b2}}{\pi D_2 b_2}\right] \text{m of air} \quad ...(17.9)$$

• Total pressure developed by a centrifugal fan,

$$(p_T)_{fan} = \rho_a \times H = \frac{\rho_a \times C_{w2} \times C_{b2}}{g}, \text{ mm of water} \quad ...[\text{From eqn. } (iii)]$$

$$= \rho_a \times C_{w2} \times C_{b2} \text{ N/m}^2$$

$$= \frac{\rho_a}{g}\left[(C_{b2})^2 - \frac{Q \cot \alpha_2 \times C_{b2}}{\pi D_2 b_2}\right], \text{ mm of water}$$

where, ρ_a = Mass density of air, kg/m³.

$$\cot \alpha_2 = \frac{C_{b2} - C_{w2}}{C_{f2}} \quad ...(\text{From eqn. } (iv))$$

or, $$C_{w2} = C_{b2} - C_{f2} \cot \alpha_2$$

For radial blades: $\alpha_2 = 90°$, therefore, $C_{w2} = C_{b2}$

For forward-curved blades: $\alpha_2 > 90°$, therefore, $C_{w2} < C_{b2}$

For backward-curved blades: $\alpha_2 < 90°$, therefore, $C_{w2} < C_{b2}$

— As $(p_T)_{fan} = \rho_a \times C_{w2} \times C_{b2}$, N/m^2, therefore, for a given velocity (C_{b2}), $(p_T)_{fan}$ is *greatest for a centrifugal fan having forward-curved blades.*

17.7 SPECIFIC SPEED OF A CENTRIFUGAL FAN

The **specific speed** (N_s) *of a centrifugal fan is defined as the speed of a geometrically similar fan which would deliver 1 m^3 of air per second against a head of 1 m of air*:

Let, D = Diameter of impeller,

b = Width of impeller,

C_f = Velocity of flow,

Q = Total quantity of air flowing through the fan,

N = Speed of the impeller, and

H = Head developed by the fan.

$$Q = \pi D b\, C_f, \quad \text{or,} \quad Q \propto D b C_f \qquad ...(i)$$

Also, $$D \propto b \qquad ...(ii)$$

$\therefore$ $$Q \propto D^2 C_f \qquad ...(iii) \text{ [From } (i) \text{ and } (ii)]$$

Further, $$C_b = \frac{\pi DN}{60}, \quad \text{or,} \quad C_b \propto DN \qquad ...(iv)$$

Also, $$C_b \propto C_f \propto \sqrt{H} \qquad ...(v)$$

$\therefore$ From eqns. (*iv*) and (*v*), we have:

$$DN \propto \sqrt{H} \quad \text{or} \quad D \propto \frac{\sqrt{H}}{N} \qquad ...(vi)$$

Substituting the value of D in eqn. (*iii*), we get:

$$Q = \frac{H}{N^2} \times C_f \quad \text{or} \quad Q \propto \frac{H}{N^2} \sqrt{H} \propto \frac{(H)^{3/2}}{N^2}$$

or, $$Q = \frac{K(H)^{3/2}}{N^2} \qquad ...(vii)$$

where, K is the constant of proportionality.

According to the definition, if Q = 1 m^3/s and H = 1 m, then N = N_s. Substituting these values in eqn. (*vii*), we have:

$$1 = \frac{K(1)^{3/2}}{(N_s)^2}, \quad \text{or,} \quad K = N_s^2$$

From eqn. (*vii*), we have:

$$Q = \frac{(N_s)^2 (H)^{3/2}}{N^2}, \quad \text{or,} \quad N_s = \frac{N\sqrt{Q}}{(H)^{3/4}} \qquad ...(17.10)$$

17.8 FAN SIMILARITY LAWS

For any series of geometrically similar fans and for any point on the characteristic curves the following *relationships hold:

$$\frac{Q}{ND^3} = \text{constant} \qquad .....(17.11)$$

$$\frac{g}{N^2D^2} = \text{constant} \qquad ...(17.12)$$

$$\frac{(p_T)_{fan}}{N^2D^2\rho} = \text{constant} \qquad ...(17.13)$$

$$\frac{P}{N^3D^5\rho} = \text{constant} \qquad ...(17.14)$$

$$\eta = \text{constant} \qquad ...(17.15)$$

where, Q = Volume rate of flow,

H = Fan total head,

$(p_T)_{fan}$ = Fan total pressure,

P = Power consumed,

N = Rotational speed,

D = Diameter of impeller,

g = Acceleration due to gravity, and

ρ = Density of the fluid.

The above dimensionless relationships, in any consistent system of units, hold good and the co-relations are known as **Fan Laws**.

Then, (*i*) $Q \times N$ ...(17.16)

(*ii*) $Q \times D^3$ ...(17.17)

It may be noted that the *laws are valid for geometrically similar fans and for the same point on the characteristic curve.*

- The following dimensionless design parameters are also used:

$$\textit{Volume coefficient, } \phi = \frac{4Q}{\pi D^2 C} \qquad ...(17.18)$$

$$\textit{Pressure coefficient, } \psi = \frac{2\,(p_T)_{fan}}{\rho C_b^2} \qquad ...(17.19)$$

$$\textit{Power coefficient, } \lambda = \frac{\phi\psi}{\eta} \qquad ...(17.20)$$

where, C_b = Peripheral velocity of impeller in consistent system of units employed.

* The similarly laws are explained in details in the author's popular book on "Fluid Mechanics and Hydraulic Machines."

17.9 FAN AND SYSTEM CHARACTERISTIC

System characteristic is *the change in pressure loss or resistance with the change in flow rate*. It is independent of the fan used in that system.

The system resistance (R) or pressure loss (p_L) of any fixed system varies as square of the flow rate, i.e.,

$$R , \quad \text{or,} \quad p_L = KQ^2$$

When R or p_L is plotted against Q, a curve (a parabola) such as OA, as shown in Fig. 17.9, is obtained; the curve is usually known as a *system characteristic curve*.

The operation of a fan in conjunction with a particular system must fulfil the condition that the loss of total pressure in the system at a given volume flow must be equal to the total pressure developed by the fan (i.e., fan total pressure) at the same volume. This condition is satisfied by the point of intersection (point D) of system characteristic curve OA and the fan performance curve BC. This point of intersection is called the *operating point* of that particular fan in that particular system.

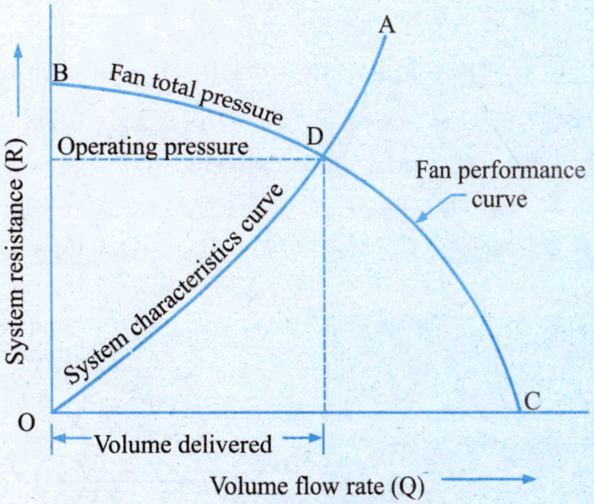

Fig. 17.9. Fan and system characteristic.

17.10 FANS IN SERIES AND IN PARALLEL

17.10.1 Fans in Series

Fans are said to be connected in *series* when the outlet of first fan is connected to the inlet of second fan and outlet of the second fan is connected to the inlet of the third fan and so on, as shown in Fig. 17.10.

When the fans are connected in *series*:

(*i*) *Volume flow rate* (Q) through each fan is *same*, i.e.,

$$Q = Q_1 = Q_2 = Q_3$$

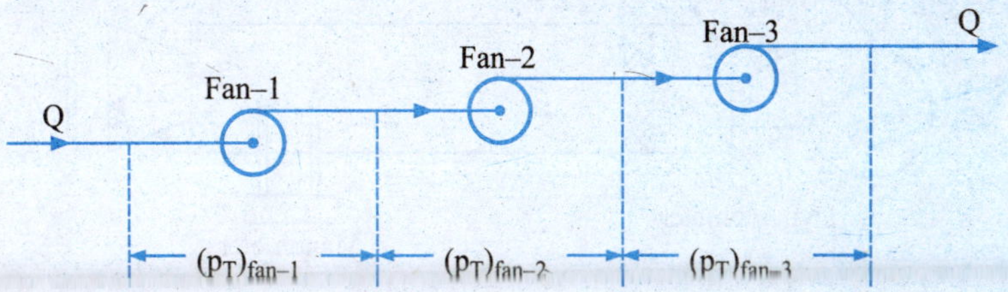

Fig. 17.10. Fans in series.

(*ii*) The overall fan total pressure $(p_T)_{fan}$ is equal to the *sum of the fan total pressures developed by the individual* fans *i.e.*,

$$(p_T)_{fan} = (p_T)_{fan-1} + (p_T)_{fan-2} + (p_T)_{fan-3}$$

17.10.2 Fans in Parallel

Fig. 17.11 shows fans connected in *parallel*.

In this case:

(*i*) The *fan total pressure* $[(p_T)_{fan}]$ of each fan is same, i.e.,

$$(p_T)_{fan-1} = (p_T)_{fan-2} = (p_T)_{fan-3}$$

(*ii*) The total volume delivered (Q) is equal to the *sum of the volumes delivered by the individual fans*, i.e.,

$$Q = Q_1 + Q_2 + Q_3.$$

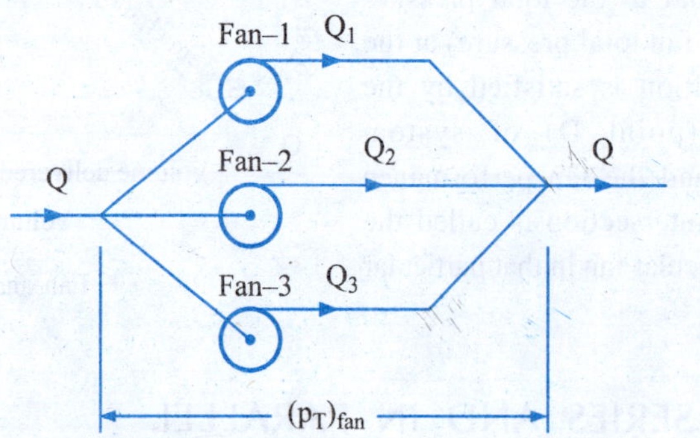

Fig. 17.11. Fans in parallel.

17.11 FAN PERFORMANCE

In order to establish the fan performance and to make fan-performance curves, the manufacturers adopt a test set-up. Such a set-up in its simplest form is shown in Fig 17.12.

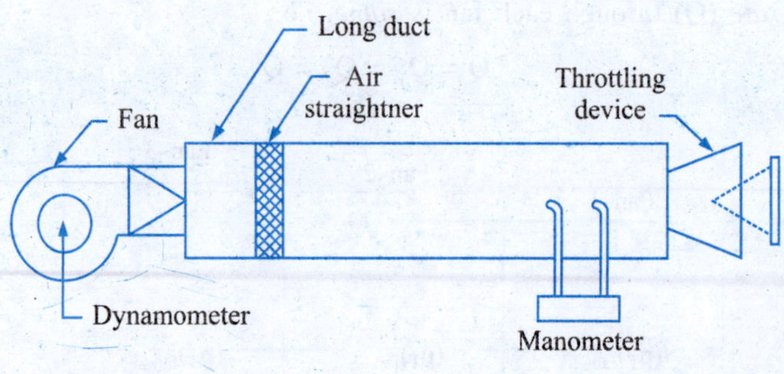

Fig. 17.12. Fan testing.

— The fan to be tested is connected to a long duct having a "*throttling*" *device* attached to its free end. The air flow can be adjusted by adjusting the opening of the throttling device.

— The "*honey-combed air straightner*" is provided to straighten out the air flow in the duct.

— After the straightner, "*manometers*" are connected to the duct *for obtaining steady readings of the total and static pressures.*

— A "*dynamometer*" is attached to the fan shaft to indicate the power consumed by the fan when it is in operation.

When the fan is run with the throttling device fully opened, air flows in the duct without experiencing any resistance, i.e., no static pressure at all. So the total pressure reading on only velocity pressure and air quantity handled by the fan is known as '*Wide open cmh*' (Wocmh). On increasing the fan speed the Wocmh will increase and consequently the velocity pressure will go up and vice-versa. The quantity of air is calculated by converting velocity pressure to velocity and multiplying it with the area of the cross-section of the duct.

The static pressure can be varied by adjusting the throttling device. The *throttling device is progressively adjusted from the fully open to fully closed position and readings of velocity pressure (total pressure-static pressure) and power consumed are taken and air quantities calculated for increasing steps of static pressures.*

By *repeating the above procedure for a number of fan speeds*, air quantities and power consumed at various static pressures and fan speeds can be obtained. These values of air quantities, static pressure, speed and power, all plotted in a graph sheet, constitute the "*fan performance curves*". The dynamometer indicates the power required at the fan shaft and it is the brake power (B.P.) of the fan. The power of the fan motor is higher than the B.P. to take of care of drive-losses, motor efficiency etc.

When the throttling device is fully closed, the air flow in the duct ceases completely, and so no velocity pressure will be indicated. Only static pressure will be shown and this is known as '*Block Tight-static Pressure*' (BTSP).

The use of fan performance curves can be made in predicting the behaviour of the fan in the air-distribution system, for which it is selected. This is done by superimposing the system resistance curves on the fan curves.

The performance curves for the various types of fans are shown in Fig. 17.13 to Fig. 17.16.

- Fig. 17.13 shows the performance curves for a centrifugal fan with forward-curved blades. These fans *require an ever-increasing amount of power as the air volume is increased.* However, this type of fan *provides greater static pressure for a given blade-tip velocity than the other types.* Inspite of this advantage, such a fan is *commonly used in air-conditioning system.*
- Fig. 17.14 shows the centrifugal fans with backward-curved blades which *require maximum power.* These fans are said to have *non-overloading power characteristic.*
- Fig. 17.15 shows the performance curves for a centrifugal fan with radial blades.
- Fig. 17.16 shows that the *axial-flow fans also have non-over-loading power characteristics.*

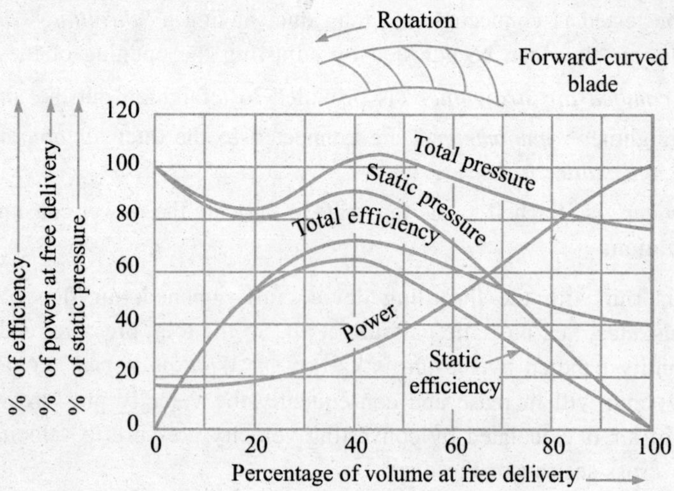

Fig. 17.13. Performance curves for a centrifugal fan with forward-curved blades.

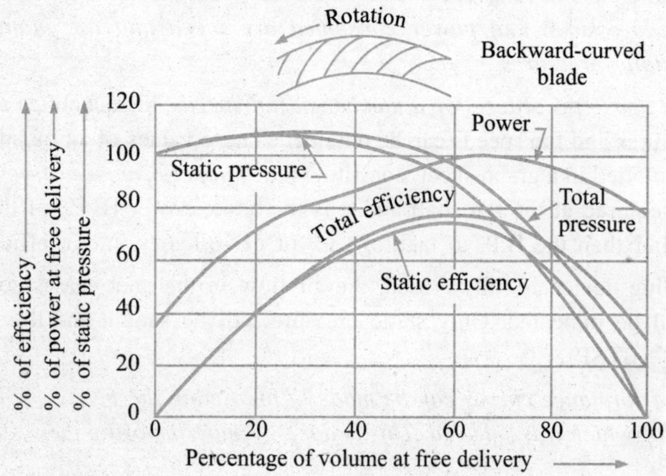

Fig. 17.14. Performance curves for a centrifugal fan with backward-curved blades.

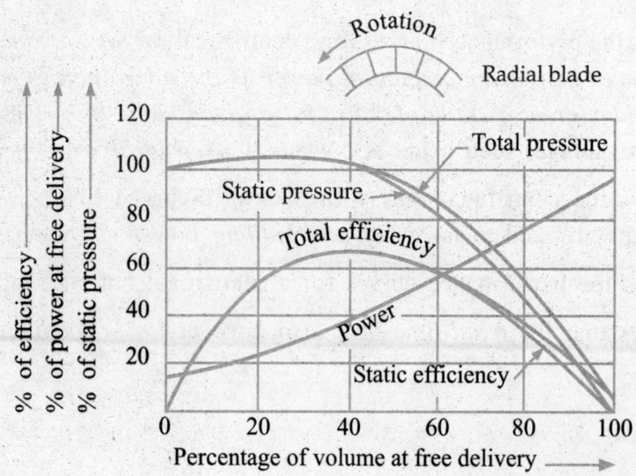

Fig. 17.15. Performance curves for a centrifugal fan with radial blades.

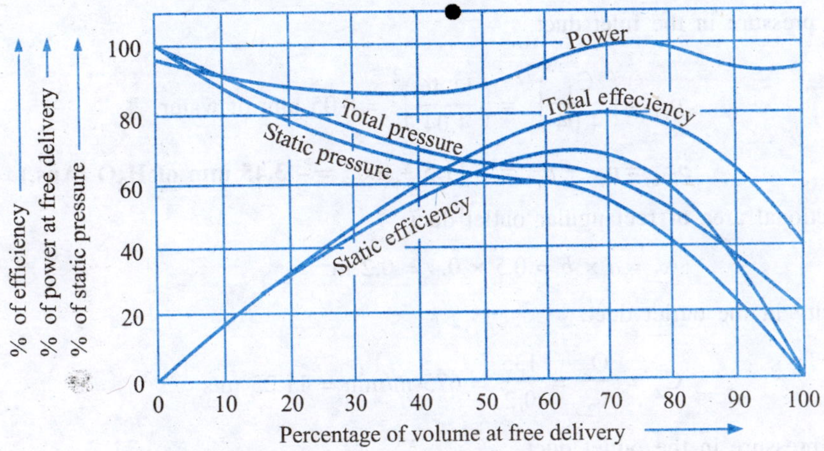

Fig. 17.16. Performance curves for an axial-flow air foil-type fan.

WORKED EXAMPLES

Examples 17.1 *The following data relate to a centrifugal fan:*

Diameter of circular inlet duct .... *0.5 m*

Size of the rectangular duct .... *0.5 m × 0.4 m*

Static pressure at the fan inlet .... *– 11.5 mm*

Static pressure at the fan outlet .... *24 mm of water*

Quantity of air delivered .... *135 m³/min*

Power absorbed .... *1.2 kW*

Assuming standard air density in both ducts and compressibility factor as 1, determine:

(*i*) *Total pressure at fan inlet and outlet,*

(*ii*) *Fan total pressure and fan static pressure, and*

(*iii*) *Fan total efficiency and fan static efficiency.*

Solution: *Given*: D = 0.5 m; a = 0.5 m; b = 0.4 m; p_{s1} = – 11.5 mm of water;

p_{s2} = 24 mm of water; Q = 135 m^3/min; B.P = 1.2 kW;

K_P = 1.

(*i*) **Total pressure at fan inlet and outlet:**

Let, p_{T1} = Total pressure of fan inlet, and

p_{T2} = Total pressure at fan outlet.

Cross-sectional area of circular inlet duct,

$$A_1 = \frac{\pi}{4} D^2 = \frac{\pi}{4} \times (0.5)^2 = 0.1963 \text{ m}^2$$

∴ Velocity of air in the inlet duct,

$$C_1 = \frac{Q}{A_1} = \frac{135}{0.1963} = 687.72 \text{ m/min} = 11.46 \text{ m/s}$$

Velocity pressure in the inlet duct,

$$p_{v1} = \left(\frac{C_1}{4.04}\right)^2 = \left(\frac{11.46}{4.04}\right)^2 = 8.05 \text{ mm of water}$$

$\therefore$ $$\boldsymbol{p_{T1}} = p_{s1} + p_{v1} = -11.5 + 8.05 = \mathbf{-3.45 \text{ mm of } H_2O \text{ (Ans.)}}$$

Cross-sectional area of rectangular outlet duct,

$$A_2 = a \times b = 0.5 \times 0.4 = 0.2 \text{ m}^2$$

$\therefore$ Velocity in the outlet duct,

$$C_2 = \frac{Q}{A_2} = \frac{135}{0.2} = 675 \text{ m/min} = 11.25 \text{ m/s}$$

Velocity pressure in the outlet duct,

$$p_{v2} = \left(\frac{C_2}{4.04}\right)^2 = \left(\frac{11.25}{4.04}\right)^2 = 7.75 \text{ mm of water}$$

$\therefore$ $$\boldsymbol{p_{T2}} = p_{s2} + p_{v2} = 24 + 7.75 = \mathbf{31.75 \text{ mm of water (Ans.)}}$$

(*ii*) **Fan total pressure and fan static pressure:**

Fan total pressure, $(\boldsymbol{p_T})_{\mathbf{fan}} = p_{T2} - p_{T1}$

$$= 31.75 - (-3.45) = \mathbf{35.2 \text{ mm of water (Ans.)}}$$

Fan static pressure, $(p_s)_{fan} = (p_T)_{fan} - p_{v2}$

$$= 35.2 - 7.75 = \mathbf{27.45 \text{ mm of } H_2O \text{ (Ans.)}}$$

(*iii*) **Fan total efficiency and fan static efficiency:**

Total fan air power,

$$(P_{fan})_{total} = \frac{9.81\, Q\, (p_T)_{fan} \times K_P}{60} \text{ watts} \qquad \text{...[Eqn. (17.2)]}$$

$$= \frac{9.81 \times 135 \times 35.2 \times 1}{60} = 776.9 \text{ W, or, } 0.7769 \text{ kW}$$

$$(P_{fan})_{static} = \frac{9.81\, Q\, (p_s)_{fan} \times K_P}{60} \text{ watts} \qquad \text{...[Eqn. (17.3)]}$$

$$= \frac{9.81 \times 135 \times 27.45 \times 1}{60} = 605.9 \text{ W, or, } 0.6059 \text{ kW}$$

$\therefore$ Fan total efficiency, $$(\eta_{fan})_{total} = \frac{(P_{fan})_{total}}{B.P} = \frac{0.7769}{1.2} = 0.6474, \text{ or, } \mathbf{64.74\% \text{ (Ans.)}}$$

Fan static efficiency, $$(\eta_{fan})_{static} = \frac{(P_{fan})_{static}}{B.P.} = \frac{0.6059}{1.2} = 0.5049, \text{ or, } \mathbf{50.49\% \text{ (Ans.)}}$$

Example 17.2 *The following data refer to a fan in which air enters the impeller with a small whirl component in the direction of impeller rotation, but the relative velocity meets the blade tangentially.*

Speed of the fan	*1000 r.p.m*
Quantity of air delivered	*135 m^3/min*
Impeller diameter	*0.75 m*
Diameter at the blade inlet	*0.5 m*
Impeller width at inlet and outlet	*170 mm and 120 mm*
The blade (backward-curved) angles	*25° and 52° with the tangents at the inlet and outlet*
Losses at inlet, in the impeller and in the casing	*72 per cent of the velocity head at impeller outlet*
Velocity head at the fan discharge	*10 per cent of the velocity head at impeller outlet.*

(*i*) *Draw the inlet and outlet velocity triangles and determine the theoretical total head produced by the impeller.*

(*ii*) *Fan static pressure in mm of water.*

Take the mass density of air to be 1.2 kg/m^3 and neglect the effect of blade thickness and interblade circulation.

Solution: *Given:* $N = 1000$ r.p.m.; $Q = 135\ m^3/min = 2.25\ m^3/s$; $D_2 = 0.75$ m;

$D_1 = 0.5$ m; $b_1 = 170$ mm $= 0.17$ m; $b_2 = 120$ mm $= 0.12$ m;

$\alpha_1 = 25°$; $\alpha_2 = 52°$; $\rho_a = 1.2\ kg/m^3$.

(*i*) **Theoretical total head developed by the impeller:**

Let, H = Theoretical total head developed by the impeller.

The inlet and outlet velocity triangles are shown in Fig. 17.17.

The blade velocity at inlet,

$$C_{b1} = \frac{\pi D_1 N}{60} = \frac{\pi \times 0.5 \times 1000}{60} = 26.18 \text{ m/s}$$

The blade velocity at outlet,

$$C_{b2} = \frac{\pi D_2 N}{60} = \frac{\pi \times 0.75 \times 1000}{60} = 39.27 \text{ m/s}$$

Quantity of air delivered,

$$Q = \pi D_1 b_1 C_{f1} = \pi D_2 b_2 C_{f2}$$

∴ Velocity of flow at inlet,

$$C_{f1} = \frac{Q}{\pi D_1 b_1} = \frac{2.25}{\pi \times 0.5 \times 0.17} = 8.43 \text{ m/s, and}$$

Velocity of flow at outlet,

$$C_{f1} = \frac{Q}{\pi D_2\, b_2} = \frac{2.25}{\pi \times 0.75 \times 0.12} = 7.96 \text{ m/s}$$

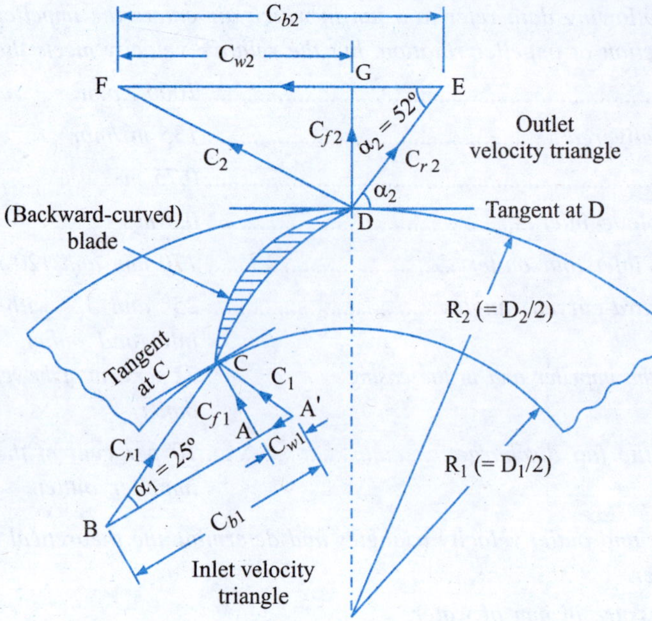

Fig. 17.17

From the *inlet velocity triangle*, we have:

$$\frac{AB}{AC} = \cot \alpha_1 , \quad \text{or,} \quad (C_{b1} - C_{w1}) = C_{f1} \cos 25°$$

or, $\qquad (26.18 - C_{w1}) = 8.43 \cot 25°$

or, $\qquad C_{w1} = 8.1 \text{ m/s}$

From the *outlet velocity triangle*, we have:

$$\frac{GE}{GD} = \cot \alpha_2 = \cot 52° , \quad \text{or,} \quad (C_{b2} - C_{w2}) = C_{f2} \cot 52°$$

or, $\qquad (39.27 - C_{w2}) = 7.96 \cot 52°$

$\therefore \qquad C_{w2} = 33.05 \text{ m/s}$

The theoretical total head developed by the impeller,

$$H = \frac{C_{w2} C_{b2} - C_{w1} C_{b1}}{g} \text{ m of air} \qquad \text{...[Eqn. (17.7)]}$$

$$= \frac{33.05 \times 39.27 - 8.1 \times 26.18}{9.81} = \mathbf{110.68 \text{ m of water (Ans.)}}$$

(*ii*) **Fan static pressure, $(p_{s2})_{fan}$:**

Total pressure developed by the impeller or fan,

$$(p_T)_{fan} = \rho_a H = 1.2 \times 110.68 = 132.82 \text{ mm of water}$$

$$= 132.82 \times 9.81 = 1302.96 \text{ N/m}^2 \qquad (\because 1 \text{ mm of water} = 9.81 \text{ N/m}^2)$$

From the *outlet velocity triangle*, absolute velocity of air leaving the blade,

$$C_2 = \sqrt{(C_{w2})^2 + (C_{f2})^2} = \sqrt{(33.05)^2 + (7.96)^2} = 34 \text{ m/s}$$

$\therefore$ Velocity head or velocity pressure at the impeller outlet,

$$H_{v2} = \frac{\rho_a (C_2)^2}{2} = \frac{1.2 \times (34)^2}{2} = 693.6 \text{ N/m}^2$$

$$= \frac{693.6}{9.81} = 70.7 \text{ mm of water}$$

$$\left[\textit{Alternatively} : H_{v2} = \left(\frac{C_2}{4.04}\right)^2 = \left(\frac{34}{4.04}\right)^2 \simeq 70.7 \text{ mm of water} \right]$$

Losses at inlet, in the impeller and in the casing

$= 72\%$ of velocity head at outlet ...(Given)

$= 0.72\ H_{v2} = 0.72 \times 70.7 = 50.9$ mm of water

Velocity head or velocity pressure at the fan discharge

$= 10\%$ of velocity head at outlet

$= 0.1 \times H_{v2} = 0.1 \times 70.7 = 7.07$ mm of water

Applying Bernoulli's equation to fan inlet and outlet, we have:

Total pressure at fan inlet + Total pressure developed by the impeller

= Losses + Total pressure at fan outlet

Since the total pressure at the fan inlet is zero and the total pressure at the fan outlet $[(p_T)_{fan}]$ is the sum of static pressure at fan outlet or fan static pressure $(p_{s2})_{fan}$ and the velocity pressure at fan outlet, therefore,

$$0 + 132.82 = 50.9 + (p_{s2})_{fan} + 7.07$$

$\therefore$ $(p_{s2})_{fan}$ **= 74.85 mm of water (Ans.)**

Example 17.3 *A fan 650 mm in diameter is running at 1200 r.p.m., delivering 120 m³/min of air at 14°C against 72 mm of water of total pressure. If its total efficiency is 82 per cent, determine the volume of air delivered, total pressure developed and power consumed, if*

(a) *Air temperature is 40°C,*

(b) *Air temperature is 40°C and the fan speed is increased to 1500 r.p.m., and*

(c) *Conditions are same as in (b) but a 500 mm diameter, geometrically similar fan is used.*

Solution: *Given:* $D_1 = 650$ mm $= 0.65$ m; $N_1 = 1200$ r.p.m.; $Q_1 = 120$ m³/min;

$t_1 = 14°C$; $(p_{T1})_{fan} = 72$ mm of water;

$(\eta_{fan})_{total} = 82\%$; $t_2 = 40°C$.

Let us firstly find the densities at temperatures 14°C and 40°C.

Let, ρ_{a1} = Density of air at 14°C, and

ρ_{a2} = Density of air at 40°C.

For the constant barometric pressure, the density of air at 40°C is given by:

$$\rho_{a2} = \rho_{a1} \left(\frac{273 + t_1}{273 + t_2}\right)$$

$$\therefore \quad \frac{\rho_{a1}}{\rho_{a2}} = \frac{273 + t_2}{273 + t_1} = \frac{273 + 40}{273 + 14} = 1.09$$

(*a*) **When air temperature is 40°C:**

Let, Q_2 = Volume of air delivered,

$(p_{T2})_{fan}$ = Total pressure developed, and

P_2 = Power consumed.

$$\frac{Q_1}{N_1 (D_1)^3} = \frac{Q_2}{N_2 (D_2)^3} \qquad \text{...[Eqn. (17.11)]}$$

Since there is no change in N and D, therefore,

$$Q_2 = Q_1 = \mathbf{120\ m^3/min\ (Ans.)}$$

Now, $$\frac{(p_{T1})_{fan}}{(p_{T2})_{fan}} = \frac{\rho_{a1}}{\rho_{a2}} \qquad \text{..(D and N unchanged)} \qquad \text{...[Eqn. (17.13)]}$$

$$\therefore \quad (p_{T2})_{fan} = (p_{T1})_{fan} \times \frac{\rho_{a2}}{\rho_{a1}}$$

$$= 72 \times \frac{1}{1.09} = 66.055 \text{ mm of water}$$

$$= 66.055 \times 9.81 = \mathbf{648\ N/m^2\ (Ans.)}$$

Power consumed, $$P_2 = \frac{Q_2 \times (p_{T2})_{fan}}{\eta} = \frac{(120/60) \times 648}{0.82} = 1580 \text{ W or } \mathbf{1.58\ kW\ (Ans.)}$$

(*b*) **When the temperature is 40°C and the speed in increased to 1500 r.p.m.:**

Let, Q_3 = Volume of air delivered,

$(p_{T3})_{fan}$ = Total pressure developed,

N_3 = New (increased) speed = 1500 r.p.m. (Given), and

P_3 = Power consumed.

Now, $$\frac{Q_3}{N_3} = \frac{Q_2}{N_2} \qquad \text{...[Eqn. (17.11)]}$$

$$\therefore \quad Q_3 = Q_2 \times \frac{N_3}{N_2} = 120 \times \frac{1500}{1200} = \mathbf{150\ m^3/min\ (Ans.)} \qquad (\because N_2 = N_1)$$

Also, $$\frac{(p_{T3})_{fan}}{(N_3)^2} = \frac{(p_{T2})_{fan}}{(N_2)^2} \qquad \text{...(D and } \rho \text{ unchanged)} \qquad \text{...[Eqn. (17.13)]}$$

$$\therefore \quad (p_{T3})_{fan} = (p_{T2})_{fan} \times \left(\frac{N_3}{N_2}\right)^2 = 6.055 \left(\frac{1500}{1200}\right)^2 = \mathbf{103.2\ mm\ of\ water}$$

and $$\frac{P_3}{(N_3)^3} = \frac{P_2}{(N_2)^3} \qquad \text{...(D and } \rho \text{ unchanged)} \qquad \text{...[Eqn. (17.14)]}$$

$$\therefore \quad P_3 = P_2 \left(\frac{N_3}{N_2}\right)^3 = 1.58 \times \left(\frac{1500}{1200}\right)^3 = \mathbf{3.08\ kW\ (Ans.)}$$

(*c*) **Conditions are same as in (*b*) but a 500 mm diameter, geometrical similar fan is used:**

Let, Q_4 = Volume of air delivered,

$(p_{T4})_{fan}$ = Total pressure developed,

D_4 = New impeller diameter = 500 mm or 0.5 m, and

P_4 = Power consumed.

Now, $$\frac{Q_4}{(D_4)^3} = \frac{Q_3}{(D_3)^3} \quad \text{(N unchanged)} \quad \text{...[Eqn. (17.11)]}$$

$$\therefore \quad Q_4 = Q_3 \times \left(\frac{D_4}{D_3}\right)^3 = 150 \times \left(\frac{0.5}{0.65}\right)^3 = \mathbf{68.28\ m^3/min\ (ans.)}$$

Also, $$\frac{(p_{T4})_{fan}}{(D_4)^2} = \frac{(p_{T3})_{fan}}{(D_3)^2} \quad \text{...(N and } \rho \text{ unchanged)} \quad \text{...[Eqn. (17.13)]}$$

$$\therefore \quad (p_{T4})_{fan} = (p_{T3})_{fan} \times \left(\frac{D_4}{D_3}\right)^2 = 103.2 \left(\frac{0.5}{0.65}\right)^2 = \mathbf{61.06\ mm\ of\ water\ (Ans.)}$$

and, $$\frac{P_4}{(D_4)^5} = \frac{P_3}{(D_3)^5} \quad \text{(N and } \rho \text{ unchanged)} \quad \text{...[Eqn. (17.14)]}$$

$$\therefore \quad P_4 = P_3 \left(\frac{D_4}{D_3}\right)^5 = 2.47 \left(\frac{0.5}{0.65}\right)^5 = \mathbf{0.665\ kW\ (Ans.)}$$

HIGHLIGHTS

1. A *fan* is a rotodynamic machine which propels air or other gas continuously.
2. A *centrifugal* or *radial flow* fan is one in which air enters the impeller axially and is discharged radially from the impeller. When air flows parallel to the axis of impeller, it is called an *axial-fan*.
3. *Total fan air power*, based on fan total pressure $[(p_T)_{fan}]$ is given by:

$$(P_{fan})_{total} = \frac{9.81\, Q\, (p_T)_{fan} \times K_P}{60} \text{ watts}$$

where, Q = Total quantity of air flowing at the fan inlet, m³/min, and

K_P = Compressibility coefficient.

Static fan air power based on fan static pressure $[(p_s)_{fan}]$ is given by:

$$(P_{fan})_{static} = \frac{9.81\, Q\, (p_s)_{fan} \times K_P}{60} \text{ watts}$$

4. *Total fan efficiency* $(\eta_{fan})_{total}$ is the ratio of the total fan air power to the driving power (or brake power) required at the fan shaft. It is also called "*mechanical efficiency*" of the fan.

Static fan efficiency $(\eta_{fan})_{static}$ is defined as the ratio of the static fan air power to the input or brake power.

5. The *specific speed* (N_s) of a centrifugal fan is defined as the speed of a geometrically similar fan which would deliver 1 m³ of air per second against a head of 1 m of air.
6. *System characteristic* is the change in pressure loss or resistance with the change in flow rate.

7. *When fans are in series:*

$$Q = Q_1 = Q_2 = Q_3 \ldots$$

$$(p_T)_{fan} = (p_T)_{fan-1} + (p_T)_{fan-2} + (p_T)_{fan-3} + \ldots$$

When fans are in parallel:

$$Q = Q_1 + Q_2 + Q_3 + \ldots$$

$$(p_T)_{fan} = (p_T)_{fan-1} = (p_T)_{fan-2} = (p_T)_{fan-3} = \ldots$$

OBJECTIVE TYPE QUESTIONS

Choose the Correct Answer:

1. The axial flow fans are particularly suitable for handling
(*a*) large volumes of air at relatively low pressures
(*b*) small volumes of air at relatively low pressures
(*c*) large volumes of air at relatively high pressures
(*d*) small volumes of air at relatively high pressures.

2. The capacity of a fan is .... cube of the impeller diameter.
(*a*) directly proportional to (*b*) inversely proportional to
(*c*) either (*a*) or (*b*) (*d*) none of the above.

3. In axial flow fans,
(*a*) the air flows parallel to the axis of impeller
(*b*) the air flows perpendicular to the axis of impeller
(*c*) the air may flow either parallel or perpendicular to the axis of impeller
(*d*) none of the above.

4. The power of a fan is directly proportional to
(*a*) $D^{2.5}$ (*b*) $D^{3.5}$ (*c*) D^4 (*d*) D^5.
where D = Impeller diameter.

5. The air at inlet of a fan is .... atmospheric pressure.
(*a*) above (*b*) below (*c*) equal to (*d*) any of these.

6. The fan total pressure is the algebraic .... between the total pressure at the fan outlet and the total pressure at the fan inlet.
(*a*) sum (*b*) difference (*c*) either of these (*d*) none of these.

7. The power of a fan is directly proportional to
(*a*) N (*b*) $N^{2.5}$ (*c*) N^3 (*d*) $N^{4.5}$
where, N = Fan speed.

8. The air guide vanes are sometimes installed in axial flow fans ir order to
(*a*) increase the static pressure (*b*) eliminate spiral flow of discharge air
(*c*) reduce high frequency sound generation (*d*) all of the above.

9. Two fans that are of different sizes, but have the same basic shape to their performance curves, are called
(*a*) base fans (*b*) tube axial fans
(*c*) geometrically similar fans (*d*) multi-stage fans.

ANSWERS

1. (*a*)	**2.** (*a*)	**3.** (*a*)	**4.** (*d*)	**5.** (*b*)
6. (*b*)	**7.** (*c*)	**8.** (*b*)	**9.** (*c*).	

THEORETICAL QUESTIONS

1. What is a 'fan'?

2. What is the function of a fan in an air-conditioning system?

3. How are fans classified?

4. Describe briefly the following types of centrifugal fans:
(*i*) Forward-curved fans, (*ii*) Radial blade fan,
(*iii*) Backward-curved fans.

5. Explain briefly the following three types of axial-flow fans:
(*i*) Propeller fans, (*ii*) Tube axial fans, and
(*iii*) Vane axial fans.

6. Define the following:
(*i*) Fan static pressure, (*ii*) Fan velocity pressure, and
(*iii*) Total fan pressure.

7. Explain briefly the following:
(*i*) Total fan air power, and (*ii*) Static fan air power.

8. Define the following:
(*i*) Total fan efficiency, and (*ii*) Static fan efficiency.

9. Define specific speed of a centrifugal fan. Derive its expression also.

10. What do you understand by a geometrically similar fan?

11. What is "system characteristic"?

12. Discuss briefly the procedure for establishing the fan performance to make fan-performance curves.

UNSOLVED EXAMPLES

1. The following data relate to a centrifugal fan:
Diameter of circular inlet duct .... 0.45 m
Size of the rectangular duct .... 0.45 m × 0.375 m
Static pressure at the fan inlet .... –12.5 mm of water
Static pressure at the fan outlet .... 25 mm of water
Quantity of air delivered .... 115 m^3/min
Power absorbed .... 1 kW
Assuming standard air density in both ducts and compressibility factor as 1, determine:
(*i*) Total pressure at fan inlet and outlet.

(*ii*) Fan total pressure and fan static pressure.

(*iii*) Fan total efficiency and fan static efficiency.

[**Ans.** (*i*) – 3.7 mm of water; 32.8 mm of water; (*ii*) 36.5 mm of water, 28.7 mm of water; (*iii*) 68.63%, 53.96%]

2. The following data refer to a fan in which air enters the impeller with a small whirl component in the direction of impeller rotation, but the relative velocity meets the blade tangentially.

Speed of the fan .... 960 r.p.m.

Quantity of air delivered .... 2 m^3/s

Impeller diameter .... 0.7 m

Diameter at the blade inlet .... 0.48 m

Impeller width at inlet and outlet .... 160 mm and 110 mm

The blade (backward-curved) angles .... 22.5° and 50° with the tangents at the inlet and outlet

Losses at inlet, the impeller and in the casing .... 70 per cent of the velocity head at impeller outlet

Velocity head at the fan discharge .... 10 per cent of the velocity head at impeller outlet.

(*i*) Draw the inlet and outlet velocity triangles and determine the theoretical total head produced by the impeller.

(*ii*) Fan static pressure in mm of water.

Take mass density of air to be 1.2 kg/m^3 and neglect the effect of blade thickness and interblade circulation.

[**Ans.** (*i*) 91.76 mm of water; (*ii*) 67.44 mm of water]

❑❑❑

Chapter 18

Load Estimation

18.1 INTRODUCTION

- *The primary requirement of cooling or heating equipment is that these must be able to remove or add heat at the rate at which it is produced, or removed and maintain the given comfort conditions in the room.*
- The proportion of sensible heat to latent heat decides the slope of '*sensible heat ratio*' or '*enthalpy humidity difference ratio*' and hence also the condition at which the conditioned air must enter the room *because it lies on the line drawn from the room condition at the slope of above ratios.*
- The cost of the equipment will be quite high if it is designed for maximum heating or cooling loads. The owner may be prepared to tolerate some uncomfort for a short time to save capital cost investment. If the equipment is designed for average loads, there may be long periods of uncomfort and the equipment may soon become unpopular. Hence the *capacity equipment must be estimated at a value, which wisely accounts for the physical and economical comfort of the owner.* It is only the exact and detailed estimation of heating and cooling loads which helps in this decision.

  The ***Estimation of load*** *involves the following variables:*

  1. Magnitude and direction of wind velocity.
  2. Outside humidity and temperature.
  3. Nature of construction, materials used.
  4. Orientation of openings, windows and doors.
  5. Periods of occupancy and the number of persons in the room, activities of the persons etc.

- One has to be satisfied with the estimation under assumed conditions keeping the variables fixed and allowances must be made for, when the actual conditions are different from those assumed. Thus to make precise load calculation is unfortunately not simple and some of the sources of the load are difficult to predict and evaluate. For a practising engineer it is not possible to make detailed calculations and therefore he must consult latest editions of 'Heating, ventilation and air conditioning guide' and make quick easy estimates.

18.2 COOLING–LOAD ESTIMATE

For air-conditioning the cooling-load can be classified as follows:

1. **Room load**—which falls on the *room directly.*
2. **Total load**—which falls on the *air-conditioning apparatus.*

1. **Room Load**

(*a*) Room sensible heat (RSH):

(*i*) Solar and transmission heat gain through walls, roof, etc.

(*ii*) Solar and transmission heat gain through glass.

(*iii*) Transmission gain through partition walls, ceiling, floor etc.

(*iv*) Infiltration.

(*v*) Internal heat gain from people, power, lights, appliances etc.

(*vi*) Additional heat gain not accounted above, safety factor etc.

(*vii*) Supply duct heat gain, supply duct leakage loss and fan power.

(*b*) Room Latent Heat (RLH):

(*i*) Infiltration.

(*ii*) Internal heat from people, steam, appliances etc.

(*iii*) Vapour transmission.

(*iv*) Additional heat gain not accounted above, safety factor etc.

(*v*) Supply duct leakage loss.

It may be noted that if we add '*by-passed air load*' to room sensible heat, we shall get '*effective room sensible heat*' (ERSH). Similarly adding 'by-passed air load' to room latent heat we shall get '*effective room latent heat*' (ERLH).

2. **Grand total load** (on air-conditioning apparatus)

(*a*) Sensible heat:

(*i*) Effective room sensible heat.

(*ii*) Sensible heat of the outside air that is not by-passed.

(*iii*) Return duct heat gain, return duct leakage gain, dehumidifier pump power and dehumidifier and piping losses.

Total sensible heat (TSH) is obtained by adding items (*i*) to (*iii*).

(b) Latent heat

(i) Effective room latent heat.

(ii) Latent heat of outside air which is not by-passed.

(iii) Return duct leakage gain.

Total latent heat (TLH) is obtained by adding items (i) to (iii).

Grand total heat (GTH) = Total sensible heat (TSH) + Total latent heat (TLH)

i.e., $$GTH = TSH + TLH \quad ...(18.1)$$

18.3 HEATING-LOAD ESTIMATE

Heating-load estimate is prepared *on the basis of 'maximum probable heat loss' of the room or space to be heated.* Following points should be considered while making heat-load calculations:

1. **Transmission heat loss.** The transmission heat loss from walls, roof, etc., is calculated *on the basis of just the outside and inside temperature difference.*
2. **Solar radiation.** Normally there is no solar radiation present and hence no solar heat gain at the time of the peak load which normally occurs in the early morning hours.
3. **Internal heat gains.** The heating requirement is reduced due to internal heat gains from occupants, lights, motors and machinery etc.

18.4 SOLAR RADIATION

- The *solar radiation intensity* normal to the sun's rays incident upon a plane surface situated in the limits of the earth's atmosphere, varies with the time of the year as the distance of the earth from the sun changes. Its value when the earth is at its mean distance from the sun is called the '*solar constant*'. The normal value of the solar constant is assumed as 5045 kJ/m^2-h. Radiation received at the surface of the earth is much less because, much of it, while passing through earth's atmosphere, is scattered and absorbed by dust and vapour particles and the gases in the atmosphere.
- The solar heat reaches part of the earth's surface in the form of two radiations:

1. **Beam or direct radiation.** The part of the sun's radiation which travels through the atmosphere and *reaches the earth's surface directly* is called *Beam or direct radiation*. It *is maximum when the surface is normal to the sun's rays.* The intensity of the radiation can be increased or decreased by *changing the orientation of the surface.*
2. **Diffuser sky radiation.** A large part of the sun's radiation is *scattered, reflected back into space and absorbed by the earth's atmosphere.* A part of this radiation is *re-radiated* and reaches the earth's surface uniformly from all directions. It is called *diffuse or sky radiation.* It *does not normally change with orientation of the surface.*

The total solar radiation reaching a surface is equal to the *sum of the direct and diffuse radiations.*

18.5 SOLAR HEAT GAIN THROUGH GLASS

- Glass which is major material of most buildings, provides the most direct route for entry of solar radiation. For these reasons, the proper estimation of heat gain through glass is necessary.
- *Heat transmitted through a glass surface depends on the wavelength of radiation and physical and chemical characteristics of the glass.* Part of the radiation is *absorbed*, part is *reflected* and the rest is *transmitted.* Glass is opaque to the radiant energy emitted from sources below 200°C. Thus glass has high *transmittivity for short wavelength and low transmitivity for long wave length radiation.*
- *Direct solar heat gain* can be *reduced* by using different types of glass, glass construction and shades as given below:

    (*i*) Double pan glass reduces the solar heat by 10% to 20%.

    (*ii*) Special heat absorbing glass reduces the solar heat by 25%.

    (*iii*) Stained glass can reduce it upto 65% depending upon its colour.

    (*iv*) Shading devices installed on the outside of windows reduce sun load upto 15%.

    (*v*) Ventilation blinds and curtain shades reduce it by 30% to 35%.

18.6 HEAT THROUGH BUILDING STRUCTURES (THERMAL BARRIERS)

Refer to Fig. 18.1. Transfer of heat by conduction (Q) from outside to inside under steady state due to the temperature difference is given by the equation:

$$Q = UA\,(t_0 - t_i) \quad ...(18.2)$$

where, Q = Heat transfer rate,

U = Overall co-efficient of heat transfer,

A = Area, m^2,

t_0 = Outside temperature, and

t_i = Inside temperature.

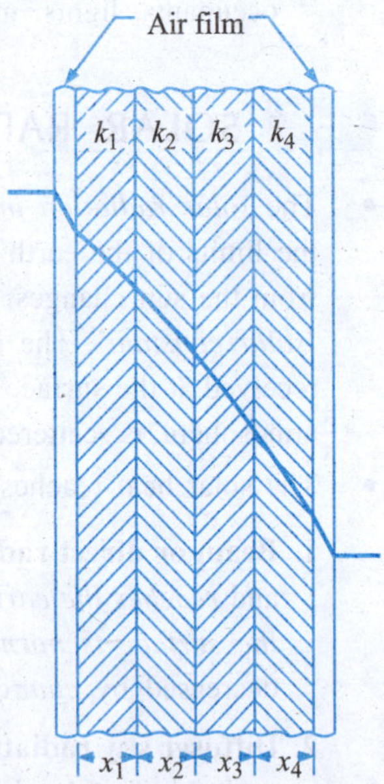

Fig. 18.1

Overall thermal resistance,

$$\frac{1}{U} = \frac{1}{f_0} + \frac{x_1}{k_1} + \frac{x_2}{k_2} + ... + \frac{1}{f_i} \quad ...(18.3)$$

where, f_0 = Film co-efficient on outside wall,

x = Thickness of material,

K = Conductivity of material, and

f_i = Film co-efficient on inside wall.

Subscripts 1, 2 etc., refer to successive layers of materials which make up the walls.

The film *co-efficient varies directly with wind velocity.*

While considering the steady state conduction, the outdoor air temperature should be assumed to be the mean average during the day. This assumption of steady state is justified only in cases where the temperature difference inside and outside is very large and fluctuation of outside air temperature is small fraction of the temperature difference between inside and outside. For the cases *where fluctuation in periodic variation is large compared to temperature difference inside and outside, the equivalent temperature difference should be used in eqn. (18.2).*

Loss or gain of heat from adjacent rooms through partitions and from the ground through floor may be calculated by using the following equation:

$$Q = UA\,(t' - t_i) \quad \text{...(18.4)}$$

where, U and A refer to the separating partition or the floor and t' refers to the air temperature of adjacent room or ground.

Table 18.1. Overall coefficient of heat transmission (U) for structures with a wind velocity of 24 km/h outside

Structure	*Overall coefficient of heat transmission (U) in W/m² K*
Brick wall, 20 cm, bare	2.84
Brick wall, 20 cm, plaster one side on brick	2.61
Brick wall, 20 cm, plaster one side on metal lath-furred	1.82
Brick wall, 40 cm, bare	1.59
Brick wall, 40 cm, plaster one side on brick	1.53
Brick wall, 40 cm, plaster one side on metal lath-furred	1.20
Hollow tile, stucco exterior, 20 cm, bare	2.27
Hollow tile, stucco exterior, 20 cm plaster on metal lath-furred	1.59
Hollow tile, stucco exterior, 30 cm bare	1.70
Hollow tile, stucco exterior 30 cm, plaster on metal lath-furred	1.25
Cinder blocks, 24 cm bare	2.38
Cinder blocks 20 cm, plaster one side on metal lath-furred	1.59
Concrete blocks, 20 cm bare	3.18
Clap-board frame construction, plaster on wood lath	1.42
Wood shingle frame construction, plaster on wood lath	1.42
Stucco frame construction, plaster on wood lath	1.70
Brick veneer frame construction, plaster on wood lath	1.53

Table 18.2. Thermal conductivity (k) of building materials.

Material	*Thermal conductivity (k) in W/m K*
Brick, low density	8.66
Brick, high density	15.93
Cement mortar	20.82
Cement plaster, typical	20.82
Concrete, typical	20.82

Material	*Thermal conductivity (k) in W/m K*
Concrete, typical fibre gypsum, 87.5% gypsum, 12.5% wood chips	1.61
Sand and gravel	21.86
Lime stone	18.72
Cinder	8.50
Cinder, boiler (12.5 mm to 19 mm)	2.13
Tile or terazzo, typical	20.82
Asbestos building board	4.71
Gypsum, between layers of heavy paper	2.44
Gypsum, plaster, typical	5.72
Wood across grain, typical	1.73
Balsa	0.657
Maple, hard, 16% moisture	2.0
Pine, yellow short leaf, 16% moisture	1.45

Overall heat transfer coefficients for structures, thermal conductivity of building materials and insulating materials, conductance of building materials and film or surface conductance for air film and air spaces are given in Tables 18.1 to 18.5 respectively.

Table 18.3. Thermal conductivity (k) of insulating materials

Material	*Thermal conductivity (k) in W/m K*
Asbestos packed	2.812
Asbestos loose	1.852
Asbestos, paper, thin layers organic binder	0.85
Asbestos mill board	1.454
Asbestos wood	4.85
Asphalt roofing (felt)	1.212
Balsa	0.607
Cement wood (saw dust and Portland cement)	1.683
Charcoal (hard woods) coarse	0.617
Cork board, typical	0.468 to 0.486
Cork regranulated, coarse	0.536
Cork regranulated	0.468
Cotton	0.676
Diatomaceous earth	0.460
Eel grass	0.412
Glass wool, high grade	0.388 to 0.460
Glass wool, commercial grade	0.444
Hair felt, not compressed	0.444
Insulation boards, various fibres	0.553 to 0.657
Kapok, loosely packed	0.416
Planer shavings, various woods	0.692
Rock wool	0.468 to 0.486
Rubber, expanded	0.364
Saw dust, various woods	0.710

Table 18.4. Conductance of building materials and construction for thickness indicated

Material	*Conductance in W/m² K*
Plaster board 9 mm thick	21.17
Plaster board 12.5 mm thick	16.14
Roofs:	
Asphalt, composition or prepared roofing	36.85
Shingles, asbestos	34.05
Shingles, asphalt	36.65
Shingles, slate	58.96
Metal lath and plaster, total thickness 12.5 mm	24.96
Wood lath and plaster, total thickness 18 mm	14.20
For shealthing 25 mm and building paper	4.88
For shealthing 25 mm building paper and yellow pine lap siding	2.83
For shealthing 25 mm building paper and stucco	4.65
Air space over 18 mm faced ordinary building materials	6.15
Typical hollow clay tile, 10 cm	5.67
Typical hollow clay tile, 15 cm	3.64
Typical hollow clay tile, 20 cm	3.40
Typical hollow clay tile, 25 cm	3.29
Typical hollow clay tile, 30 cm	2.277
Typical hollow clay tile, 40 cm	1.762
Concrete block, sand and gravel, typical, 20 cm	5.675
Concrete block, sand and gravel, typical, 30 cm	4.547
Concrete block, cinder aggregate, typical, 20 cm	3.404
Concrete block, cinder aggregate, typical, 30 cm	3.006
Surface coefficient, outside for 24 km/h wind	34.05

Table 18.5. Film or surface conductance of air film and air spaces

Material	*Surface position*	*Thermal conductance (f) in W/m² K*
Air film (surface)		
Still air (f_i) heat flowing up	Horizontal	9.25
	Sloping, 45°	9.1
Still air, heat flowing down	Horizontal	6.13
	Sloping 45°	7.5
Still air, heat flowing horizontal	Vertical	8.3
Wind, 24 km/h (f_0)	Any position	34.0
Wind, 12 km/h (f_0)	Any position	22.7

Material	*Surface position*	*Thermal conductance (f) in W/m² K*
Air spaces bounded by structural material		
Heat flowing up	Horizontal, 19 to 100 mm	6.7
	Sloping 45°, 19 to 100 mm	6.3
Heat flowing horizontal	Vertical, 19 to 100 mm	5.85
Heat flowing down	Sloping 45°, 19 to 100 mm	5.5
	Horizontal, 19 mm	5.56
	Horizontal, 38 mm	4.94
	Horizontal, 100 mm	4.6

Note: The above values of thermal conductance are for a mean temperature of 10° C to 32°C.

18.7 INFILTRATION

- Both sensible and latent heat losses result from the *leakage* of outside air into the building through cracks and openings, displacement of an equal volume of room air. The infiltration depends on the length and width of cracks around the windows, doors, walls, partitions and floors. The nature of cracks depends upon the building construction workmanship and the building age and deterioration.
- In order to estimate air leakage through *window* and *door cracks* the following two methods may be used:

 1. Air change method.
 2. Crack length method.

1. **Air change method**

In this method, the infiltration is based on the volume of the room and is given by the number of times air changes to fill the volume of the room. The sensible heat loss/cooling load due to such air changes and other infiltrations can be calculated as follows:

$$Q = \rho\, Vc\,(t_0 - t_i) \qquad ...(18.5)$$

where, ρ = Density of air, kg/m³,

V = Rate of air infiltrations or changes in m³/h

c = Specific heat of air (= 1.026 kJ/kg-°C),

t_0 = Outside air temperature, and

t_i = Inside air temperature.

- *Air change method is purely arbitrary and based on the engineering practice. It assumes that infiltration air replaces the air in the room in a certain time. It should be used only when it is not possible to use crack method due to uncertainty of the crack length. Air change method is convenient for reception halls and vestibules, where air change is very frequent.*

2. **Crack length method**

- Crack length method is *more accurate and logical.* This method gives volume of air infiltrated per metre of crack length.
- *The crack method is used in case of windows and the air change method is more convenient to use in the case of doors.* It is usually used where *greater accuracy is required.*

The amount of infiltrated air ($V_{inf.}$) through windows and walls may be found as follows:

$$V_{inf} = \frac{L \times B \times H \times A_c}{60} \text{ m}^3/\text{min}$$

where, L = Room length, m,
B = Room breadth, m,
H = Room height, m, and
A_C = Air changes per hour.

The total room infiltration air for an entire building is taken one-half of the above calculated value because infiltration takes place on the windward side of a building.

The Table 18.6 given below, shows the number of air changes per hour for a variety of room types and exposures.

Table 18.6. Number of air changes per hour

S. No.	Type of room or building	Number of air changes per hour (A_C)
1.	*Rooms with no windows or outside doors*	0.5 to 0.75
2.	*Rooms, one wall exposed*	1
3.	*Rooms, two walls exposed*	1.5
4.	*Rooms, three walls exposed*	2
5.	*Rooms, four walls exposed*	2
6.	*Entrance halls*	2 to 3
7.	*Reception halls*	2
8.	*Bath rooms*	2

For each person passing through a door leading to the outside or to an unconditioned space, the values given in the following table for the door infiltration should be added to the infiltration air through windows and walls, in order to find the total building infiltration.

Table 18.7. Infiltrated air

S. No.	Usage of door	Infiltrated air for 1.8 metres revolving door in m^3 per person per passage	
		Freely revolving door	Door equipped with brake
1.	Infrequent	2.5	2
2.	Average	2	1.75
3.	Heavy	1.5	1.25

- For rooms with weather stripped windows or storm sash, use 1/2 of the values given in Table 18.6 but it should never be less than 0.5.
- The figures in Table 18.7 are based on the assumption that there is no wind pressure and the swinging doors are in use in one wall only. Any swinging door in other walls should be kept closed to ensure air-conditioning in accordance with the recommended standards.
- The door infiltration for a 0.9 m swinging door be taken as 3 m^3/min.

Chimney or stack effect

Due to differences between temperatures and humidities, differences in the densities of air between the outside and inside of the buildings are produced. Consequently pressure differences occur causing flow of air known as ***Chimney or stack effect***. When the inside temperature is lower than the outside, the stack effect produces positive inside pressure at lower levels and negative inside pressure at higher levels. As a result, the outward flow of air takes place at lower levels and the inward flow at higher levels, with the neutral zone in the middle. The reverse is true when inside is at a higher temperature than the outside. Thus in '*summer*' *infiltration is at the top and exfiltration at the bottom.* Similarly in '*winter*' *infiltration is at the bottom and exfiltration at the top.*

18.8 INTERNAL HEAT GAINS

The sensible and latent heat gains due to occupants, lights, appliances, machines, piping etc., within the conditioned space form the components of *internal heat gains*.

18.8.1 Heat Load of Occupants

The occupants in a conditioned space give out heat at a metabolic rate that more or less depends on their rate of working. The relative proportion of the sensible and latent heats given out, however, depends on the ambient dry bulb temperature (DBT). The lower the DBT, the greater the heat given out as sensible heat. The usual problem in calculating the occupancy load lies in the *estimation of the exact number of people present*. Heat gain from occupants (in watts) is given in Table 18.8.

Table 18.8. Heat gain from occupants (in watts)

Degree of activity (1)	*Typical application* (2)	*Total heat Adults-Male* (3)	*Total heat adjusted* (4)	*Sensible heat* (5)	*Latent heat* (6)
1. Seated-at rest	Theatre-Matinee	114	97	53	44
	Theatre-Evening	114	102	57	45
2. Seated-very light work	Offices, Hotels, Apartments	132	117	57	67
3. Moderately active office work	Offices, Hotels, Apartments	139	132	59	73
4. Standing, light work, walking slowly	Department store, Retail store	161	132	59	73
5. Walking, seated, standing, walking slowly	Drug store, Bank	161	146	58	88

Degree of activity (*1*)	*Typical application* (2)	*Total heat Adults-Male* (*3*)	*Total heat adjusted* (*4*)	*Sensible heat* (*5*)	*Latent heat* (*6*)
6. Sedentary work	Restaurant	144	161	64	97
7. Light bench work	Factory	234	220	64	155
8. Moderate dancing	Dance hall	264	249	72	177
9. Walking 4.8 km/h, Moderately heavy work	Factory	293	293	88	205
10. Bowling	Bowling alley	440	425	136	288
11. Heavy work	Factory	440	425	136	288

Note: The above values are based on 27°C dry bulb temperature. For 25°C room dry bulb, the total heat remains the same, but the sensible heat values should be increased approximately 10 per cent and the latent heat values decreased accordingly.

18.8.2 Electrical Load

Electrical load consists of lights, heaters, rheostats and other heat dissipating electrical devices.

Lighting load. Electrical lights generate a sensible heat equal to the amount of the electrical power consumed. Most of the energy is liberated as heat and the rest as light which also eventually becomes heat after multiple reflections. As a rough calculation, one may use the lighting load equal to 33.5 W/m^2 to produce a lighting standard of 540 lumens/m^2 in an office space. After the wattage is known the heat gain may be calculated as follows:

Incandescent : Q = total watts,

Fluorescent : $Q = 1.25 \times$ total watts.

Appliance load:

- Both sensible and latent heats are contributed by most appliances. The latent heat produced depends on the function the appliances perform, such as drying, cooking etc.
- *Gas appliances* produce additional moisture as a product of combustion. Such loads can be considerably reduced by providing properly designed loads with a positive exhaust system of suction over the appliances.
- *Electrical motors* contribute sensible heat to conditioned space. A part of the power input is directly converted into heat due to the inefficiency of the motor and is disspated through the frame of the motor. This power is (Input) (1-motor efficiency). The rest of the power input is utilised by the driven mechanism for doing work which may or may not result in a heat gain to the space. It depends whether the energy input goes outside or to the conditioned space.

Heat gain from appliances without hoods (in watts) is given in Table 18.9.

Table 18.9. Heat gain from appliances without hoods (in watts)

Appliances	Electrical		Gas	
	Sensible	Latent	Sensible	Latent
(1)	(2)	(3)	(4)	(5)
1. Coffee brewer, $2\frac{1}{4}$ litres	264	64	396	103
2. Coffee warmer, $2\frac{1}{4}$ litres	68	18	117	29
3. Coffee brewer with tank, 20 litres	1406	352	2110	528
4. Coffee urn, $13\frac{1}{2}$ litres	645	436	733	733
5. Coffee urn, $22\frac{1}{2}$ litres	996	674	1143	1143
6. Egg boiler, 2 cups	352	234	—	—
7. Food warmer with plate warmer, per sq m of top surface	1140	1140	—	—
8. Food warmer only per sq m of top surface	650	1140	2756	1401
9. Fry kettle per sq m of fry area	1026	1465	1758	1172
10. Griddle, per sq m of fry area	880	470	—	—
11. Grill, meat, per sq m of fry area	1378	733	2930	733
12. Grill, sand which, per sq m of fry area	790	205	—	—
13. Hair dryer, blower type	674	117	—	—
14. Hair dryer, helmet type	547	98	—	—
15. Toaster, 360 slices per hour	1495	380	2257	967
16. Stoves, short order				
open top, per sq m top	—	—	1230	1230
closed top, per sq m top	—	—	967	967
Fry top, per sq m top	—	—	1055	1055
17. Permanent wave machine	250	44	—	—
18. Sterilizer instrument for physicians	190	352	—	—

18.8.3 Product Load

The term 'product' as used here means anything whose temperature is reduced by the refrigerating equipment. It includes perishable commodities such as food stuff and other items such as welding rods, plastics, medicines and liquids of all types. The quantity of heat to be given up by the product to the refrigerating space depends upon:

(*i*) Initial state of the product at the time of entering;

(*ii*) Final state of product it attains after storing;

(*iii*) Nature of product;

(*iv*) Weight of product and its specific heat above and below freezing;

(*v*) Freezing temperature;

(*vi*) Latent heat of the fusion.

18.8.4 Process Load

For each industrial air-conditioning process there is a specific procedure for calculating the cooling and heating loads. The requirements for the process may involve the control of one or more of the following factors:

(*i*) Regain of moisture content by hygroscopic materials, such as cotton, silk tobacco, etc., and the accompanying heat liberated;

(*ii*) Drying load;

(*iii*) Rate of chemical and biochemical reactions;

(*iv*) Rate of crystallization, freezing, freeze-drying etc.; and

(*v*) Sensible cooling load.

18.9 SYSTEM HEAT GAINS

The system heat gain is the heat gain (or loss) of an air-conditioning system. It includes the following:

1. Supply air duct heat gain and leakage loss.
2. Heat gain from air-conditioning fan.
3. Return air duct heat and leakage gain.
4. Heat gain from dehumidifier pump and piping.

The sensible heat gain should be initially estimated and then included in the total heat load for the air-conditioning plant. The same should be checked after the whole plant has been designed.

Note: *'Safety factor' is strictly a factor of probable error in the estimation of the load. For the purpose, additional 5 per cent heat should be added to the room sensible and latent heats.*

18.10 DIFFERENT FACTORS TO BE CONSIDERED IN LOAD ESTIMATION SHEET FOR "COMFORT" APPLICATION

The different *factors* to be considered in load estimation sheet for "*comfort*" application are :

1. Load estimate sheets are designed for limited application with respect to building type, construction, internal heat gains and exposure. For application of the form, manufacturers recommendations should be carefully followed.
2. Application of load forms require estimate of time of peak loads or range of time of peak loading.
3. Commercial load forms are based on significant internal heat gains and relatively lower external heat gain. Do not use residential forms for commercial applications and vice-versa.

4. FW sunlit windows, doors and walls:
   (*a*) Determine the exposure at peak load time.
   (*b*) In calculating the solar load, orientation of walls is critical.
   (*c*) Walls—use wall area (total) exposed to the sun during the day.
      (*i*) Light masonry means 200 mm or less.
      (*ii*) Heavy masonry means greater than 200 mm.
      (*iii*) "Insulated wall" must contain 50 mm or more of insulation. "Insulated ceilings" must contain 100 mm or more of insulation.
5. Transmission gains for walls, windows, doors and partitions:
   (*a*) Area to include all windows exposed to outside air.
   (*b*) Area for wall is total net wall exposed to outside air. Walls should be separated into sunlit and shaded portions.
   (*c*) Partitions may be considered as shaded walls if no factor is listed. Partitions exposed to unusual conditions such as kitchen, boiler room etc. may be handled by multiplying partition factors by 2, if no special allowance factor is listed.
   (*d*) Shaded wall areas may be considered as north wall area.
6. Internal heat gains:
   (*a*) Many estimate forms include infiltration allowance together with occupancy load.
   (*b*) Allowance for infiltration is dependent on door, window type, occupying, ventilation air quantity and exhaust air quantity.
   (*c*) Evaluate occupancy load carefully.
   (*d*) Adequate exhaust and hoods over appliances may reduce the loads from these appliances by as much as 50%.
   (*e*) Careful consideration should be given to lighting load. Frequently this is a major internal load particularly in office space or brilliantly lighted stores.

18.11 DESIGN OF A COLD STORAGE

A "**Cold storage**" *is a low temperature refrigeration application.* To keep the size of the refrigeration small, and power consumption low, it is necessary to construct the cold storage such that the *cooling load is minimized.* Accordingly, 10 to 15 cm thick insulation is applied on the *walls and ceiling*. The common insulating materials are *expanded polysterene* (thermocole, with $k = 0.0037$ W/in °C) and *poly-methane foam* (PUF, with $k = 0.0173$ W/m°C). *Due to its lower thermal conductivity PUF is preferable to thermocole.*

— These days manufacturers are making *sandwich PUF panels* also which are used to assemble pre-fabricated cold stores. The panels are in the form of a sandwich consisting of two steel plates with core of a rigid PUF in between. Foaming is performed in a hydraulic press. PUF is injected in a mould under high pressure. The foaming agent is a CFC free substance, say propane. The value of overall heat transfer coefficient, U, of a 10 cm PUF panel is 0.2 W/m^2°C.

The flooring can also be provided with a PUF slab, and the requisite water-proofing and RCC finish will have to be done.

For a cold storage, the ***load calculation*** *is separated into the following main sources of heat for a given 24-hour period* :

(*i*) Transmission load,

(*ii*) Air change (infiltration) load,

(*iii*) Miscellaneous loads like electric motors, lighting, people, etc.,

(*iv*) Product load, sensible, and

(*v*) Product load, respiration.

The average ***air change***, per 24 hours for storage rooms above 0°C due to the door openings and infiltration may range from 10 for small cold rooms of 90 m^3 volume to 1 for larger cold storages of about 900 m^3 volume.

Although most of the load in refrigerated room is due to transmission, air change and product, the loads due to *lighting, motors people working should not be overlooked.*

- The **product load** consists of the following *three* separate components:

  (*i*) ***Sensible.*** This is the amount of heat that must be removed to bring the product to storage temperature. It depends on specific heat of product.

  (*ii*) ***Latent***. This is the amount of heat that must be removed to freeze the product. It depends on latent heat of fusion. And it is applicable in the case of frozen food storage. *Example*: 2500 kJ/kg of heat removal is required for the hardening of ice cream.

  (*iii*) ***Respiration.*** Fresh fruits and vegetables are alive. They undergo changes even when stored at low temperatures, releasing heat of respiration.

  When product load is calculated on per hour basis then a correction factor for taking into account pull down time other than 24 hours must be applied as $\dfrac{24 \text{ hours}}{\text{pull-down time}}$. A safety factor of 10 per cent is also added to allow for any omissions and inaccuracies.

- *When load in kW has been determined, the condensing unit is selected first to give a capacity greater than the cooling/freezing load. The fan-coil unit is then selected to balance the capacity of the condensing unit.*

- The nature of the product governs the choice of relative humidity to be maintained in the room. The relative humidity, in turn, determines the choice of refrigeration temperature. *Higher the humidity, higher the refrigeration temperature*. The table below gives recommended temperature differences between the air in the storage room and the saturated suction temperature of the refrigerant in the forced convection evaporator, depending on room relative humidity.

Product type	*RH*	*Temp. Difference*	*Remarks*
Fruits, Vegetables, flowers, etc.	90%	4-5°C	Minimum moisture loss
General storage, packed products.	80-85%	6°C	Products require lower RH
Pharmaceuticals, beer, potatoes, onions, tough-skin fruits, etc.	65-80%	7-9°C	Products require moderate RH
Prep/cutting rooms, candy, etc.	50-65%	9-12°C	Products not affected by RH

- *In cold storage applications, it is recommended that load be divided into multiple units to provide for some capacity control, since the refrigeration system has to be selected for the worst conditions which may occur even for one day in the year.*
- *Important considerations in the selection and location of fan-coil unit are*:
  (*i*) Air should be directed to move where there is heat gain, *e.g.*, towards walls and ceiling.
  (*ii*) Total supply of air is around 40 to 80 air changes per hour. This is so because the dehumidified rise is very small.

Special precautions to be taken for insulation of a cold storage wall:

(1) The insulating material must have high resistance to internal chemical activity as a result of exposure to surrounding conditions.

(2) It must be strong enough to withstand the pressure coming on them.

(3) Moisture absorbing insulation deteriorate rapidly. Hence material that resists absorption either as free water or water vapour should be selected. For this purpose, a vapour barrier of metal foil, paper, or plastic film should be placed on the lived-in side of the insulation. This will prevent moist air inside the cold storage from destroying the effectiveness of the insulation.

(4) Materials that provide a medium for vermin infestation should be excluded as insulation.

WORKED EXAMPLES

Example 18.1 *The following data relate to a conference room for seating 80 persons.*

Inside design conditions .... *22°C DBT, 55% RH*

Outside design conditions .... *38° DBT, 28°C WBT*

Sensible and latent heat loads per person .... *75 W and 45 W respectively*

Lights and fans loads .... *12000 W*

Sensible heat gain through glass, walls, ceiling etc. .... *12000 W*

Air infiltration .... *18 m³/min*

Fresh air supply .... *80 m³/min.*

By-pass factors of the coils *0.1*

If two-third of recirculated room air and one-third of fresh air are mixed before entering the cooling coils, determine:

(*i*) *Apparatus dew point,*

(*ii*) *Grand total heat load, and*

(*iii*) *Effective room sensible heat factor.*

Solution: *Given:* Number of persons = 80; t_{db2} = 22°C; ϕ_2 = 55%; t_{db1} = 38°C; t_{wb1} = 28°C;

Q_S per person = 75 W; Q_L per person = 45 W;

$(Q_S)_{L\&F}$ = 12000 W = 12 kW; $(Q_S)_{G,W,C\ etc.}$ = 12000 W or 12 kW; $V_1 = 18\ m^3/min$;

$V_F = 80\ m^3/min$; BF = 0.1.

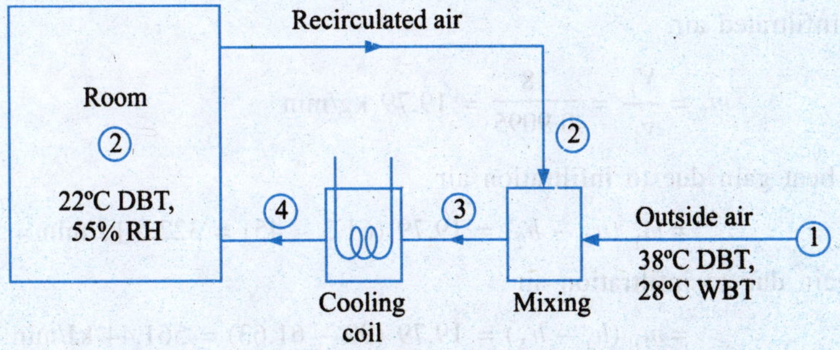

Fig. 18.2. Flow diagram for the air-conditioning system.

The flow diagram for the air-conditioning system is shown in Fig. 18.2. The various points on the psychrometric chart as shown in Fig. 18.3. are marked as discussed below:

- Locate point 1 at the intersection of 38°C DBT and 28°C WBT lines.
- Locate point 2 at the intersection of 22°C DBT line and 55% RH curve.
- Locate point A by drawing vertical and horizontal lines through points 1 and 2 respectively.

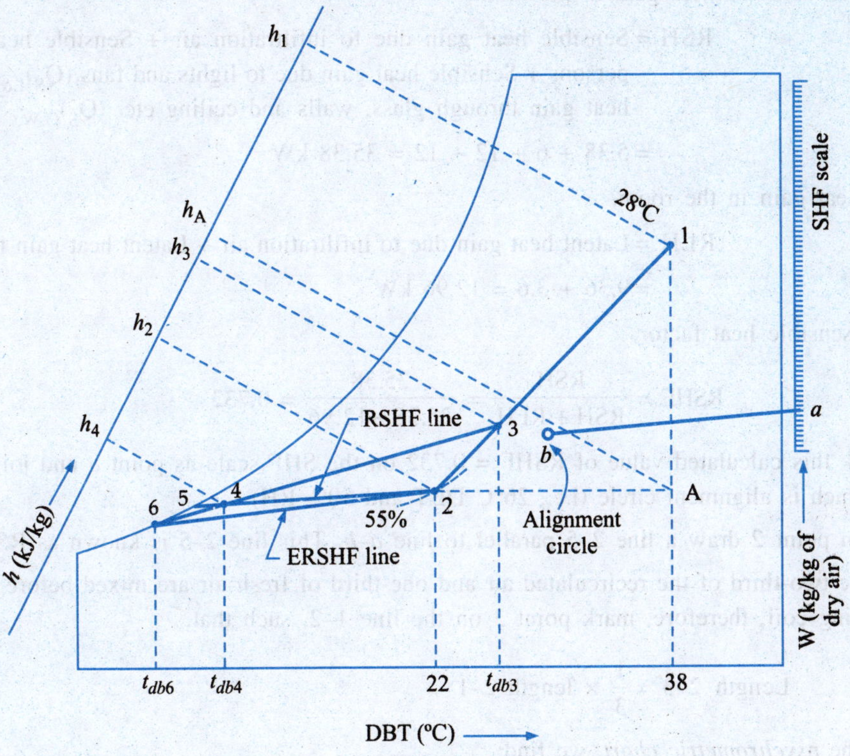

Fig. 18.3

From the *psychrometric chart*, we find:

$$v_1 = 0.9095 \text{ m}^3/\text{kg of dry air}$$

$$h_1 = 90 \text{ kJ/kg of dry air}$$

$$h_2 = 45 \text{ kJ/kg of dry air}$$

$$h_A = 61.3 \text{ kJ/kg of dry air.}$$

The mass of infiltrated air,

$$m_1 = \frac{V_1}{v_1} = \frac{18}{0.9095} = 19.79 \text{ kg/min}$$

∴ Sensible heat gain due to infiltration air

$$= m_1 (h_A - h_2) = 19.79 (61.3 - 45) = 322.6 \text{ kJ/min} = 5.38 \text{ kW}$$

Latent heat gain due to infiltration air

$$= m_1 (h_1 - h_A) = 19.79 (90 - 61.63) = 561.44 \text{ kJ/min} = 9.36 \text{ kW}$$

Total sensible heat gain from persons

$$= Q_S \text{ per person} \times \text{No. of persons}$$

$$= 75 \times 80 = 6000 \text{ W} = 6 \text{ kW}$$

Total latent heat gain from persons

$$= 45 \times 80 = 3600 \text{ W} = 3.6 \text{ kW}$$

∴ Total sensible heat gain in the room,

RSH = Sensible heat gain due to infiltration air + Sensible heat gain from persons + Sensible heat gain due to lights and fans $(Q_S)_{L\&F}$ + Sensible heat gain through glass, walls and ceiling etc. $(Q_S)_{G,W,C \text{ etc.}}$

$$= 5.38 + 6 + 12 + 12 = 35.38 \text{ kW}$$

Total heat gain in the room,

RLH = Latent heat gain due to infiltration air + Latent heat gain from persons

$$= 9.36 + 3.6 = 12.96 \text{ kW}$$

Room sensible heat factor,

$$\text{RSHF} = \frac{\text{RSH}}{\text{RSH} + \text{RLH}} = \frac{35.38}{35.38 + 12.96} = 0.732$$

- Mark this calculated value of RSHF = 0.732 on the SHF scale as point *a* and join with point *b* which is alignment circle (i.e., 26°C DBT and 50% RH).
- From point 2 draw a line 2–5 parallel to line *a–b*. This line 2–5 is known as RSHF line.
- Since two-third of the recirculated air and one-third of fresh air are mixed before entering the cooling coil, therefore, mark point 3 on the line 1–2, such that,

$$\text{Length } 2\text{–}3 = \frac{1}{3} \times \text{length } 2\text{–}1$$

From the *psychrometric chart*, we find:

$$t_{db3} = 27.6°\text{C}$$

(*i*) Apparatus dew points (ADP):

Let, t_{db6} = Apparatus dew point.

Then, $$\text{BF} = 0.1 = \frac{t_{db4} - t_{db6}}{t_{db3} - t_{db6}} = \frac{t_{db4} - t_{db6}}{27.6 - t_{db6}}$$

By trial and error method, we find:

$$t_{db4} = 9.6°C, \quad t_{db6} = \mathbf{7.6°C\ (Ans.)}$$

(*ii*) **Grand total heat load:**

We know that outside air total heat,

$$\text{OATH} = 0.02\ V_F\ (h_1 - h_2) = 0.02 \times 80\ (90 - 45) = 72\ \text{kW}$$

∴ Grand total heat,

$$\text{GTH} = \text{RSH} + \text{RLH} + \text{OATH} = 35.38 + 12.96 + 72 = \mathbf{120.34\ kW\ (Ans.)}$$

(*iii*) **Effective room sensible heat factor:**

- Join point 6 with point 2, the line 6–2 is known as effective room sensible heat factor line and ERSH factor is found to be **0.69 (Ans.)**

Example 18.2 *An office for seating 30 occupants is to be maintained at 22°C DBT and 55% RH. The outdoor conditions are 36°C DBT and 27°C WBT. The various loads in the office are:*

Solar heat gain .... *8500 W*

Sensible heat gain per occupant .... *83 W*

Latent heat gain per occupant .... *100 W*

Lighting load .... *2500 W*

Sensible heat load from other sources .... *12000 W*

Infiltration load .... *15 m³/min*

Assuming 40 per cent fresh air and 60 per cent of recirculated air passing through the evaporator coil and the by-pass factor of 0.12, determine:

(i) Dew point temperature of the coil; and

(ii) Capacity of the plant.

Solution: Given: Number of occupants = 30; t_{db2} = 22°C; ϕ_2 = 55% RH; t_{db_1} = 36°C; t_{wb_1} = 23°C; $(Q_S)_S$ = 8500 W or 8.5 kW; Q_S per occupant = 83 W; Q_L per occupant = 100 W; $(Q_S)_L$ = 2600 W or 2.6 kW; $(Q_S)_{\text{other sources}}$ = 12000 W = 12 kW; V_1 = 15 m³/min; BF = 0.15.

The flow diagram for the given air-conditioning system is shown in Fig. 18.4. The various points on the psychrometric chart, as shown in Fig. 18.4, are marked as discussed below:

- Locate point 1 at the intersection of 36°C DBT and 27°C WBT lines.
- Locate point 2 at the intersection of 22°C DBT line and 55% RH curve.
- Locate point A by drawing vertical and horizontal lines through points 1 and 2 respectively.

From *psychrometric chart*, we find:

$$v_1 = 0.902\ \text{m}^3/\text{kg of dry air}$$
$$h_1 = 85.2\ \text{kJ/kg of dry air}$$
$$h_2 = 45\ \text{kJ/kg of dry air}$$
$$h_1 = 59.6\ \text{kJ/kg of dry air.}$$

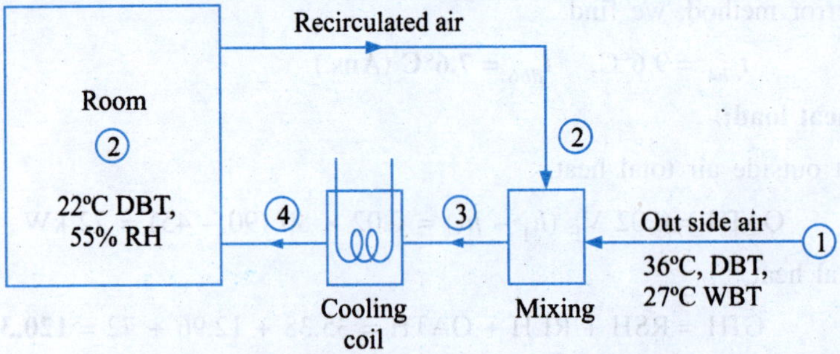

Fig. 18.4. Flow diagram of the air-conditioning system.

Mass of infiltrated air at point 1,

$$m_1 = \frac{V_1}{v_1} = \frac{15}{0.902} = 16.63 \text{ kg/min}$$

Sensible heat gain due to infiltration air

$$= m_1 (h_A - h_2) = 16.63 (59.6 - 45) = 242.8 \text{ kJ/min} = 4.05 \text{ kW}$$

Latent heat gain due to infiltration air

$$= m_1 (h_1 - h_A) = 16.63 (85.2 - 59.6) = 425.73 \text{ kJ/min} = 7.1 \text{ kW}$$

Total sensible heat gain from the occupants

$$= Q_S \text{ per occupant} \times 30$$

$$= 83 \times 30 = 2490 \text{ W} \simeq 2.5 \text{ kW}$$

Total latent heat gain from the occupants

$$= 100 \times 30 = 3000 \text{ W} = 3 \text{ kW}$$

Total sensible heat gain in the room/office,

RSH = Solar heat gain + Sensible heat gain due to infiltration air + Sensible heat gain from occupants + Sensible heat gain due to lighting + Sensible heat gain from other sources

$$= 8.5 + 4.05 + 2.5 + 2.6 + 12 = 29.65 \text{ kW, and}$$

Total latent heat gain in the room/office

RLH = Latent heat due to infiltration air + Latent heat gain from occupants

$$= 7.1 + 3 = 10.1 \text{ kW}$$

$\therefore$ Room sensible heat factor, $RSHF = \frac{RSH}{RSH + RLH}$

$$= \frac{29.65}{29.65 + 10.1} = 0.746$$

- Mark this calculated value of RSHF = 0.746 on SHF scale as point *a* and join with point *b* which is alignment circle (i.e., 26°C DBT and 50% RH) as shown in Fig. 18.5.
- From point 2, draw a line 2–5 parallel to the line *a–b*. The line 2–5 is known as RSHF line.
- Since 40% of fresh air (i.e., outside air) and 60% of recirculated air is passed through the evaporated coil, therefore mark the mixing point 3 on the line 1–2, such that,

Length 2–3 = 0.4 × Length 2–1.

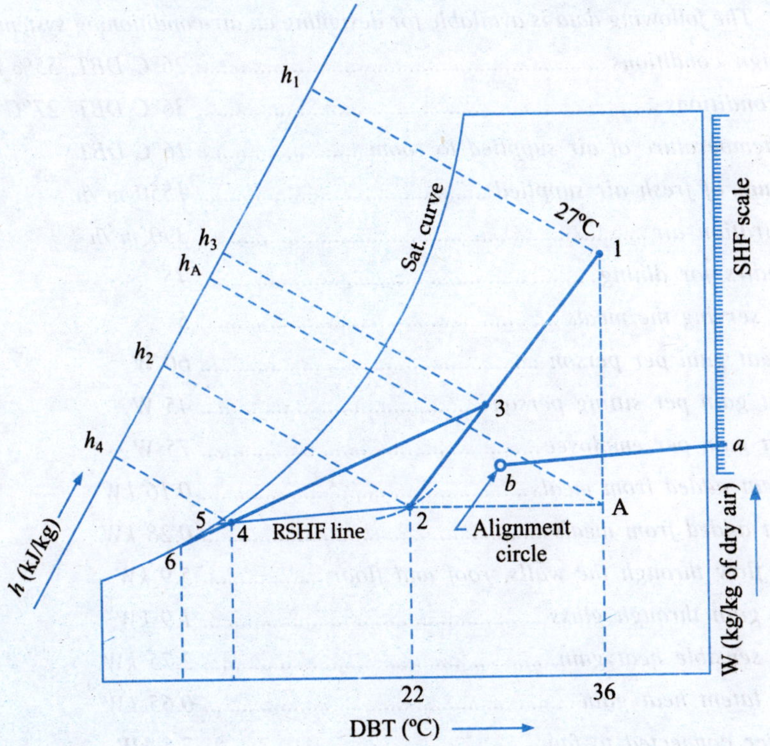

Fig. 18.5

From *psychrometric chart* : $t_{db3} = 27.6°C$

(*i*) Dew point temperature of the coil:

Let, t_{db6} = Dew point temperature of the coil.

Then, $$BF = 0.15 = \frac{t_{db4} - t_{db6}}{t_{db3} - t_{db6}} = \frac{t_{db4} - t_{db6}}{27.6 - t_{db6}}$$

By trial and error we find:

$t_{db4} \simeq 10.7°C$; $t_{db6} \simeq$ **7.8°C (Ans.)**

(*ii*) Capacity of the plant:

From the *psychrometric chart*, we find:

$h_4 = 29.5$ kJ/kg of dry air, and

$h_3 = 61$ kJ/kg of dry air.

Mass of air entering the room,

$$m_a = \frac{\text{Total room heat}}{\text{Total heat sermoved}}$$

$$= \frac{RSH + RLH}{(h_2 - h_4)} = \frac{29.65 + 10.1}{(45 - 29.5)} = 2.56 \text{ kg/s}$$

∴ Capacity of the plant

$$= m_a (h_3 - h_4) = \frac{2.56\,(61 - 29.5) \times 3600}{14000} = \textbf{20.74 tonnes (Ans.)}$$

Example 18.3 *The following data is available for designing an air-conditioning system for a restaurant:*

Inside design conditions	*26°C DBT, 55% RH*
Outdoor conditions	*36°C DBT, 27°C WBT*
Minimum temperature of air supplied to room	*16°C DBT*
Total amount of fresh air supplied	*1550 m³/h*
Total infiltration air	*390 m³/h*
Seating chairs for dining	*45*
Employees serving the meals	*5*
Sensible heat gain per person	*60 W*
Latent heat gain per sitting person	*45 W*
Latent heat gain per employee	*75 W*
Sensible heat added from meals	*0.16 kW*
Latent heat added from meals	*0.28 kW*
Total heat flow through the walls, roof and floor	*5.9 kW*
Solar heat gain through glass	*1.9 kW*
Equipment sensible heat gain	*2.75 kW*
Equipment latent heat gain	*0.65 kW*
Motor power connected to fan	*7.5 kW*

If the fan is installed before the conditioner, determine:

(*i*) *Amount of air delivered to the room in m³/h,*

(*ii*) *Percentage of recirculated air,*

(*iii*) Refrigeration load on the coil in tonnes of refrigeration, and

(*iv*) Dew point temperature of the cooling coil and by-pass factor.

Solution: *Given:* $t_{db2} = 26°C$; $\phi_2 = 55\%$; $t_{db1} = 36°C$ DBT; $t_{wb1} = 26°C$;

$t_{db4} = 16°C$; $V_F = 1550\ m^3/h$; $V_1 = 390\ m^3/h$; Dining chairs = 45;

Employees = 5; Q_S per person = 60 W; Q_L per person = 45 W;

Q_S per employee = 75 W; $(Q_S)_{meals} = 0.16$ kW; $(Q_L)_{meals} = 0.28$ kW;

$(Q_S)_{W,R\&F} = 5.9$ kW; $(Q_S)_{glass} = 1.9$ kW; $(Q_S)_{eqt.} = 2.75$ kW; $(Q_L)_{eqt} = 0.65$ kW; $Q_M = 7.5$ kW

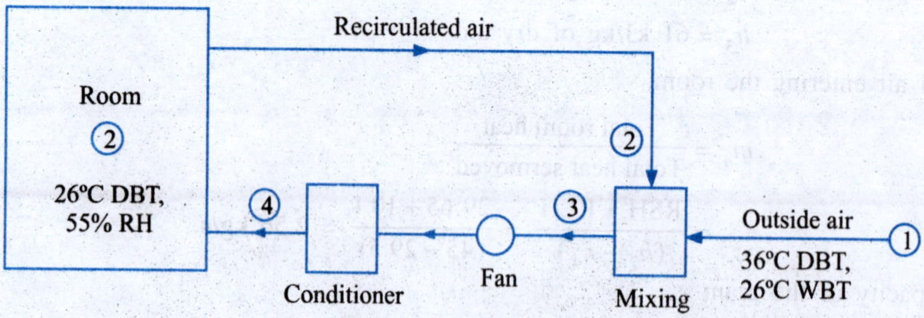

Fig. 18.6. Flow diagram for air-conditioning system of the restaurant.

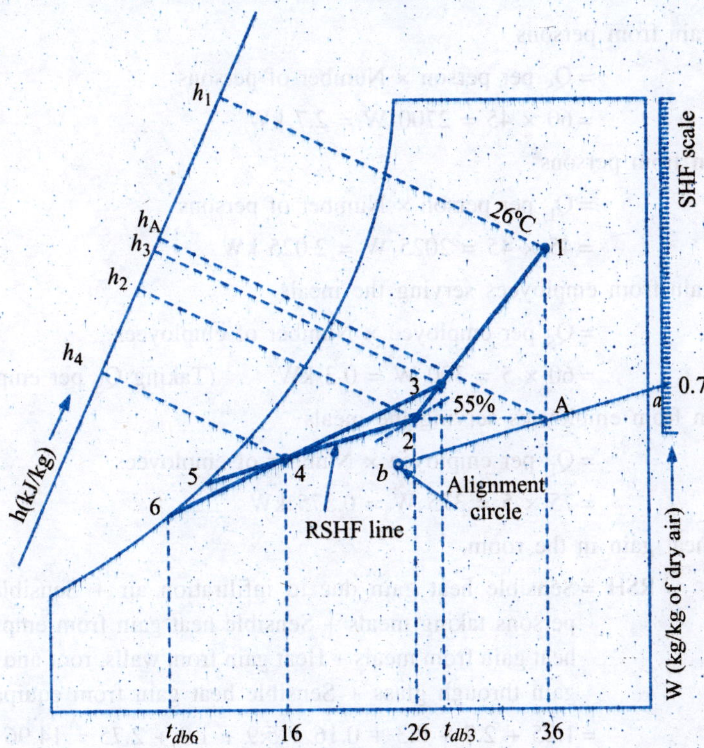

Fig. 18.7

The flow diagram for the air-conditioning system of the restaurant is shown in Fig. 18.6. The various points on the psychrometric chart, as shown in Fig. 18.7, are marked and discussed henceforth:

- Locate point 1 at the intersection of 36°C DBT and 27°C WBT lines.
- Locate point 2 at the intersection of 26°C DBT line and 55% RH curve.
- Locate point A by drawing vertical and horizontal lines from points 1 and 2 respectively.

From the *psychrometric chart*, we find:

$$v_1 = 0.902 \text{ m}^3\text{/kg of dry air,}$$

$$h_1 = 85.2 \text{ kJ/kg of dry air,}$$

$$h_2 = 55.6 \text{ kJ/kg of dry air, and}$$

$$h_A = 66 \text{ kJ/kg of dry air.}$$

Mass of infiltrated air, $m_1 = \dfrac{V_1}{v_1} = \dfrac{390}{0.902} = 432.37$ kg/h

$\therefore$ Sensible heat gain due to infiltration air

$$= m_1 (h_A - h_2) = 432.37 (66 - 55.6) = 4496.6 \text{ kJ/h}$$

$$= \frac{4496.6}{3600} = 1.25 \text{ kW, and}$$

Latent heat gain due to infiltration air

$$= m_1 (h_1 - h_A) = 432.37 (85.2 - 66) = 8301.5 \text{ kJ/h}$$

$$= \frac{8301.5}{3600} = 2.31 \text{ kW}$$

Sensible heat gain from persons

$= Q_S$ per person × Number of persons

$= 60 \times 45 = 2700$ W = 2.7 kW

Latent heat gain from persons

$= Q_L$ per person × Number of persons

$= 45 \times 45 = 2025$ W = 2.025 kW

Sensible heat gain from employees serving the meals

$= Q_S$ per employee × Number of employees

$= 60 \times 5 = 300$ W = 0.3 kW (Taking Q_S per employee = 60 W)

Latent heat gain from employees serving the meals

$= Q_L$ per employee × Number of employees

$= 75 \times 5 = 375$ W = 0.375 kW

Total sensible heat gain in the room,

RSH = Sensible heat gain due to infiltration air + Sensible heat gain from persons taking meals + Sensible heat gain from employees + Sensible heat gain from meals + Heat gain from walls, roof and floor + Solar heat gain through glass + Sensible heat gain from equipment

$= 1.25 + 2.7 + 0.3 + 0.16 + 5.9 + 1.9 + 2.75 = 14.96$ kW

Total latent heat gain in the room,

RLH = Latent heat gain due to infiltration air + Latent heat gain from persons taking meals + Latent heat gain from employees serving the meals + Latent heat gain from meals + Latent heat gain from equipment

$= 2.31 + 2.025 + 0.375 + 0.28 + 0.65 = 5.64$ kW

$$\therefore \text{ Room sensible heat factor, RSHF} = \frac{\text{RSH}}{\text{RSH} + \text{RLH}} = \frac{14.96}{14.96 + 5.64} = 0.726$$

- Mark the calculated value of RSHF = 0.726 on the SHF scale as point *a* and join with point *b* which is alignment circle (i.e., 26°C DBT and 50% RH), as shown in Fig. 18.7.
- From point 2 draw a line 2–5 parallel to the line *a–b*. This line 2–5 is known as *RSHF line*.
- Draw a vertical line through 16°C DBT (minimum temperature of air supplied to the room) which cuts RSHF line at point 4.

From the *psychrometric chart*, we find:

$v_4 = 0.832$ m^3/kg of dry air;

$h_4 = 41.3$ kJ/kg of dry air.

(*i*) Amount of air delivered to the room in m^3/h:

Mass of air delivered to the room,

$$m_a = \frac{\text{Total room heat}}{\text{Total heat removed}} = \frac{\text{RSH} + \text{RLH}}{(h_2 - h_4)} = \frac{14.96 + 5.64}{(55.6 - 41.3)}$$

$= 1.44$ kg/s = 5184 kg/h

∴ Amount of air delivered to the room in m³/h,

$$= V_a = m_a \times v_4 = 5184 \times 0.832 = \mathbf{4313\ m^3/h\ (Ans.)}$$

(*ii*) **Percentage of recirculated air:**

Mass of fresh air supplied,

$$m_F = \frac{V_F}{v_1} = \frac{1550}{0.902} = 1718.4 \text{ kg/h}$$

∴ Mass of recirculated air $= m_a - m_F = 5184 - 1718.4 = 3465.6$ kg/h

∴ Percentage of recirculated air

$$= \frac{3465.6}{5184} = 0.668 \text{ or } \mathbf{66.8\%\ (Ans.)}$$

(*iii*) **Refrigeration load on the cooling coil:**

Since the recirculated air at point 2 is 66.8%, therefore, the fresh air supplied to the room at point 1 is 33.2%. The mixing of recirculated air and fresh air before entering to the cooling coil is marked at point 3 on the psychrometric chart, such that,

Length 2–3 = 0.332 × Length 2–1.

From *psychrometric chart*, we find:

$$h_3 = 65 \text{ kJ/kg of dry air}$$

$$t_{db3} = 29.2°C$$

∴ Refrigeration load on the cooling coil

$$= m_a\,(h_3 - h_4) + \text{Heat added by fan motor}$$

$$= 1.44\,(65 - 41.3) + 7.5 = 41.63 \text{ kW}$$

$$= \frac{41.63 \times 3600}{14000} = \mathbf{10.7\ tonnes\ (Ans.)}$$

(*iv*) **Dew point temperature of the cooling coil and by-pass factor:**

- Joint point 3 with point 4 and produce the line to intersect the saturation curve at point 6.

From *psychrometric chart*, we find:

$$t_{dp} = t_{db6} = \mathbf{12.2°C\ (Ans.)}$$

By-pass factor, $$\text{BF} = \frac{t_{db4} - t_{db6}}{t_{db3} - t_{db6}} = \frac{16 - 12.2}{29.2 - 12.2} = \mathbf{0.2235\ (Ans.)}$$

Example 18.4. *The following data is available for designing an air-conditioning system for a hall:*

Inside conditions .... *23°C DBT, 65% RH*

Outdoor conditions .... *37°C DBT, 28°C WBT*

Sensible heat load in the room .... *45.5 kW*

Latent heat load in the room .... *11.5 kW*

Total infiltration air .... *1150 m³/h*

Appraratus dew point .... *9°C*

Quantity of recirculated air from the hall *60%*

If the quantity of recirculated air is mixed with the conditioned air after the cooling coil, determine:

(i) The condition of air leaving the conditioner coil and before mixing with the recirculated air;

(ii) The condition of air before entering the hall;

(iii) The mass of air-entering the cooler;

(iv) The mass of total air passing through the hall;

(v) The by-pass factor of the cooling coil;

(vi) The refrigeration load on the cooling coil in tonnes of refrigeration.

Solution: *Given:* $t_{ab2} = 23°C$; $\phi_2 = 65\%$; $t_{db1} = 37°C$; $t_{wb1} = 28°C$; $Q_S = 45.5$ kW; $Q_L = 11.5$ kW; $V_1 = 1150$ m³/h; ADP = 9°C

(*i*) The condition of air leaving the conditioner coil and before mixing with the recirculated air:

The flow diagram for the air-conditioning system of a hall are shown in Fig. 18.8. The various points on the psychrometric chart, as shown in Fig. 18.9., are marked as discussed below:

- Locate point 1 at the intersection of 37°C DBT and 28°C WBT lines.
- Locate point 2 at the intersection of 23°C DBT line and 65% RH curve.
- Locate point A by drawing vertical and horizontal times from points 1 and 2 respectively.

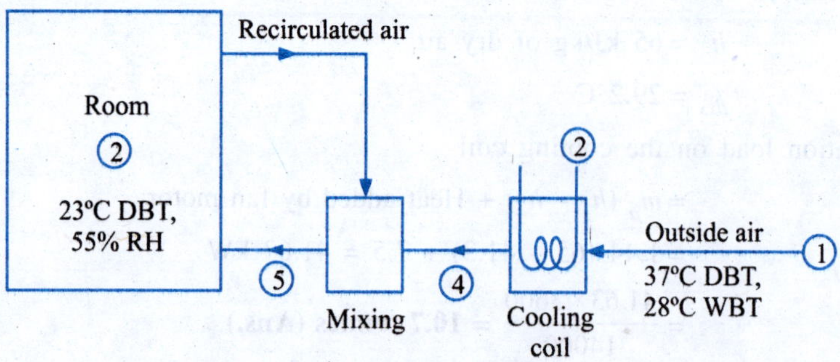

Fig. 18.8. Flow diagram for the air-conditioning system of the hall.

From the *psychrometric chart*, we find:

$$v_1 = 0.9079 \text{ m}^3/\text{kg of dry air},$$

$$h_1 = 90 \text{ kJ/kg of dry air},$$

$$h_2 = 52.3 \text{ kJ/kg of dry air, and}$$

$$h_A = 67.3 \text{ kJ/kg of dry air.}$$

Mass of infiltrated air,

$$m_1 = \frac{V_1}{v_1} = \frac{1150}{0.9079} = 1266.66 \text{ kg/h}$$

∴ Sensible heat due to infiltrated air

$$= m_1 (h_A - h_2) = 1266.66 (67.3 - 52.3) = 19000 \text{ kJ/h}$$

$$= \frac{19000}{3600} = 5.28 \text{ kW, and}$$

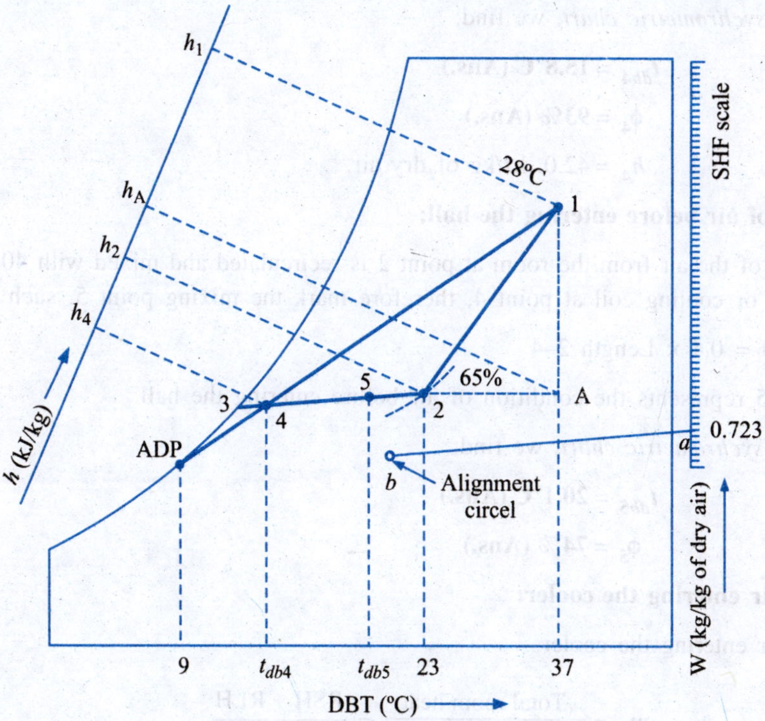

Fig. 18.9

Latent heat due to infiltrated air

$$= m_1\,(h_1 - h_A) = 1266.66\,(90 - 67.3) = 28753.2 \text{ kJ/h}$$

$$= \frac{28753.2}{3600} = 7.99 \text{ kW}$$

∴ Total sensible heat gain in the room,

$$\text{RSH} = 45.5 + 5.28 = 50.78 \text{ kW, and}$$

Total latent heat gain in the room,

$$\text{RLH} = 11.5 + 7.99 = 19.49 \text{ kW}$$

Room sensible heat factor,

$$\text{RSHF} = \frac{\text{RSH}}{\text{RSH} + \text{RLH}} = \frac{50.78}{50.78 + 19.49} = 0.723$$

- Mark this calculated value of RSHF = 0.723 on the SHF scale as point *a* and join with point *b* which is alignment circle (i.e., 26°C DBT and 50% RH), as shown in Fig. 18.9.
- From point 2 draw a line 2–3 parallel to the line *a–b*.

  This line 2–3 is known as *RSHF line*.
- Draw the line joining the points 1 and ADP = 9°C which intersects the line 2–3 at point 4. This point 4 represents the condition of air leaving the conditioner or cooling coil and before mixing the recirculated air.

From the *psychrometric chart*, we find:

$$t_{db4} = \mathbf{15.8°C\ (Ans.)}$$

$$\phi_4 = \mathbf{93\%\ (Ans.)}$$

$$h_4 = 42.0.\ \text{kJ/kg of dry air.}$$

(*ii*) **Condition of air before entering the hall:**

Since 60% of the air from the room at point 2 is recirculated and mixed with 40% of air leaving the conditioner or cooling coil at point 4, therefore mark the mixing point 5, such that,

Length 2–5 = 0.4 × Length 2–4

The point 5 represents the condition of air before entering the hall.

From the *psychrometric chart*, we find:

$$t_{db5} = \mathbf{20.1°C\ (Ans.)}$$

$$\phi_5 = \mathbf{74\%\ (Ans.)}$$

(*iii*) **Mass of air entering the cooler:**

Mass of air entering the cooler,

$$m_C = \frac{\text{Total room heat}}{\text{Total heat removed}} = \frac{\text{RSH + RLH}}{(h_2 - h_4)}$$

$$= \frac{50.78 + 19.49}{(52.3 + 42.0)} = 6.82\ \text{kg/s} = \mathbf{24552\ kg/h\ (Ans.)}$$

(*iv*) **Mass of total air passing through the hall:**

Let m_H = Mass of total air passing through the hall. Since the total air (m_H) passing through the hall consists of air passing through the cooler and the recirculated air from the hall (i.e., 60% m_H), therefore,

$$24552 + 0.6\ m_H = m_H$$

$$\therefore \quad m_H = \mathbf{61380\ kg/h\ (Ans.)}$$

(*v*) **By-pass factor of the cooling coil, BF:**

$$\text{BF} = \frac{t_{db4} - \text{ADP}}{t_{db1} - \text{ADP}} = \frac{15.8 - 9}{37 - 9} = \mathbf{0.243\ (Ans.)}$$

(*vi*) **Refrigeration load on the cooling coil:**

Refrigeration load on the cooling coil

$$= \frac{m_C\,(h_1 - h_4)}{14000}\ \text{tonnes}$$

$$= \frac{24552\,(90 - 42)}{14000} = \mathbf{84.2\ tonnes\ (Ans.)}$$

Example 18.5. *It is required to design an air-conditioning system for a restaurant with the following data:*

Inside design conditions .... *24°C DBT, 55% RH*

Outdoor conditions .... *39°C DBT, 28°C WBT*

Infiltrated air .... *16 m³/min*

No. of occupants *30*

Sensible heat gain per person .... *60 W*

Latent heat gain per person .... *60 W*

Internal lighting load .... *20 lamps of 100 W; 12 fluorescent tubes of 60 W*

Solar heat gain through walls, roof and floor .... *6.1 kW*

Solar heat gain through glass .... *5.6 kW*

Sensible heat gain from other sources .... *12.2 kW*

By-pass factor of the coil .... *0.2.*

If 25 per cent fresh air and 75 per cent recirculated air is mixed and passed through the conditioner coil, determine:

(*i*) *The amount of total air required in m³/h,*

(*ii*) *The dew point temperature of the coil,*

(*iii*) *The condition of supply air to the room, and*

(*iv*) *The capacity of the conditioning plant.*

Solution: *Given:* t_{db2} = 24°C; ϕ_2 = 55%; t_{db2} = 39°C; t_{wb2} = 28°C; V_1 = 16 m³/min;
Number of occupants = 30; Q_S per person = 60 W; Q_L per person = 60 W;
$(Q_S)_{L\&T} = 20 \times 100 + 12 \times 60 = 2720$ W = 2.72 kW; $(Q_S)_{W,\,R\,\&\,F}$ = 6.1 kW;
$(Q_S)_G$ = 5.6 kW; $(Q_S)_{other\ sources}$ = 12.2 kW; BF = 0.2.

The flow diagram for the air-conditioning system of the restaurant is shown in Fig. 18.10. The various points on the psychrometric chart, as shown in Fig. 18.10, are marked as discussed below:

- Locate point 1 at the intersection of 39°C DBT and 28°C WBT lines.
- Locate point 2 at the intersection of 24°C DBT line and 55% RH curve.
- Locate point A by drawing vertical and horizontal lines from points 1 and 2 respectively.

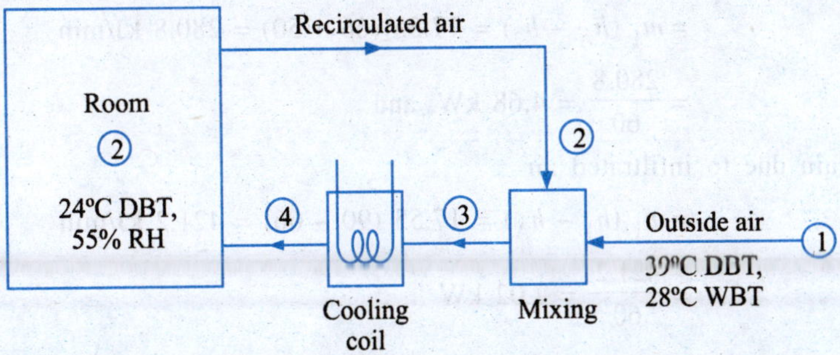

Fig. 18.10. Flow diagram for the air-conditioning system of the restaurant.

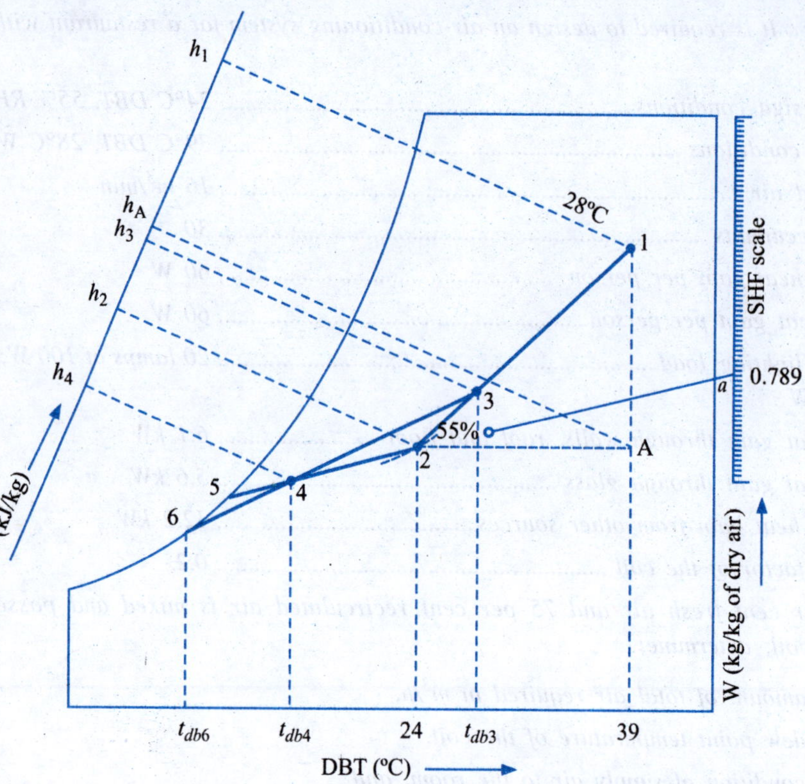

Fig. 18.11

From the *psychrometric chart*, we find:

$$v_1 = 0.9115 \text{ m}^3\text{/kg of dry air},$$

$$h_1 = 90 \text{ kJ/kg of dry air},$$

$$h_2 = 50 \text{ kJ/kg of dry air, and}$$

$$h_A = 66 \text{ kJ/kg of dry air.}$$

Mass of infiltrated air,

$$m_1 = \frac{V_1}{v_1} = \frac{16}{0.9115} = 17.55 \text{ kg/min}$$

$\therefore$ Sensible heat gain due to infiltrated air

$$= m_1 (h_A - h_2) = 17.55 (66 - 50) = 280.8 \text{ kJ/min}$$

$$= \frac{280.8}{60} = 4.68 \text{ kW, and}$$

Latent heat gain due to infiltrated air

$$= m_1 (h_1 - h_A) = 17.55 (90 - 66) = 421.2 \text{ kJ/min}$$

$$= \frac{421.2}{60} = 7.02 \text{ kW}$$

Total sensible heat gain from occupants

$= Q_S$ per person × Number of occupants

$= 60 \times 30 = 1800$ W = 1.8 kW, and

Total latent heat gain from occupants

$= 60 \times 30 = 1800$ W = 1.8 kW

Sensible heat gain from internal lighting = 2.72 kW

Sensible heat gain in the room,

RSH = Sensible heat gain due to infiltrated air + Sensible heat gain from occupants + Sensible heat gain from internal lighting + Solar heat gain through walls, roof and floor + Sensible heat gain through glass + Sensible heat gain from other sources

$= 4.68 + 1.8 + 2.72 + 6.1 + 5.6 + 12.2 = 33.1$ kW, and

Latent heat gain in the room,

RLH = Latent heat gain due to infiltrated air + Latent heat gain from occupants

$= 7.02 + 1.8 = 8.82$ kW

∴ Room sensible heat factor,

$$\text{RSHF} = \frac{\text{RSH}}{\text{RSH} + \text{RLH}} = \frac{33.1}{33.1 + 8.82} = 0.789$$

- Mark this calculated value of RSHF = 0.784 on the SHF scale as point *a* and join with point *b* which is alignment circle (i.e., 26°C DBT and 50% RH), as shown in Fig. 18.11.
- From point 2, draw a line 2–5 parallel to the line *a–b*. This line 2–5 is known as *RSHF line.*
- Since 25% of fresh air and 75% of recirculated air is mixed and passed through the conditioner coil, therefore, mark the mixing point 3 on the line 1–2 such that;

Length 2–3 = 0.25 × Length 2–1.

From the psychrometric chart, we find:

$t_{db3} = 27.6°\text{C}$,

$h_3 = 60$ kJ/kg of dry air.

Mark point 4 on the 2–5 and point 6 on the saturation curve as follows:

$$\text{BF} = 0.2 = \frac{t_{db4} - t_{db6}}{t_{db3} - t_{db6}} = \frac{t_{db4} - t_{db6}}{27.6 - t_{db6}}$$

By trial and error, we find:

$t_{db4} = 14.4°\text{C}$; $t_{db6} = 11.2°\text{C}$;

$h_4 = 37.5$ kJ/kg of dry air.

$v_4 = 0.8267$ m^3/kg of dry air

(*i*) Amount of total air required in m^3/h:

Let, V_a = Amount of total air required.

We know that, $$m_a = \frac{\text{Total room heat}}{\text{Total heat removed}} = \frac{\text{RSH} + \text{RLH}}{(h_2 - h_4)}$$

$$= \frac{33.1 + 8.82}{(50 - 37.5)} = 3.354 \text{ kg/s} = 12074.4 \text{ kg/h}$$

$\therefore$ $$v_a = m_a \times v_4 = 12074.4 \times 0.8267 = \mathbf{9982 \ m^3/h \ (Ans.)}$$

(*ii*) **Dew point temperature of the coil:**

Dew point temperature of the coil (already found above),

$$t_{db6} = \mathbf{11.2°C \ (Ans.)}$$

(*iii*) **Condition of supply air to the room:**

The point 4, as shown in Fig. 18.11, represents the condition of supply air to the room. At this point 4,

$$t_{db4} = \mathbf{14.4°C; \ (Ans.)}$$

$$\phi_4 = \mathbf{91\% \ (Ans.)}$$

(*iv*) **Capacity of the conditioning plant:**

Capacity of the conditioning plant

$$= \frac{m_a (h_3 - h_4)}{14000} \text{ tonnes}$$

$$= \frac{12074.4 (60 - 37.5)}{14000} = \mathbf{19.4 \ tonnes \ (Ans.)}$$

Example 18.6. *It is required to design a cold storage for storing 450 tonnes of vegetables with the following available data:*

Inside design conditions .... *19°C DBT, 60% RH*

Outdoor conditions .... *36°C DBT, 28°C WBT*

Infiltrated air .... *180 m³/h*

Fresh air supplied from outside .... *4500 m³/h*

No. of persons working in the cold storage *20*

Sensible heat gain through glass .... *5.5 kW*

Sensible heat gain through walls and ceilings .... *10.8 kW*

Water contents of the vegetables *74%*

Loss of water content .... *0.01 per cent per hour*

Heat from equipment and reaction heat of vegetables .... *3.1 kW*

If the air-conditioning is achieved by first cooling and dehumidifying and then heating, and the temperature of air entering the room is not to exceed 16°C, determine:

(*i*) *Amount of recirculated air, if the recirculated air is mixed with fresh air before entering the cooling coil;*

(*ii*) *Capacity of the cooling coil in tonnes of refrigeration and its by-pass factor if dew point temperature of the coil is 6°C; and*

(*iii*) *Capacity of the heating coil in kW.*

Solution: *Given:* m_{veg} = 450 tonnes; t_{db2} = 19°C; ϕ_2 = 60%; t_{db1} = 36°C; t_{wb2} = 28°C; V_1 = 180 m^3/h; V_F = 4500 m^3/h; Number of persons = 20; $(Q_S)_G$ = 5.5 kW; $(Q_S)_{W\&C}$ = 10.8 kW;

$$W_v = 74\% = 0.74; \quad W_{Loss} = 0.01\% \text{ per hour}; \quad (Q_S)_{E\&RV} = 3.1 \text{ kW}.$$

The flow diagram for the air-conditioning of the cold storage is show in Fig. 18.12. The various points on the psychrometric chart, as shown in Fig. 18.13, are marked as discussed below:

- Locate point 1 at the intersection of 36°C DBT and 28°C WBT lines.
- Locate point 2 at the intersection of 19°C DBT line and 60% RH curve.
- Locate point A by drawing vertical and horizontal lines through points 1 and 2 respectively.

From the *psychrometric chart*, we find:

$$v_1 = 0.9046 \text{ m}^3\text{/kg of dry air,}$$
$$h_1 = 90 \text{ kJ/kg of dry air,}$$
$$h_2 = 39.2 \text{ kJ/kg of dry air, and}$$
$$h_A = 57.1 \text{ kJ/kg of dry air.}$$

Mass of infiltrated air at point:

$$m_1 = \frac{V_1}{v_1} = \frac{180}{0.9046} = 198.98 \text{ kg/h}$$

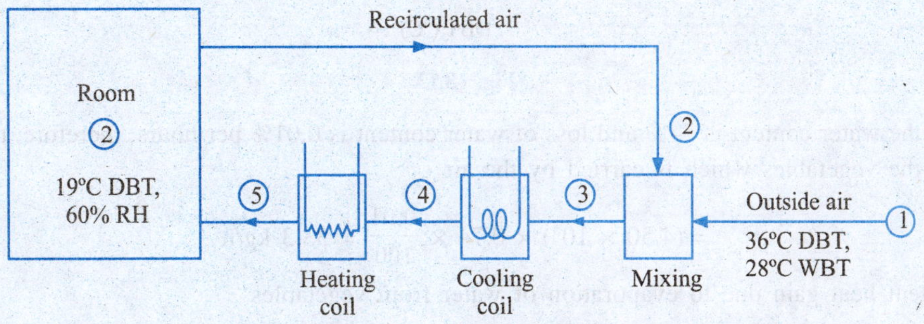

Fig. 18.12. Flow diagram for air-conditioning system of the cold storage.

∴ Sensible heat due to infiltrated air

$$= m_1 (h_A - h_2) = 198.98 (57.1 - 39.2) = 3561.7 \text{ kJ/h} = 0.99 \text{ kW}$$

Taking the sensible and latent heat gains per person and 240 kJ/h and 200 kJ/h respectively, we have:

Total sensible heat gain from the persons

$$= 240 \times 20 = 4800 \text{ kJ/h} = 1.33 \text{ kW, and}$$

Total latent heat gain from the persons

$$= 200 \times 20 = 4000 \text{ kJ/h} = 1.11 \text{ kW}$$

Latent heat due to infiltrated air

$$= m_1 (h_1 - h_A) = 198.98 (90 - 57.1) = 6546.4 \text{ kJ/h} = 1.82 \text{ kW}$$

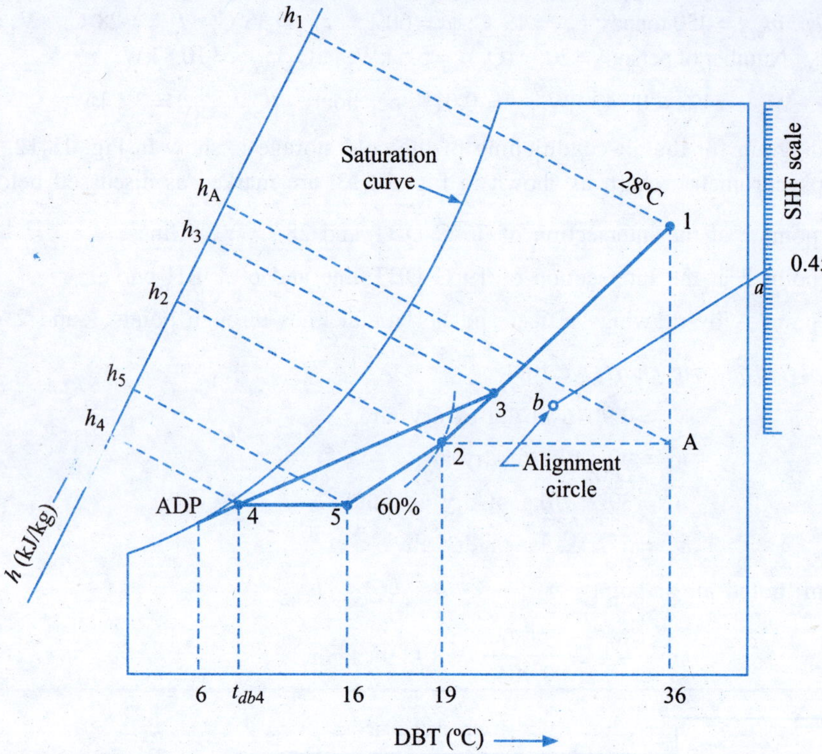

Fig. 18.13

Since the water content is 74% and loss of water content is 0.01% per hours, therefore, total water loss from the vegetables which is carried by the air

$$= (4.50 \times 10^3) \times 0.74 \times \frac{0.01}{100} = 33.3 \text{ kg/h}$$

∴ Latent heat gain due to evaporation of water from vegetables

$$= 33.3 \times 2500 = 83250 \text{ kJ/h} = 23.13 \text{ kW} \text{ ...(Taking } h_{fg} \text{ for water} = 2500 \text{ kJ/kg)}$$

Total sensible heat gain in the cold storage,

RSH = Sensible heat gain due to infiltrated air + Sensible heat gain from the persons + Sensible heat gain from glass + Sensible heat gain through walls and ceiling + Heat from equipment and reaction heat of vegetables

$$= 0.99 + 1.33 + 5.5 + 10.8 + 3.1 = 21.72 \text{ kW}$$

Total latent heat gain in the cold storage,

RLH = Latent heat gain due to infiltrated air + Latent heat gain from persons + Latent heat gain due to evaporation of water from vegetables.

$$= 1.82 + 1.11 + 23.13 = 26.06 \text{ kW}$$

Room sensible heat factor,

$$RSHF = \frac{RSH}{RSH + RLH} = \frac{21.72}{21.72 + 26.06} = 0.454$$

- Mark this calculated value of RSHF = 0.454 on the SHF scale as point *a* and join with point *b* which is alignment circle (i.e., 26°C DBT, 50% RH).
- From point 2, draw a line parallel to *a-b*, intersecting the vertical line drawn through 16°C DBT (i.e., temperature of air entering the room) at point 5. This line 2 – 5 is known as *RSHF line*. The *point 5 represents the condition of air entering the room.*

From the *psychrometric chart*, we find:

$$h_5 = 33.1 \text{ kJ/kg of dry air}$$

(*i*) **Amount of recirculated air:**

Total mass of air passing through the room,

$$m_a = \frac{\text{Total room heat}}{\text{Total heat removed}} = \frac{\text{RSH + RLH}}{(h_2 - h_5)}$$

$$= \frac{21.72 + 26.06}{(39.2 - 33.1)} = 7.83 \text{ kg/s} = 28188 \text{ kg/h}$$

Mass of fresh air supplied from outside

$$m_F = \frac{V_F}{v_1} = \frac{4500}{0.9046} = 4974.6 \text{ kg/h}$$

∴ Mass of recirculated air

$$= m_a - m_F = 28188 - 4974.6 = 23213.4 \text{ kg/h}$$

Percentage of recirculated air

$$= \frac{m_R}{m_a} \times 100 = \frac{23213.4}{28188} \times 100 = \mathbf{82.35\% \text{ (Ans.)}}$$

(*ii*) **Capacity of the cooling coil and its by-pass factor:**

- Since 82.35% of the total air passing through the room at point 2 is recirculated and mixed with 17.65% of air at point 1, therefore mark the mixing point 3 on the psychrometric chart, such that

  Length 2–3 = 0.1765 + Length 2–1.

- Now draw a line joining points 3 and ADP = 6°C intersecting the constant specific humidity line drawn through point 5, at point 4. The *point 4 represents the condition of air leaving the cooling coil.* The line 4–5 represents the heating of air.

From the *psychrometric chart*,

$$t_{db3} = 22°\text{C}; \; t_{db4} = 9.2°\text{C}$$

$$h_3 = 48.5 \text{ kJ/kg of dry air}$$

$$h_4 = 26.2 \text{ kJ/kg of dry air.}$$

∴ Cooling capacity of the coil

$$= \frac{m_a (h_3 - h_4)}{14000} = \frac{28188 (48.5 - 26.2)}{14000} = \mathbf{44.9 \text{ tonnes (Ans.)}}$$

By-pass factor, $$\text{BF} = \frac{t_{db4} - \text{ADP}}{t_{db3} - \text{ADP}} = \frac{9.2 - 6}{22 - 6} = \mathbf{0.2 \text{ (Ans.)}}$$

(*iii*) **Capacity of heating coil:**

From the *psychrometric chart*, we find:

$$h_5 = 33.1 \text{ kJ/kg of dry air}$$

∴ Capacity of the heating coil

$$= m_a (h_5 - h_4) = \frac{28188}{3600} (33.1 - 26.2) = \textbf{54.03 kW (Ans.)}$$

Example 18.7. *In a space to be air-conditioned the heat gains are as follows:*

Heat gains through walls, roof and floor	=	*10 kW*
Occupancy	:	*40 persons each with 75 W sensible and 65 W latent heat loads*
Lighting	:	*1.5 kW*
Ventilation	:	*0.33 m³/min per person*
Infiltration	:	*150 m³/h*
Outdoor conditions	:	*43°C DBT and W = 0.017 kg/kg of dry air*
Indoor conditions	:	*25°C DBT and W = 0.01 kg/kg of dry air*

$Q_S = 0.0204 \times cmm \times \Delta t$ *kW*

$Q_L = 50\ cmm \times \Delta W$ *kW*

cmm is volume flow rate in m³/min.

(*i*) *Determine RSH, RLH, OASH and OALH assuming 10% leakage if whole of duct is outside A/C space. In this case 20% of duct is outside. Heat gain due to fan is 5% of RSH and safety factor = 5%.*

(*ii*) *The cooled air enters the room at 15°C. Determine the supply air volume flow rate and by-pass factor if ADP = 10°C.* **(Punjab University)**

Solution: *Given:* $(Q_S)_{W,\ R\&F}$ = 10 kW; Number of person = 40,

Q_S per person = 75 W; Q_L per person = 65 W, $(Q_S)_L$ = 1.5 kW;

Ventilation V_o = 0.33 m³/min per person; Infiltration = 150 m³/h;

Outdoor conditions : 43°C DBT and W = 0.017 kg/kg of dry air; Indoor conditions = 25°C DBT and W = 0.01 kg/kg of dry air;

$Q_S = 0.0204 \times \text{cmm} \times \Delta t$ kW; $Q_L = 50\ \text{cmm} \times \Delta W$ kW; ADP = 10°C.

(*i*) **RSH, RLH, OASH and OATH :**

$$\text{Infiltration} = \frac{150}{60} = 2.5 \text{ m}^3\text{/min}$$

$$\text{Ventilation} = 0.33 \times 40 = 13.2 \text{ m}^3\text{/min}$$

Room sensible heat, **RSH** $= (Q_S)_{W,\ R\&C}$ + Sensible heat gain from the persons + $(Q_S)_L$ + Q_S (i.e., $0.0204 \times \text{cmm} \times \Delta t$)

$$= 10 + \frac{40 \times 75}{1000} + 1.5 + 0.0204 \times 2.5 \times (43 - 25)$$

$$= \textbf{15.418 kW (Ans.)}$$

Room latent heat, **RLH** = Latent heat gain from the persons + Q_L (i.e., 50 × cmm × ΔW)

$$= \frac{40 \times 65}{1000} + 50 \times 2.5 \times (0.017 - 0.01)$$

$$= \mathbf{3.475\ kW\ (Ans.)}$$

It is assumed that,

(*a*) The fan is located before the conditioner, and the heat gain due to fan is 5% (given) of RSH, after including the leakage load and the safety factor.

(*b*) Since leakage load is entirely sensible, it is calculated on the basis of RSH only.

(*c*) Leakage load and safety factor are considered as a part of room load for the purpose of analysis. Since 20% of the duct is outside, the leakage load will be 0.2 × 0.1 (= 0.02) of the RSH.

$$\text{Then total RSH} = 15.418 \times 1.02 \times 1.05$$

$$= 16.5127 \text{ kW}$$

$$\text{and total RLH} = 3.475 \times 1.05 = 3.6488 \text{ kW}$$

Fan load (entirely sensible)

$$= 0.05 \times 16.5127 = 0.8256 \text{ kW}$$

$$\textbf{OASH} = \text{RSH + Fan load + Ventilation sensible load}$$

$$= 16.5127 + 0.8256 + 0.0204 \times 13.2 \times (43 - 25)$$

$$= \mathbf{22.185\ kW\ (Ans.)}$$

$$\textbf{OALH} = \text{RLH + Ventilation latent load}$$

$$= 3.6488 + 50 \times 13.2 \times (0.017 - 0.01)$$

$$= \mathbf{8.2688\ kW\ (Ans.)}$$

(*ii*) **Supply air volume flow rate and by-pass factor:**

Let V_s = Supply air rate, m³/min.

The schematic arrangement, and the psychrometric chart showing the processes are shown in Figs. 18.14 and 18.15 respectively.

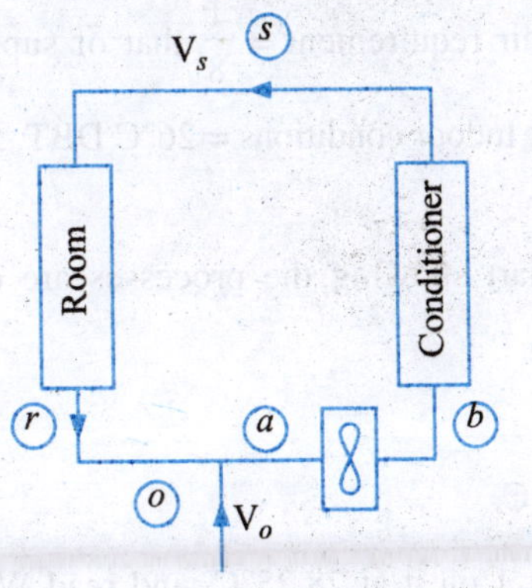

Fig. 18.14. Schematic arrangement.

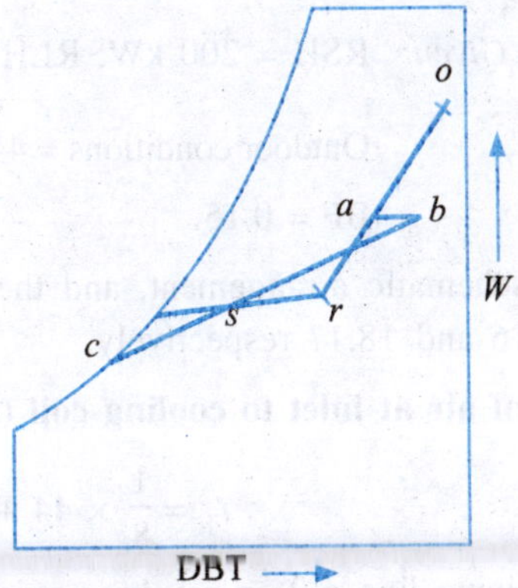

Fig. 18.15

Now, $\quad \text{RHS} = 0.0204 \times V_s\,(t_r - t_s)$

or $\quad 16.5127 = 0.0204 \times V_s\,(25 - 15)$

$\therefore \quad \mathbf{V_s = 80.94\ m^3/s}$ **(Ans.)**

$$t_a = \frac{V_0 t_0 + (V_s - V_0)\,t_r}{V_s} = \frac{13.2 \times 43 + (80.94 - 13.2) \times 25}{80.94}$$

$$= 27.9°\text{C}$$

$$\text{Fan load} = 0.8256 = 0.0204 \times V_s\,(t_b - t_a)$$

$$= 0.0204 \times 80.94\,(t_b - 27.9)$$

$$t_b = 28.4°\text{C}$$

$$\text{By-pass factor} = \frac{t_s - t_c}{t_b - t_c} = \frac{15 - 10}{28.4 - 10} = \mathbf{0.27}\ \textbf{(Ans.)}$$

Check: $\quad \text{OASH} = 0.0204 \times V_s\,(t_b - t_s)$

$$= 0.0204 \times 80.94\,(28.4 - 15)$$

$= 22.13$, which agrees fairly with above value.

Hence the assumptions made are justified.

Example 18.8. *RSH and RLH for an air-conditioned space are 200 kW and 30 kW respectively. Fresh air requirement is 1/8 that of supply air. Outdoor and indoor conditions are 44°C DBT, 30% RH and 26°C DBT, 50% RH respectively, By-pass factor of cooling coil is 0.15.*

(*i*) *Determine the state of air at inlet to cooling coil* (t_1, W_1).

(*ii*) *Set up a relation between cooling coil outlet air state* (t_2, W_2) *by using ratio of RSH and RLH [call it Equation (1)].*

(*iii*) *Set up relations between* t_2 *and* t_{ADP} *[Equation (2)] and* W_2 *and* W_{ADP} *[Equation (3)].*

(*iv*) *Solve the three equations by iteration and find the supply air volume flow rate.*

Take: $\quad Q_S = 0.0204 \times V \times \Delta t$ **(Banglore University)**

$$Q_L = 50 \times V \times \Delta W$$

Solution: *Given:* RSH = 200 kW; RLH = 30 kW; Fresh air requirement = $\frac{1}{8}$ that of supply air;

Outdoor conditions = 44°C DBT, 30% RH; Indoor conditions = 26°C DBT, 50% RH;

BF = 0.15.

The schematic arrangement, and the psychrometric chart showing the processes are depicted in Figs. 18.16 and 18.17 respectively.

(*i*) State of air at inlet to cooling coil (t_1, W_1)

$$t_1 = \frac{1}{8} \times 44 + \frac{7}{8} \times 26 = 28.25°\text{C}.$$

By joining line *r*–0 on psychrometric, chart, mark point 1 on it at 28.25°C, and read $W_1 = 0.0116$ kg/kg of air.

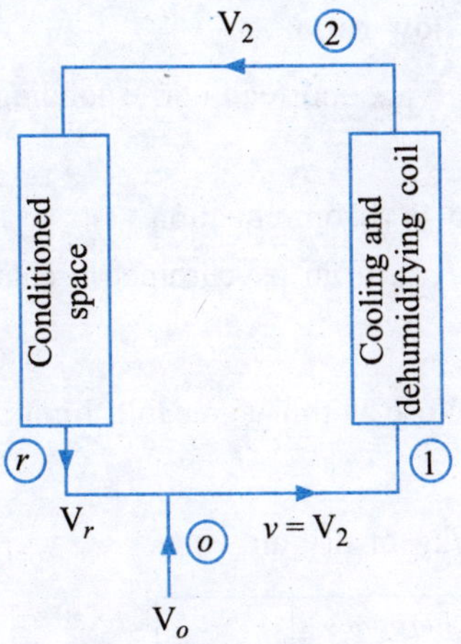

Fig. 18.16

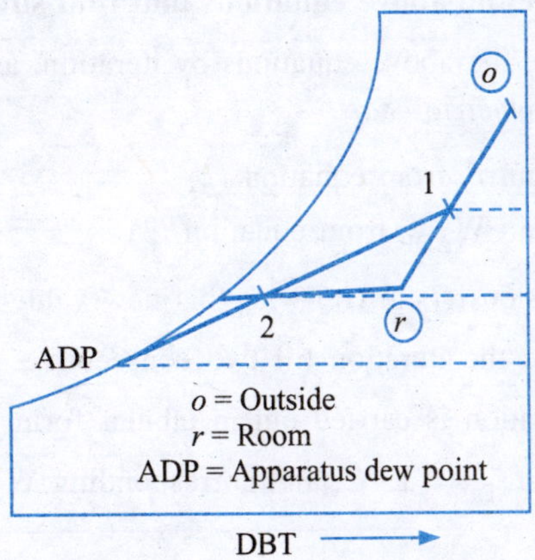

Fig. 18.17

Hence state of air at inlet to cooling coil is **28.25°C DBT, 0.0116 kg water vapour/kg of dry air. (Ans.)**

(*ii*) **Relation between cooling coil outlet air state (t_2, W_2):**

From *psychrometric chart,*

$$W_r = 0.0107 \text{ kg/kg of dryair.}$$

$$\text{Room SHF} = \frac{\text{RSH}}{\text{RSH} + \text{RLH}} = \frac{0.0204\, V_2 (t_r - t_2)}{0.0204\, V_2 (t_r - t_2) + 50\, V_2 (W_r - W_2)}$$

or,

$$\frac{200}{200 + 30} = \frac{0.0204 (26 - t_2)}{0.0204 (26 - t_2) + 50 (0.0107 - W_2)}$$

This can be solved for W_2 as:

$$\mathbf{W_2} = 0.0107 - \frac{26 - \mathbf{t_2}}{16340} \textbf{ (Ans.)} \qquad ...(1)$$

(*iii*) **Relations between t_2 and ADP and W_2 and W_{ADP}**

$$\text{BF} = \frac{t_2 - t_{ADP}}{t_1 - t_{ADP}}, \quad \text{or,} \quad 0.15 = \frac{t_2 - t_{ADP}}{28.25 - t_{ADP}}$$

This can be solved for t_2 as:

$$\mathbf{t_2 = 0.85\, t_{ADP} + 4.2375 \text{ (Ans.)}} \qquad ...(2)$$

Also,

$$\text{B} = \frac{W_2 - W_{ADP}}{W_1 - W_{ADP}}, \quad \text{or,} \quad 0.15 = \frac{W_2 - W_{ADP}}{0.0116 - W_{ADP}}$$

This can be solved for W_{ADP} as:

$$\mathbf{W_{ADP}} = \frac{W_2 - 1.74 \times 10^{-3}}{0.85} \textbf{ (Ans.)} \qquad ...(3)$$

(*iv*) **To solve the above equations and find supply air volume flow ratio:**

To solve the above equations by iteration, assume a value of t_{ADP}, and read corresponding W_{ADP} from *psychrometric chart:*

(*i*) Obtain t_2 from equation (2)
(*ii*) Obtain W_2 from equation (1)
(*iii*) Obtain W_{ADP} from equation (3)
(*iv*) Read t_{ADP} from psychrometric chart

Repeat steps (*i*) to (*iv*) with this new value of t_{ADP}.

Continue the iteration till two consecutive values of t_2 are close within agreeable limits.

The iteration is carried out in tabular form as below:

Assume t_{ADP} = 13°C, and corresponding W_{ADP} = 0.0904 kg/kg of dry air

	Iterations 1st	2nd	3rd
t_{ADP} (from psychrometric chart)	13	13.8	13.8
$t_2 = 0.85\ t_{ADP} + 4.2375$	15.23	15.97	15.97
$W_2 = 0.0107 - (26\ t_2)/16340.047$	0.0100	0.010047	
$W_{ADP} = \dfrac{W_2 - 1.74 \times 10^{-3}}{0.85}$	0.0097	0.0097	

Hence, $t_2 = 15.97°C$

$$RSH = 0.0204 \times V_2 \times (t_r - t_2)$$

or, Supply air rate, V_2, is given by:

$$V_2 = \frac{200}{0.0204\,(26 - 15.97)}$$

$$= \mathbf{977.46\ m^3/min\ (Ans.)}$$

Example 18.9. *It is required to air-condition a laboratory 27 m × 18 m × 4 m, the 27 m wall faces north. The following data is available:*

Two doors in north wall .... 2.5 m × 3 m each

Four glass windows in south wall .... 2 m × 1.5 m each

Four glass windows in east wall .... 2 m × 1.5 m each

Four glass windows in west wall .... 2 m × 1.5 m each

Lighting load .... 20 W fluorescent per m² floor area

Infiltration .... One air change

Solar heat gain factors (SHGF) for south, east and west glass....... 140, 45 and 340 W/ m² respectively.

The overall heat transfer coefficients for walls, roof, floor, door and windows .... 2.4, 1.8, 2.8, 1.4, 5.8 W/m² K respectively

The corrected equivalent temperature differences for north, south, east, west walls, roof and floor .... 11, 14, 11, 16, 19, 2.4°C respectively

Number of persons 80

Sensible and latent heat loads per person .... 70 W and 50 W respectively

Ventilation requirements per person .... 0.28 m³/min

Outdoor conditions .... 42°C DBT and 0.026 kg/kg of dry air of humidity ratio

Indoor conditions .... 24°C DBT and 0.009 kg/kg of dry air of humidity ratio

Using a factor of 1.25 for fluorescent light, determine room sensible heat load and room latent heat load with 4% safety factor, 4.5% for fan power, 1% leakage of supply air and 0.4% heat leakage to supply air duct.

Solution: *Given:* L = 27 m; B = 18 m; H = 4 m; Door size = 2.5 m × 3m; Window size = 2 m × 1.5 m;

Lighting load = 20 W fluorescent per m^2 floor area; Number of air changes = 1;

SHFG for south, east and west glass = 140, 45 and 340 W/m^2 respectively;

U_{wall} = 2.4 W/m^2 K, U_{roof} = 1.8 W/m^2 K; U_{floor} = 2.8 W/m^2 K; U_{door} = 1.4 W/m^2 K;

$U_{windows}$ = 5.8 W/m^2 K; t_e for north wall = 11°C; t_e for south wall = 14°C;

t_e for east wall = 11°C; t_e for west wall = 16°C; t_e for roof = 19°C; t_e for floor = 2.4°C;

Number of persons = 80; Q_S per person = 70 W; Q_L per person = 50 W;

v_1 = 0.28 m^3/min person; t_{db1} = 42°C; W_1 = 0.026 kg/kg of dry air; t_{db2} = 24°C;

W_2 = 0.009 kg/kg of dry air; Allowance for fluorescent light = 1.25;

Safety factor = 4%; Factor for fan power = 4.5%;

Factor for leakage of supply air 1%; Factor for heat leakage to supply air duct = 0.4%.

The plan of the laborary room is shown in Fig. 18.18.

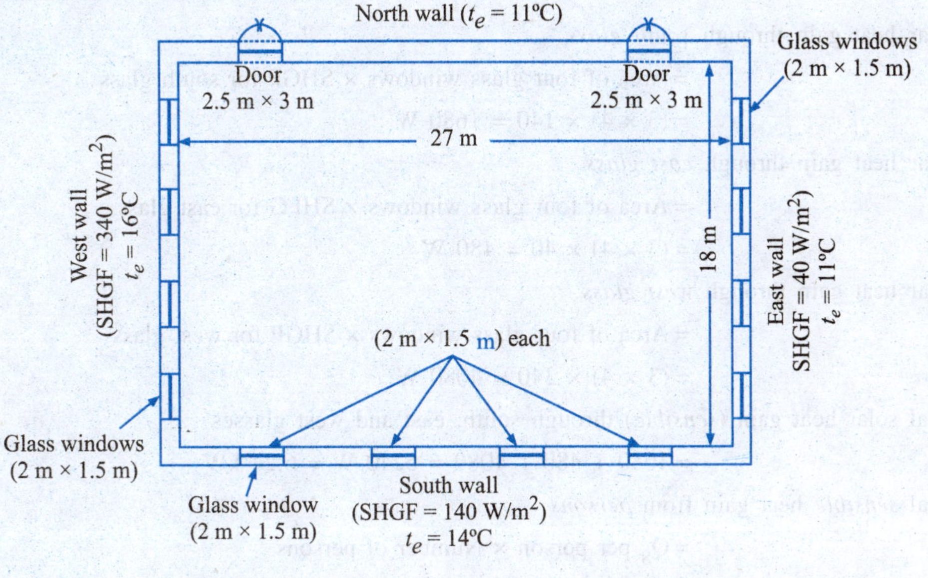

Fig. 18.18

The details of sensible heat gain from various sources is given in Table 18.8.

Table 18.8. Estimation of sensible heat gain

Area of one door = 2.5 × 3 = 7.5 m^2; Area of one window = 2 × 1.5 = 3 m^2

Source	*Overall heat transfer coefficient (U) W/m^2 K*	*Area (A)* m^2	*Equivalent temperature difference* (t_e)°C	*Sensible heat gain = U.A. t_e* W
1. North wall	2.4	27 × 4 – 2 × 7.5 = 93	11	2.4 × 93 × 11 = 2455
2. South wall	2.4	27 × 4 – 4 × 3 = 96	14	2.4 × 96 × 14 = 3226
3. East wall	2.4	18 × 4 – 4 × 3 = 60	11	2.4 × 60 × 11 = 1584
4. West wall	2.4	18 × 4 – 4 × 3 = 60	16	2.4 × 60 × 16 = 2304
5. Roof	1.8	27 × 18 = 486	19	1.8 × 486 × 19 = 16621
6. Floor	2.8	27 × 18 = 486	2.4	2.8 × 486 × 2.4 = 3266
7. Doors (2 Nos.) in north wall	1.4	2 × 7.5 = 15	11 (Same as north wall	1.4 × 15 × 11 = 231
8. Windows				
(*a*) South wall (4 Nos)	5.8	3 × 4 = 12	*18	5.8 × 12 × 18 = 1253
(*b*) East wall (4 Nos)	5.8	3 × 4 = 12	18	5.8 × 12 × 18 = 1253
(*c*) West wall (4 Nos)	5.8	3 × 4 = 12	18	5.8 × 12 × 18 = 1253 Total = 33446 W = 33.446 kW

Solar heat gain through *south glass*

= Area of four glass windows × SHGF for south glass

= (3 × 4) × 140 = 1680 W

Solar heat gain through *east glass*

= Area of four glass windows × SHFG for east glass

= (3 × 4) × 40 = 480 W

Solar heat gain through *west glass*

= Area of four glass windows × SHGF for west glass

= (3 × 4) × 340 = 4080 W

Total solar heat gain (*sensible*) through south, east and west glasses

= 1680 + 480 + 4080 = 6240 W = 6.24 kW

Total *sensible* heat gain from *persons*

= Q_S per person × Number of persons

= 70 × 80 = 5600 W = 5.6 kW

* For glass windows, the equivalent temperature difference is given as

$$t_e = t_{db1} - t_{db2} = 42 - 24 = 18°C$$

Total *latent* heat gain from *persons*

$$= Q_L \text{ per person} \times \text{Number of persons}$$

$$= 50 \times 80 = 4000 \text{ W} = 4 \text{ kW}$$

Amount of *infiltrated air,*

$$V_1 = \frac{L \times B \times H \times \text{Number of air changes}}{60}$$

$$= \frac{27 \times 18 \times 4 \times 1}{60} = 32.4 \text{ m}^3/\text{min} \quad [\because \text{Number of air change} = 1 \text{ ...Given}]$$

$\therefore$ Sensible heat gain due to infiltration air

$$= 0.02044 \text{ V}_{inf.} (t_{ab1} - t_{db2}) \quad \text{...[Eqn. 12.23]}$$

$$= 0.02044 \times 32.4 (42 - 24) = 11.92 \text{ kW, and}$$

Latent heat gain due to infiltration air

$$= 50 \text{ V}_{inf.} (W_1 - W_2) \quad \text{...[Eqn. 12.27]}$$

$$= 50 \times 32.4 (0.026 - 0.009) = 27.54 \text{ kW}$$

Volume of ventilation or outside air,

$$V_{inf.} = 0.28 \text{ m}^3/\text{min/person} \times \text{Number of persons}$$

$$= 0.28 \times 80 = 22.4 \text{ m}^3/\text{min}$$

$\therefore$ Outside air sensible heat,

$$\text{OASH} = 0.02044 \text{ V}_{vent.} (t_{db1} - t_{db2})$$

$$= 0.02044 \times 22.4 (42 - 24) = 8.24 \text{ kW, and}$$

Outside air latent heat,

$$\text{OALH} = 50 \text{ V}_{vent.} (W_1 - W_2)$$

$$= 50 \times 22.4 (0.026 - 0.009) = 19.04 \text{ kW}$$

Sensible heat gain due to lighting

$$= \text{Total wattage of lights} \times \text{Use factor} \times \text{Allowance factor}$$

$$= 20 \times (27 \times 18) \times 1 \times 1.25 = 12150 \text{ W} = 12.15 \text{ kW} (\because \text{Use factor} = 1)$$

Assuming that the *fan is placed **before** the conditioner* (i.e., neglecting factor of 4.5% for fan power), the total room sensible heat (RSH) is to be increased by: 4% (safety factor) + 1% (for leakage of supply air) + 0.4% (for heat leakage to duct) = 5.4%

$\therefore$ Total room sensible heat,

$$\textbf{RSH} = 1.054 \left[\begin{array}{l}\text{Heat gain from walls, roof, floor and windows (from table 18.8)} \\ + \text{solar heat gain through glasses} + \text{Sensible heat gain from persons} \\ + \text{Sensible heat gain due to infiltration air} + \text{Sensible heat gain due} \\ \text{to ventilation (OASH)} + \text{Sensible heat gain due to lighting.}\end{array}\right]$$

$$= 1.054 [33.446 + 6.21 + 5.6 + 11.92 + 8.24 + 12.15] = \textbf{81.78 kW (Ans.)}$$

The total room latent heat (RLH) is to be increased by : 4% (safety factor) + 1% (for leakage of supply air) = 5%

∴ Total room latent heat,

$$\textbf{RLH} = 1.05 \left[\begin{array}{l}\text{Latent heat gain from persons } + \text{ Latent heat gain due to} \\ \text{infiltration air } + \text{ Latent heat due to ventilation (OALH)}\end{array}\right]$$

$$= 1.05\ [4 + 27.54 + 19.04] = \textbf{53.11 kW (Ans.)}$$

If the *fan is located **after** the conditioner,* then 4.5% fan power should be added to RSH; *RLH will, however, remain unaffected.*

$$\therefore \quad \textbf{RSH} = (1.054 + 0.045)\ (33.446 + 6.24 + 5.6 + 11.92 + 8.24 + 12.15)$$

$$= \textbf{85.28 kW (Ans.)}$$

HIGHLIGHTS

1. The estimation of load involves the following variables:
   (*i*) Magnitude and direction of wind velocity.
   (*ii*) Outside humidity and temperature.
   (*iii*) Nature of construction, materials used.
   (*iv*) Orientation of openings, windows and doors.
   (*v*) Periods of occupancy and number of persons in the room, activities of the persons etc.
2. For air-conditioning the cooling load can be classified as follows:
   (*i*) Room load—which falls on the *room directly*.
   (*ii*) Total load—which falls on the *air-conditioning apparatus.*
3. The solar heat reaches any part of the earth's surface in the form of two radiations:
   (*ii*) Beam or direct radiation.
   (*ii*) Diffuse or sky radiation.
4. Direct solar heat gain can be reduced by using different types of glass, construction and shades.
5. Transfer of heat by conduction (Q) from outside to inside under steady state due to the temperature difference is given by the equation:

   $$Q = UA\ (t_0 - t_i)$$

   and 'overall thermal resistance'.

   $$\frac{1}{U} = \frac{1}{6} + \frac{x_1}{k_1} + \frac{x_2}{k_2} + \ldots + \frac{1}{f_i}$$

6. In order to estimate air leakage through *window and door cracks, the* following two methods may be used:
   (*i*) Air change method.
   (*ii*) Crack length method.

   Air change method is purely arbitrary and based on the engineering practice. The crack length method is used in case of windows and the air change method is more convenient to use in the case of doors.
7. The sensible and latent heat gains due to occupants, lights, appliances machines, pipes etc., within the conditioned space form the components of *internal heat gains.*

OBJECTIVE TYPE QUESTIONS

Fill in the blanks:

1. The primary requirement of .... or .... equipment is that these must be able to remove or add heat at the rate at which it is produced or removed, and maintain the given comfort conditions in the room.
2. The .... of equipment must be estimated at a value, which wisely accounts for the physical and economical comfort of the owner.
3. For air-conditioning the cooling load can be classified as .... load and .... load.
4. The normal value of solar constant is assumed as .... kJ/m^2-h.
5. The solar heat reaches any part of the earth's surface in the form of two radiations .... and ....
6. Complete the equation : Q = ... A (..............)
7. The film co-efficient varies directly with ....
8. Estimation of air leakage through window and door cracks can be done by two methods .... and ....
9. .... method is purely arbitrary and based on engineering practice.
10. .... method is used in case windows.

ANSWERS

1. Cooling, heating;	**2.** Capacity;	**3.** Room, total;	**4.** 5045;
5. Beam, diffuse;	**6.** U, t_0, t_p;	**7.** Wind velocity;	
8. Air change, crack length;		**9.** Air change;	**10.** Crack length.

THEORETICAL QUESTIONS

1. List the variables which are involved in the estimation of load.
2. Enumerate and explain the components of cooling-load estimate.
3. What points should be considered while making heat load calculations?
4. Write a short note on 'solar radiation'.
5. Write down the procedure for calculating heat gains through building structures.
6. What do you mean by the term 'infiltration'? Explain briefly how air leakage through window and door cracks can be estimated?
7. Write a short note on 'chimney or stack effect'.
8. Enumerate and explain the components of internal heat gains.
9. List the components of system heat gains.

UNSOLVED EXAMPLES

1. The following data relate to a conference room for seating 100 persons.

Inside design conditions .... 22°C DBT, 60% RH

Outside design conditions .... 40°C DBT, 27° WBT

Sensible and latent heat loads per person .... 80 W and 50 W respectively

Lights and fans loads .... 15000 W

Air infiltration .... 20 m^3/min

Sensible heat gain through glass, walls, ceiling etc. .... 15000 W

Fresh air supply .... 100 m^3/min

By-pass factor of the coils .... 0.1.

If two-third of recirculated room air and one-third of fresh air are mixed before entering the cooling coils, determine:

(*i*) Apparatus dew point;

(*ii*) Grand total heat load; and

(*iii*) Effective room sensible heat factor. **[Ans.** (*i*) 11.5°C; (*ii*) 131.72 kW; (*iii*) 0.77]

2. An office for seating 25 occupants is to be maintained at 24°C DBT, 50% RH. The outdoor conditions are 34°C DBT, 28°C WBT. The various loads in the office are:

Solar heat gain .... 9.12 kW

Sensible heat gain per occupant .... 0.09 kW

Latent heat gain per occupant .... 0.105 kW

Lighting load .... 2.3 kW

Sensible heat load from other sources .... 11.63 kW

Infiltration load .... 14 m^3/min

Assuming 40 per cent fresh air and 60 per cent of recirculated air passing through the evaporator and the by-pass factor of 0.15, determine:

(*i*) Dew point temperature of the coil; and

(*ii*) Capacity of the plant. **[Ans :** (*i*) 7.3°C, (ii) 18.8 tonnes]

3. The following data is available for designing an air-conditioning system for a restaurant:

Inside design conditions = 27°C DBT, 55% RH; Outdoor conditions = 35°C DBT, 26°C DBT; Minimum temperature of air supplied to room = 17°C DBT; Total amount of fresh air supplied = 1600 m^3/h; Total infiltration air = 400 m^3/h; Seating chairs for dining = 50; Employees serving the meals = 5; Sensible heat gain per person = 58 W; Latent heat gain per sitting person = 44 W; Latent heat gain per employee = 76 W; Sensible heat added from meals = 0.17 kW; Latent heat added from meals = 0.3 kW; Total heat flow through the walls, roof and floor = 6.2 kW, Solar heat gain through glass = 2 kW; Equipment sensible heat gain = 2.9 kW; Equipment latent heat gain = 0.7 kW; Motor power connected to the fan = 7.5 kW.

If the fan is installed before the conditioner, determine:

(*i*) Amount of air delivered to the room in m^3/h;

(*ii*) Percentage of recirculated air;

(*iii*) Refrigeration load on the coil in tonnes of refrigeration; and

(*iv*) Dew point temperature of the cooling coil and by-bass factor.

[Ans. (*i*) 4745 m^3/h, (*ii*) 68.6%, (*iii*) 9.9 tonnes, (*iv*) 0.161]

4. The following data refer to a hall to be air-conditioned:

Inside conditions = 24°C DBT, 60% RH; Outdoor conditions = 38°C DBT, 28°C WBT; Sensible heat load in the room = 46.4 kW; Latent heat load in the room = 11.6 kW; Total infiltration air = 1200 m^3/h; Apparatus dew point = 10°C; Quantity of recirculated air from the hall = 60 per cent:

If the quantity of recirculated air is mixed with the conditioned air after the cooling coil, find: (*i*) The condition of air leaving the conditioner coil and before mixing with the recirculated air, (*ii*) The condition of air before entering the hall, (*iii*) The mass of air entering the cooler, (*iv*) The mass of total air passing through the hall, (*v*) The by-pass factor of the cooling coil, and (*vi*) The refrigeration load on the cooling coil.

[**Ans.** (*i*) 16°C; 90% RH; (*ii*) 20.8°C; 70% RH, (*iii*) 23400 kg/h; (*iv*) 58500 kg/h; (*v*) 0.214; (*vi*) 79.56 tonnes]

5. It is required to design a cold storage for storing 500 tonnes of vegetables with the following available data:

Inside design conditions = 20°C DBT and 60% RH; Outdoor conditions = 35°C DBT and 28°C WBT; Infiltrated air = 200 m^3/h; Fresh air supplied from outside = 4400 m^3/h; Number of person working in the cold storage = 25; Sensible and latent heat gain per person = 250 kJ/h and 210 kJ/h respectively; Sensible heat gain through glass = 5.8 kW; Sensible heat gain through walls and ceiling = 11.6 kW; Water contents of the vegetables = 76%; Loss of water content = 0.01 per cent per hour; Heat from equipment and reaction heat of vegetables = 3.5 kW. If the air-conditioning is achieved by first cooling and dehumidifying and then heating and the temperature of air entering the room is not to exceed 15°C, determine:

(*i*) Amount of recirculated air, if the recirculated air is mixed with fresh air before entering the cooling coil;

(*ii*) Capacity of the cooling coil in tonnes of refrigeration and its by-pass factor if the dew point temperature of the coil is 5°C;

(*iii*) Capacity of the heating coil in kW. [**Ans.** (*i*) 73%, (*ii*) 41.14 tonnes, (*iii*) 38 kW]

❑❑❑

Chapter 19

Measuring Instruments—Air-Conditioning

19.1 MEASUREMENT OF HUMIDITY

The instrument used to measure humidity of air is called the *hygrometer*. Various types of hygrometers are: 1. Psychrometer 2. Dew point hygrometer 3. Organic hygrometer 4. Electrolytic hygrometer 5. Chemical and miscellaneous hygrometers.

1. **Psychrometer:** Refer to article

2. **Dew point hygrometer:**

- A dew point hygrometer consists of a bright mirrored surface placed in air and cooled at the back by evaporation of ether or dry ice. The surface temperature at which visible condensation appears on the surface gives the dew point of the air around it. The surface temperature can be measured by thermometer immersed in fluid in contact with the back of the surface or by means of thermocouple (soldered to the surface). Instead of relying on visible indication of condensation, the modern method is to use a photronic cell and a light source to detect dew or frost on the surface.
- Dew point hygrometers are *useful for dew points near the boiling point* (viz., *flue gases*).

3. **Organic hygrometers:**

- These hygrometers make use of organic materials such as human hair, wood and paper which elongate with increase in humidity of surrounding air. Due to change in dimension of material some motion takes place which acts through a linkage and causes a pointer to move across an indicating dial or moves a pen across a chart of a recorder or actuates a pneumatic, electric or electronic control mechanism. Since these organic materials are not reliable in consistent reproduction of their action over any length of time, therefore *these devices should be calibrated initially and frequently recalibrated.*

4. **Electrolytic hygrometers:**

- The Dunmore and Weaver are the electrolytic hygrometers and are based on the *principle of change in electrical resistance of an electrolytic film such as that of lithium chloride.*
  - — In *Dunmore hygrometer* there is a dual winding of small wire on an insulating tube and the assembly is coated with an electrolytic film. The humidity of surrounding air is found by calibrating the electrical resistance of this film.
  - — *Weaver hygrometer* consists of a threaded plug with a central electrode which is insulated from the plug except for the electrolytic film. The plug is exposed to the gas whose humidity is to be measured. The resistance of the film is measured/determined. The plug is then exposed to the standard gas whose humidity is known and the resistance of the film is adjusted to the previous value by changing the pressure of the standard gas. The humidity of the test gas is then calculated. This type of hygrometer is *useful for measurement of humidity of air in closed vessels.*

5. **Chemical and miscellaneous hygrometer:**

- The chemicals (sorbents) such as *sulphuric acid, lithium chloride or silica gel* are used to measure directly the humidity of a known weight of air sample by extracting the water vapour and weighing it.

  Dry ice may be used to freeze water vapour from a known air stream for humidity measurement.

  Thermal conductivity method may be used to measure low humidity for air at high temperature.

19.2 MEASUREMENT OF INFILTRATION

- The *procedure of infiltration measurement is as follows:*

A tracer gas (such as hydrogen) is introduced into a room and thoroughly mixed with room air. The concentration of tracer gas is continuously reduced due to infiltration. The concentration of tracer gas is a function of the thermal conductivity of the mixture. A *thermal conductivity meter* will give the concentration which is calculated at regular time intervals. The rate of infiltration is found by using the relation:

$$I = \frac{V(2.718C_0 - C)}{t} \quad ...(19.1)$$

where, I = Infiltration rate cmm,

V = Volume of room in m^3,

C_0 = Percent initial concentration of tracer gas,

C = Per cent final concentration tracer gas after 't' minutes, and

t = Time in minutes.

19.3 MEASUREMENT OF PRESSURE

The following methods are used to measure pressure:

1. Pressure gauges
2. Manometers
3. Aneroid barometer.

1. **Pressure gauge:**

- A *Bourdon tube* is used in most of the pressure gauges. This tube is metal tube of oval cross-section curved along its length to form almost a complete circle. The free end of the tube is closed and is connected to the pointer by suitable linkages. The tube tends to straighten when connected to the vessel under pressure.
- The *'pressure gauge' is calibrated by means of a dead weight tester whilst a 'vacuum gauge' is calibrated by mercury manometers.*
- The gauge is set to read the pressure normally used accurately.

2. **Manometers:**

- *Manometers* (in the form of U-tubes) are the primary instruments used for the measurement of low pressures and for calibration of other instruments. The legs of the U-tubes are partially filled with a liquid which may be *mercury for measurement of comparatively higher pressures or water or paraffin for low pressures.* The tubes must be chemically clean. To minimise the depression of the meniscus through the effects of surface tension they must have a minimum bore diameter form 0.5 to 1.25 cm.
- *Draft gauges* (or inclined manometers) are employed to measure the pressure difference of a few cm of water. One of the tubes of such gauges in inclined to a slope not less than 1 : 10 so as to get upto a magnification of 10 in the reading. The other tube is in form of a reservoir, the level of the liquid in which does not appreciably change. A liquid lighter than water may be used to get greater magnification. The gauge should be accurately levelled and calibrated for the liquid at different temperatures as its density varies with its temperature.
- Very sensitive micro-manometers are used to measure very low pressure to the tune of 0.0025 cm of water *column.*
- In order to measure pressure in the surrounding atmosphere a simple barometer consisting of a tube more than 75 cm long containing mercury and inverted in a mercury cup may be used. In the aneroid barometer, the variations in pressure deflect the thin surface of a diaphragm. Aneroid barometers of high class can read up to 0.005 cm of mercury.

19.4 MEASUREMENT OF AIR FLOW

Invariably, in air-conditioning, it is required to measure the flow of air through duct work and conditioned spaces. The velocity of air may be measured by using following methods:

1. Pitot tube
2. Propeller or Rotating vane anemometer
3. Deflecting vane meter
4. Hot wire resistance type anemometer
5. Kata thermometer
6. Test smoke
7. Orifices and nozzles.

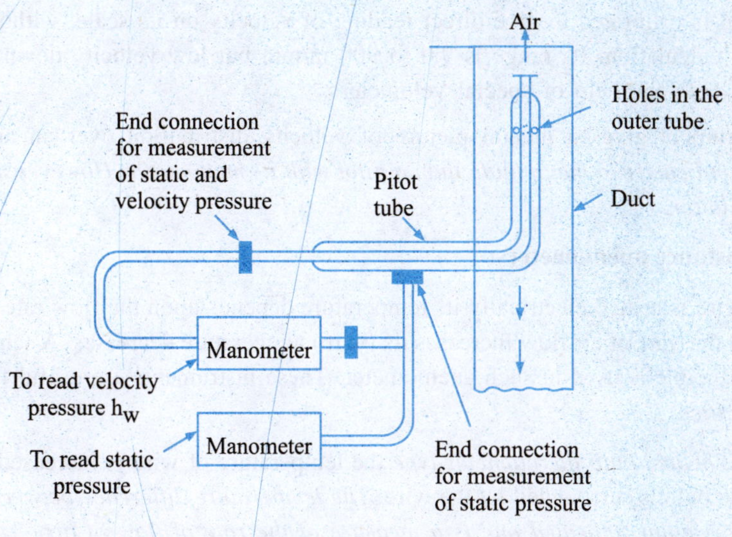

Fig. 19.1. Pitot tube.

1. **Pitot tube:**

- Pitot tube, in conjunction with some form of manometer, is used to measure the air velocity. Various types of Pitot tubes are available. Fig. 19.1 shows a schematic representation of standard N.P.L. type. There are two tubes. The *outer tube* has a number of 7 small 0.0965 cm diameter holes and is closed at end points. It is used to *determine static pressure*. The inner tube has its chamfered end facing the air flow. The static holes are arranged on the parallel portion of the outer tube. For measurement of velocity of air the connections to the manometer are made as shown in fig.19.1. If the resulting reading is h_w metre of water, then

$$V = 8400\sqrt{\frac{h_w}{w_a}} \quad \text{...(19.2)}$$

where, V = Velocity of air m/min, past the tube, and

w_a = Density of air, kgf/m^3.

— For measuring lower velocities (say less than 5 m/s) a *micro-manometer must be used.*

2. **Rotating vane anemometer:**

- A rotating vane anemometer is useful for measuring air velocities less than 30 m/min at grilles.
- It consists of a wheel which is driven by velocity energy of the wind. This wheel is connected to a set of recording dials which are calibrated to read linear metre of air. The time is noted with the help of a watch. When the velocities are low (below 60 m/min.) the instrument does not remain sensitive due to friction drag in the mechanism.

3. **Deflecting vane anemometer:**

- This instrument is also called *Velometer*.
- Here the energy of air is at the downstream opening. The vane is caused to move as the passing air exerts pressure. The vane movement is resisted by a hair spring and a damping magnet. The

instrument is calibrated to give direct reading of velocity on its scale, without the necessity of timing or calculation. Its range is 9.0 to 900 m/min but low velocity down to 3m/min can be measured with the help of special velometers.

- This instrument gives a mental picture of velocity distribution over an air way. It is *more portable and easier to manipulate than a pitot with its manometer. However this is less accurate than the latter.*

4. **Hot wire resistance anemometers:**

- When a wire is heated electrically its temperature depends upon the flow rate of the surrounding air. When the rate of air flow increases its (wire) temperature decreases. A sensing element based on this principle is used in such anemometer. These instruments are useful for *measurement of low velocities.*

— In the *heated bulb thermometer type* the temperature of wire is increased by a thermometer whose bulb is surrounded by the wire. *The temperature difference between this thermometer and a similar unheated one is a measure of the rate of flow of air.*

— In the heated *thermocouple anemometer,* the heated wire surrounds a thermocouple and hence *e.m.f. difference between this and unheated thermocouple is a measure of air flow.*

5. **Kata thermometer:**

- When it is required to measure low velocities in open free spaces Kata thermometer proves useful.
- It is an alcohol thermometer with a very large silvered bulb and with a stem marked only with divisions. For the low Kata these readings are 35ºC and 37.7°C and for high Kata these are 51.6ºC and 54.4ºC.

The *procedure* of taking a reading with this instrument is as follows:

(*i*) The instrument is heated in a water bath to a temperature until the alcohol has risen sufficiently above the upper mark on them.

(*ii*) The instrument is then taken out and its bulb dried.

(*iii*) The time taken for the alcohol to fall from the upper to the lower mark in the stem is noted.

(*iv*) The time noted above is then converted into the velocity by calculating it from the constant of Kata thermometer or by reference to a chart supplied.

- This instrument can also be employed to measure the cooling power of the surrounding air.

6. **Smoke.** Smoke is very useful in the *qualitative study of air motion.* It may be obtained from:

(*i*) Titanium tetrachloride, or

(*ii*) Burning a mixture of potassium chloride with sugar, or

(*iii*) Mixing fumes from aqua ammonia and sulphuric acid to from a white cloud.

7. **Orifices and nozzles.** By introducing orifice or nozzle in the air line the flow in a duct can be measured. After measuring the pressure differences across the device the flow can be calculated as follows:

$$Q = 8400 \times C_d A \sqrt{\frac{h_w}{w_a}} \quad \text{...(9.3)}$$

where, Q = Air flow, cm

A = Area of orifice or nozzle, m^2,

h_w = Pressure drop across orifice or nozzle, m of water,

w_a = Air density, kg_f/m^3, and

C_d = Co-efficient of discharge

= 0.6 ... for orifice approx.

= 0.95 ... for nozzle approx.

19.5 MEASUREMENT OF TEMPERATURE

The following instruments are used for the *temperature measurement:*

1. Thermometers
2. Thermocouples
3. Resistance thermometers
4. Pyrometers.

1. **Thermometers:**

- The thermometers made of glass and having liquid mercury in its bulb can have range from – 4.4.°C (freezing point of mercury) to 538°C (softening of glass).
- The range of thermometers using alcohol varies from –34°C to 120°C.
- In case the temperature of the surrounding surface is much different from the temperature to be measured, the thermometer bulb must be effectively shielded or aspirated by a high velocity air stream so that the effects of radiation from surrounding surfaces are minimised.

2. **Thermocouples:**

- A *thermocouple* may be used to determine surface temperatures and temperature of inaccessible locations by indicators located at convenient points.
- When two dissimilar wires such as copper v/s constantan are joined together by soldering or twisting, an e.m.f. exists between the two wires which depends upon the temperature of the junction; this is known as "*See beck effect*".

Fig. 19.2 shows two conductors '1' and '2' joined together either by soldering, brazing, welding or twisting; an e.m.f. exists between the two junctions maintained at different temperatures. The thermoelectric power of a thermocouple is often expressed as

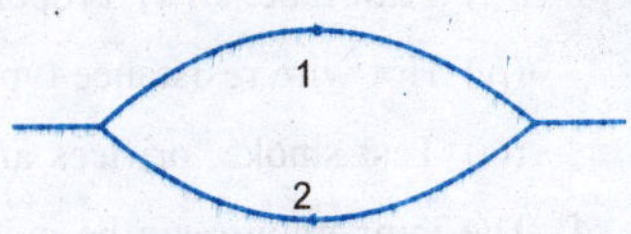

Fig. 19.2. Thermocouple.

$$= \frac{d}{dt}(E_1 - E_2) \text{ microvolts/°C} \qquad ...(9.4)$$

where, E_1 = Microvolts produced in conductor '1',

E_2 = Microvolts produced in conductor '2', and

t = Temperature measured in °C.

3. **Resistance thermometers:**

- These instruments work on the principle that, "The *electrical resistance of a wire depends upon its temperature*." Thus these are similar in action to the thermocouples except that the latter require a cold junction.
- Fig. 19.3. shows a schematic diagram of an elementary circuit of resistance thermometer. The galvanometer consists of two opposing coils. The deflection of the galvanometer depends upon the net effect of two coils.

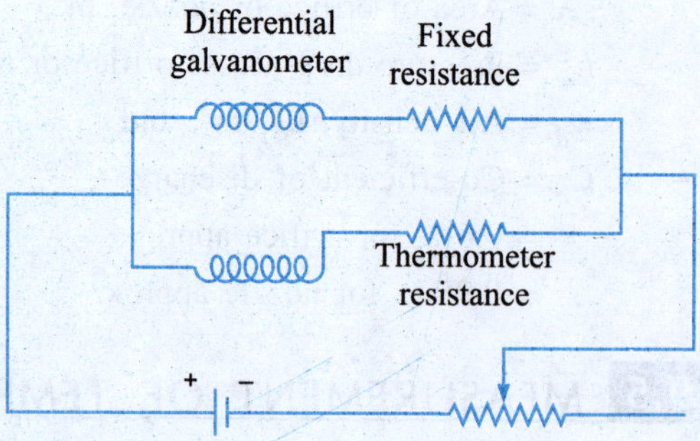

Fig. 19.3. Elementary circuit of resistance thermometer.

4. **Pyrometers:**

Pryometer is an instrument which is used to *measure high temperatures*. These are of two types:

(*i*) Radiation pyrometers.

(*ii*) Optical pyrometers.

(*i*) **Radiation pyrometers**. In these instruments the radiation from the hot body falls on a thermopile and thereby raises its temperature. The rise in temperature causes rise in e.m.f. of the themopile which can then indicate the temperature.

(*ii*) **Optical thermometers.** Here the brightness of a narrow spectral band of the heated surface is matched with the brightness of the filament of a special electric lamp by *adjusting the voltage supplied to the filament*. The surface temperature of the heated body depends upon the voltage.

HIGHLIGHTS

1. The instrument used to measure humidity of air is called the *hygrometer.*
2. The following methods are used to measure pressure:

   (*i*) Pressure gauges; (*ii*) Manometers; (*iii*) Aneroid barometer.
3. The velocity of air may be measured by using following methods:

   (*i*) Pitot tube; (*ii*) Propeller or rotating vane anemometer; (*iii*) Deflecting vane meter;

   (*iv*) Hot wire resistance type anemometer; (*v*) Kata thermometer;

   (*vi*) Test smoke, orifices and nozzels.
4. The temperature can be measured by the following instruments:

   (*i*) Thermometers; (*ii*) Thermocouples; (*iii*) Resistance thermometers; (*iv*) Pyrometers.

OBJECTIVE TYPE QUESTIONS

Fill in the Blanks or Say 'Yes' or 'No'

1. ________ is an instrument used to measure humidity of air.
2. ________ hygrometers are useful for dew points near the boiling point.
3. ________ hygrometers are based on the principle of change in electrical resistance of an electrolytic film (such as that of lithium chloride).

4. Weaver hygrometer is used for measurement of humidity of air in closed vessels.
5. ________ conductivity method may be used to measure low humidity for air at high temperature.
6. A ________ tube is used in most of the pressure gauges.
7. The 'pressure gauge' is calibrated means of a dead weight tester whilst a 'vacuum gauge' is calibrated by mercury manometers.
8. Aneroid barometers of high class can read upto ________, of mercury.
9. Very sensitive micro-manometers are used to measure very low pressure to the time of 0.0025 cm of water column.
10. Inclined manometer are used to measure the pressure difference of cm of water.
11. In inclined manometers a liquid lighter than water may be used to get magnification.
12. Pitot tube in conjuction with some form of ________, is used to measure the air velocity.
13. Deflecting vane anemometer is also called ________.
14. Hot-wire resistance anemometers are useful to measurement of ________ air velocities.
15. A ________ may be used to determine surface temperatures and temperature of inaccessible locations by indicators located at convenient points.

ANSWERS

1. Hygrometer	**2.** Dew point	**3.** Electrolytic	**4.** Yes	**5.** Thermal
6. Bourdon	**7.** Yes	**8.** 0.005 cm	**9.** Yes	**10.** a few
11. greater	**12.** manometer	**13.** Velometer	**14.** low	**15.** thermocouple.

THEORETICAL QUESTIONS

1. What is an 'hygrometer'?
2. Enumerate the various types of hygrometers and explain any two of them.
3. Explain briefly the following hygrometers:
   (*i*) Dew point hygrometer
   (*ii*) Electrolytic hygrometers.
4. How can infiltration be measured?
5. Explain briefly any two of the following methods of pressure measurement:
   (*i*) Pressure gauges
   (*ii*) Manometers
   (*iii*) Aneroid barometer.
6. Enumerate the methods which can be used for the measurement of air flow.
7. Explain briefly with a neat sketch the pitot tube method of measuring the velocity of air.
8. Name the instrument which are used for the measurement of temperature.
9. Explain briefly any two of the following methods of temperature measurement:
   (*i*) Thermometers
   (*ii*) Resistance thermometers
   (*iii*) Pyrometers

❑❑❑

PART III: MISCELLANEY

Chapters:

Chapter 20

Thermal Insulation

20.1 INTRODUCTION

- *Any material that retards or offers resistance to the flow of heat through it is known as* '**Thermal insulator**'. Thermal insulation however cannot stop the heat flow completely, but it can only retard or reduce the rate of heat flow through it. Insulating materials have extremely low thermal conductivities.
- By using insulating materials, the heat flow rates from the air-conditioned buildings (from building to outside for winter air-conditioning, and from outside to building for summer air-conditioning) can be considerably reduced. Properly applied insulation helps in economic running of refrigerating plants as well as heating plants.
- The *selection* of insulating materials for domestic refrigerators, cabinets, brine pipe lines, refrigerant pipe line, and cold storage rooms, steam carrying pipes, ducts carrying the hot air and for boilers is based on their ability to retard the heat flow. The selection of an insulator for a particular purpose depends upon number of required properties of insulating materials. The *selection is also done on the basis of economic and structural consideration.*

20.2 FUNCTIONS OF THERMAL INSULATORS

In a refrigeration system, the primary function of good heat insulating material is isolate the warm surroundings from the cold refrigerated region. These materials, commonly known as 'Thermal insulating materials' are mostly non-metallic and have a basic structure in which there are numerous cells containing air or gases.

The thermal insulating materials *may perform one or more of the following* **functions**.

1. To retard the flow of heat from surroundings.
2. To make possible better control of temperature in the refrigerated space.
3. To minimise temperature fluctuations.
4. To prevent water vapour condensation.
5. To minimise water vapour transmission.
6. To give better surface finish.
7. To add to the strength of walls and ceilings etc.
8. To absorb vibrations and reduce noise.
9. To reduce the spreading of the fire and flames in case of fire hazard.

20.3 PROPERTIES OF HEAT INSULATING MATERIALS

The properties of insulating materials can be broadly divided into following *four categories*:

1. Physical properties
2. Chemical properties
3. Mechanical properties
4. Thermal properties.

1. Physical properties. These properties include density, porosity, moisture resistance, co-efficient of thermal expansion, low odour level, inflammability, compactness and surface tension.

An insulating material should have the following *physical properties:*

(*i*) It should have *low density* (to decrease its weight).

(*ii*) It *should not possess its own odour nor it should pick up odour of other particles placed in the refrigerated space.*

(*iii*) It *should resist moisture penetration and should not deteriorate when the moisture collects on it.*

(*iv*) It *should repel vermin and insects.*

(*v*) It *should not be flammable* otherwise it may prove dangerous in the presence of electric wiring etc.

2. Chemical properties. Chemical properties enable insulation to retain its own structure in hostile environment and also prevent it from being corrosive to substances. An insulation material which is to be used in *chemical industry must possess adequate resistance to acids and corrosive gases.*

3. Mechanical properties. The important mechanical properties required generally are, *high abrasion resistance and hardness, good compressive strength.* The properties depend on the type of the insulating materials used.

4. Thermal properties. Thermal properties of insulating material are the most important properties. The are discussed below:

(*i*) **Thermal conductivity.** The ability of a material to retard the flow of heat is given by its thermal conductivity. It is a property of homogeneous material which changes with the variation of density. It should be low in case of a thermal insulator.

(*ii*) **Heat capacity.** It is essentially measure of the heat absorbed for a certain temperature rise and should be concordant with the use. *Furnace insulations must have low thermal capacity for quick response to temperature changes required.*

(*iii*) **Thermal diffusivity.** It is measure of the rate of progression of temperature rise in an insulating material when certain heat energy is available at one of its surfaces.

(*iv*) **Combustibility.** It is characterised by flash point, smoke index, fire point etc. To reduce fire hazard a low combustibility is necessary.

Requirements of thermal insulating material:

Summarily the main requirements or a *desirable properties of a good heat insulating materials* are:

1. Thermal stability
2. Chemical stability
3. Physical stability
4. Low thermal conductivity
5. Resistance to moisture
6. Low specific heat
7. Low specific gravity
8. Odourless
9. Resistance to vibration and shock
10. Non-inflammability
11. Porous and fibrous texture
12. Economical in its initial cost.

20.4 CLASSIFICATION AND DESCRIPTION OF THERMAL INSULATING MATERIALS

The heat insulating materials may be classified as follows:

1. Organic heat insulators:

(*i*) Cork (*ii*) Cotton wool
(*iii*) Rubber foam (*iv*) Sawdust
(*v*) Rice husk (*vi*) Polystrene
(*vii*) Urethane (*viii*) Phenotherm etc.

2. Inorganic heat insulators:

(*i*) Glass wool (*ii*) Slag wool
(*iii*) Ceramic products (*iv*) Asbestos etc.

The cork was considered to be the best insulation material for a long time. With the advent of glass wool, mineral wool, expanded polystyrene, polyurethane, etc., cork is rarely used these days. The physical properties of commonly used insulation materials are given in Table 20.1.

Table 20.1. Physical properties of common insulating materials

S. No.	Insulating material	Density kg/m^3	Thermal conductivity $kcal/h/m^2/°C$	Temp. range of application °C	Permeance $gm/m^2/24$ hrs per mm Hg*	Water absorption % by wt in 24 hrs.
1.	*Polyurethane*	30-32	0.015	110 to (–) 200	3.3	
2.	*Phenotherm*	30–34	0.016	130 to (–) 196	1.3	2
3.	*Expanded polystyrene (Thermocole)*	15–30	0.025 to 0.028	75 to (–) 200	0.4	0.3
4.	*Resin bonded fibreglass*	24–28	0.032	150 to (–) 30	8.5	2.3
5.	*Cork*	120–160	0.036	90 to (–) 150	1.7	4.0
6.	*Foam concrete*	320	0.062	125 to (+) 5	5.6	13.5

* Difference in the partial pressure of water vapour on either side of the material.

The type and form available and the applications of various insulations are as under :

(*i*) ***Cork:***

- Compressed and moulded into a rigid block.
- Light but strong, can be cut easily with a saw, resists water but allows relatively high rate of water vapour transmission.

(*ii*) ***Glass wool:***

- Available as semi-rigid resin bonded slabs/sheets of different densities-higher density gives strength and lower conductivity but allows vapour transmission.
- Available with foil or other coverings.

(*iii*) ***Expanded polystyrene (thermocole):***

- Can be cut easily with a saw, light weight, allows relatively low vapour transmission.
- Available as a rigid board, beads, moulded into shape for pipe/curved surface covering.

(*iv*) ***Polyenrethane:***

- Available as a rigid board, flexible board, liquid can be sprayed on surfaces and allowed to foam.
- Can be used for in situ applications.

(*v*) ***Phenotherm:***

- Can be easily cut with a saw.
- Available as slabs with different facings, and as per preformed pipe sections.

(*vi*) ***Wood shaving/saw dust:***

- Can easily settle down, fairly high conductivity, absorbs moisture/water.
- Needs good supporting compartment.

Insulation thickness required:

The thickness of insulation required depends on:

(*i*) Thermal conductivity of the insulating material selected, and

(*ii*) Difference between the surrounding and storage temperature.

Transmission of heat increases the storage temperature and so prolongs the running time of the refrigeration plant required to bring down the temperature. *A thicker insulation reduces the heat transmission thereby reducing the running time of the plant and its power consumption.*

- It may be noted that *it is necessary to make an economic analysis of the initial cost of the insulation material, cost for laying, annual interest on the investment life of the insulation vis-a-vis saving in power consumption and then only finally decide the insulation thickness.*

  Description of some commonly used heat insulating materials is given below:

1. Cork:

- It is derived from the *bark of oak trees*. It is *ground, sized and baked in moulds.* When ground and baked, the natural resin in the cork binds the material into homogeneous mass which can be pressed into flexible sheets or boards etc. It is available in the form *of granulated cork, slab cork, and regranulated baked cork.*
- The structure of cork consists of an aggregation of minute air vessel, provided with thin, strong wall, so that if material is compressed it behaves more like a gas than an elastic solid; unlike the behaviour of spring, which exerts a pressure proportionate to the linear amount of compression. Cork, when compressed, exerts a pressure which increases in a more rapid manner and varies, approximately, *inversely as the volume*.

Properties: Following are the *properties of cork*:

1. Light in colour.
2. Porous in structure.
3. Specific gravity is about 0.24.
4. Not affected by moisture.
5. Thermal conductivity is low.
6. Can be easily compressed.
7. Resilient and reasonably elastic when dry.

Uses: Following are the *uses of cork*:

1. Cork sheets and boards are used for insulating walls and ceilings, both against heat and cold and also as a sound insulator.
2. Used as a non-conducting covering for pipes carrying steam or hot water.
3. Used as non-conducting material for scientific apparatuses.
4. Used in refrigeration and cold storage insulation.
5. Also used for bottle stoppers, vibration pads and floats for rafts and fish nets.

2. Glass wool:

- Glass wool is produced by *blowing high pressure jets of steam or air on molten streams of glass at a high temperature.* Molten glass is violently scattered in all directions, to give this product.
- Glass wool is a *form of fibrous glass* with short and fine fibres, scattered in various directions.
- It is available in the form of *loose fibres, mats, rigid quilts, or semi-rigid slabs or blocks* etc.

Properties:

1. Fibrous in structure.
2. Light in weight.
3. Has good tensile and dielectric strength.
4. Low thermal conductivity.
5. Quite durable.
6. Acts as an excellent insulating material because of the presence of large pockets of air in it.
7. Not affected by low temperatures and has been used successfully at temperatures as low as –212°C.

Glass fibres have the following *characteristics* :

(*i*) Do not catch fire.
(*ii*) Not easily affected by heat.
(*iii*) Not spoiled by insects and moisture.

Uses:

1. Mostly used for insulation of *pipes, bends, valves* etc.
2. Used for panel insulation for all types of industrial equipment.
3. Can be used for thermal and sound insulation of aircrafts.
4. *Glass wool blocks* can be used in the *construction of partition walls for thermal insulation purposes.*
5. Used in boilers, ovens, cylinder or pipe insulation.

3. Rock wool:

- It is produced from flint rock containing some calcareous matter. In the absence of such a natural rock, flint and lime are mixed in the requisite proportions and melted in a furnace at temperature of about 1700°C. This molten material is then formed into small globules by means of a steam jet. These globules are then drawn into very fine fibres by hurling them in a large container. These fibres of wool are then formed into boards or blankets (to be used as insulators). It can also be pressed, rolled and secured between fabric of wire-netting of brass or copper.
- It is available in the following forms:

Loose fibres, matresses, mats, boards or felts, rigid or semi-rigid slabs, quilts.

Properties:

1. Soft and flexible.
2. Resilient and woody consistency.
3. Heat and sound proof (due to the presence of millions of minutes dead air cells).
4. Specific gravity is about 0.48.

Uses:

1. Employed for heat and sound insulation purposes.
2. Also used as an electric insulator.

4. Slag wool:

- It is an aggregate of fine filaments of slag produced by blowing air through a stream of blast furnace slag.
- It is available in the form of loose fibres.

Uses : It is used for *heat insulation in high temperature furnaces.*

5. Asbestos:

- Asbestos is a mineral fibre composed of hydrous silicate of magnesia with a small amount of iron oxide and alumina.
- Asbestos sheets or boards consist of natural asbestos fibres mixed with a binding agent (usually cement) and then rolled in the form of sheets or boards. These are available in the market under the trade name **'Salamander'**.

Properties:

1. White, grey or brown in colour.
2. Flexible and can resist high temperature.
3. Fire-proof.
4. Unaffected by acids and fumes.
5. Resistant to corrosion and vermin effect.
6. Excellently resists heat and electricity.

Uses: Employed for heat and sound insulation of buildings. Also used for insulation of furnaces.

6. Thermocole:

Thermocole is one of the trade names of *Polystrene* (Refer Plastic foams also). This product was developed (in U.S.A.) during second world war. It was made by *direct extrusion of the foam from raw materials.*

Properties:

1. It has a very attractive, natural, snow white colour.
2. Very light in weight (density : 150 to 300 N/m^3) The foam is very light because it contains over 98% (by volume) air, trapped in 3 to 6 millions closed cells per litre.

3. Compressive strength = 0.07 to 0.1 MN/m^2; Cross–breaking strength = 0.14 to 0.18 MN/m^2.
4. Very low value of thermal conductivity.
5. Highly resistant to moisture.
6. Odourless, chemically stable and resistant to fungus attack.
7. Fully resistant to water, salt, soaps, bleaching agents and HCI (35%), HNO_3 (upto 50%), H_2SO_4 (upto 95%), caustic soda, caustic potash, strong ammonia, alcohols and silicon oil.
8. Not resistant to organic solvents like benzene, paint thinners and saturated aliphatic hydrocarbons like petroleum and gasoline.
9. Very good shock-protecting properties.
10. Ability of being moulded into well-fitting, contoured cases.

Uses:

1. Thermocole (with operational range of –200°C to 80°C) is an excellent material for cold insulation in refrigerators, cold storages, air-conditioning, chilled pipe lines and chemical processes.
2. It is used for industrial insulation and insulation for buildings, against extremes of climate.
3. In the form of specially made flexible sheets, thermocole can be used on intermediate concrete floors in multistorey buildings, to *reduce impact sound transmission.*
4. It is used for *packing* electronic goods like transistors, radios, tape recorders and calculating machines, clocks, medicine bottles, cameras etc.
5. It is also used for air-dropped packaging, decorative and gifts packaging and edge protecting packaging.

7. Reflecting paper:

- Reflecting paper (also known as *building paper*) is a strong tough paper which is lined with aluminium or copper foil on the exposed side, which reflects back heat waves coming from a source and this keeps the walls and the enclosed rooms cool.
- Sometimes, reflecting coatings of *varnishes, paraffins, gums or synthetic resins* are applied to various grades of paper of fibrous materials.

Properties:

1. Strong and tough in nature.
2. Heat-resistant.
3. Possesses adequate dielectric strength.

Uses: Used for *heat insulation purposes.*

Plastic foams:

In modern air-conditioning buildings *foamed plastics* are finding rapidly growing applications. *Key consideration in selecting a plastic foam as insulating material are thermal conductivity, compressive strength, water vapour transmission, dimensional stability and method of installation.*

1. Polystyrene:

- These foams can be supplied in any desired shape for large production runs.
- Principally, the material is available in the form of foamed sheet or *pipe covering and is used:*
  - (*i*) *As thermal insulation in cold storage rooms, tanks and vessels.*
  - (*ii*) As a core for sandwich panels, encased in materials ranging from aluminium and steel to plywood.
  - (*iii*) As temporary structural formwork for placing and curing permanent shell material. The formwork then remains bonded to the permanent shell as insulation and vapour barrier, and as an interior finishing base.

2. Urethane:

- It is known as super insulation in air-conditioning industry.
- These materials are somewhat more expensive than polystyrene but because of economy of labour in their application they are the predominant plastic foam used as thermal insulant.
- Rigid urethane foam is increasingly specified for insulating ducts, pipes, walls, roofs, basements and certain walls in wide variety of new and existing structures.
- It is the *only insulating material* that can be sprayed as a liquid to reach virtually inaccessible areas.
- Urethane foam possesses high thermal resistance to the heat flow, can lower heating or cooling costs of the buildings particularly in electrically heated buildings where effective insulation is required to make operative cost competitive with other fuel system.

3. Phenolics:

- It is probably the most flame resistant cellular material developed so far.
- Phenolic foam is readily bonded to a wide range of materials like concrete, glass, steel etc., by means of commercially available adhesives.
- Owing to their high thermal resistance, phenolic foams are recommended as *insulating materials for flat roofs*. They are also used as cores for sandwich panels.

20.5 PRECAUTIONS DURING HANDLING AND LAYING OUT OF THERMAL INSULATION

While handling and laying out thermal insulation, the following *precautions* should be taken:

1. Surfaces to be insulated, such as walls, ceilings, ducts, pipes, etc. should be *thoroughly cleaned* and brushed to remove dirt, debris, cement lumps, etc. The surface is then coated, with solvent free anticorrosive paint and the insulation is laid. The surfaces should be dry before applying paint and insulation.
2. Pipe-lines and pressure vessels should be *thoroughly pressure tested* before laying out the insulation.
3. The couplings provided on chilled water lines for purge cocks, pressure gauges, thermowells, etc. should be *long enough to clear the thickness* of insulation by at least 12 mm.
4. *Coat* the inside surface, ends and edges of the pipe sections and slabs of insulation *with sealing and adhesive compounds.*

5. *Joints* (on pipe section and plain slabs) should be *staggered*.
6. Insulation materials, particularly polystrene, polyurethane, fibreglass, etc. can get easily damaged during transportation and handling. Such damaged portions are the weakest link in insulation. So care should be taken to prevent such damages.
7. *Openings* on walls in cold storages for pipes and electrical conduits should be *made*, once the pipes are laid and insulated, the openings should be closed before starting the insulation work on the walls.

HIGHLIGHTS

1. Any material that retards or offers resistance to the flow of heat through it is known as '*Thermal insulator*'.
2. In a refrigerated system, the primary function of a good heat insulating material is to isolate the warm surroundings from the cold refrigerated region.
3. The main requirements of a thermal insulator material are:

   Thermal stability; Chemical stability; Physical stability; Low thermal conductivity; Resistance to moisture; Low specific heat; Low specific gravity; Odourless; Resistance to vibration and shock; Non-flammability; Porous and fibrous texture; Economical in its initial cost.

OBJECTIVE TYPE QUESTIONS

Fill in the Blanks or Say "Yes" or "No":

1. .... is a material the retards or offers resistance to the flow of heat through it:
2. Thermal insulating materials are mostly non-metallic.
3. The ability of a material to retard the flow of heat is given by its ....
4. To reduce fire hazard a low .... is necessary.
5. Cork is derived from the bark of .... trees.
6. .... wool is produced by blowing high pressure jets of steam or air on molten stream of glass at a high temperature.
7. Glass wool is fibrous in structure.
8. .... wool is produced from flint rock containing same calcareous matter.
9. Rock wool is used for heat insulation in high temperature furnaces.
10. Thermocole is the one of the trade names of polystyrene.
11. .... is known as super insulation in air-conditioning industry.
12. .... is the only insulating material that can be sprayed as a liquid to reach virtually inaccessible areas.

ANSWERS

1. Thermal insulator.	**2.** Yes.	**3.** Thermal conductivity.	**4.** combustibility.
5. oak.	**6.** Glass	**7.** Yes	**8.** Rock
9. Yes	**10.** Yes	**11.** Urethane	**12.** Urethane

THEORETICAL QUESTIONS

1. Define a 'Thermal insulator'.
2. What are the functions of thermal insulators?
3. Explain briefly the properties of thermal insulators.
4. What are the desirable properties of a thermal insulating material?
5. Give the classification of thermal insulating materials.
6. Explain briefly any three of the following insulating materials:
   Cork, Cotton wool, Polyeurathene, Slag wool, Ceramic products.

❑❑❑

Applications of Refrigeration and Air-conditioning

Chapter 21

21.1 FOOD PRESERVATION

21.1.1 Introduction

The various varieties of foods which are used by the human being are drawn from the plants and animal kingdoms. In order to extend the life of the food it is necessary to preserve foods in storage or during transportation.

The **food preservation** *may be defined as the state in which the food may be retained over a period of time without being contaminated by pathogenic organisms or chemicals and without losing its colour, texture, flavour and nutrition value.*

Following are the **advantages** *of food preservation:*

(*i*) Addition of variety to the food.

(*ii*) Self life of food is enhanced.

(*iii*) Food supply is increased.

(*iv*) Wastage of food is reduced.

21.1.2 Factors Contributing to Food Spoilage

All type of foods, either plant or animal, contain basically three major molecular components namely *carbohydrates*, *proteins and fats*. These chemical compounds provide vitamins and minerals which are essential to safeguard the body against diseases. The *destruction of one of these components causes the spoilage of the food.*

Following are the *agents* while help activate the spoilage of food:

1. Enzymes
2. Micro-organisms

1. Enzymes:

- Enzymes are present in all substances or develop within them depending upon environmental conditions.
- These are of different types identification of which depends largely upon the type of environment and their resulting chemical action. *Examples:* Diastase, cytase, Trypsin, Pepsin, Zymase.
- They convert the food ingredients into digestible elements which cannot remain intact for a long duration of time.
- The growth of enzymes can be checked or reduced by keeping the food stuff in very cold environment. They can also be destroyed by *pasteurisation* (a method adopted to destroy micro-organisms by heating the commodities to the desired temperatures e.g., for milk the pasteurization temperature is 375 K).

2. Micro-organisms:

These undetectable living organisms are present in the surroundings. They grow very rapidly under favourable temperature and humidity conditions.

They divided into the following three sub-groups:

(*i*) Bacteria

(*ii*) Yeasts

(*iii*) Molds.

(*i*) ***Bacteria:***

- They are single cell c ganisms and are very rapid in reproduction but their life cycle is hardly few minutes. The bacteria size is of the order 10^{-5} mm.
- They are found in large number in various substances such as air, water, soil, animals, milk, eggs, butter etc.
- These are divided into two categories namely:
  - — Those *causing spoilage;*
  - — Those *employed for the food preservation.*
- The bacteria may be *destroyed by sunlight, ultra-violet rays, extreme heat and by the use of certain chemical substances.*

(*ii*) ***Yeasts:***

- Micro-organisms of size about 10^{-4} mm are called *yeasts.*
- They are one-celled plants of fungus group.
- Their growth is accelerated in the presence of air, water and warm atmosphere.
- Their action is more rapid in acidic conditions therefore they spoil the jams and syrups quickly.
- Generally all types of yeast will be destroyed when subjected to a temperature of 100°C and their activity will be stopped under low temperatures, but they remain in food for a considerably long periods as their cells are hard.

(*iii*) ***Molds:***

- These are class of plants that thrive in a damp and dark environment and is enhanced by the stagnant air.

  Examples. Rhizopus nigricans, Penicillium Italicium, Aspergillus flavour etc.
- They are larger in size than bacteria and yeast and more complex in their structure.
- They become apparent by the cottony, thread-like structure which can be seen by naked eye.
- Oxygen is essential for the growth of the molds. Thus the presence of CO_2 helps reduce the mold formation.
- The most common molds are non-poisnous. All molds are bitter in taste.
- *Mostly all molds can be destroyed with heating at 60°C and they can be made inactive by deep freezing.*

Other causes of food spoilage:

1. ***Desiccation.*** *Desiccation or dehydration is process of loss of moisture from the surface of the product by evaporation into the surrounding air.* All foods (fruits and vegetables) containing water lose their water content of the surrounding air. The effect is loss of weight and also the taste. The loss of water by evaporation is rapid in dry air.
2. ***Absorbers of odours.*** Some foods like meat and butter absorb the odours given off by the other foods and spoilage starts and is accelerated. Some foods absorb gases also given by other food and the fresh foods get destroyed eventually.
3. ***Bruising.*** It is caused by impact or squeezing which is common in packing the fruits and vegetables. This helps the other organisims to act rapidly removing the cover of fruits which is provided as protection to the inside food content.

21.1.3 Methods of Food Preservation

The various methods of food preservation are:

1. Heat-processing
2. Dehydration
3. Chemical preservation
4. Oils and spices
5. Canning
6. Pasteurization

1. Heat-processing:

- All food-spoilage agencies as enzymes, bacteria, molds and yeasts are generally destroyed when subjected to high temperatures over a period of time.
- The product temperature is raised to a level fatal to all spoilage agents and is maintained at this level until they are all destroyed. The product is then sealed in sterilized, air tight containers.

2. Dehydration:

- It is the process of removing moisture from the product. This method was used earlier for preservation of vegetables and still it is used in rural areas using heat of sun for dehydration. The taste of the foods is not preserved by this method.
- A variety of dehydrated products are available in the market e.g., dried milk, dehydrated soups, instant coffee etc.

3. Chemical preservation:

- Several micro-organisms cannot grow in the presence of salt and sugar. *The salt and sugar are used as chemical preservatives.* The high presence of salt in some foods (pickels) will not allow the micro-organisms to grow and it also maintains its food value.
- *Vinegar* is also widely used for food preservation because its acetic content prevents the development of spoilage agents.
- This method can be employed for all types of food.

4. Oils and spices:

- The oils and spices along with salt and sugar provide a medium that resists the activity of the micro-organisms in food. Moreover, they *improve the flavour of the food being preserved.*
- Spices such as chillis, fenugreek, mustard and pepper are used in pickling. When oil is used in pickling the top layer of oil prevents the micro-organism in the air from coming into contact with the food.

5. Canning:

- In this method of food preservation, the food is preserved in *sealed containers, usually after the application of heat, through steam under pressure.* During the process, some of the micro-organisms are destroyed and the rest are rendered inactive; the enzymes are also inactivated.

6. Pasteurization:

- This method is generally employed on large scale to protect milk (used for the preparation of milk products like cheese, butter and ice cream) against bacterial action.
- The pasteurization may be brought about by the "holding process" or "*high temperature short time method*".
  - — In the *holding process*, the milk is heated to at least 62°C and kept at that temperature for at least 30 minutes.
  - — In the *high temperature short time method,* the milk is heated to 70°C and kept at that temperature for at least 15 seconds.
- Pasteurization does *not* change the colour or flavour of milk.

21.1.4 Methods of Food Freezing

Food freezing is used to keep the perishable food stuff, vegetables, meat, egg, fish etc., for a longer period.

The temperatures provided for preservation, may be above freezing or below freezing temperatures of the foods, depending upon the period of preservation. *For longer periods, the temperatures below freezing are always used and the range of freezing temperature for different foods lies between –15°C to –25°C.*

A few methods of food freezing are enumerated and discussed below:

1. Slow or sharp freezing
2. Quick freezing
   (*i*) Immersion freezing
   (*ii*) Indirect contact freezing
   (*iii*) Air-blast freezing

1. **Slow or sharp freezing:**

- In this method (now almost obsolete), the food products are placed in a low temperature room and allowed to freeze slowly in still air by *natural convection*. The temperature maintained in the sharp freezers ranges from –15°C to –30°C.
- This method is used for poultry products and fishes.
- Owing to its slow rate of cooling, this method is not satisfactory for consumer's food.

2. **Quick freezing:**

Quick freezing is brought about by *forced circulation* of cold air necessary for freezing the foods in storage. *The difference between quick and sharp freezing is only size of the crystals formed within the foods during freezing.*

(*i*) ***Immersion freezing:***

- In this method the food products are immersed into low temperature liquids. Since the liquids are good conductors of heat and give high heat transfer coefficients, therefore the heat transfer is rapid and product is completely frozen in a very short time.
- The liquid mediums used *must be non-toxic* and *should not produce any bad effects on immersed foods.*

  The common liquids used for this purpose are: *Sodium chloride, brine, sugar brine and propylene glycol.*
- This method produces a *thin coating of ice* on the surface of the product resulting in prevention of dehydration of unpackaged products during the storage period.

The only **disadvantage** of *this method is the penetration of salt or sugar into the product.*

The method is commonly used for freezing fishery and poultry products at sea.

(*ii*) ***Indirect contact freezing:***

- In this method, the food is freezed by direct contact with metal surface cooled by any of the refrigerants as NH_3, R-12, R-22 or cold brine.
- The plate freezer is a common example of this category. They are available in different sizes ranging from 100 to 1200 kg loading capacity.

Plate freezers are used for freezing fillets, steaks, shrimp fish sticks, fish blocks, meat, fish, fruits etc.

The *main advantage* of this method of food freezing is that foods preserved in plate freezers *maintain their taste and colour as they not come in direct contact with the cooling medium.*

The only **disadvantage** of this method is that the ice crystals formed on the surface of the food product are of *little bigger size* than immersion freezing method.

(*iii*) ***Air-blast freezing:***

- In this method, the food products are freezed by the *contact with cold air*.

  This is widely used freezing method *since it provides excellent quality of the food among all the types.*
- These freezers are designed in the form of tunnels and take full advantage of high heat transfer coefficient. The cold air at very low temperature is obtained by circulating air over the evaporator coil, eliminating the use of brine. The rate of cooling is determined by the refrigeration capacity, heat transport by circulating air and the heat transfer characteristics of products in the packaged form.
- The temperature of –20°C to –40°C are commonly used for this method of freezing. The velocity varies considerably from 30 m to 1200 m/min. Dehydration of product which is known as *freezer burn* may occur in freezing unpacked fish in blast freezer unless the velocity of air is kept to about 160 m/min, and the period of exposure to the air is controlled.

This method of freezing is mainly used for freezing fishery products (e.g., shrimp, fish fillets, steaks etc.) in small packages.

Advantages of "*quick freezing*" over "*slower sharp freezing*":

The following are the advantages of quick freezing over slow or sharp freezing:

1. Since the time required for freezing is very less, therefore, there is no chance for separation of water in the form of pure ice and diffusion of salt from the food surface cannot take place.
2. Owing to the rapid rate of crystal formation, the *cellular expansion is greatly reduced and consequently the rupture of membranes and loss of valuable nutritions fluids are prevented.*
3. During freezing the decomposition is not possible as the freezing period is very less, thus growth of yeast and mold during freezing is *not allowed.*

21.1.5 Freeze Drying

It is possible to sublime ice from frozen material under proper conditions. When this process is applied to the dehydration of food, or biological materials, it is called as *freeze-drying* (the process of freeze-drying is based on the simple low of physics that water can exist only as solid or vapour in a very high vacuum — the actual process that takes place during freeze-drying is described as '*dehydration by sublimation*').

The rate of dehydration in the freeze-drying of solid materials may be governed either by rate of vapour diffusion or rate of heat transfer to the ice surface, whichever is limiting. There are two processes which occur in the freeze-drying of food materials which *possess a definite structure*. One involves the *transfer of water vapour from the ice surface through the dried layer, and the other involves heat transfer from the exterior to the ice surface.*

The dehydration takes place in a vacuum chamber where water vapour is pumped out of the chamber as it accumulates. The pressure in the chamber must be maintained below 5 mm Hg. If the chamber pressure approaches this value, the ice phase will melt, and sublimation of ice will cease. The heat for subliming the ice is usually supplied through heating plates which sandwich the material being dried, or it may be supplied through radiations from the surrounding chamber. Usually the rate of dehydration is limited by the transfer of heat to the ice surface. Attempts have been made to employ *microwave heating* in order to improve heat transfer through the dried layer.

21.1.6 Food Processing/Preservation by Refrigeration

The most significant application of refrigeration is in food preservation whether processing or storage. Candy manufacture, bakery products, fruits and vegetables etc. are a few examples of food processing and preservation.

1. **Candy manufacture:**

- In the candy manufacture, air-conditioning is an essential requirement. Proper control of atmosphere results in:
  — increased production,
  — lower production costs,
  — increased quality, and
  — uniformity of products.
- In candy manufacture, most of the materials are sensitive to temperature or relative humidity, or both. A properly controlled atmosphere at optimum conditions for the process then becomes a primary means of controlling the solidifying, crystallization, physical properties, quality and appearance of the finished product. Each application requires careful analysis to assure success for specific requirement.
- The latent heat of freezing of usual grades of chocolates used in candy manufacture, for average purposes, varies from 70 to 90 kJ/kg. An average value for specific heat may be taken as 2.34 kJ/kg°C before freezing and 1.26 kJ/kg°C after freezing chocolate bars are cast in metal moulds which are precooled to 15°C. In order that there is no condensation of moisture on the product, the dew point temperature of cold room in which solidification takes place and the packing room should be below 15°C.

2. **Milk and butter:**

(*i*) ***Milk.*** Milk preservation includes *blending, processing, packaging and distribution to the customers.*

- Dairy plants receive milk from different rural places and from different types of cattles. This milk is usually mixed, processed and then blended to produce a milk of uniform quality of required fat percentage.
- After the collection of milk from difference sources, *pasteurization is the first main step* in the milk preservation. All pathogenic types of bacteria and nearly all other objectionable organisms are killed by proper pasteurization.

Batch pasteurisation is accomplished by filling a vat almost full with raw milk which is heated upto abut 62°C and held for 30 minutes. Then the milk is drawn from the vat and cooled before it is passed to the bottle. The heating of vat is carried out either by *hot water or steam from* its outer surface. Then the hot milk coming out is *cooled to 4.4°C* with the help of chilled water or direct refrigerant before filling the bottle.

(*ii*) ***Butter.*** The butter prepared from the cream removed from the milk is stored at –17.8 to –33°C for long time storage. The butter is covered with a special paper or foil to prevent the absorption of odours, oxidation of fat on the surface and shrinkage in weight due to evaporation.

3. **Meat products:**

- Refrigeration systems in common use in carcass chilling and holding rooms are operated with ammonia at a system pressure corresponding to –25°C as the primary agent, and are of three general types:

  (*i*) Chilled brine spray; (*ii*) Sprayed coil; (*iii*) Dry coils.

- The average specific heat of carcass is 3.0 kJ/kg°C. The carcass temperature should be reduced to 1.5°C in a short-time after slaughter. The room temperature should be as low as –3.5°C. Refrigeration is also required for chilling water for washing the product. The *recommended storage temperature is –29°C for storage life.* The quality of the product suffers severely at temperature above –9°C.

4. **Poultry products:**

- *Chilling* and *freezing* are used to preserve poultry meat. *Chilling requires a lower capital investment whereas freezing gives quality and flexibility of operation.*
- *Liquid chilling by the use of flaked ice* is *more popular* for the processing and transportation of poultry meat. *Air chilling leads to dehydration at the surface.*
- Ready-to-cook birds are frozen in air blast freezers, with air temperatures ranging from –24°C to –40°C and air velocities of 1.5 to 5 m/s.

5. **Fishery product:**

Refrigeration is required in several ways in the processing, preservation and transportation of fresh and frozen fish and their products. Care begins from the time catching in fishing boats and trawlers.

- These days *freezing of fish* is preferred particularly for very valued varities of fish. *Individual quick freezing* (IQF) methods are used for these products.
- *Blast freezing* of fishery products is generally done in small rooms or tunnels in which cold air is circulated by one or more fans over an evaporator and around the product to be frozen. The necessary refrigerating effect is furnished by a refrigerant such as NH_3, a halocarbon, or brine flowing through pipe coil evaporator.
- *Plate freezers* are employed for large sized packages.

6. **Fruits and vegetables:**

At the time of harvest, maturity of fruit or vegetable is directly related to the refrigerated storage life and quality of the product. For any given product, almost invariably, there is a maturity which will prove to be best-suited for refrigerated storage. Under-mature product will not ripen or develop good quality during refrigerated cold storage, and over-mature product quickly deteriorates during cold storage.

- "*Controlled atmosphere storage*" is the practice of modifying the atmosphere in the refrigerated storage room by reducing the oxygen level to 3% to 5% and increasing the CO_2 level to 3% to 10%. The *controlled atmosphere drastically reduces the product respiration and also arrests deterioration of the product, substantially increasing the storage life.*

  Vegetables like beans, peas, carrots etc., are found suitable for freezing. They are partially cooked prior to freezing. *Air blast freezers are used for bulk freezing of these products.*

- "*Dehydro-freezing*" is a process *involving both dehydration* and *freezing*. Fruits and vegetables such as apples, apricots etc., are dehydro-frozen by partially drying the product before freezing. This process makes possible *substantial savings in the cost of shipping and storing food without sacrifice of quality as compared with frozen equivalent.*

21.2 COLD STORAGE

The following main *advantages* accure by cold storage of fruits, food stuffs and vegetables:

1. Substances such as potatoes, butter etc., can be stored when their supply is plentiful but which can be sold during the period of short supply.
2. Due to reduction of spoilage great saving is effected.
3. Transportation of perishable commodities from distant places is made possible.

The cold stores are of two types:

(*a*) *Long-term ware-houses* with the product in frozen or unfrozen state.

(*b*) *Short-term ware-houses* or retail storage with production usually not frozen.

Foods are *quickly frozen* to get the following *advantages* : (*i*) It limits the growth of bacteria and wild growth during the freezing period. (*ii*) It reduces the size of ice crystals formed. (*iii*) It reduces the separation of water from the cell. Quick freezing is accomplished by the use of either cold air blast, brine spray or contact evaporators or double contact plate freezers.

Types of Cooling Plant for Cold Storage:

The types of cooling plant for storage are as follows:

1. Brine coils placed parallel to and near the centre of the ceiling.
2. Unit conditioner with the condensing unit outside the store.
3. Small ceiling mounted units.

1. Brine coils placed parallel to and near the centre of the ceiling. In this arrangement/system a central brine pump supplies chilled brine to these coils situated in various rooms of a large cold storage central plant. Thermal air circulation from coils to product is set up. No fan is used.

2. Unit conditioner with the condensing unit outside the store. In this type of cooling plant the cooling coil is placed inside the store and is supplied either with direct refrigerant or else with chilled brine. Room air enters the coil at the bottom, passes over the coils and is blown in the room as chilled air passes either through louvers or a central ceiling duct. If the vapour load is very high, brine sprays are used instead of cooling coil surface. For defrosting the outer surface of the coils:

(*i*) a hot refrigeration gas is passed through the coils for direct refrigerant coils; or

(*ii*) brine is sprayed over the outer surface of brine coils; or

(*iii*) refrigerant or brine is turned off for sometime for stores above 0°C.

3. Small ceiling mounted units. These units consists of the cooling coil backed by an electric fan. The fan blows the chilled air horizontally or vertically down.

A typical plant for small cold storage is shown in the Fig. 21.1.

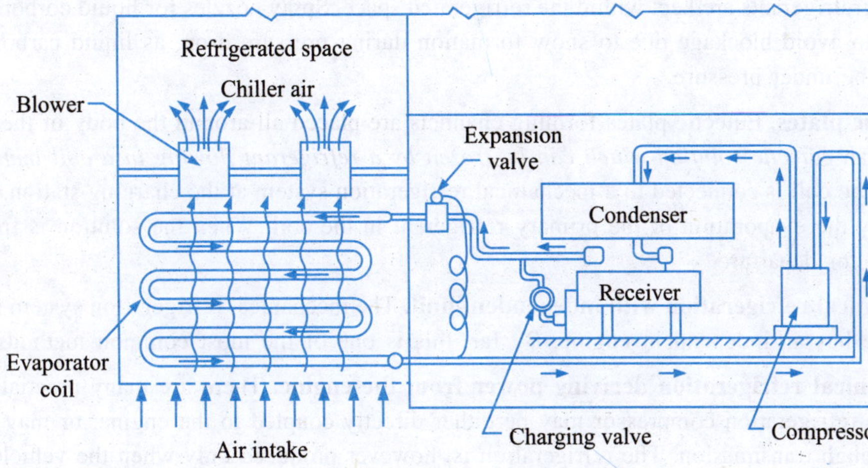

Fig. 21.1. Cold storage plant.

21.3 REFRIGERATION METHODS FOR TRUCKS AND TRAILORS

Methods of refrigeration used in refrigerated trucks and trailors are enumerated and discussed below:

1. Using water.
2. Water ice in bunker with forced air circulation.
3. Using dry ice.
4. Using liquid nitrogen or liquid carbondioxide spray.
5. Eutectic plates.
6. Mechanical refrigeration with independent unit.
7. Mechanical refrigeration deriving power from the engine.

1. Using water ice. The top of the product can be suitably iced. It is a satisfactory method for short distances and for some products only.

2. Water ice in bunker with forced air circulation. Ice bunker is fitted in front of an insulated vehicle. Air is sucked over ice by the blower (which is driven from the engine). A mixture of ice and salt can also be used for lower temperature upto –10°C.

3. Using dry ice. Dry ice is used in many small retail trucks for the delivery of frozen food and ice-cream. The dry ice blocks are kept in the ceiling and the cooling is by natural convection. If forced circulation is employed, dry ice may be placed in bunkers, just like water ice.

In the use of dry ice, care *must* be taken to avoid burns due to low temperature and suffocation due to lack of oxygen.

4. Using liquid nitrogen or liquid carbondioxide spray. Liquid nitrogen has a normal boiling point of –196°C. The is a very low temperature from the point of view of economy. Hence liquid nitrogen is used only if its is available as a by product.

In both liquid nitrogen and liquid carbondioxide cooling, the refrigerant is introduced into the vehicle through nozzles, the flow being controlled by thermostats and solenoid valves. The storage vessels for the refrigerants are kept within the refrigerated space. Spray nozzles for liquid corbondioxides are designed to avoid blockage due to snow formation during pressure drop, as liquid carbondioxide vessel has to be under pressure.

5. Eutectic plates. Eutectic plates forming channels are placed all around the body of the vehicle. They contain an *eutectic solution which can be frozen by a refrigerant flowing in a coil immersed in the solution*. The coil is connected to a mechanical refrigeration system at the charging station. Cooling is produced by the evaporation of the primary refrigerant in the coil. When the solution is frozen the truck is ready for departure.

6. Mechanical refrigeration with independent unit. The mechanical refrigeration system mounted on the vehicle has an independent engine. By far, this is one of the most common methods.

7. Mechanical refrigeration deriving power from the engine. There are many practial designs available. The refrigeration compressor may be either directly coupled to the engine, or may take off the power through transmission. The refrigeration is, however produced only when the vehicle engine is in operation.

21.4 DOMESTIC REFRIGERATOR

21.4.1 Construction and Working

Refrigerators, these days, are becoming the common item for *household use, vendor's shop, hotels, motels, offices, laboratories, hospitals, chemist and druggist shops, studios etc.* They are manufactured in different sizes to meet the needs of various groups of people. They are usually rated with *internal gross volume and the freezer volume*. The freezer space is meant to preserve perishable products at a temperature much below 0°C such as fish, meat, chicken etc., and to produce ice and ice-cream as well. The refrigerators in India are available in different sizes of various makes, i.e., 90, 100, 140, 200, 250, 380 litres of gross volume. The freezers are usually provided at top portion of the refrigerator space occupying around one-tenth to one-third of the refrigerator volume. In some refrigerators, freezers are provided at the bottom.

A domestic refrigerator consists of the following *two* main parts:

1. The refrigeration system.
2. The insulated cabinet.

Fig. 21.2 shows a flow diagram of a typical refrigeration system used in a domestic refrigerator. A simple domestic refrigerator consists of a hermetic compressor placed in the cabinet base. The condenser is installed at the back and the evaporator is placed inside the cabinet at the top.

The **working of the refrigerator** is as follows:

- The low pressure and low temperature refrigerant vapour (usually R–12) is drawn through the suction line to the compressor. The *accumulator* provided between the *suction line* and the evaporator collects liquid refrigerant coming out of the evaporator due to incomplete evaporation, if any, prevents it from entering the compressor. The *compressor* then compresses the refrigerant vapour to a high pressure and high temperature. The compressed vapour flows through the *discharge line* into condenser (vertical natural draft, wire-tube type).
- In the *condenser* the vapour refrigerant at high pressure and at high temperature is condensed to the liquid refrigerant at high pressure and low temperature.

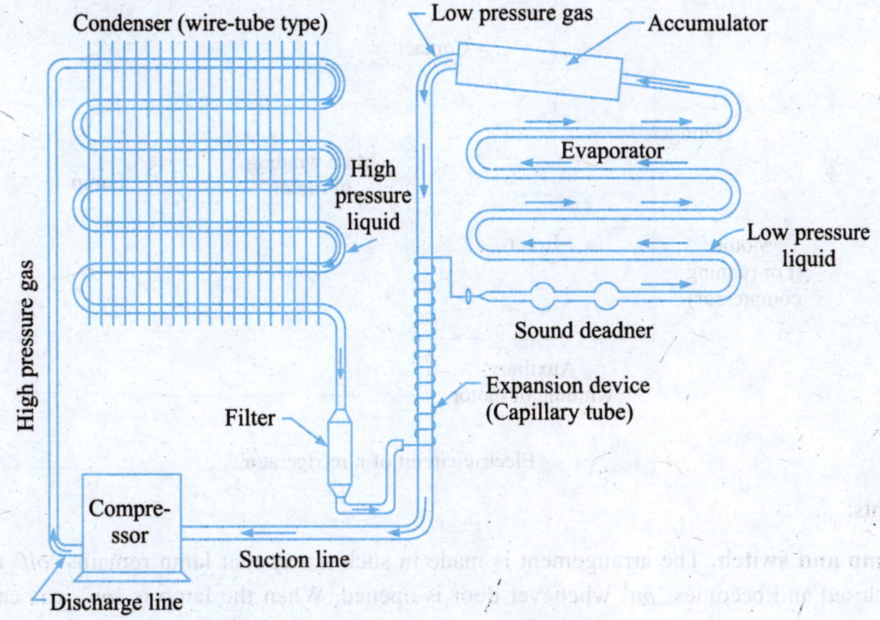

Fig. 21.2. Domestic refrigerator.

- The high pressure liquid refrigerant then flows through the *filter* and then enters the *capillary tube* (expansion device). The capillary tube is attached to the suction line as shown in Fig. 21.2. The warm refrigerant passing through the capillary tube give some of its heat to cold suction line vapour. This increases the heat absorbing quality of the liquid refrigerant slightly and increases the superheat of vapour entering the compressor.

  The capillary tube expands the liquid refrigerant at high pressure to the liquid refrigerant at low pressure so that a measured quantity of liquid refrigerant is passed into the *evaporator*.
- In the *evaporator* the liquid refrigerant gets evaporated by absorbing heat from the container/articles placed in the evaporative chamber and is sucked back into the compressor and the cycle is repeated.

21.4.2 Electrical Circuit of a Refrigerator

Fig. 21.3 shows a schematic diagram of electric circuit of a refrigerator.

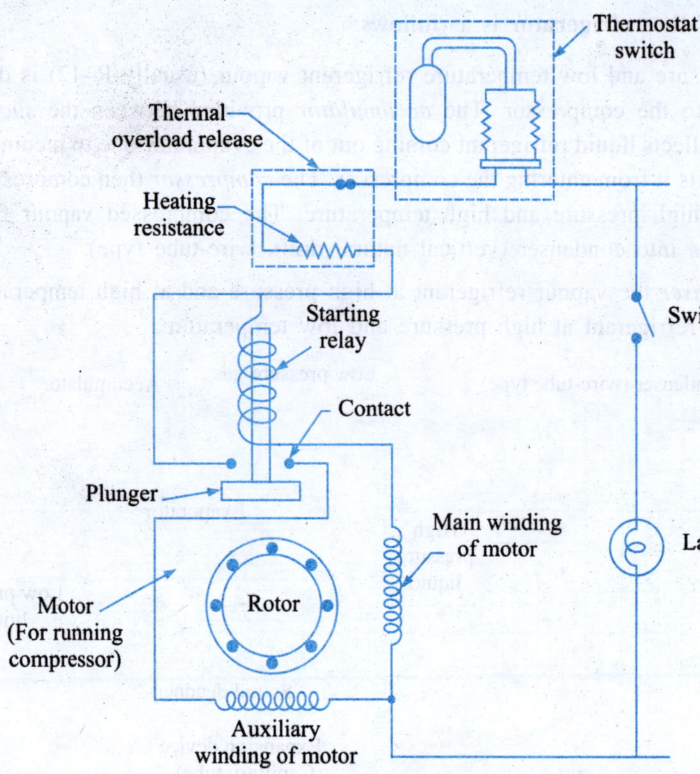

Fig. 21.3. Electric circuit of a refrigerator.

Components:

1. **Lamp and switch.** The arrangement is made in such a way that lamp remains '*off*' as the door is closed and becomes '*on*' whenever door is opened. When the lamp is '*on*' it is easy to trace the commodities placed in the refrigerator.
2. **Thermostat switch.** A thermostat switch maintains a requisite temperature in the refrigerator.
   Freezer. Temperature remains between –7°C to 5°C approx.
   Remaining part of the refrigerator. Temperature remains between 7°C to 15°C.
3. **Thermal overload release.** This component is a protective device for compressor motor unit. It operates when temperature of the compressor rises beyond a certain value or excessive current flows in the motor; under such conditions the bimetallic strips disconnect the supply to the motor.
4. **Starting relay.** A starting relay starts the motor by the putting starting winding/auxiliary winding of split phase induction motor across the supply.
5. **Electric motor.** Electric motor used is single phase induction motor, split phase type. It is a fractional horse power induction motor; its size depends upon the capacity of the refrigerator.

Working:

When electric supply is given to the refrigerator, current passes through the thermostat switch, thermal overload release, coil of starting relay and main winding of the motor. To start with, when the motor is at rest it draws a very heavy current. When this heavy current flows through the coil of the staring relay, the coil gets energised and it pulls up the plunger, short circuiting the contacts and putting auxiliary winding also in the circuit. Now since both the main winding and auxiliary winding are energised, motor starts running. When the motor gains normal speed, the current drawn by the main winding of motor becomes normal. At normal current plunger in the coil of starting relay cannot remain pulled and it is released down, opening the contacts, thus auxiliary winding gets out of circuit. The main function of starting relay is to put auxiliary winding in the circuit at the time of starting the motor and to disconnect it when the motor gains normal speed. In case starting relay fails to close, motor will not start (as explained above). But once it closes and it fails to open, then either thermal overload shall trip out or fuse shall be blown off.

21.4.3 Maintenance of Domestic Refrigerator

In order to achieve proper results in the maintenance of the refrigerator, the following points should be taken into consideration:

1. Refrigerator must be levelled properly.
2. Refrigerator should be operated on 230 volts, 50 Hz supply. If the voltage fluctuates widely it is advisable to switch off the refrigerator otherwise compressor motor may burn out.
3. Refrigerator should be close to an electrical outlet as overload or lighting extension cords may reduce the line voltage to a dangerously low level.
4. Refrigerator should not be exposed to direct rays of sun because this will add to the load of operation.
5. There should be a gap of at least 25 cm between the condenser (back of the refrigerator) and the wall.
6. Refrigerator should be operated without load at least two hours before loading it.
7. Dust of condenser should be removed periodically and make sure that no material covers the portion.
8. If there is urgent need of ice turn the knob of thermostat to 5 or more and turn the baffle tray deflection to the out position. Don't forget to change these to their normal position afterwards.
9. Defrosting should be done periodically as frost is a barrier which hinders cooling and adds to the load operation.
10. Never use harsh cleaners to hasten the defrosting process by striking at or chipping off the ice as it will result in leakage of gas.
11. Baffle tray should be kept out when it is desired to collect the dripping water from freezer chest at the time of defrosting.
12. Moist food should be placed in a closed container or wrapped in polythene or aluminium foil.
13. To safeguard against electric shock the refrigerator should be properly earthed with green wire and operated only on three point plugs. *Improper earthing may result in fatal accidents.*

21.5 WATER COOLERS

Water is one of the most needed thing for a man. In summer season cold water gives life to a thirsty man. At 10°C water is most refreshing. Thus cooling of water in summer becomes necessary. *Water coolers* are used to produce cold water at about 7°C to 13°C. The temperature of water is controlled with the help of a thermostatic switch.

Water cooler may be classified as follows:

1. *Instantaneous type water coolers:*
   (*i*) Bottle type cooler.
   (*ii*) Pressure type cooler.
   (*iii*) Self-controlled or remote type cooler.
2. *Storage type water coolers.*

1. Instantaneous Type Water Coolers:

In this type of coolers the cooling coil is *wrapped round the pipe line such that by the time water reaches the tank it is cooled to desired temperature*. The description of various types of instantaneous type water coolers is given below:

(*i*) **Bottle type cooler.** In this type water to be cooled is stored in a bottle or reservoir. For filling glass tumblers or containers faucet or similar means are provided. The dripping water from the faucet is collected in the waste water basin or water drip as shown in Fig. 21.4. Its usual *size is 25 litres* and is *suitable for places where plumbing installations is expensive and drains are available.*

(*ii*) **Pressure type cooler.** Refer Fig. 21.5. Here water is supplied under pressure. For filling glass tumblers or containers faucets or similar means are provided. A valve is employed to control an appropriate flow of water or projected stream of water from a bubbler. An arrangement should be made to collect water and allow complete collection of water spreading from the bubbler. The temperature of waste water is low, it is used for cooling the supply water by passing through a pipe coil wrapped round the drainage line. By doing so, the cooling load for cooler is reduced. Since the water is supplied under pressure the cold water can be obtained from the top mounted at any height of the cooler. In case of bottle type, faucet has to be at a height upto which syphoned water can be obtained from the tank of the cooler. The *refrigeration system is usually mounted at the bottom structure of the cooler body and a cooling coil is wrapped round the water tank, to ensure good surface contact between the evaporative tube and tank, either the tank surface is corrugated to accommodate pipe or pipes are secured using soft solder to give metal contact.* Sometimes a helical or U-type coil is immersed in the water tank. Although this arrangement *gives high heat transfer from water to the coil yet formation of undesirable salt due to chemical reaction between water contaminant and the copper surface proves to be a great disadvantage in this system.*

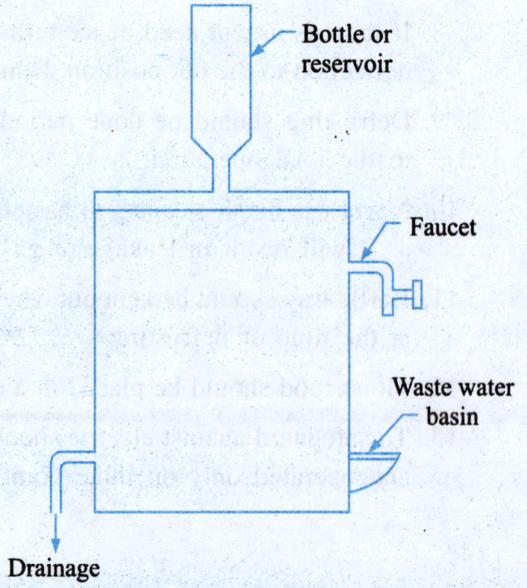

Fig. 21.4. Bottle type cooler.

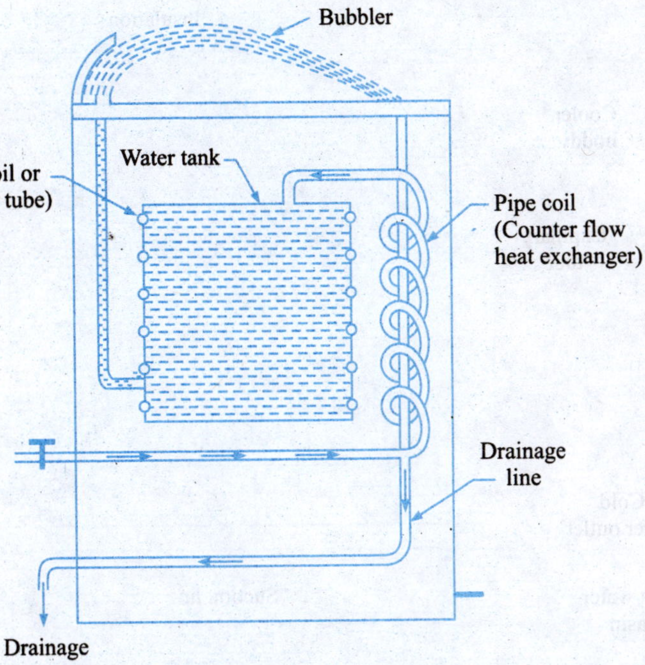

Fig. 21.5. Pressure type cooler.

(*iii*) **Self-contained or remote type.** Refer Fig. 21.6. This type of water cooler employs mechanical refrigeration system and is a factory assembled unit. A remote cooler cools the water which is supplied to the desired drinking place (away from the system). It is quite a useful unit since it does not require extra space near the place of work.

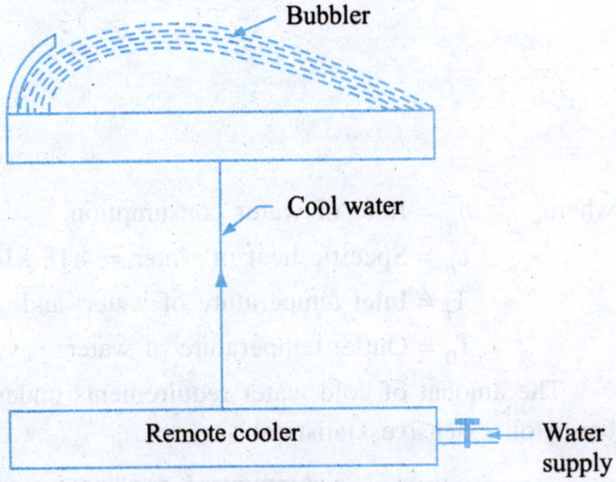

Fig. 21.6. Remote type cooler.

2. **Storage Type Water Cooler:**

Such type of coolers are used where continuous supply of water is not available. Fig. 21.7 shows a schematic storage type water cooler which is self explanatory. Here water is filled in the storage tank and level of the water is kept same by the use of a float valve. The storage tank is surrounded by an evaporator coil through which flows a low pressure liquid refrigerant which takes away the heat of water and thus makes it cold. When the water attains desired temperature the thermostat operates and disconnects the power supply to the motor. The motor used is *capacitor-start capacitor run single-phase induction motor.*

Capacity of Water Coolers

In order to calculate the cooling load for the water cooler (Q) the following relation may be used:

$$Q = m_w \times c_p (T_i - T_0) \quad ...(27.1)$$

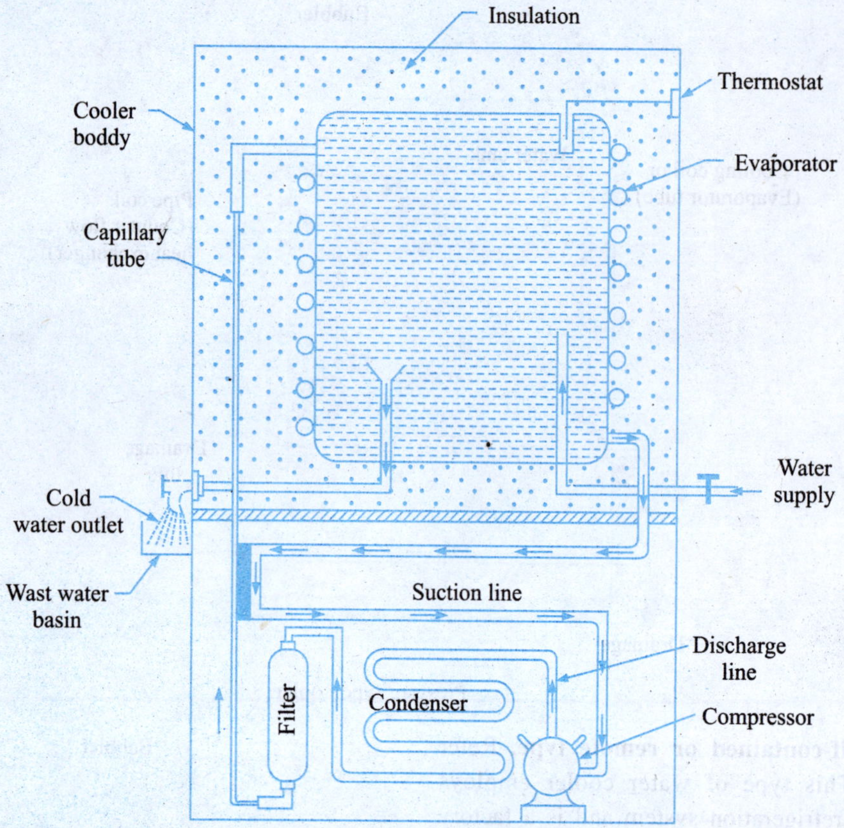

Fig. 21.7. Storage type water cooler.

where, m_w = Rate of water consumption,

c_p = Specific heat of water = 4.18 kJ/kg K,

T_i = Inlet temperature of water, and

T_0 = Outlet temperature of water.

The amount of cold water requirements under various conditions is given in Table 21.1 (Figures based on extensive statistical survey).

Table 21.1. Amount of drinking water requirements under various conditions

S. No.	***Applications***	***Temperature in °C***	***Amount of water used including drainage in litres per hour***
1.	*Restaurants*	7 – 10	0.5 per person
2.	*Schools*	7 –10	0.56 per student
3.	*Offices*	7 – 10	0.56 per person
4.	*Stores*	7 – 10	4.5 per 100 customers
5.	*Hotels*	7 – 10	0.4 per room
6.	*Hospitals*	7 – 10	0.4 per bed and per attendant
7.	*Theatres*	7 – 10	4.5 per 100 seats
8.	*Light duty manufacturing shop*	10 – 13	0.9 per person
9.	*Heavy duty manufacturing shop*	10 – 13	1.1 per person

- **R-12** is the most common refrigerant upto 1 tonne refrigeration capacity and **R-22** for 2 tonnes refrigeration capacity, and appropriate combinations for units of larger size.
- While estimating the amount of water consumption, the amount of wastage of cold water should be included.
- For ensuring insignificant heat transfer through insulation, usually heavy insulation around 4 cm to 6 cm thick glass wool or thermocole is provided.

Example 21.1. *The following data refer to a water cooler for drinking water:*

Temperature of drinking water required = 11°C

Number of workers = 450

Duty time = 6 hours

Temperature of the water available = 28°C

Heat transfer through insulation = 4% *of the total heat load*

Calculate:

(*i*) *Refrigeration tonnage of the water cooler.*

(*ii*) *Total water consumption per day.*

Take drinking water requirement per person as 1.1 litre per hour.

Solution: *Given:* $T_0 = 11 + 273 = 284$ K; $T_i = 28 + 273 = 301$ K

Number of workers = 450; Heat transfer = 4% of total heat load

Drinking water requirement per person = 1.1 litre per hour.

(*i*) **Refrigeration tonnage of the water cooler:**

Cold water supply per hour,

$$m_w = 1.1 \times 450 = 495 \text{ litres/h} = 495 \text{ kg/h} \qquad (\because \text{Sp. gr. of water} = 1)$$

Cooling load for the water cooler,

$$Q = m_w \times c_p \times (T_i - T_0)$$

$$= 495 \times 4.18 \times (301 - 284) = 35174.7 \text{ kJ/h}$$

Since the heat transfer through insulation is 4% of the total heat load, therefore, total cooling load,

$$Q_T = 0.04\, Q_T + Q \quad \text{or} \quad 0.96\, Q_T = Q$$

$$\therefore \quad Q_T = \frac{Q}{0.96} = \frac{35174.7}{0.96} = 36640.3 \text{ kJ/h}$$

$$= \frac{36640.3}{14000} = \textbf{2.62 tonnes of refrigeration (Ans.)}$$

(*ii*) **Water consumption per day:**

Water consumption per day (for 6 hours duty time)

$$= 495 \times 6 = \textbf{2970 litres/day (Ans.)}$$

21.6 DESERT COOLER

21.6.1 Principle of Working

As the name implies, *it is suitable for place where the humidity is quite low and temperature quite high.* These conditions are in conformity with *desert areas*. Hence the coolers are called **desert coolers**. The principle on which a desert cooler works is **'evaporative cooling'**. Evaporative cooling is a *process in which sensible heat is removed and moisture added to the air.* When air passes through a spray of water it gives up heat to water, some of the water evaporates and picks up heat from the air equivalent to its latent heat. The vapour thus formed are carried along in air stream. In this way *air is cooled and humidified.*

21.6.2 Construction and Working of a Desert Cooler

A desert cooler [Refer Fig. 21.8 (*a*), (*b*)], consists of following *main parts*:

1. Blower/Fan
2. Water recirculating pump
3. Water wetted pads
4. Water tank
5. Float valve
6. Water traps (3 in No.)

The water is filled in the sump of the cooler from water supply mains, the level of which is controlled by a float valve. A water pump lifts the water and supplies it at the top of the cooler to the water distribution system which consists of small branches of a copper pipe or so equipped with orifices which deliver equal amount of water to the troughs which in turns supply water to the wetted pads. The water which drops back from the pads is recirculated. The pump may be made of brass, strainless steel or even plastic.

The blower pulls the air through the wetted pads and deliver it to space to be cooled through an opening in the fourth side of the cabinet of desert cooler. The *air which is sucked through the pads is cooled by the principle of evaporative cooling.* The blower gives adequate velocity to the air before it is delivered to the spaces to be cooled.

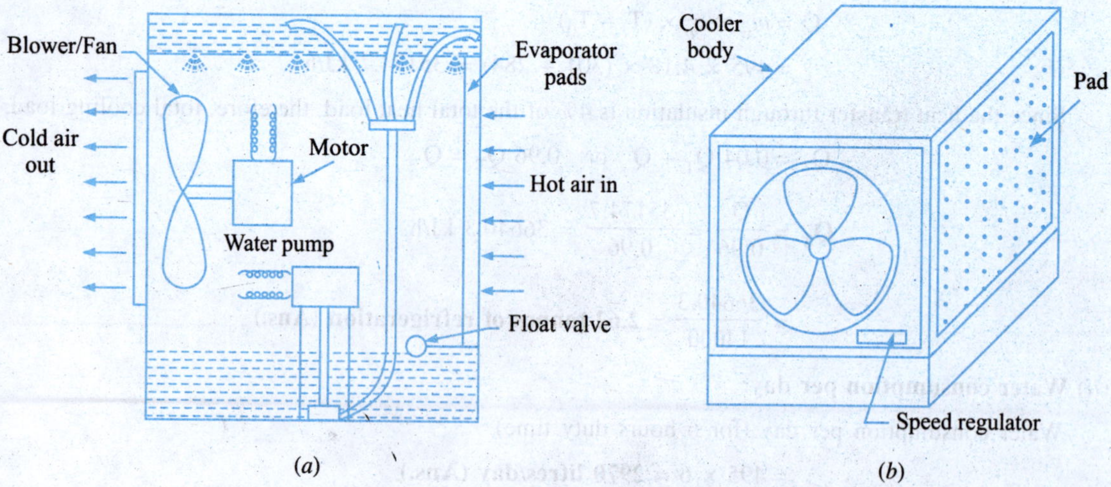

Fig. 21.8. Desert cooler.

To have long life of the desert cooler and better performance, pads should be changed every year and holes for water distribution system should be cleaned. The tank should be cleaned just after the season and coated with the corrosion resisting paint like red oxide and some other paints.

The efficiency of desert cooler is given by:

$$\eta_{\text{desert}} = \frac{t_{db} - t_{dbg}}{t_{db} - t_{wb}}$$

where, t_{db} = Dry bulb temperature of ambient air,

t_{dbg} = Temperature of air leaving the grille, and

t_{wb} = Wet bulb temperature of ambient air.

According to *Indian Standard Institution* the velocity across the pad should be around 40 m/min. The desert cooler efficiency should be around 75%. Though the high efficiency of desert cooler corresponds to cooler air, the *large humidity causes discomfort.* The desert cooler efficiency above 80% is not desirable. For good performance the efficiency of a desert cooler should not fall below 70%.

21.6.3 Comparison between a 'Desert Cooler' and a 'Window Type Air-conditioner'

S. No.	*Particulars*	*Desert cooler*	** Window type air-conditioner*
1.	*Initial cost*	Low	Quite high.
2.	*Main components*	Motor and fan	Complete refrigeration unit: Compresor, condenser, evaporator and controls etc.
3.	*Humidity control*	Not possible	Possible.
4.	*Power required per unit cooling*	More	Less.
5.	*Utility*	Can be used only to cool the air and that too in the dry season.	Can be used for both cooling . and heating purpose.
6.	*Location*	Inside or in the window of the room.	Fitted either in the window or wall.

* For details refer to Art. 15.1.1.

21.7 ICE SYSTEM OF AIR-CONDITIONING

Fig. 21.9. shows schematic representation of ice system of air-conditioning. In this system cooling is done by melting ice. This is a practical air-conditioning method for churches, theatres, and public halls that have short operating hours and relatively high peak loads. Since the investment in mechanical refrigeration equipment is expensive for short period, ice fills the need. A small quantity of ice in a water-cooling tank can release refrigeration at a fast rate.

21.8 ALL THE YEAR-ROUND AIR-CONDITIONER

An all the year round air conditioner maintains the desired conditions irrespective of season temperature variation of the ambient air. It is possible to convert simple air-conditioner into the year round air-conditioner simply by incorporating four valves as shown in the Fig. 21.10.

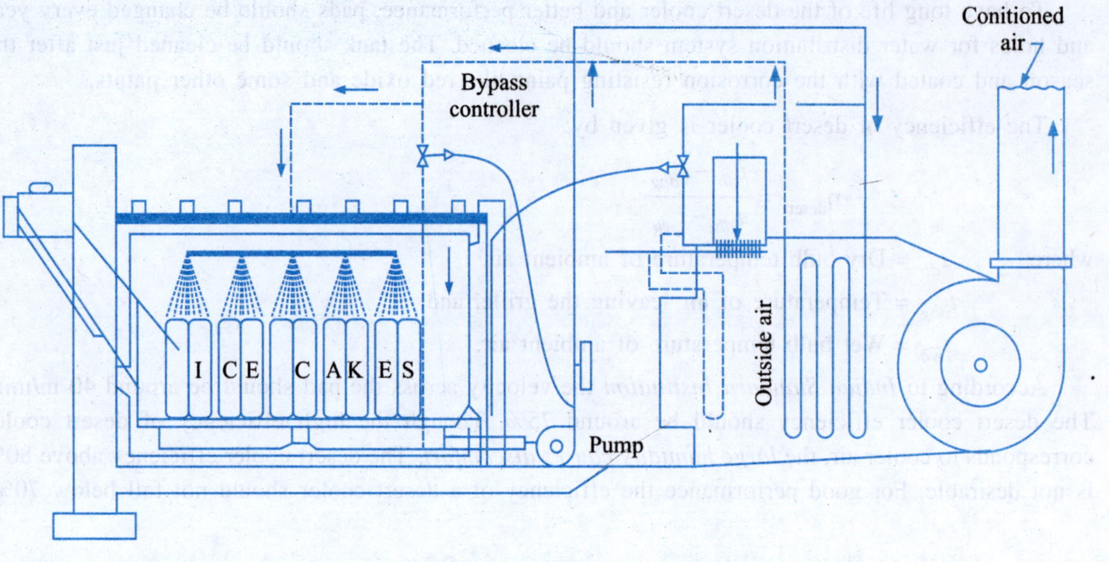

Fig. 21.9. Ice system of air-conditioning.

Summer season : Valves A and B — 'open'

Valves C and D — 'closed'

Winter season : Valves C and D — 'open'

Valves A and B — 'closed'

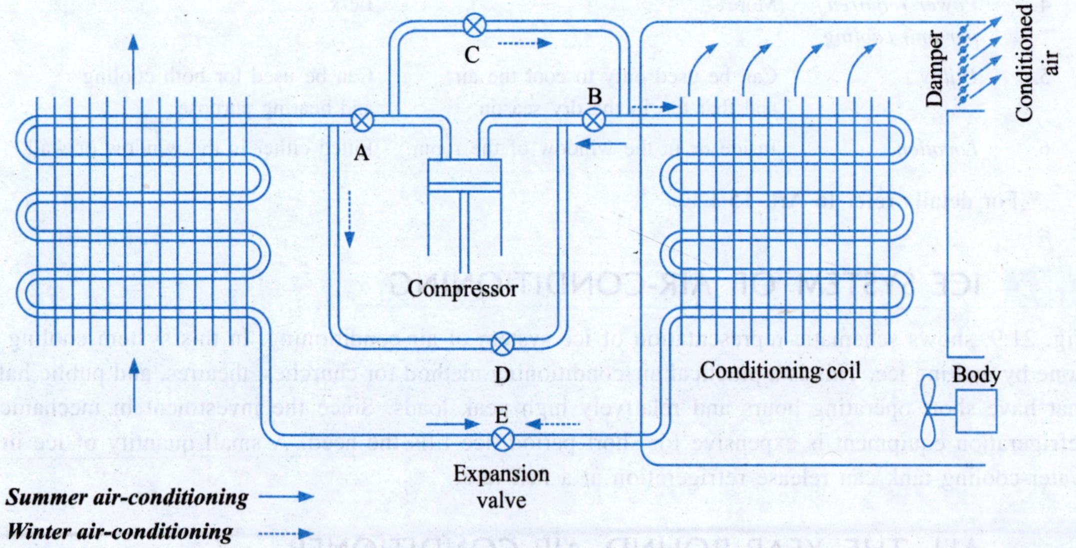

Fig. 21.10. Year-round air-conditioner.

21.9 YEAR-ROUND ABSORPTION AIR-CONDITIONER

Absorption type year-round air-conditioner uses *Lithium Bromide* solution in water which acts as absorbent and water vapour as refrigerant.

To supply heat to the conditioned space in winter, the water vapour at high pressure generated from generator on application of heat to the generator *by-passes the condenser and then led to the coil, where it is condensed to water by rejection of its latent heat to the room air*. The condensate is then taken by gravity to the generator.

In summer, the water vapour from generator is condensed in the condenser. The water then expands to a lower pressure through an expansion valve and then passes through the aforesaid coil which now acts as evaporator. The water there evaporates completely by extracting heat from the room air. The resulting low pressure vapour is then absorbed in the absorber. The solution from the absorber is then led by gravity to the generator for driving off the vapour absorbed. The cycle is then repeated.

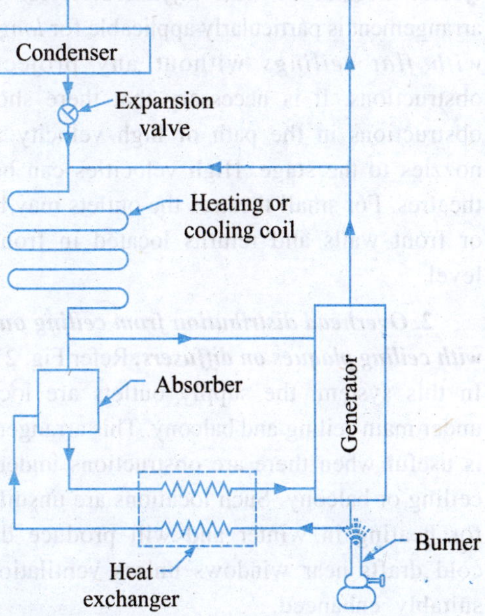

Fig. 21.11. Year-round absorption air-conditioner.

Ref. Fig. 21.11. It shows a diagrammatic arrangement of water and lithium bromide circuit for gas fired or steam heated year-round air-conditioner. The circulation is by gravity and by thermal differential in the system. There are no moving parts. These units are built in sizes of 3 to 5 tonnes and are used in residential and commercial applications.

21.10 AIR-CONDITIONING OF THEATRES

In all theatres central system of air-conditioning is employed. The cooling load mainly consists of *occupancy* and *amount of outdoor air used.*

The *outdoor air* required per occupant is 0.14 cmm. without smoking and about 0.177 cmm with smoking.

The *occupancy load consists* of men, women and children at rest. On an average, the amount of heat released in kJ/h is 420 per male occupant, 370 per female occupant and 30 per child. The heat released per occupant may therefore be taken as 370 kJ/h.

The *design condition of air* inside is 25.5% to 26.5°C and 55% R.H. which may have apparatus dew point slightly in excess of 11.5°C.

The ventilation air per occupant is 0.42 and 0.51 cmm for air delivered horizontally from the near or side wall outlets and about 0.7 cmm for air delivered from the ceiling outlets with downward diffusion.

When the weather is mild arrangement should be provided for providing full outdoor air without recirculation and conditioning in order to reduce the operating costs.

On *an average one tonne of refrigeration is required per 15 to 20 seats.*

Types of air distribution: The following *two* types of air distribution is found in theatre application:

1. *Horizontal distribution from near wall outlets with ejector nozzles or wall diffusers.* Refer Fig. 21.12. This arrangement is particularly applicable for *long narrow theatres with flat ceilings* without any projecting beams or obstructions. It is necessary that there should not be any obstructions in the path of high velocity air from ejector nozzles to the stage. High velocities can be used for large theatres. For small theatres the outlets may be located inside or front walls and returns located in front walls at floor level.

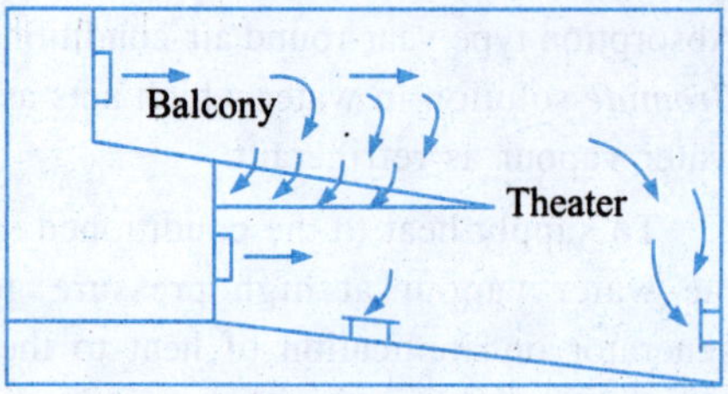

Fig. 21.12

2. *Overhead distribution from ceiling outlets with ceiling plaques on diffusers.* Refer Fig. 21.13. In this system, the supply outlets are located under main ceiling and balcony. This arrangement is useful when there are obstructions under the ceiling or balcony. Such locations are unsuitable for heating in winter and will produce down cold drafts near windows unless ventilation is suitably enhanced.

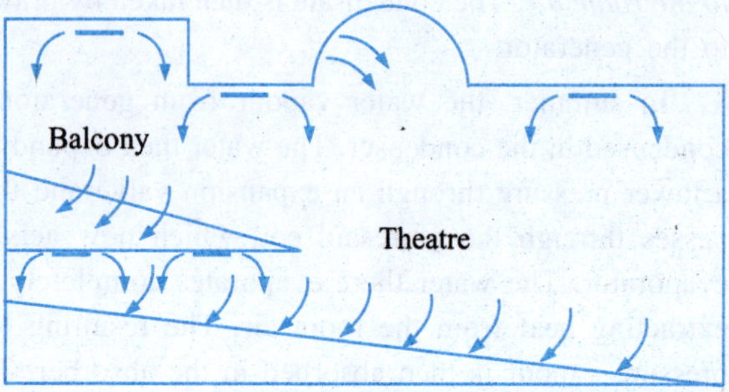

Fig. 21.13

Refrigerant and components:

The central plant *may use R–11 as refrigerant with centrifugal compressor.*

The *condenser* may be:

(*i*) Tube and shell type with cooling water supply cooled by cooling towers, or

(*ii*) Evaporative type, or

(*iii*) Air-cooled.

The air may be conditioned by:

(*i*) Direct expansion cooling coil, or

(*ii*) Cooling coil supplied with chilled water, or

(*iii*) Air washer using chilled water sprays.

It may be noted that the temperature of supply air from outlets in the theatre must be such as to maintain a temperature difference between room and supply air not exceeding 8 to 14°C otherwise it will create cold drafts causing shock to audience.

21.11 MANUFACTURE OF ICE

Fig. 21.14 shows the layout of an ice plant. The ice used for commercial purposes is produced by *freezing potable water* in standard cans placed in rectangular tanks which are filled with *chilled brine.* In order to enhance the heat transfer from the water agitators are employed to keep the brine solution in constant motion. The brine temperature is maintained at –10°C to –11°C by the refrigeration plant. The ammonia gas is used as refrigerant since it has excellent thermal properties, produces very high refrigerating effect per kg of refrigerant and has low specific volume of the refrigerant (in vapour state).

Working of ice plant:

- The high temperature, high pressure NH_3 vapours leaving the compressor are condensed in a condenser (may be of shell and tube type or evaporative type).
- The condensed liquid NH_3 is collected in the receiver and then expanded through the expansion valve. Due to expansion, the pressure of the liquid NH_3 is considerably reduced.

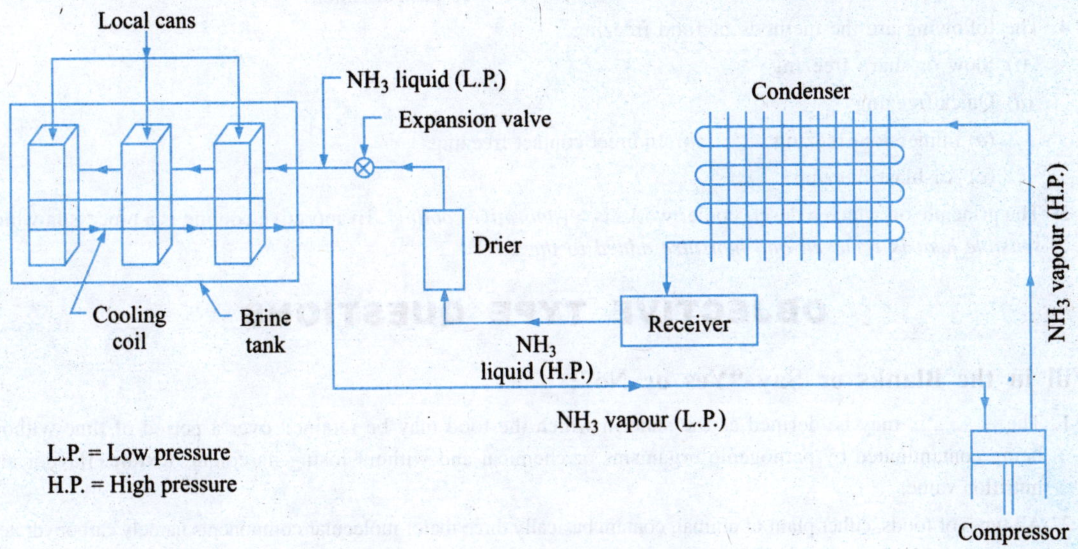

Fig. 21.14. Layout of an ice plant.

- The liquid NH_3 then passes through the evaporator coils surrounding a brine tank in which brine solution is filled. The low pressure liquid NH_3 absorbs heat from the brine solution, equivalent to its latent heat of evaporation, geths converted to vapour state and is once again fed to compressor to complete the cycle.

The following points are *worth noting:*

- Generally NaCl (sodium chloride) or CaCl (calcium chloride) are used as brine. NaCl is preferably used because of its low cost and being less injurious.
- The freezing line is dependent upon the temperature and extent of brine and water agitation.
- In order to get clear transparent ice, water in the can is agitated using low pressure air through the tubes suspended from the top.
- The ice cans are fabricated from galvanised steel sheets and are given chromium treatment to prevent corrosion due to chemical reactions.

HIGHLIGHTS

1. The *food preservation* may be defined as the state in which the food may be retained over a period of time without being contaminated by pathogenic organisms or chemicals and without losing its colour, texture, flavour and nutrition value.

2. Fool spoilage is activated by the following agents:
   (*i*) Enzymes; (*ii*) Micro-organisms
3. The various methods of food preservation are:
   (*i*) Heat processing (*ii*) Dehydration
   (*iii*) Chemical preservation (*iv*) Oils and spices
   (*v*) Canning (*vi*) Pasteurization.
4. The following are the methods of food freezing:
   (*i*) Slow or sharp freezing
   (*ii*) Quick freezing:
   (*a*) Immersion freezing (*b*) Indirect contact freezing
   (*c*) Air-blast freezing.
5. The principle on which a desert cooler works is '*evaporative cooling*'. Evaporative cooling is a process in which *sensible heat is removed and moisture added to the air.*

OBJECTIVE TYPE QUESTIONS

Fill in the Blanks or Say "Yes or No":

1. The .... may be defined as the state in which the food may be retained over a period of time without being contaminated by pathogenic organisms or chemical and without losing its colour, texture, flavour and nutrition value.
2. All types of foods, either plant or animal, contain basically three major molecular components namely carbohydrates, proteins and.............. .
3. Pasteurization is a method adopted to destroy micro-organisms by heating the commodities to the desired temperatures.
4. The bacteria cannot be destroyed by extreme heat.
5. Micro-organisms of size about 10^{-4} mm are called ....
6. Mostly all molds can be destroyed with heating at 60°C and they can be made inactive by deep freezing.
7. .... is the process of loss of moisture from the surface of the products by evaporation into the surrounding air.
8. The salt and sugar are used as chemical preservatives.
9. In .... freezing method, the food products are placed in a low temperature room and allowed to freeze slowly in still air by natural convection.
10. .... freezing is brought about by forced circulation of cold air necessary for freezing the foods in storage.
11. Immersion freezing method produces a thin coating of ice on the surface of the product.
12. In .... freezing method the food products are freezed by the contact with cold air.
13. Milk preservation includes blending, processing, packaging and distribution to the customers.
14. Chilling and freezing are used to preserve poultry meat.
15. Dehydro-freezing is a process involving both dehydration and ....
16. At 10°C water is most refreshing.
17. The usual size of a 'bottle type cooler' is .... litres.
18. .... type water coolers are used where continuous supply of water is not available.
19. The principle on which a desert-cooler works is ....
20. In air-conditioning of theatres, the cooling load mainly consists of occupancy and amount of outdoor air used.

ANSWERS

1. Food preservation.	**2.** Fats.	**3.** Yes	**4.** No.
5. Yeasts.	**6.** Yes.	**7.** Desiccation or dehydration	
8. Yes	**9.** Sharp	**10.** Quick	**11.** Yes
12. Air-blast	**13.** Yes	**14.** Yes	**15.** Freezing
16. Yes	**17.** 25	**18.** Storage	
19. Evaporative cooling	**20.** Yes.		

THEORETICAL QUESTIONS

1. What do you understand by "Food preservation"?

2. What are the advantages of food preservation?

3. State the factors which contribute to food spoilage.

4. Explain briefly the following agents which help activate the spoilage of food:

(*i*) Enzymes; (*ii*) Micro-organisms.

5. What do you mean by the term 'Desiccation'?

6. Enumerate and explain briefly three commonly used methods of food preservation.

7. Explain briefly the following methods of food freezing:

(*i*) Slow or sharp freezing;

(*ii*) Quick freezing.

8. Describe briefly the following methods of quick freezing:

(*i*) Immersion freezing;

(*ii*) Indirect contact freezing; and

(*iii*) Air-blast freezing.

9. Write a short note on "Freeze drying".

10. Explain briefly how the following foods are processed/preserved:

(*i*) Milk and butter;

(*ii*) Meat products;

(*iii*) Poultry products;

(*iv*) Fishery products.

11. State the main advantages which accrue by cold storage of fruits, food stuffs and vegetables.

12. Describe briefly with a neat sketch a cold storage plant.

13. Give briefly the methods of refrigeration used in refrigerated trucks and trailors.

14. Explain the construction and working of a domestic refrigerator, with the help of a neat sketch.

15. Draw a schematic diagram of electric circuit of a refrigerator and explain its working.

16. Write short note on 'Maintenance of domestic refrigerator'.

17. Explain briefly different types of water coolers.

18. Describe with a neat sketch the construction and working of a storage type water cooler.

19. Explain the principle of working of a desert cooler.

20. Describe briefly with the help of a neat sketch, the construction and working of a desert cooler.

21. Give the comparison between a 'Desert cooler' and a 'Window air-conditioner'.

22. Write a short note on "Ice system of air-conditioning".

23. Explain briefly the following with neat sketches:

(*i*) All the year round air-conditioner;

(*ii*) Year-round absorption air-conditioner.

24. Explain briefly the 'air-conditioning of theatres'.

25. Give the layout of an ice plant and explain briefly how ice is manufactured.

❑❑❑

Maintenance and Repair of Commercial and Domestic Equipment

Chapter 22

22.1 MAINTENANCE OF A NEW INSTALLATION—SAMPLE SCHEDULING

1. **Daily check:**

- Level of oil
- Pressure of oil
- Shaft seal leakage
- Suction pressure
- Discharge pressure
- Tension of V-belts.

2. **General instructions for maintenance:**

- Evacuate the compressor before carrying out maintenance.
- Purge the compressor after maintenance.

3. **Running operation:**

- Clean oil strainers thoroughly after *half an hour running* after initial start.
- Clean oil strainers after every *four hours* running until they remain clean.
- Clean refrigerant suction strainer after *eight hours* running for a period of one week.
- Clean refrigerant suction strainer after 24 hours' running until strainer remains practically clean.

22.2 COMPRESSOR REPAIR AND CHECKING THE EFFICIENCY

Compressor repair should be carried out only when it is made sure that the other parts of the system are trouble free. Minor repairs of the system can ordinarily be made without removing the compressor. When a compressor is worn out considerably it gives out lot of noise, its main bearing journals are worn out. In such a case it is better to use a new or reprocessed/rebuilt compressor than to attempt repairs particularly when the time and materials are of greater consideration. For the average engineer it is usually more economical to get the repairs done from a repairing shop than to make the major repairs himself.

The *following procedure may be followed to check the efficiency of a compressor:*

- Back seat the suction and discharge service valves and then remove the plugs.
- Attach a pressure gauge to the discharge valve and a compound gauge to the suction valve gauge port.
- Check both the valves.
- Start the compressor and throttle the discharge valve until pressure head of 9 bar is maintained.
- Close the suction valve slowly and when it is completely closed note the vacuum on the compound gauge.
- The *compressor may be considered as operating satisfactorily if the compressor produces and also maintains a vacuum of 50 cm or more.*
- Check the discharge valve leakage only after the compressor has been running long enough with the suction valve closed to remove any refrigerant which may be in the crankcase just to avoid the false reading.
- The suction will be leaky if it is unable to produce a 50 cm vacuum under continuous running test and abnormal high suction pressure during operation.
- Repair/replace both the suction and discharge valves if found defective.
- Faulty suction reeds located in the piston may be replaced, if removable. If not the entire piston must be replaced.
- In the compressor if there are loose or broken parts, replace the entire unit. The major repairs should be carried out only in large compressor where it is not possible to have complete replacement.

22.3 DESCALING OF A CONDENSER

The descaling of a condenser should be carried out only when it is firmly ascertained that there is a considerable scaling in the condenser. The scale formation prevents heat transfer from the refrigerant to water and thereby result in high discharge pressure. High head pressure affects the performance of the equipment adversely. The following **procedure** is adopted to descale the condenser:

- Drain out the water from the condenser after isolating it from rest of the water line.
- Connect the suction of the pump with separate tank by rubber hose and discharge the pump to condenser water inlet with suitable connectors and rubber hose.
- Connect the condenser return line back to the tank.

  Fill the tank with raw water.

  Start the pump and circulate the water for ten minutes through the condenser and make sure that water does not leak from any of the joints or hoses etc. Make sure that adequate water is available in the plant room for cleaning the condenser etc. after acid circulation.
- Add acid to the tank water at 1 : 6 ratio (of maximum 8% concentration). Circulate the solution for 6 to 8 hours and drain out the acid thereafter. Flush the system with fresh water. Then circulate washing soda in the tank for half an hour. Flush the system with fresh water once gain.

It may be noted that acid circulation should always be done by a separate pump after isolating the condenser from the existing pump and cooling tower. Acid *should never be poured in cooling tower.*

22.4 PURGING OR REMOVING AIR FROM THE SYSTEM

The following **methods** may be used to remove air from the evaporator and lines.

First Method. In this method the refrigerant in the receiver is caused to force air out. The main requirement in this method is that there must be sufficient pressure in the receiver and sufficient refrigerant to fill the system. This method entails the following *procedure.*

- Open the liquid valve (compressor service valve being assumed closed) which will cause the refrigerant to push the air ahead of it down to the suction service valve, which is closed.
- The flare nut on the suction line at the valve is loosened and *air is allowed to escape*. The nut should be kept loose until a decided hiss of the escaping vapour is heard.

This method is the *least satisfactory method and is used only on small installations.*

Second Method. In this method *air is pumped with compressor.* Its major drawback is that it draws air and moisture from the lines and the evaporator through the factory dehydrated compressor. The amount of moisture that will remain in the compressor and the rest of the system can be removed with dehydrator.

This method may be employed on small as well as large systems.

Third Method. This method *makes use of a separate vacuum pump to purge/remove the air and is best of the three method.* Here condensing unit is not started.

Refer to Fig. 22.1. The *procedure* for this method is as follows:

- Close the suction valve and remove the gauge port plug.

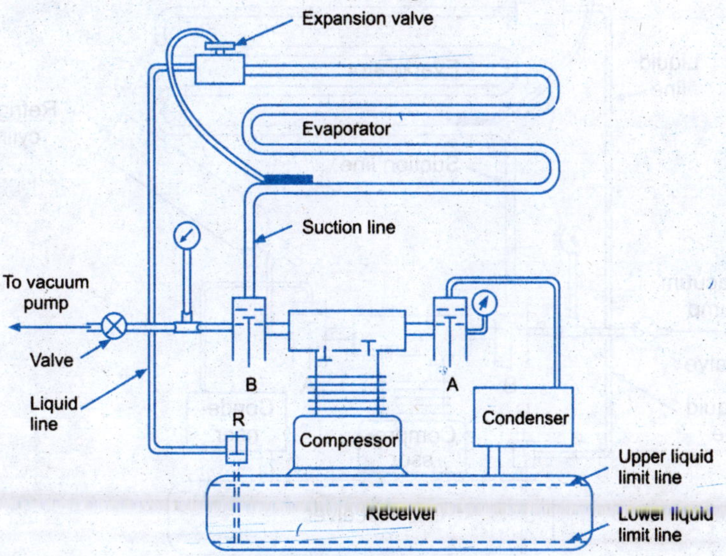

Fig. 22.1

- Screw a pipe flare connector into the gauge port and connect to a vacuum pump (which is then connected to compound gauge and a valve between the gauge and the pump).
- Start the vacuum pump and open the suction valve half way. Run the pump till highest vacuum is obtained.
- Close the valve and stop the pump. *Air evacuation will be satisfactory if vacuum gets maintained.*

22.5 CHARGING THE SYSTEM

The system may be charged with the refrigerant through the high or low sides. When changing is done on high side the refrigerant is *introduced in liquid form but when charging* is done on the low side the refrigerant should be added in vapour form in order to prevent possible damage to the compressor. Normally it is more convenient to add the refrigerant to the low side on small units or when small amounts are to be added to large machines. The refrigerant should be added in small quantities which are proportional to the size of the system. *Addition of more refrigerant than the required quantity is neither economical nor advised since it will occupy so much space in the receiver that there will not be room for the refrigerant which is in the low pressure side should pumping down the necessary.*

22.5.1 Charging through the Suction Valve (Low Side)

This charging method is normally used on small installations which need a few kilograms of refrigerants. Refer Fig. 22.2. *The system is fully evacuated and charged as follows:*

- Back seat the suction valve B and discharge the valve A.
- Connect the suction line to suction valve gauge port. Attach a gauge to the discharged valve and open half turn.
- Connect the other end of the charging line to a refrigerant cylinder (which should stand upright).

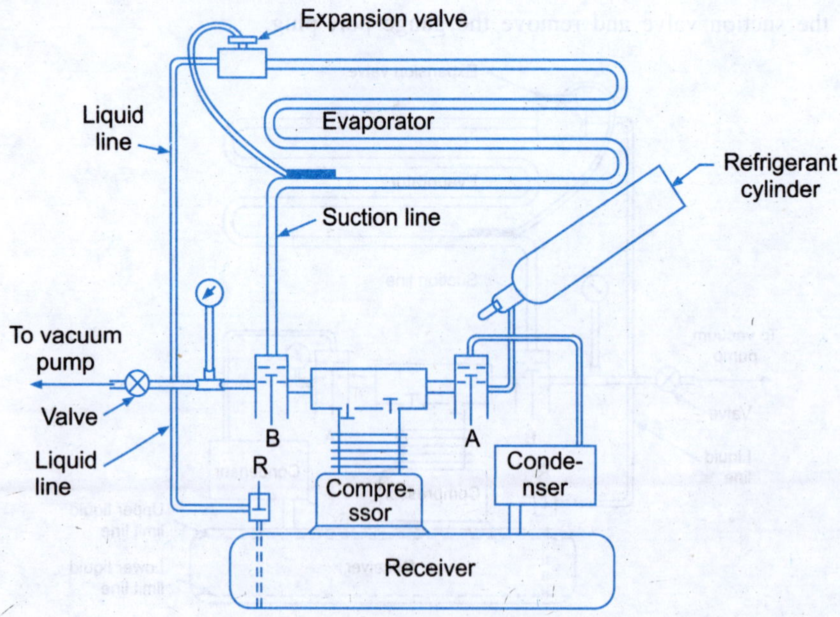

Fig. 22.2

- Open the cylinder slightly and loosen the flare nut at the compressor end. Air will be removed by this operation. Tighten the nut when sound of escaping gas is heard.
- Turn the suction valve to close the suction line, thus drawing the gas directly from the cylinder by compressor.
- Start the compressor to draw the requisite quantity of refrigerant. Make sure that the suction pressure does not exceed 2 bar gauge.
- Close the cylinder valve and allow the compressor to run for sufficient time to reduce the pressure in the charging line to zero bar gauge.
- Stop the compressor and back seat the suction valve. Detach the charging line and attach compound gauge. Turn the valve one half turn. The system is now ready for testing and normal operation.

22.5.2 Charging through Discharge Valve

This method of charging is employed under conditions when it is impracticable or impossible to add refrigerant by other methods. It is the *least used method.*

Procedure of charging. Refer Fig. 22.2.

- Back seat the discharge and suction valves. Plug the gauge port of each valve.
- Attach the compound gauge to the suction valve and turn in the valve one-half turn. Install the pipe-to-flare adapter to the discharge valve port and connect the charging line to the connector.
- Connect the other end of the charging line to the cylinder. Remove the air by loosening flare nut. Tighten the nut. Turn in the valve A about half way.
- Place the cylinder bottom side up (since the refrigerant will flow into the condenser in liquid form).
- After the refrigerant flows into the condenser, back seat the discharge valve, close the cylinder valve and remove the charging line.
- Attach the pressure gauge and turn in the valve one-half turn. Open the liquid valve. Now the system is ready for testing and operation.

22.5.3 Charging through the Charging Valve

Procedure of charging through the charging valve normally located in liquid line between the receiver and the filter or dehydrator is as follows:

- Invert the cylinder as shown in Fig. 22.3. Attach a charging line with the valve C.
- Remove/purge the air from the line by opening the cylinder valve slightly and loosening the flare nut at the charging valve C.
- Close the king valve R and open the valve C and the charging cylinder valve.
- Start the compressor adding the refrigerant in the system.
- To stop charging close the charging cylinder valve and operate the compressor for a few minutes, then close the valve C and disconnect the line.
- Open the king valve R and resume normal operation.

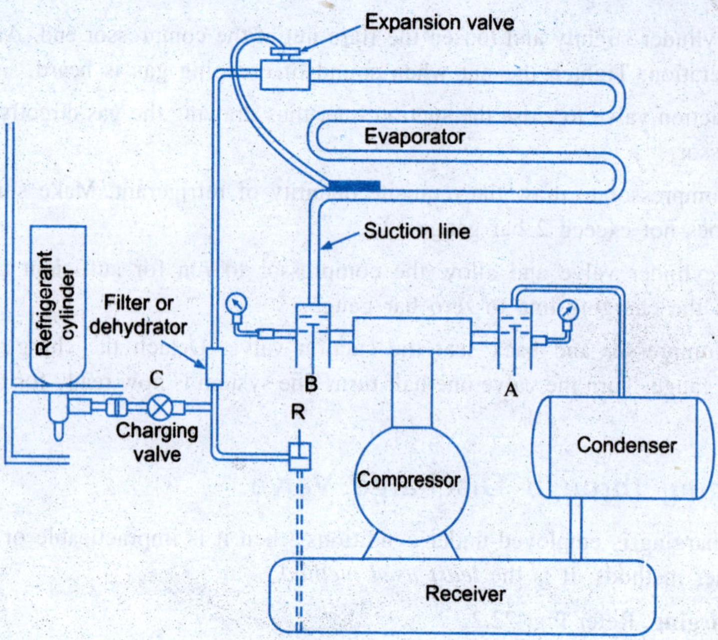

Fig. 22.3

This method is normally used in *large machines. It is the safest and fastest method of charging an installation.*

22.6 ADDING OIL TO THE COMPRESSOR

(a) When oil plug is not provided:

Procedure: Refer to Fig. 22.4.

- Add/pour some quantity of the correct grade of oil into a clean, dry glass container. The amount of oil should be more than that which is to be added.
- Connect an oil charging line containing a hand shut-off valve to the suction service valve gauge port and place the other end into the oil.

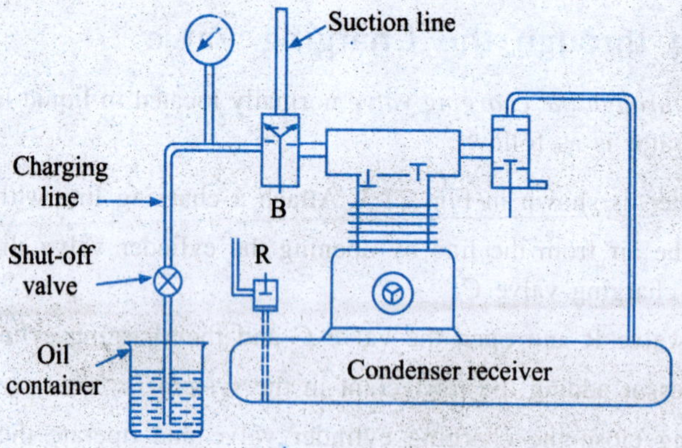

Fig. 22.4

- Close the shut-off valve and turn the service valve all the way in.
- Lift the end of the tubing out of the oil and open the shut-off valve slightly to purge the line. Reinsert the tubing into the oil.
- With the shut-off valve closed operate the compressor until a vacuum of 40 cm to 50 cm is obtained. The vacuum should hold after the compressor is stopped. If it does not repeat the pumping down process.
- Stop the compressor, open the valve and allow the desired amount of oil to be drawn into the crankcase. Do not permit the end of the tubing to be out of the oil at any time while the oil is flowing.
- Close the shut-off valve and observe the suction pressure. If it is not 0.14 bar gauge open the service valve slightly until the pressure reaches above atmospheric. The oil level may now be checked by whatever method is recommended for the particular compressor.
- To return to normal operation be sure that the shut-off valve is closed. Back seat the valves and remove the oil charging line and the gauge.

(b) When the oil plug is provided:

Procedure:

- With the suction service valve back-seated attach a compound gauge.
- Turn the valve all the way in and start the compressors. If necessary hold the control switch closed to keep the machine gauge.
- Operate the compressor until a vacuum will hold when it is stopped.
- Open the suction valve slightly until the pressure comes up to 0.14 bar gauge and then close again.
- Loosen the oil pug carefully and remove it. The desired amount of oil may then be poured into the crankcase. By working quickly air will be prevented from entering the system. Replace the oil plug.
- Open the service valve, remove the gauge and restore to normal operation.

22.7 REPLACEMENT OF THE SEALED UNIT OF A REFRIGERATOR

Procedure:

- Check whether the evaporator of a new sealed unit is to be installed from side or the back of the refrigerator body. If the back of the reingerator cabinet is sealed then the old evaporator has to go through the front i.e., through the door opening. If the old evaporator has a plate of size larger than itself at its back bolted to the cabinet, removal of the old evaporator is possible from the rear side of the cabinet.
- Declamp the thermal bulb of the thermostatic switch from the evaporator surface and detach the wire connections from the compressor.
- Expose the suction and liquid line tubing including capillary. If the tubing is visible on the back of the cabinet nothing is to be done on it but if it is on the front it is necessary to remove one or more plastic strips put at front side edges of the refrigerator cabinet.

- Unbolt the sealed compressor from the base.
- Remove the evaporator mounting screw and the entire unit will be ready for removal.
- Pull out the freezer and rest it on some crate taking care that tubing is not kinked.
- Remove the compressor from its original position (sliding of compressor and crate to be done carefully).
- Install the new sealed unit carefully.
- Reclamp the thermal bulb of the thermostatic switch and attach the wires of compressor motor. The unit is now ready for use.

22.8 TO ASSEMBLE A REFRIGERATOR UNIT

Following points should be considered while *assembling a refrigerator unit:*

- Before assembling a refrigerator *clean* all the components like evaporator, condenser and connecting tubes with *carbon tetrachloride.*
- Flush thoroughly the above parts with dry air or commercial grade nitrogen gas to make them dry from inside and to remove all the traces of washing chemicals.
- Prior to assembly, it is recommended that, at least condenser coil and evaporator should be individually dehydrated with a minimum period of 4 hours. Dehydration should be carried out by heating these parts simultaneously. These parts should preferably be flushed with dry air or with nitrogen after dehydration.
- After brazing joints sweep the system with dry nitrogen. Install the compressor.
- It is recommended that process of dehydration and vacuumisation be followed together/ simultaneously. This process is recommended even though individual parts are dehydrated. This is specially necessary in order to avoid capillary freezing problems usually encountered in such low temperature applications (due to poor system of dehydration).
- Change the refrigerant carefully and accurately (since the total amount of refrigerant to be charged in the refrigerator is quite small).

22.9 TROUBLES IN A REFRIGERATOR

The following *troubles* may be found in a refrigerator:

1. Defective compressor.
2. Chocked capillary.
3. Incorrect charge of refrigerant.

1. **Defective compressor.** A very little cooling effect is produced by a compressor which does not pump adequately. The evaporator is covered with a thin film of frost. Place your hand on the evaporator surface for 2 or 3 seconds, then examine the surface. If all of the frost melts where the evaporator was touched, it indicates that the compressor may be defective, this needs that sealed system may be checked up thoroughly.

2. **Chocked capillary.** The capillary may be chocked due to *moisture in the system, kinked tubing or foreign particles lodged in the lines.* The evaporator will then have no frost formation. Moisture

in the system will usually freeze at the outlet of the capillary tube resulting in continuous running of the compressor without any cooling effect or frosting on the evaporator. If the capillary tube is kinked the flow of refrigerant will stop and the evaporator will be free from frost. Due to presence of foreign particles as well in the capillary tube flow of refrigerant will stop. Under such conditions it is preferable to replace the capillary tube.

3. **Incorrect charge of refrigerant.** The system may be *undercharged or overcharged*. If the system is undercharged (i.e., little refrigerant is present) the *symptoms are*:

(*i*) Suction pressure will be low.

(*ii*) The compressor will run continuously.

(*iii*) Head pressure will be lower that the normal.

In order to *rectify the fault* the system should be checked for leakage and after leaking points are sealed evacuate and recharge the entire system.

The symptoms for an *overcharged system* are:

(*i*) A frost back condition on the suction line to the extent that it can be noticed on the line back to the compressor unit.

(*ii*) High suction and discharge pressures.

In order to rectify the fault, the system may be evacuated and recharged using refrigerant in proper amount.

THEORETICAL QUESTIONS

1. Give sample scheduling with regard to maintenance of a new installation.

2. When should compressor repair be carried out?

3. List the procedural steps which may be followed to check the efficiency of a compressor.

4. How is descaling of a condenser carried out?

5. Discuss the methods which may be used to remove air from the evaporator and lines.

6. Explain briefly the following methods of charging the system:

(*i*) Charging through the suction valve (low side)

(*ii*) Charging through the discharge valve.

(*iii*) Charging through the charging valve.

7. Describe the procedure for adding oil to the compressor in the following cases:

(*i*) When oil plug is not provided;

(*ii*) When oil plug is provided.

8. List the points which should be considered while assembling a refrigerator unit.

9. Enumerate the troubles which may be found in a refrigerator. Also mention the possible remedies for the same.

❑❑❑

Refrigeration and Air-Conditioning Tools

Chapter 23

23.1 LEST OF REFRIGERATION AND AIR-CONDITIONING TOOLS

A job carried out carelessly is dangerous and has no place in refrigeration and air-conditioning work. Tools if properly chosen will not only save time but will, in addition ensure a neat, dependable and satisfactory job which will result in mutual benefits to the serviceman and customer alike. There are several types of tools available in the market. Some of them are so poorly constructed that they are expensive at any price. A good tool warrants a fair price, because it is properly designed and will give service under operating conditions. A list of refrigeration and air-conditioning tools commonly used is given below:

1. Adjustable wrench; 2. Allen key set; 3. Ampere meter; 4. Blow lamp; 5. Box spanner; 6. Brushes; 7. Centre punch; 8. Charging line; 9. Chisel set; 10. Double-ended spanner; 11. Drilling machines; 12. Drills; 13. Feeler gauge; 14. Files; 15. Flaring tools; 16. Flexible charging line; 17. Gas charging stand; 18. Gas cylinder; 19. Gauge manifold; 20. Gauges: (*i*) Compound gauge, (*ii*) Pressure gauge; 21. Hacksaw frame; 22. Halide leak detector; 23. Hammers; 24. Hand shut-off valve; 25. Hollow punch; 26. Knife; 27. Magnifying glass; 28. Mallet; 29. Measuring tape; 30. Multimeter; 31. Oil gun; 32. Phase tester; 33. Pinch-off tool; 34. Pipe benders; 35. Pipe wrenches; 36. Pliers; 37. Pulley puller; 38. Ratchet wrench; 39. Refacing tool; 40. Revolution counter; 41. Ring spanner set; 42. Scissors; 43. Screw driver; 44. Screw wrench; 45. Seal alignment set; 46. Seal set puller; 47. Sealed unit servicing kit; 48. Set of die and tap; 49. Sling psychrometer; 50. Soldering iron; 51. Spring balance; 52. Spring benders; 53. Steel scale; 54. Swaging tool; 55. Thermometer; 56. Thickness gauge; 57. Try square; 58. Tube cutter; 59. Vernier calliper; 60. Voltmeter; 61. Vice.

23.2 APPLICATIONS OF TOOLS

Description/Applications of some of important tools is given below:

1. Pressure Gauges:

Pressure gauge indicate pressure in the system. Most commonly used gauges in refrigeration and air-conditioning installations are:

(*a*) High pressure gauge

(*b*) Compound pressure gauge.

(a) High pressure gauge:

Refer to Fig. 23.1. As the name suggests, this gauge is employed on the high side of the system. It is equipped with dial for measuring pressure from 0 to 300 lbs/square inch.

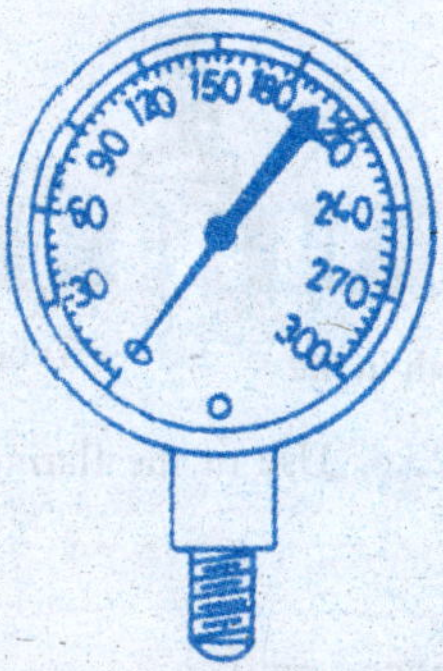

Fig. 23.1. High pressure gauge.

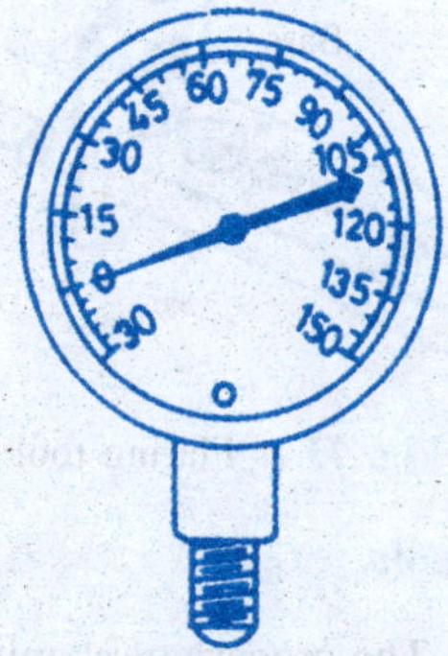

Fig. 23.2. Compound pressure gauge.

(b) Compound pressure gauge:

Refer Fig. 23.2. A compound pressure gauge can measure pressures both above and below the atmospheric pressure. *It is used on low side of the system.* A standard compound gauge's dial is graduated to record a pressure range from 30 inches of vacuum to 150 pounds per square inch. Gauge used in refrigeration service should be graduated to read approximately double the actual working pressure.

2. Combination Gauge Set:

A combination gauge set is one of the handiest aids which a service man uses while servicing small units which are not provided with gauges.

It consists essentially of two tee valves and a special tee as shown in Fig. 23.3. The tee valves are constructed so that the valve stem works only on the leg attached to the tee. Therefore, opening or closing the stem affects only this one opening; the other two openings of the valve remain open at all times. This unit can be used to obtain pressure readings on *both high and low sides* when (*i*) adding oil (*ii*) adding or recharging the refrigerant, (*iii*) purging off an excess of refrigerant or air, (*iv*) pumping the refrigerant from one side to the other.

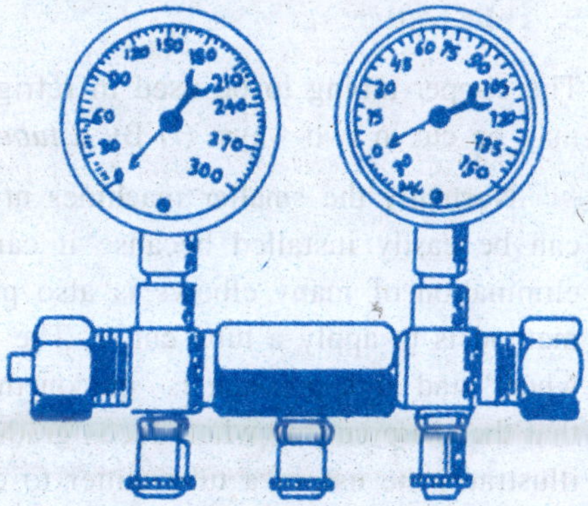

Fig. 23.3. Combination gauge set.

3. Flaring Tools:

Refer Fig. 23.4. A flaring tool consists of two bars held together by a wing nut and bolt; the bars are provided with holes for the various sizes of tubing (usually varying from 3/16" (38 mm) to 5/8" (125 mm). A yoke containing the forming die is slipped on the bars and the handle is rotated to produce a flare. This tool finds a wide use because of its simplicity, ease of performing the flaring operation; furthermore the flares produced by it are uniformly excellent in producing a tight seal.

Prior to application of pressure with the forming tool, place a drop of oil on the end of the tube. Make sure that only refrigeration oil is used (any other type oil, if used, may enter the tube and give trouble by reacting with the refrigerant at a later date).

Fig. 23.5 shows the proper use of the flaring tool.

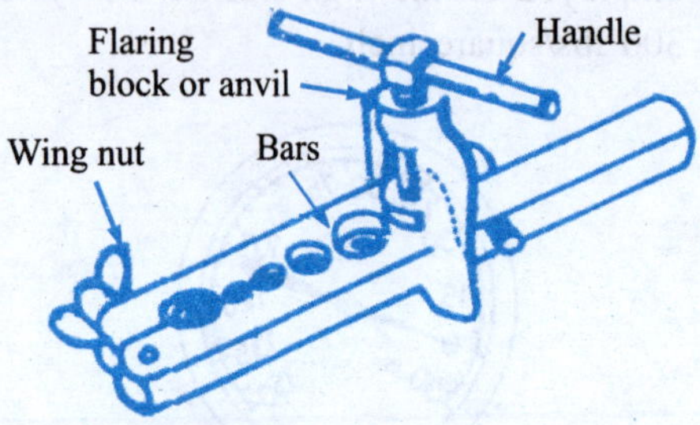

Fig. 23.4. Flaring tool.

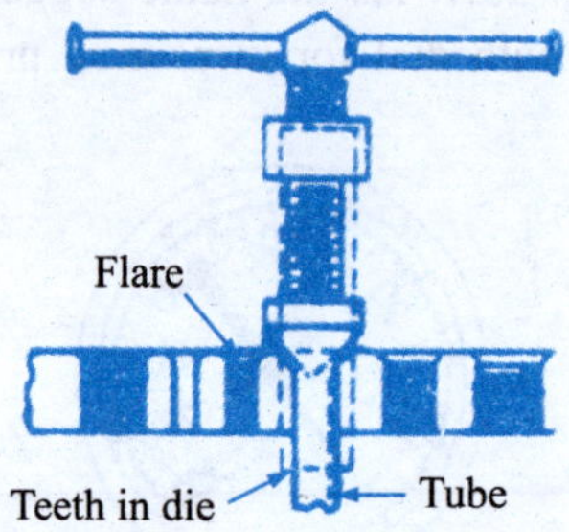

Fig. 23.5. Use of the flaring tool.

4. Swaging Tool:

Refer Fig 23.6. The process of shaping copper tubing so that two pieces may be joined without the use of a fitting is called *swaging* and the tool used for this process is known as *swaging tool*. The swaging tool can be of Punch type or Screw type. To carry out the swaging operation, hold the tube in the flaring block tool and set bar in vice, the tubing to be extended above the bar a distance of its diameter plus 3 mm and clamp. Now a specially designed punch is hammered into the tube to a depth equal to its original diameter. Now remove the punch and insert other piece of tubing in this swaged portion. Apply heat to the joint and add solder to make a proper joint.

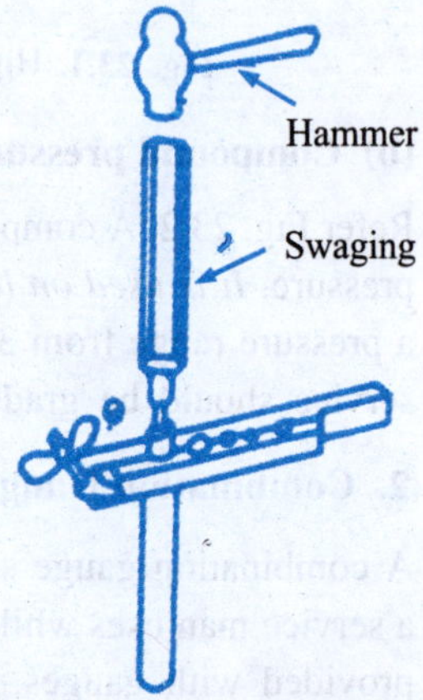

Fig. 23.6. Swaging tool.

5. Tube Cutter:

The copper tubing to be used in refrigeration and air-conditioning system may be cut in two ways: (*i*) By a *tube cutter* or (*ii*) By a *hacksaw*.

Normally the smaller machines utilise soft copper tubing, this tubing can be easily installed because it can be bent around obstructions and elimination of many elbows is also possible. While cutting soft copper tubing an easy and quick method is to apply a tube cutter. The tube to be cut is held in between the 'Roller' and the 'cutter wheel' and high pressure is put on the lead screw. Revolve the cutter slowly around the tubing so that the sharp cutting wheel feeds gradually into the tubing, making a clean right angle cut. Fig. 23.7 illustrates the use of a tube cutter to cut tubing.

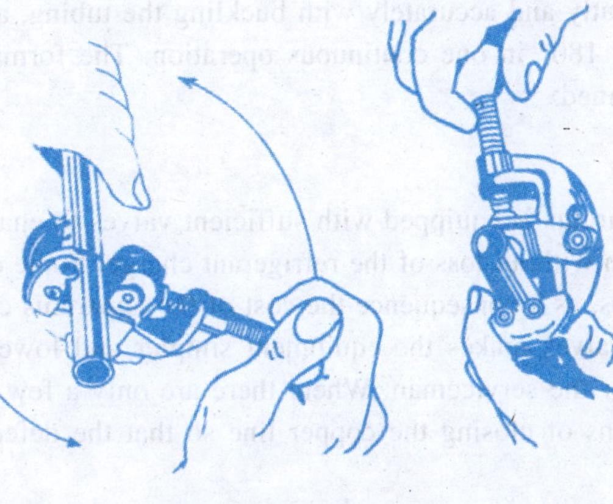

Fig. 23.7. Tube cutter in use.

Hard copper tubing can be cut by a hacksaw when it is held in the flaring block. Fig. 23.8 shows the method of cutting with hacksaw in a flaring block.

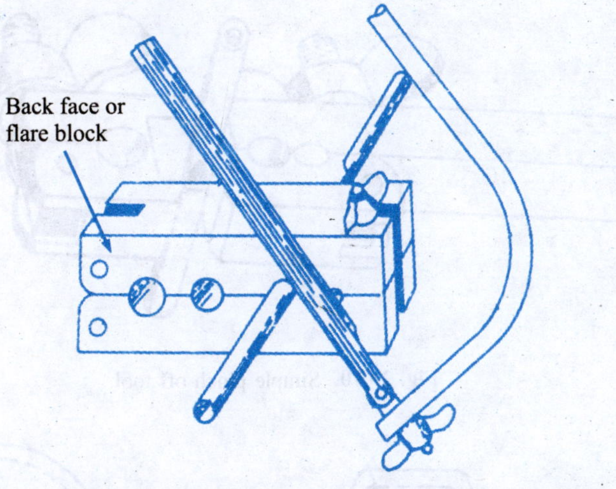

Fig. 23.8. Method of cutting tubing with hacksaw in a flaring block.

6. Tube Bender:

For carrying out bending operation several special tools are available. While using *spring type bender* it should be borne in mind that the tubing should be bent somewhat farther than it is required and then bend up to the proper angle. Thus the spring is loosened and can be easily removed. Fig 23.9 shown spring type bender.

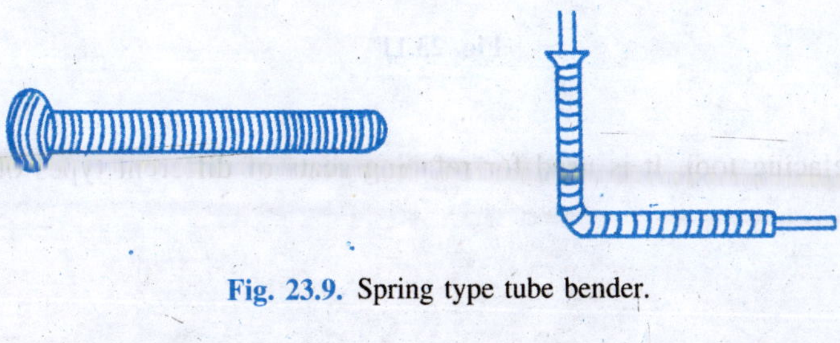

Fig. 23.9. Spring type tube bender.

In order to form bends neatly and accurately with buckling the tubing, a lever type cutter is used. These tools form bends up to 180° in one continuous operation. The forming wheel is calibrated to show the degree of bend attained.

7. Pinch-off Tools:

Usually most of the large apparatus is equipped with sufficient valves to enable any portion or section to be closed off and worked on without loss of the refrigerant charge. Some of the smaller and simpler systems have only a few valves, as a consequence the cost of the apparatus can be kept to a minimum. Although the elimination of valves makes the equipment simpler and lower priced yet it makes the servicing some what harder for the serviceman. Where there are only a few or no valves in a system, the pich-off tool offers a means of closing the copper line so that the defective or worn part can be removed or replaced.

Fig. 23.10 shows a simple pinch-off tool. The pinching tool jaws are designed in such a way that the tuning is pinched shut put not mashed (Fig. 23.11 *a*, *b*). The round opening is used to round up the pinched section when the repair or replacement has been completed.

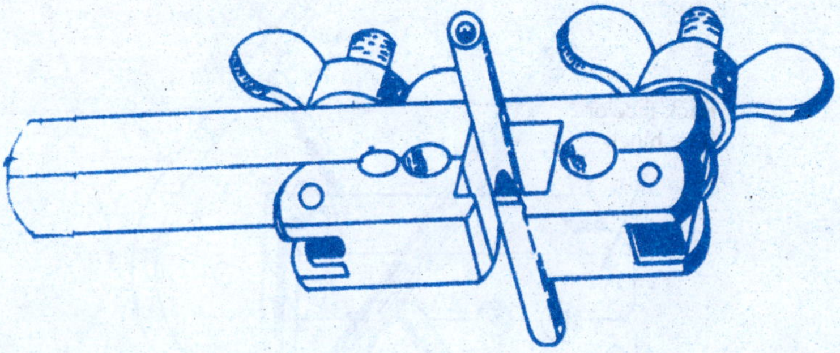

Fig. 23.10. Simple pinch-off tool.

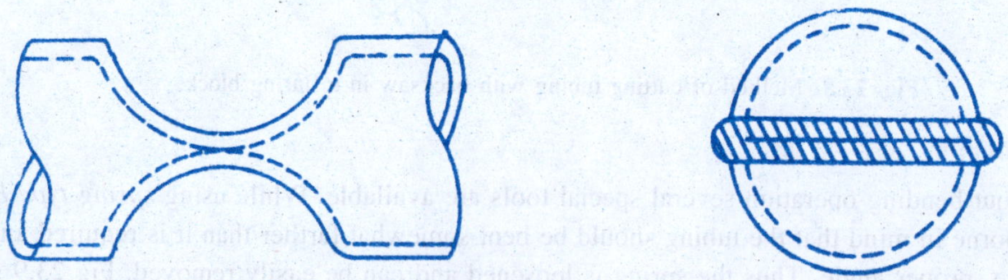

(*a*) (*b*)

Fig. 23.11

8. Refacing Tool:

Fig. 23.12 shows a refacing tool. It is used for refacing seats of different types of connectors.

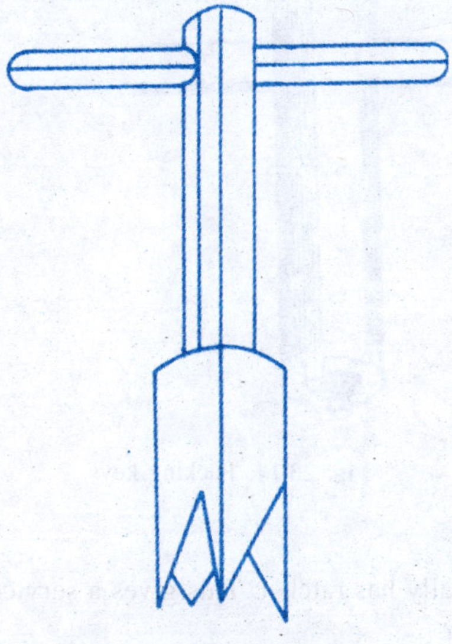

Fig. 23.12. Refacing tool.

9. Valve Keys:

Some refrigeration apparatus is equipped with valves which are without handles. The stems are provided with a square end, so that a handle can be applied or a key fitted to the valve stem as shown in Fig. 23.13.

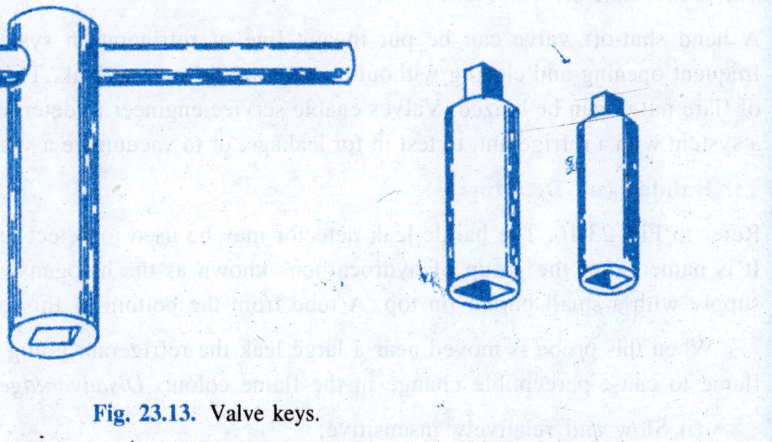

Fig. 23.13. Valve keys.

10. Packing Keys:

Some valves are packed with such a material that serves to prevent the refrigerant from leaking out from the stem. A packing nut can be tightened by means of a wrench. When tightening external packing nuts, use a 12-point wrench if one is available, or make use of the proper open-end wrench. The packing nut is generally the internal type and is provided with a slot across its face for tightening.

The stem of the valve projects through this nut and makes it necessary to use a special gland or packing nut key, such as shown in Fig. 23.14. This key is hollow in the centre, permitting the valve stem to extend through the tool, and is provided with two teeth that engage the slot in the packing nut.

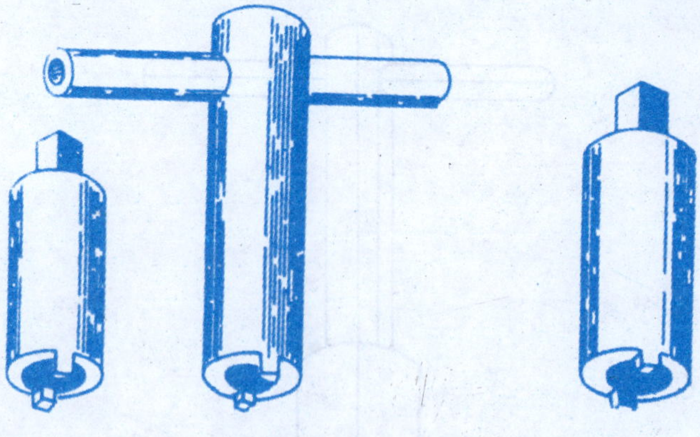

Fig. 23.14. Packing keys.

11. Ratchet Wrench:

Refer to Fig. 23.15. This tool usually has ratchet. This gives a service engineer rapid control for slight opening or closing of valve stem.

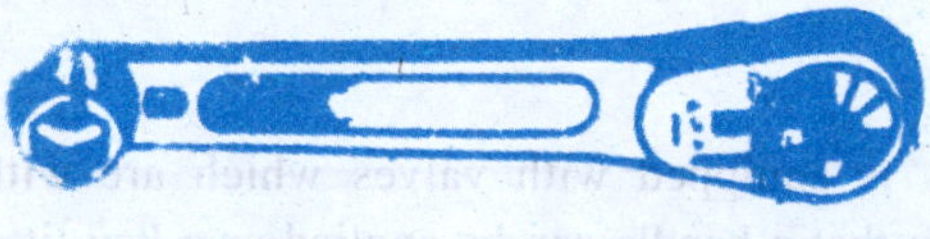

Fig. 23.15. Ratchet wrench.

12. Hand Shut-off Valve:

A hand shut-off valve can be put in any line of refrigeration system. These valves can withstand frequent opening and closing without an external or internal leak. These can be installed with the help of flare nut or can be brazed. Valves enable service engineer to determine operation pressure, to charge a system with a refrigerant, to test in for leakage, or to vacuumize a system depending upon its location.

13. Halide Leak Detector:

Refer to Fig. 23.16. The halide leak detector may be used to detect leaks in Freon refrigerant system. It is named after the group of hydrocarbons known as the halogens. It consists of a combustible gas supply with a small burner on top. A tube from the bottom of this burner forms the probe.

When this probe is moved near a large leak the refrigerant going into the tube combines with the flame to cause perceptible change in the flame colour. *Disadvantages* of using this detector are:

(*i*) Slow and relatively insensitive;

(*ii*) In a brightly lighted area it is almost impossible to locate smaller leaks;

(*iii*) If the area is darkened, a smaller leak can be located; however, the change in colour of the flame is difficult to interpret because other gases and dust cause the flame to change colour.

14. Electronic Type Leak Detector:

As compared to halide leak detector, an electronic type leak detector is *more sensitive, faster responding* and is capable of detecting a leak even though the surrounding air may be contaminated.

It consists of an internal pump which draws air and leaking halogen tracer gases through a probe and hose as the probe is passed over leaking joits, gaskets, welds etc. The air gas mixture is drawn

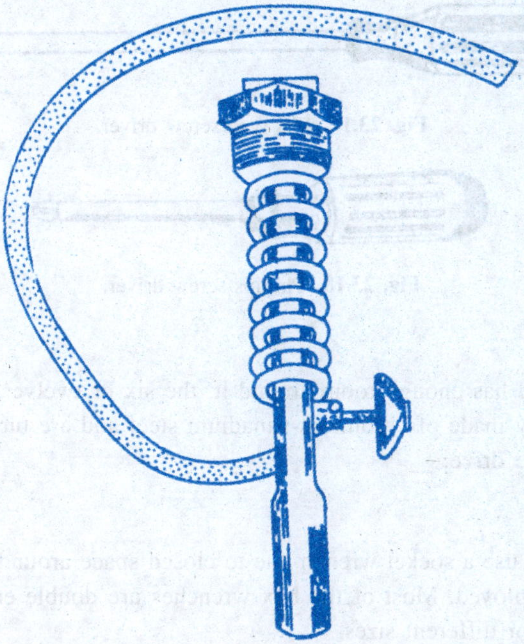

Fig. 23.16. Halide leak detector.

between two electrodes in the control unit—the *ion-emitter* and *collector*. The presence of halogen gas in the inter-electrode space causes an increase in interelectrode conductance that is proportional to the amount of halogen gas present, which results in greater positive ion current. In the electronic detector, this emission current is electronically amplified and the amplifier output causes the leak indicating lamp to flash. In order to adjust the sensitivity to detect leaks a knob on the control unit is provided to do so. This detector is specially designed for use in atmosphere containing a relatively high concentration of halogen gas.

To detect leaks by this type of detector, following procedure is followed: (*i*) Plug the test detector into an AC outlet (220/110 V). (*ii*) Turn the sensitivity knob to the right and allow a 1-minute warm up; (*iii*) Check the operation by turning the sensitivity knob quickly form one position to the anothe. This should light the probe lamp, (*iv*) Probe for leaks, starting from maximum sensitivity; (*v*) If the probe lamp lights twice for each leak, reduce the sensitivity; (*vi*) recheck the leaks for confirmation.

15. Screw Drives:

In refrigeration service work screw drivers are employed to loosen and tighten the screws. A screw drive primarily consists of a handle, ferrule and the blade. The length of a screw driver is measured from the blade tip to the handle. Handles are not measured.

Several types of screw drivers are used in refrigeration and air conditioning field out of which following two are most important: 1. Flat type and 2. Phillips type. The flat is a standard screw driver (Fig. 23.17) being very similar to chisel. The tip of the screw drive is applied on the job and head is moved to remove or insert the screw where as the *phillips screw driver* (Fig. 23.18) is made with a specially shaped tip to deal with cross slot screws.

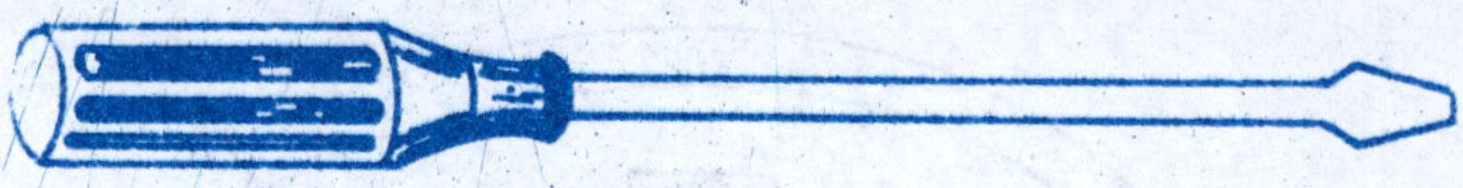

Fig. 23.17. Flat type screw driver.

Fig. 23.18. Phillips screw driver.

16. Socket Wrenches:

Where the nut or bolt head has enough room around it, the six or twelve point wrench is best used. These sockets are generally made of chromium-vanadium steel and are turned by handles that have a 1/4, 3/8 or 1/2 inch square drive.

17. Box Wrenches:

Where it is not possible to use a socket wrench due to closed space around the nut or bolt head a box wrench is successfully employed. Most of the box wrenches are double ended (Fig. 23.19); both the ends may be of the same or different sizes.

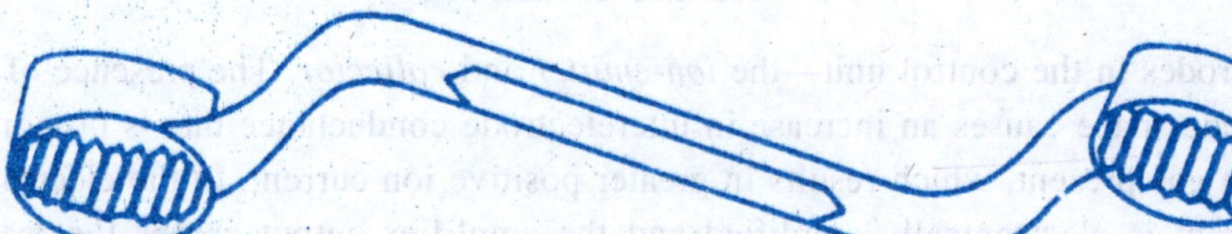

Fig. 23.19. Box wrench.

18. Adjustable Wrenches:

Refer to Fig. 23.20. These wrenches are used when odd sizes of nuts and bolts heads are to be handled. As these wrenches are not so strong (as other solid types), so their use should be restricted when forces used are of large magnitudes. In order to ensure that they are not cracked or bent out of shape, the jaw and adjustment screw must be checked up periodically.

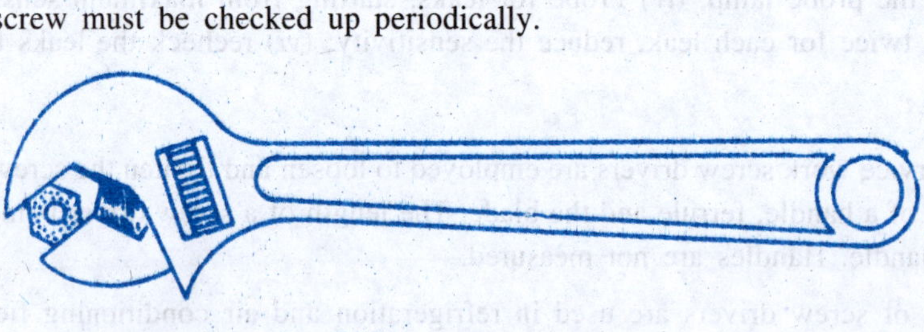

Fig. 23.20. Adjustable wrench.

19. Oper Ended Wrench/Spanners:

These wrenches are the non-adjustable type with openings in each end. This type of wrench can slide on the nut or bolt head from the side. These wrenches are available in sets, each set containing about ten wrenches with openings of different sizes. These are used in close spaces on unions and other places where the socket wrench and box wrench cannot be used. Fig. 23.21 shows a typical double ended wrench/spanner.

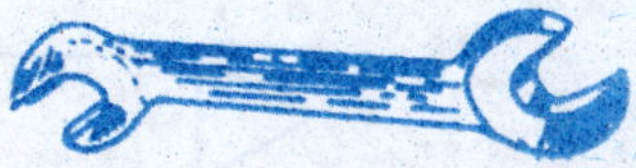

Fig. 23.21. Double ended wrench/spanner.

20. Pliers:

Pliers of different designs are available, each for a particular purpose. For example: (*i*) '*Cutting pliers*' are used for skinning a cable insulation and cutting wires when working on the wiring of a refrigeration equipment; (*ii*) '*Round nose pliers*' are used to shape wire into loops and to bend sheet metal edges; (*iii*) '*Slip nose plier*' and '*duck bill pliers*' are used in hard to reach places.

Figs. 23.22, 23.23 and 23.24 show some of the commonly used pliers.

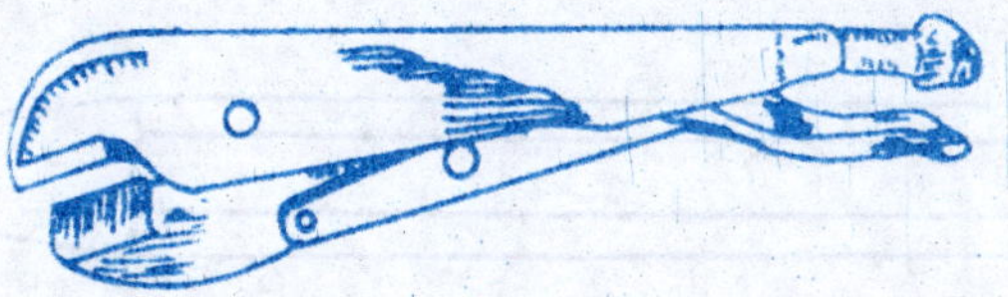

Fig. 23.22

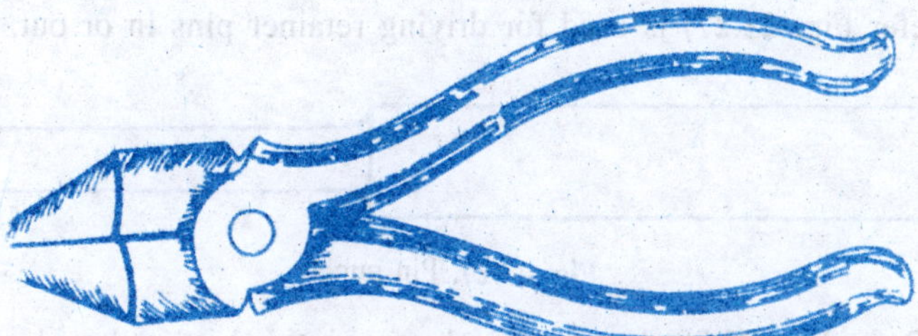

Fig. 23.23

Fig. 23.24

21. Punches:

The following four types of punches are used for carrying out different jobs.

1. Centre punch
2. Drift punch
3. Pin punch
4. Prick punch.

The *centre punch* (Refer Fig. 23.25) is made of steel and the point is hardened. The angle can vary for 60° to 90°. One of the main uses of a centre punch is to mark the centres of holes for drilling. For this purpose a much deeper mark is required than is necessary for simply marking lines. Such punches may also be used to make alignment marks on part before dismantling.

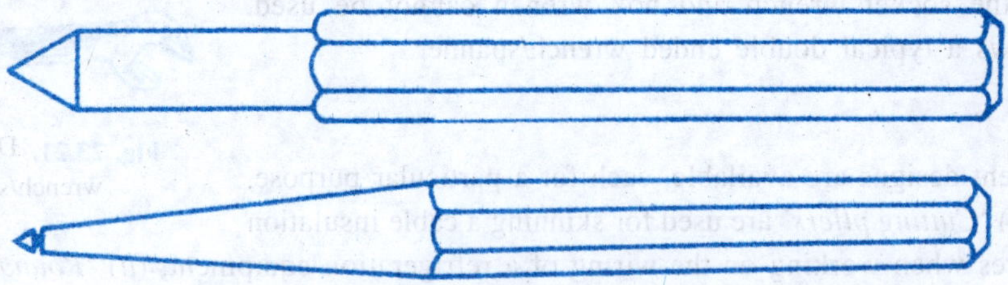

Fig. 23.25. Centre punch.

A *drift punch* (Refer Fig. 23.26) is employed to line up holes in the mating surfaces and drive out the keys.

Fig. 23.26. Drift punch.

A *pin punch* (Refer Fig. 23.27) is used for driving retainer pins in or out.

Fig. 23.27. Pin punch

A *prick punch* (Refer Fig. 23.28) has a long, sharp point and is used exclusively for layout work.

Fig. 23.28. Prick punch.

22. Vice:

A vice is employed to hold parts for drilling, filing and assembling. The most useful type is a vice with parallel jaws (Refer to Fig. 23.29). It should be fixed firmly on the bench at the front and preferably near one corner. Usually the jaws are roughened somewhat like a file. This gives a good grip but it also damages the surface of the part held in it. For some purposes, therefore it is desirable to cover the jaws with what are called clamps.

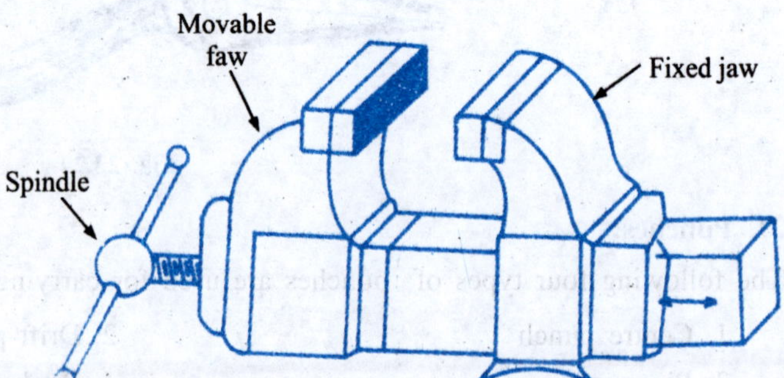

Fig. 23.29. Vice.

23. Hacksaw:

Refer to Fig. 23.30. It consists of a metal bow frame and handle with some sort of screw adjuster, usually at the end of a frame. This has a small peg on one side and a similar peg at the handle end. On these pegs is fitted the saw blade which is a thin springy steel strip with teeth along side. Although it can be bent a good deal the saw blades breaks quite easily if mised. To fit a new blade the adjuster must be turned so as to shorten the distance between the two pegs.

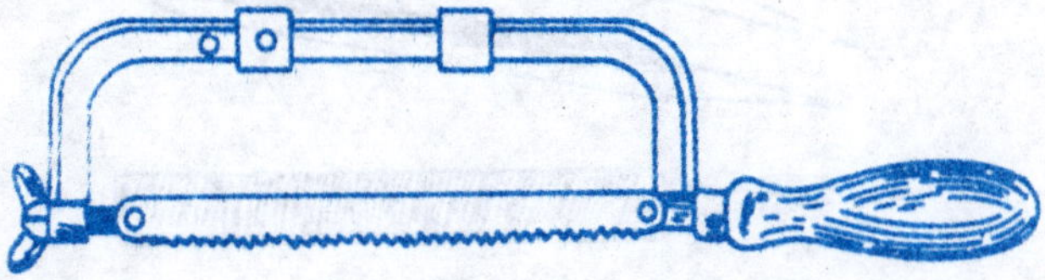

Fig. 23.30. Hacksaw.

The hacksaws are used to cut rods, strips, bars and pipes into desired lengths.

24. Hammer:

Hammer is a striking tool frequently used in bench work. Its size is indicated by weight. These are classified as follows:

1. *Hard*:
   (*a*) Ball peen
   (*b*) Cross peen
   (*c*) Straight peen
   (*d*) Sledge.
2. *Soft*:
   (*a*) Raw hide
   (*b*) Mallet
   (*c*) Brass
   (*d*) Plastic
   (*e*) Hard rubber.

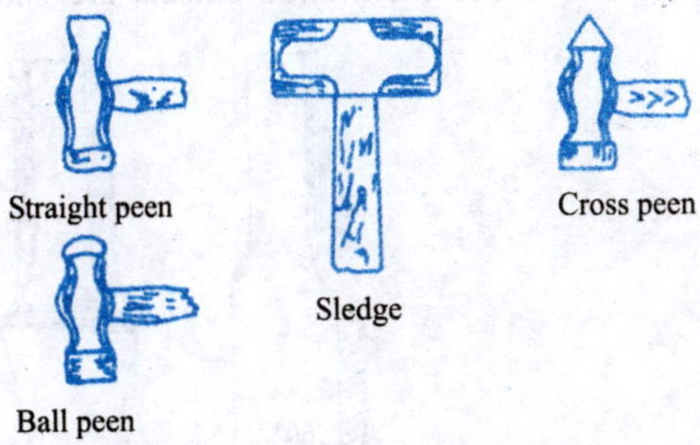

Fig. 23.31. Commonly used hammers.

Fig. 23.31 shows some of the commonly used hammers.

25. Files:

Files are used to abrade or smooth other substances such as metal, wood etc. These are classified on the basis of size, cut of teeth and shapes as follows : (*i*) Flat file (used for general work) ; (*ii*) Round file (used to enlarge the holes or circular openings and to file concave surface, (*iii*) Square file (used to file key ways, slots etc.) (*iv*) Three square file (used to file internal angles and clean out corners).

Fig. 23.32 shows some files.

Fig. 23.32. Some files.

26. Chisels:

Chisels are used for cutting or clipping metal. These are made of good grade tool steel (carbon steel) with hardened cutting edge. They are classified as given below:

(*i*) Flat chisel (*ii*) Cross-cut chisel (*iii*) Round nose chisel

(*iv*) Diamond point chisel (*v*) Side chisel.

The above mentioned chisels are shown in Fig. 23.33.

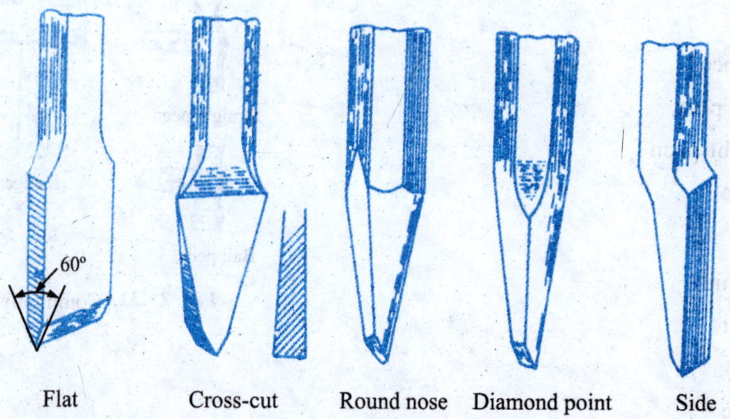

Fig. 23.33. Chisels.

27. Drills:

Drill is a tool used for the production of round holes in metal or other solid material.

Fig. 23.34 shows the leading characteristics of a twist drill.

In the Fig. 23.35 are shown different types of drills.

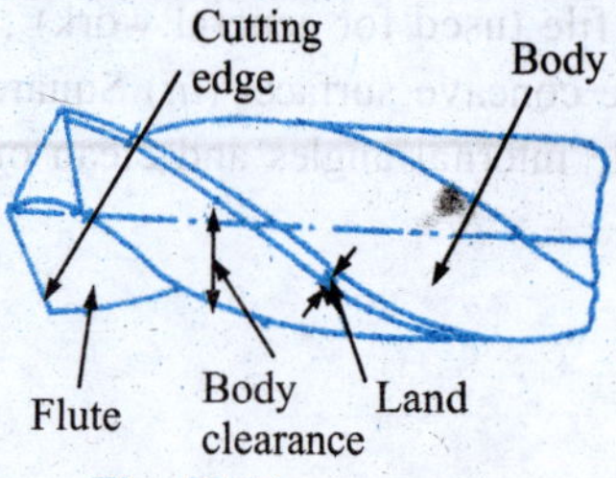

Fig. 23.34. Twist drill.

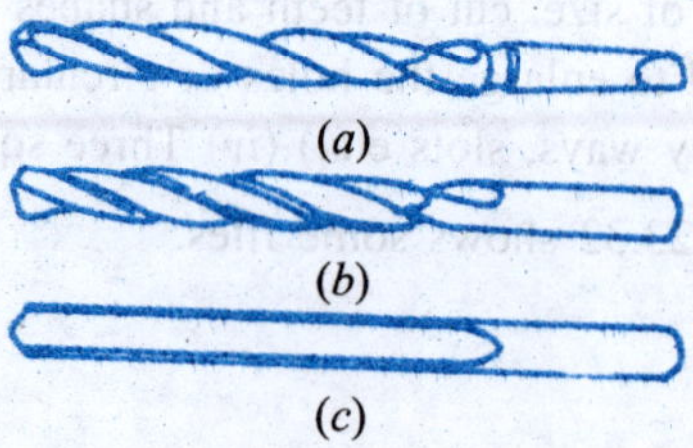

Fig. 23.35. Types of drills.

28. Taps:

A screw tap is a tool used in producting a screw thread in a hole, i.e., an internal thread. A tap is a tool which has threads like a bolt and three or four flutes cut across the threads. Taps are made of carbon steel or high speed steel hardened and tempered. The flutes provide cutting edges as well as passage for oil to reach the cutting point and for removal of chips. The square shank is for the tap handle which rotates the screw.

The taps are classified as under:

(*i*) Hand taps (*ii*) Machine screw taps
(*iii*) Tapered taps (*iv*) Pulley taps (*v*) Machine taps.

29. Dies:

A screw die is a tool used when threading cylindrical pieces beyond the capacity of a die plate. Dies are more efficient than die plates or die nuts and are made in many different designs both for hand and machine use.

Refer to Fig. 23.36.

(*a*), (*b*) : Stock and dies—the older, heavier adjustable.

(*c*), (*d*) : Non-adjustable and adjustable circular dies.

(*e*), (*f*), (*g*) : Inserted chasers.

Dies will cut only in one direction of rotation, so it is no good, after running down the thread once with stocks and dies, to tighten up the dies still further in the hope that more will be removed while winding the stocks back.

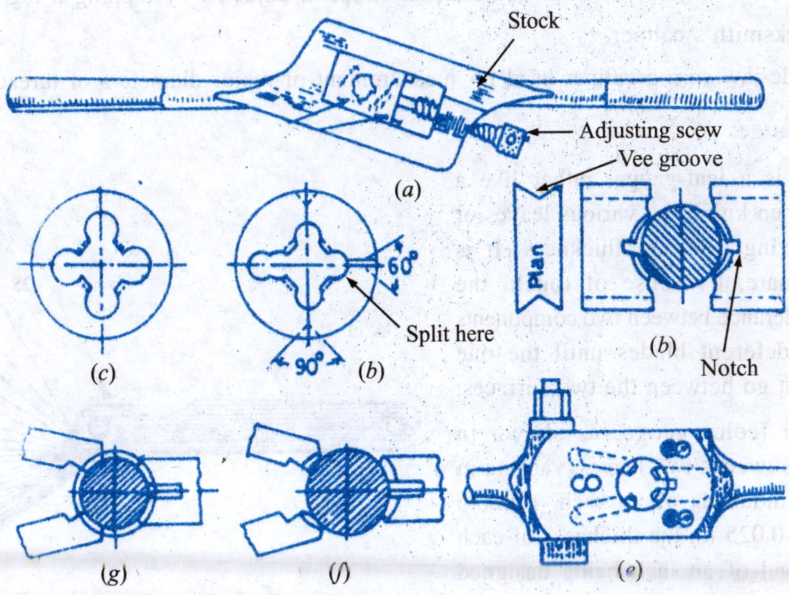

Fig. 23.36. Dies.

30. Dividers:

These are two-legged instruments of compass-like form, sometimes called *points*. They are used for measuring distances between points, for scribing arcs or circles on metals, for transferring lengths from scales, etc. Dividers may be ordinary firm joint or spring type. For smaller and more accurate jobs spring dividers are preferred. In this pattern the legs have a spring joint and the necessary adjustment is obtained by a knurled nut and screw (Refer Fig. 23.37).

Fig. 23.37. Divider.

31. Caliper:

Caliper is an instrument used for measuring distance between or over surfaces, or for comparing dimensions of work pieces with such standards as plug gauges, graduated rules, etc.

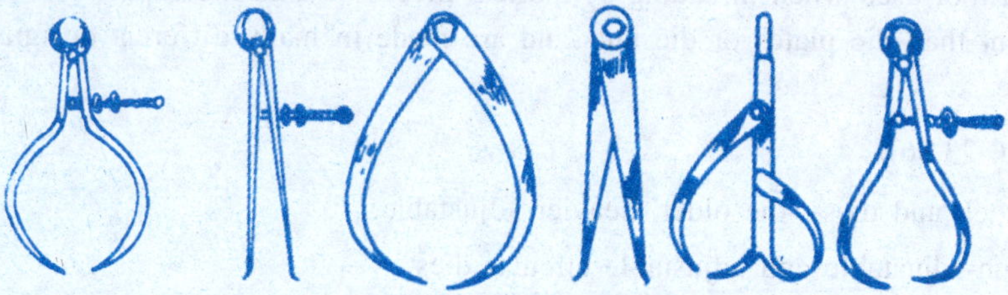

Fig. 23.38. Calipers.

Fig. 23.38 shows the various types of calipers in use:

(*a*) and (*b*) : external and internal calipers with adjusting screws.

(*c*) and (*d*) : firm-joint external and internal calipers, adjusted by tapping a leg or the joint.

(*e*) : Blacksmith's caliper.

(*f*) : Wide-jaw spring caliper used for measurement of major diameters of threaded work.

32. Feeler Gauge:

Feeler gauge is a leaf-gauge, rather like a multi-bladed pen knife, the various leaves or blades all having different thickness. It is used to estimate, by sense of touch, the clearance or tolerance between two components by inserting deferent blades until the one which will just go between the two surfaces.

A typical feeler gauge, as shown in Fig. 23.39, has twenty-two leaves, varying in thickness by thousnds of an inch, running from 0.004 to 0.025 in, the thickness of each leaf in thousand of an inch being designed by the number engraved upon it. The leaves may be used singly or in combination with

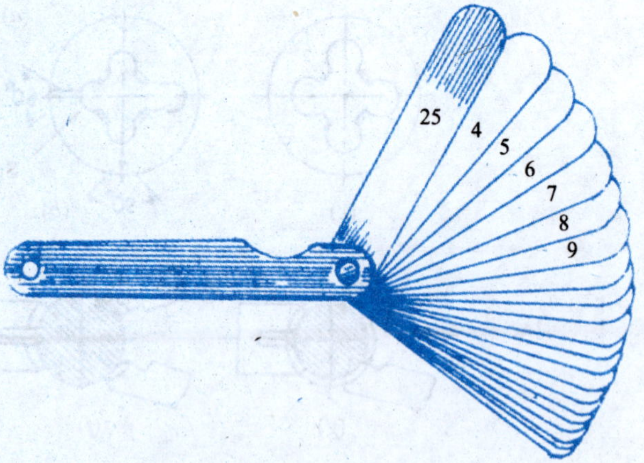

Fig. 23.39. Feeler gauge.

others, and any thickness within their limits can easily be obtained. Feeler gauges are also made in metre sizes form 5/100 to 1 mm thickness.

23.3 SAFETY PRECAUTIONS

While working with tools and equipment in a refrigeration and air-conditioning shop the following ***precautions*** must be observed:

- Always keep the shop neat and tidy.
- Do not wear loose dress.
- Do not temper with the safety devices provided on the machines.
- Bend the tubing as large as possible.
- Do not use files without handles (as the tang may injure the hand).
- Always wear goggles while drilling, otherwise the flying chips may cause serious injury to eye.
- 'Mushroom' heads should be ground from chisels and punches as these tools are likely to fly when struck with hammer thus causing serious injury to the operator.
- Do not use emery cloth to clean tubing prior to soldering as it may leave an oily deposit.
- During the use of silver brazing materials it should be ensured that the work place is properly ventilated since sometimes silver brazing materials contain cadmium which give out poisonous fumes when heated.
- To detect leaks in tubing circuits, use carefully low or medium pressure CO_2 or N_2, O_2 should never be used.
- Blunt edged tools must never be used.
- Where ever a dangerous thing is placed, mark caution sign.
- Always 'crack' service valves and cylinder valves before opening. This gives quick control to the flow gases if there is any danger.
- Keep everything connected with refrigerating mechanism thoroughly dry.
- Cylinders containing refrigerant should never be filled more that 85% of their capacity as they are likely to burst due to hydrostatic pressurre if overfilled.

OBJECTIVE TYPE QUESTIONS

Fill in the Blanks:

1. High pressure gauge is applied on the........... side of the system.
2. A................ pressure gauge can measure pressure both above and below the atmospheric pressure.
3. The process of shaping copper tubing so that two pieces may be joined without the use of a fitting is called..............
4. .... tool is used for refacing the seats of different types of connectors.
5. The .... leak detector may be used to detect leaks in Freon refrigerant system.
6. Where it is not possible to use a socket wrench due to closed space around the nut or bolt head a .... wrench is successfully employed.

7. .... wrenches are used when odd sizes of nuts and bolts heads are to be handled.
8. .... pliers are used for skinning a cable insulation and cutting wires.
9. A.................. punch is used for driving retainer pins in or out.
10. .... are used to abrade or smooth other substance such as metal, wood etc.

ANSWERS

1. High;	**2.** Compound;	**3.** Swaging;	**4.** Refacing;
5. Halide;	**6.** Box;	**7.** Adjustable;	**8.** Cutting;
9. Pin;	**10.** Files.		

THEORETICAL QUESTIONS

1. Enumerate a list refrigeration and air-conditioning tools.
2. What does a pressure gauge indicate? What is the different between a 'high' pressure gauge ' and a 'compound' pressure gauge?
3. Explain briefly any *three* of the following:
   (*i*) Swaging tool (*ii*) Tube cutter (*iii*) Tube bender (*iv*) Pinch-off tools
   (*v*) Refacing tool.
4. Write short notes on any three of the following:
   (*i*) Valve keys (*ii*) Hand shut-off valve (*iii*) Screw divers (*iv*) Pliers
   (*v*) Vice (*vi*) Hacksaw.
5. Explain briefly the following:
   (*i*) Halide leak detector (*ii*) Electronic type leak detector.

❑❑❑

Universities' Questions (Latest) with "Solutions"

Chapter 24

SECTION-A: SHORT ANSWER QUESTIONS

Q. 1. *What are the limitations of Carnot cycle?*

Ans. Carnot cycle is theoretically most efficient when working between given fixed temperature limits. But it *cannot* be used in practice *because machine has to run at high speed during adiabatic process followed by extremely slow speed at isothermal process and such variation of speed is not possible to attain.*

Q. 2. *In vapour compression refrigeration system what happens to power requirement if the throttling valve was replaced by an insentropic turbine?*

Ans. Theoretically the power requirement should decrease, but in practice the work obtainable by expanding liquid in the turbine would *not be sufficient* even to overcome the mechanical fraction loss in the turbine.

Q. 3. *What are the three primary factors of air on which human comfort depends?*

Ans. Temperature, humidity and air motion.

Q. 4. *When we buy a cold canned drink from vending machine on a humid and hot day, dew forms on the can, compare the temperature of the drink with dew-point temperature of the surrounding air.*

Ans. Temperature of the drink is *less* than the dew-print temperature of the surrounding air.

Q. 5. *What are the three common sources of irreversibilities which make actual vapour compression cycle differ from the ideal one?*

Ans. Throttling in expansion valve, wire drawing in suction and delivery valve, and non-isentropic compression.

Q. 6. *What is the thumb rule of the percentage of improvement of C.O.P. in vapour compression refrigeration system, with each °C raising of evaporating temperature or lowering of the condensing temperature?*

Ans. Each °C *raising evaporating temperature* will *raise* the C.O.P. by about 2.5%. Each °C lowering of *condensing temperature* will *raise* the C.O.P. by about 2%.

Q. 7. *Sketch reverse Brayton cycle and corresponding state points on a T s diagram showing irreversibility.*

Ans. Reverse Brayton cycle is shown in given Fig. 1.

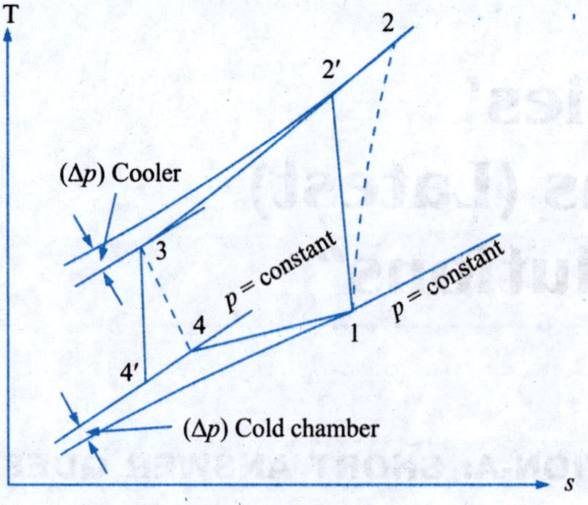

Fig. 1. Reverse Brayton cycle

Q. 8. *Define entrainment process as applied to steam jet refrigeration system.*

Ans. The high velocity vapour from the nozzle of the ejector imparts momentum to the vapour from the flash chamber, and thereby the flash vapour moves along with the motive steam through the ejector. The process of momentum exchange between the flash chamber vapour and the motive steam, and the carry over to the former by the latter is termed as *entrainment.*

Q. 9. *In cooling load calculation, name the load which depends upon volume of air-conditioned space.*

Ans. Infiltration load.

Q. 10. *Classify 'cooling load'* used for summer air-conditioning as per 'ASHRAE'.

Ans. Four categories:

(*i*) Wall heat gain; (*ii*) Air changes;

(*iii*) Product load;

(*iv*) Miscellaneous (including electrical energy dissipation such as light, fan and heaters).

Q. 11. *Name three methods of duct design.*

Ans. Velocity reduction method, equal friction method, and static regain method.

Q. 12. *Why 'aspect-ratio' is considered as very important factor in duct design?*

Ans. The cost of duct, the pressure drop, and heat gain/loss depends upon aspect ratio. Hence aspect ratio is considered as very important factor in duct design. As the *aspect ratio increases, heat gain/loss by the duct, the pressure drop, and the cost of duct also increase.*

Q. 13. *Consider a refrigeration system using R-12 as the working fluid. If this refrigerator is to operate in the environment of 32°C, what is the minimum pressure to which refrigerant should be compressed? Why?*

Ans. 7.5 bar, being saturation temperature for R-12 at 30°C.

Q. 14. *How does the C.O.P. of cascade refrigeration system, compare to that of a simple vapour compression cycle operating between the same pressure limits?*

Ans. The C.O.P. of cascade system is *less* than that of a simple vapour compression system operating between the same pressure limits.

Q. 15. *Why are semiconductor materials preferable to metals in thermoelectric refrigerator?*

Ans. To obtain high performance of thermoelectric refrigeration system, materials with high value of *Z* (figure of merit) are required, i.e. having high seebeck coefficient, high electrical conductivity, and low thermal conductivity. Metals do not have seebeck coefficient and low thermal conductivity. Only semiconductors have all the above three characteristics.

Q. 16. *A thermoelectric refrigerator removes heat from a refrigerated space at – 5 °C at a rate of 130 W and rejects it to an environment at 20°C. Determine the maximum C.O.P. this thermoelectric refrigerator can have, and minimum required input.*

Ans. $$(\text{C.O.P.})_{max} = \frac{[273+(-5)]}{[20-(-5)]} = 10.75$$

$$(\text{Power input})_{input} = \frac{\text{RE}}{(\text{C.O.P.})_{max}} = \frac{130}{10.75} = 12.127 \text{ W.}$$

Q. 17. *In absorption refrigeration cycles, why is the fluid in the absorber cooled and the fluid in the generator heated.*

Ans. When refrigerant is absorbed in the condenser, heat is evolved. To keep the solution cool (so that it can absorb more refrigerant) the absorber is to be cooled. The fluid in the generator is heated to evolve the high pressure refrigerant so that it can be condensed in the condenser.

Q. 18. *Moist air is passed through a cooling section where it is cooled and dehumidified. How do (a) the specific humidity, and (b) relative humidity of air change?*

Ans. (*a*) Specific humidity is reduced.

(*b*) Relative humidity is increased.

Q. 19. *When are the dry bulb and dew-point temperatures of air identical?*

Ans. For saturated air.

Q. 20. *How does the air motion in the vicinity of the human body affect human comfort.*

Ans. Air motion in the vicinity of human body increases the rate of heat loss from the body by increasing the convective film coefficient. It also increases the heat loss by evaporation (drying of sweat). Both these effects decrease the Efficient Temperature (ET) of the surrounding air, and provide comfort to the human body.

Q. 21. *Compare enzymes and micro-organisms which are responsible for food spoilage.*

Ans. *Enzymes* are *internal* agents which are inherent in all organic materials. They are complex protein like chemical substances capable of bringing about chemical changes in organic materials.

Micro-organisms are *external* agents which grow in and on the surfaces of the foodstuff. They are a group of minute plants and animals of microscopic and sub-microscopic sizes.

Q. 22. *Define Energy Efficiency Ratio (EER).*

Ans. EER of a particular cooling device is the ratio of output cooling to input electrical power in a given operating point. The EER is related to the C.O.P. with the primary difference being that of the C.O.P. of a cooling device is limitless. The cooling load and the electrical power need to run the device are both measured using the same limits. Therefore, a C.O.P. is universal and can be used in any system of units. However, the *C.O.P.* is an *instantaneous* measure, whereas *EER* is *averaged* over a duration of time; the time duration considered in several hours of constant conditions for EER.

Q. 23. *What is the effect of sub-cooling of liquid on C.O.P. of a refrigerating system?*

Ans. The sub-cooling results in increase of C.O.P. provided no further energy has to be spent to obtain the extra cold coolant required.

Q. 24. *What is Dry Air Rated Temperature (DART)?*

Ans. The dry air rated temperature is the temperature in terms of which the comparison of different types of air-refrigeration systems is made.

Q. 25. *What is a secondary refrigerant? Give two examples.*

Ans. Secondary refrigerants are those circulating substances which are first cooled with the help of the primary refrigerants and are then employed for cooling purposes. These refrigerants cool substances by absorption of their sensible heat.

Examples: Ice, carbon dioxide etc.

Q. 26. *What is the function of a flash chamber used in compound vapour compression refrigeration system?*

Ans. When the high pressure liquid refrigerant from the condenser passes through the expansion valve, it gets evaporated partially. This partial evaporation of liquid refrigerant is known as 'Flash'. This formed vapour, which cannot produce any refrigerating effect, is bypassed around the evaporator and supplied directly to the compressor through a *flash chamber*. The flash chamber is located between the evaporator and the expansion valve. Flash chamber is an insulated container and separates the liquid and vapour due to 'centrifugal effect'.

Q. 27. *What is humidity ratio?*

Ans. It is the ratio of mass of water vapour per unit mass of dry air in the mixture of vapour and air. It is generally expressed as grams of water per kg of air.

Q. 28. *Define room sensible heat factor (RSHF) line.*

Ans. The straight line joining the supply air condition (conditioned air supplied to the room) to room design condition (condition of air required in room) represents RSHF line.

Q. 29. *How will you define effective temperature?*

Ans. Effective temperature is defined as a sensory index that combines into a single factor the effects of temperature, humidity and air movement on human comfort in a noise-free pure air environment. It corresponds to dry-bulb temperature of the saturated air at which a given percentage of people feel comfortable. For example, at 21°C (and RH = 100%) and air movement 8 m/min.) most people feel comfortable.

Q. 30. *How wet bulb temperature changes during sensible cooling process?*

Ans. During sensible cooling process wet bulb temperature decreases.

Q. 31. *Name various methods to produce refrigeration effect.*

Ans. The various refrigeration systems are:

(*i*) Ice refrigeration; (*ii*) Air refrigeration system;

(*iii*) Vapour compression refrigeration system;

(*iv*) Vapour absorption refrigeration system;

(*v*) Special refrigeration systems:

(*a*) Absorption refrigeration system; (*b*) Cascade refrigeration systems; (*c*) Mixed refrigeration system; (*d*) Vortex tube refrigeration system; (*e*) Thermoelectric refrigeration; (*f*) Steam jet refrigeration.

Q. 32. *Why do we call a closed air refrigeration system a dense air refrigeration system?*

Ans. In the closed air refrigeration system, pressure is maintained above atmosphere throughout, and as such, it is known as dense air system.

Q. 33. *What is hermetically sealed compressor?*

Ans. A hermetically sealed compressor is one in which the motor and compressor are inside a welded steel dome which cannot be removed easily for internal servicing. In this type of compressor the power required per tonne of refrigeration is *less* than that of the open type.

Q. 34. *Name the equipment in vapour absorption system which replaces the compressor in vapour compression system.*

Ans. In vapour absorption system, the compressor is replaced by an absorber, a pump, a generator and a pressure-reducing valve.

Q. 35. *What is 'boot strap' with reference to aircraft refrigeration systems?*

Ans. The term 'boot strap' as used in air cycle refrigeration system indicates a system in which the pressure of the working fluid is raised to a higher level than the main compressor by a separate compressor before expanding in cooling turbine.

Q. 36. *What do you mean by primary refrigerants? Give examples.*

Ans. Primary refrigerants are those working medium or heat carriers which directly take part in the refrigeration system and cool the substance by absorption of latent heat.

Examples: NH_3, CO_2, Feon group etc.

Q. 37. *List six psychometrics properties of air.*

Ans. DBT, WBT, DPT, Humidity ratio, Relative humidity and Degree of saturation.

Q. 38. *Write five engineering applications of 'cryogenics'?*

Ans. (*i*) Expansion fitting; (*ii*) Cryobiology (this term relates to the effects of low temperatures on materials of biological origins); (*iii*) Semen preservation; (*iv*) Cryosurgery; (*v*) Space research; Underground power lines.

Q. 39. *What are the lubrication methods used in a refrigeration system?*

Ans. In a refrigeration system, the only parts which need lubrication are compressors and bearings of motors. The two methods used for lubrication of these parts are:

(*i*) Splash lubrication, and (*ii*) Pressure lubrication.

Q. 40. *Differentiate between Natural and Mechanical refrigeration.*

Ans. The process of *natural cooling* that does not require any mechanical means or work input are said to follow natural refrigeration e.g. Ice making by Noctornal cooling; evaporative cooling, cooling by salt solutions.

In Noctornal cooling method, a thin layer of water, a few mm thick, is kept in shallow earthen trays and exposed to cloudless night sky. The trays are insulated by compacted hay. The trays exposed to loose heat by radiation and by early morning hours the water in the trays freezes to ice.

Mechanical refrigeration/cooling depends upon the evaporation of liquid refrigerant and its circuit includes the equipment namely evaporator, compressor, condenser and expansion valve.

Q. 41. *Mention the advantage of cascade refrigeration system over multiple vapour compression refrigeration system.*

Ans. Cascade refrigeration system permits the use of two different refrigerants, with thermodynamic properties favourable for the two temperature ranges, thus each refrigerant cycle is comparatively simple.

In this system each cycle is separately operating, there is no difficulty of lubrication. Oil from one compressor cannot wander to another compressor as it generally happens in multi-stage systems.

Q. 42. *What is room sensible heat factor (RSHF)?*

Ans. RSHF is the ratio of room sensible heat (RSH) to the sum of room sensible heat and the room latent heat (RLH). In other words,

$$RSHF = \frac{RSH}{RSH + RLH}$$

The air which is to be supplied to the conditioned space must have the capacity to offset simultaneously both of the room sensible heat and room latent heat loads.

Q. 43. *What is the effect of delivery pressure on a vapour compression system?*

Ans. The refrigerating effect is decreased and work required is increased. The net effect is to *reduce* the refrigerating capacity of the system (with the same amount of refrigerant flow and the C.O.P.)

Q. 44. *What is electrolux refrigerator system?*

Ans. This type of refrigerator is also called "Three fluids absorption system". The three fluids used are NH_3, H_2 and H_2O.

The principle involved makes use of the properties of gas-vapour mixtures. If a liquid is exposed to an inert atmosphere, it will evaporate until the atmosphere is saturated with vapour of the liquid.

Q. 45. *What is humid specific heat?*

Ans. Specific heat of wet air is called humid specific heat.

Q. 46. *Compare centrifugal and axial flow fans.*

Ans. A "centrifugal or radial flow fan" is one in which air enters the impeller axially and is discharged radially from the impeller. These fans are used for the air-conditioning and ventilation purposes.

An "axial-flow fan" is one in which the air flows parallel to the axis of impeller. These fans are particularly suitable for handling large volumes of air at relatively low pressure. They are never employed for duct air-conditioning system because they are incapable of developing high pressure.

Q. 47. *Explain the physical significance of by-pass factor with reference to a heating coil.*

Ans. When air passes over a dry surface which is at a temperature greater than its (air) dry bulb temperature, it undergoes 'sensible heating'. During such a process, the specific humidity remains constant but the dry bulb temperature rises and approaches that of the surface. The extent to which it approaches at mean effective surface temperature of the coil is conveniently expressed in terms of the equivalent 'by-pass factor'.

The 'by-pass factor' for the process is defined as the ratio of the difference between the mean surface temperature of the coil and the leaving air temperature to the difference between the mean surface and entering air temperature.

Q. 48. *What do you mean by 'standard rating of a refrigeration machine'.*

Ans. The rating of a refrigeration machine is obtained by refrigerating effect or amount of heat extracted in a given time from a body. The rating of the refrigeration machine is given by unit of refrigeration known as **'standard tonne of refrigeration'** which is defined as the refrigerating effect (RE) produced by the melting of 1 tonne of ice from and at 0°C in 24 hours. Since the latent heat of fusion of ice is 336 kJ/kg, the refrigerating effect of 336 × 1000 kJ/kg in 24 hours is rated as 1 tonne i.e.

$$\text{A tonne of refrigeration} = \frac{336 \times 1000}{24} = 14000 \text{ kJ/h}$$

Q. 49. *Differentiate between refrigerator and heat pump.*

Ans. A 'refrigerator' is a machine which is used to remove the heat from the body at low temperature and then transfer this heat to the body at high temperature. It operates between the temperature of surroundings and temperature *below* that of surroundings.

A 'heat pump' is a machine whose objective is to heat a medium which may already be warmer than surroundings. It operates between the temperature of surroundings and a temperature *above* that of surroundings.

Q. 50. *What is Energy Performance Ratio (EPR)? How is it related to 'Coefficient of Performance (C.O.P.)?*

Ans. C.O.P. for heat pump is called EPR. It is the ratio of heat rejected to the hot-body per cycle to the work done per cycle.

$$\text{EPR [i.e., (C.O.P.)}_{\text{heat pump}}] = 1 + (\text{C.O.P.})_{\text{ref.}}$$

Q. 51. *Define degree of saturation.*

Ans. The 'degree of saturation' is defined as the ratio of mass of water vapour associated with unit mass of *dry air* and mass of water vapour associated with unit mass of *dry saturated air*.

Q. 52. *Write the name of the refrigerant to be used in a ice plant.*

Ans. Ammonia (NH_3).

Q. 53. *What are the essential qualities of a good thermoelectric material?*

Ans. (*i*) High electrical conductivity, (*ii*) Low thermal conductivity, and (*iii*) High seeback coefficient.

Q. 54. *What is a pulse tube refrigeration system?*

Ans. This is a simple but fascinating method of obtaining low temperatures and the problem of lubrication at low temperature is basically absent as there are no moving parts. In this system sudden expansion and release of gas are employed to get the refrigeration effect.

This method of obtaining low temperatures and moderate pressure and without the use of moving parts can be used to construct compact and reliable miniature refrigerators.

Q. 55. *How are refrigerants designated?*

Ans. The general chemical formula for a compound derived from a "saturated hydrocarbon" is given as:

$$C_aH_bF_cCl_d$$

where, $b + c + d = 2a + 2$,

and, a = Number of carbon atoms

b = Number of hydrogen atoms

c = Number of fluorine atoms, and

d = Number of chlorine, atoms.

The complete designation number of the refrigerant is given by:

$$R(a - 1)(b + 1)\,c$$

Q. 56. *What is a dehydrator?*

Ans. A dehydrator/drier is a device employed to remove moisture from a refrigerant. It is generally placed in a liquid line. The drying material used is called the dessicant.

Q. 57. *What is 'vasomotor control' in relation to 'human comfort'?*

Ans. This control regulates the blood supply to the skin. It acts by causing vasodilation of the peripheral blood vessels, if environment temperature is increased. The thermal conductance of the skin is increased, thus minimising body heating.

If environment temperature is decreased the control causes vasocontraction of peripheral blood vessels. This decreases blood circulation and reduces body cooling.

Q. 58. *What are the requirements of good room air distribution.*

Ans. The primary requirement of good room air distribution is to create a proper combination of temperature, humidity and air motion in occupied zone which is normally at 1.8 m above the floor level. The maximum variation in temperature in a single room should not be more than 1°C and within rooms 2°C. The desirable air velocity is 7.5 to 9 m/min.

Q. 59. *What are the variables which are involved in the estimation of load?*

Ans. (*i*) Magnitude and direction of used velocity; (*ii*) Outside humidity and temperature; (*iii*) Nature of construction, materials used; (*iv*) Orientation of openings, windows and doors; (*v*) Period of occupancy and the number of persons in the room, activities of the person etc.

Q. 60. *What are the various methods of food presentation?*

Ans. (*i*) Heat-processing; (*ii*) Dehydration; (*iii*) Chemical preservation; (*iv*) Oils and spices; (*v*) Canning; (*vi*) Pasteurization.

SECTION-B: QUESTIONS WITH "SOLUTIONS"

Q. 1. (*a*) *What is 'standard commercial tonne of refrigeration'?*

(*b*) *A carnot refrigerator is required to maintain a region at a temperature of –32°C. If it needs 1.25 kW per tonne of refrigeration, determine:*

(*i*) *C.O.P. of the refrigerator;*

(*ii*) *Higher temperature of the cycle;*

(*iii*) *The heat delivered and C.O.P. when this device is used as heat pump.*

Solution: (*a*) Standard commercial *tonne* of refrigeration is defined as the *refrigerating effect (RE) produced by the melting of 1 tonne of ice at 0°C in 24 hours.*

Since the latent heat of fusion of ice is 336 kJ/kg, the refrigerating effect of 336 × 1000 kJ/kg in 24 hours is rated as *one tonne;* i.e.,

$$\text{A } \textit{tonne} \text{ of } \textit{refrigeration} = \frac{336 \times 1000}{24} = 14000 \text{ kJ/h}$$

Note: *Ton of refrigeration (TR).* A ton of refrigeration is basically and American unit of RE. It is originated from the rate at which heat is required to be removed to freeze *one ton of water from and at 0°C.* Using American units this is equal to removal of 200 BTU of heat per minute, and in MKS units it is adopted as 50 kcal/min. or 3000 kcal/hour. In SI units its conversion is rounded off to 3.5 kJ/s (kW) or 210 kJ/min.

(*1 ton of refrigeration may be taken equal to 0.9 tonne)*

(*b*) *Given:* T_2 = 273 + (–32) = 241 K; Power required per tonne of refrigeration = 1.25 kW.

(i) C.O.P. of Carnot refrigerator:

$$\text{C.O.P.}_{\text{(Carnot ref.)}} = \frac{\text{Heat absorbed}}{\text{Work done}} = \frac{1 \text{ tonne}}{1.25} = \frac{14000 \text{ kJ/h}}{1.25 \times 60 \times 60 \text{ kJ/h}} = 3.11 \textbf{ (Ans.)}$$

(ii) Higher temperature of the cycle, T_1: Refer to Fig. 1.

$$\text{C.O.P.}_{\text{(Carnot ref.)}} = \frac{T_2}{T_1 - T_2}$$

i.e.

$$3.11 = \frac{241}{T_1 - 241}$$

∴

$$T_1 = \frac{241}{3.11} + 241 = 318.5 \text{ K}$$

$$= 318.5 - 273 = \mathbf{45.5°C \ (Ans.)}$$

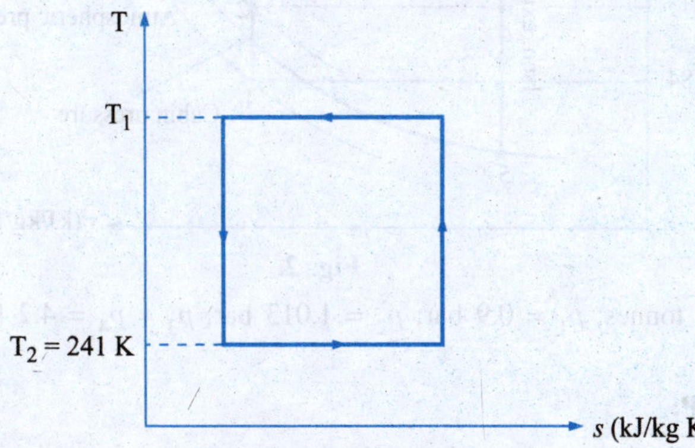

Fig. 1. Reversed Carnot cycle.

(iii) Heat delivered as heat pump and C.O.P.$_{\text{(heat pump)}}$:

$$\text{Heat delivered as heat pump} = \text{Heat absorbed} + \text{Work done}$$

$$= \frac{14000}{60} + 1.25 \times 60 = \mathbf{308.33 \text{ kJ/min (Ans.)}}$$

$$\text{C.O.P.}_{\text{(heat pump)}} = \frac{\text{Heat delivered}}{\text{Work done}} = \frac{308.33}{1.25 \times 60} = \mathbf{4.11 \ (Ans.)}$$

(Also, C.O.P.$_{\text{(heat pump)}}$ = 1 + C.O.P.$_{\text{(ref.)}}$ = 1 + 3.11 = 4.11)

Q. 2. *(a) What are the main considerations involved (in order of importance) in an aircraft refrigeration?*

(b) Following data relate to a simple air cooled system used for an aeroplane:

Load = 8 tonnes; The atmospheric pressure and temperature = 0.9 bar and 11°C respectively, during ramming pressure increases to 1.013 bar; The temperature of air reduced in the heat exchanger by 57°C; The pressure in the cabin = 1.01 bar; The temperature of air leaving the cabin = 24°C; The pressure of the compressed air = 4.2 bar.

Assume that all the expansions and compressions are isentropic.

Determine:

(i) Power required to take the load of cooling in the cabin;

(ii) C.O.P. of the system.

Solution: *(a)* The main considerations in order of merit are:

(i) Weight, *(ii)* Space, *(iii)* Operating power.

(b) Refer to Fig. 2.

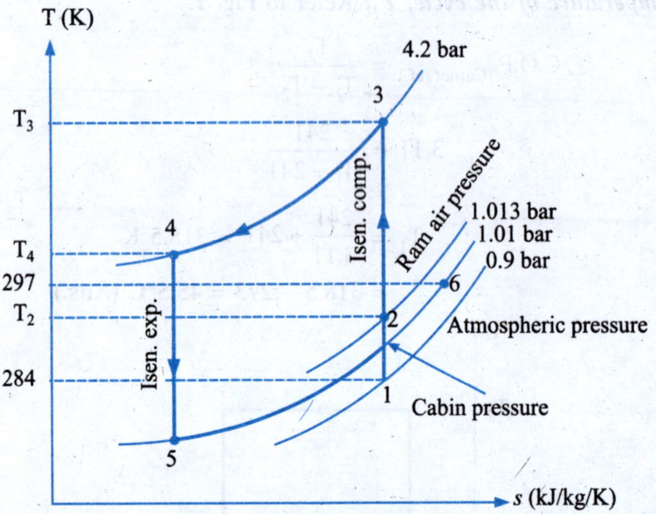

Fig. 2.

Given: Load, Q = 8 tonnes; p_1 = 0.9 bar; p_2 = 1.013 bar; $p_3 = p_4$ = 4.2 bar; $p_5 = p_6$ = 1.01 bar, T_6 = 273 + 24 = 297 K

(i) Power required P:

Let, T_2 = Temperature of air at the end of ramming or entering the main compressor,

T_3 = Temperature of air leaving the compressor,

T_4 = Temperature of air leaving the heat exchanger, and

T_5 = Temperature of air leaving the cooling turbine.

We know that,

$$\frac{T_2}{T_1} = \left(\frac{p_2}{p_1}\right)^{\frac{\gamma-1}{\gamma}} = \left(\frac{1.013}{0.9}\right)^{\frac{1.4-1}{1.4}} = 1.034$$

$$\therefore \qquad T_2 = T_1 \times 1.034 = 284 \times 1.034 = 293.6 \text{ K}$$

Similarly, $$\frac{T_3}{T_2} = \left(\frac{p_3}{p_2}\right)^{\frac{\gamma-1}{\gamma}} = \left(\frac{4.2}{1.013}\right)^{\frac{1.4-1}{1.4}} = 1.5$$

$\therefore$ $$T_3 = T_2 \times 1.5 = 293.6 \times 1.5 = 440 \text{ K} = 167°\text{C}$$

Since the temperature of air is reduced by 52°C in the heat exchanger, therefore, temperature of air leaving the heat exchanger,

$$T_4 = (167 - 57) + 273 = 383 \text{ K}$$

We know that, $$\frac{T_5}{T_4} = \left(\frac{p_5}{p_4}\right)^{\frac{\gamma-1}{\gamma}} = \left(\frac{1.01}{4.2}\right)^{\frac{1.4-1}{1.4}} = 0.665$$

$\therefore$ $$T_5 = T_4 \times 0.665 = 383 \times 0.665 = 254.7 \text{ K}$$

The mass of air (m_a) required for refrigeration purposes,

$$m_a = \frac{14000\ Q}{c_p(T_6 - T_5)} \qquad \text{...[Eqn. (2.11)]}$$

$$= \frac{14000 \times 8}{1(297 - 254.7)} = 2647.7 \text{ kg/h}$$

(Taking c_p for air = 1 kJ/kg K)

Power required to take the load of cooling in the cabin,

$$P = \frac{m_a \times c_p \times (T_3 - T_2)}{3600} \text{ kW}$$

$$= \frac{2647.7 \times 1 \times (440 - 293.6)}{3600} = \mathbf{107.7\ kW\ (Ans.)}$$

***(ii)* C.O.P. of the system:**

$$\text{C.O.P.} = \frac{14000\ Q}{P \times 3600}$$

$$= \frac{14000 \times 8}{107.7 \times 3600} = \mathbf{0.289\ (Ans.)}$$

Q. 3. (*a*) *Enumerate the various parts of a simple vapour compression system.*

(*b*) *The following data relate to a refrigeration machine, operating on a simple vapour compression system, which is required to produce at 0°C from water at 18°C:*

Consider temperature = 298 K; Evaporator temperature = 268 K; Relating efficiency of the machine = 52%; Freon-12 (refrigerant) circulated through the system per min. = 6.2 kg, enters the compressor with a dryless fraction of 0.62; Specific heat of water = 4.187 kJ/kg K; Latent heat of ice = 335 kJ/kg.

The table of properties of Freon-12 is given below:

Temperature (K)	*Liquid heat (kJ/kg)*	*Latent heat (kJ/kg)*	*Entropy of liquid (kJ/kgK)*
298	*59.7*	*138.0*	*0.2232*
268	*31.4*	*154.0*	*0.1251*

Calculate the amount of ice produced in 24 hours.

Solution: (*a*) The various parts of a vapour compression refrigeration system are:

(*i*) Compressor; (*ii*) Discharge line (or hot-gas line);
(*iii*) Condenser; (*iv*) Receiver tank;
(*v*) Liquid line; (*vi*) Expansion valve;
(*vii*) Evaporator; (*viii*) Suction line.

(*b*) Refer to Fig. 3. *Given:* m = 6.2 kg/min; $\eta_{relative}$ = 52%; x_2 = 0.62;
c_{pw} = 4.187 kJ/kg K; Latent heat of ice = 335 kJ/kg

Amount of ice produce in 24 hours, m_{ice}:

$$h_{f2} = 31.4 \text{ kJ/kg};\ h_{fg2} = 154.0 \text{ kJ/kg};\ h_{f3} = 59.7 \text{ kJ/kg};$$

$$h_{fg3} = 138 \text{ kJ/kg};\ h_{f4} = 59.7 \text{ kJ/kg}$$

$$h_2 = h_{f2} + x_2\, h_{fg2} = 31.4 + 0.62 \times 154.0 = 126.9 \text{ kJ/kg}$$

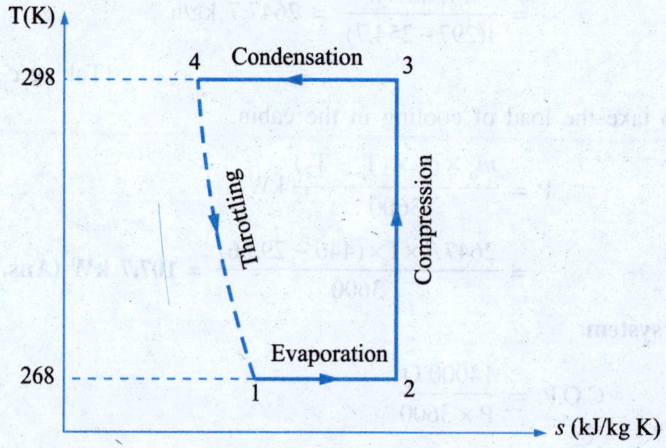

Fig. 3.

For isentropic compression 2-3,

$$s_3 = s_2$$

$$s_{f3} + x_3 \frac{h_{fg3}}{T_3} = s_{f2} + x_2 \frac{h_{fg2}}{T_2}$$

$$0.2232 \times x_3 + \frac{138}{298} = 0.1251 + 0.62 \times \frac{154}{268} = 0.4814$$

$$\therefore \quad x_3 = (0.4814 - 0.2232) \times \frac{298}{138} = 0.5575$$

Now, $h_3 = h_{f3} + x_3 h_{fg3} = 59.7 + 0.5575 \times 138 = 136.6$ kJ/kg

Also, $h_1 = h_{f4} = 59.7$ kJ/kg

$$\text{Theoretical C.O.P.} = \frac{R_4}{W} = \frac{h_2 - h_1}{h_3 - h_2} = \frac{126.9 - 59.7}{136.6 - 126.9} \simeq 6.93$$

Actual C.O.P. = $\eta_{relative} \times (\text{C.O.P.})_{theoretical} = 0.52 \times 6.93 = 3.6$

Heat *extracted* from 1 kg of water at 18°C for the formation of 1 kg of ice at 0°C

$$= \underbrace{1 \times 4.187 \times (18-0)}_{\text{Sensible heat}} + \underbrace{335}_{\text{Latent heat}} = 410.37 \text{ kJ/kg}$$

Let, m_{ice} = Mass of ice formed in kg/min.

$$(\text{C.O.P.})_{actual} = 3.6 = \frac{R_{n(actual)}}{W} = \frac{m_{ice} \times 410.37 \text{ (kJ/min)}}{m(h_3 - h_2)}$$

$$= \frac{m_{ice} \times 410.37 \text{ (kJ/min)}}{6.2\,(136.6 - 126.9)\text{ (kJ/min)}}$$

$$\therefore \quad m_{ice} = \frac{3.6 \times 6.2\,(136.6 - 126.9)}{410.37} = 0.5276 \text{ kg/min.}$$

$$= \frac{0.5276 \times 60 \times 24}{1000} \text{ tonne (in 24 hours)} = \mathbf{0.76 \text{ tonne (Ans.)}}$$

Q. 4. *The following data relate to a two-single vapour compression system:*

Condensing temperature = 40°C; Flash intercooler temperature = 15°C; Evaporating temperature = – 15°C, Refrigerant used = R-717 (NH_3); Refrigerant flow through H.P. side = 0.38 kg/s; Condensate is sub-cooled in condenser by 5°C; Compression efficiencies = 82% each.

Determine the following:

(i) *Flow rate through H.P. side;*

(ii) *Tonnes of refrigeration;*

(iii) *Total power input;*

(iv) *C.O.P.;*

(v) *Heat transfer to condenser.*

Solution: Refer to Fig. 4(*a* and *b*).

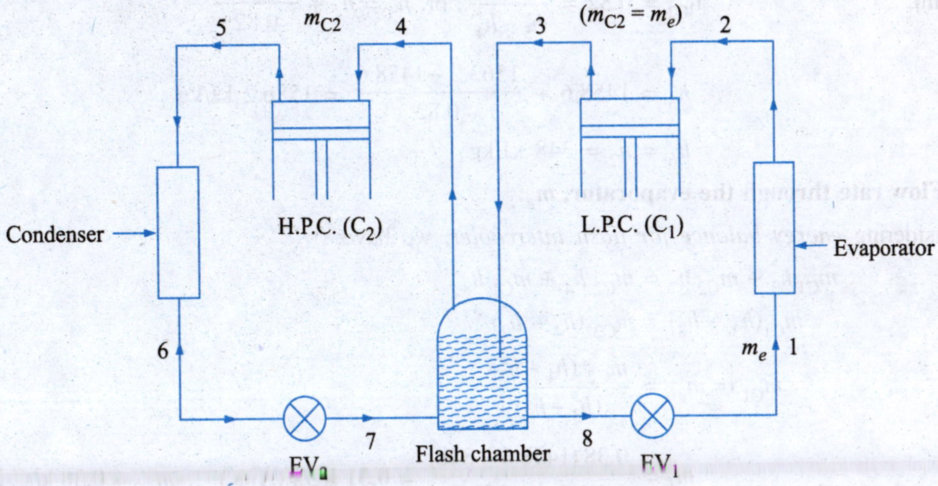

L.P.C. = Low compressor; H.P.C. = High pressure compressor,
EV_1, EV_2 = Expansion valves

(*a*) Schematic arrangement.

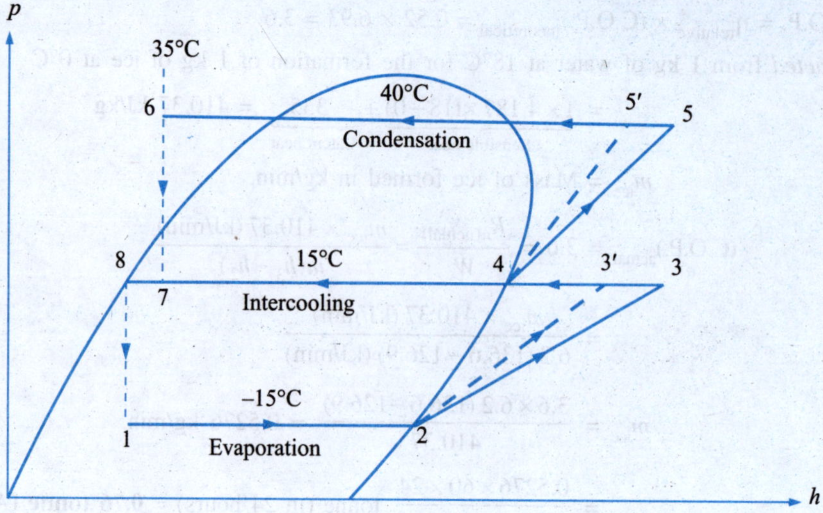

(*b*) *p-h* diagram.

Fig. 4.

Assume suction vapour in each compressor is dry and saturated.

The various values read from property tables/*p-h* diagram, for ammonia (R-717), are as follows:

h_2 = 1426.6 kJ/kg; $h_{3'}$ = 1579 kJ/kg; h_4 = 1458.6 kJ/kg;

$h_{5'}$ = 1563.2 kJ/kg; $h_8 = h_1$ = 251.4 kJ/kg.

Now, $$\eta_{C1} = 0.82 = \frac{h_{3'} - h_2}{h_3 - h_2}, \quad \text{or,} \quad h_3 = h_2 + \frac{h_{3'} - h_2}{0.82}$$

$$\therefore \quad h_3 = 1426.6 + \frac{1579 - 1426.6}{0.82} = 1612.4 \text{ kJ/kg}$$

Again, $$\eta_{C2} = 0.82 = \frac{h_{5'} - h_4}{h_5 - h_4}, \text{ or, } h_5 = h_4 + \frac{h_{5'} - h_4}{0.82}$$

$$\therefore \quad h_5 = 1458.6 + \frac{1563.2 - 1458.6}{0.82} = 1586.2 \text{ kJ/kg}$$

$$h_6 = h_7 = 348 \text{ kJ/kg}$$

(*i*) Flow rate through the evaporator, m_e:

Considering *energy balance for flash intercooler,* we have:

$$m_{C1} h_3 + m_{C2} h_7 = m_{C2} h_4 + m_{C1} h_8$$

or, $$m_{C1}(h_3 - h_8) = m_{C2}(h_4 - h_7)$$

$$\therefore \quad m_{C1} (= m_e) = \frac{m_{C2}(h_4 - h_7)}{(h_3 - h_8)}$$

$$\therefore \quad m_e = \frac{0.38\,(1458.6 - 348)}{(1612.4 - 251.4)} = \mathbf{0.31 \text{ kg/s (Ans.)}} \quad (\because m_{C2} = 0.38 \text{ k/s ... Given})$$

(*ii*) **Tonnes of refrigeration, Q:**

$$Q = \frac{m_e(h_2 - h_1) \times 3600}{14000}$$

$$= \frac{0.31\,(1426.6 - 251.4) \times 3600}{14000} = \textbf{93.7 tonnes (Ans.)}$$

(*iii*) **Total power input, P:**

$$P = P_{C1} + P_{C2} = m_{C1}(h_3 - h_2) + m_{C2}\,(h_5 - h_4)$$

$$= 0.31(1612.4 - 1426.6) + 0.38(1586.2 - 1458.6)$$

$$= \textbf{106.1 kW (Ans.)}$$

(*iv*) **C.O.P.:**

$$\text{C.O.P.} = \frac{R_n\,(\text{kW})}{W\,(\text{kW})} = \frac{m_e(h_2 - h_1)}{m_{C1}(h_3 - h_2) + m_{C2}(h_5 - h_4)}$$

$$= \frac{0.31(1426.6 - 251.4)}{106.1} = \textbf{3.43 (Ans.)}$$

[Alternatively: $$\text{C.O.P.} = \frac{RE\,(\text{kW})}{\text{Power input (kW)}} = \frac{14000\,Q/3600}{P}$$

$$= \frac{(14000 \times 93.7/3600)}{106.1} = 3.43$$]

(*v*) **Heat transfer to condenser:**

$$\text{Heat transfer to condenser} = m_{C2}(h_5 - h_6)$$

$$= 0.38(1586.2 - 348) = \textbf{470.5 kJ/s (Ans.)}$$

Q. 5. (*a*) *Give the comparison between vapour absorption and vapour compression refrigeration systems.*

(*b*) *In a 15 tonnes absorption refrigeration system the heating in generator is carried out by using steam at 3 bar and 80% dry. The refrigeration temperature is –9°C. The condensation of the refrigerant is carried out at 28°C using cooling water.*

Determine:

(*i*) *Maximum possible C.O.P. of the system;*

(*ii*) *Quantity of steam required per hour to run the plant if the steam leaves the generator as saturated water at the same pressure. Assume relative C.O.P. = 0.42.*

Solution: (*a*) Refer to Art. 5.6.

(*b*) *Given:* Q = 15 tonnes; p = 3 bar; x = 0.8; $T_E = -9 + 273 = 264$ K, $T_C = 28 + 273 = 301$ K; $(\text{C.O.P.})_{relative} = 0.42$

(*i*) **Maximum, C.O.P., (C.O.P.)$_{max}$**: From steam tables, corresponding to 3 bar, the saturation temperature of steam, $T_G = 133.5 + 273 = 406.5$ K

$$\therefore \quad (\text{C.O.P.})_{max} = \left[\frac{T_E}{(T_C - T_E)}\right]\left[\frac{T_G - T_C}{T_G}\right] \quad \text{...[Eqn. (5.1)]}$$

$$= \left[\frac{264}{(301 - 264)}\right]\left[\frac{(406.5 - 301)}{406.5}\right] = \textbf{1.85 (Ans.)}$$

(*ii*) **Quality of steam required per hour:**

$$(\text{C.O.P.})_{actual} = 0.42 \times 1.85 = 0.777$$

$$\therefore \quad \text{Actual heat supplied} = \frac{\text{Refrigeration load}}{(\text{C.O.P.})_{actual}}$$

$$= \frac{15 \times 14000}{0.777} = 270270 \text{ kJ/h}$$

h_{fg} at 3 bar (from steam tables)= 2163.2 kJ/kg

∴ Quality of steam required per hour

$$= \frac{\text{Actual heat required}}{x \times h_{fg}}$$

$$= \frac{270270}{0.8 \times 2163.2} \simeq \mathbf{156.2 \text{ (Ans.)}}$$

Q. 6. (*a*) *Explain briefly the construction and working of a 'Vortex tube'.*

(*b*) *In a thermoelectric cooling system each couple has total electric resistance = 0.0028 Ω, total thermal conductance = 0.027 W/K and thermoelectric power = 410 × 10^{-6} V/K. T_h = 320 K, T_C = 280 K, I = 19 A.*

Determine:

(*i*) *Figure of merit.*

(*ii*) *Cooling capacity in tonnes of refrigeration if 120 couples are used.*

Solution: (*a*) Refer to Article 8.2.

(*b*) *Given:* $R = 0.0028\ \Omega$; $T_h = 320$ K; $T_C = 280$ K, $U = 0.027$ W/K; $\alpha_{ab} = 410 \times 10^{-6}$ V/K; $I = 19$ A

(*i*) **Figure of merit,** *z***:**

$$Z = \frac{(\alpha_{ab})^2}{UR} \qquad \text{...[Eqn. (8.18)]}$$

$$= \frac{(410 \times 10^{-6})^2}{0.027 \times 0.0028} = \mathbf{0.002223 \text{ (Ans.)}}$$

(*ii*) **Cooling capacity with 120 couples:**

Heat absorbed at the cooling junction *per couple*,

$$Q_c = \alpha_{ab} \cdot I \cdot T_c - \frac{1}{2} I^2 R - U(T_h - T_c) \qquad \text{...[Eqn. (8.13)]}$$

$$= 410 \times 10^{-6} \times 19 \times 280 - \frac{1}{2} \times (9^2 \times 0.0028 - 0.027(320 - 280)$$

$$= 2.1812 - 0.5054 - 1.08 = 0.5958 \text{ W}$$

∴ Cooling capacity for 120 couples

$$= \frac{(0.5958 \times 3600/1000) \times 120}{14000} = \mathbf{0.01838 \text{ tonne (Ans.)}}$$

Q. 7. (*a*) *Define the following:*

(i) Saturated air; (ii) Dew point temperature (DPT); (iii) Relative humidity (RH).

(*b*) *0.0038 kg of water vapour per kg of atmospheric air is removed and temperature of air after removing the water vapour becomes 22°C. Determine: (i) Relative humidity, and (ii) Dew point temperature.*

Assume that condition of atmospheric air is 30°C, 60% RH and pressure is 1.0132 bar.

Solution: (*a*) (*i*) ***Saturated air***. Moist air is said to be saturated when its condition is such that it can *co-exist in natural equilibrium* with an associated condensed moisture phase presenting a flat surface to it. For a given temperature, a given quantity of air can be saturated with a fixed quantity of moisture.

(*ii*) ***Dew point temperature (DPT).*** It is the *temperature to which air must be cooled at constant pressure in order to cause condensation of any of its water vapour*. It is equal to steam table saturation temperature corresponding to the actual partial pressure of water in the air (t_{dp}).

(*iii*) ***Relative humidity.*** It is the *ratio of partial pressure of water vapour in the mixture to the saturated partial pressure at the dry bulb temperature, expressed as percentage.*

(*b*) Corresponding to 30°C, from steam tables, p_{vs} = 0.0425 bar.

∴ Relative humidity (RH), $$\phi = \frac{p_v}{p_{vs}}$$

i.e., $$0.60 = \frac{p_v}{0.0425}$$

∴ $$p_v = 0.60 \times 0.0425 = 0.0255 \text{ bar}$$

Also the specific humidity,

$$W = \frac{0.622\, p_v}{p_t - p_v} = \frac{0.622 \times 0.0255}{1.0132 - 0.0255} = 0.016 \text{ kg/kg of day air}$$

The specific humidity after removing 0.0038 kg of water becomes,

0.0160 – 0.0038 = 0.0122 kg/kg of dry air and the temperature t_{db} is given as 22°C.

The partial pressure of water vapour, p_v, at this condition can be calculated as follows:

$$W = \frac{0.622\, p_v}{p_t - p_v}$$

$$0.0122 = \frac{0.622\, p_v}{1.0132 - p_v}$$

or $$0.0122\,(1.0132 - p_v) = 0.622\, p_v$$

or $$0.01236 - 0.0122\, p_v = 0.622\, p_v$$

or $$p_v = \frac{0.01236}{(0.622 + 0.0122)} = 0.0195 \text{ bar}$$

Corresponding to 22°C, from steam tables, p_{vs} = 0.0264 bar

(*i*) **Relative humidity,** $\phi = \frac{p_v}{p_{vs}} = \frac{0.195}{0.0264}$ = 0.73 or **73.9% (Ans.)**

(*ii*) **Dew point temperature,** t_{dp}: Corresponding to 0.0195 bar, from steam tables, $t_{dp} \simeq$ **17°C (Ans.)**

Q. 8. *It is required to design an air-conditioning plant for a small office room for the following winter conditions:*

Outdoor conditions = 12°C DBT and 10°C WBT; Required conditions = 20°C DBT and 60% RH; Amount of air circulation = 0.32 m³/min/person; Seating capacity of office = 50. The required condition is achieved first by heating and then by adiabatic humidification.

Determine: (i) Heating capacity of the coil in kW and the surface temperature required if the by pass factor of the coil is 0.42; (ii) The capacity of the humidifier.

Solution: Refer to Fig. 5.

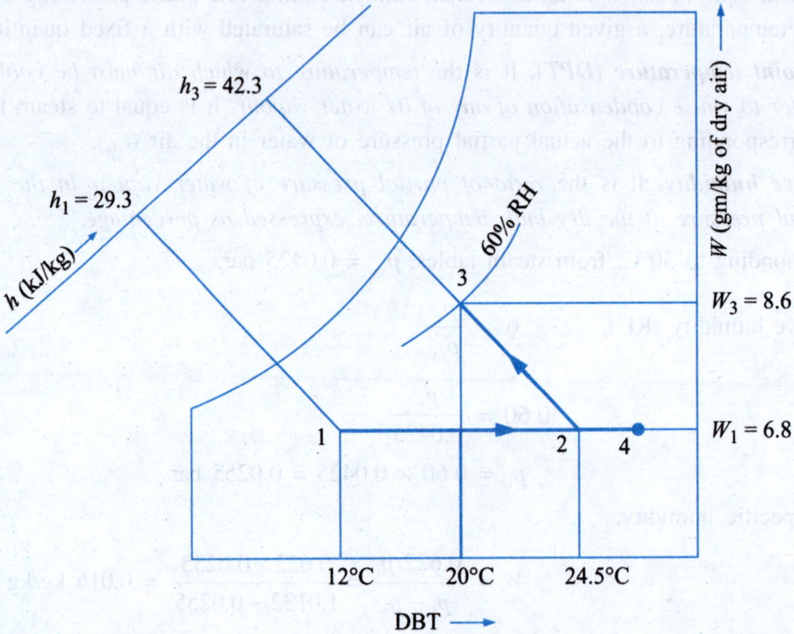

Fig. 5

- Locative the points '1' (12°C DBT, 10°C WBT) and '3' (20°C DBT, 60% RH) on the psychrometric chart.
- Draw a constant enthalpy line through '3' and constant specific humidity time through '1'.
- Locate the point '2' where the above two lines intersect.

From psyschrometric chart:

$h_1 = 29.3$ kJ/kg; $h_2 = h_3 = 42.3$ kJ/kg; $t_{db_2} = 24.5$°C; $v_{s1} = 0.817$ m³/kg

The mass of air circulated per minute,

$$m_a = \frac{0.32 \times 50}{0.817} = 19.58 \text{ kg/min}$$

(*i*) Heating capacity of the heating coil

$$= m_a (h_2 - h_1) = 23.5\ (42.3 - 29.3) = \mathbf{305.5\ kJ/min} \quad \textbf{(Ans.)}$$

Coil surface temperature, t_{db4}:

The by-pass factor (BF) of heating coil is given by:

$$\text{BF} = \frac{t_{db_4} - t_{db_2}}{t_{db_4} - t_{db_1}}$$

$$0.42 = \frac{t_{db_4} - 24.5}{t_{db_4} - 12}$$

$\therefore$ $$0.42(t_{db_4} - 12) = t_{db4} - 24.5$$

$$0.42\, t_{db_4} - 5.04 = t_{db4} - 24.5$$

i.e., $$t_{db4} = \mathbf{33.55°C\ (Ans.)}$$

(*ii*) **Capacity of the humidifier:**

$$\text{Capacity of the humidifier} = \frac{m_a(W_3 - W_1)}{1000} \times 50 \text{ kg/h}$$

$$= \frac{19.58\,(8.6 - 6.8)}{1000} \times 50 = \mathbf{1.762\ kg/h\ (Ans.)}$$

Q. 9. (*a*) *What points should be considered while making heat-load calculations?*

(*b*) *Explain briefly the following two methods for estimating air leakage through the window and door cracks.*

Solution: (*a*) Refer to Art. 18.3.

(*b*) Refer to Art. 18.7.

Q. 10. (*a*) *What are the advantages of food preservation?*

(*b*) *Explain with a neat diagram the working of an "All the year-round air-conditioner".*

Solution: (*a*) Refer to Art. 21.1.1.

(*b*) Refer to Art. 21.8.

Q. 11. (*a*) *What are the applications of refrigeration?*

(*b*) *The following data refer to a cold storage plant which is required to store 15 tonnes of fish:*

The temperature of the fish when supplied = 26°C; Storage temperature of fish required = –7°C; Specific heat of fish above freezing point = 2.93 kJ/kg °C; Specific heat of fish below freezing point = 1.25 kJ/kg °C; Freezing point of fish = –3°C; Latent heat of fish = 232 kJ/kg.

If the cooling is achieved within 10 hours, find out: (i) Capacity of the refrigerating plant; (ii) Reversed Carnot cycle C.O.P. between this temperature range; (iii) If the actual C.O.P. is 35% of the Carnot C.O.P., determine the power required to run the plant.

Solution: (*a*) The main applications of refrigeration are:

(*i*) Ice making;

(*ii*) Transportation of foods above and below freezing;

(*iii*) Industrial air-conditioning;

(*iv*) Comfort air-condition;

(*v*) Medical and surgical aids;

(*vi*) Processing food products and beverages;

(*vii*) Oil refining and synthetic rubber manufacturing;

(*viii*) Manufacturing and treatment of metals.

(*b*) Heat removed in 10 hours from each kg of fish

$$= 1 \times 2.93 \times [26 - (-3)] + 223 + 1 \times 1.25\,[-3 - (-7)]$$

$$= 312.97 \text{ kJ/kg}$$

Heat removed by the plant/hour

$$= \frac{312.97 \times (25 \times 1000)}{10} = 782425 \text{ kJ/h}$$

(*i*) **Capacity of the refrigerant plant:**

$$\text{Capacity of the refrigerant plant} = \frac{782425}{14000} = \mathbf{55.88 \text{ tonnes (Ans.)}}$$

(*ii*) **C.O.P. of reversed Carnot cycle:**

$$\text{C.O.P.} = \frac{T_2}{T_1 - T_2} = \frac{(-7+273)}{(26+273)-(-7+273)} = \mathbf{8.06 \text{ (Ans.)}}$$

(*iii*) **Power required:**

Actual C.O.P. = 0.35 × 8.06 = 2.82

But, actual C.O.P. $= \frac{R_n}{W}$

or $$2.82 = \frac{782425}{W} \text{ kJ/h}$$

or $$W = \frac{782425}{2.82} = 277455 \text{ kJ/h} = 77.07 \text{ kJ/s}$$

∴ Power required to run the plant = **77.07 kW (Ans.)**

Q. 12. (*a*) *Show the simple vapour compression cycle on (a) p-h chart.*

(*b*) *A refrigerator operating on standard vapour compression cycle has a C.O.P. of 6.2 and is driven by a 45 kW compressor. The enthalpies of saturated liquid and saturated vapour refrigerant at the operating condensing temperature of 35°C are 69.55 kJ/kg and 201.45 kJ/kg, respectively. The saturated refrigerant vapour leaving the evaporator has an enthalpy of 187.55 kJ/kg. If the c_p of refrigerant vapour is 0.6155 kJ/kg °C, calculate the refrigerant temperature at compressor discharge.*

Solution: (*a*) Refer to Art. 3.6.

(*b*) Refer to Fig. 6. *Given:* C.O.P. = 6.2; W = 45 kW; $h_{3'}$ = 201.45 kJ/kg; $h_{f4} = h_1$ = 69.55 kJ/kg; h_2 = 187.53 kJ/kg; c_p = 0.6155 kJ/kg °C.

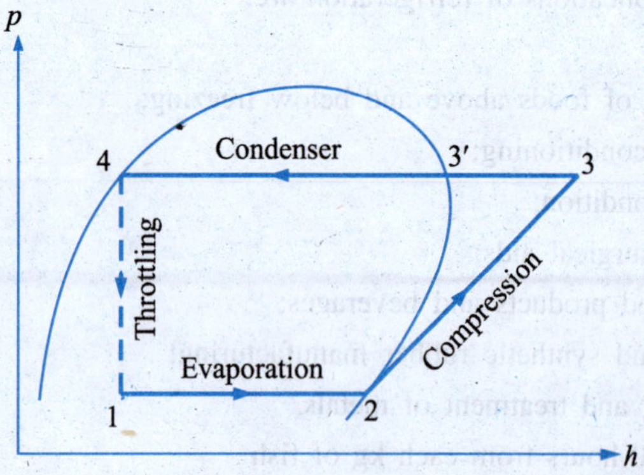

Fig. 6.

Refrigerant temperature t_3:

Refrigerating capacity, $RE = 45 \times \text{C.O.P.} = 45 \times 6.2 = 279$ kW

Heat extracted per kg of refrigerant

$$= (h_2 - h_1) = 187.53 - 69.55 = 118 \text{ kJ/kg}$$

Refrigerant flow rate $= \dfrac{279}{118} = 2.36$ kg/s

Compressor power = 45 kW (= 45 kJ/s)

Heat input per kg $= \dfrac{45 \text{ (kJ/s)}}{2.36 \text{ (kg/s)}} = 19.07$ kJ/kg

Enthalpy of vapour after compression

$$= h_2 + 19.07 = 187.53 + 19.07 = 206.6 \text{ kJ/kg}$$

Superheat $= 206.6 - h_3' = 206.6 - 201.45 = 5.15$ kJ/kg

But, $5.15 = 1 \times c_p(t_3 - t_{3'}) = 1 \times 0.6155 \times (t_3 - 35)$

$\therefore$ $t_3 = \dfrac{5.15}{0.6155} + 35 = $ **43.37°C (Ans.)**

Q. 13. (*a*) *What do you mean by cryogenics?*

(*b*) *Explain with a neat sketch a cascade refrigeration system.*

Solution: (*a*) Refer to Art. 7.1.

(*b*) Refer to Art. 7.3.

Q. 14. (*a*) *Give the comparison between 'thermoelectric refrigeration' and 'vapour compression system'.*

(*b*) *The pressure and temperature of air in a room are 1.0132 bar and 25°C respectively. If the relative humidity is found to be 45%, determine:*

(i) The partial pressure of water vapour; (ii) The drew point temperature; (iii) The specific volume of each component; (iv) The specific humidity.

Solution: (*a*) Refer to Art. 8.1.6.

(*b*) *Given:* $p_v = 1.0132$; $t_{db} = 25°\text{C}$, $\phi = 45\%$

(*i*) **The partial pressure of water vapour, p_v:**

From steam tables, corresponding to 25°C, $p_{vs} = 0.0317$ bar.

Also, $\phi = \dfrac{p_v}{p_{vs}}$

or, $0.45 = \dfrac{p_v}{0.0317}$

$\therefore$ $p_v = 0.45 \times 0.0317 = 0.0143$ bar

(*ii*) **The dew point temperature, t_{dp}:**

The dew point temperature, corresponding to 0.0143 bar $\simeq$ **12 °C (Ans.)**

(*iii*) **The specific volume of each component, v_v, v_a:**

For vapour, $v_v = \dfrac{R_v T}{p_v} = \dfrac{462 \times (25 + 273)}{0.0143 \times 10^5} = $ **96.28 m³/kg (Ans.)**

For air, $v_a = \dfrac{R_a T}{p_a} = \dfrac{287 \times (25 + 273)}{(1.0132 - 0.0143) \times 10^5}$ = **0.856 m³/kg Ans.**

[$\because R_v = 462$ J/kg, $R_a = 287$ J/kg and $p_t = p_a + p_v$]

(*iv*) **The specific humidity, W:**

$$W = \frac{m_v}{m_a} = \frac{v_a}{v_v} \quad \text{...[Eqn. (12.5)]}$$

$$= \frac{0.856}{96.28} = \textbf{0.00889 kg/kg of dry air (Ans.)}$$

Q. 15. (*a*) *Explain briefly how the body reacts to changes in temperature of human body.*

(*b*) *Atmospheric air at 36°C and 30% relative humidity passes through an evaporative cooler. If the final temperature of air is 19°C, how much water is added per kg of dry air and what is the final relative humidity?*

Solution: (*a*) Refer to Art. 13.1.

(*b*) ***At 36°C:*** p_{vs} = 0.0595 bar; h_{g1} = 2567.1 kJ/kg; and $p_v = \phi \times p_{vs}$ = 0.3 × 0.0595 = 0.01785 bar

At 19°C: p_{vs} = 0.0220 bar; h_{g2} = 2536.3 kJ/kg

$$\therefore \quad W_1 = \frac{0.622 \times 0.01785}{1.0132 - 0.01785} = 0.01115 \text{ kg/kg of dry air}$$

Since enthalpy remains constant during the process,

$$c_p\, t_{db1} + W_1 hg_1 = c_p\, t_{db2} + W_2\, h_{g2}$$

$$1.005 \times 36 + 0.01115 \times 2567.1 = 1.005 \times 19 + W_2 \times 2536.3$$

or $$64.80 = 19.09 + 2536.3\, W_2$$

or $$W_2 = \frac{64.80 - 19.09}{2536.3} = 0.01774 \text{ kg/kg of dry air.}$$

$\therefore$ **Amount of water added** $= W_2 - W_1$

$= 0.01774 - 0.01115 =$ **0.00659 kg/kg of dry air (Ans.)**

Also, $$0.00659 = \frac{0.622\, p_{v_2}}{1.0132 - p_{v_2}}$$

or, $$0.00659\,(1.0132 - p_{v2}) = 0.622\, p_{v2}$$

or, $$0.00668 - 0.00659\, p_{v2} = 0.622\, p_{v2}$$

or, $$p_{v2} = 0.01063 \text{ bar}$$

Final relative humidity $= \dfrac{0.01063}{0.0220} = 0.483$ or **48.3% (Ans.)**

Q. 16. *The following data relate to an office for seating 30 occupants.*

The conditions to be maintained in the office = 22 °C DBT and 55% RH; The outdoor conditions = 36 °C DBT and 27 °C DBT; Solar heat gain = 8500 W; Sensible heat gain per occupant = 83 W; Latent heat gain per person = 100 W; Lighting load = 2500 W; Sensible heat load from other sources = 12000 W; Infiltration load = 15 m³/min.

Assuming 40% fresh air and 60% of recirculated air passing through the evaporator coil and the by-pass factor 0.12, determine the following:

(i) Dew point temperature of the coil; and (ii) Capacity of the plant.

Solution: Given: Number of occupants = 30; t_{db2} = 22°C; ϕ_2 = 55% RH; t_{db1} = 36°C; t_{wb1} = 23°C; $(Q_S)_S$ = 8500 W or 8.5 kW; Q_S per occupant = 83 W; Q_L per occupant = 100 W; $(Q_S)_L$ = 2600 W or 2.6 kW; $(Q_S)_{other\ sources}$ = 12000 W = 12 kW; V_1 = 15 m^3/min; BF = 0.15.

The flow diagram for the given air-conditioning system is shown in Fig. 7. The various points on the psychrometric chart, as shown in Fig. 8, are marked as discussed below:

- Locate point 1 at the intersection of 36°C DBT and 27°C WBT lines.
- Locate point 2 at the intersection of 22°C DBT line and 55% RH curve.
- Locate point A by drawing vertical and horizontal lines through points 1 and 2 respectively.

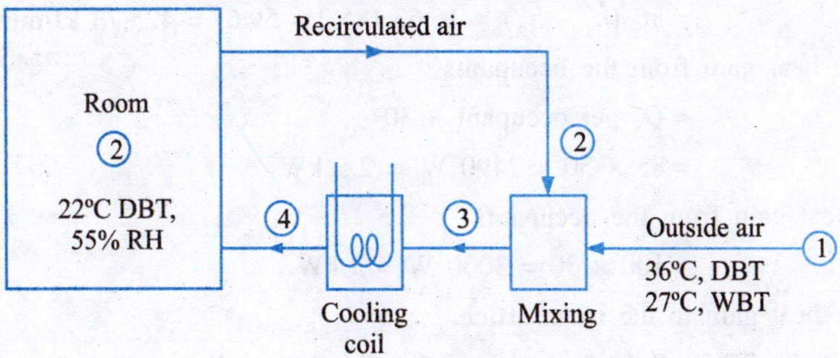

Fig. 7. Flow diagram of the air-conditioning system

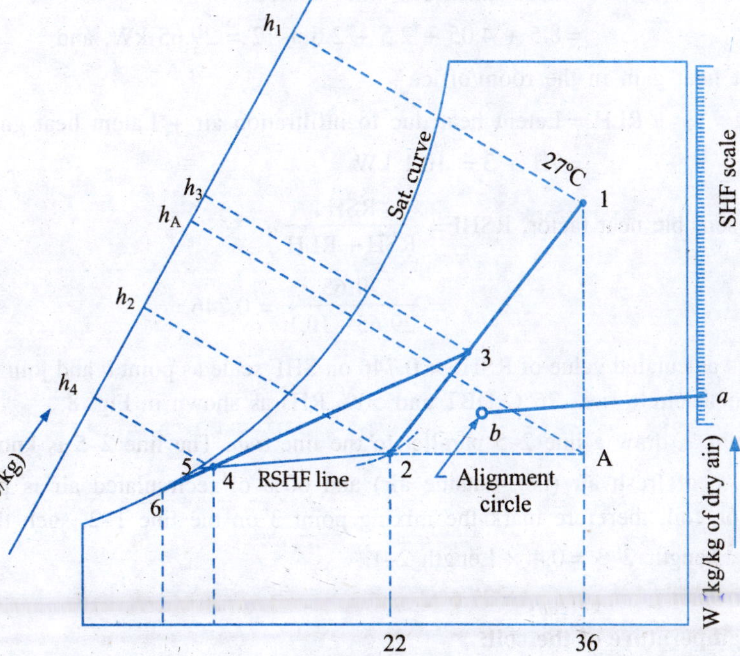

Fig. 8.

From *psychrometric chart*, we find:

$$v_1 = 0.902 \text{ m}^3\text{/kg of dry air}$$
$$h_1 = 85.2 \text{ kJ/kg of dry air}$$
$$h_2 = 45 \text{ kJ/kg of dry air}$$
$$h_1 = 59.6 \text{ kJ/kg of dry air}$$

Mass of infiltrated air at point 1,

$$m_1 = \frac{V_1}{v_1} = \frac{15}{0.902} = 16.63 \text{ kg/min}$$

Sensible heat gain due to infiltration air

$$= m_1 (h_A - h_2) = 16.63 (59.6 - 45) = 242.8 \text{ kJ/min} = 4.05 \text{ kW}$$

Latent heat gain due to infiltration air

$$= m_1 (h_1 - h_A) = 16.63 (85.2 - 59.6) = 425.73 \text{ kJ/min} = 7.1 \text{ kW}$$

Total sensible heat gain from the occupants

$$= Q_S \text{ per occupant} \times 30$$
$$= 83 \times 30 = 2490 \text{ W} \simeq 2.5 \text{ kW}$$

Total latent heat gain from the occupants

$$= 100 \times 30 = 3000 \text{ W} = 3 \text{ kW}$$

Total sensible heat gain in the room/office,

RSH = Solar heat gain + Sensible heat gain due to infiltration air + Sensible heat gain from occupants + Sensible heat gain due to lighting + Sensible heat gain from other sources

$$= 8.5 + 4.05 + 2.5 + 2.6 + 12 = 29.65 \text{ kW, and}$$

Total latent heat gain in the room/office

RLH = Latent heat due to infiltration air + Latent heat gain from occupants

$$= 7.1 + 3 = 10.1 \text{ kW}$$

$\therefore$ Room sensible heat factor, $\text{RSHF} = \dfrac{\text{RSH}}{\text{RSH} + \text{RLH}}$

$$= \frac{29.65}{29.65 + 10.1} = 0.746$$

- Mark this calculated value of RSHF = 0.746 on SHF scale as point *a* and join with point *b* which is alignment circle (i.e., 26°C DBT and 50% RH) as shown in Fig. 8.
- From point 2, draw a line 2–5 parallel to the line *a–b*. The line 2–5 is known as RSHF line.
- Since 40% of fresh air (i.e., outside air) and 60% of recirculated air is passed through the evaporator coil, therefore mark the mixing point 3 on the line 1–2, such that,

Length 2–3 = 0.4 × Length 2–1.

From *psychrometric chart*: $t_{db3} = 27.6°\text{C}$

(*i*) **Dew point temperature of the coil:**

Let, t_{db6} = Dew point temperature of the coil.

Now, $\quad BF = 0.15 = \dfrac{t_{db4} - t_{db6}}{t_{db3} - t_{db6}} = \dfrac{t_{db4} - t_{db6}}{27.6 - t_{db6}}$

By trial and error we find:

$$t_{db4} \simeq 10.7°C; \quad \mathbf{t_{db6} \simeq 7.8°C \ (Ans.)}$$

(*ii*) **Capacity of the plant:**

From the *psychrometric chart*, we find:

$$h_4 = 29.5 \text{ kJ/kg of dry air, and}$$

$$h_3 = 61 \text{ kJ/kg of dry air.}$$

Mass of air entering the room,

$$m_a = \frac{\text{Total room heat}}{\text{Total heat sermoved}}$$

$$= \frac{\text{RSH} + \text{RLH}}{(h_2 - h_4)} = \frac{29.65 + 10.1}{(45 - 29.5)} = 2.56 \text{ kg/s}$$

$\therefore$ Capacity of the plant

$$= m_a (h_3 - h_4) = \frac{2.56\,(61 - 29.5) \times 3600}{14000} = \mathbf{20.74 \ tonnes \ (Ans.)}$$

Q. 17. (*a*) *What are the various types of air-refrigeration systems employed for aircrafts.*

(*b*) *A refrigerating system operates on the reversed Carnot cycle. The higher temperature of the refrigerant in the system is 32°C and the lower temperature is –14°C. The capacity is to be 10 tonnes. Neglecting losses, determine:*

(*i*) *C.O.P.;* (*ii*) *Heat rejected from the system per hour;* (*iii*) *Power required.*

Solution: (*a*) Various types of air refrigeration systems employed for aircrafts are:

1. Basic or simple air-refrigeration system:
   (*i*) Without evaporative cooling.
   (*ii*) With evaporative cooling.
2. Boot-strap air-refrigeration system:
   (*i*) Without evaporative cooling.
   (*ii*) With evaporative cooling.
3. Regenerative air-refrigeration system.
4. Reduced ambient air refrigeration system.

(*b*) *Given*: $T_1 = 32 + 273 = 305$ K; $T_2 = -14 + 273 = 259$ K; Capacity = 10 tonnes.

(*i*) **C.O.P.:**

$$\text{C.O.P.} = \frac{T_2}{T_1 - T_2} = \frac{259}{(305 - 259)} = \mathbf{5.63 \ (Ans.)}$$

(*ii*) **Heat rejected from the system per hour:**

$$\text{C.O.P.} = \frac{\text{Refrigerating effect}}{\text{Work input}}$$

$$5.63 = \frac{10 \times 14000 \text{ (kJ/h)}}{\text{Work input}}$$

$$\therefore \quad \text{Work input} = \frac{10 \times 14000}{5.63} = 24867 \text{ kJ/h}$$

Thus, heat rejected per hour = Refrigerating effect/hour + work input/hour

= 10 × 14000 + 24867 = **164867 kJ/h (Ans.)**

(*iii*) **Power required:**

$$\text{Power required} = \frac{\text{Work input/hour}}{60 \times 60} = \frac{24867}{60 \times 60} = \textbf{6.91 kW (Ans.)}$$

Q. 18. (*a*) *Discuss the effect of the following on the performance of a vapour compression system:*

(*i*) *Effect of superheating;*

(*ii*) *Effect of sub-cooling of liquid.*

(*b*) *What are the advantages of compound compression with intercooler over single-stage compression?*

Solution: (*a*) Refer to Art. 3.7.

(*b*) Refer to Art. 4.2.

Q. 19. *Discuss the arrangement used for producing low temperatures by adiabatic demagnetisation of a paramagnetic salt.*

Solution: Refer to Art. 8.5.

Q. 20. (*a*) *What is a secondary refrigerant?*

(*b*) *Among R-11, R-12, R-21 and R-22 which are best suited for centrifugal compressors and why?*

(*c*) *Explain automatic expansion valve.*

Solution: (*a*) Refer to Art. 9.2.

(*b*) R-11 and R-21 are best suited for centrifugal compressors due to *low evaporation pressures and large volumes to be handled.* Also big size air-conditioning plants using R-12 where large volume is to be handled employ centrifugal compressors.

(*c*) Refer to Art. 10.5.3.

Q. 21. *In a laboratory test, psychrometer recorded dry bulb temperature as 32°C and wet bulb temperature 26°C. Calculate the following:*

(*i*) *Vapour pressure;* (*ii*) *Relative humidity;*

(*iii*) *Specific humidity;* (*iv*) *Degree of saturation;*

(*v*) *Dew point temperature;* (*vi*) *Enthalpy of mixture.*

Take barometric pressure as 1.0132 bar.

Solution: *Given:* t_{db} = 32°C; t_{wb} = 26°C; p_t = 1.0132 bar.

From steam tables, corresponding to 26°C,

$$(p_{vs})_{wb} = 0.0336 \text{ bar}$$

(*i*) **Vapour pressure, (p_v):**

$$p_v = (p_{vs})_{wb} - \frac{[p_t - (p_{vs})_{wb}](t_{db} - t_{wb})}{1527.4 - 1.3\, t_{db}} \quad \text{...[Eqn. (12.1)]}$$

$$= 0.0336 - \frac{[1.0132 - 0.0336](32 - 26)}{1527.4 - 1.3 \times 32}$$

$$= 0.0336 - \frac{5.8776}{1485.8} = \mathbf{0.0296\ bar\ (Ans.)}$$

(*ii*) **Relative humidity, ϕ:**

$$\phi = \frac{p_v}{p_{vs}}$$

From steam tables, corresponding to 32°C, p_{vs} = 0.0476 bar

$$\therefore \quad \phi = \frac{0.0296}{0.0476} = 0.622 \text{ or } \mathbf{62.2\%\ (Ans.)}$$

(*iii*) **Specific humidity, W:**

$$W = \frac{0.622\, p_v}{p_t - p_v} = \frac{0.622 \times 0.0296}{(1.0132 - 0.0296)} = \mathbf{0.0187\ kg/kg\ of\ dry\ air\ (Ans.)}$$

(*iv*) **Degree of saturation, μ:**

$$\mu = \frac{p_v(p_t - p_{vs})}{p_{vs}(p_t - p_v)} \quad \text{...[Eqn. (12.7)]}$$

$$= \frac{0.0296\,(1.0132 - 0.0476)}{0.0476\,(1.0132 - 0.0296)} = \mathbf{0.61\ (Ans.)}$$

(*v*) **Dew point temperature, t_{dp}:**

t_{dp} corresponding to p_v (i.e., 0.0296 bar), from steam tables $\simeq$ **24°C (Ans.)**

(*vi*) **Enthalpy of mixture *h*:**

$$h = 1.005\, t_{db} + W(2500 + 1.88\, t_{db}) \quad \text{...[Eqn. (12.12b)]}$$

$$= 1.005 \times 32 + 0.0187\,(2500 + 1.88 \times 32) = \mathbf{80.03\ kJ/kg\ of\ dry\ air\ (Ans.)}$$

Q. 22. *Enumerate various methods of air-conditioning duct design and explain in detail the "Equal friction method".*

Solution: Refer to Art. 16.1.7.2.

Q. 23. (*a*) *List of the variables which are involved in the estimation of load.*

(*b*) *Enumerate and explain the components of internal heat gains.*

Solution: (*a*) Refer to Art. 18.1.

(*b*) Refer to Art. 18.8.

GATE and UPSC Examinations' Questions (Latest-Selected) with Answers/Solutions

Chapter 25

A. CONVENTIONAL QUESTIONS WITH ANSWERS/SOLUTIONS

(**Note:** 1 ton = 210 kJ/min 12600 kJ/h, 1 tonne = 14000 kJ/h i.e. 1 ton = 0.9 tonne)

Q. 25.1. *Calculate all the psychrometric properties of air at 1 bar and 25°C dbt and 15°C wbt. The following properties of water may be assumed:*

Temp. (°C)	*Sat. pressure (bar)*	*Sp. volume of vapour* (m^3/kg)	*Enthalpy (kJ/kg)*	
			Sat. liq.	*Sat. vap.*
25	0.03166	43.40	–	2547.3
15	0.017307	77.98	62.94	–
10	0.01078	–	–	–
0	0.01002	–	–	–

The following expression may be used, if necessary:

$$p_v = (p_{vsat})_{t_{wb}} - \frac{(p - p_{vsat})(dbt - wbt) \times 1.8}{2854 - 1.325\,(1.8 t_{db} + 32)} \qquad \textbf{(IES)}$$

Solution: *Given:* p = 1 bar, t_{db} = 25°C, t_{wb} = 15°C

$$p_v = (p_{vsat})_{t_{wb}} - \frac{(p - p_{vsat})\,(t_{db} - t_{wb}) \times 1.8}{2854 - 1.325\,(1.8 \times t_{db} + 32)} \qquad \text{...(Given)}$$

$$= 0.017307 - \frac{(1 - 0.03166)\,(25 - 15) \times 1.8}{2854 - 1.325\,(1.8 \times 25 + 32)}$$

$$= 0.017307 - \frac{17.43}{2751.97} = 0.01097 \text{ bar}$$

Psychometric properties of air:

(*i*) **Specific humidity, W:**

$$W = 0.622 \times \left(\frac{p_v}{p - p_v}\right) = 0.622 \times \left(\frac{0.01097}{1 - 0.01097}\right)$$

= **0.006899 kJ/kg of dry air (Ans.)**

(*ii*) **Relative humidity, ϕ:**

$$\phi = \frac{p_v}{p_{vsat}} = \frac{0.01097}{0.03166} = 0.3465 \text{ or } \mathbf{34.65\%\ (Ans.)}$$

(*iii*) **Degree of saturation, μ:**

$$\mu = \frac{p_v}{p_{vsat}}\left(\frac{p - p_{vsat}}{p - p_v}\right) \qquad \text{...[Eqn. (12.7)]}$$

$$= \frac{0.01097}{0.03166}\left(\frac{1 - 0.03166}{1 - 0.01097}\right) = 0.3392 \quad \textbf{(Ans.)}$$

(*iv*) **Enthalpy of moist air, *h*:**

$$h = 1.005\ t_{db} + W(2500 + 1.88 t_{db}) \qquad \text{...[Eqn. 12.12(b)]}$$

$$= (1.005 \times 25) + 0.006899\ (2500 + 1.88 \times 25)$$

$$= \mathbf{42.697\ kJ/kg\ of\ dry\ air\ (Ans.)}$$

Q. 25.2. *An R-12 vapour compression plant producing 10 tons of refrigeration operates with condensing and evaporating temperatures of 35°C and –10°C respectively. A suction line heat exchanger is used to subcool the saturated liquid leaving the condenser. Saturated vapour leaving the evaporator is superheated in the suction line heat exchanger to the extent that a discharge temperature of 60°C is obtained after isentropic compression. Calculate:*

(*i*) *The subcooling achieved in the heat exchanger.*

(*ii*) *The refrigerant flow rate in kg/s.*

(*iii*) *The cylinder dimensions of the two-cylinder compressor, if the speed is 900 rpm, stroke-to-bore ratio is 1 : 1 and the volumetric efficiency is 80%.*

(*iv*) *The COP of the plant, and*

(*v*) *The power required to drive the compressor in kW.*

Draw the cycle on p-h diagram.

You may use the following table:

Temperature (°C)	*Specific volume (m^3/kg) vapour (v_g)*	*Specific enthalpy kJ/kg*		*Specific entropy kJ/kgK*	
		Liquid (h_f)	*Vapour (h_g)*	*Liquid (s_f)*	*Vapour (s_g)*
–10	*0.0766*	*190.88*	*347.13*	*0.9660*	*1.5600*
+35	*0.0206*	*233.50*	*365.92*	*1.1139*	*1.5419*

Average specific heat of desuperheating vapour at condenser = 0.796 kJ/kgK.

Average specific heat of vapour between evaporator outlet and compression suction = 0.658 kJ/kgK. **(IES)**

Solution: Refer to Fig. 25.1.

Given: $TR = 10$ tons; $T_1\ (= T_{2'}) = -10 + 273 = 263$ K; $T_{3'} = 35 + 273 = 308$ K; $T_3 = 60 + 273 = 333$ K;

$\eta_{vol.} = 80\%$; $N = 900$ rpm, $\frac{L}{D} = 1$

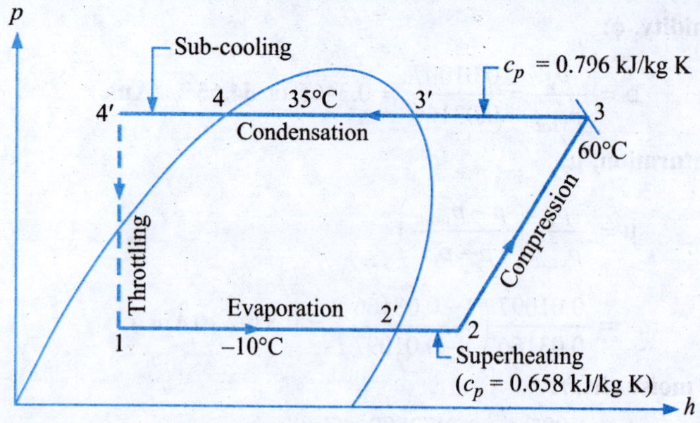

Fig. 25.1

From table, we have: $v_{2'} = 0.0766$ m³/kg; $h_{2'} = 347.13$ kJ/kg; $s_{2'} = 1.56$ kJ/kgK; $h_4 = 233.5$ kJ/kg; $s_4 = 1.1139$ kJ/kgK; $v_{3'} = 0.0206$ m³/kg; $h_{3'} = 365.92$ kJ/kg; $s_{3'} = 1.5419$ kJ/kgK

Now, $s_3 = s_{3'} + c_p \ln\left(\frac{T_3}{T_{3'}}\right) = 1.5419 + 0.796 \ln\left(\frac{333}{308}\right) = 1.604$ kJ/kgK

For isentropic compression, $s_2 = s_3$

or, $$1.604 = s_{2'} + c_p \ln\left(\frac{T_2}{T_{2'}}\right) = 1.56 + 0.658 \ln\left(\frac{T_2}{263}\right)$$

or, $$\ln\left(\frac{T_2}{263}\right) = (1.604 - 1.56)/0.658 = 0.06687 \text{ or } T_2 = 281 \text{ K}$$

Now, $$h_2 = h_{2'} + c_p \Delta t = 347.13 + 0.658\ (281 - 263) = 358.9 \text{ kJ/kg}$$

and, $$h_3 = h_{3'} + c_p (T_3 - T_{3'}) = 365.92 + 0.796\ (333 - 308) = 385.82 \text{ kJ/kg}$$

(*i*) **Subcooling achieved:**

In regenerative heat exchanger applying energy balance, we get:

$$h_4 - h_{4'} = h_2 - h_{2'}$$

or, $$h_{4'} = h_4 - (h_2 - h_{2'}) = 233.5 - (358.9 - 347.13) = 221.7 \text{ kJ/kg}$$

Subcooling achieved $= h_4 - h_{4'} = 233.5 - 221.7 =$ **11.8 kJ/kg (Ans.)**

(*ii*) **Refrigerant flow rate, $\dot{m}$:**

Refrigerating effect, RE = 10TR = 10 × 3.5 = 35 kW (∵ 1 ton = 3.5 kJ/s (kW))

or, $\dot{m}\ (h_{2'} - h_1) = 35$

or, $\dot{m}\ (h_{2'} - h_{4'}) = 35$

or, $$\dot{m} = \frac{35}{h_{2'} - h_{4'}} = \frac{35}{(347.13 - 221.7)} = \mathbf{0.28\ kg}\ \ \textbf{(Ans.)}$$

(*iii*) **Cylinder dimensions, *L*, *D*:**

Now, $$\frac{v_{2'}}{T_{2'}} = \frac{v_2}{T_2} \text{ or } v_2 = \frac{v_{2'} T_2}{T_{2'}} = \frac{0.0766 \times 281}{263} = 0.0818 \text{ m}^3/\text{kg}$$

∴ Actual refrigerant intake/stroke $= \dot{m} \times v_2 = 0.28 \times 0.0818 = 0.0229 = 9$ m³/s

Also, Volumetric efficiency, $\eta_{vol.} = \dfrac{\text{Actual refrigerant intake/stroke}}{\text{Swept volume}}$

or, $$\frac{80}{100} = \frac{0.0229}{\left(\frac{\pi}{4} D^2 \times L\right) \times \left(\frac{N}{60}\right)} = \frac{0.0229 \times 60}{2 \times \frac{\pi}{4} \times D^3 \times 900} \quad (\because L = D)$$

or, $$D^3 = \frac{0.0229 \times 60}{\left(\frac{80}{100}\right) \times \frac{\pi}{4} \times 900} = 0.002429 \text{ m}^3$$

or, $D = \mathbf{0.1344}$ **m, or D = L = 13.44 cm (Ans.)**

(*iv*) **COP:** $$\text{COP} = \frac{h_{2'} - h_1}{h_3 - h_2} = \frac{347.13 - 221.7}{385.82 - 358.9} = \mathbf{4.66 \text{ (Ans.)}} \quad (\because h_1 = h_{4'})$$

(*v*) **Power required, P:**

Power required to drive the compressor,

$$P = \dot{m}\ (h_3 - h_2) = 0.28\ (385.82 - 358.9) = \mathbf{7.54 \text{ kW (Ans.)}}$$

Q. 25.3. *A dense air refrigeration machine operating on Bell-Coleman cycle operates between 3.4 bar and 17 bar. The temperature of air after the cooler is 15°C and after the refrigerator is 6°C. If the refrigeration capacity is 6 tonnes, calculate:*

(*i*) *Temperature after compression and expansion.*

(*ii*) *Air circulation per min.*

(*iii*) *Work of compressor and expander.*

(*iv*) *Theoretical COP.*

(*v*) *Rate of water circulation required in the cooler in kg/min if the rise in temperature is limited to 30°C.* **(GATE)**

Solution: Refer to Fig. 25.2.

Given: $p_1\ (= p_4) = 17$ bar; $p_2\ (= p_3) = 3.4$ bar; $T_1 = 15 + 273 = 288$ K; $T_3 = 6 + 273 = 279$ K; Refrigeration capacity = 6 tonnes.

(*i*) **Temperatures, T_4, T_2:**

Now, $$\frac{T_4}{T_3} = \left(\frac{p_4}{p_3}\right)^{\frac{\gamma-1}{\gamma}} = \left(\frac{17}{3.4}\right)^{\frac{1.41-1}{1.41}} = 1.597$$

or, $$\frac{T_4}{279} = 1.584 \text{ or } T_4 = 279 \times 1.597 = 445.6 \text{ K} \quad \textbf{(Ans.)}$$

Again, $$\frac{T_2}{T_1} = \left(\frac{p_2}{p_1}\right)^{\frac{\gamma-1}{\gamma}} = \left(\frac{3.4}{17}\right)^{\frac{1.41-1}{1.41}} = 0.626$$

or, $$\frac{T_2}{288} = 0.626 \text{ or } \boldsymbol{T_2} = 288 \times 0.626 = \mathbf{180.3 \text{ K}} \quad \textbf{(Ans.)}$$

(*ii*) **Air circulation per min., m_{air}:**

Heat extracted (or refrigerating effect per kg of air circulated) in cold chamber

$$Q = c_p\ (T_3 - T_2) = 1.005 \times (279 - 180.3) = 99.19 \text{ kJ/kg}$$

$$\therefore m_{air} = \frac{\text{Refrigeration capacity}}{\text{Heat extracted in chamber}}$$

$$= \frac{6 \times (14000/60)}{99.19} = \mathbf{14.11 \text{ kg/min Ans.}} \quad [\because 1 \text{ tonne} = 14000 \text{ kJ/h and } 1 \text{ ton} = 12600 \text{ kJ/h}]$$

(*iii*) **Work of compressor and expander W_{comp}/W_{exp}:**

$$W_{comp.} = m_{air} \times \frac{\gamma}{\gamma - 1} \times \frac{R}{J}(T_4 - T_3)$$

$$= \frac{14.11}{60} \times \frac{1.41}{1.41 - 1} \times \frac{0.287}{1} (445.6 - 279)$$

$$= \mathbf{38.67\ kW\ (Ans.)}$$

$$W_{exp.} = m_{air} \times \frac{\gamma}{\gamma - 1} \times \frac{R}{J}(T_1 - T_2)$$

$$= \frac{14.11}{60} \times \frac{1.41}{1.41 - 1} \times \frac{0.287}{1} (288 - 180.3)$$

$$\approx \mathbf{25\ kW\ (Ans.)}$$

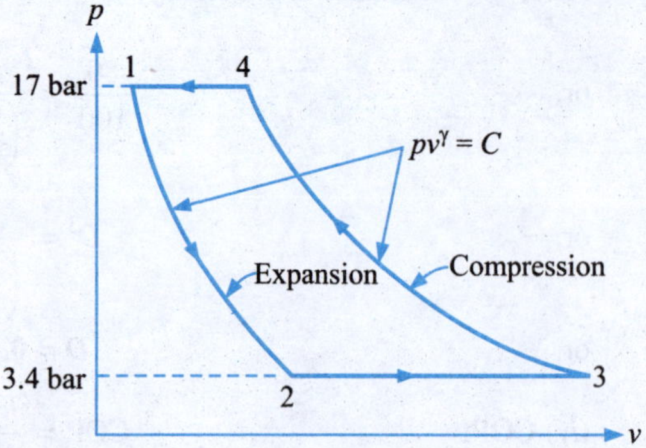

Fig. 25.2

(*iv*) **Theoretical COP:**

$$\text{COP} = \frac{\text{Heat extracted}}{\text{Net work done}}$$

Net work done = Heat received – Heat extracted

$= c_p (T_4 - T_1) - 99.19 = 1.005 (445.6 - 288) - 99.19 = 59.2$ kJ/kg

$$\therefore \qquad \text{COP} = \frac{99.19}{59.2} = \mathbf{1.67\ (Ans.)}$$

(v) **Rate of water circulation required, m_w:**

We know, $\quad m_w\, c_{pw}\, (\Delta T_w) = m_{air} \times (c_{pa} \times \Delta T_a)$

or $\quad m_w \times 4.187 \times 30 = 14.11 \times 1.005\ (445.6 - 288) = 2234.85$

or $$m_w = \frac{2234.85}{4.187 \times 30} = \mathbf{17.79\ kg/min\ (Ans.)}$$

Q. 25.4. *A compressor having a stroke volume of 500 c.c. runs at 500 r.p.m. and works with CO_2 refrigerant. The evaporator and condenser temperatures are –15°C and 25°C, respectively. The condenser liquid is subcooled to 15°C before passing on to the expansion valve. Assume wet compression with an initial quality x of CO_2 as 0.9. Assuming isentropic compression and volumetric efficiency of the compressor as 0.85, calculate the following: (i) The COP, (ii) The power required by the compressor (kW), and (iii) The refrigerating capacity of the compressor in tonnes of refrigeration.*

The following data for CO_2 may be used:

t°C	*p, bar*	*v, m³/kg*		*h, kJ/kg*		*s, kJ/kg K*	
		Liquid	*Vapour*	*Liquid*	*Vapour*	*Liquid*	*Vapour*
– 15	22.88	0.00101	0.0166	49.62	322.86	0.1976	1.2567
15	50.92	0.00130	0.0063	127.75	308.08	0.4697	1.0959
25	64.32	0.00147	0.0042	164.17	283.63	0.5903	0.9912

Take specific heat c_p of CO_2 gas = 2.4 kJ/kg K. **(IES)**

Solution: Refer to Fig. 25.3.

Given: $V_s = 500\ cm^3 = 500 \times 10^{-6}\ m^3$;

$N = 500$ r.p.m.; $x_2 = 0.9$; $\eta_{vol.} = 0.85$;

$c_{pg} = 2.4$ kJ/kg K.

Assuming single-acting, the compressor displacement $= 500 \times 10^{-6} \times 500 = 0.25\ m^3/min$.

Specific volume at compressor inlet,

$$v_2 = x_2 v_{g2} + (1 - x_2)\, v_{f2}$$
$$= 0.9 \times 0.0166 + (1 - 0.9) \times 0.00101$$
$$= 0.01504\ m^3/kg$$

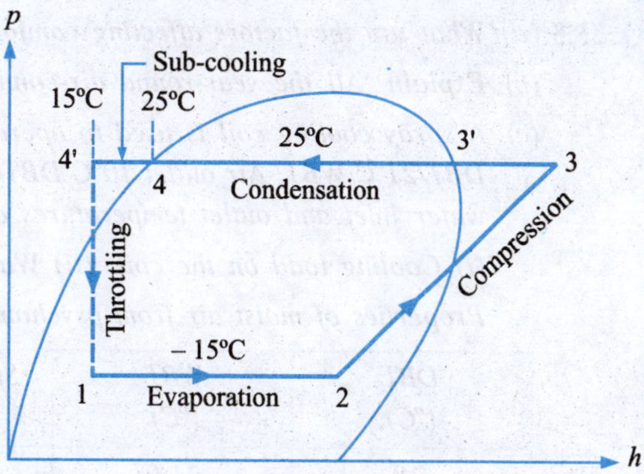

Fig. 25.3

Mass flow rate of refrigerant,

$$m = \frac{0.25 \times 0.85}{0.01504} = 14.129 \text{ kg/min}$$

For isentropic compression process 2-3, we have:

$$s_2 = s_3$$

$$s_{f2} + x_2\,(s_{g2} - s_{f2}) = s_{g3} + c_{pg} \ln\left(\frac{t_3 + 273}{25 + 273}\right)$$

$$0.1976 + 0.9\,(1.2567 - 0.1976) = 0.9912 + 2.14 \ln\left(\frac{t_3 + 273}{298}\right)$$

or, $$t_3 = 45.5°C$$

$$\therefore \quad h_3 = h_{g3'} + c_{pg}\,(45.5 - 25) = 283.63 + 2.4 \times 20.5 = 332.83 \text{ kJ/kg}$$

$$h_2 = h_{f2} + x_2\,(h_{g2} - h_{f2}) = 49.62 + 0.9\,(322.86 - 49.62) = 295.5 \text{ kJ/kg}$$

$$h_{f4'} = h_1 = 127.75 \text{ kJ/kg}$$

$$\text{Ref. effect/kg} = h_2 - h_1 = 295.5 - 127.75 = 167.75 \text{ kJ/kg}$$

$$\text{Work/kg} = h_3 - h_2 = 332.83 - 295.5 = 37.33 \text{ kJ/kg}$$

(*i*) **C.O.P.:** $$\text{C.O.P.} = \frac{h_2 - h_1}{h_3 - h_2} = \frac{167.75}{37.33} = \mathbf{4.49} \quad \textbf{(Ans.)}$$

(*ii*) **The power required by the compressor, P:**

$$P = \frac{14.129}{60}\,(h_3 - h_2) = \frac{14.129}{60} \times 37.33 = \mathbf{8.79\ kW} \quad \textbf{(Ans.)}$$

(*iii*) **Refrigerating capacity:**

$$\text{Refrigerating capacity} = \frac{m(h_2 - h_1) \times 60}{14000} = \frac{14.129 \times 167.75 \times 60}{14000}$$

$$= \mathbf{10.16\ tonnes} \quad \textbf{(Ans.)}$$

Q. 25.5. *(a) What are the factors affecting comfort air conditioning?*

(b) Explain 'All the year-round air-conditioner'.

(c) A spray cooling coil is used to operate under the following conditions – Air inlet 28°C DBT/21°C WBT, Air outlet 10°C DBT/6°C WBT. Total air flow rate 2000 m³/min. Chilled water inlet and outlet temperatures are 7°C and 12°C. Find:

(i) Cooling load on the coil; (ii) Water flow rate through the coil.

Properties of moist air from psychometric chart are as follows:

DBT (°C)	*WBT (°C)*	*Sp. humidity (gm/kg)*	*Sp. vol. (m³/kg)*	*Enthalpy (kJ/kg)*
28	*21.0*	*12.95*	*0.87*	*61.0*
10	*6.0*	*4.2*	*0.81*	*21.0*

(IES)

Solution: (*a*) Factors affecting comfort air-conditioning are:

(*i*) Climate and seasonal difference; (*ii*) Duration of occupancy;

(*iii*) Clothing; (*iv*) Age and sex;

(*v*) Activity; (*vi*) Radiant heat;

(*vii*) Shock effect.

(*b*) Refer to Article **21.8**.

(*c*) *Given:* Air inlet: 28°C DBT/21°C WBT; Air outlet: 10°C DBT/6°C WBT; Total air flow rate m_a = 2000 m³/min; Inlet and outlet temperatures of chilled water; t_i = 7°, t_o = 12°C.

(*i*) **Cooling load on coil:**

$$\text{Cooling load on coil} = m_a (h_1 - h_2) = \left(\frac{2000}{60} \times \frac{1}{0.87}\right) \times (61 - 21) = \mathbf{1532.57\ kW\ (Ans.)}$$

(*ii*) **Water flow rate, m_w:**

$$m_w\, c_p\, (t_o - t_i) = 1532.57$$

$$m_w \times 4.187\, (12 - 7) = 1532.57$$

or,
$$m_w = \frac{1532.57}{4.187(12-7)} = \mathbf{73.2\ kg/s\ (Ans.)}$$

Q. 25.6. *A Bell-Coleman refrigeration system is used to produce 10 tons of refrigeration. The cooler and refrigerator pressures are 4.2 bars and 1.4 bar. Air is cooled in the cooler to 45°C and temperature of air at the inlet of the compressor is –20°C. For an ideal cycle, calculate COP, mass of air circulated/min, theoretical piston displacement of compressor and power required per ton of refrigeration. Assume c_p for air as 1.005 kJ/kg-K. Find the cylinder dimensions if the compressor is single-acting single-cylinder with L/D ratio of 1.2 and runs at 600 rpm.*

(IES)

Solution: Refer to Fig. 25.4.

Given. TR = 10 tons; $p_1 = p_4$ = 4.2 bars, $p_2 = p_3$ = 1.4 bar; T_1 = 45 + 273 = 318 K; T_3 = –20 + 273 = 253 K; c_p = 1.005 kJ/kg K; *L*/*D* = 1.2; *N* = 600 rpm.

COP (ideal cycle):

$$\text{COP} = \frac{1}{(r_p)^{\gamma-1/\gamma} - 1} \qquad \text{...[Eqn. (2.6)]}$$

$$= \frac{1}{\left(\frac{4.2}{1.4}\right)^{\frac{1.41-1}{1.41}} - 1} = \mathbf{2.66\ (Ans.)}$$

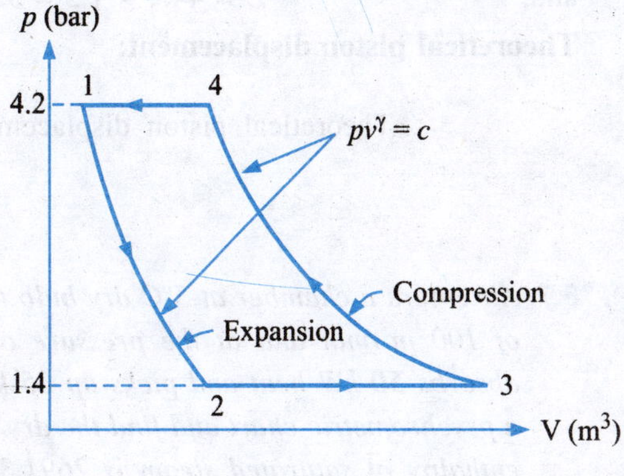

Fig. 25.4

Mass of air circulated/min, m_a:

$$\frac{T_4}{T_3} = \left(\frac{p_4}{p_3}\right)^{\frac{\gamma-1}{\gamma}}$$

or, $$\frac{T_4}{253} = \left(\frac{4.2}{1.4}\right)^{\frac{1.41-1}{1.41}}$$

or, $$T_4 = 253 \times (3)^{\frac{0.41}{1.41}} = 348.2 \text{ K}$$

and, $$\frac{T_1}{T_2} = \left(\frac{p_1}{p_2}\right)^{\frac{\gamma-1}{\gamma}} \text{ or } \frac{318}{T_2} = \left(\frac{4.2}{1.4}\right)^{\frac{1.41-1}{1.41}} = 1.376$$

or, $$T_2 = \frac{318}{1.376} = 231.1 \text{ K}$$

Now, refrigeration capacity = 10 × 3.5 = 35 kW = $m_a\, c_p\, (T_3 - T_2)$

$$\therefore \quad m_a = \frac{35}{c_p (T_3 - T_2)} = \frac{35}{1.005 \times (253 - 231.1)} \times 60 = \mathbf{95.4\ kg/min\ (Ans.)}$$

Power required per ton:

$$\text{Power required} = m_a\, c_p\, [(T_4 - T_3) - (T_1 - T_2)]$$

$$= \frac{95.4}{60} \times 1.005\, [(348.2 - 253) - (318 - 231.1)] = 13.26 \text{ kW}$$

$$\therefore \text{Power required per ton} = \frac{13.26}{10} = \mathbf{1.326\ kW/ton\ (Ans.)}$$

Cylinder dimensions, *L*, *D*:

Now, $$\rho_3 = \frac{p_3}{RT_3} = \frac{1.4 \times 10^5}{287 \times 253} = 1.928 \text{ kg/m}^3$$

Assuming volumetric efficiency 100%, we get:

$$\frac{\pi}{4} D^2 L \times N = \frac{\dot{m}_a}{\rho_3}$$

or, $$\frac{\pi}{4} D^2 \times 1.2D \times \left(\frac{600}{60}\right) = \frac{(95.4/60)}{1.928} = 0.8247$$

or $$D^3 = \frac{0.8247 \times 4 \times 60}{\pi \times 1.2 \times 600} \quad \text{or} \quad D = 0.444 \text{ m or } 44.4 \text{ cm} \textbf{ (Ans.)}$$

and, $$L = 44.4 \times 1.2 = \mathbf{53.28 \text{ cm (Ans.)}}$$

Theoretical piston displacement:

$$\text{Theoretical piston displacement} = \frac{\pi}{4} D^2 L$$

$$= \frac{\pi}{4} \times (44.4)^2 \times 53.28 = \mathbf{82493.5 \text{ cm}^3 \text{ (Ans.)}}$$

Q. 25.7. *Air enters a chamber at 5°C dry bulb temperature and 2.5°C wet bulb temperature at the rate of 100 m³/min and at the pressure of 1 bar. While passing through the chamber, the air absorbs 50 kW heat and picks up 35 kg/h of saturated steam at 110°C. Show the process on a psychrometric chart and find the dry and wet bulb temperatures of the leaving air. At 110°C, enthalpy of saturated steam is 2691.3 kJ/kg.* **(GATE)**

Solution: Refer to Fig. 25.5.

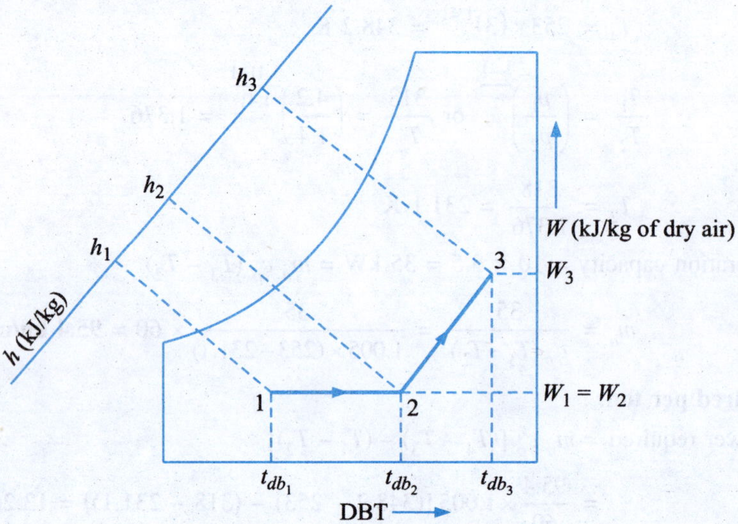

Fig. 25.5

t_{db_3}, t_{wb_3} :

Process 1-2: It is a '*sensible heating*' process in which heat absorbed, $Q = 50$ kW

i.e. $$m_a (h_2 - h_1) = 50 \quad ...(i)$$

$$W_1 = 0.0035 \text{ (from psychrometric chart)}$$

Specific volume, $$v_{s1} = 0.792 \text{ m}^3/\text{kg}$$

$$h_1 = 14 \text{ kJ/kg}$$

Volume of flow rate = 100 m³/min. = 1.67 m³/s

$\therefore$ Mass of air flow rate, $$m_a = \frac{1.67}{v_{s1}} = \frac{1.67}{0.792} = 2.11 \text{ kg/s}$$

Inserting the values in Eqn. (*i*), we get:

$$2.11\ (h_2 - 14) = 50$$

or, $$h_2 = \frac{50}{2.11} + 14 = 37.7 \text{ kJ/kg}$$

Again, at *state* 2 : W_2 (= W_1) = 0.0035 kg/kg of dry air; h_2 = 37.76 kJ/kg

After '*humidification*' by steam injection,

$$W_3 = W_2 + \frac{m_s}{m_a} = 0.0035 + \frac{(35/3600)}{2.11} = 0.0081 \text{ kg/kg of dry air}$$

Now, Enthalpy of leaving air, $h_3 = h_2 + \frac{m_s}{m_a}(h_g)$

$$= 37.76 + \frac{(35/3600)}{2.11} \times 2691.3 = 50.16 \text{ kJ/kg}$$

At state 3: From psychrometric chart, corresponding to W_3 = 0.0081 kg/kg of dry air, h_3 = 50.16 kJ/kg, we have:

$$t_{db_3} = 31°\text{C};\ t_{w_3} = 18.3°\text{C (Ans.)}$$

Q. 25.8. *A simple air-refrigeration system is employed in an aeroplane. The main compressor delivers the air at 5.2 bar and 210°C. The air taken from compressor is passed through a heat exchanger, cooled with the help of ram air so that the temperature of air leaving the heat exchanger is 50°C and pressure is 4.7 bar. The cooling turbine drives the exhaust fan which is used to force the ram air through the heat exchanger. The air leaving the heat exchanger passes through the cooling turbine and then supplied to cabin at 1 bar. The pressure loss between the cooling turbine and cabin is 0.2 bar. If the rate of flow of air through the cooling turbine is 18 kg/min, determine the following:*

(*i*) *The temperature of the air leaving the expander,*

(*ii*) *The power delivered to the ram air which is passed through the heat exchanger, and*

(*iii*) *The refrigeration load in tonnes when the temperature of the air leaving the cabin is limited to 24°C.*

Assume that the isentropic efficiency of the cooling turbine is 76 per cent and there is no loss of heat from air between the cooling turbine and cabin. Take c_p = 1 kJ/kg K; γ = 1.4. **(IES)**

Solution: *Given:* p_3 = 5.2 bar;

T_3 = 210 + 273 = 483 K;

T_4 = 50 + 273 = 323 K;

p_4 = 4.7 bar;

T_6 = 24 + 273 = 297 K;

p_6 = 1 bar;

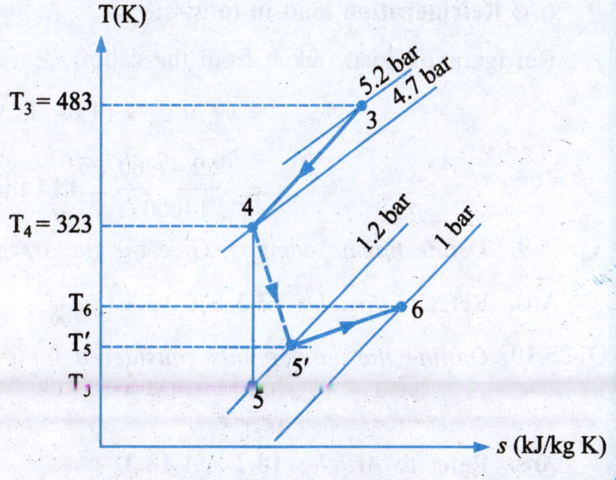

Fig. 25.6. T-*s* diagram

p_5 = 1.2 bar ; m_a = 18 kg/min;

η_T = 76%;

c_p = 1 kJ/kg K; γ = 1.4.

The various processes on T-*s* diagram are shown in Fig. 25.6:

- The point 3 represents the air delivered from the compressor to heat exchanger;
- The point 4 shows the condition of air leaving the heat exchanger;
- The vertical line 4-5 represents the isentropic expansion of air in the cooling turbine;
- The curve 4-5′ shows the actual expansion of air in the cooling turbine due to internal friction;
- The line 5′-6 represents the refrigeration process.

(*i*) The temperature of air leaving the expander, $T_{5'}$:

For isentropic expansion 4-5, we have:

$$\frac{T_5}{T_4} = \left(\frac{p_5}{p_4}\right)^{\frac{\gamma-1}{\gamma}} = \left(\frac{1.2}{4.7}\right)^{\frac{1.4.1}{1.4}} = 0.677$$

$$\therefore \quad T_5 = T_4 \times 0.677 = 323 \times 0.677 = 218.7 \text{ K}$$

$$\text{Now,} \quad \eta_T = 0.76 = \frac{T_4 - T_{5'}}{T_4 - T_5} = \frac{323 - T_{5'}}{323 - 218.7}$$

$$\therefore \quad T_{5'} = 323 - 0.76\,(323 - 218.7) = \mathbf{243.7\ K} \quad \textbf{(Ans.)}$$

(*ii*) Power delivered to the ram air which is passed through the heat exchanger:

Power delivered to the ram air which is passed through the heat exchanger

$$= m_a \times c_p \times (T_4 - T_{5'}) = 18 \times 1 \times (323 - 243.7) = 1427.4 \text{ kJ/min}$$

$$= \frac{1427.4}{60} = \mathbf{23.79\ kW} \quad \textbf{(Ans.)}$$

(*iii*) Refrigeration load in tonnes:

Refrigeration load (taken from the cabin)

$$= m_a \times c_p \times (T_6 - T_{5'}) = 18 \times 1 \times (297 - 243.7) = 959.4 \text{ kJ/min}$$

$$= \frac{959.4 \times 60}{14000} = \mathbf{4.11\ tonnes} \quad \textbf{(Ans.)} \qquad (\because 1 \text{ tonne of ref.} = 14000 \text{ kJ/h})$$

Q. 25.9. *Define human comfort. Give the factors governing optimum effective temperature.*

Ans. Refer to *Articles* **13.3** and **13.4**. **(GATE)**

Q. 25.10. *Outline the main points considered for cooling load estimate and heating load estimate.* **(IES)**

Ans. Refer to *Articles* **18.2** and **18.3**.

Q. 25.11. *Discuss in detail the factors affecting the performance of vapour compression refrigeration.* **(IES)**

Ans. Refer to *Article* **3.7.**

Q. 25.12. *What is the purpose of undercooling in vapour compression refrigeration system?* **(GATE)**

Ans. Refer to *Article* **3.7** (4).

Q. 25.13. *(i) List the vapour refrigeration controls.*

(ii) What is metabolism? **(IES)**

Ans. (*i*) Refer to *Article* **10.1.**

(*ii*) Refer to *Article* **13.1.**

Q. 25.14. *With the help of a heat sketch, discuss the working principle of a practical vapour absorption system.* **(IES)**

Ans. Refer to *Article* **5.3.**

Q. 25.15. *Briefly describe the various methods of air-conditioning duct design.* **(GATE)**

Ans. Refer to *Article* **16.17.**

Q. 25.16. *Discuss the measurement of wet bulb temperature by means of wick-covered bulb of a thermometer.* **(IES)**

Ans. Refer to *Article* **12.5.**

B. MULTIPLE-CHOICE QUESTIONS WITH ANSWERS AND "EXPLANATIONS"

(1 ton = 0.9 tonne) [1 ton = 12600 kJ/h; 1 tonne = 14000 kJ/h]

Q. 1. A heat pump for domestic heating operates between a cold system at 0ºC and the hot system at 60ºC. What is the minimum electric power consumption if the heat rejected is 80000 kJ/h?

(*a*) 2 kW (*b*) 3 kW (*c*) 4 kW (*d*) 5 kW.

Ans. (c)

Explanation: Refer to Fig. 1.

$$(\text{COP})_{\text{heat pump}} = \frac{T_2}{T_1 - T_2} = \frac{273}{333-273} = 4.55$$

Also, $\text{COP} = \dfrac{Q_2}{W}$ or $Q_2 = 4.55\ W$

Now, $Q_1 = Q_2 + W = 4.55\ W + W = 5.55\ W$

or, $\dfrac{80000}{3600} = 5.55\ W$

or, $W = \dfrac{80000}{3600 \times 5.55} = 4\ \text{kW}$ **(Ans.)**

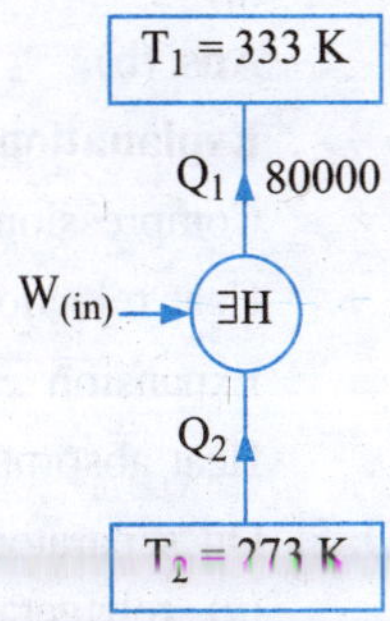

Fig. 1

Q. 2. A Carnot refrigerator requires 1.5 kW/ton of refrigeration to maintain a region at a temperature of –30ºC. The COP of the Carnot refrigerator is

(*a*) 1.42 (*b*) 2.33 (*c*) 2.87 (*d*) 3.26.

Ans. (b)

Explanation:

$$1 \text{ TR} = 210 \text{ kJ/min} = \frac{210}{60} = 3.5 \text{ kW}$$

$$\therefore \quad (\text{COP})_{\text{carnot}} = \frac{R_n}{W} = \frac{3.5}{1.5} = 2.33 \textbf{ (Ans.)}$$

Q. 3. A refrigerator working on a reversed Carnot cycle has a COP of 4. If it works as a heat pump and consumes 1 kW, the heating effect will be

(*a*) 1 kW (*b*) 4 kW (*c*) 5 kW (*d*) 6 kW.

Ans. (c)

Explanation:

$$(\text{COP})_{\text{heat pump}} = (\text{COP})_{\text{ref.}} + 1 = 4 + 1 = 5$$

$$\text{or,} \quad 5 = \frac{\text{Heating effect}}{W_{\text{in}}}$$

$$\text{or,} \quad \text{Heating effect} = 5 \times W_{\text{in}} = 5 \times 1 = 5 \text{ kW} \textbf{ (Ans.)}$$

Q. 4. Match List-I (Process) with List-II (Type) for Bell Coleman or Joule or Reverse Brayton cycle for gas cycle refrigeration and select the correct answer using the codes given below the lists:

List-I	List-II
A. Compression	1. Isobaric
B. Heat rejection	2. Isothermal
C. Expansion	3. Isentropic
D. Heat absorption	4. Isobaric

Codes:

	A	B	C	D
(*a*)	3	1	4	2
(*b*)	3	1	3	1
(*c*)	3	2	3	2
(*d*)	3	1	2	2.

Ans. (b)

Explanation: Refer to Fig. 2.

Compression ... Isentropic (3-4)

Heat rejection ... Isobaric (4-1)

Expansion ... Isentropic (1-2)

Heat absorption ... Isobaric (2-3) **(Ans.)**

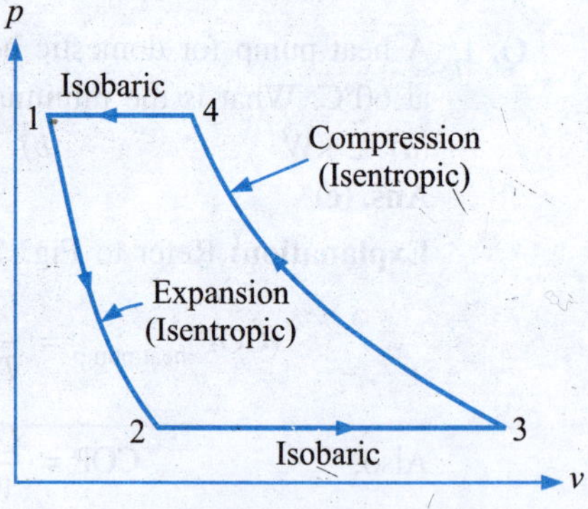

Fig. 2 Bell Coleman Cycle

Q. 5. Oil separator is not required in refrigeration system if

(*a*) refrigerant and oil are immiscible at all the pressures and temperatures.

(*b*) refrigerant and oil are immiscible at condensation pressure and temperature.

(*c*) refrigerant and oil are miscible at all pressures and temperatures.
(*d*) refrigerant and oil are miscible at condensation pressures and temperatures.

Ans. (b)

Explanation: In refrigeration system, *oil separator* is *not* required when refrigerant and oil are *miscible* at all pressures and temperatures because they do not get separated in evaporator fouling does not take place.

Q. 6. Consider the following statements in respect of absorption refrigeration and vapour compression refrigeration systems:

1. The former runs on low grade energy.
2. The pumping work in the former is negligible since specific volume of strong liquid solution is small.
3. The latter uses an absorber while former uses a generator.
4. The liquid pump alone replaces compressor of the latter.

Which of these statements are correct?

(*a*) 1 and 2 (*b*) 1 and 3 (*c*) 1 and 4 (*d*) 2 and 4.

Ans. (a)

Explanation: In absorption refrigeration system, both absorber and generator are used. Compressor is replaced by an absorber, a pump, generator and a pressure reducing value. **(Ans.)**

Q. 7. Moist air enters the cooling coil with mass flow rate of 10 kg-dry air/s at dry bulb temperature of 30ºC and humidity ratio of 0.017 kg vapour/kg dry air. It leaves the cooling coil at dry bulb temperature of 16ºC and humidity ratio of 0.008 kg vapour/kg dry air. If specific heat of humid air is 1.02 kJ/kg da K and latent heat of water vapour is 25000 kJ/kg vapour. The sensible and latent heat transfer of cooling coil are, respectively

(*a*) 140 kW and 25000 kW (*b*) 142.8 kW and 2.25 kW
(*c*) 142.8 kW and 225 kW (*d*) 225 kW and 142.8 kW.

Ans. (c)

Explanation:

Sensible heat, $Q_s = m_a\, c_{pa}\, \Delta t$

or, $Q_s = 10 \times 1.02 \times (30 - 16) = 142.8$ kW **(Ans.)**

Latent heat, $Q_L = m_a\, h_{fg}\, \Delta W$

$= 10 \times 2500\,(0.017 - 0.008)$

$= 225$ kW **(Ans.)**

Q. 8. If the volume of moist air with 50% relative humidity is isothermally reduced to half its original volume, then relative humidity of moist air becomes

(*a*) 25% (*b*) 60% (*c*) 75% (*d*) 100%.

Ans. (d)

Explanation:

Relative humidity, $\phi = \dfrac{p_v}{p_{vs}}$, Also, $pV = RT$

$$\therefore \quad \frac{\phi_1}{\phi_2} = \frac{RT/V_1}{RT/(V_1/2)} = \frac{RT/V_1}{2RT/V_1} \quad \therefore \frac{1}{2} \text{ or, } \phi_2 = 2\phi_1 = 2 \times 50 = 100\% \text{ (Ans.)}$$

Q. 9. An air-conditioned room has length, width and height of 20 m, 30 m and 4 m respectively. The infiltration is assumed to be one air change. The outdoor and indoor dry bulb temperatures are 40ºC and 25ºC respectively. The sensible heat load due to infiltration is

(*a*) 734 kW (*b*) 12.24 kW (*c*) 0.204 kW (*d*) 10 kW.

Ans. (b)

Explanation: The sensible heat load due to infiltration, SHL

$$= 0.0204 \times V \times \Delta t$$

$$= 0.0204 \times \left[1 \times \frac{20 \times 30 \times 4}{60}\right] \times (40 - 25)$$

$$= 12.24 \text{ kW } \textbf{(Ans.)}$$

Q. 10. Which of the following properties increase(s) during sensible heating of air-water vapour mixture?

1. Relative humidity
2. Humidity ratio
3. Wet bulb temperature
4. Specific enthalpy of air-vapour mixture

Select the correct answer from the codes given below:

(*a*) 1 and 2 (*b*) 3 only
(*c*) 2 and 3 (*d*) 3 and 4.

Ans. (d)

Explanation: Refer to Fig. 3.

During sensible heating of air-water vapour mixture, wet bulb temperature and specific enthalpy of air-vapour mixture increase. **(Ans.)**

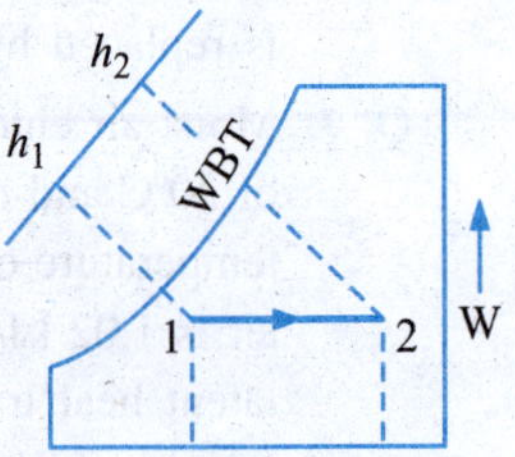

Fig. 3

Q. 11. In a one ton capacity water cooler, water enters at 30ºC at the rate of 200 litres per hour. The outlet temperature of water will be (sp. heat of water = 4.18 kJ/kgK)

(*a*) 3.5ºC (*b*) 6.3ºC (*c*) 23.7ºC (*d*) 15ºC.

Ans. (d)

Explanation:

As, $1\ TR = 3.5$ kJ/s, therefore:

$3.5 \times 3600 = 200 \times 4.18 \times (30 - t)$, where t is the outlet temperature of water

or, $$t = 30 - \frac{3.5 \times 3600}{200 \times 4.18} = 15^\circ\text{C} \quad \textbf{(Ans.)}$$

Q. 12. When the wet and dry bulb temperatures are identical, which of the following statements is/ are true?

1. Air is fully saturated.
2. Dew point temperature is reached.
3. Humidity ratio is unity.
4. A partial pressure of vapour equals total pressure.

Select the correct answer from the code given below:

(*a*) 1 only (*b*) 1 and 2 (*c*) 3 and 4 (*d*) 1, 2, 3 and 4.

Ans. (b)

Explanation: When wet bulb and dry bulb temperatures are equal and air is fully saturated, dew point temperature is reached. **(Ans.)**

Q. 13. A heat engine with 30% efficiency drives a refrigerator of COP 5. What would be the net heat input to the engine for each MW of heat removed in the refrigerator?

(*a*) 66.67 kJ (*b*) 600 kJ (*c*) 666.67 kJ (*d*) 6600 kJ.

Ans. (c)

Explanation: *Given:* η_{HE} = 30% or 0.3; $(COP)_{Ref.}$ = 5; Heat removed = 1 MW or 1000 kW

We know that: $\eta_{HE} = \frac{W}{Q_1}$ or $0.3 = \frac{W}{Q_1}$

Also, $(COP)_{ref.} = 5 = \frac{1000}{W}$ or $W = 200$ kW

$\therefore$ $0.3 = \frac{200}{Q_1}$ or $Q_1 = \frac{200}{0.3} = 666.67$ kJ/s **(Ans.)**

Q. 14. In a vapour compressor refrigeration system, the refrigeration capacity is 2100 kJ/min and heat rejection factor is 1.2. What will respectively be the heat rejected from the condenser and COP?

(*a*) 5040 kJ/min and 5 (*b*) 2520 kJ/min and 5

(*c*) 2520 kJ/min and 4 (*d*) 5040 kJ/min and 4.

Ans. (b)

Explanation:

$$\text{Heat rejection factor} = \frac{\text{Heat rejected in condenser } (Q_R)}{\text{Refrigering effect } (RE)}$$

or, $1.2 = \frac{Q_R}{2100}$ or $Q_R = 1.2 \times 2100 = 2520$ kJ/min **(Ans.)**

Now, $Q_R = W_{in} + RE$ or $2520 = W_{in} + 2100$ or $W_{in} = 420$ kJ/min **(Ans.)**

$\therefore$ $COP = \frac{RE}{W_{in}} = \frac{2100}{420} = 5$ **(Ans.)**

Q. 15. In a domestic refrigerator periodic defrosting is required because frosting

(*a*) causes corrosion of materials (*b*) reduces heat extraction

(*c*) overcools food stuff (*d*) partially blocks refrigerant flow.

Ans. (b)

Explanation: In a domestic refrigerator periodic defrosting in required because frosting *reduces heat extraction*. When air is cooled *below* its dew point temperature (DPT), it gives up moisture which deposits on nearest colder surface. As the colder surface in a refrigerated area is *evaporator*, therefore, the moisture deposits on the evaporator surface. When the temperature of the evaporator fall below 0°C, the moisture deposited on the evaporator surface freezes and forms a coating of *frost*. If this frost is not removed periodically, it acts as an insulator and reduces heat extraction. **(Ans.)**

Q. 16. Air at 35°C DBT and 25°C dew point temperature passes through the water shower whose temperature is maintained at 20°C. What is the process involved?

(*a*) Cooling and humidification (*b*) Sensible cooling

(*c*) Cooling and dehumidification (*d*) Heating and humidification.

Ans. (c)

Explanation: The process involved is *'cooling and dehumidification'*, as the temperature of water shower is *less* than dew point temperature and consequently vapour starts condensing. **(Ans.)**

Q. 17. Atmospheric air at dry bulb temperature of 15ºC enters a heating coil whose surface temperature is maintained at 40ºC. The air leaves the heating coil at 25ºC. What will be the by-pass factor of the heating coil?

(*a*) 0.376 (*b*) 0.4 (*c*) 0.6 (*d*) 0.67

Ans. (c)

Explanation: Refer to Fig. 4.

$$\text{Bypass factor, BF} = \frac{t_{db_3} - t_{db_2}}{t_{db_3} - t_{db_1}}$$

where, t_{db_1} = Temp. of inlet air = 15ºC,

t_{db_2} = Exit temp. of air = 25ºC, and

t_{db_3} = Temp. of heating coil = 40ºC

$$\therefore \quad \text{BF} = \frac{40-25}{40-15} = 0.6 \text{ (Ans.)}$$

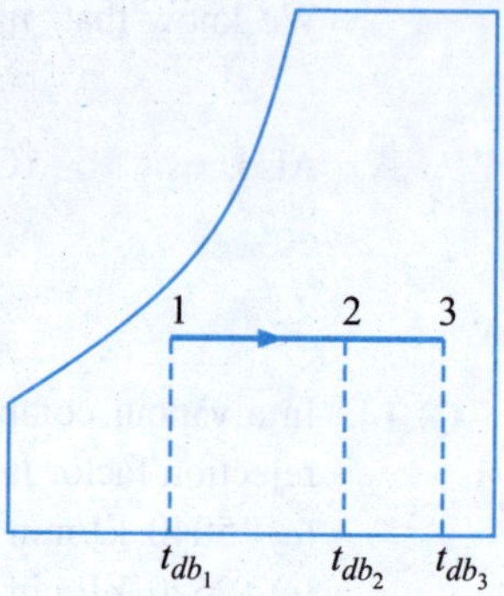

Fig. 4

Q. 18. Velocity of air passing through a rectangular duct and a circular duct is same. Which one of the following is the correct expression for the equivalent diameter of the circular duct in respect of a rectangular duct for the same pressure loss per unit length? (*a* and *b* are the length and breadth of the rectangular duct cross-section)

(*a*) $\frac{a+b}{ab}$ (*b*) $\frac{2ab}{a+b}$ (*c*) $\frac{2a}{a-b}$ (*d*) $\frac{2b}{a+b}$.

Ans. (b)

Explanation: Equivalent diameter of circular duct, $D_{eq} = \frac{2ab}{a+b}\left(=\frac{2a}{\frac{a}{b}+1}\right)$

where $\frac{a}{b}$ is known as *"Aspect. ratio"*.

• In case of rectangular ducts the aspect ratio should not be greater than 8 in any case. **(Ans.)**

Q. 19. Which one of the following is correct?

(*a*) Effective temperature is the index which correlates the combined effects of air dry bulb temperature, air humidity and air movement upon human comfort.

(*b*) The value of effective temperature in winter and summer should be same for human comfort.

(*c*) Effective temperature and wet bulb temperature are the same.

(*d*) The value of the effective temperature should be higher in winter than in summer for comfort.

Ans. (a)

Explanation: *'Effective temperature'* is defined as a sensory index that combines into a single factor the effects of temperature, humidity and air movement on human comfort in a noise free pure air environment. The effective temperature corresponds to the dry-bulb temperature of the saturated air at which a given percentage of people feel comfortable. For example at 21°C, RH it = 100% and air movement 8 m/min, most people feel comfortable. **(Ans.)**

Q. 20. Which one of the following are normally desired comfort conditions in an air-conditioning system?

(*a*) 25ºC DBT and 50% RH (*b*) 22ºC DBT and 90% RH
(*c*) 15ºC BDT and 75% RH (*d*) 15ºC DBT and 40% RH

Ans. (a)

Explanation: Normally desired comfort conditions in an air-conditioning system are 25ºC DBT and 50% RH. **(Ans.)**

Q. 21. In an air-conditioning plant the refrigeration load on the coil is 100 TR. The mass and enthalpy of air leaving the coil are 420 kg/min and 40 kJ/kg respectively. The enthalpy of the air at the inlet to the coil under these condition is

(*a*) 80 kJ/kg (*b*) 90 kJ/kg (*c*) 100 kJ/kg (*d*) 102.5 kJ/kg.

Ans. (b)

Explanation: Load on the coil = $100\ TR = 100 \times 210 = 21000$ kJ/min

$$\text{Refrigeration capacity} = m_a\,(h_i - h_f)$$

or, $$21000 = 420\,(h_i - 40)$$

or, $$h_i = \frac{21000}{420} + 40 = 90 \text{ kJ/kg } \textbf{(Ans.)}$$

Q. 22. In a 0.5 TR capacity water cooler, water enters at 30ºC and leaves at 15ºC. What is the actual water flow rate?

(*a*) 50 litres/hour (*b*) 75 litres/hour (*c*) 100 litres/hour (*d*) 125 litres/hour.

Ans. (c)

Explanation: $$Q = 0.5 \times (210/60) = 1.75 \text{ kW}$$

Also, $$Q = 1.75 = \dot{m}\ c\ \Delta t \text{ or } 1.75 = \dot{m} \times 4.187 \times (30 - 15)$$

or, $$\dot{m} = \frac{1.75}{4.187 \times 15} = 0.02786 \text{ kg/s}$$

$$= 0.02786 \times 3600 \text{ litres/hour}$$

$$= 100 \text{ litres/hour} (1 \text{ hour}/1 \text{ kg of water} = 1 \text{ litre}) \textbf{ (Ans.)}$$

Q. 23. A refrigerator based on reversed Carnot cycle works between two such temperatures that the ratio between the low and high temperatures is 0.8. If a heat pump is operated between same temperature range, then what would be its COP?

(*a*) 2 (*b*) 3 (*c*) 4 (*d*) 5.

Ans. (d)

Explanation: $$\frac{T_L}{T_H} = 0.8 \text{ (Given)}; \quad (\text{COP})_{\text{heat pump}} = \frac{T_H}{T_H - T_L} = \frac{1}{1 - \dfrac{T_L}{T_H}}$$

$$= \frac{1}{1 - 0.8} = 5 \textbf{ (Ans.)}$$

Q. 24. In an Electrolux refrigerator

(*a*) Ammonia is absorbed in water (*b*) Ammonia is absorbed in hydrogen

(*c*) Hydrogen is evaporated in ammonia (*d*) Ammonia is evaporated in hydrogen.

Ans. (a)

Explanation: In an electrolux refrigerator water is used as an *absorbent* because it has ability to *absorb ammonia* refrigerant. **(Ans.)**

Q. 25. Which one of the following is responsible for the operation of a thermostatic expansion valve?

(*a*) Pressure changes in evaporator (*b*) Temperature changes in evaporator

(*c*) Degree of superheat in evaporator (*d*) Degree of subcooling in evaporator.

Ans. (c)

Explanation: A "thermostatic expansion value" is a throttling device which works automatically, maintaining *proper and correct liquid flow* as per the requirements of the load on the evaporator. This valve is also called a 'constant superheat valve' because it *maintains a constant degree of superheat of the vapour refrigerant* at the end of the evaporator coil *by controlling the flow of liquid refrigerant through the evaporator.*

Q. 26. In a vapour compression refrigeration plant, the refrigerant leaves the evaporator at 195 kJ/kg and the condenser at 65 kJ/kg. For 1 kg/s of refrigerant, the refrigeration effect is

(*a*) 70 kW (*b*) 100 kW (*c*) 130 kW (*d*) 160 kW.

Ans. (c)

Explanation: RE (Refrigerating effect) $= \dot{m}_{\text{refrigerant}} \; (h_{\text{evaporator}} - h_{\text{condenser}})$

$= 1 \times (195 - 65) = 130$ kW **(Ans.)**

Q. 27. When the wet bulb and dry bulb temperatures are equal, which of the following statements is/are correct?

1. Air is fully saturated.
2. Dew point temperature is reached.
3. Partial pressure of vapour equals to the total pressure.
4. Humidity ratio is 100%.

Select the correct answer using the code given below:

(*a*) 1 and 2 (*b*) 1 only (*c*) 1, 2 and 4 (*d*) 2 and 3

Ans. (a)

Explanation: When the wet bulb temperature (WBT) and dry bulb temperature (DBT) are *equal* the air is fully saturated and dew point temperature (DPT) is reached. **(Ans.)**

Q. 28. In order to have a low by-pass factor of a cooling coil, the fin spacing and the number of tube rows should be

(*a*) Wide apart and high, respectively (*b*) Wide apart and low, respectively

(*c*) Close and high, respectively (*d*) Close and low, respectively.

Ans. (c)

Explanation: A cooling coil will have a low by-pass factor (BF) when the fin spacing is close (i.e., coils are closely spaced) and the number of tube rows is high. **(Ans.)**

Q. 29. An air-conditioned room of volume 10 m^3 has infiltration of air equivalent to 3 air changes per hour. Density of air is 1.2 kg/m^3, specific heat c_p is 1 kJ/kgK and temperature difference between room and ambient air is 20 K. What is the sensible heat load due to infiltrated air?

(*a*) 60 kJ/h (*b*) 12 kJ/h (*c*) 0.45 kW (*d*) 0.2 kW.

Ans. (d)

Explanation: Sensible heat load due to infiltrated air,

$$Q_s = 0.0204 \times V \times \Delta T = 0.0204 \times \frac{10 \times 3}{60} \times 20 \approx 0.2 \text{ kW } \textbf{(Ans.)}$$

Q. 30. A heat pump is used to heat a house in the winter and then reversed to cool the house in the summer. The inside temperature of the house is to be maintained at 20ºC. The heat transfer through the house wall is 7.9 kJ/s and the outside temperature in winter is 5ºC. What is the minimum power (approximate) required to drive the heat pump?

(*a*) 40.5 W (*b*) 405 W (*c*) 42.5 W (*d*) 425 W.

Ans. (b)

Explanation: Refer to Fig. 5.

$$T_1 = 20 + 273 = 293 \text{ K};$$

$$T_2 = 5 + 273 = 278 \text{ K}; \; Q = 7.9 \text{ kJ/s}$$

$$(\text{COP})_{\text{heat temp.}} = \frac{T_1}{T_1 - T_2} = \frac{293}{293 - 278} = 19.5$$

Also, $$(\text{COP})_{\text{heat pump}} = \frac{Q_1}{W}$$

or, $$19.5 = \frac{7.9}{W}$$

or, $$W = \frac{7.9}{19.5} \times 1000 = 405 \text{ watts } \textbf{(Ans.)}$$

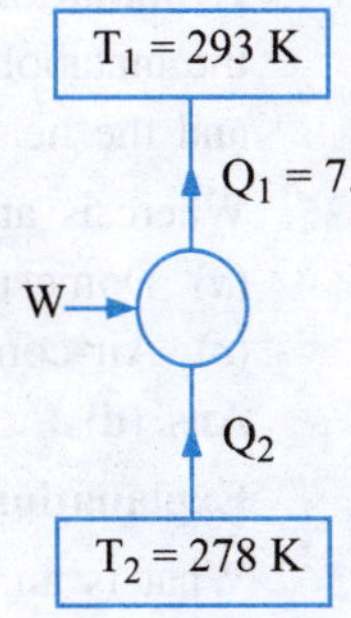

Fig. 5

Q. 31. The leaks in a refrigeration system using freon are detected by

(*a*) A halide torch, which on detection produces greenish flame lighting

(*b*) Sulphur sticks, which on detection give white smoke

(*c*) Using reagents

(*d*) Sensing reduction in pressure.

Ans. (a)

Explanation: In a refrigeration system using freon, *'halide torch'* may be used to detect the leak of freon vapour. The halide torch detector (acetylene) produces nearly colourless flame.

- In case 'freon' has escaped into the room in such a large quantity that the detector burns with green flame continuously, then a soap method must be used. **(Ans.)**

Q. 32. The wet bulb depression is zero, when relative humidity is equal to

(*a*) 100% (*b*) 60% (*c*) 40% (*d*) Zero.

Ans. (a)

Explanation: Wet bulb depression = Dry bulb temp. (DBT) – Wet bulb temp. (WBT).

When the relative humidity is 100%

(i.e., air is saturated), the wet bulb depression is zero since then DBT is equal to WBT. **(Ans.)**

Q. 33. Consider the following statements:

Air washer can work as

1. Humidifier only
2. Dehumidifier only
3. Filter only

Which of these statements is/are correct?

(*a*) only 1 (*b*) only 2 and 3 (*c*) only 1 and 3 (*d*) 1, 2 and 3.

Ans. (d)

Explanation: In an air washer, air flows through a spray of water and during this flow air may be cooled or heated, humidified or dehumidified, or simply adiabatically saturated, depending on the mean surface temperature of water. The air washer can work as filter also. **(Ans.)**

Q. 34. A human body feels comfortable when the heat produced by the metabolism of human body is equal to

(*a*) Heat dissipated to the surroundings (*b*) Heat stored in the human body

(*c*) Sum of (*a*) and (*b*) (*d*) Difference of (*a*) and (*b*).

Ans. (c)

Explanation: A human body feels comfortable thermodynamically when the heat produced by the metabolism of human body is equal to the sum of the heat dissipated to the surroundings and the heat stored in human body by raising the temperature of the body tissues. **(Ans.)**

Q. 35. Where is an air refrigeration cycle generally employed?

(*a*) Domestic refrigerators (*b*) Commercial refrigerators

(*c*) Air-conditioning (*d*) Gas liquefaction.

Ans. (d)

Explanation: An air refrigeration cycle is widely used for *gas liquefaction*. **(Ans.)**

Q. 36. What is an azeotrope?

(*a*) A non-halogenic refrigerant. (*b*) A refrigerant dissolved in alcohol

(*c*) A mixture of refrigerants without phase separation

(*d*) An eco-friendly refrigerant.

Ans. (c)

Explanation: The refrigerants belonging to 'Azeotropes' group consist of mixtures of different substances. These substance *cannot be separated into components by distillation.* They possess *fixed thermodynamic properties and do not undergo any separation with changes in temperature and pressure.* An azeotrope behaves like a simple substance. **(Ans.)**

Q. 37. Environmental protection agencies advise against the use of Chloro-fluoro-carbon refrigerants because these react with

(*a*) Water vapour and cause acid rain (*b*) Plants and cause green house effect

(*c*) Oxygen and cause its depletion (*d*) Ozone layer and cause its depletion.

Ans. (d)

Explanation: Environmental protection agencies advise against the use of Chloro-fluoro-carbon (CFC) refrigerants because they react with ozone layer and cause its depletion. **(Ans.)**

Q. 38. What is hunting of thermostatic expansion valve?

(*a*) A variation of evaporator load with degree of superheat

(*b*) A variation in pressure of the evaporator with variation of load

(*c*) Alternate over-feeding and starving of refrigerant flow in the evaporator

(*d*) This is not used in connection with expansion valve.

Ans. (c)

Explanation: The alternate over-feeding and starving of refrigerant flow in the evaporator is termed as 'hunting of thermostatic expansion valve'. **(Ans.)**

Q. 39. Match List-I (Devices) with List-II (Processes undergone by air) and select the correct answer using the code given below the lists:

List-I	*List-II*
A. Cooling tower	1. Heating
B. Air coolers	2. Cooling and dehumidification
C. Evaporator coil	3. Cooling and humidification
D. Air cooled condenser	4. Heating and humidification

Codes:

	A	B	C	D
(*a*)	2	1	4	3
(*b*)	4	3	2	1
(*c*)	2	3	4	1
(*d*)	4	1	2	3.

Ans. (b)

Explanation:

4. Heating and humidification	A. Cooling tower
3. Cooling and humidification	B. Air coolers
2. Cooling and dehumidification	C. Evaporator coil
1. Heating	D. Air cooled condenser **(Ans.)**

Q. 40. In vapour absorption refrigeration system heating in generator is done at 177ºC, refrigeration in evaporator at –3ºC and cooling in condenser at 27ºC. Then what will be the maximum COP of the system?

(*a*) 1.5 (*b*) 2.0 (*c*) 3.0 (*d*) 4.0.

Ans. (c)

Explanation: *Given:* T_G = 177ºC = 450 K; T_C = 27ºC = 300 K; T_E = –3 + 273 = 270 K

We know: $$(\text{COP})_{max} = \left(\frac{T_E}{T_C - T_E}\right)\left(\frac{T_G - T_G}{T_G}\right)$$

or, $$(\text{COP})_{max} = \left(\frac{270}{300 - 270}\right)\left(\frac{450 - 300}{450}\right) = 3$$ ...[Eqn. (5.1)] **(Ans.)**

Q. 41. In conventional refrigerants what is the element responsible for ozone depletion?

(*a*) Chlorine (*b*) Fluorine (*c*) Carbon (*d*) Hydrogen.

Ans. (b)

Explanation: In conventional refrigerants, the element responsible for ozone depletion is *fluorine*. **(Ans.)**

Q. 42. The refrigerating efficiency, that is the ratio of actual COP to reversible COP of a refrigeration cycle is 0.8, the condenser and evaporator temperatures are 51ºC and –30ºC respectively. If cooling capacity of the plant is 2.4 kW then what is the work requirement?

(*a*) 1.00 kW (*b*) 1.33 kW (*c*) 1.25 kW (*d*) 2.008 kW.

Ans. (a)

Explanation: *Given:* T_E = 51ºC = 324 K; T_E = –30ºC = 243 K; *RE* = 2.4 kW

$$\text{Actual COP} = 0.8 \times \left(\frac{243}{324 - 243}\right) = 2.4$$

Also, $$\text{Actual COP} = \frac{RE}{W} \quad \text{or, } 2.4 = \frac{2.4}{W} \quad \text{or } W = 1 \text{ kW } \textbf{(Ans.)}$$

Q. 43. The atmospheric air at 760 mm of Hg, dry bulb temperature 15ºC and wet bulb temperature 11ºC enters a heating coil whose temperature is 41ºC. If the by-pass factor of the heating coil is 0.5, then what will be the dry bulb temperature of the air leaving the coil?

(*a*) 28ºC (*b*) 29ºC (*c*) 30ºC (*d*) 26ºC.

Ans. (a)

Explanation: Refer to Fig. 6.

$$BF = 0.5 = \frac{41 - t_{db}}{41 - 15}$$

or $t_{db} = 41 - 0.5\ (41 - 15) = 28$ºC **(Ans.)**

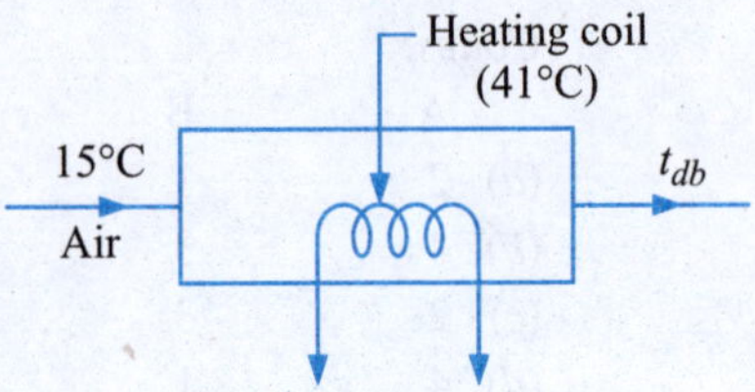

Fig. 6

Q. 44. The humidity ratio of atmospheric air at 28ºC DBT and 760 mm of Hg is 0.016 kg of vapour per kg of dry air. What is the partial pressure of water vapour?

(*a*) 2.242 kN/m^2 (*b*) 2.535 kN/m^2 (*c*) 3.535 kN/m^2 (*d*) 4.242 kN/m^2.

Ans. (b)

Explanation: Humidity ratio, $W = 0.622 \times \left(\frac{p_v}{p_t - p_v}\right)$

or, $$0.016 = 0.622 \times \left(\frac{p_v}{101.32 - p_v}\right)$$

($\because p_t = 1.0132$ bar $= 101.32$ kN/m^2)

or, $$\frac{0.016}{0.622} = \frac{p_v}{101.32 - p_v} \quad \text{or } 0.02572 = \frac{p_v}{101.32 - p_v}$$

or, $$0.02572\ (101.32 - p_v) = p_v \quad \text{or, } 2.606 - 0.02572\ p_v = p_v$$

or, $$1.02572\ p_v = 2.606 \quad \text{or} \quad p_v = 2.54 \text{ kN/m}^2 \textbf{ (Ans.)}$$

Q. 45. Operating temperature of a cold storage is –2ºC. From the surrounding at ambient temperature of 40ºC heat leaked into the cold storage is 30 kW. If the actual COP of the plant is 1/10 of the maximum possible COP, then what will be the power required to pump out the heat to maintain the cold storage temperature at –2ºC?

(*a*) 1.90 kW (*b*) 3.70 kW (*c*) 20.28 kW (*d*) 46.50 kW.

Ans. (d)

Explanation: $$\text{Actual COP} = \frac{RE}{W}$$

or, $$\frac{1}{10}\left(\frac{271}{313 - 271}\right) = \frac{30}{W} \quad \text{or } W = 46.5 \text{ kW } \textbf{(Ans.)}$$

Q. 46. Air enters a rectangular duct measuring 30 × 40 cm with a velocity of 8.5 m/s and a temperature of 40°C. Kinematic viscosity of the air is 16.95 × 10^{-6} m^2/s. What will be the Reynolds number?

(*a*) 1.72 × 10^5 (*b*) 2.58 × 10^5 (*c*) 0.86 × 10^5 (*d*) 0.72 × 10^5.

Ans. (c)

Explanation: *Given:* $a = 30$ cm, $b = 40$ cm, $U = 8.5$ m/s; $\nu = 16.95 \times 10^{-6}$ m^2/s

$$\text{Equivalent diameter, } D_{eq} = \frac{2ab}{a+b} = \frac{2 \times 30 \times 40}{(30+40)} = 34.28 \text{ cm}$$

$$\therefore \quad \text{Raynolds number, } Re = \frac{U D_{eq}}{\nu} = \frac{8.5 \times (34.28 \times 10^{-2})}{16.95 \times 10^{-6}} = 1.72 \times 10^5 \textbf{ (Ans.)}$$

Q. 47. As an index of comfort, the temperature of saturated air at which a person would experience the same feeling of comfort as experienced in the actual unsaturated environment is called the

(*a*) Comfort temperature (*b*) Effective temperature

(*c*) Wet bulb temperature (*d*) Soothing temperature.

Ans. (b)

Explanation: *'Effective temperature'* is the temperature of saturated air at which a person would experience the same feeling of comfort of as experienced in the actual unsaturated environment. **(Ans.)**

Q. 48. A certain air has DBT 35ºC and DPT of 20ºC, the corresponding saturation pressure of water being 5.628 kPa and 2.33 kPa respectively. When the atmospheric pressure is assumed as 1.0132 bar, the specific humidity of air will be

(*a*) 2.5×10^{-3} (*b*) 7.8×10^{-3} (*c*) 14.7×10^{-3} (*d*) 25×10^{-3}.

Ans. (c)

Explanation: *Given:* $t_{db} = 35$ºC, $t_{dp} = 20$ºC; $p_v = 2.33$ kPa; $p_t = 1.0132$ bar or 101.32 kPa.

$$\text{Specific humidity, } W = 0.622\left(\frac{p_v}{p_t - p_v}\right) = 0.622 \times \left(\frac{2.33}{101.32 - 2.33}\right)$$

$$= 14.64 \times 10^{-3} \text{ kg/kg of air } \textbf{(Ans.)}$$

Q. 49. A refrigerator operates between the temperatures of –23ºC and 27ºC. If one TR = 3.5 kW, the minimum power required per TR to operate the refrigerator is

(*a*) 0.5 kW (*b*) 0.7 kW (*c*) 0.9 kW (*d*) 1.0 kW.

Ans. (b)

Explanation: $T_1 = 27 + 273 = 300$ K; $T_2 = -23 + 273 = 250$ K; 1 TR (RE) = 3.5 kW

Also, $$(COP)_{Ref.} = \frac{T_2}{T_1 - T_2} = \frac{250}{300 - 250} = 5$$

$$(COP)_{Ref.} = \frac{R_E}{W_{in}} \quad \text{or} \quad 5 = \frac{3.5}{W_{in}} \quad \text{or } W_{in} = 0.7 \text{ kW } \textbf{(Ans.)}$$

Q. 50. If air flows over a cooling coil, dehumidification of air will take place if the coil surface temperature is below the following of the entering air

(*a*) Wet bulb temperature (*b*) Dry bulb temperature

(*c*) Dew point temperature (*d*) Adiabatic saturation temperature.

Ans. (c)

Explanation: If air flows over a cooling coil, 'dehumidification of air' will take place if the coil surface temperature is *below* the *'dew point temperature'* of the entering air. **(Ans.)**

Q. 51. A condenser of a refrigeration system rejects heat at a rate of 120 kW, while the compressor of the system consumes a power of 30 kW. The COP of the system will be

(*a*) 1/4 (*b*) 1/3 (*c*) 3 (*d*) 4.

Ans. (c)

Explanation: Refer to Fig. 7.

Given: $Q_{rejected} = 120$ kW; $W_{in} = 30$ kW

Now, $Q_{absorbed} + W_{in} = Q_{rejected}$ or $Q_{absorbed} + 30 = 120$

or, $Q_{absorbed} = 120 - 30 = 90$ kW

$\therefore$ $(COP)_{ref.} = \dfrac{Q_{absorbed}}{W_{in}} = \dfrac{90}{30} = 3$ **(Ans.)**

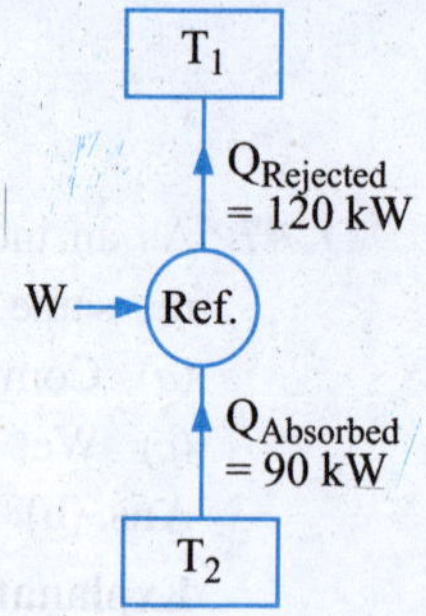

Fig. 7

Q. 52. In an aircraft refrigeration system, the pressure at the cooling turbine outlet is equal to

(*a*) Ambient pressure (*b*) Cabin pressure
(*c*) Compressor inlet pressure (*d*) Evaporator pressure.

Ans. (b)

Explanation: In an aircraft refrigeration system, the pressure at the cooling turbine outlet is equal to *cabinet pressure*. **(Ans.)**

Q. 53. Lowest COP is of vapour

(*a*) Compression cycle with superheated vapour
(*b*) Compression cycle with dry compression
(*c*) Compression cycle with wet compression
(*d*) Absorption cycle.

Ans. (d)

Explanation: Lowest COP is of *'vapour absorption cycle'* since it *operates on 'heat energy'*. **(Ans.)**

Q. 54. For a 2-stage air compressor for the maximum efficiency of compressor, consider the following statements:

1. $p_2 = \dfrac{1}{2}(p_1 + p_3)$
2. $p_2 = \sqrt{p_1 \cdot p_3}$
3. Intercooling between the stages is complete
4. Work is equally shared by the two stages

Which of these statements are correct?

(*a*) 1 and 4 only (*b*) 1, 3 and 4 (*c*) 2, 3 and 4 (*d*) 2 and 3 only.

Ans. (c)

Explanation: The efficiency of a 2-stage compressor will be maximum when $p_2 = \sqrt{p_1 p_3}$, the intercooling between the two stages is complete and work is equally shared between the two stages. **(Ans.)**

Q. 55. The following readings refer to a vapour compression refrigerator:
The enthalpy per kg of refrigerant flow, from *p-h* chart
at inlet to compressor = 1500 kJ/kg
at outlet to compressor = 1700 kJ/kg
at exit of condenser = 300 kJ/kg
The COP of the refrigerator is

(*a*) 3 (*b*) 4 (*c*) 5 (*d*) 6.

Ans. (d)

Explanation: $W_{in} = 1700 - 1500 = 200$ kJ/kg; $RE = 1500 - 300 = 1200$ kJ/kg

$\therefore$ $COP = \dfrac{RE}{W_{in}} = \dfrac{1200}{200} = 6$ **(Ans.)**

Q. 56. During the sensible cooling process, specific humidity
(*a*) remains constant (*b*) increases
(*c*) decreases (*d*) unpredictable.

Ans. (a)

Explanation: Refer to Fig. 8.

During the sensible cooling process, specific humidity (W) remains *constant*. **(Ans.)**

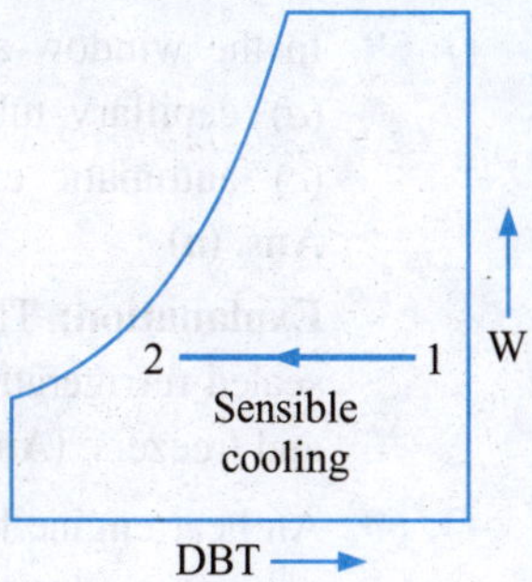

Fig. 8

Q. 57. An industrial heat pump operates between the temperatures of 27ºC and –13ºC. The rates of heat addition and heat rejection are 750 W and 1000 W, respectively. The COP for the heat pump is
(*a*) 7.5 (*b*) 6.5 (*c*) 4.0 (*d*) 3.0.

Ans. (c)

Explanation: Refer to Fig. 9.

$$T_1 = 27 + 273 = 300 \text{ K}$$

$$T_2 = -13 + 273 = 260 \text{ K}$$

Given: Heat addition = 750 W; Heat rejection = 1000 W

$$W_{in} + \text{heat added} = \text{Heat rejected}$$

or, $$W_{in} = 1000 - 750 \text{ or, } W_{in} = 250 \text{ W}$$

$$\text{COP} = \frac{\text{Heat rejected}}{W_{in}} = \frac{1000}{250} = 4 \textbf{ (Ans.)}$$

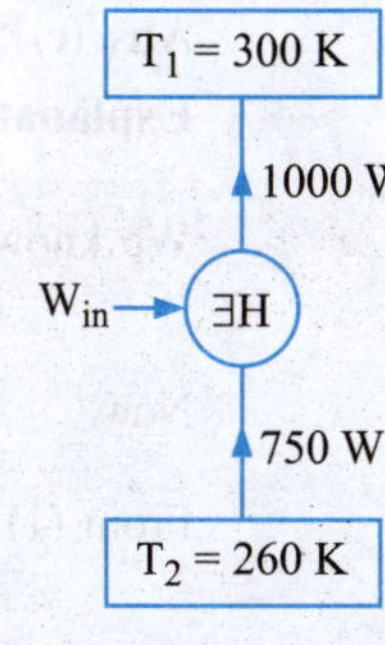

Fig. 9

Q. 58. A refrigerator based on ideal vapour compression cycle operates between the temperature limits of –20ºC and 40ºC. The refrigerant enters the condenser as saturated vapour and leaves as saturated liquid. The enthalpy and entropy values for saturated liquid and vapour at these temperatures are given in the table below:

*T*ºC	h_f *(kJ/kg)*	h_g *(kJ/kg)*	s_f *(kJ/kg K)*	s_g *(kJ/kg K)*
–20	20	180	0.07	0.7366
40	80	200	0.3	0.67

If refrigerant circulation rate is 0.025 kg/s, the refrigeration effect is equal to
(*a*) 2.1 kW (*b*) 2.5 kW (*c*) 3.0 kW (*d*) 4.0 kW.

Ans. (a)

Explanation: Refer to Fig. 10.

Given: $h_3 = 200$ kJ/kg; $s_2 = 0.67$ kJ/kg K; $h_1 = h_{f4} = 80$ kJ/kg

First calculating quality (x) of vapour, we have:

$$s_3 = s_2$$

or, $$s_3 = 0.07 + x\,(0.7366 - 0.07)$$

or, $$0.67 = 0.07 + 0.6666x \text{ or, } x = 0.90$$

Enthalpy at point 2,

$$h_2 = 20 + 0.90\,(180 - 20)$$
$$= 20 + 0.90 \times 160 = 164 \text{ kJ/kg}$$

Refrigerant effect (R_n) = $m\,(h_2 - h_1) = 0.025\,(164 - 80) = 2.1$ kW **(Ans.)**

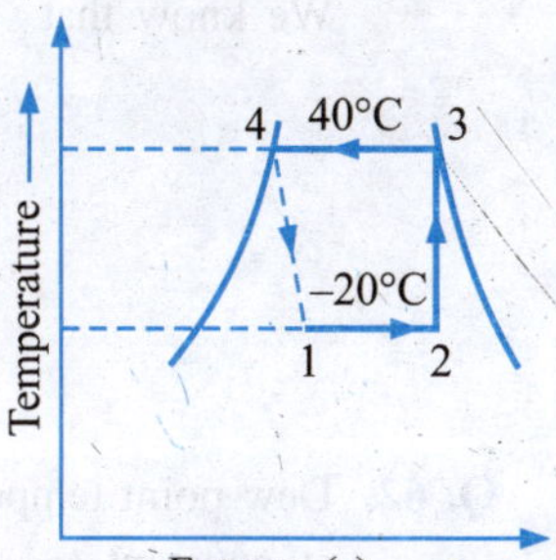

Fig. 10

Q. 59. In the window air conditioner, the expansion device used is

(*a*) capillary tube (*b*) thermostatic expansion valve
(*c*) automatic expansion valve (*d*) float valve.

Ans. (a)

Explanation: The capillary tube is used as an expansion device in small capacity hermetic sealed refrigeration units such as in domestic refrigerators, water coolers, room air-conditioners and freezers. **(Ans.)**

Q. 60. An heat engine having an efficiency of 70% is used to drive a refrigerator having a coefficient of performance of 5. The energy absorbed from low temperature reservoir by the refrigerator for each kJ of energy absorbed from high temperature source by the engine is

(*a*) 0.14 kJ (*b*) 0.71 kJ (*c*) 3.5 kJ (*d*) 7.1 kJ.

Ans. (c)

Explanation: Refer to Fig. 11.

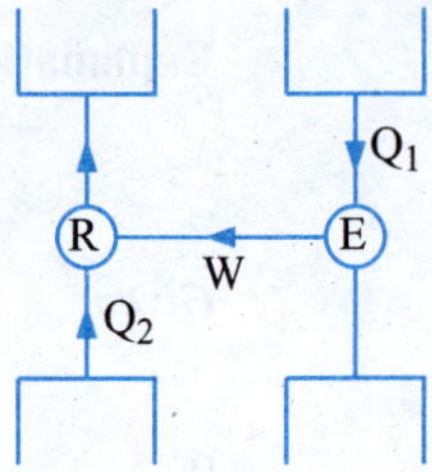

Fig. 11

We know that: $$R_{\text{engine}} = \frac{W}{Q_1} = 0.7 \quad ...(1)$$

Now, $$(\text{COP})_R = \frac{Q_2}{W} = 5 \quad ...(2)$$

From (1) and (2), we get:

$$0.7\, Q_1 = \frac{Q_2}{5} \quad \text{or,} \quad \frac{Q_2}{Q_1} = 0.7 \times 5 = 3.5$$

Energy absorbed from low temperature reservoir by the refrigerator for each kJ of energy absorbed from high temperature source by the engine = 3.5 kJ **(Ans.)**

Q. 61. Dew point temperature of air at one atmospheric pressure (1.013 bar) is 18°C. The air dry bulb temperature is 30°C. The saturation pressure of water at 18°C and 30°C are 0.02062 bar and 0.04241 bar respectively. The specific heats of air and water vapour respectively are 1.005 and 1.88 kJ/kgK and the latent heat of vaporization water of 0°C is 2500 kJ/kg. The specific humidity (kg/kg of dry air) and enthalpy (kJ/kg of dry air) of this moist air respectively, are

(*a*) 0.01051, 52.64 (*b*) 0.01291, 63.15 (*c*) 0.01481, 78.60 (*d*) 0.01532, 81.40.

Ans. (b)

Explanation: *Given:* $p_t = 1.013$ bar; $p_v = 0.02062$ bar (at dew point)

We know that

$$\text{Specific humidity} = \frac{0.622\, p_v}{p_t - p_v} = \frac{0.622 \times 0.02062}{1.013 - 0.02062} = 0.01292 \text{ kg/kg of dry air}$$

$$\text{Enthalpy, } h = 1.005\, t_{db} + W\,(2500 + 1.88\, t_{db})$$
$$= 1.005 \times 30 + 0.01292\,(2500 + 1.88 \times 30)$$
$$= 63.17 \text{ kJ/kg of dry air } \textbf{(Ans.)}$$

Q. 62. Dew point temperature is the temperature at which condensation begins when the air is cooled at constant

(*a*) volume (*b*) entropy (*c*) pressure (*d*) enthalpy.

Ans. (c)

Explanation: Refer to Fig. 12.

If a sample (A) of unsaturated air, containing superheated water vapour, is cooled at constant pressure, the partial pressure (p_v) of each constituent remains constant until the water vapour reaches the saturated state as shown by point B. At this point B, the first drop of dew will be formed and hence the temperature at point B is called *dew point temperature*. **(Ans.)**

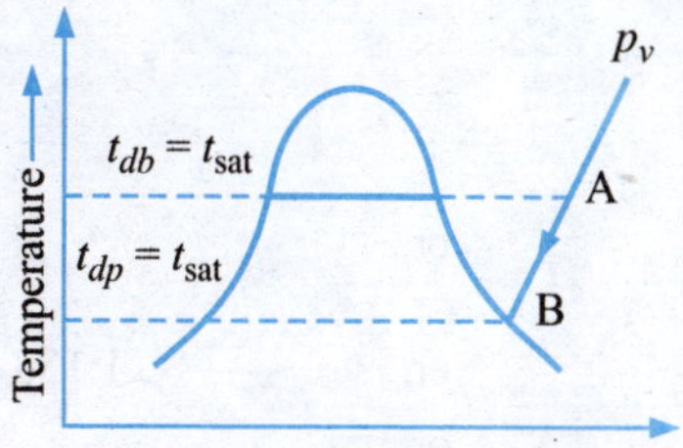

Fig. 12

Q. 63. Atmospheric air at a flow rate of 3 kg/s (on dry air basis) enters a cooling and dehumidifying coil with an enthalpy of 85 kJ/kg of dry air and a humidity ratio of 19 grams/kg of dry air. The air leaves the coil with an enthalpy of 43 kJ/kg of dry air and a humidity ratio of 8 grams/kg of dry air. If the condensate water leaves the coil with an enthalpy of 67 kJ/kg, the required cooling capacity of the coil in kW is

(*a*) 75.0 (*b*) 123.8 (*c*) 128.2 (*d*) 159.0.

Ans. (c)

Explanation:

$$W_1 = 19 \text{ grams/kg of dry air} = 19 \times 10^{-3} \text{ kg/kg of dry air}$$

$$W_2 = 8 \text{ grams/kg of dry air}$$

Hence, at inlet, mass of water vapour,

$$m_{v1} = 19 \times 10^{-3} \times (3 \text{ kg/s}) = 57 \times 10^{-3} \text{ kg/s.}$$

At outlet mass of water vapour

$$m_{v2} = 8 \times 10^{-3} \times (3 \text{ kg/s}) = 24 \times 10^{-3} \text{ kg/s.}$$

Hence, mass of water condensed

$$= (57 - 24) \times 10^{-3} \text{ kg/s.} = 33 \times 10^{-3} \text{ kg/s}$$

Required cooling capacity = Change in enthalpy of condensed water + change in enthalpy of dry air

$$= (67 \text{ kJ/kg}) \times 33 \times 10^{-3} \text{ kg/s} + (85 \text{ kJ/kg} - 43 \text{ kJ/kg}) \times 3 \text{ kg/s}$$

$$= 128.2 \text{ kW } \textbf{(Ans.)}$$

Q. 64. Moist air at a pressure of 100 kPa is compressed to 500 kPa and then cooled to 35°C in an aftercooler. The air at the entry to the aftercooler is unsaturated and becomes just saturated at the exit of the aftercooler. The saturation pressure of water at 35°C is 5.628 kPa. The partial pressure of water vapour (in kPa) in the moist air entering the compressor is closest to

(*a*) 0.57 (*b*) 1.13 (*c*) 2.26 (*d*) 4.52.

Ans. (b)

Explanation: At the exit of aftercooler saturation pressure of water at 35°C = 5.628 kPa.

$\therefore$ The air becomes saturated, $\dfrac{p_v}{p_{vs}} = 1$

or, $p_v = 5.628$ kPa.

Now,

$$W = 0.622\left(\frac{p_v}{p_t - p_v}\right) = 0.622\left(\frac{5.628}{500 - 5.628}\right)$$

$$= 7.0809 \times 10^{-3} \text{ kg/kg dry air}$$

$$\therefore \quad 7.0809 \times 10^{-3} = 0.622\left(\frac{p_v}{p - p_v}\right)$$

$$\text{or,} \quad 0.01138 = \left(\frac{p_v}{100 - p_v}\right)$$

$$\text{or,} \quad 1.138 - 0.01138\, p_v = p_v$$

$$\text{or,} \quad 1.01138\, p_v = 1.138 \text{ or, } p_v \simeq 1.13 \text{ kPa } \textbf{(Ans.)}$$

Q. 65. Air (at atmospheric pressure) at a dry bulb temperature of 40ºC and wet bulb temperature of 20ºC is humidified in an air washer operating with continuous water recirculation. The wet bulb depression (*i.e.*, the difference between the dry and wet bulb temperatures) at the exit is 25% of that at the inlet. The dry bulb temperature at the exit of the air washer is closest to

(*a*) 10ºC (*b*) 20ºC (*c*) 25ºC (*d*) 30ºC.

Ans. (c)

Explanation: Air (at atmospheric pressure) DBT = 40ºC

$$(\text{Wet bulb depression})_{exit} = 25\% \ (\text{wet bulb depression})_{inlet}$$

$$t_{db} - 20 = 0.25\ (20)$$

$$(t_{db})_{exit} = 25°\text{C } \textbf{(Ans.)}$$

Q. 66. In an ideal vapour compression refrigeration cycle, the specific enthalpy of refrigerant (in kJ/kg) at the following states is given as:

Inlet of condenser : 283

Exit of condenser : 116

Exist of evaporator : 232

The COP of this cycle is

(*a*) 2.27 (*b*) 2.75 (*c*) 3.27 (*d*) 3.75.

Ans. (a)

Explanation: Refer to Fig. 13.

Inlet of condenser, $h_3 = 283$ kJ/kg

Exit of condenser, $h_4 = 116$ kJ/kg

Exit of evaporator, $h_2 = 232$ kJ/kg

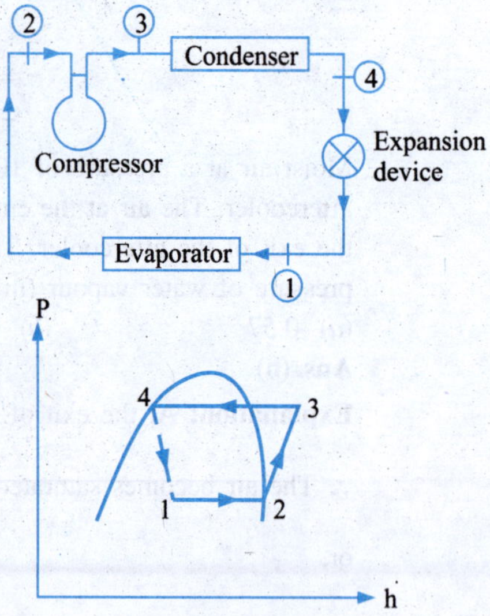

Fig. 13

$$\text{Now,} \quad \text{COP} = \frac{\text{Refrigeration Effect (R.E.)}}{\text{Work Input}}$$

$$\text{COP} = \frac{h_2 - h_1}{h_3 - h_2}$$

$$\text{COP} = \frac{h_2 - h_4}{h_3 - h_2} \qquad (\because h_1 = h_4)$$

($h_4 = h_1$ because 4 to 1 isenthalpic throttling process)

$$\text{COP} = \frac{232 - 116}{283 - 232} = \frac{116}{51}$$

$$= 2.275 \textbf{ (Ans.)}$$

Q. 67. A moist air sample has dry bulb temperature of 30°C and specific humidity of 11.5 g water vapour per kg dry air. Assume molecular weight of air as 28.93. If the saturation vapour pressure of water at 30°C is 4.24 kPa and the total pressure is 90 kPa, then the relative humidity (in %) of air sample is

(*a*) 50.5 (*b*) 38.5 (*c*) 56.5 (*d*) 68.5.

Ans. (b)

Explanation: Relative humidity, $\phi = \frac{p_v}{p_{vs}}$

$$\text{Specific humidity, } W = 0.622\left(\frac{p_v}{p_t - p_v}\right)$$

$$11.5 \times 10^{-3} = 0.622\left(\frac{p_v}{90 - p_v}\right) \quad \text{or, } p_v = 1.63378 \text{ kPa}$$

$$\text{R.H., } \phi = \frac{p_v}{p_{vs}} = \frac{1.63378}{4.24} = 0.385 = 38.5\% \textbf{ (Ans.)}$$

Q. 68. If a mass of moist air in an airtight vessel is heated to a higher temperature, then

(*a*) specific humidity of the air increases

(*b*) specific humidity of the air decreases

(*c*) relative humidity of the air increases

(*d*) relative humidity of the air decreases.

Ans. (d)

Explanation: Refer to Fig. 14.

Process is shown in psychrometric chart.

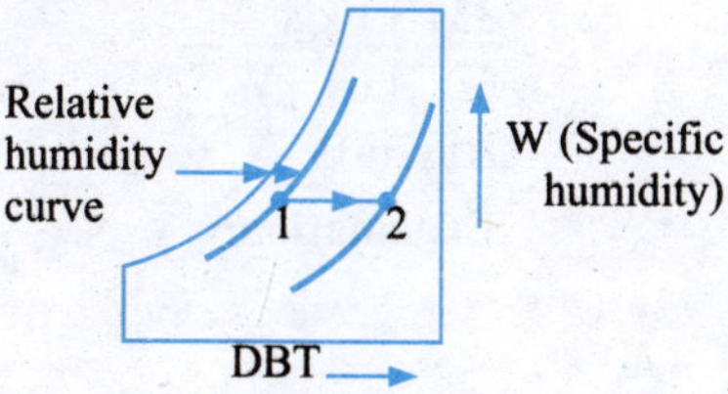

Fig. 14

Q. 69. A Carnot refrigerator requires 1.3 kW per tonne of refrigeration to maintain a region at low temperature of –38°C. Determine:

(*i*) COP of Carnot refrigerator, (*ii*) Higher temperature of the cycle.

(*a*) 1.22, 20.6°C (*b*) 2.54, 26.6°C (*c*) 2.99, 40.6°C (*d*) 3.99, 60.6°C.

Ans. (c)

Explanation: Refer to Fig. 15

$$T_2 = 273 - 38 = 235 \text{ K}$$

Power required per tonne of refrigeration = 1.3 kW

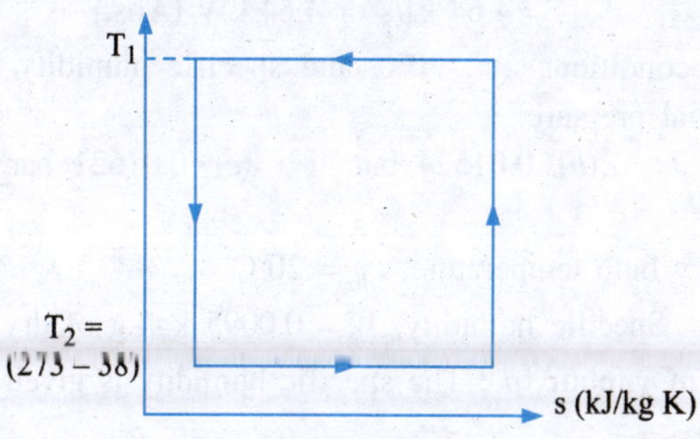

Fig. 15

(*i*) **COP of Carnot refrigerator:**

$$COP_{(Carnot\ ref.)} = \frac{\text{Heat absorbed}}{\text{Work done}}$$

$$= \frac{1\ \text{tonne}}{1.3} = \frac{14000\ \text{kJ/h}}{1.3 \times 60 \times 60\ \text{kJ/h}} = 2.99\ \textbf{(Ans.)}$$

(*ii*) **Higher temperature of the cycle, T_1:**

$$COP_{(Carnot\ ref.)} = \frac{T_2}{T_1 - T_2}$$

i.e.,
$$2.99 = \frac{235}{T_1 - 235}$$

$\therefore$
$$T_1 = \frac{235}{2.99} + 235 = 313.6\ \text{K}$$

$$= 313.6 - 273 = 40.6°\text{C}\ \textbf{(Ans.)}$$

Q. 69. Find the least power of a perfect reversed heat engine that makes 400 kg of ice per hour at –8ºC from feed water at 18ºC. Assume specific heat of ice as 2.09 kJ/kg K and latent heat 334 kJ/kg.

(*a*) 4.64 kW (*b*) 5.64 kW (*c*) 6.64 kW (*d*) 7.64 kW.

Ans. (a)

Explanation: $T_1 = 18 + 273 = 291$ K; $T_2 = -8 + 273 = 265$ K

$$COP = \frac{T_2}{T_1 - T_2} = \frac{265}{291 - 265} = 10.19$$

Heat absorbed per kg of water (to form ice at 8ºC, *i.e.*, 265 K),

$$R_n = 1 \times 4.18\ (291 - 273) + 334 + 1 \times 2.09 \times (273 - 265)$$

$$= 425.96\ \text{kJ/kg}.$$

Also,
$$COP = \frac{\text{Net refrigerating effect}}{\text{Work done}} = \frac{R_n}{W}$$

i.e.,
$$10.19 = \frac{425.96 \times 400}{W}$$

$\therefore$
$$W = \frac{425.96 \times 400}{10.19} = 16720.7\ \text{kJ/h}$$

$$= 4.64\ \text{kJ/s or } 4.64\ \text{kW}\ \textbf{(Ans.)}$$

Q. 70. The atmospheric conditions are: 20ºC and specific humidity of 0.0095 kg/kg of dry air. Calculate the partial pressure.

(*a*) 0.01224 bar (*b*) 0.01524 bar (*c*) 0.01621 bar (*d*) 0.01723 bar.

Ans. (b)

Explanation: Dry bulb temperature, $t_{db} = 20$ºC

Specific humidity, $W = 0.0095$ kg/kg of dry air.

Partial pressure of vapour, p_v: The specific humidity is given by,

$$W = \frac{0.622\ p_v}{p_t - p_v}$$

$$0.0095 = \frac{0.622\, p_v}{1.0132 - p_v}$$

$$0.0095\,(1.0132 - p_v) = 0.622\, p_v$$
$$0.009625 - 0.0095\, p_v = 0.622\, p_v$$
$$p_v = 0.01524 \text{ bar. } \textbf{(Ans.)}$$

Q. 71. A heat pump working on a reversed Carnot cycle has a COP of 5. If it works as a refrigerator taking 1 kW of work input, the refrigerating effect will be

(*a*) 1 kW (*b*) 2 kW (*c*) 3 kW (*d*) 4 kW.

Ans. (d)

Explanation: $(\text{COP})_{\text{heat pump}} = 5 = \dfrac{\text{Heat rejected}}{\text{Work done}}$ or Heat rejected = 5 × Work done

But, Heat rejected = Work done + refrigerating effect

or 5 × work done = Work done + refrigerating effect

∴ 4 × work done = refrigerating effect

Hence, refrigerating effect = 4 × 1 = 4 kW **(Ans.)** (∵ Work done = 1 kW ... given)

Q. 72. A refrigerating system operating on reversed Brayton refrigeration cycle is used for maintaining 250 K. If the temperature at the end of constant pressure cooling is 300 K and rise in the temperature of air in the refrigerator is 50 K, then the network of compression will be (assume air as the working substance with c_p as 1 kJ per kg per°C)

(*a*) 250 kJ/kg (*b*) 200 kJ/kg (*c*) 50 kJ/kg (*d*) 25 kJ/kg.

Ans. (d)

Explanation: Reversed Brayton refrigerating cycle is shown in the Fig. 16.

Network of compression = $(h_2 - h_1) - (h_3 - h_4)$

Also, $\dfrac{T_2}{T_1} = \dfrac{T_3}{T_4}$ or $T_2 = \dfrac{T_3 \times T_1}{T_4} = \dfrac{300 \times 250}{200} = 375$ K

Net work = $c_p\,[(375 - 250) - (300 - 200)] = 25$ kJ/kg **(Ans.)** (∵ $c_p = 1$ ... given)

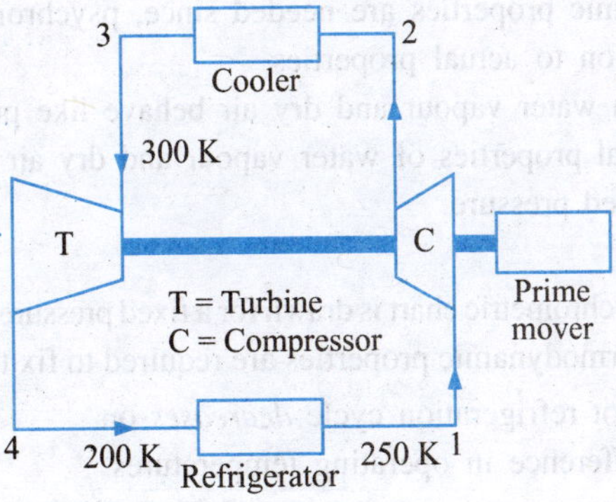

Fig. 16

Q. 73. In a vapour compression refrigeration plant, the refrigerant leaves the evaporator at 195 kJ/kg and the condenser at 65 kJ/kg. For every kg of refrigerant the plant can supply per second, a cooling load of

(*a*) 70 kW (*b*) 100 kW (*c*) 130 kW (*d*) 160 kW.

Ans. (c)

Explanation: Refer to Fig. 17.

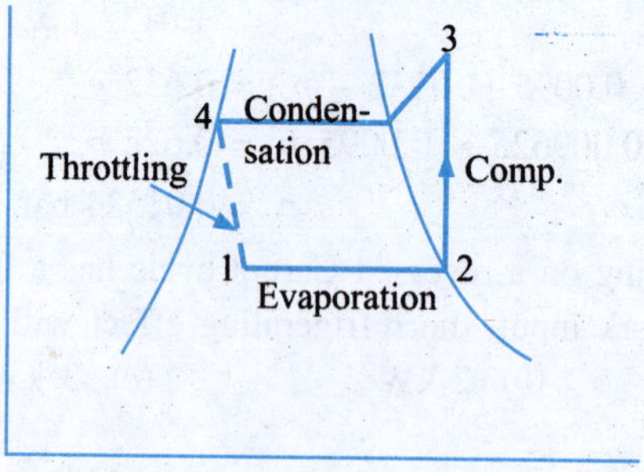

Fig. 17

$h_2 = 195$ kJ/kg, $h_4 = h_1 = 65$ kJ/kg (throttling process)

$\therefore$ Refrigerating effect $= h_2 - h_1 = 195 - 65 = 130$ kJ/kg

Hence, for every kg of refrigerant the plant can supply *per second* a cooling load of 130 kW. **(Ans.)**

Q. 74. A reversible engine has an ideal thermal efficiency of 30%. When it is used as a refrigerating machine with all other conditions unchanged, the coefficient of performance will be

(*a*) 3.33 (*b*) 3.00 (*c*) 2.33 (*d*) 1.33.

Ans. (c)

Explanation:
$$\eta_{carnot} = \frac{T_1 - T_2}{T_1} = 0.3 \quad \text{or} \quad 1 - \frac{T_2}{T_1} = 0.3$$

or,
$$\frac{T_2}{T_1} = 0.7 \quad \text{or} \quad T_2 = 0.7\, T_1$$

$$(\text{COP})_{ref.} = \frac{T_2}{T_1 - T_2} = \frac{0.7\, T_1}{T_1 - 0.7\, T_1} = 2.33 \; \textbf{(Ans.)}$$

Q. 75. Moist air is a mixture of dry and water vapour. Hence three independent intrinsic thermodynamic properties are required to fix its thermodynamic state. While using psychrometric chart, however, only two thermodynamic properties are needed since, psychrometirc chart

(*a*) is an approximation to actual properties

(*b*) assumes that both water vapour and dry air behave like prefect gases

(*c*) is drawn for actual properties of water vapour and dry air

(*d*) is drawn for a fixed pressure.

Ans. (d)

Explanation: Since psychrometric chart is drawn for a fixed pressure (standard atmospheric pressure), therefore, only two thermodynamic properties are required to fix thermodynamic state. **(Ans.)**

Q. 76. The COP of the Carnot refrigeration cycle *decreases* on

(*a*) decreasing the difference in operating temperatures

(*b*) keeping the upper temperature constant and increasing the lower temperature

(*c*) increasing the upper temperature and keeping the lower temperature constant

(*d*) increasing the upper temperature and decreasing the lower temperature.

Ans. (d)

Explanation: $(COP)_{carnot\ ref.} = \frac{T_2}{T_1 - T_2}$. From this relation it is evident that COP will decrease if upper temperature T_1 is increased and T_2 is decreased. Hence (*d*) *is the correct choice.*

Q. 77. In a psychrometric process, the sensible heat added is 30 kJ/sec and the latent heat added is 20 kJ/sec. The sensible heat factor for the process will be

(*a*) 0.3 (*b*) 0.6 (*c*) 0.67 (*d*) 1.5.

Ans. (b)

Explanation: Sensible heat factor (SHF) $= \frac{\text{Sensible heat (SH)}}{\text{Sensible heat (SH) + latent heat (LH)}}$

$$= \frac{30}{30 + 20} = 0.6 \textbf{ (Ans.)}$$

Q. 78. Consider the following statements:

Moisture should be removed from refrigerants to avoid:

1. compressor seal failure
2. freezing at the expansion valve
3. restriction to refrigerant flow
4. corrosion of steel parts

Of these statements

(*a*) 1, 2, 3 and 4 are correct (*b*) 1 and 2 are correct
(*c*) 2, 3 and 4 are correct (*d*) 1, 3 and 4 are correct.

Ans. (c)

Explanation: The compressor seal *cannot fail due to moisture,* all the other conditions do occur due to presence of moisture. **(Ans.)**

Q. 79. Consider the following statements:

1. Practically all common refrigerants have approximately the same COP and power requirement.
2. Ammonia mixes freely with lubricating oil and this helps lubrication of compressors.
3. Dielectric strength of refrigerants is an important property in hermetically sealed compressor units.
4. Leakage of ammonia can be detected by halide torch method.

Of these statements:

(*a*) 1, 2 and 4 are correct (*b*) 2 and 4 are correct
(*c*) 1, 2 and 3 are correct (*d*) 1 and 3 are correct.

Ans. (d)

Explanation: All common refrigerants like F_{11}, F_{12}, F_{22}, NH_3 etc. have approximately the same COP ranging from 4.76 to 5.09 and H.P./ton varies from 0.99 to 1.01.

The electric resistance of the refrigerant is an important factor when it is used in hermetically sealed unit where the motor is exposed to the refrigerant. **(Ans.)**

Q. 80. For large tonnage (more than 200 tons) air-conditioning applications, which one of the following types of compressors is recommended?

(*a*) Reciprocating (*b*) Rotating (*c*) Centrifugal (*d*) Screw.

Ans. (c)

Explanation: The reason why centrifugal compressors are used for large tonnage is that *they can handle large volumes of refrigerant, also the part load efficiency of this kind is higher.* **(Ans.)**

Q. 81. A refrigerating machine working on reversed carnot cycle takes out 2 kW per minute of heat from the system while working between temperature limits of 300 K and 200 K. COP and power consumed by the cycle will be respectively

(*a*) 1 and 1 kW (*b*) 1 and 2 kW (*c*) 2 and 1 kW (*d*) 2 and 2 kW.

Ans. (c)

Explanation:

$$\text{COP} = \frac{T_2}{T_1 - T_2} = \frac{200}{300 - 200} = 2$$

$$\text{Power consumed} = \frac{R_n \text{ (net refrigerating effect)}}{\text{COP}} = \frac{2 \text{ kW}}{2} = 1 \text{ kW (Ans.)}$$

❑❑❑

APPENDICES

Table A.1. Physical properties of various industrial materials

Materials	*Thermal conductivity kJK/m-h-°C*	*Temp. t °C*	*Density kg/m³*	*Specific heat c_p kJ/kg-K*	*Temp. T (K)*
Bakelite varnish	1.0467	—	1400	—	—
Brick red, machine moulded	2.7634	0	1800	0.8792	273
Card board	0.502- 1.256	20	—	1.3073	293
Concrete dry	3.0146	20	1600	0.8374	293
Concrete reinforced	5.5687	20	2200	0.8374	293
Concrete slag	2.5122	20	1500	0.7955	293
Cork sheets	0.1507- 0.1926	80	147-198	1.7585	363
Cotton wool	0.1507	30	80	—	303
Glass plate	2.805- 3.182	0-100	2550	0.7787	273-373
Plywood	1.256- 1.465	20-25	—	2.7215	233-298
Woolen material	0.1884	—	240	—	—

Table A.2. Minimum design working pressure for evaporator

Refrigerant	*Chemical formula*	*Symbol*	*Design pressure bar (gauge)*
Trichloromonofluoromethance	CCl_3F	R-11	2.07
Dichlorodifluoromethance	CCl_2F_2	R-12	9.65
Monochlorodifluoromethane	$CHClF_2$	R-22	10.34
Dichlorotetrafluoromethane	$CClF_2CClF_2$	R-114	3.45
Ammonia	NH_3	R-717	10.34

Table A.3. Allowable S values of stress for refrigeration pipe

S. No.	*Metal and type of joint*	*Stress (bar)*
1.	Seamless Steel Grade A	830
2.	Seamless Steel Grade B	1030
3.	Lap Welded Steel	620
4.	Butt Welded Steel	470
5.	Electric Resistance Welded Steel	700
6.	Lap Welded Wrought Iron	550
7.	Butt Welded Wrought Iron	410
8.	Seamless Red Brass	480
9.	Copper	410

Table A.4. The amount of drinking water required under various conditions

Applications	*Temp. °C*	*Amount of water used including drainage (l = litre)*
Restaurants	7-10	0.5 *l*/h per person
Schools	7-10	0.56 *l*/h per student
Offices	7-10	0.56 *l*/h per person
Stores	7-10	4.5 *l*/h per 100 customers
Hotels	7-10	0.4 *l*/h per room
Hospitals	7-10	0.4 *l*/h per bed and 0.4 *l*/h per attendant
Theatres	7-10	4.5 *l*/h per 100 seats
Light duty manufacturing shop	10-13	0.9 *l*/h per person
Heat duty manufacturing shop	10-13	1.1 *l*/h per person

Table A.5. Minimum test pressure for leak testing in refrigerant system

S. No	*Refrigerant*			*Minimum pressure bars**	
	Name	*Symbol*	*Chemical formula*	*High pressure side*	*Low pressure side*
1.	Ammonia	R-717	NH_3	21.7	11.4
2.	Butane	R-600	C_4H_{10}	77.56	3.8
3.	Carbondioxide	R-744	CO_2	108.5	72.3
4.	Dichlorodifluoromethane	R-12	CCl_2F_2	17.2	10.7
5.	Dichloroethylene	R-1130	$CH\ Cl = CH\ Cl$	2.17	2.17
6.	Dichloromethane (Carren No.1) (Methylene chloride)	R-30	$CH_2\ Cl_2$	2.17	2.17
7.	Dichloromonofluoromethane	R-21	$CH\ Cl_2F$	5.8	3.77
8.	Dichlorotetrafluoroethane	R-114	$C_2\ Cl_3\ F_4$	4.46	4.46
9.	Ethane	R-170	$C_2\ H_6$	86.8	49.3
10.	Ethyl chloride	R-160	$C_2\ H_5\ Cl$	5.15	4.46
11.	Ethylene	R-1150	$C_2\ H_4$	111.3	86.8
12.	Isobutane	R-601	$(CH_3)_3\ CH$	9.97	5.8
13.	Methyl chloride	R-40	$CH_3\ Cl$	15.49	8.68
14.	Methyl formate	R-611	$HCOO\ CH_3$	4.46	4.46
15.	Monochlorodifluoromethane	R-22	$CH\ Cl\ F_2$	22.8	11.4
16.	Propane	R-290	$C_2\ H_8$	22.8	11.4
17.	Sulphur-dioxide	R-764	SO_2	12.73	6.36
18.	Trichloromono flurometháne	R-11	$C\ Cl_3\ F$	2.17	2.17
19.	Trichlorotrifluoroethane	R-113	$C_2\ Cl_3\ F_3$	2.17	2.17

* Pressures are absolute = (p gauge + 1.013 bar)

Table A.6. Overall heat transfer coefficients U (kJ/m^2-h-K) for liquid coolers

S. No.	Type of evaporator	Minimum	Maximum
1.	Flooded shell-and-plain-tube (water to refrigerants 12, 22 and 717)	2700	3800
2.	Flooded shell-and-finned-tube (water to refrigerant 12 or 22)	1900	2900
3.	Flooded shell-and-plain-tube (brine to refrigerant 717)	940	2090
4.	Flooded shell-and-plain-tube (brine to refrigerant 12 or 22)	630	1900
5.	Dry expansion, shell-and-plain-tube (water to R-12, 22, or 717) (Refrigerant in tube)	1670	3350
6.	Dry expansion, shell-and-internal finned-tubes (Water to R-12 or 22) (Refrigerant in tubes)	3350	5200
7.	Dry expansion, shell-and-plain-tube (brine to R-12, 22 or 717) (refrigerant in tubes)	1250	2900
8.	Dry expansion, shell-and-internal finned tubes (non-salt brines to R-12 or 22)	2090	3550
9.	Shell-and-plain-tube coil (water in shell) (R-12, 22, or 717 in coil)	210	520
10.	Baudelot cooler, dry expansion (refrigerant 12 or 22 to water)	2090	4200
11.	Baudelot cooler, dry expansion (refrigerant 717 to water)	1250	3140
12.	Baudelot cooler, dry expansion (refrigerant 12 or 22 to water)	1250	2500
13.	Double pipe cooler (refrigerant 717 to water)	1050	3140
14.	Double pipe cooler (refrigerant 717 to brine)	1050	2600
15.	Spray type shell-and-type water coolers (refrigerant 11, 12, 22, 113 or 717)	3140	2500
16.	Tank-and-agitator, coil type water cooler (flooded, refrigerant 717)	1670	2600
17.	Tank-and-agitator, coil type water cooler (flooded, refrigerant 12 or 22)	1250	2090
18.	Tank, ammonia (R-717) to brine cooling, coils between can in ice tank	310	840
19.	Tank, high velocity raceway type (refrigerant 717 to brine)	1670	2300

Table A.7. Specific heat of liquid R-22, kJ/kg-K
(Valid from-17 °C to 55 °C)

$$c_{pf} = 0.2843 + 0.0032826\ T,\ T \text{ in K}$$

Table A.8. The physical properties of vapour at atmospheric pressure

Temp. °C	Thermal conductivity $k \times 10^3$ kJ/m-h.-C	Dynamic viscosity $\mu \times 10^6$ kg-s/m^2	Pr	Kinematic viscosity $\nu \times 10^4$ m^2/sec.	Specific heat c_p kJ/kg-K	Temp. K
			Sulphur dioxide, SO_2			
0	30.146	1.23	0.874	4.14	0.6071	273
100	44.382	1.64	0.863	7.51	0.6615	373
200	59.874	2.04	0.856	11.8	0.7117	473
			Ammonia (NH_3)			
0	75.785	0.954	0.908	12.2	2.0432	273
100	122.260	1.33	0.852	23.2	2.2191	373
200	175.854	1.70	0.818	38.0	2.3991	473
300	235.728	2.10	0.812	56.4	2.5833	573

Table A.9. Dynamic viscosity of R-22

t	C	– 40	– 20	0	20	40	60	80
	(cp)	0.351	0.295	0.262	0.238	0.221	0.207	0.197

Table A.10. Minimum odour-free air requirement to eject objectionable body-odours

Type of occupant	*Space per person*	*Odour free air supply*
Sendentary adults of average socio-economic status	m^3	m^3
	2.83	0.71
	5.66	0.453
	8.50	0.34
	14.1	0.20
Labourers	5.66	0.65
Grade school children of average socio-economic status	2.83	0.82
	5.66	0.594
	8.5	0.481
	14.1	0.311
Labourers Grade school children of average lower socio-economic status	5.66	1.08
Private grade school	2.83	0.623

Table A.11. Recommended fouling factors for use in calculating heat transfer rates (m^2-h-°C-/kJ)

Application	*Non-Ferrous tubes*	*Ferrous tubes*
Brine coolers (brine side)		
Inhibited salt brines	0.000025	0.00005
Non-inhibited salt brines	0.00005	0.0001
Solvent brines (methylene chloride, fluorinated hydrocarbon refrigerants)	None	None
Water coolers (water side)		
Recirculated water, closed systems	0.000025	0.00005
Recirculated water, open system	0.00005	0.0001
Non-circulated water	Use 1/2 of that allowed for condenser service but in no case than 0.000025	

Table A.12. Ventilation standards

Application	Smoking	Ventilation (m^3/s per person) $\times 10^{-2}$		Ventilation $m^3/s/m^2$ of floor area
		Recommended	Minimum	Minimum ×1000 +
Apartment, average...	Some...	9.5	4.7	1.00
Apartment, Deluxe...	Some...	14.2	12.0	1.676
Banking space...	Occasional	4.7	3.5	
Barber shops...	Considerable	7.1	4.7	
Brokers parlers	Occasional	4.7	3.5	
Brokers' board rooms...	Very heavy	24.0	9.5	
Cocktail bars	Heavy	19.0	12.0	
Corridors (Supply or exhaust)	—	—	—	1.27
Department stores	None	3.5	2.4	0.254
Directors' room	Extreme	24	14.2	
Drug stores†	Considerable	4.7	3.5	
Factories**	None	4.7	3.5	0.508
Five and ten cent stores	None	3.5	2.4	
Funeral parlers	None	4.7	3.5	
Garages††	—	—	—	5.08
Hospitals operating rooms	None	—	—	10.16
Hospitals private rooms	None	14.2	12.0	1.676
Hospitals wards	None	9.5	4.7	
Hotel rooms	Heavy	14.2	12.0	1.676
Kitchens, Restaurant	—	—	—	20.32
Kitchens, Residence	—	—	—	10.16
Laboratories†	Some	9.5	7.1	
Meeting rooms	Very heavy	24	14.2	6.35
Offices, General	Some	7.1	4.7	
Private	None	12.0	7.1	1.27
Private	Considerable	14.2	12	1.27
Restaurant cafeteria†	Considerable	5.7	4.7	
Dining Room†	Considerable	7.1	0.7	
School Rooms	None	—	—	
Shop, Retail	None	4.7	3.6	
Theatre**	None	3.6	2.4	
Theatre	Some	7.1	4.7	
Toilets** (exhaust)	—	—	—	10.16

* When minimum is used take the larger of two.

** See local codes which may govern.

† May be governed by exhaust.

†† May be governed by special sources of contamination of local codes.

All outside air recommended to overcome explosion hazard of anesthetics.

Table A.13. Infilteration load

Room volume m^3	Number of air changes per 24 h		Number of air changes per 24 h long storage	Room volume m^3 long storage
	Average load	Heavy		
5	22.6	35.45	12.65	55
10	20.5	32.26	11.26	60
15	19.25	30.44	10.69	70
20	18.25	28.76	9.92	80
25	17.37	27.27	8.92	90
30	16.60	26.04	8.26	100
35	15.91	25.10	7.22	120
40	15.29	24.45	6.44	140
45	14.74	24.08	6.13	150
50	14.25	23.99	5.85	160
55	13.80		5.37	180
60	13.40		4.99	200
65	13.03		4.42	240
70	12.70		4.30	250
75	12.40		4.01	280
80	12.12		3.86	300
85	11.87		3.55	350
90	11.67		3.38	400
95	11.43		3.19	450
			3.08	500
			2.94	600
			2.86	700
			2.82	800
			2.80	1000
			2.80	1500
			2.80	2000
			2.75	2500
			2.73	3000
			2.73	

Table A.14. Thermal energy release from human body

Degree of activity (At t_{db} = 26.50 °C, t_{wb} = 25.50 °C)	*Energy release kJ/h (Male, adult)*
Seated at rest	419
Seated very light work	469
Moderately active office work	502
Standing light work	586
Light bench work factor	837
Walking; moderately heavy work	1047
Heavy work factor	1570
Seriously reading or thinking	1570

Table A.15. Storage conditions and properties of perishable products

Products	Storage temp.°C		Optimum humidity		Freezing point °C	Specific heat kJ/kg		Latent heat of fusion kJ/kg	Storage period months	Chilling	
	Retail whole sale	Ware house storage	Gravity air circulation	Forced air circulation		Above freezing	Below freezing			Time h	Rate factor
Apples	2 →5	– 1→ 0	85	88	– 2.2	3.768.	2.051	285	8	–	–
Bananas	13 → 13.5	13 → 13.5	80	80	– 3 → 1	3.768	–	–	10 days	12	0.1
Grapes	2 → 5	1 → 0.0	80	85	– 0.2	3.768	2.554	285	3 to 6	20	0.8
Oranges	5 → 7	i → 1.0	80	85	2.8	3.768	1.967	293	1 → 6	22	0.7
Carrots	2 → 5	0 → 1.0	90	95	– 1.5	3.600	1.884	293	2	24	0.8
Potatoes	2 → 10	4 → 6	85	90	+ 1.5	3.600	1.967	264	4 to 6	24	1.0
Onions	10 → 16	0 → 1.0	70	75	– 1	3.768	2.135	293	6	24	0.8
Mangoes	– –	0.5 → 2.0	85	85	0.0	3.768	1.926	311	5 to 6	–	–
Fish (Green)	– 10 → 6	– 15 → 12	80	80	– 2.0	3.182	1.716	234	10 days	2	1.0
Egg (Frozen)	– 5 → 7	1 → 0.5	85	85	– 3.0	3.182	1.674	232	12	24	0.67
Butter	7 → 10	– –	80	80	25→21.0	2.679	–	–	10 days	–	–

Usage factor, kJ/m^3 K–24 h

Table A.16. Heat emission from commodities

Commodities	*Temp. °C*	*kJ ton-24 h*
Apples	0.0	628
Bananas (Green)	12	3475
Carrot	0	2093
	15	8374
Lemons	0	586
	27	6699
Oranges	0	712
	27	8370
Potatoes	0	2510
	15	6280
Tomatoes	0	1050
	15	5860

Table A.17. Usage heat gain, kJ per 24 h-m^3 of the interior volume °C

Storage capacity m^3		*Services*	
	Average	*Heavy*	*Long storage*
0.5	82	92	
0.75	58	76.2	
1.0	46	66.8	
1.5	35.6	55.2	
2.0	30.2	48.0	
2.5	27.3	42.4	
3.0	25.4	39.4	
5.0	22.5	36.0	
10.0	20.3	31.6	
15.0	19.0	29.8	
20.0	18.0	28.5	
30.0	16.4	26.2	
40.0	15.0	24.2	
50.0	13.8		13.5
60	13.2		11.8
75			11.2
100			8.32
125			7.05
150			6.25
200			5.10
250			4.23
300			3.75
400			3.35
500			3.14
600			2.90
750			2.80
1000			2.80
1500			2.75
2000			2.73
3000			2.70

Table A.18. Physical properties of liquids at saturation conditions

Temperature °C	*Density* kg/m^3	*Kinematic viscosity* × 10^4 m^2/s	*Thermal diffusivity* × 10^6 m^2/h	*Prandtl number* Pr	*Thermal conductivity* kW/m°C	*Specific heat* c_p kJ/kg °C
			Water H_2O		R-718	
0	1002	1.788	471	13.600	552.4	4.216
20	1000	1.006	515	7.020	597.8	4.178
40	995	0.657	544	4.340	628.0	4.178
60	985	0.478	559	3.020	651.3	4.183
80	974	0.364	589	2.220	668.7	4.195
100	961	0.293	605	1.740	680.4	4.216
120	945	0.247	615	1.445	856.0	4.250
140	928	0.213	621	1.241	683.8	4.283

Table A.19. Allowances for solar R-Roof, EW-East Wall, W-West and S-South

Radiation	*EW*	*SW*	*WW*	*FR*
Dark coloured tar roofing, Slate roof	4.4	3.0	4.4	11
Medium colour wood, brick, cement, gray	3.3	2.2	3.3	8.3
Light coloured, white stone white paint	2.2	2.1	2.2	5

Table A.20. Wet and dry bulb temperature relative humdity (Dry bulb temp., Wet bulb temp. in °C)

Dry bulb °C	*Difference ($t_{db} - t_{wb}$) in °C*									
	0.5	*1*	*1.5*	*2*	*2.5*	*3*	*4*	*5*	*6*	*7*
– 1	90	79	69	59	49	39	20	1.0	–	–
0	90	81	71	61	52	44	25	7	–	–
1	90	81	73	64	55	47	29	13	–	–
2	91	82	73	64	57	49	33	17	1.0	–
5	92	84	76	68	61	53	38	24		
10	93	87	81	74	68	62	53	38	27	16
15	94	89	84	78	73	68	58	48	39	30
20	95	91	86	81	77	73	64	56	48	40
25	96	92	88	84	80	76	68	61	54	47
30	96	93	89	85	82	78	72	65	59	53
36	96	93	90	87	84	81	75	69	63	58
40	96	94	91	88	85	82	76	71	66	61
46	97	94	91	89	86	83	78	73	68	64
50	97	95	92	89	87	82	79	74	70	65

REFRIGERATION AND AIR-CONDITIONING TABLES AND CHARTS

Table 1. Saturated Trichloromonofluoromethane ($CC1_3F$), R-11

Datum at – 40°C, $h_f = 0$, $s_f = 0$

Saturation temp., °C	Saturation pressure, bar	Specific volume, m^3/kg		Specific enthalpy, kJ/kg			Specific entropy, kJ/kg K	
		Liquid	Vapour	Liquid	Vapour	Latent	Liquid	Vapour
(t)	(p)	(v_f)	(v_g)	(h_f)	(h_g)	(h_{fg})	(s_f)	(s_g)
– 40	0.05096	0.000617	2.7625	0.00	203.48	203.48	0.0000	0.8730
– 38	0.05942	0.000619	2.5360	1.65	204.48	202.83	0.00068	0.8701
– 36	0.06787	0.000621	2.3094	3.30	205.48	202.18	0.0137	0.8671
– 34	0.07633	0.000623	2.0827	4.95	206.48	201.53	0.0206	0.8642
– 32	0.08478	0,000624	1.8562	6.60	207.48	200.88	0.0274	0.8613
– 30	0.09323	0.000626	1.6296	8.25	208.48	200.23	0.0343	0.8583
– 28	0.13045	0.000627	1.4448	9.90	209.48	199.58	0.0410	0.8554
– 26	0.11586	0.000629	1.3123	11.55	210.49	198.94	0.0477	0.8533
– 24	0.12828	0.000631	1.1798	13.20	211.49	198.29	0.0544	0.8508
– 22	0.14292	0.000633	1.0722	14.85	212.50	197.65	0.0611	0.8483
– 20	0.15869	0.000634	0.9770	16.51	213.50	196.99	0.0678	0.8462
– 18	0.17414	0.000636	0.8793	18.19	214.52	196.33	0.0741	0.8441
– 16	0.19280	0.000638	0.8045	19.82	215.53	195.71	0.0808	0.8420
– 14	0.21260	0.000639	0.7345	21.49	216.54	195.05	0.0875	0.8403
– 12	0.23390	0.000641	0.6713	23.17	217.55	194.38	0.0938	0.8387
– 10	0.25848	0.000643	0.6191	24.83	218.56	193.73	0.1009	0.8366
– 8	0.28306	0.000645	0.5671	26.50	219.57	193.07	0.1063	0.8349
– 6	0.30280	0.000647	0.5194	28.17	220.59	192.42	0.1126	0.8332
– 4	0.33967	0.000649	0.4801	29.85	221.61	191.76	0.1189	0.8315
– 2	0.36983	0.000615	0.4409	31.54	222.63	191.09	0.1252	0.8303
0	0.40360	0.000653	0.4069	33.22	223.64	190.42	0.1315	0.8286
1	0.42190	0.000653	0.3920	34.06	224.14	190.08	0.1344	0.8280
2	0.44021	0.000654	0.3770	34.90	224.65	189.75	0.1374	0.8274
3	0.45852	0.000655	0.3620	35.74	225.15	189.41	0.1436	0.8267
4	0.47683	0.000656	0.3471	36.59	225.65	189.06	0.1436	0.8261
5	0.49720	0.000657	0.3340	37.43	226.16	188.73	0.1465	0.8254
6	0.51917	0.000658	0.3225	38.28	226.66	188.38	0.1498	0.8248
7	0.54118	0.000659	0.3109	39.13	227.17	188.04	0.1526	0.8242
8	0.56317	0.000660	0.2994	39.98	227.68	187.70	0.1562	0.8236
9	0.58517	0.000661	0.2878	40.83	228.18	187.35	0.1586	0.8229
10	0.60717	0.000662	0.2763	41.68	228.69	187.07	0.1616	0.8223
11	0.3320	0.000663	0.2673	42.53	229.20	186.67	0.1646	0.8219
12	0.65921	0.000664	0.2584	43.39	229.71	186.32	0.1675	0.8215
13	0.68523	0.000665	0.2495	44.24	230.21	185.97	0.1706	0.8210
14	0.71124	0.000666	0.2405	45.10	230.72	185.62	0.1738	0.8206

Contd....

t	p	v_f	v_g	h_f	h_g	h_{fg}	s_f	s_g
15	0.73727	0.000667	0.2316	45.95	231.22	185.27	0.1767	0.8200
16	0.76552	0.000669	0.2235	46.81	231.73	184.92	0.1796	0.8194
17	0.79655	0.000670	0.2163	47.67	232.23	184.56	0.1825	0.8170
18	0.82759	0.000671	0.2092	48.54	232.73	184.19	0.1855	0.8186
19	0.85862	0.000672	0.2021	49.40	233.24	183.84	0.1885	0.8180
20	0.88966	0.000673	0.1950	50.26	233.74	183.48	0.1913	0.8173
21	0.92070	0.000674	0.1878	51.12	234.24	183.12	0.1943	0.8169
22	0.95625	0.000675	0.8121	51.99	234.74	182.75	0.1972	0.8165
23	0.99237	0.000676	0.1764	52.85	235.24	182.39	0.2003	0.8163
24	1.02496	0.000677	0.1708	53.72	235.75	182.03	0.2031	0.8160
25	1.06117	0.000678	0.1651	54.59	236.25	181.66	0.2061	0.8157
26	1.10074	0.000679	0.1595	55.46	236.75	181.29	0.2089	0.8152
27	1.13880	0.000680	0.1542	56.32	237.25	180.93	0.2120	0.8150
28	1.17917	0.000681	0.1495	57.20	237.76	180.56	0.2148	0.8148
29	1.22273	0.000682	0.1452	58.07	238.25	180.18	0.2177	0.8144
30	1.26070	0.000684	0.1401	58.94	238.76	179.82	0.2206	0.8140
31	1.30665	0.000685	0.1362	59.82	239.25	179.43	0.2234	0.8135
32	1.34821	0.000686	0.1316	60.70	239.75	179.05	0.2261	0.8131
33	1.39568	0.000687	0.1278	61.57	240.24	178.67	0.2290	0.8128
34	1.44414	0.000689	0.1242	62.45	240.73	178.28	0.2320	0.8126
35	1.49276	0.000690	0.1205	63.32	214.23	177.91	0.2349	0.8123
36	1.54130	0.000691	0.1168	64.20	214.72	177.52	0.2378	0.8119
37	1.58983	0.000692	0.1131	65.07	242.22	177.15	0.2406	0.8116
38	1.64000	0.000693	0.1096	65.95	242.72	176.77	0.2433	0.8114
39	1.69571	0.000694	0.1066	66.84	234.20	176.66	0.2462	0.8112
40	1.75145	0.000696	0.1037	67.74	243.69	175.95	0.2490	0.8110
41	1.80718	0.000697	0.1007	68.63	244.18	175.55	0.2518	0.8108
42	1.86292	0.000698	0.0977	69.52	244.67	175.15	0.2546	0.8106
43	1.91866	0.000699	0.0947	70.41	245.16	174.75	0.2574	0.8104
44	1.97952	0.000700	0.0921	71.31	245.65	174.34	0.2600	0.8102
45	2.04300	0.000701	0.0896	72.20	246.14	173.94	0.2029	0.8100
46	2.10641	0.000702	0.0871	73.08	246.63	173.55	0.2658	0.8098
47	2.16980	0.000704	0.0847	73.98	247.12	173.14	0.2686	0.8096
48	2.23310	0.000705	0.0823	74.87	247.60	172.73	0.2713	0.8094
49	2.29670	0.000706	0.0798	75.76	248.07	172.31	0.2741	0.8092
50	2.36000	0.000707	0.0774	76.67	248.53	171.86	0.2768	0.8090

Table 2. Saturated Dichlorodifluoromethane (CCl_2F_2), R-12

Datum at $-40°C$, $h_f = 0$, $s_f = 0$

Saturation temp., °C	*Saturation pressure, bar*	*Specific volume, m³/kg*		*Specific enthalpy, kJ/kg*			*Specific entropy, kJ/kg K*	
		Liquid	*Vapour*	*Liquid*	*Vapour*	*Latent*	*Liquid*	*Vapour*
(t)	*(p)*	(v_f)	(v_g)	(h_f)	(h_g)	(h_{fg})	(s_f)	(s_g)
– 100	0.01185	0.000600	10.1951	– 51.84	142.00	193.84	– 0.2567	0.8628
– 95	0.01864	0.000604	6.6231	– 47.56	144.22	191.78	– 0.2323	0.8442
– 90	0.02843	0.000608	4.4206	– 43.28	146.46	189.74	– 0.2086	0.8273
– 85	0.04254	0.000613	3.0531	– 39.00	148.73	187.73	– 0.1856	0.8122
– 80	0.06200	0.000617	2.1519	– 34.72	151.02	185.74	– 0.1631	0.7985
– 75	0.08826	0.000622	1.5462	– 30.42	153.32	183.74	– 0.1412	0.7861
– 70	0.12298	0.000627	1.1314	– 26.12	155.63	181.75	– 0.1198	0.7665
– 65	0.16807	0.000632	0.8421	– 21.81	157.96	179.77	– 0.0988	0.7648
– 60	0.22665	0.000637	0.6401	– 17.48	160.29	177.77	– 0.0783	0.7558
– 55	0.30052	0.000643	0.4930	– 13.14	162.62	175.76	– 0.0581	0.7475
– 50	0.39237	0.000648	0.3845	– 8.78	164.95	173.73	– 0.0384	0.7401
– 45	0.50512	0.000654	0.3035	– 4.39	167.27	171.66	– 0.0190	0.7334
– 40	0.64190	0.000660	0.2422	0.00	169.60	169.60	0.0000	0.7274
– 38	0.70460	0.000663	0.2221	1.76	170.52	168.76	0.0075	0.7251
– 36	0.77196	0.000665	0.2040	3.53	171.44	167.91	0.0149	0.7230
– 34	0.84421	0.000667	0.1877	5.31	172.36	167.05	0.0224	0.7209
– 32	0.92776	0.000670	0.1729	7.08	173.28	166.20	0.0298	0.7190
– 30	1.00441	0.000673	0.1596	8.86	174.20	165.34	0.0371	0.7171
– 28	1.09311	0.000676	0.1475	10.64	175.11	164.47	0.0444	0.1753
– 26	1.18778	0.000678	0.1364	12.43	176.02	163.59	0.0516	0.7135
– 24	1.28858	0.000681	0.1265	14.22	176.93	162.71	0.0588	0.7118
– 22	1.39581	0.000683	0.1173	16.02	177.83	161.81	0.0660	0.7102
– 20	1.50972	0.000686	0.1090	17.82	178.73	160.91	0.0731	0.7087
– 18	1.63104	0.000689	0.1014	19.62	179.63	160.01	0.0801	0.7073
– 16	1.75963	0.000692	0.0944	21.43	180.53	159.10	0.0871	0.7059
– 14	1.89575	0.000695	0.0880	23.23	181.42	159.19	0.0941	0.7045
– 12	2.04605	0.000698	0.0821	25.05	182.31	157.26	0.1010	0.7032
– 10	2.19172	0.000701	0.0767	26.87	183.19	156.32	0.1080	0.7019
– 8	2.35272	0.000704	0.0717	28.70	184.06	155.36	0.1148	0.7007
– 6	2.52244	0.000707	0.0672	30.53	184.94	154.41	0.1217	0.6996
– 4	2.70116	0.000710	0.0630	32.37	185.80	153.43	0.1285	0.6986
– 2	2.88921	0.000713	0.0591	34.20	186.67	152.47	0.1352	0.6975
0	3.08690	0.000717	0.0555	36.05	187.53	151.48	0.1420	0.6965
1	3.18974	0.000719	0.0538	36.98	187.95	150.97	0.1453	0.6961
2	3.29513	0.000720	0.0521	37.90	188.38	150.48	0.1489	0.6956

Contd....

t	p	v_f	v_g	h_f	h_g	h_{fg}	s_f	s_g
3	3.40310	0.000721	0.0505	38.83	188.81	149.98	0.1521	0.6951
4	3.51367	0.000723	0.0490	39.76	189.23	149.47	0.1553	0.6947
5	3.62690	0.000725	0.0475	40.69	189.65	148.96	0.1586	0.6943
6	3.74280	0.000727	0.0461	41.62	190.07	148.45	0.1620	0.6938
7	3.86141	0.000729	0.0447	42.56	190.49	147.93	0.1653	0.6933
8	3.98283	0.000730	0.0434	43.50	190.91	147.41	0.1686	0.6929
9	4.10702	0.000732	0.0422	44.43	191.32	146.89	0.1719	0.6925
10	4.23407	0.000734	0.0410	45.37	191.74	146.37	0.1752	0.6921
11	4.36442	0.000736	0.0398	46.31	192.15	145.84	0.1784	0.6917
12	4.49763	0.000738	0.0386	47.26	192.56	145.30	0.1817	0.6913
13	4.63386	0.000739	0.0375	48.20	192.97	144.77	0.1850	0.6909
14	4.77312	0.000741	0.0365	49.15	193.38	144.23	0.1883	0.6906
15	4.91545	0.000743	0.0355	50.10	193.79	143.69	0.1915	0.6902
16	5.06087	0.000745	0.0345	51.05	194.19	143.14	0.1948	0.6898
17	5.20942	0.000747	0.0335	52.00	194.59	142.59	0.1981	0.6894
18	5.36117	0.000749	0.0326	52.95	194.99	142.04	0.2013	0.6891
19	5.51614	0.000751	0.0317	53.91	195.38	141.47	0.2046	0.6888
20	5.67441	0.000753	0.0308	54.87	195.78	140.91	0.2078	0.6884
21	5.83635	0.000756	0.0300	55.83	196.17	140.34	0.2110	0.6881
22	6.00171	0.000758	0.0292	56.79	196.56	139.77	0.2143	0.6878
23	6.17050	0.000759	0.0284	57.75	196.96	139.21	0.2174	0.6875
24	6.34269	0.000761	0.0276	58.73	197.34	138.61	0.2207	0.6872
25	6.51840	0.000764	0.0269	59.70	197.73	138.03	0.2239	0.6868
26	6.69765	0.000766	0.0262	60.67	198.11	137.44	0.2271	0.6865
27	6.88048	0.000768	0.0255	61.65	198.50	136.85	0.2303	0.6862
28	7.06704	0.000770	0.0248	62.63	198.87	136.24	0.2335	0.6859
29	7.25738	0.000772	0.0241	63.61	199.25	135.64	0.2368	0.6856
30	7.45103	0.000775	0.0235	64.59	199.62	135.03	0.2400	0.6853
31	7.64903	0.000777	0.0230	65.58	199.99	134.41	0.2431	0.6850
32	7.85089	0.000779	0.0225	66.57	200.36	133.79	0.2463	0.6847
33	8.05662	0.000782	0.0218	67.56	200.73	133.17	0.2495	0.6845
34	8.26621	0.000784	0.0212	68.56	201.09	132.53	0.2527	0.6842
35	8.48000	0.000786	0.0207	69.56	201.45	131.89	0.2559	0.6839
36	8.69766	0.000789	0.0202	70.55	201.80	131.25	0.2591	0.6836
37	8.91904	0.000792	0.0196	71.55	202.16	130.61	0.2623	0.6833
38	9.14483	0.000794	0.0191	72.56	202.51	129.95	0.2654	0.6830
39	9.37497	0.000796	0.0186	73.57	202.86	129.29	0.2685	0.6828
40	9.60897	0.000799	0.0182	74.59	203.20	128.61	0.2718	0.6825
41	9.84793	0.000802	0.0177	75.61	203.54	127.93	0.2750	0.6822
42	10.09131	0.000804	0.0173	76.62	203.87	127.25	0.2782	0.6820
43	10.33862	0.000807	0.0168	77.65	204.21	126.56	0.2814	0.6817
44	10.59021	0.000810	0.0164	78.68	204.55	125.87	0.2846	0.6814

Contd....

t	p	v_f	v_g	h_f	h_g	h_{fg}	s_f	s_g
45	10.84655	0.000813	0.0160	79.71	204.87	125.16	0.2878	0.6811
46	11.10758	0.000815	0.0156	80.75	205.19	125.44	0.2909	0.6808
47	11.37304	0.000818	0.0153	81.79	205.51	123.72	0.2941	0.6805
48	11.64290	0.000821	0.0149	82.83	205.83	123.00	0.2973	0.6802
49	11.91724	0.000824	0.0146	83.88	206.14	122.26	0.3005	0.6800
50	12.19655	0.000827	0.0142	84.94	206.45	121.51	0.3037	0.6797

Table 3. Saturated Monochlorodifluoromethane ($CHClF_2$), R-22

Datum at – 40°C, $h_f = 0$, $s_f = 0$

Saturation temp., °C	Saturation pressure, bar	Specific volume, m^3/kg		Specific enthalpy, kJ/kg			Specific entropy, kJ/kg K	
(t)	(p)	Liquid (v_f)	Vapour (v_g)	Liquid (h_f)	Vapour (h_g)	Latent (h_{fg})	Liquid (s_f)	Vapour (s_g)
– 100	0.02009	0.000643	8.3412	– 63.45	203.73	267.18	– 0.3144	1.2293
– 95	0.03150	0.000647	5.4344	– 58.14	206.19	264.33	– 0.2843	1.2004
– 90	0.04792	0.000652	3.6381	– 52.87	208.64	261.51	– 0.2250	1.1736
– 85	0.07731	0.000656	2.5204	– 47.61	211.11	258.72	– 0.2269	1.1489
– 80	0.10393	0.000661	1.7816	– 42.40	213.60	256.00	– 0.1989	1.1267
– 75	0.14759	0.000666	1.2842	– 37.17	216.11	253.28	– 0.1721	1.1066
– 70	0.20517	0.000672	0.9420	– 31.93	218.62	250.55	– 0.1461	1.0874
– 65	0.27965	0.000677	0.7037	– 26.68	221.16	247.84	– 0.1206	1.0702
– 60	0.37448	0.000683	0.5351	– 21.42	223.67	245.09	– 0.0959	1.0543
– 55	0.49621	0.000689	0.4131	– 16.13	226.18	242.31	– 0.0712	1.0396
– 50	0.64758	0.000696	0.3229	– 10.81	228.69	239.50	– 0.0473	1.0262
– 45	0.83241	0.000702	0.2556	– 5.40	231.20	236.60	– 0.0234	1.0137
– 40	1.05586	0.000709	0.2049	0.00	233.67	233.67	0.000	1.0024
– 38	1.15862	0.000712	0.1882	2.20	234.65	232.45	0.0096	0.9982
– 36	1.26910	0.000715	0.1728	4.40	235.65	231.25	0.0188	0.9940
– 34	1.38731	0.000719	0.1590	6.58	236.60	230.02	0.0280	0.9902
– 32	1.51324	0.000712	0.1465	8.83	237.60	228.77	0.0373	0.9860
– 30	1.64690	0.000724	0.1353	11.05	238.55	227.50	0.0464	0.9283
– 28	1.78938	0.000727	0.1253	13.29	239.52	226.23	0.0557	0.9785
– 26	1.94069	0.000731	0.1161	15.58	240.48	224.90	0.0645	0.9747
– 24	2.10207	0.000734	0.1077	17.77	241.41	223.64	0.0733	0.9710
– 22	2.27448	0.000738	0.1000	19.99	242.33	222.34	0.0821	0.9676
– 20	2.45793	0.000741	0.0930	22.21	243.25	221.04	0.0908	0.9638
– 18	2.65310	0.000744	0.0865	24.47	244.17	219.70	0.0996	0.9605
– 16	2.85903	0.000748	0.0806	26.72	245.08	218.36	0.1080	0.9576
– 14	3.07876	0.000752	0.0752	28.94	245.96	217.02	0.1164	0.9538
– 12	3.31172	0.000756	0.0701	31.16	246.84	215.68	0.1248	0.9508
– 10	3.55793	0.000759	0.0655	33.40	247.72	214.32	0.1336	0.9479
– 8	3.81321	0.000763	0.0612	35.66	248.60	212.94	0.1419	0.9454
– 6	4.09172	0.000767	0.0573	37.92	249.46	211.54	0.1503	0.9425
– 4	4.37972	0.000771	0.0536	40.19	250.30	210.11	0.1591	0.9396
– 2	4.68317	0.000775	0.0503	42.52	251.14	208.62	0.1675	0.9370
0	5.00207	0.000779	0.0471	44.94	251.97	207.03	0.1763	0.9345
1	5.16841	0.000 781	0.0457	46.16	252.37	260.21	0.1809	0.9333
2	5.33917	0.000783	0.0443	47.38	252.77	205.39	0.1855	0.9320

Contd....

t	p	v_f	v_g	h_f	h_g	h_{fg}	s_f	s_g
3	5.51434	0.000785	0.0429	48.64	253.17	204.53	0.1901	0.9303
4	5.69352	0.000787	0.0416	49.91	253.56	203.65	0.1943	0.9291
5	5.87621	0.000790	0.0403	51.16	253.95	202.79	0.1989	0.9278
6	6.06276	0.000792	0.0391	52.41	254.33	201.92	0.2035	0.9266
7	6.25393	0.000794	0.0379	53.68	254.71	201.03	0.2077	0.9253
8	6.44993	0.000797	0.0368	54.94	255.08	200.14	0.2123	0.9241
9	6.65034	0.000799	0.0357	56.22	255.44	199.22	0.2169	0.9228
10	6.85517	0.000801	0.0346	57.52	255.81	198.29	0.2211	0.9216
11	7.06621	0.000803	0.0336	58.80	256.15	197.35	0.2257	0.9203
12	7.27724	0.000805	0.0326	60.07	256.48	196.41	0.2303	0.9190
13	7.49793	0.000808	0.0316	61.38	256.83	195.45	0.2345	0.9178
14	7.72552	0.000810	0.0307	62.72	257.17	194.45	0.2391	0.9165
15	7.95517	0.000813	0.0298	64.02	257.50	193.48	0.2437	0.9152
16	8.18758	0.000815	0.0289	65.32	257.81	192.48	0.2483	0.9140
17	8.42552	0.000817	0.0281	66.63	258.13	191.50	0.2529	0.9127
18	8.63241	0.000820	0.0273	67.95	258.44	190.49	0.2571	0.9115
19	8.91793	0.000822	0.0266	69.27	258.73	189.46	0.2617	0.9102
20	9.17241	0.000825	0.0258	70.59	259.00	188.41	0.2663	0.9089
21	9.43310	0.000827	0.0251	71.93	259.29	187.36	0.2709	0.9077
22	9.69931	0.000830	0.0244	73.31	259.58	186.27	0.2755	0.9065
23	9.97103	0.000833	0.0237	74.66	259.85	185.19	0.2801	0.9052
24	10.2482	0.000835	0.0230	76.04	260.11	184.07	0.2847	0.9039
25	10.53103	0.000838	0.0224	77.39	260.38	182.99	0.2889	0.9027
26	10.81931	0.000841	0.0218	78.79	260.64	181.85	0.2935	0.9014
27	11.11310	0.000844	0.0212	80.16	260.89	180.73	0.2981	0.9002
28	11.41241	0.000846	0.0206	81.54	216.12	179.58	0.3023	0.8989
29	11.69034	0.000849	0.0200	82.96	216.37	178.41	0.3069	0.8977
30	12.03448	0.000852	0.0194	84.38	261.60	177.22	0.3115	0.8964
31	12.35103	0.000855	0.0189	85.77	216.81	176.04	0.3161	0.8948
32	12.67310	0.000858	0.0184	87.17	262.02	174.85	0.3207	0.8935
33	13.00070	0.000861	0.0179	88.57	262.22	173.65	0.3249	0.8922
34	13.33793	0.000864	0.0174	90.00	262.40	172.40	0.3295	0.8909
35	13.68276	0.000867	0.0169	91.43	262.58	171.15	0.3337	0.8893
36	14.03034	0.000870	0.0165	92.85	262.74	169.89	0.3338	0.8880
37	14.38207	0.000874	0.0161	94.24	262.88	168.64	0.3429	0.8868
38	14.74345	0.000877	0.0156	95.63	263.00	167.37	0.3471	0.8851
39	15.11103	0.000881	0.0152	97.03	263.13	166.10	0.3517	0.8838
40	15.48965	0.000884	0.0148	98.44	263.21	164.77	0.3563	0.8822
41	15.86827	0.000887	0.0144	99.82	263.29	163.47	0.3601	0.8809
42	16.25793	0.000891	0.0140	101.24	263.38	162.14	0.3651	0.8797
43	16.65380	0.000894	0.0137	102.68	263.48	160.80	0.3693	0.8780
44	17.05517	0.000898	0.0133	104.12	263.58	159.46	0.3739	0.8767

Contd....

t	p	v_f	v_g	h_f	h_g	h_{fg}	s_f	s_g
45	17.46552	0.000902	0.0130	105.58	263.67	158.09	0.3785	0.8755
46	17.88414	0.000906	0.0126	107.04	263.76	156.72	0.3827	0.8738
47	18.30827	0.000910	0.0123	108.51	263.87	155.36	0.3871	0.8723
48	18.73793	0.000914	0.0119	109.98	263.93	153.95	0.3915	0.8709
49	19.12414	0.000918	0.0116	111.42	263.99	152.57	0.3959	0.8694
50	19.61380	0.000922	0.0113	112.86	264.05	151.91	0.4003	0.8680

Table 4. Saturated Methyl Chloride (CH_3Cl), R-40

Datum at – 40°C, $h_f = 0$, $s_f = 0$

Saturation temp., °C	Saturation pressure, bar	Specific volume, m^3/kg		Specific enthalpy, kJ/kg			Specific entropy, kJ/kg K	
		Liquid	Vapour	Liquid	Vapour	Latent	Liquid	Vapour
(t)	(p)	(v_f)	(v_g)	(h_f)	(h_g)	(h_{fg})	(s_f)	(s_g)
– 40	0.47434	0.000971	0.7950	0.00	443.48	443.48	0.0000	1.9026
– 38	0.52793	0.000974	0.7310	2.98	444.66	441.68	0.0126	1.8917
– 36	0.58152	0.000977	0.6669	5.97	445.85	439.88	0.0255	1.8808
– 34	0.63793	0.000981	0.6064	8.95	447.03	438.08	0.0377	1.8703
– 32	0.70434	0.000984	0.5590	11.95	448.22	436.27	0.0502	1.8603
– 30	0.77069	0.000988	0.5114	14.95	449.41	434.46	0.0628	1.8498
– 28	0.84345	0.000991	0.4692	17.96	450.57	432.61	0.0749	1.8402
– 26	0.88517	0.000994	0.4336	20.98	451.73	430.75	0.0871	1.8310
– 24	1.00483	0.000998	0.3979	24.00	452.87	428.87	0.0992	1.8213
– 22	1.09655	0.001001	0.3679	27.03	454.04	427.01	0.1114	1.8126
– 20	1.18896	0.001004	0.3383	30.08	455.17	425.09	0.1235	1.8033
– 18	1.31545	0.001009	0.3080	33.81	456.56	422.75	0.1382	1.7929
– 16	1.40165	0.001013	0.2906	36.19	457.43	421.24	0.1474	1.7862
– 14	1.51972	0.001016	0.2700	39.24	458.57	419.33	0.1595	1.7782
– 12	1.64207	0.001020	0.2506	42.30	459.69	417.39	0.1712	1.7698
– 10	1.77380	0.001023	0.2329	45.38	460.76	415.38	0.1830	1.7619
– 8	1.90760	0.001027	0.2179	48.47	461.80	413.33	0.1947	1.7544
– 6	2.06138	0.001031	0.2027	51.58	462.83	411.24	0.2065	1.7464
– 4	2.22690	0.001035	0.1898	54.69	463.83	409.14	0.2177	1.7389
– 2	2.38900	0.001039	0.1769	57.79	464.83	407.04	0.2290	1.7309
0	2.55931	0.001043	0.1650	60.89	465.80	404.91	0.2408	1.7238
1	2.65055	0.001045	0.1597	62.46	466.29	403.83	0.2466	1.7200
2	2.74400	0.001048	0.1546	64.02	466.77	402.75	0.2520	1.7129
3	2.84014	0.001050	0.1497	65.58	467.25	401.67	0.2579	1.7163
4	2.93862	0.001052	0.1450	67.15	467.72	400.57	0.2638	1.7098
5	3.04000	0.001054	0.1404	68.72	468.19	399.47	0.2696	1.7058
6	3.14414	0.001056	0.1360	70.29	468.67	398.38	0.2749	1.7024
7	3.25048	0.001058	0.1317	71.86	469.12	397.26	0.2803	1.6991
8	3.35931	0.001060	0.1277	73.43	469.57	396.14	0.2860	1.6957
9	3.47131	0.001062	0.1237	75.00	470.04	395.04	0.2916	1.6924
10	3.58552	0.001065	0.1200	76.60	470.50	393.90	0.2973	1.6890
11	3.70283	0.001067	0.1164	78.16	470.94	392.78	0.3029	1.6857
12	3.82386	0.001069	0.1129	79.75	471.40	391.65	0.3086	1.6828
13	3.94690	0.001072	0.1096	81.33	471.83	390.50	0.3140	1.6794
14	4.07380	0.001074	0.1063	82.91	472.27	389.36	0.3194	1.6760

Contd....

t	p	v_f	v_g	h_f	h_g	h_{fg}	s_f	s_g
15	4.20276	0.001076	0.1032	84.50	472.71	388.21	0.3251	1.6727
16	4.33641	0.001079	0.1002	86.09	473.14	387.05	0.3308	1.6698
17	4.47265	0.001081	0.0973	87.68	473.57	385.89	0.3362	1.6664
18	4.61048	0.001083	0.0945	89.28	474.00	384.72	0.3416	1.6631
19	4.75386	0.001085	0.0918	90.89	474.43	383.54	0.3471	1.6601
20	4.89724	0.001087	0.0892	92.48	474.85	382.37	0.3525	1.6572
21	5.04776	0.001090	0.0867	94.08	476.25	381.17	0.3578	1.6539
22	5.19828	0.001092	0.0842	95.68	475.64	379.96	0.3631	1.6509
23	5.35434	0.001094	0.0819	97.29	476.04	378.75	0.3685	1.6476
24	5.51200	0.001096	0.0796	98.89	476.43	377.54	0.3739	1.6446
25	5.67380	0.001099	0.0774	100.49	476.83	376.34	0.3793	1.6417
26	5.83890	0.001101	0.0753	102.11	477.20	375.09	0.3848	1.6388
27	6.00648	0.001103	0.0733	103.72	477.57	373.85	0.3903	1.6358
28	6.17903	0.001105	0.0713	105.33	477.94	372.61	0.3956	1.6329
29	6.35227	0.001108	0.0694	106.94	478.31	371.37	0.4009	1.6300
30	6.53103	0.001111	0.0676	108.55	478.69	370.14	0.4061	1.6275
31	6.71352	0.001114	0.0658	110.17	479.03	368.86	0.4114	1.6248
32	6.90096	0.001116	0.0641	111.79	479.38	367.59	0.4167	1.6218
33	7.08903	0.001119	0.0624	113.41	479.74	366.33	0.4220	1.6191
34	76.28414	0.001121	0.0608	115.03	480.08	365.05	0.4274	1.6164
35	7.48276	0.001124	0.0592	116.65	480.41	363.76	0.4328	1.6136
36	7.68414	0.001126	0.0577	118.27	480.75	362.48	0.4382	1.6108
37	7.88896	0.001128	0.0562	119.90	481.08	361.18	0.4435	1.6082
38	8.09586	0.001131	0.0548	121.53	481.41	359.88	0.4487	1.6055
39	8.31000	0.001133	0.0534	123.16	481.74	358.58	0.4438	1.6029
40	8.52414	0.001136	0.0520	124.79	482.06	357.27	0.4589	1.6003
41	8.74758	0.001139	0.0508	126.43	482.38	355.95	0.4642	1.5976
42	8.97103	0.001142	0.0495	128.07	482.69	354.62	0.4694	1.5950
43	9.20172	0.001144	0.0483	129.71	483.00	353.29	0.4746	1.5923
44	9.43448	0.001147	0.0472	131.34	483.31	351.96	0.4797	1.5897
45	9.67496	0.001150	0.0461	132.98	483.60	350.62	0.4848	1.5872
46	9.92172	0.001153	0.0450	134.63	483.88	349.25	0.4900	1.5848
47	10.16827	0.001156	0.0439	136.28	484.17	347.89	0.4952	1.5823
48	10.41517	0.001159	0.0429	137.93	484.46	346.53	0.5003	1.5798
49	10.66441	0.001162	0.0418	139.58	484.74	345.16	0.5055	1.5774
50	10.93380	0.001165	0.0409	141.23	485.01	343.78	0.5106	1.5750
51	11.20317	0.001168	0.0400	142.88	485.27	342.39	0.5158	1.5726
52	11.47255	0.001172	0.0391	144.55	485.53	340.98	0.5209	1.5702
53	11.74193	0.001175	0.0381	146.21	485.80	339.59	0.5260	1.5677
54	12.01131	0.001178	0.0372	147.86	486.06	338.20	0.5311	1.5653
55	12.29448	0.001181	0.0364	149.53	486.31	335.78	0.5362	1.5630
56	12.58869	0.001184	0.0356	151.19	486.55	335.36	0.5413	1.5606

Contd....

t	p	v_f	v_g	h_f	h_g	h_{fg}	s_f	s_g
57	12.88290	0.001187	0.0348	152.86	486.78	333.92	0.5463	1.5583
58	13.17710	0.001190	0.0340	154.53	487.01	332.48	0.5514	1.5560
59	13.47131	0.001193	0.0332	156.20	487.25	331.05	0.5564	1.5536
60	13.76552	0.001197	0.0324	157.86	487.48	329.62	0.5615	1.5513

Table 5. Saturated Ammonia (NH_3), R-717

Datum at –40°C, $h_f = 0$, $s_f = 0$

Saturation temp., °C	Saturation pressure, bar	Specific Volume, m^3/kg		Specific enthalpy, kJ/kg			Specific entropy, kJ/kg K	
		Liquid	Vapour	Liquid	Vapour	Latent	Liquid	Vapour
(t)	(p)	(v_f)	(v_g)	(h_f)	(h_g)	(h_{fg})	(s_f)	(s_g)
– 50	0.40896	0.001426	2.6281	– 44.43	1373.27	1417.70	– 0.1943	6.1603
– 48	0.45972	0.001431	2.3565	– 35.44	1376.80	1412.24	– 0.1551	6.1192
– 46	0.51600	0.001436	2.1177	– 26.60	1380.20	1406.80	– 0.1157	6.0789
– 44	0.57710	0.001441	1.9062	– 17.81	1383.31	1401.12	– 0.0769	6.0394
– 42	0.64455	0.001446	1.7196	– 8.97	1386.67	1395.64	– 0.0384	6.0008
– 40	0.71793	0.001451	1.5537	0.00	1390.02	1390.02	0.0000	5.9631
– 38	0.79384	0.001456	1.4077	8.97	1393.13	1384.16	0.0381	5.9262
– 36	0.88607	0.001462	1.2775	17.81	1393.35	1378.54	0.0758	5.8900
– 34	0.98096	0.001467	1.1614	26.84	1399.51	1372.67	0.1131	5.8545
– 32	1.08165	0.001472	1.0574	35.68	1402.48	1366.80	0.1506	5.8198
– 30	1.19586	0.001477	0.9644	44.66	1405.60	1360.94	0.1876	5.7856
– 28	1.31724	0.001483	0.8820	53.68	1408.53	1354.85	0.2242	5.7521
– 26	1.44790	0.001488	0.8069	62.61	1411.45	1348.84	0.2607	5.7195
– 24	1.58841	0.001494	0.7397	71.73	1414.39	1342.66	0.2970	5.6872
– 22	1.73986	0.001500	0.6793	80.76	1417.28	1336.52	0.3330	5.6556
– 20	1.90276	0.001505	0.6244	89.78	1420.02	1330.24	0.3684	5.6244
– 18	2.07807	0.001511	0.5750	98.76	1422.72	1323.96	0.4043	5.5939
– 16	2.26551	0.001517	0.5303	107.83	1425.28	1317.45	0.4397	5.5639
– 14	2.46634	0.001523	0.4896	116.95	1427.88	1310.93	0.4747	5.5356
– 12	2.68069	0.001529	0.4526	126.16	1430.54	1304.38	0.5096	5.5055
– 10	2.90896	0.001536	0.4189	135.37	1433.05	1297.68	0.5443	5.4770
– 8	3.15365	0.001541	0.3884	144.35	1435.33	1290.98	0.5789	5.4487
– 6	3.41380	0.001548	0.3604	153.56	1437.93	1284.37	0.6139	5.4210
– 4	3.69062	0.001554	0.3348	162.77	1440.02	1277.25	0.6473	5.3940
– 2	3.98427	0.001561	0.3113	171.98	1442.17	1270.19	0.6812	5.3670
0	4.29586	0.001567	0.2898	181.20	1444.45	1263.25	0.7151	5.3405
1	4.45848	0.001571	0.2798	185.80	1445.49	1259.69	0.7321	5.3277
2	4.62662	0.001574	0.2702	190.40	1446.54	1256.14	0.7487	5.3145
3	4.79931	0.001578	0.2610	195.17	1447.59	1252.42	0.7653	5.3017
4	4.97682	0.001582	0.2521	199.85	1448.63	1248.78	0.7818	5.2888
5	5.15862	0.001585	0.2436	204.46	1449.56	1245.10	0.7989	5.2765
6	5.34745	0.001589	0.2354	209.06	1450.49	1241.43	0.8154	5.2638
7	5.54007	0.001592	0.2276	213.73	1451.54	1237.81	0.8320	5.2513
8	5.73820	0.001595	0.2201	218.50	1452.54	1234.04	0.8487	5.2389
9	5.94186	0.001598	0.2128	223.11	1453.39	1230.28	0.8652	5.2266

Contd....

t	p	v_f	v_g	h_f	h_g	h_{fg}	s_f	s_g
10	5.15103	0.001603	0.2060	227.72	1464.22	1226.50	0.8814	5.2141
11	5.36641	0.001607	0.1992	232.53	1455.30	1222.77	0.8979	5.2020
12	6.58731	0.001610	0.1928	237.15	1456.11	1218.96	0.9142	5.1900
13	5.81420	0.001614	0.1866	241.92	1456.96	1215.04	0.9307	5.1780
14	7.04717	0.001517	0.1807	246.60	1457.80	1211.20	0.9470	51659
15	7.28276	0.001621	0.1751	251.44	1458.63	1207.19	0.9634	5.1542
16	7.53104	0.001624	0.1696	256.14	1469.47	1203.33	0.9794	5.1421
17	7.78138	0.001628	0.1643	259.88	1460.24	1200.36	0.9956	5.1302
18	8.03862	0.001633	0.1592	256.54	1460.92	1195.38	0.0118	5.1186
19	8.30551	0.001637	0.1543	270.35	1461.75	1191.40	1.0280	5.1073
20	8.57241	0.001641	0.1496	275.16	1462.60	1187.44	1.0442	5.0956
21	8.85172	0.001645	0.1450	279.77	1463.21	1183.44	1.0604	5.0843
22	9.13655	0.001648	0.1407	284.56	1463.84	1179.28	1.0763	5.0729
23	9.42690	0.001652	0.1365	289.37	1464.63	1175.26	1.0924	5.0616
24	9.72690	0.001656	0.1324	294.19	1465.33	1171.14	1.1083	5.0503
25	10.02760	0.001661	0.1284	298.90	1465.84	1166.94	1.1242	5.0391
26	10.34069	0.001665	0.1246	303.82	1466.59	1162.77	1.1402	5.0279
27	10.66137	0.001669	0.1210	308.63	1467.22	1158.59	1.1563	5.0170
28	10.99172	0.001673	0.1174	313.45	1467.85	1154.40	1.1721	5.0061
29	11.32758	0.001678	0.1140	318.26	1468.45	1150.19	1.1879	4.9951
30	11.66896	0.001682	0.1107	323.08	1468.87	1145.79	1.2037	4.9842
31	11.01655	0.001686	0.1075	327.89	1469.50	1141.61	1.2195	4.9733
32	12.37517	0.001691	0.1045	332.71	1469.94	1137.23	1.2350	4.9624
33	12.74482	0.001695	0.1015	337.52	1470.36	1132.84	1.2508	4.9517
34	13.12137	0.001700	0.0987	342.48	1470.92	1128.44	1.2664	4.9409
35	13.50345	0.001704	0.0960	347.50	1471.43	1123.93	1.2821	4.9302
36	13.89379	0.001709	0.0932	352.29	1471.70	1119.41	1.2978	4.9196
37	14.29517	0.001713	0.0907	357.25	1472.19	1114.94	1.3135	4.9091
38	14.70620	0.001718	0.0882	362.12	1472.45	1110.33	1.3290	4.8985
39	15.12276	0.001723	0.0857	367.11	1472.87	1105.76	1.3445	4.8885
40	15.54483	0.001727	0.0834	371.93	1473.30	1101.37	1.3600	4.8774
41	15.98551	0.001732	0.0811	376.95	1473.50	1096.55	1.3754	4.8669
42	16.43172	0.001738	0.0789	381.79	1473.70	1091.91	1.3908	4.8563
43	16.88344	0.001742	0.0768	386.76	1473.91	1087.15	1.4065	4.8461
44	17.34482	0.001747	0.0747	391.79	1474.12	1082.33	1.4221	4.8357
45	17.81724	0.001752	0.0727	396.81	1474.33	1077.52	1.4374	4.8251
46	18.30070	0.001757	0.0708	401.84	1474.54	1072.70	1.4528	4.8147
47	18.79518	0.001763	0.0689	406.86	1474.68	1067.82	1.4683	4.8045
48	19.29931	0.001768	0.06070	411.89	1474.76	1062.87	1.4835	4.7938
49	19.80965	0.001773	0.0652	416.91	1474.84	1057.93	1.4990	4.7834
50	20.33103	0.001779	0.0636	421.94	1474.92	1052.92	1.5118	4.7732

Table 6. Saturated Carbon dioxide (CO_2), R-744

Datum at – 40°C, $h_f = 0$, $s_f = 0$

Saturation temp., °C	Saturation pressure, bar	Specific Volume, m^3/kg		Specific enthalpy, kJ/kg			Specific entropy, kJ/kg K	
		Liquid	Vapour	Liquid	Vapour	Latent	Liquid	Vapour
(*t*)	(*p*)	(v_f)	(v_g)	(h_f)	(h_g)	(h_{fg})	(s_f)	(s_g)
– 40	10.05517	0.000898	0.03821	0.00	320.52	320.52	0.0000	1.3754
– 38	10.84965	0.000904	0.03577	3.77	320.86	317.09	0.0161	1.3654
– 36	11.64413	0.000911	0.03332	7.54	321.19	313.65	0.0322	1.3553
– 34	12.46676	0.000918	0.03100	11.32	321.51	310.19	0.0483	1.3453
– 32	13.38786	0.000925	0.02906	15.17	321.76	306.59	0.0641	1.3356
– 30	14.30896	0.000932	0.02712	19.02	322.01	302.99	0.0800	1.3259
– 28	15.28855	0.000939	0.02535	22.95	322.22	299.27	0.0956	1.3163
– 26	16.34124	0.000947	0.02379	26.97	322.39	295.42	0.1109	1.3067
– 24	17.39393	0.000955	0.02222	30.99	322.56	291.57	0.1264	1.2971
– 22	18.54262	0.000963	0.02084	35.06	322.72	287.66	0.1419	1.2879
– 20	19.73931	0.000971	0.01957	39.17	322.89	283.72	0.1574	1.2787
– 18	20.93600	0.000980	0.01830	43.27	323.06	279.79	0.1733	1.2692
– 16	22.22758	0.000990	0.01716	47.45	322.89	275.44	0.1888	1.2605
– 14	23.58345	0.001000	0.01611	51.73	322.61	270.88	0.2060	1.2523
– 12	25.01131	0.001010	0.01513	56.32	322.58	266.26	0.2260	1.2464
– 10	25.54069	0.001021	0.01426	60.85	322.24	261.39	0.2432	1.2372
– 8	28.07007	0.001032	0.01338	65.37	321.91	256.54	0.2600	1.2280
– 6	29.66069	0.001043	0.01256	70.06	321.54	251.48	0.2771	1.2192
– 4	31.37379	0.001056	0.01182	75.08	321.12	246.04	0.2954	1.2101
– 2	33.08689	0.001069	0.01110	80.10	320.71	240.61	0.3135	1.2010
0	34.91034	0.001082	0.01041	85.27	320.01	234.74	0.3312	1.1909
1	35.86620	0.001090	0.01010	87.91	319.55	231.64	0.3400	1.1854
2	35.82207	0.001098	0.00979	90.54	319.09	228.55	0.3488	1.1799
3	37.77800	0.001106	0.00947	93.18	318.63	225.45	0.3575	1.1744
4	38.73380	0.001113	0.00916	95.82	318.11	222.35	0.3662	1.1690
5	39.75034	0.001122	0.00887	98.55	317.57	219.02	0.3754	1.1633
6	40.81545	0.001131	0.00860	101.36	316.86	215.60	0.3849	1.1575
7	41.88055	0.001140	0.00834	104.16	316.14	211.98	0.3944	1.1516
8	42.94565	0.001149	0.00807	106.97	315.43	208.46	0.4039	1.1458
9	44.01076	0.001158	0.00780	109.77	314.72	204.95	0.4134	1.1400
10	45.07586	0.001167	0.00753	112.58	314.01	201.43	0.4229	1.1342
11	46.25517	0.001178	0.00729	115.55	312.79	197.24	0.4329	1.1274
12	47.43448	0.001190	0.00705	118.52	311.58	193.06	0.4430	1.1205
13	48.61379	0.001202	0.00682	121.49	310.36	188.87	0.4534	1.1137
14	49.79310	0.001213	0.00658	124.47	309.15	184.68	0.4635	1.1068

Contd....

t	p	v_f	v_g	h_f	h_g	h_{fg}	s_f	s_g
15	50.97241	0.001225	0.00634	127.44	307.94	180.50	0.4735	1.1000
16	52.20580	0.001238	0.00611	130.68	306.41	175.79	0.4843	1.0921
17	53.50676	0.001254	0.00590	134.05	304.48	170.43	0.4954	1.0830
18	54.80772	0.001270	0.00569	137.48	302.55	165.07	0.5065	1.0738
19	56.10870	0.001286	0.00547	140.92	300.63	159.71	0.5176	1.0647
20	57.40965	0.001302	0.00526	144.35	298.70	154.35	0.5286	1.0556
21	58.71062	0.001318	0.00505	147.78	296.78	149.00	0.5397	1.0465
22	60.12745	0.001345	0.00482	151.96	293.29	141.33	0.5539	1.0332
23	61.55876	0.001374	0.00460	156.23	289.60	133.37	0.5681	1.0194
24	62.99007	0.001404	0.00438	160.50	285.92	125.42	0.5832	1.0056
25	64.42138	0.001433	0.00416	164.77	282.23	117.46	0.5978	0.9918
26	65.85270	0.001462	0.00394	169.04	278.55	109.51	0.6124	0.9781
27	67.31930	0.001501	0.00371	174.08	274.16	100.08	0.6289	0.9617
28	68.85655	0.001560	0.00347	180.64	268.37	87.73	0.6490	0.9402
29	70.39380	0.001619	0.00323	187.19	262.58	75.39	0.6691	0.9187
30	71.93103	0.001678	0.00299	193.75	256.79	63.04	0.6892	0.8972
31	73.53103	0.002160	0.00216	225.62	225.62	0.00	0.7913	0.7913

Table 7. Saturated Sulphur dioxide (SO_2), R-764

Datum at – 40° C, $h_f = 0$, $s_f = 0$

Saturation temp., °C	Saturation pressure, bar	Specific volume, m^3/kg		Specific enthalpy, kJ/kg			Specific entropy, kJ/kg K	
		Liquid	Vapour	Liquid	Vapour	Latent	Liquid	Vapour
(t)	(p)	(v_f)	(v_g)	(h_f)	(h_g)	(h_{fg})	(s_f)	(s_g)
– 40	0.21628	0.000652	1.4012	0.00	415.44	415.44	0.0000	1.7820
– 38	0.24594	0.000654	1.2694	2.45	416.52	414.07	0.0101	1.7715
– 36	0.27561	0.000656	1.1375	4.90	417.61	412.71	0.0203	1.7610
– 34	0.30731	0.000659	1.0143	7.38	418.68	411.30	0.0305	1.7505
– 32	0.34578	0.000661	0.9211	9.93	419.64	409.71	0.0409	1.7404
– 30	0.38432	0.000663	0.8280	12.49	420.62	408.13	0.0514	1.7303
– 28	0.42757	0.000665	0.7462	15.09	421.56	406.47	0.0619	1.7203
– 26	0.47673	0.000667	0.6794	17.75	422.45	404.70	0.0726	1.7106
– 24	0.52590	0.000669	0.6125	20.42	423.33	402.91	0.0833	1.7008
– 22	0.58345	0.000671	0.5575	23.13	424.17	401.04	0.0941	1.6912
– 20	0.64517	0.000673	0.5090	25.88	424.93	399.05	0.1049	1.6818
– 18	0.70693	0.000676	0.4606	28.63	425.73	397.10	0.1158	1.6723
– 16	0.77683	0.000678	0.4199	31.44	426.46	395.02	0.1267	1.6630
– 14	0.85324	0.000680	0.3841	34.24	427.13	392.89	0.1376	1.6538
– 12	0.93517	0.000682	0.3518	37.06	427.74	390.68	0.1485	1.6447
– 10	1.02345	0.000685	0.3227	39.89	428.31	388.42	0.1594	1.6356
– 8	1.11820	0.000687	0.2966	42.76	428.86	386.10	0.1703	1.6267
– 6	1.21972	0.000689	0.2729	45.63	429.36	383.73	0.1812	1.6177
– 4	1.32758	0.000692	0.2509	40.50	429.80	381.30	0.1919	1.6088
– 2	1.44289	0.000695	0.2319	51.39	430.21	378.82	0.2028	1.6000
0	1.56620	0.000697	0.2148	54.26	430.54	376.28	0.2135	1.5912
1	1.63076	0.000699	0.2061	55.73	430.71	374.28	0.2189	1.5868
2	1.69782	0.000701	0.1983	57.19	430.86	373.67	0.2242	1.5824
3	1.76783	0.000702	0.1910	58.65	430.99	372.34	0.2295	1.5781
4	1.83785	0.000703	0.1837	60.11	431.11	371.00	0.2349	1.5737
5	1.91103	0.000704	0.1768	61.56	431.22	369.66	0.2402	1.5694
6	1.98676	0.000705	0.1704	63.02	431.31	368.29	0.2455	1.5651
7	2.06248	0.000707	0.1640	64.48	431.40	366.92	0.2509	1.5608
8	2.14276	0.000708	0.1581	65.94	431.48	365.54	0.2561	1.5564
9	2.22496	0.000710	0.1524	67.40	431.54	364.14	0.2614	1.5522
10	2.30689	0.000711	0.1467	68.87	431.61	362.74	0.2667	1.5479
11	2.39627	0.000712	0.1417	70.34	431.65	361.31	0.2719	1.5436
12	2.48565	0.000714	0.1366	71.80	431.68	359.88	0.2771	1.5393
13	2.57655	0.000715	0.1317	73.27	431.70	358.43	0.2823	1.5350
14	2.67289	0.000716	0.1273	74.72	431.69	356.97	0.2875	1.5308

Contd....

t	p	v_f	v_g	h_f	h_g	h_{fg}	s_f	s_g
15	2.67924	0.000718	0.1228	76.18	431.68	355.50	0.2927	1.5266
16	2.86910	0.000719	0.1186	77.64	431.66	354.02	0.2978	1.5224
17	2.97338	0.000720	0.1146	79.09	431.62	352.53	0.3030	1.5182
18	3.07765	0.000722	0.1106	80.55	431.58	351.03	0.3082	1.5140
19	3.18673	0.000723	0.1069	82.01	431.52	349.51	0.3132	1.5097
20	3.29820	0.000725	0.1033	83.46	431.46	348.00	0.3183	1.5055
21	3.40965	0.000726	0.0997	84.92	431.39	346.47	0.3233	1.5013
22	3.52910	0.000728	0.0965	86.38	431.29	344.91	0.3283	1.4971
23	3.64951	0.000729	0.0933	87.83	431.19	343.36	0.3334	1.4929
24	3.77092	0.000731	0.0902	89.29	431.08	341.79	0.3384	1.4887
25	3.90027	0.000732	0.0873	90.74	430.94	340.20	0.3434	1.4846
26	4.02965	0.000734	0.0844	92.19	430.80	338.61	0.3484	1.4805
27	4.16207	0.000735	0.0817	93.64	430.65	337.01	0.3532	1.4764
28	4.30124	0.000737	0.0791	95.08	430.48	335.40	0.3680	1.4721
29	4.44027	0.000738	0.0765	96.53	430.31	333.78	0.3629	1.4677
30	4.58276	0.000740	0.0740	97.97	430.12	332.15	0.3677	1.4635
31	4.73172	0.000741	0.0718	99.41	429.91	330.50	0.3725	1.4593
32	4.88069	0.000743	0.0695	100.86	429.70	328.84	0.3773	1.4551
33	5.03641	0.000745	0.0673	102.29	429.47	327.18	0.3821	1.4509
34	5.19407	0.000746	0.0653	103.71	429.22	325.51	0.3867	1.4467
35	5.35172	0.000748	0.0632	105.13	428.98	323.85	0.3914	1.4425
36	5.52353	0.00749	0.0612	105.56	428.70	322.14	0.3961	1.4383
37	5.69534	0.000751	0.0593	107.98	428.43	320.45	0.4007	1.4341
38	5.87100	0.000753	0.0575	109.40	428.14	318.74	0.4053	1.4299
39	6.06019	0.000754	0.0558	110.81	427.81	317.00	0.4098	1.4256
40	5.24938	0.000756	0.0542	112.21	427.48	315.27	0.4143	1.4214
41	5.43856	0.000758	0.0526	113.62	427.15	313.53	0.4188	1.4172
42	5.62775	0.000759	0.0510	115.03	426.82	311.79	0.4233	1.4129
43	6.81693	0.000761	0.0493	116.43	426.46	310.06	0.4278	1.4087
44	7.05520	0.000763	0.0478	117.83	426.11	308.28	0.4322	1.4044
45	7.31793	0.000765	0.0465	119.22	425.71	306.49	0.4365	1.4002
46	7.58080	0.000767	0.0451	120.61	425.32	304.71	0.4409	1.3959
47	7.84360	0.000769	0.0438	122.00	424.92	302.92	0.4452	1.3916
48	8.10640	0.000771	0.0424	123.39	424.52	301.13	0.4495	1.3874
49	8.36598	0.000773	0.0411	124.78	424.11	299.33	0.4538	1.3831
50	8.59986	0.000775	0.0401	126.13	423.60	297.47	0.4578	1.3787

Table 8. Psychrometric properties of air at standard barometric pressure of 760 mm, Hg. or 1.01325 bar

	Properties of water and steam (Taken from ISI Steam Tables)					*Properties of Dry Air at 760 mm. Hg. (ASHRAE)*		*Properties of Mixture of Dry Air and Saturated Steam i.e. (1 + W) kg. at Total Pressure of 760 mm. Hg. (ASHRAE)*		
Temperature	*Saturation pressure of water and steam, p_s*		*Enthalpy of saturated water, h_f*	*Enthalpy of saturated steam, h_g*	*Specific volume of saturated steam, v_g*	*Specific volume, v_a*	*Enthalpy, h_a*	*Volume of mixture (1 + W)kg per kg. of dry air, v_{as}*	*Enthalpy of mixture (1 + W)kg per kg. of dry air, h_{as}*	*Specific humidity per kg. of dry air, W*
t°C	*mm. Hg.*	*bar*	*kJ/kg.*	*kJ/kg.*	*m³/kg.*	*m³/kg.*	*kJ/kg.*	*m³/kg.*	*kJ/kg.*	*kJ/kg.*
0	4.580	0.006108	0.00	2501.6	206.31	0.7734	0.000	0.7781	9.473	0.003789
1	4.924	0.006566	4.17	2503.4	192.61	0.7762	1.006	0.7813	11.204	0.004076
2	5.291	0.007055	8.39	2505.2	179.92	0.7791	2.012	0.7845	12.982	0.004381
3	5.681	0.007575	12.60	2507.1	168.17	0.7819	3.018	0.7878	14.811	0.004707
4	6.097	0.008129	16.80	2508.9	157.27	0.7848	4.024	0.7911	16.696	0.005054
5	6.539	0.008718	21.01	2510.7	147.16	0.7876	5.030	0.7944	18.639	0.005424
6	7.010	0.009345	25.21	2512.6	137.79	0.7904	6.036	0.7978	20.644	0.005818
7	7.510	0.010012	29.41	2514.4	129.06	0.7933	7.041	0.8012	22.713	0.006237
8	8.041	0.010720	33.60	2516.2	120.97	0.7961	8.047	0.8046	24.852	0.006683
9	8.605	0.011472	37.80	2518.1	113.44	0.7990	9.053	0.8081	27.064	0.007157
10	9.204	0.012270	41.99	2519.9	106.43	0.8018	10.059	0.8116	29.352	0.007661
11	9.839	0.013116	46.19	2521.7	99.909	0.8046	11.065	0.8152	31.724	0.008197
12	10.512	0.014014	50.38	2523.6	93.835	0.8075	12.071	0.8188	34.179	0.008766
13	11.225	0.014965	54.57	2525.4	88.176	0.8103	13.077	0.8225	36.726	0.009370
14	11.982	0.015973	58.75	2527.2	82.900	0.8132	14.084	0.8262	39.370	0.010012
15	12.781	0.017039	62.94	2529.1	77.978	0.8160	15.090	0.8300	42.113	0.010692
16	13.628	0.018168	67.13	2530.9	73.384	0.8188	16.096	0.8338	44.963	0.011413
17	14.524	0.019362	71.31	2532.7	69.095	0.8217	17.102	0.8377	47.926	0.012178
18	15.470	0.020624	75.50	2534.5	65.087	0.8245	18.108	0.8414	51.008	0.012989
19	16.470	0.021957	79.68	2536.4	61.341	0.8274	19.114	0.8457	54.216	0.013848

Contd....

Temperature	Properties of water and steam (Taken from ISI Steam Tables)					Properties of Dry Air at 760 mm. Hg. (ASHRAE)		Properties of Mixture of Dry Air and Saturated Steam i.e. (1 + W) kg. at Total Pressure of 760 mm. Hg. (ASHRAE)		
	Saturation pressure of water and steam, p_s		Enthalpy of saturated water, h_f	Enthalpy of saturated steam, h_g	Specific volume of saturated steam v_g	Specific volume, v_a	Enthalpy h_a	Volume of mixture (1 + W)kg per kg. of dry air, v_{as}	Enthalpy of mixture (1 + W)kg. per kg. of dry air, h_{as}	Specific humidity per kg. of dry air, W
°C	mm. Hg.	bar	kJ/kg.	kJ/kg.	m^3/kg.	m^3/kg.	kJ/kg.	m^3/kg.	kJ/kg.	kJ/kg.
20	17.53	0.023366	83.86	2538.2	57.838	0.8302	20.121	0.8498	57.555	0.014758
21	18.64	0.024853	88.04	2540.0	54.561	0.8330	21.127	0.8540	61.035	0.015721
22	19.82	0.026422	92.23	2541.8	51.492	0.8359	22.133	0.8583	64.660	0.016741
23	21.06	0.028076	96.41	2543.6	48.619	0.8387	23.140	0.8627	68.440	0.017821
24	22.37	0.029821	100.59	2545.5	45.926	0.8416	24.146	0.8671	72.385	0.018963
25	23.75	0.031660	104.77	2547.3	43.402	0.8444	25.153	0.8717	76.500	0.020170
26	25.50	0.033597	108.95	2549.1	41.034	0.8472	26.159	0.8764	80.798	0.021448
27	26.73	0.035636	113.13	2550.9	38.813	0.8501	27.165	0.8811	85.285	0.022798
28	28.34	0.037782	117.31	2552.7	36.728	0.8529	28.172	0.8860	89.976	0.024226
29	30.03	0.040040	121.48	2554.5	34.769	0.8558	29.179	0.8910	94.878	0.025735
30	31.81	0.042415	125.66	2556.4	32.929	0.8586	30.185	0.8962	100.006	0.027329
31	33.69	0.044911	129.84	2558.2	31.199	0.8614	31.192	0.9015	105.369	0.029014
32	35.65	0.047534	134.02	2560.0	29.572	0.8643	32.198	0.9069	110.979	0.030793
33	37.72	0.050288	138.20	2561.8	28.042	0.8671	33.205	0.9125	116.857	0.032674
34	39.89	0.053180	142.38	2563.6	26.601	0.8700	34.212	0.9183	123.011	0.034660
35	42.17	0.056216	146.56	2565.4	25.245	0.8728	35.219	0.9242	129.455	0.036756
36	44.55	0.059400	150.74	2567.2	23.967	0.8756	36.226	0.9303	136.209	0.038971
37	47.06	0.062739	154.92	2569.0	22.763	0.8785	37.233	0.9366	143.290	0.041309
38	49.69	0.066240	159.09	2570.8	21.627	0.8813	38.239	0.9431	150.713	0.043778
39	52.44	0.069908	163.27	2572.6	20.557	0.8842	39.246	0.9498	158.504	0.046386
40	55.31	0.073750	167.45	2574.4	19.546	0.8870	40.253	0.9568	166.683	0.049141
41	58.33	0.077773	171.63	2576.2	18.592	0.8898	41.261	0.9640	175.265	0.052049
42	61.49	0.081986	175.81	2577.9	17.692	0.8927	42.268	0.9714	184.275	0.055119
43	64.80	0.086391	179.99	2579.7	16.841	0.8955	43.275	0.9792	193.749	0.058365
44	68.25	0.091001	184.17	2581.5	16.036	0.8983	44.282	0.9872	203.699	0.061791

INDEX

S

T

V

W

❑❑❑

Win Prizes!

ATTENTION READERS

We request you for your frank assessment regarding some of the aspects of the book given as under:

A Textbook of REFRIGERATION AND AIR-CONDITIONING

By:

Er. R.K. RAJPUT

Please fill up the given space in neat capital letters.

(i) What topics of your syllabus that are important from you examination point of view are not covered in the book ?

....

....

....

....

(ii) What are the chapters wherein the treatment of the subject matter is not systematic or organised or updated ?

....

....

....

....

(iii) Have you come across misprints/mistakes/factual inaccuracies in the book? Please specify the chapters and the page numbers.

....

....

....

....

(iv) What other books on the same subject have you found/heard better that the present book? Please specify in terms of price and quality.

....

....

....

....

(v) Further suggestions and comments for the improvement of the book:

....

....

....

....

(PS: If need be please attach a separate sheet to write your views.)

Other Details :

(i) Who recommended you the book ? (Please tick in the box near the option relevant to you.)

☐ Teachers ☐ Friends ☐ Bookseller

(ii) Name of the recommending teacher, his designation and address.

....

....

....

....

(iii) Name and address of the bookseller you purchased the book form :

....

....

....

....

(iv) Name and address of your institution (Please mention the University or Board, as the case may be)

....

....

....

....

(v) Your name and complete postal address :

....

....

....

....

(vi) Course you are enrolled in :

....

....

....

....

The best assessment will be awarded each month. The award will be in the form of our publications as decided by the Editorial Board.

Please mail the filled up coupon at your earliest to:

S.K. Kataria & Sons®

4885/109, Prakash Mahal, Dr. Subhash Bhargav Lane,
Opposite Delhi Medical Association, Daryaganj, New Delhi-110002
Phones: +91-11-23243489, 43551243; Mobile: +91-9871775858
e-mail: katariabook@gmail.com; katariabooks@yahoo.com
Website: www.skkatariaandsons.com